# MANTLE DYNAMICS

# Treatise on Geophysics

# MANTLE DYNAMICS

Editor-in-Chief

**Professor Gerald Schubert**

*Department of Earth and Space Sciences and Institute of Geophysics and Planetary Physics,*
*University of California Los Angeles, Los Angeles, CA, USA*

Volume Editor

**Dr. David Bercovici**

*Yale University, New Haven, CT, USA*

## ELSEVIER

AMSTERDAM • BOSTON • HEIDELBERG • LONDON • NEW YORK • OXFORD
PARIS • SAN DIEGO • SAN FRANCISCO • SINGAPORE • SYDNEY • TOKYO

Elsevier B.V.
Radarweg 29, 1043 NX Amsterdam, the Netherlands

First edition 2009

**British Library Cataloguing in Publication Data**
A catalogue record for this book is available from the British Library

**Library of Congress Control Number:** 2009929985

ISBN: 978-0-444-53456-9

For information on all Elsevier publications
visit our website at elsevierdirect.com

Printed and bound in Spain

09 10 11 12 10 9 8 7 6 5 4 3 2 1

# Contents

# Preface

Geophysics is the physics of the Earth, the science that studies the Earth by measuring the physical consequences of its presence and activity. It is a science of extraordinary breadth, requiring 10 volumes of this treatise for its description. Only a treatise can present a science with the breadth of geophysics if, in addition to completeness of the subject matter, it is intended to discuss the material in great depth. Thus, while there are many books on geophysics dealing with its many subdivisions, a single book cannot give more than an introductory flavor of each topic. At the other extreme, a single book can cover one aspect of geophysics in great detail, as is done in each of the volumes of this treatise, but the treatise has the unique advantage of having been designed as an integrated series, an important feature of an interdisciplinary science such as geophysics. From the outset, the treatise was planned to cover each area of geophysics from the basics to the cutting edge so that the beginning student could learn the subject and the advanced researcher could have an up-to-date and thorough exposition of the state of the field. The planning of the contents of each volume was carried out with the active participation of the editors of all the volumes to insure that each subject area of the treatise benefited from the multitude of connections to other areas.

Geophysics includes the study of the Earth's fluid envelope and its near-space environment. However, in this treatise, the subject has been narrowed to the solid Earth. The *Treatise on Geophysics* discusses the atmosphere, ocean, and plasmasphere of the Earth only in connection with how these parts of the Earth affect the solid planet. While the realm of geophysics has here been narrowed to the solid Earth, it is broadened to include other planets of our solar system and the planets of other stars. Accordingly, the treatise includes a volume on the planets, although that volume deals mostly with the terrestrial planets of our own solar system. The gas and ice giant planets of the outer solar system and similar extra-solar planets are discussed in only one chapter of the treatise. Even the *Treatise on Geophysics* must be circumscribed to some extent. One could envision a future treatise on Planetary and Space Physics or a treatise on Atmospheric and Oceanic Physics.

Geophysics is fundamentally an interdisciplinary endeavor, built on the foundations of physics, mathematics, geology, astronomy, and other disciplines. Its roots therefore go far back in history, but the science has blossomed only in the last century with the explosive increase in our ability to measure the properties of the Earth and the processes going on inside the Earth and on and above its surface. The technological advances of the last century in laboratory and field instrumentation, computing, and satellite-based remote sensing are largely responsible for the explosive growth of geophysics. In addition to the enhanced ability to make crucial measurements and collect and analyze enormous amounts of data, progress in geophysics was facilitated by the acceptance of the paradigm of plate tectonics and mantle convection in the 1960s. This new view of how the Earth works enabled an understanding of earthquakes, volcanoes, mountain building, indeed all of geology, at a fundamental level. The exploration of the planets and moons of our solar system, beginning with the Apollo missions to the Moon, has invigorated geophysics and further extended its purview beyond the Earth. Today geophysics is a vital and thriving enterprise involving many thousands of scientists throughout the world. The interdisciplinarity and global nature of geophysics identifies it as one of the great unifying endeavors of humanity.

The keys to the success of an enterprise such as the *Treatise on Geophysics* are the editors of the individual volumes and the authors who have contributed chapters. The editors are leaders in their fields of expertise, as distinguished a group of geophysicists as could be assembled on the planet. They know well the topics that had to be covered to achieve the breadth and depth required by the treatise, and they know who were the best of their colleagues to write on each subject. The list of chapter authors is an impressive one, consisting of geophysicists who have made major contributions to their fields of study. The quality and coverage achieved by this group of editors and authors has insured that the treatise will be the definitive major reference work and textbook in geophysics.

Each volume of the treatise begins with an 'Overview' chapter by the volume editor. The Overviews provide the editors' perspectives of their fields, views of the past, present, and future. They also summarize the contents of their volumes and discuss important topics not addressed elsewhere in the chapters. The Overview chapters are excellent introductions to their volumes and should not be missed in the rush to read a particular chapter. The title and editors of the 10 volumes of the treatise are:

Volume 1: Seismology and Structure of the Earth

> Barbara Romanowicz
> University of California, Berkeley, CA, USA
> Adam Dziewonski
> Harvard University, Cambridge, MA, USA

Volume 2: Mineral Physics

> G. David Price
> University College London, UK

Volume 3: Geodesy

> Thomas Herring
> Massachusetts Institute of Technology, Cambridge, MA, USA

Volume 4: Earthquake Seismology

> Hiroo Kanamori
> California Institute of Technology, Pasadena, CA, USA

Volume 5: Geomagnetism

> Masaru Kono
> Okayama University, Misasa, Japan

Volume 6: Crust and Lithosphere Dynamics

> Anthony B. Watts
> University of Oxford, Oxford, UK

Volume 7: Mantle Dynamics

> David Bercovici
> Yale University, New Haven, CT, USA

Volume 8: Core Dynamics

> Peter Olson
> Johns Hopkins University, Baltimore, MD, USA

Volume 9: Evolution of the Earth

> David Stevenson
> California Institute of Technology, Pasadena, CA, USA

Volume 10: Planets and Moons

> Tilman Spohn
> Deutsches Zentrum für Luft-und Raumfahrt, GER

In addition, an eleventh volume of the treatise provides a comprehensive index.

The *Treatise on Geophysics* has the advantage of a role model to emulate, the highly successful *Treatise on Geochemistry*. Indeed, the name *Treatise on Geophysics* was decided on by the editors in analogy with the geochemistry compendium. The *Concise Oxford English Dictionary* defines treatise as "a written work dealing formally and systematically with a subject." Treatise aptly describes both the geochemistry and geophysics collections.

The *Treatise on Geophysics* was initially promoted by Casper van Dijk (Publisher at Elsevier) who persuaded the Editor-in-Chief to take on the project. Initial meetings between the two defined the scope of the treatise and led to invitations to the editors of the individual volumes to participate. Once the editors were on board, the details of the volume contents were decided and the invitations to individual chapter authors were issued. There followed a period of hard work by the editors and authors to bring the treatise to completion. Thanks are due to a number of members of the Elsevier team, Brian Ronan (Developmental Editor), Tirza Van Daalen (Books Publisher), Zoe Kruze (Senior Development Editor), Gareth Steed (Production Project Manager), and Kate Newell (Editorial Assistant).

G. Schubert
*Editor-in-Chief*

.

# Contributors

D. Bercovici
*Yale University, New Haven, CT, USA*

A. Davaille
*IPGP and CNRS, Paris, France*

G. Ito
*University of Hawaii, Honolulu, HI, USA*

C. Jaupart
*Institut de Physique du Globe, Paris, France*

S. D. King
*Virginia Tech, Blacksburg, VA, USA*

S. Labrosse
*Institut de Physique du Globe, Paris, France*

A. Limare
*IPGP and CNRS, Paris, France*

J.-C. Mareschal
*GEOTOP-UQAM-McGill, Montreal, QC, Canada*

L. N. Moresi
*Monash University, Clayton, VIC, Australia*

E. M. Parmentier
*Brown University, Providence, RI, USA*

N. M. Ribe
*Institut de Physique du Globe, Paris, France*

Y. Ricard
*Université de Lyon, Villeurbanne, France*

P. J. Tackley
*Institut für Geophysik, ETH Zurich, Switzerland*

P. E. van Keken
*University of Michigan, Ann Arbor, MI, USA*

D. A. Yuen
*University of Minnesota, Minneapolis, MN, USA*

S. J. Zhong
*University of Colorado, Boulder, CO, USA*

# 1 Mantle Dynamics Past, Present, and Future: An Introduction and Overview

**D. Bercovici**, Yale University, New Haven, CT, USA

## 1.1 Introduction

Much of what we refer to as geology, or more accurately geological activity on Earth, is due to the simple act of our planet cooling to space. What allows this activity to persist over the lifetime of the solar system is that the major and most massive portion of the planet, namely the mantle, is so large, moves so slowly, and cools so gradually that it sets the pace of cooling for the whole Earth. If the Earth's other components, such as the crust and core, were allowed to lose heat on their own, their small size or facile motion would have allowed them to cool rapidly and their activity would have ceased eons ago.

For this reason, the study of the dynamics of the mantle, both its evolution and circulation, is critical to our understanding of how the entire planet functions. Processes from plate tectonics and crustal evolution to core freezing and hence the geodynamo are governed, and in many ways driven, by the cooling of the mantle and the attendant phenomenon of mantle convection, wherein hot buoyant material rises and cold heavy material sinks.

## 1.2 A Historical Perspective on Mantle Dynamics

To some extent the development of the field of mantle dynamics is most closely linked with the history of theories of continental drift and plate tectonics. Although mantle convection was invoked to provide a driving mechanism for continental (or plate) motions, the hypothesis that the mantle flows and circulates predates even that of continental drift (see Schubert *et al.*, 2001, chapter 1). As discussed recently by England *et al.* (2007), John Perry used the notion of mantle convection in 1895 to refute the estimate for the age of the Earth given by his former mentor William Thomson (Lord Kelvin). However, a great deal of progress on understanding mantle convection also comes, obviously, from the general study of the physics of thermal convection, not specifically applied to the mantle.

Histories of plate tectonics (or continental drift) are in abundance (e.g., Menard, 1986; Hallam, 1987) and the recent text on mantle dynamics by Schubert *et al.* (2001) gives an excellent summary of the history of the development of mantle convection theory in conjunction with plate tectonics. However, the historical context and personalities associated with some of the steps in this development are important to understand in terms of how the field evolved, and to some extent how science in general has been done and is done now. Thus, rather than merely repeat other historical summaries here, we will focus instead on the contributions (pertaining primarily to mantle convection) and professional and personal histories of some of the leading names in the development of the theories of thermal convection and mantle dynamics. Roughly keeping with the structure of this volume, this section concentrates on the origins of the physics, theory, and systematic experiments of convection by visiting Benjamin Thompson (Count Rumford), John William Strutt (Lord Rayleigh), and Henri Claude Bénard. This will be followed by reviewing the lives of some of the pioneers of the quantitative analysis of mantle convection as a driving force of 'continental drift' namely Arthur Holmes, Anton Hales, and Chaim Pekeris, and then two leading proponents of convection and its association with the modern theory of seafloor spreading, subduction, and plate tectonics, Harry Hammond Hess and Stanley Keith Runcorn. Apart from hopefully providing an in-depth perspective on the origins of the science of mantle convection, this survey also reveals the rather fascinating historical ties many of these famous characters had with one another; for example that Rayleigh had become a professor in the institution that Rumford established, that Holmes had studied under Rayleigh's son, and that Chaim Pekeris was intimately involved with the birth of the state of Israel that Rayleigh's brother-in-law Arthur Balfour helped create.

### 1.2.1 Benjamin Thompson, Count Rumford (1753–1814)

Benjamin Thompson is perhaps one of the more colorful and complex characters in the history of science (**Figure** 1). He was simultaneously a brilliant observationalist, an egotistical opportunist, and a dedicated social reformer and champion of the poor. His role as spy against the rebelling American colonies on behalf of the British gives him a dubious role in American (although not European) history in that one of the fathers of thermodynamics also played a role not unlike that of Benedict Arnold.

Rumford is primarily known for his work on the theory of heat as motion – leading eventually to the kinetic theory and thermodynamics – and for working to debunk the caloric theory of heat. Histories of convection will often note that Rumford is credited

**Figure 1** Benjamin Thompson, Count Rumford (1753–1814) (Smithsonian Institution Libraries Digital Collection).

as possibly being the first to observe convection; in fact, the study of the mass transport of heat was a significant part of his overall body of work (Brown, 1967), and he wrote an important article on convection in 1797, although the use of the word 'convection' was not coined until much later, by Prout in 1834 (see Schubert *et al.*, 2001).

Benjamin Thompson was born in Woburn Massachusetts in 1753 to a line of Thompsons that can be traced back to a James Thompson who arrived 10 years after the landing of the *Mayflower* (1620), along with eventual Massachusetts Governor John Winthrop. Thompson's father and grandfather were reasonably wealthy farmers, but his father died young when Benjamin was less than 2 years old. The family farm was inherited by Benjamin's uncle who appears to have treated his nephew well with a significant income, a portion of land, and a high-quality education. As with all the subjects of our histories here, Thompson was a brilliant student, displaying talents in mechanics and natural philosophy; however, he was also known for being a somewhat spoiled child at his family's farm.

Thompson left school at age 13 for an apprenticeship in retail, but continued his studies independently in engineering, mathematics, medicine, experimental philosophy, along with French, fencing, music, and draftsmanship. He also carried out independent experiments in science, including astronomy, engineering, anatomy, and nearly electrocuted himself trying to repeat Benjamin Franklin's experiments on thunderstorm electrification.

At age 18, Thompson set out to generate much-needed income and turned to tutoring the children of local wealthy families, which led to his being invited by the Reverend Timothy Walker of Concord, New Hampshire, to run a school in his village. Concord was originally known as Rumford and it is from this town that Thompson was to derive his name upon being ennobled. Thompson courted The Reverend Walker's daughter, Sarah, who had earlier married a much older wealthy landowner who died after one year of marriage. Less than a year after Sarah was widowed, Thompson married her in 1772 when he was 19. Thompson's new wealthy wife facilitated his connections with the British governing class, in particular by dressing him up in a fine hussar uniform and parading him about Boston, where he made such an impression on Governor John Wentworth that he was given a British major's commission in the 2nd New Hampshire Regiment. In 1774, Thompson and his wife had a daughter – also Sarah.

By this time hostilities between the Colonies and the Crown had been mounting; these included the Stamp Act and subsequent riots (1765), the Boston Massacre (1770), and the Boston Tea Party (1773), which was followed by both the relocation of the Massachusetts capital from Boston to Salem and the passing of a series of acts, called, by the colonists, the Coercive or Intolerable Acts. These events had by now led to the First Continental Congress in Philadelphia in September 1774, which demanded repeal of these acts, as well as calling on civil disobedience and the buildup of local militia called the Minutemen. Eventually war broke out near Boston in April 1775, at Lexington and Concord.

Benjamin Thompson's acceptance of a British commission in the 2nd New Hampshire Regiment was thus problematic for several reasons. While British officers resented the presence among their ranks of an inexperienced schoolmaster, the people of Concord and Woburn regarded him as a traitor and he had to face two trials for "being unfriendly to Liberty," in 1774 and 1775, both of which were weathered without formal charges. Nevertheless, Thompson left for Boston in 1775 to offer his services to the British Army, in particular to gather intelligence on rebels by various and nefarious means. In March 1776, the British evacuated Boston and Thompson left with them, abandoning his wife and daughter; he was never to see his wife again, and his daughter not again for 20 years.

In England, Thompson worked for the Secretary of State for the Colonies, and he rose rapidly to other prominent positions in the administration of colonial rule. During this time, around 1778, his studies of force and heat associated with gunpowder explosions and ballistics of large guns led to his election as Fellow of the Royal Society in 1781.

Thompson also briefly went back to the American colonies in 1781 to take up command in the King's American Dragoons in New York. With the end of the war shortly thereafter, Thompson returned to England as a professional soldier and colonel, and went to the continent as a soldier of fortune in 1783, landing in Bavaria, eventually to become the advisor to Elector Karl Theodor. As a consequence of his new employment, he was knighted by the British King George III in 1784 to secure his loyalty to England. It was in Bavaria, with the financial and technical backing of his court position, that Thompson did much of his scientific work, especially on insulating properties of materials and on transport of heat by fluids (i.e., convection). He also did much in terms of military, educational, social, and economic reform in Bavaria, all for which he was rewarded by being made a Count of the Holy Roman Empire in 1792; he assumed the title of Count Rumford, adopting the original name of the New Hampshire village where his rapid climb began.

Rumford's frenetic activity and modern innovations met with great resistance by the established order who eventually held sway with the Elector and caused Rumford to be removed back to London as a Minister to the English Court. However, due to a miscommunication with the British Crown, he arrived jobless in London. Unable to remain idle, Rumford established the Royal Institution in London which employed natural philosophers like Thomas Young and Humphrey Davy to perform experiments and give public lectures. However, his dominating style and large ego led to battles with institute managers and he eventually left both the Royal Institution and England in 1802. He flitted for political reasons between Bavaria and France, until the invasion of Bavaria by Austria in 1805 sent him permanently to France, where he hoped to gain favor with Napoleon Bonaparte. In France, he married the widow of the "father of chemistry" Antoine Lavoisier, who had been sent to the guillotine at the height of the French Revolution in 1794. However, in France, Rumford fought scientifically with Laplace and Lagrange and separated from Madame Lavoisier de Rumford. He died in August 1814, and was buried in the village of Auteuil.

Rumford was of course known widely for his philosophic papers on the nature of heat and mounted perhaps the most coherent assault on the caloric theory. The caloric theory is – as with many failed theories – frequently explained with hindsight as an absurdity. However, it is important to understand that it was a reasonably sophisticated physical and mathematical theory that made distinct quantitative predictions about the nature of both heat and matter that were in fact borne out by experiments. Caloric was considered the medium or 'fluid' that transported and retained heat within matter. But more specifically, caloric was thought to provide the repulsive force – by providing a caloric atmosphere around each atom – within matter to keep it from collapsing under the influence of gravitational attraction. The mathematical predictions of the caloric theory were borne out by the experiments of Dulong and Petit on, for example, thermal expansion. Similarly, predictions about dependence of specific heat on temperature, phase changes (in which caloric is either absorbed or released), adiabatic heating under pressurization (wherein caloric is squeezed out), and heat conduction as driven by caloric potential, all verified the caloric theory. Indeed, the notion of caloric substance remains a relic in our discussion of heat as 'flowing', or existing as quantities with certain densities such as specific heat or latent heat (Brown, 1967).

Thus, Rumford's assault on caloric theory was not on a ludicrous model that could be easily dispatched by simple thought experiments. However, it did conflict with certain observations, a few key ones Thompson was responsible for. Most notably, these included his work on frictional heating, in particular his observation that during work on boring out cannons, heat is created with no evident source of caloric from other matter; and also the propagation of heat in a vacuum by radiation (as caloric was thought to exist only in proximity to atoms).

*Rumford's observations of thermal convection.* Rumford's work on thermal convection was a rather famous paper that was published in multiple venues in 1797 (Thompson, 1797; Brown, 1967) during his tenure in Bavaria. Much of this work involved the investigation of the insulating properties of matter, which he was investigating as part of an effort to improve cold-weather clothing for field soldiers, as well as the poor and underclassed. But this work was also relevant in his attack on caloric theory in that it showed that heat could be transported by the motion of mass itself, and if such mass flux did not occur the material was an insulator (or nearly so). This also prompted him to note that the internal motion of fluid particles also provided a ready source of heat and this was a precursor to his theory of heat as particle motion.

Rumford's most explicit observation of convection itself occurred while examining the "communication of Heat" by taking oversized thermometers filled with liquids (including alchohol derived from wine) and subjecting them to heating. Upon leaving one thermometer in a window to cool, Rumford noticed rapid fluid motion that was well delineated by dust particles (which had been introduced because he had let the tubes remain open for two years without cleaning) illuminated by the sunlight:

> I observed an appearance which surprised me, and at the same time interested me very much indeed. I saw the whole mass of the liquid in the tube in a most rapid motion, running swiftly in two opposite directions, up and down at the same time (Thompson, 1797).

On closer inspection, Rumford noticed a regular circulation of upwelling along the axis of the tube and downwelling along the cooler glass boundary. He further found that dousing the tube with ice water hastened the motion, but that the circulation eventually ceased as the entire thermometer reached room temperature.

Rumford reasoned that heat in liquids and gases was only carried by particle motion and to prove this he contrived experiments in which the fluid was subjected to thermal gradients but the particle motion was obstructed or 'embarrassed'. He compared heat transport in water, when pure and when mixed with substances such as eider (duck) down and stewed apples, which would "impair its fluidity." Rumford concluded that, "Heat is propagated in water in *consequence* of internal motions or that it is transported or carried by the particles of that liquid...." The manner in which heat transport can be inhibited by impaired fluidity led Rumford to speculate on the role of convective heat transport in nature, in particular that impaired fluidity was God's design, for example observing that

> ... when we advert to the additional increased viscosity of the [tree] sap in winter, and to the almost impenetrable covering for confining Heat which is formed by the bark, we shall no longer be at a loss to account for the preservation of trees during winter ... (Thompson, 1797)

Rumford had in fact noted an important feature of convection in many natural systems, viz. the concept of self-regulation of heat during convective heat transport. This is something considered also in mantle convection, in particular that high temperatures lead to low viscosity and rapid convective expulsion of heat, while low temperatures lead to greater effective insulation and eventual buildup of heat.

Count Rumford was in every sense an experimentalist and observationalist, but his contributions to the study of heat marked the initiation of the modern theory of classical thermodynamics, kinetic theory, and heat transport by particle motion both microscopically and through thermal convection.

Further reading, including sources for this brief history, are Brown (1967) and Brown (1999).

### 1.2.2 John William Strutt, Lord Rayleigh (1842–1919)

The contributions that John William Strutt (**Figure 2**) made to various fields of physics is evident in the number of phenomena named for him, including

**Figure 2** John William Strutt, Third Baron Rayleigh (1842–1919) (AIP Emilio Segre Visual Archives, Physics Today Collection).

Rayleigh scattering, Rayleigh–Jeans criterion, Rayleigh waves, and, within the study of convection, the Rayleigh number. His greatest legacy is perhaps his contribution to the theory of waves in solids and fluids; however, he received the Nobel Prize in Physics for the discovery and isolation of atmospheric argon, which was essentially experimental chemistry.

Although Strutt was one of the few noble scientists actually born into nobility (as opposed to say Benjamin Thompson who was ennobled in his own life), the Strutts were not from ancient noble stock. The family could be traced back to 1660, where it was known for milling corn with water-driven mills, a business that eventually established the family's financial standing. In 1761, John Strutt (Rayleigh's great-grandfather) purchased what was to become the family estate of Terling Manor, and where Rayleigh would eventually build his laboratory in which he did much of his experimental work. John Strutt's oldest surviving son Joseph Holden was a colonel in the West Essex Militia, and a member of parliament (like his father); for his public services, he was offered a peerage by George III. However, for not entirely clear reasons, other than an apparent conviction not to accept personal honors, Joseph Strutt declined the peerage and asked that it be bestowed on his wife Lady Charlotte

(Fitzgerald) Strutt, whom he had married in 1789. When she died in 1836, the peerage went to Joseph's only son John James, who became the Second Baron Rayleigh, during his father's own lifetime. The name Rayleigh was in fact after a small market town in Essex (and possibly for no particularly deep reason other than its noble sound). John James married Clara Vicars in 1842 (nearly 30 years his junior); in biographies of Rayleigh, it is usually claimed that his scientific and mathematical talent came not from the Strutts but from the Vicars, who boasted several members of the Royal Engineers and direct descendancy from the brother of physicist Robert Boyle. In November of the year of their marriage, John William Strutt, oldest son and heir to the barony, was born, prematurely. As a child he did not speak properly till past the age of 3, prompting his grandfather Colonel Strutt to remark that "That child will either be very clever or an idiot" (Strutt, 1968).

John William suffered poor health throughout his youth and had to be withdrawn from school twice to be tutored privately. He entered Trinity College Cambridge in 1861 where he studied rigorous mathematics with, for example, Sir George Stokes, the Lucasian Professor of Mathematics, and graduated with honors in 1865. Immediately after graduating Strutt opted not to go on the traditional grand tour of the Continent, but rather visited the United States, which was then in the throes of post-Civil War Reconstruction. Upon his return to America, he purchased experimental equipment to set up his own laboratory. There was at the time no formal university physics laboratory in Cambridge; although much experimental work had already been done by the likes of Michael Faraday and Humphrey Davy, most of this was outside the university (in particular at the Royal Institution, which, as noted above, was founded by Count Rumford). It was not until 1871 that James Clerk Maxwell was named the First Cavendish Professor of experimental physics, and not till 1873 that the Cavendish labs were built.

Strutt's first paper in 1869 was on Maxwell's theory of electromagnetism, for which he received much encouragement from Maxwell himself. In 1871, he married Evelyn Balfour, the sister of Arthur James Balfour. (Arthur James Balfour became British Prime Minister from 1902 to 1905 and was the author of the Balfour Declaration (1917) in which Britain formally supported the creation of a Jewish homeland in Palestine during the partitioning of the Ottoman Empire after World War I.) However, soon after his marriage, Strutt had a seious bout of rheumatic fever,

which prompted a lengthy tour of Egypt; it was during these travels that he did much of his work on his famous book, *The Theory of Sound*. After his return in 1873, he assumed the title of Third Baron Rayleigh on the death of his father, and took up residence in Terling where he built his laboratory in which he did much of his life's scientific work. His book on sound was published in 1877 and 1878 (in 2 volumes), an achievement that emphasized his lifelong fascination with sound and wave theory in general. During these early years, he also continued to work on sound, electomagnetism, light, as well as his theory for light scattering and the cause of the sky's blue color.

James Clerk Maxwell's untimely death in 1879, at the age of 48, left the Cavendish Experimental Chair vacant. The first Cavendish Chair had been offered to Sir William Thomson (later Lord Kelvin) in 1871, but he had refused in order to stay in Glasgow, and his decision remained even upon being reoffered the job in 1879. Rayleigh was then offered the position and reluctantly decided to take it because he needed the money given that revenue from Terling had dropped off during the agricultural depression of the late 1870s. Indeed, Rayleigh only held the position until 1884 after which his economic situation improved and he returned to his residence and work in Terling. His brief tenure as the Cavendish Professor was the only academic position he ever held. However, during his time in Cambridge, he carried out a vigorous program of experimental instruction and research on electrical standards, now having access to lab assistants and more facilities. Also, in his years at Cambridge, he developed a close acquaintance with Sir William Thomson.

After resigning his professorship and returning to Terling, Rayleigh became Secretary of the Royal Society for the next 11 years (1885–96). In 1887, he became professor of natural philosophy at the Royal Institution (see Section 1.2.1), at which he gave over a hundred lectures until he left it in 1905.

One of the findings that Rayleigh is most noted for is his discovery of argon in the atmosphere. Rayleigh had first suggested the presence of an unknown atmospheric gas because of an apparent anomaly in the density of atmospheric nitrogen relative to that extracted from compounds. Later, he did careful and difficult experiments separating argon from atmospheric gas in 1895. William Ramsay, following up on Rayleigh's earlier suggestion, also conducted experiments to extract argon. Some controversy ensued over whether Ramsay had a right to follow Rayleigh's suggestion and infringe on his research. Nevertheless, the two shared priority of the discovery for which they each received the Nobel Prize in 1904, Rayleigh's in physics and Ramsay's in chemistry.

Rayleigh became President of the Royal Society from 1905–08, and took on numerous public service roles, chief of which was Chancellor of Cambridge University in 1908. He published his complete papers up to 1910 in 5 volumes; the remaining papers from 1911–19 were published posthumously under the editorship of his son Robert John (who was himself a prominent physicist; see Section 1.2.4) in 1920. Rayleigh was author of approximately 450 papers, working essentially up to his death in June 1919; indeed three of his papers were still in review or in press on the day of his death.

*Rayleigh's linear theory of convection.* Lord Rayleigh's paper on convection (Strutt, 1916) was written very late in his life, in 1916, only three years before his death. He had, however, worked sporadically on the stability of fluid flows since at least the 1880s. Rayleigh's paper on convection was inspired by Henri Bénard's experiments which had taken place 15 years earlier (Section 1.2.3). In his 1916 paper, Rayleigh described Bénard's experiments, recognizing two phases, which were an initial transient phase in which an irregular or semiregular pattern is established (with polygons of four to seven sides) and then a second phase of stable and regular polygonal patterns. In a footnote, he also commented that Bénard was perhaps unaware of the work of James Thomson (Lord Kelvin's older brother), who did fundamental work on evaporative convection finding similar polygonal patterns (see Berg *et al.*, 1966). Rayleigh's theory was in fact focused on the first phase, or the transient onset; he did not recognize in the second phase what was later termed 'exchange of stabilities' in that the state to which the perturbations are moving the system is itself stable and nonoscillatory (see Chandrasekhar, 1961). Rayleigh followed a first-order pertubation analysis, although in a rather informal manner. He explored the effect on stability of various individual parameters (e.g., viscosity (considering both inviscid and viscous cases), thermal diffusivity, sign of imposed temperature gradient, gravity, and thermal expansivity), but did not recognize dynamic similarity and dependence on a few dimensionless numbers (*see* Chapter 4). Although he did not pose the dimensionless number called the Rayleigh number, it is easily recognizable in his solution for the minimum critical temperature gradient or 'density drop' across the layer (Strutt, 1916, equations (44) and (46)). Rayleigh also commented on the degeneracy of the linear problem (i.e., mode selection only in

terms of wave number squared). However, he also recognized that the basic differential operators in the equations allowed wave number pairs that permitted regular polyhedral patterns such as squares and rolls, although he admitted that while hexagons and triangles were obvious, the problem was not immediately tractable (it was, however, later solved analytically; see Chandrasekhar (1961)). Bénard's hexagonal patterns were cause for some confusion in the comparison of theory and experiment, and they were later inferred to be characteristic of systems where vertical symmetry of the convecting fluid layer is broken (e.g., with temperature-dependent viscosity; see Manneville (2006) and Section 1.2.3).

Lord Rayleigh's paper is of course one of the great seminal works in the study of convection, not just for establishing the theoretical framework of the problem and developing the concept of marginal stability, but also for inferring pattern selection at convective onset. Further reading and sources for this history can be found in Strutt (1968) and Lindsay (1970).

### 1.2.3 Henri Claude Bénard (1874–1939)

While the phenomeon of thermal convection had been observed at least as early as the accounts of Count Rumford (Section 1.2.1). Henri Bénard (**Figure 3**) is widely recognized for having done the first systematic quantiaive examination of natural cellular convection. Even given all the problems later found with his experimental results, he is justifiably recognized as the father of experimental studies of convection, which later inspired decades

**Figure 3** Henri Claude Bénard (1874–1939). From Wesfried J (2006) Scientific biography of Henri Bénard (1874–1939). In: Mutabazi I, Wesfried J, and Guyon E (eds.) *Dynamics of Spatio-Temporal Cellular Structure – Henri Bénard Centenary Review*, pp.9–37. New York: Springer. (fig 2.3, p.15), with kind permission of Springer Science and Business Media.

of work not only on convection but also on self-organizaton and critical phenomena (Wesfried, 2006; Manneville, 2006).

Henri Claude Bénard was born in Lieurey, a small village in Normandy, in October of 1874. His father was a financial investor who died when Bénard was young. After attending schools in Caen and later Paris, Bénard was, in 1894, accepted into the highly competitive Ecole Normale Supérieure de Paris, one of the French Grand Ecole's, in a year with about 5% acceptance rate in the sciences. While at the ENS-Paris, he studied with some very notable classmates, for example, physicist Paul Langevin and mathematician Henri Lebesgue, and was witness to the groundswell of outrage over the Dreyfus affair. In 1897, he received the *agrégé de physique* (now termed *agrégation*, whereby state teaching credentials are obtained) and began work in the department of experimental physics in the Collége de France, which is unattached to any university and does not grant degrees, but provides open public lectures (similar to the Royal Institution). While there, Bénard was assistant to Eleuthére Mascart and Marcel Brillouin, who were, respectively, grandfather and father of the physicist Leon Brillouin (Wesfried, 2006). While at the Collége de France, he worked on the rotation of polarized light through sugar solutions, which gave him training in the use of optics for measurements of fluid motion; this specific work also led to the second topic of his PhD thesis, which was then required for the French doctorate.

The experiments on convection for which Bénard is well known today (Bénard, 1900, 1901) were the primary part of his PhD thesis, but the topic was arrived at much by accident (see below). In defense of his thesis on 15 March 1901, his committee found the work satisfactory but less than inspiring, criticizing Bénard because, while it was innovative (mostly interesting in its application of optical methods) and above average, it was disappointing in that it provided no general theoretical development to explain the experiments. The report on his defense stated that "though Bénard's main thesis was very peculiar, it did not bring significant elements to our knowledge. The jury considered that the thesis should not be considered as the best of what Bénard could produce" (Wesfried, 2006).

After his thesis in 1900, he briefly settled in Paris, got married (but had no children), and was shortly thereafter appointed to the Faculty of Sciences in Lyon (1902). In Lyon, Bénard carried out much of his well-known work on vortex shedding around

bluff (prismatic) bodies, and further developed his ingenious employment of cinematography in laboratory experiments. In 1910, he was appointed to the Faculty of Science at the University of Bordeaux in physics under the department head Pierre Duhem.

With the outbreak of war in 1914, and as a former student of the Ecole Normale, which carried with it state obligations, Bénard entered the military and was made an officer. He was put on a military scientific commission wherein he worked on problems relating to food refrigeration under transport, and, later, on the military use of optics (e.g., using polarized light for tracking ships and submarines, and for improving periscopes).

After the war, Bénard returned to science, and in 1922 he moved to Paris as professor at the Sorbonne University. In 1928, he became the President of the French Physical Society. In 1929, he participated in the development of the Institute of Fluid Mechanics, and in 1930 became professor of experimental physics. For the next decade he continued to work with various students on convection and vortex shedding. He died on 29 March 1939, slightly shy of 65.

*Bénard's experiments on convection* Henri Bénard's first observations of cellular convective motion came in 1898, at the Collége de France, while trying to make a coherer of solid dielectrics. (A coherer is a loose, often granular, agglomeration of conductors or semi-nconductors whose conductivity is affected by impingement of radio waves.) In the preparation, he noticed a polygonal pattern in melted paraffin that had graphite dust in it. From there, Bénard detoured into a painstaking and systematic study of convection, in particular the difficult task of finding the onset of convection as near to conductive stability as possible. To carry out this task meant eliminating minute thermal fluctuations imposed at the boundaries, and so Bénard constructed an apparatus comprised of a metal container with steam circulation in its walls to provide a nearly uniform isothermal bottom boundary; the top however was exposed to air. Bénard observed different patterns of convection cells which had polygonal structure of four to seven sides (see also Section 1.2.2, as well as Chapter 3), but were predominatly hexagaonal. He termed these cells *tourbillons cellulaires* or cellular vortices, although we now refer to them as Bénard cells or convective cells. Bénard was able to measure closed streamlines of particle flow, and observed an initial transient state of polygon formation, settling down to a stable hexagonal configuration after some time, which allowed him to measure cell sizes accurately. He also observed convective rolls (or *tourbillons en bandes*, what Rayleigh later referred to as striped vortices), which occurred at low heat flux, as well as high-heat-flux turbulent 'vortex worms'. What is highly notable is the large number of innovative optical techniques Bénard used and developed. He not only introduced the use of cinematography, but also particle trajectories, interference fringes due to light reflected off hills and valleys on the convectively warped free surface, and also light transmission across the fluid layer, which is essentially the same as the shadowgraph technique commonly used today (*see* Chapter 3). The combination of these optical effects in fact allowed him to estimate isotherms with (quite impressively) 0.1°C contours.

Years later, in 1916, Rayleigh analyzed Bénard's experiments (see Section 1.2.2), but assumed free-slip top and bottom boundaries (which was an analytically tractable configuration of boundary conditions). Rayleigh's work confirmed some of the patterns observed by Bénard, but not the critical conditions for convective onset (what we now call the critical Rayleigh number) due to inappropriate boundary conditions. Because of World War I, Bénard was not aware of Rayleigh's work until the late 1920s, well after Rayleigh's death. In the late 1920s and early 1930s, Bénard compared his experiments to the theoretical work of Sir Harold Jeffreys who had repeated Rayleigh's stability analysis but with the appropriate mixed boundary conditions of a free-slip top and a no-slip bottom (see Wesfried, 2006). Bénard found the patterns predicted in Jeffrey's work matched some of his experiments, but the conditions for convective onset still did not agree, implying a significant disparity between theory and Bénard's experiments.

Bénard was aware at the time of his first experiments of problems inherent with an open surface, but he was mainly concerned with the fact that some working fluids, especially volatile ones such as alcohol, experienced evaporative convection, which is often coincident with Marangoni (surface tension driven) convection (see Berg *et al.*, 1966). Bénard thus used fats or oils (i.e., spermacetti) and wax that had higher melting temperature and thus lower vapor pressures. Block (1956), however, repeating Bénard's experiments, later suggested that thermally induced surface tension gradients, rather than thermal buoyancy, were the cause for the observed motions and surface deflection; surface tension-driven convection, now known as Marangoni–Bénard convection, was formally developed by Pearson (1958).

Throughout his life, Bénard continued to make analogies between the cells of his experiments and natural ones, in some cases incorrectly (e.g., Taylor–Couette rolls). However, he correctly advocated the cellular cause of solar granulation. Moreover, he promoted the idea that cloud streets were due to longitudinal convective rolls aligned parallel to wind; he directed experiments on convection in a tilted layer, and ones with a moving top boundary, to find alignment of convection rolls (see Wesfried (2006) and references therein), essentially identical to what it is referred to in mantle dynamics as Richter rolls. Additional reading and historical sources for this brief history are Wesfried (2000) and Mannville (2006).

### 1.2.4   Arthur Holmes (1890–1965)

The prospect of convective heat transport in the mantle was alluded to in the early to mid-nineteenth century (e.g., Schubert *et al.*, 2001; England *et al.*, 2007). However, Arthur Holmes (**Figure 4**) can rightly be considered one of the founders of the physical theory of mantle convection as it pertains to the driving mechanism of continental drift. Even so, Holmes is

**Figure 4**   Arthur Holmes (1890–1965). Used with permission from the Arthur Holmes Isotope Geology Laboratory at Durham University.

still perhaps most well known for championing the science of radiometric dating to infer the age of the Earth, and for establishing the geologic timescale (Lewis, 2000).

Arthur Holmes was born in Gateshead, in the northeast of England, in 1890. As a precocious teenage student, he was strongly influenced by a teacher who introduced him to both physics and geology through the writings of William Thomson (Lord Kelvin) and the Swiss geologist Edward Suess. At the age of 17 he went to London to study physics at the Royal College of Science, which was then being absorbed into the new Imperial College London, but he then changed directions to geology. However, his background and interest in physics would serve his geological ventures throughout his life. In his final year at Imperial College he studied the novel and exciting phenomenon of radioactivity, and the prospect of radioactive dating of rocks, under the new young professor Robert John Strutt, the son of Lord Rayleigh. Strutt had himself been involved in a vigorous and public debate over the age of the Earth with his father's old friend, Lord Kelvin (Lewis, 2000).

The debate over the age of the Earth is traditional historical fare in gesoience (see Lewis, 2000). Suffice it to say that the controversy swirled about both Bishop Usher's biblically inferred age of 6 ky and Lord Kelvin's cooling age of 20 My, neither of which could be reconciled with geologists' observations of sedimentation and speciation rates that required the Earth to be no less than 100 My old. However, the discovery of radioactivity in rocks and the resolution of radioactive half-lives suggested a nearly direct measure of rock ages. (It also provided a heat source for keeping the Earth from having to cool from a recent molten state.) Several scientists attempted to develop the technique of radiometric dating, including R. J. Strutt and Ernest Rutherford. But, in the end, Yale scientist Bertram Boltwood identified lead as the final product of the uranium–radium decay series; since lead was a nonvolatile and thus a nonleaking daughter product – unlike helium – its concentration could be reliably measured. Boltwood used this decay series to infer the age of rocks in both Connecticut and Ceylon (now Sri Lanka) and determined ages of these rocks between 535 My and 2.2 Gy and published these results in 1907 (see Turekian and Narendra, 2002). Using Boltwood's method, Holmes similarly dated Devonian rocks and arrived at an age of 370 My; his results were presented (in his absence) at the Royal Society in 1911.

By 1911, Holmes was already in Mozambique where he had taken a job in mineral prospecting because his scholarly stipend in London was not enough to live on. However, he was there only 6 months, having fallen seriously ill with malaria. He returned to London to take a job as demonstrator in the Imperial College where he continued to push the geochronological methods. In 1913, at the age of 23, Holmes published his first book, *The Age of the Earth*, where he calculated the planet's age to be 1.6 Gy old, which was less than Boltwood's estimate. Although geologists were, on the whole, relieved that the 6 ky and 20 My ages were proven wrong, many still held fast to the 100 My date and were reluctant to accept the radiometrically inferred ages, probably because many did not understand the new physical principles of radioactivity (Lewis, 2000, 2002).

Holmes married Margaret (Maggie) Howe in 1914 and he continued working as a demonstrator in the Imperial College through World War I. After the birth of his son, Norman, in 1918, Holmes found that he was unable to support his small family on his demonstrator's salary, and thus took a job prospecting for oil with a company in Burma in 1920. However, the company soon went bankrupt. Moreover, Holmes' 4-year-old son Norman fell ill with dysentery and, despite available medical attention, died in 1922. Holmes and his wife soon returned to England, impecunious and grieving. His misfortunes continued, however, and he could not find a position until 1924 (in the meantime running a curio shop for income), when he was offered a professorship at the University of Durham to build a new geology department.

Holmes remained at Durham for nearly 20 years. In that time, he produced his now famous papers on mantle convection as the cause for continental drift (see below) and wrote his text *Principles of Physical Geology*, the first edition of which appeared in 1944. However, also during those years, Holmes' wife Maggie died (in 1938), but he was soon thereafter remarried to Doris Reynolds, a fellow geologist.

Holmes left Durham in 1943 to assume the Regius Professorship at the University of Edinburgh where he remained until retirement in 1956. While at Edinburgh, he continued to work on refining the age of the Earth and developing the geologic timescale, in addition to other geological pursuits. He received various high honors (e.g., the Wollaston and Penrose medals in 1956; and Vetlesen prize in 1964), primarily for his work on the geologic timescale, not on continental drift and convection. In the early 1960s, evidence for plate tectonics was mounting, especially with the Vine–Matthews work in 1963 on seafloor spreading. Holmes' sense of vindication is evident in his revised text, which was published the year he died, in 1965.

*Holmes and mantle convection.* By the late 1920s and early 1930s, the debate over the age of the Earth had given way to a new controversy over the theory of continental drift. As is well known, the German meteorologist Alfred Wegner had proposed his idea based largely on geographical evidence (Wegener, 1924; Hallam, 1987). But he also proposed that continents plowed through oceanic crust like ships, and the driving force was due to centrifugal effects. Although Sir Harold Jeffreys had done some of the most fundamental work on convection theory (see Schubert *et al.* (2001) and Section 1.2.3), he had argued that there were no available forces sufficient to deform the Earth's crust during continental drift. Holmes, on the other hand, supported the idea of continental drift but, along with Bull (1931), proposed that subsolidus convection in the mantle – powered by heat production from radioactive decay – was instead the driving mechanism for continental breakup, seafloor formation (not spreading), crustal accumulation at convergence zones, and continental drift (Holmes, 1931, 1933). Holmes' ideas of subsolidus convection were similar to present-day understanding. Some notable differences his model has with contemporary mantle convection theory are that Holmes believed that mantle flow would establish jets and prevailing winds as in the atmosphere, although these in fact arise through the combination of convection and planetary rotation; as discussed in Chapter 2, the effects of rotation are not significant in mantle circulation. Moreover, Holmes also proposed that seafloor formation was associated with deep and active mantle upwellings, which was later proved to be unlikely even as early as the 1960s (see Section 1.2.7). Holmes' theories of convection were, like Wegener's theory of continental drift, rebuffed and ignored, although Holmes continued to teach these ideas while at Durham and Edinburgh; indeed, his famous text contains a final chapter discussing his view of continental drift and convection.

Further reading and sources for this section can be found in Hallam (1987), Lewis (2000, 2002), and Schubert *et al.* (2001).

## 1.2.5   Anton Hales and Chaim Pekeris

Arthur Holmes is largely seen as a visionary in being the champion of convection as the driving mechanism for continental drift. However, near the same time as his first papers, two important papers on mantle convection were published, that is, by Anton Hales (Hales, 1936) and Chaim Pekeris (Pekeris, 1935). Both of these used modern fluid dynamic theory to estimate not only the conditions for convection, as Holmes had done, but also to calculate the finite-amplitude velocity and stresses of convective currents, and to compare them with predictions from gravity observations. Not only were their theories fluid dynamically sophisticated, but their predictions were borne out 30 years later in measurements of plate motions. Moreover, their respective papers were precursors to the modern analysis of how convection is reflected in gravity, geoid, and topography. Thus these two authors warrant some discussion. Both Hales and Pekeris were perhaps better known for their lifelong contributions outside of mantle dynamics (e.g., seismology) but they both played important roles in the growth of geophysics in the twentieth century.

### 1.2.5.1   Anton Linder Hales (1911–2006)

Anton Hales (**Figure 5**) was born in Mossel Bay, in the Cape Province of South Africa in March 1911. As with all our historical subjects, Hales showed an early aptitude and talent for science and graduated from the University of Capetown with a BSc in physics

**Figure 5**   Anton L. Hales (1911–2006). Used with permission from the Department of Geosciences of the University of Texas at Dallas.

and mathematics, at the age of 18, and a MSc at 19. He then, in 1931 at the age of 20, took up a post as a junior lecturer in mathematics at the University of Witwatersrand in Johannesberg (Lilley, 2006). However, after only one more year, he received a a scholarship to study at Cambridge. Although intending to study quantum mechanics, he was convinced by Sir Basil Schonland (a senior lecturer in physics at the University of Capetown) that this was the "wrong choice" (Lambeck, 2002) and that he should instead study geophysics. Thus, while in Cambridge, Hales studied with Sir Harold Jeffreys and interacted with Keith Bullen, and received his BA in mathematics from St. John's College in Cambridge in 1933.

Hales returned to South Africa where he resumed his post as junior lecturer and eventually senior lecturer in applied mathematics at the University of Witwatersrand. While there he also carried out research primarily in seismology for which he received his PhD from the University of Capetown in 1936. That year Hales married Marjorie Carter with whom he had two sons, James and Peter (Lilley, 2006).

At the outbreak of World War II, Hales scientific career was put on hold and he served as an engineering officer in the North African campaign. After the war, Hales left his lecturer position to become a senior researcher at the Bernard Price Institute for Geophysical Research at the University of Witwatersrand where he worked on development of seismic and gravity measurement methods. However, in 1949, he left BPI for a professorship in applied mathematics and became the head of the mathematics department at the University of Capetown. During these years, he went briefly to Cambridge (1952) to receive a masters degree. In 1954, he returned to Witwatersrand as director of BPI and professor of geophysics. While director of BPI, he continued to push for development of geophysical methods, including paleomagnetism, and he was involved with some of the first measurements anticipating plate tectonic motions by looking at pole paths in South Africa, which began to convince Hales of the validity, after all, of continental drift (Lambeck, 2002). It was also during his time at BPI, in 1957, that Marjorie, his wife of 21 years, died.

In 1962, Hales left South Africa for the United States to become the founding director of the Geoscience Division for the Southwest Center for Advanced Studies, later to become the University of Texas at Dallas. That year he also married Denise Adcock with whom he had two more sons, Mark and Colin. While in Texas, Hales continued to build a powerful research institute and made major contributions to seismic studies of the crust and upper mantle.

In 1973, at the age of 62, Hales was convinced (most notably by Ted Ringwood and John Jaeger (Lambeck, 2002)) to move to the Australian National University (ANU) to become the founding director of the Research School of Earth Sciences (RSES) where he served until 1978. Hales was an active and unique director in that he minimized departmental structure and bureacracy and pushed his scientific staff to work globally rather than on regional Australian studies, thereby establishing the school's reputation as one of the world's foremost Earth science institutes. Moreover, under his directorship, the sensitive high resolution ion microprobe (SHRIMP)-then a very new and expensive technological advance in geochemical analysis – was developed.

Hales retired from ANU in 1978 and returned to the University of Texas as Professor of Geophysics. He retired from the University of Texas shortly thereafter in 1982, and returned to ANU and RSES to resume his position as Emeritus Professor until 2002. By the time of his retirement, he had been made a Fellow of the Royal Society of South Africa, the American Geophysical Union (AGU), and the Australian Academy of Sciences; in 2003, he was given the Centenary Medal from the Australian government. Hales stayed in and around Canberra for the remainder of his life. However, in 2004, his son Mark from his second marriage was tragically killed in a car accident (Lilley, 2006), only two years before Anton Hales himself passed away, in December of 2006, at the age of 95.

*Hale's and convection in the mantle.* Anton Hale's work on the viability of mantle convection occurred during his doctoral studies but was unrelated to his dissertation research; his paper on the subject was published in 1935 before receiving his PhD. The problem he examined, suggested to him by Harold Jeffreys, concerned the plausibility of convection with regard to whether the buoyant stresses driven by a mantle of a certain viscosity and heat input were consistent with those inferred by the gravity anomalies measured during the famous submarine gravity surveys of F. A. Vening Meinesz. To calculate convective stresses, Hales estimated convective velocities by equating mid-mantle advective heat transport with the conductive surface heat transport necessary to remove the net radiogenic heat production. Hales' relationship for his velocities could be shown to yield very plausible tectonic velocities, as mentioned by Jeffreys himself in the later editions of his famous text (Jeffreys, 1959). (Similarly, as shown in Bercovici (2003), a simple balance of the net advective heat extraction by slabs of a given mean temperature anomaly against the known mantle cooling rate predicts quite readily slab velocities of 10 cm yr$^{-1}$.) Hales used these velocities to estimate the stresses of convective currents, and compared these with the stresses necessary to support mass anomalies inferred from gravity measurments. Although Jeffreys no doubt believed that convective stresses in a stiff mantle would be far in excess of those predicted by gravity, Hales showed that the stress estimates were in fact very close and easily permitted a convecting mantle. Jeffreys 'communicated' (i.e., sponsored) his former student's findings to the Royal Astronomical Society for publication (Hales, 1936).

Further reading, including sources for this brief history, are Lambeck (2002) and Lilley (2006).

### 1.2.5.2 Chaim Leib Pekeris (1908–93)

Chaim Leib Pekeris (**Figure 6**) was born in Lithuania, in June 1908 in the town of Alytus, where his father was

**Figure 6**  Chaim L. Pekeris (1908–93). From Gilbert F (2004) Chaim Leib Pekeris, June 15, 1908–February 24, 1993. *Biographical Memoirs* 85: 217–231. National Academy of Sciences. Reprinted with permission from the National Academies Press Copyright 2004, National Academy of Sciences.

a baker. He was the oldest of five siblings, and had two brothers and two sisters. As a youth, he was (unsurprisingly) precocious in mathematics, and was apparently teaching high school math by age 16. In the 1920s, he and his two brothers emigrated to the United States with the help of family and friends already in America (Gillis, 1995; Gilbert, 2004). The three Pekeris brothers became American citizens and continued their education in the US. In contrast, one of his sisters moved to Palestine in 1935 as a Zionist. The remaining sister and their parents were, however, later murdered by anti-Semites in Alytus during the Holocaust.

Chaim Pekeris entered the Massachusetts Institute of Technology (MIT) in 1925 to study meteorology, and obtained his BSc in 1929. He stayed on for graduate work with Carl-Gustave Rossby and obtained his doctorate in 1933. During his graduate career, he was also a Guggenheim Fellow and studied meteorology in Oslo. After finishing graduate studies, he became an assistant geophysicist in the Department of Geology at MIT, and from there rapidly transitioned away from meteorology. He also received a Rockefeller Foundation Fellowship in 1934, and at the same time was married to Leah Kaplan.

Pekeris had been hired at MIT by Louis Slichter who had himself been hired by MIT to establish a geophysics program. Pekeris was his first hire and the second was Norman Haskell, a new PhD from Harvard. The combination of Pekeris and Haskell during the early 1930s made important contributions to the burgeoning problem of continental drift and mantle flow. As is well known, Haskell was to perform the first analysis of Fennoscandian uplift from which the viscosity of the mantle was initially estimated (Haskell, 1937). In conjunction with Haskell's work, Pekeris did a 'hydrodynamic' analysis of thermal convection in the Earth's mantle, which led to an accurate first-order estimate of mantle flow and velocities (see below).

Chaim Pekeris also worked on free oscillations, but in particular of stellar atmospheres, although this laid the foundation for his study of free oscillations of the Earth for which he was perhaps best known (Gilbert, 2004). He further studied pulse propagation and inverse problems in sonar sounding and was promoted to associate geophysicist in 1936. From 1941 to 1945 Pekeris worked for the Division of War Research at the Hudson Laboratories of Columbia University, again studying the propagation of acoustic pulses and waves; he continued there as director of the Mathematical Physics Group (1945–50), and had a joint appointment at the Institute for

Advanced Study at Princeton. For his war research, he was given the title of honorary admiral.

After the war, Pekeris continued his scientific research in several areas, including microwaves, atomic physics, and explosive sound propagation through a fluid–fluid interface, which led to a seminal paper on normal modes and dispersion. He also produced the first theoretical derivation for the critical Reynolds number for onset of instability in pipe flow (Gilbert, 2004).

In addition to his scientific work, Pekeris was involved with assisting in the 'birth' and stability of the new state of Israel by aiding in the transfer of US military surplus to Palestine (Gilbert, 2004). In 1950, Chaim Weizmann, the first President of Israel, convinced Pekeris to move to the Weizmann Institute of Science in Rehovot, Israel, to be the founding chair of the Department of Applied Mathematics.

At that time conditions for the university in Rehovot were extreme since Israel was in an almost constant state of war, with the students and many faculty in the army. The situation likely contributed to Pekeris' pragmatic approach of pursuing development of applied math and physics through a computational effort (Gillis, 1995). Thus, part of his negotiation to come to the Weizman Institute was to be given funds to build one of the world's first digital computers for scientific studies, the Weizmann automatic computer (WEIZAC), which was completed in 1955 and whose design was based on the von Neumann machine. The WEIZAC's first use was to solve Laplace's equations for Earth's ocean tides for realistic continental boundaries; this was a major accomplishment showing the power of computers to turn theory into 'modeling' (Gillis, 1995).

Pekeris continued to build and recruit for the applied mathematics department and mentored many graduate students who went on in science and other faculty positions. He was also involved in establishing the first Israeli geophysical survey which led to the discovery of oil in Israeli territory. In 1952, he was elected to the US National Academy of Sciences. While in Israel he continued to work on atomic physics (on the ground states of helium), as well as wave propagation and free oscillations in both stars and the Earth, and he was able to test his free oscillations theory with data from the giant 1960 Chile earthquake. He and his students continued development of computing synthetic seismograms from generalized ray theory, which was computationally intensive and required a computing upgrade from WEIZAC to the more powerful GOLEM series of Israeli-developed 'supercomputers'. He continued to work on all these

problems of atomic phyiscs, seismology, tides, free oscillations, and hydrodynamics until his retirement at the age of 65 in 1973, which was also the year that his wife Leah passed away. Even after retirement he continued to do research, for example, publishing again on the physics of ocean tides in 1978. Near and during this postretirement time, he was recognized for his lifelong contributions and was elected to various societies and given prizes such as the Vetlesen prize (1973), the Gold Medal from the Royal Astronomical Society (1980), and the Israel prize (1981). In 1990, the Mathematical Geophysics meeting was held in Jerusalem in honor of Pekeris and his contributions. In February of 1993, Chaim Pekeris died as a result of injuries from a fall in his home in Rehovot. The following year the Weizmann Institute of Science began the annual Pekeris Memorial Lecture.

*Pekeris' model of mantle convection.* Chaim Pekeris's paper on the viability of mantle convection (Pekeris, 1935) was in many ways one of the first truly sophisticated analyses of mantle convection. His basic hypothesis was to examine the convective circulation caused by lateral thermal gradients associated with the difference between a warm subcontinental mantle and a cooler suboceanic mantle, both estimated from crustal thickness and mantle heat production. Pekeris also used Haskell's value for mantle viscosity, and calculated that the convective velocities were of the order of $1\,\mathrm{cm\,yr^{-1}}$ which is a perfectly plausible tectonic velocity. However, Pekeris' theory also laid the groundwork for other effects, including accounting for convection in a deep spherical shell, the effect of convection on distortion of the surface, and the net effect of convective density anomalies and surface deflections on the gravity field, thereby predating modern analysis of geoid and topography by 50 years. Pekeris' paper was published 5 months before Anton Hale's paper, although they were both received at the same time (December 1935) and were similarly communicated by Harold Jeffreys to the Royal Astronomical society. As with Hales, Pekeris was not to examine mantle dynamics again since it is likely that this field was perceived as too speculative and without promise.

Further reading, including sources for this brief history, are Gilbert (2004) and Gillis (1995).

### 1.2.6  Harry Hammond Hess (1906–69)

Harry Hess (**Figure** 7) was one of the giants of the plate tectonics revolution, not only through his legacy of careful seagoing observations, but in his

**Figure 7**    Harry H. Hess (1906–69). From Shagham R, Hargraues RB, Morgan WJ, Van Houten FB, Burk CA, Holland HD, Hollister LC (eds.) (1973) *Studies in Earth and Space Sciences: A Memoir in Honor of Harrg Hanmond Hess. The Geological Society of America Memoir 132.* Boulder, CO: The Geological Society of America.

landmark paper (Hess, 1962) hypothesizing the essence of seafloor spreading and subduction. Hess was also (along with Keith Runcorn; see Section 1.2.7) one of the leading proponents, at the dawn of the plate tectonics revolution, for the mantle-convection driving mechanism of plate motion.

Harry Hammond Hess was born in New York City in May 1906. At the age of 17, he entered Yale University to study electrical engineering, but switched to geology and, despite purported failures at mineralogy, graduated with a BS in 1927.

After graduating from Yale, he worked as a mineral prospector in northern Rhodesia for two years. He then returned to the US where he started graduate school at Princeton. For his PhD he worked on ultramafic peridotite, thought to be part of the mantle. During graduate school he also took part in some of F. A. Vening Meinesz' submarine gravity surveys, particularly of the West Indies island arc, which later inspired him for further marine surveys while in the navy. Hess

finished his graduate studies and received his PhD in 1932. After graduation and brief appointments at Rutgers and the Geophysical Laboratory at Carnegie, he returned to Princeton where he joined the faculty in 1934 and was to remain there for the rest of his life, other than brief visiting professorships at the University of Capetown from 1949 to 1950 (where he likely interacted with Anton Hales), and Cambridge in 1965.

To continue the submarine activities he started with Vening Meinesz, Hess arranged in the 1930s a commission as an officer (lieutenant) in the US Navy reserve. After the attack on Pearl Harbor in December of 1941, he was called to active duty. He was first involved with enemy submarine detection in the North Atlantic and developed a technique for locating German submarines. He volunteered for hazardous duty to complete the submarine detection program by joining the submarine decoy vessel USS *Big Horn*. He was later made commanding officer of the USS *Cape Johnson*, which was a transport ship. Hess was involved with some of the major island-hopping landings of the Pacific theatre, such as the landings at the Marianas, Leyte, Linguayan, and Iwo Jima. However, Hess also continuously ran his ship's sonar echo sounder to detect bathymetry en route to these landings, thereby collecting seafloor profiles all across the North Pacific. These surreptitious surveys led to the discovery of 'guyots', flat-topped seamounts, which Hess named after Swiss geographer Arnold Guyot who founded the Princeton department. Hess inferred these guyots to be islands that had been eroded to sea level but had become submarine due to seafloor migration, hence providing one of the first major clues of seafloor mobility. Such seafloor mapping continued under the newly formed Office of Naval Research, and this led to the discovery of the mid-ocean ridge system, which was also found to have in many instances rift-shaped valleys running along its length.

After the war Hess returned to Princeton and initiated and directed the massive multinational Princeton Carribean Research Project which set out to perform a comprehensive exploration of Carribean geology, producing more than 30 PhD's in the process. Hess was made department chair in 1950, a position he kept for 16 years, and during which led a large expansion of the Princeton department. In 1952, he was elected to the National Academy (the same year as Chaim Pekeris' election) and also served on various national advising committees. In the late 1950s, he and Walter Munk initiated the Mohole Project to drill through the ocean crust into the mantle, and he continued his work on Mohole through to 1966, leading to technical breakthroughs that paved the way for the Deep Sea Drilling Project.

Harry Hess continued to serve his department and as national advisor through the 1960s. He was involved with development of the national space program and was on a special panel appointed to analyze rock samples brought back from the Moon by Apollo 11. He received numerous honors and awards: apart from National Academy membership, he also was given the Penrose Medal in 1966, elected to the American Academy of Arts and Sciences in 1968, and in 1969, just months before his death, he was given an honorary doctorate by Yale.

In late August 1969, Hess was chairing a meeting of the Space Science Board of the National Academy in Woods Hole Massachusetts, which was discussing the scientific objectives of lunar exploration only a month since the amazingly successful Apollo 11 mission. During the meeting on August 25, Hess suffered a fatal heart attack. He was buried in Arlington National Cemetery. He was pothumously given the NASA Distinguished Public Service award, and the AGU established, as one of its primary awards, the Hess medal in his honor (Dunn, 1984).

*Hess, plate tectonics, and mantle convection.* In 1960, Harry Hess wrote an internal report to the Office of Naval Research propounding his hypothesis of sea-floor spreading. Years of experience in seafloor surveying led him to believe that mid-ocean ridges had rift valleys; his knowledge of sedimentology and petrology also made evident to him that neither sea-floor sediments nor fossils were ever more than a few hundred million years old. Hess, was essentially convinced by Wegener's observations of continental breakup, and he proposed the idea that the ocean crust diverged away from linear ridges of volcanic activity, later known as seafloor spreading, and that sediments were swept into trenches. He also proposed that ocean crust was subducted, but continental crust tended to scrape off sea sediments to make mountain ranges. In 1962, Hess republished the same paper in a peer-reviewed volume (Hess, 1962), and this paper became one of the most famous and highly cited papers in geoscience, and was a landmark in the plate tectonics revolution. However, evidence and verification of the idea did not come until the Vine–Matthews study of magnetic seafloor lineations in 1963. The idea that the seafloor traveled like a conveyor belt essentially satisfied the main objections to continental drift.

At the same time that Hess proposed seafloor spreading he also concluded, in the same paper, that the driving mechanism for surface motions was mantle convection,

based on observations of gravity anomalies and studies of peridotites, which being ultramafic were assumed to upwell from the mantle. His convection postulate was not based on fluid dynamical analysis but on a synthesis of observations and physical intuition about the ramifications of seafloor spreading and seafloor destruction at trenches. His intuition about the importance of subduction as convection was largely correct, although he, like Holmes, assumed that ridges involved active upwelling (which he also thought carried much more water than is presently known to occur) and that ridges were henced pried apart by convection (Hess, 1962).

Additional reading and historical sources for this brief history are Buddington (1973), Leitch (1978), and Dunn (1984).

### 1.2.7   Stanley Keith Runcorn (1922–95)

Keith Runcorn (**Figure 8**) is a unique figure in that he was at the forefront of major discoveries in paleo-magnetism that led to the modern theory of plate tectonics, and was also an early and vigorous proponent for the theory of mantle convection, not just as a driving mechanism for continental drift, but also in other planets, particularly the Moon.

**Figure 8**   S. Keith Runcorn (1922–95). American Geophysical Union, courtesy, AIP Emilio Segre Visual Archives.

Stanley Keith Runcorn was born in Lancashire, England, in November 1922, and was educated there as a youth. While an intellectually active young man, he was more interested in history and geography, although he was eventually persuaded by his father and school headmaster to pursue science, particularly astronomy (Collinson, 1998). At 18, in 1940, he went to Cambridge to study electrical engineering and graduated in 1943. He subsequently joined a telecommunications research firm in Worcestershire to work on radar during the remainder of World War II.

After the war, in 1946, Runcorn took an assistant lectureship in physics at Manchester University to initially work on cosmic radiation, but then moved to study stellar and planetary magnetic fields. The first dynamo theories by Elsasser and Bullard had just been proposed, and Runcorn's first work was to test the core-origin of the geomagnetic field by examing the variation of horizontal field strength in coal mines, which he found to increase with depth thereby lending support to the core-origin hypothesis; this work led to Runcorn's PhD degree in 1949 (see Collinson, 1998).

In 1950, Runcorn moved to the Cambridge Department of Geodesy and Geophysics and worked on remanent magnetism of rocks of different ages to infer polar wander paths for both Great Britain and North America, which were found to differ. After testing and eliminating the possibility of large uncertainty in the data, Runcorn became convinced that the variation in paths was due to continental drift, and this set of observations in the end formed one of the cornerstones for the advent of plate tectonics (see Girdler, 1998).

In 1956, Runcorn left Cambridge for the Chair of Physics at King's College, University of Durham at Newcastle upon Tyne. He was to remain there until his retirement from the British system in 1988. He quickly established a geophysics program within the Physics Department (largely by moving Cambridge colleagues with him) and continued to work on and foster paleomagnetic studies, with considerable field work of his own in the Western United States. During this time, he also established his well-known reputation for extensive travel and departmental absence, earning the facetious title of "Theoretical Professor of Physics" (Collinson, 1998).

It was during his time in Newcastle that Runcorn began his work on mantle convection (see below), which also led to his growing-core hypothesis that had implications for the change of length of day. These ideas prompted Runcorn to move toward the study of coral growth rings, which demonstrated that

he had no fear of moving into a different field, even if it was biology. He also continued work through the 1960s on ocean currents and tides, but then eventually landed on the study of the Moon, which, given the advent of space exploration and the Apollo program, was a field he would pursue for the remainder of his life. Runcorn continued to work on ideas of convection and geomagnetism in the Moon as well as on various other topics including Chandler wobble, Jupiter's rotation, and Mars geodesy. In the late 1980s, before his retirement, he also became interested in variations in the gravitational constant $G$, but this pursuit was eventually found to be fruitless. Runcorn retired from Newcastle in 1988 and took a part-time Chair at the University of Alaska at Fairbanks, while also keeping a position in physics at Imperial College London. During his life, Runcorn had been elected a Fellow of the Royal Society (1965), received the Gold Medal of the Royal Astronomical Society (1984), the Fleming Medal of the AGU (1983), Vetlesen Prize (1971), Wegener Medal (1987), and several other honors.

In 1995, on the way to the AGU meeting in San Francisco, Keith Runcorn stopped off in San Diego to give a seminar and discuss Galileo Orbiter results with a colleague. On December 5, he was found dead, the victim of a violent homicide, the perpetrator of which was later apprehended and found guilty of first degree murder and sentenced to imprisonment (*Imperial College Reporter*, 1996, *Nature*, 1997).

*Runcorn and mantle convection.* Not long after his move to Cambridge in 1956, Keith Runcorn became interested in mantle convection as a driving mechanism for continental drift. In two papers in *Nature* in 1962 (Runcorn, 1962a, 1962b), he established the theoretical arguments for subsolidus creep in the mantle under buoyancy stresses and made a quantitative prediction of convective velocities similar (although perhaps unbeknown to him) to those of Hales and Pekeris. He also believed that a growing core influenced the onset of convective-driven drift, an idea that never found much traction.

Runcorn also proposed the then radical idea that the long-wavelength geoid was not frozen into the Earth but was due to mantle convection (Runcorn, 1963). While Runcorn's analysis of the geoid was not as sophisticated as modern analysis (*see* Chapters 2, 4, 8), it did lead him to infer that spreading centers were not sites of deep active upwellings, as had been inferred by Holmes and Hess, but were better explained by broad upwellings, which is a more modern and plausible view of mantle flow (*see* Chapter 7). This observation also prompted Runcorn to have little faith in the significance of the ridge-push force (Runcorn, 1974; Girler, 1998). In many ways, Keith Runcorn's views of mantle convection were well ahead of their time and have largely been borne out by the last several decades of analysis.

Additional reading and historical sources for this brief history are Collinson (1998) and Girdler (1998).

### 1.2.8   Mantle Convection Theory in the Last 40 Years

The origin of modern mantle convection theory was not only contemporaneous with the birth (or rebirth) of the theory of plate tectonics but also with the advent of the study of nonlinear convection. Up till the early 1960s, the study of linear convection, that is, the onset of convective instability from infinitesimal perturbations (*see* Chapters 2, 4) was well studied across several fields, as summarized in the classic treatise by Subramanyan Chandrasekhar (1910–95; Nobel Prize 1983) (Chandrasekhar, 1961). However, the mid-1960s witnessed increased activity in experimental studies of convection as well in theories of finite-amplitude or nonlinear convection (see Manneville, 2006). Well-controlled experiments that documented transitions in convective state (e.g., patterns) (*see* Chapter 3) provided inspiration and testing for various studies of nonlinear convection, leading to several seminal theoretical approaches such as nonlinear perturbation theory and matched asymptotic analysis (*see* Chapter 4).

Along with the plate tectonics revolution and the renewed study of the physics of convection, mantle convection theory made rapid progress and growth and achieved a rather mature level in the late 1960s and early 1970s (e.g., Turcotte and Oxburgh, 1972; Oxburgh and Turcotte, 1978) through the work of various investigators that had migrated in from physics and engineering, most notably Dan McKenzie, Jason Morgan, Donald Turcotte, and Gerald Schubert, the former two also having made major contributions to formulating the modern working theory of plate tectonics. Throughout the late 1960s and 1970s, many of the fundamental problems of convection in a solid-state mantle had been identified, although not necessarily solved. This included the fact that mantle convection was occuring in a fluid with a complex variable rheology, and that convection was occuring through multiple solid–solid phase transformations most notably at the top and bottom of the Earth's transition zone at depths of 410 and 660 km, respectively (*see* Chapters

2, 8). This work set the stage for the explosion in the field of mantle dynamics that occurred during rapid improvements in computing power and numerical methods. In particular, many seminal contributions in the numerical analysis of mantle convection were made by Ulrich Christensen and colleagues during the 1980s and 1990s. With the combination of laboratory, theoretical, and numerical analyses, along with improved observations coming from, in particular, seismology, isotope geochemistry, and mineral physics, our picture of mantle convection has progressed in the last 20 years to highly sophisticated levels of complexity and realism, although much still remains to be understood.

## 1.3   Observations and Evidence for Mantle Convection

Many of the observations that are relevant to mantle dynamics are covered in other volumes of this treatise, as well as within this volume. As discussed above in Section 1.2, the modern theory of mantle convection was motivated as the driving mechanism for continental drift and plate tectonics. The observations that in themselves inspired the resurgence of the mobile-surface theory of plate tectonics was largely from paleomagnetic studies showing relative continental motion and seafloor spreading (see Sections 1.2.6, 1.2.7). Global seismicity also gave clear delineation of the structure of plates and locations of plate boundaries, as well as outlines of subducting slabs along zones of deep earthquakes or Wadati–Benioff zones. With the advent of the space program and satellite measurements came accurate geodetic measurements of sea-surface height and hence global models of the Earth's geoid and gravity field which provide important constraints about the density structure of the mantle associated with convection.

Mantle geochemistry and petrology also provided important constraints on mantle dynamics through the analysis of magma reaching the surface from the mantle at mid-ocean ridges, ocean islands, large igneous provinces, and at subduction-related arcs. Isotopic and petrologic analyses of melting, as well as melt fractionation of trace elements, gave important information regarding the depth of melting beneath ridges and hot spots. However, the disparity between the concentration of incompatible elements in mid-ocean ridge basalts (MORBs) and ocean-island basalts (OIB), in addition to a host of other geochemical arguments involving, for example, noble gas isotopes, has been one of the driving motivations for inferences about the preservation of isolated reservoirs (e.g., layering) in the mantle. These geochemical observations, however, seem to conflict with geophysical evidence for whole-mantle stirring by sinking slabs (van der Hilst *et al.*, 1997; Grand *et al.*, 1997), and this has engendered a long-standing debate about the structure and nature of mantle convection (*see* Section 1.6.2, Chapter 10).

Our understanding of mantle dynamics also draws from observations on other planets. The lack of cratering record on Venus argues for massive resurfacing events, and giant volcanoes on Mars are evidence of extensive and deep magmatism. However, that neither of our neighboring terrestrial planets appears to have at least present-day plate tectonics continues to be one of the cornerstones in the argument that plate tectonics requires liquid water (Tozer, 1985).

## 1.4 Mantle Properties

The study of mantle convection has a boundless appetite for information on the properties of the convecting medium. Indeed, what caused the theory of continental drift to be marginalized for decades was the material property argument by Sir Harold Jeffreys that the Earth was too strong to permit movement of continents through ocean crust. Later measurements of mantle viscosity by postglacial rebound were an important key element in recognizing that the mantle is fluid on long timescales (*see* Chapter 2).

Fluid dynamics is rife with dimensionless numbers and one of the most important such numbers in the study of convection is the Rayleigh number (*see* Chapter 2). The Rayleigh number defines the vigor of convection in terms of the competition between gravitationally induced thermal buoyancy that acts to drive convective flow, and the dissipative or resistive effects of both fluid viscosity, which retards convective motion, and thermal diffusion which acts to diminish thermal anomalies. The Rayleigh number is written generically as

$$Ra = \frac{\rho g \alpha \Delta T d^3}{\kappa \mu} \qquad [1]$$

where $\rho$ is density, $g$ is gravitational acceleration, $\alpha$ is the thermal expansivity, $\Delta T$ is the typical temperature

contrast from the hottest to coldest parts of the fluid layer, $d$ is the dimension of the layer such as the layer thickness, $\kappa$ is thermal diffusivity, and $\mu$ is dynamic viscosity (*see* Chapter 2).

An estimate of the mantle Rayleigh number by itself requires knowledge of the material responses to various inputs such as heat and stress. The response to heat input involves heat capacity, thermal expansivity $\alpha$, and heat conduction or thermal diffusion ($\kappa$) (*see* Chapter 2). The density structure inferred from high-pressure and temperature experiments, and seismology, constrain how mantle density $\rho$ responds to pressure changes as upwellings and downwellings traverse the mantle, undergoing simple compression or decompression, as well as solid–solid phase transitions; both effects can have either stabilizing or destabilizing effects on mantle currents (Chapters 2, 8, 9).

One of the most important factors within the Rayleigh number and in the overall study of mantle convection is viscosity $\mu$. That the mantle is viscous at all was one of the key elements in determining the viability of the mantle convection hypothesis. That continents were inferred to be in isostatic balance clearly implied that they are floating in a fluid mantle; but isostatic equilibrium does not indicate how fluid the mantle is since it gives no information about how long it takes for the floating continents to reach an isostatic state. However, measurements of this approach to isostasy could be taken by examining postglacial rebound, that is, the uplift of high-latitude continental masses such as Scandinavia and Canada, following the melting of the glacial ice caps after the end of the last ice age (Haskell, 1937). From these analyses came one of the most crucial and well-known material properties: the average viscosity of the mantle of $\mu = 10^{21}$ Pa s. Today, the analysis of the mantle's response to changing loads (e.g., melting ice caps or a decrease in Earth's rotation rate) is done through an increasingly sophisticated combination of geodetic satellite and field analyses, which further refine the viscosity structure of the mantle (*see* Chapters 2, 4).

The viscosity of the mantle is so large that it is called a slowly moving or creeping fluid and thus does not suffer the complexities of classical turbulence (Chapter 2). However, one of the greatest of all complexities in mantle dynamics is associated with the various exotic rheological behaviors of mantle rocks, which are almost exclusively inferred from laboratory experiments. Mantle viscosity is well known to be a strong function of temperature, and the dramatic increase in viscosity toward the surface leads to various conundra about how plate tectonics forms or functions at all and/or how subduction zones can ever initiate from such a cold strong lithosphere (*see* Chapters 2, 3, 8). The mantle's rheology is also complicated by the various deformation mechanisms it can assume, prevalently diffusion creep at 'low' stress, and dislocation creep at higher stress, although other creep and slip mechanisms are also possible (*see* Chapter 2). In diffusion creep, viscosity is a function of mineral grain size and this effect can induce dramatic changes if grain-growth or grain-reduction mechanisms exist. In dislocation creep, viscosity is non-Newtonian and is a function of stress itself, thereby undergoing pseudo-plastic behavior in which the material softens the faster it is deformed. These and many more complexities (see Section 1.6.4) continue to keep mantle convection a rich field.

## 1.5    Questions about Mantle Convection We Have Probably Answered

### 1.5.1    Does the Mantle Convect?

The existence of this entire volume, as well as an enormous body of literature on mantle convection, would seem to obviate this question. However, less than 50 years ago, the notion that the mantle convects was not entirely accepted. The physical requirements or conditions for a fluid mantle to convect (i.e., a sufficiently high Rayleigh number) could be inferred from material property measurements; these indeed imply that the Rayleigh number is perhaps a million times what is needed to just barely convect at all, and thus should be convecting vigorously (*see* Chapter 2).

However, the direct observation of a convecting mantle is most closely linked to those that verify plate tectonics and to seismic imaging of the Earth's interior. Plate spreading and subduction, and the 'creation' and 'destruction' cycle of lithosphere that they represent, demand vertical transfer from the surface into the mantle and vice versa. That heat flow and bathymetry measurements show the lithosphere going from a hot ridge to a cold trench is not only evidence that ridges and trenches are the expression of upwellings and downwellings, respectively, but that the cooling lithosphere is nothing more than a convective thermal boundary layer (*see* Chapters 2, 4). Lastly, seismology has given us

not only the deep-earthquake trace of the cold sinking slab along the Wadati–Benioff zone but also tomographic images of these same slabs sinking deep into the mantle (van der Hilst *et al.*, 1997; Grand *et al.*, 1997). There remains little if any doubt that the mantle convects, but it is important to remember what the first-order evidence is for this conclusion, especially considering that the mantle is more inaccessible to direct observation than are distant galaxies.

### 1.5.2    Is the Mantle Layered at 660 km?

The average structure of the mantle in terms of density and elastic properties was shown by seismological studies (augmented by mineral physics) to contain discontinuities, most notably at about 410 and 660 km depths, the latter one being considerably more distinct. The presence of a strong discontinuity in density at 660 km depth suggested that the lower mantle was a denser and perhaps isolated and sluggish layer convecting separately from the upper mantle. This layered-convection argument was reinforced by observations that deep earthquakes along subducting slabs seem to cease around 700 km, and that some tomographic images show slabs stalling at this depth. This view also fits well with the geochemical inferences that MORBs and OIBs must be coming from different layers (the former from the upper mantle above 660 km, the latter from the lower mantle below 660 km; see also Section 1.6.2 and Chapter 10). The convergence of seismology and geochemistry toward a single model of a mantle layered into compositionally distinct regions was indeed a compelling argument. The prospect of mantle convection existing in two layers separated at 660 km depth was a prevalent theme in the study of mantle convection for several decades starting in the late 1960s.

However, high-pressure mineral physics experiments indicated that mantle discontinuities are most likely associated with solid–solid phase transitions, not compositional changes. The major upper-mantle component olivine was shown to undergo a change to a spinel structure called wadsleyite at 410 km depth; wadsleyite itself undergoes a less dramatic transition to ringwoodite at around 510 km, and then, at 660 km depth, ringwoodite changes to a combination of perovskite and magnesiowüstite. Studies of convection in the presence of such phase changes indicated that they might impede convection temporarily but not indefinitely (*see* Chapters 2, 8); and indeed seismic tomographic studies using body waves showed that many slabs do indeed penetrate this boundary and sink well into the lower mantle (van der Hilst *et al.*, 1997; Grand *et al.*, 1997).

In the end, the predominant evidence points to the mantle not being layered with an impermeable boundary at 660 km depth. This of course leads to other conundra, especially with regard to explaining geochemical observations, which has thus inspired several variants of deep-mantle layering and ways of isolating reservoirs or chemical components (*see* Chapter 10).

### 1.5.3 What Are the Driving Forces of Tectonic Plates?

Upon the widespread acceptance of the plate tectonic model, considerable effort was put forward to make direct estimates of the plate-driving forces. This effort was perhaps best represented by the seminal work of Forsyth and Uyeda (1975), although considerable work has followed since then (Jurdy and Stefanick, 1991). Several plate forces have been coined, most notably the two driving forces of ridge push and slab pull. Which of these two forces is dominant was the subject of some debate (Hager and O'Connell, 1981; Jurdy and Stefanick, 1991). However, perhaps the most compelling evidence for slab pull is the profound correlation, shown in the original paper by Forsyth and Uyeda (1975), between the fraction of convergent (trench) boundary and plate velocity (**Figure 9**); essentially no other meaningful correlation was shown between any other plate characteristic and plate velocity. This correlation showed that the fastest plates (of order 10 cm yr$^{-1}$) have the most amount of slab connected to them, while the slowest plates (of order 1 cm yr$^{-1}$) have little or no slab connected to them, which demonstrated rather conclusively that slab pull is the driving force of plate tectonics, because, quite simply, for a plate to move it needs a slab (Chapter 8). From a mantle convection perspective this inference fits very well with the modern picture of cold downwelling or slab-dominated mantle circulation. That is, the Earth's mantle is primarily driven by the surface cooling of a mantle that is more or less uniformly heated by radioactive decay of uranium, thorium, and (at one time) potassium, and is also losing primordial or fossil heat (*see* Chapter 6); such a configuration of distributed heat production and surface heat loss typically leads to convection dominated by cold downwelling currents, which are synonymous with subducting slabs (*see* Chapter 8). Most importantly of all, the estimate of plate-driving forces are thus completely reconcilable with and even subsumed

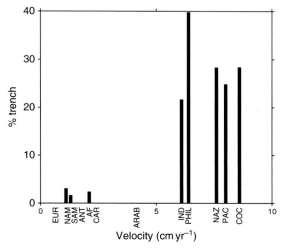

**Figure 9** A histogram of percent trench length (percent that a plate boundary is comprised of subduction zone) versus plate velocity for the major plates. (Initials on the abscissa are for European (EUR), N. American (NAM), S. American (SAM), Antarctic (ANT), African (AF), Caribbean (CAR), Arabian (ARAB), Indo-Australian (IND), Philippine (PHIL), Nazca (NAZ), Pacific (PAC), and Cocos (COC) plates.) Adapted from Forsyth D and Uyeda S (1975) On the relative importance of the driving forces of plate motion. *Geophysical Journal of the Royal Astronomical Society* 43: 163–200 after Bercovici D (2003) The generation of plate tectonics from mantle convection. *Earth and Planetary Science Letters* 205: 107–121, Elsevier.

by the theory of mantle convection. Moreover, this inference leads to the important conclusion that the tectonic plates are not so much driven by convection, they are convection. The plates are cooling thermal boundary layers that are both driven by and become slabs, which are in themselves convective downwellings; the plates are thus convection.

## 1.6 Major Unsolved Issues in Mantle Dynamics

### 1.6.1 Energy Sources for Mantle Convection

The discovery of radioactive elements at the turn of the nineteenth century was a key discovery in many regards, including providing evidence that the Earth has internal sources of heat. This was also the key argument to refute Lord Kelvin's estimate for the age of the Earth since he assumed it was cooling freely (without heat sources) from an initially molten state (see Section 1.2.4). Early estimates of the concentration of radioactive elements inside the Earth are based on heat flow and geochemical measurements of crustal rocks, in conjunction with cooling models (*see* Chapters 4, 6); these arguments tended to point toward a high enough concentration of radioactive elements in the mantle to account for as much as 70–80% of the heat flow out of the Earth to be due to radiogenic heating (Schubert *et al.*, 2001). This satisfied the condition that the Earth has been cooling relatively slowly over its 4.5 Gy lifetime. However, recent estimates of radioactive element abundances from the study of chondrites (thought to be representative of planetary building blocks) possibly suggest somewhat less radiogenic heating (*see* Chapter 6). If radiogenic heating is small, then more of the Earth's heat flow is from loss of fossil heat (as with Lord Kelvin's assumptions; see also England *et al.* (2007); this would demand rapid cooling from a recently excessively hot or even molten state, unless the physical process of convection was itself somehow very different in the past in order for the mantle to retain its heat (*see* Chapter 6). Alternatively, if chondritic estimates of these radiogenic sources are very wrong, then it implies significant problems with the chondritic model for planetary composition and thus for our understanding of Earth's formation. Either possibility leads to many intriguing directions for future inquiry.

### 1.6.2   Is the Mantle Well Mixed, Layered, or Plum Pudding?

As mentioned above in Sections 1.3 and 1.5.2, there are various lines of geochemical evidence-ranging from the disparity between trace-element abundances in MORBs and those in OIBs, sources and reservoirs of noble gases, source and origin of continental crust, etc. – that suggest that the mantle has isolated reservoirs, such as distinct layers. Although layering of the mantle at 660 km depth has probably been eliminated, the motivation still persists to reconcile the geochemical inference of an unmixed mantle with the geophysical evidence of a well-stirred mantle. The problem remains unsolved, although various models and solutions have been proposed, in particular deep layering, or mechanisms for keeping the mantle poorly mixed (the 'plum pudding' model). This issue comprises the major thrust of Chapter 10, and is touched upon in Chapter 9, as well as elsewhere, notably.

### 1.6.3   Are There Plumes?

Convective plumes are relatively narrow cylindrical upwellings typical of convection and especially in fluids with strongly temperature-dependent viscosity, such as mantle rocks. The existence of plumes in the mantle was proposed by Jason Morgan (*see* Chapter 9) to explain anomalous intraplate volcanism such as at Hawaii. That hot spots appeared to be more or less immobile relative to plates suggested a deep origin, and it has often been supposed that plumes emanate from the most obvious heated boundary in the mantle, the core–mantle boundary. However, deep plumes have eluded direct observation by seismic methods, other than with recent forefront techniques that still remain controversial (Montelli *et al.*, 2004). To some extent, evidence for the existence of plumes is still circumstantial and thus they remain the subject of ongoing debate. The subject of plumes and melting anomalies is discussed in Chapter 9.

### 1.6.4   Origin and Cause of Plate Tectonics

As noted already, the modern theory of mantle convection was motivated to provide the driving mechanism for plate tectonics. We also argued above (Section 1.5.3) that plate tectonics is indeed convection. However, there still remains no unified theory of mantle dynamics and plate tectonics, wherein plate tectonics arises naturally and self-consistently from mantle convection. There are some aspects of plate tectonics that are reasonably well explained by convective theory, in particular the existence of cold planar downwellings akin to subducting slabs (*see* Chapter 8). But still many first-order questions remain. Just with regard to subduction itself, there is still no widely accepted theory of how a subduction zone and sinking slab initiates from a thick cold and, by all appearances, immobile lithosphere (Chapters 2, 3, 8); lithospheric instabilities and sublithospheric small-scale convection are easily generated (*see* Chapter 7), but a widely accepted mechanism for getting the entire stiff cold lithosphere (from surface to base) to bend and sink remains elusive. Also, asymmetric subduction (only one plate subducts at a trench) is not easily obtained with convection theory, although such asymmetries are likely associated with disparity between oceanic and continental lithosphere.

However, many other issues remain in the problem of 'plate generation'. Although mid-ocean ridges are associated with upwelling from the mantle, all evidence points to the upwelling being shallow and the spreading being passive; that is, ridges are pulled apart by slabs at a distance, rather than pried apart by a deep upwelling (*see* Chapter 7). How such passive upwelling occurs in a convection calculation is not universally understood, although self-consistent convection calculations with near-surface melting do a reasonably good job of predicting the formation of passive ridges (Tackley, 2000b; see **Figure 10**).

One of the long-standing problems in understanding the plate-like features of mantle convection is the generation of toroidal motion, which involves strike-slip shear and spin of plates (Hager and O'Connell, 1979; O' Connell *et al.*, 1991; Dumoulin *et al.*, 1998). Toroidal motion is enigmatic because it is not directly generated by convective forces but must arise by the coupling of buoyancy driven flow and large viscosity variability (*see* Chapter 4). The dependence of toroidal flow on rheological effects also links it closely to the generation of narrow weak plate boundaries separated by broad strong plates. Plate-like toroidal flow and plate-like structures have been shown to both require severe velocity-weakening mechanisms that are well beyond even the reasonably complex viscous creep rheologies typical of the mantle (**Figure 11**); see, for example, the review by Bercovici (2003). The focusing of

deformation associated with plate boundary formation and the strike-slip form of toroidal flow are classified as a natural occurrence of shear localization. Although some creep rheologies (showing plastic behavior) can generate some plausible localization, their effect is instantaneous in that the weak zones only persist as long as they are being deformed, whereas actual plate boundaries have long lives even if inactive and can hence be reactivated (Gurnis *et al.*, 2000). The mechanisms for such localization are thus likely to involve 'state' variables that will grow under a rapidly deforming state, but decay away slowly after deformation ceases. Temperature is a simple analogy of such a state variable in that thermal anomalies can be generated by frictional dissipation and cause weakening, but will diffuse away gradually after forcing stops. More plausible but more exotic shear-localizing state variables might be defect and microcrack density, or as has been proposed by Bercovici and Ricard (2005), the most effective mechanism may be grain-size reduction through damage (**Figure 12**).

Regardless of their successes, none of these shear-localizing mechanisms have yet to explain the role of water as is assumed to exist (Section 1.3). Moreover, essentially all plate generation models have been designed to address present-day instanteneous plate

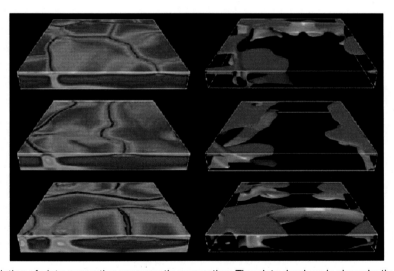

**Figure 10** A simulation of plate generation over mantle convection. The plate rheology is viscoplastic and the viscosity reduction associated with melting is parametrized into the model, leading to exceptional plate-like behavior and apparent passive spreading (i.e., narrow spreading centers not associated with any deep upwelling). The right panels show surfaces of constant temperature, which here are dominated by cold downwellings; the left panels show the viscosity field (red being high viscosity and blue low viscosity). Different rows show different times in the simulation. After Tackley P (2000b) Self-consistent generation of tectonic plates in time-dependent, three-dimensional mantle convection simulations. Part 2: strain weakening and asthenosphere. (G³) 1 (doi: 10.1029/2000GC000,43). American Geophysical Union.

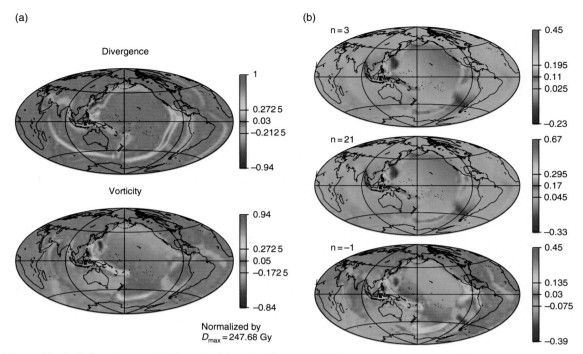

**Figure 11**  A shallow-layer model of mantle–lithosphere flow uses the Earth's present-day divergence field as a source–sink field (upper left) to drive motion; various non-Newtonian rheologies for the lithosphere are examined to see which best recovers the present-day strike-slip or vertical vorticity field (lower left). Power-law rheologies are characterized by the power-law index $n$ such that strain-rate goes as stress$^n$. Positive power indices of $n = 3$ (typical of mantle rocks) or even $n = 21$ (closer to viscoplasticity) are insufficient to recover the vertical vorticity field (upper two right panels). A more exotic rheology with power-law index of $n = -1$, which allows for stick-slip or velocity weakening behavior, is much more successful at reproducing the Earth's vorticity field. From Bercovici D (1995) A source-sink model of the generation of plate tectonics from non-newtonian mantle flow. *Journal of Geophysical Research* 100: 2013–2030. American Geophysical Union.

motions, and none have even begun to address plate motion changes, and plate growth and shrinkage, although some effort has been made to understand the convective forces that can cause plate motion changes (e.g., Lowman *et al.*, 2003). Much still remains to be examined in the basic and important problem of plate generation, and aspects of it are discussed throughout this Treatise, in particular Chapters 2, 3, 4, 8. Several reviews on plate generation can also be found in the literature, in particular those of Bercovici (2003), Bercovici *et al.* (2000), and Tackley (2000a).

## 1.7  Burgeoning and Future Problems in Mantle Dynamics

### 1.7.1  Volatile Circulation

The interaction of the ocean and atmosphere with the mantle has in fact been an important and fertile field of study for the last few decades, although it has largely been the province of mantle petrology and geochemistry. However, questions of how much the mantle entrains, returns, and stores various important volatiles such as water and carbon dioxide remain important and largely unanswered (Williams and Hemley, 2001; Karato, 2003; Hirschmann, 2006). The relevance of the problem is manifold, but can be summarized perhaps in two themes. First, the ingestion of volatiles by the mantle affects its convective circulation and thus both the thermal and chemical evolution of the mantle, mainly because of the rheological effects of volatiles (which tend to weaken rocks), as well as their tendency to facilitate melting and hence chemical and isotopic fractionation. Second, how the mantle ingests, stores, and releases various volatiles controls the evolution of the oceans and atmosphere, both in their size (or mass), as well as composition (since different volatiles are likely to be entrained and stored differently).

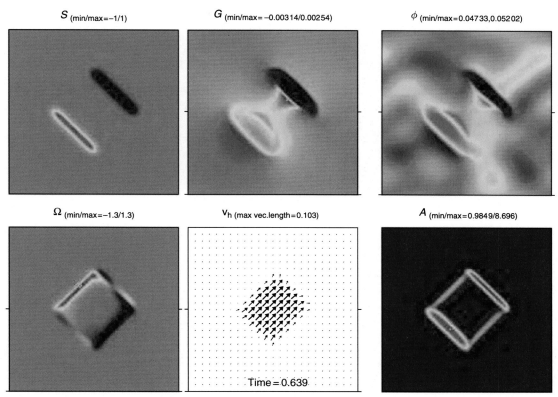

**Figure 12** A simple source–sink model of shallow two-phase flow with a shear localizing damage mechanism. Damage *per se* involves transfer of deformational work to the creation of surface energy on interfaces by void and/or grain boundary generation in the continuum. In the case shown, all damage is focused on grain size reduction. The panel meanings are indicated by symbols where $S$ is the imposed divergence rate (i.e., the source-sink field) that drives flow; $G$ is the dilation rate due to void formation; $\phi$ is void volume fraction; $\Omega$ is vertical vorticity or rate of strike-slip shear; $v_h$ is horizontal velocity; and $\mathcal{A}$ is the 'fineness' or inverse grain size. This particular calculation shows that fineness-generating, or grain size-reducing, damage is very effective at creating localized fault-like strike-slip zones in vorticity $\Omega$, and solid-body-like translation in the velocity field $v_h$. Adapted from Bercovici D and Richard Y (2005) Tectonic plate generation and two-phase damage: Void growth versus grainsize reduction. *Journal of Geophysical Research* 110(B03): 401 (doi: 10.1029/2004JB003,181). American Geophysical Union.

The flux of volatiles into the mantle is primarily through subduction zones (Peacock, 1990; Hirschmann, 2006). Crustal rocks entering a subduction zone are necessarily hydrated by virtue of being submarine. Whether the lithosphere as a whole is hydrated is questionable since it tends to be dried during the process of mantle melting and formation of crust at ridges. Absorption of carbon dioxide is first by dissolution in seawater to form an acidic solution that reacts with calcium-rich sediments to produce carbonates (Falkowski *et al.*, 2000), which are then entrained by subduction. However, the quantification of volatile entrainment through subduction zones is problematic for many reasons; in the case of water, it is difficult to make an accurate estimate of how much gets carried by slabs, how much continues to be carried down after devolatilization through arc

magmatism, and which hydrous silicate phases in slabs are capable of carrying water to significant mantle depths (Williams and Hemley, 2001; Hirschmann, 2006).

Storage of volatiles in the mantle is also the subject of much debate. Although solid rocks cannot absorb volatiles in great concentrations, the mass of the mantle is so large that it could conceivably hold several to tens of world ocean masses. Moreover, the region of the mantle between the phase transitions at 410 and 660 km depths – called the transition zone – is known for having anomalous solubility of at least water and while it is only a tenth of the thickness of the mantle, it could hold much more water than the upper mantle (above 410 km) and lower mantle (below 660 km) combined. The transfer of anomalous water from the transition zone through convection could possibly cause deep melting, as has been proposed since the early 1990s (e.g., Inoue and Sawamoto, 1992; Inoue, 1994); this melting might have an effect on trace-element circulation and led to the appearance of layering in whole-mantle convection (Bercovici and Karato, 2003; Karato *et al.*, 2006). How much volatile mass is eventually returned to the oceans and atmospheres through volcanism is perhaps better constrained since most nonarc volcanic output occurs at mid-ocean ridges whose basalts are relatively dry. The ingestion of water by slabs and meager water return at ridges suggests that the mantle is on the whole absorbing oceans, although it is equally possible that little water is taken into the mantle beyond arc volcanism (Dixon *et al.* (2002); *see* Chapter 8; and Hirschmann (2006). The balance of volatiles between the oceans and atmosphere is an ongoing debate since it relies on various feedback mechanims, some of which are still not well articulated. Volatiles will tend to reduce mantle viscosity and enhance convective vigor, but whether this enhances ingestion or output of water is not entirely known and similar models can lead to very different conclusions, such as complete mantle degassing in a brief time (McGovern and Schubert, 1989), or drainage of the oceans into the mantle over billions of years (Bounama *et al.*, 2001). However, evidence that ocean masses have remained more or less constant over billions of years (see Hirschmann, 2006) implies a steady-state exchange of the mantle with the oceans and atmosphere that has either a trivial explanation (i.e., there is no exchange), or requires a self-regulating feedback mechanism. Such mechanisms might involve the ocean–mantle contact area that governs the size (depth and breadth) of the world's oceans, and thus the size of ocean basins, plate boundary lengths, and plate sizes. To simply state that subduction zones and ridges must ingest and eject an equal amount of water is not an explanation since the mechanisms of absorption at subduction zones and release at ridges are vastly different and it would be fortuitous if they could balance each other for any given plate or ocean-size configuration.

### 1.7.2    Mantle Convection, Water, and Life

The presence of liquid water on Earth is likely a necessary condition for plate tectonics, and an obvious necesary condition for the existence of life as we know it. A major question still remains as to whether all three are linked; that is, that plate tectonics has allowed liquid water and life, or possibly even whether life has influenced plate tectonics. The volcanic return of subducted carbon dioxide, which is removed from the atmosphere by oceans and seafloor sediments, likely sustains a greenhouse state that keeps the surface temperature sufficiently high for water to remain liquid, and hence permit plate tectonics. Moreover, the aborption of carbon dioxide by ocean–sediment reactions, which prohibit buildup of $CO_2$ and thus a possible runaway greenhouse, is probably also tectonically controlled by the continuous exhumation of calcium-rich minerals through mountain building (Walker *et al.*, 1981). Thus plate tectonics possibly plays a role in keeping the Earth at the right temperature for liquid water (and life) to exist, and thus for plate tectonics itself to persist (Ward and Brownlee, 2000). Whether or not life is a mere passive player in this balance remains an open question.

## 1.8    Summary and Context of the Rest of This Volume

This volume on mantle dynamics is designed to follow two themes: (1) how is mantle convection studied, and (2) what do we understand about mantle dynamics to date. The first four chapters following this overview are thus concerned with pedagogical reviews of the physics of mantle convection (Chapter 2), laboratory studies of the fluid dynamics of convection relevant to the mantle (Chapter 3), theoretical analysis of mantle dynamics (Chapter 4), and numerical analysis and methods of mantle convection (Chapter 5). The subsequent chapters concentrate on leading issues of mantle convection itself, which include the energy budget of the mantle (Chapter 6), the upper mantle and lithosphere in and near the spreading center (mid-ocean ridge) environment (Chapter 7), the dynamics of subducting slabs (Chapter 8), hot spots, melting anomalies, and mantle plumes (Chapter 9), and lastly geochemical mantle dynamics and mixing (Chapter 10).

The physics of mantle convection is extensively multidisciplinary since it involves not only fluid mechanics, but also gravitational potential theory (to understand not only self-gravitation of our massive convecting medium but how the shape of the geoid or sea surface is affected by and informs us about convection), seismology and mineral physics (to understand the thermodynamic state and properites of the mantle such as density, thermal expansivity, and solid–solid phase transitions), material science (to understand rheology and deformation mechanisms and transport phenomena), and geochemistry, petrology, and complex multiphase, multicomponent flows (to understand chemical mixing and mantle melting). Chapter 2 begins by elucidating the basic physics of convection universal to any fluid system, and then proceeds to discuss the complexities associated with the mantle, including, for example, compressibility, phase changes, viscoelasticity, silicate rheology, etc. Many of the future research topics involving mantle complexities cannot be described with single-component and single-phase fluid mechanics; thus some care is given to develop an introduction to complex fluids entailing multicomponent systems undergoing mixing and chemical transport, as well as deformable multiphase media.

As discussed above in Section 1.2.3, the initiation of the modern study of convection is attributed to the first systematic experimental studies of Henri Bénard. Thus, Chapter 3 involves a survey of laboratory methods for studying convection, and for approaching the particular complexities of mantle convection itself. Convection in the mantle is, of all geophysical flows, perhaps most easily scaled to laboratory conditions since the ratio of viscosity to the cube of layer thickness is easily preserved between the mantle and the laboratory model. Moreover, laboratory models do not suffer the assumptions made in physical theory to obtain a closed system, the simplifying approximations of analytic theory to obtain a mathematical solution, or the limitations of numerical resolution in computer models; thus the need for studying convection in real materials is paramount. However, the mantle is an exotic fluid in that it is undoubtedly chemically inhomogeneous, has phase changes, is heated and cooled internally, and has rheology sensitive to various state variables such as temperature, pressure, and stress or grain size, and these effects are all difficult to reproduce in the laboratory and with available laboratory fluids. Thus, progress in laboratory models is both vital for exploring new physics and testing theories, but remains an extremely challenging field. Chapter 3 thus not only discusses the methods for creating and observing convection experiments, but also reviews the considerable progress in incorporating the complexities of the mantle itself into these experiments.

The birth of convection as a field of physics is also associated with Lord Rayleigh's seminal theoretical work on convective instability (Section 1.2.2). In the last 40–50 years, the theoretical analysis of convection in both the linear and especially nonlinear regimes has burgeoned, not only because of its relevance to planetary and stellar atmospheres and interiors, but because convection is the classic paradigm of a nonlinear dynamical system undergoing chaotic behavior and self-organization (e.g., Nicolis, 1995). Thus, Chapter 4 surveys the wealth of theoretical analyses on convection itself, as well as on individual theories relating to features of convective circulation. Thus the initiation of mantle plumes is examined through stability (Rayleigh–Taylor) theory, and fully developed plumes and mantle diapirs through simpler fluid dynamical models. Similarly, subducting slabs are examined through, for example, the theory of bending viscous sheets. Finally, convection is studied holistically through a review of weakly nonlinear perturbation theories which predict three-dimensional patterns at convective onset, to matched-asymptotic boundary layer theories for strongly supercritical convection.

Progress in the study of mantle dynamics in the last 20 years has been driven by the rapid increase in computational power, as well as ever-growing sophistication in numerical models. Numerical methods are invariably the most versatile of all methods of study since they can examine strongly nonlinear convection without excessive simplifying assumptions, and can incorporate any relevant physics that can at least be articulated mathematically. Thus, Chapter 5 reviews the numerical analysis of mantle dynamics by surveying the leading methods that are employed today. Classic methods of numerical modeling include finite-difference, finite-volume, and spectral methods, all of which are outlined in Chapter 5; but perhaps the most powerful and versatile method is finite elements and this is given special attention. Some classic examples of how numerical analysis is used to attack key problems are also discussed; these include the omnipresent mantle complexities of thermochemical convection, phase changes, and non-Newtonian rheology.

With the major tools of studying mantle convection surveyed in detail, the volume progresses to topics specific to the mantle itself. An obvious and key issue of mantle dynamics is the energy source for convection, which remains an active and at times controversial issue, and this is reviewed in Chapter 6. The energy budget of the Earth relies on various important quantities that are unfortunately not easily constrained. The loss of heat through the Earth's surface is a first-order observation but is problematic in that the measurements are difficult to make (e.g., measurements of conduction through the lithosphere are easily contaminated by the effects of hydrothermal circulation) and global coverage of heat flow is difficult given the large variation in both crustal thickness and properties between and within oceans and continents. Second, the estimate of the heat sources inside the Earth is also problematic since it involves understanding the composition of the bulk Earth, which involves measurements of radioactive element concentrations from various sources such as continental crust, oceanic basalts, and chondritic meteorites, but of course never directly from the mantle itself. Finally, basic understanding of convection from theoretical, numerical, and analytical models (Chapters 3, 4, 5) allows construction of thermal histories of how the Earth cools and has evolved under the action of convection.

Chapters 7, 8, and 9 concentrate on some of the key individual features of convective currents. Chapter 7 treats the problem of convection in the upper mantle, its interaction with mid-ocean ridges, and the oceanic lithosphere. It is widely recognized that the plates and lithosphere are the dominant convective thermal boundary layer of the mantle as it is cooled from above. How the oceanic lithosphere forms following complex magmatic processes at ridges can play a key role in the entire convective cycle, since, for example, melting at ridges can cause dehydration and strengthening of the lithosphere. How this same lithosphere transmits heat and thickens as it moves away from mid-ocean ridges is further an important feature of mantle convection. However, small-scale convection (as well as hot spot activity) in the sublithospheric asthenosphere can have a profound effect on how heat is transferred into the lithosphere. Moreover, plate motions can affect the structure of this small-scale convection, typically by aligning convective rolls with plate motion. Similar to small-scale convection is the convective instability of the lithosphere itself, which can lead to delamination and further change of lithospheric structure and cooling.

Chapter 8 continues with the plate and lithosphere as it enters subduction zones and becomes what is clearly the driving force for mantle convection, that is, subducting slabs. The chapter first reviews the basic mechanism and energy release of slab descent. It then proceeds to examine evidence for slab structure from seismological observations, which can then be used to constrain dynamic models of sinking slabs as they descend through the subduction zone wedge environment and through mantle phase transformations. Phase changes have a particularly important effect on slabs because they can not only impede slab transfer into the lower mantle, but the slow kinetics inside cold slabs might cause

metastable phases to exist, which potentially cause fine-scale buoyant stresses that could influence slab deformation and seismicity. One of the primary constraints on slab-related convection and mantle structure involves the geoid and topographic signature of subduction zones; in particular, the geoid over subduction zones is generally positive which suggests that the cold positive mass anomaly of the slab is supported by a viscosity increase with depth rather than by a downward deflection of the surface. Finally, as discussed above, the volatile flux into the Earth is entirely coupled to subduction and slab dynamics, and this issue is necessarily discussed in detail.

To contrast the discussion of cold slabs, Chapter 9 concerns anomalously hot mantle, melting anomalies, and the mantle's other primary convecting current, the elusive mantle plume. The chapter reviews first the observations relating to hot spots and melting, including discussion of age-progression volcanism (which has been the observational foundation of the stationary hot-spot model); the topographic structure of hot-spot swells (which suggest a larger mantle structure than just the volcanic edifice itself); the signature of large igneous provinces (thought to be associated with large starting plumes and/or the initiating of continental rifting); and the geochemical, petrological, and seismological evidence for the depth of origin of hot spots. Next, the process of hot-spot melting and volcanism, the formation of plume swells, and the fluid dynamics of deep-mantle plumes are examined and evaluated.

The final chapter in this volume, Chapter 10, involves the ongoing problem of interpreting observations of geochemical heterogeneity from the perspective of mantle convection theory. The chapter first reviews the observations and evidence for mantle heterogeneity, starting with its origin during early segregation of the mantle, crust, and core, to its present state. As discussed in Section 1.6.2, much of the evidence for current heterogeneity involves analysis of trace (incompatible) elements in MORBs and OIBs, as well as budgets of noble gases, and these primary constraints are explained and reviewed as well. Following the survey of observations, the chapter examines models of mixing, trace-element transport, and layering in the mantle to address the fate, scale, and isotopic signature of dispersed heterogeneities, as well as the stability of reservoirs of large-scale chemical heterogeneity. The problem of chemical evolution and heterogeneity remains one of the biggest unsolved problem in mantle dynamics and thus recent and future directions in this field are also discussed.

In the end, this volume is designed to give both a classical and state-of-the-art introduction to the methods and science of mantle dynamics, as well as a survey of leading order problems (both solved and unsolved) and our present understanding of how the mantle works.

## Acknowledgments

Thanks are due to Ondrej Sramek for his assistance, to Gerald Schubert for his comments and suggestions, and also to John Ferguson for the photograph of Anton Hales, and Graham Pearson for the photograph of Arthur Holmes.

## References

Bénard H (1900) Les tourbillons cellulaires dans une nappe liquide. *Revue Générale des Sciences Pures et Appliquées* 11: 1261–1271 and 1309–1328.

Bénard H (1901) Les tourbillons cellulaires dans une nappe liquide transportant de la chaleur par convection en régime permanent. *Annales de Chimie et de Physique* 23: 62–144.

Bercovici D (1995) A source-sink model of the generation of plate tectonics from non-newtonian mantle flow. *Journal of Geophysical Research* 100: 2013–2030.

Bercovici D (2003) The generation of plate tectonics from mantle convection. *Earth and Planetary Science Letters* 205: 107–121.

Bercovici D and Karato S (2003) Whole mantle convection and the transition-zone water filter. *Nature* 425: 39–44.

Bercovici D and Ricard Y (2005) Tectonic plate generation and two-phase damage: Void growth versus grainsize reduction. *Journal of Geophysical Research* 110(B03): 401 (doi:10.1029/2004JB003,181).

Bercovici D, Ricard Y, and Richards M (2000) The relation between mantle dynamics and plate tectonics: A primer. In: Richards MA, Gordon R, and van der Hilst R (eds.) *Geophysical Monograph Series, vol. 121: History and Dynamics of Global Plate Motions*, pp. 5–46. Washington, DC: American Geophysical Union.

Berg J, Acrivos A, and Boudart M (1966) Evaporative convection. *Advances in Chemical Engineering* 6: 61–123.

Block M (1956) Surface tension as the cause for bénard and surface deformation in a liquid film. *Nature* 178: 650–651.

Bounama C, Franck S, and von Bloh W (2001) The fate of Earth's ocean. *Hydrology and Earth System Sciences* 5: 569–575.

Brown G (1999) *Scientist, Soldier, Statesman, Spy: Count Rumford – The Extraordinary Life of a Scientific Genius*. Stroud: Sutton Publishing Ltd.

Brown S (1967) *Benjamin Thompson – Count Rumford: Count Rumford on the Nature of Heat*. New York: Pergamon Press.

Buddington A (1973) Memorial to Harry Hammond Hess - 1906–1969. *Geological Society of America Memorials* 1: 18–26.

Bull A (1931) The convection current hypothesis of mountain building. *Geological Magazine* 58: 364–367.

Chandrasekhar S (1961) *Hydrodynamic and Hydromagnetic Stability*. New York: Oxford University Press.

Collinson D (1998) The life and work of S. Keith Runcorn, F.R.S. *Physics and Chemistry of the Earth* 23: 697–702.

Davis GF (1999) *Dynamic Earth*. Cambridge: Cambridge University Press.

Dixon JE, Leist L, Langmuir C, and Schilling J-G (2002) Recycled dehydrated lithosphere observed in plume-influenced mid-ocean-ridge basalt. *Nature* 420: 385–389.

Dumoulin C, Bercovici D, and Wessel P (1998) A continuous plate-tectonic model using geophysical data to estimate plate margin widths, with a seismicity based example. *Geophysical Journal International* 133: 379–389.

Dunn D (1984) Harry Hammond Hess (1906–1969) Essay for AGU Honors, American Geophysical Union, URL http://www.agu.org/inside/awards/hess2.html.

England P, Molnar P, and Richter F (2007) John Perrys neglected critique of Kelvins age for the Earth: A missed opportunity in geodynamics. *GSA Today* 17: 4–9.

Falkowski P, Scholes RJ, Boyle E, *et al.* (2000) The global carbon cycle: A test of our knowledge of earth as a system. *Science* 290: 291–296.

Forsyth D and Uyeda S (1975) On the relative importance of the driving forces of plate motion. *Geophysical Journal of the Royal Astronomical Society* 43: 163–200.

Gilbert F (2004) Chaim Leib Pekeris, June 15, 1908–February 24, 1993. *Biographical Memoirs*, 85: 217–231. National Academy of Sciences.

Gillis J (1995) Chaim L. Pekeris, 15 June 1908-25 February 1993. *Proceedings of the American Philosophical Society* 139: 179–181.

Girdler R (1998) From polar wander to dynamic planet: A tribute to Keith Runcorn. *Physics and Chemistry of the Earth* 23: 709–713.

Grand S, van der Hilst R, and Widiyantoro S (1997) Global seismic tomography: A snapshot of convection in the Earth. *GSA Today* 7: 1–7.

Gurnis M, Zhong S, and Toth J (2000) On the competing roles of fault reactivation and brittle failure in generating plate tectonics from mantle convection. In: Richards MA, Gordon R, and van der Hilst R (eds.) *Geophysical Monograph Series, vol. 121: History and Dynamics of Global Plate Motions*, pp. 73–94. Washington, DC: American Geophysical Union.

Hager B and O'Connell R (1979) Kinematic models of large-scale flow in the Earth's mantle. *Journal of Geophysical Research* 84: 1031–1048.

Hager B and O'Connell R (1981) A simple global model of plate dynamics and mantle convection. *Journal of Geophysical Research* 86: 4843–4867.

Hales A (1936) Convection currents in the Earth. *Monthly Notices of the Royal Astronomical Society Geophysics Supplement* 3: 372–379.

Hallam A (1987) Alfred Wegener and the hypothesis of continental drift. In: Gingerich O (ed.) Readings from Scientific American: *Scientific Genius and Creativity*, pp. 77–85. New York: W. H. Freeman.

Haskell N (1937) The viscosity of the asthenosphere. *American Journal of Science* 33: 22–28.

Hess H (1962) History of ocean basins. In: Engeln A, James H, and Leonard B (eds.) *Petrologic Studies - A Volume in Honor of A. F. Buddington*, pp. 599–620. New York: Geological Society of America.

Hirschmann MM (2006) Water, melting, and the deep earth $H_2O$ cycle. *Annual Review of Earth and Planetary Sciences* 34: 62953.

Holmes A (1931) Radioactivity and Earth movements. *Geological Society of Glasgow Transactions* 18: 559–606.

Holmes A (1933) The thermal history of the Earth. *Journal of the Washington Academy of Sciences* 23: 169–195.

Imperial College Reporter (1996) Murder charge, *In Brief*, Issue 22.

Inoue T (1994) Effect of water on melting phase relations and melt composition in the system $Mg_2SiO_4 – MgSiO_3 – H_2O$ up to 15 GPa. *Physics of the Earth and Planetary Sciences* 85: 237–263.

Inoue T and Sawamoto H (1992) High pressure melting of pyrolite under hydrous condition and its geophysical implications. In: Syono Y and Manghani M (eds.) *High Pressure Research: Application to Earth and Planetary Sciences*, pp. 323–331. Washington DC: American Geophysical Union.

Jeffreys H (1959) *The Earth, Its Origin, History and Physical Constitution*, 4th edn. Cambridge, UK: Cambridge University Press.

Jurdy D and Stefanick M (1991) The forces driving the plates: Constraints from kinematics and stress observations. *Philosophical Transactions of the Royal Society of London Series A* 337: 127–138.

Karato S (2003) Mapping water content in the upper mantle. In: Eiler J (ed.) *AGU Monograph, vol. 138: Subduction Factory*, pp. 289–313. Washington, DC: American Geophysical Union.

Karato S, Bercovici D, Leahy G, Richard G, and Jing Z (2006) The transition zone water filter model for global material circulation: where do we stand? In: Jacobsen S and van der Lee S *AGU Monograph Series, vol. 168: Earths Deep Water Cycle*, pp. 289–313. Washington, DC: American Geophysical Union.

Lambeck K (2002) Professor Anton Hales – Terrestrial and Planetary Scientist, Interviews with Australian Scientists Program, Australian Academy of Science, URL www.science.org.au/scientists/hales.htm.

Leitch A (1978) *A Princeton Companion*. Princeton, NJ: Princeton University Press.

Lewis C (2000) *The Dating Game - One Man's Search for the Age of the Earth*. Cambridge, UK: Cambridge University Press.

Lewis C (2002) Arthur Holmes: An Ingenious Geoscientist. *GSA Today* 12: 16–17.

Lilley T (2006) Obituary for Professor Anton Linder Hales, Research School of Earth Science, Australian National University, URL http://wwwrses.anu.edu.au/admin/index.php?p=hales_obituary.

Lindsay R (1970) *Lord Rayleigh – The Man and His Work*. New York: Pergamon Press.

Lowman J, King S, and Gable C (2003) The role of the heating mode of the mantle in periodic reorganizations of the plate velocity field. *Geophysical Journal International* 152: 455–467.

Manneville P (2006) Rayliegh-Bénard convection: Thirty years of experimental, theoretical, and modeling work. In: Mutabazi I, Wesfried J, and Guyon E (eds.) *Dynamics of Spatio-Temporal Cellular Structure - Henri Bénard Centenary Review*, pp. 41–65. New York: Springer.

McGovern P and Schubert G (1989) Thermal evolution of the Earth: Effects of volatile exchange between atmosphere and interior. *Earth and Planetary Science Letters* 96: 27–37.

Menard H (1986) *The Ocean of Truth. A Personal History of Global Tectonics*. Princeton, NJ: Princeton University Press.

Montelli R, Nolet G, Dahlen FA, Masters G, Engdahl ER, and Hung S-H (2004) Finite-frequency tomography reveals a variety of plumes in the mantle. *Science* 303: 338–343.

Nature (1997) News in brief: Kick-boxer jailed for death of geophysicist. *Nature* 389: 657.

Nicolis G (1995) *Introduction to Nonlinear Science*. Cambridge: Cambridge University Press.

O'Connell R, Gable C, and Hager B (1991) Toroidal-poloidal partitioning of lithospheric plate motion. In: Sabadini R and Boschi E (eds.) *Glacial Isostasy, Sea Level and Mantle Rheology*, pp. 535–551. Norwell, MA: Kluwer Academic.

Oxburgh E and Turcotte D (1978) Mechanisms of continental drift. *Reports on Progress in Physics* 41: 1249–1312.

Peacock S (1990) Fluid processes in subduction zones. *Science* 248: 329–337.

Pearson J (1958) On convection cells induced by surface tension. *Journal of Fluid Mechanics* 4: 489–500.

Pekeris C (1935) Thermal convection in the interior of the Earth. *Monthly Notices of the Royal Astronomical Society, Geophysics Supplement* 3: 343–367.

Runcorn S (1962a) Convection currents in the Earth's mantle. *Nature* 195: 1248–1249.

Runcorn S (1962b) Towards a theory of continental drift. *Nature* 193: 311–314.

Runcorn S (1963) Satellite gravity measurements and convection in the mantle. *Nature* 200: 628–630.

Runcorn S (1974) On the forces not moving lithospheric plates. *Tectonophysics* 21: 197–202.

Schubert G, Turcotte D, and Olson P (2001) *Mantle Convection in the Earth and Planets*. Cambridge, UK: Cambridge University Press.

Shagham R, Hargraves RB, Morgan WJ, Van Houten FB, Burk CA, Holland HD, and Hollister LC (eds.) (1973) *Studies in Earth and Space Sciences: A Memoir in Honor of Harry Hanmond Hess. The Geological Society of America Memoir 132*. Boulder, CO: The Geological Society of America.

Strutt John William (Lord Rayleigh) (1916) On convective currents in a horizontal layer of fluid when the higher temperature is on the under side. *Philosophical Magazine* 32: 529–546.

Strutt Robert John (Fourth Baron Rayleigh) (1968) *Life of John William Strutt, Third Baron Rayleigh*. Madison, WI: University of Wisconsin Press.

Tackley P (2000a) The quest for self-consistent generation of plate tectonics in mantle convection models. In: Richards MA, Gordon R, and van der Hilst R *Geophysical Monograph Series, vol. 121: History and Dynamics of Global Plate Motions*, pp. 47–72. Washington, DC: American Geophysical Union.

Tackley P (2000b) Self-consistent generation of tectonic plates in time-dependent, three-dimensional mantle convection simulations. Part 2: strain weakening and asthenosphere. $(G^3)$ 1 (doi:10.1029/2000GC000,43).

Thompson Benjamin (Count Rumford) (1797) The propagation of heat in fluids. *Bibliotheque Britannique (Science et Arts)* 90: 97–200 (reprinted in Brown, 1967).

Tozer D (1985) Heat transfer and planetary evolution. *Geophysical Survey* 7: 213–246.

Turcotte D and Oxburgh E (1972) Mantle convection and the new global tectonics. *Annual Review of Fluid Mechanics* 4: 33–66.

Turekian KK and Narendra BL (2002) Earth sciences. In: Altman S (ed.) *Science at Yale*, pp. 157–170. New Haven, CT: Yale University.

van der Hilst R, Widiyantoro S, and Engdahl E (1997) Evidence for deep mantle circulation from global tomography. *Nature* 386: 578–584.

Walker J, Hayes P, and Kasting J (1981) A negative feedback mechanism for the long-term stabilization of Earth's surface temperature. *Journal of Geophysical Research* 86: 9776–9782.

Ward P and Brownlee D (2000) *Rare Earth*. New York: Copernicus - Springer Verlag.

Wegener A (1924) *The Origin of Continents and Oceans*. London: Methuen and Co.

Wesfried J (2006) Scientific biography of Henri Bénard (1874-1939). In: Mutabazi I, Wesfried J, and Guyon E (eds.) *Dynamics of Spatio-Temporal Cellular Structure – Henri Bénard Centenary Review*, pp. 9–37. New York: Springer.

Williams Q and Hemley R (2001) Hydrogen in the deep Earth. *Annual Review of Earth and Planetary Sciences* 29: 365–418.

# 2 Physics of Mantle Convection

**Y. Ricard**, Université de Lyon, Villeurbanne, France

## 2.1 Introduction

In many text books of fluid dynamics, and for most students, the word 'fluid' refers to one of the states of matter, either liquid or gaseous, in contrast to the solid state. This definition is much too restrictive. In fact, the definition of a fluid rests in its tendency to deform irrecoverably. Basically, any material that appears as elastic or nondeformable, with a crystalline structure (i.e., belonging to the solid state) or with a disordered structure (e.g., a glass, which from a thermodynamic point of view belongs to the liquid state) can be deformed when subjected to stresses for a long enough time.

The characteristic time constant of the geological processes related to mantle convection, typically $10 \, \text{My}$ $(3 \times 10^{14} \, \text{s})$, is so long that the mantle, although stronger than steel and able to transmit seismic shear waves, can be treated as a fluid. Similarly, ice, which is the solid form of water, is able to flow from mountain tops to valleys in the form of glaciers. A formalism that was developed for ordinary liquids or gases can therefore be used in order to study the inside of planets. It is not the equations themselves, but their parameters (viscosity, conductivity, spatial dimensions, etc.) that characterize their applicability to mantle dynamics.

Most materials can therefore behave like elastic solids on very short time constants and like liquids at long times. The characteristic time that controls the appropriate rheological behavior is the ratio between viscosity, $\eta$, and elasticity (shear modulus), $\mu_R$, called the Maxwell time $\tau_M$ (Maxwell, 1831–79):

$$\tau_M = \frac{\eta}{\mu_R} \qquad [1]$$

The rheological transition in some materials like silicon putty occurs in only a few minutes; a silicon ball can bounce on the floor, but it turns into a puddle when left on a table for tens of minutes. The transition time is of the order of a few hundred to a few thousand years for the mantle (see Section 2.3.2). Phenomena of a shorter duration than this time will experience the mantle as an elastic solid while most tectonic processes will experience the mantle as an irreversibly deformable fluid. Surface loading of the Earth by glaciation and deglaciation involves times of a few thousands years for which elastic aspects cannot be totally neglected with respect to viscous aspects.

Although the word 'convection' is often reserved for flows driven by internal buoyancy anomalies of thermal origin, in this chapter, we will more generally use 'convection' for any motion of a fluid driven by internal or external forcing. Convection can be kinematically forced by boundary conditions or induced by density variations. The former is 'forced' and the latter is 'free convection' which can be of compositional or thermal origin. We will however, mostly focus on the aspects of thermal convection (or Rayleigh–Bénard convection (Rayleigh, 1842–1919; Bénard, 1874–1939; see Chapter 1) when the fluid motion is driven by thermal anomalies and discuss several common approximations that are made in this case. We know, however, that many aspects of mantle convection can be more complex and involve compositional and petrological density anomalies or multiphase physics. We will therefore review some of these complexities.

The physics of fluid behavior, like the physics of elastic media, is based on the general continuum hypothesis. This hypothesis requires that quantities like density, temperature, or velocity are defined everywhere, continuously and at 'points' or infinitesimal volumes that contain a statistically meaningful number of molecules so that these quantities represent averages, independent of microscopic molecular fluctuations. This hypothesis seems natural for ordinary fluids at the laboratory scale. We will adopt the same hypothesis for the mantle although we know that it is heterogeneous at various scales and made of compositionally distinct grains.

## 2.2   Conservation Equations

The basic equations of this section can be found in more detail in many classical text books (Batchelor, 1967; Landau and Lifchitz, 1980). We will only emphasize the aspects that are pertinent for Earth's and terrestrial mantles.

### 2.2.1   General Expression of Conservation Equations

Let us consider a fluid transported by the velocity field **v**, a function of position **X**, and time $t$. There are two classical approaches to describe the physics in this deformable medium. Any variable $A$ in a flow can be considered as a simple function of position and time, $A(\mathbf{X}, t)$, in a way very similar to the specification of an electromagnetic field. This is the Eulerian point of view (Euler, 1707–1783). The second point of view is traditionally attributed to Lagrange (Lagrange, 1736–1813). It considers the trajectory that a material element of the flow initially at $\mathbf{X}_0$ would follow $\mathbf{X}(\mathbf{X}_0, t)$. An observer following this trajectory would naturally choose the variable $A(\mathbf{X}(\mathbf{X}_0, t), t)$. The same variable $A$ seen by an Eulerian or a Lagrangian observer would have very different time derivatives. According to Euler the time derivative would simply be the rate of change seen by an observer at a fixed position, that is, the partial derivative $\partial/\partial t$. According to Lagrange, the time derivative, noted with $D$, would be the rate of change seen by an observer riding on a material particle

$$\frac{\mathrm{D}A}{\mathrm{D}t} = \left( \frac{\mathrm{d}A(\mathbf{X}(\mathbf{X}_0, t), t)}{\mathrm{d}t} \right)_{\mathbf{X}_0} = \sum_{i=1,3} \frac{\partial A}{\partial X_i} \frac{\partial X_i}{\partial t} + \frac{\partial A}{\partial t} \quad [2]$$

where the $X_i$ are the coordinates of **X**. Since **X** is the position of a material element of the flow, its partial time derivative is simply the flow velocity **v**. The Lagrangian derivative is also sometimes material called the derivative, total derivative, or substantial derivative.

The previous relation was written for a scalar field $A$ but it could easily be applied to a vector field **A**.

The Eulerian and Lagrangian time derivatives are thus related by the symbolic relation

$$\frac{D}{Dt} = \frac{\partial}{\partial t} + (\mathbf{v} \cdot \nabla) \qquad [3]$$

The operator $(\mathbf{v} \cdot \nabla)$ is the symbolic vector $(v_1 \partial/\partial x_1, v_2 \partial/\partial x_2, v_3 \partial/\partial x_3)$ and a convenient mnemonic is to interpret it as the scalar product of the velocity field by the gradient operator $(\partial/\partial x_1, \partial/\partial x_2, \partial/\partial x_3)$. The operator $(\mathbf{v} \cdot \nabla)$ can be applied to a scalar or a vector. Notice that $(\mathbf{v} \cdot \nabla) \mathbf{A}$ is a vector that is neither parallel to $\mathbf{A}$ nor to $\mathbf{v}$.

In a purely homogeneous fluid, the flow lines are not visible and the mechanical properties are independent of the original position of fluid particles. In this case the Eulerian perspective seems natural. On the other hand, a physicist describing elastic media can easily draw marks on the surface of deformable objects and flow lines become perceptible for him. After an elastic deformation, the stresses are also dependent on the initial equilibrium state. The Lagrangian perspective is therefore more appropriate. We will mostly adopt the Eulerian perspective for the description of the mantle. However when we discuss deformation of heterogeneities embedded in and stirred by the convective mantle, the Lagrangian point of view will be more meaningful (see Section 2.5.1.7).

A starting point for describing the physics of a continuum are the conservation equations. Consider a scalar or a vector extensive variable (i.e., mass, momentum, energy, entropy, number of moles) with a density per unit volume $A$, and a virtual but fixed volume $\Omega$ enclosed by the surface $\Sigma$. This virtual volume is freely crossed by the flow. The temporal change of the net quantity of $A$ inside $\Omega$ is

$$\frac{d}{dt} \int_\Omega A \, dV = \int_\Omega \frac{\partial A}{\partial t} \, dV \qquad [4]$$

Since $\Omega$ is fixed, the derivative of the integral is the integral of the partial time derivative.

The total quantity of the extensive variable $A$ in a volume $\Omega$ can be related to a local production, $H_A$ (with units of $A$ per unit volume and unit time), and to the transport (influx or efflux) of $A$ across the interface. This transport can either be a macroscopic advective transport by the flow or a more indirect transport, for example, at a microscopic diffusive level. Let us call $\mathbf{J}_A$ the total flux of $A$ per unit surface

area. The conservation of $A$ can be expressed in integral form as

$$\int_\Omega \frac{\partial A}{\partial t} \, dV = -\int_\Sigma \mathbf{J}_A \cdot d\mathbf{S} + \int_\Omega H_A \, dV \qquad [5]$$

where the infinitesimal surface element vector $d\mathbf{S}$ is oriented with the outward unit normal; hence, the minus sign associates outward flux with a sink of quantity $A$. Equation [5] is the general form of any conservation equation. When the volume $\Omega$, surface $\Sigma$, and flux $\mathbf{J}_A$ are regular enough (in mathematical terms when the volume is compact, the surface piecewise smooth, and the flux continuously differentiable), we can make use of the divergence theorem

$$\int_\Sigma \mathbf{J}_A \cdot d\mathbf{S} = \int_\Omega \nabla \cdot \mathbf{J}_A \, dV \qquad [6]$$

to transform the surface integral into a volume integral. The divergence operator transforms the vector $\mathbf{J}$ with Cartesian coordinates $(\mathcal{J}_1, \mathcal{J}_1, \mathcal{J}_2)$ (Descartes, 1596–1650) into the scalar $\nabla \cdot \mathbf{J} = \partial \mathcal{J}_1/\partial x_1 + \partial \mathcal{J}_2/\partial x_2 + \partial \mathcal{J}_3/\partial x_3$, scalar product of the symbolic operator $\nabla$ by the real vector $\mathbf{J}$ (notice the difference between the scalar $\nabla \cdot \mathbf{J}$ and the operator $\mathbf{J} \cdot \nabla$). Since the integral eqn [5] is valid for any virtual volume $\Omega$, we can deduce that the general differential form of the conservation equation is

$$\frac{\partial A}{\partial t} + \nabla \cdot \mathbf{J}_A = H_A \qquad [7]$$

A similar expression can be used for a vector quantity $\mathbf{A}$ with a tensor flux $\mathbf{J}$ and a vector source term $\mathbf{H_A}$. In this case, the divergence operator converts the second-order tensor with components $\mathcal{J}_{ij}$ into a vector whose Cartesian components are $\Sigma_{j=1,3} \partial \mathcal{J}_{ij}/\partial x_j$.

We now apply this formalism to various physical quantities. Three quantities are strictly conserved: the mass, the momentum, and the energy. This means that they can only change in a volume $\Omega$ by influx or efflux across the surface $\Sigma$. We must identify the corresponding fluxes but no source terms should be present (in fact, in classical mechanics the radioactivity appears as a source of energy) (see also Section 2.2.5.3). One very important physical quantity is not conserved – the entropy – but the second law of thermodynamics insures the positivity of the associated sources.

## 2.2.2 Mass Conservation

The net rate at which mass is flowing is

$$\mathbf{J}_\rho = \rho\mathbf{v} \qquad [8]$$

Using either the Eulerian or the Lagrangian time derivatives, mass conservation becomes

$$\frac{\partial\rho}{\partial t} + \nabla\cdot(\rho\mathbf{v}) = 0 \qquad [9]$$

or

$$\frac{\mathrm{D}\rho}{\mathrm{D}t} + \rho\nabla\cdot\mathbf{v} = 0 \qquad [10]$$

In an incompressible fluid, particles have constant density, and so in the particle frame of reference, the Lagrangian observer does not see any density variation and $\mathrm{D}\rho/\mathrm{D}t = 0$. In this case, mass conservation takes the simple form $\nabla\cdot\mathbf{v} = 0$. This equation is commonly called the continuity equation although this terminology is a little bit vague.

Using mass conservation, a few identities can be derived that are very useful for transforming an equation of conservation for a quantity per unit mass to a quantity per unit volume. For example for any scalar field $A$,

$$\frac{\partial(\rho A)}{\partial t} + \nabla\cdot(\rho A\mathbf{v}) = \rho\frac{\mathrm{D}A}{\mathrm{D}t} \qquad [11]$$

and for any vector field $\mathbf{A}$,

$$\frac{\partial(\rho\mathbf{A})}{\partial t} + \nabla\cdot(\rho\mathbf{A}\otimes\mathbf{v}) = \rho\frac{\mathrm{D}\mathbf{A}}{\mathrm{D}t} \qquad [12]$$

where $\mathbf{A}\otimes\mathbf{v}$ is a dyadic tensor of components $A_i v_j$.

## 2.2.3 Momentum Conservation

### 2.2.3.1 General momentum conservation

The changes of momentum can be easily deduced by balancing the changes of momentum with the body forces acting in the volume $\Omega$ and the surface force acting on its surface $\Sigma$, that is, Newton's second law (Newton, 1642–1727). The total momentum is

$$\int_\Omega \rho\mathbf{v}\,\mathrm{d}V \qquad [13]$$

and its variations are due to

- advective transport of momentum across the surface $\Sigma$,
- forces acting on this surface, and
- internal body forces.

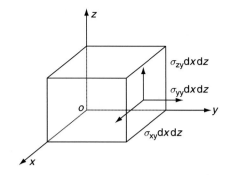

**Figure 1** The force per unit area applied on a surface directed by the normal vector $\mathbf{n}_i$ is by definition $\underline{\boldsymbol{\sigma}}\cdot\mathbf{n}_i$. The component of this force along the unit vector $\mathbf{e}_j$ therefore $\mathbf{e}_j\cdot\underline{\boldsymbol{\sigma}}\cdot\mathbf{n}_i$.

The momentum conservation of an open, fixed volume, can therefore be expressed in integral form as

$$\int_\Omega \frac{\partial(\rho\mathbf{v})}{\partial t}\,\mathrm{d}V = -\int_\Sigma \rho\mathbf{v}(\mathbf{v}\cdot\mathrm{d}\mathbf{S})$$
$$+ \int_\Sigma \underline{\boldsymbol{\sigma}}\cdot\mathrm{d}\mathbf{S} + \int_\Omega \mathbf{F}\,\mathrm{d}V \qquad [14]$$

The tensor $\underline{\boldsymbol{\sigma}}$ corresponds to the total stresses applied on the surface $\Sigma$ (see **Figure 1**). Our convention is that $\sigma_{ij}$ is the $i$-component of the force per unit area across a plane normal to the $j$-direction. The term $\mathbf{F}$ represents the sum of all body forces, and in particular the gravitational forces $\rho\mathbf{g}$ (we will not consider electromagnetic forces).

Using the divergence theorem (for the first term on the right-hand side, $\rho\mathbf{v}(\mathbf{v}\cdot\mathrm{d}\mathbf{S})$ can also be written $\rho(\mathbf{v}\otimes\mathbf{v})\cdot\mathrm{d}\mathbf{S}$) and the equality [12], the differential form of momentum conservation becomes

$$\rho\frac{\mathrm{D}\mathbf{v}}{\mathrm{D}t} = \nabla\cdot\underline{\boldsymbol{\sigma}} + \mathbf{F} \qquad [15]$$

It is common to divide the total stress tensor into a thermodynamic pressure $-P\underline{\mathbf{I}}$ where $\underline{\mathbf{I}}$ is the identity stress tensor, and a velocity-dependent stress $\underline{\boldsymbol{\tau}}$. The relationship between the tensor $\underline{\boldsymbol{\tau}}$ and the velocity field will be discussed later in Section 2.3.2. Without motion, the total stress tensor is thus isotropic and equal to the usual pressure. In most geophysical literature, it has been assumed that the velocity-dependent tensor has no isotropic component, that is, it is traceless $\mathrm{tr}(\underline{\boldsymbol{\tau}}) = 0$. In this case the thermodynamic pressure $P$ is the average isotropic stress, $\mathrm{tr}(\underline{\boldsymbol{\sigma}}) = -3P$, which is not the hydrostatic pressure (see Section 2.3.2 for more details). The velocity-dependent stress tensor $\underline{\boldsymbol{\tau}}$ is thus also the deviatoric stress tensor.

As $\nabla \cdot (P\underline{\mathbf{I}}) = \nabla P$, momentum conservation [14], in terms of pressure and deviatoric stresses, is

$$\rho \frac{D\mathbf{v}}{Dt} = -\nabla P + \nabla \cdot \underline{\boldsymbol{\tau}} + \mathbf{F} \qquad [16]$$

This equation is called the Navier–Stokes equation (Navier, 1785–1836; Stokes, 1819–1903) when the stress tensor is linearly related to the strain rate tensor and the fluid incompressible (see Section 2.3.2).

### 2.2.3.2 Inertia and non-Galilean forces

In almost all studies of mantle dynamics the fact that the Earth is rotating is simply neglected. It is however worth discussing this point. Let us define a reference frame of vectors $\mathbf{e}_i$ attached to the solid Earth. These vectors rotate with the Earth and with respect to a Galilean frame such that

$$\frac{d\mathbf{e}_i}{dt} = \boldsymbol{\omega} \times \mathbf{e}_i \qquad [17]$$

where $\boldsymbol{\omega}$ is the angular velocity of Earth's rotation. A point of the Earth, $\mathbf{X} = \Sigma_i X_i \mathbf{e}_i$, has a velocity in the Galilean frame:

$$\left(\frac{d\mathbf{X}}{dt}\right)_{\text{Gal}} = \sum_{i=1,3}\left[\left(\frac{dX_i}{dt}\right)\mathbf{e}_i + X_i\left(\frac{d\mathbf{e}_i}{dt}\right)\right]$$
$$= \mathbf{v}_{\text{Earth}} + \boldsymbol{\omega} \times \mathbf{X} \qquad [18]$$

where $\mathbf{v}_{\text{Earth}}$ is the velocity in the Earth's frame, and by repeating the derivation, an acceleration $\boldsymbol{\gamma}_{\text{Gal}}$ in a Galilean frame (Galileo, 1564–1642)

$$\boldsymbol{\gamma}_{\text{Gal}} = \boldsymbol{\gamma}_{\text{Earth}} + 2\boldsymbol{\omega} \times \mathbf{v} + \boldsymbol{\omega} \times (\boldsymbol{\omega} \times \mathbf{X}) + \frac{d\boldsymbol{\omega}}{dt} \times \mathbf{X} \qquad [19]$$

In this well-known expression, one recognizes on the right-hand side, the acceleration in the non-Galilean Earth reference frame, the Coriolis (Coriolis, 1792–1843), centrifugal and Poincaré accelerations (Poincaré, 1854–1912).

To quantify the importance of the three first acceleration terms (neglecting the Poincaré term), let us consider a characteristic length scale (the Earth's radius, $a = 6371$ km), and mantle velocity (the maximum plate tectonic speed, $U = 10$ cm yr$^{-1}$) and let us compare the various acceleration terms. One immediately gets

$$\frac{\text{inertia}}{\text{Coriolis}} = \frac{U}{2\omega a} = \frac{1}{2.9 \times 10^{11}} \qquad [20]$$

$$\frac{\text{Coriolis}}{\text{gravitational force}} = \frac{2\omega U}{g} = \frac{1}{2.1 \times 10^{13}} \qquad [21]$$

$$\frac{\text{centrifugal}}{\text{gravitational force}} = \frac{\omega^2 a}{g} = \frac{1}{291} \qquad [22]$$

Thus, the inertial term is much smaller than the Coriolis term (this ratio is also known as the Rossby number (Rossby, 1898–1957)), which is itself negligible relative to gravitational force. Even if we argue that a more meaningful comparison would be between the whole Coriolis force $2\rho\omega U$ and the lateral variations of the gravitational force $\delta\rho g$ (this ratio would be the inverse of the Eckman number, (Eckman, 1874–1954)), inertia and Coriolis accelerations still play a negligible role in mantle dynamics. Neglecting inertia means that forces are instantaneously in balance and that changes in kinetic energy are negligible since inertia is the time derivative of the kinetic energy. We can perform a simple numerical estimate of the mantle kinetic energy. The kinetic energy of a lithospheric plate (a square of size 2000 km, thickness 100 km, velocity 5 cm yr$^{-1}$, and density 3000 kg m$^{-3}$) is 1.67 kJ, which is comparable to that of a middle size car (2000 kg) driven at only 4.65 km h$^{-1}$!

The centrifugal term is also quite small but not so small (1/291 of gravitational force). It controls two effects. The first is the Earth's flattening with an equatorial bulge of 21 km (1/300 of Earth's radius) which is a static phenomenon that has no interactions with convective dynamics. The second effect is the possibility that the whole planet rotates along an equatorial axis in order to keep its main inertial axis coincident with its rotational axis (Spada et al., 1992a; Ricard et al., 1993b). This rotational equilibrium of the Earth will not be discussed here (see, e.g., Chandrasekhar (1969) for the static equilibrium shape of a rotating planet and, e.g., Munk and MacDonald (1960) for the dynamics of a deformable rotating body).

We neglect all the acceleration terms in the following but we should remember that in addition to the convective motion of a nonrotating planet, a rotation of the planet with respect to an equatorial axis is possible. This motion documented by paleomagnetism is called True Polar Wander (Besse and Courtillot, 1991).

### 2.2.3.3 Angular momentum conservation

The angular momentum per unit mass $\mathbf{J} = \mathbf{X} \times \mathbf{v}$ obeys a law of conservation. This law can be obtained in two different ways. First, as we did for mass and momentum conservation, we can express the balance of angular momentum in integral form. In the absence of intrinsic angular momentum sources, its variations are due to

- advective transport of angular momentum across the surface $\Sigma$,
- torque of forces acting on this surface, and
- torque of internal body forces.

The resulting balance is therefore

$$\int_\Omega \frac{\partial(\rho \mathbf{J})}{\partial t}\,dV = -\int_\Sigma \rho\mathbf{J}(\mathbf{v}\cdot d\mathbf{S}) + \int_\Sigma \mathbf{X}\times(\underline{\boldsymbol{\sigma}}\cdot d\mathbf{S}) \\ + \int_\Omega \mathbf{X}\times\mathbf{F}\,dV \qquad [23]$$

The only difficulty to transform this integral form into a local equation is with the integral involving the stress tensor. After some algebra, eqn [23] becomes

$$\rho\frac{D\mathbf{J}}{Dt} = \mathbf{X}\times\boldsymbol{\nabla}\cdot\underline{\boldsymbol{\sigma}} + \mathbf{X}\times\mathbf{F} + \mathcal{T} \qquad [24]$$

where the torque $\mathcal{T}$ is the vector $(\tau_{zy}-\tau_{yz},\,\tau_{xz}-\tau_{zx},\,\tau_{yx}-\tau_{xy})$. A second expression can be obtained by the vectorial multiplication of the momentum equation [15] by $\mathbf{X}$. Since

$$\mathbf{X}\times\frac{D\mathbf{v}}{Dt} = \frac{D\mathbf{J}}{Dt} - \frac{D\mathbf{X}}{Dt}\times\mathbf{v} = \frac{D\mathbf{J}}{Dt} - \mathbf{v}\times\mathbf{v} = \frac{D\mathbf{J}}{Dt} \qquad [25]$$

we get

$$\rho\frac{D\mathbf{J}}{Dt} = \mathbf{X}\times\boldsymbol{\nabla}\cdot\underline{\boldsymbol{\sigma}} + \mathbf{X}\times\mathbf{F} \qquad [26]$$

which differs from [24] by the absence of the torque $\mathcal{T}$. This proves that in the absence of sources of angular momentum, the stress (either $\underline{\boldsymbol{\sigma}}$ or $\underline{\boldsymbol{\tau}}$) must be represented by a symmetrical tensor,

$$\underline{\boldsymbol{\sigma}} = \underline{\boldsymbol{\sigma}}^{t}, \quad \underline{\boldsymbol{\tau}} = \underline{\boldsymbol{\tau}}^{t} \qquad [27]$$

where $[\,]^{t}$ denotes tensor transposition.

## 2.2.4  Energy Conservation

### 2.2.4.1  First law and internal energy
The total energy per unit mass of a fluid is the sum of its internal energy, $\mathcal{U}$, and its kinetic energy (this approach implies that the work of the various forces is separately taken into account; another approach that we use in Section 2.2.5.3, adds to the total energy the various possible potential energies and ignores forces). In the fixed volume $\Omega$, the total energy is thus

$$\int_\Omega \rho\left(\mathcal{U} + \frac{v^2}{2}\right)dV \qquad [28]$$

A change of this energy content can be caused by

- advection of energy across the boundary $\Sigma$ by macroscopic flow,
- transfer of energy through the same surface without mass transport, by say diffusion or conduction,
- work of body forces,
- work of surface forces, and
- volumetrically distributed radioactive heat production

Using the divergence theorem, the balance of energy can therefore be written as

$$\frac{\partial}{\partial t}\left(\rho\left(\mathcal{U} + \frac{v^2}{2}\right)\right) = -\boldsymbol{\nabla}\cdot\left(\rho\left(u + \frac{v^2}{2}\right)\mathbf{v} + \mathbf{q} + P\mathbf{v} - \underline{\boldsymbol{\tau}}\cdot\mathbf{v}\right) \\ + \mathbf{F}\cdot\mathbf{v} + \rho H \qquad [29]$$

where $\mathbf{q}$ is the diffusive flux, $H$ the rate of energy production per unit mass, and where the stresses are divided into thermodynamic pressure and velocity dependent stresses.

This expression can be developed and simplified by using [11] and the equations of mass and momentum conservation, [9] and [16] to reach the form

$$\rho\frac{D\mathcal{U}}{Dt} = -\boldsymbol{\nabla}\cdot\mathbf{q} - P\boldsymbol{\nabla}\cdot\mathbf{v} + \underline{\boldsymbol{\tau}}:\boldsymbol{\nabla}\mathbf{v} + \rho H \qquad [30]$$

The viscous dissipation term $\underline{\boldsymbol{\tau}}:\boldsymbol{\nabla}\mathbf{v}$ is the contraction of the two tensors $\underline{\boldsymbol{\tau}}$ and $\boldsymbol{\nabla}\mathbf{v}$ (of components $\partial v_i/\partial x_j$). Its expression is $\sum_{ij}\tau_{ij}\partial v_i/\partial x_j$.

### 2.2.4.2  State variables
The internal energy can be expressed in terms of the more usual thermodynamic state variables, namely, temperature, pressure, and volume. We use volume to follow the classical thermodynamics approach, but since we apply thermodynamics to points in a continuous medium, the volume $V$ is in fact the volume per unit mass or $1/\rho$. We use the first law of thermodynamics which states that during an infinitesimal process the variation of internal energy is the sum of the heat $\delta Q$ and reversible work $\delta W$ exchanged. Although irreversible processes occur in the fluid, we assume that we can adopt the hypothesis of a local thermodynamic equilibrium.

The increments in heat and work are not exact differentials: the entire precise process of energy exchange has to be known to compute these increments, not only the initial and final stages. Using either a $T - V$ or $T - P$ formulation, we can write

$$\delta Q = C_V dT + l dV = C_P dT + h dP \qquad [31]$$

where $C_P$ an $C_V$ are the heat capacities at constant pressure and volume, respectively, and $h$ and $l$ are two other calorimetric coefficients necessary to account for heat exchange at constant temperature. For fluids the reversible exchange of work is only due to the work of pressure forces

$$\delta W = -P\,dV \qquad [32]$$

This implies that only the pressure term corresponds to an energy capable to be stored and returned without loss when the volume change is reversed. On the contrary, the stresses related to the velocity will ultimately appear in the dissipative, irrecoverable term of viscous dissipation. This point will be further considered in Section 2.3.2 about rheology.

Thermodynamics states that the total variations of energy, $d\mathcal{U} = \delta Q + \delta W$, enthalpy, $d\mathcal{H} = d\mathcal{U} + d(PV)$, or entropy, $d\mathcal{S} = \delta Q / T$, are exact differentials and $\mathcal{U}$, $\mathcal{H}$, and $\mathcal{S}$ are potentials. This means that the net change in energy (enthalpy, entropy) between an initial and a final state depends only on the initial and final states themselves, and not on the intermediate stages. This implies mathematically that the second partial derivatives of these potentials with respect to any pair of variables are independent of the order of differentiation. Using these rules a large number of relations can be derived among the thermodynamic coefficients and their derivatives. These are called the Maxwell relations and are discussed in most thermodynamics textbooks (e.g., Poirier, 1991). We can in particular derive the values of $l$ (starting from $d\mathcal{U}$ and $d\mathcal{S}$ in $T - V$ formulation) and $h$ (starting from $d\mathcal{H}$ and $d\mathcal{S}$ in $T - P$ formulation),

$$l = \alpha T K_T \quad \text{and} \quad h = -\frac{\alpha T}{\rho} \qquad [33]$$

In these expressions for $l$ and $h$, we introduced the thermal expansivity $\alpha$ and the isothermal incompressibility $K_T$,

$$\alpha = \frac{1}{V}\left(\frac{\partial V}{\partial T}\right)_P = -\frac{1}{\rho}\left(\frac{\partial \rho}{\partial T}\right)_P$$

$$K_T = -V\left(\frac{\partial P}{\partial V}\right)_T = \rho\left(\frac{\partial P}{\partial \rho}\right)_T \qquad [34]$$

The thermodynamic laws and differentials apply to a closed deformable volume $\Omega(t)$. This corresponds to the perspective of Lagrange. We can therefore interpret the differential symbols 'd' of the thermodynamic definitions [31] of [32] as Lagrangian derivatives 'D'.

Therefore, in total, when the expressions for $l$ and $h$ are taken into account, [33], and when the differential symbols are interpreted as Lagrangian derivatives, the change of internal energy, $d\mathcal{U} = \delta Q + \delta W$, can be recast as

$$\frac{D\mathcal{U}}{Dt} = C_V\frac{DT}{Dt} + (\alpha T K_T - P)\frac{\nabla \cdot \mathbf{v}}{\rho} \qquad [35]$$

or

$$\frac{D\mathcal{U}}{Dt} = C_P\frac{DT}{Dt} - \frac{\alpha T}{\rho}\frac{DP}{Dt} - P\frac{\nabla \cdot \mathbf{v}}{\rho} \qquad [36]$$

In these equations we also have replaced the volume variation using mass conservation [9]

$$\frac{DV}{Dt} = \frac{D(1/\rho)}{Dt} = -\frac{1}{\rho^2}\frac{D\rho}{Dt} = \frac{\nabla \cdot \mathbf{v}}{\rho} \qquad [37]$$

### 2.2.4.3    Temperature

We can now employ either thermodynamic relation [35] or [36], in our conservation equation deduced from fluid mechanics, [30], to express the conservation of energy in terms of temperature variations

$$\rho C_P\frac{DT}{Dt} = -\nabla \cdot \mathbf{q} + \alpha T\frac{DP}{Dt} + \underline{\boldsymbol{\tau}} : \nabla\mathbf{v} + \rho H$$

$$\rho C_V\frac{DT}{Dt} = -\nabla \cdot \mathbf{q} + \alpha T K_T \nabla \cdot \mathbf{v} + \underline{\boldsymbol{\tau}} : \nabla\mathbf{v} + \rho H \qquad [38]$$

Apart from diffusion, three sources of temperature variations appear on the right-hand side of these equations. The last term $\rho H$ is the source of radioactive heat production. This term is of prime importance for the mantle, mostly heated by the decay of radioactive elements like $^{235}U$, $^{238}U$, $^{236}Th$, and $^{40}K$. Altogether these nuclides generate about $20 \times 10^{12}$ W (McDonough and Sun, 1995). Although this number may seem large, it is in fact very small. Since the Earth now has about $6 \times 10^9$ inhabitants, the total natural radioactivity of the Earth is only $\sim 3$ kW/person, not enough to run the appliances of a standard kitchen in a developed country. It is amazing that this ridiculously small energy source drives plate tectonics, raises mountains, and produces a magnetic field. In addition to the present-day radioactivity, extinct radionucleides, like that of $^{36}Al$ (with a half life of 0.73 My), have played an important role in the initial stage of planet formation (Lee et al., 1976).

The viscous dissipation term $\underline{\boldsymbol{\tau}} : \nabla\mathbf{v}$ converts mechanical energy into a temperature increase. This term explains the classical Joule experiment (equivalence between work and heat (Joule, 1818–89), in which the potential energy of a load (measured in joules) drives a propeller in a fluid and dissipates the mechanical energy as thermal energy (measured in calories).

The remaining source term, containing the thermodynamic coefficients ($\alpha$ or $\alpha K_T$ in (38)) cancels when the fluid is incompressible (e.g., when $\alpha = 0$ or when $\boldsymbol{\nabla} \cdot \mathbf{v} = 0$). This term is related to adiabatic compression and will be discussed in Section 2.4.2.2.

### 2.2.4.4 Second law and entropy

We now consider the second law of thermodynamics and entropy conservation. Assuming local thermodynamic equilibrium, we have $d\mathcal{U} = T d\mathcal{S} - P dV$. Using the equation of conservation for the internal energy $\mathcal{U}$, [30], and expressing the volume change in terms of velocity divergence, [37], we obtain

$$\rho T \frac{D\mathcal{S}}{Dt} = -\boldsymbol{\nabla} \cdot \mathbf{q} + \underline{\boldsymbol{\tau}} : \boldsymbol{\nabla} \mathbf{v} + \rho H \qquad [39]$$

To identify the entropy sources, we can express this equation in the form of a conservation equation (see [7]),

$$\frac{\partial(\rho\mathcal{S})}{\partial t} = -\boldsymbol{\nabla} \cdot \left(\rho\mathcal{S}\mathbf{v} + \frac{\mathbf{q}}{T}\right) - \frac{1}{T^2}\mathbf{q} \cdot \boldsymbol{\nabla} T$$
$$+ \frac{1}{T}\underline{\boldsymbol{\tau}} : \boldsymbol{\nabla}\mathbf{v} + \frac{1}{T}\rho H \qquad [40]$$

The physical meaning of this equation is therefore that the change of entropy is related to a flux of advected and diffused entropy, $\rho\mathcal{S}\mathbf{v}$ and $\mathbf{q}/T$, and to three entropy production terms, including from radiogenic heating.

A brief introduction to the general principles of nonequilibrium thermodynamics will be given in Section 2.5.1.4. Here, we simply state that the second law requires that in all situations, the total entropy production is positive. When different entropy production terms involve factors of different tensor orders (tensors, vectors, or scalars), they must separately be positive. This is called the Curie principle (Curie, 1859–1906) (see, e.g., de Groot and Mazur (1984) and Woods (1975)). It implies that

$$-\mathbf{q} \cdot \boldsymbol{\nabla} T \geq 0 \quad \text{and} \quad \underline{\boldsymbol{\tau}} : \boldsymbol{\nabla}\mathbf{v} \geq 0 \qquad [41]$$

The usual Fourier law (Fourier, 1768–1830) with a positive thermal conductivity $k > 0$,

$$\mathbf{q} = -k\boldsymbol{\nabla} T \qquad [42]$$

satisfies the second law.

When the conductivity $k$ is uniform, the thermal diffusion term of the energy equation $-\boldsymbol{\nabla} \cdot \mathbf{q}$ becomes $k\nabla^2 T$, where $\nabla^2 = \boldsymbol{\nabla} \cdot \boldsymbol{\nabla}$ is the scalar Laplacian operator (Laplace, 1749–1827). Instead of a thermal conductivity, a thermal diffusivity $\kappa$ can be introduced:

$$\kappa = \frac{k}{\rho C_P} \qquad [43]$$

(in principle, isobaric and isochoric thermal diffusivities should be defined). In situations with uniform conductivity, without motion and radioactivity sources, the energy equation [38] becomes the standard diffusion equation

$$\frac{\partial T}{\partial t} = \kappa \nabla^2 T \qquad [44]$$

The relation between stress and velocity satisfying [41] will be discussed in detail in Section 2.3.2. We will show that the relationship

$$\underline{\boldsymbol{\tau}} = 2\eta\left(\underline{\dot{\boldsymbol{\epsilon}}} - \frac{1}{3}\boldsymbol{\nabla} \cdot \mathbf{v}\right) \qquad [45]$$

is appropriate for the mantle, where the strain rate tensor $\underline{\dot{\boldsymbol{\epsilon}}}$ is defined by

$$\underline{\dot{\boldsymbol{\epsilon}}} = \frac{1}{2}\left([\boldsymbol{\nabla}\mathbf{v}] + [\boldsymbol{\nabla}\mathbf{v}]^t\right) \qquad [46]$$

Using this relation and assuming $\eta$ uniform, the divergence of the stress tensor that appears in the momentum conservation equation has the simple form

$$\boldsymbol{\nabla} \cdot \underline{\boldsymbol{\tau}} = \eta\nabla^2\mathbf{v} + \frac{\eta}{3}\boldsymbol{\nabla}(\boldsymbol{\nabla} \cdot \mathbf{v}) \qquad [47]$$

where the vectorial Laplacien $\nabla^2\mathbf{v}$ is $\boldsymbol{\nabla}(\boldsymbol{\nabla} \cdot \mathbf{v}) - \boldsymbol{\nabla} \times (\boldsymbol{\nabla} \times \mathbf{v})$. This relationship suggests a meaningful interpretation of the viscosity. The momentum equation [16] can be written as

$$\frac{\partial\mathbf{v}}{\partial t} = \frac{\eta}{\rho}\nabla^2\mathbf{v} + \text{other terms} \ldots \qquad [48]$$

forgetting the other terms, a comparison with the thermal diffusion equation [44] shows that the kinematic viscosity, $\nu$, defined by

$$\nu = \frac{\eta}{\rho} \qquad [49]$$

should rather be called the momentum diffusivity; it plays the same role with respect to the velocity as thermal diffusivity does with respect to temperature.

### 2.2.5 Gravitational Forces

#### 2.2.5.1 Poisson's equation

In this chapter, the only force is the gravitational body force. The gravity is the sum of this gravitational body force and the centrifugal force already discussed (see Section 2.2.3.2). The gravitational force per unit mass is the gradient of the gravitational potential $\psi$, a solution of Poisson's equation (Poisson, 1781–1840), that is,

$$\mathbf{g} = -\boldsymbol{\nabla}\psi \quad \text{and} \quad \nabla^2\psi = 4\pi G\rho \qquad [50]$$

where $G$ is the gravitational constant. In the force term that appears in the momentum equation [16], $\mathbf{F} = \rho\mathbf{g}$, the gravitational force per unit mass should be in agreement with the distribution of masses: the Earth should be self-gravitating.

### 2.2.5.2  Self-gravitation

When dealing with fluid dynamics at the laboratory scale, the gravitational force can be considered as constant and uniform. The gravitational force is related to the entire distribution of mass in the Earth (and the Universe) and is practically independent of the local changes in density in the experimental environment. Therefore, at the laboratory scale, it is reasonable to ignore Poisson's equation and to assume that $\mathbf{g}$ is a uniform and constant reference gravitational field.

Inside a planet, the density can be divided into an average depth-dependent density, $\rho_0(r)$, the source of the reference depth-dependent gravitational field, $\mathbf{g}_0(r)$, and a density perturbation $\delta\rho$, the source a gravitational perturbation $\delta\mathbf{g}$. The force term, $\mathbf{F}$, is therefore to first order $\rho_0\mathbf{g}_0 + \delta\rho\mathbf{g}_0 + \rho_0\delta\mathbf{g}$; it is tempting to assume that each term in this expression is much larger than the next one and hopefully that only the first two terms are of importance (neglecting the second term would suppress any feed back between density perturbations and flow). Practically, this assumption would imply consideration of the total density anomalies but only the depth-dependent gravitational field. Solving Poisson's equation to compute the perturbed gravitational field would thus be avoided. We can test the above idea and show that unfortunately, the third term, $+\rho_0\delta\mathbf{g}$, may be of the same order as the second one (Ricard *et al.*, 1984; Richards and Hager, 1984; Panasyuk *et al.*, 1996). To perform this exercise we have to introduce the spherical harmonic functions $Y_{lm}(\theta, \phi)$. These functions of latitude $\theta$ and longitude $\phi$ oscillate on a sphere just like two-dimensional (2-D) sinusoidal functions on a plane. Each harmonic function changes sign $l - m$ times from north to south pole, and $m$ times over the same angle (180°) around the equator. The degree $l$ can thus be interpreted as corresponding to a wavelength of order $2\pi a/l$, where $a$ is radius. Spherical harmonics constitute a basis for functions defined on the sphere and are also eigenfunctions of the angular part of Laplace's equation which facilitates the solution of Poisson's equation.

Let us consider a density anomaly $\delta\rho = \sigma\delta(r-a)\times Y_{lm}(\theta, \phi)$ at the surface of a sphere of radius $a$ and uniform density $\rho_0$ ($\delta(r-a)$ is the Dirac delta

function (Dirac, 1902–84), $\sigma$ has unit of $\mathrm{kg\,m^{-2}}$. This mass distribution generates inside the planet the radial gravitational perturbation field of

$$\delta g = 4\pi G\sigma \frac{l}{2l+1} \left(\frac{r}{a}\right)^{l-1} Y_{lm}(\theta, \phi) \qquad [51]$$

We can compare the terms $\rho_0\langle\delta g\rangle$ and $g_0\langle\delta\rho\rangle$, both averaged over the planet radius. For a uniform planet, the surface gravitational force per unit mass is $g_0 = 4/3\pi G\rho_0 a$. Since $\langle\delta\rho\rangle = \sigma Y_{lm}(\theta, \phi)/a$, we get

$$\frac{\rho_0\langle\delta g\rangle}{g_0\langle\delta\rho\rangle} = \frac{3}{2l+1} \qquad [52]$$

This estimate is certainly crude and a precise computation taking into account a distributed density distribution could be done. However, this rule of thumb would remain valid. At low degree the effect of self-gravitation $\rho_0\delta\mathbf{g}$ is about 50% of the direct effect $\delta\rho\mathbf{g}_0$ and reaches 10% of it only near $l \sim 15$. Self-gravitation has been taken into account in various models intended to explain the Earth's gravity field from mantle density anomalies (*see also* Chapter 4). Some spherical convection codes seem to neglect this effect, although it is important at the longest wavelengths.

### 2.2.5.3  Conservative forms of momentum and energy equations

In the general remarks on conservation laws in Section 2.2.1, we wrote that conserved quantities like mass, momentum, and energy can only be transported but do not have production terms (contrary to entropy). However, in the momentum conservation [16] and in the energy conservation [29], two terms, $\rho\mathbf{g}$ and $\rho\mathbf{g}\cdot\mathbf{v}$, appear as sources (we also said that the radioactive term $\rho H$ appears because the classical physics does not identify mass as energy and vice versa. A negligible term $-\rho H/c^2$, where $c$ is the speed of light, should, moreover, be present in the mass conservation).

It is interesting to check that our equations can be recast into an exact conservative form. An advantage of writing equations in conservative form is that it is appropriate to treat with global balances, interfaces, and boundaries (see Section 2.2.6). We can obtain conservative equations by using Poisson's relation and performing some algebra (using $\mathbf{g}\cdot\nabla\mathbf{g} = \nabla\mathbf{g}\cdot\mathbf{g}$, since $\mathbf{g} = -\nabla\psi$)

$$\rho\mathbf{g} = -\frac{1}{4\pi G}\nabla\cdot\left(\mathbf{g}\otimes\mathbf{g} - \frac{1}{2}g^2\mathbf{I}\right) \qquad [53]$$

$$\rho \mathbf{g} \cdot \mathbf{v} = -\rho \frac{D\psi}{Dt} - \frac{1}{4\pi G} \mathbf{\nabla} \cdot \left( \mathbf{g} \frac{\partial \psi}{\partial t} \right) - \frac{1}{8\pi G} \frac{\partial g^2}{\partial t} \quad [54]$$

If we substitute these two expressions in the momentum and the energy conservation equations, [16] and [29], we obtain the conservative forms

$$\frac{\partial(\rho \mathbf{v})}{\partial t} = -\mathbf{\nabla} \cdot \left( \rho \mathbf{v} \otimes \mathbf{v} + P\underline{\mathbf{I}} - \underline{\boldsymbol{\tau}} \right.$$
$$\left. + \frac{1}{4\pi G} \mathbf{g} \otimes \mathbf{g} - \frac{1}{8\pi G} g^2 \underline{\mathbf{I}} \right) \quad [55]$$

$$\frac{\partial}{\partial t} \left( \rho \left( \mathcal{U} + \psi + \frac{v^2}{2} \right) + \frac{g^2}{8\pi G} \right)$$
$$= -\mathbf{\nabla} \cdot \left( \rho \mathbf{v} \left( \mathcal{U} + \psi + \frac{v^2}{2} \right) + \mathbf{q} \right.$$
$$\left. + P\mathbf{v} - \underline{\boldsymbol{\tau}} \cdot \mathbf{v} + \frac{\mathbf{g}}{4\pi G} \frac{\partial \psi}{\partial t} \right) + \rho H \quad [56]$$

When the gravitational force is time independent, a potential $\psi$ can simply be added to the kinetic and internal energies to replace the work of gravitational forces. When gravitational force and its potential are time-dependent (due to mass redistribution during convection, segregation of elements, etc.), two new terms must be added; a gravitational energy proportional to $g^2$ and a gravitational flux proportional to $\mathbf{g}\partial\psi/\partial t$ (this is equivalent to the magnetic energy proportional to $B^2$ where $\mathbf{B}$ is the magnetic induction, in tesla (Tesla, 1856–1943), and to the Poynting vector of magneto-hydrodynamics (Poynting, 1852–1914)).

In a permanent or in a statistically steady regime, the time-dependent terms of energy equation [56] can be neglected and the equation can then be integrated over the volume of the Earth. The natural assumption is that the Earth's surface velocities are perpendicular to the Earth's surface normal vector and that the surface is either stress free or with no horizontal velocity (we exclude the case of convection forced by imposing a nonzero surface velocity). Using the divergence theorem to transform the volume integral of the divergence back to a surface integral of flux, most terms cancel and all that remains is

$$\int_{\Sigma} \mathbf{q} \cdot d\mathbf{S} = \int_{\Omega} \rho H \, dV \quad [57]$$

The surface flux in a statistically steady regime is simply the total radiogenic heat production.

It is surprising at first that viscous dissipation does not appear in this balance. To understand this point, we can directly integrate the energy equation written in terms of temperature [38],

$$\int_{\Omega} \rho C_P \frac{DT}{Dt} dV = -\int_{\Sigma} \mathbf{q} \cdot d\mathbf{S} + \int_{\Omega} \left( \alpha T \frac{DP}{Dt} + \underline{\boldsymbol{\tau}} : \mathbf{\nabla v} \right) dV$$
$$+ \int_{\Omega} \rho H \, dV \quad [58]$$

On the right-hand side, the first and last terms cancel each other out by [57]. On the left-hand side, we can use [11] to replace $\rho DT/Dt$ by $\partial(\rho T)/\partial t + \mathbf{\nabla} \cdot (\rho \mathbf{v} T)$. The modest assumptions that $C_P$ is a constant and that the temperature is statistically constant lead to

$$\int_{\Omega} \left( \alpha T \frac{DP}{Dt} + \underline{\boldsymbol{\tau}} : \mathbf{\nabla v} \right) dV = 0 \quad [59]$$

The total heat production due to dissipation is balanced by the work due to compression and expansion over the convective cycle (Hewitt et al., 1975). This balance is global, not local. Dissipation occurs mostly near the boundary layers of the convection and compressional work is done along the downwellings and upwellings of the flow.

### 2.2.6 Boundary and Interface Conditions

#### 2.2.6.1 General method

A boundary condition is a special case of an interface condition when certain properties are taken as known on one side of the interface. Sometimes the properties are explicitly known (e.g., the three velocity components are zero on a no slip surface), but often an interface condition simply expresses the continuity of a conserved quantity. To obtain the continuity conditions for a quantity $A$, the general method is to start from the conservation equation of $A$ in its integral form (see [5]). We choose a cylindrical volume $\Omega$ (a pill-box) of infinitely small radius $R$ where the top and bottom surfaces are located at a distance $\pm \epsilon$ from a discontinuity surface (see **Figure 2**). We choose two Cartesian axis $Ox$ and $Oy$, we call $\mathbf{n}$ the upward unit vector normal to the interface and $\mathbf{t}$, a radial unit vector, normal to the cylindrical side of the pill-box, and $\theta$ is the angle between $\mathbf{t}$ and $Ox$. If we now make the volume $\Omega(\epsilon)$ shrink to zero by

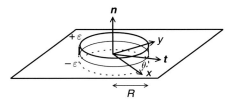

**Figure 2** The pillbox volume used to derive the interface and boundary conditions.

decreasing $\epsilon$ at constant $R$, the volume integrals of the time dependent and the source terms will also go to zero (unless the source term contains explicit surface terms like in the case of surface tension, but this is irrelevant at mantle scales). Since the surface of the pillbox $\Sigma(\epsilon)$ remains finite, we must have

$$\lim_{\epsilon \to 0} \int_{\Sigma(\epsilon)} \mathbf{J}_A \cdot d\mathbf{S} = 0 \qquad [60]$$

(the demonstration is here written for a vector flux, but is easily extended to tensor flux). This condition can also be written as

$$\pi R^2 \left[\mathbf{J}_A\right] \cdot \mathbf{n} + R \int_{-\epsilon}^{+\epsilon} \int_0^{2\pi} \mathbf{J} \cdot \mathbf{t} \, dz \, d\theta \qquad [61]$$

where $[X]$ is the jump of $X$ across the interface, sometimes noted $X^+ - X^-$. In most cases, the second term goes to zero with $\epsilon$ because the components of $\mathbf{J}$ are bounded, or is exactly zero when the flux is not a function of $\theta$ (since the double integral becomes the product of an integral in $z$ times $\int_0^{2\pi} \mathbf{t} \, d\theta = 0$). In these cases, the boundary condition for $A$ becomes

$$\left[\mathbf{J}_A\right] \cdot \mathbf{n} = 0 \qquad [62]$$

At an interface, the normal flux of $A$ must therefore be continuous. However in some cases, for example, when $\mathbf{J}$ varies with $x$ and $y$ but contains a $z$-derivative, the second term may not cancel and this happens in the case of boundaries associated with phase changes.

### 2.2.6.2  Interface conditions in the 1-D case and for bounded variables

Using the mass, momentum, energy, and entropy conservations in their conservative forms in (9), (55), (56), and (40) and assuming for now that no variable becomes infinite at an interface, the interface conditions in the reference frame where the interface is motionless are

$$[\rho \mathbf{v}] \cdot \mathbf{n} = 0$$

$$[\boldsymbol{\tau}] \cdot \mathbf{n} - [P]\mathbf{n} = 0$$

$$[\rho \mathbf{v} \mathcal{U} + \mathbf{q} - \boldsymbol{\tau} \cdot \mathbf{v} + P\mathbf{v}] \cdot \mathbf{n} = 0 \qquad [63]$$

$$\left[\rho \mathbf{v} \mathcal{S} + \frac{\mathbf{q}}{T}\right] \cdot \mathbf{n} = 0$$

(the gravitational force per unit mass and its potential are continuous). In these equations, we neglected the inertia and the kinetic energy terms in the second and third equations of [63] as appropriate for the mantle. When these terms are accounted for (adding $[-\rho \mathbf{v} \otimes \mathbf{v}] \cdot \mathbf{n}$ to the second equation and

$[\rho \mathbf{v} v^2/2] \cdot \mathbf{n}$ to the third), these equations are known as Hugoniot–Rankine conditions (Hugoniot, 1851–87; Rankine, 1820–72).

On any impermeable interfaces where $\mathbf{v} \cdot \mathbf{n} = 0$, the general jump conditions [63] without inertia imply that the heat flux, $[\mathbf{q}] \cdot \mathbf{n}$, the entropy flux $[\mathbf{q}/T] \cdot \mathbf{n}$ (and therefore the temperature $T$), and the stress components $[\boldsymbol{\tau}] \cdot \mathbf{n} - [P]\mathbf{n}$ are continuous. In 3-D, four boundary conditions are necessary on a surface to solve for the three components of velocity and for the temperature. The temperature (or the heat flux) can be imposed and, for the velocity and stress, either free slip boundary conditions ($\mathbf{v} \cdot \mathbf{n} = 0$, which is the first condition of [63] and $(\boldsymbol{\tau} \cdot \mathbf{n}) \times \mathbf{n} = 0$), or no slip boundary conditions ($\mathbf{v} = 0$), are generally used.

### 2.2.6.3  Phase change interfaces

Mantle minerals undergo several phase transitions at depth and at least two of them, the olivine $\rightleftharpoons$ wadsleyite and the ringwoodite $\rightleftharpoons$ perovskite + magnesiowustite transitions around 410 and 660 km depth, respectively, are sharp enough to be modeled by discontinuities. Conditions [63] suggest that $[\rho \mathbf{v}] \cdot \mathbf{n} = 0$ and $[\boldsymbol{\tau} \cdot \mathbf{n} - P\mathbf{n}] = 0$ and these seem to be the conditions used in many convection models. However as pointed by Corrieu et al. (1995), the first condition is correct, not the second one. The problem arises from the term in $\partial v_z/\partial z$ present in the rheological law [100] that becomes infinite when the material is forced to change its density discontinuously. To enforce the change in shape that occurs locally, the normal horizontal stresses have to become infinite and therefore their contributions to the force equilibrium of a pillbox do not vanish when the pillbox height is decreased.

To derive the appropriate interface condition we have to consider again [61] where $\mathbf{J}_A$ is substituted by $\boldsymbol{\sigma}$. The only terms may be unbounded on the interface are $\sigma_{xx}$, $\sigma_{yy}$, and $\sigma_{zz}$. Omitting the other stress components, that would make no contribution to the interface condition when $\epsilon$ goes to zero, the stress continuity becomes

$$\pi R^2 [\boldsymbol{\sigma}] \cdot \mathbf{n} + \mathbf{e}_x R \int_{-\epsilon}^{+\epsilon} \int_0^{2\pi} \sigma_{xx} \cos \theta \, dz \, d\theta +$$

$$R\mathbf{e}_y \int_{-\epsilon}^{+\epsilon} \int_0^{2\pi} \sigma_{yy} \sin \theta \, dz \, d\theta = 0 \qquad [64]$$

Since $R$ is small, we can replace the stresses on the cylindrical side of the pillbox by their first-order expansions, for example, $\sigma_{xx} = \sigma_{xx}(0) + (\partial \sigma_{xx}/\partial x) \times$

$R\cos\theta + (\partial\sigma_{xx}/\partial_y)R\sin\theta$ and perform the integration in $\theta$. After simplification by $\pi R^2$, one gets

$$[\boldsymbol{\sigma}] \cdot \mathbf{n} + \mathbf{e}_x \frac{\partial}{\partial x} \int_{-\epsilon}^{+\epsilon} \sigma_{xx}\,\mathrm{d}z + \mathbf{e}_y \frac{\partial}{\partial y} \int_{-\epsilon}^{+\epsilon} \sigma_{yy}\,\mathrm{d}z = 0 \quad [65]$$

This expression already demonstrates the continuity of $\sigma_{zz}$. Using $\sigma_{xx} = \sigma_{zz} + 2\eta\partial v_x/\partial x - 2\eta\partial v_z/\partial z$, and assuming that the viscosity remains uniform, we see that

$$\lim_{\epsilon\to 0}\int_{-\epsilon}^{\epsilon}\sigma_{xx}\,\mathrm{d}z = -2\eta\lim_{\epsilon\to 0}\int_{-\epsilon}^{\epsilon}\frac{\partial v_z}{\partial z}\,\mathrm{d}z = -2\eta[v_z] \quad [66]$$

The same result holds for the $\sigma_{yy}$ term. Since $v_z$ is discontinuous, forcing a sudden change in volume implies a discontinuity of the tangential stresses. The boundary conditions are thus

$$[\tau_{xz}] - 2\eta\frac{\partial}{\partial x}[v_z] = [\tau_{yz}] - 2\eta\frac{\partial}{\partial y}[v_z] = [\tau_{zz} - P] = [v_x]$$
$$= [v_y] = [\rho v_z] = 0 \quad [67]$$

When the kinetic energy is neglected, and the viscous stresses are much smaller than the pressure term, which are two approximations valid for the mantle, the last two boundary conditions are, assuming continuity of temperature,

$$\left[\rho\mathbf{v}\left(\mathcal{U} + \frac{P}{\rho}\right)\right] \cdot \mathbf{n} + [\mathbf{q}] \cdot \mathbf{n} = 0$$
$$[\rho\mathbf{v}\mathcal{S}] \cdot \mathbf{n} + \frac{1}{T}[\mathbf{q}] \cdot \mathbf{n} = 0 \quad [68]$$

The diffusive flux $\mathbf{q}$ can be eliminated from these two equations. Since $\rho\mathbf{v}$ is continuous and remembering that $\mathcal{U} + P/\rho$ is the enthalpy $\mathcal{H}$, we simply recognize the Clapeyron condition, which is latent heat release,

$$\Delta\mathcal{H} = T\Delta\mathcal{S} \quad [69]$$

where the enthalpy and entropy jumps, $[\mathcal{H}]$ and $[\mathcal{S}]$, were replaced by their more traditional notations, $\Delta\mathcal{H}$ and $\Delta\mathcal{S}$. The heat flux is discontinuous across an interface,

$$\Delta\mathcal{H}\rho\mathbf{v} \cdot \mathbf{n} + [\mathbf{q}] \cdot \mathbf{n} = 0 \quad [70]$$

and the discontinuity amounts to the enthalpy released by the mass flux that has undergone a chemical reaction or a phase change.

### 2.2.6.4 Weakly deformable surface of a convective cell

When a no slip condition is imposed at the surface, both normal and shear stresses are present at the boundary. These stresses, according to the second interface condition [63], must balance the force $-\boldsymbol{\tau} \cdot \mathbf{n} + P\mathbf{n}$ exerted by the fluid. This is reasonable for a laboratory experiment with a fluid totally enclosed in a tank whose walls are rigid enough to resist fluid traction. However in the case of free slip boundary conditions, it may seem strange that by imposing a zero vertical velocity, a finite normal stress results at the free surface. It is therefore worth discussing this point in more detail.

The natural boundary conditions should be that both the normal and tangential stresses applied on the free deformable surface, $z = h(x, y, t)$, of a convective fluid are zero

$$(\boldsymbol{\tau} \cdot \mathbf{n} - P\mathbf{n})_{\text{on } z = h} = 0 \quad [71]$$

(neglecting atmospheric pressure). In this expression the topography $h$ is unknown and the normal, computed at the surface of the planet, is $\mathbf{n} = (\mathbf{e}_z - \boldsymbol{\nabla}_H h)/\sqrt{1 + |\boldsymbol{\nabla}h|^2}$ where $\mathbf{e}_z$ is the unit vector along $z$, opposite to gravity.

The variation of topography is related to the convective flow and satisfies

$$\frac{\partial h}{\partial t} + \mathbf{v}_H^0 \cdot \boldsymbol{\nabla}_H h - v_z^0 = 0 \quad [72]$$

This equation expresses the fact that a material particle on the surface remains always on it. In this expression $\mathbf{v}_H^0$ and $v_z^0$ are the horizontal and vertical velocity components at the surface of the planet. We will see in Section 2.4 that lateral pressure and stress variations are always very small compared to the average pressure (this is because in most fluids, and in the mantle, the lateral density variations remain negligible compared to the average density). This implies that the surface topography is not much affected by the internal dynamics and remains close to horizontal, $|\boldsymbol{\nabla}_H h| \ll 1$. Boundary condition [71] and topography advection [72] can therefore be expanded to first order to give

$$(\boldsymbol{\tau} \cdot \mathbf{e}_z - P\mathbf{e}_z)_{\text{on } z = 0} \simeq -\rho_0 g_0 h\mathbf{e}_z \quad [73]$$

$$\frac{\partial h}{\partial t} = v_z^0 \quad [74]$$

where we again make use of the fact that the total stress remains close to hydrostatic, that is, $\mathbf{n} \cdot \boldsymbol{\tau} \cdot \mathbf{n} \ll P$ ($\rho_0$ and $g_0$ are the surface values of density and gravity). To first order, the stress boundary condition on a weakly deformable top surface is therefore zero shear stress but with a time-dependent normal stress related to the surface topography and vertical velocity.

The convection equations with these boundary conditions could be solved but this is not always useful. Since the boundary conditions involve both displacement $h$ and velocity $v_z^0$, the solution is akin to an eigenvalue problem. It can be shown that for an internal density structure of wavelength $\lambda$, $v_z^0$ goes to zero in a time of order $\bar{\eta}/\rho_0 g_0 \lambda$, where $\bar{\eta}$ is the typical viscosity of the underlying liquid over the depth $\lambda$ (Richards and Hager, 1984). For the Earth, this time is the characteristic time of postglacial rebound and is typically a few thousand years for wavelengths of a few thousand kilometers (e.g., Spada *et al.*, 1992b).

For convection, where the characteristic times are much longer, it is thus appropriate to assume that the induced topography is in mechanical equilibrium with the internal density structure. A zero normal velocity can therefore be imposed and the resulting normal stress can be used to estimate the topography generated by the convective flow. Internal compositional interfaces can be treated in a similar manner if they are only weakly deformable (i.e., when their intrinsic density jumps are much larger than the thermal density variations). This is the case for the core–mantle boundary (CMB).

For short wavelength structures and for rapid events (e.g., for a localized thermal anomaly impinging the Earth's surface), the time for topographic equilibration becomes comparable to the timescale of internal convective processes. In this case the precise computation of a history-dependent topography is necessary and the finite elasticity of the lithosphere, the coldest part of the mantle, plays an important role (Zhong *et al.*, 1996).

## 2.3 Thermodynamic and Rheological Properties

Section 2.2 on conservation equations is valid for all fluids (although the interface conditions are mostly discussed when inertia and kinetic energy are negligible). The differences between mantle convection and core, oceanic, or atmospheric convection come from the thermodynamic and transport properties of solids that are very different from those of usual fluids. We review some basic general properties of solids in Section 2.3 and will be more specific in Section 2.6.

### 2.3.1 Equation of State and Solid Properties

The equation of state of any material (EoS) relates its pressure, density, and temperature. The equation of state of a perfect gas, $PV/T = $ constant, is well known, but irrelevant for solids. Unfortunately, there is no equation for solids based on a simple and efficient theoretical model. In the Earth mineralogical community, the third-order finite strain Birch–Murnaghan EoS seems highly favored (Birch, 1952). This equation is cumbersome and is essentially empirical. More physical approaches have been used in Vinet *et al.* (1987), Poirier and Tarantola (1998), and Stacey and Davis (2004), but it seems that for each solid, the EoS has to be obtained experimentally.

In the simplest cases, the density varies around $\rho_0$ measured at temperature $T_0$ and pressure $P_0$ as

$$\rho = \rho_0 \left( 1 - \alpha(T - T_0) + \frac{P - P_0}{K_T} \right) \quad [75]$$

where the thermal expansivity $\alpha$ and incompressibility $K_T$ have been defined in [34]. This expression is a first-order expansion of any EoS. Equation [75] can however be misleading if one forgets that the parameters $\alpha$ and $K_T$ are not independent but must be related through Maxwell relations (e.g., their definitions [34] imply that $\partial(\alpha\rho)/\partial P = -\partial(\rho/K_T)/\partial T$).

Equation [75] can be used for a very simple numerical estimate that illustrates an important characteristics of solid Earth geophysics. Typically for silicates $\alpha \sim 10^{-5}\,\mathrm{K}^{-1}$, $K_T \sim 10^{11}\,\mathrm{Pa}$, while temperature variations in the mantle, $\Delta T$, are of a few 1000 K with a pressure increase between the surface and the core, $\Delta P$, of order of $10^{11}\,\mathrm{Pa}$. This indicates that the overall density variations due to temperature differences are negligible compared to those due to pressure differences ($\alpha\Delta T << 1$ but $\Delta P/K_T \sim 1$). In planets, to first order, the radial density is only a function of pressure, not of temperature. This is opposite to most liquid or solid laboratory experiments, where the properties are usually controlled by temperature.

A very important quantity in the thermodynamics of solids is the Grüneisen parameter (Grüneisen, 1877–1949)

$$\Gamma = \frac{\alpha K_T}{\rho C_V} = \frac{1}{\rho C_V}\left(\frac{\partial P}{\partial T}\right)_V \quad [76]$$

The Grüneisen parameter is dimensionless, does not vary much through the mantle ($\Gamma$ is typically around

1 within 50%), and can reasonably be considered as independent of the temperature. An empirical law (Anderson, 1979) relates $\Gamma$ with the density

$$\Gamma = \Gamma_0 \left(\frac{\rho_0}{\rho}\right)^q \qquad [77]$$

where $q$ is around 1. The Grüneisen parameter can also be related to the microscopic vibrational properties of crystals (Stacey, 1977). At high temperature, above the Debye temperature (Debye, 1884–1966), all solids have more or less the same heat capacity at constant volume. This is called the Dulong and Petit rule (Dulong, 1785–1838; Petit, 1791–1820). At high $T$, each atom vibrates and the thermal vibrational energy is equipartioned between the three dimensions of space (degrees of freedom) which leads to $C_{Vm} = 3R$ per mole of atoms, independent of the nature of the solid ($R$ is the gas constant). Assuming that the mantle is made of pure forsterite $Mg_2SiO_4$ that contains seven atoms for a molar mass of 140 g, its heat capacity at constant volume is therefore close to $C_{Vm} = 21R = 174.56 \, J \, K^{-1} \, mol^{-1}$ or $C_V = 1247 \, J \, K^{-1} \, kg^{-1}$. The approximate constancy of $C_V$ and the fact that $\Gamma$ is only a function of $\rho$ [77], allow us to integrate [76]:

$$P = F(\rho) + \alpha_0 K_T^0 (T - T_0) \left(\frac{\rho}{\rho_0}\right)^{1-q} \qquad [78]$$

where $\alpha_0$ and $K_T^0$ are the thermal expansivity and incompressibility at standard conditions and where $F(\rho)$ is a density-dependent integration constant. A rather simple but acceptable choice for the function $F(\rho)$, at least for mantle dynamicists, is the Murnaghan EoS (Murnaghan, 1951) at constant $T$ that allows us to write an EoS for solids of the form

$$P = \frac{K_T^0}{n} \left[\left(\frac{\rho}{\rho_0}\right)^n - 1\right] + \alpha_0 K_T^0 (T - T_0) \left(\frac{\rho}{\rho_0}\right)^{1-q} \qquad [79]$$

with an exponent $n$ of order of 3. This equation could easily be used to derive any thermodynamic property like $\alpha \, (P, T)$ or $K_T \, (P, T)$. This equation has been used implicitly in various models of mantle convection (e.g., Glatzmaier, 1988; Bercovici et al., 1989a, 1992). An important consequence of this EoS assuming $q \sim 1$ is that $\alpha K_T$ is more or less constant and that

$$K_T \sim K_T^0 \left(\frac{\rho}{\rho_0}\right)^n, \quad \alpha \sim \alpha_0 \left(\frac{\rho}{\rho_0}\right)^{-n} \qquad [80]$$

In the mantle, the incompressibility increases and the thermal expansion decreases significantly with depth. The geophysical consequences are further discussed in Section 2.6.5.1.

Two other thermodynamic equalities can also be straightforwardly deduced by chain rules of derivatives and will be used in the following. A relation between the two heat capacities $C_P$ and $C_V$ of the energy equations [38] can be derived from the two expressions for heat increments, [31] and the definition of $h$, [33],

$$C_P - C_V = \frac{\alpha T}{\rho} \left(\frac{\partial P}{\partial T}\right)_V \qquad [81]$$

The same expressions for heat increments, [31] and the $l$ and $h$ definitions, [33], imply that for an adiabatic transformation (when $\delta Q$ and $dS$ are zero),

$$\left(\frac{\partial P}{\partial T}\right)_S = \frac{\rho C_P}{\alpha T} \quad \text{and} \quad \left(\frac{\partial V}{\partial T}\right)_S = -\frac{C_V}{\alpha K_T T} \qquad [82]$$

Equations [81]–[82] take simpler forms when the Grüneisen parameter [76] and the adiabatic compressibility defined by

$$K_S = \rho \left(\frac{\partial P}{\partial \rho}\right)_S \qquad [83]$$

are used; they are

$$\frac{C_P}{C_V} = \frac{K_S}{K_T} = 1 + \Gamma \alpha T \qquad [84]$$

Since $\Gamma \sim 1$ and since $\alpha T << 1$, the two heat capacities are basically equal. It seems safer to assume that $C_V$ is constant (the Dulong and Petit rule) and infer $C_P$ from it. The incompressibility $K_S$ is defined similarly to $K_T$ but at constant entropy. The theory of elastic waves introduces this parameter that can be obtained from seismic observations

$$K_S = \rho \left(v_p^2 - \frac{4}{3} v_s^2\right) \qquad [85]$$

where $v_p$ and $v_s$ are the p and s, compressional and shear seismic, wave velocities. This important parameter provides a connection between geodynamics and seismology.

## 2.3.2 Rheology

In Section 2.2.3, no assumption is made on the rheology of the fluid, that is, on the relation between the stress tensor and the flow itself. In contrast, the discussion of energy conservation (Section 2.2.4) relies on the assumption that the pressure-related work is entirely recoverable [32]; as a consequence, the work of the deviatoric stresses ends up entirely as a dissipative term, hence a source of entropy. In a real fluid, this may be wrong for two reasons: part of the deviatoric stresses may be recoverable and part of the isotropic work may not be recoverable. In the first case, elasticity may be present, in the second case, bulk viscosity.

### 2.3.2.1   Elasticity

On a very short timescale, the mantle is an elastic solid in which compressional and shear waves propagate (e.g., Kennett, 2001). In an elastic solid, the linear strain tensor,

$$\underline{\epsilon}^e = \frac{1}{2}(\nabla \mathbf{u} + [\nabla \mathbf{u}]^t) \qquad [86]$$

where $\mathbf{u}$ is the displacement vector (this is valid for small deformations (see, e.g., Malvern, 1969; Landau and Lifchitz, 2000) for the large deformation case) is linearly related to the stress tensor,

$$\sigma^e_{ij} = A^e_{ijkl}\epsilon^e_{kl} \qquad [87]$$

where $\underline{\mathbf{A}}^e$ is the fourth-rank stiffness tensor. Since both the stress and the strain tensors are symmetric and because of the Maxwell thermodynamic relations for internal energy (including the elastic energy), $\partial^2\mathcal{U}/\partial\epsilon_{ij}\partial\epsilon_{kl} = \partial^2\mathcal{U}/\partial\epsilon_{kl}\partial\epsilon_{ij}$, the elastic tensor is invariant to permutations of $i$ and $j$, $k$ and $l$, $ij$ and $kl$. This leaves in the most general case of anisotropy, 21 independent stiffness coefficients (Malvern, 1969). In crystals, this number decreases with the number of symmetries of the unit cell. For isotropic elastic solids, only two parameters are needed, the incompressibility $K$ and the rigidity $\mu_R$ and the elastic behavior satisfies

$$\underline{\sigma}^e = K\,\mathrm{tr}(\underline{\epsilon}^e)\underline{\mathbf{I}} + 2\mu_R\left(\underline{\epsilon}^e - \frac{1}{3}\mathrm{tr}(\underline{\epsilon}^e)\underline{\mathbf{I}}\right) \qquad [88]$$

where $\mathrm{tr}(\underline{\epsilon}^e) = \nabla \cdot \mathbf{u}$.

Two remarks can be made on this rapid presentation of elasticity which are more deeply developed in textbooks of mechanics (e.g., Malvern, 1969; Landau and Lifchitz, 2000) or of seismology (e.g., Dahlen and Tromp, 1998). First, the expression [88] assumes that the displacement vector is computed from an initial situation where the solid is perfectly stress free, that is, $\underline{\sigma}^e = 0$ when $\underline{\epsilon}^e = 0$. In practical problems, only incremental displacements with respect to an initial prestressed state are known and $\underline{\sigma}^e$ has to be understood as a variation of the stress tensor. Second, temperature variations are associated with changes in elastic stresses and the incompressibility $K$ takes these variations into account. The incompressibility should be $K = K_S$ for rapid adiabatic seismic waves

and $K = K_T$ for isothermal variations. The other elastic parameters that are often introduced, Poisson's ratio, Young's modulus (Young, 1773–1829), Lamé's parameters (Lamé, 1795–1870), are simple functions of incompressibility and rigidity. Since the term proportional to $\mu_R$ is traceless, eqn [88] leads to, $\mathrm{tr}(\underline{\sigma}^e) = 3K\,\mathrm{tr}(\underline{\epsilon}^e)$, the rheology law can also be written in terms of compliance (i.e., getting $\underline{\epsilon}^e$ as a function of $\underline{\sigma}^e$):

$$\underline{\epsilon}^e = \frac{1}{9K}\mathrm{tr}(\underline{\sigma}^e)\underline{\mathbf{I}} + \frac{1}{2\mu_R}\left(\underline{\sigma}^e - \frac{1}{3}\mathrm{tr}(\underline{\sigma}^e)\underline{\mathbf{I}}\right) \qquad [89]$$

In these equations, the trace of the stress tensor can also be replaced by the pressure definition

$$\mathrm{tr}(\underline{\sigma}^e) = -3P \qquad [90]$$

The momentum equation [15] remains valid in a purely elastic solid (except that the advective transport is generally neglected, $D/Dt \sim \partial/\partial t$), but the discussion of energy conservation and thermodynamics is different for elastic and viscous bodies. The work the elastic stress is entirely recoverable: a deformed elastic body returns to its undeformed shape when the external forces are released. The internal energy change due to the storage of elastic stress is $\delta W = V\underline{\sigma}_e : \mathrm{d}\underline{\epsilon}^e$ instead of $\delta W = -P\mathrm{d}V$ and this is provided by the deformation work term $\underline{\tau} : \nabla\mathbf{v}$, which is therefore nondissipative. Thus, for an elastic body, the temperature equations [38] and the entropy equation [39] hold but with the $\underline{\tau} : \nabla\mathbf{v}$ source term removed.

### 2.3.2.2   Viscous Newtonian rheology

On a very long timescale, it is reasonable to assume that the internal deviatoric stresses become eventually relaxed and dissipated as heat. This is the assumption that we have implicitly made and that is usual in fluid mechanics. Since the dissipative term is $\underline{\tau} : \nabla\mathbf{v} = \underline{\tau} : \underline{\dot{\epsilon}}^v$ and must be positive according to the second law, this suggests a relationship between velocity-related stresses and velocity derivatives such that the total stress tensor has the form

$$\sigma^v_{ij} = -P\delta_{ij} + A^v_{ijkl}\dot{\epsilon}^v_{kl} \qquad [91]$$

$\delta_{ij}$ being the Kronecker symbol (Kronecker, 1823–91). Except for the time derivative, the only formal difference between this expression and [87] is that pressure exists in a motionless fluid but is always associated with deformation in an elastic solid.

Using the same arguments as for the elastic case, the viscous rheology in the isotropic case can therefore be written in term of stiffness

$$\underline{\sigma}^v = (-P + \zeta \mathrm{tr}(\underline{\dot{\epsilon}}^v))\mathbf{I} + 2\eta\left(\underline{\dot{\epsilon}}^v - \frac{1}{3}\mathrm{tr}(\underline{\dot{\epsilon}}^v)\mathbf{I}\right) \quad [92]$$

where $\mathrm{tr}(\underline{\dot{\epsilon}}^v) = \nabla \cdot \mathbf{v}$. Using $\mathrm{tr}(\underline{\sigma}^v) = 3(-P + \zeta \mathrm{tr}(\underline{\dot{\epsilon}}^v))$, the rheology can also be expressed in term of compliance

$$\underline{\dot{\epsilon}}^v = \frac{1}{9\zeta}(3P + \mathrm{tr}(\underline{\sigma}^v))\mathbf{I} + \frac{1}{2\eta}\left(\underline{\sigma}^v - \frac{1}{3}\mathrm{tr}(\underline{\sigma}^v)\mathbf{I}\right) \quad [93]$$

The two parameters $\eta$ and $\zeta$ are positive according to the second law and are called the shear and bulk viscosities. When they are intrinsic material properties (i.e., independent of the flow itself), the fluid is called linear or Newtonian. The hypothesis of isotropy of the rheology is probably wrong for a mantle composed of highly anisotropic materials (see Karato, 1998) but only a few papers have tried to tackle the problem of anisotropic viscosity (Christensen, 1997a; Muhlhaus et al., 2004).

Since $\mathrm{tr}(\underline{\sigma}^v)/3 = -P + \zeta\nabla \cdot \mathbf{v}$, the isotropic average of the total stress is not the pressure term, unless $\zeta\nabla \cdot \mathbf{v} = 0$. Therefore, part of the stress work, $\boldsymbol{\tau}: \underline{\dot{\epsilon}}^v$, during isotropic compaction could be dissipated in the form of the heat source $\zeta(\nabla \cdot \mathbf{v})^2$. A density-independent bulk viscosity allows an infinite compression under a finite isotropic stress. The bulk viscosity parameter $\zeta$ is generally only introduced to be immediately omitted and we will do the same. However, using [92] with $\zeta = 0$ but keeping $\nabla \cdot \mathbf{v} \neq 0$ does not seem valid since it would remove all resistance to isotropic compression. We will see that considering $\zeta = 0$ in [92] is formally correct although the real physical explanation is more complex: elastic stresses must be present to provide a resistance to isotropic viscous compression. The bulk viscosity, or some equivalent concept, is however necessary to handle two phase compression problems (McKenzie, 1984; Bercovici et al., 2001a) (see Section 2.5.2).

### 2.3.2.3  Maxwellian visco-elasticity

To account for the fact that the Earth behaves elastically on short time constants and viscously at long times, it is often assumed that under the same stress, the deformation has both elastic and viscous components. By summing the viscous compliance equation [93] with the time derivative of the elastic compliance equation, [89] and in the case of an infinite bulk viscosity $\zeta$, we get

$$\underline{\dot{\epsilon}} = \frac{1}{9K}\mathrm{tr}(\underline{\dot{\sigma}})\mathbf{I} + \frac{1}{2\eta}\left(\underline{\sigma} - \frac{1}{3}\mathrm{tr}(\underline{\sigma})\mathbf{I}\right)$$
$$+ \frac{1}{2\mu_R}\left(\underline{\dot{\sigma}} - \frac{1}{3}\mathrm{tr}(\underline{\dot{\sigma}})\mathbf{I}\right) \quad [94]$$

where $\underline{\sigma} = \underline{\sigma}^v = \underline{\sigma}^e$ and $\underline{\epsilon} = \underline{\epsilon}^v + \underline{\epsilon}^e$. This time-dependent rheological law is the constitutive law of a linear Maxwell solid.

A few simple illustrations of the behavior of a Maxwellian body will illustrate the physical meaning of eqn [94] (see also Chapter 4). First, we can consider the case where stress and strain are simple time-dependent sinusoidal functions with frequency $\omega$ (i.e., $\underline{\sigma} = \underline{\sigma}_0 \exp(i\omega t)$ and $\underline{\epsilon} = \underline{\epsilon}_0 \exp(i\omega t)$). The solution to this problem can then be used to solve other time-dependent problems by Fourier or Laplace transforms. Equation [94] becomes

$$\underline{\epsilon}_0 = \frac{1}{9K}\mathrm{tr}(\underline{\sigma}_0)\mathbf{I} + \frac{1}{2\mu_R}\left(1 - \frac{i}{\omega\tau}\right)\left(\underline{\sigma}_0 - \frac{1}{3}\mathrm{tr}(\underline{\sigma}_0)\mathbf{I}\right) \quad [95]$$

where $\tau = \eta/\mu_R$ is the Maxwell time, [1]. This equation can be compared to [89], and shows that the solution of a visco-elastic problem is formally equivalent to that of an elastic problem with a complex elastic rigidity. This is called the correspondence principle.

We can also solve the problem of a purely 1-D Maxwellian body (only $\sigma_{zz}$ and $\epsilon_{zz}$ are nonzero), submitted to a sudden load $\sigma_{zz} = \sigma_0 H(t)$ (where $H$ is the Heaviside distribution (Heaviside, 1850–1925)), or to a sudden strain $\epsilon_{zz} = \epsilon_0 H(t)$. The solutions are, for $t \geq 0$,

$$\epsilon_0 = \frac{1}{3k\mu_R}\sigma_0 + \frac{1}{3\eta}\sigma_0 t \quad [96]$$

and

$$\sigma_0 = 3k\mu_R \exp\left(-k\frac{t}{\tau}\right)\epsilon_0 \quad [97]$$

respectively, with $k = 3K/(3K + \mu_R)$. In the first case, the finite elastic deformation is followed by a steady flow. In the second case, the initial elastic stresses are then dissipated by viscous relaxation over a time constant, $\tau/k$. This time constant is different from the Maxwell time constant as both deviatoric and nondeviatoric stresses are present. For mantle material the time $\tau/k$ would however be of the same order as the Maxwell time constant $\tau$ (in the mid-mantle, $K \sim 2\mu_R \sim 200$ GPa).

From eqn [94], we can now understand what rheology must be used for a compressible viscous mantle. For phenomena that occur on time constants much larger than the Maxwell time, the deviatoric stresses can only be supported by the viscosity. As a typical viscosity for the deep mantle is in the range $10^{19}$–$10^{22}$ Pa s (see Sections 2.4 and 2.6), the appropriate Maxwell times are in the range 30 yr–30 kyr, much shorter than those of convection. By contrast, the isotropic stress remains only supported by elasticity in the approximation where the bulk viscosity $\zeta$ is infinitely large. The appropriate rheology for mantle convection is therefore given by

$$\dot{\underline{\epsilon}} = \frac{1}{9K}\mathrm{tr}(\dot{\underline{\sigma}})\mathbf{I} + \frac{1}{2\eta}\left(\underline{\sigma} - \frac{1}{3}\mathrm{tr}(\underline{\sigma})\mathbf{I}\right) \qquad [98]$$

This equation is simultaneously a rheology equation for the deviatoric stress and an EoS for the isotropic stress. Using $P = -\mathrm{tr}(\underline{\sigma})/3$, the stress tensor verifies

$$\underline{\sigma} = -P\mathbf{I} + 2\eta\left(\dot{\underline{\epsilon}} - \frac{\dot{P}}{3K}\mathbf{I}\right) \qquad [99]$$

This equation is intrinsically a visco-elastic equation, that can be replaced by a purely viscous equation plus an EoS

$$\underline{\sigma} = -P\mathbf{I} + 2\eta\left(\dot{\underline{\epsilon}} - \frac{1}{3}\mathrm{tr}(\dot{\underline{\epsilon}})\mathbf{I}\right) \qquad [100]$$

$$\mathrm{tr}(\epsilon) = \frac{P}{K} \qquad [101]$$

Equation [100] is therefore the appropriate limit of eqn [92] for slow deformation, when $\zeta = +\infty$ and when isotropic compaction is resisted by the elastic stresses.

The use of a Maxwell visco-elastic body to represent the mantle rheology on short timescale remains however rather arbitrary. Instead of summing the elastic and viscous deformations for the same stress tensor, another linear viscoelastic body could be obtained by partitioning the total stress into elastic and viscous components for the same strain rate. Instead of having the elasticity and the viscosity added like a spring and a dashpot in series (Maxwell rheology), this Kelvin–Voigt rheology would connect in parallel a viscous dashpot with an elastic spring (Kelvin, 1824–1907; Voigt, 1850–1919). Of course, further degrees of complexity could be reached by summing Maxwell and Voigt bodies, in series or in parallel. Such models have sometimes be used for the Earth but the data that could support or dismiss them are scarce (Yuen *et al.*, 1986).

### 2.3.2.4  Nonlinear rheologies

Even without elasticity and bulk viscosity, the assumption of a linear Newtonian rheology for the mantle is problematic. The shear viscosity cannot be a direct function of velocity since this would contradict the necessary Galilean invariance of material properties. However, the shear viscosity could be any function of the invariants of the strain rate tensor. There are three invariants of the strain rate tensor; its trace (but $\mathrm{tr}(\dot{\underline{\epsilon}}) = 0$, for an incompressible fluid), its determinant and the second invariant $I_2 = \sqrt{\dot{\underline{\epsilon}} : \dot{\underline{\epsilon}}}$ (where as in [30], the double dots denote tensor contraction). The viscosity could therefore be a function of $\det(\dot{\underline{\epsilon}})$ and $I_2$.

The main mechanisms of solid-state deformation pertinent for mantle conditions (excluding the brittle and plastic deformations) are either diffusion creep or dislocation creep (see Poirier (1991)). In the first case, finite deformation is obtained by summing the migrations of individual atoms exchanging their positions with crystalline lattice vacancies. In crystals, the average number of lattice vacancies $C$ varies with pressure, $P$, and temperature, $T$, according to Boltzmann statistics (Boltzmann, 1844–1906),

$$C \propto \exp\left(-\frac{PV}{RT}\right) \qquad [102]$$

($V$ is the atomic volume, $R$ the gas constant). A mineral is composed of grains of size $d$ with an average concentration of lattice vacancies $C_0$. A first-order expansion of [102] indicates that a gradient of vacancies of order $|\nabla C| \propto (C_0/d)(\tau V/RT)$ appears under a deviatoric stress $\tau$, due to the difference in stress regime between the faces in compression and the faces in extension ($\tau V \ll RT$) (see Poirier, 1991; Ranalli, 1995; Turcotte and Schubert, 1982; Schubert *et al.*, 2001). This induces a flux of atoms (number of atoms per unit surface and unit time)

$$\mathcal{J} \propto D\frac{C_0}{d}\frac{\tau}{RT} \qquad [103]$$

where $D$ is a diffusion coefficient. This flux of atoms goes from the grain faces in compression to the grain faces in extension. Along the direction of maximum compression, each crystal grain shortens by a quantity $\delta d$ which corresponds to a total transport of $d^2\delta d/V$ atoms. These atoms can be transported in a time $\delta t$ by the flux $\mathcal{J}_V$ across the grain of section $d^2$ (with volume diffusion $D_V$). They can also be transported by grain boundary flux $\mathcal{J}_B$ (with grain boundary diffusion $D_B$) along the grains interfaces through a surface $hd$ ($h$ being the thickness of the grain boundary), according to

$$\frac{d^2 \delta d}{V} \sim \mathcal{J}_D d^2 \delta t \quad \text{or} \quad \frac{d^2 \delta d}{V} \sim \mathcal{J}_B hd \delta t \quad [104]$$

As $\dot{\epsilon} = (\delta d / \delta t)/d$, the previous equations lead to the stress–strain rate relationship

$$\dot{\underline{\epsilon}} \propto \frac{V}{d^2 RT} \left( D_V + D_B \frac{h}{d} \right) \underline{\tau} \quad [105]$$

This diffusion mechanisms lead to a Newtonian rheology but with a grain-size dependence of the viscosity; $\eta \propto d^2$ for Nabarro–Herring creep with diffusion inside the grain (Nabarro, 1916–2006; Herring, 1914) and $\eta \propto d^3$ for Coble grain-boundary creep (Coble, 1928–92). The viscosity is also very strongly $T$-dependent not so much because of the explicit factor $T$ in [105], but because diffusion is a thermally activated process, $D \propto \exp(-E_{\text{diff}}/RT)$, where $E_{\text{diff}}$ is an activation enthalpy of diffusion.

In the case of dislocation creep, lines or planar imperfections are present in the crystalline lattice and macroscopic deformation occurs by slip motion along these imperfections, called dislocations. Instead of the grain size $d$ for diffusion creep, the mean spacing $d_d$ between dislocations provides the length scale. This distance is often found to vary as $1/I_2$. Therefore, instead of a diffusion creep with a viscosity in $d^n$ the resulting rheology is rather in $I_2^{-n}$ and is also thermally activated with an activation energy $E_{\text{dis}}$. Dislocation creep leads to a nonlinear regime where the equivalent viscosity varies with the second invariant with a power $-n$, where $n$ is typically of order 2,

$$\dot{\underline{\epsilon}} \propto I_2^n \exp(-E_{\text{dis}}/RT) \underline{\tau} \quad [106]$$

(this relationship is often written, in short, $\dot{\underline{\epsilon}} \propto \underline{\tau}^m$ with a stress exponent $m$ of order 3 but $\underline{\tau}^m$ really means $I_2^{m-1} \underline{\tau}$).

In general, for a given stress and a given temperature, the mechanism with the smallest viscosity (largest strain rate) prevails. Whether linear (grain size dependent), or nonlinear (stress dependent), viscosities are also strongly dependent upon temperature, pressure, melt content, water content, mineralogical phase, and oxygen fugacity (e.g., Hirth and Kolhstedt (1996)). In Section 2.6.3 we will further discuss the rheological mechanisms appropriate for the Earth.

## 2.4 Physics of Convection

The complex and very general system of equations that we have discussed in Sections 2.2 and 2.3 can be used to model an infinite number of mantle flow situations. Mantle flow can sometimes be simply modeled as driven by the motion of plates (some examples are discussed in Chapters 5, 7 and 8). It can also be induced by compositional density anomalies (some examples are discussed in Chapter 10). However, a fundamental cause of motion is due to the interplay between density and temperature and this is called thermal convection.

The phenomenon of thermal convection is common to all fluids (gas, liquid, and creeping solids) and it can be illustrated by simple experiments (*see* Chapter 3). The simplest can be done using water and an experimental setup called the shadowgraph method. Parallel light enters a transparent fluid put in a glass tank and is deflected where there are refractive index gradients due to temperature variations in the fluid. A pattern of bright regions and dark shadows is formed on a screen put on the other side of the tank. From this shadowgraph the structure of the temperature pattern can be qualitatively assessed (see examples of shadowgraphs in Tritton (1988)).

### 2.4.1 Basic Balance

From a simple thought experiment on thermal convection, we can derive the basic dynamic balance of convection. Let us consider a volume of fluid, $\Omega$, of characteristic size $a$, in which there is a temperature excess $\Delta T$ with respect to the surrounding fluid. The fluid is subject to a gravity $\mathbf{g}$, it has an average density $\rho$ and thermal expansivity $\alpha$. The volume $\Omega$, because of its anomalous temperature, experiences an Archimedian force, or buoyancy (Archimedes around 287–212 BC) given by

$$\mathbf{F} = -c_1 a^3 \rho \alpha \Delta T \mathbf{g} \quad [107]$$

($c_1$ is a constant taking into account the shape of $\Omega$, e.g., $c_1 = 4\pi/3$ for a sphere). If the volume $\Omega$ is in a fluid of viscosity $\eta$, it will sink or rise with a velocity given by Stokes law (Stokes, 1819–1903)

$$\mathbf{v}_s = -c_1 c_2 \frac{a^2 \rho \alpha \Delta T \mathbf{g}}{\eta} \quad [108]$$

($c_2$ is a drag coefficient accounting for the shape of $\Omega$, i.e., $c_2 = \pi/6$ for a sphere).

During its motion, the volume $\Omega$ exchanges heat by diffusion with the rest of the fluid and the diffusion equation [44] tells us that a time of order

$$t_d = c_3 \frac{\rho C_p a^2}{k} \qquad [109]$$

is needed before temperature equilibration with its surroundings. During this time, the fluid parcel travels the distance $l = v_s t_d$.

A natural indication of the possibility that the parcel of fluid moves can be obtained by comparing the distance $l$ to the characteristic size $a$. When $l >> a$, that is, when the fluid volume can be displaced by several times its size, motion will be possible. On the contrary when $l << a$, thermal equilibration will be so rapid or the Stokes velocity so slow that no motion will occur.

The condition $l >> a$, when $v_s$ and $t_d$ are replaced by the above expressions, depends on only one quantity, the Rayleigh number $Ra$

$$Ra = \frac{\rho^2 \alpha \Delta T g a^3 C_P}{\eta k} = \frac{\alpha \Delta T g a^3}{\kappa \nu} \qquad [110]$$

in terms of which motion occurs when $Ra >> 1$ (assuming that $c_1 c_2 c_3 \sim 1$). The Rayleigh number compares the driving mechanism (e.g., the Archimedian buoyancy) to the two resistive mechanisms, the diffusion of heat, represented by $\kappa$ (see [43]), and the diffusion of momentum, represented by $\nu$ (see [49]).

This simple balance suggests that a large nondimensional number $Ra$ favors fluid motion. How large $Ra$ needs to be is a question that we cannot address at this moment but it will be discussed in Section 2.4.4. Convection lifts hot fluid and causes cold fluid to sink (assuming $\alpha > 0$, which is true for most fluids and for the mantle). A convective system will rapidly reach an equilibrium where all thermal heterogeneities are swept up or down (if $Ra$ is large) or thermally equilibrated (if $Ra$ is small), unless a forcing mechanism continuously injects new cold parcels at the top and new hot parcels at the bottom. This can be done by cooling the top surface or heating the bottom one. When a fluid is heated from the side, a lateral temperature anomaly is constantly imposed and the liquid lateral thermal equilibration is prevented. The fluid remains in motion regardless of the amplitude of the imposed temperature anomaly.

### 2.4.2   Two Simple Solutions

#### 2.4.2.1   The diffusive solution
Trying to directly and exactly solve the mass, momentum, energy, and Poisson's equations and accounting for a realistic EoS would certainly be a formidable task. This complex system of equations has however two rather obvious but opposite solutions.

A steady and motionless solution is indeed possible. The assumption $\partial / \partial t = 0$ and $\mathbf{v} = 0$ satisfies the mass equation [9], the momentum equation [16] when the pressure is hydrostatic,

$$0 = -\nabla P + \rho \mathbf{g} \qquad [111]$$

and the energy equation [38] when the temperature is diffusive (using the Fourier law [42]),

$$\nabla \cdot (k \nabla T) + \rho H = 0 \qquad [112]$$

Solving analytically for the hydrostatic pressure and the diffusive temperature is trivial when $H$, $k$, and $\rho$ are uniform. For example, choosing a depth $z$ positive downward, we get

$$P = \rho g z, \quad T = T_0 + \Delta T \frac{z}{h} + \frac{1}{2} \rho H z (h - z) \qquad [113]$$

across a conductive solution with $T = T_0$, $P = 0$ at $z = 0$, and $T = T_0 + \Delta T$ at $z = h$. Computing analytically the conductive solution remains feasible, but could be quite cumbersome if one introduces a realistic EoS and computes gravity in agreement with the density distribution using Poisson's equation [50]. In Section 2.4.4 we will understand why the fluid does not necessarily choose the diffusive solution.

#### 2.4.2.2   The adiabatic solution
The previous diffusive solution was obtained for a steady motionless situation. However, the opposite situation where the velocities are very large also corresponds to a rather simple situation. The energy equation [38] can also be written as

$$\rho C_V T \left( \frac{\mathrm{D} \ln T}{\mathrm{D}t} - \Gamma \frac{\mathrm{D} \ln \rho}{\mathrm{D}t} \right) = -\nabla \cdot \mathbf{q} + \underline{\tau} : \nabla \mathbf{v} + \rho H \qquad [114]$$

or

$$\rho C_P T \left( \frac{\mathrm{D} \ln T}{\mathrm{D}t} - \frac{\alpha}{\rho C_P} \frac{\mathrm{D}P}{\mathrm{D}t} \right) = -\nabla \cdot \mathbf{q} + \underline{\tau} : \nabla \mathbf{v} + \rho H \qquad [115]$$

The right-hand sides of these equations were previously shown to be equal to $\rho T \mathrm{D}\mathcal{S}/\mathrm{D}t$ in [39]. If we decrease the viscosity in a fluid, the convective velocity increases. The advection terms, $\mathbf{v} \cdot \nabla T$, $\mathbf{v} \cdot \nabla \ln \rho$, and $\mathbf{v} \cdot \nabla P$ become then much larger than the time dependent, diffusion, and radioactive production terms. With a low viscosity, the fluid becomes also unable to sustain large stresses. As a consequence, when convection is vigorous enough, the fluid should evolve toward a situation where

$$(\nabla \ln T)_S - \Gamma(\nabla \ln \rho)_S = 0$$
$$(\nabla \ln T)_S - \frac{\alpha}{\rho C_P}(\nabla P)_S = 0 \qquad [116]$$

Since such equations imply that the entropy is exactly conserved, $DS/Dt = 0$, this equilibrium is called the adiabatic equilibrium. We added a subscript $[\ ]_S$ to denote the isentropic state.

Notice that, since the Grüneisen parameter is only density dependent, see [77], density and temperature are simply related along the adiabat. For example, if the Grüneisen parameter is a constant, $\Gamma_0$ (using $q = 0$ in [77]), the first of eqns [116] implies

$$T = T_0 \left(\frac{\rho}{\rho_0}\right)^{\Gamma_0} \qquad [117]$$

where $T_0$ and $\rho_0$ are two reference values. This equation implies that the adiabatic temperature increases by a factor 1.72 (e.g., from 1300 to 2230 K) from the asthenosphere ($\rho_0 \sim 3200 \text{ kg m}^{-3}$) to the CMB ($\rho \sim 5500 \text{ kg m}^{-3}$), if we assume $\Gamma_0 = 1$.

### 2.4.2.3 Stability of the adiabatic gradient

When a fluid is compressed, it heats up and cools down when decompressed. This is the same physics that explains why atmospheric temperature decreases with altitude. Of course, this adiabatic effect is vanishingly small in laboratory experiments, but not always in nature.

If a parcel of fluid is rapidly moved up or down along $z$ by a distance $a$, it changes its temperature adiabatically, by the quantity $a(dT/dz)_S$. However, the surrounding fluid will be at the temperature $a(dT/dz)$ where $dT/dz$ is just the temperature gradient, not necessarily adiabatic, of the fluid at rest. We can define $\Delta T_{na}$ as the nonadiabatic temperature: $\Delta T_{na} = a(dT/dz - (dT/dz)_S)$. The parcel being warmer or colder than the surroundings will rise or sink with a Stokes velocity that, rather than [108] will be of order

$$\mathbf{v}_s = -c_1 c_2 \frac{a^2 \rho \alpha \Delta T_{na} \mathbf{g}}{\eta} \qquad [118]$$

Since $z$ is depth, $dz$ is positive along $\mathbf{g}$, the adiabatic gradient is positive and the fluid parcel locally unstable when the gradient in the surrounding fluid is larger (superadiabatic) than the adiabatic gradient. On the contrary, a subadiabatic gradient is stable with respect to convection. It is therefore not the total temperature difference between the top and bottom of the fluid that drives motion, but only its nonadiabatic part.

To compare the Stokes velocity with the thermal equilibration time, we need to introduce a modified Rayleigh number

$$Ra = \frac{\alpha \Delta T_{na} g_0 a^3}{\kappa \nu} \qquad [119]$$

This number is based on the nonadiabatic temperature difference in excess of the adiabatic variation imposed over the height $a$.

We have shown that inside a convective cell, the thermal gradient should be superadiabatic. Superadiabaticity is the source of convective instability, but vigorous convective stirring involves largely rapid adiabatic vertical motion; so much of convecting fluid is indeed adiabatic while most of superadiabaticity is bound up in the thermal boundary layers where the vertical motion goes to zero. This mechanism suggests that an adiabatic reference background should not be such a bad assumption for a convective fluid.

This adiabaticity hypothesis should, however, not be taken too literally (Jeanloz and Morris, 1987). In most numerical simulations, the resulting averaged geotherm can be far (a few hundred kelvins) from adiabatic (Bunge et al., 2001). First, radioactive heating, dissipation, and diffusion are never totally negligible, second, even if each fluid parcel follows its own adiabatic geotherm, the average geotherm may not correspond to any particular adiabat.

### 2.4.3 Approximate Equations

#### 2.4.3.1 Depth-dependent reference profiles

We assume that the thermodynamic state is not far from an hydrostatic adiabat; thus, we choose this state as a reference and rewrite the equations of fluid dynamics in term of perturbations to this state (see also Jarvis and McKenzie, 1980; Glatzmaier, 1988; Bercovici et al., 1992). We denote all the reference variables with an overbar. We choose a reference hydrostatic pressure given by

$$\nabla \bar{P} = \bar{\rho}\bar{\mathbf{g}} \qquad [120]$$

and adiabatic temperature and densities obeying [116]

$$\nabla \bar{T} = \frac{\bar{\alpha}\bar{\mathbf{g}}}{\bar{C}_P} \bar{T}$$

$$\nabla \bar{\rho} = \frac{\bar{\alpha}\bar{\mathbf{g}}}{\bar{C}_P \Gamma} \bar{\rho}$$

[121]

where all the parameters are computed along the reference geotherm and where $\bar{\mathbf{g}}$ has been solved using Poisson's equation [50] with the reference density $\bar{\rho}$.

The reference parameters are depth dependent and usually, even for a simple EoS cannot be analytically obtained. They can however be computed numerically from [116] assuming that the Grüneisen parameter is only $\rho$-dependent. Using the EoS [79], all the thermodynamic quantities become functions of depth only, so that the reference profile can be obtained by quadratures.

### 2.4.3.2   Nondimensionalization

As a principle, the validity of equations cannot depend on the units in which the quantities are expressed. The laws of physics can only relate dimensionless combinations of parameters (e.g., Barenblatt, 1996). This is fundamental in fluid dynamics where a large number of quantities appear in the equations (*see also* Chapter 4). A necessary starting point is therefore to rephrase any fluid dynamics problem involving $N$-dimensional parameters in term of $M$ dimensionless quantities ($N - M \leq 0$ is the number of independent physical dimensions of the problem).

One cannot perform the nondimensionalization of the convection equations using the variable reference profiles. We must therefore introduce constant parameters, with indices $[\ ]_0$, corresponding to some typical or mantle-averaged values of the depth-dependent reference values. We introduce, for example, $\alpha_0$, $\rho_0$, $g_0$, or $C_P^0$.

The adiabatic equations [121] impose the natural scale for temperature variation, $\bar{C}_P/(\bar{\alpha} || \bar{\mathbf{g}} ||)$. This scale varies with depth but should not be too different from the value computed with the constant parameters (with indices 0). We therefore introduce a nondimensional number, the dissipation number $D_0$,

$$D_0 = \frac{\alpha_0 g_0 a}{C_P^0}$$

[122]

that compares the natural scale of temperature variations with the layer thickness, $a$. The dissipation number $D_0$ is around 0.5 for the Earth's mantle. For any laboratory experiment this number would be infinitely small, only geophysical or astrophysical

problems have large dissipation numbers. The hydrostatic and adiabatic reference profiles satisfy approximately

$$\frac{\mathrm{d}\bar{T}}{\mathrm{d}z} \sim D_0 \frac{\bar{T}}{a} \quad \text{and} \quad \frac{\mathrm{d}\bar{\rho}}{\mathrm{d}z} \sim \frac{D_0}{\Gamma_0} \frac{\bar{\rho}}{a}$$

[123]

A zero dissipation number leads to uniform reference temperature and pressure, as the adiabatic compression effects are not large enough to affect these quantities.

From top to bottom, the reference temperature increases adiabatically by $\Delta T_S$ while a total temperature jump $\Delta T_S + \Delta T_{\mathrm{na}}$ is imposed. Only the excess nonadiabatic temperature $\Delta T_{\mathrm{na}}$ is really useful to drive convection. The term driving convective instability is thus the dimensionless quantity $\epsilon$,

$$\epsilon = \alpha_0 \Delta T_{\mathrm{na}}$$

[124]

which is always a small number (all the necessary numerical values are listed in **Table 1**).

All these preliminaries have been somewhat lengthy but we are now ready to nondimensionalize the various equations. Then, we will get the approximate equations (Jarvis and Mckenzie, 1980), by simply taking into account that $\epsilon << 1$ (anelastic equations) or more crudely that both $\epsilon << 1$ and $D_0 << 1$ (the so-called Boussinesq equations, Boussinesq, 1842–1929). The mathematical formulation is heavy because of the number of symbols with or without overbars, tildes or indices, but is straightforward. The reader may jump directly to the results of Section 2.4.3.3.

To nondimensionalize the equations we can use the quantities $a$ and $\Delta T_{\mathrm{na}}$; we also need a characteristic velocity and pressure. Following our discussion of the basic force balance in Section 2.4.1, we use a typical Stokes velocity, $V_0 = a^2 g_0 \rho_0 \alpha_0 \Delta T_{\mathrm{na}}/\eta_0$, time $a/V_0$, and pressure $P_0 = \eta_0 V_0/a$. Using the definitions of $D_0$, [122], $\Gamma_0$, [76] and $\epsilon$, [124], we perform the following change of variables:

$$\mathbf{v} = \epsilon \frac{a K_T^0}{\eta_0} \frac{D_0}{\Gamma_0} \frac{C_P^0}{C_V^0} \tilde{\mathbf{v}}$$

$$P = \bar{P} + \epsilon K_T^0 \frac{D_0}{\Gamma_0} \frac{C_P^0}{C_V^0} \tilde{P}$$

$$\alpha_0 T = \alpha_0 \bar{T} + \epsilon \tilde{T}$$

[125]

$$\nabla = \frac{1}{a} \tilde{\nabla}$$

$$\frac{\partial}{\partial t} = \epsilon \frac{K_T^0}{\eta_0} \frac{D_0}{\Gamma_0} \frac{C_P^0}{C_V^0} \frac{\partial}{\partial \tilde{t}}$$

**Table 1** Typical parameter values for numerical models of mantle convection.

| | | Mantle | Core | |
|---|---|---|---|---|
| Size | $a$ | $3 \times 10^6$ | $3 \times 10^6$ | m |
| Dyn. viscosity | $\eta_0$ | $10^{21}$ | $10^{-3}$ | Pa s |
| Heat capacity | $C_P^0$ ou $C_V^0$ | 1000 | 700 | J K$^{-1}$ kg$^{-1}$ |
| Density | $\rho_0$ | 4000 | 11000 | kg m$^{-3}$ |
| Heat cond. | $k_0$ | 3 | 50 | W m$^{-1}$ K$^{-1}$ |
| Expansivity | $\alpha_0$ | $2 \times 10^{-5}$ | $10^{-5}$ | K$^{-1}$ |
| Temperature excess | $\Delta T_{na}$ | 1500 | 1? | K |
| Radiactivity prod. | $H$ | $7 \times 10^{-11}$ | 0? | W kg$^{-1}$ |
| Gravity | $g_0$ | 9.8 | 5 | m s$^{-2}$ |
| Incompressibility | $K_T^0$ | $10^{11}$ | $10^{12}$ | Pa |
| Kin. viscosity | $\nu = \eta_0/\rho_0$ | $2.5 \times 10^{17}$ | $9.1 \times 10^{-8}$ | m$^2$ s$^{-1}$ |
| Thermal diff. | $\kappa = k_0/(\rho_0 C_P^0)$ | $7.5 \times 10^{-7}$ | $6.5 \times 10^{-6}$ | m$^2$ S$^{-1}$ |
| Driving term | $\epsilon = \alpha_0 \Delta T_{na}$ | $3.0 \times 10^{-2}$ | $1.0^{-5}$ | |
| Dissip. number | $D_0$ | 0.59 | 0.21 | |
| Grüneisen par. | $\Gamma$ | 0.50 | 1.3 | |
| Rayleigh | $Ra$ | $4.2 \times 10^7$ | $2.2 \times 10^{27}$? | |
| Intern. Rayleigh | $Ra_H$ | $2.4 \times 10^{10}$ | 0? | |
| Prandtl | $Pr$ | $3.3 \times 10^{23}$ | $1.4 \times 10^{-2}$ | |
| Reynolds | $Re = Ra/Pr$ | $1.3 \times 10^{-15}$ | $1.6 \times 10^{29}$ | |
| Mach | $M$ | $5.0 \times 10^{-15}$ | $2.3 \times 10^{-16}$ | |
| | $RaPrM^2$ | $3.5 \times 10^{-2}$ | $1.6 \times 10^{-6}$ | |

To emphasize the drastic differences between the highly viscous mantle and a real liquid (in which shear waves do not propagate), we added estimates for the core assuming that core convection is so efficient that only 1 K of nonadiabatic temperature difference can be maintained across it. Notice that with only 1 K of temperature difference, the Rayleigh number of the fluid core would already reach $10^{27}$!

The EoS [75] can then be expanded at first order for a thermodynamic state close to the hydrostatic adiabatic reference

$$\rho = \bar{\rho} + \frac{\bar{\rho}}{\bar{K}_T}(P - \bar{P}) - \bar{\alpha}\bar{\rho}(T - \bar{T}) \qquad [126]$$

After nondimensionalization, the EoS becomes

$$\rho = \bar{\rho}\left(1 + \epsilon\left(\frac{K_T^0}{\bar{K}_T}\frac{D_0}{\Gamma_0}\frac{C_P^0}{C_V^0}\widetilde{P} - \frac{\bar{\alpha}}{\alpha_0}\widetilde{T}\right)\right) \qquad [127]$$

As $C_P^0 \sim C_V^0$, $K_T \sim K_T^0$, $\alpha \sim \alpha_0$, $\widetilde{P} \sim \widetilde{T} \sim 1$, and $\epsilon \ll 1$, it shows that the density remains close to the reference profile within terms of order $\epsilon$, $\rho = \bar{\rho}(1 + O(\epsilon))$.

### 2.4.3.3 Anelastic approximation

To approximate the equations of convection, we perform the change of variables, [125], and use the fact that $\epsilon$ is small (see also Schubert *et al.*, 2001). The choice of the reference state leads to the cancellation of the terms $O(1)$ of the mass, momentum, and energy equations, [9], [16], and [38].

At order $O(\epsilon)$, the first terms that remain in each equation are

$$\widetilde{\nabla} \cdot \left(\frac{\bar{\rho}}{\rho_0}\widetilde{\mathbf{v}}\right) = 0$$

$$\frac{Ra}{Pr}\frac{D\widetilde{\mathbf{v}}}{D\widetilde{t}} = -\widetilde{\nabla}\widetilde{P} + \widetilde{\nabla} \cdot \widetilde{\boldsymbol{\tau}} + \frac{\bar{\rho}}{\rho_0}\frac{\bar{\mathbf{g}}}{g_0}\frac{K_T^0}{\bar{K}_T}\frac{D_0}{\Gamma_0}\frac{C_P^0}{C_V^0}\widetilde{P}$$
$$\qquad\qquad - \frac{\bar{\rho}}{\rho_0}\frac{\bar{\mathbf{g}}}{g_0}\frac{\bar{\alpha}}{\alpha_0}\widetilde{T}$$

$$\frac{\bar{\rho}}{\rho_0}\frac{\bar{C}_P}{C_P^0}\frac{D\widetilde{T}}{D\widetilde{t}} = \frac{1}{Ra}\widetilde{\nabla} \cdot \left[\frac{\bar{k}}{k_0}\widetilde{\nabla}\left(\frac{\bar{T}}{\Delta T_{na}} + \widetilde{T}\right)\right] \qquad [128]$$
$$\qquad + \frac{\bar{\alpha}}{\alpha_0}\frac{\bar{\rho}}{\rho_0}\frac{\bar{g}}{g_0}D_0\widetilde{T}\widetilde{v}_g + \frac{\bar{\rho}}{\rho_0}\frac{1}{Ra}\frac{\rho_0 H a^2}{k_0 \Delta T}$$
$$\qquad + D_0\widetilde{\boldsymbol{\tau}} : \widetilde{\nabla}\widetilde{\mathbf{v}}$$

In the last equation, $\widetilde{v}_g$ is the component of the velocity field is along the radial reference gravity $\bar{\mathbf{g}} \cdot \widetilde{\mathbf{v}} = \bar{g}\widetilde{v}_g$ This new set of equations constitutes the equation of fluid dynamics in the anelastic approximation. The term 'anelastic' comes from the fact that the propagation of sound waves is impossible since the term in $\partial\rho/\partial t$ is neglected.

In eqns [128], we introduced the Rayleigh, $Ra$, and Prandtl, $Pr$, numbers

$$Ra = \frac{\alpha_0 \Delta T_{na} \rho_0^2 g_0 a^3 C_P^0}{\eta_0 k_0} = \frac{\alpha_0 \Delta T_{na} g_0 a^3}{\nu_0 \kappa_0} \quad [129]$$

$$Pr = \frac{\eta_0 C_P^0}{k_0} = \frac{\nu_0}{\kappa_0} \quad [130]$$

The physical meaning of the Rayleigh number as a measure of convective vigor has already been discussed. The Prandtl number (Prandtl, 1875–1953) compares the two diffusive processes: namely the diffusion of momentum and heat.

In the momentum equation, $\widetilde{\tau} = \underline{\tau} a/(\eta_0 V_0)$ is the nondimensionalized stress tensor which in the Newtonian case (without bulk viscosity) becomes

$$\widetilde{\underline{\tau}} = \frac{\eta}{\eta_0} \left( \widetilde{\nabla}\widetilde{v} + [\widetilde{\nabla}\widetilde{v}]' \right) - \frac{2}{3} \frac{\eta}{\eta_0} \nabla \cdot \widetilde{v} \quad [131]$$

The formalism is already so heavy that we have not included the self-gravitational term. This term is not negligible at long wavelengths (see Section 2.2.5.2). To account for this term we should have added on the right-hand side of second equation of [128] a term $-\bar{\rho}/\rho_0 \widetilde{\nabla}\widetilde{\psi}$, where the perturbed gravitational potential due to the departure of the density from the reference profile satisfies Poisson's equation

$$\widetilde{\nabla}^2 \widetilde{\psi} = 3 \frac{\bar{\rho}}{<\rho>} \left( \frac{K_T^0}{\bar{K}_T} \frac{D_0}{\Gamma_0} \frac{C_P^0}{C_V^0} \widetilde{P} - \frac{\bar{\alpha}}{\alpha_0} \widetilde{T} \right) \quad [132]$$

where $<\rho>$ is the average density of the Earth.

### 2.4.3.4  Dimensionless numbers

The ratio $Ra/Pr$ is also called the Grashof number $Gr = \alpha g \Delta T_{na} a^3/\nu^2$ (Grashof, 1826–1893). This number can also be written as $V_0 a/\nu$ and could be called the Reynolds number, $Re$, of the flow (Reynolds, 1842–1912). Using one or the other names depends on the quantities that are best known. For example, if the velocity $V$ is a parameter imposed by a boundary condition, using it to perform the nondimensionalization and speaking in terms of Reynolds number would be more natural than using the Grashof number. If a thermal structure is imposed by a velocity boundary condition (e.g., by the thickening of the oceanic lithosphere with age), it would seem natural to introduce a Péclet number $Va/\kappa$ (Péclet, 1793–1857), which is nothing more than the Rayleigh number of the flow if the velocity is imposed by the internal dynamics.

The Rayleigh and Prandtl numbers can be estimated in different ways. In fact, the only difficult parameter to know is the viscosity. In most textbooks the value of $10^{21}$ Pa s, first proposed by Haskell (1937), is given with a unanimity that hides very large uncertainties and most probably a large geographical variability. The mantle viscosity and its depth dependence can be constrained by postglacial rebound, geoid, true polar wander, change of flattening of the Earth, and plate force balance models or extrapolated from laboratory measurements. An increase of viscosity with depth between one or two orders of magnitudes is likely, with an asthenosphere significantly less viscous ($10^{19}$ Pa s) at least under oceanic plates and a lower mantle probably around $10^{22}$ Pa s (see details in Section 2.6.1). It is impossible to give justice to all the papers on this subject but some geodynamic estimates of mantle viscosity can be found in, for example, Peltier (1989), Sabadini and Yuen (1989), Lambeck and Johnston (1998), or Ricard et al. (1993a). Whatever the value of the real viscosity, the ratio $Ra/Pr$ is so small that inertia plays no role in the mantle and the left-hand side of the momentum equation in the anelastic approximation [128] can safely be set to zero (see numerical values in **Table 1**).

We can also introduce the Rayleigh and Prandtl number in the EoS [127] and define a new number, $M$,

$$\rho = \bar{\rho} \left( 1 + \frac{K_T^0}{\bar{K}_T} Ra Pr M^2 \widetilde{P} - \frac{\bar{\alpha}}{\alpha_0} \widetilde{T}\epsilon \right) \quad [133]$$

This number is the Mach number $M = v_D/v_M$, ratio of the velocity of thermal diffusion, $v_D = \kappa_0/a$, to the velocity $v_M = \sqrt{K_T/\rho}$ (Mach, 1838–1916). This last velocity is very close to the bulk velocity $v_\phi = \sqrt{K_S/\rho}$ (see [85]). The anelastic approximation is sometimes called small Mach number approximation. The Earth's Mach number is indeed of order $10^{-15}$ since thermal diffusion is much slower than the sound speed. However, the anelastic approximation really requires a small $Ra Pr M^2$ and this quantity is just $\epsilon D_0 C_P^0/\Gamma_0 C_V^0$, that is, of order $\epsilon = 10^{-2}$ (Bercovici et al., 1992). A planet could have a very low Mach number but so large a Prandtl number that the anelastic approximation would not be valid. Similarly, a planet can have a low Reynolds number (creeping convection) with a very large Rayleigh number (chaotic convection).

### 2.4.3.5  Boussinesq approximation

As was expected from Section 2.2.5.3, where we had shown that the dissipation and the adiabatic terms balance each other in a statistical steady-state regime, these terms are proportional to the same dissipation number $D_0$. Although $D_0$ is not so small, most of the physics of mantle convection, except for the additional adiabatic temperature gradient, is captured with models where $D_0$ is arbitrarily set to zero. This is due to the fact that the mantle viscosity is so high that a huge $\Delta T_{na}$ is maintained across the top and bottom boundary layers. The other sources and sinks of energy and density anomaly (viscous dissipation and adiabatic heating) are therefore always small (Jarvis and Mckenzie, 1980; Glatzmaier, 1988; Bercovici et al., 1992; Tackley, 1996).

In the Boussinesq approximation, $D_0 = 0$, the reference density and temperature become constants according to [123] and the EoS [127] indicates that the density becomes only a function of temperature (when the reference temperature is uniform we can also choose $C_P = C_V = \bar{C}_P = \bar{C}_V$). The nonadiabatic temperature increase $\Delta T_{na}$ becomes simply the total temperature increase $\Delta T$. This approximation is of course excellent for laboratory scale experiments where effectively $D_0 \ll 1$ and the fluid dynamics equations become

$$\widetilde{\boldsymbol{\nabla}} \cdot \widetilde{\mathbf{v}} = 0$$

$$\frac{Ra}{Pr}\frac{D\widetilde{\mathbf{v}}}{D\widetilde{t}} = -\widetilde{\boldsymbol{\nabla}}\widetilde{P} + \widetilde{\boldsymbol{\nabla}} \cdot \left( \frac{\eta}{\eta_0}\left( \widetilde{\boldsymbol{\nabla}}\widetilde{\mathbf{v}} + \left[ \widetilde{\boldsymbol{\nabla}}\widetilde{\mathbf{v}} \right]' \right) \right)$$

$$- \frac{\bar{\mathbf{g}}}{g_0}\frac{\bar{\alpha}}{\alpha_0}\widetilde{T} \qquad [134]$$

$$\frac{D\widetilde{T}}{D\widetilde{t}} = \frac{1}{Ra}\widetilde{\boldsymbol{\nabla}} \cdot \left[ \frac{\bar{k}}{k_0}\widetilde{\boldsymbol{\nabla}}\widetilde{T} \right] + \frac{1}{Ra}\frac{\rho_0 Ha^2}{k_0 \Delta T}$$

Here again as in the mantle, $Gr = Ra/Pr \ll 1$, the inertia in the momentum equation [134] can be neglected. The self-gravitational term $-\bar{\rho}/\rho_0 \widetilde{\boldsymbol{\nabla}}\widetilde{\psi}$ should be added to the momentum equation (second equation of [134]) for large-scale simulations, the gravitational potential being solution of [132] where only the thermal part of the density variations needs to be taken into account.

The physical behavior of a large Rayleigh number is obvious in [134]. When $Ra \to \infty$, the temperature becomes a purely advected and conserved quantity, $D\widetilde{T}/D\widetilde{t} = 0$.

### 2.4.3.6   Internal heating

In the nondimensionalization, we assumed that the nonadiabatic temperature $\Delta T_{na}$ and the radioactive sources are two independent quantities. Of course, in the case where the mantle is only heated from within, the excess temperature is not anymore a free

parameter but must result from the properties of the flow itself. In the nondimensionalization, we can replace $\Delta T_{na}$ by $\rho_0 Ha^2/k_0$ in such a way that the radioactive heat source of the anelastic or Boussinesq energy equations, [128] or [134], are simply $1/Ra$. This choice requires the introduction of a somewhat different Rayleigh number, the internally heated Rayleigh number (Roberts, 1967)

$$Ra_{\mathrm{H}} = \frac{\alpha_0 H \rho_0^3 g_0 a^5 C_P^0}{\eta_0 k_0^2} \qquad [135]$$

### 2.4.3.7   Alternative forms

Using the formulation of internal energy in terms of $C_V$, we would have reached the equivalent anelastic energy equation

$$\frac{\bar{\rho}}{\rho_0}\frac{\bar{C}_V}{C_P^0}\frac{D\widetilde{T}}{D\widetilde{t}} = \frac{1}{Ra}\widetilde{\boldsymbol{\nabla}} \cdot \left[ \frac{\bar{k}}{k_0}\widetilde{\boldsymbol{\nabla}}\left( \frac{\bar{T}}{\Delta T} + \widetilde{T} \right) \right]$$

$$+ \frac{\bar{\alpha}}{\alpha_0}\frac{\bar{\rho}}{\bar{\rho}_0}\frac{\bar{g}}{g_0}\frac{\bar{K}_T}{\bar{K}_S}D_0\widetilde{T}\widetilde{v}_z$$

$$+ \frac{\bar{\rho}}{\rho_0}\frac{1}{Ra}\frac{\rho_0 Ha^2}{k_0 \Delta T} + D_0\widetilde{\boldsymbol{\tau}} : \widetilde{\boldsymbol{\nabla}}\widetilde{\mathbf{v}} \qquad [136]$$

where the reference incompressibility $\bar{K}_S$ is the incompressibility measured along the reference adiabatic profile

$$\bar{K}_S = \bar{\rho}\left( \frac{\partial P}{\partial \rho} \right)_S = \bar{\rho}\frac{||\boldsymbol{\nabla}\bar{P}||}{||\boldsymbol{\nabla}\bar{\rho}||} = \frac{\bar{\rho}^2 \bar{g}}{||\boldsymbol{\nabla}\bar{\rho}||} \qquad [137]$$

This relationship is given here as a definition of $\bar{K}_S$, and this incompressibility is built from a theoretical hydrostatic and adiabatic model. However if the real Earth is indeed hydrostatic and adiabatic, then this relationship [137] relates the density gradient of the real Earth, $||\boldsymbol{\nabla}\rho(r)||$, to a seismological observation $K_S(r)/\rho(r)$. This is the important Bullen hypothesis (Bullen, 1940) used to build the reference density of the Earth (e.g., Dziewonski and Anderson, 1981).

### 2.4.3.8   Change of nondimensionalization

We used a Stokes velocity to nondimensionalize the equations. We have therefore introduced a velocity $V_0$ of order of $300\,\mathrm{m\,yr}^{-1}$ and a time $a/V_0 = 10\,000$ years (see **Table 1**). This is certainly very fast and short compared with geological scales. Most physical and geophysical textbooks (e.g., Schubert et al., 2001) use instead a diffusive time $t_D = \rho C_P a^2/k_0$ and velocity $a/t_D$. This is perfectly valid but **Table 1** shows that the diffusive time and velocity amount to $t_D = 400$ billion years and $V_D = 7 \times 10^{-6}\, \text{m\,yr}^{-1}$. These values are, on the contrary, very slow and long compared with geological scales. A nondimensionalization using a diffusive time scale leads to the anelastic equations

$$\tilde{\boldsymbol{\nabla}} \cdot \left( \frac{\bar{\rho}}{\rho_0} \tilde{\mathbf{v}} \right) = 0$$

$$\frac{1}{Pr} \frac{D\tilde{\mathbf{v}}}{D\tilde{t}} = -\tilde{\boldsymbol{\nabla}} \tilde{P} + \tilde{\boldsymbol{\nabla}} \cdot \tilde{\tau} + \frac{\bar{\rho}}{\rho_0} \frac{\bar{\mathbf{g}}}{g_0} \frac{K_T^0}{\bar{K}_T} \frac{D_0}{\Gamma_0} \frac{C_P^0}{C_V^0} \tilde{P}$$

$$\qquad - \frac{\bar{\rho}}{\rho_0} \frac{\bar{\mathbf{g}}}{g_0} \frac{\bar{\alpha}}{\alpha_0} Ra\, \tilde{T}$$

$$\frac{\bar{\rho}}{\rho_0} \frac{\bar{C}_P}{C_P^0} \frac{D\tilde{T}}{D\tilde{t}} = \tilde{\boldsymbol{\nabla}} \cdot \left[ \frac{\bar{k}}{k_0} \tilde{\boldsymbol{\nabla}} \left( \frac{\bar{T}}{\Delta T} + \tilde{T} \right) \right] \qquad [138]$$

$$\qquad + \frac{\bar{\alpha}}{\alpha_0} \frac{\bar{\rho}}{\rho_0} \frac{\bar{\mathbf{g}}}{g_0} D_0 \tilde{T} \tilde{v}_g + \frac{\bar{\rho}}{\rho_0} \frac{\rho_0 H a^2}{k_0 \Delta T}$$

$$\qquad + \frac{D_0}{Ra} \tilde{\tau} : \tilde{\boldsymbol{\nabla}} \tilde{\mathbf{v}}$$

and the Boussinesq equations,

$$\tilde{\boldsymbol{\nabla}} \cdot \tilde{\mathbf{v}} = 0$$

$$\frac{1}{Pr} \frac{D\tilde{\mathbf{v}}}{D\tilde{t}} = -\tilde{\boldsymbol{\nabla}} \tilde{P} + \tilde{\boldsymbol{\nabla}} \cdot \left( \frac{\eta}{\eta_0} \left( \tilde{\boldsymbol{\nabla}} \tilde{\mathbf{v}} + \left[ \tilde{\boldsymbol{\nabla}} \tilde{\mathbf{v}} \right]' \right) \right)$$

$$\qquad - \frac{\bar{\mathbf{g}}}{g_0} \frac{\bar{\alpha}}{\alpha_0} Ra\, \tilde{T} \qquad [139]$$

$$\frac{D\tilde{T}}{D\tilde{t}} = \tilde{\boldsymbol{\nabla}} \cdot \left[ \frac{\bar{k}}{k_0} \tilde{\boldsymbol{\nabla}} \tilde{T} \right] + \frac{1}{Ra} \frac{\rho_0 H a^2}{k_0 \Delta T}$$

Notice that the $Ra$ number appears in different places than in [128] or [134]. Of course, after their appropriate changes of variables, the dimensional solutions are the same.

### 2.4.4 Linear Stability Analysis for Basally Heated Convection

To understand why the diffusive solution is not necessarily the solution chosen by the fluid, the standard way to test the stability of a solution is what physicists call a study of marginal stability (see also Chapter 4). It consists of substituing into the basic equations a known solution plus an infinitely small perturbation and checking whether or not this perturbation amplifies, decreases, or propagates. It is only if the perturbation decreases in amplitude, that the tested solution is stable.

We use the Boussinesq approximation, with constant viscosity and conductivity, neglecting inertia, and without internal heating. The nondimensionalized

equations [134] (the tilde sign has been omitted for simplicity) can then be written as

$$\boldsymbol{\nabla} \cdot \mathbf{v} = 0$$

$$-\boldsymbol{\nabla} P + \nabla^2 \mathbf{v} - T \mathbf{e}_z = 0 \qquad [140]$$

$$\frac{\partial T}{\partial t} + \mathbf{v} \cdot \boldsymbol{\nabla} T = \frac{1}{Ra} \nabla^2 T$$

($\mathbf{e}_z$ is the normal vector directed along $\mathbf{g}$). The steady diffusive nondimensional temperature solution is $T = z$, and we test a solution of the form $T = z + \delta T$. The temperature boundary condition $T = 0$ on top and $T = 1$ at the bottom requires that $\delta T$ vanishes for $z = 0$ and $z = 1$. As in the diffusive case the velocity is zero, the velocity induced by $\delta T$ will be infinitely small $\delta \mathbf{v}$. In the nonlinear term, we can approximate $\mathbf{v} \cdot \boldsymbol{\nabla} T = \delta \mathbf{v} \cdot \boldsymbol{\nabla}(z + \delta T)$ by $\delta v_z = v_z$. With this approximation, the equations are linear and we can find a solution in the form of a plane wave.

For a fluid confined between $z = 0$ and $z = 1$ and unbounded in the $x$-direction, a solution $\delta T = \theta(t) \sin(\pi z)\sin(kx)$ is appropriate and satisfies the boundary conditions. This solution is 2-D, has a single mode in the $z$-direction, and is periodic in $x$ with wavelength $\lambda = 2\pi/k$. More complex patterns could be tried but the mode we have chosen would destabilize first (see Chapter 4). It is then straightforward to deduce that for such a thermal anomaly, the energy equation imposes a vertical velocity

$$v_z = -\left( \dot{\theta} + \frac{(k^2 + \pi^2)}{Ra} \theta \right) \sin(\pi z)\sin(kx) \qquad [141]$$

From mass conservation the $x$-component of the velocity must be

$$v_x = -\frac{\pi}{k} \left( \dot{\theta} + \frac{(k^2 + \pi^2)}{Ra} \theta \right) \cos(\pi z)\cos(kx) \qquad [142]$$

This flow has a vertical component that vanishes on the top and bottom surfaces where the horizontal component is maximum. The choice of the temperature structure corresponds to free-slip velocity conditions. When the velocity and the temperature are introduced in the momentum equation, the time evolution of the temperature perturbation is found:

$$\dot{\theta} = \theta \left( \frac{k^2}{(\pi^2 + k^2)^2} - \frac{(\pi^2 + k^2)}{Ra} \right) \qquad [143]$$

For any wave number $k$, a small enough Rayleigh number corresponds to a stable solution, $\dot{\theta}/\theta < 0$. When the Rayleigh number is increased, the

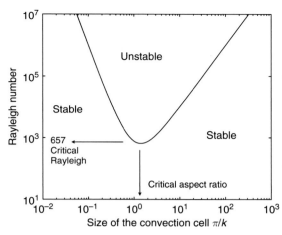

**Figure 3** Critical Rayleigh number as a function of the half wavelength π/k (the size of the convection cells). Above this curve, convection occurs with a whole range of unstable wavelengths. Below this curve, the conductive temperature is stable since temperature perturbations of any wavelength decrease. When the Rayleigh number is increased, the first unstable wavelength corresponds to a convection cell of aspect ratio √2 and a critical Rayleigh number of 657.

temperature component of wave number $k$ becomes unstable at the threshold Rayleigh number

$$Ra = \frac{(\pi^2 + k^2)^3}{k^2} \qquad [144]$$

This $Ra(k)$ curve is plotted in **Figure 3** This curve has a minimum when

$$k = \frac{\pi}{\sqrt{2}}, \quad Ra_c = \frac{27}{4}\pi^4 \sim 657 \qquad [145]$$

What can be interpreted as the size of one convective cell is $\pi/k$ since one wavelength corresponds to two contrarotating cells. When $Ra \gtrsim Ra_c$, the steady diffusive solution becomes marginally unstable and a convective cell with an aspect ratio, width over height, of $\sqrt{2}$, develops.

A Rayleigh number of 657 is the critical Rayleigh number for convection heated from below with free-slip boundary conditions. As soon as $Ra > Ra_c$, there is a wave number interval over which convection begins. Of course, when convection grows in amplitude, the marginal stability solution becomes less and less pertinent as the assumption that $\delta v \cdot \nabla \delta T << \delta v \cdot \nabla z$ becomes invalid.

### 2.4.5 Road to Chaos

In Cartesian geometry, when the Rayleigh number reaches its critical value, convection starts, and forms

rolls. When the Rayleigh number is further increased, complex series of convection patterns can be obtained, first stationary, then periodic, and finally, chaotic (*see* Chapter 3). Using the values of **Table 1**, the critical Rayleigh number of the mantle would be attained for a nonadiabatic temperature difference between the surface and the CMB of only 0.025 K! The mantle Rayleigh number is several orders of magnitude higher than critical and the mantle is in a chaotic state of convection.

**Figure 4** shows a stationary convection pattern at $Ra = 10^5$ and three snapshots of numerical simulation of convection at higher Rayleigh number. The color scale has been chosen differently in each panel to emphasize the thermal structures that decrease in length scale with $Ra$. This view is somewhat misleading since all the thermal anomalies become confined in a top cold boundary layer and in a hot bottom one at large Rayleigh numbers. Most of the interior of the cell becomes just isothermal (or adiabatic when anelastic equations are used). The various transitions of convection as the Rayleigh number increases will be discussed in other chapters of this treatise (*see* e.g., Chapters 3, 4 and 5).

## 2.5 Introduction to Physics of Multicomponent and Multiphase Flows

The mantle is not a simple homogeneous material. It is made of grains of variable bulk composition and mineralogy and contains fluids, magma, and gases. Discussion of multicomponent and multiphase flows could deal with solids, liquids, or gases, include compressibility or not, and consider elastic, viscous, or more complex rheology. For each combination of these characteristics a geophysical application is possible. Here we will restrict the presentation to viscous creep models (i.e., without inertia), where the various components are treated with continuous variables (i.e., each component is implicitly present everywhere). We do not consider approaches where the various components are separated by moving and deformable interfaces. Our presentation excludes cases where the problem is to match properties at macroscopic interfaces between regions of different but homogeneous compositions (e.g., Manga and Stone, 1993).

We will focus on two cases. First, when all the components are perfectly mixed in variable proportions. This corresponds to the classical chemical approach of multiple components in a solution. This will provide some tools to understand mantle

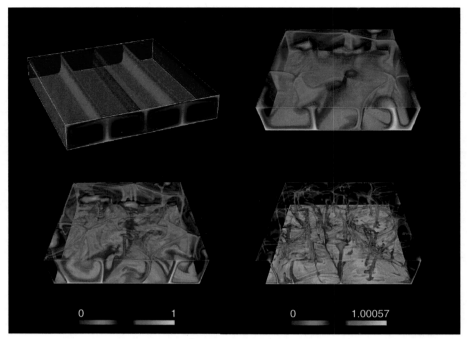

**Figure 4** Convection patterns of a fluid heated from below at Rayleigh number $10^5$, $10^6$, $10^7$, $10^8$. The temperature color bars range from 0 (top boundary) to 1 (bottom boundary). The Boussinesq approximation was used (numerical simulations by F. Dubuffet). The increase in Rayleigh number corresponds to a decrease of the boundary layer thicknesses and the width of plumes. Only in the case of the lowest Rayleigh number (top left) is the convection stationary with cells of aspect ratio $\sim \sqrt{2}$ as predicted by marginal stability. For higher Rayleigh number, the patterns are highly time dependent.

phase transitions and the physics of chemical diffusion and mixing. We will be rather formal and refer the applications and illustrations to other chapters of this treatise (e.g., Chapters 7 and 10). Our goal is to explain why and when the advection diffusion equation can be used in mantle dynamics. The irreversible thermodynamics of multicomponent flows is discussed in various classical books (e.g., Haase, 1990; de Groot and Mazur, 1984). However as usual with geophysical flows, the mantle has many simplifications and a few complexities that are not necessarily well documented in these classical textbooks.

The second case will be for two phase flows in which the two phases are separated by physical interfaces which are highly convolved and with spatial characteristics much smaller than the typical size of geodynamic models. This is typically the case where magma can percolate through a compacting matrix. This approach was used to model melt extraction and core–mantle interaction (McKenzie, 1984; Scott and Stevenson, 1984). Magma migration has also been treated in a large number of publications where solid and magma are considered as separated in

studies of dike propagation through hydraulic fracturing, (e.g., Lister and Kerr, 1991), or where fusion is parametrized in some way (e.g., Ito *et al.*, 1999; Choblet and Parmentier, 2001). We do not discuss these latter approaches.

### 2.5.1 Fluid Dynamics of Multicomponent Flows in Solution

#### 2.5.1.1 Mass conservation in a multicomponent solution

If we want to study the evolution of major or trace element concentration in the convecting mantle, we can consider the mantle, instead of a homogeneous fluid, as a solution of various components $i$ in volumetric proportions $\phi_i$ (with $\Sigma_i \phi_i = 1$) having the densities $\rho_i$ and velocities $\mathbf{v}_i$ (and later, thermal expansivities $\alpha_i$, heat capacities $C_P^i$, etc.).

Using a mass balance very similar to what we had discussed for a homogeneous fluid, we obtain mass conservation equations of the form

$$\frac{\partial(\phi_i \rho_i)}{\partial t} + \boldsymbol{\nabla} \cdot (\phi_i \rho_i \mathbf{v}_i) = \Gamma_i \qquad [146]$$

where $\Gamma_i$ is the rate of mass production of component $i$. This rate of mass production is zero if no reactions produce the component $i$.

In the fluid, the average density is

$$\bar{\rho} = \sum \phi_i \rho_i \qquad [147]$$

and various average velocities can be defined (weighted by the mass, the volume, the number of moles of each component $i$). In this section, we introduce the barycentric velocity, $\mathbf{v}_b$ (velocity of the center of mass), defined by

$$\mathbf{v}_b = \frac{\sum \phi_i \rho_i \mathbf{v}_i}{\bar{\rho}} \qquad [148]$$

The average mass conservation can be obtained by summing the equations of component conservation [146],

$$\frac{\partial \bar{\rho}}{\partial t} + \nabla \cdot (\bar{\rho} \mathbf{v}_b) = 0 \qquad [149]$$

since the sum of the rates of mass production is zero:

$$\sum_i \Gamma_i = 0 \qquad [150]$$

In eqn [146], instead of the various component velocities $\mathbf{v}_i$, we can introduce the barycentric velocity $\mathbf{v}_b$ and the diffusive flux of the component $i$ with respect to this average flow,

$$\frac{\partial (\phi_i \rho_i)}{\partial t} + \nabla \cdot (\phi_i \rho_i \mathbf{v}_b) = -\nabla \cdot \mathbf{J}_i + \Gamma_i \qquad [151]$$

where we define the diffusive flux, $\mathbf{J}_i$, by

$$\mathbf{J}_i = \phi_i \rho_i (\mathbf{v}_i - \mathbf{v}_b) \qquad [152]$$

By definition of the barycentric velocity [148], the sum of the diffusive flows just cancels out:

$$\sum_i \mathbf{J}_i = 0 \qquad [153]$$

A diffusive transport is nothing else than an advective transfer with respect to the average barycentric velocity. We will show later in simple cases that the diffusive transports are driven by concentration gradients (Woods, 1975; de Groot and Mazur, 1984; Haase, 1990).

If we introduce the mass fraction $C_i = \phi_i \rho_i / \bar{\rho}$ (in kg of $i$ per kg of mixture), we can easily show from [149] and [151] that

$$\bar{\rho} \frac{DC_i}{Dt} = -\nabla \cdot \mathbf{J}_i + \Gamma_i \qquad [154]$$

where the Lagrangian derivative is defined with the barycentric velocity:

$$\frac{D}{Dt} = \frac{\partial}{\partial t} + \mathbf{v}_b \cdot \nabla \qquad [155]$$

### 2.5.1.2 Momentum and energy in a multicomponent solution

In a multicomponent solution, all constituents are present at each point and they are all locally submitted to the same pressure and stresses. We assume that the viscous stress is simply related to $\mathbf{v}_b$ and we neglect inertia as appropriate for the mantle. Newton's second law (here, simply the balance of forces) can be applied to the barycenter and implies

$$\nabla \cdot \underline{\tau} - \nabla P + \bar{\rho} \mathbf{g} = 0 \qquad [156]$$

where the only force is due to the (constant) gravity. The momentum equation thus remains identical to that of a fluid with uniform composition and without inertia [16].

Since there is only one momentum equation for $i$ components, the $i - 1$ other velocity equations will be found by using the constraints of the laws of thermodynamics and in particular the positivity of the entropy source. To derive the energy conservation, we perform the standard balance to account for all the energy exchanges in a volume $\Omega$ and across its surface $\Sigma$. Instead of the one component equation [30], we have to sum up various contributions and we get

$$\sum_i \frac{\partial (\phi_i \rho_i \mathcal{U}_i)}{\partial t} = -\nabla \cdot \left( \sum_i \phi_i \rho_i \mathcal{U}_i \mathbf{v}_i \right.$$
$$\left. + P \sum_i \phi_i \mathbf{v}_i + \mathbf{q} - \underline{\tau} \cdot \mathbf{v}_b \right)$$
$$+ \mathbf{g} \cdot \sum_i \phi_i \rho_i \mathbf{v}_i + \bar{\rho} \bar{H} \qquad [157]$$

In this expression, we recognize the temporal changes in energy ($\mathcal{U}_i$ is the component internal energy per unit mass, the kinetic energies are neglected), the bulk energy flux, the pressure work, the thermal diffusion, the viscous stress work, the gravity work, and the radioactivity production ($\bar{\rho} \bar{H} = \Sigma_i \phi_i \rho_i H_i$). The various $\phi_i$ come from the assumption that each component $i$ is present in proportion $\phi_i$ in both the volume $\Omega$ and its surface $\Sigma$. We assume that thermal diffusion acts equally for each component and that the surface work of the stress tensor is only related to the barycentric velocity.

Using the definition of the barycentric velocity [148], of the diffusive fluxes, [152], of the momentum conservation, [156], and using $\Sigma_i \phi_i = 1$, the energy expression can be simplified to

$$\sum_i \phi_i \rho_i \frac{D\mathcal{H}_i}{Dt} = -\nabla \cdot \mathbf{q} - \sum_i \mathbf{J}_i \cdot \nabla\mathcal{H}_i + \frac{DP}{Dt} \\ - \sum_i \Gamma_i \mathcal{H}_i + \underline{\underline{\tau}} : \nabla\mathbf{v}_b + \bar{\rho}H \quad [158]$$

where $\mathcal{H}_i$ are the component enthalpies:

$$\mathcal{H}_i = \mathcal{U}_i + \frac{P}{\rho_i} \quad [159]$$

The enthalpy variation for each component $i$ can be expressed as a function of the state variables $P$ and $T$. From $d\mathcal{H}_i = \delta Q_i + V_i dP$ and the expression of the exchanged heat [31], we can write

$$\rho_i \frac{D\mathcal{H}_i}{Dt} = \rho_i C_P^i \frac{DT}{Dt} + (1 - \alpha_i T) \frac{DP}{Dt} \quad [160]$$

Finally, the expression for the temperature evolution is

$$\bar{\rho}\bar{C}_P \frac{DT}{Dt} = -\nabla \cdot \mathbf{q} - \sum_i \mathbf{J}_i \cdot \nabla\mathcal{H}_i + \bar{\alpha}T\frac{DP}{Dt} \\ - \sum_i \Gamma_i \mathcal{H}_i + \underline{\underline{\tau}} : \nabla\mathbf{v}_b + \bar{\rho}H \quad [161]$$

where the average heat capacity and thermal expansivity are $\bar{C}_P = \Sigma_i \phi_i \rho_i C_P^i / \bar{\rho}$ and $\bar{\alpha} = \Sigma_i \phi_i \alpha_i$.

Compared to the homogeneous case [38], two new heat source terms are present, the enthalpy exchange through chemical reactions, $\Sigma_i \Gamma_i \mathcal{H}_i$, and the enthalpy redistribution by component diffusion, $\Sigma_i \mathbf{J}_i \cdot \nabla\mathcal{H}_i$.

### 2.5.1.3 Entropy conservation in a multicomponent solution

Entropy conservation is essential for deriving the expressions of the diffusive fluxes. The general expression of entropy conservation [7] is

$$\sum_i \frac{\partial(\phi_i \rho_i \mathcal{S}_i)}{\partial t} = -\nabla \cdot \mathbf{J}_\mathcal{S} + H_\mathcal{S} \quad [162]$$

where $\mathbf{J}_\mathcal{S}$ and $H_\mathcal{S}$ are the yet unknown entropy flux and source. The entropy of the various components take into account their specific entropies as well as their configurational entropies or mixing entropies due to the dispersion of the component $i$ in the solution. Introducing the barycentric velocities and the diffusive fluxes, this equation can be recast as

$$\sum_i \phi_i \rho_i \frac{D\mathcal{S}_i}{Dt} = \nabla \cdot \left( \sum_i \phi_i \rho_i \mathcal{S}_i \mathbf{v}_b + \sum_i \mathcal{S}_i \mathbf{J}_i - \mathbf{J}_\mathcal{S} \right) \\ + H_\mathcal{S} - \mathcal{S}_i \Gamma_i - \mathbf{J}_i \cdot \nabla\mathcal{S}_i \quad [163]$$

However, a second expression of the entropy conservation can be obtained from the enthalpy conservation, [158]: using $d\mathcal{H}_i = T d\mathcal{S}_i + V_i dP$, which in our case can be expressed as

$$\rho_i \frac{D\mathcal{H}_i}{Dt} = \rho_i T\frac{D\mathcal{S}_i}{Dt} + \frac{DP}{Dt} \quad [164]$$

we derive

$$\sum_i \phi_i \rho_i T\frac{D\mathcal{S}_i}{Dt} = -\nabla \cdot \mathbf{q} - \sum_i \mathbf{J}_i \cdot \nabla\mathcal{H}_i \\ - \sum_i \Gamma_i \mathcal{H}_i + \underline{\underline{\tau}} : \nabla\mathbf{v}_b + \bar{\rho}H \quad [165]$$

A comparison of the two expressions for the entropy conservation, [163] and [165], allows us to identify the total entropy flux

$$\mathbf{J}_\mathcal{S} = \frac{\mathbf{q}}{T} + \mathbf{v}_b \sum_i \phi_i \rho_i \mathcal{S}_i + \sum_i \mathcal{S}_i \mathbf{J}_i \quad [166]$$

and the entropy sources

$$TH_\mathcal{S} = -\left( \frac{\mathbf{q}}{T} + \sum_i \mathcal{S}_i \mathbf{J}_i \right) \cdot \nabla T - \sum_i \mathbf{J}_i \cdot \nabla\mu_i \\ - \sum_i \Gamma_i \mu_i + \underline{\underline{\tau}} : \nabla\mathbf{v}_b + \bar{\rho}H \quad [167]$$

where we introduced the chemical potentials $\mu_i = \mathcal{H}_i - T\mathcal{S}_i$. The total entropy flux, [166], is related to thermal diffusion and to advection and chemical diffusion of component entropies.

In [167], the gradients of chemical potential and temperature are not independent as the chemical potential gradients implicitly include the temperature gradient, so that alternative expressions can be found. For example, using

$$\nabla\mu_i = T\nabla\frac{\mu_i}{T} + \mu_i \frac{\nabla T}{T} \quad [168]$$

and $\mu_i = \mathcal{H} - T\mathcal{S}_i$, the entropy source [167] can be written as

$$TH_\mathcal{S} = -\left( \mathbf{q} + \sum_i \mathcal{H}_i \mathbf{J}_i \right) \frac{\nabla T}{T} - T\sum_i \mathbf{J}_i \cdot \nabla\frac{\mu_i}{T} \\ - \sum_i \Gamma_i \mu_i + \underline{\underline{\tau}} : \nabla\mathbf{v}_b + \bar{\rho}H \quad [169]$$

We can also introduce the gradient of $\mu$ at constant temperature $\nabla_T\mu$ as

$$\nabla_T\mu_i = \nabla\mu_i + \mathcal{S}_i \nabla T \quad [170]$$

which leads to

$$TH_S = -\frac{1}{T}\mathbf{q} \cdot \boldsymbol{\nabla}T - \sum_i \mathbf{J}_i \cdot \boldsymbol{\nabla}_T\mu_i$$
$$- \sum_i \Gamma_i\mu_i + \underline{\boldsymbol{\tau}} : \boldsymbol{\nabla}\mathbf{v}_b + \bar{\rho}\bar{H} \qquad [171]$$

This last equation has the advantage of separating the temperature contribution, $\boldsymbol{\nabla}T$, from the compositional contribution, $\boldsymbol{\nabla}_T\mu_i$ ($\boldsymbol{\nabla}_T\mu_i$ varies mostly with composition as composition can change over very short distances; however, this term is also related to pressure variations) (see de Groot and Mazur, 1984).

### 2.5.1.4 Advection–diffusion equation and reaction rates

Among the entropy sources, only terms involving similar tensorial ranks can be coupled in an isotropic medium, according to Curie's principle. The positivity of the entropy production imposes three conditions: coupling tensors, vectors, and scalars.

$$\underline{\boldsymbol{\tau}} : \boldsymbol{\nabla}\mathbf{v}_b \geq 0, \quad -\mathbf{q} \cdot \frac{\boldsymbol{\nabla}T}{T} - \sum_i \mathbf{J}_i \cdot \boldsymbol{\nabla}_T\mu_i \geq 0,$$
$$- \sum_i \Gamma_i\mu_i \geq 0 \qquad [172]$$

The first term relates tensors and we have already discussed its implications for the rheology in Section 2.3.2.

The second term relates vectors and we assume, in agreement with the general principle of nonequilibrium thermodynamics (de Groot and Mazur, 1984), that a matrix a phenomenological matrix $M$ relates the thermodynamic fluxes $\mathbf{J} = \mathbf{J}_1 \ldots \mathbf{J}_i \ldots \mathbf{q}$ to the thermodynamic forces $\mathbf{X} = -\boldsymbol{\nabla}_T\mu_1 \ldots -\boldsymbol{\nabla}_T\mu_i \ldots -\boldsymbol{\nabla}T/T$:

$$\begin{pmatrix} \mathbf{J}_1 \\ \mathbf{J}_2 \\ \ldots \\ \mathbf{q} \end{pmatrix} = - \begin{pmatrix} m_{11} & m_{12} & \ldots & m_{1q} \\ m_{21} & m_{22} & \ldots & m_{2q} \\ \ldots & \ldots & \ldots & \ldots \\ m_{q1} & m_{q2} & \ldots & m_{qq} \end{pmatrix} \begin{pmatrix} \boldsymbol{\nabla}_T\mu_1 \\ \boldsymbol{\nabla}_T\mu_2 \\ \ldots \\ \boldsymbol{\nabla}T/T \end{pmatrix} \qquad [173]$$

This linear relationship implies that the term of vectorial rank (with superscript $v$), in the entropy source, $TH_S^{(v)}$ appears as

$$TH_S^{(v)} = \mathbf{X}^t M \mathbf{X} = \mathbf{X}^t \frac{M + M^t}{2} \mathbf{X} \qquad [174]$$

According to the second law of thermodynamics, the symmetric part of the matrix $M$, $(M + M^t)/2$, must be positive definite, that is, the right-hand side of eqn [174] must be positive for any vectors $\mathbf{X}$.

At microscopic scale, a process and its reverse occur at the same rate. A consequence, known as the Onsager reciprocal relations, is the existence of symmetry or antisymmetry between $m_{ij}$ and $m_{ji}$ (Onsager, 1903–76). A general discussion can be found in, for example, de Groot and Mazur (1984) or Woods (1975). When the forces are even functions of the velocities and in the absence of magnetic field, the matrix $M$ must be symmetric. As $\boldsymbol{\nabla}T/T$ and $\boldsymbol{\nabla}_T\mu_i$ are even functions, as independent of the velocities, $m_{ij} = m_{ji}$.

In the general case, the transport of heat by concentration gradients (the Dufour effect (Dufour, 1832–92)) or the transport of concentration by temperature gradients (the Soret effect (Soret, 1827–90)) are possible. In many situations these cross-effects are small and we will assume that the matrix $M$ does not couple thermal and compositional effects (the last row and column of $M$ are zero except for $m_{qq}/T = k$, the thermal conductivity). In some case, however, when the chemical potential changes very rapidly with temperature, it becomes impossible to neglect the coupling between chemical diffusion and temperature variations. In this case it may be safer to consider a formalism where the thermodynamic force that drives the chemical diffusion of the component $i$ is $T\boldsymbol{\nabla}(\mu_i/T)$ rather than $\boldsymbol{\nabla}_T\mu_i$ (see also Richard et al. (2006)).

Even without coupling between thermal and compositional effects, chemical diffusion in a multicomponent system remains difficult to discuss in the most general case (the positive definiteness of a symmetric $i$ by $i$ matrix is not a very strong constraint). We therefore restrict our study to a simple - two-component system where

$$\begin{pmatrix} \mathbf{J}_1 \\ \mathbf{J}_2 \end{pmatrix} = - \begin{pmatrix} m_{11} & m_{12} \\ m_{21} & m_{22} \end{pmatrix} \begin{pmatrix} \boldsymbol{\nabla}_T\mu_1 \\ \boldsymbol{\nabla}_T\mu_2 \end{pmatrix} \qquad [175]$$

For such a simple case, the sum of the fluxes must cancel (see [153]), and since the Onsager relations impose the symmetry of the matrix, the coefficients $m_{ij}$ must verify

$$m_{11} + m_{21} = m_{12} + m_{22} = m_{12} - m_{21} = 0 \qquad [176]$$

Only one coefficient, for example, $m_{11}$, can be freely chosen, and the fluxes can be written as

$$\mathbf{J}_1 = m_{11}\boldsymbol{\nabla}_T(\mu_2 - \mu_1) \qquad [177]$$

$$\mathbf{J}_2 = m_{11}\boldsymbol{\nabla}_T(\mu_1 - \mu_2) \qquad [178]$$

and the second law requires $m_{11} > 0$. If the component 1 is in small quantity (the solute) and the component 2 is in large quantity (the solvent, with $\boldsymbol{\nabla}(\mu_2) = 0$), we can easily track the evolution of solute concentration $C_1$. Its chemical diffusion flux is $\mathbf{J}_1 = -m_{11}\boldsymbol{\nabla}_T\mu_1$ and according to [154], its concentration satisfies

$$\rho\left(\frac{\partial C}{\partial t} + \mathbf{v}_b \cdot \boldsymbol{\nabla}C\right) = \boldsymbol{\nabla} \cdot (m\boldsymbol{\nabla}_T\mu) + \Gamma \qquad [179]$$

where the subscripts 1 have been omitted.

For a solute the chemical potential is a standard chemical potential $\mu_0$ plus a mixing term expressing the entropy gain (configurational entropy associated with the increased disorder) made by dispersing the solute into the solvent, of the form $RT\log a(C)$ (for crystalline solids, the activity $a(C)$ of the mixing term can be complex since it depends on the number and multiplicity of crystallographic sites (Spear, 1993), but we just need to know that it is related to $C$). In a domain where the standard chemical potential and the average density are uniform, the advection–diffusion equation is obtained:

$$\frac{\partial C}{\partial t} + \mathbf{v}_b \cdot \boldsymbol{\nabla}C = \boldsymbol{\nabla} \cdot (D\boldsymbol{\nabla}C) + \frac{\Gamma}{\rho} \qquad [180]$$

with a diffusion coefficient $D = m/\bar{\rho}(\partial\mu/\partial C)$, most likely $T$-dependent. The negative linear relationship between chemical diffusion and concentration gradient, $\mathbf{J} = -\bar{\rho}D\boldsymbol{\nabla}C$ is called the first Fick's law (Fick, 1829–1901).

When a component is present in two domains separated by a compositional interface, its standard chemical potential $\mu_0$ is generally discontinuous. In this case the gradient of the chemical potential at constant $T$, $\boldsymbol{\nabla}_T\mu$, is a mathematical distribution that contains a term $\boldsymbol{\nabla}_T\mu_0$, infinite on the compositional interface. This discontinuity drives an infinitely fast diffusion of the solute component across the interface until the equilibrium $[\mu] = [\mu_0 + RT\log a(C)] = 0$. The concentration ration of $C$ (or partition coefficient of $C$) must therefore verify

$$\frac{a(C)^+}{a(C)^-} = \exp\left(-\frac{\mu_0^+ - \mu_0^-}{RT}\right) \qquad [181]$$

where $[\ ]^+$ and $[\ ]^-$ denote the values on the two sides of the discontinuity. This equation corresponds to the general rule of chemical equilibrium.

The last entropy source in [172] relates two scalars (production rates and chemical potentials). In a mixture of $i$ components involving $k$ stable atomic species, the conservation of these atomic species implies that only $r = i - k$ linearly independent reactions exist. Let $n_i^j$ be the stoichiometric coefficient of the component $i$ in the $j$th $= (1, \ldots, r)$ chemical reaction with reaction rate, $\Gamma_j$. We can express $\Gamma_i$ as

$$\Gamma_i = \sum_{j=1,\ldots,r} n_i^j\Gamma_j \qquad [182]$$

and the second law imposes

$$-\sum_{j=1,\ldots,r} \Gamma_j\sum_i n_i^j\mu_i \geq 0 \qquad [183]$$

The positivity of the entropy source is satisfied if the kinetic rates of the $j$th $= 1,\ldots,r$ chemical reaction are proportional to the their chemical affinities, $-\Delta G_j = -\sum n_i^j\mu_i$, with positive reaction rate factors $\mathcal{R}_j$:

$$\Gamma_j = -\mathcal{R}_j\Delta G_j \qquad [184]$$

Chemical reaction rates are very rarely simply proportional to the affinities and the $\mathcal{R}_j$ are likely some complex, but positive, functions of $P$, $T$ and concentrations $C_i$. In the case of exact thermodynamic equilibrium, $\sum_i n_i^j\mu_i = 0$, the second law is of course satisfied.

In the same way as we defined the affinity $-\Delta G_j$ of the reaction $j$, we can define its enthalpy $\Delta\mathcal{H}_j = \sum_i n_i^j\mathcal{H}_i$ (see de Groot and Mazur (1984)). The enthalpy exchange term of the energy equation [161], $\sum_i\Gamma_i\mathcal{H}_i$, can also be written $\sum_j\Gamma_j\Delta\mathcal{H}_j$, which represents the products of the reaction rates and the reaction enthalpies. Various phase changes take place in the mantle, most notably at 410 and 660 km depth. Their effects on mantle convection have been studied by various authors and will be discussed in Section 2.6.6.

### 2.5.1.5 Conservation properties of the advection–diffusion equation

We now make the hypothesis that the evolution of concentration of a solute in the convective fluid is controlled by the advection–diffusion equation [180], and that this solute is not involved in any chemical reaction, $\Gamma = 0$. For simplicity, we assume that the barycentric flow is incompressible ($C$ can therefore be a concentration per unit volume or per unit mass) and the diffusion coefficient $D$ is a constant. The fluid and the solute cannot escape the domain $\Omega$; the normal velocity and normal diffusive flux are thus zero on the boundaries of the domain, that is, $\mathbf{v} \cdot \mathbf{n} = 0$ and $\boldsymbol{\nabla}C \cdot \mathbf{n} = 0$ on the surface $\Sigma$ with normal vector $\mathbf{n}$.

First, it is obvious that when integrated over the total domain $\Omega$ and with the divergence theorem, the advection diffusion equation [180] implies

$$\frac{\mathrm{d}}{\mathrm{d}t}\int_\Omega C\,\mathrm{d}V = -\int_\Sigma (C\mathbf{v} - D\boldsymbol{\nabla}C) \cdot \mathrm{d}\mathbf{S} = 0 \qquad [185]$$

The initial heterogeneity does not disappear; it is just redistributed through time.

To understand how the heterogeneity is redistributed, we can express the evolution of the concentration variance. Multiplying [180] by $2C$, we get after some algebra

$$\frac{DC^2}{Dt} = D\nabla^2 C^2 - 2D|\nabla C|^2 \qquad [186]$$

This expression when integrated over the closed volume $\Omega$ implies that

$$\frac{d}{dt}\int_\Omega C^2 \, dV = -2D \int_\Omega |\nabla C|^2 \, dV \qquad [187]$$

Note that the actual variance is $C^2 - \bar{C}^2$ where $\bar{C}$ is the average concentration, but The $\bar{C}^2$ term makes no contribution to [187]. Since the right-hand side is always negative, the variance must continuously decrease until $|\nabla C| = 0$ which corresponds to a state of complete homogenization.

The concept of mixing is associated with the idea of homogenization where the concentration variance decreases with time. Since the average mixing rate is proportional to the diffusion $D$, [187], we note, however, that a nondiffusive flow does not homogenize at all. A diffusive flow just stirs the heterogeneities. In other terms, if the initial concentration is either $C = 1$ or $C = 0$, a perfect homogenization is achieved after a time $t$ if the concentration is everywhere $C = \bar{C}$, the average concentration. When there is no diffusion, the initial heterogeneity is stirred and stretched, but the local concentrations remain, for all time, either $C = 1$ or $C = 0$, but never an intermediate value (see **Figure 5**).

In the case of the Earth's mantle, the solid-state diffusion coefficients are all very low ($D = 10^{-19}$ m$^2$ s$^{-1}$ for uranium, $D = 10^{-13}$ m$^2$ s$^{-1}$ for helium (see **Table 2**)) and many studies have totally neglected chemical diffusion. We see, at face value, that these models are not really homogenizing, only stirring the heterogeneities. Without diffusion a chemical heterogeneity (e.g., a piece of subducted oceanic crust) will forever remain the same petrological heterogeneity, only its shape will change.

Since the mixing rate is related to the compositional gradient [187], we should discuss the evolution of this gradient. We multiply [180] by the operator $2\nabla C \cdot \nabla$ to obtain

$$\frac{D|\nabla C|^2}{Dt} = -2\nabla C \cdot \underline{\dot{\epsilon}} \cdot \nabla C$$
$$+ 2D\left[\nabla \cdot \left(\nabla^2 C\nabla C\right) - \left(\nabla^2 C\right)^2\right] \qquad [188]$$

**Figure 5** An initial heterogeneity (top) is introduced at $t = 0$ into a time-dependent convection cell. Without diffusion, $D = 0$ (bottom left), the heterogeneity is stirred by convection and then stretched on the form of thin ribbons. However, the variance of the heterogeneity concentration remains constant. It is only with diffusion, $D \neq 0$ (bottom right), that a real homogenization occurs with a decrease of the heterogeneity variance.

which can be integrated as

$$\frac{d}{dt}\int_\Omega |\nabla C|^2 \, dV$$
$$= -2\int_\Omega \nabla C \cdot \underline{\dot{\epsilon}} \cdot \nabla C \, dV - 2D \int_\Omega \left(\nabla^2 C\right)^2 \, dV \qquad [189]$$

The rate of gradient production is related to the flow properties through the strain rate tensor $\underline{\dot{\epsilon}}$ and to the diffusion. The diffusion term is negative and decreases the sharpness of compositional gradients.

The term related to the flow properties through the strain tensor (first term on the right-hand side of [189]), could in principle be either positive or negative. However as time evolves, this term must become positive. The strain rate tensor has locally three principal axes and three principal strain rates, the sum of them being zero since the flow is incompressible. The stretched heterogeneities become elongated along the direction of the maximum

**Table 2** Homogenization times for helium and uranium assuming an heterogeneity of initial thickness $2l_0 = 7$ km and a strain rate of $5 \times 10^{-16}$ s$^{-1}$

|  | **Uranium** | **Helium** |
|---|---|---|
| $D$ (m$^2$ s$^{-1}$) | $10^{-19}$ | $10^{-13}$ |
| $t_0$ (Ma) | $3.88 \times 10^{13}$ | $3.88 \times 10^7$ |
| $t_L$ (Ma) | $3.60 \times 10^5$ | $3.60 \times 10^3$ |
| $t_T$ (Ma) | 1920 | 1490 |

principal strain rate and the concentration gradients reorient themselves along the minimum, and negative, principal strain rate. The term under the first integral on the right-hand side of [189] is thus of order of $|\dot{\epsilon}_{min}| \, |\nabla C|^2$ ($\dot{\epsilon}_{min}$ is the local, negative eigenvalue of the strain rate tensor). Stirring is thus the source of production of concentration gradient.

We can now understand the interplay between advection and diffusion. Even when the diffusion coefficient $D$ is vanishingly small in [189], the stirring of the flow by convection will enhance the concentration gradients until the average diffusion term, proportional to the concentration gradients, will become large enough (see [187]) for a rapid decrease of the concentration variance. We illustrate this behavior in the next two sections by choosing a simple expression for the strain rate and computing the evolution of concentration through time.

### 2.5.1.6   Laminar and turbulent stirring

The efficiency of mixing, mostly controlled by stirring, is therefore related to the ability of the flow to rapidly reduce the thickness of heterogeneities (Olson et al., 1984). In this section we set aside diffusion and discuss the stirring properties of a flow (see also Chapter 10). Let us consider a vertical piece of heterogeneity of width $2d_0$, height $2L$ ($L \ll d_0$) in a simple shear flow $v_x = \dot{\epsilon} z$. Its top and bottom ends are at $(0, L_0)$, $(0, -L_0)$ and they will be advected to $(\dot{\epsilon} t L_0, L_0)$, $(-\dot{\epsilon} t L_0, -L_0)$ after a time $t$. As the heterogeneity length increases as $2L_0(1 + \dot{\epsilon}^2 t^2)$, mass conservation implies that its half-width $d(t)$ decreases as

$$d(t) = \frac{d_0}{\sqrt{1 + \dot{\epsilon}^2 t^2}} \qquad [190]$$

Such flows, in which heterogeneities are stretched at rate $\sim 1/t$, are called flows with laminar stirring. They are not very efficient in enhancing the diffusion because they do not increase the concentration gradients, typically of order $1/d(t)$, fast enough.

On the contrary, in a pure shear flow, $v_z = \mathrm{d}z/\mathrm{d}t = z\dot{\epsilon}$, the length of the heterogeneity would increase as $L = L_0 \exp(\dot{\epsilon} t)$ and its width would shrink as

$$d(t) = d_0 \exp(-\dot{\epsilon} t) \qquad [191]$$

Such a flow is said to induce turbulent stirring. This is unfortunate terminology because turbulent stirring can occur in a creeping flow with $Re = 0$. Mantle convection is not turbulent but it generates turbulent stirring.

Chaotic mixing flows have globally turbulent stirring properties and the qualitative idea that highly time-dependent convection with high Rayleigh number mixes more efficiently than low Rayleigh number convection is often true (Schmalzl et al., 1996). However, steady 3-D flows can also induce turbulent mixing. This surprising phenomenon called Lagrangian chaos is well illustrated for some theoretical flows (Dombre et al., 1986) and for various simple flows (Ottino, 1989; Toussaint et al., 2000). For example, in a steady flow under an oceanic ridge offset by a transform fault, the mixing is turbulent (Ferrachat and Ricard, 1998).

### 2.5.1.7   Diffusion in Lagrangian coordinates

In Section 2.5.1.5 we discussed the mixing properties from an Eulerian viewpoint. We can also understand the interplay between diffusion and stretching (stirring) by adopting a Lagrangian viewpoint (Kellogg and Turcotte, 1987; Ricard and Coltice, 2004), that is, by solving the advection–diffusion equations in a coordinate frame that follows the deformation.

Let us consider a strip of thickness $2l_0$ with an initial concentration $C_0$ embedded in a infinite matrix of concentration $C_\infty$. In the absence of motion, the solution of the advection–diffusion equation [180] can be expressed using the error function and the time-dependent concentration $C(x, t)$ is given by

$$\frac{C(x, t) - C_\infty}{C_0 - C_\infty} = \frac{1}{2}\left[ \mathrm{erf}\left(\frac{l_0 - x}{2\sqrt{Dt}}\right) + \mathrm{erf}\left(\frac{l_0 + x}{2\sqrt{Dt}}\right) \right] \qquad [192]$$

where $x$ is a coordinate perpendicular to the strip and is zero at its center.

The concentration at the center of the strip ($x = 0$) is

$$\frac{C(0, t) - C_\infty}{C_0 - C_\infty} = \mathrm{erf}\left(\frac{l_0}{2\sqrt{Dt}}\right) \qquad [193]$$

and the concentration decreases by a factor of about 2 in the diffusive time

$$t_0 \sim \frac{l_0^2}{D} \qquad [194]$$

(erf(1/2) is not far from 1/2). The time needed to homogenize a 7-km thick piece of oceanic crust introduced into a motionless mantle is extremely long (see **Table 2**); even the relatively mobile helium would be frozen in place since the Earth formed as it would only have migrated around 50 cm.

However, this idea of a $\sqrt{t}$ diffusion is faulty since the flow stirs the heterogeneity and increases compositional gradients (see [189]) which in turn accelerates the mixing process (e.g., [187]). Assuming that the problem remains 2-D enough so that diffusion only occurs perpendicular to the deforming heterogeneity, let $l(t)$ be its thickness. The velocity perpendicular to the strip would locally be at first order

$$v_x = \frac{x}{l(t)} \frac{\mathrm{d}}{\mathrm{d}t} l(t) \qquad [195]$$

(each side of the strip at $x = \pm l(t)$ moves at $\pm \mathrm{d}l(t)/\mathrm{d}t$).

We can choose as a new space variable $\widetilde{x} = x l_0/l$ $(t)$ in such a way that the Lagrangian coordinate inside the strip, $\widetilde{x}$, will vary between the fixed values $-l_0$ and $l_0$. The diffusion equation becomes

$$\left(\frac{\partial C}{\partial t}\right)_{\widetilde{x}} = D\left(\frac{l_0}{l(t)}\right)^2 \frac{\partial^2 C}{\partial \widetilde{x}^2} \qquad [196]$$

where the partial time derivative is now computed at constant $\widetilde{x}$. We see that the advection diffusion equation has been turned into a pure diffusive equation where the diffusivity $D$ has been replaced by $D(l_0/l(t))^2$. This equivalent diffusivity is larger than $D$ and increases with time as $l(t)$ decreases.

To solve analytically equation [196], it is appropriate to rescale the time variable by defining $\widetilde{t} = F(t)$ with

$$F(t) = \int_0^t (l_0/l(u))^2 \, \mathrm{d}u \qquad [197]$$

and the resulting advection–diffusion equation in Lagrangian coordinates becomes the simple diffusion equation with constant diffusivity. Its solution is given by [192] where $t$ and $x$ are replaced by $\widetilde{t}$ and $\widetilde{x}$. For example, the concentration at the center of the deformable strip varies like

$$\frac{C(0, t) - C_\infty}{C_0 - C_\infty} = \mathrm{erf}\left(\frac{l_0}{2\sqrt{DF(t)}}\right) \qquad [198]$$

and the concentration diminishes in amplitude by a factor of 2 after a time $l$ that satisfies

$$F(t) \sim \frac{l_0^2}{D} \qquad [199]$$

To perform a numerical application let us consider that the flow is either a simple shear [190], or a pure shear deformation, [191]. Computing $F(t)$ from eqn

[197] is straightforward and, assuming $\dot{\epsilon}t \gg 1$, we get from [199], the homogenization times

$$t_L \sim \frac{3^{1/3} l_0^{2/3}}{\dot{\epsilon}^{2/3} D^{1/3}} \qquad [200]$$

and

$$t_T \sim \frac{1}{2\dot{\epsilon}} \log \frac{2 l_0^2 \dot{\epsilon}}{D} \qquad [201]$$

respectively. For the same oceanic crust of initial thickness 7 km, we get homogenization times of about 1.49 billion years for He and 1.92 billion years for U if we use the pure shear mechanism and assume rather arbitrarily that $\dot{\epsilon} = 5 \times 10^{-16} \, \mathrm{s}^{-1}$ (this corresponds to a typical plate velocity of 7 cm yr$^{-1}$ over a plate length of 5000 km). Although He and U have diffusion coefficients six orders of magnitude apart, their residence times in a piece of subducted oceanic crust may be comparable.

The use of tracers to simulate the evolution of chemical properties in the mantle is our best method since solid-state diffusion is too slow to be efficiently accounted for in a numerical simulations (e.g., van Keken *et al.*, 2002; Tackley and Xie, 2002). However, by using tracers, we do not necessarily take into account that some of them may represent points that have been stretched so much that their initial concentration anomalies have completely diffused into the background. In other words, even if diffusion seems negligible, it will erase all heterogeneities after a finite time that is mostly controlled by the stirring properties of the flow.

## 2.5.2 Fluid Dynamics of Two Phase Flows

Up to now, in Section 2.5.1, all components were mixed in a single phase. However, another important geophysical application occurs when the multicomponents belong to different phases. This case can be illustrated with the dynamics of partial melt in a deformable compacting matrix. Partial melts are obviously present under ridges and hot spots, but they may also be present in the middle and deep mantle (Williams and Garnero, 1996; Bercovici and Karato, 2003) and they were certainly more frequent when the Earth was younger. We discuss the situation where two phases, fluid and matrix, can interact. In contrast to Section 2.5.1, where the proportion and velocity of each component in solution was defined everywhere at a microscopic level, in a partial melt aggregate, the local velocity at a microscopic level is either the velocity of a matrix grain, $\tilde{\mathbf{v}}_m$, or the interstitial velocity of the melt, $\tilde{\mathbf{v}}_f$.

We assume that the two phases are individually homogeneous, incompressible, and with densities $\rho_f$ and $\rho_m$. They have Newtonian rheologies with viscosities $\eta_f$ and $\eta_m$. They are isotropically mixed and connected. Their volume fractions are $\phi$ (the porosity) and $1 - \phi$. The rate of magma melting or freezing is $\Delta\Gamma$ (in $kg\,m^{-3}\,s^{-1}$). Although the two phases have very different physical properties, we will require the equations to be material invariant until we need to use numerical values. This means that swapping $f$ and $m$, $\phi$ and $1 - \phi$, and $\Delta\Gamma$ and $-\Delta\Gamma$ must leave the equations unchanged. This rule is both a physical requirement and a strong guidance in establishing the general equations (Bercovici *et al.*, 2001a).

We make the hypothesis that there is a mesoscopic size of volume $\delta V$ which includes enough grains and interstitial fluid that averaged and continuous quantities can be defined. Classical fluid dynamics also has its implicit averaging volume $\delta V$ that must contain enough atoms that quantum effects are negligible, but what is needed here is a much larger volume. This averaging approach remains meaningful because the geophysical macroscopic phenomenon that we want to understand (say, melting under ridges) has characteristic sizes large compared to those of the averaging volume (say, a few $cm^3$) (Bear, 1988).

To do the averaging, we define at microscopic level a function $\theta$ that takes the value 1 in the interstitial fluid and the value 0 in the matrix grain. Mathematically, this function is rather a distribution and it has a very convoluted topology. From it, we can define first the porosity (volume fraction of fluid) $\phi$, then, the fluid and matrix-averaged velocities, $\mathbf{v}_f$ and $\mathbf{v}_m$ (Bercovici *et al.*, 2001a) by

$$\phi = \frac{1}{\delta V}\int_{\delta V}\theta\,dV \qquad [202]$$

$$\phi\mathbf{v}_f = \frac{1}{\delta V}\int_{\delta V}\theta\tilde{\mathbf{v}}_f\,dV,$$
$$(1-\phi)\mathbf{v}_m = \frac{1}{\delta V}\int_{\delta V}(1-\theta)\tilde{\mathbf{v}}_m\,dV \qquad [203]$$

### 2.5.2.1   Mass conservation for matrix and fluid

Having defined the average quantities, the derivation of the two mass conservation equations is fairly standard (McKenzie, 1984; Bercovici *et al.*, 2001a). They are

$$\frac{\partial\phi}{\partial t} + \nabla\cdot[\phi\mathbf{v}_f] = \frac{\Delta\Gamma}{\rho_f} \qquad [204]$$

$$-\frac{\partial\phi}{\partial t} + \nabla\cdot[(1-\phi)\mathbf{v}_m] = -\frac{\Delta\Gamma}{\rho_m} \qquad [205]$$

We get the same equations as in [146] except that we refer to $\phi$, $1 - \phi$, and $\Delta\Gamma$ instead of $\phi_1$, $\phi_2$, and $\Gamma_1$. When averaged, the mass conservation equations of two separated phases takes the same form as the mass conservation equations of two components in a solution.

We define an average and a difference quantity for any general variable $q$, by

$$\bar{q} = \phi q_f + (1-\phi)q_m, \quad \Delta q = q_m - q_f \qquad [206]$$

The velocity $\bar{\mathbf{v}}$ is volume averaged and is different from the barycentric velocity [148], $\mathbf{v}_b = (\phi\rho_f\mathbf{v}_f + (1-\phi)\rho_m\mathbf{v}_m)/\bar{\rho}$. By combining the fluid and matrix mass conservation equations, we get the total mass conservation equation

$$\frac{\partial\bar{\rho}}{\partial t} + \nabla\cdot(\bar{\rho}\mathbf{v}_b) = 0 \qquad [207]$$

(as before [149]), and the time rate of change in volume during melting

$$\nabla\cdot\bar{\mathbf{v}} = \Delta\Gamma\frac{\Delta\rho}{\rho_f\rho_m} \qquad [208]$$

### 2.5.2.2   Momentum conservation of matrix and fluid

Total momentum conservation, that is, the balance of the forces applied to the mixture, is

$$\nabla\cdot\bar{\underline{\boldsymbol{\tau}}} - \nabla\bar{P} + \bar{\rho}\mathbf{g} = 0 \qquad [209]$$

We have considered that the only force is due to gravity, although surface tension between the two phases could also be introduced (Bercovici *et al.*, 2001a). In this equation, $\bar{P}$, $\bar{\underline{\boldsymbol{\tau}}}$, and $\bar{\rho}$ are the average pressure, stress, and density. The equation is not surprising and looks identical to its counterpart for a multicomponent solution [156]. However, the average pressure and stresses, $\bar{P} = \phi P_f + (1 - \phi)\,P_m$ and $\bar{\underline{\boldsymbol{\tau}}} = \phi\underline{\boldsymbol{\tau}}_f + (1-\phi)\underline{\boldsymbol{\tau}}_m$, are now the sum of two separate contributions, from two separate phases having most likely very different rheologies and different pressures. Hypothesizing that the two phases may feel the same pressure does not rest on any physical justification and certainly cannot hold if surface tension is present. We will show later that their pressure difference controls the rate of porosity change.

We split the total momentum equation into two equations, one for the fluid and one for the matrix

$$-\boldsymbol{\nabla}[\phi P_\mathrm{f}] + \phi\rho_\mathrm{f}\mathbf{g} + \boldsymbol{\nabla}\cdot[\phi\boldsymbol{\tau}_\mathrm{f}] + \mathbf{h}_\mathrm{f} = 0 \quad [210]$$

$$-\boldsymbol{\nabla}[(1-\phi)P_\mathrm{m}] + (1-\phi)\rho_\mathrm{m}\mathbf{g} \\ + \boldsymbol{\nabla}\cdot[(1-\phi)\boldsymbol{\tau}_\mathrm{m}] + \mathbf{h}_\mathrm{m} = 0 \quad [211]$$

where $\mathbf{h}_\mathrm{f}$ and $\mathbf{h}_\mathrm{m}$ satisfy $\mathbf{h}_\mathrm{f} + \mathbf{h}_\mathrm{m} = 0$ and represent the interaction forces acting on the fluid and on the matrix, across the interfaces separating the two phases. Because of the complexities of the interfaces, these two interaction forces must be parametrized in some way.

The simplest contribution to the interfacial forces that preserves Galilean invariance is a Darcy-like term $c\Delta\mathbf{v} = c(\mathbf{v}_\mathrm{m} - \mathbf{v}_\mathrm{f})$ (Drew and Segel, 1971; McKenzie, 1984) (Darcy, 1803–58). The interaction coefficient $c$ is related to permeability which is itself a function of porosity (Bear, 1988). A symmetrical form compatible with the usual Darcy term is (see Bercovici et al., 2001a)

$$c = \frac{\eta_\mathrm{f}\eta_\mathrm{m}}{k_0\left[\eta_\mathrm{f}(1-\phi)^{n-2} + \eta_\mathrm{m}\phi^{n-2}\right]} \quad [212]$$

where the permeability of the form $k_0\phi^n$ was used (usually $n \sim 2-3$). Assuming $n = 2$ and $\eta_\mathrm{m} \gg \eta_\mathrm{f}$, the interaction coefficient becomes a constant, $c = \eta_\mathrm{f}/k_0$.

In the absence of gravity and when the pressures are uniform and equal, no motion should occur even in the presence of nonuniform porosity. In this situation where $\Delta\mathbf{v} = \boldsymbol{\tau}_\mathrm{f} = \boldsymbol{\tau}_\mathrm{m} = 0$ and where $P$ is uniform, $P = P_\mathrm{f} = P_\mathrm{m}$, the force balances are $-P\boldsymbol{\nabla}\phi + \mathbf{h}_\mathrm{f} = -P\boldsymbol{\nabla}(1-\phi) + \mathbf{h}_\mathrm{m} = 0$. Therefore, the interface forces $h_\mathrm{f}$ and $h_\mathrm{m}$ must also include $P\boldsymbol{\nabla}\phi$ and $P\boldsymbol{\nabla}(1-\phi)$ when the two pressures are equal. This led Bercovici and Ricard (2003) to write the interaction terms

$$\mathbf{h}_\mathrm{f} = c\Delta\mathbf{v} + P_\mathrm{f}\boldsymbol{\nabla}\phi + \omega\Delta P\boldsymbol{\nabla}\phi \\ \mathbf{h}_\mathrm{m} = -c\Delta\mathbf{v} + P_\mathrm{m}\boldsymbol{\nabla}(1-\phi) + (1-\omega)\Delta P\boldsymbol{\nabla}\phi \quad [213]$$

with $0 \leq \omega \leq 1$. These expressions verify $\mathbf{h}_\mathrm{f} + \mathbf{h}_\mathrm{m} = 0$, are Galilean and material invariant, and allow equilibrium of a mixture with nonuniform porosity but uniform and equal pressures (see also Mckenzie (1984)).

At microscopic level, the matrix–melt interfaces are not sharp discontinuities but correspond to layers (called 'selvedge' layers) of disorganized atom distributions. The coefficient $0 < \omega < 1$ controls the partitioning of the pressure jump (and potentially of the surface tension) between the two phases (Bercovici and Ricard, 2003) and represents the fraction of the volume-averaged surface force exerted on

the fluid phase. The exact value of $\omega$ is related to the microscopic behavior of the two phases (molecular bond strengths and thickness of the interfacial selvage layers) and measures the extent to which the microscopic interface layer is embedded in one phase more than the other. The only general physical constraints that we have are that $\omega$ must be zero when the fluid phase disappears (when $\phi = 0$) and when the fluid phase becomes unable to sustain stresses (when $\eta_\mathrm{f} = 0$). A symmetrical form like

$$\omega = \frac{\phi\eta_\mathrm{f}}{\phi\eta_\mathrm{f} + (1-\phi)\eta_\mathrm{m}} \quad [214]$$

satisfies these conditions.

To summarize, general expressions for the equations of fluid and matrix momentum conservation are (Bercovici and Ricard, 2003)

$$-\phi[\boldsymbol{\nabla}P_\mathrm{f} - \rho_\mathrm{f}\mathbf{g}] + \boldsymbol{\nabla}\cdot[\phi\boldsymbol{\tau}_\mathrm{f}] + c\Delta\mathbf{v} + \omega\Delta P\boldsymbol{\nabla}\phi = 0 \quad [215]$$

$$-(1-\phi)[\boldsymbol{\nabla}P_\mathrm{m} - \rho_\mathrm{m}\mathbf{g}] + \boldsymbol{\nabla}\cdot[(1-\phi)\boldsymbol{\tau}_\mathrm{m}] - c\Delta\mathbf{v} \\ + (1-\omega)\Delta P\boldsymbol{\nabla}\phi = 0 \quad [216]$$

The relationship between stress and velocities does not include an explicit bulk viscosity term (Bercovici et al., 2001a), and for each phase $j$ the deviatoric stress is simply

$$\boldsymbol{\tau}_j = \eta_j\left(\boldsymbol{\nabla}\mathbf{v}_j + [\boldsymbol{\nabla}\mathbf{v}_j]^t - \frac{2}{3}\boldsymbol{\nabla}\cdot\mathbf{v}_j\mathbf{I}\right) \quad [217]$$

where $j$ stands for $f$ or $m$. There is no difference in constitutive relations for the isolated component and the component in the mixture.

### 2.5.2.3 Energy conservation for two-phase flows

In the case where surface energy and entropy exist on interfaces the conservation of energy deserves more care (Sramek et al., 2007). Otherwise the global conservation is straightforward and can be expressed by the following equation where the left-hand side represents the temporal change of internal energy content in a fixed control volume and the right-hand side represents the different contributions to this change, namely internal heat sources, loss of energy due to diffusion, advection of energy, and rate of work of both surface and body forces:

$$\frac{\partial}{\partial t}[\phi\rho_\mathrm{f}\mathcal{U}_\mathrm{f} + (1-\phi)\rho_\mathrm{m}\mathcal{U}_\mathrm{m}] \\ = \bar{\rho}\bar{H} - \boldsymbol{\nabla}\cdot\mathbf{q} - \boldsymbol{\nabla}\cdot[\phi\rho_\mathrm{f}\mathcal{U}_\mathrm{f}\mathbf{v}_\mathrm{f} + (1-\phi)\rho_\mathrm{m}\mathcal{U}_\mathrm{m}\mathbf{v}_\mathrm{m}] \\ + \boldsymbol{\nabla}\cdot[-\phi P_\mathrm{f}\mathbf{v}_\mathrm{f} - (1-\phi)P_\mathrm{m}\mathbf{v}_\mathrm{m} + \phi\mathbf{v}_\mathrm{f}\cdot\boldsymbol{\tau}_\mathrm{f} \\ + (1-\phi)\mathbf{v}_\mathrm{m}\cdot\boldsymbol{\tau}_\mathrm{m}] + \phi\mathbf{v}_\mathrm{f}\cdot\rho_\mathrm{f}\mathbf{g} \\ + (1-\phi)\mathbf{v}_\mathrm{m}\cdot\rho_\mathrm{m}\mathbf{g} \quad [218]$$

The last equation is manipulated in the standard way using the mass and momentum equations. Because the two phases are incompressible, their internal energies are simply $d\mathcal{U}_f = C_f dT$ and $d\mathcal{U}_m = C_m dT$. After some algebra we get

$$\phi \rho_f C_f \frac{D_f T}{Dt} + (1-\phi)\rho_m C_m \frac{D_m T}{Dt}$$
$$= -\nabla \cdot \mathbf{q} - \Delta P \frac{D_\omega \phi}{Dt} + \Delta \mathcal{H} \Delta \Gamma + \Psi + \bar{\rho}\bar{H} \quad [219]$$

where $\Psi$ is the rate of deformational work

$$\Psi = \phi \nabla \mathbf{v}_f : \underline{\tau}_f + (1-\phi)\nabla \mathbf{v}_m : \tau_m + c(\Delta v)^2 \quad [220]$$

It contains the dissipation terms of each phase plus a term related to the friction between the two phases. The fundamental derivatives are defined by

$$\frac{D_j}{Dt} = \frac{\partial}{\partial t} + \mathbf{v}_j \cdot \nabla \quad [221]$$

where $\mathbf{v}_j$ is to be substituted with the appropriate velocity $\mathbf{v}_f$, $\mathbf{v}_m$, or $\mathbf{v}_\omega$ with

$$\mathbf{v}_\omega = \omega \mathbf{v}_f + (1-\omega)\mathbf{v}_m \quad [222]$$

In contrast to Section 2.5.1 it would not make much sense to try to keep the equations in terms of an average velocity like $\mathbf{v}_b$ plus some diffusion terms. Here the two components may have very different velocities and material derivatives.

Since $\omega$ represents a partitioning of pressure jump, it is not surprising to find the velocity $\mathbf{v}_\omega$ (included in $D_\omega/Dt$) in the work term related to this pressure jump. Associated with this partitioning factor, we can also introduce interface values, $q^\omega$, that we will use later. Any quantity $\Delta q = q_m - q_f$ can also be written as $(q_m - q^\omega) - (q_f - q^\omega)$. When the property jump is embedded entirely in the matrix ($\omega = 0$), there should be no jump within the fluid and we must have $q^\omega = q_f$. Reciprocally, when $\omega = 1$, we should have $q^\omega = q_m$. This prompts us to define interface values by

$$q^\omega = (1-\omega)q_f + \omega q_m \quad [223]$$

Notice in the expressions of the interface velocity $\mathbf{v}_\omega$, [222], and interface value, $q^\omega$ [223], that $\omega$ and $1 - \omega$ are interchanged.

The right-hand side of [219] contains two new expressions in addition to the usual terms (heat production, diffusion and deformational work). The first term includes the changes in porosity $D_\omega \phi / Dt$ times the difference in pressures between phases, $\Delta P$. The other term contains the difference in the specific

enthalpies $\Delta \mathcal{H} = \mathcal{H}_m - \mathcal{H}_f$ where the enthalpy of phase $j$ is defined by $\mathcal{H}_j = \mathcal{U}_j + P_j / \rho_j$. A similar term was found for components reacting in a solution [158].

### 2.5.2.4  *Entropy production and phenomenological laws*

Entropy conservation is needed to constrain the pressure jump between phases and the melting rate. Starting from entropy conservation and following Sramek *et al.* (2007)

$$\frac{\partial}{\partial t}[\phi \rho_f \mathcal{S}_f + (1-\phi)\rho_m \mathcal{S}_m] = -\nabla \cdot \mathbf{J}_\mathcal{S} + H_\mathcal{S} \quad [224]$$

where $\mathbf{J}_\mathcal{S}$ is the total entropy flux and $H_\mathcal{S}$ is the internal entropy production; we compare the energy and the entropy equations [219] and [224] taking into account that, for each incompressible phase, $dS_j = C_j dT/T = du_j/T$. After some algebra, one gets

$$\mathbf{J}_\mathcal{S} = \phi \rho_f \mathcal{S}_f \mathbf{v}_f + (1-\phi)\rho_m \mathcal{S}_m \mathbf{v}_m + \frac{\mathbf{q}}{T} \quad [225]$$

$$TH_\mathcal{S} = -\frac{1}{T}\mathbf{q} \cdot \nabla T - \Delta P \frac{D_\omega \phi}{Dt} + \Delta \mu \Delta \Gamma$$
$$+ \Psi + \bar{\rho}\bar{H} \quad [226]$$

where we have introduced the difference in chemical potentials between the two phases

$$\Delta \mu = \Delta \mathcal{H} - T\Delta S \quad [227]$$

and $\Delta S = \mathcal{S}_m - \mathcal{S}_f$ is the change in specific entropies.

Following the standard procedure of nonequilibrium thermodynamics, we choose $\mathbf{q} = -k\nabla T$ and we assume that there is a linear relationship between the two thermodynamic fluxes, $D_\omega \phi / Dt$ and $\Delta \Gamma$, and the two thermodynamic forces $-\Delta P$ and $\Delta \mu$ since they have the same tensorial rank. We write

$$\begin{pmatrix} D_\omega \phi / Dt \\ \Delta \Gamma \end{pmatrix} = \begin{pmatrix} m_{11} & m_{12} \\ m_{21} & m_{22} \end{pmatrix} \begin{pmatrix} -\Delta P \\ \Delta \mu \end{pmatrix} \quad [228]$$

The matrix of phenomenological coefficients $m_{ij}$ is positive definite and symmetrical by Onsager's theorem, $m_{12} = m_{21}$ (see Sramek *et al.*, 2007). For a $2 \times 2$ matrix, it is rather simple to show that the positivity implies, $m_{22} > 0$, $m_{11} > 0$, and $m_{11}m_{22} - m_{12}^2 > 0$ (positivity of the determinant).

The form of the phenomenological coefficients $m_{ij}$ can be constrained through thought experiments. First using mass conservations [204] and [205] and the definitions of $\mathbf{v}_\omega$ and $\rho^\omega$, [222] and [223], we can combine equation [228] to get

$$\Delta P = -\frac{m_{22}}{m_{11}m_{22} - m_{12}^2} \times \left[(1-\omega)(1-\phi)\mathbf{\nabla}\cdot\mathbf{v}_m - \omega\phi\mathbf{\nabla}\cdot\mathbf{v}_f \right.$$
$$\left. + \left(\frac{\rho^\omega}{\rho_f\rho_m} - \frac{m_{12}}{m_{22}}\right)\Delta\Gamma\right] \qquad [229]$$

In the limiting case where the two phases have the same density $\rho_f = \rho_m = \rho^\omega$, melting can occur with no motion, $\mathbf{v}_m = \mathbf{v}_f = 0$, [229] should therefore predict the equality of pressure between phases, $\Delta P = 0$. In this case we must choose $m_{12}/m_{22} = 1/\rho_f = 1/\rho_m$. Let us consider now a situation of homogeneous isotropic melting where the melt has such a low viscosity that it cannot sustain viscous stresses and cannot interact with the solid by Darcy terms. For such an inviscid melt, $\omega = 0$, $\rho^\omega = \rho_f$, and $\mathbf{v}_\omega = \mathbf{v}_m$. In this case, since the melt can escape instantaneously, the matrix should not dilate, $\mathbf{\nabla}\cdot\mathbf{v}_m = 0$, and thus the two pressures should also be the same, $\Delta P = 0$. In this situation all the terms in eqn [229] are 0 except for the term proportional to $\Delta\Gamma$. Thus, in the general case,

$$\frac{m_{12}}{m_{22}} = \frac{\rho^\omega}{\rho_f\rho_m} \qquad [230]$$

Using this condition and introducing two positive coefficients, $\zeta = m_{22}/(m_{11}m_{22} - m_{12}^2)$, and $\mathcal{R} = m_{22}$, we can recast eqns [228] as

$$\Delta P = -\zeta\left[\frac{D_\omega\phi}{Dt} - \frac{\rho^\omega}{\rho_f\rho_m}\Delta\Gamma\right] \qquad [231]$$

$$\Delta\Gamma = \mathcal{R}\left[\Delta\mu - \frac{\rho^\omega}{\rho_f\rho_m}\Delta P\right] \qquad [232]$$

The first equation establishes a general relation controlling the pressure drop between phases. The coefficient $\zeta$ that links the pressure jump between the two phases to the porosity changes in excess of the melting rate is in fact equivalent to a bulk viscosity as introduced in Section 2.3.2 (see also Section 2.5.2.5). The physical requirement that the two phase mixture should have the incompressible properties of either the matrix or the fluid when $\phi = 0$ or $\phi = 1$ imposes a porosity dependence to $\zeta$ with $\lim_{\phi\to 0}\zeta(\phi) = \lim_{\phi\to 1}\zeta(\phi) = +\infty$. Simple micromechanical models (e.g., Nye, 1953; Bercovici et al., 2001a) allow us to estimate the bulk viscosity as

$$\zeta = K_0\frac{(\eta_f + \eta_m)}{\phi(1-\phi)} \qquad [233]$$

The dimensionless constant $K_0$ accounts for grain/pore geometry and is of $O(1)$.

A more general, but more hypothetical, interpretation of the entropy positivity could argue that some deformational work, $\Psi$, might affect the pressure drop of [231]. This hypothesis led to a damage theory developed in Bercovici et al. (2001a, 2001b), Ricard and Bercovici (2003), and Bercovici and Ricard (2003, 2005). Here we assume that the system remains close enough to mechanical equilibrium that damage does not occur.

The second equation [232] controls the kinetics of the melting/freezing and by consequence defines the equilibrium condition. In the case of mechanical equilibrium, when there is no pressure drop between the two phases, the melting rate cancels when there is equality of the chemical potentials of the two single phases. In case of mechanical disequilibrium ($\Delta P \neq 0$), the chemical equilibrium does not occur when the two chemical potentials are equal. We define a new effective chemical potential,

$$\mu_i^* = \mathcal{U}_i + \frac{P^\omega}{\rho_i} - T\mathcal{S}_i \qquad [234]$$

where $i$ stands for 'f' or 'm' and write the kinetic equation [232],

$$\Delta\Gamma = \mathcal{R}\Delta\mu^* \qquad [235]$$

Chemical equilibrium imposes the equality of the effective potentials on the interface, at the pressure $P^\omega$ at which the phase change effectively occurs.

Using [231] and [232], we can show that the entropy production is indeed positive and given by

$$TH_{\mathcal{S}} = k\frac{1}{T}|\mathbf{\nabla}T|^2 + \frac{\Delta P^2}{\zeta} + \frac{(\Delta\mu^*)^2}{\mathcal{R}} + \Psi + \bar{\rho}\bar{H} \qquad [236]$$

Chemical relaxation and bulk compression are associated with dissipative terms.

### 2.5.2.5 Summary equations

For convenience we summarize the governing equations when the matrix is much more viscous than the fluid phase ($\eta_f \ll \eta_m$) as typical for melting scenarios, which implies that $\underline{\underline{\tau}}_f = 0$, $\omega = 0$, $\rho^\omega = \rho_f$, $P^\omega = P_f$, and $\mathbf{v}_\omega = \mathbf{v}_m$.

The mass conservations equations are [204] and [205]. The equation of conservation of momentum for the fluid phase is

$$-\phi[\mathbf{\nabla}P_f + \rho_f\mathbf{g}z] + c\Delta\mathbf{v} = 0 \qquad [237]$$

This is Darcy's law with $c = \eta_f\phi^2/k(\phi)$, where $k$ is the permeability (often varying as $k_0\phi^n$ with $n = 2$ or 3). The second momentum equation could be the matrix momentum [211] or a combined force-difference (or action–reaction) equation

$$-\boldsymbol{\nabla}[(1-\phi)\Delta P] + (1-\phi)\Delta\rho\mathbf{g} + \boldsymbol{\nabla}\cdot[(1-\phi)\boldsymbol{\tau}_m]$$
$$-\frac{c\Delta\mathbf{v}}{\phi} = 0 \qquad [238]$$

where the deviatoric stress in the matrix is given by

$$\boldsymbol{\tau}_m = \eta_m\left(\boldsymbol{\nabla}\mathbf{v}_m + [\boldsymbol{\nabla}\mathbf{v}_m]^T - \frac{2}{3}\boldsymbol{\nabla}\cdot\mathbf{v}_m\underline{\mathbf{I}}\right) \qquad [239]$$

and the pressure jump between phases, [231], becomes

$$\Delta P = -\zeta(1-\phi)\boldsymbol{\nabla}\cdot\mathbf{v}_m \qquad [240]$$

if an equivalent bulk viscosity is used, or

$$\Delta P = -K_0\eta_m\frac{\boldsymbol{\nabla}\cdot\mathbf{v}_m}{\phi} \qquad [241]$$

from the micromechanical model, [233], of Bercovici et al. (2001a).

The action–reaction equation [238] can also be written in a different way for example by elimination of $\Delta\mathbf{v}$ taken from the Darcy equilibrium [237],

$$-\boldsymbol{\nabla}P_f + \boldsymbol{\nabla}\cdot\left[(1-\phi)\boldsymbol{\tau}_m^*\right] + \bar{\rho}\mathbf{g} = 0 \qquad [242]$$

where $\boldsymbol{\tau}_m^*$ includes the $\Delta P$ term and is defined by

$$\boldsymbol{\tau}_m^* = \eta_m\left(\boldsymbol{\nabla}\mathbf{v}_m + [\boldsymbol{\nabla}\mathbf{v}_m]^T - \frac{2}{3}\boldsymbol{\nabla}\cdot\mathbf{v}_m\mathbf{I}\right)$$
$$+ \zeta(1-\phi)\boldsymbol{\nabla}\cdot\mathbf{v}_m\mathbf{I} \qquad [243]$$

This shows that if the pressure is defined everywhere as the fluid pressure, then it is equivalent to use for the matrix a rheology, (see [93]), with a bulk viscosity $(1-\phi)\zeta\sim\zeta$ (McKenzie, 1984). This analogy only holds without surface tension between phases (Bercovici et al., 2001a; Ricard et al., 2001).

The rate of melting is controlled by

$$\Delta\Gamma = \mathcal{R}\left(\Delta\mu + P_f\left(\frac{1}{\rho_m} - \frac{1}{\rho_f}\right) - T\Delta S\right) \qquad [244]$$

and the energy equation is

$$\rho_f\phi C_f\frac{D_f T}{Dt} + \rho_m(1-\phi)C_m\frac{D_m T}{Dt} - \Delta\mathcal{H}\Delta\Gamma$$
$$= \bar{\rho}\bar{H} - \boldsymbol{\nabla}\cdot\mathbf{q} + \frac{\Delta\Gamma^2}{\mathcal{R}} + K_0\eta_m\frac{1-\phi}{\phi}(\boldsymbol{\nabla}\cdot\mathbf{v}_m)^2 + \Psi \qquad [245]$$

where we have assumed the relation [241].

These equations have been used by many authors with various levels of approximation. The most benign have been to replace $1-\phi$ by 1. Most authors have also considered the bulk viscosity $\zeta$ as a porosity-independent parameter, (e.g., McKenzie, 1984;

Scott and Stevenson, 1984; Richter and Mckenzie, 1984; Ribe, 1985a, 1985b; Scott and Stevenson, 1986, 1989; Stevenson, 1989; Kelemen et al., 1997; Choblet and Parmentier, 2001; Spiegelman et al., 2001; Spiegelman and Kelemen, 2003; Katz et al., 2004). This overestimates the possibilities of matrix compaction at low porosity. Porosity-dependent parameters have been explicitly accounted for in other papers (e.g., Fowler, 1985; Connolly and Podladchikov, 1998; Schmeling, 2000; Bercovici et al., 2001a; Ricard et al., 2001; Rabinowicz et al., 2002). The melting rates are sometimes arbitrarily imposed (e.g., Turcotte and Morgan, 1992) or solved assuming univariant transition (Fowler, 1989; Sramek et al., 2007). Surface tension is added in Riley et al. (1990) and Hier-Majumder et al. (2006). Similar equations have also been used to describe the interaction between iron and silicates near the CMB (Buffett et al., 2000), the formation of dendrites (Poirier, 1991) or the compaction of lava flows (Massol et al., 2001).

## 2.6 Specifics of Earth's Mantle Convection

In this last section we discuss various aspects of physics unique to large-scale mantle convection. We leave the problems of partial melting to Chapters 7 and 9. We are aware of the impossibility to be exhaustive but most of the important points are more deeply developed in other chapters of the treatise (see also the books by Schubert et al. (2001) and Davies (1999)).

### 2.6.1 A mantle without Inertia

The most striking difference between mantle convection and most other forms of convection is that inertia is totally negligible. This is because the Prandtl number is much larger than the (already very large) Rayleigh number. This implies that the mantle velocity flow satisfies

$$\boldsymbol{\nabla}\cdot(\bar{\rho}\mathbf{v}) = 0$$
$$-\boldsymbol{\nabla}P + \boldsymbol{\nabla}\cdot\boldsymbol{\tau} + \delta\rho\mathbf{g} + \bar{\rho}\delta\mathbf{g} = 0$$
$$\nabla^2\psi = 4\pi G\delta\rho \qquad [246]$$
$$\delta\mathbf{g} = -\boldsymbol{\nabla}\psi$$

in agreement with [128]. In this set of equations we kept the self-gravitation term as appropriate at long wavelengths. If the internal loads $\delta\rho$ are knowns, the flow can be computed independently of the temperature equation. This time-independent system has been used by many authors to try to infer the mantle flow properties.

### 2.6.1.1 Dynamic models

The system of eqn [246] can be solved analytically for a depth-dependent viscosity, when variables are expressed on the basis of spherical harmonics (see Hager and Clayton (1989) and also Chapter 4). Various possible surface observables (geoid height or gravity free air anomalies, velocity divergence, amplitude of deviatoric stress at the surface, surface dynamic topography, CMB topography, etc.) can be expressed on the basis of spherical harmonics with components $O_{lm}$. Through [246], they are related to the spherical harmonics components of the internal density variations $\delta\rho_{lm}(r)$ by various degree-dependent Green's functions (Green, 1793–1841)

$$O_{lm} = \int G_l^O(r)\delta\rho_{lm}(r)\,dr \qquad [247]$$

(*see* Chapter 4 for analytical details). The Green's functions $G_l^O(r)$ can be computed from the averaged density and viscosity profiles.

Before seismic imaging gave us a proxy of the 3-D density structure of the mantle, various theoretical attempts have tried to connect models of mantle convection to plate velocities (Hager and O'Connell, 1979, 1981), to the Earth's gravity field (or to the geoid, proportional to $\psi$), to the lithospheric stress regime, or to the topography (Runcorn, 1964; Parsons and Daly, 1983; Lago and Rabinowicz, 1984; Richards and Hager, 1984; Ricard et al., 1984).

An internal load of negative buoyancy induces a downwelling flow that deflects the Earth's surface, the CMB, and any other internal compositional boundaries, if they exist. The amount of deflection corresponds to the usual isostatic rule for a load close to an interface: the weight of the induced topography equals at first order the mass of the internal load. The total gravity anomaly resulting from a given internal load is affected by the mass anomalies associated with the flow-induced boundary deflections as well as by the load itself. Due to the deflection of the Earth's surface, the geoid perturbation induced by a dense sinking anomaly is generally negative (e.g., free air gravity has a minimum above a dense load). However, when the mantle viscosity increases significantly with depth, by one to two orders of magnitude, a mass anomaly close to the viscosity increase induces a larger CMB deformation and a lower surface deformation. The resulting gravity anomaly corresponds to a geoid high. The fact that cold subduction zones correspond to a relative geoid high suggests a factor $\geq 30$ viscosity increase around the upper–lower mantle interface (Lago and Rabinowicz, 1984; Hager et al., 1985). Shallow anomalies and anomalies near the CMB, being locally compensated, do not contribute to the long-wavelength gravity field. The lithospheric stress field, like the geoid, is affected by mid-mantle density heterogeneities. The surface deflection induced by a deep-seated density anomaly decreases with the depth of this anomaly but even lower-mantle loads should significantly affect the surface topography.

### 2.6.1.2 Mantle flow and postglacial models

As soon as seismic tomography started to image the mantle structures, these seismic velocity anomalies have been used to further constrain the mantle viscosity. The fact that the geoid and the seismic velocity anomalies are positively correlated around the transition zone but negatively in the deep mantle heterogeneities suggests a viscosity larger than 10 but not too large (less than 100), otherwise the mantle would be everywhere positively correlated with gravity (Hager et al., 1985; Hager and Clayton, 1989; King and Hager, 1994). The same modeling approach, assuming a proportionality between seismic velocity anomalies and density variations, was also used to match the observed plate divergence (Forte and Peltier, 1987), the plate velocities (Ricard and Vigny, 1989; Ricard et al., 1991), and the lithospheric stresses (Bai et al., 1992; Lithgow-Bertelloni and Guynn, 2004). The initial Boussinesq models were extended to account for compressibility (Forte and Peltier, 1991).

Joint inversions of gravity with postglacial rebound were also performed to further constrain the mantle viscosity profile. The viscosity increase required by subduction was initially thought to be too large to reconcile with postglacial rebound (Peltier, 1996). The various approaches (time dependent for the postglacial models and time independent for the geoid models) seem to have converged to a standard viscosity profile with a significant increase with depth (Mitrovica and Forte, 1997). Whether this viscosity increase occurs across a discontinuity (at the upper–lower mantle interface, or deeper) or as a gradual increase is probably beyond the resolution of these approaches (**Figure 6**).

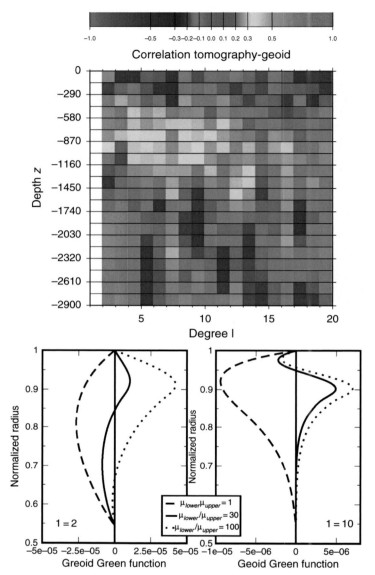

**Figure 6** Correlations between gravity and the synthetic tomographic Smean model (Becker and Boschi, 2002) as a function of degree *l* and normalized radius (top). The seismic velocities, proxy of the density variations, are positively correlated with gravity around the upper–lower mantle interface (warm colors) but negatively correlated, near the surface and in the deep lower mantle (cold colors). Geoid Green functions for degree 2 (bottom left) and degree 10 (bottom right) and three possible viscosity increases between upper and lower mantle. The geoid Green function for a uniform viscosity (dashed line) is everywhere negative and all the anomalies around the upper–lower mantle would induce a gravity signal opposite to that observed. A too large viscosity increase (a factor 100 for the dotted lines) cannot explain the rather good negative correlation between lower-mantle anomalies and the geoid at long wavelength. A moderate increase (a factor 30 for the solid line) leads to the best fit as the sign of the Green functions is everywhere that of the observed density–gravity correlations. The different Green functions are computed for an incompressible mantle, with a lithosphere, 10 times more viscous than the upper mantle.

Although these dynamic models explain the observed geoid, they require surface topography of order 1 km, induced by mantle convection and called dynamic topography (in contrast to the isostatic topography related to crustal and lithospheric density variations). Its direct observation, from the Earth's topography corrected for isostatic crustal contributions, is difficult and remains controversial (e.g., Colin and Fleitout, 1990; Kido and Seno, 1994; Lestunff and Ricard, 1995; Lithgow-Bertelloni and Silver, 1998).

### 2.6.1.3 Time-dependent models

The thermal diffusion in the mantle is so slow that even over 100–200 My it can be neglected in some long wavelength global models. Equations [246] can thus be solved by imposing the known paleo-plate velocities at the surface and advect the mass anomalies with the flow without solving explicitly the energy equation. This forced-convection approach has shown that the deep mantle structure is mostly inherited from the Cenozoic and Mesozoic plate motion (Richards and Engebretson, 1992; Lithgow-Bertelloni and Richards, 1998). From plate paleoslab reconstructions only, a density models can be obtained that gives a striking fit to the observed geoid and is in relative agreement with long-wavelength tomography (Ricard *et al.*, 1993a). This approach was also used to study the hot-spot fixity (Richards, 1991; Steinberger and O'Connell, 1998), the sea level changes (Lithgow-Bertelloni and Gurnis, 1997) or the polar wander of the Earth (Spada *et al.*, 1992a; Richards *et al.*, 1997).

## 2.6.2 A Mantle with Internal Heating

When the top and bottom boundary conditions are the same (i.e., both free-slip or both no-slip), purely basally heated convection in a Cartesian box leads to a perfectly symmetric system. We could simultaneously reverse the vertical axis and the color scale of **Figure 4** and get temperature patterns that are also convective solutions. The convective fluid has a near adiabatic core and the temperature variations are confined into two boundary layers, a hot bottom layer and a cold top layer. The thicknesses of these two boundary layers and the temperature drops across them, are the same. The mid-depth temperature is simply the average of the top and bottom temperatures. Instabilities develop from the bottom layer (hot rising plumes) and the cold layer (cold downwelling plumes). They have a temperature hotter or colder than the depth-dependent average temperature. They are active structures driven by their intrinsic positive or negative buoyancy. The Earth's mantle has however a large number of characteristics that break the symmetry between upwellings and downwellings.

What is probably the major difference between mantle convection and purely basally heated convection is that the Earth is largely powered by radiogenic heating from the decay of uranium, thorium, and potassium. Convection purely heated from within is depicted in **Figure 7**. In the extreme case where the fluid is entirely heated from within, the fluid has no hot bottom boundary layer. There are only concentrated downwelling currents sinking from the top cold boundary layer. The downwellings are active as they are moved by their own negative buoyancy. To compensate for the resulting downwelling flow, the background is rising passively, that is, without

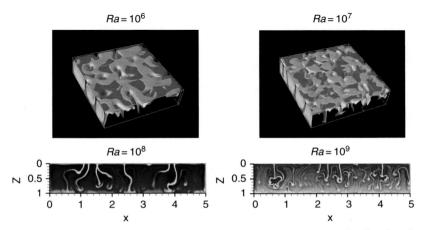

**Figure 7** Convection patterns of a fluid entirely heated from inside at Rayleigh number $10^6$, $10^7$, $10^8$, $10^9$ (simulations ran by Fabien Dubuffet). Cold finger-like instabilities are sinking from the top boundary layer, and spread on the bottom boundary layer. No active upwellings are present (compare with convection patterns for a fluid heated from below, **Figure 4**).

being pushed up by a positive buoyancy (Bercovici *et al.*, 2000). In the case of basal heating, any plume leaving the top or bottom boundary layer travels adiabatically (neglecting diffusion and shear heating). However, in the case of internal heating, while the rapid downwellings remain close to adiabatic, the radioactive decay can accumulate heat during the slow upwellings. This heating is opposite to the adiabatic cooling and the average temperature in an internally heated system remains more homogeneous and with a significant subadiabatic gradient (Parmentier *et al.*, 1994).

The Earth's mantle is however not in such an extreme situation. Some heat flow is extracted across the CMB from the molten iron outer core. This basal heat flux drives active upwellings (hot spots). The ratio of the internal radioactive heat to the total heat extracted at the Earth's surface is called the Urey number (Urey, 1951). Geochemical models of mantle composition (McDonough and Sun, 1995; Rudnick and Fountain, 1995) imply that about 50% of the surface heat flux is due to mantle and core cooling and only 50% or even less (Lyubetskaya and Korenaga, 2007), to radioactive decay. Generally, geophysicists have difficulties with these numbers as they seem to imply a too large mantle temperature in the past (Davies, 1980; Schubert *et al.*, 1980). From convection modeling of the Earth's secular cooling, they often favor ratios of order of 80% radioactive and 20% cooling, although the complex properties of the lithosphere may allow to reconcile the thermal history of the Earth with a low radiogenic content (Korenaga, 2003; Grigne *et al.*, 2005) (*see* Chapter 6). The basal heat flux at the CMB represents the core cooling component, part of the total cooling rate of the Earth. The secular cooling and the presence of internal sources tend to decrease the thickness of the hot bottom layer compared to that of the cold top layer, increase the active role of downwellings (the subducting slabs), and decrease the number or the strength of the active upwellings (the hot spots).

### 2.6.3   A Complex Rheology

We have shown that the rheological laws of crystalline solids may be linear or nonlinear, depending on temperature, grain size, and stress level. Various deformation mechanisms (grain diffusion, grain boundary diffusion, dislocation creep, etc.) act simultaneously. The equivalent viscosity of each individual mechanism can be written in the form

$$\eta = A I_2^{-n} d^m \exp\left(\frac{E^* + P V^*}{RT}\right) \qquad [248]$$

where $E^*$ and $V^*$ are the activation energy and volume, $P$ and $T$ the pressure and temperature, $R$ the perfect gas constant, $d$ the grain size, $m$ the grain size exponent, $I_2$ the second stress invariant, and $n$ a stress exponent (Weertman and Weertman, 1975; Ranalli, 1995). The multiplicative factor $A$ varies with water content, melt content and mineralogy. In general, the composite rheology is dominated by the mechanism leading to the lowest viscosity.

In **Figure 8**, we plot as a function of temperature, and for various possible grain sizes (0.1 mm, 1 mm, 1 cm) the stress rate at which the strain rate predicted for the diffusion and dislocation mechanisms are the same (see [105] and [106]). The data correspond to dry upper mantle (Karato and Wu, 1993). Low stress and temperature favor diffusion creep while high stress and high temperature favor dislocation creep. Below the lithosphere, in the upper mantle or at least in its shallowest part, nonlinear creep is likely to occur. As depth increases, the decrease in the average deviatoric stress favors a diffusive regime with a Newtonian viscosity. The observation of rheological

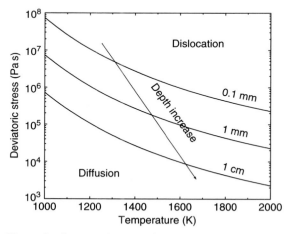

**Figure 8**  Creep regime map for dry olivine. High deviatoric stress or temperature favor a dislocation mechanism. A decrease in the grain size favors diffusion. In the upper mantle the stress and temperature conditions tend to bring the creep regime from dislocation to diffusion at depth ($n = 2.5$, $m = 0$, and $E^* = 540\,\text{kJ mol}^{-1}$ for dislocation creep, $n = 0$, $m = 2.5$, and $E^* = 300\,\text{kJ mol}^{-1}$ for diffusion creep).

parameters at lower-mantle conditions are more difficult but the lower mantle should mostly be in diffusive linear regime except the zones of intense shear around subductions (McNamara et al., 2001).

The lateral variations of viscosity due to each separate parameter, stress exponent, temperature, water content, or grain size can potentially be very large. Surprisingly, attempts to deduce these variations directly from geodynamic observations have not been very successful. Attempts to explain the Earth's gravity from internal loads do not seem to require lateral viscosity variations in the deep mantle (Zhang and Christensen, 1993). Near the surface, viscosity variations are present, at least between continental and oceanic lithosphere (Ricard et al., 1991; Cadek and Fleitout, 2003). The gain in fitting the Earth's gravity or postglacial rebound data with 3-D viscosity models remains rather moderate compared to the complexities added to the modeling (Gasperini et al., 2004) and most models of mantle viscosity are restricted to radial profiles (Mitrovica and Forte, 1997). Even the modeling of slabs, their curvatures, and their stress patterns do not really require that they are much stiffer than the surrounding mantle (Tao and O'Connell, 1993).

### 2.6.3.1 Temperature dependence of viscosity

Mantle viscosity is a strong function of temperature and the cold lithosphere seems to have a viscosity of order $10^{25}$ Pa s (Beaumont, 1978), four to six orders of magnitude stiffer than the asthenosphere. The activation energy $E^*$ is typically from 300 to 600 kJ mol$^{-1}$ (Drury and FitzGerald, 1998), the lowest values being for diffusion creep. This implies a factor $\sim 10$ in viscosity decrease for a 100 K temperature increase (using $T \sim 1600$ K). The effect of temperature dependence of viscosity on the planform of convection was recognized experimentally using oils or syrups (Booker, 1976; Richter, 1978; Nataf and Richter, 1982; Weinstein and Christensen, 1991). In the case of a strongly temperature-dependent viscosity, the definition of the Rayleigh number is rather arbitrary as the maximum, the minimum or some average viscosity can be chosen in its definition. Another nondimensional number (e.g., the ratio of viscosity variations, $\eta_{max}/\eta_{min}$) must be known to characterize the convection.

Not surprisingly, two extreme regimes are found. For a viscosity ratio lower than about 100, the convection pattern remains quite similar to convection with uniform viscosity. On the other hand, if the viscosity of the cold boundary layer (the lithosphere) is more than 3000 times that of the underlying asthenosphere, the surface becomes stagnant (Solomatov, 1995). Below this immobile lid, the flow resembles convection below a rigid top surface (Davaille and Jaupart, 1993). In between, when the viscosity ratios are in the range 100–3000, the lithosphere deforms slowly and in this sluggish regime, the convection cells have large aspect ratios.

Convection with temperature-dependent viscosity has been investigated by various authors (Parmentier et al., 1976; Christensen, 1984b; Tackley et al., 1993; Trompert and Hansen, 1998b; Kameyama and Ogawa, 2000). Since the top boundary layer is stiffer than the bottom boundary layer, the top boundary layer is also thicker than the bottom one. This impedes surface heat removal, eases the heat flux across the bottom boundary layer, and raises the average mantle temperature. Convection patterns computed with $T$-dependent viscosity remain however quite far from Earth-like convection. The major difference is that when the $T$-dependence is too strong, the surface freezes and becomes immobile while on the real Earth, the lithosphere is highly viscous but broken into tectonic plates separated by weak boundaries. Without mechanisms other than a simple $T$-dependence of viscosity, the Earth would be in a stagnant-lid regime.

Various modelers have thus tried to use $T$-dependent rheologies but have imposed a plate-like surface velocity. This has been very useful to understand the initiation of subduction (Toth and Gurnis, 1998), the interaction of slabs with the phase changes in the transition zone (Christensen, 1996, 1997b), and the relationship between subduction and gravity (King and Hager, 1994). These numerical experiments, mostly intended to model slabs, compare satisfactorily with laboratory experiments (Kincaid and Olson, 1987; Guilloufrottier et al., 1995).

To conclude this brief section on temperature dependence of viscosity, we discuss the general concept of self-regulation of planetary interiors (Tozer, 1972). If a planet were convecting too vigorously, it would lose more heat than radioactively produced. It would therefore cool down until the viscosity is large enough to reduce the heat transfer. On the contrary, a planet convecting too slowly would not extract its radioactive energy, and would heat up until the viscosity is reduced sufficiently (*see also* Chapter 6). The internal temperature of planets is mostly controlled to the activation energy (or rather enthalpy) of the viscosity (assuming that planets have similar amount of heat sources). To first order, large and small terrestrial rocky planets probably have the same internal temperatures.

### 2.6.3.2  Depth dependence of viscosity

The activation volume of the viscosity is typically around $10^{-5}$ m$^3$ mol$^{-1}$. Extrapolating to CMB conditions, this suggests a large viscosity increase throughout the mantle. However, measurements of viscosity at both high $T$ and $P$ conditions are very difficult. The viscosity increase by a factor 30–100 suggested by geodynamics (see Section 2.6.1) is probably a constrain as robust as what can be deduced from mineralogic experiments.

The effect of a depth-dependent viscosity on the planform of convection has been studied by, for example, Gurnis and Davies (1986), Cserepes (1993), or Dubuffet *et al.* (2000). At least two important geodynamic observations can be explained by an increase of viscosity with depth. One is the relative stability of hot spots. A sluggish lower mantle where convection is decreased in intensity by a larger viscosity (and also by a smaller expansivity and a potentially larger thermal conductivity as discussed in Section 2.6.5.3) favors the relative hot-spot fixity (Richards, 1991; Steinberger and O'Connell, 1998). A second consequence is a depth dependence of the wavelengths of the thermal heterogeneities. A viscosity increase (together with the existence of plates and continents that impose their own wavelengths; see Section 2.6.7) induces the existence of large-scale thermal anomalies at depth (Bunge and Richards, 1996). A slab crossing a factor 30–100 viscosity increase should thicken by a factor of order 3–5 (Gurnis and Hager, 1988). This thickening is observed in tomographic models (van der Hilst *et al.*, 1997) and can be inferred from geoid modeling (Ricard *et al.*, 1993a).

### 2.6.3.3  Stress dependence of viscosity

Starting from Parmentier *et al.* (1976) the effect of a stress-dependent viscosity has been studied by Christensen (1984a), Malevsky and Yuen (1992), van Keken *et al.* (1992), and Larsen *et al.* (1993), assuming either entirely nonlinear or composite rheologies (where deformation is accommodated by both linear and nonlinear mechanisms). At moderate

Rayleigh number, the effect of a nonlinear rheology is not very significant. In fact, the nonlinearity in the rheology is somewhat opposed to the temperature dependence of the rheology. As shown by Christensen (1984a), a $T$-dependent, nonlinear rheology with an exponent $n \sim 3$ leads to convection cells rather similar to what would be obtained with a linear rheology and an activation energy divided by $\sim n$. Convection with both nonlinear and $T$-dependent rheology looks more isoviscous than convection with only stress-dependent or only $T$-dependent, rheologies.

At large Rayleigh number, however, nonlinear convection becomes more unstable (Malevsky and Yuen, 1992) and the combination of nonlinear rheology, $T$-dependent rheology, and viscous dissipation can accelerate the rising velocity of hot plumes by more than an order of magnitude (Larsen and Yuen, 1997).

### 2.6.3.4  Grain size dependence of viscosity

The viscosity law [248] can also be a function of the grain size $d$ with an exponent $m$ of order 3 in the diffusion regime (Karato *et al.*, 1986) (when the rheology is linear in terms of stress, it becomes nonlinear in terms of grain size). There is a clear potential feedback interaction between deformation, grain size reduction by dynamic recrystallization, viscosity reduction, and further localization (Jaroslow *et al.*, 1996). Grain size reduction is offset by grain growth (e.g., the fact that surface energy drives mass diffusion from small grains to larger grains) which provides an effective healing mechanism (Hillert, 1965). A grain size-dependent viscosity has been introduced into geodynamic models (e.g., Braun *et al.*, 1999; Kameyama *et al.*, 1997). The effect is potentially important in the mantle and even more important in the lithosphere.

### 2.6.4  Importance of Sphericity

An obvious difference between the convection planform in a planet and in an experimental tank is due to the sphericity of the former. In the case of purely basally heated convection, the same heat flux (in a statistical sense) has to be transported through the bottom boundary layer and the top boundary layer. However as the CMB surface is about 4 times smaller than the top surface, this implies a 4-times larger thermal gradient through the bottom boundary layer than across the lithosphere. A bottom boundary layer thicker than the top boundary layer reinforces the upwelling hot instabilities with respect to the downgoing cold instabilities. Sphericity affects the average temperature and the top and bottom boundary layer thickness in a way totally opposite to the effects of internal sources (see Section 2.6.2) or $T$-dependent viscosity (see Section 2.6.3). Although numerically more difficult to handle, spherical convection models are more and more common (Glatzmaier, 1988; Bercovici *et al.*, 1989a, 1989b, 1992; Tackley *et al.*, 1993; Bunge *et al.*, 1997; Zhong *et al.*, 2000).

### 2.6.5 Other Depth-Dependent Parameters

#### 2.6.5.1 Thermal expansivity variations

The thermal expansivity varies with depth, as predicted by the EoS [79], from which we can easily deduce that

$$\alpha = \frac{\alpha_0}{(\rho/\rho_0)^{n-1+q} + \alpha_0(T - T_0)} \qquad [249]$$

It decreases with both temperature and density, and thus with depth. The expansivity varies from $\sim 4 \times 10^{-5}$ K$^{-1}$ near the surface to $\sim 8 \times 10^{-6}$ K$^{-1}$ near the CMB (Chopelas and Boehler, 1992). This diminishes the buoyancy forces and slows down the deep mantle convection (Hansen *et al.*, 1993). Like the increase of viscosity with depth, a depth-dependent thermal expansivity broadens the thermal structures of the lower mantle, and suppresses some hot instabilities at the CMB. On the other hand, hot instabilities gain buoyancy as they rise in the mantle, which favors their relative lateral stationarity. In addition to its average depth dependence, the temperature dependence of the expansivity also affects the buoyancy of slabs (Schmeling *et al.*, 2003).

#### 2.6.5.2 Increase in average density with depth

To take into account compressibility and the depth dependence of density, the Boussinesq approximation has been replaced by the anelastic approximation in several studies. Such investigations have been carried out by Glatzmaier (1988), Bercovici *et al.* (1992) and since extended to higher Rayleigh numbers (e.g., Balachandar *et al.*, 1992, 1993; Zhang and Yuen, 1996).

One of the difficulties with compressible fluids is that the local criterion for instability (see Section 2.4.2.3) is related to the adiabatic gradient. Depending on assumptions about the curvature of the reference geotherm with depth (the slope of the adiabatic gradient), part of the fluid can be unstable while the other part is stable. Assuming a uniform adiabatic gradient does not favor the preferential destabilization of either the upper or the lower mantle. On the other hand, assuming that the reference temperature increases exponentially with depth (i.e., taking the order of magnitude equations [123] as real equalities) would lead to an easier destabilization of the top of the mantle than of its bottom as a much larger heat flux would be carried along the lower mantle adiabat. In the real Earth, the adiabatic gradient (in K km$^{-1}$) should decrease with depth (due to the decrease in expansivity $\alpha$ with depth insufficiently balanced by the density increase, see [121]). Since less heat can be carried out along the deep mantle adiabat, compressibility should favor the destabilization of the deep mantle.

Compressible convection models generally predict downgoing sheets and cylindrical upwellings reminiscent of slabs and hot spots (Zhang and Yuen, 1996). Viscous dissipation is positive (as an entropy-related source) but maximum just below the cold boundary layer and just above the hot boundary layer, where rising or sinking instabilities interact with the layered structures of the boundary layers. On the contrary the adiabatic source heats the downwellings and cools the upwellings. On average, it reaches a maximum absolute value in the mid-mantle. Locally, viscous dissipation and adiabatic heatings can be larger than radiogenic heat production although integrated over the whole mantle and averaged over time, the adiabatic and dissipative sources cancel out (see [59]).

#### 2.6.5.3 Thermal conductivity variations

The thermal conductivity of a solid is due to two different effects. First, a hot material produces blackbody radiation that can be absorbed by neighboring atoms. This radiative transport of heat is probably a minor component since the mean free path of photons in mantle materials is very small. Second, phonons, which are collective vibrations of atoms, are excited and can dissipate their energies by interacting with other phonons, with defects and grain boundaries. The free paths of phonons being larger, they are the main contributors to the thermal conductivity.

According to Hofmeister (1999), thermal conductivity should increase with depth by a factor $\sim$2–3. The recent observations of phase transitions in the bottom of the lower mantle should also be associated with another conductivity increase (Badro *et al.*, 2004). This is one more effect (with the viscosity increase and the thermal expansivity decrease) that should decrease the deep mantle convective vigor. It also broadens the thermal anomalies, increases the average mantle temperature, and thins the bottom boundary layer (Dubuffet *et al.*, 1999).

### 2.6.6 Thermo-Chemical Convection

Except in Section 2.5, a simple negative relationship was assumed between density variations and temperature variations, through the thermal expansivity, $\Delta\rho = -\alpha\rho\Delta T$. However, in the mantle several sources of density anomalies are present. The density in the mantle varies with the temperature $T$ for a given mineralogical composition, or phase content, symbolized by the symbol $\phi$ (e.g., for a given proportion of oxides and perovskite in the lower mantle). The mineralogical phase content is itself a function of the bulk elemental composition $\chi$ (e.g., of the proportion of Mg, Fe, O, etc., atoms) and evolves with pressure and temperature to maintain the Gibbs energy minimum. Therefore, the variations of density in the mantle at a given pressure have potentially three contributions that can be summarized as

$$\Delta\rho = \left(\frac{\partial\rho}{\partial T}\right)_\phi \Delta T + \left(\frac{\partial\rho}{\partial\phi}\right)_T \left(\frac{\partial\phi}{\partial T}\right)_\chi \Delta T$$
$$+ \left(\frac{\partial\rho}{\partial\phi}\right)_T \left(\frac{\partial\phi}{\partial\chi}\right)_T \Delta\chi \qquad [250]$$

The first term on the right-hand side is the intrinsic thermal effect computed assuming a fixed mineralogy; we have already discussed this term. The second term is a thermochemical effect. The density is a function of the mineralogical composition controlled at uniform pressure and elemental composition, by the temperature variations. This effect is responsible for a rise in the 410 km deep interface and it deepens the 660 km interface in the presence of cold downwellings (Irifune and Ringwood, 1987). The last term is the intrinsic chemical effect (related to variations of the mineralogy due to changes in the elemental composition at constant temperature). The three contributions have very similar amplitudes and none of them is negligible (Ricard *et al.*, 2005).

The effect of the second term has been rather well studied (Schubert *et al.*, 1975; Christensen and Yuen, 1984; Machetel and Weber, 1991; Peltier and Solheim, 1992; Tackley *et al.*, 1993; Tackley, 1995; Christensen, 1996). Phase changes in cold downgoing slabs occur at shallower depth in the case of exothermic phase changes and at greater depth for endothermic phase changes (the ringwoodite to oxides plus perovskite phase change at 660 km depth is endothermic, all the important other phase changes of the transition zone are exothermic). These sources of anomalies and their signs are related to the Clapeyron slope of the phase transitions. The existence of latent heat release during phase change (see [161]) is a secondary and minor effect. The recent discovery of a phase transformation in the deep lower mantle (Murakami *et al.*, 2004) (the postperovskite phase) suggests that part of the complexities of the D″ layer are related to the interaction between a phase change and the hot boundary layer of the mantle (Nakagawa and Tackley, 2006).

The fact that below the normal 660 km depth interface there is a region where slabs remain in a low-density upper-mantle phase instead of being transformed into the dense lower-mantle phase is potentially a strong impediment to slab penetration. The idea that this effect induces a layering of convection at 660 km or a situation where layered convection is punctuated by large 'avalanche' events dates back to Ringwood and Irifune (1988) and was supported by numerical simulations in the 1990s (e.g., Machetel and Weber, 1991; Honda *et al.*, 1993; Tackley, 1995). It seems however that the importance of this potential effect has been reduced in recent simulations with more realistic Clapeyron slopes, phase diagrams (taking into account both the pyroxene and garnet phases), thermodynamic reference values (the phase change effect has to be compared with thermal effects and thus an accurate choice for the thermal expansivity is necessary), and viscosity profiles.

The last contribution to the density anomalies are related to variations in chemical composition (*see* Chapter 10). There are indications of large-scale depth and lateral variations of Fe or Si contents in the mantle (Bolton and Masters, 2001; Saltzer *et al.*, 2004). A large well-documented elemental differentiation is between the oceanic crust (poor in Mg, rich in Al and Si) and the mantle.

The oceanic crust in its high-pressure eclogitic facies is ~5% denser than the average mantle density in most of the mantle except in the shallowest 100 km of the lower mantle where it is lighter (Irifune and Ringwood, 1993). In the deepest mantle it is not yet totally clear whether the eclogite remains denser, neutrally buoyant or even slightly lighter than the average mantle (e.g., Ricolleau *et al.*, 2004). Thermochemical simulations starting with the pioneering paper of Christensen and Hofmann (1994) show the possibility of a partial segregation of oceanic crust during subduction, forming pyramidal piles on the CMB. These results have been confirmed by, for example, Tackley (2000b) and Davies (2002). These compositional pyramids may anchor the hot spots (Jellinek and Manga, 2002; Davaille *et al.*, 2002). The presence of a petrologically dense component of the source of hot spots also seems necessary to explain their excess temperature (Farnetani, 1997).

Not only present-day subduction can generate compositional anomalies in the mantle. Geochemists have often argued for a deep layer of primitive material. This layer should be intrinsically denser to resist entrainment by convection. The stability of such a layer has been discussed by various authors (Davaille, 1999; Kellogg *et al.*, 1999; Tackley and Xie, 2002; LeBars and Davaille, 2002; Samuel and Farnetani, 2003). Numerical simulations of thermo-chemical convection are certainly going to replace the thermal convection models in the next years. They will help to bridge the gap between geochemical observations and convection modeling (Coltice and Ricard, 1999; van Keken *et al.*, 2002).

### 2.6.7 A Complex Lithosphere: Plates and Continents

The lithosphere is part of the convection cell, and plate tectonics and mantle convection cannot be separated. The fact that the cold lithosphere is much more viscous and concentrates most of the mass heterogeneities of the mantle, makes it behaving to some extent like a membrane on top of a less viscous fluid. This suggests some analogy between mantle convection and what is called Marangoni convection (Marangoni, 1840–1925). Marangoni convection (Nield, 1964) is controlled by temperature-dependent surface tension on top of thin layers of fluids.

The Earth's mantle is certainly not controlled by surface tension, and Marangoni convection, strictly speaking, has nothing to do with mantle convection. However, the equations of thermal convection with cooling from the top and with a highly viscous lithosphere can be shown to be mathematically related (through a change of variables) to those of Marangoni convection (Lemery *et al.*, 2000). There are large differences between mantle convection and surface driven convection but this analogy has sometimes been advocated as a 'top-down' view of the mantle dynamics (Anderson, 2001). More classically, the interpretation of plate cooling in terms of ridge-push force (Turcotte and Schubert, 1982), or the analysis of tectonic stresses using thin sheet approximations (England and Mckenzie, 1982) belong to the same approach that emphasizes the importance of the lithosphere as a stress guide and as a major source of density anomalies.

Due to the complexities of the lithosphere properties, the boundary condition at the surface of the Earth is far from being a uniform free-slip condition. Both continents and tectonic plates impose their own wavelengths and specific boundary conditions on the underlying convecting asthenosphere. Of course, the position of the continents and the number and shape of the plates are themselves consequences of mantle convection. The plates obviously organize the large-scale flow in the mantle (Hager and O'connell, 1979; Ricard and Vigny, 1989). They impose a complex boundary condition where the angular velocity is piecewise constant. The continents with their reduced heat flow (Jaupart and Mareschal, 1999) also impose a laterally variable heat flux boundary condition.

Convection models with continents have been studied numerically (Gurnis and Hager, 1988; Grigné and Labrosse, 2001, Coltice *et al.*, 2007) and experimentally (Guillou and Jaupart, 1995). Continents with their thick lithosphere tend to increase the thickness of the top boundary layer and the temperature below them (see **Figure 9**). Hot rising currents are predicted under continents and downwellings are localized along continental edges. The existence of a thick and stable continental root must be due to a chemically lighter and more viscous subcontinental lithosphere (Doin *et al.*, 1997). The ratio of the heat flux extracted across continents compared to that extracted across oceans increases with the Rayleigh number. This suggests that the continental geotherms were not much different in the past when the radiogenic sources were larger; it is mostly the oceanic heat flux that was larger (Lenardic, 1998). Simulating organized plates

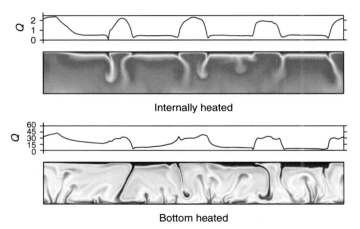

Internally heated

Bottom heated

**Figure 9** Convection patterns in the presence of four continents. The total aspect ratio is 7, the continents are defined by a viscosity increase by a factor $10^3$ over the depth 1/10. The viscosity is otherwise constant. The Rayleigh number based on the total temperature drop (bottom panels) or on the internal radioactivity (top panels) is $10^7$. The downwellings are localized near the continent margins. A large difference in heat flux is predicted between oceans and continents. In the case of bottom heating, hot spots tend to be preferentially anchored below continents where they bring an excess heat. This tends to reduce the surface heat flux variations.

self-consistently coupled with a convective mantle has been a very difficult quest. The attempts to generate plates using $T$-dependent or simple nonlinear rheologies have failed. Although in 2-D some successes can be obtained in localizing deformation in plate-like domains, (Schmeling and Jacoby, 1981; Weinstein and Olson, 1992; Weinstein, 1996), they are obtained with stress exponents (e.g., $n \geq 7$) that are larger than what can be expected from laboratory experiments ($n \sim 2$). The problems are however worst in 3-D. Generally, these early models do not predict the important shear motions between plates that is observed (Christensen and Harder, 1991; Ogawa *et al.*, 1991).

Some authors have tried to mimic the presence of plates by imposing plate-like surface boundary conditions. These studies have been performed in 2-D and 3-D (Ricard and Vigny, 1989; Gable *et al.*, 1991; King *et al.*, 1992; Monnereau and Quéré, 2001). Although they have confirmed the profound effect of plates on the wavelengths of convection, on its time dependence and on the surface heat flux, these approaches cannot predict the evolution of surface plate geometry. **Figure 10** illustrates the organizing effect of plates in spherical, internally heated compressible convection with depth-dependent viscosity (Bunge and Richards, 1996). To obtain a self-consistent generation of surface plates, more complex rheologies that include brittle failure, strain

softening, and damage mechanisms must be introduced (e.g., Bercovici, 1993, 1995; Moresi and Solomatov, 1998; Auth *et al.*, 2003). The existence of plates seems also to require the existence of a weak sublithospheric asthenosphere (Richards *et al.*, 2001). In the last years, the first successes in computing 3-D models that spontaneously organize their top boundary layer into plates have been reached (Tackley, 1998, 2000c, 2000d, 2000e; Trompert and Hansen, 1998a; Stein *et al.*, 2004). Although the topological characteristics of the predicted plates and their time evolution may be still far from the observed characteristics of plate tectonics, and often too episodic (stagnant-lid convection punctuated by plate-like events), a very important breakthrough has been made by modelers (see **Figure 11**).

The Earth's plate boundaries keep the memory of their weakness over geological times (Gurnis *et al.*, 2000). This implies that the rheological properties cannot be a simple time-independent function of stress or temperature but has a long-term memory. The rheologies that have been used to predict plates in convective models remain empirical and their interpretation in terms of microscopic behavior and damage theory remains largely to be done (Bercovici and Ricard, 2005). Reviews on the rapid progress and the limitations of self-coherent convection models can be found in Bercovici *et al.* (2000), Tackley (2000a), and Bercovici (2003).

Free-slip surface

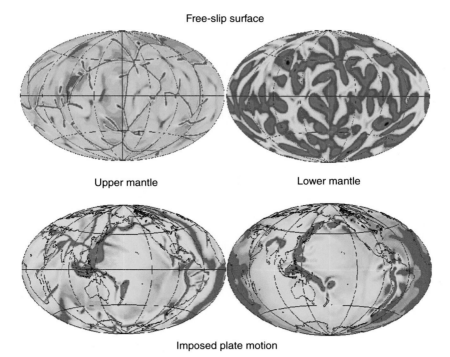

Upper mantle           Lower mantle

Imposed plate motion

**Figure 10** Spherical compressible internally heated convection models where the viscosity increases with depth (simulations by Peter Bunge). In the first row, a uniform free-slip condition on top has been used. In the second row, the present-day observed plate motion is imposed at the surface. The left column shows the temperature field in the middle of the upper mantle, the right column in the middle of the lower mantle. The figure summarizes various points discussed in the text: the presence of linear cold downwellings, the absence of active upwellings in the absence of basal heating, and the enlargement of thermal structure in the more viscous lower mantle (top row). Although the modeling is not self-consistent (i.e., the presence of plates and the constancy of plate velocities are totally arbitrary), it is clear that the presence of plates can change radically the convection patterns (compare top and bottom rows).

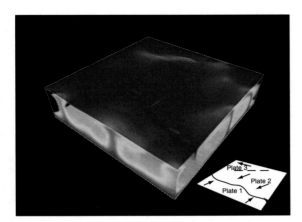

**Figure 11** Convection models with self-coherent plate generation (Stein *et al.*, 2004). Snapshot of the temperature field for a model calculation in a box of aspect ratio 4. The viscosity is temperature, pressure, and stress dependent. The flow pattern reveals cylindrical upwellings and sheet-like downflow. Three plates have formed sketched in the small plot.

# References

Anderson OL (1979) Evidence supporting the approximation $\gamma\rho = \text{const.}$ for the gruneisen parameter of the earth lower mantle. *Journal of Geophysical Research* 84: 3537–3542.

Anderson DL (2001) Geophysics – Top-down tectonics? *Science* 293: 2016–2018.

Auth C, Bercovici D, and Christensen UR (2003) Two-dimensional convection with a self-lubricating, simple-damage rheology. *Geophysical Journal International* 154: 783–800.

Badro J, Rueff JP, Vanko G, Monaco G, Fiquet G, and Guyot F (2004) Electronic transitions in perovskite: Possible nonconvecting layers in the lower mantle. *Science* 305: 383–386.

Bai WM, Vigny C, Ricard Y, and Froidevaux C (1992) On the origin of deviatoric stresses in the lithosphere. *Journal of Geophysical Research* 97: 11729–11737.

Balachandar S, Yuen DA, and Reuteler D (1992) Time-dependent 3-dimensional compressible convection with depth-dependent properties. *Geophysical Research Letters* 19: 2247–2250.

Balachandar S, Yuen DA, and Reuteler D (1993) Viscous and adiabatic heating effects in 3-dimensional compressible convection at infinite prandtl number. *Physics of Fluids A-fluid Dynamics* 5: 2938–2945.

Barenblatt GI (1996) *Scaling*. Cambridge, UK: Cambridge University Press.

Batchelor (1967) *An Introduction to Fluid Dynamics*. Cambridge, UK: Cambridge University Press.

Bear J (1988) *Dynamics of Fluids in Porous Media*. New York: Dover publishers.

Beaumont C (1978) Evolution of sedimentary basins on a viscoelastic lithosphere – Theory and examples. *Geophysical Journal of the Royal Astronomical Society* 55: 471–497.

Becker TW and Boschi L (2002) A comparison of tomographic and geodynamic mantle models. *Geochemistry Geophysics Geosystems* 3: 1003 (doi:10.1029/2001GC000,168).

Bercovici D (1993) A simple-model of plate generation from mantle flow. *Geophysical Journal International* 114: 635–650.

Bercovici D (1995) A source-sink model of the generation of plate-tectonics from non-newtonian mantle flow. *Journal of Geophysical Research-Solid Earth* 100: 2013–2030.

Bercovici D (2003) The generation of plate tectonics from mantle convection. *Earth and Planetary Science Letters* 205: 107–121.

Bercovici D and Karato S (2003) Whole-mantle convection and the transition-zone water filter. *Nature* 425: 39–44.

Bercovici D and Ricard Y (2003) Energetics of a two-phase model of lithospheric damage, shear localization and plate-boundary formation. *Geophysical Journal International* 152: 581–596.

Bercovici D and Ricard Y (2005) Tectonic plate generation and two-phase damage: Void growth versus grain size reduction. *Journal of Geophysical Research* 110: B03401.

Bercovici D, Ricard Y, and Richards M (2000) The relation between mantle dynamics and plate tectonics: A primer. In: Richards MA, Gordon R, and van der Hilst R (eds.) *Geophysical Monograph Series, vol. 121: History and Dynamics of Global Plate Motions*, pp. 5–46. Washington, DC: AGU.

Bercovici D, Ricard Y, and Schubert G (2001a) A two-phase model of compaction and damage. Part 1: General theory. *Journal of Geophysical Research* 106: 8887–8906.

Bercovici D, Ricard Y, and Schubert G (2001b) A two-phase model for compaction and damage. Part 3: Applications to shear localization and plate boundary formation. *Journal of Geophysical Research* 106: 8925–8939.

Bercovici D, Schubert G, and Glatzmaier GA (1989a) 3-dimensional spherical-models of convection in the Earths mantle. *Science* 244: 950–955.

Bercovici D, Schubert G, and Glatzmaier GA (1992) 3-dimensional convection of an infinite-prandtl-number compressible fluid in a basally heated spherical-shell. *Journal of Fluid Dynamics* 239: 683–719.

Bercovici D, Schubert G, Glatzmaier GA, and Zebib A (1989b) 3-dimensional thermal-convection in a spherical-shell. *Journal of Fluid Mechanics* 206: 75–104.

Besse J and Courtillot V (1991) Revised and synthetic apparent polar wander paths of the African, Eurasian, North-America and Indian plates, and true polar wander since 200 ma. *Journal of Geophysical Research* 96: 4029–4050.

Birch F (1952) Elasticity and constitution of the Earth's interior. *Journal of Geophysical Research* 57: 227–286.

Bolton H and Masters G (2001) Travel times of P and S from the global digital seismic networks: Implications for the relative variation of P and S velocity in the mantle. *Journal of Geophysical Research* 106: 13527–13540.

Booker JR (1976) Thermal-convection with strongly temperature-dependent viscosity. *Journal of Fluid Dynamics* 76: 741–754.

Braun J, Chery J, Poliakov A, et al. (1999) A simple parameterization of strain localization in the ductile regime due to grain size reduction: A case study for olivine. *Journal of Geophysical Research* 104: 25167–25181.

Buffett BA, Garnero EJ, and Jeanloz R (2000) Sediments at the top of Earth's core. *Science* 290: 1338–1342.

Bullen KE (1940) The problem of the Earth's density variation. *Seismological Society of America Bulletin* 30: 235–250.

Bunge HP, Ricard Y, and Matas J (2001) Non-adiabaticity in mantle convection. *Geophysical Research Letters* 28: 879–882.

Bunge HP and Richards MA (1996) The origin of large scale structure in mantle convection: Effects of plate motions and viscosity stratification. *Geophysical Research Letters* 23: 2987–2990.

Bunge HP, Richards MA, and Baumgardner JR (1997) A sensitivity study of three-dimensional spherical mantle convection at 10(8) Rayleigh number: Effects of depth-dependent viscosity, heating mode, and an endothermic phase change. *Journal of Geophysical Research* 102: 11991–12007.

Cadek O and Fleitout L (2003) Effect of lateral viscosity variations in the top 300 km on the geoid and dynamic topography. *Geophysical Journal International* 152: 566–580.

Chandrasekhar S (1969) *Ellipsoidal Figures of Equilibrium*. New Haven, CT: Yale Press.

Choblet G and Parmentier EM (2001) Mantle upwelling and melting beneath slow spreading centers: Effects of variable rheology and melt productivity. *Earth and Planetary Science Letters* 184: 589–604.

Chopelas A and Boehler R (1992) Thermal expansivity in the lower mantle. *Geophysical Research Letters* 19: 1983–1986.

Christensen U (1984a) Convection with pressure-dependent and temperature-dependent non-newtonian rheology. *Geophysical Journal of the Royal Astronomical Society* 77: 343–384.

Christensen UR (1984b) Heat-transport by variable viscosity convection and implications for the Earth's thermal evolution. *Physics of the Earth and Planetary Interiors* 35: 264–282.

Christensen UR (1996) The influence of trench migration on slab penetration into the lower mantle. *Earth and Planetary Science Letters* 140: 27–39.

Christensen UR (1997a) Some geodynamical effects of anisotropic viscosity. *Geophysical Journal of the Royal Astronomical Society* 91: 711–736.

Christensen UR (1997b) Influence of chemical buoyancy on the dynamics of slabs in the transition zone. *Journal of Geophysical Research* 102: 22435–22443.

Christensen UR and Harder H (1991) 3-d convection with variable viscosity. *Geophysical Journal International* 104: 213–226.

Christensen UR and Hofmann A (1994) Segregation of subducted oceanic crust in the convecting mantle. *Journal of Geophysical Research* 99: 19867–19884.

Christensen UR and Yuen DA (1984) The interaction of a subducting lithospheric slab with a chemical or phase-boundary. *Journal of Geophysical Research* 89: 4389–4402.

Colin P and Fleitout L (1990) Topography of the ocean-floor – thermal evolution of the lithosphere and interaction of deep mantle heterogeneities with the lithosphere. *Geophysical Research Letters* 17: 1961–1964.

Coltice N and Ricard Y (1999) Geochemical observations and one layer mantle convection. *Earth and Planetary Science Letters* 174: 125–137.

Coltice N, Phillips BR, Bertrand H, Ricard Y, and Rey P (2007) Global warming of the mantle at the origin of flood basalts over supercontinents. *Geology* 35: 391–394.

Connolly JAD and Podladchikov YY (1998) Compaction-driven fluid flow in viscoelastic rock. *Geodinamica Acta* 11: 55–84.

Corrieu V, Thoraval C, and Ricard Y (1995) Mantle dynamics and geoid green-functions. *Geophysical Journal International* 120: 516–523.

Cserepes L (1993) Effect of depth-dependent viscosity on the pattern of mantle convection. *Geophysical Research Letters* 20: 2091–2094.

Dahlen FA and Tromp J (1998) *Theoretical Global Seismology*. Princeton, NJ: Princeton University Press.

Davaille A (1999) Two-layer thermal convection in viscous fluids. *Journal of Fluid Dynamics* 379: 223–253.

Davaille A, Girard F, and LeBars M (2002) How to anchor hotspots in a convecting mantle? *Earth and Planetary Science Letters* 203: 621–634.

Davaille A and Jaupart C (1993) Transient high-rayleigh-number thermal-convection with large viscosity variations. *Journal of Fluid Dynamics* 253: 141–166.

Davies GF (1980) Thermal histories of convective Earth models and constraints on radiogenic heat-production in the Earth. *Journal of Geophysical Research* 85: 2517–2530.

Davies GF (1999) *Dynamic Earth Plates, Plumes and Mantle Convection*, 458p. Cambridge, UK: Cambridge University Press.

Davies GF (2002) Stirring geochemistry in mantle convection models with stiff plates and slabs. *Geochimica et Cosmochimica Acta* 66: 3125–3142.

de Groot SR and Mazur P (1984) *Non-Equilibrium Thermodynamics*. New York: Dover Publications.

Doin MP, Fleitout L, and Christensen U (1997) Mantle convection and stability of depleted and undepleted continental lithosphere. *Journal of Geophysical Research* 102: 2771–2787.

Dombre T, Frisch U, Greene JM, Henon M, Mehr A, and Soward AM (1986) Chaotic streamlines in the abc flows. *Journal of Fluid Dynamics* 167: 353–391.

Drew DA and Segel LA (1971) Averaged equations for 2-phase flows. *Studies in Applied Mathematics* 50: 205–220.

Drury MR and FitzGerald JD (1998) Mantle rheology: Insights from laboratory studies of deformation and phase transition. In: Jackson I (ed.) *The Earth's Mantle: Composition, Structure, and Evolution*, pp. 503–559. Cambridge, UK: Cambridge University Press.

Dubuffet F, Rabinowicz M, and Monnereau M (2000) Multiple scales in mantle convection. *Earth and Planetary Science Letters* 178: 351–366.

Dubuffet F, Yuen DA, and Rabinowicz M (1999) Effects of a realistic mantle thermal conductivity on the patterns of 3-D convection. *Earth and Planetary Science Letters* 171: 401–409.

Dziewonski AM and Anderson D (1981) Preliminary reference Earth model. *Physics of the Earth and Planetary Interiors* 25: 297–356.

England P and Mckenzie D (1982) A thin viscous sheet model for continental deformation. *Geophysical Journal of the Royal Astronomical Society* 70: 295–321.

Farnetani CG (1997) Excess temperature of mantle plumes: The role of chemical stratification across d″. *Geophysical Research Letters* 24: 1583–1586.

Ferrachat S and Ricard Y (1998) Regular vs, chaotic mantle mixing. *Earth and Planetary Science Letters* 155: 75–86.

Forte AM and Peltier WR (1987) Plate tectonics and aspherical earth structure – the importance of poloidal-toroidal coupling. *Journal of Geophysical Research* 92: 3645–3679.

Forte AM and Peltier WR (1991) Viscous flow models of global geophysical observables. Part 1: Forward problems. *Journal of Geophysical Research* 96: 20131–20159.

Fowler AC (1985) A mathematical model of magma transport in the asthenosphere. *Geophysical and Astrophysical Fluid Dynamics* 33: 63–96.

Fowler AC (1989) Generation and creep of magma in the Earth. *SIAM Journal on Applied Mathematics* 49: 231–245.

Gable CW, Oconnell RJ, and Travis BJ (1991) Convection in 3 dimensions with surface plates – generation of toroidal flow. *Journal of Geophysical Research* 96: 8391–8405.

Gasperini P, Forno GD, and Boschi E (2004) Linear or non-linear rheology in the Earth's mantle: The prevalence of power-law creep in the postglacial isostatic readjustment of laurentia. *Geophysical Journal International* 157: 1297–1302.

Glatzmaier GA (1988) Numerical simulation of mantle convection: Time-dependent, three dimensional, compressible, spherical shell. *Geophysical and Astrophysical Fluid Dynamics* 43: 223–264.

Grigné C and Labrosse S (2001) Effects of continents on Earth cooling: Thermal blanketing and depletion in radioactive elements. *Geophysical Research Letters* 28: 2707–2710.

Grigne C, Labrosse S, and Tackley PJ (2005) Convective heat transfer as a function of wavelength: Implications for the cooling of the Earth. *Journal of Geophysical Research-Solid Earth* 110: B03409.

Guillou L and Jaupart C (1995) On the effect of continents on mantle convection. *Journal of Geophysical Research* 100: 24217–24238.

Guilloufrottier L, Buttles J, and Olson P (1995) Laboratory experiments on the structure of subducted lithosphere. *Earth and Planetary Science Letters* 133: 19–34.

Gurnis M and Davies GF (1986) Numerical study of high rayleigh number convection in a medium with depth-dependent viscosity. *Geophysical Journal of the Royal Astronomical Society* 85: 523–541.

Gurnis M and Hager BH (1988) Controls of the structure of subducted slabs. *Nature* 335: 317–321.

Gurnis M, Moresi L, and Mueller RD (2000) Models of mantle convection incorporating plate tectonics: The Australian region since the Cretaceous. In: Richards MA, Gordon R, and van der Hilst R (eds.) *The History and Dynamics of Global Plate Motions*, pp. 211–238. Washongton, DC: American Geophysical Union.

Haase R (1990) *Thermodynamics of Irreversible Processes*. New York: Dover Publications.

Hager BH and Clayton RW (1989) Constraints on the structure of mantle convection using seismic observation, flow models and the geoid. In: Peltier WR (ed.) *Mantle Convection, Plate Tectonics and Global Dynamics*, pp. 657–763. New York: Gordon and Breach Science Publishers.

Hager BH, Clayton RW, Richards MA, Comer RP, and Dziewonski AM (1985) Lower mantle heterogeneity, dynamic topography and the geoid. *Nature* 313: 541–546.

Hager BH and O'Connell RJ (1979) Kinematic models of large-scale flow in the Earth's mantle. *Journal of Geophysical Research* 84: 1031–1048.

Hager BH and O'Connell RJ (1981) A simple global model of plate tectonics and mantle convection. *Journal of Geophysical Research* 86: 4843–4867.

Hansen U, Yuen DA, Kroening SE, and Larsen TB (1993) Dynamic consequences of depth-dependent thermal expansivity and viscosity on mantle circulations and thermal structure. *Physics of the Earth and Planetary Interiors* 77: 205–223.

Haskell NA (1937) The viscosity of the asthenosphere. *American Journal of Science* 33: 22–28.

Hewitt JM, McKenzie DP, and Weiss NO (1975) Dissipative heating in convective flows. *Journal of Fluid Dynamics* 68: 721–738.

Hier-Majumder S, Ricard Y, and Bercovici D (2006) Role of grain boundaries in magma migration and storage. *Earth and Planetary Science Letters* 248: 735–749.

Hillert M (1965) On theory of normal and abnormal grain growth. *Acta Metallurgica* 13: 227–235.

Hirth G and Kolhstedt DL (1996) Water in the oceanic upper mantle: Implications for rheology, melt extraction and the evolution of the lithosphere. *Earth and Planetary Science Letters* 144: 93–108.

Hofmeister AM (1999) Mantle values of thermal conductivity and the geotherm from phonon lifetimes. *Science* 283: 1699–1706.

Honda S, Yuen DA, Balachandar S, and Reuteler D (1993) 3-dimensional instabilities of mantle convection with multiple phase-transitions. *Science* 259: 1308–1311.

Irifune T and Ringwood A (1993) Phase transformations in subducted oceanic crust and buoyancy relationships at depths of 600–800 km in the mantle. *Earth and Planetary Science Letters* 117: 101–110.

Irifune T and Ringwood AE (1987) Phase transformations in a harzburgite composition to 26 GPa: Implications for dynamical behaviour of the subducting slab. *Earth and Planetary Science Letters* 86: 365–376.

Ito G, Shen Y, Hirth G, and Wolfe CJ (1999) Mantle flow, melting, and dehydration of the Iceland mantle plume. *Earth and Planetary Science Letters* 165: 81–96.

Jaroslow GE, Hirth G, and Dick HJB (1996) Abyssal peridotite mylonites: Implications for grain-size sensitive flow and strain localization in the oceanic lithosphere. *Tectonophysics* 256: 17–37.

Jarvis GT and Mckenzie DP (1980) Convection in a compressible fluid with infinite prandtl number. *Journal of Fluid Dynamics* 96: 515–583.

Jaupart C and Mareschal JC (1999) The thermal structure and thickness of continental roots. *Lithos* 48: 93–114.

Jeanloz R and Morris S (1987) Is the mantle geotherm subadiabatic. *Geophysical Research Letters* 14: 335–338.

Jellinek AM and Manga M (2002) The influence of a chemical boundary layer on the fixity, spacing and lifetime of mantle plumes. *Nature* 418: 760–763.

Kameyama M and Ogawa M (2000) Transitions in thermal convection with strongly temperature-dependent viscosity in a wide box. *Earth and Planetary Science Letters* 180: 355–367.

Kameyama M, Yuen DA, and Fujimoto H (1997) The interaction of viscous heating with grain-size dependent rheology in the formation of localized slip zones. *Geophysical Research Letters* 24: 2523–2526.

Karato SI (1998) Seismic anisotropy in the deep mantle, boundary layers and the geometry of mantle convection. *Pure and Applied Geophysics* 151: 565–587.

Karato SI, Paterson MS, and gerald JDF (1986) Rheology of synthetic olivine aggregates – influence of grain-size and water. *Journal of Geophysical Research* 91: 8151–8176.

Karato SI and Wu P (1993) Rheology of the upper mantle: A synthesis. *Science* 260: 771–778.

Katz RF, Spiegelman M, and Carbotte SM (2004) Ridge migration, asthenospheric flow and the origin of magmatic segmentation in the global mid-ocean ridge system. *Geophysical Research Letters* 31: L15605.

Kelemen PB, Hirth G, Shimizu N, Spiegelman M, and Dick HJB (1997) A review of melt migration processes in the adiabatically upwelling mantle beneath oceanic spreading ridges. *Philosophical Transactions of the Royal Society of London Series A – Mathematical Physical and Engineering Sciences* 355: 283–318.

Kellogg LH, Hager BH, and van der Hilst RD (1999) Compositional stratification in the deep mantle. *Science* 283: 1881–1884.

Kellogg LH and Turcotte DL (1987) homogeneization of the mantle by convective mixing and diffusion. *Earth and Planetary Science Letters* 81: 371–378.

Kennett BLN (2001) *The Seismic Wavefield*. Cambridge, UK: Cambridge University Press.

Kido M and Seno T (1994) Dynamic topography compared with residual depth anomalies in oceans and implications for age-depth curves. *Geophysical Research Letters* 21: 717–720.

Kincaid C and Olson P (1987) An experimental-study of subduction and slab migration. *Journal of Geophysical Research* 92: 13832–13840.

King SD and Hager BH (1994) Subducted slabs and the geoid. Part 1: Numerical experiments with temperature-dependent viscosity. *Journal of Geophysical Research* 99: 19843–19852.

King SD, Gable CW, and Weinstein SA (1992) Models of convection-driven tectonic plates – A comparison of methods and results. *Geophysical Journal International* 109: 481–487.

Korenaga J (2003) Energetics of mantle convection and the fate of fossil heat. *Geophysical Research Letters* 30: 1437.

Lago B and Rabinowicz M (1984) Admittance for convection in a layered spherical shell. *Geophysical Journal of the Royal Astronomical Society* 77: 461–482.

Lambeck K and Johnston P (1998) The viscosity of the mantle. In: Jackson I (ed.) *The Earth's Mantle: Composition, Structure, and Evolution*, pp. 461–502. Cambridge, UK: Cambridge University Press.

Landau L and Lifchitz E (1980) *An Introduction to Fluid Dynamics*. Oxford: Pergamon Press.

Landau L and Lifchitz E (2000) *Theory of Elasticity Butterworth-Heinemann*.

Larsen TB, Malevsky AV, Yuen DA, and Smedsmo JL (1993) Temperature-dependent Newtonian and non-Newtonian convection – Implications for lithospheric processes. *Geophysical Research Letters* 20: 2595–2598.

Larsen TB and Yuen DA (1997) Fast plumeheads: Temperature-dependent versus non-Newtonian rheology. *Geophysical Research Letters* 24: 1995–1998.

LeBars M and Davaille A (2002) Stability of thermal convection in two superimposed miscible viscous fluids. *Journal of Fluid Dynamics* 471: 339–363.

Lee T, Papanastassiou DA, and Wasserburg GL (1976) Demonstration of $^{26}$mg excess in allende and evidence for al$^{26}$. *Geophysical Research Letters* 3: 41–44.

Lemery C, Ricard Y, and Sommeria J (2000) A model for the emergence of thermal plumes in Rayleigh–Benard convection at infinite Prandtl number. *Journal of Fluid Mechanics* 414: 225–250.

Lenardic A (1998) On the partitioning of mantle heat loss below oceans and continents over time and its relationship to the Archaean paradox. *Geophysical Journal International* 134: 706–720.

Lestunff Y and Ricard Y (1995) Topography and geoid due to lithospheric mass anomalies. *Geophysical Journal International* 122: 982–990.

Lister JR and Kerr RC (1991) Fluid-mechanical models of crack-propagation and their application to magma transport in dykes. *Journal of Geophysical Research – Solid Earth Planets* 96: 10049–10077.

Lithgow-Bertelloni C and Gurnis M (1997) Cenozoic subsidence and uplift of continents from time-varying dynamic topography. *Geology* 25: 735–738.

Lithgow-Bertelloni C and Guynn JH (2004) Origin of the lithospheric stress field. *Journal of Geophysical Research* 109, doi: 10.1029/2003JB002,467.

Lithgow-Bertelloni C and Richards MA (1998) Dynamics of cenozoic and mesozoic plate motions. *Reviews of Geophysics* 36: 27–78.

Lithgow-Bertelloni C and Silver PG (1998) Dynamic topography, plate driving forces and the African superswell. *Nature* 395: 269–272.

Lyubetskaya T and Korenaga J (2007) Chemical composition of Earth's primitive mantle and its variance 1 Method and results. *Journal of Geophysical Research* 112, doi 10:1029/2005JB004223.

Machetel P and Weber P (1991) Intermittent layered convection in a model mantle with an endothermic phase change at 670 km. *Nature* 350: 55–57.

Malevsky AV and Yuen DA (1992) Strongly chaotic non-Newtonian mantle convection. *Geophysical Astrophysical Fluid Dynamics* 65: 149–171.

Malvern L (1969) *Prentice-Hall Series in Engineering of the Physical Sciences: Introduction to the Mechanics of a Continuum Medium*. Saddle River, NJ: Prentice-Hall.

Manga M and Stone MA (1993) Buoyancy-driven interactions between 2 deformable viscous drops. *Journal of Fluid Mechanics* 256: 647–683.

Massol H, Jaupart C, and Pepper DW (2001) Ascent and decompression of viscous vesicular magma in a volcanic conduit. *Journal of Geophysical Research* 106: 16223–16240.

McDonough WF and Sun SS (1995) The composition of the Earth. *Chemical Geology* 120: 223–253.

McKenzie D (1984) The generation and compaction of partially molten rock. *Journal of Petrology* 25: 713–765.

McNamara AK, Karato SI, and van Keken PE (2001) Localisation of dislocation creep in the lower mantle: Implications for the origin of seismic anisotropy. *Earth and Planetary Science Letters* 191: 85–99.

Mitrovica JX and Forte AM (1997) Radial profile of mantle viscosity: Results from the joint inversion of convection and postglacial rebound observables. *Journal of Geophysical Research* 102: 2751–2769.

Monnereau M and Quéré S (2001) Spherical shell models of mantle convection with tectonic plates. *Earth and Planetary Science Letters* 184: 575–588.

Moresi L and Solomatov V (1998) Mantle convection with a brittle lithosphere: thoughts on the global tectonic styles of the Earth and Venus. *Geophysical Journal International* 133: 669–682.

Muhlhaus HB, Moresi L, and Cada M (2004) Emergent anisotropy and flow alignment in viscous rocks. *Pure and Applied Geophysics* 161: 2451–2463.

Munk WH and MacDonald GJF (1960) *The Rotation of the Earth: A Geophysical Discussion*. Cambridge, UK: Cambridge University Press.

Murakami T, Hirose K, Kawamura K, Sata N, and Ohishi Y (2004) Post-perovskite phase transition in MgSiO$_3$. *Science* 304: 855–858.

Murnaghan FD (1951) *Finite Deformation of an Elastic Solid*. New York: John Wiley and Sons.

Nakagawa T and Tackley PJ (2006) Three-dimensional structures and dynamics in the deep mantle: Effects of post-perovskite phase change and deep mantle layering. *Geophysical Research Letters* 33: L12S 11 (doi:1029/2006GL025719).

Nataf HC and Richter FM (1982) Convection experiments in fluids with highly temperature-dependent viscosity and the thermal evolution of the planets. *Physics of the Earth and Planetary Interiors* 29: 320–329.

Nield DA (1964) Surface tension and buoyancy effects in cellular convection. *Journal of Fluid Dynamics* 19: 341–352.

Nye JF (1953) The flow law of ice from measurements in glacier tunnels, laboratory experiments and the jungfraufirn borehole experiment. *Proceedings of the Royal Society of London Series A – Mathematical and Physical Sciences* 219: 477–489.

Ogawa M, Schubert G, and Zebib A (1991) Numerical simulations of 3-dimensional thermal-convection in a fluid with strongly temperature-dependent viscosity. *Journal of Fluid Dynamics* 233: 299–328.

Olson P, Yuen DA, and Balsiger D (1984) Mixing of passive heterogeneties by mantle convection. *Journal of Geophysical Research* 89: 425–436.

Ottino JM (1989) *The Kinematics of Mixing: Stretching, Chaos and Transport*. New-York: Cambridge University Press.

Panasyuk S, Hager B, and Forte A (1996) Understanding the effects of mantle compressibility on geoid kernels. *Geophysical Journal International* 124: 121–133.

Parmentier EM, Sotin C, and Travis BJ (1994) Turbulent 3-d thermal-convection in an infinite prandtl number, volumetrically heated fluid – Implications for mantle dynamics. *Geophysical Journal International* 116: 241–251.

Parmentier EM, Turcotte DL, and Torrance KE (1976) Studies of finite-amplitude non-newtonian thermal convection with application to convection in Earth's mantle. *Journal of Geophysical Research* 81: 1839–1846.

Parsons B and Daly S (1983) The relationship between surface topography, gravity anomaly and the temperature structure of convection. *Journal of Geophysical Research* 88: 1129–1144.

Peltier WR (1989) Mantle viscosity. In: Peltier WR (ed.) *Mantle Convection: Plate Tectonics and Global Geodynamics*, pp. 389–478. New York: Gordon and Breach Science Publishers.

Peltier WR (1996) Mantle viscosity and ice-age ice sheet topography. *Science* 273: 1359–1364.

Peltier WR and Solheim LP (1992) Mantle phase transition and layered chaotic convection. *Geophysical Research Letters* 19: 321–324.

Poirier JP (1991) *Introduction to the Physics of the Earth's Interior*. Cambridge, UK: Cambridge University Press.

Poirier JP and Tarantola A (1998) A logarithmic equation of state. *Physics of the Earth and Planetary Interiors* 109: 1–8.

Rabinowicz M, Ricard Y, and Gregoire M (2002) Compaction in a mantle with a very small melt concentration: Implications for the generation of carbonatitic and carbonate-bearing high alkaline mafic melt impregnations. *Earth and Planetary Science Letters* 203: 205–220.

Ranalli G (1995) *Rheology of the Earth*. Dodrecht: Kluwer academic publishers.

Ribe NM (1985a) The deformation and compaction of partial molten zones. *Geophysical Journal of the Royal Astronomical Society* 83: 487–501.

Ribe NM (1985b) The generation and composition of partial melts in the Earth's mantle. *Earth and Planetary Science Letters* 73: 361–376.

Ricard Y and Bercovici D (2003) Two-phase damage theory and crustal rock failure: The theoretical 'void' limit, and the prediction of experimental data. *Geophysical Journal International* 155: 1057–1064.

Ricard Y, Bercovici D, and Schubert G (2001) A two-phase model for compaction and damage. Part 2: Applications to compaction, deformation, and the role of interfacial surface tension. *Journal of Geophysical Research* 106: 8907–8924.

Ricard Y and Coltice N (2004) Geophysical and geochemical models of mantle convection: Successes and future challenges. In: Sparks RSJ and Hawkesworth CJ (eds.) *Geophysical Monograph Series, vol. 150: The state of the planet: Frontiers and challenges in geophysics*, pp. 59–68. Washington, DC: American Geophysical Union.

Ricard Y, Doglioni C, and Sabadini R (1991) Differential rotation between lithosphere and mantle – A consequence of lateral mantle viscosity variations. *Journal of Geophysical Research* 96: 8407–8415.

Ricard Y, Fleitout L, and Froidevaux C (1984) Geoid heights and lithospheric stresses for a dynamic Earth. *Annales Geophysicae* 2: 267–286.

Ricard Y, Richards M, Lithgow-Bertelloni C, and Stunff YL (1993a) A geodynamic model of mantle density heterogeneity. *Journal of Geophysical Research* 98: 21895–21909.

Ricard Y, Spada G, and Sabadini R (1993b) Polar wander of a dynamic Earth. *Geophysical Journal International* 113: 284–298.

Ricard Y and Vigny C (1989) Mantle dynamics with induced plate-tectonics. *Journal of Geophysical Research* 94: 17543–17559.

Ricard Y, Vigny C, and Froidevaux C (1989) Mantle heterogeneities, geoid and plate motion: a monte carlo inversion. *Journal of Geophysical Research* 94: 13739–13754.

Ricard Y, Mattern E, and Matas J (2005) Synthetic tomographic images of slabs from mineral physics. In: van der Hilst RD, Bass J, Matas J, and Trampert J (eds.) *AGU Monograph: Earth's Deep Mantle: Structure, Composition, and Evolution.* pp. 285–302. Washington, DC: American Geophysical Union.

Richard G, Bercovici D, and Karato SI (2006) Slab dehydration in the Earth's mantle transition zone. *Earth and Planetary Science Letters* 251: 156–167.

Richards M (1991) Hotspots and the case for a high viscosity lower mantle. In: Sabadini R and Boschi E (eds.) *Glacial Isostasy, Sea-Level and Mantle Rheology*, pp. 571–587. Dordrecht: Kluwer Academic Publishers.

Richards MA and Engebretson DC (1992) Large-scale mantle convection and the history of subduction. *Nature* 355: 437–440.

Richards MA and Hager BH (1984) Geoid anomalies in a dynamic Earth. *Journal of Geophysical Research* 89: 5987–6002.

Richards MA, Ricard Y, Lithgow-Bertelloni C, Spada G, and Sabadini R (1997) An explanation for Earth's long-term rotational stability. *Science* 275: 372–375.

Richards MA, Yang WS, Baumgardner JR, and Bunge HP (2001) Role of a low-viscosity zone in stabilizing plate tectonics: Implications for comparative terrestrial planetology. *Geochemistry Geophysics Geosystems* 2, doi:10.1029/2000GC000115.

Richter FM (1978) Experiments on the stability of convection rolls in fluids whose viscosity depends on temperature. *Journal of Fluid Dynamics* 89: 553–560.

Richter FM and Mckenzie D (1984) Dynamical models for melt segregation from a deformable matrix. *Journal of Geology* 92: 729–740.

Ricolleau A, Fiquet G, Perillat J, et al. (2004) The fate of subducted basaltic crust in the Earth's lower mantle: An experimental petrological study. in *Eos Transaction AGU, Fall* Fall Meeting, vol. 85(47), abstract U33B-02.

Riley GN, Kohlstedt DL, and Richter FM (1990) Melt migration in a silicate liquid-olivine system – An experimental test of compaction theory. *Geophysical Research Letters* 17: 2101–2104.

Ringwood AE and Irifune T (1988) Nature of the 650-km seismic discontinuity – Implications for mantle dynamics and differentiation. *Nature* 331: 131–136.

Roberts PH (1967) Convection in horizontal layers with internal heat generation Theory. *Journal of Fluid Mechanics* 30: 33–49.

Rudnick RL and Fountain DM (1995) Nature and composition of the continental-crust – A lower crustal perspective. *Reviews of Geophysics* 33: 267–309.

Runcorn SK (1964) Satellite gravity measurements and the laminar viscous flow model of the Earth's mantle. *Journal of Geophysical Research* 69: 4389.

Sabadini R and Yuen DA (1989) Mantle stratification and long-term polar wander. *Nature* 339: 373–375.

Saltzer RL, Stutzmann E, and van der Hilst RD (2004) Poisson's ratio in the lower mantle beneath Alaska: Evidence for compositional heterogeneity. *Journal of Geophysical Research* 109: B06301.

Samuel H and Farnetani CG (2003) Thermochemical convection and helium concentrations in mantle plumes. *Earth and Planetary Science Letters* 207: 39–56.

Schmalzl J, Houseman GA, and Hansen U (1996) Mixing in vigorous, time-dependent three-dimensional convection and application to Earth's mantle. *Journal of Geophysical Research* 101: 21847–21858.

Schmeling H (2000) Partial Melting and melt migration in a convecting mantle. In: Bagdassarov N, Laporte D, and Thompson AB (eds.) *Physics and Chemistry of Partially Molten Rocks*, pp. 141–178. Dordrecht: Kluwer Academic Publisher.

Schmeling H and Jacoby WR (1981) On modeling the lithosphere in mantle convection with non-linear rheology. *Journal of Geophysics - Zeitschrift Fur Geophysik* 50: 89–100.

Schmeling H, Marquart G, and Ruedas T (2003) Pressure- and temperature-dependent thermal expansivity and the effect on mantle convection and surface observables. *Geophysical Journal International* 154: 224–229.

Schubert G, Stevenson D, and Cassen P (1980) Whole planet cooling and the radiogenic heat-source contents of the Earth and Moon. *Journal of Geophysical Research* 85: 2531–2538.

Schubert G, Turcotte D, and Olson P (2001) *Mantle Convection in the Earth and Planets*. Cambridge UK: Cambridge University Press.

Schubert G, Yuen DA, and Turcotte DL (1975) Role of phase-transitions in a dynamic mantle. *Geophysical Journal of the Royal Astronomical Society* 42: 705–735.

Scott DR and Stevenson DJ (1984) Magma solitons. *Geophysical Research Letters* 11: 9283–9296.

Scott DR and Stevenson DJ (1986) Magma ascent by porous flow. *Journal of Geophysical Research* 91: 9283–9296.

Scott DR and Stevenson DJ (1989) A self-consistent model of melting, magma migration and buoyancy-driven circulation beneath mid-ocean ridges. *Journal of Geophysical Research* 94: 2973–2988.

Solomatov VS (1995) Scaling of temperature-dependent and stress-dependent viscosity convection. *Physics of Fluids* 7: 266–274.

Spada G, Ricard Y, and Sabadini R (1992a) Excitation of true polar wander by subduction. *Nature* 360: 452–454.

Spada G, Sabadini R, Yuen DA, and Ricard Y (1992b) Effects on postglacial rebound from the hard rheology in the transition zone. *Geophysical Journal International* 109: 683–700.

Spear FS (1993) *Metamorphic Phase Equilibria and Pressure–Temperature–Time Paths*, 799p. Washington, DC: Mineralogical Society of America.

Spiegelman M and Kelemen PB (2003) Extreme chemical variability as a consequence of channelized melt transport. *Geochemistry Geophysics Geosystems* 4: 1055.

Spiegelman M, Kelemen PB, and Aharonov E (2001) Causes and consequences of flow organization during melt transport: The reaction infiltration instability in compactible media. *Journal of Geophysical Research* 106: 2061–2077.

Sramek O, Ricard Y, and Bercovici D (2007) Simultaneous melting and compaction in deformable two phase media. *Geophysical Journal International* 168: 964–982.

Stacey FD (1977) Application of thermodynamics to fundamental Earth physics. *Geophysical Surveys* 3: 175–204.

Stacey FD and Davis PM (2004) High pressure equations of state with applications to the lower mantle and core. *Physics of the Earth and Planetary Interiors* 142: 137–184.

Stein C, Schmalzl J, and Hansen U (2004) The effect of rheological parameters on plate behaviour in a self-consistent model of mantle convection. *Physics of the Earth and Planetary Interiors* 142: 225–255.

Steinberger B and O'Connell RJ (1998) Advection of plumes in mantle flow: Implications for hotspot motion, mantle viscosity and plume distribution. *Geophysical Journal International* 132: 412–434.

Stevenson DJ (1989) Spontaneous small-scale melt segregation in partial melts undergoing deformation. *Geophysical Research Letters* 16: 1067–1070.

Tackley P (1995) On the penetration of an endothermic phase transition by upwellings and downwellings. *Journal of Geophysical Research* 100: 15477–15488.

Tackley PJ (1996) Effects of strongly variable viscosity on three-dimensional compressible convection in planetary mantles. *Journal of Geophysical Research – Solid Earth* 101: 3311–3332.

Tackley PJ (1998) Self-consistent generation of tectonic plates in three-dimensional mantle convection. *Earth and Planetary Science Letters* 157: 9–22.

Tackley P (2000a) The quest for self-consistent incorporation of plate tectonics in mantle convection. In: Richards MA, Gordon R, and van der Hilst R (eds.) AGU *Geophysical Monograph Series, vol.121: History and Dynamics of Global Plate Motions*, pp. 47–72. Washington, DC: AGU.

Tackley PJ (2000b) Mantle convection and plate tectonics: Toward an integrated physical and chemical theory. *Science* 288: 2002–2007.

Tackley PJ (2000c) Mantle convection and plate tectonics: Toward an integrated physical and chemical theory. *Science* 288: 2002–2007.

Tackley PJ (2000d) Self consistent generation of tectonic plates in time-dependent, three dimensional mantle convection simulations. Part 1: Pseudoplastic yielding. *Geochemistry Geophysics Geosystems* 1(8), doi: 10.1029/2000GC000036.

Tackley PJ (2000e) Self consistent generation of tectonic plates in time-dependent, three dimensional mantle convection simulations. Part 2: Strain weakening and asthenosphere. *Geochemistry Geophysics Geosystems* 1(8), doi: 10.1029/2000GC000043.

Tackley PJ, Stevenson DJ, Glatzmaier GA, and Schubert G (1993) Effects of an endothermic phase-transition at 670 km depth in a spherical model of convection in the Earth's mantle. *Nature* 361: 699–704.

Tackley PJ and Xie SX (2002) The thermochemical structure and evolution of Earth's mantle: constraints and numerical models. *Philosophical Transactions of the Royal Society of London Series A-Mathematical Physical and Engineering Sciences* 360: 2593–2609.

Tao WC and O'Connell RJ (1993) Deformation of a weak subducted slab and variation of seismicity with depth. *Nature* 361: 626–628.

Toth J and Gurnis M (1998) Dynamics of subduction initiation at preexisting fault zones. *Journal of Geophysical Research* 103: 18053–18067.

Toussaint V, Carriere P, Scott J, and Gence JN (2000) Spectral decay of a passive scalar in chaotic mixing. *Physics of Fluids* 12: 2834–2844.

Tozer DC (1972) The present thermal state of the terrestrial planets. *Physics of the Earth and Planetary Interiors* 6: 182–197.

Tritton DJ (1988) *Physical Fluid Dynamics*. Oxford, UK: Oxford University Press.

Trompert R and Hansen U (1998a) Mantle convection simulations with rheologies that generate plate-like behaviour. *Nature* 395: 686–689.

Trompert RA and Hansen U (1998b) On the rayleigh number dependence of convection with a strongly temperature-dependent viscosity. *Physics of Fluids* 10: 351–360.

Turcotte DL and Morgan JP (1992) The physics of magma migration and mantle flow beneath a mid-ocea ridge. In: Morgan JP, Blackmann DK, and Simpson JM (eds.)

*Mantle flow and Melt Generation at Mid-Ocean Ridges*, pp. 155–182. New York: AGU.

Turcotte DL and Schubert G (1982) *Geodynamics: Applications of Continuum Physics to Geological Problems*, pp. 1–450 New York: Wiley.

Urey HC (1951) The origin and development of the Earth and other terrestrial planets. *Geochimica et Cosmochimica Acta* 1: 209–277.

van der Hilst RD, Widiyantoro S, and Engdahl ER (1997) Evidence for deep mantle circulation from global tomography. *Nature* 386: 578–584.

van Keken PE, Hauri EH, and Ballentine CJ (2002) Mantle mixing: The generation, preservation, and destruction of chemical heterogeneity. *Annual Review of Earth and Planetary Sciences* 30: 493–525.

van Keken P, Yuen DA, and Vandenberg A (1992) Pulsating diapiric flows – Consequences of vertical variations in mantle creep laws. *Earth and Planetary Science Letters* 112: 179–194.

Vinet P, Ferrante J, Rose JH, and Smith JR (1987) Compressibility of solids. *Journal of Geophysical Research* 92: 9319–9325.

Weertman J and Weertman JR (1975) High-temperature creep of rock and mantle viscosity. *Annual Review of Earth and Planetary Sciences* 3: 293–315.

Weinstein SA (1996) Thermal convection in a cylindrical annulus with a non-Newtonian outer surface. *Pure and Applied Geophysics* 146: 551–572.

Weinstein SA and Christensen U (1991) Convection planforms in a fluid with a temperature-dependent viscosity beneath a stress-free upper boundary. *Geophysical Research Letters* 18: 2035–2038.

Weinstein SA and Olson PL (1992) Thermal-convection with non-Newtonian plates. *Geophysical Journal International* 111: 515–530.

Williams Q and Garnero EJ (1996) Seismic evidence for partial melt at the base of Earth's mantle. *Science* 273: 1528–1530.

Woods LC (1975) *The Thermodynamics of Fluid Systems*. Oxford: Clarendon Press.

Yuen D, Sabadini RCA, and Gasperini P (1986) On transient rheology and glacial isostasy. *Journal of Geophysical Research* 91: 1420–1438.

Zhang SX and Christensen U (1993) Some effects of lateral viscosity variations on geoid and surface velocities induced by density anomalies in the mantle. *Geophysical Journal International* 114: 531–547.

Zhang SX and Yuen DA (1996) Various influences on plumes and dynamics in time-dependent, compressible mantle convection in 3-D spherical shell. *Physics of the Earth and Planetary Interiors* 94: 241–267.

Zhong SJ, Gurnis M, and Moresi L (1996) Free surface formulation of mantle convection. Part 1: Basic theory and implication to plumes. *Geophysical Journal International* 127: 708–718.

Zhong SJ, Zuber MT, Moresi L, and Gurnis M (2000) Role of temperature-dependent viscosity and surface plates in spherical shell models of mantle convection. *Journal of Geophysical Research* 105: 11063–11082.

# 3 Laboratory Studies of Mantle Convection

**A. Davaille and A. Limare**, Institut de Physique du Globe, Paris, France

## 3.1 Introduction

Because laboratory experiments are crucial for exploring new physics and testing theories, they have long played a central role in investigations of thermal convection and mantle dynamics. This chapter is devoted to laboratory experiments considered as tools for understanding the physics that governs mantle dynamics. We shall review both the techniques employed to run well-controlled experiments and acquire quantitative data, and the results obtained.

Investigations of mantle dynamics and of thermal convection have long been closely intertwined. Indeed, the mantle is cooled from above, heated from within by radioactive elements, and heated from below by the core, which loses its heat through the mantle (*see* Chapter 6). The emergence of mantle convection models was dictated by the failure of static, conductive, and/or radiative thermal history models to account for the mantle temperature regime, the Earth energy budget, and the Earth's lateral surface motions. In the late 1930s, Arthur Holmes, among others, hypothesized that thermal convection in the Earth's mantle provided the necessary force to drive continental motions (Holmes, 1931). Convection, which transports heat by material advection, is the only physical mechanism capable of explaining these observations. The force driving advection is gravity, whereby material lighter than its environment rises while denser material sinks. Such density anomalies can be produced by differences in composition and/or temperature. In the simple configuration of a plane layer (**Figure** 1), the former will give rise to Rayleigh–Taylor instabilities, while the latter will generate 'Rayleigh–Bénard instabilities.

Rayleigh–Bénard convection develops when a plane layer of fluid is heated from below and cooled from above (**Figure** 1). It has been identified as a major feature of the dynamics of the oceans, the atmosphere, and the interior of stars and planets. First identified by Count Rumford (1798), the phenomenon was observed several times in the nineteenth century (e.g., Thomson, 1882). However, it was the carefully controlled and quantitative laboratory experiments of Henri Bénard (1900, 1901) that focused the interest of the scientific community on the problem. Bénard studied the patterns of convection developing in thin layers with a free upper surface. He was interested in the influence of viscosity on the pattern and use several fluids, including spermaceti and paraffin. He determined quantitatively the characteristic length scales of the patterns, the deformation of the interface, and the direction of flow within the fluid. Although we now know that the beautiful hexagonal patterns he observed (**Figure** 1(c)) were due to the temperature dependence of surface tension (Pearson, 1958), it was these experiments which motivated Lord Rayleigh to apply hydrodynamic stability theory to thermal convection in the absence of surface tension (Rayleigh, 1916). When the thermal convection experiments were carried out correctly, they were found to be very well predicted by Rayleigh's theory (e.g., Schmidt and Milverton, 1935; Silveston, 1958). For more on the history of these investigations, *see* Chapter 1.

Thermal convection in an isoviscous fluid is characterized by two parameters. The Rayleigh number $Ra(H, \Delta T)$ compares the driving thermal buoyancy forces to the resisting effects of thermal diffusion and viscous dissipation across the whole system:

$$Ra(H, \Delta T) = \frac{\alpha g \Delta T H^3}{\kappa \nu} \qquad [1]$$

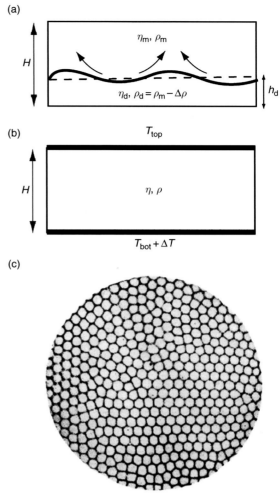

**Figure 1** (a) Rayleigh–Taylor instabilities develop when a layer of lighter fluid is introduced under a layer of denser fluid. (b) Rayleigh–Bénard instabilities develop when a layer of fluid of density $\rho$ and viscosity $\eta$ is heated from below and cooled from above. (c) Bénard cells in spermaceti. From Bénard H (1901) Les tourbillons cellulaires dans une nappe liquide transportant la chaleur par convection en régime permanent. *Annales de Chimie et de Physique* 23: 62–144.

where $H$ is the depth of the layer, $\Delta T$ the temperature difference applied across it, $g$ the gravitational acceleration, $\alpha$ the thermal expansivity, $\kappa$ the thermal diffusivity, and $\nu = \eta/\rho$ the kinematic viscosity. Convection starts when $Ra$ exceeds a critical value (Rayleigh, 1916; *see* Chapters 2 and 4), and exhibits a sequence of transitions toward chaos as $Ra$ increases. The second parameter is the Prandtl number, $Pr$, the ratio of the diffusivity of momentum and that of heat:

$$Pr = \nu/\kappa \qquad [2]$$

When $Pr \gg 1$, the fluid motion stops as soon as the heat source disappears, that is, inertial effects are negligible compared to viscous effects. This is the case for the Earth's mantle, where $Pr > 10^{23}$.

Numerous theoretical, laboratory, and numerical studies in the last 50 years have been devoted to characterizing thermal convection as a function of $Ra$ and $Pr$. Laboratory experiments have proved especially useful for determining patterns and characterizing the high $Ra$ regime. For high $Pr$ fluids, isoviscous convection has been studied for $Ra$ up to $10^9$, which includes the range of values estimated for the mantle ($10^6$–$10^8$).

However, mantle dynamics is much more complicated than isoviscous convection, in particular due to the complex rheology of mantle material (Chapter 2). For example, the generation of Plate Tectonics and its coexistence with hot spots is not reproduced in isoviscous fluids. Studies have therefore focused on progressively more complicated systems, either taking a global view (e.g., convection with temperature-dependent rheology, or internal heating) or a more local view to study some particular mantle feature (such as plumes or subduction). Analog laboratory experiments have been extensively used because of four advantages: (1) since they let Nature solve the equations, they can explore new phenomena for which such equations do not yet exist. (2) They can also explore ranges of parameters, or geometries, where the equations are too non-linear to be solved analytically or numerically. (3) They are inherently three-dimensional (3-D). (4) They can usually be run in the appropriate range of mantle parameters (which is not the case for the atmosphere or the oceans).

This chapter is organized as follows: Sections 3.2 and 3.3 are devoted to experimental setups, fluids, measurements, and visualization methods. Mantle dynamics on geological time scales is dominated by 'fluid' behavior, so that we can generally use liquids around room temperature, and fluid mechanics techniques. With the development of computer power and lasers, it has become possible in the last years to measure the temperature, velocity, concentration, and deformation fields in experimental tanks. Sections 3.4–3.7 focus on gravitational instabilities (Rayleigh–Taylor instabilities and Rayleigh–Bénard convection). Sections 3.8–3.11 describes the laboratory experiments related to more specific mantle features – plumes, mixing, accretion, and subduction – which are also described more fully in other chapters of this volume.

## 3.2  Experimental Setups and Fluids

### 3.2.1  Designing an Experiment: Scaling

The goal of any fluid mechanics modeling is to determine 'scaling laws', or functional relations that link certain parameters of interest and the various other parameters on which they depend. These scaling laws then make it possible to predict the behavior of similar systems, such as the mantle in geodynamics. Hence, there is no question of building a miniature Earth in the laboratory. A phenomenon has to be selected, and a simplified laboratory model is then constructed, where only a few parameters can vary and in a controlled manner. Each individual experiment aims at describing, through quantitative measurements, the behavior of the system for a given set of control parameters. By varying systematically the values of these parameters, a database is constituted. Scaling laws derived from fundamental physical principles can then be tested against the experimental data.

However, the results of a laboratory experiment will be applicable to other natural systems, such as the mantle, only if the dynamic similarity between the scaled-down system (laboratory, computer) and the natural system is respected. Dynamic similarity can be viewed as a generalization of the concept of geometrical similarity (*see* Chapter 4), and requires the following:

1. Similar boundary conditions (mechanical, thermal, geometry).
2. Similar rheological laws. In other words, the mechanical equation of state that relates differential flow stress to strain rate should differ only in the proportionality constant (e.g., Weijermars and Schmeling, 1986).
3. Similar balances between the different forces or operative physical effects. $Ra$ and $Pr$ reflect such balances.

Dynamic similarity gives birth to a set of dimensionless parameters. When the governing equations are known, they can be nondimensionalized to make the relevant dimensionless parameters appear explicitly in the equations and/or the initial and boundary conditions. When the equations are not known, one may use the Buckingham-$\Pi$ theorem, which is a consequence of the fundamental principle that the validity of a physical law cannot depend on the units in which it is expressed. According to this theorem, if a given experiment is described by $N$-dimensional parameters of which $M$ have independent physical dimensions, then the experiment can be completely described by $(N - M)$-independent nondimensionless combinations of the dimensional parameters (*see* Chapter 4).

Besides $Ra$, $Pr$, and ratios that characterize the variations of physical properties such as viscosity within the experimental tank, other important dimensionless numbers for geodynamics include:

1. the Reynolds number $Re$, which compares the advection of momentum by the fluid motion and the viscous diffusion of momentum:

$$Re = \frac{UH}{\nu} \qquad [3]$$

where $U$ is the characteristic velocity in the system. For the mantle, $Re \sim 10^{-19}$.

2. the Peclet number, $Pe$, when the flow is forced by a boundary velocity $U$ (such as a plate velocity). $Pe$ compares the heat transport by advection and the heat transport by conduction:

$$Pe = \frac{Ud}{\kappa} \qquad [4]$$

where $d$ is for example the thickness of the plate.

All systems with the same dimensionless parameter (say $Ra$) will behave in the same way, irrespective of their size. However, the timescale and/or the distance over which the phenomenon occurs will depend on the system size. This is how convection in the laboratory on a scale of hours can be analogous to convection in the mantle over geological times. Dynamical similarity and scaling analysis are therefore essential to analog laboratory modeling. Their principles and techniques are discussed in Chapter 4.

### 3.2.2 Experimental Fluids

Except when focusing on lithospheric processes such as accretion or subduction, laboratory experiments usually assume that mantle material flows like a Newtonian fluid, with a linear relation between stress and strain rate. This is especially true for convection experiments, because non-Newtonian fluids are usually more difficult to characterize, and to handle.

In the mantle $Pr \sim 10^{24}$. This is not possible to obtain in the laboratory, but because the dynamics becomes independent of $Pr$ for $Pr > 100$ (see Section 3.5), the use of fluids with $Pr > 100$ is adequate to ensure the dominance of viscous over inertial effects.

Another consideration is the temperature dependence of viscosity. Solid-state creep is thermally activated and therefore mantle material has a highly temperature-dependent viscosity. The viscosity ratio across the lithosphere where the temperature difference is the highest reaches $10^7$. A number of experiments have investigated in detail the influence of a strongly temperature-dependent viscosity on convection. The viscosity of liquids always depends on temperature. The silicone oils have the smallest, with a change by no more than a factor of 3 over 50°C. Sugar syrups have a much stronger temperature dependence, with Golden Syrup (from Tate and Lyle) the winner, showing a $10^6$ viscosity change between 60°C and −20°C (**Figure 2**).

Silicone oils are available in different grades (with viscosity at 20°C between 10 mP s and $10^3$ Pa s), and should be used when negligible temperature-related viscosity variations are required in an experiment. Thickeners in aqueous solutions have also been

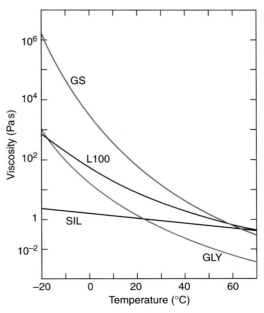

**Figure 2** Viscosity as a function of temperature. SIL, Silicone oil 47V1000; GLY, Glycerol; GS, Golden Syrup; L100, polybutene.

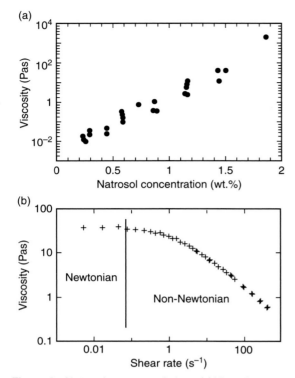

**Figure 3** Natrosol aqueous solutions. (a) Viscosity as a function of shear rate (measured with a Haake rotative viscometer) for 1.4 wt.% Natrosol. There is a Newtonian plateau for low shear rates. (b) Viscosity as a function of the Natrosol concentration. From Davaille A (1999a) Two-layer thermal convection in viscous fluids *Journal of Fluid Mechanics* 379: 223–253.

used (e.g., Tait and Jaupart, 1989; Davaille, 1999a, 1999b; Namiki, 2003). By adding less than 2 wt.% of thickener (ex: Hydroxy-ethylcellulose, trade name Natrosol), the viscosity of water can be multiplied by $10^7$ (**Figure 3(a)**). Although the resulting fluid is shear-thinning (**Figure 3(b)**), there is a Newtonian plateau at the low shear rates that typically obtain in convection laboratory experiments.

Well-controlled experiments require a precise knowledge of the physical properties of the fluids used. Thermal conductivity and specific heat do not change much from one batch to another, and one can generally rely on the values given by the manufacturer. But it is advisable to measure density (and thermal expansion if needed) and viscosity prior to any new experiments. Moreover, because of the temperature dependence of viscosity, experiments should prefererably be run in a temperature-controlled laboratory. This will also ensure that electronic measurements do not drift with time.

### 3.2.3 Experimental Setup for Convection

A thermal convection experiment should minimize heat losses to allow reliable heat flux determination, and should also allow good flow visualization (pattern, thermal structure, etc.). Unfortunately, these two requirements cannot be perfectly met at the same time. We shall therefore present three commonly used experimental setups.

#### 3.2.3.1 Horizontal pattern visualization

The convecting fluid is bounded between two glass plates; the bottom is heated from below by hot water flushing along its outer surface, while the top is cooled by cold water flushed along the top, as shown in **Figure 4(a)** (e.g., Busse and Whitehead, 1971; Richter and Parsons, 1975; Whitehead and Parsons, 1978; White, 1988; Weinstein and Olson, 1990; Weinstein and Christensen, 1991). An alternative is to use as the bottom heater a metal plate whose upper surface is polished to a mirror finish (e.g., Chen and Whitehead, 1968; Heutmaker and Gollub, 1987). The horizontal heat exchangers should be carefully leveled since imperfect horizontal alignment modifies convection patterns and heat transfer (e.g., Namiki and Kurita, 2002; Chilla *et al.*, 2004). Precautions also have to be taken to eliminate lateral inhomogeneities due to the side walls. In Busse and Whitehead's setup (**Figure 4(a)**), the area where observations on convection were made was bounded on the sides by walls of 2″-thick polyvinylchloride, whose thermal conductivity is close to that of the working fluid (silicone oil). This provided a working area 80 cm × 80 cm. Outside these walls was another convecting region approximately 20 cm in width so that the temperature gradients on both sides of the walls were similar. Visualization was done using the shadowgraph technique (see Section 3.3), whereby collimated light (i.e., with parallel rays) is directed vertically up through the tank (**Figure 4(a)**). In the convecting fluid, the index of refraction of light depends on temperature such that light rays diverge away from hot regions and converge toward colder regions. The projection of the resulting rays onto a white screen thus show hot zones as dark shadows and cold zones as concentration of brightness.

#### 3.2.3.2 Heat flux determination

The dependence of the global heat transport on convection characteristics is one of the fundamental

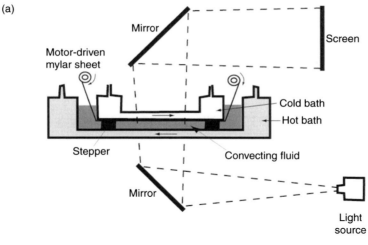

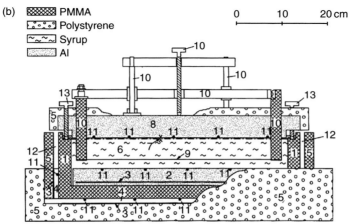

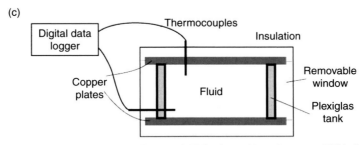

**Figure 4** (a) Experimental setup from Richter and Parsons (1975) adapted from Busse and Whitehead (1971). Thermostated water is flowed in the top and bottom glass assemblages. A mylar sheet can be introduced just underneath the top cold plate and driven by a motor at constant speed. (b) Convection apparatus from Giannandrea and Christensen (1993). 1, convection tank; 2, heating plate; 3, heating foil; 4, guard heaters; 5, thermal insulation; 6, working fluid; 7, oil-filled gap; 8, cooling block; 9, nickel wire; 10, device for vertical displacement of the Ni-wire; 11, thermocouples; 12, air gap; 13, adjusting screw. (c) 'hybrid' experimental setup. The insulation windows can be removed for visualization.

questions of Rayleigh–Bénard studies (see Section 3.4). The most accurate measurement of the global heat flow is obtained through the measurement of the electric power needed to heat the bottom plate (e.g., Schmidt and Milverton, 1935; Malkus, 1954; Silveston, 1958; Giannandrea and Christensen, 1993; Brown et al., 2005; Funfschilling et al., 2005). It requires special design of the experimental tank in order to minimize the heat losses from the heating plate to the bottom (e.g., **Figure 4(b)**). A second experimental problem is the influence of the side walls on the heat transport in the fluid. This problem is significantly reduced when the experimental tank aspect ratio is large (e.g., Giannandrea and Christensen, 1993), and by choosing a wall of relatively small thermal conductivity (e.g., plexiglas) compared to that of the fluid, plus another insulation in low conductivity material such as polystryrene foam. Even with that, sides losses must be estimated, and models have been derived to correct for them (e.g., Ahlers, 2001). A third problem is the effect of the finite conductivity of the top and bottom plates on the heat transport by the fluid (e.g., Chilla *et al.*, 2004; Verzicco, 2004): the less conductive the plates, the more the heat transport in the fluid is diminished, and the effect is a function of *Ra*. A correction has been proposed, which fits well the data for turbulent convection (Verzicco, 2004; Brown *et al.*, 2005). In practice, the best conductor available is copper ($k = 319 \, \mathrm{W \, m^{-1} \, K^{-1}}$), followed by aluminum ($k = 161 \, \mathrm{W \, m^{-1} \, K^{-1}}$). So good heat flux determination precludes the visualization of the planform because of the metal plates, and also visualization from the side because of the insulation.

### 3.2.3.3   Hybrid solution at high Pr and Ra

It is of course necessary to correlate heat flux measurements with the geometry of convection. Moreover, modern visualization techniques can give *in situ* measurements of the velocity, temperature, and concentration fields (see Section 3.3), but they require transparent sides. Therefore, a number of studies have used a 'lighter' setup (**Figure 4(c)**; e.g., Olson, 1984; Jaupart and Brandeis, 1986; Davaille and Jaupart, 1993; Weeraratne and Manga, 1998). The bottom and top heat exhangers are made or copper or aluminum, and are regulated either by electric heating or by circulation of thermostated water. The tank sides are made of plexiglas or glass. The whole tank is insulated with styrafoam, but the insulation on the sides can be removed for visualization. In this configuration, it is advisable to run the experiments in conditions such that the mean temperature inside the experimental cell is close to ambient temperature, to minimize heat losses.

Last, to obtain high Rayleigh numbers ($10^6$–$10^9$) in high *Pr* viscous fluids, we need fluids thicknesses between at least 10 and 50 cm. It is then no longer possible to have large aspect ratios. It therefore will always be a possibility that the flow will be influenced by the mechanical boundary conditions (zero velocity) at the side walls. We shall discuss this question in more detail later.

### 3.2.3.4   Moving boundaries and free-slip boundaries

To impose a moving upper boundary (e.g., to simulate plate tectonics), a mylar sheet (**Figure 4(a)**) is usually introduced just below the upper plate and slowly driven by a step-motor (e.g., Richter and Parsons, 1975; Kincaid *et al.*, 1995, 1996; Jellinek *et al.*, 2003).

Because the mantle is bounded by the liquid core at the bottom, and oceans or atmosphere at the top, its mechanical outer boundary conditions are free slip. Moreover, analytical and numerical models usually are best resolved with free-slip boundary conditions. To obtain the latter in the laboratory, thin layers of a fluid much less viscous (at least a 1000 times) than the working fluid must be introduced between the heat exchangers and the high *Pr* convecting fluid. For a bottom free-slip condition, thin layers of mercury (e.g., Solomon and Gollub, 1991) or salted water (e.g., Jellinek *et al.*, 2002; Jellinek and Manga, 2002, 2004) may be used; while a thin layer of conductive oil may be used to obtain a free-slip upper surface (e.g., Giannandrea and Christensen, 1993). However, since these layers have a finite conductivity, they will modify the thermal boundary conditions as discussed above.

### 3.2.3.5   Centrifuge

Some experiments have been run in centrifuges (e.g., Ramberg, 1972; Nataf *et al.*, 1984; Weiermars, 1988), where the model is subjected to a centrifugal force able to produce accelerations up to $20\,000g$. This approach decreases the experimental time required and allows the use of very viscous materials (with $Pr > 10^6$) at Rayleigh numbers up to $6 \times 10^5$. A detailed description of the experimental setup is given by Nataf *et al.*, (1984) and Weijermars (1988). However, this technique is heavy to implement. With the apparent confirmation (in part because of the early centrifuge experiments) that convective dynamics becomes independent of *Pr* when $Pr > 100$, the technique has not been used extensively.

## 3.3 Measurements and Visualization Techniques

### 3.3.1 Patterns

Laboratory experiments have been used extensively to determine convective and/or mixing patterns. One then seeks quantitative information on the morphology, wavelength, and evolution (plume ascent rate, blob stretching, etc.) of a particular feature in the experimental fluid.

#### 3.3.1.1 Dye

*3.3.1.1.(i) One-shot visualization* Patterns and motions in two-fluid experiments such as compositional plumes (**Figure 5**, Olson and Singer, 1985, Section 3.6), Rayleigh–Taylor instabilities (e.g., Whitehead and Luther, 1975; see Section 3.4), mixing (e.g., Ottino, 1989; see Section 3.9) or two-layer convection (e.g., Olson, 1984; Olson and Kincaid, 1991; Davaille, 1999; see Section 3.9) are easily observed by the addition of dye to one of the transparent fluids. Most commonly used dyes are fluoresceine, rhodamine, or supermarket food dye. The latter is very easily visualized through illumination with a white light (**Figure 5**), while the former are most spectacular using laser sheets (e.g., Tsinober *et al.*, 1983; Fountain *et al.*, 2000). In mixing experiments, the dye can be continuously injected by seringes (e.g., Fountain *et al.*, 2000; Kerr and Mériaux, 2004). In all cases, the dye quantity is so small that it does not change the physical properties of the fluids.

The fluid initially at rest can also be marked by dye streaks that allow subsequent fluid motions to be recorded by their distortion. Griffiths (1986) visualized the motions of initially cold fluid due to the passage of a thermal by injecting (from a seringe) before the run at various heights a number of horizontal lines of the same fluid containing a small concentration of dye (**Figure 6(a)**). Using very viscous putties, Weijermars (1988, 1989) imprinted black grids on some cross-sections prior to experiments in a centrifuge, which recorded the subsequent deformation (**Figure 6(b)**).

*3.3.1.1.(ii) Electrochemical technique* In water and aqueous solutions, an electrochemical technique using thymol blue, a pH indicator, can be employed (Baker, 1966). Thymol blue is either blue or yellow-orange depending upon whether the pH of the solution is greater or less than 8. Approximately, 0.01% by weight of thymol blue is added to the water and the solution is titrated to the end-point with sodium hydroxyde. Then, by drop addition of hydrochloric acid, the solution is made orange in color. When a small DC voltage ($\sim$10–20 V) from a drycell source is impressed between a pair of electrodes situated within such a fluid, H+ ions are removed from solution at the negative electrode, with a corresponding change in color from orange to blue. The dye thus created is neutrally buoyant and faithfully follows the motion of the fluid. Sparrow *et al.* (1970) used the bottom plate of the experimental cell as the negative electrode itself, while the positive electrode was a large copper sheet situated adjacent to the fluid but well removed from the bottom plate. They hereby obtained the first photographs of thermals rising from a heated horizontal surface (**Figure 7**). Haramina and Tilgner (2004) recently used the same method to image coherent structures in boundary

(a)          (b)          (c)

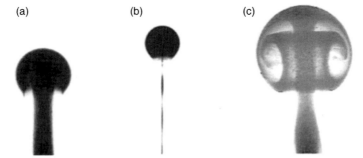

**Figure 5** Morphology of starting plumes when the dyed plume material is injected into a transparent medium. (a, b) Purely compositional plumes (a) diapiric plume, more viscous than its surroundings; (b) cavity plume, less viscous than its surroundings. (c) Starting thermal plume: the injected dyed material is hotter and less viscous. (a, b) From Olson P and Singer H (1985). Creeping plumes. *Journal of Fluid Mechanics* 158: 511–531. (c) From Laudenbach N and Christensen UR (2001). An optical method for measuring temperature in laboratory models of mantle plumes. *Geophysical Journal International* 145: 528–534.

(a)   (b)

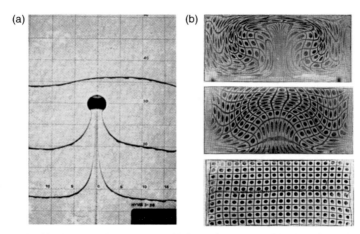

**Figure 6**   Fluid initially marked by horizontal dye streaks. (a) Rise of a pocket of hot fluid (so-called 'a thermal'); (b) rise of hot fluid continuously released from a hot source ('hot plume'). The gravity has been accelerated 2000 times in a centrifuge. Time increases from bottom to top pictures. (a) From Griffiths RW (1986) Thermals in extremely viscous fluids, including the effects of temperature-dependent viscosity. *Journal of Fluid Mechanics* 166: 115–138. (b) From Weijermars R (1988). New laboratory method for analyzing deformation and displacement in creeping fluid: Examples from stokes flow and a thermal plume. *Journal of Geophysical Research* 93: 2179–2190.

**Figure 7**   Thermal boundary layer instabilities in water visualized by an electrochemical technique. Adapted from Sparrow EM, Husar RB, and Goldstein RJ (1970). Observations and other characteristics of thermals. *Journal of Fluid Mechanics* 41: 793–800.

layers of Rayleigh–Bènard convection at very high Rayleigh numbers.

### 3.3.1.2   Shadowgraph and schlieren

Since the refractive index of a fluid depends on temperature and composition (e.g., salt or sugar content), convective features in a transparent medium will alter the refractive index distribution, also called schlieren (from the german; sometimes spelled 'shlieren' in the literature). From the deformation of the optical wave front, one can therefore deduce information on the refractive index, hence on the temperature or compositional field. The optical methods are divided into two groups: the shadow and schlieren techniques using the deflection of light in the measurement media (e.g., Settles, 2001),

and interference methods based on differences of length of the optical paths, that is, on the phase (e.g., Hauf and Grigull 1970). In the following, we shall refer to a 'schlieren object', when considering the influence of this object on light deflection, and to a 'phase object' when considering its influence on the optical path length.

Both shadow and schlieren techniques are whole-field integrated optical systems that project line-of-sight information onto a viewing screen or camera focal plane. They use light intensity and light ray identification, so that only a good white light source is required. They are more appropriate for two-dimensional (2-D) or axisymmetric phenomena since they integrate the information along the light ray path, but are still qualitatively useful for any phenomenon. Both techniques have a long history: R. Hooke developed a schlieren method as early as 1672 and the first published shadowgram of a turbulent plume over a flame by Marat dates back to 1783. Settles (2001) gives a comprehensive review of these two techniques.

### 3.3.1.2.(i)   Shadowgraph

Whithout the object present in the field of view, the light source illuminates the screen uniformly. With the object present, some rays are refracted, bent, and deflected from their original paths, producing a shadow. The optical inhomogeneities of the object redistribute the screen illuminance. The illuminance level responds to the second spatial derivative or Laplacian of the

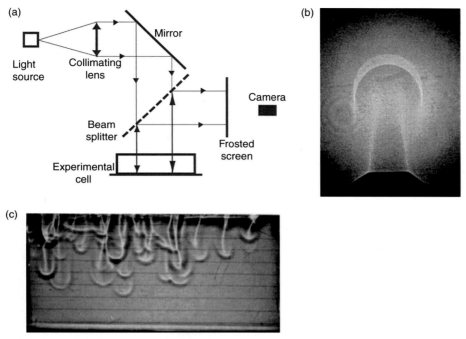

**Figure 8** Shadowgraph. (a) Alternative setup to **Figure 4(a)** when the bottom plate of the cell is a mirror. (b) Hot thermal plume out of a small heat source. (c) cold thermal instabilities from a cold horizontal plate. From Jaupart C and Brandeis G (1986) The stagnant bottom layer of convecting magma chambers. *Earth and Planetary Science Letters* 80: 183–199. Cold material focuses light and appears bright, whereas hot material acts as a divergent lens and appears dark.

refractive index. Best results are obtained when the cell is illuminated by a (quasi-) parallel source of light (**Figures 4(a)** and **8(a)**). Due to the negative slope of the temperature dependence of the refractive indices of most liquids, the hot material acts as a diverging lens and appears dark (**Figure 8(b)**), while the cold material acts as a focusing lens and appears light (**Figure 8(c)**). Shadowgraph techniques are easy (and cheap!) to implement and allow the visualization of large objects. Since they image strong temperature gradients well, they have been used to determine convective patterns and measure their characteristic wavelength (e.g., Chen and Whitehead, 1968; Heutmaker and Gollub, 1987; see Section 3.5), measure thermal boundary layer thicknesses (Olson *et al.*, 1988), or follow the evolution of laminar thermal plumes (e.g., Shlien, 1976; Moses *et al.*, 1993). However, the whole image cannot be interpreted quantitatively in terms of temperature.

***3.3.1.2.(ii) Schlieren*** The schlieren image is a conjugated optical image of the schlieren object formed by a lens or a mirror. The method requires a knife-edge or some other sharp cutoff of the refracted light. **Figure 9(a)** presents the diagram of a simple schlieren system with a knife-edge placed at the focus of the second lens. Adding an object will bend light rays away from their original paths. The refracted rays miss the focus of the optical system. The upward-deflected ray brightens a point on the screen, but the downward-deflected ray hits the knife-edge causing a dark point against a bright background. So a vertical gradient of the refractive index is converted into an amplitude difference. In general, the illuminance level of the schlieren image responds to the first derivative of the refractive index on a direction perpendicular to the knife-edge. **Figures 9(b)** and **9(c)** show the schlieren image of a hot plume obtained with a horizontal and vertical knife edge, respectively. **Figure 9(d)** shows the schlieren image of the same plume using a circular spatial filter of 0.5 mm diameter. The bright line in the center of the image and contours represent the zones with zero refractive index gradient.

For weak disturbances, the schlieren technique has a much higher sensitivity than the shadowgraph. It has been used to visualize the onset of thermal convection above a heated horizontal plate (Schmidt and Milverton, 1935), 2-D boundary

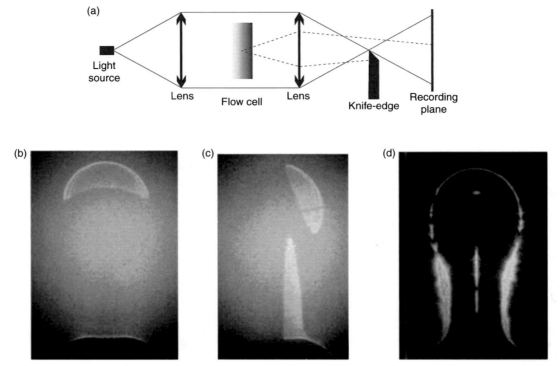

**Figure 9**   Schlieren. (a) Optical setup. (b–d) Pictures of a thermal plume out of a small heat source obtained with (b) a horizontal knife edge, (c) a vertical knife edge, and (d) a 0.5 mm diameter pin hole.

layers along vertical walls (e.g., Hauf and Grigull, 1970), and shock waves (e.g., Settles, 2001). Another advantage over shadowgraph is the 1:1 image correspondence with the object of study. Moreover, 3-D study of the phase object is also possible by using a multisource system (Hanenkamp and Merzkirch, 2005).

Other optical methods can be derived from schlieren techniques. In the so-called 'lens-and-grid' techniques, an array of light/dark stripes is used as a background grid. This simple method, also called background grid distortion, can be used when a large field-of-view is necessary with no need for high sensitivity. The sensitivity can be improved by adding another grid between the focusing lens and the image plane (**Figure 10**). The Moiré method can be seen as variant of lens-and-grid method with grids on either sides of the schlieren object. Dalziel *et al.* (2000) contributed to this method by simulating electronically the second grid in the image capture device. Another variant called Moiré shearing interferometry (or Talbot interferometer) was used for mapping phase objects such as candle flames (Lohmann and Silva, 1972).

## 3.3.2   Temperature and Heat Flow Measurements

Since the precise characterization of thermal convection requires the quantitative knowledge of the temperature field, temperature measurements are a major step in data acquisition. Local measurements have long been provided by thermocouples or thermistors embedded within the flow, but with the risk of disturbing it. The last 30 years have seen the development of noninvasive methods. Laser beam scan and interferometry techniques can provide accurate temperature structure of 2-D or axisymmetric features. More recently, the use of thermochromic liquid crystals has allowed for the first time the accurate determination of a fully 3-D, time-dependent temperature field.

### 3.3.2.1   Local temperature measurements

Point temperature measurements can be obtained electronically using the variation with temperature of the electrical resistance of metals (Pt resistance thermometers, thermistors) or using the

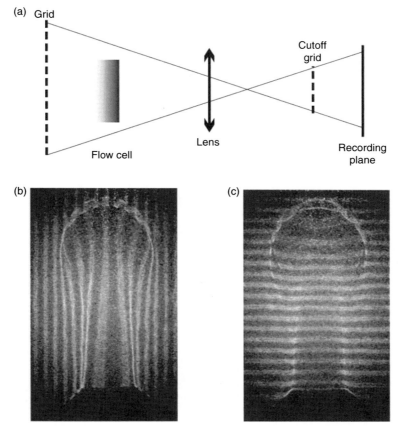

**Figure 10** Lens-and-grid method. (a) Optical setup. (b, c) Same thermal plume as **Figure 9(b)** obtained with (b) a vertical grid and (c) an horizontal grid.

thermoelectric effect at the junction between two metals (thermocouples).

Pt-thermometers exploit the increase with temperature of the electrical resistance of platinum. The most widely used sensor is the $100\,\Omega$ or $1000\,\Omega$ platinum resistance thermometer. They are the most accurate and stable sensors over a long time period. However, they are expensive and do not come in diameters smaller than a few millimeters. So they are generally used to calibrate all the other temperature sensors in a lab, but are not used directly in the experimental cell.

Thermistors are made from certain metal oxides whose resistance decreases with increasing temperature. Their behavior is highly nonlinear, which limits their useful temperature span, and they are less stable than Pt-thermometers. However, they are cheaper, and can be produced in very small designs (0.1 mm) with a fast response and low thermal mass. The measurement of temperature then requires an electric power supply and a voltmeter. Properly calibrated, thermistors can give a precision of $0.002\,°C$ over a $10\,°C$ range. They are used in high Rayleigh number convection experiments in water and lower viscosity fluids (e.g., Castaing et al., 1989; Niemela et al., 2000; Chilla et al., 2004).

Thermocouples are based on the thermoelectric effect: the junction between two different metals produces a voltage which increases with temperature. In order for this thermal voltage to produce a flow of current, the two metals must also be connected together at the other end so that a closed circuit is formed. With different temperatures at the junctions the voltages generated are different and a current flows. A thermocouple can thus only measure temperature differences between the measurement junction and the reference junction. The latter is at a known temperature, which is nowadays usually that of a commercial electronic ice-point cell. The temperature range of thermocouples is bigger than for thermistors and their behavior is more linear. However, the voltage produced by the thermoelectric effect is very small. For the commonly used type 'K'and type 'E', it amounts, respectively, to 40 and 60 µV per degree celsius, so that a six-digit voltmeter will allow temperature measurements with an accuracy of ±0.025°C. Their response time is about 2 s, that is, much shorter than the typical timescale of instabilities in viscous fluids.

The temperature sensors can be introduced permanently in the experimental tank either one by one at different locations (e.g., Sparrow *et al.*, 1970; Davaille and Jaupart, 1993; Weeraratne and Manga, 1998), or on given vertical or horizontal profiles (e.g., Guillou and Jaupart, 1995; Le Bars and Davaille, 2002, 2004a). The latter allows one to follow the temperature profiles through time but the number of sensors in a 2-mm-diameter probe is limited to 14. Alternatively, the sensors can be mounted on a stepping motor, and moved vertically to measure the vertical temperature profile (**Figure 4(b)**). With viscous fluids, a large volume of fluid is carried along each time the probe is moved to a new depth. Therefore, the probe must be kept several minutes at the same height in order for the system to equilibrate, and the vertical temperature profiles are determined from the time-averaged measurements at each depth (e.g., Olson, 1984; Giannandrea and Christensen, 1993; Matsumoto *et al.*, 2006).

It is also possible to measure directly an horizontal average of the temperature field by stretching a set of very thin platinum wires (0.2 mm diameter) horizontally across the experimental cell (**Figure 8(c)**; Jaupart and Brandeis, 1986; Davaille and Jaupart, 1993; Giannandrea and Christensen, 1993). Because the wire resistance varies as a function of temperature, these wires are operated in the same way as thermistors, within a circuit made of a stable precision tension generator and a precision resistance. This setup is very delicate and time consuming to calibrate, but allows one to measure every second the horizontally averaged temperature with a 0.1°C accuracy with a six-digit voltmeter. This method was used to study penetrative convection in constant-viscosity fluids (Jaupart and Brandeis, 1986) and in strongly temperature-dependent viscosity fluids (Davaille and Jaupart, 1993, 1994).

The advantages of local temperature sensors are their high precision and the ability to obtain long time series (high-frequency sampling and/or long run) at low cost. However, the sensors are introduced within the fluid, which imposes a zero velocity condition for the flow on their surface. They therefore

always perturb the flow locally. For nonsteady flow and thin probes, this perturbation remains negligible- and the probes do not preferentially focus convective features on themselves (e.g., Davaille and Jaupart, 1993). However, for steady flow (like low Rayleigh number convection), upwellings or downwellings can be locked on the temperature probes. In such cases, one should use only removable temperature sensors mounted on step motors, and/or the nonintrusive methods to which we turn now.

### 3.3.2.2  Deflection of a light beam

In the cases of 2-D or axisymmetrical refractive index distributions, it is possible to recover the temperature field by scanning the test zone with a laser beam and recording its deflection. Nataf and *et al.*, (1988) and Rasenat *et al.* (1989) applied this method to determine the type of coupling in two-layer convection at low Rayleigh numbers. Laudenbach and Christensen (2001) described the method in detail for the axisymmetric case (**Figure 11**) and applied it to thermal plume conduit and solitary waves.

The same treatment can be applied to the lens-and-grid methods described in Section 3.3.1.2, since the ray displacement can be calculated from the grid distortion provided that the structure is 2-D or axisymmetrical. The spatial resolution of the method then depends on the spatial frequency of the grid lines.

### 3.3.2.3  Interferometry

Compared to shadowgraph and schlieren methods, interferometric methods offer more detailed information about the phase object (e.g., Yang, 1989), since they play with the light path, but they usually require laser light and a good optical control. They have been used for pattern determination (e.g., Nataf *et al.*, 1981, 1984, 1988; Jaupart *et al.*, 1984; Kaminski and Jaupart, 2003) as well as for quantitative measurement of the temperature field (e.g., Gebhart *et al.*, 1970; Shlien and Boxman, 1981).

### 3.3.2.3.(i)  Mach–Zender interferometry  The
most common two-beam interferometer is the Mach–Zender interferometer (**Figure 12(a)**; Hauf and Grigull, 1970). The phase object is placed in one of the legs of the interferometer formed by two beam-splitter mirrors and two total reflecting mirrors. For a 2-D object, the interference fringes can be interpreted as isothermal lines. It has been used for boundary-layer dynamics along vertical walls or in double-diffusive systems (e.g., Lewis *et al.*, 1982), and

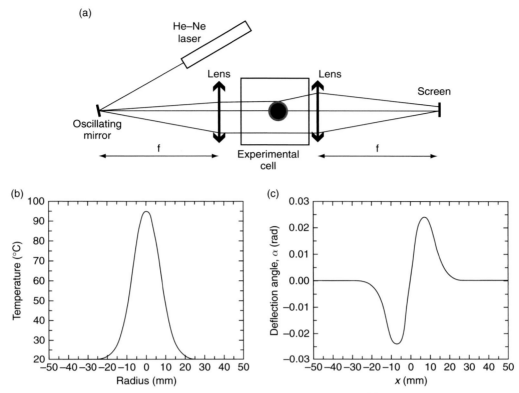

**Figure 11** Deflection of a laser beam by a thermal plume. (a) Experimental setup. (b) Radial temperature structure of the plume. (c) Beam deflection angle. Adapted from Laudenbach N and Christensen UR (2001). An optical method for measuring temperature in laboratory models of mantle plumes. *Geophysical Journal International* 145: 528–534.

steady laminar plumes above a horizontal line heat source (**Figure 12(b)**; Gebhart *et al.*, 1970). For a 3-D object, the phase difference between the two optical paths is integrated over the longitudinal length of the object, and the interferogram is not readily interpretable (Merzkirch, 1993). In a cylindrical symmetry the radial temperature distribution can be obtained rather simply by an Abel transformation. If the refractive index has no symmetry, tomography algorithms and reconstruction techniques have to be applied. The Mach–Zender interferometer was used to measure the temperature field of axisymmetric, laminar thermal plumes in liquids (**Figure 12(c)**; e.g., Boxman and Shlien, 1978; Shlien and Boxman, 1979, 1981; Chay and Shlien 1986) and for turbulent mixing of salt solutions (Boxman and Shlien, 1981). More recently, Qi *et al.* (2006) used the Michelson interferometer, more often used to measure surface displacement (see Section 3.3.5), to measure the temperature field of 3-D axisymmetrical flames.

Holographic interferometry was introduced in flow visualization by Heflinger *et al.* (1966) and Tanner (1966). Holography has opened a new dimension for two-beam interferometry: the reference and test beams can be separated in time rather than in space. In a holographic interferometer two consecutive exposures are taken through the field of interest, usually the first exposure without object and the second in the presence of an object of varying temperature. The double exposed plate is developed and placed in the holographic reference beam for reconstruction. The pattern which becomes visible after this reconstruction is equivalent to the one obtained by a Mach–Zender interferometer (Merzkirch, 1993; Hauf and Grigull, 1970). The main advantage of holographic interferometry over 'classical' interferometers is that, since the geometrical path of the signal in the two exposures is identical, the quality of the optical components is not crucial. The eventual optical disturbances and impurities in the test section cancel out.

(a)

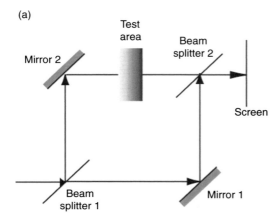

(b)

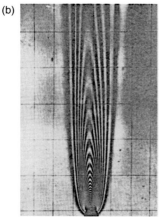

(c)

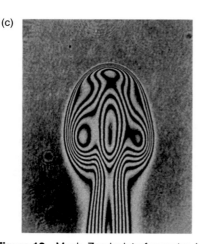

**Figure 12**   Mach–Zender interferometry. (a) Set up. (b) 2-D laminar plume from a heat line. From Gebhart B, Pera W, and Schorr A (1970) Steady laminar natural convection plumes above a horizontal line heat source. *International Journal of Heat and Mass Tranfer* 13: 161–171. (c) Axisymmetric laminar plume from a point source in water. From Shlien DJ and Boxman RL (1981) Laminar starting plume temperature field measurement. *International Journal of Heat and Mass Transfer* 24: 919–930.

### 3.3.2.3.(ii)   Differential interferometry

Differential interferometry (also called shearing interferometry) is a method to measure derivatives of light phase distortions. In practice, this can be done in two ways. One possibility is to send two beams slightly displaced through a phase object and to put them back together on the camera. The best way for doing this is to use two Wollaston prisms and polarizers (Oertel and Bühler, 1978; Nataf *et al.*, 1981). Another possibility is to divide one wave into two identical beams behind the object and to put them on a camera with a small displacement. This can be done for instance by using one Wollaston prism (Sernas and Fletcher, 1970; Small *et al.*, 1972). Another variant is the use of a Mach–Zender interferometer with the phase under investigation placed outside (Pretzel *et al.*, 1993). In this arrangement, using a very compact Mach–Zender interferometer minimized its inherent sensitivity to vibrations. Another simple way of separating into two beams is to use an optical flat plate tilted at a certain angle (**Figure 13**; Jaupart *et al.*, 1984; Kaminski and Jaupart, 2003). Extinction lines follow constant horizontal temperature gradients. Vertical extinction lines away from the plume correspond to light beams through uniform background.

The main advantage of differential interferometry is its variable sensitivity: carrier fringe orientation and frequency can be chosen separately. It is therefore possible to visualize with the same apparatus the flow configuration in two fluids with very different temperature dependence of the refractive index (e.g., Oertel and Buhler, 1978). Moreover, the separation between the object and the interferometer facilitates the investigation of large and complicated objects. The evaluation of the interferogram can be done by

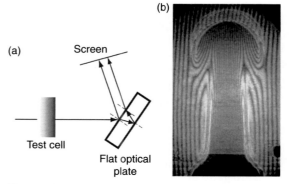

(a)

(b)

**Figure 13**   Differential interferometry. (a) Setup. (b) Thermal plume in silicone oil V5000.

Fourier analysis, and the result, being the first derivative of the integral phase shift caused by the object, is obtained with high accuracy. For radially symmetrical objects, the spatial distribution of the refraction index can be obtained easily because the Abel inversion formula is reduced to a simple integration, which can be done more reliably than a differentiation. In that respect, differential interferometry results are less noisy than Mach–Zender interferometry, where the spatial distribution of the refractive index is obtained through numerical differentiation of the optical path difference. More details on the digital processing of interferograms can be found in the monograph by Yang (1989).

Speckle interferometry can also enter into this category (Merzkirch, 1995). In double-exposure speckle photography, two speckle patterns are recorded on the same photographic plate. Between the two exposures the scattering plate (ground glass) is shifted to obtain the interferometric fringes. After photographic development, the specklegram is scanned by a laser beam. By measuring the Young's fringe spacing and orientation, it is possible to measure the two components of the displacement and convert them into deflection angles (resembling for this reason the shadowgraph and schlieren methods). In addition, precise multiprojection speckle photography allow the reconstruction of a 3-D temperature field using computer tomography (Asseban et al., 2000). Among the most important advantages of speckle photography are its spatial resolution (about 0.2 mm) and the possibility to collect a great amount of experimental information from a single specklegram. Large amounts of data can be processed and analyzed statistically. The potential of analyzing spatial characteristics of turbulent flows was demonstrated (Merzkirch et al., 1998).

### 3.3.2.4 Isotherms: thermochromic liquid crystals

The use of thermochromic liquid crystals (TLCs) allows one to visualize the temperature field on a 2-D-plane in the fluid flow without perturbing it (Rhee et al., 1984; Dabiri and Gharib, 1991; Willert and Gharib, 1991). Liquid crystals are mesomorphic phases which present peculiar properties due to the presence of some degree of anisotropy (Chandrasekhar, 1977). One particular class of these mesophases, *chiral nematics (cholesterics)*, have a structure that undergoes an helical distortion, and because of their periodic structure, they generate Bragg reflections at optical wavelengths. The

pitch of the wavelength of the Bragg-reflected light depends on the temperature $T$ (de Gennes and Prost, 1993). Thus, the color of the material can change drastically over a temperature interval of a few degrees. When the liquid crystals are illuminated by white light, their color changes with increasing temperature from colorless to red at low temperatures, passes through green and blue to violet, and turns colorless again at high temperatures. It was first used as paint on a surface to determine convective patterns qualitatively (e.g., Chen and Chen, 1989; Lithgow-Bertelloni et al., 2001).

Subsequently, it became possible to mix the TLC slurry directly within the fluid and to illuminate the tank cell on cross-sections (**Figure 14(a)**). One of the main applications of this technique has been the study of aqueous turbulent flows (Solomon and Gollub, 1990, 1991; Gluckman et al., 1993; Moses et al., 1993; Dabiri and Gharib, 1996; Park et al., 2000; Pottebaum and Gharib, 2004). The use of this method to measure the temperature field quantitatively requires a high-precision color CCD camera and a very precise calibration of the color of the liquid crystals particles against the true temperature. Moreover, the total temperature range accessible with one particular TLC slurry is usually around 2–3°C. This technique has lately been further extended to the joint measurement of temperature and velocity in a 3-D field (Kimura et al., 1998; Fujisawa and Funatani, 2000; Ciofalo et al., 2003). An uncertainty analysis performed by Fujisawa and Hashizume (2001) gives an error less than 0.1°C, or 5% of the total imposed temperature difference, for a calibration method based on a hue-saturation-intensity approach (Fujisawa and Adrian, 1999).

Convection in conditions analogous to those of the mantle involves viscous fluids (see Section 3.2) and typical temperature heterogeneities ~10–25°C. Therefore, the use of one TLC slurry is not enough to image the whole temperature field. The use of white light and a single TLC was used to determine the influence of mechanical (Namiki and Kurita, 1999, 2001) and thermal boundary heterogeneities (Matsumoto et al., 2006) on thermal convection. The type of TLC was chosen to image well the hot instabilities. This permitted quantification of the convective pattern, but not accurate measurement of the temperature field. Illuminating several TLC slurries by a single laser wavelength plane sheet and recording the images with a high precision black and white CCD camera (**Figure 14(a)**), Davaille et al.

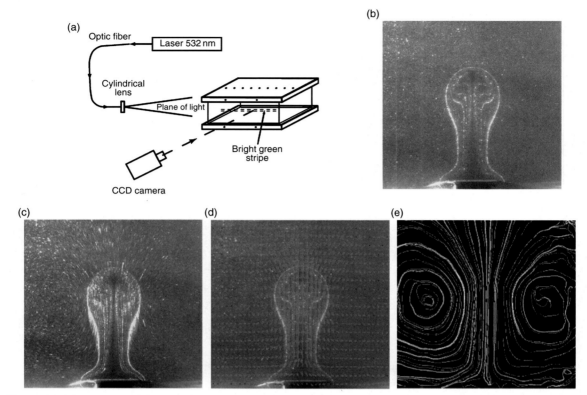

**Figure 14**  Temperature and velocity field of a thermal plume out of a localized heat source. (a) Setup. (b) Isotherms. (c) Streaklines taken over 10 s. (d) Velocity field calculated by PIV. (e) Streamlines calculated from the velocity field.

(2006) demonstrated that it was possible to obtain isotherms and local temperature gradients. The laser beam was expanded into a beam using a cylindrical lens. Each slurry brightens over a different temperature subrange and therefore generates a horizontal bright line (**Figure 14(b)**). Each stripe presents a finite thickness: although each TLC responds at a given wavelength to a precise temperature, the polymeric capsules enclosing them introduce a scatter around this value. On a plot of the intensity as a function of depth, or temperature, each stripe corresponds therefore to a peak whose maximum defines the value of the 'isotherm', and whose thickness gives a measure of the local temperature gradient. After calibration, the light intensity maximum gives the 'isotherm' temperature with a precision of $\pm 0.1\,^{\circ}$C, and the half width of the bright lines gives the local temperature gradient. This method was used to study two-layer convection (Le Bars and Davaille, 2002, 2004a, 2004b; Kumagai et al., 2007), plumes arising from a viscous thermal boundary layer (Davaille and Vatteville, 2005) and plumes from a point heat source (Vatteville, 2004).

Since viscous fluid motions are slow, it is possible to scan the experimental cell with the laser sheet (by means of an oscillating mirror driven by a galvanometer) to obtain the 3-D structure of the temperature field. This method can be combined with laser-induced fluorescence (LIF) and particle image velocimetry (PIV) to obtain simultaneously the composition (see Section 3.3.3) and velocity fields (see Section 3.3.4).

### 3.3.2.5  Heat flow measurements

The dependence of the global heat transport on convection characteristics is one of the fundamental questions of Rayleigh–Bénard studies (see Section 3.5).

The most accurate measurement of the global heat flow is obtained through the measurement of the electric power needed to heat the bottom plate (e.g., Schmidt and Milverton, 1935; Malkus, 1954; Silveston, 1958; Giannandrea and Christensen, 1993; Brown et al., 2005; Funfschilling et al., 2005). With careful design of the experimental tank in order to minimize the heat losses (cf. Section 3.2.3), late results have reached a precision of 0.1% (e.g., Funfschilling *et al.*, 2005).

The heat flow can also be deduced from the first derivative of the measured temperature profiles, either locally (using temperature probes or isotherms) or horizontally averaged (using platinum wires). This method requires a precise knowledge of thermal conductivity (especially as a function of temperature), and temperature measurements at several different heights within the thermal boundary layers close to the horizontal plates to resolve the steep gradients there. This becomes increasingly difficult as the Rayleigh number increases and the boundary layers become thinner, with the danger of underestimating the heat flux. For Rayleigh numbers up to $10^8$, a precision of 5% has been obtained (e.g., Davaille and Jaupart, 1993).

### 3.3.3 Composition

Measurement of the time evolution of composition is required in studies on entrainment, mixing and two-layer convection. It usually amounts to measuring a tracer dilution. Point-based techniques use either *in situ* probes (e.g., conductivity), or extraction of samples by suction from various points in the flow field. Tait and Jaupart (1989) used conductivity measurements to study the mushy crystallization of ammonium chloride in viscous solutions. Davaille (1999) studied entrainment processes in two-layer convection by periodic sampling of the fluid layers, measuring the salinity of each sample using the dependence of the refractive index on the salt concentration, and a food dye concentration by a UV absorption technique. However, these point-based techniques have some major drawbacks: the flow may be disturbed by the probes, the number of measurement points is very limited, and extraction techniques can yield only time-averaged concentrations and cannot capture their instantaneous fluctuations.

Concentration profiles averaged over the experimental cell (e.g., Olson, 1984) or over one of its dimension (e.g., Solomon and Gollub, 1988a, 1988b, 1991) can be deduced from the optical absorption by the cell of an extended white light beam (e.g., Olson 1984), or a laser beam (e.g., Solomon and Gollub, 1988a, 1988b, 1991).

The advent of LIF techniques in the 1970s enabled simultaneous capture of the entire instantaneous tracer (fluorescent dye) concentration field over a planar sampling area (laser sheet), with a experimental set up similar to **Figure 14**. LIF is a nonintrusive technique which has been applied to turbulent jet flows (e.g., Villermaux and Innocenti, 1999), laminar mixing (e.g., Ottino *et al.*, 1988; Fountain *et al.*, 2000), and two-layer convection (e.g., Kumagai *et al.*, 2007). Since fluoresceine dye shifts the laser light frequency, it is possible to use LIF and TLCs to measure composition and temperature simultaneously, by recording the images with two different filters, one for the laser frequency and one for the fluoresceine (**Figure 15**; Kumagai *et al.*, 2007). To obtain a 3-D field, the laser beam can be swept through the flow at high speed and images captured with a synchronized camera.

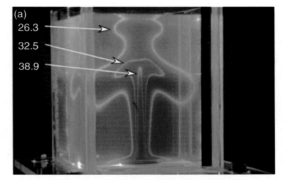

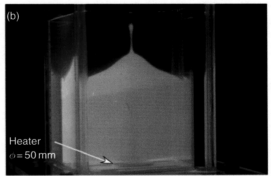

**Figure 15** Interaction of a thermal plume with a density stratification. The denser layer has been dyed with fluoresceine. Two-dimensional visualization when the tank is illuminated with a laser sheet (532nm): (a) isotherms (26.3, 32.5, and 38.9°C) and (b) compositional fields (LIF method). These two images were taken by a 3-CCD camera at almost same time, but using two different optical filters, a band-pass filter for temperature and a cut-off filter for composition. From Kumagai I, Davaille A, and Kurita K (2007). On the fate of mantle thermal plumes at density interface. *Earth and Planetary Science Letters* 254: 180–193.

### 3.3.4   Fluid Motions

#### 3.3.4.1   *Local measurements*

It is easy to follow the displacement of an interface defined by a grid (e.g., Weijermars, 1988), or by abrupt gradients either in refraction index (e.g., visualized by shadowgraph or by interferometry) or in dye concentration. These techniques have been used to measure the rising velocity of thermals (Griffiths, 1986) or plumes (e.g., Olson and Singer, 1985; Griffiths and Campbell, 1990; Moses *et al.*, 1993; Coulliette and Loper, 1995; Kaminski and Jaupart, 2003).

'Hot wire' probes can also be used whereby the local flow velocity is measured by sensing the rate of cooling of fine, electrically heated wires. However, their use is usually confined to turbulent or high-speed flows of low viscosity fluids. Morevover, as an intrusive method, it suffers of the same drawbacks as temperature sensors (cf. Section 3.3.2.1).

*Laser Doppler velocimetry* (LDV) allows one to measure continuously in time and at a given position in space up to the three components of the velocity of tracer particles (e.g., Adrian, 1983; Merzkirch, 2000). The method is based on the optical Doppler effect and requires seeding of the working fluid with micrometer-size particles. Incident light is scattered by the moving particles, and the frequency of the scattered radiation is Doppler-shifted. Since this frequency shift is relatively small, its detection requires the use of monochromatic incident laser light. Maps of the spatial dependence of the velocity field can be obtained by translating the cell apparatus over the LDV system. Solomon and Gollub (1988a, 1988b, 1991) applied this technique to measure the velocity field in steady quasi-2-D Rayleigh–Bénard convection.

#### 3.3.4.2   *2D and 3D field measurements*

One of the oldest techniques for measuring velocities in a fluid is seeding particles (e.g., hollow glass beads, aluminum flakes, and tiny air bubbles) into the fluid and illuminating a cross-section of the flow cell by a thin sheet of light. The foreign particles should be small and have densities as close to that of the fluid as possible, in order to follow passively the local flow.

There are then several ways to describe the flow. The trajectory of a single fluid particle over time defines a 'pathline'. A 'streakline' connects all the fluid elements that have passed through a given point (**Figure 14(c)**). 'Streamlines' are tangential to the flow directions at a given time (**Figure 14(e)**).

In steady flow, pathlines, streaklines, and streamlines are identical, but usually not in time-dependent flow.

The availability of high-power laser sources together with fast digital processors led to the development of sophisticated whole-field velocimetry techniques such as PIV and particle tracking velocimetry (PTV). Both techniques provide a quantification of the velocity field over the entire plane. The optical setup is the same as for the visualization of isotherms (**Figure 14(a)**). PTV tracks individual particles in subsequent images, while PIV follows a group of particles through statistical correlation of local windows of the image field from two sequential images (Adrian, 1991; Raffel *et al.*, 1998; Merzkirch, 2000). From the known time difference and the measured displacement the velocity is calculated.

PTV is convenient when the seeding is sparse, and/or near boundaries, since each particle trace is analyzed individually. The latter operation is time consuming. Moreover, as the particles are present at random locations in the fluid, the velocity estimators are found at random locations in the flow, too.

The PIV scheme, on the contrary, requires a high particle density, and calculates the 'mean' displacement of particles in a small region of the image (the interrogation window) by cross-correlation of the transparency signal of the same window in two subsequent images. It therefore removes the problem of identifying individual particles, which is often associated with tedious operations and large errors in the detection of particle pairs. Its spatial resolution is uniform (function of the interrogation window size) over the whole image. In geodynamics, this technique has recently been used to study subduction (Kincaid and Griffiths, 2003, 2004) and thermal convection (Davaille and Vatteville, 2005; Kumagai *et al.*, 2007).

Current research aims at combining the resolution of PTV with the fast algorithm of PIV. Funiciello *et al.* (2006) recently developed such a feature tracking (FT) method to map mantle flow during retreating subduction.

PIV has also been used simultaneously with TLCs in two ways. When the temperature difference is small and only one TLC slurry is needed to determine a continuous temperature field, the velocity is measured using the same TLC particles as Lagrangian flow tracers (e.g., Dabiri and Gharib, 1991; Park *et al.*, 2000; Ciofalo *et al.*, 2003). When several TLC slurries are used to visualize isotherms, the fluid can also be seeded uniformly with $10\,\mu m$ size particles (hollow glass or latex) to calculate the

PIV (**Figure 14**; Davaille and Vatteville, 2005). 3-D determination of the two fields is possible by scanning the experimental tank with the laser sheet (e.g., Ciofalo *et al.*, 2003). The velocity field can also be obtained in a volume by stereoscopy, whereby two cameras record the particle displacements from different angles.

### 3.3.4.3 Stress and strain rate

Stress and strain rate can be determined from the displacement and velocity fields measured by the methods just described; indeed, this is the most common method.

However, one can also use optical 'streaming birefringence'. This effect, already known to Mach (1873) and Maxwell (1873), is the property of certain liquids to become birefringent under the action of shear forces in a flow. Such fluids consist mainly of elongated and deformable molecules (e.g., polymers) or contain elongated, solid, crystal-like particles in solution. At rest, these particules are randomly distributed and the fluid is optically isotropic. The shear forces in a flow cause the particles to align in a preferential direction, which renders the fluid anisotropic. This phenomenon is also well known in solids (and used in photoelasticity), and in the mantle is responsible for seismic anisotropy. Light propagation in such a medium is directionally dependent: an incident light wave separates in the birefringent liquid into two linearly polarized components whose planes of polarization are perpendicular to each other, and which propagate with different phase velocities. Different values of the refraction index are therefore assigned to the two components. They are out of phase when leaving the birefringent liquid, and this difference in optical path can be measured by interferometry.

Although theoretically most fluids should show this effect, the best results have been obtained by using Milling Yellow dye dissolved in aqueous solutions (for its physico-chemical properties; see Swanson and Green (1969) and Pindera and Krishnamurthy (1978)). Milling Yellow solutions can be strongly non-Newtonian, depending on concentration and temperature. One special concentration is particularly interesting, since its flow curve is similar to that of human blood (Schmitz and Merzkirch, 1981).

Interferometric fringes, 'isochromates', can then be obtained with a Max–Zender interferometer or with a polariscope (**Figure 16**). However, a theoretically based flow-optic relation that would allow

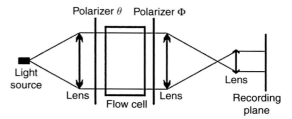

**Figure 16** Streaming birefringence. Polariscope setup. From Merzkirch W (1989) Streaming birefringence. In: Yang W-J (ed.) *Handbook of Flow Visualization*, pp. 177–180. Taylor and Francis.

quantitative determination of the flow from the refraction index distribution does not yet exist. Empirical relations concern only small values of the maximum strain rate, or small Reynolds numbers (creeping flow). For 2-D or axisymmetric situations, a linear relationship between the refractive index differences and the strain rate has been successfully used to determine the flow in pipes of varying cross-sections (e.g., Schmitz and Merzkirch, 1981). Horsmann and Merzkirch (1981) developed a flow-optic relationship which applies to the general 3-D case. More details on the technique can be found in Merzkirch (1989).

### 3.3.5 Surface Displacement

On Earth, hot spots are usually located on top of broad topographic swells (*see* Chapter 9). Experiments on the interaction between a rising buoyant plume and the lithosphere, commonly proposed to explain hot-spot volcanism, therefore involved measurement of the upper surface displacement. Two techniques were used. Olson and Nam (1986) measured the amplitude of the surface topography above the axis of a rising buoyant drop using a high-frequency induction coil proximity probe (**Figure 17(a)**). They were able to follow the temporal evolution of the surface elevation with a sensitivity of 0.02 mm. Using a Michelson interferometer, Griffiths *et al.* (1989) could reconstruct from the fringe pattern both the shape and width of the swell, in addition to its height (**Figure 17(b)**). Fringes in the interferogram are topographic contours with a vertical distance apart of half the laser wavelength (i.e., 0.633 μm for a He–Ne laser). The surface height at any position of the swell is therefore given by the number of fringes.

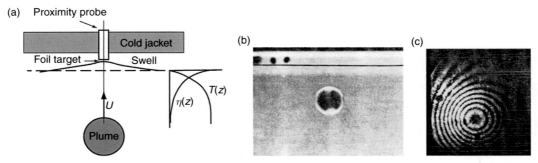

**Figure 17** Measurements of surface displacements above a rising drop. (a) Proximity probe. (b) Side view of the rising drop. (c) Holographic interferometry on the free surface. The center of the interference fringes corresponds to the surface highest point and is centered on the drop vertical axis. (a) Adapted from Olson and Nam (1986). (c) From Griffths RW, Gurnis M, and Eitelberg G (1989) Holographic measurements of surface topography in laboratory models of mantle hotspots. *Geophysical Journal* 96: 1–19.

## 3.4   Rayleigh–Taylor Instabilities

Whenever light fluid underlies a heavier fluid in a field of gravity, the interface between them is inherently unstable to small perturbations (**Figure 1(a)**). This is the classic form of the so-called Rayleigh–Taylor instability (hereafter RTI), first studied theoretically by Rayleigh (1883) and later by Taylor (1950) (*see* Chapter 4). RTIs have been used to model a number of geophysical processes, including the formation and distribution of salt domes (e.g., Nettleton, 1934; Selig, 1965; Biot and Ode, 1965; Ribe, 1998), the emplacement of gneissic domes and granitic batholiths (Fletcher, 1972), instability of continental lithosphere beneath mountain belts (Houseman and Molnar, 1997), subduction of oceanic lithosphere (Canright and Morris, 1993), the temporal and spatial periodicity of volcanic activity in a variety of geological settings, namely island arcs (Marsh and Carmichael, 1974; Fedotov, 1975; Marsh, 1979; Kerr and Lister, 1988), continental rifts (Mohr and Wood, 1976; Bonatti, 1985; Ramberg and Sjostrom, 1973), Iceland (Sigurdsson and Sparks, 1978), and mid-ocean ridges (Whitehead *et al.*, 1984; Shouten *et al.*, 1985; Crane, 1985; Whitehead, 1986; Kerr and Lister, 1988), segregation and mixing in the early history of Earth's core and mantle (Jellinek *et al.*, 1999), and the initiation of instabilities deep in the mantle (Ramberg, 1972; Whitehead and Luther, 1975; Stacey and Loper, 1983; Loper and Eltayeb, 1986; Ribe and de Valpine, 1994; Kelly and Bercovici, 1997; Bercovici and Kelly, 1997).

Laboratory experiments have proved to be powerful tools for studying the development and morphology of RTI. We focus here on the case where the density field is nondiffusing, and the fluids highly viscous, so that the effects of compositional diffusion, inertia, and surface tension, can be neglected. Then, the nature of overturning only depends on the geometry of the boundaries, the viscosities, and densities of the fluids and the layer depths. A large number of experimental studies using a variety of materials have been performed. For the experiments using the most viscous fluids (e.g., silicone putties), the effective gravity was enhanced up to 800*g* by spinning in a centrifuge (e.g., Ramberg, 1967, 1968; Jackson and Talbot, 1989). Early experiments with putty and other non-Newtonian fluids have been extensively photographed and compared with geological formations (esp. salt domes) by Nettleton (1934, 1943), Parker and McDowell (1955), and Ramberg (1967, 1972). However, there was no intercomparison between these laboratory experiments and theory due to the unknown rheology of the laboratory materials. We shall restrict the discussion below to experiments with Newtonian fluids, easier to control.

Whitehead and Luther (1975) performed experiments under normal gravity and showed that the morphology of the rising pockets of light fluid depends on the viscosity ratio between the two fluids: more viscous instabilities develop in finger-like 'diapirs', while less viscous instabilities develop into mushroom-shaped 'cavity plume' (**Figure 18**). The latter case is of particular interest for the initiation of instabilities deep in the mantle since hot mantle material is probably also less viscous (e.g., Stacey and Loper, 1983). The characteristic spacing $\lambda$ and growth rate $\sigma$ of RTI strongly depend on the viscosity of the materials (Selig, 1965; Whitehead and Luther, 1975; Canright and Morris, 1993; Bercovici

(a)                    (b)

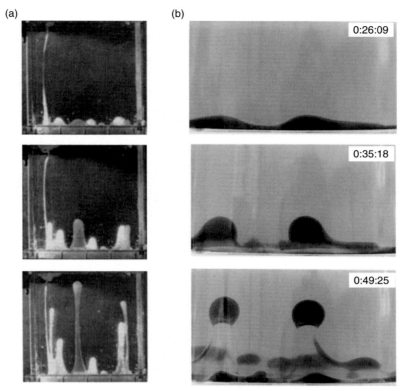

**Figure 18** Development of Rayleigh–Taylor instabilities. (a) Fingers develop when the instabilities are more viscous than the ambient fluid; (b) cavity plumes when the instabilities are less viscous. (a) From Whitehead JA and Luther DS (1975) Dynamics of laboratory diapir and plume models. *Journal of Geophysical Research* 80: 705–717. (b) From Bercovici D and Kelly A (1997). Nonlinear initiation of diapirs and plume heads. *Physics of the Earth and Planetary Interiors* 101: 119–130.

and Kelly, 1997; Ribe, 1998), as well as on the geometry of their source (Kerr and Lister, 1988; Lister and Kerr, 1989; Wilcock and Whitehead, 1991). For a thin denser lower layer of thickness $h_d$ and viscosity $\nu_d$ underlying an infinite layer of viscosity $\nu_m$, they scale as:

$$\lambda \sim h_d \; F(\gamma) \qquad [5a]$$

and

$$\sigma \approx \frac{\Delta \rho g h_d}{\rho \nu} G(\gamma) \qquad [5b]$$

where $g$ is the gravity acceleration, $\gamma = \nu_d/\nu_m$ and $v = \max(\nu_d, \; \nu_m)$. Equation [5b] shows that the instability growth rate is limited by the fluid with the larger viscosity. The functions $F$ and $G$ depend on the viscosity ratio and the geometry of the system. For a thin plane more viscous layer, linear stability analysis shows that $F(\gamma) \sim \gamma^{1/5}$ and $G(\gamma) \sim 1$, while for a thin plane less viscous layer, $F(\gamma) \sim \gamma^{-1/3}$ and $G(\gamma) \sim \gamma^{1/3}$ (Whitehead and Luther, 1975; Selig,

1965). Laboratory measurements agree with these scalings (**Figure 19**), which also confirms the ability of the linearized equations to predict the dominant wavelength of the developed instability (Whitehead and Luther, 1975). However, for instabilities developing from a cylinder in a denser fluid, the RTI spacing and growth rate still depend on the cylinder size but are independent of the viscosity ratio (**Figure 19**; Kerr and Lister, 1988; Lister and Kerr, 1989). For a thin layer of light fluid embedded in denser fluid, linear stability again predicts a fastest growing wavelength much greater than the thickness of the low-viscosity layer, which agrees with the experimental results (**Figure 19**; Lister and Kerr, 1989; Wilcock and Whitehead, 1991). Moreover, the displacement of the lower interface can reach a significant fraction of that of the upper interface. This leads to entrainment of the substratum into the light diapir (Ramberg, 1972; Wilcock and Whitehead, 1991). For very thin layers, a second instability is also observed at a scale much greater than the

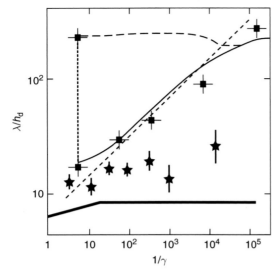

**Figure 19** RTI wavelength normalized by the thin layer depth as the inverse of the viscosity ratio. The dashed line represents the infinite plane layer solution $l/h_d \sim \gamma^{-1/3}$ (eqn [5a]). Dots are experimental data for a thin layer embedded in more viscous fluid; the dotted line connects observations from an experiment exhibiting a bimodal wavelength instability; characteristic wavelengths predicted by the numerical model are shown in dot-dashed and thin solid lines (Wilcock and Whitehead, 1991). Stars represent observations for a cylinder and the thick solid line shows the theoretical prediction (Kerr and Lister, 1988).

characteristic wavelength; it arises from perturbations predominantly involving thickening rather than translation of the buoyant layer (Wilcock and Whitehead, 1991).

Bercovici and Kelly (1997) described both theoretically and experimentally the evolution of the interface from the initial linear instability to the formation and lift-off of cavity plumes. This involves nonlinear feedback mechanisms between the growth of the instability and the draining of the low-viscosity channel, whereby the proto-plume can stall for a long period of time before it separates and begins its ascent. The radius of the cavity plume $a$ and its trailing conduit $R$ now scale as

$$a \sim h_d \gamma^{-2/9} \quad \text{and} \quad R \sim h_d \qquad [6]$$

which shows that the radius of the head will be significantly larger than the conduit radius.

After RTI lift-off, the plumes may not always continue to rise along a vertical line as solitary bodies. Depending on their relative size, and the vertical and horizontal separation between them, they can become attracted to one another, clustering

and even coalescing (Kelly and Bercovici, 1997). Combining theoretical, numerical, and laboratory results, Manga (1997) further showed that attraction between plumes was enhanced by plumes' ability to deform (*see* Chapter 4).

Once the instabilities have started, the light fluid will gather at the top of the box, where it will remain since the density configuration is now stable. An initially unstable compositional stratification will therefore go unstable only once. In that respect, the Rayleigh–Taylor instability appears as an essential process of segregation of two intermingled materials.

But the 'one-shot' character of RTI can be altered when the lighter fluid is continuously released from the bottom through a diffusive interface (Loper *et al.*, 1988; Jellinek *et al.*, 1999) or through melting (Jellinek *et al.*, 1999). By introducing a silk membrane on the interface between a thin water layer and a heavy viscous corn syrup layer, Loper and McCartney (1986) and Loper *et al.*, (1988) observed intermittent RTI generation. When the lighter fluid is less viscous, the RTI morphology is 3-D as already described, whereby it is sheetlike when the lighter fluid is more viscous (Jellinek *et al.*, 1999). Moreover, for high release flux (or high Reynolds number), the RTI can entrain on its way up a significant amount of the ambient denser fluid, and the final stratification is significantly altered by mixing compared to the initial stage (Jellinek *et al.*, 1999).

More commonly, episodicity can be sustained indefinitely when the density field is diffusing. This is the case in thermal convection, to which we turn now.

## 3.5    Simple Rayleigh–Benard Convection Studies

In buoyancy-driven flows, the exact governing equations are intractable. Some approximation is needed, and the simplest one which admits buoyancy is the Oberbeck–Boussinesq approximation (*see* Chapter 2). It assumes that

1. In the equations for the rate of change of momentum and mass, density variations may be neglected except when they are coupled to the gravitational acceleration in the buoyancy force. In practice, it requires $\alpha \Delta T \ll 1$.
2. All other fluid properties can be considered as constant over the experimental cell.
3. Viscous dissipation is negligible.

In this section, we focus on cases where the Oberbeck–Boussinesq approximation is valid. Since mantle material has a high $Pr$, emphasis will be on results using high $Pr$ fluids. The convective flows have usually been determined using planforms visualized by shadowgraph, heat flow measurements, or time series of temperature measurements within the experimental cell.

### 3.5.1 Convection at Relatively Low $Ra(Ra_c \leq Ra \leq 10^6)$

As predicted by linear stability theory (Rayleigh, 1916; Chandrasekhar, 1961), convection should set in above a critical value $Ra_c$, which depends on the boundary conditions (*see* Chapter 4). **Figure 20** shows a regime diagram of thermal convection as a function of Rayleigh and Prandtl numbers when both the top and bottom boundaries are rigid (zero velocity on the boundary) and isothermal. Rayleigh–Bénard convection exhibits a sequence of transitions toward chaos as $Ra$ increases. These transitions can also be seen on the measurements of the heat flow $Q_S$ extracted by the system (**Figure 21**), compared to the conductive heat flux $k\Delta T/H$. Their ratio defines the Nusselt number

$$Nu = \frac{Q_S}{k.\Delta T/H} \qquad [7]$$

It is equal to unity for conduction and exceeds unity as soon as convection starts. It was by measuring $Q_S$ as a function of $\Delta T$ that Schmidt and Silverton (1935) determined experimentally for the first time the critical Rayleigh number for the onset of convection. They confirmed the predicted theoretical value of 1708 with an accuracy better than 10%. Among others, Silveston (1958) extended the measurements to a wider range of fluids (**Figure 21**) and showed that, as predicted, the value of $Ra_c$ does not depend on $Pr$. However, this is not the case for the subsequent transitions between patterns at higher $Ra$ (**Figure 20**), for which there is a clear difference between low $Pr$ fluids (e.g., water, air, mercury, compressed gases like helium) and fluids with $Pr \geq 100$.

For $Pr \geq 60$, laboratory experiments performed with silicone oils revealed the following sequence of patterns in the convective planform as $Ra$ increases (Busse and Whitehead, 1971; Krishnamurti, 1970a, 1970b; Richter and Parsons, 1975): 2-D rolls (**Figure 22(a)**) are stable only for Rayleigh numbers less than 13 times critical. For $Ra$ greater than this, a second set of rolls perpendicular to the original ones grows, and the new planform is rectangular. In this so-called 'bimodal' regime (**Figure 22(b)**), the upwellings assume the geometry of two adjacent sides of a square, while the downwellings form the

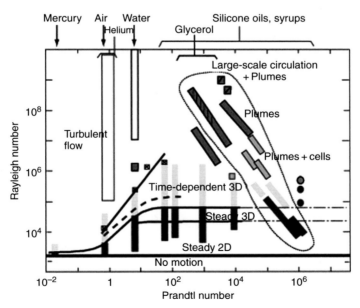

**Figure 20** Regimes diagram of thermal convection. Data compiled from Krishnamurti (1970a, 1970b, 1979), Whitehead and Parsons (1978), Nataf *et al.* (1984), Guillou and Jaupart (1995), Zhang *et al.* (1997), Manga and Weeraratne (1999), and Xi *et al.* (2004). The square bars show the range of experiments performed. The experiments performed with a viscosity contrast within the tank greater than 10 are enclosed in the dotted line. The three round dots show experiments in centrifuge by Nataf *et al.* (1982).

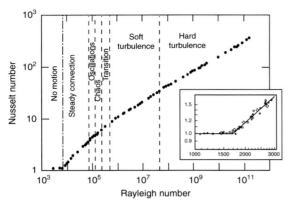

**Figure 21** Nusselt number as a function of Rayleigh number mesured in He. The different transitions observed are indicated. Adapted from Heslot F, Castaing B, and Libchaber A (1987). Transitions to turbulence in helium gas. *Physical Review A* 36: 5870–5873. The inset shows the data from Silveston (1958) around the convection onset.

(a)

(b)

(c)

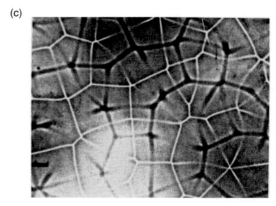

**Figure 22** Shadowgraphs of the different stable cellular patterns in isoviscous convection at high *Pr*: (a) rolls, $Ra = 20 \times 10^3$; (b) bimodal, Ra = $64 \times 10^3$ Instabilities of convection rolls in high Prandtl number fluid. *Journal of Fluid Mechanics* 47: 305–320); (c) spoke, $Ra = 1.4 \times 10^5$. (a) From Busse FH and Whitehead JA (1971) Instabilities of convection rolls in high Prandtl number fluid. *Journal of Fluid Mechanics* 47: 305–320 and (c) From Richter FM and Parsons B (1975). On the interaction of two scales of convection in the mantle. *Journal of Geophysical Research* 80: 2529–2541.

other two sides of the square. At *Ra* about 100 times critical, a new planform develops through a collective instability: the corners of the bimodal rectangles join, and the resulting planform is time dependent and resembles spokes radiating out of the central upwellings or downwellings (multimodal regime, **Figure 22(c)**). The two transitions in patterns are correlated with kinks in the *Nu-Ra* curves (e.g., Malkus, 1954; Willis and Deardorff, 1967; Krishnamurti, 1970a, 1970b), and the *Ra* value at each transition does not vary systematically with *Pr* (Busse and Whitehead, 1971; Krishnamurti, 1970a, 1970b). However, the transition toward the time-dependent spoke pattern is not sharp, but there is a domain in *Ra* where both time-dependent and stable cells can coexist. Hence, the pattern will depend strongly on the initial conditions. Under controlled initial conditions, Whitehead and Parsons (1978) observed stationary bimodal convection up to $Ra = 7.6 \times 10^5$, but an oscillating spoke pattern if starting from random conditions. At $Pr = 10^6$ (using a centrifuge), Nataf *et al.*, (1984) observed steady motions at $Ra = 0.95 \times 10^5$, but time-dependent instabilities at $Ra = 6.3 \times 10^5$. When the Oberbeck–Boussineq approximation applies, cold and hot currents have symmetric characteristics, even for time-dependent spokes. The horizontally averaged thermal structure is symmetric relative to the temperature at the mean depth (**Figure 23**).

For intermediate and low *Pr*, the pattern for a Boussinesq fluid always consists of straight rolls above but close to onset. However, the bifurcation toward stationary bimodal flow is replaced by an oscillatory secondary instability (e.g., Busse, 1978, 1989), and quite rapidly by chaos, then turbulence, as *Ra* increases further (**Figure 21**).

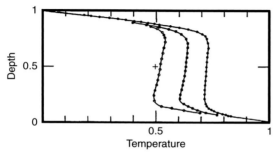

**Figure 23** Nondimensional temperature as a function of depth. (o), Silicone oil, $\gamma = 3$, $Ra = 2.16 \times 10^5$; L-100, $\gamma = 22$, $Ra = 1.36 \times 10^5$; L-100, $\gamma = 750$, $Ra = 4.95 \times 10^5$. In all but the last case, data near the boundaries have been omitted. From Richter FM, Nataf H-C., and Daly SF (1983). Heat transfer and horizontally averaged temperature of convection with large viscosity variations. *Journal of Fluid Mechanics* 129: 173–192.

### 3.5.2 Patterns and Defects

The study of convection planforms and pattern selection is a very rich and fundamental field in itself, and has motivated a large body of theoretical (*see* Chapter 4) and experimental work. Recent reviews on the subject include papers by Busse (1978, 1989), Geitling (1998), Bodenschatz *et al.* (2000), and Ahlers (2005). Experiments are usually done in thin fluid layers with large aspect ratios (up to 100), and visualized by shadowgraph. We will mention here only a few striking results.

#### 3.5.2.1 Generation of an initial prescribed pattern

To determine the domain of stability of the different patterns and follow the formation and evolution of the defects, the convective pattern must be initiated in a well-controlled way. The method developed by Chen and Whitehead (1968) consists in placing a grid made up of alternating blocked and clear areas over the top transparent channel. A 300–500 W incandescent lamp then is shone down through the pattern, so that the fluid lying below is slightly radiatively heated in the desired pattern. After a certain time ($\sim$1 h), the temperature of the top bath is decreased and the temperature of the bottom bath increased at equal rates (typically $2°\text{min}^{-1}$) until the desired Rayleigh number is reached. It is important to change the baths at the same rates in order to exclude asymmetric or hexagonal modes of flow. Convection then starts with the location of the rising limbs of the cells being controlled by the places which were heated.

#### 3.5.2.2 Patterns for $Pr \geq 100$

Chen and Whitehead (1968) and Busse and Whitehead (1971) studied experimentally the stability of rolls for a silicone oil with $Pr = 100$. Rolls of known aspect ratio (roll width over tank thickness) were initiated, and allowed to evolve. The results were compared with the theory of Busse (1967a) which predicts a balloon-shaped region of stable rolls (**Figure 24**). The zigzag instability (**Figure 25(a)**) occurs when the aspect ratio is greater than 1 and tends to reduce it to 1. The cross-roll instability (**Figure 25(b)**) occurs when the aspect ratio is much less than 1, or when the Rayleigh number exceeds about $2 \times 10^4$: the initial rolls are replaced by two sets of rolls with different wavelength at right angles to each other. The pattern then evolves toward one set of rolls with increased aspect ratio (in the first case) or to the bimodal mode (in the second case). When two sets of rolls of

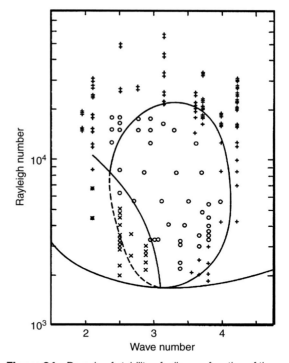

**Figure 24** Domain of stability of rolls as a function of the wave number and the Rayleigh number ('Busse's balloon'). Solid lines are the theoretical predictions (Busse, 1967a). Experimental results by Busse and Whitehead (1971): (o), stable rolls; (x), zigzag instability; (+), cross-roll instability leading to rolls; (++), cross-roll instability leading to bimodal convection; (+++), cross-roll instability inducing transient rolls with subsequent local processes. From Busse FH and Whitehead JA (1971). Instabilities of convection rolls in high Prandtl number fluid. *Journal of Fluid Mechanics* 47: 305–320.

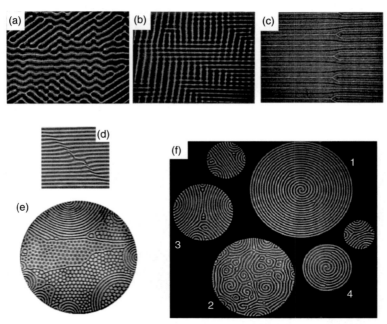

**Figure 25**    Patterns and defects. (a) Zig-zag, (b) cross-roll, (c) pinching instabilities, (d) skewed-varicose instability; (e) coexisting up- and downflow hexagons together with rolls; (f) two-armed spiral patterns in cells 1 and 4. Spiral defect chaos in cell 2. (a, b, c) From Busse FH and Whitehead JA (1971). Instabilities of convection rolls in high Prandtl number fluid. *Journal of Fluid Mechanics* 47: 305–320; (d) From Plapp BB (1997) *Spatial-Pattern Formation in Rayleigh–Bénard Convection.* PhD Thesis, Cornell University, Ithaca, New York; (e) Assenheimer M, Steinberg V (1996) Observation of coexisting upflow and downflow hexagons in Boussinesq Rayleigh–Bénard convection. *Physical Review Letters* 76: 756–759; (f) from Plapp BB, Egolf DA, Bodenschatz E, and Pesch W (1998). Dynamics and selection of giant spirals in Rayleigh-Bénrad convection. *Physical Review Letters* 81: 5334–5337.

different wavelengths are combined, local rearrangement occurs through the pinching instability (**Figure 25(c)**). The agreement between the experimental results and the theory is quite remarkable (**Figure 24**; *see also* Chapter 4).

### 3.5.2.3 *Patterns for Pr < 100*

Patterns and defects have been extensively studied more recently, using primarily compressed gases as the working fluid (small cell sizes can be used and *Pr* can be tuned between 0.17 and 30), sensitive shadowgraph visualization coupled to digital image analysis, and quantitative heat flow measurements (for a review, see Bodenschatz *et al.* (2000)). **Figure 25** shows examples of the patterns observed. Above but close to onset, the pattern for a Boussinesq fluid always consists of straight rolls, possibly with some defects induced by the sidewalls. At $Pr \sim 1$, the skewed-varicose instability arises to transform rolls into rolls of larger wavelength (**Figure 25(d)**). When non-Boussinesq conditions prevail, hexagons develop instead of rolls (**Figure 25(e)**). Further above onset, spiral-defect chaos (Morris *et al.*, 1993) occurs in systems with *Pr*

of order 1 or less (**Figure 25(f)**). This state is a bulk property and is not induced by side-walls. On the other hand, spirals occur in circular cells (**Figure 25**).

### 3.5.3    Thermal Convection at High Rayleigh Numbers($10^5 \leq Ra$)

At high *Ra*, the convective pattern becomes disorganized and flow is generated locally by thermal boundary layer (TBL) instabilities (Elder, 1968), which develop into plumes (**Figures 7** and **8**). Their thermal anomalies are easily recorded by thermocouples located within the fluid (e.g., Townsend, 1957; Deardorff *et al.*, 1969; Sparrow *et al.*, 1970; Castaing *et al.*, 1989; Davaille and Jaupart, 1993; Weeraratne and Manga, 1998; Lithgow-Bertelloni *et al.*, 2001; **Figure 28**).

### 3.5.3.1    *Thermal boundary layer instabilities*

Plume generation in fully developed convection is a cyclic process in which the TBLs grow by thermal diffusion, become unstable when they reach a critical thickness $\delta_c$ such that $Ra(\Delta T, \delta\square) = Ra_c$, and then empty themselves rapidly into plumes, at which point the cycle begins again (**Figure 26**). The characteristic timescale $\tau_c$ for this process is the time required for the growing TBL to become unstable, and is (Howard, 1964)

$$\tau_c = \frac{H^2}{\pi\kappa}\left(\frac{Ra_c}{Ra(H, \Delta T)}\right)^{2/3} \qquad [8]$$

Because $Ra \sim H^3$ (eqn [1]), $\tau_c$ is independent of the layer depth $H$: the two TBLs do not interact anymore. This phenomenological model (Howard, 1964) and the scaling [8] is confirmed by laboratory experiments (**Figures 26** and **27**; e.g., Sparrow *et al.*, 1970; Davaille and Jaupart, 1993; Davaille, 1999a; Manga *et al.*, 2001; Davaille and Vatteville, 2005). Experiments further show that the typical spacing between plumes is 3–6 times $\delta_c$ (e.g., Sparrow *et al.*, 1970; Tamai and Asaeda, 1984; Asaeda and Watanabe, 1989; Davaille *et al.*, 2002; Jellinek and Manga, 2004). Moreover, the plumes seem connected

by a network of 'ridges' near the base of the layer (Tamai and Asaeda, 1984; Asaeda and Watanabe, 1989). This morphology was confirmed by detailed numerical studies (e.g., Houseman, 1990; Christensen and Harder, 1991; Tackley, 1998; Trompert and Hansen, 1998a; Parmentier and Sotin, 2000). Plumes have a head-and-stem structure, and both head and stem diameters decrease with increasing $Ra$ (Lithgow-Bertelloni *et al.*, 2001). Plumes can eventually detach from the TBL once they have drained it (**Figure 26**). They are transient features with a lifetime comparable to $\tau_c$ (Davaille and Vatteville, 2005). Since the two TBLs do not interact, the global heat flow should not depend on $H$ anymore, implying

$$Nu \sim Ra^{1/3} \qquad [9]$$

This scaling was verified by Globe and Dropkin (1959) and Guillou and Jaupart (1995) using silicone oils for $Ra$ up to $Ra \sim 2 \times 10^7$. Goldstein *et al.*, (1990), who employed an electrochemical analog to

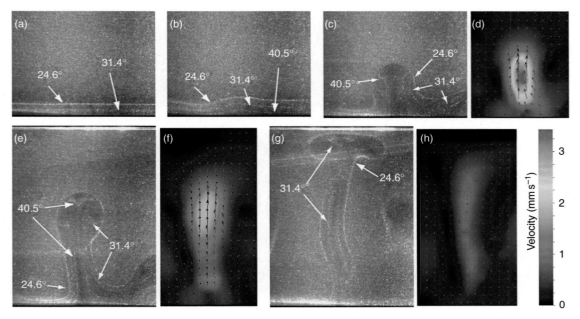

**Figure 26**  Thermal boundary layer instabilities in a layer of sugar syrup, initially at 21°C and suddenly heated from below at 53°C ($Ra = 1.7 \times 10^6$). (a) Negative of the picture taken at $t = 300$ s. The isotherms appear as white lines. The TBL is growing by conduction from the lower boundary. (b) $t = 400$ s. The TBL becomes unstable. (c) $t = 460$ s. A thermal plume, outlined by the 24.6°C isotherm, rises from the TBL, and the TBL begins to shrink. (d) Corresponding velocity field deduced from PIV. The color background represents the velocity magnitude. (e) $t = 500$ s. The plume is well developed and the TBL is emptying itself in it. (f) Corresponding velocity field. (g) $t = 600$ s. The plume head has reached the upper boundary and begins to spread under it. The TBL has nearly disappeared and the conduit is disconnected from its source. (h) Corresponding velocity field. From Davaille A and Vatteville J (2005). On the transient nature of mantle plumes. *Geophysical Research Letters* 32, doi:10.1029/2005GL023029.

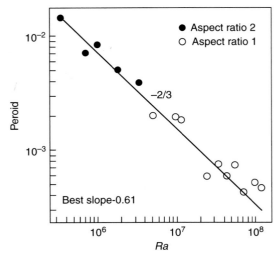

**Figure 27**  Periodicity of the TBL instabilities as a function of Rayleigh number. The solid line represents Howard's scaling in $Ra^{-2/3}$. From Manga M and Weeraratne D (1999). Experimental study of non-Boussinesq Rayleigh–Bénard convection at high Rayleigh and Prandtl numbers. *Physics of Fluids* 10: 2969–2976.

convection to reach $Ra \sim 10^{12}$ with $Pr = 2750$, recovered also (9). This regime of plumes has sometimes been named 'soft turbulence'.

### 3.5.3.2  Soft and hard turbulence at low Pr

Steady-state experiments by Heslot *et al.* (1987) and Castaing *et al.* (1989) in helium ($0.6 \leq Pr \leq 1.7$) indicated that the spectrum of temperature fluctuations changes above $Ra = 4 \times 10^7$, defining a new dynamical regime called 'hard turbulence'. This transition is also seen on the $Nu$–$Ra$ relationship (**Figure 21**), which follows the 1/3 power law [9] for $Ra = 4 \lesssim 10^7$ (the 'soft turbulence' regime), but changes to a 2/7 power law for $Ra \geq 4 \times 10^7$. On the other hand, Katsaros *et al.* (1977) recovered the 1/3 power law [9] for $Ra$ up to $10^9$ during the transient cooling from above of a water tank. Castaing *et al.* (1989) suggested that, in steady state, the dynamics of boundary layer instabilities were affected by the establishement of a large-scale circulation over the whole tank. Since the experiments were done in a container with aspect ratio 1, the large-scale circulation could have been caused by the wall effects. However, the establishment of a large-scale circulation had already been observed by Krishnamurti and Howard (1981) for larger aspect ratio. Since these experiments, much theoretical and experimental work has been done to confirm or disprove, extend,

and explain the new regime (for recent reviews, see Siggia (1994), Grossmann and Lohse, (2000); Chavanne *et al.* (2001)). It was found that a large-scale forced flow can decrease the convective heat flow (Solomon and Gollub, 1991). For aspect ratios smaller than 1, several regimes seem to coexist, due to the geometry of the large-scale circulation which can switch from one cell to several superimposed cells (e.g., Chilla *et al.*, 2004). In the search for the ultimate turbulent regime, $Ra \sim 10^{17}$ has now been obtained using cryogenic He (Niemela *et al.*, 2000). However, from $10^6$ to $10^{17}$, these authors found neither a 2/7 nor a 1/3 power law, but a 0.30 power-law. Following Kraichman (1962), Grossmann and Lohse (2000) proposed a systematic theory to predict all the different asymptotic regimes as a function of $Ra$ and $Pr$. To be able to compare accurately theory and experiments, special care is now taken to estimate heat losses from the sides (e.g., Alhers, 2001) and from the finite conductivity of the bottom and top plates (e.g., Brown *et al.*, 2005; see Section 3.2.3), as well as the $Pr$ variations. However, it seems that the 'ultimate' regim has not been found yet.

### 3.5.3.3  Large-scale (cellular) circulation and plumes at high Pr

To obtain high Rayleigh number in high $Pr$ fluids requires large temperature differences (typically $\geq 20\,^{\circ}$C) and large depths (typically $\geq 20$–30 cm). The aspect ratios of these experiments are therefore small (between 1 and 3 at most). Even for fluids with very weak temperature dependence of viscosity (e.g., silicone oils), the viscosity can no longer be treated as constant (**Figures 2 and 20**). Its value is therefore taken at the mean of the boundary temperatures, ($T_{bot} + T_{top}$)/2.

Weeraratne and Manga (1998), Manga and Weraratne (1999), and Manga *et al.*, (2001) determined the different styles of time-dependent convection from temperatures recorded on the tank mid-plane (**Figure 28**). For steady flow ($Ra \leq 10^5$), these remain constant. When the flow first becomes unsteady, temperature fluctuations with large amplitude and long period appear (**Figure 28(a)**). As $Ra$ increases, the amplitude of long-period fluctuations decreases and small short-period fluctuations appear (e.g., $Ra \sim 5 \times 10^6$, **Figure 28(b)**). At still higher $Ra$, the large-amplitude fluctuations disappear completely and only the short-period fluctuations remain (**Figure 28(c)**). As mushroom-shaped plumes were

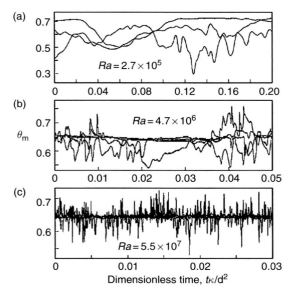

**Figure 28** Temperature fluctuations $\theta_m = (T-T_{top})/\Delta T$ at three thermocouples located at mid-height in a tank of convecting corn syrup. Time is normalized by the diffusive timescale $d^2/k$. (a) Unsteady convection dominated by large-scale flow ($Ra = 2.5 \times 10^5$). (b) Increased plume activity along with large-scale flow ($Ra = 4.7 \times 10^6$). (c) Plume dominated convection with short period fluctuations ($Ra = 5.5 \times 10^7$). From Weeraratne D and Manga M (1998) Transitions in the style of mantle convection at high Rayleigh numbers. *Earth and Planetary Science Letters* 160: 563–568.

observed in this regime, the small-scale fluctuations were interpreted as their signature, and the long-period signal as the signature of large-scale (cellular) flow. From $Ra \sim 10^7$ to $2 \times 10^8$, no large-scale circulation was observed. Such circulation was also absent from transient experiments where the fluid was continuously heated from below (Lithgow-Bertelloni et al., 2001) or cooled from above (Davaille and Jaupart, 1993), and the other boundaries were insulated. Using glycerol (Zhang et al., 1997, 1998) and dipropylene glycol (Xi et al., 2004) in aspect ratio 1 cells, a different picture was found for $Ra \sim 10^8$–$10^9$: a large-scale circulation develops which entrains the plumes as soon as they have formed. Measuring the velocity field by PIV, Xi et al. (2004) demonstrated that the flow becomes organized because of the plumes, which interact through their velocity boundary layers typically as soon as they have risen a distance comparable to their spacing: plume clustering then occurs, as in Rayleigh–Taylor instabilities (Bercovici and Kelly, 1997) and in experiments on the interaction between two thermal plumes (Moses

et al., 1993). In this regime where plumes and large-scale circulation coexist, the exponent of the $Nu$–$Ra$ power law becomes again closer to 2/7 (Xi et al., 2004). It would now be interesting to run the same experiments with larger aspect ratio to determine if the large-scale circulation always spans the whole tank, or if it has its own wavelength.

### 3.5.4 Convective Regime in an Isoviscous Mantle

From **Figure 20** and eqn [1], one can determine the convective regime of a mantle layer as a function of its thickness and average viscosity (**Figure 29**). Plumes will develop in the upper mantle (of thickness 660 km) only if its viscosity is lower than $10^{21}$ Pa s. For an average mantle viscosity of $10^{22}$ Pa s, plumes will be generated only if the mantle layer thickness exceeds 2000 km. Convective motions originating at the CMB and developing over the whole mantle thickness should therefore take the form of plumes (Parmentier et al., 1975; Loper and Stacey, 1983). However, those plumes are transient phenomena. According to eqn [8], $\tau_c \sim 10$ My, 40 My and 200 My for dynamic viscosities $\nu_m = 10^{19}$ Pa s, $5 \times 10^{20}$ Pa s, and $10^{22}$ Pa s, respectively. Such recurrence times are probably also upper bounds on plume lifetime. The corresponding values of the critical TBL thickness $\delta_c \sim H(Ra_c/Ra)^{1/3}$ for the same viscosities are 31, 62, and 140 km, respectively, which

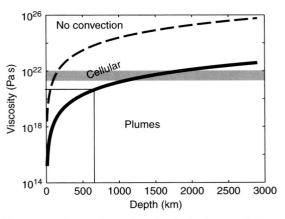

**Figure 29** Convective regime developing in a mantle layer as a function of its thickness and viscosity. The pattern is cellular for $Ra \geq Ra_c = 650$, and in plumes for $Ra \geq 10^6$. In gray the average viscosity inferred for the whole mantle (Kolhstedt, volume 2). The thin gray line delimits the upper mantle. The calculation has been done from eqn [1] taking $\kappa = 10^{-6}$ m$^2$ s$^{-1}$, $\alpha = 2 \times 10^{-5}$ K$^{-1}$ and $\Delta T = 3000$ K.

implies a plume spacing 100–840 km in the mantle. Given the 2900 km thickness of the mantle, plume clustering could occur (e.g., Schubert *et al.*, 2004). Moreover, from **Figure 26**, we expect that a plume will eventually detach from the hot TBL from which it originated. This implies that the absence of a plume conduit in a tomographic image need not indicate the absence of a plume.

However, if isoviscous convection predicts plumes, it does not predict plates, which are the main signature of convection on the Earth's surface (e.g., Chapters 1 and 2). Part of the difficulty in determining the convective pattern in the mantle resides in the complexity of mantle rheology, which depends on temperature (strongly), pressure, partial melting (strongly), water content (strongly), chemistry and strain rate. The rheology of the lithosphere is especially difficult to quantify since it comprises a brittle skin as well as a viscous lower part. The convective pattern driven by the 'cold' upper-mantle boundary appears to be cellular, although time dependent. Inverting **Figure 29** for this regime, it would therefore indicate an 'effective' mantle viscosity greater than $10^{23}$ Pa s, that is, much lower than the viscosity of the cold lithosphere, but at least an order of magnitude higher than the measured bulk mantle viscosity. This asymmetry in mantle convection could be caused by its mixed heating mode (internal and bottom heating), and by its rheology, in particular the temperature dependence of viscosity. We now turn to the experiments which have been devoted to these two effects.

## 3.6   Temperature-Dependent Viscosity

A strong temperature dependence of the viscosity breaks the symmetry in the convection cell: cold instabilities are now more viscous than the fluid in the bulk of the tank, while the hot instabilities are less viscous. According to what has been observed in RTI (Section 3.4, **Figure 18**), their morphology should change. **Figure 30** shows that this is indeed the case: hot TBL instabilities are mushroom shaped, while cold instabilities become diapirs, which can fold on the bottom plate before spreading (**Figure 30(b)**). Material in the cold TBL also flows less easily and rapidly than in the rest of the fluid, which will diminish the efficiency of convection and the heat transport. The system is now characterized

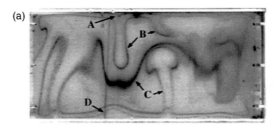

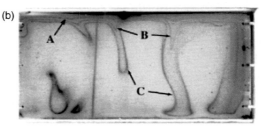

**Figure 30**   Thermal convection in a strongly temperature-dependent viscosity fluid (glucose syrup) cooled from above and heated from below. The black lines are the isotherms (A: 10.5°C; B: 31.4°C; C: 24.6°C; D: 40.5°C). (a) Hot and cold TBL instabilities ($Ra = 4.7 \times 10^{6}$; $\gamma = 116$). (b) Cold TBL for sugar syrup cooled from above only ($Ra = 1.2 \times 10^{7}$; $\gamma = 1580$).

by $Ra$ (based on the viscosity at the mean of the boundary temperatures), $Pr$, and the viscosity ratio $\gamma = \eta(T_{\text{top}})/\eta(T_{\text{bot}})$.

### 3.6.1   Rigid Boundaries

#### 3.6.1.1   Onset of convection
Stengel *et al.*, (1982) studied theoretically and experimentally the onset of convection for $\gamma \lesssim 3\ 400$. $Ra_{\text{c}}$ remains constant at low $\gamma$, increases at moderate $\gamma$ to a maximum for $\gamma \sim 3000$, and then decreases. According to Stengel *et al.* (1982), this occurs because convection begins first in a (hotter) less viscous sublayer where the local Rayleigh number is maximum. Richter *et al.* (1983) increased the range of viscosity ratio to $10^{5}$, and found similar results. Moreover, finite-amplitude disturbances are seen to grow at Rayleigh numbers below critical, as predicted by Busse (1967b) for fluids with weakly temperature-dependent viscosity.

#### 3.6.1.2   Patterns and regimes
Richter (1978) explored the stability of rolls, using glycerine and L100 (polybutene), for $Ra \lesssim 25 \times 10^{3}$ and $\gamma \lesssim 20$. For large $\gamma$, hexagons were found to be the stable pattern. White (1988) used Golden Syrup

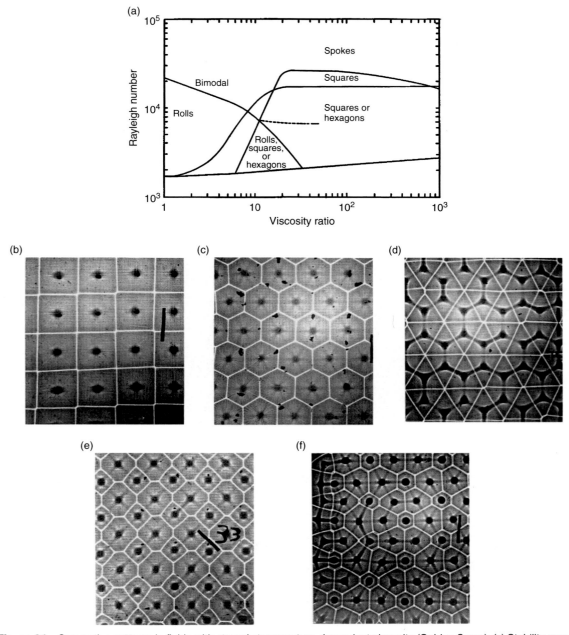

**Figure 31** Convective patterns in fluids with strongly temperature-dependent viscosity (Golden Syrup). (a) Stability map for rolls, hexagons, and squares. There is a large amount of hysteresis where the realized planform depends on what pattern was originally induced. (b) Squares. (c) Hexagons. (d) Triangles. (e, f) Mosaic instabilities. From White DB (1988) The planforms and onset of convection with a temperature-dependent viscosity fluid? *Journal of Fluid Mechanics* 191: 247–286.

to study the stability of patterns for $\gamma \leq 1000$ and $Ra \leq 63 \times 10^3$ (**Figure 31**). Besides rolls and hexagons, a new planform of squares (**Figure 31(b)**) was found to be stable at large viscosity variations, and pattern instabilities developed in a mosaic form (**Figure 31(e)** and **31(f)**). At fixed wave number,

rolls are always unstable as soon as $\gamma > 40$. Moreover, squares break down to a spoke pattern around $Ra = 20\,000$, that is much sooner than in the constant viscosity case (transition at $Ra \sim 10^5$). The wavelength of squares is also observed to decrease with increasing $\gamma$. Temperature profiles through the layer (e.g., **Figure 22**) revealed that this shift is associated with the development of a thick, cold boundary layer, and an increase of the interior temperature (White, 1988; Richter *et al.*, 1983). The temperature profiles allowed the calculation of the convective heat flow and viscosity profile, which showed that most of the temperature variations across the tank occur in the cold TBL, and that the top part of the latter becomes stagnant (zero convective heat flow). Hence, convection develops in a sublayer where the viscosity ratio reaches at most 100 (Richter *et al.*, 1983). This stagnant-lid regime has been observed for *Ra* up to $10^8$ (Davaille and Jaupart, 1993).

Global heat flow measurements were carried out by Booker (1976), Booker and Stengel (1978), and Richter *et al.* (1983). For the entire range of parameters ($10^4 \leq Ra \leq 6 \times 10^5$ and $\gamma < 10^5$), the heat transfer differs little from that of a uniform-viscosity fluid when the Rayleigh number is defined with the viscosity at the mean of the boundary temperatures. The relationship between Nusselt number and supercriticality ($Ra/Ra_c(\gamma)$) is even more remarkable since it is independent of $\gamma$ and indistinguishable from the results of Rossby (1969) for constant viscosity:

$$Nu = 1.46\left(\frac{Ra}{Ra_c(\gamma)}\right)^{0.281} \qquad [10]$$

Subsequent high-quality heat flow measurements by Giannandrea and Christensen (1993) agree with this law. Weeraratne and Manga (1998), Manga and Weraratne (1999), and Manga *et al.* (2001) extended the range of *Ra* up to $2 \times 10^8$ for $\gamma = 1800$. We already described in the previous section the sequence of regimes which they observed for $\gamma = 170$, that is, in absence of a conductive lid (**Figure 20**). Their heat flux data agree with eqn [10] for *Ra* up to $2 \times 10^6$ (**Figure 32**). For higher *Ra* and higher $\gamma$, the heat flux is less than that predicted by [10] and by numerical calculations at infinite *Pr*. This trend was attributed to inertial effects (Manga and Weeraratne, 1999).

### 3.6.1.3  *Characteristics of the stagnant lid regime*

Davaille and Jaupart (1993, 1994) studied transient high *Ra* convection in a layer of Golden Syrup suddenly cooled from above. With this setup, the viscosity contrast across the cold TBL, $\gamma_{TBL}$, ranged between 2 and $10^6$. For $\gamma > 100$, convection developed only in the

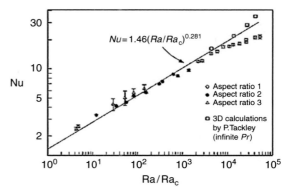

**Figure 32** Nusselt number as a function of Rayleigh number. From Manga M and Weeraratne D (1999) Experimental study of non-Boussinesq Rayleigh–Bénard convection at high Rayleigh and Prandtl numbers. *Physics of Fluids* 10: 2969–2976.

lower part of the TBL, and the upper part remained stagnant (**Figure 33**). The Rayleigh number and all the scaling laws therefore used the viscosity of the well-mixed interior, since it would be meaningless to characterize the flow with a viscosity corresponding to a temperature well within the stagnant fluid. At the onset of convection, the viscosity contrast across the unstable region is about 3; in fully developed convection, it reaches a typical value of 10. The temperature difference $\Delta T_{eff}$ across the unstable part of the lithosphere depends only on the interior temperature $T_m$ and on a temperature scale controlled by the rheology $\Delta T_c$ (e.g., Morris, 1982; Morris and Canright, 1984; Davaille and Jaupart, 1993):

$$\Delta T_c(T_m) = \left|\frac{\eta(T_m)}{(\partial\eta/\partial T)(T_m)}\right| \qquad [11]$$

Laboratory results then give $\Delta T_{eff} = 2.24\Delta T_c$. A necessary condition for the existence of a stagnant lid is that the applied temperature difference exceeds a threshold value equal to $\Delta T_{eff}$. All the dynamics of the system are then determined locally in the unstable part of the TBL. Following once again the phenomenological model of Howard (1964), the heat flux can therefore be written as

$$Q_s = -Ck(\alpha g/\kappa\nu(T_m))^{1/3}\Delta T_c^{4/3} \qquad [12]$$

where the constant $C = 0.47 \pm 0.03$ is determined experimentally. This model also predicts well the convective onset time and the amplitude of the temperature fluctuations (Davaille and Jaupart, 1993).

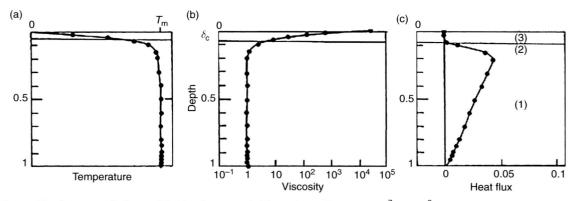

**Figure 33** Structure of a layer of Golden Syrup cooled from above ($Ra = 2.5 \times 10^7$; $\gamma = 10^6$). (a) Horizontally averaged temperature profile. (b) Corresponding viscosity profile. (c) Convective heat flux calculated from the temperature time series. Its value is 0 in the topmost viscous part of the fluid layer (stagnant lid). From Davaille A and Jaupart C (1993) Transient high-Rayleigh number thermal convection with large viscosity variations. *Journal of Fluid Mechanics* 253: 141–166.

Subsequent numerical and experimental work also recovered the functional forms of [11] and [12], with $\Delta T_{\text{eff}} = 2.2–2.6 \times \Delta T_c$ and $C = 0.4–0.9$ (e.g., Grasset and Parmentier, 1998; Trompert and Hansen, 1998a; Dumoulin *et al.*, 1999; Solomatov and Moresi, 2000; Manga *et al.*, 2001; Gonnermann *et al.*, 2004). In the case of steady Rayleigh–Bénard convection at high *Ra*, the temperature difference in the hot TBL has been found to scale as $(1.1–1.3) \times \Delta T_c$ (Solomatov and Moresi, 2000; Manga *et al.*, 2001), which together with [11] allows one to predict the temperature of the well-mixed interior (Solomatov, 1995; Manga *et al.*, 2001). Moreover, since the cold and hot unstable thermal boundary layers have different characteristics, plumes are released from these layers with different frequencies, and cold more viscous downwellings carry a greater temperature anomaly than their hot counterparts. Measurements show that whereas there is a single frequency for cold plume formation, hot plumes form with multiple frequencies and in particular may be triggered by cold sinking plumes (Schaeffer and Manga, 2001).

### 3.6.2 Plate Tectonics in the Laboratory?

Mantle convection is characterized by mobile plates on its top surface, which have no chance to be recovered in experiments with a rigid top boundary since the latter imposes a zero-fluid velocity on the surface. Moreover, 2-D numerical simulations had shown that the heat transport strongly depends on the mobility of the surface boundary layer (Christensen, 1985), which is controlled by the viscosity contrast. Giannandrea and Christensen (1993) and Weinstein

and Christensen (1991) therefore carried out experiments in syrups in a large aspect ratio tank (**Figure 39**) to study the effect of a stress-free boundary (**Figure 4(b)**). They observed two regimes. For $\gamma \leq 1000$, the surface layer is mobile. At $Ra = 10^5$, the morphology of downwellings changes, from the spokes obtained with a rigid boundary, to a dendritic network of descending sheets, with a wavelength more than 3 times that of the spoke pattern (**Figure 34**). The existence of this 'whole-layer' or 'sluggish lid' mode with an increased wavelength is in agreement with earlier predictions (Stengel *et al.*, 1982; Jaupart and Parsons, 1985) and 3-D numerical calculations (Ogawa *et al.*, 1991). The Nusselt number drops by 20% for viscosity contrasts between 50 and 5000. For $\gamma \geq 1000$, a stagnant lid forms on the top of an actively convecting region; and both the Nusselt number and the size of the convection cells are nearly identical for both rigid and free boundaries.

**Figure 35** shows the regime diagram established from experimental, numerical, and theoretical analysis (Solomatov, 1995). With a viscosity contrast across the lithosphere $\sim 10^7$, the Earth should be in the stagnant-lid regime, that is, a one-plate planet like Mars. So, if temperature-dependent viscosity is clearly a key ingredient for plate formation, this ingredient alone is not sufficient to generate 'plate tectonics' convection. To make plate tectonics work, a failure mechanism is needed and several options have recently been proposed (see Bercovici *et al.*, 2000; Bercovici, 2003; *see* Chapters 2 and 5). The simplest is probably pseudo-plastic yielding (Moresi and Solomatov, 1998; Trompert and Hansen, 1998b; Tackley, 2000; Stein *et al.*, 2004; Grigné *et al.*, 2005): moving plates and thin

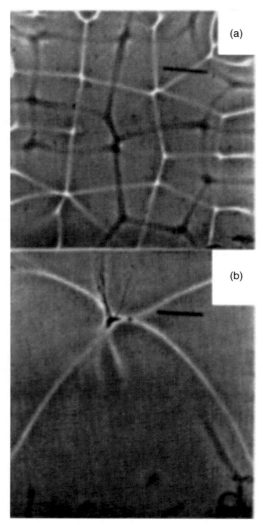

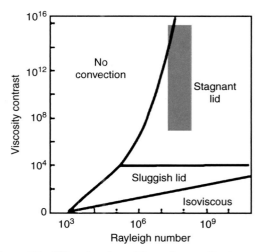

**Figure 35** Different regims of thermal convection in a strongly temperature-dependent fluid with free-slip boundaries. From Solomatov S (1995) Scaling of temperature- and stress-dependent viscosity convection. *Physics of Fluids* 7: 266–274. The gray area is the mantle parameters range.

**Figure 34** Shadowgraphs of the convection planforms found for the case of (a) rigid boundaries and (b) a stress-free upper boundary. From Weinstein SA and Christensen UR (1991) Convection planforms in a fluid with a temperature-dependent viscosity beneath a stress-free upper boundary. *Geophysical Research Letters* 18: 2035–2038.

weak boundaries appear but subduction remain symmetric. Gurnis *et al.*, (2000) also pointed out the importance of lithosphere 'memory': the lithosphere can support dormant weak zones (faults or rifts) over long time periods, but those weak zones can be preferentially reactivated to become new plate boundaries. Another ingredient could be damage (e.g., Bercovici *et al.*, 2001), which introduces some memory in the rheology and allows development over time of weak zones. This approach has been used with some success (e.g., Ogawa, 2003) but a physical understanding of the damage process from the grain to the lithosphere scale

is still lacking. Anyway, experimentalists are still looking for a laboratory fluid presenting the 'right' kind of rheology to allow plate tectonics. However, the study of the stagnant-lid regime is still relevant for the dynamics of plate cooling, to which we turn now.

### 3.6.3 Stagnant-Lid Regime and Lithosphere Cooling

The oceanic lithosphere thickens at it moves away from a mid-ocean ridge (*see* Chapter 7). It has been proposed that cold thermal instabilities develop in its lower part. Such 'small-scale convection' (SSC) was first invoked to explain the flattening of heat flux and bathymetry of the oceans at old ages (Parsons and Sclater, 1977; Parsons and McKenzie, 1978) and later several other phenomena occurring on different time and length scales, such as small-scale (150–500 km) geoid anomalies in the central Pacific (e.g., Haxby and Weissel, 1986) and central Indian (Cazenave *et al.*, 1987) oceans, differences in subsidence rates (Fleitout and Yuen, 1984; Buck, 1987; Eberle and Forsyth, 1995), ridge segmentation (e.g., Sparks and Parmentier, 1993; Rouzo *et al.*, 1995), delamination of the lithosphere under hot spots (Sleep, 1994; Moore *et al.*, 1998; Dubuffet *et al.*, 2000), and patterns of anisotropy beneath the Pacific ocean (Nishimura and Forsyth, 1988; Montagner and Tanimoto, 1990; Davaille and Jaupart, 1994; Ekström and Dziewonski, 1998).

Davaille and Jaupart (1994) applied the scaling laws derived from their experimental results, using for the mantle a Newtonian creep law with activation enthalpy $H$:

$$\nu(T) = \mu_r \exp\left(\frac{H}{RT}\right) \qquad [13]$$

For $H$ varying between 150 and 500 kJ mol$^{-1}$ (the value depends on the creep mechanism and the water content), $\Delta T_{eff}$ ranges between 310°C and 90°C according to [11] and [13]. Since small-scale instabilities develop when the local Rayleigh number based on the characteristics of the unstable part of the lithosphere (temperature, viscosity, thickness) exceeds a critical value, the onset of convection depends only on the thermal structure of the lithosphere, and on the rheology of mantle material. Following Davaille and Jaupart (1994), the onset time can be written as

$$\tau_c = \frac{a}{\kappa}\left(\frac{\alpha g \Delta T_{eff}}{\kappa \nu_m}\right)^{-2/3}$$
$$\times \left[1 + \frac{b}{f(\Delta T/\Delta T_{eff})}\left(\frac{\Delta T}{\Delta T_{eff}} - 1\right)\right]^2 \qquad [14]$$

where $\nu_m$ is the viscosity of the asthenosphere at the mantle temperature $T_m$, $a = 51.84$ and $b = 0.3013$ are two constants determined from the laboratory experiments, and $\Delta T$ is the temperature drop across the lithosphere. The function $f$ depends on the cooling model and for a half-space conductive cooling is given in **Figure 36(a)** . For a thin stagnant lid (or $\Delta T/\Delta T_{eff} < 3$), $f \sim 1$. This approximation applies well to laboratory experiments and was therefore adopted by Davaille and Jaupart (1994; thin line labeled DJ94 on **Figure 36(b)**). However, as pointed out by subsequent studies (Dumoulin et al., 2001; Zaranek and Parmentier, 2004), $\Delta T/\Delta T_{eff} > 4$ for the mantle lithosphere and the approximation $f \sim 1$ leads to overestimate the SSC onset time (**Figure 36(b)**) compared to the prediction of [14] without the approximation (thick black line on **Figure 36(b)**). In the last years, several numerical studies, using either Arrhenius or exponential viscosity laws, have allowed to extend the range of $\Delta T/\Delta T_{eff}$ studied (Choblet and Sotin, 2000; Dumoulin et al., 2001; Korenaga and Jordan, 2003; Huang et al., 2003; Zaranek and Parmentier, 2004), and to develop new scaling laws. The discrepancies between the different data sets reveal the relative importance of the onset time measurement, the type of viscosity

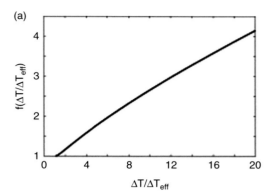

(a)

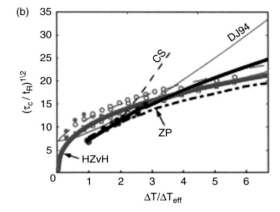

(b)

**Figure 36** (a) Function $f(x)$ described in the text (eqn [7]) as a function of $\Delta T/\Delta T_{eff}$. (b) Square root of the dimensionless SSC onset time $\tau_c/t_R$ ($t_R = \kappa^{-1} . \alpha g \Delta T/(\kappa \nu)^{-2/3}$) as a function of $\Delta T/\Delta T_{eff}$. The black disks represent the laboratory data of Davaille and Jaupart (1994). The black thick solid line was calculated with eqn [7], and the thin line labeled DJ94 with the $f \sim 1$ approximation. The thin dashed black line labeled CS shows Choblet and Sotin (2000) calculations, the blue thick line labeled HZvH, Huang et al. (2003), and the black dash-dot line labeled ZP, Zaranek and Parmentier's (2004). In red, data and scaling laws by Korenaga and Jordan (2003). Squares used an exponential law, and circles an Arrhenius law.

law (10–20% of change only), and the type of perturbation (initial-time/at-all-times, numerical-noise/finite-size-perturbation) introduced numerically which will grow toward convective instabilities. The latter has the strongest influence (Zaranek and Parmentier, 2004): the longest onset times are obtained when the numerical perturbations initially introduced are smallest (Choblet and Sotin, 2000). The trend of the laboratory experiments seems to compare best with numerical experiments which introduce thermal noise around $10^{-3} \times \Delta T$ at all times (**Figure 36(b)**). Then, in the parameter range relevant for the Earth's mantle

$(\Delta T / \Delta T_{\mathrm{eff}} \geq 4)$, numerical studies agree with [14] within 25%. So, if the magnitude of the temperature perturbations in the lithospheric mantle is similar to the laboratory one, an SSC onset time between 10 and 100 My is predicted, depending on the asthenospheric mantle temperature.

## 3.7 Complications: Internal Heating, Continents and Moving Plates

### 3.7.1 Internal Heating

We have so far focused on the dynamics of fluid layers uniformly heated from below and cooled from above. However, the Earth's mantle is not only heated along the core–mantle boundary by the hotter molten iron core, but is also partly heated from within by radiogenic decay of uranium, thorium, and potassium distributed throughout the mantle (*see* Chapter 6). Thermal convection with both internal and bottom heating is characterized by two dimensionless numbers: $Ra$ for the bottom heating given by eqn [1] and an equivalent Rayleigh number for the purely internally heated case, defined as

$$Ra_{\mathrm{H}} = \frac{\alpha g H^5 Q_{\mathrm{H}}}{\nu k \kappa} \qquad [15]$$

where $Q_{\mathrm{H}}$ is the volumetric rate of internal heat generation.

### 3.7.1.1 *Purely internally heated fluid*

The heat sources are simulated by the Ohmic dissipation of an electric current passed through an electrolyte fluid layer. Electrodes are placed either on each of the horizontal boundaries (e.g., Carrigan, 1982), or on two of the side walls of the chamber (e.g., Kulacki and Goldstein, 1972). Polarization effects in the fluid are eliminated by applying an alternating current. In all the experiments using this technique, the bottom of the tank is insulated (zero heat flux) and the top of the tank is cooled by a thermostated bath. One therefore expect only one TBL in the system, just below the cold plate.

Most of the experiments that have been performed used aqueous solutions. Tritton and Zarraga (1967) and Schwiderski and Schab (1971) visualized the planform of convection from marginal stability up to the transition to turbulence. Quasi-steady planforms were observed with downflow in the centers of the cells for $Ra_{\mathrm{H}} \leq 80 Ra_{\mathrm{c}}$. For larger $Ra$, cells broke up and turbulent motions appeared. Kulacki and

Goldstein (1972) extended the range up to $675 Ra_{\mathrm{c}}$, well into the turbulent regime. Kulacki and Nagle (1975) focused on heat flow measurements as a function of $Ra_{\mathrm{H}}$. For the turbulent regime $(1.5 \times 10^5 \leq RaH \leq 2.5 \times 10^9)$, their data are consistent with a power law $Nu \sim 0.305\, Ra^{0.239}$. As for the bottom heated case, transitions in the $Nu = f(Ra_{\mathrm{H}})$ curves were observed at each change of the convective pattern. Thermal fluctuations were also found to be maximum just outside the upper cold TBL and no bottom hot TBL was observed.

Carrigan (1982) used glycerine $(Pr \sim 3000)$ for $10^5 \leq Ra_{\mathrm{H}} \leq 10^6$. He observed cold spoke patterns up to $Ra_{\mathrm{H}} = 108 Ra_{\mathrm{c}}$, and a superposition of a large-scale circulation and cold individual plumes for larger $Ra$.

### 3.7.1.2 *Bottom- and internally-heated fluid*

Following Krishnamurti (1968a), Weinstein and Olson (1990) demonstrated the equivalence between thermal convection in a plane layer of fluid with uniformly distributed heat sources between boundaries with constant temperatures, and thermal convection in a plane layer of fluid with boundary temperatures that are spatially uniform but vary in time at a rate $\mathrm{d}T_{\mathrm{top}}/\mathrm{d}t$. The effective volumetric internal heat generation is then

$$Q_{\mathrm{H}} = -\rho C_{\mathrm{p}} \mathrm{d}T_{\mathrm{top}}/\mathrm{d}t \qquad [16]$$

This technique is simpler and more flexible to use than the electrolytic one, since it can be used with any fluid and allows the fluid to be heated from below as well as internally. Krishnamurti (1968b) performed experiments with silicone oils $(Pr \sim 200)$ near the onset of convection. Using high Prandtl number oils $(Pr \sim 9 \times 10^3)$, Weinstein and Olson (1990) studied the evolution of the convective pattern and its time dependence as the proportion of internal heating increases (**Figure 37**). At $Ra = 1.5 \times 10^5$, in the absence of internal heat generation, a spoke pattern is observed, in which connected spokes of ascending and descending flows are on average of equal strength. With internal heat generation, the buoyancy becomes concentrated into the cold downwellings, and the amount of time dependence increases. Hot active instabilities will reappear if there exists a bump on the hot bottom boundary with a height comparable to (or greater than) the TBL thickness (Namiki and Kurita, 2001). In this case, hot plumes will be anchored on the bump. So although an internally heated mantle favors

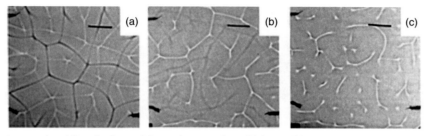

**Figure 37** Shadowgraphs of the planform of thermal convection with internal heat generation for $Ra = 1.5 \times 10^5$ (spoke pattern). (a) $RaH/Ra = 0$; light and dark lines have the same intensity as cold and hot currents have the same strength. (b) $RaH/Ra = 6$; the dark lines are fading as the downwellings become weaker. (c) $RaH/Ra = 18$; no hot upwellings are seen anymore. From Weinstein SA and Olson P (1990) Planforms in thermal convection with internal heat sources at large Rayleigh and Prandtl numbers. *Geophysical Research Letters* 17: 239–242.

cold downwellings, hot plumes can still develop, provided that the bottom of the mantle, or the $D''$ layer, has some topography.

### 3.7.2 Inhomogeneous Cooling/Continents

Continental lithosphere is thicker than its oceanic counterpart, and does not subduct. Its crust is also richer in radiocative elements. Moreover, heat flow beneath continents is significantly lower than under oceans (e.g., Jaupart *et al.*, 1998), which tends to show that continents are more insulating. Hence, continents induce a lateral heterogeneity in both the mechanical and thermal conditions at the upper boundary of the mantle. The experimental study of Rossby (1965) showed that any lateral heterogeneity in cooling or heating will generate an horizontal large-scale flow. Elder (1967) was probably the first to suggest that the continents, by acting as an insulating lid, could generate a new class of lateral motions. Later studies have therefore focused on characterizing and quantifying the influence of these lateral heterogeneities on mantle thermal convection.

Whitehead (1972, 1976) investigated both theoretically and experimentally the conditions under which a heater mounted on a mobile float could become unstable to drifting motion. For cellular convection, the drift was found of the same magnitude as the characteristic convective velocity.

Guillou and Jaupart (1995) investigated the thermal effect of a fixed continent on the convective pattern and heat flow, using silicone oils and Golden Syrup for $Ra$ ranging between critical and $10^6$. The continent was modeled as an insulating material embedded into a cold copper plate (**Figure 38**), and its size and shape were varied. They observed the generation of deep hot instabilities localized under the continent and the generation of cold instabilities outside or on its edges (**Figure 38**). The presence of a continent therefore shields part of the underlying mantle from subduction (Gurnis, 1988), and also induces a large-scale convection pattern oceanward over distances up to 3 times the continent's size (Guillou and Jaupart, 1995). Combining experiments, theory, and simulations, Lenardic *et al.* (2005) studied the influence of partial insulation on mantle heat extraction. Since partial insulation leads to increased internal mantle temperature and decreased viscosity, it allows, in turn, for the more rapid overturn of oceanic lithosphere and increased oceanic heat flux. For ratios of continental to oceanic surface area lower than 0.4, global mantle heat flow could therefore remain constant or even increase as a result. Wenzel *et al.* (2004) used the same kind of setup with two-layer convection. They suggested that, on Mars, the combination of crustal dichotomy (the martian crust is much thicker, therefore insulating, in the southern hemisphere than in the northern hemisphere) and two-layer convection would lead to the formation of a large-scale upwelling under the southern highlands that appeared early and endured for billions of years.

Zhang and Libchaber (2000) and Zhong and Zhang (2005) replaced the rigid cold boundary by a free surface cooled by air, which allowed the insulating raft to move freely in response to the convection (**Figure 39**). The experimental fluid was water, with Rayleigh numbers ranging from $10^7$ to $4 \times 10^8$. Like Guillou and Jaupart (1995), they observed that the insulating raft distorts the local heat flux and induces a coherent flow over the experimental cell, which in turn moves the plate. In their confined geometry, the plate can be driven to a periodic motion even under the action of chaotic convection. The period of

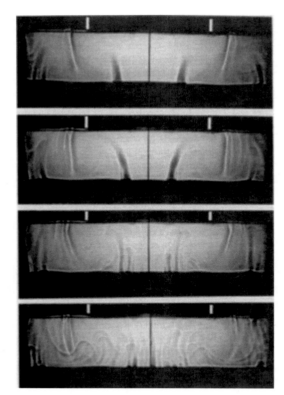

**Figure 38** Insulating continents from Guillou and Jaupart (1995). Markers along the upper boundary indicate the extent of insulating material representing continental lithosphere. Time progresses from top to bottom. Note upwelling plumes beaneath the interior and descending sheets near the margins of the insulator.

oscillation depends on the coverage ratio and on the Rayleigh number (**Figure 39**).

Nataf *et al.* (1981) investigated the influence of the sinking cold oceanic lithosphere on the neighboring subcontinental convection and on the generation of instabilities in the horizontal boundary layers by performing experiments with cooling from both the top plate and one of the sides. The system dynamics are then characterized by two Rayleigh numbers: $Ra$ based on the vertical temperature difference, and a lateral Rayleigh number based on the lateral temperature difference ($Ra_{lat}$). The lateral cooling induces a large roll with axis parallel to the cold wall and descending flow next to it. At constant $Ra$, the width of the large roll is proportional to $Ra_{lat}^{1/2}$, while for constant $Ra_{lat}/Ra$, it scales as $\sim Ra^{1/6}$. In the upper mantle such large-scale circulation could become 5 times wider than deep. The experiments also show that hot instabilities cannot develop directly beneath the cold downwelling flow, because the spreading of the latter impedes the growth of the hot TBL.

(a)

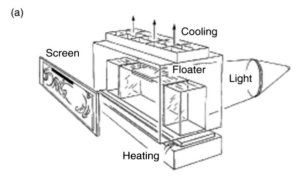

(b)

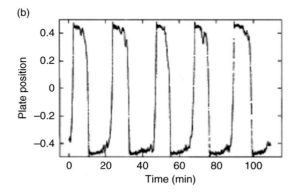

(c)

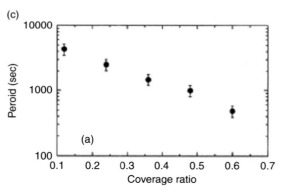

**Figure 39** Mobile insulating continent. (a) experimental setup. (b) Continent position as a function of time. (c) Continental drift periodicity as a function of the coverage ratio (length of the floater divided by tank length). From Zhang J and Libchaber A (2000) Periodic boundary motion in thermal turbulence. *Physical Review Letters* 84: 4361–4364.

### 3.7.3 Moving Top Boundary

We just saw that any large-scale circulation strongly influences the generation of convective features of smaller extent. The large-scale motion of plates is therefore expected to influence all motions within the mantle, especially small-scale convection under the cooling lithosphere, and the distribution of hot spots.

Following the theoretical study of Richter (1973), Richter and Parsons (1975) studied the influence of a moving top boundary on the convective pattern for $Ra$ between $3.3 \times 10^3$ (rolls) and $1.4 \times 10^5$ (spokes). A mylar sheet was introduced just below the top cold boundary and pulled at constant velocity $U$ (**Figure 4**). Besides $Ra$, another system now controls the system dynamics, the Peclet number $Pe = UH/\kappa$, which compares the heat transport by advection and by diffusion. It ranges between 30 and 720 in the experiments. Initially, the top boundary was at rest until the convective pattern becomes quasi-steady. For the lower $Ra$, rolls orthogonal to the spreading direction were initiated using the technique of Chen and Whitehead (1968). Then the velocity of the mylar was set to a finite value and the convective pattern followed by shadowgraph. Over the whole $Ra$ range, the pattern was observed to evolve toward rolls parallel to the mylar velocity. Scaling laws show that it would probably take 70 My for the geometry of convection to reorganize into rolls parallel to the spreading direction over the whole thickness of the upper mantle.

Experiments of Kincaid et al. (1996), Jellinek et al. (2002, 2003), and Gonnermann et al. (2004) studied how a plume-dominated free convective flow interacts with a large-scale passive flow driven by plate motions. Kincaid et al. (1996) focused on the interaction of mid-ocean ridges and convection in the upper mantle. The three other papers extended this study to the whole mantle, higher $Ra$, and fluids with strongly temperature dependent viscosity. Jellinek et al. (2002, 2003) performed laboratory experiments using corn syrup heated from below and driven at its surface by a ridge-like velocity boundary condition (**Figure 40(a)**). They documented three different flow regimes, depending on the ratio $V$ of the spreading rate to the typical plume rise velocity and the ratio of the interior viscosity to the viscosity of the hottest fluid $\gamma$. When $V \ll 1$ and $\gamma \geq 1$, the plume distribution is random and unaffected by the large-scale flow. The opposite extreme is represented by $V > 10$ and $\gamma > 100$, where plume formation is suppressed entirely and the large-scale flow carries all the heat flux. Interaction of the two flows occurs for intermediate values of $V$; established plume channels are now advected along the bottom boundary and focused toward an upwelling beneath the ridge. The basal heat flux and the TBL thickness depend on $Pe$ and $\gamma$. In particular, since cold, and therefore also more viscous, fluid is rapidly transported from the top to the bottom of the layer, the viscosity contrast across the lower TBL increases,

(a)
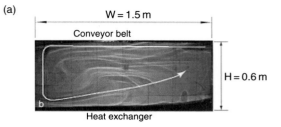

(b)
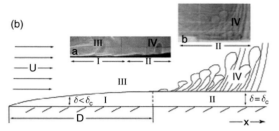

**Figure 40**   (a) Experimental setup and shadowgraph at $Ra \sim 5 \times 10^6$. $Pe \sim 10^4$ and $\gamma \sim 100$. Note the increasing thickness of the TBL in the downstream direction. (b) Schematic diagram depicting the four convecting regions. The TBL is not thick enough in region I to develop plumes in III, while TBL instabilities develop in regions II and IV. From Gonnermann HM, Jellinek AM, Richards MA, and Manga M (2004) Modulation of mantle plumes and heat flow at the core mantle boundary by plate-scale flow: Results from laboratory experiments. *Earth and Planetary Science Letters* 226: 53–67.

generating hot plumes with a bigger head to tail ratio than without plate circulation (Nataf, 1991; Jellinek et al., 2002). Running more detailed experiments, Gonnermann et al. (2004) proposed a conceptual framework (**Figure 40(b)**) whereby the tank bottom can be divided in four regions. The large-scale flow impedes the growth of the hot TBL where the cold and more viscous downwelling spreads (Nataf et al., 1981), and therefore the hot TBL is not thick enough in region I to develop plumes in III. In region II, the TBL becomes thick enough for hot instabilities to develop in IV. Scalings were derived for the lateral variations of the bottom heat flux and the spatial extent of plume suppression. Applied to the Earth's mantle, they suggest that $\sim 30\%$ of the core heat flux could be due to increased heat flux from plate-scale flow. Furthermore, the core heat flux is expected to vary laterally by at least a factor of 2. Last, these results explain well why hot spots seem confined to about a half of the surface of the Earth, away from ancient subducted slabs, and preferentially close to mid-ocean ridges (Weinstein and Olson, 1989).

## 3.8    Close-Up on Plumes: Plumes from a Point Source of Buoyancy

Since the mid-1970s, much of the best experimental work in mantle dynamics has been devoted to plumes from point sources of buoyancy. While plumes in the mantle probably develop primarily as instabilities of thermal boundary layers (e.g., Parmentier *et al.*, 1975; Loper and Stacey, 1983), such instabilities are time dependent and can be difficult to quantify (see Section 3.5). As a first step, therefore, it makes sense to investigate the simpler case of an isolated laminar 'starting plume' rising from a point source of buoyancy whose strength is constant in time. Such plumes are easily studied and photographed in the laboratory, and have given us some of the most beautiful images in the field of experimental geodynamics (e.g., Whitehead and Luther, 1975; Olson and Singer, 1985; Griffiths, 1986; Campbell and Griffiths, 1990). Indeed, these images have been so influential that they now constitute a sort of 'standard model' of a mantle plume as a large, bulbous cavity (the head) trailed by a narrow conduit (the tail) connecting it with its source (**Figure 5**). According to this model, the arrival of a plume head at the base of the lithosphere produces massive flood basalts, while the trailing conduit generates the subsequent volcanic track (Richards *et al.*, 1989). The model successfully explains important features of several prominent hot spots, including the volume of the topographic swell (e.g., Davies, 1988; Olson, 1990; Sleep, 1990) and the volume ratio between flood basalts and island chain volcanism (Olson and Singer, 1985; Richards *et al.*, 1989), as discussed in more detail in Chapter 9.

### 3.8.1    Compositionally Buoyant Plumes

The early experiments on starting plumes in viscous fluids (Whitehead and Luther, 1975; Olson and Singer, 1985) were done by injecting compositionally buoyant fluid (typically corn syrup) at a constant rate from a pipe at the bottom of a large reservoir of a second fluid. The great advantage of this experimental setup is its simplicity: experiments can be performed under ambient laboratory conditions, with no temperature control required. Whitehead and Luther (1975) investigated the evolution of starting plumes for short times after the beginning of injection, during which the buoyant fluid forms a quasi-spherical ball that grows until it is large enough to 'lift off' from the injector (**Figure 5**). They

proposed that the ball lifts off when its buoyant ascent speed (as predicted by Stokes law; *see* Chapter 4) exceeds the rate of increase of its radius, which occurs at a critical time $t_{sep}$ and radius $a_{sep}$ given by

$$t_{sep} = \left(\frac{4\pi}{3Q}\right)^{1/4}\left(\frac{\nu_m}{g^*}\right)^{3/4}, \qquad a_{sep} = \left(\frac{3Q}{4\pi}\right)^{1/3} t_{sep}^{1/3} \quad [17]$$

where $Q$ is the volumetric injection rate, $\nu_m$ is the viscosity of the ambient fluid, $g^* = g\,(\rho_m - \rho_p)/\rho = g\,\Delta\rho/\rho$ is the reduced gravitational acceleration, and $\rho_m$ and $\rho_p$ are the densities of the ambient and injected fluids, respectively. Both the time and radius of separation increase with increasing viscosity of the ambient fluid. For the large viscosities typical of the mantle, plume heads must be quite large (radius > 100 km) to separate from their source (Whitehead and Luther, 1975).

After liftoff, the morphology and dynamics of starting plumes depend strongly on the viscosity ratio $\gamma = \nu_p/\nu_m$, where $\nu_p$ is the viscosity of the injected fluid (Whitehead and Luther, 1975; Olson and Singer, 1985). When the injected fluid is more viscous ($\gamma > 1$), the plume has roughly the shape of a cylinder (**Figure 5(a)**) whose length and radius increase with time. Olson and Singer (1985) called this a 'diapiric' plume. In the more geophysically realistic limit $\gamma << 1$, the ascending plume is a 'cavity plume' comprising a large head and a thin trailing conduit (**Figure 5(b)**). The rate of change of the volume $V$ of the head is just the injected flux $Q$ less the flux required to build the (lengthening) stem, or

$$\frac{dV}{dt} = Q - \pi R_c^2 W \qquad [18]$$

where $R_c \equiv (8\nu_p Q/\pi g^*)^{1/4}$ is the radius of the conduit required to carry the flux $Q$ by Poiseuille (cylindrical pipe) flow and $W$ is the velocity of the head predicted by Stokes's law. For long times, a steady state ($dV/dt = 0$) is achieved in which the head reaches a terminal velocity $W_{term}$ and radius $a_{term}$ (Whitehead and Luther, 1975):

$$W_{term} = \left(g^* Q/8\pi v_p\right)^{1/2}, \qquad a_{term} = \left(3W_{term}\nu_m/g^*\right)^{1/2} \quad [19]$$

More recent experiments have shown that under certain conditions compositionally buoyant cavity plumes can entrain the denser ambient fluid through which they rise (Neavel and Johnson, 1991; Kumagai, 2002). Kumagai (2002) identified two distinct regimes: a 'vortex ring' regime for $\gamma \lesssim \bar{1}$ in which layers of entrained fluid form a scroll-like pattern inside the plume head, and a 'chaotic stirring' regime for $\gamma > 100$

(see Section 3.10). Structures of the vortex ring type also occur in thermal starting plumes (Shlien, 1976; Shlien and Boxman, 1981; Tanny and Shlien, 1985; Griffiths and Campbell, 1990; **Figure 7(c)**).

The stem of a chemically buoyant plume can also support solitary waves, waves of large amplitude that propagate upwards without change in shape. Their existence was first demonstrated experimentally by Olson and Christensen (1986) and Scott *et al.* (1986), who generated the waves by increasing impulsively the volume flux $Q$ injected at the bottom of a vertical conduit surrounded by a fluid with a much greater viscosity. Whitehead and Helfrich (1988) subsequently showed that the flow within such waves exhibits recirculation along closed streamlines when viewed in a frame traveling with the wave, and suggested that this might provide a mechanism for transporting deep material rapidly to the surface with little contamination. Subsequent numerical experiments (Schubert *et al.*, 1989) showed that thermal plume stems could also support solitary waves, and Laudenbach and Christensen (2001; **Figure 41**) performed an experimental study of solitary waves of

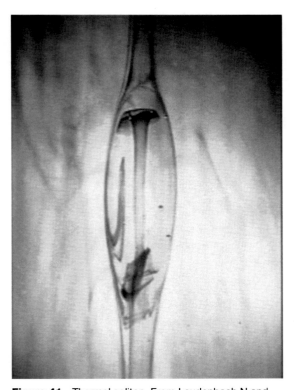

**Figure 41** Thermal soliton. From Laudenbach N and Christensen UR (2001) An optical method for measuring temperature in laboratory models of mantle plumes. *Geophysical Journal International* 145: 528–534.

hot fluid in a vertical conduit. Solitary waves on the stems of thermal plumes have been invoked to explain episodic magma production at weak hot spots and surges of activity at stronger hot spots (Laudenbach and Christensen, 2001) and the formation of the V-shaped ridges on the Reykjanes ridge south of Iceland (Albers and Christensen, 2001; Ito, 2001).

### 3.8.2 Thermal Plumes

Because the buoyancy force that causes plumes to rise in the Earth's mantle is, to a large degree, thermal in origin, numerous laboratory investigations of thermal plumes in viscous fluids have been carried out since the mid-1980s. The first experiments on laminar starting plumes (Shlien and Thompson, 1975; Shlien, 1976) were not motivated by geophysical applications, and were performed using water and a narrow electrode as a heater. Experiments in this configuration typically used interferometry and particle tracking to recover the temperature (**Figure 12(c)**) and velocity fields (e.g., Shlien and Boxman, 1979; Tanny and Shlien, 1985; Chay and Shlien, 1986). By contrast, investigations of plumes in more viscous fluids have until now mostly focused on more global features such as the volume and ascent rate of the plume head. Because the viscosity of many experimental fluids decreases strongly with increasing temperature, hot thermal starting plumes generally are cavity plumes comprising a large head and a narrow stem. In some experiments, however, the heat source is suddenly turned off to produce an isolated head ('thermal') that rises without a trailing conduit. The discussion below begins with this latter case (**Figure 6(a)**), and then turns to the dynamics of steady plume stems and thermal starting plumes (**Figures 42** and **43**).

#### 3.8.2.1 Thermals

Griffiths (1986) conducted a laboratory study of thermals generated by injecting a known volume of heated polybutene oil into a large tank of the same oil at room temperature. He observed that a thermal initially rises at constant speed through a distance of the order of its diameter, and then slows down while increasing in volume. Griffiths proposed that the enlargement occurs because diffusion of heat away from the thermal warms a thin boundary layer of ambient fluid around it, which is then advected back to the thermal's trailing edge and entrained into it. The total buoyancy of the thermal is therefore constant, and is proportional to the Rayleigh number $Ra = \alpha g \Delta T_0 V_0 / \kappa \nu_m$, where $V_0 = \pi a_0^3$ is the thermal's

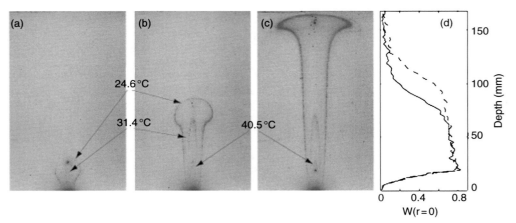

**Figure 42** Thermal and velocity structures of a plume issued from a point-source heater (with $P = 3$ W), deduced from the pictures (a)–(c) taken at the following times: (a) $t = 60$ s, (b) $t = 120$ s, (c) $t = 300$ s. The isotherms appear as black lines on the negatives. The temperature decreases in the stem and the head as height increases. (d) Velocity profiles along the axis $r = 0$ for $t = 120$ s, measured with a PIV technique. The velocity decreases with increasing height along the stem. The fluid is sugar syrup, whose viscosity depends on temperature. At 20°C, it is 6 Pa s, and on the heater (54°C) it is 0.3 Pa s. From Vatteville J (2004) Etude expérimentale des panaches: Structure et evolution temporelle. Mémoire de DEA, ENS Lyon/IPG Paris.

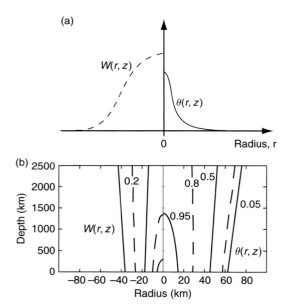

**Figure 43** Thermal plume stem at high Prandtl number. (a) Sketch of the radial temperature (solid line) and vertical velocity (dashed line) profiles for constant viscosity: the plume thermal anomaly is embedded into the velocity boundary layer. (b) Model of Olson et al. (1993) for strongly temperature-dependent viscosity in the mantle: the high velocity conduit (left) is embedded in a 'thermal halo' (right). Temperature and velocity are normalized by their maximum value on the axis.

initial volume and $\Delta T_0$ its initial temperature excess. Griffiths (1986) suggested that for times greater than $\tau_G \sim Ra^{-1/2} a_0^2/\kappa$, the thermal's volume $V$ and velocity $W$ have the self-similar forms

$$V = \frac{\pi C_1^3}{6} Ra^{3/4}(\kappa t)^{3/2}, \qquad W = \frac{f}{2\pi C_1} Ra^{3/4}(\kappa t)^{-1/2} \quad [20]$$

where $f = O(1)$ is a function of the viscosity contrast between the thermal and the ambient fluid and $C_1 \sim 1.0$ is a constant determined experimentally. Laboratory experiments using viscous oils (Griffiths, 1986a) and corn syrup (Coulliette and Loper, 1995) are consistent with [20]. In the mantle, thermals with radii >350 km would reach the lithosphere before being influenced by thermal diffusion in typically 20 My and their ascent speeds could reach 10–20 cm yr$^{-1}$.

### 3.8.2.2  The stem

When a starting plume rises from a steady point source of heat, the lower part of its stem quickly achieves a steady-state structure that is independent of the dynamics of the ascending head above. To understand this structure, Batchelor (1954) studied analytically a model in which a plume rises from a point source of heat of strength $P$ in an infinite fluid with constant viscosity $\nu$, thermal diffusivity $\kappa$, specific heat $C_p$, and thermal expansion coefficient $\alpha$. Simple scaling arguments suggest that the vertical velocity $W$, the temperature anomaly $\Delta T$, and the stem radius $R$ scale with the height $z$ above the source as

$$W \sim \left(\frac{g\alpha P}{\pi\rho C_p \nu}\right)^{1/2}, \qquad \Delta T \sim \frac{P}{\kappa\rho C_p z}$$

$$R \sim \left(\frac{\rho C_p \nu\kappa^2}{g\alpha P}\right)^{1/4} z^{1/2} \quad [21]$$

The vertical velocity is constant, but thermal diffusion causes the temperature anomaly to decrease with height as $z^{-1}$ and the stem radius to increase as $z^{1/2}$. The scalings [21] are valid to within multiplicative functions of the Prandtl number $Pr = \nu/\kappa$, which have been determined numerically and/or analytically (Fuji, 1963; Worster, 1986; Vasquez et al., 1996). Experiments in water ($Pr \sim 7$) verify the scalings [22] and agree with the complete functional form of Fuji (1963) if corrections are made for the finite size of the heat source and for downward heat losses from it (Shlien and Boxman, 1979; Tanny and Shlien, 1985).

The scaling [21] is no longer valid in the limit of infinite $Pr$, because the model problem studied by Batchelor (1954) has no solution if the fluid is infinite and inertia is totally absent (Stokes paradox; see Chapter 7). However, a solution does exist if the heat source is located on a horizontal (rigid or free) boundary, and has a vertical velocity $W$ given by [21] multiplied by a function that increases logarithmically with height (Whittaker and Lister, 2006).

The dynamics are very different when the viscosity depends strongly on temperature, as is the case in the mantle. **Figure 43** shows isotherms (a–c) and profiles of vertical velocity on the axis (d) at different times for a thermal plume above a heater in corn syrup (Vatteville, 2004). The lower portion 20 mm $\leq z \leq 70$ mm of the plume quickly reaches a steady state, as shown by the similarity of the isotherms and velocity profiles for $t = 120$ s and 300 s in this height range. Moreover, the vertical velocity decreases strongly with height, in contrast with the constant-viscosity case. Similar results were found by Laudenbach (2001) along the stem of a plume of heated corn syrup, which was injected at the bottom of a tank filled with the same fluid at room temperature.

These observations are well explained by Olson et al.,'s (1993) approximate solution of the boundary-layer equations for the steady flow above a point source of heat in a fluid whose viscosity depends exponentially on temperature. The upwelling velocity $W$ on the axis and the temperature anomaly $\Delta T$ vary with height $z$ as

$$W = \left(\frac{g\alpha P}{4\pi\rho C_p \nu_0(z)}\right)^{1/2}$$
$$\Delta T = \Delta T_0 \exp\left(\frac{-4\pi k \Delta T_r z}{P}\right) \quad [22]$$

where $\nu_0(z)$ is the viscosity on the plume centerline, $\Delta T_0$ is the temperature anomaly at the base of the plume, $k$ is the thermal conductivity, and $\Delta T_r$ is the temperature required to change the viscosity by a factor $e$. The expression [22] for $W$ is identical to the constant-viscosity expression [21] except for the variability of the centerline viscosity $\nu_0(z)$. The temperature anomaly decreases exponentially with height, and the corresponding increase in viscosity causes the upwelling velocity $W$ also to decrease with height, in qualitative agreement with the experimental observations. The differences relative to the constant-viscosity case are due to the fact that temperature-dependent viscosity concentrates the upwelling in the hottest central part of the plume stem, so that the radius $\delta_T$ of the thermal anomaly is wider than the radius $\delta_w$ of the upwelling region (the 'conduit' proper) even for $Pr \gg 1$ (**Figure 42**). For typical mantle parameter values, plumes with large viscosity variations ($\geq 1000$) would have $\delta_w \sim 25$–30 km and $\delta_T \sim 60$ km, while plumes with low viscosity variations ($\leq 100$) are typically twice as broad.

### 3.8.2.3 Thermal starting plumes

A thermal starting plume is essentially an ascending thermal connected by a stem to a source of buoyancy (**Figure 5**). Thus, the plume head can grow both by entrainment of ambient fluid and by the addition of plume fluid through the stem. Unfortunately, no reliable scaling laws have yet been found to describe the evolution of the size and speed of the plume head in this case, and different studies give conflicting results.

Griffiths and Campbell (1990) investigated thermal starting plumes by injecting hot glucose syrup with a temperature excess $\Delta T_0$ into a cooler reservoir of the same fluid at a constant volumetric rate $Q$. They proposed that the volume $V$ of the plume head evolves as

$$\frac{dV}{dt} = Q + C_2(\kappa g\alpha\Delta T_0 Qt/\nu_m)^{1/2} V^{1/3} \quad [23]$$

where $\nu_m$ is the viscosity of the cold ambient fluid and $C_2$ is an empirical entrainment constant. For large times the second (entrainment) term in [23] is dominant, implying

$$V = \left(\frac{4C_2}{9}\right)^{3/2} \beta\kappa^{3/4} t^{9/4} \quad [24]$$

where $\beta = (\alpha g \Delta T_0 Q / \nu_m)^{3/4}$. The corresponding velocity of the head, calculated from Stokes's law for a relatively inviscid drop, is

$$W = \frac{1}{8} \left( \frac{6}{\pi} \right)^{2/3} C_2^{-1/2} \beta (\kappa t)^{1/4} \qquad [25]$$

On the basis of experiments with thermals, Griffiths and Campbell (1990) state that $1 \leq C_2 \leq 4$. Equation [24] describes well the experimental data for intermediate times, although at larger times the cap growth rate becomes lower than predicted and is better described by the law $V \sim t^{3/2}$ [20] for an isolated thermal (**Figure 44**).

Experiments on plumes produced by localized heating without fluid injection (Moses *et al.*, 1993; Coulliette and Loper, 1995; Kaminski and Jaupart, 2003; Vatteville, 2004) show a different behavior: the velocity of the head increases rapidly for a short time and then attains a nearly constant value $W_{\text{term}}$. To explain this, Coulliette and Loper (1995) proposed a modified version of the Griffiths and Campbell (1990) model that accounted for incomplete entrainment of the hot thermal boundary layer surrounding the head. However, the resulting theory involves empirical constants whose values vary by factors of 2–5 among experiments. A possible cause of the difficulty may be the theory's neglect of the flux required to build the lengthening stem, which is important for compositional plumes (Whitehead and Luther, 1975; Olson and Singer, 1985).

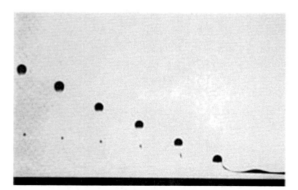

**Figure 44**    Volume of a starting plume as a function of time. The circles represent the experimental data points from Griffiths and Campbell (1990). In dashed lines, the $t^{1/3}$ law for a growing Stokes sphere (Whitehead and Luther 1975), the $t^{9/4}$ growth model of Griffiths and Campbell (1990), and the $t^{3/2}$ law due to the conductive growth of the plume cap.

### 3.8.3    Interaction of Plumes with a Large-Scale Flow

Because deep-seated plumes must traverse the convecting mantle on their way to the surface, it is important to investigate how they interact with an ambient large-scale flow. This interaction was first studied by Skilbeck and Whitehead (1978), who injected compositionally buoyant fluid into a shear flow generated by rotating the top surface of a tank containing another fluid. As the rising plume head is swept horizontally away from the source, the trailing conduit is tilted until it becomes gravitationally unstable at a tilt angle $\sim 30°$ from the vertical. The instability forms new, smaller cavity plumes whose spacing is independent of the volume flux $Q$ (Whitehead, 1982). Olson and Singer (1985) performed similar experiments by towing the source at constant speed $U$ at the bottom of a tank of motionless fluid. The characteristic spacing $\gamma$ and volume $V$ of the new cavity plumes generated by the instability were found to scale as $\lambda \approx 12(\nu_m U/g^*)^{1/2}$ and $V \approx 12(\nu_m/g^* U)^{1/2} Q$, respectively, although these relationships break down for small ($<0.3$) and large ($>10$) values of the dimensionless parameter $(g^* Q / \nu_m U^2)^{1/2}$. Skilbeck and Whitehead (1978) suggested that this instability might account for the discrete character of volcanic island chains.

Richards and Griffiths (1988) studied experimentally the interaction between a compositionally buoyant plume conduit and a large-scale shear flow, for low conduit tilt angles at which the instability documented by Skilbeck and Whitehead (1978) does not occur. They showed that the conduit's shape can be calculated by describing its motion as a vector sum of the shear velocity and a modified vertical Stokes velocity $v_s$ for individual conduit elements. For a simple linear shear flow, the resulting steady-state shape is a parabola. In response to a sudden change in the shear flow, the lateral position $x(z)$ of such a conduit adjusts to a new steady-state position in a time $z/v_s$, where $z$ is the height. Griffiths and Richards (1989) applied this theory to Hawaii, concluding that the sharpness of the bend in the Hawaiian–Emperor chain at 43 Ma implies that the deflection of the Hawaiian plume before the change in plate motion was $<200$ km.

Kerr and Mériaux (2004) carried out a comprehensive laboratory study of thermal plumes ascending in a shear flow in a fluid with temperature-dependent viscosity. Plumes were observed to rise initially with a constant velocity that is independent of the centerline viscosity $\nu_p$. Subsequently, the ambient shear tilts the plume conduits, and in some cases significant cross-stream circulation and entrainment is seen. However, the tilted conduits never develop Rayleigh–Taylor-type gravitational instabilities of the kind observed on tilted compositional conduits, perhaps because thermal diffusion is fast enough to suppress their growth (Richards and Griffiths, 1988).

### 3.8.4 Plume–Lithosphere Interactions

Because the interaction of plumes with the lithosphere is responsible for most of the geochemical and geophysical signatures observed at hot spots, the dynamics of this process has been the focus of numerous experimental studies. The more general case of the interaction of plumes with density and/or viscosity boundaries within the mantle will be treated together with two-layer convection in Section 3.9.

#### 3.8.4.1 Interaction with a stagnant lithosphere

To understand the formation of bathymetric swells by plumes, Olson and Nam (1986) performed laboratory experiments on the deformation of an initially spherical drop of buoyant low-viscosity fluid rising toward a high-viscosity boundary layer at the cooled free surface of a fluid with strongly temperature-dependent viscosity. They measured the amplitude of the surface topography as a function of time using a proximity probe (Section 3.3.5; **Figure 17**), and found that the typical evolution comprises a phase of rapid uplift followed by slower subsidence that corresponds to stagnation and lateral spreading of the drop below the cold boundary layer. The maximum swell height increased with the drop volume, but decreased with increasing lithosphere thickness. Griffiths *et al.* (1989) confirmed these results using holographic interferometry (**Figure 17**), and extended the study to larger viscosity and buoyancy contrasts between the lithosphere and the underlying mantle. They found the maximum swell height to be independent of the viscosity contrast, but a decreasing function of the density contrast.

Using experiments performed under isothermal conditions, Griffiths and Campbell (1991) determined scaling laws for the radius of a drop spreading beneath a solid plate and the thickness of the 'squeeze film' above the drop as functions of time. They observed that the gravitationally unstable

interface between the drop and the squeeze film eventually broke up via a Rayleigh–Taylor instability with a complex lateral planform dominated by short-wavelength (~drop thickness) upwellings and downwellings.

Olson *et al.* (1988) studied the penetration of mantle plumes through colder and more viscous lithosphere, whereas Jurine *et al.* (2005) investigated how thermal plumes deform and penetrate a compositionally distinct lithosphere. The amplitude and shape of the penetration was found to depend mostly on the ratio *B* between the compositional and thermal density contrasts (buoyancy number; see Section 3.9), while the timescale of the interface displacement is governed primarily by the viscosity contrast. When applied to the continental lithosphere, the experimentally derived scaling laws show that plume penetration through Archean lithosphere requires thermal anomalies larger than 300 K, corresponding to buoyancy fluxes in excess of those determined today for the Hawaiian plume. But because the experimental plumes are observed to interact differently with lithospheres of different ages and/or compositions, no single 'typical' plume signature can be identified.

#### 3.8.4.2 Interaction with a moving or rifting lithosphere

While the interaction of a plume with a stationary lithosphere is relatively easy to study in the laboratory, analog modeling of a moving lithosphere poses serious challenges, including the difficulty of controlling the temperature on the moving boundary and the unwanted influence of the return flow the boundary motion induces (see Section 3.7). Here two cases must be distinguished: interaction of the plume with an intact lithosphere moving at constant speed relative to it ('plume–plate interaction' or PPI), and interaction of the plume with a rifting lithosphere representing a mid-ocean ridge ('plume–ridge interaction' or PRI). To date there exist no experimental studies of PPI, which has been investigated solely with analytical and numerical methods (e.g., Olson, 1990; Ribe and Christensen, 1994, 1999). By contrast, PRI has been successfully modeled in the laboratory.

The first case to be studied was that of a 'ridge-centered' plume rising directly beneath a spreading ridge, Iceland being the main natural example. Feighner and Richards (1995) investigated this case by injecting compositionally buoyant fluid from a pipe beneath the boundary between two diverging mylar sheets at the top of a tank of corn syrup. They observed that the buoyant fluid spreads beneath the mylar sheets to form a thin layer with a steady 'bow-tie' shape in mapview. They found that the width of the layer along the ridge scales as $(Q/U)^{1/2}$, where $Q$ is the volumetric injection rate and $U$ is the half-spreading rate, in agreement with the predictions of numerical models based on lubrication theory (Ribe *et al.*, 1995) and the tracking of chemically buoyant tracer particles (Feighner *et al.*, 1995.) This work was extended by Kincaid *et al.* (1995) to the interaction of a thermal plume with a mylar-sheet spreading ridge located at some distance from it. The mantle was modeled using a concentrated sucrose solution, producing a highly viscous upper boundary layer (the 'lithosphere') that cooled and thickened as it moved away from the ridge. Kincaid *et al.* (1995) observed that the plume material can be guided toward the ridge along the sloping base of the lithosphere, confirming the suggestion of Schilling (1991) that an off-axis plume may communicate thermally and chemically with a spreading ridge.

## 3.9   Inhomogeneous Fluids: Mixing and Thermochemical Convection

These studies have arisen from two fundamental questions in mantle dynamics:

1. How can the dispersion observed in geochemical data be interpreted in dynamical terms? Since geochemical isotopic signatures modify neither the density nor the physical properties (rheology, thermal expansion, and thermal diffusion) of mantle materials, they constitute passive tracers of mantle flow (e.g., Allègre, 1987). The question therefore relates to the more general problem of convective mixing of passive tracers.

2. What is the convective pattern in the mantle? Can mantle convection be stratified? There are numerous sources of compositional heterogeneities in the mantle, such as slab remnants (e.g., Olson and Kincaid, 1991; Christensen and Hofmann, 1994), delaminated continental material, relics of a primitive mantle (Gurnis and Davies, 1986; Tackley, 1998; Kellogg *et al.*, 1999) enriched, for example, in iron (Javoy 1999), or chemical reactions with or infiltration from the core (e.g., Hansen and Yuen, 1988). So, how does a density heterogeneity interact with thermal convection?

Both questions involve time-dependent 3-D phenomena occuring on a wide range of spatial scales, abrupt gradients of physical properties (e.g., density and viscosity), and intricate physical phenomena (stirring, entrainment) for which even the equations remain to be found. Laboratory experiments have therefore proved to be a good tool to tackle these problems. A more complete review of mixing in the mantle is given by Tackley in Chapter 10.

### 3.9.1   Mixing of Passive Tracers

An impurity is said to be a passive tracer when it does not modify the flow field transporting it. This condition requires the impurity either to be neutrally buoyant, or to have a characteristic length scale very small compared to that of the flow. Mixing of an impurity involves two phenomena: (1) stirring, that is, stretching and folding of the heterogeneity until its length scale becomes small enough for chemical diffusion to act, followed by (2) diffusive erasure of the heteregeneity (e.g., Reynolds, 1894; Nagata, 1975; Ottino, 1989). Given the small values of chemical diffusion coefficients in the mantle (typically $10^{-15}\,\mathrm{cm^2\,s^{-1}}$; Hofmann and Hart, 1978), the latter process will act only over a few centimeters on geological timescales. Stirring is therefore the process of greater interest for geodynamics (*see* Chapters 2 and 10).

Our current understanding of mixing has significantly improved in the last 20 years, due to the establishment of a clear connection between chaos and stirring (e.g., Aref, 1984). Essential to this breakthrough were careful laboratory experiments in prototypical flows, first in 2-D, time periodic flows (e.g., Ottino *et al.*, 1988) then in 3-D flows (e.g., Fountain *et al.*, 1998). In particular, studies in simple periodic 2-D flows have helped visualize chaos (**Figure 45**): unmixed regions translate, stretch, and contract periodically, and represent the primary obstacle to efficient mixing. Particle trajectories in chaotic regions separate exponentially in time, and material filaments are continuously stretched and folded by means of horseshoes. The details seen in **Figure 45** would have been impossible to see in computer simulations since numerical diffusion is greater than the molecular diffusion in the fluid experiments. However, a special effort has been made in mixing studies to couple to the extent possible simple lab experiments and numerical simulations of the same flow, since the latter give much more ready access to the velocity field and

**Figure 45** Mixing produced by discontinuous cavity flow. The top wall moves from left to right for half a period and then stops, then the bottom wall moves from right to left and stops, and so on. Visualization is provided by a fluorescent tracer dissolved in glycerine excited by a longwave ultraviolet light of 365 nm. Folding and an island of unmixed material are clearly seen. From Ottino JM, Leong CW, Rising H, and Swanson PD (1988) Morphological structures produced by mixing in chaotic flows. *Nature* 333: 419–426.

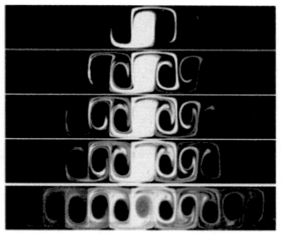

**Figure 46** Transport of uranine dye along an array of convection cells oscillating with a amplitude of 0.12. Time (from the top): 1, 2, 3, 4, 10 oscillations. From Solomon TH, Tomas S and Warner (1996) The Role of Lobes in Chaotic Mixing of Miscible and Immiscible Impurities. *Physical Review Letters* 77: 2682–2685.

other quantities such as the time-dependent stretching field. Recent theoretical work has provided a solid connection between maxima of the stretching field and invariant manifolds of the flow (Haller and Yang, 2000; Haller, 2001). In a flow which can be described analytically, the comparison between computations of stretching and dye visualization shows that the dye spreads over regions that have experienced large stretching. The use of high-resolution PIV technique in laboratory experiments now allows us to calculate the *in situ* stretching field (e.g., Voth *et al.*, 2002). This gives a powerful new insight into the geometrical structure which underlies mixing, even for nonanalytical flow field.

The more specific study of mixing by thermal convection in the laboratory has been confined to 2-D and weakly 3-D patterns. Solomon and Gollub (1988a, 1988b) studied the passive transport of methylene blue and latex spheres by steady 2-D Rayleigh–Bénard convection. The convective flow was monitored by laser Doppler velocimetry and the transport by optical absorption techniques. The transport in this system is determined by the interplay between advection of tracer particles along the streamlines within convection rolls and diffusion of these particles between adjacent rolls. They found that the transport on long space and timescales could be modeled as a diffusive process with an effective

diffusion coefficient $D^* \sim DPe^{1/2} = D(Wd/\pi D)^{1/2}$, where $D$ is the molecular diffusion, $W$ the characteristic velocity of the flow, and $Pe$ the Peclet number. When the 2-D convection is time periodic, the local effective diffusion coefficient becomes independent of the molecular diffusion but depends linearly on the local amplitude of the roll oscillation. The basic mechanism of transport is chaotic advection in the vicinity of oscillating roll boundaries (Solomon and Gollub, 1988b). Solomon *et al.* (1996) verified that $D^*$ is related to the areas of the lobes which are carrying tracers between two cells (**Figure 46**). For weakly time-dependent and 3-D flows, Solomon and Mezic (2003) observed completely uniform mixing when the pattern oscillation period was close to typical circulation times, as predicted by 'singularity-diffusion' theory.

### 3.9.2 Convection in an Initially Stratified Fluid

The most commonly studied situation comprises two superposed fluid layers of different composition, density, and viscosity (**Figure 47**) across which a temperature difference $\Delta T$ is applied. Since convection occurs in the Earth's mantle on large scales (>>1 km), and surface tension acts on millimeter scales, we shall confine our review to the case of two miscible fluids. When the viscosity depends on temperature, $\nu_m$ and $\nu_d$ are taken to be the values at

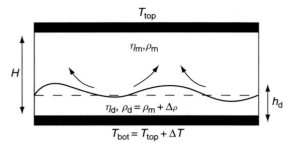

**Figure 47**  Sketch of two-layer convection. The denser layer is initially at the bottom.

the mean temperature of each layer. In cases where the thermal expansion coefficient also depends on temperature (Le Bars and Davaille, 2004a, 2004b), it is taken to be the value at the temperature of the layer interface. Accordingly, the system dynamics are characterized by five dimensionless numbers:

- the Rayleigh number, $Ra(H, \Delta T)$ defined using eqn [1] with $\nu = \max(\nu_d, \nu_m)$;
- the Prandtl number $Pr = \nu/\kappa$;
- the viscosity ratio $\gamma$;
- the depth ratio $a = h_d/H$; and
- the buoyancy ratio, $B = \Delta\rho_X/\rho\alpha\Delta T$, ratio of the stabilizing chemical density anomaly

$\Delta\rho$ to the destabilizing thermal density anomaly.

### 3.9.2.1  Regime diagrams

Early work used linear stability analysis to investigate the stability of various two-layer systems as a

function of Rayleigh number and density contrast (Richter and Johnson, 1974; Busse, 1981; Sotin and Parmentier, 1989). The presence of a deformable interface offers the possibility of Hopf bifurcations: since it introduces an additional degree of freedom, overstability can occur in the form of oscillatory interfacial instability (Richter and Johnson, 1974; Renardy and Joseph, 1985; Renardy and Renardy, 1985; Rasenat *et al.*, 1989; Le Bars and Davaille, 2002; Jaupart *et al.*, 2007). This overstable mode was observed experimentally for the first time by Le Bars and Davaille (2002) (**Figure 48**). At first, numerical studies of two-layer convection at finite amplitude have typically focused on the mechanism of coupling (thermal *vs.* mechanical) between the layers, often treating the interface as impermeable (e.g., Richter and McKenzie, 1981; Kenyon and Turcotte, 1983; Christensen and Yuen, 1984; Cserepes and Rabinowicz, 1985; Boss and Sacks, 1986; Cserepes *et al.*, 1988). For a long time, the highly nonlinear convective regimes associated with high Rayleigh numbers or entrainment at the interface were accessible only to laboratory experiments.

Using aqueous glycerol solutions, Richter and McKenzie (1981) showed that two-layer convection is stable when the ratio of chemical to thermal buoyancy $B$ exceeds unity, but they did not address the issue of partial mixing. Olson (1984) used aqueous sugar solutions to study two-layer convection in the penetrative regime $B \sim 1$ when the viscosity ratio between the two layers is lower than 10. His results

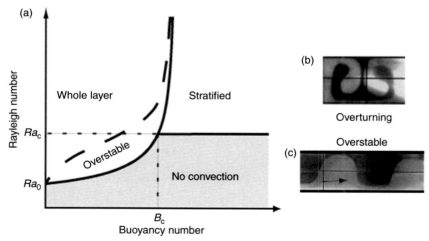

**Figure 48**  (a) Marginal stability curves separating the different convective regimes. The curves here were calculated for the case $\gamma = 6.7$ and $a = 0.5$, where $Ra_0 = 5430$, $Ra_c = 38\,227$ and $B_c = 0.302$. For $B < B_c$, the whole-layer regime develops under the form of (b) overturn for $\gamma$ around 1 and/or low $B$ (here $B = 0.048$; $Ra = 6.7 \times 10^3$; $a = 0.44$; $\gamma = 1.1$), or (c) traveling waves at low Rayleigh number (here $B = 0.20$; $Ra = 1.8 \times 10^4$; $a = 0.5$; $\gamma = 6.7$). From Le Bars M and Davaille A (2002) Stability of thermal convection in two superimposed miscible viscous fluids. *Journal of Fluid Mechanics* 471: 339–363.

established that entrainment at low Reynolds number can indeed occur, and that viscous stresses can substitute for inertial instabilities as a mixing mechanism. Nataf *et al.* (1988) demonstrated that the mechanism of coupling between immiscible fluid layers depends on conditions at the interface (e.g., surface tension and impurities), and that the oscillatory interfacial mode is stabilized by surface tension, as predicted by theory.

Numerous studies now exist on the subject and the domain of parameters investigated is therefore large, with $10^2 < Ra < 10^9$, $10^{-2} < \gamma < 6 \times 10^4$, $0.05 < a < 0.95$, $0.04 < B < 10$. The experiments have been run with constant viscosity (e.g., aqueous Natrosol solutions) or temperature-dependent viscosity fluids, (e.g., sugar or corn syrups) and with constant or temperature-dependent thermal expansion. The different regimes observed in laboratory experiments (Richter and McKenzie, 1981; Olson, 1984; Olson and Kincaid, 1991; Davaille, 1999a, 1999b; Davaille *et al.*, 2002; Le Bars and Davaille, 2002, 2004a; Jellinek and Manga, 2002, 2004; Cottrell *et al.*, 2004; Jaupart *et al.*, 2007), but also in numerical simulations (Schmeling, 1988; Tackley, 1998, 2002; Kellogg *et al.*, 1999; Montague and Kellogg, 2000; Hansen and Yuen, 2000; Samuel and Farnetani, 2002; McNamara and Zhong, 2004) are shown in **Figure 49** and **Table 1** and described in more detail below.

For $B < 0.03$, chemical density heterogeneities are negligible and the system behaves as a single homogeneous fluid (**Figure 49(a)**). For larger values of B and $Ra > 10^5$, two scales of convection coexist: compositionally homogeneous thermal plumes generated at the outer boundaries, and large-scale thermochemical instabilities that involve both fluids.

*Whole-layer regime and unstable doming.* When $B$ is less than a critical value $B_c \sim 0.4$ (the exact value depending on the viscosity and depth ratios; Le Bars and Davaille, 2002; Jaupart *et al.*, 2007), the stable compositional stratification can be overcome by thermal buoyancy. The interface between the layers then becomes unstable, and convection occurs over the whole depth. An important characteristic of the interfacial instability is the 'spouting' direction (Whitehead and Luther, 1975), defined as the direction (up or down) in which finite-amplitude perturbations grow superexponentially to form dome-like structures (**Figures 49(e)** and **49(f)**). The theory of the Rayleigh–Taylor instability shows that spouting occurs in the direction of the

layer with the lower 'resistance' (Ribe, 1998). Thus, the lower layer will spout into the upper one only if

$$b_d < \frac{H}{1 + \gamma^{-1/3}} \qquad [26]$$

otherwise, spouting is downward (Le Bars and Davaille, 2004a). Condition [26] implies, for example, that a less viscous layer will spout into a more viscous overlying mantle only if the former is much thinner. As for the case of purely compositional plumes (**Figures 5** and **18**), the morphology of thermochemical instabilities depends on the viscosity ratio $\gamma$. Suppose for definiteness that [26] is satisfied, so that spouting is upward. Then when the lower layer is more viscous ($\gamma > 1$), large cylindrical diapirs form (**Figure 49(f)**), and secondary plumes can develop above them in the upper layer (Davaille, 1999b). When $\gamma < 1$, by contrast, purely thermal instabilities first develop within the lower layer, and then merge to form large cavity plume heads (**Figure 49(e)**). In both cases, if the lower layer is thin, the plumes empty it before they reach the upper boundary, and become disconnected from the lower boundary as in the purely thermal case (**Figure 26**). If the layer is thicker and/or $0.2 < B < B_c$, the plumes reach the upper boundary before disconnecting from the lower (**Figure 49(e)**). They then begin to cool and lose their thermal buoyancy, and eventually collapse back to the bottom, whereupon the cycle begins again. When $\gamma > 5$ or $\gamma < 0.2$, each layer retains its identity over several pulsations. Since thermal buoyancy must overcome the stable compositional stratification before driving convection, the temperature anomaly $\theta$ carried by thermochemical instabilities is the slave of the compositional field (Le Bars and Davaille, 2004a) and is

$$\theta = (0.98 \pm 0.12)\frac{\Delta\rho_X}{\alpha\rho} \qquad [27]$$

Moreover, the diameter, wavelength, velocity, and cyclicity of the domes are mainly controlled by the more viscous upper layer since it retards motion over the whole depth (e.g., Whitehead and Luther, 1975; Olson and Singer, 1985; Herrick and Parmentier, 1994).

*Stratified regime: anchored hot spots, bumps, ridges, and piles.* For $B > B_c$, thermal buoyancy cannot overcome the stable compositional stratification. Convection remains stratified, and a TBL forms at the interface from which long-lived thermochemical plumes are

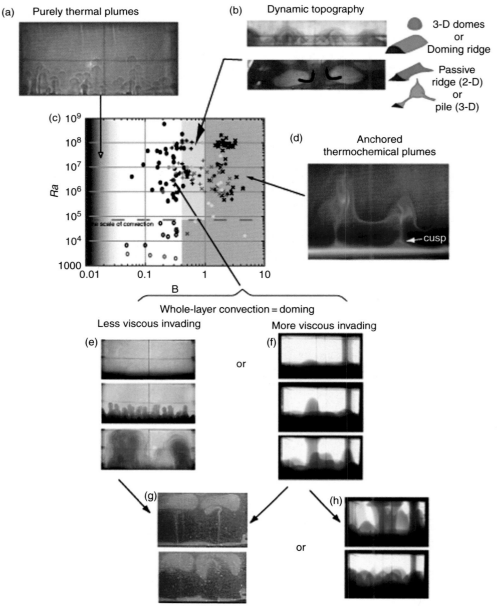

**Figure 49** Regimes of thermochemical convection as a function of Rayleigh number and buoyancy number. Parts (a)–(h) are explained in details in **Table 1**, and in the text. Black circles: whole-layer convection (open circles for experiments close to marginal stability.), '+': stratified convection and anchored plumes with a deformed interface (light shaded area.), 'x': stratified convection with a flat interface, '*': stratified convection and anchored plumes with a non-convecting thinner layer. Laboratory experiments are shown in red (Richter and McKenzie, 1981); dark blue (Olson and Kincaid, 1991); yellow (Jellineck and Manga, 2004); and black (Davaille, 1999a, 1999b; Davaille et al., 2002; Le Bars and Davaille, 2002, 2004a). Numerical calculations are shown in gray (Schmeling, 1988), light blue (Tackley, 1998, 2002), brown (Kellogg et al., 1999; Montague and Kellogg, 2000), purple (Hansen and Yuen, 2000), green (Samuel and Farnetani, 2002), and pink (McNamara and Zhong, 2004).

generated (**Figure 49(d)**). These plumes do not have a well-defined head, and the thermal anomaly they carry is weak, proportional to the temperature difference across the unstable part of the TBL above the interface (Christensen, 1984; Farnetani, 1997; Tackley, 1998). They entrain a thin filament (at most 5% of the total plume volume) of the denser bottom layer by viscous coupling and locally deform

**Table 1** Convective regime and upwellings morphology as a function of $B$, viscosity ratio $\gamma$, layer depth ratio $a$, and internal Rayleigh number of the denser bottom layer $Ra_d$

| $B$ | Regime | $\gamma = \eta_d/\eta_m$ | $Ra_d$ | $a = h_d/H$ | Upwellings morphology | Figure |
|---|---|---|---|---|---|---|
| <0.03 | 1-layer | | | | Thermal plumes | 49a |
| $0.03 < B < B_c \sim 0.4$ | Whole layer | | | | Active domes and passive ridges | |
| | | | $<Ra_c \sim 1000$ | | Passive ridges = return flow to downwellings | 49c |
| | | <1 | $>Ra_c$ | $a < a_c$ $a_c = 1/(1+\gamma^{-1/3})$ | Active hot upwellings • Cavity plumes (or 'mega-plumes') through collection of small thermal instabilities • detach from hot bottom boundary (HBB) | 49e 49g |
| | | | $>Ra_c$ | $a > a_c$ | Passive ridges • return flow to cold more viscous downwellings | 49c |
| | | >1 | $>Ra_c$ | | • Active hot diapirs detach from HBB if a < 0.3 and B < 0.2 continuous fingers from HBB otherwise • Secondary plumes on top of domes | 49f 49g 49h |
| | | $1/5 < \gamma < 5$ | | | Overturning = immediate stirring after first instabilities | |
| | | $\gamma < 1/5$ or $\gamma > 5$ | | | Pulsations = two layers retain their identity for several doming cycles | 49f + 49h |
| $\rightarrow B_c$ | 2-layers | | | | • Stratified convection above and below interface • Anchored hot thermochemical plumes arise from TBL at the interface. | 49b–d 49d |
| $B > 1$ | | | | | Nearly flat interface | |
| $B_c < B < 1$ | | | | | Dynamic topography, does not reach the upper boundary | 49b–c |
| | | <1 | | | Passive ridges (2D) or piles (3D) formed in response to cold viscous downwellings | 49c |
| | | >1 | $<Ra_c$ | | Passive ridges (2D) or piles (3D) | 49c |
| | | | $Ra_c < Ra_d < 10^4$ | | Upwelling ridges | 49b |
| | | | $>10^4$ | | Upwelling domes, or superplumes | 49b |

For depth- or temperature-dependent properties, the viscosities are taken at the averaged temperature of each layer, and $\alpha$ is taken at the interface. We focus on high global $Ra$ (>$10^6$).
Note that if $Ra_d < Ra_c$, the denser layer cannot convect on his own.

the interface into cusps (**Figure 49(d)**; Davaille, 1999b; Jellinek and Manga, 2002, 2004). The interfacial topography and temperature anomaly serves in turn to anchor the plumes (Davaille 1999a, 1999b; Namiki and Kurita, 1999; Davaille *et al.*, 2002; Jellinek and Manga, 2002, 2004; Matsumo *et al.*, 2006), which persist until the chemical stratification disappears through entrainment.

When $B_c < B < 1$, thermal buoyancy is sufficient to maintain local thermochemical 'bumps' on the interface (**Figures 49(b)** and **49(c)**), whose maximum height increases with increasing $Ra$ and decreasing $B$ (Le Bars and Davaille, 2004a; McNamara and Zhong, 2004; Cottrell *et al.*, 2004; Jaupart *et al.*, 2007). If the lower layer is thin enough, it can break up and form stable 'ridges' and 'piles' (Hansen and Yuen, 1988; Tackley 1998, 2002; **Table 1**). The maximum height of the thermochemical bumps is the height of 'neutral' buoyancy, where local compositional buoyancy is just balanced by local thermal buoyancy (LeBars and Davaille, 2004a; Kumagai *et al.*, 2007). Upon reaching this level, the dome can, depending on its internal Rayleigh number, either stagnate or collapse. This generates very complicated morphologies (**Figure 50**).

*Relevance of the experiments to the deep mantle: pressure dependence of B.* The laboratory experiments span well the parameter space of the mantle, except in one respect: the depth dependence of the physical properties. This can be a problem since in the mantle, $B$ could vary with pressure, henceforth depth, because of two effects: (1) the thermal expansion coefficient $\alpha$

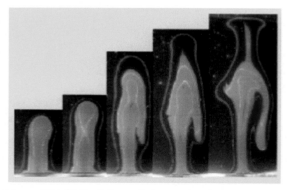

**Figure 50** 'Failing' dome: A thermochemical dome was generated on a flat heater with a constant power. Initially the orange denser material was in a thin layer at the bottom of the tank. The green lines are the isotherms. Courtesy of Ichiro Kumagai and Kei Kurita.

varies with depth (e.g., Tackley, 1998; Montague and Kellogg, 2000; Hansen and Yuen, 2000), and/or (2) the density contrast of compositional origin $\Delta\rho_X$ varies with depth since the two materials have different compressibilities (Tan and Gurnis, 2005; Samuel and Bercovici, 2006).

1. The interface stability is determined by the ability of the thermal density contrast across the interface to overcome the compositional density contrast. The former should therefore be calculated using the value of $\alpha$ at the interface depth, and we do not expect $B_c$ to vary significantly (see discussion in LeBars and Davaille (2004a)). This is indeed confirmed by **Figure 48**, where both the numerical simulation results using a depth-dependent $\alpha$ and laboratory experiments collapse on the same diagram. In the mantle, $\alpha$ decreases with increasing depth, which, for a given compositional density contrast, has two effects on thermochemical convection. First, the regime of a given two-layer system will change with the location of the interface (Davaille, 1999b), from doming in the mid-mantle to stratified just above the CMB. Second, any dome starting deep in the mantle will gain buoyancy as it rises, which might enhance its velocity (LeBars and Davaille, 2004b).

2. Now, Si-enriched (i.e., denser) compositions would also induce a lower compressibility of the dense anomaly compared to the surrounding mantle, implying an increase of the compositional density excess with decreasing depth (Samuel and Bercovici, 2006; Tan and Gurnis, 2005). Therefore, although Si-enriched thermochemical domes could fully develop and rise toward the surface if $B < B_c$ initially, upon reaching the level of neutral buoyancy they would stagnate and spread under it. If this level is significantly higher than the initial interface, large and well-formed plume heads could therefore stagnate at various depths of the lower mantle (Samuel and Bercovici, 2006). If the neutral buoyancy level is around the initial interface depth, tabular less viscous piles with steep vertical walls could form (Tan and Gurnis, 2005). This is the only way to generate less viscous active ridges or piles.

*Stability of the continental lithosphere.* Combining stability analysis and experiments, Cottrell *et al.* (2004) and Jaupart *et al.* (2007) studied the stability of the cooling continental lithosphere (*see* Chapter 7), a thin, compositionally lighter, but also more viscous, layer on top of an (quasi)-infinite mantle. For this combination of transient cooling from above, small depth ratio $a$ and large viscosity ratio $\gamma$, two types of oscillatory instabilities are obtained, depending on the value of $B$ (Jaupart *et al.*, 2007). In particular, for $0.275 < B < {\sim}0.5$, oscillations are vertical and the lithosphere could grow by entrainment of chemically denser material from the asthenosphere. These studies further suggest that the lithosphere could have developed in a state near that of instability with different thicknesses depending on its intrinsic buoyancy.

*Diversity of mantle upwellings.* As described by Ito and van Keken in Chapter 9, hot spots and mantle upwellings present a large range of different signatures, which are difficult to reconcile with one single origin: there are probably different kinds of hot spots and upwellings in the mantle. According to **Table 1** and **Figure 49**, this diversity could be explained by the coexistence of the different regimes of thermochemical convection in a compositionally heterogeneous mantle which convects today primarily on one layer (e.g., Davaille, 1999; Davaille *et al.*, 2005). According to Section 3.7.3, these thermochemical upwellings would develop primarily away from downwellings (e.g., Gonnermann *et al.*, 2004; McNamara and Zhong, 2004), that is, in two mantle 'boxes' separated by the circum-Pacific subduction zones (e.g., Davaille *et al.*, 2002; McNamara and Zhong, 2005).

Then, long-lived hot spots could be due to plumes anchored on the topography of a dense ($B > 0.5$) basal layer, which could also develop piles and ridges. Scaling laws based on laboratory experiments (Davaille, 1999a, 1999b; Davaille *et al.*, 2002; Jellinek and Manga, 2004) suggest that plumes anchored in D″ could develop for $\Delta\rho_X > 0.6\%$, and that they could survive hundreds of millions of years, depending on the spatial extent and magnitude of the anchoring chemical heterogeneity (Davaille *et al.*, 2002; Gonnermann *et al.*, 2002; Jellinek and Manga, 2002, 2004). Such plumes could therefore produce much longer-lived hot spots than the transient plumes observed experimentally in isochemical convection (**Figure 27**). However, plume and piles longevity does not necessarily imply spatial fixity: plumes and their anchors could be advected by large-scale flow associated with strong downwellings such as subducting plates (Tan *et al.*, 2002; Davaille *et al.*, 2002; McNamara and Zhong, 2004).

On the other hand, 'superswells', hot spot clusters and large seismic slow anomalies in the deep mantle could be due to doming of a compositionally denser layer with $B < 0.5$. Then, according to [27], thermochemical instabilities with temperature anomalies 300–500 K could be produced at the base of the mantle by compositional density heterogeneities of 0.3–0.6%. If viscosity depends only on temperature, the instabilities would then be 10–3000 times less viscous than the bulk mantle, and should therefore have the form of cavity plumes. According to [26], the thickness of the dense less viscous layer from which they come must be less than 200–700 km. The scaling laws of Le Bars and Davaille (2004a, 2004b) predict velocities ${\sim}5$–$20\,\mathrm{cm\,yr}^{-1}$, radius ${\sim}500$–$1000\,\mathrm{km}$, spacing ${\sim}2000$–$3500\,\mathrm{km}$, and cyclicity ${\sim}100$–$200\,\mathrm{My}$ for cavity plumes at the base of the mantle. Moreover, heat flux could vary by as much as 200% both laterally and spatially on those same time and length scales (Davaille *et al.*, 2003; Namiki and Kurita, 2003; Le Bars and Davaille, 2004b). Note that the spacing predicted being much smaller than the typical size of a mantle 'box', several instabilities should be observed in each box.

One fundamental question now remains: how might the large-scale circulation created by plate tectonics influence the different regims described above, and especially their time dependence? Can cyclic doming be maintained in such a flow? Experiments including plate-scale motions are now needed.

### 3.9.2.2   Entrainment and mixing

*Entrainment from an interface.* Olson's experiments (1984) on two-layer convection established that entrainment at low Reynolds number can indeed occur, and that viscous stresses can substitute for inertial instabilities as a mixing mechanism. The entrainment occurs in two steps: first, the thermal heterogeneities at the interface induce circulations in the two layers; then the viscous drag due to those convective motions becomes sufficient to overcome the negative buoyancy forces due to the stable chemical density gradient, and thin tendrils of material are entrained (**Figure 51(a)**; Sleep, 1988; Lister, 1989). The entrainment rate and the filament thickness depend on the intensity of convection, its geometry, the viscosity ratio, and the buoyancy ratio: the more stable the fluid, the harder it is to entrain. A significant entrainment corresponds also to a cusp height greater than a critical value comparable to the thickness of the thermal boundary layer above the interface (Jellinek and Manga, 2004). Scaling laws have been derived (Davaille 1999a, 1999b; Davaille (a)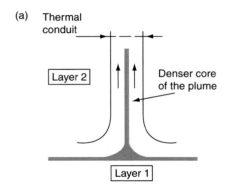

(b)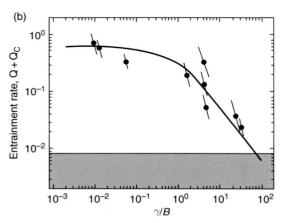

**Figure 51**   (a) Sketch of a thermochemical plume. (b) Volumetric entrainment rate of a thermochemical plume divided by the convective scale as a function of $g/B$. The circles represent the experimental measurements, while the solid line stands for the theoretical scaling. The shaded area shows the domain where salt diffusion becomes important. From Davaille A, Girard F, and Le Bars M (2002) How to anchor plumes in a convecting mantle? *Earth and Planetary Science Letters* 203: 62–634.

(a)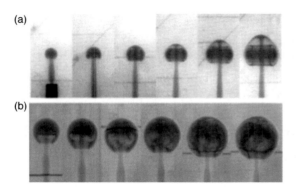

(b)

(c)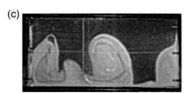

**Figure 52**   Entrainment in compositional plumes (a, b) and thermochemical domes (c) by viscous coupling. (a) Low viscosity ratio: stirring within the plume head follows a vortex ring structure. (b) high viscosity ratio: stirring within the head is chaotic. (c) Mixing pattern in whole layer thermochemical convection. The lower layer, initially denser and more viscous, is dyed with fluoresceine. The experimental tank is illuminated by a laser sheet. After three cycles, the two layers are still not mixed, but one can distinguish filaments and blobs of dark, light fluid within the viscous domes. The thin filaments were entrained by viscous coupling while the blobs were encapsulated when the viscous dome collapsed upon cooling. Since the blobs are lighter, they are bursting up through the viscous fluid as Rayleigh–Taylor instabilities. ($B = 0.30$; $Ra = 2.7 \times 10^7$; $a = 0.3$; $\gamma = 175$). (b) From Kumagai I (2002) On the anatomy of mantle plumes: Effect of the viscosity ratio on entrainment and stirring. *Earth and Planetary Science Letters* 198: 211–224. (c) From Davaille A, Le Bars M, and Carbonne C (2003) Thermal convection in a heterogeneous mantle. *Comptes Rendus De L Academie Des Sciences Géosciences* 335(1): 141–156.

*et al.*, 2002; Jellinek and Manga, 2002, 2004) which explain the data well (e.g., **Figure 51(c)**).

*Stirring.* The mixing pattern in thermochemical convection (**Figure 52**) is very complex, and still poorly characterized since convection creates compositional heterogeneities with two different typical sizes and topologies: (1) thin filaments generated by mechanical entrainment across the interface, leading to the development of 'marble cake' structures in both reservoirs, even though they can remain dynamically separated for a long time and (2) domes, and blobs encapsulated within them, are generated by instabilities.

*Experimental limitation: species diffusion.* The geodynamical irrelevance of surface tension effects requires the use of miscible fluids in the laboratory. Therefore,

chemical diffusion becomes a problem when the ratio of chemical to thermal diffusivity (the Lewis number) $Le = D/\kappa$ is too small. In the mantle, this ratio is $>10^9$. In the laboratory experiments reported here, it was always greater than 1000. Even then, chemical diffusion becomes important when convection is slow (cf. Section 3.9.1), and the mechanism of entrainment across the interface then switches to advection-enhanced diffusion (see the discussion in Davaille (1999a), and gray area in **Figure 51(b)**). Another problem arises when three-component solutions are used (e.g., water + salt + Natrosol) in which the two dissolved species (e.g., salt and Natrosol) do not diffuse at the same speed. Then, if the more viscous layer is initially the upper layer (i.e., the layer with no salt is the layer with the most Natrosol), salt-Natrosol fingers develop, even before the onset of thermal convection. This phenomenon, well known in thermohaline convection, changes drastically the physics and the entrainment rate at the interface. This led Davaille (1999a, 1999b) to discard those experiments. This could also explain the results of Namiki (2003), who observed, using Natrosol solutions, that the interfacial entrainment rate when the more viscous layer was on the top was different than when it was at the bottom.

### 3.9.3   Interaction of a Plume with a Density and/or Viscosity Interface

This situation is of interest for the mantle since the 660-km depth seismic discontinuity corresponds to the phase transition from the spinel phase at low pressure to perovskite at higher pressure. Although this endothermic phase transition probably cannot stratified mantle convection, it is sufficient to delay flow (*see* Chapters 2 and 5). Moreover, the perovskite phase is also 3–100 times stiffer than the upper mantle phases.

Laboratory experiments coupled to numerical modelling have shown that a viscosity jump alone is sufficient to modify the shape of an upwelling compositional plume, the head of which becomes elongated as it crosses the discontinuity from the stiffer lower mantle to the less viscous upper mantle (Manga *et al.*, 1993). For sufficently large viscosity contrast, the plume head can even accelerate so much in the upper mantle that it becomes disconnected from the plume tail, which aggregates at the interface and forms a second plume 'head' (Bercovici and Mahoney, 1994). If the plume is compositionally heterogeneous, segregation can occur at the interface and the two head events then have different geochemical signatures (Kumagai and Kurita, 2000). These phenomena would explain well the emplacement timing and the geochemical data at Ontong Java Plateau (Bercovici and Mahoney, 1994; Kumagai and Kurita, 2000).

Kumagai *et al.* (2007) extended the previous studies to thermal plumes and investigated the interaction between a lower-layer thermal plume and an inner interface. The interaction mode depends on the local buoyancy number ($B_L$: the ratio of the stabilizing chemical buoyancy to the plume thermal buoyancy at the interface), $Ra$, and $\gamma$. For $B_L < 0.6$, the 'pass-through' mode develops, whereby a large volume of the lower material rises through the upper layer and reaches the top surface, since the plume head has a large thermal buoyancy compared to the stabilizing

density contrast between the two layers. When $B_L > 0.6$, the 'rebirth' mode occurs, where the thermal plume ponds and spreads under the chemical boundary and secondary thermal plumes are generated from the interface. Depending on the magnitude of $B_L$, these plumes can entrain a significant amount of lower-layer material upwards by viscous coupling.

## 3.10   Mid-Ocean Ridges and Wax Tectonics

The study of ridges and accretion is challenging because it involves processes which involves both brittle fracture and ductile flow, which is quite difficult to study numerically. For example, current numerical simulations still cannot reproduce the coupled fluid–solid deformation processes responsible for microplates at mid-ocean ridges. On the other hand, published results from a wax analog model yielded the first observations of overlapping spreading centers (OSCs), morphological precursors to microplates, before their discovery on the seafloor (Oldenburg and Brune, 1972). Two main questions have been tackled with laboratory experiments: the ability of a buoyant mantle upwelling to break the crust, and the morphology of ridges.

### 3.10.1   Can a Buoyant Mantle Upwelling Generate Sufficient Stress to Rupture a Brittle Crust?

Using surface layers of various brittle and visco-elastic materials, Ramberg and Sjostrom (1973) studied whether a relatively stiff crustal layer could break in tension above a buoyant diapir and become displaced laterally in a manner simulating the breakup of Pangea and the wandering continents. For better visualization of the deformation, the material was initially dyed in different colored layers. The model was then centrifuged for several minutes at accelerations up to 3000g. All experiments show the breakup of the brittle layer by the diapir head. However, the material rheology was not sufficently well characterized to permit a quantitative study, or to investigate the morphology of the ridges.

### 3.10.2   Morphology of Ridges

Ridge morphology has been studied using paraffin wax as an analog mantle material: solid wax simulates brittle mantle and molten material simulates the

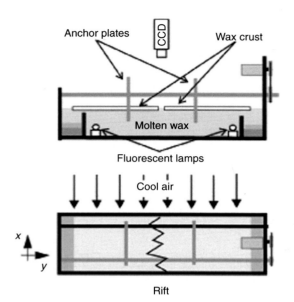

**Figure 53** Experimental setup. The tank is heated from below at constant temperature and cooled from above by a constant flow of air. Before each run, a layer of wax is allowed to grow on the surface until thermal equilibrium is reached. Two skimmers embedded in the solid wax are attached to a threaded rod which is driven by a stepping motor. The rift is initiated with a straight cut through the wax, perpendicular to the spreading direction. Divergence at this cut causes liquid wax to rise into the rift and solidify. Illumination from below permits to image the plate thickness at the rift from above using a video camera. From Ragnarsson R, Ford JL, Santangelo CD, and Bodenschatz E (1996) Rifts in spreading wax layers. *Physical Review Letters* 76: 3456–9.

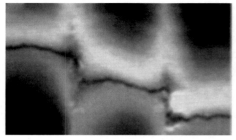

**Figure 54** Transmission image of transform faults, obtained in Shell Wax Callista 158 and with a spreading rate of 78 mm s$^{-1}$. Image size is 9 mm. From Bodenschatz E, Gemelke N, Carr J, and Ragnarsson R (1997) Rifts in spreading wax layers: An analogy to the mid-ocean rift formation. *Localization Phenomena and Dynamics of Brittle and Granular Systems*. Columbia University.

region that deforms as a viscous fluid (Oldenburg and Brune, 1972, 1975; Ragnarsson *et al.*, 1996). The molten wax was frozen at the surface by a flow of cold air. Then the solid crust was pulled apart with constant velocity and a rift was formed separating the crust into two solid plates (**Figure 53**). In this case, the solid wax thickness increases as the square root of the distance to the ridge (Oldenburg and Brune, 1975), in agreement with theory (Parker and Oldenburg, 1973).

Oldenburg and Brune (1972) first observed that a straight rift initially perpendicular to the pulling direction evolved into a pattern consisting of straight segments interrupted by faults orthogonal to the rift and parallel to the pulling direction (**Figure 54**). The ability of a wax to generate orthogonal transform faults depends on the ratio of the shear strength of its solid phase to the resistive stresses acting along the transform fault: if the latter exceed the former, a breakup of the solid near the transform fault wall should occur (Oldenburg and Brune, 1975).

Ragnarsson *et al.* (1996) and Bodenschatz *et al.* (1997) studied the phase diagram in more detail, investigating the rift structure as a function of the pulling speed for two different waxes. The regime diagram was found to depend critically on the rheology of the wax:

1. In the first case (Shell Wax 120), the solid layer could be divided in two regions because the paraffin wax undergoes a solid–solid phase transition: a colder phase, hard and brittle, where strain is mostly accomodated by crack formation during extension; and a warmer ductile phase, where strain is accomodated primarily by viscous flow. This wax gave birth to the different regimes shown in **Figure 55** (Ragnarsson *et al.*, 1996), but the orthogonal transform faults observed by Oldenburg and Brune (1972, 1975) were never observed. At slow spreading rates, the rift, initially perpendicular to the spreading direction, was stable (**Figure 55(a)**). Above a critical spreading rate, a 'spiky' rift developed with fracture zones almost parallel to the spreading direction (**Figure 55(b)**). At yet higher spreading rates a second transition from the spiky rift to a zigzag pattern occurred (**Figure 55(c)**). With further increase of the spreading rate the zigzags steepened (**Figure 55(d)**). In **Figure 55(a)** and **55(b)**, the rift was frozen, whereas in **Figures 55(c)** and **55(d)**, molten fluid reached the surface. In the zigzag regime, the angle of the interface with respect to the spreading direction can be described by a simple geometrical model, since the problem is dominated by only two velocities, the externally given spreading rate (pulling speed) and the internally selected maximal growth

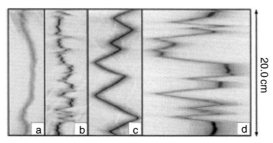

20.0 cm

**Figure 55** Different regimes as the spreading velocity increases. From Ragnarsson R, Ford JL, Santangelo CD and Bodenschatz E (1996) Rifts in spreading wax layers. *Physical Review Letters* 76 3456–3459.

velocity of the solidifying fronts. When the spreading rate exceeds the solidification speed, the interface has to grow at an angle to keep up. The experimental data on the angles agree well with this model.

2. In the second case (Shell Wax Callista 158), the solid phase is more brittle (Bodenschtaz *et al.*, 1997; Oblath *et al.*, 2004). Three distinct morphological regimes were observed. At slow spreading rates, a straight rift is stable and forms a topographic low. At moderate rates, it becomes unstable and OSCs and microplates form, evolve, and die on the ridge, which has little or no relief. At high spreading rates, the microplates lose their internal rigidity and become transform faults at the ridge (**Figure 54**), as previously described by Oldenburg and Brune (1972, 1975). The ridge corresponds to a topographic high

(Bodenschtaz *et al.*, 1997; Oblath *et al.*, 2004). Katz *et al.* (2005) focused on the formation of microplates (**Figure 56**). In wax, like on Earth, they originate from OSCs. The latter nucleate predominantly on sections of obliquely spreading rift, where the rift normal is about 45° from the spreading direction.

These experiments are very interesting for pattern formation, and show striking geometrical similarities with what is observed on the ocean floor. However, their relevance to mid-ocean ridge systems is still questionable for two reasons: first, dynamic similarity does not obtain, since the viscosity contrast between the solid and the liquid wax is much greater than the viscosity contrast across the solid lithosphere. Experiments involving wax are therefore probably more relevant to the dynamics of ice satellites (e.g., Manga and Sinton, 2004) or of lava lakes. Second, the physics of the wax experiments reported above is far from being understood. For example, the influence of latent heat effects on the dynamics of the wax systems remains unknown. To apply their results to the Earth mid-ocean ridges system, Oldenburg and Brune (1972, 1975) neglected these effects and proposed a semi-quantitative theory based on shrinkage and the mechanical properties of wax. The more recent experiments, although they very nicely showed the diversity of rift morphologies, were not able to relate them quantitatively to the wax cooling, nor to its rheology. So a quantitative

(a)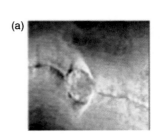

(b)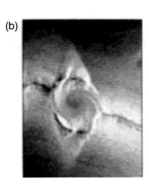

(c)

**Figure 56** Time series of images showing a growing microplate: (a) 15 s, (b) 30 s, (c) 60 s. Green arrows show the direction of spreading of the main plates and the direction of microplate rotation. In red, the spreading ridge and the pseudo-faults pair. The inner ones were generated with the kinematic model of Shouten *et al.* (1993). From Katz RF, Ragnarsson R, and Bodenschatz E (2005) Tectonic Microplates in a Wax Model of Sea Floor Spreading. *New Journal of Physics* 7, doi:10.1088/1367-2630/7/1/037.

understanding of lithosphere rifting remains to be achieved, and will require a better knowledge of fluid rheology.

## 3.11   Subduction-Related Experiments

The subduction of oceanic lithosphere is the most prominant signature of thermal convection in the Earth's mantle. Subducting plates constitute the cold downwellings of mantle convection, which sink into the mantle because of their negative buoyancy. Subduction involves a wide variety of physical and chemical processes: earthquakes, volcanism (dehydration, melting, and melting migration), phase transformations, thermal effects, mantle circulation, and plate motion. They are reviewed in detail by King in Chapter 8. We focus here on the experimental work which has been restricted to the study of the mantle-scale dynamics of subduction, namely the relations between motion of subducted slabs, mantle flow, and back-arc spreading (**Figure 57**). Laboratory models of these phenomena have three advantages: they are inherently 3-D; they do not suffer from limited computational resolution of the temperature and velocity fields; and it is easy to work with strong viscosity variations.

### 3.11.1   Ingredients for Subduction

It has been long recognized that cold plates are not produced by isoviscous convection, and that the strong temperature dependence of mantle viscosity is required (e.g., Torrance and Turcotte, 1971), as we

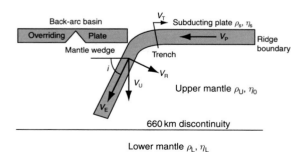

**Figure 57**   Sketch of a subduction zone with the main features studied in the laboratory. $V_P$: surface plate velocity; $V_T$: retrograde trench velocity; $V_R$: retrograde slab velocity; $V_E$: downdip velocity; $V_U$: sinking velocity; $i$: dip angle; $d$: slab thickness, $H$: upper layer thickness.

showed in Sections 3.5 and 3.6. Turner (1973) illustrated this idea using glycerine in a tank cooled from above, in which convective overturn was forced by a stream of bubbles of cold $CO_2$ released from dry ice at the bottom of the tank. Surface cooling then produced a thin viscous sheet which was driven away from regions of upward convection, and then plunged steeply into the less viscous interior, maintaining its identity as it did so. Jacoby (1970, 1973, 1976) and Jacoby and Schmeling (1982) studied the gravitational sinking of cold viscous high-density lithosphere into asthenosphere. Jacoby first studied the mode of sinking of a heavy elastic rubber sheet into water under isothermal conditions without active 'mantle' convection (Jacoby, 1973). Later, he used the same paraffin as Oldenburg and Brune (1975) to include the effects of thermal convection. The paraffin was heated from below above its melting point, and cooled from above below its melting point. It therefore formed a thin solid dense skin on top of the fluid layer. However, the skin could not be generated or melted at the same rate as that of the sinking, preventing real plate tectonics behavior (Jacoby, 1976). Moreover, surface tension prevented thick skins from sinking by themselves. However, sinking could be initiated by loading the edge of the skin or by wetting its surface with molten paraffin (**Figure 58**). The heavy skin first bends with a radius of curvature which depends on the skin thickness, then accelerates down the trench as the portion within the mantle increases. Upon interacting with the tank bottom, the skin can 'fold' and the trench can even 'roll back' in the oceanward direction (**Figure 58**). Convection cells caused by localized heating at the bottom of the tank do not perturb significantly the slab's evolution. Although these pioneering experiments were only 'semi-quantitative' (Jacoby, 1976), they already reproduced most features of subduction (down-dip motion and roll back, 3-D, time dependence, folding on the box bottom, etc.), as well as highlighting the difficulties that quantitative experiments have to overcome (subduction initiation, surface tension effects, etc.).

Kincaid and Olson (1987) designed experiments to further study the lateral migration of slabs and their penetration through the transition zone (**Figure 4**). Cold, negatively buoyant molded slabs of concentrated sucrose solution were introduced into a more dilute, two-layer sucrose solution representing the upper and lower mantle. The transition zone was modeled by a step increase in both density and viscosity. The initial setup consisted of two horizontal

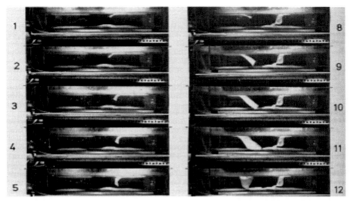

**Figure 58** Series of photographs of two sinking paraffin skins, taken at approximately 3 s intervals with a longer interval between 5 and 8. Changes in slab dip, retrograde motion, and folding on the bottom boundary are clearly seen. From Jacoby WR (1976) Paraffin model experiment of plate tectonics. *Tectonophysics* 35: 103–113.

plates separated by a trench gap, with one plate attached to a dipping slab, simulating a developing subduction zone with an overriding plate (**Figure 57**). The trailing end ('ridge') of the slab was either locked ($V_P = 0$), or free to move along with the plate. Besides the Prandtl number [2], important dimensionless numbers are the ratio of slab thickess to mantle depth $\delta/H$ and the ratios $\eta_S/\eta_U$ and $\eta_S/\eta_L$ of the slab viscosity to those of the upper and lower mantle. It was necessary to have $\eta_S/\eta_L > 10^3$ to maintain a tabular shape during subduction. No velocity is imposed and the slab sinks under its own weight. This phenomenon can be described by an effective Rayleigh number, defined in terms of the slab/upper layer density contrast $\rho_S - \rho_U$:

$$Ra_S = \frac{(\rho_S - \rho_U)gH^3}{\kappa \eta_U} \qquad [28]$$

In the experiments, $Ra_S \sim 1.2 \times 10^5$. On the other hand, the style of slab penetration through the density discontinuity depends on the ratio of the slab buoyancy in the upper and the lower mantle:

$$B_S = \frac{\rho_S - \rho_L}{\rho_S - \rho_U} \qquad [29]$$

Homogeneous fluids correspond to $B_S = 1.0$, while negative values correspond to strong stratification. In agreement with the numerical study by Christensen and Yuen (1984), Kincaid and Olson (1987) found three modes of slab penetration (**Figure 59(a)**). Regime I, in which $B_S \leq -0.2$, corresponds to strong stratification and virtually no slab penetration. In the intermediate regime II ($-0.2 \leq B_S \leq 0.5$), there is partial slab penetration and formation of a root beneath the interface. When

$B_S > 0.5$, the stratification is weak enough to permit the slab to sink into the lower layer with minor deformation (regime III). Retrograde slab motion and trench migration occurred in nearly every case, being greatest in regime I (**Figure 59(b)**), episodic in regime II, and nearly absent in regime III, especially when $\eta_L/\eta_U$ is around 1. Moreover, the episodicity of trench rollback is related to the interaction of the slab with the density stratification, and is greatest in the case of a fixed ridge. The complexity of the time-dependent slab motion, both downdip and retrograde, and its interaction with any bottom boundary, made it difficult to derive scaling laws to describe subduction. Work in the last 20 years has aimed at obtaining those scalings, using experiments where more parameters were controlled.

### 3.11.2 The Surface Story and the Initiation of Subduction

Subduction initiates within the lithosphere and produces large-scale tectonic features. The first analog modeling of tectonic processes probably dates back to 1815 when Sir James Hall attempted to model folds in geological strata, using layered beds of clay and clothes. Numerous experiments have been performed since to reproduce lithosphere and crustal dynamics. It is beyond the scope of this chapter to review them all (for recent reviews, see Schellart (2002)). We shall describe here only studies relevant to the nucleation and evolution of subduction.

Lithospheric processes necessitate taking into account the complex rheology of the lithosphere. Shemenda (1992, 1993) considered the subduction

(a)                                        (b)

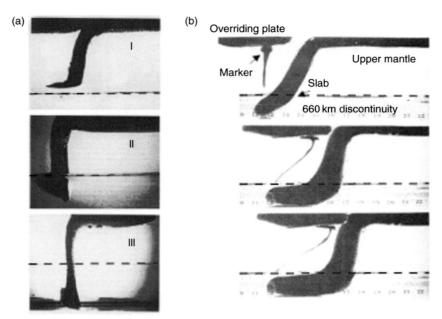

**Figure 59**    (a) Styles of slab penetration through a density discontinuity: I, slab deflection with $B_s \sim -0.2$; II, partial slab penetration and formation of a root beneath the interface with $B_s \sim 0$; III, complete slab penetration for $B_s > 0.5$. (b) Trench migration and retrograde subduction with the fixed ridge boundary condition. The dye streak indicates the flow pattern in the back arc edge. From Kincaid C and Olson P (1987) An experimental study of subduction and slab migration. *Journal of Geophysical Research* 92: 13832–13840.

of an elasto-plastic lithosphere (a mixture of powder and hydrocarbons) into a low-viscosity asthenosphere (water), driven by both horizontal gravitational sinking (i.e., plate/asthenosphere density contrast) and an horizontal compressional force driven by a piston. However, subduction did not initiate without the presence of pre-existing discontinuities. In this case, the horizontal compression produced buckling instabilities, then localization of deformation, leading to failure and eventually subduction of the lithosphere. Trench rollback was then accompanied by back-arc opening along pre-existing faults. The lithosphere has also been modeled as a succession of brittle and ductile layers of different strengths and rheologies (e.g., Davy and Cobbold, 1991). Following this approach, Pinet and Cobbold (1992) and Pubellier and Cobbold (1996) studied the consequences of oblique subduction, that is, the partitioning between down-dip motion and transverse motion along faults. They used a sand mixture to model the brittle upper crust, silicone putty for the ductile lower crust/upper mantle, and glucose syrup for the asthenosphere.

Faccenna *et al.* (1996, 1999) and Becker *et al.* (1999) use the same layered system to study the behavior of an ocean-continent plate system subjected to

compressional strain over geological timescales. Compressional stress was achieved by displacing a piston at constant velocity perpendicular to the plate margin (**Figure 60**). The coupled interface between oceanic and continental plates is found to resist a rapid surge of compressional stress and the shortening is accomodated by undulations within the ocean plate (Martinod and Davy, 1994). But if the system evolves under a low compressive strain rate (slow ridge-push), the oceanic plate becomes unstable to

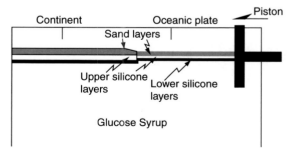

**Figure 60**    Four layers setup to study the initiation of subduction. From Faccenna C, Giardini D, Davy P, and Argentieri A (1999) Initiation of subduction at Atlantic-type margins: Insights from laboratory experiments. *Journal of Geophysical Research* 104: 2749–2766.

RTI and subduction develops (Faccenna *et al.* 1999). The trench is then first localized at the ocean-continent boundary because of the lateral heterogeneity there. The dynamics of the system can therefore be controlled by the 'buoyancy number' $F$, the ratio of driving buoyant to resisting viscous forces (e.g., Houseman and Gubbins, 1997):

$$F(\rho_S - \rho_U, \delta_S, \eta_S, U) = \frac{(\rho_S - \rho_U)g\delta_S^2}{\eta_S U} = \frac{\sigma}{U/\delta_S} \quad [30]$$

where $U$ is the imposed horizontal velocity, and $\sigma$ is the RTI growth rate for a more viscous slab (see Section 3.4). If $F(\rho_S - \rho_U, \delta_S, \eta_S, U) < 1$, the RTI is inhibited and the oceanic plate is deformed by folding. Subduction initiates if $F(\delta_S) > 1$, that is, if the negative buoyancy force of the oceanic plate exceeds the viscous resisting force of the model lithosphere. Faccenna *et al.* (1999) further show that the passive margin behavior is not sensitive to the high shear strength of the brittle layer but only to the resistance of the ductile layer. So subduction intiation is essentially a viscous fluid process. Since its growth rate $\sigma$ depends on the inverse of the slab viscosity (which is high), it is a slow process. However, once the developed slab sinks into the less viscous mantle, its style and dynamics are predominantly governed by a buoyancy number $F(\rho_S - \rho_U, H, \eta_U, U)$ based on the viscosity and the thickness of the upper mantle (Becker *et al.*, 1999). The velocity and angle of subduction and the rate of trench rollback are found to be strongly time dependent and to increase exponentially over tens of millions years before the slab interacts with the 660 km discontinuity.

### 3.11.3  Mantle Flow Induced by a Cold Rigid Slab

The use of rigid slabs enables one to maintain constant along the slab the dip angle $i$ and the slab material velocity $V_E$, and to control easily both $V_E$ and the velocity $V_R$ of retrograde motion either by varying the trench velocity $V_T$ or the time variation of the dip angle $di/dt$ (**Figure 60**). In this setup, all plate motions are kinematically determined. They can be scaled to the mantle through the Peclet number (eqn [4]) $Pe = V_E.\delta/\kappa$, which compares advective heat transport and conductive transport. For mantle subduction, $10 \leq Pe \leq 400$. This means that the slab will have reached 660 km depth before having lost much heat.

#### 3.11.3.1  Generation of seismic anisotropy

Motivated by the growing body of evidence that trench rollback could be associated with a particular pattern of seismic anisotropy, whereby the *a*-axis of olivine crystals would align parallel to the trench oceanward of the slab and turn around its edges (e.g., Russo and Silver, 1994), Buttles and Olson (1998) used small cylinders (whiskers) suspended in a viscous fluid as an analog to olivine crystals, and studied their orientation in vertical and horizontal cross-sections in the vicinity of a subducting slab. The slab was simulated by a rigid Plexiglas plate whose dip angle, down-dip motion, and rollback were controlled independently (**Figure 60**), and the mantle was modeled with Newtonian corn syrup. The whole system was isothermal. They show that trench-parallel olivine *a*-axis orientation in the seaward-side mantle is indeed controlled by the amount of slab rollback, and that orientations in the mantle wedge depend upon slab dip angle.

#### 3.11.3.2  Mantle flow and thermal evolution of the slab

Kincaid and Griffiths (2003, 2004) used the same kind of setup to study the variability of flow and temperature in subduction zones. The slab was made of a composite laminate cooled to a prescribed temperature before the experiment, and containing temperature sensors to monitor its thermal evolution as it is forced into an isothermal tank of hotter glucose syrup. The thermal boundary layers developing in the fluid were visualized by a Moiré technique and the flow field measured using tiny air bubbles and PIV (**Figure 61**). The experiments $Pe$ ranged between 80 and 450. The mantle return flow induced by pure longitudinal sinking creates two cells, one oceanward and one in the wedge (**Figure 61**), where the velocity can reach 40% of the slab downdip speed. Rollback subduction induces flow both around and beneath the sinking slab (**Figure 25**), with larger velocities in the wedge and flow focused toward the center of the plate. These 3-D flows strongly influence the thermal evolution and structure of the plate: they speed up slab reheating. Moreover, the highest slab reheating occurs along the slab centerline when there is rollback, while it occurs along the edges for pure longitudinal sinking.

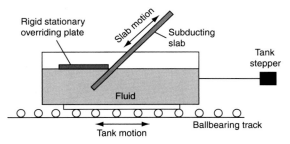

**Figure 61** Typical experimental setup allowing to control independently the trench velocity and the slab motion (velocity and dip angle). The subducting slab can be either a rigid plate mounted on a stepper (e.g., Buttles and Olson, 1998; Kincaid and Griffiths, 2003, 2004), or a viscous fluid extruded at a constant rate from a thermostated reservoir (e.g., Guillou-Frottier et al., 1995; and Griffiths et al., 1995).

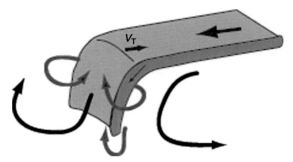

**Figure 63** Schematic sketch of subduction and slab rollback-induced flow. Down-dip motion induces two poloidal flow cells on each side of the slab (black arrows), while slab rollback expels oceanward mantle material into the wedge, creating three toroidal flow cells, two around the lateral slab edges, and one around the tip of the slab when the slab has not reached yet the bottom of the box (or the density stratification).

### 3.11.4    Coupled Mantle-Slab Fluid System

These experiments consider the interactions between the mantle and a mature, purely viscous slab. The subduction is induced either by extruding a viscous slab with constant velocity and dip angle (e.g., **Figure 61**), or by bending the tip of an horizontal slab and forcing it into the mantle until the slab sinks under its own weight (e.g., **Figure 57**).

#### 3.11.4.1    Fixed rollback velocity

These experiments were designed to further investigate the interaction of a sinking slab with a density and viscosity interface, mimicking the discontinuity at 660 km depth between the upper and lower mantles (Kincaid and Olson, 1987) (**Figures 62** and **63**). The slab velocity and initial dip angle were this time imposed. Griffiths and Turner (1988) injected tabular slabs and cylindrical plumes vertically ($V_T = 0$) onto a fluid interface under isothermal conditions and determined the conditions under

which folding of the slab or plume occurred. Mixtures of water and Golden Syrup were used to control the viscosity and density of the slab and the mantle layers. As in the experiments of folding on a solid surface (Cruickshank and Munson, 1981; **Figure 64(a)**), they found that folding occurs in a compressed slab when the ratio of its length to its width exceeds a critical value (∼6). They also determined the amount of fluid entrained within the folds in the lower layer, and when this entrainment could sufficiently reduce the buoyancy of the slab to prevent its sinking further in the lower mantle. Griffiths et al. (1995) and Guillou-Frottier et al. (1995) extended these results to retreating slabs ($V_E$ and $V_T$ constant). Griffiths et al. (1995) used again isothermal mixtures of Golden Syrup and water so that their slab was between 1.5 and 50 times more viscous than their lower mantle. Guillou-Frottier et al. (1995) used

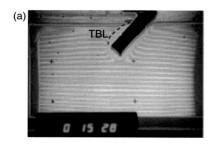

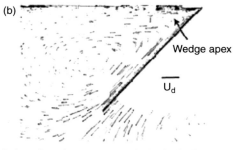

**Figure 62** Rigid slab experiments. (a, b) A cold rigid slab is introduced at constant velocity in the fluid, with no trench roll back. (a) The cold TBL developing around the slab is visualized with a Moiré technique. (b) Corner flow motions in the mantle wedge visualized with streaks. From Kincaid C and Griffiths RW (2004) Variability in flow and temperatures within mantle subduction zones. *Geochemistry, Geophysics, Geosystems* 5: Q06002 (doi:10.1029/2003GC000666).

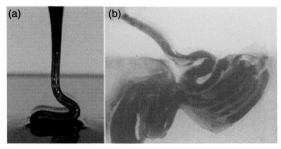

**Figure 64** (a) A sheet of viscous fluid poured on a surface can develop folds and form a pile (b) The same folding and piling behavior is observed when a viscous slab sinking in another viscous fluid encounters a viscosity and/or density interface. Here the slab viscosity is 1000 times that of the lower layer. (a) From Ribe NM (2003) Periodic folding of viscous sheets. *Physical Review E* 68: 036305; (b) from Guillou-Frottier L, Buttles J, and Olson P (1995). Laboratory experiments on structure of subducted lithosphere. *Earth and Planetary Science Letters* 133: 19–34.

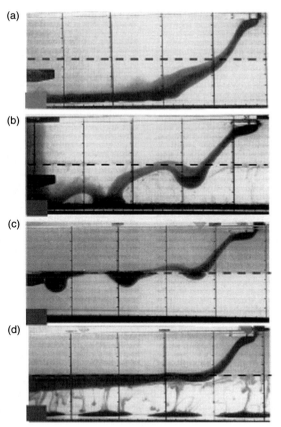

**Figure 65** Different regimes of slab interaction with a viscosity/density interface for a slab which is between 3 and 50 times more viscous than the lower layer. The trench velocity increases from (a) to (c) and the viscosity contrast from (c) to (d). From Griffiths RW, Hackney RI, and van der Hilst RD (1995). A laboratory investigation of effects of trench migration on the descent of subducted slabs. *Earth and Planetary Science Letters* 133(1–2): 1–17.

corn syrup, and extruded a very cold and viscous slab into their two-layer mantle, so that the experiments $Pe \sim 30$–150 and the slab was $10^2$–$10^4$ more viscous than the lower mantle. Both sets of experiments highlighted the essential role of trench rollback, in additon to $B_S$ and $\eta_L/\eta_U$, in the slab penetration regime. Increasing trench rollback enhances slab flattening on the interface (**Figures 59** and **65**). Increased resistance in the lower mantle also promotes folds and piles (**Figure 64(b)**), which can either stagnate in the lower mantle or sink to the bottom (Guillou-Frottier *et al.*, 1995). The amplitude of the observed folds is well predicted by numerically determined scaling laws, which also predict amplitudes (400–500 km) consistent with tomographic images of slabs beneath some subduction zones (Ribe *et al.*, 2007). On the other hand, when the slab is not too viscous, RTIs have time to develop from the slab resting on the interface, and to sink into the lower mantle (**Figure 65**; Griffiths *et al.*, 1995). When, however, the slab material reaches the depth of neutral buoyancy, it will spread at this level as a gravity current (Kerr and Lister, 1987; Lister and Kerr, 1989). Slab deformation modes therefore finally depend on three parameters: the slab to mantle viscosity ratio $\eta_S/\eta_L$, the ratio of sinking (Stokes) velocities in the upper and lower mantle ($\sim B_S\eta_U/\eta_L$), and the ratio of the slab's horizontal velocity to its vertical velocity. As the latter can change through time in Nature, a slab probably passes through a number of different deformation modes during its history.

### 3.11.4.2 Free rollback velocity

If rollback controls in part the slab deformation at depth, what happens if the trench is free to move in response to the mantle flow? A number of systematic studies have recently been devoted to the dynamics of a slab sinking under its own weight (**Figure 66(a)**). Regardless of the boundary conditions, subduction is always strongly time dependent. After initiation, subduction accelerates as the mass of slab in the mantle increases (slab-pull) while the slab dip is close to 90°. Then, subduction slows down after interaction with the 660 km interface, reaching a steady state when the slab spreads on the interface, followed by eventual penetration, accompanied by folding (Becker *et al.*, 1999; Funiciello *et al.*, 2003, 2004, 2006; Schellart, 2004a, 2004b, 2005). Besides, the amount

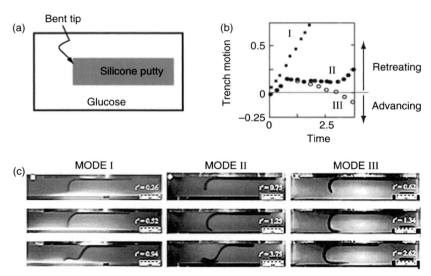

**Figure 66** (a) Experimental setup used by the Roma III group. A slab of silicone putty is spread horizontally on Glucose Syrup. Subduction is initiated by forcing its tip 3 cm into the glucose. (b) Trench motion vs. time for the three modes of subduction images in (c). From Bellahsen N, Faccenna C, and Funiciello F (2005) Dynamics of subduction and plate motion in laboratory experiments: Insights into the 'plate tectonics' behavior of the Earth. *Journal of Geophysical Research* 110: B01401 (doi:10.1029/2004JB002999).

of trench and slab rollback depends on the degree of lateral confinement of the flow (Funiciello *et al.*, 2003, 2004; Schellart, 2004a, 2005), and is much reduced in 2-D or when the side walls are within 600 km of the slab edges (Funiciello *et al.*, 2003). As in the experiments with rigid slabs (Buttles and Olson, 1998; Kincaid and Griffiths, 2003, 2004), slab rollback induces flow around the slab edges. The lateral flow in turn forces the hinge line to adopt a convex shape toward the direction of retreat (Funiciello *et al.*, 2003; Schellart, 2004a; **Figure 63**). By varying the thickness, width, viscosity, and density of the slab and mantle, Bellahsen *et al.* (2005) observed three characteristic modes of subduction for a slab connected to a plate with a free ridge (**Figures 11(b)** and **11(d)**): a retreating trench mode (mode I), a retreating trench mode following a transient period of advancing trench (mode II), and an advancing trench mode (mode III). The same modes are found when the plate velocity $V_P$ is fixed at the ridge ('ridge push'; Funiciello *et al.*, 2004; Schellart, 2005): a relatively low $V_P$ results in relatively fast hinge-retreat $V_T$ with backward sinking of the slab and a backward draping slab geometry (mode I). With increasing $V_P$, hinge migration is relatively small, resulting in subvertical sinking of the slab and a folded slab piling geometry (mode II). For very high $V_P$, the hinge migrates forward, resulting in a forward draping slab geometry. These three modes are characterized by different

partitioning of the subduction into plate and trench motion. The lithospheric radius of curvature, which depends upon plate characteristics (stiffness and thickness) and the mantle thickness, exerts a primary control on the trench behavior. Moreover, the subduction velocity results from the balance between acting and resisting forces, where lithospheric bending represents 75–95% of the total resisting forces (Bellahsen *et al.*, 2006). Martinod *et al.* (2005) studied how ridges and plateaus on the subducting lithosphere modify the subduction regime. Using feature tracking image analysis, Funiciello *et al.* (2006) further investigated the influence of plate width and mantle viscosity/density on rollback and the induced poloidal and toroidal components of the mantle circulation. The poloidal component is the response to the viscous coupling between the slab motion and the mantle, while the toroidal component is produced by lateral slab migration. The experiments show that both components are important from the beginning of the subduction process, and strongly intermittent.

The experiments have therefore established that mantle subduction is a strongly time-dependent and 3-D phenomenon. Hence, mantle flow in subduction zones cannot be correctly described by models assuming a 2-D steady-state process. Subduction dynamics involve slab-pull, ridge-push, but also trench rollback and interaction with the 660 km depth discontinuity. All these effects

explain the diversity of slab deformation and subduction history observed in the laboratory and in
Nature (e.g., Kincaid and Olson, 1987; Guillou-
Frottier *et al.*, 1995; Griffiths *et al.*, 1995; Faccenna
*et al.*, 2001; Schellart, 2004a). Despite the multiplicity of regimes, time-dependent trench rollback is
the most common occurrence, which explains well
the formation of back-arc basins (e.g., Faccenna
*et al.*, 2001). The challenge now is to obtain scaling
laws to predict the complete regime diagram of
subduction and the characteristics of each regime.
This will require including thermal effects, and
adressing the problem of lithospheric rheology.
All the experiments which have been done so far
used a slab with Newtonian rheology. However, the
true 'effective' viscosity of the lithosphere is still
controversial. In the laboratory, it has been chosen
either $10^3$–$10^4$ times more viscous than the mantle
(e.g., Olson and co-workers, Funiciello and co-
workers) so that the slab retains its tabular shape,
or only 10–100 times more viscous (e.g., Griffiths
and co-workers, Schellart), which produced somewhat different results. Another problem is to
estimate the influence of surface tension, which is
unavoidable in laboratory experiments with free
surfaces (Jacoby, 1976) and which must be
accounted for to obtain quantitative measurements
of the forces resisting subduction and laboratory-
based scaling laws for trench motion that can be
applied to the mantle.

## 3.12   Conclusions

Quantitative analog laboratory experiments have
played an indispensable role in advancing the field
of geodynamics since the earliest days of its existence as a discipline. Reflecting on this history helps
one to understand more clearly what an experimental approach has to offer. At the top of any list must
surely be the discovery of new phenomena.
Laboratory experiments have always been the
source of many of the most exciting new discoveries
in fluid mechanics generally, and geodynamics itself
offers numerous examples: a partial list would
include the convective planforms in a variable-viscosity fluid, the diversity of thermal and
compositional plumes, the entrainment in plumes,
and the different modes of subduction. A second
important contribution of the experimental
approach is to the formation of influential new concepts and models. An obvious example here is the

classical 'head plus tail' picture of a mantle plume; it
is no exaggeration to say that this model had its
origin almost entirely in the remarkable laboratory
experiments of Whitehead and Luther (1975),
Olson and Singer (1985), and Griffiths (1986a)
among others, and in the beautiful photographic
images that came from them. A third important
role that laboratory experiments can play is to test
theories arrived at using other methods: one thinks
for example of the experiments performed by
Whitehead and Luther (1975) to test the linear
theory of the Rayleigh–Taylor instability.
Experimental tests are especially important whenever the theory in question involves assumptions of
an asymptotic nature, or whose validity is not
obvious.

The rapid evolution of high-performance numerical computing in recent years has provided still
another role for experimental geodynamics: that of
verifying and benchmarking complex numerical
codes for phenomena such as convection with temperature-dependent viscosity (Busse *et al.*, 1994),
subduction (Schmeling *et al.*, 2004), and convection
with a chemical field (van Keken *et al.*, 1997). Such
benchmarks are particularly important for 3-D time-
dependent flows with large variations in material
properties, which are still difficult to treat reliably
by numerical means. By the same token, the ability of
numerical models to deliver detailed representations
of 2-D and 3-D field variables has encouraged the
development of powerful new techniques for quantitative flow visualization in the laboratory, such as the
observation of the temperature field using TLCs
(e.g., Rhee *et al.*, 1984; Davaille *et al.*, 2006), or the
determination of the velocity and stretching fields by
PIV (e.g., Adrian, 1991; Voth *et al.*, 2002; Funiciello
*et al.*, 2006).

While the future of experimental geodynamics is
impossible to predict, it is not hard to identify some
critical scientific questions where an experimental
approach is likely to be fruitful. One is the nature of
entrainment and mixing between different mantle
reservoirs, which involves phenomena occurring on
short length scales and with little diffusion that are
hard to resolve numerically. A second is the generation of plate tectonics by mantle convection, which
cannot be understood in terms of traditional
Newtonian fluid mechanics. Modeling this process
in the laboratory will require new developments in
the characterization and experimental deployment of
fluids with complex rheology.

# References

Adrian RJ (1983) Laser velocimetry. In: Goldstein RJ (ed.), *Fluid Mechanics Measurements*, pp. 155–244. Washington: Hemisphere.

Adrian RJ (1991) Particle imaging techniques for experimental fluid mechanics. *Annual Review of Fluid Mechanics* 23: 261–304.

Ahlers G (2001) Effect of sidewall conductance on heat-transport measurements for turbulent Rayleigh–Bénard convection. *Physical Review E* 62: 015303.

Ahlers G (2005) Experiments with Rayleigh–Bénard convection. In: Mutabazi I, Jose EW, and Guyon E (eds.) *Springer Tracts in Modern Physics: Dynamics of Spatio-Temporal Cellular Structures – Henri Benard Centenary Review*, pp. 67–94. New York: Springer.

Albers M and Christensen UR (2001) Channeling of plume flow beneath mid-ocean ridges. *Earth and Planetary Science Letters* 187: 207–220.

Allègre CJ (1987) Isotope geodynamics. *Earth and Planetary Science Letters* 86: 175–203.

Aref H (1984) Stirring by chaotic advection. *Journal of Fluid Mechanics* 143: 1–15.

Asaeda T and Watanabe K (1989) The mechanism of heat transport in thermal convection at high Rayleigh numbers. *Physics of Fluids A* 1: 861–867.

Asseban A, Lallemand M, Saulnier J-B, et al. (2000) Digital speckle photography and speckle tomography in heat transfer studies. *Optics and Laser Technology* 32: 583–592.

Assenheimer M and Steinberg V (1993) Rayleigh–Bénard convection near the gas–liquid critical point. *Physical Review Letters* 70: 3888–3891.

Assenheimer M and Steinberg V (1996) Observation of coexisting upflow and downflow hexagons in Boussinesq Rayleigh–Bénard convection. *Physical Review Letters* 76: 756–759.

Baker DJ (1966) A technique for the precise measurement of small fluid velocities. *Journal of Fluid Mechanics* 26: 573–575.

Batchelor GK (1954) Heat convection and buoyancy effects in fluids. *Quarterly Journal of the Royal Meteorological Society* 80: 339–358.

Becker T, Faccenna C, Giardini D, and O'Connell R (1999) The development of slabs in the upper mantle: Insights from numerical and laboratory experiments. *Journal of Geophysical Research* 104: 15207–15226.

Bellahsen N, Faccenna C, and Funiciello F (2005) Dynamics of subduction and plate motion in laboratory experiments: Insights into the 'plate tectonics' behavior of the Earth. *Journal of Geophysical Research* 110: B01401 (doi:10.1029/2004JB002999).

Bénard H (1900) Les tourbillons cellulaires dans une nappe liquide. *Revue Generale des Science Pures et Appliquees* 11: 1261–1271; 1309–1328.

Bénard H (1901) Les tourbillons cellulaires dans une nappe liquide transportant la chaleur par convection en régime permanent. *Annales de Chimie et de Physique* 23: 62–144.

Bercovici D (2003) The generation of plate tectonics from mantle convection. *Earth and Planetary Science Letters* 205: 107–121.

Bercovici D and Mahoney J (1994) Double flood basalts and plume head separation at the 660-kilometer discontinuity. *Science* 266: 1367–1369.

Bercovici D and Kelly A (1997) Nonlinear initiation of diapirs and plume heads. *Physics of the Earth and Planetary Interiors* 101: 119–130.

Bercovici D, Ricard Y, and Richards MA (2000) The relation between mantle dynamics and plate tectonics: A primer. In: Richards MA, Gordon R, and Van der Hilst R (eds.) AGU Geophysical Monograph 21: *The History and Dynamics of Global Plate Motions*, pp. 5–46. Washington DC: AGU.

Bercovici D, Ricard Y, and Schubert G (2001) A two-phase model for compaction and damage, 1. General theory. *Journal of Geophysical Research* 106: 8887–8906.

Bercovici D, Ricard Y, and Schubert G (2001) A two-phase model for compaction and damage, 3. Applications to shear localization and plate boundary formation. *Journal of Geophysical Research* 106: 8925–8940.

Biot MA and Ode H (1965) Theory of gravity instability with variable overburden and compaction. *Geophysics* 30: 213–227.

Bodenschatz E, Gemelke N, Carr J, and Ragnarsson R (1997) Rifts in spreading wax layers: An analogy to the mid-ocean rift formation. *Localization Phenomena and Dynamics of Brittle and Granular Systems*. Columbia University.

Bodenschatz E, Pesch W, and Ahlers G (2000) Recent developments in Rayleigh–Bénard convection. *Annual Review of Fluid Mechanics* 32: 709–773.

Bonatti E (1985) Punctiform initiation of seafloor spreading in the Red Sea during transition from a continental to an oceanic rift. *Nature* 316: 33–37.

Booker JR (1976) Thermal convection with strongly temperature dependent viscosity. *Journal of Fluid Mechanics* 76: 741–754.

Booker JR and Stengel KC (1978) Futher thoughts on convective heat transport in a variable viscosity fluid. *Journal of Fluid Mechanics* 86: 289–291.

Boss AP and Sacks IS (1986) High spatial resolution models of time-dependent, layered mantle convection. *Geophysical Journal of the Royal Astronomical Society* 87: 241–264.

Boxman RL and Shlien DJ (1978) Interferometric measurement technique for the temperature field of axisymmetric buoyant phenomena. *Applied Optics* 17: 2788–2793.

Boxman RL and Shlien DJ (1981) Laminar starting plume temperature field measurement. *Journal of Heat and Mass Transfer* 24: 919–931.

Brown E, Nikilaenko A, Funfschilling D, and Ahlers G (2005) Heat transport in turbulent Rayleigh–Bénard convection: Effect of finite top- and bottom-plate conductivities. *Physics of Fluids* 17: 075108.

Buck WR (1987) Analysis of the cooling of a variable viscosity fluid with applications to the Earth. *Geophysical Journal of the Royal Astronomical Society* 89: 549–577.

Busse FH (1967a) On the stability of two-dimensional convection in a layer heated from below. *Journal of Applied Mathematics and Physics* 46: 140–149.

Busse FH (1967b) The stability of finite amplitude cellular convection and its relation to an extremum principle. *Journal of Fluid Mechanics* 30: 625–635.

Busse FH (1978) Non linear properties of thermal convection. *Reports on Progress in Physics* 41: 1929–1967.

Busse FH (1981) On the aspect ratios of two-layer mantle convection. *Physics of the Earth and Planetary Interiors* 24: 320–324.

Busse FH (1989) Fundamentals of thermal convection. In: Peltier WR (ed.), *Mantle Convection: Plate tectonics and Global Dynamics*, pp. 23–95. Montreux: Gordon and Breach.

Busse FH and Whitehead JA (1971) Instabilities of convection rolls in high Prandtl number fluid. *Journal of Fluid Mechanics* 47: 305–320.

Busse FH, Christensen U, Clever R, et al. (1994) 3D convection at infinite Prandtl number in cartesian geometry – A benchmark comparison. *Geophysical and Astrophysical Fluid Dynamics* 75: 39–59.

Buttles J and Olson P (1998) A laboratory model of subduction zone anisotropy. *Earth and Planetary Science Letters* 164: 245–262.

Campbell IH and Griffiths RW (1990) Implications of mantle plume structure for the evolution of flood basalts. *Earth and Planetary Science Letters* 99: 79–93.

Canright D and Morris S (1993) Buoyant instability of a viscous film over a passive fluid. *Journal of Fluid Mechanics* 255: 349–372.

Cardin P and Nataf H-C (1991) Nonlinear dynamical coupling oberserved near the threshold of convection in a two-layer system. *Europhysics Letters* 14: 655–660.

Carrigan CR (1982) Multiple scale convection in the Earth's mantle: A three dimensional study. *Science* 215: 965–967.

Castaing B, Gunaratne G, Heslot F, et al. (1989) Scaling of hard turbulence in Rayleigh–Bénard convection. *Journal of Fluid Mechanics* 204: 1–30.

Cazenave A, Monnerau M, and Gibert D (1987) Seasat gravity undulations in the central indian ocean. *Physics of the Earth and Planetary Interiors* 48: 130–141.

Chandrasekhar S (1977) *Liquid Crystals*, 342 p. Cambridge: Cambridge University Press.

Chandrasekhar S (1961) *Hydrodynamic and Hydromagnetic Stability*. New York: Dover.

Chavanne X, Chillà F, Chabaud F, Castaing B, and Hebral B (2001) Turbulent Rayleigh–Bénard convection in gaseous and liquid He. *Physics of Fluids* 13: 1300–1320.

Chay A and Shlien DJ (1986) Scalar field measurements of a laminar starting plume cap using digital processing of interferograms. *Physics of Fluids* 29: 2358–2366.

Chen F and Chen CF (1989) Experimental investigation of convective stability in a superposed fluid and porous layer when heated from below. *Journal of Fluid Mechanics* 207: 311–321.

Chen MM and Whitehead JA (1968) Evolution of two-dimensional periodic Rayleigh convection cells of arbitrary wave-numbers. *Journal of Fluid Mechanics* 31: 1–15.

Chilla F, Rastello M, Chaumat S, and Castaing B (2004) Long relaxation times and tilit sensitivity in Rayleigh–Bénard turbulence. *European Physical Journal B* 40: 223–227.

Choblet G and Sotin C (2000) 3D thermal convection with variable viscosity: Can transient cooling be described by a quasi-static scaling law? *Physics of the Earth and Planetary Interiors* 119: 321–336.

Christensen U (1984) Instability in a hot boundary layer and initiation of thermochemical plumes. *Annales de Geophysique* 2: 311–320.

Christensen U (1985) Heat transport by variable viscosity convection II: Pressure influence, non-Newtonian rheology and decaying heat sources. *Physics of the Earth and Planetary Interiors* 37: 183–205.

Christensen U and Yuen DA (1984) The interaction of a subducting slab with a chemical or phase boundary. *Journal of Geophysical Research* 89: 4389–4402.

Christensen UR and Harder H (1991) 3-D Convection with variable viscosity. *Geophysical Journal International* 104: 213–226.

Christensen UR and Hofmann AW (1994) Segregation of subducted oceanic crust in the convecting mantle. *Journal of Geophysical Research* 99: 19867–19884.

Ciofalo M, Signorino M, and Simiano M (2003) Tomographic particle-image velocimetry and thermography in Rayleigh–Bénard convection using suspended thermochromic liquid crystals and digital image processing. *Experiments in Fluids* 34: 156–172.

Cottrell E, Jaupart C, and Molnar P (2004) Marginal stability of thick continental lithosphere. *Geophysical Research Letters* 31 (doi: 10.1029/2004GL020332).

Coulliette DL and Loper DE (1995) Experimental, numerical and analytical models of mantle starting plumes. *Physics of the Earth and Planetary Interiors* 92: 143–167.

Crane K (1985) The spacing of rift axis highs: dependence upon diapiric processes in the underlying asthenosphere? *Earth and Planetary Science Letters* 72: 405–414.

Cruickshank JO and Munson BR (1981) Viscous fluid buckling of plane and axisymmetric jets. *Journal of Fluid Mechanics* 113: 221–239.

Cserepes L and Rabinowicz M (1985) Gravity and convection in a two-layered mantle. *Earth and Planetary Science Letters* 76: 193–207.

Cserepes L, Rabinowicz M, and Rosemberg-Borot C (1988) Three-dimensional infinite Prandtl number convection in one and two layers with implications for the Earth's gravity field. *Journal of Geophysical Research* 93: 12009–12025.

Dabiri D and Gharib M (1991) Digital particle image thermometry: The method and implementation. *Experiments in Fluids* 11: 77–86.

Dabiri D and Gharib M (1996) The effects of forced boundary conditions on flow within a cubic cavity using digital particle image thermometry and velocimetry (DPITV). *Experimental Thermal and Fluid Science* 13: 349–363.

Dalziel SB, Hughes GO, and Sutherland BR (2000) Whole field density measurements by 'synthetic schlieren'. *Experiments in Fluids* 28: 322–345.

Davaille A and Jaupart C (1993) Transient high-Rayleigh number thermal convection with large viscosity variations. *Journal of Fluid Mechanics* 253: 141–166.

Davaille A and Jaupart C (1994) Onset of thermal convection in fluids with temperature-dependent viscosity: Application to the oceanic mantle. *Journal of Geophysical Research* 99: 19853–19866.

Davaille A (1999a) Two-layer thermal convection in viscous fluids. *Journal of Fluid Mechanics* 379: 223–253.

Davaille A (1999b) Simultaneous generation of hotspots and superswells by convection in a heterogeneous planetary mantle. *Nature* 402: 756–760.

Davaille A, Girard F, and Le Bars M (2002) How to anchor plumes in a convecting mantle? *Earth and Planetary Science Letters* 203: 62–634.

Davaille A, Le Bars M, and Carbonne C (2003) Thermal convection in a heterogeneous mantle. *Comptes Rendus De L Academie Des Sciences Géosciences* 335(1): 141–156.

Davaille A and Vatteville J (2005) On the transient nature of mantle plumes. *Geophysical Research Letters* 32 (doi:10.1029/2005GL023029).

Davaille A, Limare A, Vidal V, et al. (2006) Imaging isotherms in viscous fluids. *Experiments in Fluids* (submitted).

Davies GF (1988) Ocean bathymetry and mantle convection, 1. Large-scale flow and hotspots. *Journal of Geophysical Research* 93: 10467–10480.

Davy P and Cobbold PR (1991) Experiments on shortening of a 4-layer continental lithosphere. *Tectonophysics* 188: 1–25.

Deardorff JW, Willis GE, and Lilly DK (1969) Laboratory investigation of non-steady penetrative convection. *Journal of Fluid Mechanics* 35: 7–31.

de Gennes P and Prost J (1993) *The Physics of Liquid Crystals*, 2nd edn., 597 pp. Oxford: Clarendon Press.

Dubuffet F, Rabinowicz M, and Monnereau M (2000) Multiple scales in mantle convection. *Earth and Planetary Science Letters* 178: 351–366.

Dumoulin C, Doin M-P, and Fleitout L (2001) Numerical simulations of the cooling of an oceanic lithosphere above a convecting mantle. *Physics of the Earth and Planetary Interiors* 125: 45–64.

Eberle MA and Forsyth DW (1995) Regional viscosity variations, small-scale convection and slope of the depth-age$^{1/2}$ curve. *Geophysical Research Letters* 22: 473–476.

Ekström G and Dziewonski AM (1998) The unique anisotropy of the pacific upper mantle. *Nature* 394: 168–172.

Elder JW (1967) Convective self-propulsion of continents. *Nature* 214: 657–660.

Elder JW (1968) The unstable thermal interface. *Journal of Fluid Mechanics* 32: 69–96.

Faccenna C, Davy P, Brun J-P, et al. (1996) The dynamics of back-arc extension: An experimental approach to the opening of the Tyrrhenian Sea. *Geophysical Journal International* 126: 781–795.

Faccenna C, Giardini D, Davy P, and Argentieri A (1999) Initiation of subduction at Atlantic-type margins: Insights from laboratory experiments. *Journal of Geophysical Research* 104: 2749–2766.

Faccenna C, Piromallo C, Crespo-Blanc A, Jolivet L, and Rossetti F (2004) Lateral slab deformation and the origin of the western Mediterranean arcs. *Tectonics* 23: TC1012 (doi:10.1029/2002TC001488).

Farnetani CG (1997) Excess temperature of mantle plumes: The role of chemical stratification across D″. *Geophysical Research Letters* 24: 1583–1586.

Fedotov SA (1975) Mechanism of magma ascent and deep feeding channels of island arc volcanoes. *Bulletin of Volcanology* 39: 241–254.

Feighner MA and Richards MA (1995) The fluid dynamics of plume-ridge and plumeplate interactions: An experimental investigation. *Earth and Planetary Science Letters* 129: 171–182.

Feighner MA, Kellogg LH, and Travis BJ (1995) Numerical modeling of chemically buoyant plumes at spreading ridges. *Geophysical Research Letters* 22: 715–718.

Fleitout L and Yuen DA (1984) Steady-state, secondary convection beneath lithospheric plates with temperature and pressure-dependent viscosity. *Journal of Geophysical Research* 89: 9227–9244.

Fletcher RC (1972) Application of mathematical model to emplacement of mantle gneiss domes. *American Journal of Science* 272: 197–216.

Fountain GO, Khakhar DV, and Ottino JM (1998) Visualization of three-dimensional chaos. *Nature* 281: 683–686.

Fountain GO, Khakhar DV, Mezic I, and Ottino JM (2000) Chaotic mixing in a bounded three-dimensional flow. *Journal of Fluid Mechanics* 417: 265–301.

Fuji T (1963) Theory of the steady laminar natural convection above a horizontal line heat source and a point heat source. *International Journal of Heat and Mass Transfer* 6: 597–606.

Fujisawa N and Adrian R (1999) Three-dimensional temperature measurement in turbulent thermal convection by extended range scanning liquid crystal thermometry. *Journal of Visualization* 1: 355–364.

Fujisawa N and Funatani S (2000) Simultaneous measurement of temperature and velocity in a turbulent thermal convection by the extended range liquid crystal visualization technique. *Experiments in Fluids* 29: 158–165.

Fujisawa N and Hashizume Y (2001) An uncertainty analysis of temperature and velocity measured by a liquid crystal visualization technique. *Measurement Science and Technology* 12: 1235–1242.

Funiciello F, Morra G, Giardini D, and Regenauer-Lieb K (2003a) Dynamics of retreating slabs. Part 1: Insights from 2-D numerical experiments. *Journal of Geophysical Research* 108: 2206 (doi: 10.1029/2001JB000898).

Funiciello F, Faccenna C, Giardini D, and Regenauer-Lieb K (2003b) Dynamics of retreating slabs: Part 2. Insights from three-dimensional laboratory experiments. *Journal of Geophysical Research* 108: 2207 (doi:10.1029/2001JB000896).

Funiciello F, Faccenna C, and Giardini D (2004) Role of lateral mantle flow in the evolution of subduction system: Insights from 3-D laboratory experiments. *Geophysical Journal International* 157: 1393–1406.

Funiciello F, Moroni M, Piromallo C, Faccenna C, Cenedese A, and Bui HA (2006) Mapping mantle flow during retreating subduction: Laboratory models analyzed by feature

tracking. *Journal of Geophysical Research* 111: B03402 (doi:10.1029/2005JB003792).

Funfschilling D, Brown E, Nikolaenko A, and Ahlers G (2005) Heat transport by turbulent Rayleigh-Benard Convection in cylindrical cells with aspect ratio one and larger. *Journal of Fluid Mechanics* 536: 145–155.

Gebhart B, Pera LW, and Schorr A (1970) Steady laminar natural convection plumes above a horizontal line heat source. *International Journal of Heat and Mass Transfer* 13: 161–171.

Geitling AV (1998) *Rayleigh-Bénard Convection*. Singapore: World Scientific.

Giannandrea E and Christensen U (1993) Variable viscosity convection experiments with a stress-free upper boundary and implications for the heat transport in the Earth's mantle. *Physics of the Earth and Planetary Interiors* 78: 139–152.

Globe S and Dropkin D (1959) Natural-convection heat transfert in liquids confined by two horinzontal plates and heated from below. *Journal of Heat Transfer* 31: 24–28.

Gluckman B, Willaime H, and Gollub J (1993) Geometry of isothermal and isoconcentration surfaces in thermal turbulence. *Physics of Fluids A* 5: 647–651.

Goldstein RJ, Chiang HD, and See DL (1990) High Rayleigh-number convection in a horizontal enclosure. *Journal of Fluid Mechanics* 213: 111–126.

Gonnermann HM, Manga M, and Jellinek AM (2002) Dynamics and longevity of an initially stratified mantle. *Geophysical Research Letters* 29 (doi:10. 1029/2002GL01485).

Gonnermann HM, Jellinek AM, Richards MA, and Manga M (2004) Modulation of mantle plumes and heat flow at the core mantle boundary by plate-scale flow: Results from laboratory experiments. *Earth and Planetary Science Letters* 226: 53–67.

Grasset O and Parmentier EM (1998) Thermal convection in a volumetrically, infinite Prandtl number fluid with strongly temperature-dependent viscosity: Implications for planetary evolution. *Journal of Geophysical Research* 103: 18171–18181.

Griffiths RW (1986) Thermals in extremely viscous fluids, including the effects of temperature-dependent viscosity. *Journal of Fluid Mechanics* 166: 115–138.

Griffiths RW and Turner JS (1988) Folding of viscous plumes impinging on a density or viscosity interface. *Geophysical Journal* 95: 397–419.

Griffiths RW and Richards MA (1989) The adjustment of mantle plumes to changes in plate motion. *Geophysical Research Letters* 16: 437–440.

Griffths RW, Gurnis M, and Eitelberg G (1989) Holographic measurements of surface topography in laboratory models of mantle hotspots. *Geophysical Journal* 96: 1–19.

Griffiths RW and Campbell IH (1990) Stirring and structure in mantle starting plumes. *Earth and Planetary Science Letters* 99: 66–78.

Griffiths RW and Campbell IH (1991) Interaction of mante plume heads with the Earth's surface and the onset of small-scale convection. *Journal of Geophysical Research* 96: 18295–18310.

Griffiths RW, Hackney RI, and van der Hilst RD (1995) A laboratory investigation of effects of trench migration on the descent of subducted slabs. *Earth and Planetary Science Letters* 133(1–2): 1–17.

Grigné C, Labrosse S, and Tackley PJ (2005) Convective heat transfer as a function of wavelength: Implications for the cooling of the Earth. *Journal of Geophysical Research* 110: B03409 (doi:10.1029/2004JB003376).

Grossmann S and Lohse D (2000) Scaling in thermal convection: A unifying theory. *Journal of Fluid Mechanics* 407: 27–56.

Guillou L and Jaupart C (1995) On the effect of continents on mantle convection. *Journal of Geophysical Research* 100: 24217–24238.

Guillou-Frottier L, Buttles J, and Olson P (1995) Laboratory experiments on structure of subducted lithosphere. *Earth and Planetary Science Letters* 133: 19–34.

Gurnis M (1988) Large-scale mantle convection and the aggregation and dispersal of supercontinents. *Nature* 335: 695–699.

Gurnis M and Davies GF (1986) The e_ect of depth-dependent viscosity on convective mixing in the mantle and the possible survival of primitive mantle. *Geophysical Research Letters* 13: 541–544.

Gurnis M, Zhong S, and Toth J (2000) On the competing roles of fault reactivation and brittle failure in generating plate tectonics from mantle convection. In: Richards MA, Gordon R, and Van der Hilst R (eds.) AGU Geophysical Monograph 21: *The History and Dynamics of Global Plate Motions*, pp. 73–94. Washington DC: AGU.

Haller G (2001) Finding finite-time invariant manifolds in two-dimensional velocity fields. *Chaos* 10: 99–108.

Haller G and Yuan G (2000) Lagrangian coherent structures and mixing in two-dimensional turbulence. *Physica D* 147: 352–370.

Hanenkamp A and Merzkirch W (2005) Investigation of the properties of a sharp-focusing schlieren system by means of Fourier analysis. *Optics and Lasers in Engineering* 44: 159–169.

Hansen U and Yuen DA (1988) Numerical simulations of thermo-chemical instabilities at the core-mantle boundary. *Nature* 33: 237–240.

Hansen U and Yuen DA (2000) Extended-Boussinesq thermal–chemical convection with moving heat sources and variable viscosity. *Earth and Planetary Science Letters* 176: 401–411.

Haramina T and Tilgner A (2004) Coherent structures in boundary layers of Rayleigh–Bènard convection. *Physical Review E* 69: 056306.

Hauf W and Grigull U (1970) Optical methods in heat transfer. In: Hartnett JP and Irvine TF Jr. (eds.) *Advances in Heat Transfer*, pp. 133–366. New York: Academic Press.

Haxby WF and Weissel JK (1986) Evidence for small-scale convection from seasat altimeter data. *Journal of Geophysical Research* 91: 3507–3520.

Heflinger LO, Wuerker RF, and Brooks RE (1966) Holographic interferometry. *Journal of Applied Physics* 37: 642–649.

Herrick DL and Parmentier EM (1994) Episodic large-scale overturn of two-layer mantles in terrestrial planets. *Journal of Geophysical Research* 99: 2053–2062.

Heslot F, Castaing B, and Libchaber A (1987) Transitions to turbulence in helium gas. *Physical Review A* 36: 5870–5873.

Heutmaker MS and Gollub JP (1987) Wave-vector field of convective flow patterns. *Physical Review A* 35: 242–260.

Hofmann AW and Hart SR (1978) An assessment of local and regional isotopic equilibrium in the mantle. *Earth and Planetary Science Letters* 38: 44–62.

Holmes A (1931) Radioactivity and Earth movements. *Transactions of the Geological Society of Glasgow* 18: 559–606.

Horsmann M and Merzkirch W (1981) Scattered light streaming birefringence in colloidal solutions. *Rheologica Acta* 20: 501–510.

Houseman GA (1990) The thermal structure of mantle plumes: Axisymmetric or triple-junction? *Geophysical Journal International* 102: 25–43.

Houseman GA and Molnar P (1997) Gravitational (Rayleigh–Taylor) instability of a layer with non-linear viscosity and convective thinning of continental lithosphere. *Geophysical Journal International* 128: 125–150.

Houseman GA and Gubbins D (1997) Deformation of subducted oceanic lithosphere. *Geophysical Journal International* 131: 535–551.

Howard LN (1964) Convection at high Rayleigh number. In: Gortler H (ed.), *Proceedings of the 11th International Congress Applied Mechanics*, pp. 1109–1115. Berlin: Springer.

Huang J, Zhong S, and van Hunen J (2003) Controls on sublithospheric small-scale convection. *Journal of Geophysical Research* 108(B8): 2405 (doi:10.1029/2003JB002456).

Ito G (2001) Reykjanes 'V'-shaped ridges originating from a pulsing and dehydrating mantle plume. *Nature* 411: 681–684.

Jacoby WR (1970) Instability in the upper mantle and global plate movements. *Journal of Geophysical Research* 75: 5671–5680.

Jacoby WR (1973) Model experiment of plate movements. *Nature (Physical Sciences)* 242: 130–134.

Jacoby WR (1976) Paraffin model experiment of plate tectonics. *Tectonophysics* 35: 103–113.

Jacoby WR and Schmeling H (1982) On the effects of the lithosphere on mantle convection and evolution. *Physics of the Earth and Planetary Interiors* 29: 305–319.

Jackson MPA and Talbot CJ (1989) Anatomy of mushroom shaped diapirs. *Journal of Structural Geology* 11: 211–230.

Jaupart C, Brandeis G, and Allègre CJ (1984) Stagnant layers at the bottom of convecting magma chambers. *Nature* 308: 535–538.

Jaupart C and Brandeis G (1986) The stagnant bottom layer of convecting magma chambers. *Earth and Planetary Science Letters* 80: 183–199.

Jaupart C and Parsons B (1985) Convective instabilities in a variable viscosity fluid cooled from above. *Physics of the Earth and Planetary Interiors* 39: 41–32.

Jaupart C, Mareschal J-C, Guillou-Frottier L, and Davaille A (1998) Heat flow and thickness of the lithosphere in the canadian shield. *Journal of Geophysical Research* 103: 15269–15286.

Jaupart C, Molnar P, and Cottrell E (2007) Instability of a chemically dense layer heated from below and overlain by a deep less viscous fluid. *Journal of Fluid Mechanics* 572: 433–469.

Javoy M (1999) Chemical Earth models. *Comptes Rendus De L Academie Des Sciences* 329: 537–555.

Jellinek AM, Kerr RC, and Griffiths RW (1999) Mixing and compositional stratification produced by natural convection. 1. Experiments and their application to Earth's core and mantle. *Journal of Geophysical Research* 104: 7183–7201.

Jellinek AM and Manga M (2002) The influence of a chemical boundary layer on the fixity, spacing and lifetime of mantle plumes. *Nature* 41: 760–763.

Jellinek AM, Lenardic A, and Manga M (2002) The influence of interior mantle temperature on the structure of plumes: Heads for Venus, Tails for the Earth. *Geophysical Research Letters* 29 (doi:10.1029/2001GL014624).

Jellinek AM, Gonnermann HM, and Richards MA (2003) Plume capture by divergent plate motions: Implications for the distribution of hotspots, geochemistry of midocean ridge basalts, and heat flux at the core–mantle boundary. *Earth and Planetary Science Letters* 205: 367–378.

Jellinek AM and Manga M (2004) Links between long-lived hotspots, mantle plumes, D'', and plate tectonics. *Reviews of Geophysics* 42 (doi:10.1029/2003RG000144).

Jurine D, Jaupart C, Brandeis G, and Tackley PJ (2005) Penetration of mantle plumes through depleted lithosphere. *Journal of Geophysical Research* 110 (doi:10.1029/2005JB003751).

Kaminski E and Jaupart C (2003) Laminar starting pluimes in high-Prandtl-number fluids. *Journal of Fluid Mechanics* 478: 287–298.

Katz RF, Ragnarsson R, and Bodenschatz E (2005) Tectonic Microplates in a Wax Model of Sea Floor Spreading. *New Journal of Physics* 7 (doi:10.1088/1367-2630/7/1/037).

Katsaros KB, Liu WT, Businger JA, and Tillman JE (1977) Heat thermal structure in the interfacial boundary layer measured in an open tank of water in turbulent free convection. *Journal of Fluid Mechanics* 83: 311–335.

Kellogg LH, Hager BH, and van der Hilst RD (1999) Compositional stratification in the deep mantle. *Science* 283: 1881–1884.

Kelly A and Bercovici D (1997) The clustering of rising diapirs and plume heads. *Geophysical Research Letters* 24: 201–204.

Kenyon PM and Turcotte DL (1983) Convection in a two-layer mantle with a strongly temperature-dependent viscosity. *Journal of Geophysical Research* 88: 6403–6414.

Kerr RC and Lister JR (1987) The spread of subducted lithoshperic material along the mid-mantle boundary. *Earth and Planetary Science Letters* 85: 241–247.

Kerr RC and Lister JR (1988) Island arc and mid-ocean ridge volcanism, modelled by diapirism from linear source regions. *Earth and Planetary Science Letters* 88: 143–152.

Kerr RC and Mériaux C (2004) Structure and dynamics of sheared mantle plumes. *Geochemistry, Geophysics, Geosystems* 5 (doi:10.1029/2004GC000749).

Kimura I, Hyodo T, and Ozawa M (1998) Temperature and velocity measurement of 3D thermal flow using thermosensitive liquid crystals. *Journal of Visualization* 1: 145–152.

Kincaid C and Olson P (1987) An experimental study of subduction and slab migration. *Journal of Geophysical Research* 92: 13832–13840.

Kincaid C, Ito G, and Gable C (1995) Laboratory investigation of the interaction of off-axis mantle plumes and spreading centres. *Nature* 376: 758–761.

Kincaid C, Sparks D, and Detrick R (1996) The relative importance of plate-driven and buoyancy- driven flow at mid-ocean ridge spreading centers. *Journal of Geophysical Research* 101: 16177–16193.

Kincaid C and Griffiths RW (2003) Laboratory models of the thermal evolution of the mantle during rollback subduction. *Nature* 425: 58–62.

Kincaid C and Griffiths RW (2004) Variability in flow and temperatures within mantle subduction zones. *Geochemistry, Geophysics, Geosystems* 5: Q06002 (doi:10.1029/2003GC000666).

Korenaga J and Jordan TH (2003) Physics of multiscale convection in Earth's mantle: Onset of the sublithospheric convection. *Journal of Geophysical Research* 108 (doi:10.1029/2002JB001760).

Kraichman RH (1962) Turbulent thermal convection at arbitrary Prandtl number. *Physics of Fluids* 5: 1374–1389.

Krishnamurti R (1968a) Finite amplitude convection with changing mean temperature. Part 1. Theory. *Journal of Fluid Mechanics* 33: 445–455.

Krishnamurti R (1968b) Finite amplitude convection with changing mean temperature. Part 2. An experimental test of the theory. *Journal of Fluid Mechanics* 33: 457–463.

Krishnamurti R (1970a) On the transition to turbulent convection. Part 1. Transition from two to three dimensional flow. *Journal of Fluid Mechanics* 42: 295–307.

Krishnamurti R (1970b) On the transition to turbulent convection. Part 2. Transition to time-dependent flow. *Journal of Fluid Mechanics* 42: 309–320.

Krishnamurthi R (1979) Theory and experiment in cellular convection. *Proceedings of NATO Advanced Study Institute on Continental Drift and the Mechanism of Plate Tectonics*, pp. 245–257. England: University of Newcastle upon Tyne.

Krishnamurti R and Howard LN (1981) Large scale flow generation in turbulent convection. *Proceedings of the National Academy of Sciences* 78: 1981–1985.

Kulacki FA and Goldstein RJ (1972) Thermal convection in a horizontal fluid layer with volumetric heat sources. *Journal of Fluid Mechanics* 55: 271.

Kulacki FA and Nagle ME (1975) Natural convection in a horizontal fluid layer with volumetric heat sources. *Journal of Heat Transfer* 91: 204–211.

Kumagai I (2002) On the anatomy of mantle plumes: Effect of the viscosity ratio on entrainment and stirring. *Earth and Planetary Science Letters* 198: 211–224.

Kumagai I and Kurita K (2000) On the fate of mantle plumes at density interfaces. *Earth and Planetary Science Letters* 179: 63–71.

Kumagai I, Davaille A, and Kurita K (2007) On the fate of mantle thermal plumes at density interface. *Earth and Planetary Science Letters* 254: 180–193.

Laudenbach N (2001) *Experimentelle und numerische Untersuchungen zur Ausbreitung von Volumenstörungen in thermishen Plumes.* PhD Dissertation, Gottingen.

Laudenbach N and Christensen UR (2001) An optical method for measuring temperature in laboratory models of mantle plumes. *Geophysical Journal International* 145: 528–534.

Le Bars M and Davaille A (2002) Stability of thermal convection in two superimposed miscible viscous fluids. *Journal of Fluid Mechanics* 471: 339–363.

Le Bars M and Davaille A (2004a) Large interface deformation in two-layer thermal convection of miscible viscous fluids. *Journal of Fluid Mechanics* 499: 75–110.

Le Bars M and Davaille A (2004b) Whole-layer convection in an heterogeneous planetary mantle. *Journal of Geophysical Research* 109 (doi:10.1029/2003JB002617).

Lenardic A, Moresi L-N, Jellinek AM, and Manga M (2005) Continental insulation, mantle cooling, and the surface area of oceans and continents. *Earth and Planetary Science Letters* 234: 317–333.

Lewis WT, Incropera FP, and Viskanta R (1982) Interferometric study of stable salinity gradients heated from below or cooled from above. *Journal of Fluid Mechanics* 116: 411–430.

Lister JR (1989) Selective withdrawal from a viscous two-layer system. *Journal of Fluid Mechanics* 198: 231–254.

Lister JR and Kerr RC (1989) The effect of geometry on the gravitational instability of a buoyant region of viscous fluid. *Journal of Fluid Mechanics* 202: 231–254.

Lister JR and Kerr RC (1989) The propagation of two-dimensional and axisymmetric viscous gravity currents along a fluid interface. *Journal of Fluid Mechanics* 203: 215249.

Lithgow-Bertelloni C, Richards MA, Conrad CP, and Griffiths RW (2001) Plume generation in natural thermal convection at high Rayleigh and Prandtl numbers. *Journal of Fluid Mechanics* 434: 1–21.

Lohmann AW and Silva DE (1972) A Talbot interferometer with circular gratings. *Optics Communications* 4: 326–328.

Loper DE and Stacey FD (1983) The dynamical and thermal structure of deep mantle plumes. *Physics of the Earth and Planetary Interiors* 33: 305–317.

Loper DE and Eltayeb IA (1986) On the stability of the D″ layer. *Geophysical and Astrophysical Fluid Dynamics* 36: 229–255.

Loper DE and McCartney K (1986) Mantle plumes and the periodicity of magnetic field reversals. *Geophysical Research Letters* 13: 1525–1528.

Loper DE, McCartney K, and Buzyna G (1988) A model of correlated episodicity in magnetic-field reversals, climate and mass extinctions. *Journal of Geology* 88: 1–14.

Mach E (1873) *Optish-akustusche.* Prague: Calve.

McNamara AK and Zhong S (2004) Thermochemical structures within a spherical mantle: Superplumes or piles?. *Journal of Geophysical Research* 109 (doi:10.1029/2003JB002847).

McNamara AK and Zhong S (2005) Thermochemical structures beneath Africa and the Pacific Ocean. *Nature* 437: 1136–1139.

Malkus WVR (1954) Discrete transitions in turbulent convection. *Proceedings of the Royal Society of London A* 225: 185–195.

Manga M (1997) Interactions between mantle diapirs. *Geophysical Research Letters* 24: 1871–1874.

Manga M, Stone HA, and O'Connell RJ (1993) The interaction of plume heads with compositional discontinuities in the Earth's mantle. *Journal of Geophysical Research* 98: 19979–19990.

Manga M and Weeraratne D (1999) Experimental study of non-Boussinesq Rayleigh–Bénard convection at high Rayleigh and Prandtl numbers. *Physics of Fluids* 10: 2969–2976.

Manga M, Weeraratne D, and Morris SJS (2001) Boundary-layer thickness and instabilities in Bénard convection of a liquid with a temperature-dependent viscosity. *Physics of Fluids* 13: 802–805.

Manga M and Sinton A (2004) Formation of bands, ridges and grooves on Europa by cyclic deformation: Insights from analogue wax experiments. *Journal of Geophysical Research* 109: E09001 (10.1029/2004JE002249).

Marsh BD (1979) Island arc development: Some observations and speculations. *Journal of Geology* 87: 687–713.

Marsh BD and Carmichael ISE (1974) Benioff zone magmatism. *Journal of Geophysical Research* 79: 1196–1206.

Martinod J and Davy P (1994) Periodic instabilities during compression of the lithosphere 2. Analogue experiments. *Journal of Geophysical Research* 99: 12057–12070.

Matsumoto N, Namiki A, and Sumita I (2006) Influence of a basal thermal anomaly on mantle convection. *Physics of the Earth and Planetary Interiors* 157: 208–222.

Maxwell JC (1873) Double refraction of viscous fluids in motion. *Proceedings of the Royal Society of London Series A* 22: 46–47.

Merzkirch W (1987) *Flow Visualization 2nd edn.* Orlando, FL: Academic Press.

Merzkirch W (1989) Streaming birefringence. In: Yang W-J (ed.), *Handbook of Flow Visualization* pp. 177–180. Taylor and Francis.

Merzkirch W (1993) Interferometric methods. *Lecture Series – van Kareman Institute for Fluid Dynamics* 9: 1–96.

Merzkirch W (1995) Density-sensitive whole-field flow measurement by optical speckle photography. *Experimental Thermal and Fluid Science* 10: 435–443.

Merzkirch W, Vitkin D, and Xiong W (1998) Quantitative flow visualization. *Meccanica* 33: 503–516.

Merzkirch W (2000) Non-intrusive measurement techniques. *Lecture series, Van Karman Institute for Fluid Dynamics* 6: 1–37.

Morgan WJ (1971) Convection plumes in the lower mantle. *Nature* 230: 42–43.

Mohr PA and Wood CA (1976) Volcano spacing and lithospheric attenuation in the eastern rift of Africa. *Earth and Planetary Science Letters* 33: 126–144.

Montagner J-P and Tanimoto T (1990) Global anisotropy in the Upper Mantleinferred from the regionalization of phase velocities. *Journal of Geophysical Research* 95: 4797–4819.

Montague LN and Kellogg LH (2000) Numerical models of a dense layer at the base of the mantle and implications for the geodynamics of D". *Journal of Geophysical Research* 105(2000): 11101–11114.

Moore WB, Schubert G, and Tackley P (1998) Three-dimensional simulations of plume-lithosphere interaction at the Hawaiian swell. *Science* 279: 1008–1011.

Moresi L and Solomatov V (1998) Mantle ocnvection with a brittle lithosphere: thoughts on the global tectonic styles of the Earth and Venus. *Geophysical Journal International* 133: 669–682.

Morris S (1982) The effects of a strongly temperature-dependent viscosity on slow flow past a hot sphere. *Journal of Fluid Mechanics* 124: 1–25.

Morris S and Canright DR (1984) A boundary-layer analysis of Benard convection with strongly temperature-dependent viscosity. *Physics of the Earth and Planetary Interiors* 36: 355–373.

Morris SW, Bodenschatz E, Cannell DS, and Ahlers G (1993) Spiral defect chaos in large aspect ratio Rayleigh–Bénard convection. *Physical Review Letters* 71: 2026.

Moses E, Zocchi G, and Libchaber A (1993) An experimental study of laminar plumes. *Journal of Fluid Mechanics* 251: 581–601.

Nagata S (1975) *Mixing: Principles and Applications.* New York: Halsted Press.

Namiki A (2003) Can the mantle entrain D"? *Journal of Geophysical Research* 108 (doi:10.1029/2002JB002315).

Namiki A and Kurita K (1999) The influence of boundary heterogeneity in experimental models of mantle convection. *Geophysical Research Letters* 26: 1929–1932.

Namiki A and Kurita K (2001) The influence of boundary heterogeneity in experimental models of mantle convection with internal heat sources. *Physics of the Earth and Planetary Interiors* 128: 195–205.

Namiki A and Kurita K (2002) Rayleigh-Benard convection with an inclined upper boundary. *Physical Review E* 65: 056301-1–10.

Namiki A and Kurita K (2003) Heat transfer and interfacial temperature of the two-layered convection: Implications for the D"-mantle coupling. *Geophysical Research Letters* 30 (doi:10.1029/2002GL015809).

Nataf H-C, Froidevaux C, Levrat JL, and Rabinowicz M (1981) Laboratory convection experiments: effect of lateral cooling and generation of instabilities in the horizontal boundary layers. *Journal of Geophysical Research* 86: 6143–6154.

Nataf H-C, Hager BH, and Scott RF (1984) Convection experiments in a centrifuge and the generation of plumes in a very viscous fluid. *Annales de Geophysique* 2: 303–310.

Nataf H-C, Moreno S, and Cardin P (1988) What is responsible for thermal coupling in layered convection? *Journal of Physics France* 49: 1707–1714.

Nataf H-C (1991) Mantle convection, plates and hotspots. *Tectonophysics* 187: 361–377.

Neavel KE and Johnson AM (1991) Entrainment in compositionally buoyant plumes. *Tectonophysics* 200: 1–15.

Nettleton LL (1934) Fluid mechanics of salt domes. *American Association of Petroleum Geologists Bulletin* 18: 1175–1204.

Nettleton LL (1943) Recent experimental and geophysical evidence of mechanics of salt-dome formation. *American Association of Petroleum Geologists Bulletin* 27: 51–63.

Niemela JJ, Skrbek L, Sreenivasan KR, and Donelly RJ (2000) Turbulent convection at very high Rayleigh numbers. *Nature* 404: 837–840.

Nishimura CE and Forsyth DL (1988) Rayleigh wave phase velocities in the Pacific with implications for azimuthal anisotropy and lateral heterogeneities. *Geophysical Journal International* 94: 479–501.

Oblath N, Daniels K, and Bodenschatz E (2004) Spreading rate dependent morphology of mid-ocean-ridge-like dynamics. *Physical Review Letters* (submitted).

Oertel H and Buhler K (1978) A special differential interferometer used for heat convection investigations. *International Journal of Heat and Mass Transfer* 21: 1111–1115.

Ogawa M, Schubert G, and Zebib A (1991) Numerical simulations of three-dimensional thermal convection in a fluid with strongly temperature-dependent viscosity. *Journal of Fluid Mechanics* 233: 299–328.

Ogawa M (2003) Plate-like regime of a numerically modeled thermal convection in a fluid with temperature-, pressure-, and stress-history-dependent viscosity. *Journal of Geophysical Research* 108: doi:10.1029/2000JB000069.

Oldenburg DW and Brune JN (1972) Ridge transform fault spreading patterns in freezing wax. *Science* 178: 301–304.

Oldenburg DW and Brune JN (1975) Explanation for the othogonality of ocean ridges and transform faults. *Journal of Geophysical Research* 80: 2575–85.

Olson P (1984) An experimental approach to thermal convection in a two-layered mantle. *Journal of Geophysical Research* 89: 11293–11301.

Olson P and Singer H (1985) Creeping plumes. *Journal of Fluid Mechanics* 158: 511–531.

Olson P and Nam IS (1986) Formation of seafloor swells by mantle plumes. *Journal of Geophysical Research* 91: 7181–7191.

Olson P and Christensen UR (1986) Solitary wave propagation in a fluid conduit within a viscous matrix. *Journal of Geophysical Research* 91: 6367–6374.

Olson P, Schubert G, Anderson C, and Goldman P (1988) Plume formation and lithosphere erosion: a comparison of laboratory and numerical experiments. *Journal of Geophysical Research* 93: 15065–15084.

Olson P (1990) Hot spots, swells and mantle plumes. In: Ryan MP (ed.), *Magma Transport and Storage*, pp. 33–51. New York: Wiley.

Olson P and Kincaid C (1991) Experiments on the interaction of thermal convection and compositional layering at the base of the mantle. *Journal of Geophysical Research* 96: 4347–4354.

Olson P, Schubert G, and Anderson C (1993) Structure of axisymmetric mantle plumes. *Journal of Geophysical Research* 98: 6829–6844.

Ottino JM, Leong CW, Rising H, and Swanson PD (1988) Morphological structures produced by mixing in chaotic flows. *Nature* 333: 419–426.

Ottino JM (1989) *The Kinematics of Mixing: Stretching, Chaos, and Transport*, 364 pp. Cambridge: Cambridge University Press.

Park H, Dabiri D, and Gharib M (2000) Digital particle image velocimetry/thermometry and application to the wake of a heated circular cylinder. *Experiments in Fluids* 30: 327–338.

Parker TJ and McDowell AN (1951) Scale models as a guide to interpretation of salt-dome faulting. *American Association of Petroleum Geologists Bulletin* 35: 2076–2086.

Pearson JRA (1958) On convection cells induced by surface tension. *Journal of Fluid Mechanics* 19: 489–500.

Parmentier EM, Turcotte DL, and Torrance KE (1975) Numerical experiments on the structure of mantle plumes. *Journal of Geophysical Research* 80: 4417–4424.

Parmentier EM and Sotin C (2000) Three-dimensional numerical experiments on thermal convection in a very viscous fluid: Implications for the dynamics of a thermal boundary layer at high Rayleigh number. *Physics of Fluids* 12: 609–617.

Parsons B and McKenzie DP (1978) Mantle convection and the thermal structure of the plates. *Journal of Geophysical Research* 82: 4485–4495.

Parsons B and Sclater JC (1977) An analysis of the variations of ocean floor bathymetry and heat flow with age. *Journal of Geophysical Research* 82: 803–827.

Pindera JT and Krishnamurthy AR (1978) Characteristic relations of flow birefringence: Part 2. Relations in Scattered radiations. *Experimental Mechanics* 18: 41–48.

Pinet N and Cobbold PR (1992) Experimental insights into the partitioning of motion within zones of oblique subduction. *Tectonophysics* 206: 371–388.

Plapp BB (1997) *Spiral-pattern formation in Rayleigh–Bénard convection*. PhD Thesis. Cornell University, Ithaca, NY.

Plapp BB, Egolf DA, Bodenschatz E, and Pesch W (1998) Dynamics and selection of giant spirals in Rayleigh-Bénrad convection. *Physical Review Letters* 81: 5334–5337.

Pottebaum TS and Gharib M (2004) The pinch-off process in a starting buoyant plume. *Experiments in Fluids* 37: 87–94.

Pretzel G, Jäger H, and Neger T (1993) High-accuracy differential interferometry for the investigation of phase objects. *Measurement Science and Technology* 4: 649–658.

Pubellier M and Cobbold PR (1996) Analogue models for the transpressional docking of volcanic arcs in the western Pacific. *Tectonophysics* 253: 33–52.

Qi JA, Leung CW, Wong WO, and Probert SD (2006) Temperature-field measurements of a premixed butane/air circular impinging-flame using reference-beam interferometry. *Applied Energy* 83: 1307–1316.

Raffel M, Willert C, and Kompenhans J (1998) *Particle Image Velocimetry*. Springer.

Ragnarsson R, Ford JL, Santangelo CD, and Bodenschatz E (1996) Rifts in spreading wax layers. *Physical Review Letters* 76: 3456–9.

Ramberg H (1967) Model experimentation of the effect of gravity on tectonic processes. *Geophysical Journal of the Royal Astronomical Society* 14: 307–329.

Ramberg H (1968) Instability of layered systems in the field of gravity. *Physics of the Earth and Planetary Interiors* 1: 427–447.

Ramberg H (1972) Mantle diapirism and its tectonic and magmatic consequences. *Physics of the Earth and Planetary Interiors* 4: 45–60.

Ramberg H and Sjostrom H (1973) Experimental geodynamical models relating to continental drift and orogenesis. *Tectonophysics* 19: 105–132.

Ramberg H and Ghosh SK (1977) Rotation and strain of linear and planar structures in three-dimensional progressive deformation. *Tectonophysics* 40: 309–337.

Rasenat S, Busse FH, and Rehberg I (1989) A theoretical and experimental study of double-layer convection. *Journal of Fluid Mechanics* 199: 519–540.

Rayleigh L (1883) Investigations of the character of an incompressible heavy fluid of variable density. *Proceedings of the London Mathematical Society* 14: 170–177.

Rayleigh L (1916) On convection currents in a horizontal layer of fluid, when the higher temperature is on the underside. *Philosophical Magazine* 32: 529–546.

Renardy Y and Joseph DD (1985) Oscillatory instability in a Bénard problem of two fluids. *Physics of Fluids* 28: 788–793.

Renardy Y and Joseph DD (1985) Perturbation analysis of steady and oscillatory onset in a Bénard problem with two similar liquids. *Physics of Fluids* 28: 2699–2708.

Reynolds O (1894) Study of fluid motion by means of colored bands. *Nature* 50: 161–164.

Rhee H, Koseff J, and Street R (1984) Flow visualization of a recirculating flow by rheoscopic and liquid crystal techniques. *Experiments in Fluids* 2: 57–64.

Ribe NM (1998) Spouting and planform selection in the Rayleigh–Taylor instability of miscible viscous fluids. *Journal of Fluid Mechanics* 234: 315–336.

Ribe NM (2003) Periodic folding of viscous sheets. *Physical Review E* 68: 036305.

Ribe NM and Christensen U (1994) Three-dimensional modeling of plume-lithosphere interaction. *Journal of Geophysical Research* 99: 669–682.

Ribe NM and de Valpine DP (1994) The global hotspot distribution and instability of D'. *Geophysical Research Letters* 21: 1507–1510.

Ribe NM, Christensen UR, and Theissing J (1995) The dynamics of pllume-ridge interaction, 1: Ridge-centered plumes. *Earth and Planetary Science Letters* 134: 155–168.

Ribe NM and Christensen U (1999) The dynamical origin of Hawaiian volcanism. *Earth and Planetary Science Letters* 171: 517–531.

Ribe NM, Stutzmann E, Ren Y, and van der Hilst R (2007) Buckling instabilities of subducted lithosphere beneath the transition zone. *Earth and Planetary Science Letters* 254: 173–179.

Richards MA and Griffiths RW (1988) Deflection of plumes by mantle shear flow: Experimental results and a simple theory. *Geophysical Journal* 94: 367–376.

Richards MA, Duncan RA, and Courtillot VE (1989) Flood basalts and hot-spot tracks: Plume heads and tails. *Science* 246: 103–107.

Richter FM (1973) Convection and the large-scale circulation of the mantle. *Journal of Geophysical Research* 78: 8735–8745.

Richter FM (1978) Experiments on the stability of convection rolls in fluids whose viscosity depends on temperature. *Journal of Fluid Mechanics* 89: 553.

Richter FM and Johnson CE (1974) Stability of a chemically layered mantle. *Journal of Geophysical Research* 79: 1635–1639.

Richter FM and Parsons B (1975) On the interaction of two scales of convection in the mantle. *Journal of Geophysical Research* 80: 2529–2541.

Richter FM and McKenzie DP (1978) Simple plate models of mantle convection. *Journal of Geophysics* 44: 441.

Richter FM and McKenzie DP (1981) On some consequences and possible causes of layered convection. *Journal of Geophysical Research* 86: 6133–6124.

Richter FM, Nataf H-C, and Daly SF (1983) Heat transfer and horizontally averaged temperature of convection with large viscosity variations. *Journal of Fluid Mechanics* 129: 173–192.

Rossby HT (1965) On thermal convection driven by non-uniform heating form below: an experimental study. *Deep-Sea Research* 12: 9–16.

Rossby HT (1969) A study of Bénard convection with and without convection. *Journal of Fluid Mechanics* 36: 309–35.

Rouzo S, Rabinowicz M, and Briais A (1995) Segmentation of mid-ocean ridges with an axial valley induced by small-scale convection. *Nature* 374: 795–798.

Rumford C (1798) Heat is a form of motion: An experiment in Boring Cannon. *Philosophical Transactions* vol. 88.

Russo RM and Silver PG (1994) Trench-parallel flow beneath the Nazca plate from seismic anisotropy. *Science* 263: 1105–1111.

Samuel H and Farnetani CG (2002) Thermochemical convection and helium concentrations in mantle plumes. *Earth and Planetary Science Letters* 207: 39–56.

Samuel H and Bercovici D (2006) Oscillating and stagnating plumes in the Earth's lower mantle. *Earth and Planetary Science Letters* 248: 90–105.

Schaeffer N and Manga M (2001) Interaction of rising and sinking mantle plumes. *Geophysical Research Letters* 21: 765–768.

Schellart WP (2002) Analogue modelling of large-scale tectonic processes. *Journal of Virtual Explorer* 7: 1–6.

Schellart WP (2004a) Kinematics of subduction and subduction-induced flow in the upper mantle. *Journal of Geophysical Research* 109: B07401 (doi:10.1029/2004JB002970).

Schellart WP (2004b) Quantifying the net slab pull force as a driving mechanism for plate tectonics. *Geophysical Research Letters* 31: L07611 (doi:10.1029/2004GL019528).

Schellart WP (2005) Influence of the subducting plate velocity on the geometry of the slab and migration of the subdutcion hinge. *Earth and Planetary Science Letters* 231: 197–219.

Schilling J-G (1991) Fluxes and excess temperatures of mantle plumes inferred from their interaction with migrating mid-ocean ridges. *Nature* 352: 397–403.

Schmitz E and Merzkirch W (1981) A test fluid for simulating blood flows. *Experiments in Fluids* 2: 103–104.

Schmeling H (1988) Numerical models of Rayleigh-Taylor instabilities superimposed upon convection. *Bulletin, Geological Institute, University of Uppsala* 14: 95–109.

Schmeling H, Cada M, Babeyko A, et al. (2004) A benchmark comparison of subduction models. *Geophysical Research Abstracts* 6: 01484.

Schmidt RJ and Milverton SW (1935) On the instability of a fluid when heated from below. *Proceedings of the Royal Society of London* A 152: 586–594.

Schubert G, Olson P, Anderson, and Goldman P (1989) Solitary waves in mantle plumes. *Journal of Geophysical Research* 94: 9523–9532.

Schubert G, Masters G, Olson P, and Tackley P (2004) Superplumes or plume clusters? *Physics of the Earth and Planetary Interiors* 146: 147–162.

Schwiderski EW and Schwab HJA (1971) Convection experiments with electrolytically heated fluid layers. *Journal of Fluid Mechanics* 48: 703–717.

Scott DR, Stevenson DJ, and Whitehead JA (1986) Observations of solitary waves in a viscously deformable pipe. *Nature* 319: 759–761.

Selig F (1965) A theoretical prediction if salt-dome patterns. *Geophysics* 30: 633–645.

Sernas V and Fletcher LS (1970) A schlieren interferometer method for heat transfer studies. *Journal of Heat Transfer* 92: 202–204.

Settles GS (2001) *Schlieren and Shadowgraph Techniques; Visualizing Phenomena in Transparent Media*. Heidelberg: Springer-Verlag.

Shemenda AI (1992) Horizontal lithosphere compression and subduction: Constraints provided by physical modeling. *Journal of Geophysical Research* 97: 1097–1116.

Shemenda AI (1993) Subduction of the lithoshpere and back arc dynamics: Insights from physical modeling. *Journal of Geophysical Research* 98: 16167–16185.

Shlien DJ (1976) Some laminar thermal and plume experiments. *Physics of Fluids* 19: 1089–1098.

Shlien DJ and Thompson DJ (1975) Some experiments on the motion of an isolated laminar thermal. *Journal of Fluid Mechanics* 72: 35–47.

Shlien DJ and Boxman RL (1979) Temperature field measurement of an axisymmetric laminar plume. *Physics of Fluids* 22: 631–634.

Shlien DJ and Brosh A (1979) Velocity field measurements of a laminar thermal. *Physics of Fluids* 22: 1044–1053.

Shlien DJ and Boxman RL (1981) Laminar starting plume temperature field measurement. *International Journal of Heat and Mass Transfer* 24: 919–930.

Schouten J, Klitgord KD, and Whitehead JA (1985) Segmentation of midocean ridges. *Nature* 317: 225–229.

Siggia (1994) High Rayleigh number convection. *Annual Review of Fluid Mechanics* 26: 137–168.

Sigurdsson H and Sparks SRJ (1978) lateral magam flow within rifted Icelandic crust. *Nature* 274: 126–130.

Silveston PL (1958) Wärmedurchgang in waagerechten Flüssigkeitschichten. *Forschung im Ingenieurwesen* 24: 59–69.

Skilbeck JN and Whitehead JA (1978) Formation of discrete islands in linear islands chains. *Nature* 272: 499–501.

Sleep NH (1988) Gradual entrainment of a chemical layer at the base of the mantle by overlying convection. *Geophysical Journal* 95: 437–447.

Sleep NH (1990) Hotspots and mantle plumes: Some phenomenology. *Journal of Geophysical Research* 95: 6715–6736.

Sleep NH (1994) Lithospheric thinning by midplate mantle plumes and the tehrmal history of hot plume material ponded at sublithospheric depths. *Journal of Geophysical Research* 99: 9327–9343.

Small RD, Sernas VA, and Page RH (1972) Single beam schlieren interferometer using a Wollaston prism. *Applied Optics* 11: 858–862.

Solomatov S (1995) Scaling of temperature- and stress-dependent viscosity convection. *Physics of Fluids* 7: 266–274.

Solomatov VS and Moresi LN (2000) Scaling of time-dependent stagnant lid convection: Application to small-scale convection on Earth and other terrestrial planets. *Journal of Geophysical Research* 105: 21785–21817.

Solomon TH and Gollub JP (1988a) Passive transport in Steady Rayleigh-Benard Convection. *Physics of Fluids* 31, 1372–1379.

Solomon TH and Gollub JP (1988b) Chaotic Particle Transport in Time-Dependent Rayleigh–Bénard Convection. *Physical Review A* 38: 6280–6286.

Solomon TH and Gollub JP (1990) Sheared boundary layers in turbulent Rayleigh–Bénard convection. *Physical Review Letters* 64: 3102–3106.

Solomon T and Gollub J (1991) Thermal boundary layers and heat flux in turbulent convection: the role of recirculating flows. *Physical Review A* 43: 6683–6693.

Solomon TH, Tomas S, and Warner (1996) The Role of Lobes in Chaotic Mixing of Miscible and Immiscible Impurities. *Physical Review Letters* 77: 2682–2685.

Solomon TH and Mezic I (2003) Uniform resonant chaotic mixing in fluid flows. *Nature* 425: 376–380.

Sotin C and Parmentier EM (1989) On the stability of a fluid layer containing a univariant phase transition: application to planetary interiors. *Physics of the Earth and Planetary Interiors* 55: 10–25.

Sparks DW and Parmentier EM (1993) The structure of three-dimensional convection beneath oceanic spreading centers. *Geophysical Journal International* 112: 81–91.

Sparrow EM, Husar RB, and Goldstein RJ (1970) Observations and other characteristics of thermals. *Journal of Fluid Mechanics* 41: 793–800.

Stacey FD and Loper DE (1983) The thermal boundary layer interpretation of D″ and its role as a plume source. *Physics of the Earth and Planetary Interiors* 33: 45–55.

Stacey FD and Loper DE (1984) Thermal histories of the core and mantle. *Physics of the Earth and Planetary Interiors* 36: 99–115.

Stein C, Schmalz G, and Hansen U (2004) The effects of rheological parameters on plate behaviour in a self-consistent model of mantle convection. *Physics of the Earth and Planetary Interiors* 142: 225–255.

Stengel KC, Oliver DS, and Booker JR (1982) Onset of convection in a variable-viscosity fluid. *Journal of Fluid Mechanics* 120: 411–431.

Swanson WM and Green RL (1969) Colloidal suspension properties of Milling Yellow dye. *Journal of Colloid and Interface Science* 29: 161–163.

Tackley PJ (1998) Three-dimensional simulations of mantle convection with a thermo-chemical basal boundary layer: D″? In: Gurnis M, Wysession M, Knittle E, and Buffett B (eds.) *The Core-Mantle Boundary Region*, pp. 231–253. Washington, DC: AGU.

Tackley PJ (2000) Self-consistent generation of tectonic plates in time-dependent, threedimensional mantle convection simulations. 1. Pseudoplastic yielding. *Geochemistry, Geophysics, Geosystems* 1: 2000Gc000041.

Tackley PJ (2002) Strong heterogeneity caused by deep mantle layering. *Geochemistry, Geophysics, Geosystems* 3(4) 1024 doi: 10.1029/2001GC000167.

Tait S and Jaupart C (1989) Compositional convection in viscous melts. *Nature* 338: 571–574.

Tamai N and Asaeda T (1984) Sheetlike plumes near a heated bottom plate at large Rayleigh number. *Journal of Geophysical Research* 89: 727–734.

Tan E, Gurnis M, and Han L (2002) Slabs in the lower mantle and their modulation of plume formation. *Geochemistry, Geophysics, Geosystems* 3 (doi:10.1029/2001GC000238).

Tan E and Gurnis M (2005) Metastable superplumes and mantle compressibility. *Geophysical Research Letters* 32 (doi:10.1029/2005GL024190).

Tanner LH (1966) Some applications of holography in fluid mechanics. *Journal of Scientific Instruments* 43: 81–83.

Tanny J and Shlien DJ (1985) Velocity field measurements of a laminar starting plume. *Physics of Fluids* 28: 1027–1032.

Taylor G (1950) The instability of liquid surfaces when accelerated in a direction perpendicaular to their planes. *Proceedings of the Royal Society of London Series A* 201: 192–196.

Thompson PF and Tackley PJ (1998) Generation of mega-plumes from the core-mantle boundary in a compressible mantle with temperature-dependent viscosity. *Geophysical Research Letters* 25, 1999–2002.

Thomson J (1882) On a changing tesselated structure in certain liquids. *Proceedings of the Philosophical Society of Glasgow* 13: 464–468.

Torrance KE and Turcotte DL (1971) Thermal convection with large viscosity variations. *Journal of Fluid Mechanics* 47: 113–125.

Townsend AA (1957) Temperature fluctuations over a heated horizontal surface. *Journal of Fluid Mechanics* 5: 209–241.

Tritton DJ and Zarraga MM (1967) Convection in horizontal fluid layers with internal heat generation experiments. *Journal of Fluid Mechanics* 30: 21–31.

Trompert RA and Hansen U (1998) On the Rayleigh number dependence of convection with a strongly Temperature-dependent viscosity. *Physics of Fluids* 10: 351–360.

Tsinober AB, Yahalom Y, and Shlien DJ (1983) A point source of heat in a stable salinity gradient. *Journal of Fluid Mechanics* 135: 199–217.

Turner JS (1973) Convection in the mantle: a laboratory model with temperature-dependent viscosity. *Earth and Planetary Science Letters* 17: 369–374.

van Keken P (1997) Evolution of starting mantle plumes: a comparison between numerical and laboratory models. *Earth and Planetary Science Letters* 148: 1–11.

van Keken P, King S, Schmeling H, Christensen U, Neumeister D, and Doin M-P (1997) A comparison of methods for the modeling of thermochemical convection. *Journal of Geophysical Research* 102: 22477–22496.

Vasquez PA, Perez AT, and Castellanos A (1996) Thermal and electrohydrodynamics plumes. A comparative study. *Physics of Fluids* 8: 2091–2096.

Vatteville J (2004) Etude expérimentale des panaches: Structure et evolution temporelle. Mémoire de DEA, ENS Lyon/IPG Paris.

Verzicco R (2004) Effect of non-perfect thermal sources in turbulent thermal convection. *Physics of Fluids* 16: 1965–1975.

Villermaux E and Innocenti C (1999) On the geometry of turbulent mixing. *Journal of Fluid Mechanics* 393: 123–147.

Voth GA, Haller G, and Gollub JP (2002) Experimental Measurements of stretching Fields in Fluid Mixing. *Physical Review Letters* 88: 254501.1–254501.4.

Weeraratne D and Manga M (1998) Transitions in the style of mantle convection at high Rayleigh numbers. *Earth and Planetary Science Letters* 160: 563–568.

Weijermars R (1986) Flow behavior and physical chemistry of Bouncing Putties and related polymers in view of tectonics laboratory applications. *Tectonophysics* 124: 325–358.

Weijermars R and Schmeling H (1986) Scaling of Newtonian and non-Newtonian fluid dynamics without inertia for quantitative modelling of rock flow due to gravity (including the concept of rheological similarity). *Physics of the Earth and Planetary Interiors* 43: 316–330.

Weijermars R (1988) New laboratory method for analyzing deformation and displacement in creeping fluid: Examples from stokes flow and a thermal plume. *Journal of Geophysical Research* 93: 2179–2190.

Weijermars R (1989) Experimental pictures of deformation patterns in a possible model of the earth's interior. *Earth and Planetary Science Letters* 91: 367–373.

Weinstein SA and Olson PL (1989) The proximity of hotspots to convergent and divergent plate boundaries. *Geophysical Research Letters* 16: 433–436.

Weinstein SA and Olson P (1990) Planforms in thermal convection with internal heat sources at large Rayleigh and Prandtl numbers. *Geophysical Research Letters* 17: 239–242.

Weinstein SA and Christensen UR (1991) Convection planforms in a fluid with a temperature-dependent viscosity beneath a stress-free upper boundary. *Geophysical Research Letters* 18: 2035–2038.

Wenzel MJ, Manga M, and Jellinek AM (2004) Tharsis as a consequence of Mars'dichotomy and layered mantle. *Geophysical Research Letters* 31 (doi:10.1029/2003GL019306).

White DB (1988) The planforms and onset of convection with a temperature-dependent viscosity fluid? *Journal of Fluid Mechanics* 191: 247–286.

Whitehead JA (1972) Moving heaters as a model of continental drift. *Physics of the Earth and Planetary Interiors* 18: 199–212.

Whitehead JA and Luther DS (1975) Dynamics of laboratory diapir and plume models. *Journal of Geophysical Research* 80: 705–717.

Whitehead JA (1976) Convection models: Laboratory versus mantle. *Tectonophysics* 35: 215–228.

Whitehead JA and Parsons B (1978) Oberservations of convection at Rayleigh numbers up to 760000 in a fluid with large Prandtl number. *Geophysical and Astrophysical Fluid Dynamics* 9: 201–217.

Whitehead JA (1982) Instabilities of fluid conduits in a flowing Earth – Are plates lubricated by the asthenosphere? *Geophysical Journal of the Royal Astronomical Society* 70: 415–433.

Whitehead JA, Dick HHB, and Schouten H (1984) A mechanism for magmatic accretion under spreading centers. *Nature* 312: 146–148.

Whitehead JA (1986) Buoyancy driven instabilities of low viscosity zones as models of magma-rich zones. *Special Issue: Partial Melting Phenomena in the Earth and Planetary Evolution. Journal of Geophysical Research* 91: 9303–9314.

Whitehead JA and Helfrich KR (1988) Wave transport of deep mantle material. *Nature* 336: 59–61.

Whittaker RJ and Lister JR (2006) Steady axisymmetric creeping plumes above a planar boundary. Part I: A point source. *Journal of Fluid Mechanics* 567: 361–378.

Whittaker RJ and Lister JR (2006) Steady axisymmetric creeping plumes above a planar boundary. Part II: A distributed source. *Journal of Fluid Mechanics* 567: 379–397.

Wilcock WSD and Whitehead JA (1991) The Rayleigh–Taylor instability of an embedded layer of low-viscosity fluid. *Journal of Geophysical Research* 96: 12193–12200.

Willert C and Gharib M (1991) Digital particle image thermometry. *Experiments in Fluids* 10: 181–193.

Willis GE and Deardorff JW (1967) Confirmation and renumbering of the discrete heat flux transitions of Malkus. *Physics of Fluids* 10: 1861–1866.

Wilson TJ (1963) Evidence from islands on the spreading of the ocean floor. *Canadian Journal of Physics* 41: 863–868.

Worster MG (1986) The axisymmetric laminar plume: Asymptotic solution for large Prandtl number. *Studies in Applied Mathematics* 75: 139–152.

Xi H-D, Lam S, and Xia K-Q (2004) From laminar plumes to organized flows: The onset of large-scale circulation in turbulent thermal convection. *Journal of Fluid Mechanics* 503: 47–56.

Yang W-J (1989) *Handbook of Flow Visualization*. Washington, DC: Taylor and Francis.

Zaranek SE and Parmentier EM (2004) The onset of convection in fluids with strongly temperature-dependent viscosity cooled from above with implications for planetary lithospheres. *Earth and Planetary Science Letters* 224: 71–386.

Zhang J, Childress S, and Libchaber A (1997) Non-Boussinesq effect: Thermal convection with broken symmetry. *Physics of Fluids* 9: 1034–1042.

Zhang J, Childress S, and Libchaber A (1998) Non-Boussinesq effect: Asymmetric velocity profiles in thermal convection. *Physics of Fluids* 10: 1534–1536.

Zhang J and Libchaber A (2000) Periodic boundary motion in thermal turbulence. *Physical Review Letters* 84: 4361–4364.

Zhong J-Q and Zhang J (2005) Thermal convection with a freely moving top boundary. *Physics of Fluids* 17: 115105.

# 4 Analytical Approaches to Mantle Dynamics

**N. M. Ribe**, Institut de Physique du Globe, Paris, France

## 4.1  Introduction

Geodynamics is the study of how Earth materials deform and flow over long ($\geq 10^2$–$10^3$ years) time-scales. It is thus a science with dual citizenship: at once a central discipline within the Earth sciences and a branch of fluid dynamics more generally.

The basis of fluid dynamics is a set of general conservation laws for mass, momentum, and energy, which are usually formulated as partial differential equations (PDEs) (Batchelor, 1967, chapter 3). However, because these equations apply to all fluid flows, they describe none in particular, and must therefore be supplemented by material constitutive relations and initial and/or boundary conditions that are appropriate for a particular phenomenon of interest. The result, often called a model problem, is the ultimate object of study in fluid mechanics.

Once posed, a model problem can be solved in one of three ways. The first is to construct a physical analog in the laboratory and let nature do the solving. The experimental approach has long played a central role in geodynamics. Another approach is to solve the model problem numerically on a computer, using one of the methods. The third possibility, the subject of the present chapter, is to solve the problem analytically.

Admittedly, analytical approaches are most effective when the model problem at hand is relatively simple, and lack some of the flexibility of the best experimental and numerical methods. However, they compensate for this by providing a degree of understanding and insight that no other method can match. What is more, they also play a critical role in the interpretation of experimental and numerical results. For example, dimensional analysis is required to ensure proper scaling of experimental and numerical results to the Earth; and local scaling analysis of numerical output can reveal underlying laws that are obscured by numerical tables and graphical images. For all these reasons, the central role that analytical methods have always played in geodynamics is unlikely to diminish.

The purpose of this chapter is to survey the principal analytical methods of geodynamics and the major results that have been obtained using them. These methods are remarkably diverse, and require a correspondingly broad and comprehensive treatment. However, it is equally important to highlight the common structures and styles of argumentation that give analytical geodynamics an impressive unity. In this spirit, we begin with a discussion of the art of formulating geophysical model problems, focusing on three paradigmatic phenomena (heat transfer from magma diapirs, gravitational instability of buoyant layers, and plume–lithosphere interaction (PLI)) that will subsequently reappear in the course of the chapter treated by different methods. Thus, heat transfer from diapirs is treated using dimensional analysis (Sections 4.3.1 and 4.3.2), scaling analysis (Section 4.3.3), and boundary-layer (BL) theory (Section 4.7.1.2); gravitational instability using scaling analysis (Section 4.4.3), long-wave analysis (Section 4.8.3), and linear stability analysis (Section 4.9.1); and PLI using lubrication theory and scaling analysis (Sections 4.8.1 and 4.8.2). Moreover, the discussions of these and other phenomena are organized as much as possible around three recurrent themes. The first is the importance of scaling arguments (and the scaling laws to which they lead) as tools for understanding physical mechanisms and applying model results to the Earth. Examples of scaling arguments can be found in Sections 4.3.3, 4.4.3, 4.7.2, 4.8.2, 4.9.2.5, 4.9.2.6, 4.9.3.1, and 4.9.3.4. The second is the ubiquity of self-similar behavior in geophysical flows, which typically occurs in parts of the spatiotemporal flow domain that are sufficiently far from the inhomogeneous initial or boundary conditions that drive the flow to be uninfluenced by their structural details (Sections 4.4.1, 4.4.2, 4.5.3.3, 4.5.3.4, 4.7.1.1, 4.7.3, 4.8.1, 4.8.2, and 4.9.3.4). The third theme is asymptotic analysis, in which the smallness of some crucial parameter in the model problem is exploited to simplify the governing equations, often via a reduction of their dimensionality (Sections 4.4.1, 4.5.7, 4.7, 4.7.2, 4.8.1, 4.8.3, 4.8.4, 4.8.5, 4.8.6, 4.9.2.5, 4.9.2.6, 4.9.3.2, 4.9.3.3, and 4.9.3.4). While these themes by no means encompass everything the chapter contains, they can serve as threads to guide the reader through what might otherwise appear a trackless labyrinth of miscellaneous methods.

A final aim of the chapter is to introduce some less familiar methods, drawn from other areas of fluid mechanics, that deserve to be better known among geodynamicists. Examples include the use of Papkovich–Fadle eigenfunction expansions (Section 4.5.4.1) and complex variables (Section 4.5.5) for two-dimensional (2-D) Stokes flows, solutions in bispherical coordinates for 3-D Stokes flows (Section 4.5.4.3), and multiple-scale analysis of modulated convection rolls (Sections 4.9.2.5 and 4.9.2.6).

**Table 1** Frequently used abbreviations

| Abbreviation | Meaning |
| --- | --- |
| 1-D | One-dimensional |
| 2-D | Two-dimensional |
| 3-D | Three-dimensional |
| BL | Boundary layer |
| BVP | Boundary-value problem |
| CMB | Core–mantle boundary |
| LHS | Left-hand side |
| MEE | Method of eigenfunction expansions |
| MMAE | Method of matched asymptotic expansions |
| ODE | Ordinary differential equation |
| PDE | Partial differential equation |
| PLI | Plume-lithosphere interaction |
| RHS | Right-hand side |
| R-T | Rayleigh–Taylor |
| SBT | Slender-body theory |
| TBL | Thermal boundary layer |

Throughout this chapter, unless otherwise stated, Greek indices range over the values 1 and 2; Latin indices range over 1, 2, and 3; and the standard summation convention over repeated subscripts is assumed. Subscript notation (e.g., $u_i$, $e_{ij}$) and vector notation (e.g., $\mathbf{u}$, $\mathbf{e}$) are used interchangeably as convenience dictates, and the notations $(x, y, z) = (x_1, x_2, x_3)$ for Cartesian coordinates and $(u, v, w) = (u_1, u_2, u_3)$ for the corresponding velocity components are equivalent. Unit vectors are denoted by symbols $\mathbf{e}_x$, $\mathbf{e}_r$, etc. Partial derivatives are denoted either by subscripts or by the symbol $\partial$, and $\partial_i = \partial/\partial x_i$. Thus, for example,

$$T_x = \partial_x T = \partial_1 T = \frac{\partial T}{\partial x} = \frac{\partial T}{\partial x_1} \qquad [1]$$

The symbols $\Re[\ldots]$ and $\Im[\ldots]$ denote the real and imaginary parts, respectively, of the bracketed quantities. Frequently used abbreviations are listed in **Table 1**.

## 4.2 Formulating Geodynamical Model Problems: Three Case Studies

Ideally, a geodynamical model should respect two distinct criteria: it should be sufficiently simple that the essential physics it embodies can be easily understood, yet sufficiently complex and realistic that it can be used to draw inferences about the Earth. It is seldom easy to satisfy both these desiderata in a single model; and so most geophysicists tend to emphasize one or the other, according to temperament and education.

However, there is a way around this dilemma: to investigate not just a single model, but rather a hierarchical series of models of gradually increasing complexity and realism. Such an investigation – whether carried out by one individual or by many – is a cumulative one in which the initial study of a highly simplified model provides the physical understanding required to guide the formulation and investigation of more complex models. To show how this process works in practice, I have chosen three exemplary geophysical phenomena as case studies: heat transfer from mantle diapirs; buoyant instability of thermal boundary layers (TBLs); and the interaction of mantle plumes with the lithosphere. Mathematical detail and bibliographical references are kept to a minimum in order to focus on the conceptual structure of the hierarchical approach.

### 4.2.1  Heat Transfer from Mantle Diapirs

Our first example is the ascent of a hot blob or diapir of magma through the lithosphere, a possible mechanism for the formation of island-arc volcanoes (Marsh and Carmichael, 1974; Marsh, 1978). To evaluate this model, one needs to know how far the diapir can move through the colder surrounding material before losing so much of its excess heat that it solidifies. **Figure 1** illustrates a series of model problems that can be used to investigate this question.

Probably the simplest model that still retains much of the essential physics (Marsh, 1978) can be formulated by assuming that (1) the diapir is spherical and has a constant radius; (2) the diapir's interior temperature is uniform and (3) does not vary with time; (4) the ascent speed and (5) the temperature of the lithosphere far from the diapir are constants; (6) the lithosphere is a uniform viscous fluid with constant physical properties; and (7) the viscosity of the diapir is much less than that of the lithosphere. The result is the model shown in **Figure 1(a)**, in which an effectively inviscid fluid sphere with radius $a$ and temperature $\Delta T$ moves at constant speed $U$ through a fluid with constant density $\rho$, thermal diffusivity $\kappa$, and viscosity $\nu$, and constant temperature $T = 0$ far from the sphere. An analytical solution for the rate of heat transfer $q$ from the diapir (Section 4.7.1.2) can now be obtained if one makes the additional (and realistic) assumption (8) that the Peclet number $Pe \equiv Ua/\kappa \gg 1$, in which case the temperature variations around the leading hemisphere of the diapir are confined to a thin BL of

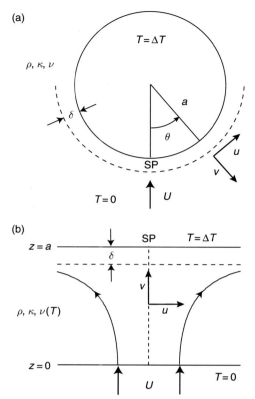

**Figure 1**  Models for the heat transfer from an ascending magma diapir. (a) Original model in spherical geometry. A sphere of radius $a$ and constant temperature $\Delta T$ is immersed in an infinite fluid with density $\rho$, kinematic viscosity $\nu$, and thermal diffusivity $\kappa$. The fluid far from the sphere moves with a constant streaming speed $U$, and its temperature is zero. The colatitude measured from the upstream stagnation point (SP) is $\theta$, and the components of the velocity in the colatitudinal and radial directions are $u$ and $v$, respectively. In the limit $Ua/\kappa \gg 1$, temperature variations in the hemisphere $\theta \leq \pi/2$ are confined to a boundary layer of thickness $\delta \ll a$. The viscosity $\nu$ may be constant (Marsh, 1978) or temperature-dependent (Morris, 1982). (b) Modified stagnation-point flow model of Morris (1982). The surface of the hot sphere is replaced by the plane $z = a$, and the steady streaming velocity $U$ is imposed as a boundary condition at $z = 0$.

thickness $\delta \ll a$ (**Figure 1(a)**). One thereby finds (see Section 4.3.3 for the derivation)

$$q \sim k_c a \Delta T Pe^{1/2} \qquad [2]$$

where $k_c$ is the thermal conductivity.

While the model just described provides a first estimate of how the heat transfer scales with the ascent speed and the radius and excess temperature of the diapir, it is far too simple for direct application to the Earth. A more realistic model can be obtained by relaxing assumptions (3) and (5), allowing the temperatures of the diapir and the ambient lithosphere to vary with time. If these variations are slow enough, the heat transfer at each instant will be described by a law of the form [2], but with a time-dependent excess temperature $\Delta T(t)$. A model of this type was proposed by Marsh (1978), who obtained a solution in the form of a convolution integral for the evolving temperature of a diapir ascending through a lithosphere with a prescribed far-field temperature $T_{\text{lith}}(t)$.

A different extension of the simple model of **Figure 1(a)**, also suggested by Marsh (1978), begins from the observation that the viscosity of mantle materials decreases strongly with increasing temperature. A hot diapir will therefore be surrounded by a thin halo of softened lithosphere, which will act as a lubricant and increase the diapir's ascent speed. The effectiveness of this mechanism depends on whether the halo is thick enough, and/or has a viscosity low enough, to carry a substantial fraction of the volume flux $\sim \pi a^2 U$ that the sphere must displace in order to move. Formally, this model is obtained by replacing the constant viscosity $\nu$ in **figure 1(a)** by one that depends exponentially on temperature as $\nu = \nu_0 \exp(-T/\Delta T_{\text{r}})$.

While this new variable-viscosity model is more realistic and dynamically richer than the original model, its spherical geometry makes analytical solution impossible except in certain limiting cases (Morris, 1982; Ansari and Morris, 1985). However, closer examination reveals that the spherical geometry is not in fact essential: all that matters is that the flow outside the softened halo varies over a characteristic length scale $a$ that greatly exceeds the halo thickness. This recognition led Morris (1982) to study a simpler model in which the flow around the sphere is replaced by a stagnation-point flow between two planar boundaries $z = 0$ and $z = a$ (**Figure 1(b)**). The model equations now admit 1-D solutions $T = T(z)$ and $\nu = \nu(z)$ for the temperature and the vertical velocity, respectively, which can be determined using the method of matched asymptotic expansions (MMAE) (Section 4.7.2) in the limit of large viscosity contrast $\Delta T/\Delta T_{\text{r}} \gg 1$ (Morris, 1982).

### 4.2.2    Plume Formation in TBLs

Our second example (**Figure 2**) is the formation of plumes via the gravitational instability of a horizontal TBL. The first step in formulating a model for this process is to choose a simple representation for the relevant physical properties (density, viscosity, and thermal diffusivity) of the fluid. Because these depend on pressure, temperature, and (possibly) chemical composition, they will vary continuously with depth in the

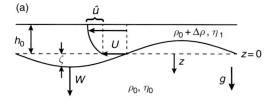

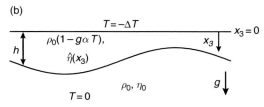

**Figure 2**   Models for plume formation in a dense/cold thermal boundary layer. (a) Rayleigh–Taylor instability of a layer of fluid with density $\rho_0 + \Delta\rho$ and viscosity $\eta_1$ above a half-space of fluid with density $\rho_0$ and viscosity $\eta_0$. The initial thickness of the dense layer is $h_0$, and the deformation of the interface is $\zeta$. The maximum values of the horizontal and vertical velocities at the interface $z = \zeta$ are $U$ and $W$, respectively, and $\hat{u}$ is the magnitude of the change in horizontal velocity across the layer. (b) Buoyant instability of a cold thermal boundary layer. The upper surface $x_3 = 0$ of a fluid half-space is held at a fixed temperature $-\Delta T$ relative to the interior. The density of the fluid varies with temperature as $\rho = \rho_0(1 - g\alpha T)$, where $\alpha$ is the thermal expansion coefficient. The thickness of the boundary layer is $h(x_1, x_2, t)$ and the viscosity within it is $\hat{\eta}(x_3)$.

TBL and with time (due to thermal diffusion). As a first approximation, however, one can model the depth distribution of the fluid properties as a nondiffusing and spatially discontinuous two-fluid configuration in which a dense layer with constant thickness $h_0$, density $\rho_0 + \Delta\rho$, and viscosity $\eta_1$ overlies a half-space with density $\rho_0$ and viscosity $\eta_0$ (**Figure 2(a)**). The case of a less-dense BL beneath a denser fluid is obtained by turning the system upside down and switching the sign of $\Delta\rho$. Because both configurations are gravitationally unstable, any small perturbation of the interface between the two fluids will grow with time via the Rayleigh–Taylor (R–T) instability. The growth rate of an infinitesimal sinusoidal perturbation with arbitrary wave number $k$ can be determined analytically (Selig, 1965; Whitehead and Luther, 1975) via a standard linear stability analysis (Section 4.9.1).

While the simple RT model embodies some of the essential physics of plume formation, it is unable to describe such crucial features as the characteristic periodicity of TBL instabilities (Howard, 1964). For this purpose, a more realistic model that incorporates a diffusing temperature field is required. **Figure 2(b)** shows such a model (Canright, 1987; Lemery et al.,

2000), in which a diffusive TBL grows away from a cold surface and subsequently becomes unstable. The density of the fluid varies linearly with temperature, and the viscosity $\hat{\eta}$ can vary as a function of the depth $x_3$ within the TBL. The additional assumptions that the wavelength of the instability greatly exceeds the thickness of the TBL and that the horizontal components of the fluid velocity are constant across it then permit an analytical reduction of the full 3-D equations to an equivalent set of 2-D equations for the horizontal velocities and the first moment of the temperature in the TBL (Section 4.8.3).

### 4.2.3 Plume–Lithosphere Interaction

PLI refers to the processes that occur after a rising mantle plume strikes the base of the lithosphere. Because the plume fluid is buoyant relative to its surroundings, it will spread beneath the lithosphere, eventually forming a thin layer or pancake whose lateral dimensions greatly exceed its thickness.

**Figure 3** shows a series of fluid dynamical models that have been used to study PLI, beginning with the kinematic model of Sleep (1987) (**Figure 3(a)**). Sleep's insight was that the flow associated with a plume rising beneath a moving plate can be regarded as the sum of two parts: a (horizontal) radial flow representing buoyant plume fluid emanating from a steady localized source at the top of the plume conduit; and an ambient mantle wind in the direction of the plate motion. Fluid from the source can travel only a finite distance upstream against the wind before being blown back downstream again, leading to the formation of a stagnation point (labeled SP in **Figure 3(a)**) at which the wind speed just equals the speed of radial outflow from the source. The stagnation streamline that passes through this point (heavy line in **Figure 3(a)**) divides the $(x, y)$ plane into an inner region containing fluid from the source, and an outer region containing fluid brought in from upstream by the wind. The stagnation streamline resembles the shape of the topographic swell around the Hawaiian island chain (Richards *et al.*, 1988).

While the kinematic model of Sleep (1987) nicely illustrates the geometry of PRI, it neglects the (driving) buoyancy force and (resisting) viscous force that control the spreading of the plume pancake. We now seek the simplest possible model that embodies these dynamics. Following the logic of Section 4.2.2, we replace the continuous variation of fluid properties by a two-fluid structure, comprising a pancake with constant viscosity $\eta$ spreading in an ambient fluid with viscosity $\eta_m$. We further assume that $\eta_m/\eta$ is not too large, so that the influence of the ambient mantle on the spreading of the plume pancake can be neglected. Another simplifying assumption (to be relaxed later) is that the plate does not move with respect to the source of plume fluid. Finally, we suppose that the strength of the source is constant. The result is the viscous gravity current model (Huppert, 1982), wherein

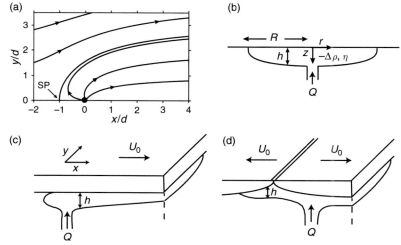

**Figure 3** Hierarchy of models for plume–lithosphere interaction. (a) Steady streamlines for the 2-D kinematic model of Sleep (1987). The source is indicated by the black circle, the heavy solid line is the stagnation streamline, and $d$ is the distance between the source and the stagnation point SP. (b) Spreading beneath a rigid surface of an axisymmetric current of viscous fluid with viscosity $\eta$ and anomalous density $-\Delta\rho$, supplied at a volumetric rate $Q$. (c) Same as (b), but beneath a rigid plate moving at speed $U_0$. (d) Same as (c), but beneath two plates separated by a spreading ridge with half-spreading rate $U_0$.

fluid with viscosity $\eta$ and anomalous density $-\Delta\rho$ relative to its surroundings is supplied at a constant volumetric rate $Q$ and spreads beneath a stationary rigid surface (**Figure 3(b)**). The analytical solution for the closely related problem of a current with constant volume $V$ is discussed in Section 4.8.1.

The next step is to generalize the gravity current model by allowing the plate to move with a constant velocity $U_0$ relative to the source (Olson, 1990). The resulting refracted plume model (**Figure 3(c)**) is in essence a dynamically self-consistent extension of the 2-D kinematic model of Sleep (1987), and is discussed further in Section 4.8.2.

As a final illustration, **Figure 3(d)** shows a further extension of the refracted plume model in which the uniform plate is replaced by two plates separated by a spreading ridge. Despite the increased complexity of this plume–ridge interaction model, analytical methods can still profitably be applied to it (Ribe *et al.*, 1995).

## 4.3 Dimensional and Scaling Analysis

The goal of studying model problems in fluid mechanics is typically to determine functional relations, or scaling laws, that obtain between certain parameters of interest and the various other parameters on which they depend. Two simple yet powerful methods that can be used for this purpose are dimensional analysis and scaling analysis.

### 4.3.1 Buckingham's Π-Theorem and Dynamical Similarity

Dimensional analysis begins from the principle that the validity of physical laws cannot depend on the units in which they are expressed. An important consequence of this principle is the Π-theorem of Buckingham (1914).

Suppose that there exists a (generally unknown) functional relationship among $N$-dimensional parameters $P_1\ P_2, \ldots, P_N$, such that

$$f_1(P_1,\ P_2, \ldots,\ P_N) = 0 \qquad [3]$$

Let $M < N$ be the number of the parameters $P_n$ which have independent physical dimensions (note that a dimensionally consistent functional relationship among the parameters $P_n$ is impossible if $M = N$). In most (but not all!) cases, $M$ is just the number of

independent units that enter into the problem, for example, $M = 3$ for mechanical problems involving units of m, kg, and s and $M = 4$ for thermomechanical problems involving temperature (units K) in addition. The Π-theorem states that the functional relationship [3] is equivalent to a relation of the form

$$f_2(\Pi_1,\ \Pi_2, \ldots,\ \Pi_{N-M}) = 0 \qquad [4]$$

where $\Pi_1,\ \Pi_2, \ldots,\ \Pi_{N-M}$ are $N - M$ independent dimensionless combinations (or groups) of the dimensional parameters $P_n$. The fact that all systems of units (SI, cgs, etc.) are equivalent requires that each dimensionless group $\Pi_i$ be a product of powers of the dimensional parameters $P_n$; no other functional form preserves the value of the dimensionless group when the system of units is changed. The function $f_2$, by contrast, can have any form. While the total number of independent groups $\Pi_i$ is fixed ($\equiv N - M$), the definitions of the individual groups are arbitrary and can be chosen as convenient. A more detailed discussion and proof of the Π-theorem can be found in Barenblatt (1996, chapter 1).

The Π-theorem is the basis for the concept of dynamical similarity, according to which two physical systems behave similarly (i.e., proportionally) if they have the same values of the dimensionless groups $\Pi_i$ that define them. The crucial point is that two systems may have identical values of $\Pi_i$ even though they are of very different size, that is, even if the values of the dimensional parameters $P_n$ are very different. Dynamical similarity is thus a natural generalization of the concept of geometrical similarity, whereby, for example, two triangles of different sizes are similar if they have the same values of the dimensionless parameters (angles and ratios of sides) that define them. Geometrical similarity is a necessary, but not a sufficient, condition for dynamical similarity.

The importance of dynamical similarity for physical modeling is that it allows results obtained in the laboratory or on a computer to be applied to another system with very different scales of length, time, etc. Its power derives from the fact that the function $f_2$ in [4] has $M$ fewer arguments than the original function $f_1$. Thus, an experimentalist or numerical analyst who seeks to determine how a target dimensional parameter $P_1$ depends on the other $N-1$ parameters need not vary all of the latter individually; it suffices to vary only $N - M - 1$ dimensionless parameters. Consequently, if the variation of a given dimensional parameter requires $\approx 10$ samplings, then use of the Π-theorem reduces the effort involved in searching the parameter space by a factor $\approx 10^M$ (Barenblatt, 1996; see also Chapter 2) By the same token, the $\Pi$-theorem makes possible a far more economical representation of experimental or numerical data. As an example, suppose that we have $N = 5$ dimensional parameters $P_1$–$P_5$ from which $N - M = 2$ independent dimensionless groups $\Pi_1$ and $\Pi_2$ can be formed. To represent our data without the help of the $\Pi$-theorem, we would need many shelves (one for each value of $P_5$), each containing many books (one for each value of $P_4$), each containing many pages (one for each value of $P_3$), each containing a plot of $P_2$ versus $P_1$. By using the $\Pi$-theorem, however, we can collapse the whole library onto a single plot of $\Pi_2$ versus $\Pi_1$.

As a simple illustration of the $\Pi$-theorem, consider again the model for heat transfer from a hot sphere (**Figure 1(a)**). Suppose that we wish to determine the radial temperature gradient $\beta$ (proportional to the local conductive heat flux) as a function of position on the sphere's surface. A list of all the relevant parameters includes the following eight (units in parentheses): $\beta(\text{K m}^{-1})$, $a(\text{m})$, $U(\text{m s}^{-1})$, $\rho(\text{kg m}^{-3})$, $\nu(\text{m}^2\,\text{s}^{-1})$, $\kappa(\text{m}^2\,\text{s}^{-1})$, $\Delta T(\text{K})$, and $\theta$ (dimensionless). However, $\rho$ can be eliminated immediately because it is the only parameter that involves units of mass: no dimensionless group containing it can be defined. The remaining parameters are $N = 7$ in number, $M = 3$ of which (e.g., $a$, $\Delta T$, and $U$) have independent units, so $N - M = 4$ independent dimensionless groups can be formed. It is usually good practice to start by defining a single group containing the target parameter ($\beta$ here.) While inspection usually suffices, one can also proceed more formally by writing the group ($\Pi_1$ say) as a product of the desired parameter ($\beta$) and unknown powers of any set of $M$ parameters with independent dimensions, for example, $\Pi_1 = \beta a^{n_1} \Delta T^{n_2} U^{n_3}$. The requirement that $\Pi_1$ be dimensionless then implies $n_1 = 1$, $n_2 = -1$, and $n_3 = 0$. Additional groups are then obtained by applying the same procedure to the remaining dimensional parameters in the list ($\kappa$ and $\nu$ in this case.) Finally, any remaining parameters in the list that are already dimensionless ($\theta$ in this case) can be used as groups by themselves. For the hot sphere, the result is

$$\frac{\beta a}{\Delta T} = \text{fct}\left(\frac{Ua}{\kappa}, \frac{Ua}{\nu}, \theta\right) \qquad [5]$$

where fct is an unknown function. The groups $Ua/\kappa \equiv Pe$, $Ua/\kappa \equiv Re$, and $\beta a/\Delta T \equiv \mathcal{N}$ are traditionally called the Peclet number, the Reynolds number, and the (local) Nusselt number, respectively (the last

to be distinguished from the global Nusselt number $Nu \equiv \int_s \mathcal{N}\, dS$ that measures the total heat flux across the sphere's surface $S$.) As we remarked earlier, the definitions of the dimensionless groups in a relation like [5] are not unique. Thus one can replace any group by the product of itself and arbitrary powers of the other groups, for example, $Pe$ in [5] by the Prandtl number $\nu/\kappa \equiv Pe/Re$. Furthermore, it often (but not always!) happens that the target parameter ceases to depend on a dimensionless group whose value is very large or very small. For example, in a very viscous fluid such as the mantle, $Re \ll 1$ because inertia is negligible, and so $Re$ no longer appears as an argument in [5].

### 4.3.2   Nondimensionalization

When the equations governing the dynamics of the problem at hand are known, another method of dimensional analysis becomes available: nondimensionalization. We illustrate this using the same example of a hot sphere.

The first step is to write down the governing equations, together with all the relevant initial and boundary conditions. Because the problem is both steady and axisymmetric, the dependent variables are the temperature $T(r, \theta)$ and the velocity $\mathbf{u}(r, \theta)$, where $r$ and $\theta$ are the usual spherical coordinates. If viscous dissipation is negligible, the governing equations and boundary conditions are (*see* Chapter 6)

$$\nabla \cdot \mathbf{u} = 0, \quad \mathbf{u} \cdot \nabla T = \kappa \nabla^2 T, \qquad [6a]$$
$$\mathbf{u} \cdot \nabla \mathbf{u} = -\rho^{-1} \nabla p + \nu \nabla^2 \mathbf{u}$$

$$T(a, \theta) - \Delta T = T(\infty, \theta) = \mathbf{u}(a, \theta)$$
$$= \mathbf{u}(\infty, \theta) - U\mathbf{e}_z = 0 \qquad [6b]$$

where $\mathbf{e}_z$ is a unit vector in the direction of the steady stream far from the sphere. The next step is to define dimensionless variables (denoted, e.g., by primes) using scales that appear in the equations and/or initial and boundary conditions. An obvious (but not unique) choice is $r' = r/a$, $T' = T/\Delta T$, $\mathbf{u}' = \mathbf{u}/U$, and $p' = a(p - p_0)/\rho \nu U$, where $p_0$ is the (dynamically insignificant) pressure far from the sphere. Substituting these definitions into [6] and immediately dropping the primes to avoid notational overload, we obtain the dimensionless BVP:

$$\nabla \cdot \mathbf{u} = 0, \quad Pe\,\mathbf{u} \cdot \nabla T = \nabla^2 T, \qquad [7a]$$
$$Re\,\mathbf{u} \cdot \nabla \mathbf{u} = -\nabla p + \nabla^2 \mathbf{u}$$

$$T(1, \theta) - 1 = T(\infty, \theta) = \mathbf{u}(1, \theta) = \mathbf{u}(\infty, \theta) - \mathbf{e}_z = 0 \qquad [7b]$$

where $Pe$ and $Re$ are the Peclet and Reynolds numbers defined in Section 4.3.1. Now because these are the only dimensionless parameters appearing in [7a] and [7b], the dimensionless temperature must have the form $T = T(r, \theta, Pe, Re)$. Differentiating this with respect to $r$ and evaluating the result on the surface $r = 1$ to obtain the quantity $\beta a / \Delta T$, we find the same result [5] as we did using the $\Pi$-theorem.

Whether one chooses to do dimensional analysis using the $\Pi$-theorem or nondimensionalization depends on the problem at hand. The $\Pi$-theorem is of course the only choice if the governing equations are not known, but its effective use then requires a good intuition of what the relevant physical parameters are. When the governing equations are known, nondimensionalization is usually the best choice, as the relevant physical parameters appear explicitly in the equations and initial/boundary conditions.

### 4.3.3 Scaling Analysis

Except when $N - M = 1$, dimensional analysis yields a relation involving an unknown function of one or more dimensionless arguments. To determine the functional dependence itself, methods that go beyond dimensional analysis are required. The most detailed information is provided by a full analytical or numerical solution of the problem, but finding such solutions is rarely easy. Scaling analysis is a powerful intermediate method that provides more information than dimensional analysis while avoiding the labor of a complete solution. It proceeds by estimating the orders of magnitude of the different terms in a set of governing equations, using both known and unknown quantities, and then exploiting the fact that the terms must balance (the definition of an equation!) to determine how the unknown quantities depend on the known.

To illustrate this, we consider once again the problem of determining the local Nusselt number $\mathcal{N}$ for the hot sphere, but now in the specific limit $Re \ll 1$ and $Pe \gg 1$. Recall that $Re$ measures the ratio of advection to diffusion of gradients in velocity ($\equiv$vorticity), while $Pe$ does the same for gradients in temperature. In the limit $Re \ll 1$, advection of velocity gradients is negligible relative to diffusion everywhere in the flow field, and $\mathbf{u}$ is given by the classic Stokes–Hadamard solution for slow viscous flow around a sphere of another fluid (Section 4.5.3.2). When $Pe \gg 1$, temperature gradients are transported by advection with negligible diffusion

everywhere except in a thin TBL of thickness $\delta(\theta) \ll a$ around the leading hemisphere where advection and diffusion are of the same order. Because radial temperature gradients greatly exceed surface-tangential gradients within this layer, the temperature distribution there is described by the simplified BL forms of the continuity and energy equations (cf. Section 4.7)

$$a \sin \theta v_r + (u \sin \theta)_\theta = 0 \qquad [8a]$$

$$a^{-1} u T_\theta + v T_r = \kappa T_{rr} \qquad [8b]$$

where $u(r, \theta)$ and $v(r, \theta)$ are the tangential ($\theta$) and radial ($r$) components of the velocity, respectively, and subscripts indicate partial derivatives. Equations [8] are obtained from [145] by setting $x = a\theta$ and $r = a \sin \theta$.

We begin by determining the relative magnitudes of the velocity components $u$ and $v$ in the BL. While these can be found directly from the Stokes–Hadamard solution, it is more instructive to do a scaling analysis of the continuity equation [8a]. Now $v_r \sim \Delta v / \delta$, where $\Delta v$ is the change in $v$ across the BL; but because $v(a, \theta) = 0$, $\Delta v = v$. Similarly, $u_\theta \sim \Delta u / \Delta \theta$, where $\Delta u$ is the change in $u$ over an angle $\Delta \theta$ of order unity from the forward stagnation point $\theta = 0$ toward the equator $\theta = \pi/2$. But because $u(r, 0) = 0$, $\Delta u = u$. The continuity equation therefore implies

$$v \sim (\delta / a) u \qquad [9]$$

We turn now to the left-hand side (LHS) of the energy equation [8b], whose two terms represent advection of temperature gradients in the tangential and radial directions, respectively. Now the radial temperature gradient $T_r \sim \Delta T / \delta$ greatly exceeds the tangential gradient $a^{-1} T_\theta \sim \Delta T / a$, but this difference is compensated by the smallness of the radial velocity $v \sim (\delta / a) u$, and so the terms representing tangential and radial advection are of the same order. The balance of advection and diffusion in the BL is therefore $a^{-1} u T_\theta \sim \kappa T_{rr}$, which together with $T_{rr} \sim \Delta T / \delta^2$ implies

$$\delta^2 \sim \kappa a / u \qquad [10]$$

It remains only to determine an expression for $u$, which depends on the ratio $\gamma$ of the viscosity of the sphere to that of the surrounding fluid. We consider the limiting cases $\gamma \ll 1$ (a traction-free sphere) and $\gamma \gg 1$ (an effectively rigid sphere). Because the fluid outside the sphere has constant viscosity, $\mathbf{u}$ varies smoothly over a length scale $\sim a$. Within the TBL, therefore, $\mathbf{u}$ can be approximated by the first term of its Taylor series expansion in the radial distance $r - a \equiv \zeta$ away from the sphere's surface. If the sphere is traction free, $u_\zeta|_{\zeta=0} = 0$, implying that $u \sim U$ is constant across the TBL to lowest order. If however the sphere is rigid, $u|_{\zeta=0} = 0$ and $u \sim (\zeta/a)U$. The tangential velocity $u$ at the outer edge $\zeta \sim \delta$ of the TBL is therefore

$$u \sim (\delta/a)^n U \qquad [11]$$

where $n = 0$ for a traction-free sphere and $n = 1$ for a rigid sphere. Substituting [11] into [10] and noting that $\mathcal{N} \sim a/\delta$, we obtain

$$\mathcal{N} \sim Pe^{1/(n+2)} f_n(\theta) \qquad [12]$$

where $f_n(\theta)$ ($n = 1$ or 2) are unknown functions. Thus when $Pe \gg 1$, $\mathcal{N} \sim Pe^{1/2}$ if the sphere is traction-free and $\mathcal{N} \sim Pe^{1/3}$ if it is rigid. $\mathcal{N}$ is greater in the former case because the tangential velocity $u$, which carries the heat away from the sphere, is $\sim U$ across the whole TBL.

## 4.4 Self-Similarity and Intermediate Asymptotics

In geophysics and in fluid dynamics more generally, one often encounters functions that exhibit the property of scale-invariance or self-similarity. As an illustration, consider a function $f(y, t)$ of two arbitrary variables $y$ and $t$. The function $f$ is self-similar if it has the form

$$f(y, t) = G(t)F\left(\frac{y}{\delta(t)}\right) \qquad [13]$$

where $F$, $G$, and $\delta$ are arbitrary functions and $\eta \equiv y/\delta(t)$ is the similarity variable. Self-similarity simply means that curves of $f$ versus $y$ for different values of $t$ can be obtained from a single universal curve $F(\eta)$ by stretching its abscissa and ordinate by factors $\delta(t)$ and $G(t)$, respectively.

Self-similarity is closely connected with the concept of intermediate asympotics (Barenblatt, 1996). In many physical situations, one is interested in the behavior of a system at intermediate times, long after it has become insensitive to the details of the initial conditions but long before it reaches a final equilibrium state. The behavior of the system at these intermediate times is often self-similar, as we now illustrate using a simple example of conductive heat transfer (Barenblatt, 1996, Section 4.2.1).

### 4.4.1 Conductive Heat Transfer

Consider the 1-D conductive heat transfer in a rod $y \in [0, L]$ in which the initial temperature is zero everywhere except in a heated segment of length $h$ centered at $y = y_0$ (**Figure 4**). The width of the heated segment is much smaller than the distance to either end of the rod, and the segment is much closer to the left end than to the right end, that is, $h \ll y_0$ and $y_0 \ll L - y_0$. The ends of the rod are held at zero temperature, and its sides are insulated.

The equation and initial/boundary conditions governing the temperature $T(y, t)$ in the rod are

$$T_t = \kappa T_{yy} \qquad [14a]$$

$$T(y, 0) = T_0(y), \quad T(0, t) = T(L, t) = 0 \qquad [14b]$$

where $T_0(y)$ is the concentrated initial temperature distribution. While it is relatively easy to solve [14a] and [14b] numerically for an arbitrary initial temperature $T_0(y)$, such an approach would not reveal the essential fact that the solution has two distinct intermediate asymptotic, self-similar stages. The first obtains long after the temperature distribution has forgotten the details of the initial distribution $T_0(y)$, but long before it feels the influence of the left boundary condition $T(0, t) = 0$, that is, for (roughly) $0.1h^2/\kappa \ll t \ll 0.1y_0^2/\kappa$. In this time range, the rod appears effectively infinite, and the integrated temperature anomaly

$$Q = \int_{-\infty}^{\infty} T(y, t)\mathrm{d}y \qquad [15]$$

is constant. The temperature $T$ can depend only on $Q$, $\kappa$, $t$, and $y - y_0$, and only three of these five parameters have independent dimensions. Applying the $\Pi$-theorem with $N = 5$ and $M = 3$, we find

$$T = \frac{Q}{(\kappa t)^{1/2}} F_1(\eta), \quad \eta = \frac{y - y_0}{(\kappa t)^{1/2}} \qquad [16]$$

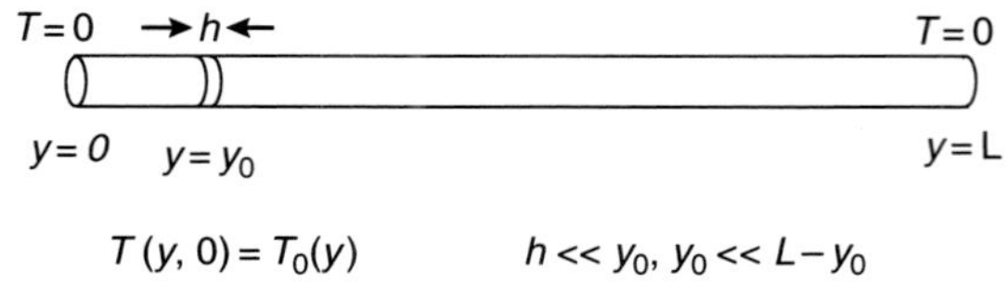

**Figure 4** Model for 1-D conductive heat transfer in a rod $y \in [0, L]$ in which the initial temperature is zero everywhere except in a heated segment of length $h$ centered at $y = y_0$. The width of the heated segment is much smaller than the distance to either end of the rod, and the segment is much closer to the left end than to the right end. Both ends of the rod are held at zero temperature, and its sides are insulated.

which is of the general self-similar form [13]. Substituting [16] into [14a] and [15], we find that $F_1$ satisfies

$$2F_1'' + \eta F_1' + F_1 = 0, \qquad \int_{-\infty}^{\infty} F_1 \, d\eta = 1 \qquad [17]$$

Upon solving [17] subject to $F_1(\pm\infty) = 0$, [16] becomes

$$T = \frac{Q}{2\sqrt{\pi \kappa t}} \exp\left[-\frac{(y-y_0)^2}{4\kappa t}\right] \qquad [18]$$

The second intermediate asymptotic stage occurs long after the temperature distribution has begun to be influenced by the left boundary condition $T(0, t) = 0$, but long before the influence of the right boundary condition $T(L, t) = 0$ is felt, or $y_0^2/\kappa \ll t \ll 0.1(L - y_0)^2/\kappa$. During this time interval, the rod is effectively semi-infinite, and the temperature satisfies the boundary conditions

$$T(0, \ t) = T(\infty, \ t) = 0 \qquad [19]$$

The essential step in determining the similarity solution is to identify a conserved quantity. Multiplying [14a] by $y$, integrating from $y = 0$ to $y = \infty$, and then taking the time derivative outside the integral sign, we obtain

$$\frac{d}{dt}\int_0^\infty yT \, dy = \kappa \int_0^\infty yT_{yy} dy \qquad [20]$$

However, the RHS of [20] is zero, as can be shown by integrating by parts, applying the conditions [19], and noting that $yT_y|_{y=\infty} = 0$ because $T_y \to 0$ more rapidly (typically exponentially) than $y \to \infty$. The temperature moment

$$M = \int_0^\infty yT \, dy \qquad [21]$$

is therefore constant; and because the initial temperature distribution is effectively a delta-function concentrated at $y = y_0$, $M = Qy_0$. Now in the time interval in question, the influence of the temperature distribution that existed at the time $\approx 0.1 y_0^2/\kappa$ when the heated region first reached the near end $y = 0$ of the rod will no longer be felt. The temperature will therefore no longer depend on $y_0$, but only on $M, \kappa, y$, and $t - t_0$, where $t_0$ is the effective starting time for the second stage, to be determined later. Applying the $\Pi$-theorem as before, we find

$$T = \frac{M}{\kappa(t-t_0)} F_2(\eta), \qquad \eta = \frac{y}{\sqrt{\kappa(t-t_0)}} \qquad [22]$$

Now substitute [22] into [14a] and [21], and solve the resulting equations subject to $F_2(0) = F_2(\infty) = 0$, whereupon [22] becomes

$$T = \frac{My}{2\sqrt{\pi}[\kappa(t-t_0)]^{3/2}} \exp\left[-\frac{y^2}{4\kappa(t-t_0)}\right] \qquad [23]$$

The final step is to determine the starting time $t_0 = -y_0^2/6\kappa$ (Barenblatt, 1996, p. 74). Because $t_0 < 0$, that is, before the rod was originally heated, it represents a virtual starting time with respect to which the behavior of the second stage is self-similar.

### 4.4.2   Classification of Self-Similar Solutions

The solutions [18] and [23] are examples of what Barenblatt (1996) calls self-similar solutions of the first kind, for which dimensional analysis (in some cases supplemented by scaling analysis of the governing equations) suffices to find the similarity variable. They are distinguished from self-similar solutions of the second kind, for which the similarity variable can only be found by solving an eigenvalue problem. We will meet some examples of these below, in the sections on viscous eddies in a corner (Section 4.5.3.4) and the spreading of viscous gravity currents (Section 4.8.1).

An example of a self-similar solution that does not fit naturally into either class is the impulsive cooling of a half-space deforming in pure shear. Suppose that the half-space $y \geq 0$ has temperature $T = 0$ initially, and that at time $t = 0$ the temperature at its surface $y = 0$ is suddenly decreased by an amount $\Delta T$. The 2-D velocity field in the half-space is $\mathbf{u} = \dot{\epsilon}(x\mathbf{e}_x - y\mathbf{e}_y)$, where $\dot{\epsilon}$ is the constant rate of extension of the surface $y = 0$ and $\mathbf{e}_x$ and $\mathbf{e}_y$ are unit vectors parallel to and normal to the surface, respectively. Given this velocity field, a temperature field $T = T(y, t)$ that is independent of the lateral coordinate $x$ is an allowable solution of the governing advection–diffusion equation $T_t + \mathbf{u} \cdot \nabla T = \kappa \nabla^2 T$, which takes the form

$$T_t - \dot{\epsilon}yT_y = \kappa T_{yy} \qquad [24]$$

subject to the conditions $T(y, 0) = T(\infty, t) = T(0, t) + \Delta T = 0$. The limit $\dot{\epsilon} = 0$ corresponds to the classic problem of the impulsive cooling of a static half-space.

Neither dimensional analysis nor scaling analysis is sufficient to determine the similarity variable, which does not have the standard power-law monomial form. However, the solution can be found via a generalized form of the familiar separation-of-variables procedure often used to solve PDEs such as Laplace's equation. Note first that the amplitude $\Delta T$ of the temperature in the half-space is a constant. This implies that the function $G(t)$ in the similarity transformation [13] must be independent of time, whence $T(y, t) = \Delta T F(y/\delta(t))$. Substituting this expression into [24] and bringing all terms involving $\delta(t)$ to the LHS, we obtain

$$\frac{\delta(\dot{\delta} + \dot{\epsilon}\delta)}{\kappa} = -\frac{F''}{\eta F'} \quad [25]$$

where dots and primes denote differentiation with respect to $t$ and $\eta \equiv y/\delta(t)$, respectively. Now the LHS of [25] is a function of $t$ only, whereas the RHS depends on $y$ through the similarity variable $\eta$. Equation [25] is therefore consistent only if both sides are equal to a constant $\lambda^2$, which is positive because $\dot{\delta} > 0$. The solutions for $\delta$ and $F$ subject to the conditions $\delta(0) = F(\infty) = F(0) - 1 = 0$ are

$$F = \mathrm{erfc}\frac{\lambda y}{\sqrt{2}\delta}, \quad \delta = \lambda\left\{\frac{\kappa}{\dot{\epsilon}}[1 - \exp(-2\dot{\epsilon}t)]\right\}^{1/2} \quad [26]$$

Evidently $\lambda$ cancels out when the solution for $\delta$ is substituted into the solution for $F$, because different values of $\lambda$ merely correspond to different (arbitrary) definitions of the thermal layer thickness $\delta$. With $\dot{\epsilon} = 0$ and the conventional choice $\lambda = \sqrt{2}$, we recover the well-known solution $\delta = 2\sqrt{\kappa t}$ for a static half-space. However, when $\dot{\epsilon} > 0$, the BL thickness approaches a steady-state value $\delta = (2\kappa/\dot{\epsilon})^{1/2}$ for which the downward diffusion of temperature gradients is balanced by upward advection, and the similarity variable involves an exponential function of time. The only reliable way to find such nonstandard self-similar solutions is the separation-of-variables procedure outlined above. But the same procedure works just as well for problems with similarity variables of standard form, and therefore will be used throughout this chapter.

In conclusion, we note that similarity transformations can also be powerful tools for reducing and interpreting the output of numerical models. As a simple example, suppose that some such model yields values of a dimensionless parameter $W$ as a function of two dimensionless groups $\Pi_1$ and $\Pi_2$. Depending on the physics of the problem, it may be possible to express the results in the self-similar form

$$W(\Pi_1, \Pi_2) = F_1(\Pi_1)F_2\left(\frac{\Pi_1}{F_3(\Pi_2)}\right) \quad [27]$$

where $F_1$–$F_3$ are functions to be determined numerically. A representation of the form [27] is not

guaranteed to exist, but when it does it provides a compact way of representing multidimensional numerical data by functions of a single variable that can be fit by simple analytical expressions. An example of the use of this technique for a problem involving five dimensionless groups is the lubrication theory model for plume–ridge interaction of Ribe and Delattre (1998).

### 4.4.3 Intermediate Asymptotics with Respect to Parameters: The R–T Instability

The concept of intermediate asymptotics is not limited to self-similar behavior of systems that evolve in time, but also applies in a more general way to functions of one or more parameters that exhibit simple (typically power-law) behavior in some asymptotically defined region of the parameter space. Because power-law scaling usually results from a simple dynamical balance between two competing effects, the identification of intermediate asymptotic limits that have this form is crucial for a physical understanding of the system in question.

The dynamical significance of intermediate asymptotic limits and the role that scaling arguments play in identifying them are nicely illustrated by the RT instability of a fluid layer with density $\rho_0 + \Delta\rho$, viscosity $\eta_1$, and thickness $h_0$ above a fluid half-space with density $\rho_0$ and viscosity $\eta_0$ (**Figure 2(a)**). The following discussion is adapted from Canright and Morris (1993).

Linear stability analysis of this problem (cf. Section 4.9.1) shows that an infinitesimal sinusoidal perturbation $\zeta = \zeta_0 \sin kx$ of the initially flat interface $z = 0$ in **Figure 2(a)** grows exponentially at a rate (Whitehead and Luther, 1975)

$$s = s_1 \frac{\gamma}{2\epsilon}\left[\frac{\gamma(C-1) + S - 2\epsilon}{\gamma^2(S + 2\epsilon) + 2\gamma C + S - 2\epsilon}\right] \quad [28]$$

where $s_1 = g\Delta\rho h_0/\eta_1$, $\gamma = \eta_1/\eta_0$, $\epsilon = h_0 k$, $C = \cos h(2\epsilon)$, and $S = \sin h(2\epsilon)$. Here we shall consider only the long-wavelength limit $\epsilon \ll 1$, for which [28] reduces to

$$\frac{s}{s_1} \sim \frac{\epsilon\gamma(2\epsilon + 3\gamma)}{2(2\epsilon^3 + 3\gamma + 6\epsilon\gamma^2)} \quad [29]$$

where the viscosity contrast $\gamma$ is arbitrary.

By noting the ranges of $\gamma(\epsilon)$ for which different pairs of terms in [29] (one each in the numerator and the denominator) are dominant, one finds that [29] has four intermediate asymptotic limits: $\gamma \ll \epsilon^3$, $\epsilon^3 \ll \gamma \ll \epsilon$, $\epsilon \ll \gamma \ll \epsilon^{-1}$, and $\gamma \gg \epsilon^{-1}$. The essential

**Table 2**   Rayleigh–Taylor instability: intermediate asymptotic limits

| Limit | $\gamma$ | $W/U$ | $\hat{u}/U$ | Balancing pressure | $s$ |
|---|---|---|---|---|---|
| 1 | $\ll \epsilon^3$ | $\epsilon^2/\gamma$ | $\epsilon/\gamma$ | $p_0 \sim \eta_0 k W$ | $\Delta\rho g/2k\eta_0$ |
| 2 | $\epsilon^3 \ll \gamma \ll \epsilon$ | $\epsilon^2/\gamma$ | $\epsilon/\gamma$ | $p_1 \sim \eta_1 \hat{u}/h_0^2 k \sim \eta_1 W/h_0^3 k^2$ | $\Delta\rho g h_0^3 k^2/3\eta_1$ |
| 3 | $\epsilon \ll \gamma \ll \epsilon^{-1}$ | $\epsilon$ | $\epsilon/\gamma$ | $p_1 \sim \eta_1 \hat{u}/h_0^2 k \sim \eta_0 W/h_0^2 k$ | $\Delta\rho g h_0^2 k/2\eta_0$ |
| 4 | $\gg \epsilon^{-1}$ | $\epsilon$ | $\epsilon^2$ | $p_1 \sim \eta_1 k U \sim \eta_1 W/h_0$ | $\Delta\rho g h_0/4\eta_1$ |

dynamics associated with each are summarized in columns 3–5 of **Table 2**. Column 3 shows the ratio of the amplitudes of the vertical ($W$) and horizontal ($U$) components of the velocity at the interface (**Figure 2(a)**). As $\gamma$ increases, the motion of the interface changes from dominantly vertical in limit 1 to dominantly horizontal in limits 3 and 4. The ratio of shear deformation to plug flow in the layer is measured by the ratio $\hat{u}/U$ (column 4), where $\hat{u}$ is the change in horizontal velocity across the layer (**Figure 2(a)**). The layer deforms mainly by shear in limits 1 and 2, and by plug flow in limits 3 and 4.

The growth rate is determined by whether the interfacial buoyancy $\sim g\Delta\rho\zeta_0$ is supported by the pressure $p_1$ in the layer or the pressure $p_0$ in the half-space. Column 5 of **Table 2** gives the expression for the pressure that balances the buoyancy in each limit, and column 6 shows the corresponding growth rate $s = W/\zeta_0$. While the pressures can be calculated directly from the analytical solution of the problem (Section 4.9.1), it is more revealing to obtain them via a scaling analysis of the horizontal component $\partial_x p = \eta\nabla^2 u$ of the momentum equation. In the half-space, the only length scale is $k^{-1}$, so that $\partial_x \sim \partial_z \sim k$. The continuity equation then implies $u \sim w$. The magnitude of $u \sim w$ is set by the larger of the two components of the velocity at the interface, namely, $u \sim [U, W]$, where $[\ldots]$ denotes the maximum of the enclosed quantities. Turning now to the layer, we note that the horizontal and vertical length scales are different, so that $\partial_x \sim k$ and $\partial_z \sim h_0^{-1}$. Moreover, $\nabla^2 u \sim [\hat{u}/h_0^2, k^2 U]$ is the sum of terms arising from the shear and plug flow components of $u$. We thereby find

$$p_1 \sim \frac{\eta_1}{h_0^2 k}\left[\hat{u}, \epsilon^2 U\right], \quad p_0 \sim \eta_0 k[U, W] \qquad [30]$$

In view of the pressure scales [30] and those for $W/U$ and $\hat{u}/U$ from **Table 2**, the essential dynamics of each of the four intermediate limits can be summarized as follows. In limit 1, the half-space feels the layer as an effectively traction-free boundary, the buoyancy is balanced by the pressure $p_0 \sim \eta_0 k W$ in the half-space, and $s$ is controlled by the half-space viscosity $\eta_0$. In limit 2, the half-space still sees the layer as traction-free, but the pressure $p_1 \sim \eta_1 \hat{u}/h_0^2 k$ induced by shear flow in the layer is nevertheless sufficient to balance the buoyancy. Because the layer deforms mostly in shear, $\hat{u}$ is related directly to $W$ via the continuity equation ($W \sim \epsilon\hat{u}$), so $s$ is controlled by the layer viscosity $\eta_1$. In limit 3, each layer feels the shear stress applied by the other. While the buoyancy is still balanced by the shear-induced pressure in the layer, the dominance of plug flow means that $\hat{u}$ and $W$ are no longer related via the continuity equation, but rather by the matching condition on the shear stress. The growth rate is therefore controlled by the half-space viscosity $\eta_0$. Finally, in limit 4 the layer feels the half-space as a traction-free boundary, the buoyancy is balanced by the pressure $p_1 \sim \eta_1 k U$ induced by plug flow in the layer, and $s$ is controlled by the layer viscosity $\eta_1$.

## 4.5 Slow Viscous Flow

Flows with negligible inertia are fundamental in the Earth's mantle, where the Reynolds number $Re \approx 10^{-20}$. A particularly important subclass of inertialess flow – variously called slow, creeping, or low Reynolds number flow – comprises flows in which the fluid is incompressible, isothermal, and has a rheology with no memory (elasticity). These conditions, while obviously restrictive, are nevertheless sufficiently realistic to have served as a basis for many important geophysical models.

### 4.5.1 Basic Equations and Theorems

The most general equations required to describe the slow viscous flows discussed in this section are (*see* Chapter 6)

$$\partial_j u_j = 0 \qquad [31a]$$

$$\partial_j \sigma_{ij} + b_i = 0 \qquad [31b]$$

$$b_i = -\rho \partial_i \chi \qquad [31c]$$

$$\nabla^2 \chi = 4\pi G \rho \qquad [31d]$$

$$\sigma_{ij} = -p + 2\eta e_{ij}, \quad e_{ij} = \frac{1}{2}\left(\partial_i u_j + \partial_j u_i\right) \qquad [31e]$$

$$\eta = \eta_0 (I/I_0)^{-1+1/n}, \quad I = \left(e_{ij} e_{ij}\right)^{1/2} \qquad [31f]$$

where $u_i$ is the velocity vector, $\sigma_{ij}$ is the stress tensor, $b_i$ is the gravitational body force per unit volume, $\chi$ is the gravitational potential, $\rho$ is the density, $p$ is the pressure, $\eta$ is the viscosity, and $e_{ij}$ is the strain-rate tensor with second invariant $I$. Equation [31a] is the incompressibility condition. Equation [31b] expresses conservation of momentum in the absence of inertia, and states that the net force (viscous plus gravitational) acting on each fluid element is zero. Equation [31d] is Poisson's equation for the gravitational potential. Equation [31e] is the standard constitutive relation for a viscous fluid. Finally, [31f] is the strain-rate-dependent viscosity for a power-law fluid (sometimes called a generalized Newtonian fluid), where $\eta_0$ is the viscosity at a reference strain rate $I = I_0$ and $n$ is the power-law exponent. A Newtonian fluid has $n = 1$, while the rheology of dry olivine deforming by dislocation creep is well described by [31f] with $n \approx 3.5$ (Bai et al., 1991). A discussion of more complicated non-Newtonian fluids is beyond the scope of this chapter; the interested reader is referred to Bird et al. (1987).

Viscous flow described by [31] can be driven either externally, by velocities and/or stresses imposed at the boundaries of the flow domain, or internally, by buoyancy forces (internal loads) arising from lateral variations of the density $\rho$. On the scale of the whole mantle, the influence of long-wavelength lateral variations of $\rho$ on the gravitational potential $\chi$ (self-gravitation) is significant and cannot be neglected. In modeling flow on smaller scales, however, one generally ignores Poisson's equation [31d] and replaces $\nabla \chi$ in [31c] by a constant gravitational acceleration $-\mathbf{g}$.

Slow viscous flow exhibits the property of instantaneity: $u_i$ and $\sigma_{ij}$ at each instant are determined throughout the fluid solely by the distribution of forcing (internal loads and/or boundary motions) acting at that instant. Instantaneity requires that the fluid have no memory (elasticity) and that acceleration and inertia be negligible; there is then no time lag between the forcing and the fluid's response to it. A corollary is that slow viscous flow is quasi-static, any time-dependence being due entirely to the time-dependence of the forcing.

The theory of slow viscous flow is most highly developed for the special case of Newtonian fluids (Stokes flow), and several excellent monographs on the subject exist (Ladyzhenskaya, 1963; Langlois, 1964; Happel and Brenner, 1991; Kim and Karrila, 1991; Pozrikidis, 1992). Relative to general slow viscous flow, Stokes flow exhibits the important additional properties of linearity and reversibility. Linearity implies that for a given geometry, a sum of different solutions (e.g., for different forcing distributions) is also a solution. It also implies that $u_i$ and $\sigma_{ij}$ are directly proportional to the forcing that generates them, and hence for example, that the force acting on a body in Stokes flow is proportional to its speed. Reversibility refers to the fact that changing the sign of the forcing terms reverses the signs of $u_i$ and $\sigma_{ij}$ for all material particles. The reversibility principle is especially powerful when used in conjunction with symmetry arguments. It implies, for example, that a body with fore-aft symmetry falling freely in any orientation in Stokes flow experiences no torque, and that the lateral separation of two spherical diapirs with different radii is the same before and after their interaction (Manga, 1997).

An important theorem concerning Stokes flow is the 'Lorentz reciprocal theorem,' which relates two different Stokes flows $(u_i, \sigma_{ij}, b_i)$ and $(u_i^*, \sigma_{ij}^*, b_i^*)$. This theorem is the starting point for the boundary-integral representation of Stokes flow derived in Section 4.5.6.4. Consider the scalar quantity $u_i^* \partial_j \sigma_{ij}$, which we manipulate as follows:

$$\begin{aligned} u_i^* \partial_j \sigma_{ij} &= \partial_j \left(u_i^* \sigma_{ij}\right) - \sigma_{ij} \partial_j u_i^* \\ &= \partial_j \left(u_i^* \sigma_{ij}\right) - \left(-p\delta_{ij} + 2\eta e_{ij}\right)\partial_j u_i^* \\ &= \partial_j \left(u_i^* \sigma_{ij}\right) - 2\eta e_{ij} e_{ij}^* \end{aligned} \qquad [32]$$

By subtracting from [32] the analogous expression with the starred and unstarred fields interchanged and setting $\partial_j \sigma_{ij} = -b_i$ and $\partial_j \sigma_{ij}^* = -b_i^*$, we obtain the differential form of the Lorentz reciprocal theorem:

$$\partial_j \left(u_i^* \sigma_{ij} - u_i \sigma_{ij}^*\right) = u_j b_j^* - u_j^* b_j \qquad [33]$$

An integral form of the reciprocal theorem is obtained by integrating [33] over a volume $V$ bounded by a surface $S$ and applying the divergence theorem, yielding

$$\int_S u_i^* \sigma_{ij} n_j \, \mathrm{d}S + \int_V b_j u_j^* \, \mathrm{d}V = \int_S u_i \sigma_{ij}^* n_j \, \mathrm{d}S$$
$$+ \int_V b_j^* u_j \, \mathrm{d}V \qquad [34]$$

where $n_j$ is the outward unit normal to $S$.

Two additional theorems for Stokes flow concern the total rate of energy dissipation

$$E = 2\eta \int_V e_{ij} e_{ij} \mathrm{d}V \qquad [35]$$

in a volume $V$. The first states that the solution of the Stokes equations subject to given boundary conditions is unique, and is most easily proved by showing that the energy dissipated by the difference of two supposedly different solutions is zero (Kim and Karrila, 1991, p. 14.) The second is the minimum dissipation theorem, which states that a solution of the Stokes equations for given boundary conditions dissipates less energy than any other solenoidal vector field satisfying the same boundary conditions (Kim and Karrila, 1991, p. 15.) Note that this theorem merely compares a Stokes flow with other flows that do not satisfy the Stokes equations. It says nothing about the relative rates of dissipation of Stokes flows with different geometries and/or boundary conditions, and therefore its use as a principle of selection among such flows is not justified.

### 4.5.2 Potential Representations for Incompressible Flow

In an incompressible flow satisfying $\nabla \cdot \mathbf{u} = 0$, only $N-1$ of the velocity components $u_i$ are independent, where $N = 3$ in general and $N = 2$ for 2-D and axisymmetric flows. This fact allows one to express all the velocity components in terms of derivatives of $N-1$ independent scalar potentials, thereby reducing the number of independent variables in the governing equations. The most commonly used potentials are the streamfunction $\psi$ (for $N = 2$) and the poloidal potential $\mathcal{P}$ and the toroidal potential $\mathcal{T}$ (for $N = 3$). Below we give expressions for the components of $\mathbf{u}$ in terms of these potentials in Cartesian $(x, y, z)$, cylindrical $(\rho, \phi, z)$, and spherical $(r, \theta, \phi)$ coordinates, together with the PDEs they satisfy for the important special case of constant viscosity.

#### 4.5.2.1  2-D and axisymmetric flows

A 2-D flow is one in which the velocity vector $\mathbf{u}$ is everywhere perpendicular to a fixed direction ($\mathbf{e}_z$ say) in space. The velocity can then be represented in terms of a streamfunction by

$$\mathbf{u} = \mathbf{e}_z \times \nabla \psi = -\mathbf{e}_x \psi_y + \mathbf{e}_y \psi_x = -\mathbf{e}_\phi \psi_\rho + \frac{\mathbf{e}_\rho}{\rho} \psi_\phi \qquad [36]$$

The PDE satisfied by $\psi$ is obtained by applying the operator $\mathbf{e}_z \times \nabla$ to the momentum equation [31b] with the constitutive law [31e]. If the viscosity is constant and $b_i = 0$, $\psi$ satisfies the biharmonic equation

$$\nabla_1^4 \psi = 0, \quad \nabla_1^2 = \partial_{xx}^2 + \partial_{yy}^2 = \rho^{-1} \partial_\rho (\rho \partial_\rho) + \rho^{-2} \partial_{\phi\phi}^2 \qquad [37]$$

An axisymmetric flow (without swirl) is one in which $\mathbf{u}$ at any point lies in the plane containing the point and some fixed axis ($\mathbf{e}_z$, say). For this case,

$$\mathbf{u} = \frac{\mathbf{e}_\phi}{\rho} \times \nabla \psi = -\frac{\mathbf{e}_z}{\rho} \psi_\rho + \frac{\mathbf{e}_\rho}{\rho} \psi_z \qquad [38a]$$

$$\mathbf{u} = \frac{\mathbf{e}_\phi}{r \sin\theta} \times \nabla \psi = \frac{\mathbf{e}_\theta}{r \sin\theta} \psi_r - \frac{\mathbf{e}_r}{r^2 \sin\theta} \psi_\theta \qquad [38b]$$

where $\psi$ is referred to as the Stokes streamfunction. The PDE satisfied by the Stokes streamfunction is obtained by applying the operator $\mathbf{e}_\phi \times \nabla$ to the momentum equation. For a fluid with constant viscosity and no body force, the result is

$$E^4 \psi = 0,$$
$$E^2 = \frac{h_3}{h_1 h_2} \left[ \frac{\partial}{\partial q_1} \left( \frac{h_2}{h_1 h_3} \frac{\partial}{\partial q_1} \right) + \frac{\partial}{\partial q_2} \left( \frac{h_1}{h_2 h_3} \frac{\partial}{\partial q_2} \right) \right] \qquad [39]$$

where $(q_1, q_2)$ are orthogonal coordinates in any half-plane normal to $\mathbf{e}_\phi$, $(h_1, h_2)$ are the corresponding scale factors, and $h_3$ is the scale factor for the azimuthal coordinate $\phi$. Thus $(q_1, q_2, h_1, h_2, h_3) = (z, \rho, 1, 1, \rho)$ in cylindrical coordinates and $(q_1, q_2, h_1, h_2, h_3) = (r, \theta, 1, r, r \sin\theta)$ in spherical coordinates. The operator $E^2$ is in general different from the Laplacian operator

$$\nabla_1^2 = \frac{1}{h_1 h_2 h_3} \left[ \frac{\partial}{\partial q_1} \left( \frac{h_2 h_3}{h_1} \frac{\partial}{\partial q_1} \right) + \frac{\partial}{\partial q_2} \left( \frac{h_1 h_3}{h_2} \frac{\partial}{\partial q_2} \right) \right] \qquad [40]$$

the two being identical only for 2-D flows ($h_3 = 1$).

#### 4.5.2.2  3-D flows

The most commonly used (but not the only) potential representation of 3-D flows in geophysics is a decomposition of the velocity $\mathbf{u}$ into poloidal and toroidal components (Backus, 1958; Chandrasekhar, 1981). This representation requires the choice of a preferred direction, which is usually chosen to be an upward vertical ($\mathbf{e}_z$) or radial ($\mathbf{e}_r$) unit vector. Relative to Cartesian coordinates, the poloidal/toroidal decomposition has the form

$$\mathbf{u} = \nabla \times (\mathbf{e}_z \times \nabla \mathcal{P}) + \mathbf{e}_z \times \nabla \mathcal{T}$$
$$= \mathbf{e}_x(-\mathcal{P}_{xz} - \mathcal{T}_y) + \mathbf{e}_y(-\mathcal{P}_{yz} + \mathcal{T}_x) + \mathbf{e}_z \nabla_1^2 \mathcal{P} \qquad [41]$$

where $\mathcal{P}$ is the poloidal potential and $\mathcal{T}$ is the toroidal potential. The associated vorticity $\boldsymbol{\omega} \equiv \nabla \times \mathbf{u}$ is

$$\boldsymbol{\omega} = \mathbf{e}_x(\nabla^2 \mathcal{P}_y - \mathcal{T}_{xz}) + \mathbf{e}_y(-\nabla^2 \mathcal{P}_x - \mathcal{T}_{yz}) + \mathbf{e}_z \nabla_1^2 \mathcal{T} \quad [42]$$

Inspection of [41] and [42] immediately reveals the fundamental distinction between the poloidal and toroidal fields: the former has no vertical vorticity, while the latter has no vertical velocity.

The PDEs satisfied by $\mathcal{P}$ and $\mathcal{T}$ are obtained by applying the operators $\nabla \times (\mathbf{e}_z \times \nabla)$ and $\mathbf{e}_z \times \nabla$, respectively, to the momentum equation [31b] and [31e]. Supposing that the viscosity is constant but retaining a body force $\mathbf{b} \equiv -g\delta\rho \mathbf{e}_z$ where $\delta\rho(\mathbf{x})$ is a density anomaly, we find

$$\nabla_1^2 \nabla^4 \mathcal{P} = \frac{g}{\eta} \nabla_1^2 \delta\rho, \quad \nabla_1^2 \nabla^2 \mathcal{T} = 0 \quad [43]$$

Note that the equation for $\mathcal{T}$ is homogeneous, implying that flow driven by internal density anomalies in a fluid with constant viscosity is purely poloidal. The same is true in a fluid whose viscosity varies only as a function of depth, although the equations satisfied by $\mathcal{P}$ and $\mathcal{T}$ are more complicated than [43]. When the viscosity varies laterally, however, the equations for $\mathcal{P}$ and $\mathcal{T}$ are coupled, and internal density anomalies drive a toroidal flow that is slaved to the poloidal flow. Toroidal flow will also be driven by any surface boundary conditions having a nonzero vertical vorticity, even if the viscosity does not vary laterally.

Because of the Earth's spherical geometry, the spherical-coordinate form of the poloidal–toroidal representation is particularly important in geophysics. However, the definitions of $\mathcal{P}$ and $\mathcal{T}$ used by different authors sometimes differ by a sign and/or a factor of $r$. Following Forte and Peltier (1987)

$$\mathbf{u} = \nabla \times (r\mathbf{e}_r \times \nabla \mathcal{P}) + r\mathbf{e}_r \times \nabla \mathcal{T}$$
$$= \mathbf{e}_\theta \left[ -\frac{1}{r}(r\mathcal{P})_{r\theta} - \frac{\mathcal{T}_\phi}{\sin\theta} \right] + \mathbf{e}_\phi \left[ -\frac{1}{r\sin\theta}(r\mathcal{P})_{r\phi} + \mathcal{T}_\theta \right]$$
$$+ \frac{\mathbf{e}_r}{r} \mathcal{B}^2 \mathcal{P} \quad [44]$$

where

$$\mathcal{B}^2 = \frac{1}{\sin\theta}\frac{\partial}{\partial\theta}\sin\theta\frac{\partial}{\partial\theta} + \frac{1}{\sin^2\theta}\frac{\partial^2}{\partial\phi^2} \quad [45]$$

Another common convention is that of Chandresekhar (1981, appendix III), who uses a poloidal potential $\Phi \equiv -r\mathcal{P}$ and a toroidal potential $\Psi \equiv -r\mathcal{T}$.

Two additional quantities of interest are the lateral divergence $\nabla_1 \cdot \mathbf{u}$ and the radial component $\mathbf{e}_r \cdot (\nabla \times \mathbf{u})$ of the vorticity, which are

$$\nabla_1 \cdot \mathbf{u} = -\mathcal{B}^2\left[\frac{1}{r^2}(r\mathcal{P})_r\right], \quad \mathbf{e}_r \cdot (\nabla \times \mathbf{u}) = \frac{\mathcal{B}^2 \mathcal{T}}{r} \quad [46]$$

The lateral divergence depends only on the poloidal component of the flow, whereas the radial vorticity depends only on the toroidal component. At the Earth's surface $r = a$, therefore, divergent and convergent plate boundaries (where $\nabla_1 \cdot \mathbf{u} \neq 0$) are associated with poloidal flow, while transform faults (where $\mathbf{e}_r \cdot (\nabla \times \mathbf{u}) \neq 0$) reflect toroidal flow.

The PDEs satisfied by $\mathcal{P}$ and $\mathcal{T}$ in a fluid of constant viscosity in spherical coordinates are obtained by applying the operators $\nabla \times (r\mathbf{e}_r \times \nabla)$ and $r\mathbf{e}_r \times \nabla$, respectively, to the momentum equation [31b] and [31e], yielding

$$\mathcal{B}^2 \nabla^4 \mathcal{P} = \frac{g}{\eta r}\mathcal{B}^2 \delta\rho, \quad \mathcal{B}^2 \nabla^2 \mathcal{T} = 0 \quad [47]$$

Because the above equation for $\mathcal{T}$ is homogeneous, the remarks following [43] apply also to spherical geometry.

### 4.5.3 Classical Exact Solutions

The equations of slow viscous flow admit exact analytical solutions in a variety of geophysically relevant geometries. Some of the most useful of these solutions are the following:

#### 4.5.3.1 Steady unidirectional flow

The simplest conceivable fluid flow is a steady unidirectional flow with a velocity $w(x, y, t)\mathbf{e}_z$, where $(x, y)$ are Cartesian coordinates in the plane normal to $\mathbf{e}_z$. If the viscosity is constant and inertia is negligible, $w$ satisfies (Batchelor, 1967)

$$\eta \nabla_1^2 w = -G \quad [48]$$

where $-G \equiv p_z$ is a constant pressure gradient and $\eta$ is the viscosity. Two special cases are of interest in geophysics. The first is the steady Poiseuille flow through a cylindrical pipe of radius $a$ driven by a pressure gradient $-G$, for which the velocity $w$ and the volume flux $Q$ are

$$w = \frac{G}{4\eta}(a^2 - r^2), \quad Q \equiv 2\pi \int_0^a rw\,dr = \frac{\pi G a^4}{8\eta} \quad [49]$$

Poiseuille flow in a vertical pipe driven by an effective pressure gradient $g\Delta\rho$ has been widely used as a model for the ascent of buoyant fluid in the conduit or tail of a mantle plume (e.g., Whitehead and Luther, 1975). The second case is that of a 2-D channel $0 \leq y \leq d$ bounded by rigid walls in which flow is driven by a combination of an applied pressure gradient $-G$ and motion of the boundary $y = d$ with speed $U_0$ in its own plane. For this case,

$$w = \frac{G}{2\eta} y (d - y) + \frac{U_0 y}{d} \qquad [50]$$

An example of a geodynamical application of [50] is the asthenosphere flow model of Yale and Phipps Morgan (1998).

### 4.5.3.2  Stokes–Hadamard solution for a sphere

Another classical result that is useful in geophysics is the Stokes–Hadamard solution for the flow in and around a fluid sphere with radius $a$ and viscosity $\eta_1$ in an unbounded fluid with viscosity $\eta_0 = \eta_1 / \gamma$ and velocity $U \mathbf{e}_z$ far from the sphere (Batchelor, 1967, pp. 230–238). The outer $(n = 0)$ and inner $(n = 1)$ streamfunctions are $\psi_n = U a^2 \sin^2 \theta f_n(r)$, where

$$f_0 = \frac{(2 + 3\gamma)\hat{r} - \gamma \hat{r}^{-1} - 2(1 + \gamma)\hat{r}^2}{4(1 + \gamma)}, \quad f_1 = \frac{\hat{r}^2 - \hat{r}^4}{4(1 + \gamma)} \qquad [51]$$

and $\hat{r} = r/a$. The drag on the sphere is

$$\mathbf{F} = 4\pi a \eta_0 U \frac{1 + 3\gamma/2}{1 + \gamma} \mathbf{e}_z \qquad [52]$$

If the densities of the two fluids differ such that $\rho_1 = \rho_0 + \Delta\rho$, the steady velocity $\mathbf{V}$ of the sphere as it moves freely under gravity is obtained by equating $\mathbf{F}$ to the Archimedean buoyancy force, yielding

$$\mathbf{V} = \frac{a^2 \mathbf{g} \Delta\rho}{3\eta_0} \frac{1 + \gamma}{1 + 3\gamma/2} \qquad [53]$$

where $\mathbf{g}$ is the gravitational acceleration. The speed of an effectively inviscid sphere $(\gamma = 0)$ is only 50% greater than that of a rigid sphere $(\gamma \to \infty)$. Equation [53] has been widely used to estimate the ascent speed of plume heads (e.g., Whitehead and Luther, 1975) and isolated thermals (e.g., Griffiths, 1986) in the mantle.

### 4.5.3.3  Models for subduction zones and ridges

Two-dimensional viscous flow in a fluid wedge driven by motion of the boundaries (corner flow) has been widely used to model mantle flow in subduction zones and beneath mid-ocean ridges. **Figures 5(a)** and **5(b)**

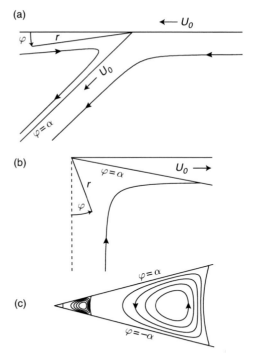

**Figure 5** Models for slow viscous flow in wedge-shaped regions. (a) subduction zone; (b) mid-ocean ridge; (c) self-similar viscous corner eddies generated by an agency far from the corner. In all models, streamlines (for Newtonian rheology) are shown by solid lines with arrows.

show the geometry of the simplest corner-flow models of these features, due respectively to McKenzie (1969) and Lachenbruch and Nathenson (1974).

The models of **Figures 5(a)** and **5(b)** admit analytical solutions for both Newtonian and power-law rheology. The boundary conditions for both models can be satisfied by a self-similar (of the first kind; cf. Section 4.4.2) streamfunction $\psi = -U_0 r F(\varphi)$, where $(r, \varphi)$ are polar coordinates. The only nonzero component of the strain-rate tensor $\mathbf{e}$ is $e_{r\theta} = U_0(F + F'')/2r$, whence the second invariant of $\mathbf{e}$ is $I \equiv (e_{ij} e_{ij})^{1/2} = \sqrt{2} e_{r\theta}$. The equation satisfied by $F$ is obtained by applying the operator $\mathbf{e}_z \times \nabla$ to the momentum equation [31b] with the power-law constitutive relation [31e] and [31f], where $\mathbf{e}_z$ is a unit vector normal to the flow plane. The result is

$$\left( \frac{d^2}{d\varphi^2} + \frac{2n-1}{n^2} \right) (F'' + F)^{1/n} = 0 \qquad [54]$$

where $n$ is the power-law index. The associated pressure, obtained by integrating the $\mathbf{e}_r$-component of the momentum equation with respect to $r$ subject to the condition of vanishing pressure at $r = \infty$ is

$p = -\eta U_0 r^{-1} (F' + F''')$. Equation [54] can be solved analytically if $n$ is any positive integer (Tovish *et al.*, 1978). The values most relevant to geophysics are $n = 1$ and $n = 3$, for which the general solutions are

$$F_1 = A_1 \sin \varphi + B_1 \cos \varphi + C_1 \varphi \sin \varphi + D_1 \varphi \cos \varphi \quad [55]$$

$$F_3 = A_3 \sin \varphi + B_3 \cos \varphi + C_3 H(\varphi, D_3) \quad [56]$$

respectively, where

$$H(\varphi, D_3) = 27 \cos \frac{\sqrt{5}}{3} (\varphi + D_3) - \cos \sqrt{5} (\varphi + D_3) \quad [57]$$

and $A_n - D_n$ are arbitrary constants for $n = 1$ and $n = 3$ that are determined by the boundary conditions. For the ridge model,

$$\{A_1, B_1, C_1, D_1\} = \frac{\{c^2, 0, 0, -1\}}{\alpha - sc} \quad [58]$$

$$\{A_3, B_3, C_3, D_3\}$$
$$= \left\{ -C_3(b_1 s + b_1' c), 0, \frac{1}{b_1' s - b_1 c}, \frac{3\pi}{2\sqrt{5}} \right\} \quad [59]$$

where $s = \sin \alpha$, $c = \cos \alpha$, $b_1 = H(\alpha, D_3)$, and $b_1' = H_\varphi(\alpha, D_3)$. For the Newtonian ($n = 1$) subduction model in the wedge $0 \le \varphi \le \alpha$,

$$\{A_1, B_1, C_1, D_1\} = \frac{\{\alpha s, 0, \alpha c - s, -\alpha s\}}{\alpha^2 - s^2} \quad [60]$$

For the power-law ($n = 3$) subduction model in $0 \le \varphi \le \alpha$, $D_3$ satisfies $b_1 - b_0 c - b_0' s = 0$ and the other constants are

$$C_3 = (b_1' + b_0 s - b_0' c)^{-1}, \quad B_3 = -b_0 C_3,$$
$$A_3 = -b_0' C_3 \quad [61]$$

where $b_0 = H(0, D_3)$, and $b_0' = H_\varphi(0, D_3)$. If needed, the solutions in the wedge $\alpha \le \varphi \le \pi$ can be obtained from [55] and [56] by applying the boundary conditions shown in **Figure 5(a)**. The solution [60] together with the corresponding one for the wedge $\alpha \le \varphi \le \pi$ was the basis for Stevenson and Turner's (1977) hypothesis that the angle of subduction is controlled by the balance between the hydrodynamic lifting torque and the opposing gravitational torque acting on the slab. Their results were extended to power-law fluids by Tovish *et al.* (1978).

#### 4.5.3.4   Viscous eddies

Another important exact solution for slow viscous flow describes viscous eddies near a sharp corner (Moffatt, 1964). This solution is an example of a self-similar solution of the second kind (Section

4.4.2), the determination of which requires the solution of an eigenvalue problem. Here we consider only Newtonian fluids; for power-law fluids, see Fenner (1975).

The flow domain is a 2-D wedge $|\varphi| \le \alpha$ bounded by rigid walls (**Figure 5(c)**). Flow in the wedge is driven by an agency (e.g., stirring) acting at a distance $\sim r_0$ from the corner. We seek to determine the asymptotic character of the flow near the corner, that is, in the limit $r/r_0 \to 0$. Because the domain of interest is far from the driving agency, we anticipate that the flow will be self-similar.

The streamfunction satisfies the biharmonic equation [37], which admits separable solutions of the form

$$\psi = r^\lambda F(\varphi) \quad [62]$$

Substituting [62] into [37], we find that $F$ satisfies

$$F'''' + [\lambda^2 + (\lambda - 2)^2] F'' + \lambda^2 (\lambda - 2)^2 = 0 \quad [63]$$

where primes denote $d/d\varphi$. For all $\lambda$ except 0, 1, and 2, the solution of [63] is

$$F = A \cos \lambda\varphi + B \sin \lambda\varphi + C \cos(\lambda - 2)\varphi + D \sin(\lambda - 2)\varphi \quad [64]$$

where $A–D$ are arbitrary constants. The solutions for $\lambda = 0$, 1, and 2 do not exhibit eddies (the solution with $\lambda = 1$ is the one used in the models of subduction zones and ridges in Section 4.5.3.3).

The most interesting solutions are those for which $\psi(r, \varphi)$ is an even function of $\varphi$. Application of the rigid-surface boundary conditions $F(\pm\alpha) = F'(\pm\alpha) = 0$ to [64] with $B = D = 0$ yields two equations for $A$ and $C$ that have a nontrivial solution only if

$$\sin 2(\lambda - 1)\alpha + (\lambda - 1)\sin 2\alpha = 0 \quad [65]$$

The only physically relevant roots of [65] are those with $\Re(\lambda) > 0$, corresponding to solutions that are finite at $r = 0$. When $2\alpha < 146°$, these roots are all complex; let $\lambda_1 \equiv p_1 + iq_1$ be the root with the smallest real part, corresponding to the solution that decays least rapidly towards the corner. **Figure 5(c)** shows the streamlines $\Re(r^\lambda F) = \text{cst}$ for $2\alpha = 30°$, for which $p_1 = 4.22$ and $q_1 = 2.20$. The flow comprises an infinite sequence of self-similar eddies with alternating senses of rotation, whose successive intensities decrease by a factor $\exp(\pi p_1/q_1) \approx 414$ towards the corner. In the limit $\alpha = 0$ corresponding to flow between parallel planes, $p_1 = 4.21$ and $q_1 = 2.26$, all the eddies have the same size ($\approx 2.78$ times the channel width), and the intensity ratio $\approx 348$.

In a geophysical context, corner eddies are significant primarily as a simple model for the tendency of forced viscous flows in domains with large aspect ratio to break up into separate cells. An example (Section 4.9.3), is steady 2-D cellular convection at high Rayleigh number, in which Stokes flow in the isothermal core of a cell is driven by the shear stresses applied to it by the thermal plumes at its ends. When the aspect ratio (width/depth) $\beta = 1$, the core flow comprises a single cell ; but when $\beta = 2.5$, the flow separates into two distinct eddies (Jimenez and Zufiria, 1987, Figure 2).

### 4.5.4 Superposition and Eigenfunction Expansion Methods

The linearity of the equations governing Stokes flow is the basis of two powerful methods for solving Stokes flow problems in regular domains: the methods of superposition and eigenfunction expansion. In both methods, a complicated flow is represented by infinite sums of elementary separable solutions of the Stokes equations for the coordinate system in question, and the unknown coefficients in the expansion are determined to satisfy the boundary conditions. In the superposition method, the individual separable solutions do not themselves satisfy all the boundary conditions in any of the coordinate directions. In the method of eigenfunction expansions (henceforth MEE), by contrast, the separable solutions are true eigenfunctions that satisfy all the (homogeneous) boundary conditions at both ends of an interval in one of the coordinate directions, so that the unknown constants are determined entirely by the boundary conditions in the other direction(s). Let us turn now to some concrete illustrations of these methods in three coordinate systems of geophysical interest: 2-D Cartesian, spherical, and bispherical.

#### 4.5.4.1 2-D flow in Cartesian coordinates

The streamfunction $\psi$ for 2-D Stokes flow satisfies the biharmonic equation [37], which has the general solution

$$\psi = f_1(x + iy) + f_2(x - iy) + (y + ix)f_3(x + iy) + (y - ix)f_4(x - iy) \qquad [66]$$

where $f_1$–$f_4$ are arbitrary functions of their (complex) arguments. However, the most useful solutions for applications are the separable solutions that

vary periodically in one direction. Setting $\psi \propto \exp ikx$ and solving [37] by separation of variables, we find

$$\psi = [(A_1 + A_2 y)\exp(-ky) + (A_3 + A_4 y)\exp ky]\exp ikx \qquad [67]$$

where $A_1$–$A_4$ are arbitrary functions of $k$. An analogous solution is obtained by interchanging $x$ and $y$ in [67], and both solutions remain valid if $k$ is complex.

The solution [67] has been widely used in geodynamics to describe flows generated by loads that vary sinusoidally in the horizontal direction (e.g., Fleitout and Froidevaux, 1983.) The representation of more general flows, however, typically requires the use of a superposition or an eigenfunction expansion. To illustrate the use of these two methods, we consider the problem of driven cavity flow in a rectangular domain $x \in [-\beta/2, \beta/2]$, $y \in [-1/2, 1/2]$ with impermeable ($\psi = 0$) boundaries (**Figure 6**). The sidewalls $x = \pm\beta/2$ and the bottom $y = -1/2$ are rigid and motionless, while the upper boundary $y = 1/2$ moves in its own plane with velocity $U(x)$ (**Figure 6**, boundary conditions a).

In the superposition method, the solution is represented as a sum of two ordinary Fourier series in the two coordinate directions. The flow is the sum of two parts that are even and odd functions of $y$,

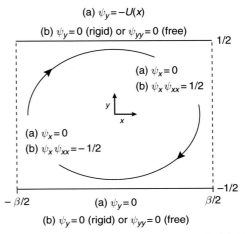

**Figure 6** Geometry and boundary conditions for (a) driven cavity flow and (b) steady cellular convection in a rectangle $x \in [-\beta/2, \beta/2]$, $y \in [-1/2, 1/2]$ with impermeable boundaries. All lengths are nondimensionalized by the height of the rectangle.

respectively, and the representation for the even part is (Meleshko, 1996)

$$\psi = \sum_{m=1}^{\infty} A_m F\left(y, \ p_m, \ 1\right) \cos p_m x$$

$$+ \sum_{n=1}^{\infty} B_n F\left(x, \ q_n, \ \beta\right) \cos q_n y \qquad [68a]$$

$$p_m = \frac{2m-1}{\beta}\pi, \qquad q_n = (2n-1)\pi \qquad [68b]$$

$$F(z, k, r) = \frac{r \tanh(rk/2)\cosh kz - 2z \sinh kz}{2 \cosh(rk/2)} \qquad [68c]$$

where $A_m$ and $B_n$ are undetermined coefficients. Due to the choice of the wave numbers $p_m$ and $q_n$ and the function $F(z,k,r)$ (which vanishes at $z = \pm r/2$), each of the series in [68a] satisfies the impermeability condition $\psi = 0$ on all boundaries. Moreover, because the trigonometric systems $\cos p_m x$ and $\cos q_n y$ are complete, the superposition [68] (with suitable choices of $A_m$ and $B_n$) can represent arbitrary distributions of tangential velocity on $y = \pm 1/2$ and $x = \pm \beta/2$.

Unlike the superposition method, the MEE makes use of so-called Papkovich–Fadle eigenfunctions $\phi(x)$ that satisfy all the homogeneous boundary conditions in the direction ($x$ in this case) perpendicular to two motionless walls. For simplicity, suppose that $U(x) = U(-x)$, so that $\psi$ is an even function of $x$. The even eigenfunctions on the canonical unit interval $x \in [-1/2, 1/2]$ are obtained by substituting $\psi = \phi(x) \exp(\lambda y)$ into the biharmonic equation [37] and solving the resulting equation for $\phi$ subject to the conditions $\phi(\pm 1/2) = \phi_x(\pm 1/2) = 0$, yielding

$$\phi_n(x) = x \sin \lambda_n x - \frac{1}{2}\tan\frac{\lambda_n}{2}\cos\lambda_n x \qquad [69]$$

where $\lambda_n$ are the first-quadrant complex roots of $\sin\lambda_n + \lambda_n = 0$. The streamfunction for the flow in the cavity with $x \in [-\beta/2, \beta/2]$ (**Figure 6**) can then be written as (Shankar, 1993)

$$\psi = \sum_{n=1}^{\infty}\left\{ A_n \Phi_n \exp\left[-\lambda_n\left(y + \frac{1}{2}\right)\right]\right.$$

$$+ \bar{A}_n \bar{\Phi}_n \exp\left[-\bar{\lambda}_n\left(y + \frac{1}{2}\right)\right]$$

$$+ B_n \Phi_n \exp\left[-\lambda_n\left(\frac{1}{2} - y\right)\right]$$

$$\left. + \bar{B}_n \bar{\Phi}_n \exp\left[-\bar{\lambda}_n\left(\frac{1}{2} - y\right)\right]\right\} \qquad [70]$$

where $\Phi_n = \phi_n(x/\beta)$, overbars denote complex conjugation, and the constants $A_n$ and $B_n$ are chosen to

satisfy the boundary conditions at $y = \pm 1/2$. The difficulty, however, is that the reduced biharmonic equation $(d^2/dx^2 + \lambda^2)^2 \phi = 0$ satisfied by the eigenfunctions $\phi_n$ is not self-adjoint, as can be seen by rewriting it in the form

$$\frac{d^2}{dx^2}\begin{pmatrix}\phi \\ -\lambda^{-2}\phi''\end{pmatrix} = \lambda^2\begin{pmatrix}0 & -1 \\ 1 & -2\end{pmatrix}\begin{pmatrix}\phi \\ -\lambda^{-2}\phi''\end{pmatrix} \qquad [71]$$

and noting that the square matrix on the RHS is not Hermitian. Consequently, the eigenfunctions $\phi_n$ are not mutually orthogonal, which makes the determination of $A_n$ and $B_n$ a nontrivial matter. One solution is to use a complementary set of adjoint eigenfunctions $\chi_m$ which are biorthogonal to the set $\phi_m$, although considerable care must then be taken to ensure convergence of the expansion (see Katopodes *et al.* (2000) for a discussion and references to the relevant literature). A cruder but very effective approach is to determine $A_n$ and $B_n$ via a numerical least-squares procedure that minimizes the misfit of the solution [70] to the boundary conditions (Shankar, 1993; Bloor and Wilson, 2006). Shankar (2005) showed how this approach can be extended to an irregular domain by embedding the latter in a larger, regular domain on which a complete set of eigenfunctions exists.

Another geophysically relevant 2-D Stokes problem that can be solved using superposition and eigenfunction expansion methods is that of the flow within the isothermal core of a vigorous (high Rayleigh number) convection cell. The boundary conditions for this case (**Figure 6**, conditions b) comprise either rigid ($\psi_y = 0$) or traction-free ($\psi_{yy} = 0$) conditions at $y = \pm 1/2$ and nonlinear sidewall conditions $\psi_x \psi_{xx} = \pm 1/2$ (derived in Section 4.9.3.2) that represent the shear stresses applied to the core by the buoyant thermal plumes. If the boundaries $y = \pm 1/2$ are traction free, the conditions $\psi = \psi_{yy} = 0$ involve only even derivatives of $\psi$, and are satisfied identically if $\psi \propto \cos q_n y$ where $q_n$ is defined by [68b]. The streamfunction that also satisfies the sidewall impermeability conditions $\psi(\pm \beta/2, y) = 0$ is (Roberts, 1979)

$$\psi = \sum_{n=0}^{\infty} A_n F(x, \ q_n, \ \beta)\cos q_n y \qquad [72]$$

where $F(x, q_n, \beta)$ is defined by [68c]. The constants $A_n$ are then determined iteratively to satisfy the boundary conditions $\psi_x \psi_{xx}(\pm \beta/2, y) = \pm 1/2$. Strictly speaking, [72] is an eigenfunction expansion, because the functions $\cos q_n y$ satisfy all the homogeneous boundary conditions at $y = \pm 1/2$. In practice, however, the term eigenfunction expansion is usually reserved for expressions like [70] that involve a sequence of complex wave numbers $\lambda_n$. No such terminological ambiguity applies for a convection cell bounded by rigid surfaces, for which the flow can be represented by the eigenfunction expansion (Roberts, 1979)

$$\psi = \sum_{n=1}^{\infty} \left[ A_n \phi_n(y) \cosh \lambda_n x + \bar{A}_n \bar{\phi}_n(y) \cosh \bar{\lambda}_n x \right] \quad [73]$$

Other examples of the use of superposition and eigenfunction expansion methods for cellular convection problems can be found in Turcotte (1967), Turcotte and Oxburgh (1967), Olson and Corcos (1980), Morris and Canright (1984), and Busse et al. (2006).

### 4.5.4.2 Spherical coordinates

Lamb (1932) derived a general solution of the equations of Stokes flow in spherical coordinates $(r, \theta, \phi)$. Because the pressure $p$ satisfies Laplace's equation, it may be expressed as a sum of solid spherical harmonics $p_l$

$$p = \sum_{l=-\infty}^{\infty} p_l, \quad p_l = r^l \sum_{m=-l}^{l} c_{lm} Y_l^m(\theta, \phi) \quad [74]$$

where $Y_l^m$ are surface spherical harmonics and $c_{lm}$ are complex coefficients. The velocity $\mathbf{u}$ can then be written

$$\mathbf{u} = \sum_{l=-\infty}^{\infty} \left[ \nabla \Phi_l + \nabla \times (\mathbf{x} \chi_l) \right]$$
$$+ \sum_{\substack{l=-\infty \\ l \neq 1}}^{\infty} \left[ \frac{(1/2)(l+3) r^2 \nabla p_l - l \mathbf{x} p_l}{\eta (l+1)(2l+3)} \right] \quad [75]$$

where $\eta$ is the viscosity, $\mathbf{x}$ is the position vector and $\Phi_l$ and $\chi_l$ are solid spherical harmonics of the form [74] but with different coefficients. The first sum in [75] is the solution of the homogeneous Stokes equations $\nabla^2 \mathbf{u} = 0$, $\nabla \cdot \mathbf{u} = 0$, whereas the second sum is the particular solution of $\nabla^2 \mathbf{u} = \nabla p / \eta$. A recent application of [75] is the solution of Gomilko et al. (2003) for steady Stokes flow driven by the motion of one of three mutually perpendicular walls that meet in a corner, a 3-D generalization of the 2-D corner-flow model (**Figure 5(a)**). The main difficulty in using Lamb's solution is in applying boundary conditions, because the elements of [75], while complete, do not form an orthogonal basis for vector functions on the surface of a sphere in the way that standard spherical

harmonics form an orthonormal basis for scalar functions. Methods for dealing with this problem are discussed in chapter 4 of Kim and Karrila (1991).

### 4.5.4.3 Bispherical coordinates

Flow in a domain bounded by nonconcentric spheres is a useful model for the interaction of one buoyant diapir or plume head with another or with a flat interface (i.e., a sphere of infinite radius). Such flow problems can often be solved analytically using bispherical coordinates $(\xi, \theta, \phi)$, which are related to the Cartesian coordinates $(x, y, z)$ by

$$(x, y, z) = \frac{(a \sin \theta \cos \phi, \ a \sin \theta \sin \phi, \ a \sinh \xi)}{\cosh \xi - \cos \theta} \quad [76]$$

where $a$ is a fixed length scale. Surfaces of constant $\xi$ are nonconcentric spheres with their centers on the axis $x = y = 0$, and $\xi = 0$ corresponds to the plane $z = 0$ (**Figure 7**). For axisymmetric (i.e., independent of $\phi$) Stokes flow, the general solution for the streamfunction $\psi$ and the pressure $p$ is (Stimson and Jeffreys, 1926)

$$\psi = (\cosh \xi - \eta)^{-3/2} \sum_{n=0}^{\infty} C_{n+1}^{-1/2}(\eta) \left[ a_n \cosh \beta_{-1} \xi \right.$$
$$\left. + b_n \sinh \beta_{-1} \xi + c_n \cosh \beta_3 \xi + d_n \sinh \beta_3 \xi \right] \quad [77a]$$

$$p = (\cosh \xi - \eta)^{-1/2} \sum_{n=0}^{\infty} \left[ A_n \exp \beta_1 \xi \right.$$
$$\left. + B_n \exp(-\beta_1 \xi) \right] P_n(\eta), \quad [77b]$$

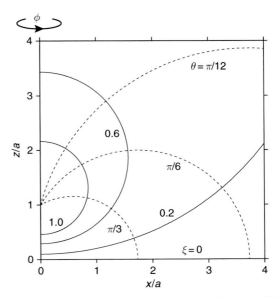

**Figure 7** Bispherical coordinates $(\xi, \theta, \phi)$ defined by [76]. Selected lines of constant $\xi$ (solid) and constant $\theta$ (dashed) are shown.

where $\eta = \cos\theta$, $\beta_m = n + m/2$, $C_{n+1}^{-1/2}$ is a Gegenbauer polynomial of order $n+1$ and degree $-1/2$, $P_n$ is a Legendre polynomial, and $a_n - d_m$, $A_m$ and $B_n$ are arbitrary constants. Koch and Ribe (1989) used a solution of the form [77] to model the effect of a viscosity contrast on the topography and gravity anomalies produced by the motion of a buoyant and deformable fluid sphere beneath a free surface of a fluid with a different viscosity. A representation of general nonaxisymmetric flows in bispherical coordinates was derived by Lee and Leal (1980), and used to determine the flow due to the arbitrary motion of a sphere near a plane wall.

### 4.5.5    The Complex-Variable Method for 2-D Flows

A powerful method for 2-D Stokes flows and analogous problems in elasticity (Muskhelishvili, 1953) is based on the Goursat representation of a biharmonic function $\psi$ in terms of two analytic functions $\phi$ and $\chi$ of the complex variable $z \equiv x + iy$:

$$\psi = \Re[\bar{z}\phi(z) + \chi(z)] \qquad [78]$$

where an overbar denotes complex conjugation. If $\psi$ is the streamfunction of a 2-D flow, then the velocity components $u = -\psi_y$ and $v = \psi_x$ are

$$v - iu = \phi(z) + z\overline{\phi'(z)} + \overline{\chi'(z)} \qquad [79]$$

the vorticity $\omega \equiv \nabla^2\psi$ and the pressure $p$ are

$$\omega + ip/\eta = 4\phi'(z) \qquad [80]$$

where $\eta$ is the viscosity and the components of the stress tensor are

$$\sigma_{xx} = -2\eta\Im[2\phi'(z) - \bar{z}\phi''(z) - \chi''(z)] \qquad [81a]$$

$$\sigma_{yy} = -2\eta\Im[2\phi'(z) + \bar{z}\phi''(z) + \chi''(z)] \qquad [81b]$$

$$\sigma_{xy} = 2\eta\Re[\bar{z}\phi''(z) + \chi''(z)] \qquad [81c]$$

The Goursat representation reduces the task of solving the biharmonic equation to one of finding two analytic functions that satisfy the relevant boundary conditions. The method is most powerful when used in conjunction with conformal mapping, which allows a flow domain with a complex shape to be mapped onto a simpler one (such as the interior of the unit circle). A remarkable example is Jeong and Moffatt's (1992) analytical solution for the formation of a cusp above a vortex dipole located at depth $d$ beneath the free surface of a viscous fluid. In the geodynamically relevant limit of zero surface tension, the surface displacement $y(x)$ satisfies $x^2y = -(y + 2d/3)^3$, which has an infinitely sharp cusp at $(x,y) = (0, -2d/3)$. Although the model is too idealized for direct application to the mantle, its dynamics are relevant to the formation of cusp-like features by entrainment in thermochemical convection (Davaille, 1999).

### 4.5.6    Singular Solutions and the Boundary-Integral Representation

The Stokes equations admit a variety of singular solutions in which the velocity and/or the pressure becomes infinite at one or more points in space. Such solutions are the basis of the boundary-integral representation, whereby a Stokes flow in a given domain is expressed in terms of surface integrals of velocities and stresses over the domain boundaries. The dimensionality of the problem is thereby reduced by one (from 3-D to 2-D or from 2-D to 1-D), making possible a powerful numerical technique – the boundary element method – that does not require discretization of the whole flow domain (Pozrikidis, 1992.)

The most useful singular solutions fall into two classes: those involving point forces, and those associated with volume sources and sinks.

#### 4.5.6.1    Flow due to point forces

The most important singular solution of the Stokes equations is that due to a point force $F_i$, or Stokeslet, applied at a position $\mathbf{x}'$ in the fluid. The velocity $u_i$ and stress tensor $\sigma_{ij}$ induced at any point $\mathbf{x}$ satisfy

$$\partial_j\sigma_{ij} = -F_i\delta(\mathbf{x} - \mathbf{x}'), \quad \partial_j u_j = 0 \qquad [82]$$

where $\delta(\mathbf{x} - \mathbf{x}') = \delta(x_1 - x_1')\delta(x_2 - x_2')\delta(x_3 - x_3')$ and $\delta$ is the Dirac delta-function. Here and throughout Section 4.5.6, vector arguments of functions are denoted using boldfaced vector notation, while other quantities are written using Cartesian tensor (subscript) notation. In an infinite fluid, the required boundary conditions are that $u_i \rightarrow 0$ and $\sigma_{ij} \rightarrow -p_0\delta_{ij}$ as $|\mathbf{x} - \mathbf{x}'| \rightarrow \infty$, where $p_0$ is the (dynamically irrelevant) far-field pressure. Because the response of the fluid is proportional to the applied force, $u_i = \mathcal{J}_{ij}F_j/\eta$ and $\sigma_{ik} = K_{ijk}F_j$, where $\eta$ is the viscosity and $\mathcal{J}_{ij}$ and $K_{ijk}$ are tensorial Green's functions representing the response to a unit force. Substituting these expressions into [82] and eliminating the arbitrary vector $F_j$, we obtain

$$\partial_i K_{ijk} = -\delta_{jk}\delta(\mathbf{x} - \mathbf{x}'), \quad \partial_i \mathcal{J}_{ij} = 0 \qquad [83]$$

The solutions of [83] can be found using a Fourier transform (Kim and Karrila, 1991, p. 33) or by reducing [83] to Poisson's equation (Pozrikidis, 1992, p. 22), and are

$$\mathcal{J}_{ij}(\mathbf{r}) = \frac{1}{8\pi}\left(\frac{\delta_{ij}}{r} + \frac{r_i r_j}{r^3}\right), \quad K_{ijk}(\mathbf{r}) = -\frac{3}{4\pi}\frac{r_i r_j r_k}{r^5} \quad [84]$$

where $\mathbf{r} = \mathbf{x} - \mathbf{x}'$ and $r = |\mathbf{r}|$. The tensor $\mathcal{J}_{ij}$ is often called the Oseen tensor. The analogs of [84] for a 2-D flow are

$$\mathcal{J}_{ij}(\mathbf{r}) = \frac{1}{4\pi}\left(-\delta_{ij}\ln r + \frac{r_i r_j}{r^2}\right), \quad K_{ijk}(\mathbf{r}) = -\frac{1}{\pi}\frac{r_i r_j r_k}{r^4} \quad [85]$$

The expression [85] for $\mathcal{J}_{ij}$ does not vanish as $r \rightarrow \infty$, which is related to the fact that a 2-D Stokes flow around an infinitely long cylinder does not exist (Stokes's paradox). Below we will see how this paradox is resolved by the presence of a boundary.

Starting from the Stokeslet solution, one can use the principle of superposition to construct an infinite variety of additional singular solutions. An example is the flow due to a force dipole, comprising a point force $F_i$ at $\mathbf{x}'$ and an equal and opposite force $-F_i$ at $\mathbf{x}' - d\mathbf{n}$, where $\mathbf{n}$ is a unit vector directed from the negative to the positive force. The associated velocity field is $u_i(\mathbf{x}) = [\mathcal{J}_{ij}(\mathbf{r}) - \mathcal{J}_{ij}(\mathbf{r} + d\mathbf{n})]F_j$. In the limit $d \rightarrow 0$ with $dF_j n_k$ fixed,

$$u_i = dF_j n_k G_{ijk}^{\mathrm{FD}} \quad [86a]$$

$$G_{ijk}^{\mathrm{FD}}(\mathbf{r}) = -\partial_k \mathcal{J}_{ij}(\mathbf{r}) \equiv \frac{1}{8\pi}\left[\frac{\delta_{ij}r_k - \delta_{ik}r_j - \delta_{jk}r_i}{r^3} + \frac{3r_i r_j r_k}{r^5}\right] \quad [86b]$$

where $G_{ijk}^{FD}$ is the force-dipole Green's function. The force-dipole moment $dn_j F_k$ is sometimes decomposed into symmetric (stresslet) and antisymmetric (rotlet) parts (Kim and Karrila, 1991, Section 4.2.5.)

### 4.5.6.2  Flows due to point sources

The basic singular solution of the second class is that associated with a volume source of strength $Q$ at $\mathbf{x}'$, which generates a spherically symmetric flow

$$u_i = \frac{Qr_i}{4\pi r^3} \quad [87]$$

The flow due to a source doublet comprising a source and sink with equal strengths $Q$ separated by a vector $d\mathbf{n}$ pointing from the sink to the source is

$$u_i = Qdn_j G_{ij}^{\mathrm{SD}},$$
$$G_{ij}^{\mathrm{SD}}(\mathbf{r}) = -\frac{1}{4\pi}\partial_j\left(\frac{r_i}{r^3}\right) \equiv \frac{1}{4\pi}\left(\frac{3r_i r_j}{r^5} - \frac{\delta_{ij}}{r^3}\right) \quad [88]$$

### 4.5.6.3  Singular solutions in the presence of a boundary

The flow produced by a point force is modified by the presence of an impermeable wall. Consider a force $F_i$ at a point $\mathbf{x}'$ located a distance $d$ from the wall, and let $\mathbf{n}$ be a unit vector normal to the wall directed towards the side containing $\mathbf{x}'$ (**Figure 8(a)**). The modified velocity can be written as $u_i = G_{ij}^{\mathrm{B}}F_j$, where $G_{ij}^{\mathrm{B}}(\mathbf{x}, \mathbf{x}')$ is a Green function that satisfies all the required boundary conditions at the wall. Its general form is

$$G_{ij}^{\mathrm{B}}(\mathbf{x}, \mathbf{x}') = \mathcal{J}_{ij}(\mathbf{x} - \mathbf{x}') + G_{ij}^{\mathrm{IM}}(\mathbf{x} - \mathbf{x}^{\mathrm{IM}}) \quad [89]$$

where $G_{ij}^{\mathrm{IM}}$ is a Green function that is singular at the image point $\mathbf{x}^{\mathrm{IM}} = \mathbf{x}' - 2d\mathbf{n}$ (**Figure 8a**).

The two limiting cases of greatest interest are traction-free and rigid walls. Because a traction-free wall is equivalent to a plane of mirror symmetry, the modified flow for this case can be constructed simply by adding a reflected Stokeslet with strength $\mathbf{R} \cdot \mathbf{F} \equiv \mathbf{F}^*$ at the image point $\mathbf{x}^{\mathrm{IM}}$, where $R_{ij} \equiv \delta_{ij} - 2n_i n_j$ is a reflection tensor that reverses the sign of the wall-normal component of a vector while leaving its wall-parallel components unchanged (**Figure 8(a)**). Therefore

$$G_{ij}^{\mathrm{IM}}(\mathbf{r}^{\mathrm{IM}}) = R_{jk}\mathcal{J}_{ik}(\mathbf{r}^{\mathrm{IM}}) \quad [90]$$

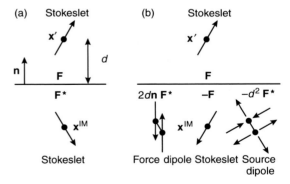

**Figure 8** Singular solutions required to describe the flow due to a point force **F** located at a point $\mathbf{x}'$ a distance $d$ above a plane wall. The boundary conditions on the wall (horizontal lines) are satisfied by adding one or more singular solutions at the image point $\mathbf{x}^{\mathrm{IM}} \equiv \mathbf{x}' - 2d\mathbf{n}$, where $\mathbf{n}$ is the unit vector normal to the wall. The strength or moment of each required singular solution is indicated, and $\mathbf{F}^*$ is the reflection of the vector $\mathbf{F}$ across the wall. (a) traction-free wall; (b) rigid wall.

where $\mathbf{r}^{IM} = \mathbf{x} - \mathbf{x}^{IM}$. If the surface is rigid, the boundary conditions can be satisfied by adding three different singular solutions at $\mathbf{x}^{IM}$: a Stokeslet with strength $-\mathbf{F}$, a force dipole with moment $2d n \mathbf{F}^*$, and a source dipole with strength $-d^2 \mathbf{F}^*$ (**Figure 8(b)**) The result is (Blake, 1971)

$$G_{ij}^{IM}(\mathbf{r}^{IM}) = -\mathcal{J}_{ij}(\mathbf{r}^{IM}) \\ + 2dR_{jl}n_k G_{ikl}^{FD}(\mathbf{r}^{IM}) - d^2 R_{jk} G_{ik}^{SD}(\mathbf{r}^{IM}) \quad [91]$$

where $G_{ij}^{FD}$ and $G_{ij}^{SD}$ are defined by [86b] and [88], respectively.

Expressions analogous to those above also apply to 2-D flow in the presence of a boundary. The basic idea is illustrated by the free-surface case, for which both [89] and [90] remain valid if $\mathcal{J}_{ij}$ has its 2-D form [85]. Now, however, the presence of the image singularity $G_{ij}^{IM}$ cancels the logarithmic divergence of $\mathcal{J}_{ij}$ at large distances from the point force, thereby resolving Stokes's paradox and rendering the problem well-posed. A rigid surface has the same effect, although the corresponding expressions for the Green's function are more complicated.

### 4.5.6.4  Boundary-integral representation

The boundary-integral representation for Stokes flow expresses the velocity $u_i$ at any point in a fluid volume $V$ bounded by a surface $S$ in terms of the velocity and traction on $S$. The starting point is the integral form [34] of the Lorentz reciprocal theorem. Let $(u_i, \sigma_{ij})$ be the flow of interest in a fluid with no distributed body forces ($b_i = 0$), and let $u_i^* \equiv \mathcal{J}_{ij}(\mathbf{x} - \mathbf{x}')F_j/\eta$ and $\sigma_{ik}^* \equiv K_{ijk}(\mathbf{x} - \mathbf{x}')F_j$ be the flow produced by a point force $b_i^* \equiv F_i\delta(\mathbf{x} - \mathbf{x}')$ at the point $\mathbf{x}'$. Substituting these expressions into [34] and dropping the arbitrary vector $F_i$, we obtain

$$\frac{1}{\eta}\int_S \mathcal{J}_{ij}(\mathbf{x}-\mathbf{x}')\sigma_{ik}(\mathbf{x})n_k(\mathbf{x})\mathrm{d}S(\mathbf{x}) - \int_V u_j(\mathbf{x})\delta(\mathbf{x}-\mathbf{x}')\mathrm{d}V(\mathbf{x}) \\ = \int_S K_{ijk}(\mathbf{x}-\mathbf{x}')u_i(\mathbf{x})n_k(\mathbf{x})\mathrm{d}S(\mathbf{x}) \quad [92]$$

where the normal $\mathbf{n}$ points out of $V$. Now

$$\int_V u_j(\mathbf{x})\delta(\mathbf{x}-\mathbf{x}')\mathrm{d}V = \chi(\mathbf{x}')u_j(\mathbf{x}') \quad [93]$$

where $\chi(\mathbf{x}') = 0$, $1/2$, or $1$ depending on whether $\mathbf{x}'$ lies outside $V$, right on $S$, or inside $V$, respectively. Substituting [93] into [92] and interchanging the

roles of $\mathbf{x}$ and $\mathbf{x}'$, we obtain the boundary-integral representation

$$\frac{1}{\eta}\int_{S'} \mathcal{J}_{ij}\sigma_{ik}n_k\mathrm{d}S' - \int_{S'} K_{ijk}u_i n_k\mathrm{d}S' = \chi(\mathbf{x})u_j(\mathbf{x}) \quad [94]$$

where the arguments of the quantities in the integrands ($\mathbf{x}'$ for $n_k$, $\sigma_{ik}$, and $u_i$; $\mathbf{x}' - \mathbf{x}$ for $\mathcal{J}_{ij}$ and $K_{ijk}$) have been suppressed for brevity and $\mathrm{d}S' = \mathrm{d}S(\mathbf{x}')$. The first integral in [94] represents the velocity due to a surface distribution of point forces with density $\sigma_{ik}n_k\mathrm{d}S'$. It is called the single-layer potential by analogy to the electrostatic potential generated by a surface distribution of electric charges. The second integral, called the double-layer potential, represents the velocity field generated by the sum of a distribution of sources and sinks and a symmetric distribution of force dipoles (Kim and Karrila, 1991, Section 4.2.4.2).

An important extension of the integral representation [94] is to the buoyancy-driven motion of a fluid drop with viscosity $\eta_2 \equiv \gamma\eta_1$ and density $\rho_2 \equiv \rho_1 + \Delta\rho$ in another fluid with viscosity $\eta_1$ and density $\rho_1$ (Pozrikidis, 1990; Manga and Stone, 1993.) Let $S$, $V_1$, and $V_2$ be the surface of the drop and the volumes outside and inside it, respectively. We begin by writing separate integral equations of the form [94] for each fluid:

$$-\frac{1}{\eta_1}\int_{S'} \mathcal{J}_{ij}\sigma_{ik}^{(1)}n_k\mathrm{d}S' + \int_{S'} K_{ijk}u_i^{(1)}n_k\mathrm{d}S' \\ = \chi_1(\mathbf{x})u_j^{(1)}(\mathbf{x}) \quad [95a]$$

$$\frac{1}{\eta_2}\int_{S'} \mathcal{J}_{ij}\sigma_{ik}^{(2)}n_k\mathrm{d}S' - \int_{S'} K_{ijk}u_i^{(2)}n_k\mathrm{d}S' \\ = \chi_2(\mathbf{x})u_j^{(2)}(\mathbf{x}) \quad [95b]$$

where $\mathbf{n}$ points out of the drop and the volume ($V_1$ or $V_2$) in which a given quantity is defined as indicated by a superscript in parentheses. Here $\chi_1(\mathbf{x}) = 0$, $1/2$, or $1$ if $\mathbf{x}$ is in $V_2$, right on $S$, or in $V_1$, respectively, and $\chi_2(\mathbf{x})$ is defined similarly but with the subscripts 1 and 2 interchanged. Now multiply [95b] by $\gamma$, add the result to [95a], and apply the matching conditions $u_j^{(1)} = u_j^{(2)} = u_j$ and $(\sigma_{ik}^{(1)} - \sigma_{ik}^{(2)})n_k = n_i\Delta\rho g_k x_k'$ on $S$, where $g_k$ is the gravitational acceleration. The result is

$$\chi_1(\mathbf{x})u_j^{(1)}(\mathbf{x}) + \gamma\chi_2(\mathbf{x})u_j^{(2)}(\mathbf{x}) - (1-\gamma)\int_{S'} K_{ijk}u_i n_k\mathrm{d}S' \\ = -\frac{g_k\Delta\rho}{\eta_1}\int_{S'} \mathcal{J}_{ij}n_i x_k'\mathrm{d}S' \quad [96]$$

For points $\mathbf{x}$ on $S$, [96] reduces to

$$\frac{1}{2}u_j(\mathbf{x}) - \frac{1-\gamma}{1+\gamma}\int_{S'} K_{ijk}u_i n_k\mathrm{d}S' \\ = -\frac{g_k\Delta\rho}{\eta_1(1+\gamma)}\int_{S'} \mathcal{J}_{ij}n_i x_k'\mathrm{d}S' \quad (\mathbf{x}\in S) \quad [97]$$

a Fredholm integral equation of the second kind for the velocity **u** of the interface. Once **u** on $S$ has been determined by solving [97], **u** at points in $V_1$ and $V_2$ can be determined if desired from [96]. A generalization of [97] to $N > 1$ interacting drops was derived by Manga and Stone (1993).

The integral equation [97], which must in general be solved numerically, has been used in geodynamics to model systems comprising distinct fluids with different viscosities. Manga and Stone (1993) solved the $N$-drop generalization of [97] using the boundary element method (Pozrikidis, 1992) to investigate the buoyancy-driven interaction between two drops in an infinite fluid with a different viscosity. Manga *et al.* (1993) used a similar method to study the interaction of plume heads with compositional discontinuities in the Earth's mantle. The deformation of viscous blobs in a 2-D cellular flow was investigated by Manga (1996), who concluded that geochemical reservoirs can persist undisturbed for relatively long times if they are 10–100 times more viscous than the surrounding mantle. Finally, Manga (1997) showed that the deformation-induced mutual interactions of deformable diapirs in a rising cloud causes the diapirs progressively to cluster, but that the rate of this clustering is probably too slow to affect significantly the lateral spacing of rising diapirs in the mantle.

### 4.5.7 Slender-Body Theory

Slender-body theory (SBT) is concerned with Stokes flow around thin rod-like bodies whose length greatly exceeds their other two dimensions. The approach takes its departure from Stokes's paradox: the fact that a solution of the equations for Stokes flow around an infinitely long circular cylinder moving steadily in an unbounded viscous fluid does not exist, due to a logarithmic singularity that makes it impossible to satisfy all the boundary conditions (Batchelor, 1967). However, the problem can be regularized in one of three ways: by including inertia, by making the length of the cylinder finite, or by making the domain bounded. SBT is concerned with the second (and by extension the third) of these possibilities.

The canonical problem of SBT is to determine the force **F** on a rod of length $2\ell$ and radius $a \ll \ell$ moving with uniform velocity **U** in a viscous fluid. The solution can be found using the MMAE, which exploits the fact that the flow field comprises two distinct regions characterized by very different length scales. The first or inner region includes points whose radial distance $\rho$ from the rod is small compared to their distance from the rod's nearer end. In this region, the fluid is not affected by the ends of the rod, and sees it as an infinite cylinder with radius $a$. The second, outer region $\rho \gg a$ is at distances from the rod that are large compared to its radius. Here, the fluid is unaffected by the rod's finite radius, and sees it as a line distribution of point forces with effectively zero thickness. The basic idea of the MMAE is to obtain two different asymptotic expansions for the velocity field that are valid in the inner and outer regions, respectively, and then to match them together in an intermediate or overlap region where both expansions must coincide.

As the details of the matching are rather complicated, we defer our discussion of the method to Section 4.2, where it will be applied in the context of BL theory. Here we simply quote the lowest-order result that the force on a cylinder whose axis is parallel to a unit vector $\mathbf{e}_z$ is

$$\mathbf{F} = -4\pi\eta\ell\epsilon[2\mathbf{U} - (\mathbf{U} \cdot \mathbf{e}_z)\mathbf{e}_z] \qquad [98]$$

where $\epsilon = (\ln 2\ell/a)^{-1}$. The force on a cylinder moving normal to its axis ($\mathbf{U} \cdot \mathbf{e}_z = 0$) is thus twice that on one moving parallel to its axis. For further details and extensions of SBT, see Batchelor (1970), Cox (1970), Keller and Rubinow (1976), and Johnson (1980). Geophysically relevant applications of SBT include Olson and Singer (1985), who used [98] to predict the rise velocity of buoyant quasi-cylindrical (diapiric) plumes. Koch and Koch (1995) used an expression analogous to [98] for an expanding ring to model the buoyant spreading of a viscous drop beneath the free surface of a much more viscous fluid. Whittaker and Lister (2006a) presented a model for a creeping plume above a planar boundary from a point source of buoyancy, in which they modeled the flow outside the plume as that due to a line distribution of Stokeslets.

### 4.5.8 Flow Driven by Internal Loads

Because inertia is negligible in the mantle, the flow at each instant is determined entirely by the distribution of internal density anomalies (loads) at that instant. This principle is the basis of a class of internal loading models in which an instantaneous mantle flow field is determined by convolving a load distribution with a Green function that represents the mantle's response to a unit load. However, it proves convenient here to define the unit load not as a point force, but rather as a surface force concentrated at a single radius and whose amplitude is proportional to a spherical harmonic $Y_l^m(\theta, \phi)$ of degree $l$ and order $m$.

Because the Stokes equations are separable in spherical coordinates, the Green function representing the response to an harmonic surface load of a given degree and order satisfies an ordinary differential equation (ODE) in the radial coordinate $r$ that can be solved analytically. This approach was pioneered by Parsons and Daly (1983) for a plane layer with constant viscosity, and was extended by Richards and Hager (1984), Ricard et al. (1984), and Forte and Peltier (1987) to spherical geometry with self-gravitation and radially variable viscosity. To illustrate the method, we sketch below Forte and Peltier's (1987) analytical derivation of the Green function for a self-gravitating mantle with uniform viscosity.

#### 4.5.8.1  Wave-domain Green functions

Our starting point is the equations [31] with $n=1$ that govern slow flow in an incompressible, self-gravitating, Newtonian mantle. Let

$$\rho = \rho_0 + \hat{\rho}(r, \theta, \phi), \quad \boldsymbol{\sigma} = \boldsymbol{\sigma}_0(r) + \hat{\boldsymbol{\sigma}}(r, \theta, \phi), \\ \chi = \chi_0(r) + \hat{\chi}(r, \theta, \phi) \qquad [99]$$

where hatted quantities are perturbations of the field variables about a reference hydrostatic state denoted by a subscript 0. Substituting [99] into [31b] and [31d] and neglecting products of perturbation quantities, we obtain

$$0 = \boldsymbol{\nabla} \cdot \hat{\boldsymbol{\sigma}} + \mathbf{g}_0 \hat{\rho} - \rho_0 \boldsymbol{\nabla} \hat{\chi} \qquad [100a]$$

$$\nabla^2 \hat{\chi} = 4\pi G \hat{\rho} \qquad [100b]$$

where $\mathbf{g}_0 = -\boldsymbol{\nabla} \chi_0$. The third term on the RHS of [100b] represents the buoyancy force acting on the internal density anomalies, and the fourth represents the additional force associated with the perturbations in the gravitational potential that they induce (self-gravitation). For consistency with other sections of this chapter, the signs of all gravitational potentials ($\chi_0$, $\hat{\chi}$, etc.) referred to below are opposite to those of Forte and Peltier (1987).

As noted in Section 4.5.2, the flow driven by internal density anomalies in a fluid with constant or depth-dependent viscosity is purely poloidal. The poloidal potential $\mathcal{P}(r, \theta, \phi)$ satisfies the first of equations [47], which remain valid in a self-gravitating mantle. We now substitute into this equation the expansions

$$\mathcal{P} = \sum_{l=0}^{\infty} \sum_{m=-l}^{l} \mathcal{P}_l^m(r) Y_l^m(\theta, \phi), \\ \hat{\rho} = \sum_{l=0}^{\infty} \sum_{m=-l}^{l} \hat{\rho}_l^m(r) Y_l^m(\theta, \phi) \qquad [101]$$

where $Y_l^m(\theta, \phi)$ are surface spherical harmonics satisfying $\mathcal{B}^2 Y_l^m = -l(l+1) Y_l^m$ and $\mathcal{B}^2$ is defined by [45]. We thereby find that $\mathcal{P}_l^m(r)$ satisfies

$$\mathcal{D}_l^2 \mathcal{P}_l^m(r) = \frac{g_0 \hat{\rho}_l^m}{\eta r}, \quad \mathcal{D}_l = \frac{d^2}{dr^2} + \frac{2}{r}\frac{d}{dr} - \frac{l(l+1)}{r^2} \qquad [102]$$

where the gravitational acceleration $g_0$ has been assumed constant (Forte and Peltier, 1987). Now define a poloidal Green's function $P_l(r, r')$ that satisfies

$$\mathcal{D}_l^2 P_l(r, r') = \delta(r - r') \qquad [103]$$

$P_l(r, r')$ represents the poloidal flow generated at the radius $r$ by an infinitely thin density contrast of spherical harmonic degree $l$ and unit amplitude located a radius $r'$. The poloidal flow due to a distributed density anomaly $\hat{\rho}_l^m(r)$ is then obtained by convolving $\hat{\rho}_l^m(r)$ with the Green's function, yielding

$$\mathcal{P}_l^m(r) = \frac{g_0}{\eta} \int_{a_2}^{a_1} \frac{\hat{\rho}_l^m(r')}{r'} P_l(r, r') dr' \qquad [104]$$

where $a_2$ and $a_1$ are the inner and outer radii of the mantle, respectively.

At all radii $r \neq r'$, the Green's function $P_l(r, r')$ satisfies the homogeneous form of [103], which has the general solution

$$P_l(r, r') = A_n r^l + B_n r^{-l-1} + C_n r^{l+2} + D_n r^{-l+1} \qquad [105]$$

where $A_n - D_n$ are undetermined constants with $n=1$ for $r' < r \leq a_1$ and $n=2$ for $a_2 \leq r < r'$. These eight constants are determined by the boundary conditions at $r=a_1$ and $r=a_2$ and by matching conditions at $r=r'$. The vanishing of the radial velocity at $r=a_1$ and $r=a_2$ requires

$$P_l(a_1, r') = P_l(a_2, r') = 0 \qquad [106]$$

and the vanishing of the shear stress requires

$$\frac{d^2 P_l}{dr^2}(a_1, r') = \frac{d^2 P_l}{dr^2}(a_2, r') = 0 \qquad [107]$$

Turning now to the matching conditions, we define $[A] \equiv A(r'+) - A(r'-)$ to be the jump in the quantity $A$ across the radius $r=r'$. Continuity of the normal and tangential velocities and the shear stress requires

$$[P_l] = \left[\frac{dP_l}{dr}\right] = \left[\frac{d^2 P_l}{dr^2}\right] = 0 \qquad [108]$$

The normal stress, however, is discontinuous at $r=r'$. By integrating [103] from $r'-$ to $r'+$ and applying [108], we find

$$\left[\frac{d^3 P_l}{dr^3}\right] = 1 \qquad [109]$$

Substitution of [105] into [106]–[109] yields eight equations for $A_n$–$D_n$, the solutions of which are

$$C_n = -\frac{B_n}{a_n^{2l+3}}$$

$$= \frac{1}{2(2l+3)(2l+1)(r')^{l-1}}$$

$$\times \frac{(a_1/a_n)^{2l+3} - (r'/a_2)^{2l+3}}{1 - (a_1/a_2)^{2l+3}} \qquad [110a]$$

$$D_n = -a_n^{2l-1}A_n$$

$$= \frac{a_n^{2l-1}}{2(4l^2-1)(r')^{l-3}} \frac{(a_1/a_n)^{2l-1} - (r'/a_2)^{2l-1}}{1 - (a_1/a_2)^{2l-1}} \qquad [110b]$$

The next step is to determine the gravitational potential anomaly. Because the flow induces deformations (dynamic topography) of the Earth's surface and of the core–mantle boundary (CMB), the total potential anomaly is

$$\hat{\chi}_l^m(r) = (\hat{\chi}_0)_l^m(r) + (\hat{\chi}_1)_l^m(r) + (\hat{\chi}_2)_l^m(r) \qquad [111]$$

where $\hat{\chi}_0$, $\hat{\chi}_1$, and $\hat{\chi}_2$ are the potentials associated with the internal load, the deformation $\hat{a}_1$ of the Earth's surface, and the deformation $\hat{a}_2$ of the CMB, respectively. To determine $\hat{\chi}_0$, we note that the general solution of [100b] is

$$\hat{\chi}(\mathbf{x}) = -G \int_V \frac{\hat{\rho}(\mathbf{x}')}{|\mathbf{x}-\mathbf{x}'|} dV' \qquad [112]$$

where $\mathbf{x}$ is the 3-D position vector and the integral is over the whole mantle. We now invoke the expansion (Jackson, 1975, p. 102)

$$|\mathbf{x}-\mathbf{x}'|^{-1} = 4\pi \sum_{l=0}^{\infty} \sum_{m=-l}^{l} \frac{1}{2l+1}$$

$$\times \frac{r_<^l}{r_>^{l+1}} Y_l^m(\theta, \phi) \bar{Y}_l^m(\theta', \phi') \qquad [113]$$

where $r_< = \min(r, r')$ and $r_> = \max(r, r')$ and an over-bar denotes complex conjugation. Substituting [113] into [112] and integrating over $\theta'$ and $\phi'$, we obtain

$$(\hat{\chi}_0)_l^m(r) = -\frac{4\pi G}{2l+1} \int_{a_2}^{a_1} (r')^2 \frac{r_<^l}{r_>^{l+1}} \hat{\rho}_l^m(r') dr' \qquad [114]$$

Expressions for $\hat{\chi}_1$ and $\hat{\chi}_2$ can be obtained from [114] by replacing $\hat{\rho}_l^m(r')$ by $(\rho_0 - \rho_w)(\hat{a}_1)_l^m \delta(r'-a_1)$ and $(\rho_c - \rho_0)(\hat{a}_2)_l^m \delta(r'-a_2)$, respectively, where $\rho_w$ is the density of seawater and $\rho_c$ is the core density. The results are

$$(\hat{\chi}_1)_l^m(r) = -\frac{4\pi G a_1}{2l+1}(\rho_0 - \rho_w)\left(\frac{r}{a_1}\right)^l (\hat{a}_1)_l^m \qquad [115a]$$

$$(\hat{\chi}_2)_l^m(r) = -\frac{4\pi G a_2}{2l+1}(\rho_c - \rho_0)\left(\frac{a_2}{r}\right)^{l+1}(\hat{a}_2)_l^m \qquad [115b]$$

The boundary deformations $\hat{a}_n$ are determined from the principle that the normal stress must be continuous across the deformed surfaces $r = a_n + \hat{a}_n \equiv r_n$. This is equivalent to the requirement that the nonhydrostatic normal stress $\hat{\sigma}_{rr}$ acting on the reference surfaces $r = a_n$ be equal to the weight of the topography there. Expanded in spherical harmonics, this condition reads

$$-\hat{p}_l^m(a_n) + 2\eta \frac{d\hat{w}_l^m}{dr}(a_n) = -g_0 \Delta \rho_n (\hat{a}_n)_l^m \qquad [116]$$

where $p$ is the nonhydrostatic pressure, $w$ is the radial velocity, $\Delta\rho_1 = \rho_0 - \rho_w$, and $\Delta\rho_2 = \rho_0 - \rho_c$. An expression for $\hat{p}$ in terms of the poloidal scalar is obtained by integrating the $\mathbf{e}_\theta$-component of the momentum equation [100a] and expanding the result in spherical harmonics:

$$\hat{p}_l^m = -\eta \frac{d}{dr}\left(r D_l \mathcal{P}_l^m\right) - \rho_0 \hat{\chi}_l^m \qquad [117]$$

Substituting [117] and $\hat{w}_l^m = -l(l+1)\mathcal{P}_l^m/r$ into [116] and applying the boundary conditions [106], we obtain

$$(\hat{a}_n)_l^m = \Delta\rho_n^{-1}\left[X_l^m(a_n) - \frac{\rho_0}{g_0}\hat{\chi}_l^m(a_n)\right] \qquad [118a]$$

$$X_l^m(r) = \frac{\eta}{g_0}\left[-r\frac{d^3}{dr^3} + \frac{3l(l+1)}{r}\frac{d}{dr}\right]\mathcal{P}_l^m(r) \qquad [118b]$$

Now by substituting [115] into [111] and using [118a], we obtain the following expression for the total gravitational potential:

$$\chi_l^m(r) = (\hat{\chi}_0)_l^m(r) - \frac{4\pi a_1 G}{2l+1}\left(\frac{r}{a_1}\right)^l \left[X_l^m(a_1) - \frac{\rho_0}{g_0}\hat{\chi}_l^m(a_1)\right]$$

$$+ \frac{4\pi a_2 G}{2l+1}\left(\frac{a_2}{r}\right)^{l+1}\left[X_l^m(a_2) - \frac{\rho_0}{g_0}\hat{\chi}_l^m(a_2)\right] \qquad [119]$$

The boundary potentials $\hat{\chi}_l^m(a_1)$ and $\hat{\chi}_l^m(a_2)$ are determined by solving the coupled equations obtained by evaluating [119] at $r = a_1$ and $r = a_2$, and are then eliminated from [119]. Next, the resulting equation for $\hat{\chi}_l^m$ is rewritten as a convolution integral using [114], [118b], [104], [105], and [110]. Finally, the result is evaluated at $r = a_1$ to obtain the surface potential

$$\hat{\chi}_l^m(a_1) = -\frac{4\pi a_1 G}{2l+1}\int_{a_2}^{a_1} G_l(r')\hat{\rho}_l^m(r')dr' \qquad [120]$$

where the Green function or kernel $G_l$ is

$$G_l(r) = \left(1 - K_1 + K_2 K_r \beta^{2l+1}\right)^{-1} \left(\frac{r}{a_1}\right)^{l+2}$$

$$\times \left\{1 - K_r \beta \left(\frac{a_2}{r}\right)^{2l+1} - E_1\left(1 - K_r \beta^{2l+2}\right)\left(\frac{a_1}{r}\right)^{l+2}\right.$$

$$\left. - E_2(1 - K_r \beta)\left(\frac{a_2}{r}\right)^{l+2}\right\} \qquad [121a]$$

$$E_n = \frac{l(l+2)}{2l+1}\left(\frac{a_n}{r}\right)^{l-2}\frac{(r/a_1)^{2l-1} - (a_2/a_n)^{2l-1}}{1 - \beta^{2l-1}}$$

$$+ \frac{l(l-1)}{2l+1}\left(\frac{a_n}{r}\right)^{l}\frac{(r/a_1)^{2l+3} - (a_2/a_n)^{2l+3}}{1 - \beta^{2l+3}} \qquad [121b]$$

$$K_n = \frac{4\pi a_n \rho_0 G}{(2l+1)g_0}, \quad K_r = \frac{K_1}{1 + K_2}, \quad \beta = \frac{a_2}{a_1} \qquad [121c]$$

**Figure 9** shows $G_l(r)$ for $l = 2$ and $l = 10$, both with (solid lines) and without ($K_1 = K_2 = 0$; dashed lines) self-gravitation. The kernels are negative at all radii because the negative gravitational potential of the deformed upper surface exceeds the positive contribution of the internal mass anomaly itself. The maximum potential anomaly is produced by loads in the mid-mantle when $l = 2$, and in the upper mantle when $l = 10$. The effect of self-gravitation is nearly a factor of 2 for $l = 2$, but only about 10% for $l = 10$.

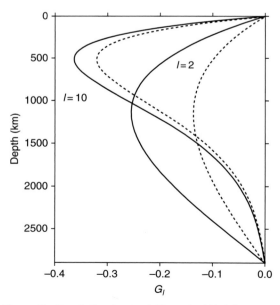

**Figure 9** Gravitational potential kernels $G_l(r)$ defined by [121] as functions of depth for spherical harmonic degrees $l = 2$ and $l = 10$, with (solid lines) and without (dashed lines) self-gravitation.

An expression analogous to [121] for a mantle comprising an upper layer $r > a_1 - z$ with viscosity $\eta_U$ and a lower layer $r < a_1 - z$ with viscosity $\eta_L$ is given by Forte and Peltier (1987). The general effect of a viscosity contrast $\eta_L/\eta_U > 1$ is to enhance the dynamic deformation of the CMB and reduce that of the upper surface. The reduced (negative) gravitational potential anomaly of the upper surface then counteracts less effectively the positive anomaly due the (sinking) load, with the result that $G_l(r)$ increases at all depths relative to the kernels for $\eta_L/\eta_U = 1$. For $\eta_L/\eta_U = 30$ and $z = 670$ km, for example, $G_{10}(r) > 0$ for all $r$, while $G_2(r)$ is positive in the upper mantle and negative in the lower mantle (Forte and Peltier, 1987, figure 17).

The simplest application of the kernel approach (e.g., Hager et al., 1985; Forte and Peltier, 1987) is to start with a load function $\hat{\rho}(r,\theta,\phi)$ estimated from seismic tomography or the global distribution of subducted lithosphere, and then to determine by repeated forward modeling the values of $\eta_L/\eta_U$ and $z$ for which geoid and other anomalies (surface divergence, surface topography, etc.) predicted by formulae like [120] best match their observed counterparts on the real Earth. A robust result of the early studies that has been confirmed by later work (e.g., Mitrovica and Forte, 2004) is that the lower mantle must be more viscous than the upper mantle by a factor $\eta_L/\eta_U \sim 10 - 100$. In many of these studies, the kernels were obtained using an alternative analytical technique, the propagator matrix method, discussed next.

### 4.5.8.2  The propagator-matrix method

The analytical Green function approach outlined in Section 4.5.8.1 for a constant-viscosity mantle can in principle be extended to models comprising any number $N$ of discrete layers with different viscosities. In practice, however, the rapidly increasing complexity of the analytical expressions limits the method to $N = 2$ (Forte and Peltier, 1987). A more efficient approach for models with multiple layers is the propagator-matrix method, whereby a flow solution is propagated from one layer interface to the next by simple matrix multiplication. The method is in fact applicable to any system of linear ODEs with constant coefficients of the form

$$\frac{d\mathbf{y}}{dz} = \mathbf{A}\mathbf{y} + \mathbf{b}(z) \qquad [122]$$

where $\mathbf{y}(z)$ is a vector of dependent variables, $\mathbf{A}$ is a constant square matrix, and $\mathbf{b}(z)$ is an inhomogeneous vector. The general solution of [122] is (Gantmacher, 1960, I, p. 120, eqn [53])

$$\mathbf{y}(z) = \mathbf{P}(z, \ z_0)\mathbf{y}(z_0) + \int_{z_0}^{z} \mathbf{P}(z, \ \zeta)\mathbf{b}(\zeta)\mathrm{d}\zeta \qquad [123a]$$

$$\mathbf{P}(z, \ z_0) = \exp[\mathbf{A}(z - z_0)] \qquad [123b]$$

where $\mathbf{y}(z_0)$ is the solution vector at the reference point $z = z_0$ and $\mathbf{P}(z, z_0)$ is the propagator matrix. The form of the solution [123] is identical to that for a scalar variable $y(z)$, except that the argument of the exponential in [123b] is now a matrix rather than a scalar quantity. Analytical expressions for functions of matrices such as [123b] are given by Gantmacher (1960, vol. I, pp. 95–110).

As an illustration, we determine the propagator matrix for a poloidal flow driven by internal density anomalies in a self-gravitating spherical shell with radially variable viscosity $\eta(r)$ and laterally averaged density $\rho(r)$ (Hager and O'Connell, 1981; Richards and Hager, 1984). As in Section 4.5.8.1, we suppose that the driving density anomaly is $\hat{\rho}_l^m(r)Y_l^m(\theta,\phi)$, where $Y_l^m$ is a surface spherical harmonic.

The first step is to transform the equations governing the flow in the mantle to the canonical form [122], where $\mathbf{A}$ – to repeat – is a constant matrix. The standard Stokes equations [31] for our model mantle do not have this form for three reasons: (1) the fluid properties $\eta(r)$ and $\rho(r)$ vary with radius; (2) the expression for the differential operator $\nabla$ in spherical coordinates involves the scale factor $r^{-1}$; and (3) a system of first-order equations cannot be written in terms of the primitive variables $u_i$ and $p$. Difficulty (1) is circumvented by dividing the mantle into $N$ discrete layers $n = 1, 2, \ldots, N$, in each of which the viscosity $\eta_n$ and the density $\rho_n$ is constant. Difficulty (2) is overcome by using a transformed radial variable $z = \ln(r/a_1)$, where $a_1$ is the outer radius of the mantle. Finally, one circumvents (3) by using the independent variables

$$\mathbf{y} = \left[\hat{u}, \ \hat{v}, \ \frac{r\hat{\sigma}_{rr}}{\eta_0}, \ \frac{r\hat{\sigma}_{r\theta}}{\eta_0}, \ \frac{\rho_0 r\hat{\chi}}{\eta_0}, \ \frac{\rho_0 r^2 \partial_r \hat{\chi}}{\eta_0}\right]^{\mathrm{T}} \qquad [124]$$

where $\hat{u}$ and $\hat{v}$ are the $\mathbf{e}_r$- and $\mathbf{e}_\theta$-components of the velocity, respectively, $\eta_0$ and $\rho_0$ are reference values of the viscosity and density, respectively, and the spherical harmonic dependence of each variable ($\hat{u}$, $\hat{\chi}_{rr}$, $\hat{\chi}$, and $\partial_r \hat{\chi} \propto Y_l^m$, $\hat{v}$ and $\hat{\sigma}_{r\theta} \propto \partial_\theta Y_l^m$) has been suppressed for clarity. The Stokes equations within each layer $n$ then takes the form [122] with

$$\mathbf{A} = \begin{bmatrix} -2 & L & 0 & 0 & 0 & 0 \\ -1 & 1 & 0 & 1/\eta^* & 0 & 0 \\ 12\eta^* & -6L\eta^* & 1 & L & 0 & \rho^* \\ -6\eta^* & 2(2L-1)\eta^* & -1 & -2 & \rho^* & 0 \\ 0 & 0 & 0 & 0 & 1 & 1 \\ 0 & 0 & 0 & 0 & L & 0 \end{bmatrix} \qquad [125a]$$

$$\mathbf{b} = [0, \ 0, \ r^2 g_0(r)\hat{\rho}(r)/\eta_0, \ 0, \ 0, \\ -4\pi r^3 G\rho_0 \hat{\rho}(r)/\eta_0]^{\mathrm{T}} \qquad [125b]$$

where $L = l(l+1)$, $\eta^* = \eta_n/\eta_0$, and $\rho^* = \rho_n/\rho_0$. Further simplification is achieved by recasting the density anomaly $\hat{\rho}(r)$ as a sum of equivalent surface density contrasts $\hat{\sigma}_n$ (units $\mathrm{kg\,m}^{-2}$) localized at the midpoints $r_n$ of the layers, according to

$$\hat{\rho}(r) = \sum_n \delta(r - r_n)\hat{\sigma}_n \qquad [126]$$

In practice, $\hat{\sigma}_n$ is different in each layer, while $\eta_n$ and $\rho_n$ may be constant over several adjacent layers. For a group of $M$ such adjacent layers bounded by the depths $z$ and $z_0$, the solution [123a] now takes the approximate form

$$\mathbf{y}(z) = \mathbf{P}(z, \ z_0)\mathbf{y}(z_0) + \sum_{m=1}^{M} \mathbf{P}(z, \ z_m)\mathbf{b}_m \qquad [127a]$$

$$\mathbf{b}_m = [0, \ 0, \ r_m g_0(r_m)\hat{\sigma}_m/\eta_0, \ 0, \ 0, \\ -4\pi r_m^2 G\rho_0 \hat{\sigma}_m/\eta_0]^{\mathrm{T}} \qquad [127b]$$

where $z_m = \ln(r_m/a_1)$. To use [127], one must apply boundary conditions at the CMB and at the Earth's surface, taking into account the dynamic topography of these boundaries (Richards and Hager, 1984, Appendix 1).

An analytical expression for $\mathbf{P}(z, z_0)$ can be written in terms of the minimal polynomial $\psi(\lambda)$ of $\mathbf{A}$. For the matrix [125a], $\psi(\lambda)$ is identical to the characteristic polynomial, and is

$$\psi(\lambda) = \prod_{i=1}^{4}(\lambda - \lambda_i)^{m_i} \qquad [128a]$$

$$\lambda_1 = l + 1, \quad \lambda_2 = -l, \\ \lambda_3 = l - 1, \quad \lambda_4 = -l - 2 \qquad [128b]$$

$$m_1 = m_2 = 2, \quad m_3 = m_4 = 1 \qquad [128c]$$

where $\lambda_1$–$\lambda_4$ are the four distinct eigenvalues of $\mathbf{A}$. Following Gantmacher (1960, I, pp. 95–102)

$$\mathbf{P}(z, \ z_0) = \sum_{k=1}^{4} \sum_{j=1}^{m_k} \alpha_{kj}(\mathbf{A} - \lambda_k \mathbf{I})^{j-1}\Psi_k \qquad [129]$$

where

$$\alpha_{kj} = \frac{1}{(j-1)!} \frac{d^{j-1}}{d\lambda^{j-1}} \left[ \frac{\exp\lambda(z-z_0)}{\psi_k(\lambda)} \right]_{\lambda=\lambda_k} \quad [130]$$

$$\psi_k(\lambda) = \frac{\psi(\lambda)}{(\lambda-\lambda_k)^{m_k}}, \quad \Psi_k = \prod_{\substack{i=1 \\ i \neq k}}^{4} (\mathbf{A} - \lambda_i \mathbf{I})^{m_i} \quad [131]$$

and $\mathbf{I}$ is the identity matrix.

## 4.6   Elasticity and Viscoelasticity

Fluid convection with a period $\tau \sim 10^{15}$ s is the extreme limit of a spectrum of deformational processes in the mantle spanning an enormous range of timescales, including postglacial rebound ($\tau \sim 10^{11}$ s), the Chandler wobble ($\tau \sim 4 \times 10^7$ s), and elastic free oscillations ($\tau \sim 10-10^3$ s). All these processes can be understood by regarding the mantle as a viscoelastic body that deforms as a fluid when $\tau \gg \tau_M$ and as an elastic solid when $\tau \ll \tau_M$, where $\tau_M$ is the Maxwell time of the material. $\tau_M$ is just the ratio of the viscosity of the material to its elastic shear modulus, and is of the order of a few hundred years ($\sim 10^{10}$ s) in the mantle.

Because the theory of viscous flow is valid only at very long-periods ($\tau \gg \tau_M$), other rheological models are required to understand phenomena with shorter periods. The two most commonly used models are the linear elastic solid (for short periods $\tau \ll \tau_M$) and the linear Maxwell solid (for intermediate periods $\tau \sim \tau_M$.) The reader is referred to Chapter 6 for a more extensive discussion of these models. Here we focus on the mathematical analogies (correspondence principles) among the viscous, elastic, and viscoelastic models and the powerful analytical techniques these analogies make possible.

### 4.6.1   Correspondence Principles

Stokes (1845) and Rayleigh (1922) demonstrated the existence of a mathematical correspondence (the Stokes–Rayleigh analogy) between small incompressible deformations of an elastic solid and slow flows of a viscous fluid. As discussed in Chapter 6, the constitutive law for a linear (Hookean) elastic solid is

$$\sigma_{ij} = Kv_{kk}\delta_{ij} + 2\mu\left(v_{ij} - \frac{1}{3}v_{kk}\delta_{ij}\right) \quad [132]$$

where $\sigma_{ij}$ is the stress tensor, $v_{ij} \equiv (\partial_i v_j + \partial_j v_i)/2$ is the linearized strain tensor, $v_i$ is the displacement vector, $K$ is the bulk modulus, and $\mu$ is the shear modulus. Alternatively, [132] can be written in terms of the Young's modulus $E$ and Poisson's ratio $\sigma$, which are related to $K$ and $\mu$ by

$$E = \frac{9K\mu}{3K+\mu}, \quad \sigma = \frac{3K-2\mu}{2(3K+\mu)} \quad [133]$$

An incompressible elastic solid corresponds to the limits $K/E \to \infty$, $\sigma \to 1/2$, and an incompressible deformation to the limit $\Delta \to 0$. As these limits are approached, however, the product $-K\Delta$ tends to a finite value, the pressure $p$. Equation [132] then becomes

$$\sigma_{ij} = -p\delta_{ij} + 2\mu v_{ij} \quad [134]$$

which is identical to the constitutive relation [31e] for a viscous fluid if the shear modulus $\mu$ and the displacement $v_i$ are replaced by the viscosity $\eta$ and the velocity $u_i$, respectively. Consequently, solutions of problems in linear elasticity and Stokes flow can be transformed into one another via the transformations

$$(v_i, \ E, \ \sigma) \to (u_i, \ 3\eta, \ 1/2), \quad (u_i, \ \eta) \to (v_i, \ \mu) \quad [135]$$

Examples of the use of [135] are described in Section 4.8.4 on thin-shell theory.

A second useful correspondence principle relates problems in linear viscoelasticity and linear elasticity. Although this principle is valid for any linear viscoelastic body (Biot, 1954), its geophysical application is usually limited to the special case of a linear Maxwell solid, for which the constitutive relation is (e.g., Peltier, 1974)

$$\dot{\sigma}_{ij} + \frac{\mu}{\eta}\left(\sigma_{ij} - \frac{1}{3}\sigma_{kk}\delta_{ij}\right) = 2\mu\left(\dot{v}_{ij} - \frac{1}{3}\dot{v}_{kk}\delta_{ij}\right) + K\dot{v}_{kk}\delta_{ij} \quad [136]$$

where dots denote time derivatives. Transforming [136] into the frequency domain using a Laplace transform as described in Chapter 6, we obtain

$$\bar{\sigma}_{ij} = K\bar{v}_{kk}\delta_{ij} + 2\bar{\mu}(s)\left(\bar{v}_{ij} - \frac{1}{3}\bar{v}_{kk}\delta_{ij}\right) \quad [137a]$$

$$\bar{\mu}(s) = \frac{\mu s}{s + \mu/\eta} \quad [137b]$$

where $s$ is the complex frequency and overbars denote Laplace transforms of the quantities beneath. Equation

[137] is identical in form to Hooke's law [132] for an elastic solid, but with a frequency-dependent shear modulus $\bar{\mu}$. Accordingly, any viscoelastic problem can be reduced to an equivalent elastic problem, which can be solved and then inverse-Laplace transformed back into the time domain to yield the solution of the original viscoelastic problem. We now turn to one of the principle geophysical applications of this procedure.

### 4.6.2 Surface Loading of a Stratified Elastic Sphere

The fundamental problem of postglacial rebound is to determine the response of a radially stratified Maxwell Earth to a time-dependent surface load $P(\theta, \phi, t)$ that represents the changing global distribution of glacial ice and meltwater. Formally, this problem can be solved by convolving the load distribution with a Green function that describes the Earth's response to an impulsive point load applied at time $t$ at a point $\mathbf{r}$ on the Earth's surface and then immediately removed. Now an impulse function in the time domain corresponds to a constant in the frequency domain. According to the correspondence principle, therefore, our problem reduces to that of determining the deformation of a stratified elastic sphere by a static point load (Longman, 1962; Farrell, 1972; Peltier, 1974). The derivation below follows Peltier (1995).

The general equations governing the deformation of a self-gravitating elastic sphere are (Backus, 1967)

$$0 = \nabla \cdot \hat{\boldsymbol{\sigma}} + \mathbf{g}_0 \hat{\rho} - \rho_0 \nabla \hat{\chi} - \nabla(\rho_0 g_0 \hat{u}) \qquad [138a]$$

$$\nabla^2 \hat{\chi} = 4\pi G \hat{\rho} \qquad [138b]$$

where $\hat{\mathbf{v}} \equiv \hat{u}\mathbf{e}_r + \hat{v}\mathbf{e}_\theta + \hat{w}\mathbf{e}_\phi$ is the displacement vector and the other symbols are the same as in the analogous equations [100] for slow viscous flow.

The elastic equations [138b] differ from their viscous analogs [100] in three ways. First, the stress tensor $\hat{\boldsymbol{\sigma}}$ in [138a] is related to the displacement vector $\hat{\mathbf{v}}$ by the elastic constitutive law [137] with a frequency-dependent shear modulus. Second, because the deformation is driven by surface loading rather than by internal density anomalies, the perturbation density $\hat{\rho} \equiv -\nabla \cdot (\rho_0 \hat{\mathbf{v}})$ is determined entirely by the requirement of mass conservation in the deformed solid. Third, [138a] contains an additional term $-\nabla(\rho_0 g_0 \hat{u})$, that corrects for the fact that the strain tensor that appears in the elastic constitutive

law is at a fixed material particle and not at a fixed point in space (Backus, 1967, p. 96).

The reduction of the governing equations [138] to solvable form proceeds as for the viscous flow equations in Section 4.5.8.2, but with a few significant differences. As in the viscous case, the goal is to reduce [138] to a sixth-order system of ODEs of the form

$$\frac{d\mathbf{y}}{dr} = \mathbf{A}\mathbf{y} \qquad [139]$$

where $\mathbf{y}$ is the unknown vector and $\mathbf{A}$ is a $6 \times 6$ matrix. Equation [139] contains no inhomogeneous vector $\mathbf{b}$ like the one in the viscous equations [122], because the loads in the elastic problem are applied at the boundaries rather than internally. Another important point is that the reference profiles of density $\rho_0(r)$ and the elastic moduli $\mu(r)$ and $K(r)$ that appear in the equations are essentially continuous functions, given a priori by a seismological reference model (e.g., PREM; Dziewonski and Anderson, 1981). Consequently, a discrete-layer solution method like the propagator-matrix technique is not practical, and it is therefore superfluous to reduce [139] to constant-coefficient form via the transformation $z = \ln(r/a_1)$.

We turn now to the definition of the variables $\mathbf{y}$. By symmetry, the deformation of the sphere can depend only on the radius $r$ and the angular distance $\theta$ from the point load. The field variables can therefore be expanded as

$$\hat{\mathbf{v}} = \sum_{l=0}^{\infty} [\hat{u}_l(r, s)P_l(\cos\theta)\mathbf{e}_r + \hat{v}_l(r, s)\partial_\theta P_l(\cos\theta)\mathbf{e}_\theta]$$

$$[140a]$$

$$\hat{\chi} = \sum_{l=0}^{\infty} \hat{\chi}_l(r, s)P_l(\cos\theta) \qquad [140b]$$

where $P_l(\cos\theta)$ are the Legendre polynomials. Now for each angular degree $l$, let

$$\mathbf{y} = \left[\hat{u}, \ \hat{v}, \ \hat{\sigma}_{rr}, \ \hat{\sigma}_{r\theta}, \ \hat{\chi}, \ \partial_r\hat{\chi} \right.$$
$$\left. + (l+1)\hat{\chi}/r + 4\pi G\rho_0\hat{u}\right]^{\mathrm{T}} \qquad [141]$$

where the subscript $l$ on each variable has been suppressed for simplicity. Apart from factors of $r$, the definition of $y_6$, and the replacement of velocities by displacements, [141] is identical to its viscous analog [124]. With the choice [141], the matrix $\mathbf{A}$ in [139] is

$$\mathbf{A} = \begin{bmatrix} -\dfrac{2\lambda}{\beta r} & \dfrac{L\lambda}{\beta r} & \dfrac{1}{\beta} & 0 & 0 & 0 \\[4pt] -r^{-1} & r^{-1} & 0 & \bar{\mu}^{-1} & 0 & 0 \\[4pt] \dfrac{4}{r}\left(\dfrac{\gamma}{r}-\rho_0 g_0\right) & -\dfrac{L}{r}\left(\dfrac{2\gamma}{r}-\rho_0 g_0\right) & -\dfrac{4\bar{\mu}}{\beta r} & \dfrac{L}{r} & -\dfrac{\rho_0 M}{r} & \rho_0 \\[4pt] \dfrac{1}{r}\left(\rho_0 g_0 - \dfrac{2\gamma}{r}\right) & \dfrac{L(\gamma+\bar{\mu})-2\bar{\mu}}{r^2} & -\dfrac{\lambda}{\beta r} & -\dfrac{3}{r} & \dfrac{\rho_0}{r} & 0 \\[4pt] -4\pi G\rho_0 r & 0 & 0 & 0 & -\dfrac{M}{r} & 1 \\[4pt] -\dfrac{4\pi M G\rho_0}{r} & \dfrac{4\pi L G\rho_0}{r} & 0 & 0 & 0 & \dfrac{l-1}{r} \end{bmatrix} \qquad [142]$$

where $L = l(l+1)$, $M = l+1$, $\beta = K + 4\bar{\mu}/3$, $\gamma = 3K\bar{\mu}/\beta$, $\lambda = K - 2\bar{\mu}/3$, and $\bar{\mu}$ is defined by [137b]. The system [139] can be solved by a standard numerical method for two-point boundary-value problems (BVPs) (e.g., shooting) subject to appropriate boundary conditions. Inversion of the resulting solution back into the time domain and its subsequent convolution with a given time-dependent surface load function are then performed numerically. The use of this procedure to infer the mantle viscosity profile $\eta(r)$ is reviewed by Peltier (1995).

## 4.7  BL Theory

Many geophysical flows occur in layers or conduits whose length greatly exceeds their thickness: examples include TBLs, subducted oceanic lithosphere, mantle plume stems, and gravity currents of buoyant plume material spreading beneath the lithosphere. In all these cases, the gradients of the fluid velocity and/or temperature across the layer greatly exceed the gradients along it, a fact that can be exploited to simplify the governing equations substantially. The classic example of this approach is BL theory, which describes the flow in layers whose thickness is controlled by a balance of diffusion and advection.

A BL is defined as a thin region in a flow field, usually adjoining an interface or boundary, where the gradients of some quantity transported by the fluid (e.g., vorticity, temperature, or chemical concentration) are large relative to those elsewhere in the flow. Physically, BLs arise when the boundary acts as a source of the transported quantity, which is then prevented from diffusing far from the boundary by strong advection. BLs thus occur when $UL/D \gg 1$, where $U$ and $L$ are characteristic velocity and length scales for the flow and $D$ is the diffusivity of the quantity in question. In classical BL theory, the transported quantity is

vorticity, the relevant diffusivity is the kinematic viscosity $\nu$, and BLs form when the Reynolds number $Re \equiv UL/\nu \gg 1$. Such BLs do not occur in the mantle, where $Re \sim 10^{-20}$. However, because the thermal diffusivity $\kappa \sim 10^{-23}\nu$, the Peclet number $UL/\kappa \gg 1$ for typical mantle flows, implying that TBLs will be present.

Although BLs are 3-D structures in general, nearly all BL models used in geodynamics involve one of the three simple geometries shown in **Figure 10**: 2-D flow (1), an axisymmetric plume (2), and axisymmetric flow along a surface of revolution (3). Let $x$ and $y$ be the coordinates parallel to and normal to the boundary (or symmetry axis), respectively, $u$ and $v$ be the corresponding velocity components, and $\delta(x)$ be the thickness of the BL.

The fundamental hypothesis of BL theory is that gradients of the transported quantity (heat in this case) along the BL are much smaller than those across it. Diffusion of heat along the layer is therefore negligible. As an illustration, consider the simplest case of a thermal BL in steady flow with negligible viscous

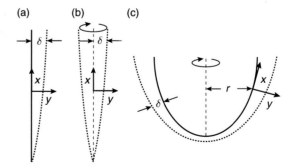

**Figure 10** Typical geometries for boundary-layer flows: (a) 2-D flow, (b) an axisymmetric plume, and (c) axisymmetric flow along a surface of revolution. Impermeable boundaries are shown by heavy lines, axes of symmetry by light dashed lines, and the edges of BLs by heavy dashed lines. The coordinates $x$ and $y$ are parallel to and normal to the boundary (or symmetry axis), respectively, $\delta(x)$ is the thickness of the BL, and $r(x)$ is the radius of the surface of revolution.

dissipation of energy. The BL forms of the continuity and energy equations are then

$$u_x + v_y = 0,$$ [143a]

$$uT_x + vT_y = \kappa T_{yy}$$ [143b]

for 2-D flow,

$$u_x + y^{-1}(yv)_y = 0$$ [144a]

$$uT_x + vT_y = \kappa y^{-1}(yT_y)_y$$ [144b]

for an axisymmetric plume, and

$$(ru)_x + rv_y = 0$$ [145a]

$$uT_x + vT_y = \kappa T_{yy}$$ [145b]

for a surface of revolution. As noted in Section 4.3.3, the terms $uT_x$ and $vT_y$ are of the same order.

### 4.7.1 Solution of the BL Equations Using Variable Transformations

A powerful technique for solving problems involving BLs is the use of variable transformations to reduce the BL equations to equations of simpler form. One of the most important of these transformations is that of Von Mises (1927), which transforms the 2-D BL equations into the classical heat conduction equation. A second useful transformation, due to Mangler (1948), relates the structure of an axisymmetric BL on a surface of revolution to that of a 2-D BL on a flat surface. After introducing both transformations, we will show how they can be used together to obtain a solution for the heat transfer from a hot sphere moving in a viscous fluid (**Figure 1(a)**).

#### 4.7.1.1 Von Mises's transformation

The essential trick involved in this transformation is to use the streamfunction $\psi$ instead of $y$ as the transverse coordinate in the BL. Denoting the streamwise coordinate by a new symbol $s \equiv x$ for clarity, we transform the derivatives in [143b] using the chain rule as

$$\frac{\partial T}{\partial x} = \frac{\partial T}{\partial s}\frac{\partial s}{\partial x} + \frac{\partial T}{\partial \psi}\frac{\partial \psi}{\partial x} \equiv \frac{\partial T}{\partial s} - v\frac{\partial T}{\partial \psi}$$ [146]

$$\frac{\partial T}{\partial y} = \frac{\partial T}{\partial s}\frac{\partial s}{\partial y} + \frac{\partial T}{\partial \psi}\frac{\partial \psi}{\partial y} \equiv u\frac{\partial T}{\partial \psi}$$ [147]

where the streamfunction is defined according to the convention $(u, v) = (\psi_y, -\psi_x)$. Equation [143b] then becomes

$$\frac{\partial T}{\partial s} = \kappa \frac{\partial}{\partial \psi}\left(u\frac{\partial T}{\partial \psi}\right)$$ [148]

Equation [148] takes still simpler forms if the surface $y = \psi = 0$ is either traction-free or rigid. Near a free surface, $u \approx U(s)$ is constant across the BL. Upon introducing a new downstream coordinate $\tau$ such that $d\tau = U(s)ds$, [148] becomes

$$\frac{\partial T}{\partial \tau} = \kappa \frac{\partial^2 T}{\partial \psi^2}$$ [149]

which is just the classical equation for diffusion of heat in a medium with constant thermal diffusivity $\kappa$. The units of the time-like variable $\tau$, however, are now those of diffusivity ($\mathrm{m^2\,s^{-1}}$) rather than of time. Near a rigid surface, $u = yf(s)$ and $\psi = y^2 f(s)/2$, implying $u \approx (2\psi f)^{1/2}$, where $f(s)$ is arbitrary. Substituting this result into [148] and introducing a new downstream variable $\tau$ such that $d\tau = (2f)^{1/2}ds$, we obtain

$$\frac{\partial T}{\partial \tau} = \kappa \frac{\partial}{\partial \psi}\left(\psi^{1/2}\frac{\partial T}{\partial \psi}\right)$$ [150]

which describes the diffusion of heat in a medium with a position-dependent thermal diffusivity $\kappa\psi^{1/2}$ as a function of a time-like variable $\tau$ having units of $\mathrm{m\,s^{-1/2}}$.

Of special interest are the self-similar solutions of [149] and [150] that exist when the wall temperature $T_0$ is constant ($= \Delta T$, say) and the upstream temperature profile is $T(0, \psi) = 0$. The solution of the free-surface equation [149] has the form $T = \Delta T F(\eta)$, where $\eta = \psi/\delta(\tau)$. Separating variables, we obtain $-F''/\eta F' = \delta\dot{\delta}/\kappa = \lambda^2$ (constant), which when solved subject to the conditions $F(0) - 1 = F(\infty) = \delta(0) = 0$ yields

$$T = \Delta T \,\mathrm{erfc}\left(\frac{\psi}{2\sqrt{\kappa\tau}}\right)$$ [151]

The solution of the rigid-surface equation [150] has the form $T = \Delta T F(\eta)$, where $\eta = \sqrt{\psi}/\delta(\tau)$. Separating variables, we find $-F''/4\eta^2 F' = \delta^2\dot{\delta}/\kappa = \lambda^2$ (constant), which when solved subject to $F(0) - 1 = F(\infty) = \delta(0) = 0$ with $\lambda^2 = 1/3$ for convenience gives

$$\frac{T}{\Delta T} = 1 - \frac{3}{\Gamma(1/3)}\left(\frac{4}{9}\right)^{1/3}\int_0^\eta \exp\left(-\frac{4}{9}x^3\right)dx,$$
$$\eta = \frac{\psi^{1/2}}{(\kappa\tau)^{1/3}}$$ [152]

where $\Gamma$ is the gamma function. Below we show how the free-surface solution [151] can be applied to the problems of heat transfer from a sphere (next subsection) and steady cellular convection (Section 4.9.3.3.)

### 4.7.1.2 Mangler's transformation

Mangler (1948) showed that the equations governing the axisymmetric BL on a surface of revolution with radius $r(x)$ (**Figure 10**) are related to the 2-D BL equations by the variable transformation

$$\bar{x} = \int_0^x (r/L)^2 \mathrm{d}x, \quad \bar{y} = \frac{r}{L}y, \quad \bar{u} = u,$$
$$\bar{v} = \frac{L}{r}\left(v + \frac{r'}{r}yu\right), \quad \bar{\psi} = L^{-1}\psi \qquad [153]$$

where the variables with and without overbars are those of the 2-D and the axisymmetric flows, respectively, $L$ is an arbitrary constant length scale, and $\psi$ is the Stokes streamfunction. The transformation [153], which applies equally to vorticity and thermal BLs, allows solutions of the 2-D BL equations to be transformed directly into solutions of the axisymmetric BL equations on a surface of revolution. To illustrate, we determine the heat flow from a traction-free isothermal sphere of radius $a$ and excess temperature $\Delta T$ moving at constant speed $U$ in a viscous fluid (**Figure 1**) by transforming the Cartesian BL solution [151], which in terms of the barred variables is

$$T = \Delta T \operatorname{erfc}\left(\frac{\bar{\psi}}{2\sqrt{\kappa\tau}}\right), \quad \tau(\bar{x}) = \int_0^{\bar{x}} \bar{u}\,\mathrm{d}\bar{x} \qquad [154]$$

For a sphere, $r(x) = a\sin(x/a) \equiv a\sin\theta$, and $L = a$ is the natural choice. The Stokes–Hadamard solution (Section 4.5.3.2) gives $\psi \approx (1/2)Uay\sin^2\theta$, and $u \approx (1/2)U\sin\theta$, and [153] implies

$$\bar{\psi} = \frac{1}{2}Uy\sin^2\theta, \quad \bar{x} = \frac{1}{2}x - \frac{a}{4}\sin\frac{2x}{a} \qquad [155]$$

The stretched downstream variable $\tau$ is therefore

$$\tau = \int_0^{\bar{x}} \bar{u}\,\mathrm{d}\bar{x}$$
$$= \int_0^x u\frac{\mathrm{d}\bar{x}}{\mathrm{d}x}\mathrm{d}x = \frac{2}{3}aU(2+\cos\theta)\sin^4\frac{\theta}{2} \qquad [156]$$

Substitution of [155] and [156] into [154] yields the temperature $T(y, \theta)$ everywhere in the BL, and the corresponding local Nusselt number is

$$\mathcal{N}(\theta) \equiv -\frac{a}{\Delta T}\frac{\partial T}{\partial y}(y=0)$$
$$= \left(\frac{3}{2\pi}\right)^{1/2}\frac{1+\cos\theta}{(2+\cos\theta)^{1/2}}Pe^{1/2} \qquad [157]$$

where $Pe = Ua/\kappa$ is the Peclet number.

**Table 3** Heat transfer from a hot sphere for $Re \ll 1$: analytical predictions

| Technique | Local Nusselt number $\mathcal{N}$ | Validity |
|---|---|---|
| Dimensional analysis | $\mathrm{fct}(Pe, \theta)$ | universal |
| Scaling analysis | $Pe^{1/2}\mathrm{fct}(\theta)$ | $Pe \gg 1$ |
| BL theory | $\left(\dfrac{3}{2\pi}\right)^{1/2}\dfrac{1+\cos\theta}{(2+\cos\theta)^{1/2}}Pe^{1/2}$ | $Pe \gg 1$, $0 < \dfrac{\pi}{2}$ |

The result [157], together with our previous treatments of the hot sphere in Sections 4.3.1, 4.3.2, and 4.3.3, shows that BL theory represents a third stage in a hierarchy of techniques (dimensional analysis, scaling analysis, BL theory) that give progressively more detailed information about the structure of the solution in the asymptotic limit of negligible inertia ($Re \ll 1$) and $Pe \to \infty$. **Table 3** summarizes the local Nusselt number $\mathcal{N}(\theta)$ for the (traction-free) sphere predicted by each of the three techniques. Note that the cost of the increasing precision of the results is a decreasing range of validity.

### 4.7.2 The MMAE

We noted in Section 4.5.7 that the MMAE is a powerful method for solving problems where the field variables exhibit distinct regions characterized by very different length scales. The method is particularly well suited for BLs, whose characteristic thickness $\delta$ is much smaller than the scale $L$ of the flow outside the BL. As an illustration of the method, consider a simple axisymmetric stagnation-point flow model for the steady temperature distribution in a plume upwelling beneath a rigid lithosphere (**Figure 11**), in which fluid with temperature $T_1$ and upward vertical velocity $-w_1$ at a depth $z = d$ ascends towards a rigid surface with temperature $T = T_0$. If viscous dissipation of energy is negligible, the pressure is (nearly) hydrostatic, and all physical properties of the fluid are constant, then $T(\mathbf{x})$ satisfies

$$\mathbf{u} \cdot \nabla T - \frac{g\alpha}{c_p}\mathbf{u} \cdot \hat{\mathbf{z}}T = \kappa\nabla^2 T \qquad [158]$$

where $\alpha$ is the thermal expansion coefficient and $c_p$ is the heat capacity at constant pressure. The three terms in [158] represent advection of temperature

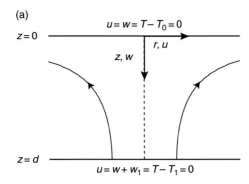

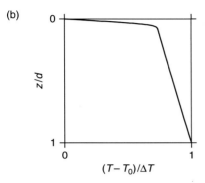

**Figure 11** Model for the temperature distribution in a steady stagnation-point flow beneath a rigid lithosphere. (a) The base of the lithosphere $z = 0$ is at temperature $T_0$, and a uniform upward vertical velocity $w_1 > 0$ is imposed at a depth $z = d$ where the temperature is $T_1$. The curved lines with arrows are streamlines. (b) Temperature as a function of depth for $\epsilon = 0.0001$, $\beta = 0.1$. and $\lambda = 2.0$.

gradients, adiabatic decompression, and thermal diffusion, respectively.

Let $\mathbf{u} = u(r, z)\mathbf{e}_r + w(r, z)\mathbf{e}_z$. Because $w(r, d) \equiv -w_1$ is constant, $w = w(z)$, whence the continuity equation implies $u = -rw'(z)/2$. Equation [158] therefore admits a 1-D solution $T = T(z)$ that satisfies

$$wT' - \frac{g\alpha}{c_p}wT = \kappa T'' \qquad [159]$$

where primes denote d/dz. By substituting $w = w(z)$ and $u = -rw'(z)/2$ into the constant-viscosity Stokes equations in cylindrical coordinates and solving the resulting equation for $w(z)$ subject to the boundary conditions shown in **Figure 11**, we find

$$\frac{w}{w_1} = 2\left(\frac{z}{d}\right)^3 - 3\left(\frac{z}{d}\right)^2 \qquad [160]$$

Upon introducing dimensionless variables $\tilde{z} = z/d$ and $\tilde{T} = (T - T_0)/(T_1 - T_0) \equiv (T - T_0)/\Delta T$ and

then immediately dropping the tildes, [159] together with [160] and the boundary conditions on $T$ become

$$\left(2z^3 - 3z^2\right)\left[T' - \beta(T + \lambda)\right] = \epsilon T'' \qquad [161a]$$

$$T(0) = T(1) - 1 = 0 \qquad [161b]$$

where

$$\epsilon = \frac{\kappa}{dw_1} \equiv Pe^{-1}, \quad \beta = \frac{g\alpha d}{c_p}, \quad \lambda = \frac{T_0}{T_1 - T_0} \qquad [162]$$

We wish to solve [161] in the limit $\epsilon \to 0$ ($Pe \to \infty$), assuming for simplicity that $\beta = O(1)$ and $\lambda = O(1)$. However, note that $\epsilon$ appears in [161a] as the coefficient of the most highly differentiated term. We therefore cannot simply set $\epsilon = 0$ in [161a], because that would reduce the order of the ODE and make it impossible to satisfy all the boundary conditions. Equations [161] therefore constitute a singular perturbation problem: the solution for small values of $\epsilon > 0$ is not a small perturbation of a solution for $\epsilon = 0$, which in any case does not exist. The resolution of this apparent paradox is that the solution exhibits a thin BL where the term $\epsilon T''$ in [161a] is important, no matter how small $\epsilon$ may be.

We therefore anticipate that the solution to [161] will comprise two distinct regions governed by different dynamics: an inner region (the BL) of dimensionless thickness $\delta \ll 1$ in which advection is balanced by diffusion, and an outer region where advection is balanced by adiabatic decompression. Consider the outer region first, and let the temperature there be $T = b(z, \epsilon)$. We seek a solution in the form of an asymptotic expansion

$$b = b_0(z) + \eta_1(\epsilon)b_1(z) + \eta_2(\epsilon)b_2(z) + \cdots \qquad [163]$$

where $\eta_n(\epsilon)$ are (as yet unknown) gauge functions that form an asymptotic sequence such that $\lim_{\epsilon \to 0} \eta_n/\eta_{n-1} = 0$. The function $\eta_0(\epsilon) = 1$ because $T = O(1)$ in the outer region. Substituting [163] into [161a] and retaining only the lowest-order terms, we obtain

$$b_0' - \beta(b_0 + \lambda) = 0 \qquad [164]$$

Because [164] is a first-order ODE, its solution can satisfy only one of the boundary conditions [161b]. Evidently this must be the condition at $z = 1$, because $z = 0$ is the upper limit of a TBL that cannot be described by [164]. The same conclusion can be reached in a more formal algorithmic way by assuming contrary to fact that the BL is at $z = 1$, solving [164] subject to the wrong boundary condition $T(0) = 0$, and finally realizing that the inner solution for the supposed

BL is unphysical because it increases exponentially upward. The lowest-order terms of the (correct) boundary condition are $h_0(1) = 1$, and the solution of [164] that satisfies this is

$$h_0 = (1 + \lambda) \exp \beta(z - 1) - \lambda \qquad [165]$$

Turning now to the inner region, the first task is to determine which term on the LHS of [161a] balances the RHS. Suppose (contrary to fact, as will soon appear) that $\beta(T + \lambda) \gg T'$ (decompression $\gg$ advection). Because $z \sim \delta$ and $z^2 \gg z^3$ in the BL, the balance $z^2 \beta(T + \lambda) \sim \epsilon T''$ implies $\delta \sim \epsilon^{1/4}$. But then $T' \gg \beta(T + \lambda)$, which contradicts our original assumption. The correct balance must therefore be $z^2 T' \sim \epsilon T''$ (advection $\sim$ diffusion), which implies $\delta \sim \epsilon^{1/3}$ and $T' \gg \beta(T + \lambda)$, consistent with our original assumption. A more physical argument that leads to the same conclusion is to note that the vertical scale length over which adiabatic decompression is significant in the mantle ($\sim 1000$ km) is much greater than a typical BL thickness.

Now that we know the thickness of the BL, we can proceed to determine its structure. To discern the thin BL distinctly, we use a sort of mathematical magnifying glass: a new stretched depth coordinate $\hat{z} = z/\delta \equiv z\epsilon^{-1/3}$ that is of order unity in the BL. Denoting the inner solution by $f(\hat{z}, \epsilon)$ and writing [161a] in terms of $\hat{z}$, we find

$$\left(2\epsilon\hat{z}^3 - 3\epsilon^{2/3}\hat{z}^2\right)\left[\epsilon^{-1/3}f' - \beta(f + \lambda)\right] = \epsilon^{1/3}f'' \qquad [166]$$

where primes denote differentiation with respect to $\hat{z}$. We now seek a solution in the form

$$f = f_0(z) + \nu_1(\epsilon)f_1(z) + \nu_2(\epsilon)f_2(z) + \cdots \qquad [167]$$

where $\nu_n(\epsilon)$ are unknown gauge functions. Substituting [167] into [166] and retaining only the lowest-order terms, we obtain

$$f_0'' + 3\hat{z}^2 f_0' = 0 \qquad [168]$$

The solution of [168] that satisfies the boundary condition $f_0(0) = 0$ is

$$f_0 = A \int_0^{\hat{z}} \exp\left(-x^3\right) dx \equiv \frac{A}{3}\gamma\left(1/3, \hat{z}^3\right) \qquad [169]$$

where $\gamma(a, x) = \int_0^x \exp\left(-t\right)t^{a-1}dt$ is the incomplete Gamma function and $A$ is an unknown constant that must be determined by matching [169] to the outer solution [165].

The most rigorous way to do the matching is to rewrite both the inner and outer expansions in terms of an intermediate variable $z_{int} = \xi(\epsilon)z$ such that $\epsilon^{-1/3} \ll \xi \ll 1$, and then to choose the values of any unknown constants ($A$ in this case) so that the two expressions agree. However, it is often possible to use a simpler matching principle, due to Prandtl, which states that the inner limit of the outer expansion must be equal to the outer limit of the inner expansion – roughly speaking, that the two expansions must match at the edge of the BL. For our problem, Prandtl's principle is

$$\lim_{\hat{z}\to\infty} f_0(\hat{z}) = \lim_{z\to 0} h_0(z), \quad \text{whence}$$
$$A = \frac{3[(1 + \lambda)\exp(-\beta) - \lambda]}{\Gamma(1/3)} \qquad [170]$$

where $\Gamma$ is the Gamma function.

The last step is to construct a composite expansion that is valid both inside and outside the BL. This is just the sum of the inner and outer expansions less their shared common part $h_0(z \to 0)$, or

$$T = h_0(z) + [(1 + \lambda)\exp(-\beta) - \lambda]\left[\frac{\gamma(1/3, z^3/\epsilon)}{\Gamma(1/3)} - 1\right] \qquad [171]$$

The procedure described above is a first-order matching that retains only the first terms in the expansions [163] and [167]. If desired, the matching can be carried out to higher order by working back and forth between the inner and outer expansions, determining the gauge functions $\eta_n(\epsilon)$ and $\nu_n(\epsilon)$ and matching at each step. Higher-order matching often requires the use of the more rigorous intermediate matching principle; for examples see Hinch (1991), Kevorkian and Cole (1996), or the somewhat less formal treatments of Nayfeh (1973) or Van Dyke (1975). For many problems, however, first-order matching suffices to reveal the essential structure and physical significance of the solution.

The MMAE has been used to solve a variety of geophysically relevant problems involving BLs, mostly with constant viscosity. Umemura and Busse (1989) studied the axisymmetric convective flow in a cylindrical container of height $d$ with free-slip boundaries, using the MMAE to match the interior flow to the central rising plume and the circumferential downwelling. They found that the vertical velocity in the plume is $w \sim (\kappa/d)(-\ln\epsilon)^{1/2}Ra^{2/3}$, where the dimensionless plume radius $\epsilon \equiv a/d$ satisfies $\epsilon(-\ln\epsilon)^{1/4} \sim Ra^{-1/6}$, $Ra \gg 1$ being the Rayleigh number. Whittaker and Lister (2006a) studied creeping plumes from a point heat source with buoyancy flux $B$ on an impermeable plane boundary by matching the BL flow to an outer flow that sees the plume as a line distribution of Stokeslets. They found that the vertical velocity $w \sim (B/\nu)^{1/2}[\ln(z/z_0)]^{1/2}$, where $z$ is the height and $z_0 = 32\pi\kappa^2\nu/B$. Whittaker and Lister (2006b) studied the dynamics of a plume above a heated disk on a plane boundary, and used the MMAE to match the flow within the plume to both an outer flow and a horizontal BL flow across the disk. The results show that the Nusselt number $N \sim Ra^{1/5}$ for a rigid boundary and $N \sim (Ra/\ln Ra)^{1/3}$ for a free-slip boundary, where $Ra \gg 1$ is defined using the disk radius. An application of the MMAE to a problem with variable viscosity (Morris, 1982) appears in the next subsection.

### 4.7.3 BLs with Strongly Variable Viscosity

In a fluid with constant viscosity and infinite Prandtl number, thermal BLs are not accompanied by vorticity BLs, because the velocity field varies on a length scale much larger than the thickness of the TBL. As a result, the velocity field within the BL can be represented by the first term of its Taylor series expansion (or its multipole expansion in the case of an axisymmetric plume; cf. Whittaker and Lister, 2006a), which greatly simplifies the task of solving the thermal BL equation. This happy state of affairs no longer obtains if the viscosity of the fluid depends on temperature, because the velocity and temperature fields are then coupled.

Most studies of variable-viscosity BLs in the geodynamics literature focus on mantle plumes in either planar (**Figure 10(a)**) or axisymmetric (**Figure 10(b)**) geometry. The basic procedure is to supplement the BL equations [143] or [144] with a simplified BL form of the vertical component of the momentum equation in which derivatives along the BL are neglected relative to those normal to it. In physical terms, this equation simply states that the buoyancy force in the plume is balanced by the lateral gradient of the vertical shear stress $\tau$, or

$$y^{-n}\partial_y(y^n\tau) + \rho g\alpha(T - T_\infty) = 0 \qquad [172]$$

where $T_\infty$ is the temperature far from the plume, $y$ is the coordinate normal to the plume, and $n = 0$ or $1$ for planar or axisymmetric geometries, respectively.

An important early study based on [172] in planar geometry was that of Yuen and Schubert (1976), who investigated the buoyant upwelling adjacent to a vertical, isothermal, and traction-free plane of a fluid with temperature-dependent Newtonian or power-

law rheology. The governing BL equations governing admit a similarity transformation of the form

$$T = T_\infty + f(\eta), \quad \psi = x^{(n+2)/(n+3)}g(\eta),$$
$$\eta = yx^{-1/(n+3)} \qquad [173]$$

where $\psi$ is the streamfunction and $n$ is the power-law exponent. The more realistic case of an axisymmetric plume from a point source of heat in a fluid with temperature-dependent viscosity was studied by Morris (1980), Loper and Stacey (1983), and Olson et al. (1993). Using a similarity transformation, Morris (1980) found that the temperature on the plume axis decreases exponentially upward with a scale height $Q/12\pi k_c\Delta T_r$, where $Q$ is the total heat flux carried by the plume, $\Delta T_r$ is the temperature change required to change the viscosity by a factor $e$, and $k_c$ is the thermal conductivity. Using a simpler approach in which the functional form of the temperature profile across the plume is assumed, Olson et al. (1993) found a scale height equal to three times that determined by Morris (1980). Hauri et al. (1994) considered a similar problem, but with an empirical superexponential temperature- and depth-dependent viscosity law.

While plumes involve free or buoyancy-driven convection, variable-viscosity BLs can also arise in situations where a large-scale background flow or wind is imposed externally (forced convection). An example is the previously introduced stagnation-point flow model for the heat transfer from a hot sphere moving in a fluid with strongly temperature-dependent viscosity (**Figure 1(b)**). Using the MMAE, Morris (1982) showed that three distinct dynamical regimes occur: a conduction limit; a Stokes limit in which the flow around the sphere resembles that in an isoviscous fluid; and a lubrication limit in which most of the volume flux is carried by the low-viscosity BL adjoining the hot plate. Solutions in spherical geometry were obtained by Morris (1982) and Ansari and Morris (1985) for the lubrication limit and an intermediate (lubrication/Stokes) limit.

## 4.8 Long-Wave Theories

Long-wave theories comprise a variety of loosely related approaches whose goal is to describe the evolving thickness or shape of a thin layer. The fundamental idea underlying these approaches is the distinction between small amplitude and small slope. To take a simple example, if a 2-D fluid layer has thickness $h_0 + \Delta h \sin kx$, its surface has small amplitude if $\Delta h/h_0 \ll 1$ and small slope (or long wavelength) if $k\Delta h \ll 1$. Long-wave theories typically exploit the fact that small slope does not imply small amplitude to derive equations governing the nonlinear (finite-amplitude) evolution of long-wave disturbances in a layer.

### 4.8.1 Lubrication Theory

Viscous flows in thin layers are common in geodynamics: examples include lava flows, the deformation of continents, and the spreading of buoyant plume material beneath the lithosphere. Such flows are described by a simplified form of the Navier–Stokes equations, called the lubrication equations because of their importance in the design of industrial lubrication bearings. To develop the theory, we return to the simple model of a viscous drop or gravity current spreading on a rigid surface sketched in **Figure 3(b)**. The lubrication equations that describe such a current can be obtained from the full Navier–Stokes equations by exploiting the fact that $h/R \equiv \epsilon \ll 1$, where $h$ and $R$ are the layer's characteristic thickness and lateral extent, respectively. The thinness of the layer has three important consequences: (1) The horizontal fluid velocity $\mathbf{u} \equiv u\mathbf{e}_x + v\mathbf{e}_y$ along the layer greatly exceeds the velocity $w$ normal to it. The flow is therefore quasi-unidirectional, and inertia is negligible. (2) Derivatives of the velocity components across the layer greatly exceed the derivatives along the layer $(\partial_z \sim h^{-1} \gg \partial_x \sim \partial_y \sim R^{-1})$. (3) The pressure gradient across the layer is approximately hydrostatic. The Navier–Stokes equations then take the simplified forms

$$\mathbf{\nabla}_1 \cdot \mathbf{u} + w_z = 0, \quad \mathbf{\nabla}_1 p = \eta \mathbf{u}_{zz}, \quad p_z = -\rho g \quad [174]$$

where $w$ is the vertical velocity and $\mathbf{\nabla}_1 = \mathbf{e}_x \partial_x + \mathbf{e}_y \partial_y$ is the horizontal gradient operator.

To illustrate the use of the lubrication equations, we determine the shape and spreading rate of an axisymmetric gravity current with constant volume $V$ (Huppert, 1982). The geometry of this situation is that of **Figure 3(b)** except that the vertical conduit supplying fluid to the current is absent. Let $r$ be the (horizontal) radial coordinate, $u(z, r, t)$ the radial component of the velocity, $h(r, t)$ the thickness of the current, and $R(t)$ its radius. In this (cylindrical) geometry, [174] take the forms

$$r^{-1}(ru)_r + w_z = 0, \quad p_r = \eta u_{zz}, \quad p_z = -\rho g \quad [175]$$

Integrating [175c] subject to $p(z = h) = 0$, we obtain

$$p = \rho g(h - z) \quad [176]$$

Next, we substitute [176] into [175b] and integrate subject to the no-slip condition on the plate ($u(0, r, t) = 0$) and vanishing traction at $z = h$. Now because the current's upper surface is nearly horizontal, the traction there $\approx \eta(u_z + w_x)$. However, $w_x \sim (h/R)^2 u_z$ in the lubrication approximation, so the condition of vanishing traction is simply $u_z(h, r, t) = 0$. The profile of radial velocity across the current is therefore

$$u = \frac{\rho g h_r}{2\eta}(z^2 - 2hz) \quad [177]$$

where $\rho g h_r$ is the radial gradient of the hydrostatic pressure that drives the flow. Next, the continuity equation [175a] is integrated across the current subject to the impermeability condition $w(0, r, t) = 0$ to obtain

$$0 = w(h, r, t) + r^{-1} \int_0^h (ru)_r \, dz \quad [178]$$

We now simplify [178] by using the kinematic surface condition to eliminate $w(h, r, t) \equiv h_t + u(h, r, t)h_r$; taking $\partial_r$ outside the integral using the standard expression for the derivative of an integral with a variable limit; and evaluating the integral using [177]. We thereby find that $h(r, t)$ satisfies the nonlinear diffusion equation

$$h_t = 4\sigma r^{-1}(rh^3 h_r)_r, \quad \sigma = \frac{\rho g}{12\eta} \quad [179]$$

where $\sigma$ is the spreadability. Conservation of the current's volume requires

$$2\pi \int_0^R rh \, dr = V \quad [180]$$

We anticipate that for long times, the current will achieve a universal self-similar shape that retains no memory of the initial shape $h(r, 0)$. Self-similarity requires that $h$ depend on the normalized radius $\eta \equiv r/R(t) \in [0, 1]$, and conservation of volume requires $hR^2 \sim V$. We therefore seek a solution of the form

$$h = \frac{V}{R^2} H\left(\frac{r}{R}\right) \quad [181]$$

Upon substituting [181] into [179] and [180], separating variables in the now-familiar way (see Section 4.4.2), and solving the resulting equations subject to the conditions $H'(0) = H(1) = R(0) = 0$, we obtain

$$H = \frac{4}{3\pi}\left(1 - \eta^2\right)^{1/3}, \quad R = \left(\frac{4096\sigma V^3 t}{81\pi^3}\right)^{1/8} \quad [182]$$

The radius of the gravity current increases as the $1/8$ power of the time.

The solution [182] is a special case of a more general class of similarity solutions of [179] studied by Gratton and Minotti (1990) using a phase-plane formalism. Rather than work with the single PDE [179], Gratton and Minotti (1990) wrote down two coupled PDEs for the thickness $h$ and the mean horizontal velocity $v$, and then used a similarity transformation to reduce them to coupled ODEs. The solutions of these ODEs can be represented as segments of integral curves on a phase plane that connect singular points representing different boundary conditions such as sources, sinks, and current fronts. Gratton and Minotti (1990) give an exhaustive catalog of the solutions thus found, including a novel second kind similarity solution (cf. Section 4.4.2) for the evolution of an axisymmetric gravity current surrounding a circular hole.

Geophysical applications of viscous gravity-current theory include the solutions of Lister and Kerr (1989b) for the spreading of 2-D and axisymmetric currents at a fluid interface, which were applied by Kerr and Lister (1987) to the spread of subducted lithosphere along the boundary between the upper and lower mantle. In the next subsection, we discuss an extension of the theory to currents spreading on moving surfaces, which have been widely used to model the interaction of mantle plumes with a moving or rifting lithosphere.

### 4.8.2   Plume–Plate and Plume–Ridge Interaction Models

The geometry of these models was introduced in Section 4.2.3, and is sketched in **Figures 3(c)** and **3(d)**. Motion of a plate (or plates) with a horizontal velocity $\mathbf{U}_0(x, y)$ generates an ambient flow $\mathbf{U}(x, y, z)$ in the mantle below, which is assumed to have uniform density $\rho$ and viscosity $\eta_m$. The plume conduit is represented as a volume source of strength $Q$ fixed at $(x, y) = (0, 0)$ that emits buoyant fluid with density $\rho - \Delta\rho$ and viscosity $\eta_p \ll \eta_m$. This fluid spreads laterally beneath the lithosphere to form a thin layer whose thickness $h(x, y, t)$ is governed by a balance of buoyancy-driven spreading and advection by the ambient mantle flow. Strictly speaking, a correct solution of such a problem requires the simultaneous determination of the flow in both fluids subject to the usual matching conditions on velocity and stress at their interface. However, useful results can be obtained via a simpler approach in which the mantle flow $\mathbf{U}(x, y, z)$ is specified *a priori* and is assumed to be unaffected by the flow in the plume layer. Olson

(1990) proposed a model of this type for a plume beneath a plate moving at constant speed $U_0$ (**Figure 3(c)**), and derived a lubrication equation for $h(x, y, t)$ assuming a uniform mantle flow $\mathbf{U}(x, y, z) = U_0 \mathbf{e}_x$. For the important special case of a steady-state plume layer, the lubrication equation is

$$U_0 h_x = \bar{\sigma}\nabla_1^2 h^4 + Q\,\delta(x)\delta(y), \quad \bar{\sigma} = \frac{\Delta\rho g}{48\eta_p} \quad [183]$$

The three terms in [183] represent advection, buoyancy-driven lateral spreading, and injection of the plume fluid, respectively. The spreadability $\bar{\sigma}$ is four times smaller than that for the gravity current (eqn [179]) because Olson (1990) applied a no-slip boundary condition $\mathbf{u}(x, y, h) = U_0\mathbf{e}_x$ at $z = h$, where the plume fluid is in contact with the much more viscous mantle.

The fundamental scales for the thickness $h$ and width $L$ of the plume fluid layer can be found via a scaling analysis of [183]. By requiring the three terms in [183] to be of the same order and noting that $\partial_x \sim \partial_y \sim \delta(x) \sim \delta(y) \sim L^{-1}$, one finds (Ribe and Christensen, 1994)

$$L \sim \frac{Q^{3/4}\bar{\sigma}^{1/4}}{U_0} \equiv L_0, \quad h \sim \left(\frac{Q}{\bar{\sigma}}\right)^{1/4} \equiv h_0 \quad [184]$$

Further insight is provided by an analytical similarity solution of [183] that is valid far downstream $(x \gg L_0)$ from the plume source. We anticipate that at these distances, the layer thickness $h$ will vary more strongly in the direction normal to the plate motion than parallel to it, so that $\nabla_1^2 h^4 \sim (h^4)_{yy}$. Equation [183] now reduces to

$$U_0 h_x = \bar{\sigma}(h^4)_{yy}, \quad U_0\int_{-L/2}^{L/2} h\,\mathrm{d}y = Q \quad [185]$$

where the source strength $Q$ now appears in an integral relation expressing conservation of the downstream volume flux. Equations [185] admit the similarity solution (Ribe and Christensen, 1994)

$$\hat{h} = \left(\frac{3}{40}\right)^{1/3}\hat{x}^{-1/5}\left(C^2 - \hat{y}^2\hat{x}^{-2/5}\right)^{1/3},$$

$$C = \left(\frac{390625}{9\pi^3}\right)^{1/10}\left[\frac{\Gamma(5/6)}{\Gamma(1/3)}\right]^{3/5} \quad [186]$$

where $\hat{h} = h/h_0$ and $(\hat{x}, \hat{y}) = (x, y)/L_0$. The width of the plume layer increases with downstream distance as $L = 2CL_0(x/L_0)^{1/5}$. This law remains valid for more realistic 3-D numerical models with moderately temperature- and pressure-dependent viscosity (Ribe and

Christensen, 1994, 1999), but must eventually break down if the mantle/plume viscosity contrast becomes too large.

The interaction of a plume with an oceanic spreading ridge has been studied using a modified version of Olson's (1990) model in which the uniform mantle flow $\mathbf{U} = U_0\mathbf{e}_x$ is replaced by a corner flow driven by surface plates diverging with a half-spreading rate $U_0$ (**Figure 3(d)**; Ribe *et al.*, 1995). The balance of the source term $Q\delta(x)\delta(y)$ with vertical advection directly beneath the ridge yields a new length scale $L_1 \sim (Q/U_0)^{1/2}$, which agrees well with the along-strike extent of the plume fluid beneath the ridge (waist width) determined by laboratory experiments (Feighner and Richards, 1995) and numerical models (Ribe *et al.*, 1995.)

An important extension of the theory for geodynamical applications is to currents in which both the buoyancy and viscosity depend on temperature. Such a theory was developed by Bercovici (1994) and Bercovici and Lin (1996), who supplemented the usual lubrication theory equations with an energy equation describing the temperature distribution inside the current. They found that variable viscosity and buoyancy strongly influence the current's shape and spreading rate, which typically no longer exhibit the self-similar behavior typical of currents with constant properties.

### 4.8.3 Long-Wave Analysis of Buoyant Instability

In convecting systems such as the Earth's mantle, plumes arise as instabilities of horizontal TBLs. A long-wave model for this process has been proposed by Lemery *et al.* (2000; henceforth LRS00), based on two assumptions: that the wavelength of the initial instability greatly exceeds the BL thickness, and that the horizontal velocity of the fluid is approximately constant across the BL. These assumptions allow the coupled 3-D dynamics of the BL and the fluid outside it to be reduced to 2-D equations for the lateral velocity at the edge of the BL and a temperature moment that describes the distribution of buoyancy within it.

The domain of the model is a fluid half-space bounded by a cold traction-free surface $x_3 = 0$ held at temperature $-\Delta T$ relative to the fluid far from it (**Figure 2(b)**) The BL occupies the depth interval $x_3 < b(x_1, x_2, t)$, where $x_\alpha$ are Cartesian coordinates parallel to the BL and $t$ is time. In the following, hatted and unhatted variables are those in the BL

and in the interior, respectively, and an argument in parentheses indicates a value of $x_3$. The (constant) viscosity of the outer fluid is $\eta_0$, and the viscosity within the BL is $\hat{\eta}(x_3)$.

The starting point is the momentum equation within the BL, namely, $-\partial_i\hat{p} + \partial_j\hat{\tau}_{ij} = \rho_0\alpha g\hat{T}\delta_{i3}$, where $\hat{\tau}_{ij}$ is the deviatoric part of the stress tensor. Taking the curl of this equation, applying the continuity equation $\hat{\tau}_{33} = -\hat{\tau}_{\gamma\gamma}$, and noting that $(\partial_{11}^2\hat{\tau}_{\alpha3}, \partial_{12}^2\hat{\tau}_{\alpha3}, \partial_{22}^2\hat{\tau}_{\alpha3}) \ll \partial_{33}^2\hat{\tau}_{\alpha3}$ in the long-wavelength approximation, we obtain

$$\partial_{33}^2\hat{\tau}_{\alpha3} + \partial_3\hat{A}_\alpha = -\rho_0\alpha g\partial_\alpha\hat{T}, \quad \hat{A}_\alpha = \partial_\alpha\hat{\tau}_{\gamma\gamma} + \partial_\gamma\hat{\tau}_{\alpha\gamma} \quad [187]$$

Physically, [187] are the lateral ($\alpha = 1$ or 2) components of the vorticity equation. Now multiply [187] by $x_3$, integrate across the BL from $x_3 = 0$ to $x_3 = b$, and take lateral derivatives outside the integral signs by neglecting the small lateral variation of the upper limit $b(x_1, x_2, t)$. The result is

$$-\hat{\tau}_{\alpha3}(b) + b\partial_3\hat{\tau}_{\alpha3}(b) + b\hat{A}_\alpha(b) - \langle\hat{A}_\alpha\rangle = \rho_0\alpha g\partial_\alpha M$$
$$[188]$$

where $\langle\rangle = \int_0^b \mathrm{d}x_3$ and

$$M = -\langle x_3\hat{T}\rangle \quad [189]$$

is the temperature moment. Now $b\partial_3\hat{\tau}_{\alpha3}(b) \ll \hat{\tau}_{\alpha3}(b)$ in the long-wave limit, and continuity of shear stress at $x_3 = b$ requires $\hat{\tau}_{\alpha3}(b) = \tau_{\alpha3}(b)$. But because the interior fluid sees the BL as a skin with zero thickness, $\tau_{\alpha3}(b) \approx \tau_{\alpha3}(0) \approx \eta_0\partial_3 u_\alpha(0)$. Moreover, the lateral velocity components are constant across the layer to lowest order and must match those in the interior fluid, requiring $\hat{u}_\alpha = u_\alpha(0)$ and $\hat{\tau}_{\alpha\beta} = \hat{\eta}(x_3)[\partial_\alpha u_\beta(0) + \partial_\beta u_\alpha(0)]$. Substituting these expressions into [188], we find

$$-\partial_3\mathbf{u} - \nabla\cdot[\sigma(2\mathbf{e} - \mathbf{I}\nabla\cdot\mathbf{u})] - \nabla(3\sigma\nabla\cdot\mathbf{u}) = \frac{\rho_0 g\alpha}{\eta_0}\nabla M \quad [190]$$

where

$$\sigma = \left\langle\frac{\hat{\eta}(x_3) - \eta_0}{\eta_0}\right\rangle \quad [191]$$

is a relative excess surface viscosity, $\mathbf{u}$ is the horizontal velocity vector, $\mathbf{e}$ is the 2-D strain-rate tensor, $\nabla$ is the 2-D gradient operator, $\mathbf{I}$ is the identity tensor, and all terms are evaluated at $x_3 = 0$. Equation [191] is an effective boundary condition that represents the influence of the BL on the interior fluid. It shows that the BL acts like an extensible skin with shear viscosity $\sigma\eta_0$ and compressional viscosity $3\sigma\eta_0$ that applies a shear stress proportional to $\nabla M$ to the outer fluid.

The next step is to determine explicitly the flow in the outer fluid that is driven by lateral variations in $M$ by solving the Stokes equations in the half-space $x_3 \geq 0$ subject to [190]. Evaluating the resulting solution at $x_3 = 0$, we obtain the closure relationship (LRS00, eqn [2.41])

$$\bar{\mathbf{u}}(\mathbf{k}) = \frac{i\mathbf{k}}{2k(1 + 2\sigma k)} \frac{\rho_0 g \alpha}{\eta_0} \bar{M}(\mathbf{k}) \qquad [192]$$

where $\mathbf{k}$ is the horizontal wave vector, $k = |\mathbf{k}|$, and overbars denote the Fourier transform. Equation [192] is only valid if $\sigma$ does not vary laterally. Finally, an evolution equation for $M$ is obtained by taking the first moment of the energy equation, yielding

$$\partial_t M + \mathbf{u} \cdot \nabla M + 2M\nabla \cdot \mathbf{u} = \kappa(\nabla^2 M + \Delta T) \qquad [193]$$

Equations [192] and [193] are three equations for $u_\alpha(x_1, x_2, t)$ and $M(x_1, x_2, t)$ that can be solved numerically subject to periodic boundary conditions in the lateral directions for a specified initial condition $M(x_1, x_2, 0)$.

An important special case of the above equations is the R–T instability of a layer with density $\rho_0 + \Delta\rho$ and viscosity $\eta_1 = \gamma\eta_0$, obtained by the transformation

$$M \rightarrow b^2\Delta\rho/2\rho_0\alpha, \quad \sigma \rightarrow b(\gamma - 1), \quad \kappa \rightarrow 0 \qquad [194]$$

A linear stability analysis of [192] and [193] can now be performed by setting $b = b_0 + \tilde{b}\exp(i\mathbf{k} \cdot \mathbf{x})\exp(st)$ and $\mathbf{u} = \bar{\mathbf{u}}\exp(i\mathbf{k} \cdot \mathbf{x})\exp(st)$ and linearizing in the perturbations $\tilde{b}$ and $\tilde{u}$. The resulting growth rate is $s/s_1 = \epsilon\gamma/2[1 + 2\epsilon(\gamma - 1)]$, where $\epsilon = b_0 k$ and $s_1 = g\Delta\rho b_0/\eta_1$, which agrees with the exact analytical expression [29] if $\gamma \gg \epsilon$. Equations [190]–[193] are therefore valid as long as the BL is not too much less viscous than the outer fluid.

The above equations also describe the finite-amplitude evolution of the R–T instability of a dense viscous layer over a passive half-space (Canright and Morris, 1993 ≡ CM93). Because the half-space is effectively inviscid ($\gamma \rightarrow \infty$), the closure law (192) is not meaningful, and the relevant equations are [190] and [193]. Rewriting these using [194] and noting that the first term in [190] (shear stress applied by the inviscid fluid) is negligible, we obtain

$$\nabla\left(\frac{g\Delta\rho}{4\eta_1}b^2 + b\nabla \cdot \mathbf{u}\right) + \nabla \cdot (b\mathbf{e}) = 0 \qquad [195]$$

$$\partial_t b + \nabla \cdot (b\mathbf{u}) = 0 \qquad [196]$$

which are just the dimensional forms of eqns [3.8] and [3.7], respectively, of CM93.

A remarkable feature of the eqns [192]–[193] and [195]–[196] is the existence of similarity solutions in which $M$ or $b$ becomes infinite at a finite time $t_s$, corresponding to the runaway escape of the plume from its source layer. The general form of the solution for $b$ or $M^{1/2}$ is

$$b \text{ or } M^{1/2} = (t_s - t)^a \text{fct}\left(\frac{x}{(t_s - t)^b}\right) \qquad [197]$$

where $x$ is the lateral or radial distance from the peak of the instability and $a$ and $b$ are exponents. CM93's solution of [195]–[196] gives $a = -1$ for a Newtonian fluid, whereas LRS00 solved [192]–[193] in the limit $\sigma \gg 1$ to find $a = -1/2$. The discrepancy appears to be due to the fact that LRS00 treated $\sigma$ as a constant, whereas $\sigma \propto b$ in the problem studied by CM93.

### 4.8.4 Theory of Thin Shells, Plates, and Sheets

A central problem in geodynamics is to determine the response of the lithosphere to applied loads such as seamounts, plate boundary forces (ridge push, slab pull, etc.), and tractions imposed by underlying mantle convection. Such problems can be solved effectively using thin-shell theory, a branch of applied mechanics concerned with the behavior of sheet-like objects whose thickness $b$ is much smaller than their typical radius of curvature $R$. This condition is evidently satisfied for the Earth's lithosphere, for which $b \approx 100\,\text{km}$ and $R \approx 6300\,\text{km}$. A further assumption of thin-shell theory is that the stresses within the shell vary laterally on a length scale $L \gg b$. Thin-shell theory therefore properly belongs to the general class of long-wave theories.

The basic idea of thin-shell theory is to exploit the smallness of $b/R$ to reduce the full 3-D dynamical equations to equivalent 2-D equations for the dynamics of the shell's mid-surface. Let the 3-D Cartesian coordinates of any point on this surface be $\mathbf{x}_0(\theta_1, \theta_2)$, where $\theta_\alpha$ are coordinates on the mid-surface itself. In the most general formulations of shell theory (e.g., Niordson, 1985), $\theta_\alpha$ are allowed to be arbitrary and nonorthogonal. Such a formulation is useful for problems involving large finite deformation, because $\theta_\alpha$ can be treated as Lagrangian coordinates. If the deformation is small, however, it makes sense to define $\theta_\alpha$ as orthogonal lines-of-curvature coordinates whose isolines are parallel to the two directions of principal curvature of the mid-surface at each point. This less elegant but more readily understandable formulation is the one used in most geodynamical applications of shell theory.

Relative to lines-of-curvature coordinates, the fundamental quantities that describe the shape of the mid-surface are the principal radii of curvature $R_\alpha$ and the Lamé parameters $A_\alpha = |\partial_\alpha \mathbf{x}_0|$, where $\partial_\alpha = \partial/\partial_{\theta_\alpha}$. For notational convenience, let $B_\alpha = 1/A_\alpha$ and $K_\alpha = 1/R_\alpha$. All vector and tensor quantities defined on the mid-surface are expressed relative to a local orthonormal basis comprising two surface-parallel unit vectors $\mathbf{d}_1$, $\mathbf{d}_2$ and a normal vector $\mathbf{d}_3$ defined by

$$\mathbf{d}_1 = B_1 \partial_1 \mathbf{x}_0, \quad \mathbf{d}_2 = B_2 \partial_2 \mathbf{x}_0, \quad \mathbf{d}_3 = \mathbf{d}_1 \times \mathbf{d}_2 \quad [198]$$

In the following, $z$ is a coordinate normal to the mid-surface $z = 0$. Moreover, the repeated subscript $\alpha$ is not summed unless explicitly indicated, and $\beta = 2$ when $\alpha = 1$ and vice versa.

The general equilibrium equations for a shell without inertia are (Novozhilov, 1959, p. 39)

$$\partial_\alpha \left( A_\beta T_\alpha \right) + \partial_\beta (A_\alpha S) + S \partial_\beta A_\alpha - T_\beta \partial_\alpha A_\beta$$
$$+ K_\alpha \left[ \partial_\alpha \left( A_\beta M_\alpha \right) - M_\beta \partial_\alpha A_\beta \right.$$
$$\left. + 2 \partial_\beta (A_\alpha H) + 2 R_\alpha K_\beta H \partial_\beta A_\alpha \right] = -A_1 A_2 P_\alpha \quad [199a]$$

$$\sum_{\alpha=1}^{2} \left\{ B_1 B_2 \partial_\alpha \left[ B_\alpha \left( \partial_\alpha \left( A_\beta M_\alpha \right) - M_\beta \partial_\alpha A_\beta \right. \right. \right.$$
$$\left. \left. \left. + \partial_\beta (A_\alpha H) + H \partial_\beta A_\alpha \right) \right] - K_\alpha T_\alpha \right\} = -P_3 \quad [199b]$$

where

$$T_\alpha = \int_{-b/2}^{b/2} \left(1 + K_\beta z\right) \sigma_{\alpha\alpha} \, dz,$$

$$S = \int_{-b/2}^{b/2} \left(1 - K_1 K_2 z^2\right) \sigma_{12} \, dz \quad [200a]$$

$$M_\alpha = \int_{-b/2}^{b/2} z \left(1 + K_\beta z\right) \sigma_{\alpha\alpha} \, dz,$$

$$H = \int_{-b/2}^{b/2} z [1 + (K_1 + K_2) z/2] \sigma_{12} \, dz \quad [200b]$$

$P_i$ is the total load vector (per unit midsurface area), and $\sigma_{\gamma\lambda}$ is the usual Cauchy stress tensor. The essential content of [199] is that a loaded shell can deform in two distinct ways: by in-plane stretching and shear, the intensity of which is measured by the stress resultants $T_\alpha$ and $S$, and by bending, which is measured by the bending moments $M_\alpha$ and $H$. In general both modes are present, in a proportion that depends in a complicated way on the mid-surface shape. Shells that deform only by extension and in-plane shear ($M_\alpha = H = 0$) are called membranes.

Equations [199] are valid for a shell of any material. To solve them, we need constitutive relations that link $T_\alpha$, $S$, $M_\alpha$, and $H$ to the displacement $v_i$ (for an elastic shell) or the velocity $u_i$ (for a fluid shell) of the mid-surface. For an elastic shell with Young's modulus $E$ and Poisson's ratio $\sigma$, these are (Novozhilov, 1959, pp. 24, 48)

$$T_\alpha = \frac{Eb}{1-\sigma^2} \left(\epsilon_\alpha + \sigma \epsilon_\beta\right),$$
$$M_\alpha \frac{Eb^3}{12(1-\sigma^2)} \left(\kappa_\alpha + \sigma \kappa_\beta\right) \quad [201a]$$

$$S = \frac{Eb}{2(1+\sigma)} \omega, \quad H = \frac{Eb^3}{12(1+\sigma)} \tau \quad [201b]$$

where

$$\epsilon_\alpha = B_\alpha \partial_\alpha v_a + B_1 B_2 v_\beta \partial_\beta A_\alpha + K_\alpha v_3 \quad [202a]$$

$$\omega = \sum_{\alpha=1}^{2} A_\beta B_\alpha \partial_\alpha \left(B_\beta v_\beta\right) \quad [202b]$$

$$\kappa_\alpha = -B_\alpha \partial_\alpha \left(B_\alpha \partial_\alpha v_3 - K_\alpha v_\alpha\right)$$
$$- B_1 B_2 \partial_\beta A_\alpha \left(B_\beta \partial_\beta v_3 - K_\beta v_\beta\right) \quad [202c]$$

$$\tau = -B_1 B_2 \partial_{12}^2 v_3 + \sum_{\alpha=1}^{2} \left\{ B_1 B_2 B_\alpha \left(\partial_\beta A_\alpha\right) \partial_\alpha v_3 \right.$$
$$\left. + K_\alpha \left(B_\beta \partial_\beta v_\alpha - B_1 B_2 u_3 \partial_\beta A_\alpha\right) \right\} \quad [202d]$$

are the six independent quantities that describe the deformation of the mid-surface: the elongations $\epsilon_\alpha$ and the changes of curvature $\kappa_\alpha$ in the two coordinate directions, the in-plane shear deformation $\omega$, and the torsional (twist) deformation $\tau$. The analogous expressions for a fluid shell are obtained by applying the transformation [135] to [201] and [202].

The above equations include all the special cases commonly considered in geodynamics. The first is that of a flat ($R_1 = R_2 = \infty$, $A_1 = A_2 = 1$) elastic plate with constant thickness $b$. The (uncoupled) equations governing the flexural and membrane modes are (Landau and Lifshitz, 1986)

$$\frac{Eb^3}{12(1-\sigma^2)} \nabla_1^4 v_3 = P_3 \quad [203a]$$

$$\frac{Eb}{2(1-\sigma^2)} \left[(1-\sigma)\nabla^2 \mathbf{v} + (1+\sigma)\nabla(\nabla \cdot \mathbf{v})\right] = -P_1 \mathbf{e}_1 - P_2 \mathbf{e}_2$$
$$[203b]$$

where $\nabla^2$, $\nabla$, and $\mathbf{v}$ are the 2D (in-plane) Laplacian operator, gradient operator, and displacement vector, respectively, $v_3$ is the normal displacement, and $\mathbf{P}$ is the 3D load vector. The quantity $Eb^3/12(1 - \sigma^2) \equiv D$ is called the flexural rigidity. Equation [203a] has been widely used to model the deformation of the lithosphere caused by topographic loading (e.g., Watts, 1978). McKenzie and Bowin (1976) generalized [203a] to a thick incompressible plate that need not be thin relative to the load wavelength.

A second important special case is that of a spherical shell with radius $R = R_1 = R_2$. If $\theta_1 = \theta$ (colatitude) and $\theta_2 = \phi$ (longitude), then $A_1 = R$ and $A_2 = R \sin \theta$. Turcotte (1974) used these equations in the membrane limit to estimate the magnitude of the stresses generated in the lithosphere when it moves relative to the Earth's equatorial bulge, and found that they may be large enough to cause propagating fractures. Tanimoto (1998) used the complete (membrane plus flexural) equations to estimate the displacement and state of stress in subducting lithosphere, and concluded that the state of stress is strongly influenced by the spherical geometry.

A third case is pure membrane flow in a flat fluid sheet. A well-known application is the thin-sheet model for continental deformation of England and McKenzie (1983), whose eqn, (16) is just the fluid version of [203b] with a power-law rheology of the form [31f] and an expression for $P_1$ and $P_2$ representing the lateral forces arising from variations in crustal thickness.

A final limiting case of interest is the finite-amplitude deformation of a 2-D Newtonian fluid sheet. By symmetry, $\partial_2 = u_2 = K_2 = \epsilon_2 = \omega = \kappa_2 = \tau = S = H = 0$ for this case. If $\theta_1 \equiv s$ is the arclength along the mid-surface, then $A_1 = A_2 = 1$. The instantaneous mid-surface velocity $\mathbf{u}(s, t)$ produced by a given loading distribution $P_i(s, t)$ is governed by the fluid analogs of [199]–[202], while the evolution of the mid-surface position $\mathbf{x}_0(s, t)$ and the sheet thickness $b(s, t)$ are described by the kinematic equations

$$\mathcal{D}_t \mathbf{x}_0 = \mathbf{u}, \quad \mathcal{D}_t b = -b\epsilon_1 \qquad [204]$$

where $\mathcal{D}_t$ is a convective derivative that follows the motion of material points on the (stretching) mid-surface (Buckmaster et al., 1975.) Ribe (2003) used these equations to determine scaling laws for the periodic buckling instability of a viscous sheet falling onto a horizontal surface. Ribe et al. (2007) subsequently showed that these scaling laws predict well the anomalous apparent widening of subducted lithosphere imaged by seismic tomography at 700–1200 km depth beneath some subduction zones (e.g., Central America, Java).

## 4.8.5 Effective Boundary Conditions From Thin-Layer Flows

The interaction of a convective flow with a rheologically distinct lithosphere can be studied using a simple extension of thin-shell theory in which the shell's dynamics is reduced to an equivalent boundary condition on the underlying flow. For simplicity, we consider a flat fluid sheet with constant thickness $b$ and laterally variable viscosity $\hat{\eta}(x_1, x_2)$, overlying a mantle whose viscosity just below the sheet is $\bar{\eta}(x_1, x_2)$. In the following, superposed hats and bars denote quantities within the sheet and in the mantle just below it, respectively.

Suppose that the sheet deforms as a membrane (cf. Section 4.8.4), so that the lateral velocities $\hat{u}_\alpha$ and the pressure $\hat{p}$ are independent of the depth $x_3$, where $x_3 = 0$ and $x_3 = -b$ are the upper and lower surfaces of the sheet, respectively. Integrating the lateral force balance $\partial_\alpha \hat{\sigma}_{\alpha\beta} + \partial_3 \hat{\sigma}_{\beta3} = 0$ across the sheet subject to the free-surface condition $\hat{\sigma}_{\beta3}|_{z=0} = 0$, we obtain

$$b\partial_\alpha \left[ -\hat{p}\delta_{\alpha\beta} + 2\hat{\eta}\hat{e}_{\alpha\beta} \right] - \hat{\sigma}_{\beta3}|_{z=-b} = 0 \qquad [205]$$

where $e_{\alpha\beta} = (\partial_\alpha u_\beta + \partial_\beta u_\alpha)/2$ is the strain-rate tensor. Now continuity of the velocity and stress at $x_3 = -b$ requires $\hat{u}_i = \bar{u}_i$, $\hat{e}_{\alpha\beta} = \bar{e}_{\alpha\beta}$, $\hat{\sigma}_{\beta3}|_{z=-b} = \bar{\sigma}_{\beta3}$, and $-\hat{p} + 2\hat{\eta}\hat{e}_{33} \equiv -\hat{p} - 2\hat{\eta}\hat{e}_{\lambda\lambda} = \bar{\sigma}_{33}$, where the continuity equation $\hat{e}_{33} = -\hat{e}_{\lambda\lambda}$ has been used. Substituting these relations into [205], we obtain an effective boundary condition that involves only mantle (barred) variables and the known viscosity $\hat{\eta}$ of the sheet:

$$b^{-1}\bar{\sigma}_{\beta3} = 2\partial_\alpha \left[ \hat{\eta} \left( \bar{e}_{\alpha\beta} + \bar{e}_{\lambda\lambda}\delta_{\alpha\beta} \right) \right] + \partial_\beta\bar{\sigma}_{33} \qquad [206]$$

Equation [206] remains valid even if the sheet has a power-law rheology [31f], because $\hat{I} \approx (\hat{e}_{\alpha\beta}\hat{e}_{\alpha\beta} + \hat{e}_{\lambda\lambda}^2)^{1/2}$ can be written in terms of the mantle variables using the matching condition $\hat{e}_{\alpha\beta} = \bar{e}_{\alpha\beta}$.

In reality, the sheet thickness is governed by the conservation law $\partial_t b = -\partial_\alpha(b\hat{u}_\alpha)$, and will therefore not in general remain constant. However, this effect is ignored in most applications. Weinstein and Olson (1992) used the 1-D version of (206) with a power-law rheology for $\hat{\eta}$ to study the conditions under which a highly non-Newtonian sheet above a vigorously convecting Newtonian fluid exhibits plate-like behavior. Ribe (1992) used the spherical-coordinate analog of [206] to investigate the generation of a toroidal component of mantle flow by lateral viscosity variations in the lithosphere.

### 4.8.6 Solitary Waves

Another flow that can be studied using a long-wave approximation is the motion of finite-amplitude solitary waves in a two-fluid system. The potential importance of such waves in geodynamics was first demonstrated theoretically using equations describing the migration of a low-viscosity melt phase in a viscous porous matrix (Richter and McKenzie, 1984; Scott and Stevenson, 1984; see also Chapter 6) Scott *et al.* (1986) and Olson and Christensen (1986) subsequently showed that similar waves can exist in a cylindrical conduit of low-viscosity fluid embedded in a fluid of higher viscosity. These waves are examples of kinematic waves (Whitham, 1974), which occur when there is a functional relation between the density or amplitude of the medium and the flux of a conserved quantity like mass; other examples include the propagation of a pulse of wastewater down a gutter and the flow of traffic on a crowded highway.

To illustrate the basic principles, we follow here the derivation of Olson and Christensen (1986) for solitary waves in viscous conduits. Consider an infinite vertical conduit with a circular cross-section of radius $R(z, t)$ and area $A(z, t)$, where $z$ is the height and $t$ is time. The conduit contains fluid with density $\rho_L$ and viscosity $\eta_L$, and is embedded in an infinite fluid with density $\rho_M \equiv \rho_L + \Delta\rho$ and viscosity $\eta_M \gg \eta_L$. Conservation of mass in the conduit requires

$$A_t + Q_z = 0 \qquad [207]$$

where $Q(z, t)$ is the volume flux and subscripts indicate partial derivatives. Because $\eta_M \gg \eta_L$, the fluid in the conduit sees the wall as rigid. The volume flux is therefore given by Poiseuille's law

$$Q = -\frac{A^2}{8\pi\eta_L} P_z \qquad [208]$$

where the nonhydrostatic pressure $P$ in the conduit is determined by the requirement that the normal stress $\sigma_{rr}$ in the radial ($r$) direction be continuous at $r = R$. Now the pressure in the matrix is hydrostatic, and the deviatoric component of $\sigma_{rr}$ in the conduit is negligible relative to that in the matrix because $\eta_M \gg \eta_L$. Continuity of $\sigma_{rr}$ therefore requires $\rho_L g z - P = \rho_M g z + 2\eta_M u_r(R)$. However, because the flow in the matrix is dominantly radial, the continuity equation requires $u \propto r^{-1}$, whence $u_r(R) = -u(R)/R = -A_t/2A$ and

$$P = -\Delta\rho g z + \eta_M A^{-1} A_t \qquad [209]$$

Substituting [209] into [208] and using [207], we obtain

$$Q = \frac{A^2}{8\pi\eta_L}\left[\Delta\rho g + \eta_M\left(A^{-1}Q_z\right)_z\right] \qquad [210]$$

The nonlinear coupled equations [207] and [210] admit finite-amplitude traveling wave solutions wherein $A = A(z - ct)$ and $Q = Q(z - ct)$, where $c$ is the wave speed. Substitution of these forms into [207] and [210] yields a dispersion relation for $c$ as a function of the minimum and maximum amplitudes of the wave (Olson and Christensen, 1986, eqn, (22)). The most interesting case is that of an isolated solitary wave that propagates without change of shape along an otherwise uniform conduit with $A = A_0$ and $Q = \Delta\rho g A_0^2 / 8\pi\eta_L \equiv Q_0$. In the limit of large amplitude, the velocity of such a wave is

$$c = 2c_0\ln(A_{max}/A_0) \qquad [211]$$

where $c_0 \equiv Q_0/A_0$ is the average Poiseuille velocity in the conduit far from the wave and $A_{max}$ is the maximum cross-sectional area of the wave. The speed of the wave thus increases with its amplitude.

The mathematical properties of solitary waves in deformable porous media were further investigated by Barcilon and Richter (1986), who concluded that such waves are probably not solitons, that is, that they do not possess an infinite number of conservation laws. Whitehead and Helfrich (1986) showed that the equations describing solitary waves in both porous media and fluid conduits reduce to the Korteweg–de Vries equation in the limit of small amplitude. Geophysical applications of conduit solitary waves include Whitehead and Helfrich's (1988) suggestion that such waves might transport deep-mantle material rapidly to the surface with little diffusion or contamination. Schubert *et al.* (1989) subsequently showed numerically that solitary waves can also propagate along the conduits of thermal plumes in a fluid with temperature-dependent viscosity. Such solitary waves have been invoked to explain the origin of the 'V-shaped' topographic ridges on the ocean floor south of Iceland (Albers and Christensen, 2001; Ito, 2001).

## 4.9 Hydrodynamic Stability and Thermal Convection

Not every correct solution of the governing equations of fluid mechanics can exist in nature or in the laboratory. To be observable, the solution must also be stable, that is, able to maintain itself against the small disturbances or perturbations that are ubiquitous in any physical environment. Whether this is the case can be determined by linear stability analysis, wherein one solves the linearized equations that govern infinitesimal perturbations of a solution of interest (the basic state) to determine the conditions under which these perturbations grow (instability) or decay (stability). Typically, one expands the perturbations in normal modes that satisfy the equations and boundary conditions, and then determines which modes have a rate of exponential growth with a positive real part. However, linear stability analysis describes only the initial growth of perturbations, and is no longer valid when their amplitude is sufficiently large that nonlinear interactions between the modes become important. Various nonlinear stability methods have been developed to describe the dynamics of this stage.

Here we illustrate these methods for two geodynamically important instabilities: the R–T instability of a buoyant layer, and Rayleigh–Bénard convection between isothermal surfaces. The Section concludes with a discussion of how similar methods have been applied to thermal convection in more complex and realistic systems.

### 4.9.1   R–T Instability

We have already encountered the R–T instability in Section 4.4.3, where it served as an example of how intermediate asymptotic limits of a function can be identified and interpreted. We now outline the linear stability analysis that leads to the expression [28] for the growth rate of infinitesimal perturbations.

The model geometry is shown in **Figure 2(a)**. The flow in both fluids satisfies the Stokes equations $\nabla p = \eta \nabla^2 \mathbf{u}$, which must be solved subject to the appropriate boundary and matching conditions. Let the velocities in the two layers be $\mathbf{u}_n \equiv (u_n, v_n, w_n)$. The vanishing of the normal velocity and the shear stress at $z = -b_0$ requires

$$w_1(-b_0) = \partial_z u_1(-b_0) = \partial_z v_1(-b_0) = 0 \qquad [212]$$

where the arguments $x_1$, $x_2$, and $t$ of the variables have been suppressed. The velocity must vanish at $z = \infty$, which requires

$$u_2(\infty) = v_2(\infty) = w_2(\infty) = 0 \qquad [213]$$

Finally, continuity of the velocity and traction at the interface requires

$$[\mathbf{u} \cdot \mathbf{t}] = [\mathbf{u} \cdot \mathbf{n}] = [\mathbf{t} \cdot \boldsymbol{\sigma} \cdot \mathbf{n}] = [\mathbf{n} \cdot \boldsymbol{\sigma} \cdot \mathbf{n}] + g\Delta\rho\zeta = 0 \qquad [214]$$

where $[\ldots]$ denotes the jump in the enclosed quantity from fluid 1 to fluid 2 across the interface $z = \zeta$, $\sigma$ is the nonhydrostatic stress tensor,

$$\mathbf{n} = \left(1 + |\boldsymbol{\nabla}_1\zeta|^2\right)^{-1/2}(\mathbf{e}_z - \boldsymbol{\nabla}_1\zeta) \qquad [215]$$

is the unit vector normal to the interface, and $\mathbf{t}$ is any unit vector tangent to the interface. Because the hydrostatic pressure gradients $-\mathbf{e}_z g\rho_n$ in the two fluids are different, the nonhydrostatic part of the normal stress jumps by an amount $-g\Delta\rho\zeta$ across the interface even though the total normal stress is continuous there. The final relation required is the kinematic condition

$$\partial_t\zeta + \mathbf{u}(\zeta) \cdot \boldsymbol{\nabla}_1\zeta = w(\zeta) \qquad [216]$$

which expresses the fact that the interface $z = \zeta$ is a material surface.

Because the Stokes equations are linear, the nonlinearity of the problem resides entirely in the conditions [214] and [216]. To linearize them, we expand all field variables in Taylor series about the undisturbed position $z = 0$ of the interface and eliminate all terms of order $\zeta^2$ and higher, noting that $\mathbf{u} \propto \zeta$ because the flow is driven entirely by the buoyancy associated with the disturbance of the interface. The resulting linearized matching and kinematic conditions are

$$[u] = [v] = [w] = [\sigma_{xz}] = [\sigma_{yz}] = [\sigma_{zz}] + g\Delta\rho\zeta = 0 \qquad [217]$$

$$\partial_t\zeta = w(0) \qquad [218]$$

To lowest order in $\eta$, the flow is purely poloidal (Ribe, 1998) and can be described by two poloidal potentials $\mathcal{P}_n$ (one for each of the layers $n = 0$ and 1) that satisfy $\nabla_1^2\nabla^4\mathcal{P}_n = 0$ (see Section 4.5.2). Let the (exponentially growing) deformation of the interface be $\zeta = \zeta_0 f(x, y)e^{st}$, where $s$ is the growth rate and $f$ is a planform function satisfying $\nabla_1^2 f = -k^2 f$, where $k \equiv |\mathbf{k}|$ is the magnitude of the horizontal wave vector $\mathbf{k}$. Then the solutions for $\mathcal{P}_n$ must have the form

$$\mathcal{P}_n = P_n(z)f(x, y)e^{st} \qquad [219]$$

Substitution of [219] into [43], [212], [213], and [217] yields the two-point BVP

$$\left(D^2 - k^2\right)^2 P_n = 0 \qquad [220a]$$

$$P_1(-b_0) = D^2P_1(-b_0) = P_2(\infty) = DP_2(\infty) = 0 \quad [220b]$$

$$[P] = [DP] = \left[\eta(D^2 + k^2)P\right]$$
$$= \left[\eta D(3k^2 - D^2)P\right] - g\Delta\rho\zeta_0 = 0 \quad [220c]$$

where $D = \mathrm{d}/\mathrm{d}z$. The general solution of [220a] is

$$P_n = (A_n + B_n z)\mathrm{e}^{-kz} + (C_n + D_n z)\mathrm{e}^{kz} \quad [221]$$

Substitution of [221] into [220b] and [220c] yields eight algebraic equations which can be solved for the constants $A_n - D_n$. The growth rate is then determined from the transformed kinematic condition $\zeta_0 s = -k^2 P_1(0)$, yielding [28].

The R–T instability has been used to model a variety of mantle processes. The instability of a thin layer beneath an infinite fluid half-space was studied by Selig (1965) and Whitehead and Luther (1975), and used by the latter as a model for the initiation of mantle plumes. Lister and Kerr (1989a) studied analytically the instability of a rising horizontal cylinder of buoyant fluid, motivated in part by suggestions that the R–T instability in this geometry might explain the characteristic spacing of island-arc volcanoes (Marsh and Carmichael, 1974) and of volcanic centers along mid-ocean ridges (Whitehead *et al.*, 1984). They found that the growth rate $s$ and the most unstable wavelength are independent of the ambient fluid/cylinder viscosity ratio $\gamma$ when $\gamma \gg 1$, unlike flat layers for which $s \propto \gamma^{1/3}$. Canright and Morris (1993) performed a detailed scaling analysis of the instability for two layers of finite depth (see Section 4.4.3), and studied the nonlinear evolution of a Newtonian or power-law layer above an effectively inviscid half-space as a model for the initiation of subduction (see Section 4.8.3.) Ribe (1998) used a weakly nonlinear analysis to study planform selection and the direction of superexponential growth (spouting) in the R–T instability of a two-layer system. Finally, analytical solutions of the R–T instability with more complicated rheological and density structures have been used to model the delamination of the lowermost lithosphere (Conrad and Molnar, 1997; Houseman and Molnar, 1997; Molnar and Houseman, 2004).

### 4.9.2  Rayleigh–Bénard Convection

Rayleigh–Bénard (R–B) convection in a fluid layer is the paradigmatic case of a pattern-forming instability. The classic R–B configuration (**Figure 12**) comprises fluid with constant kinematic viscosity $\nu$ and thermal diffusivity $\kappa$ confined between horizontal planes $z =$

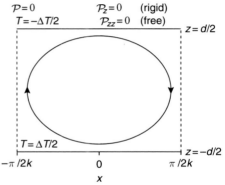

**Figure 12**  Geometry of Rayleigh–Bénard convection. Fluid with constant kinematic viscosity $\nu$, thermal diffusivity $\kappa$, and thermal expansivity $\alpha$ is confined between horizontal planes $z = \pm d/2$ held at temperatures $\mp\Delta T/2$. The characteristic horizontal wave number of the convection pattern is $k$. For the special case of 2-D rolls, the roll axis is parallel to the $y$- direction. The boundary conditions on the poloidal scalar $\mathcal{P}$ for free and rigid boundaries are indicated above the top boundary, and subscripts denote partial differentiation.

$\pm d/2$ held at temperatures $\mp\Delta T/2$. The fluid density depends on temperature as $\rho = \rho_0(1 - \alpha T)$, where $\alpha$ is the coefficient of thermal expansion. The planes $z = \pm d/2$ may be either rigid or traction-free, and $k$ is the characteristic horizontal wave number of the convection pattern. In the rest of this subsection, all variables will be nondimensionalized using $d$, $d^2/\kappa$, $\kappa/d$, and $\Delta T$ as scales for length, time, velocity, and temperature, respectively, and $\{x_1, x_2, x_3\} \equiv \{x, y, z\}$.

The basic state is a motionless ($u_i = 0$) layer with a linear (conductive) temperature profile $T = -z$. Because the viscosity is constant, buoyancy forces generate a purely poloidal flow

$$u_i = \mathcal{L}_i\mathcal{P}, \quad \mathcal{L}_i = \delta_{i3}\nabla^2 - \partial_i\partial_3 \quad [222]$$

where $\mathcal{P}$ is the poloidal potential defined by [41]. In the geodynamically relevant limit of negligible inertia, the equations satisfied by $\mathcal{P}$ and the perturbation $\theta$ of the conductive temperature profile are

$$\nabla^4\mathcal{P} = -Ra\,\theta \quad [223a]$$

$$\theta_t + \mathcal{L}_j\mathcal{P}\partial_j\theta - \nabla_1^2\mathcal{P} = \nabla^2\theta \quad [223b]$$

where $\nabla_1^2 = \partial_{11}^2 + \partial_{22}^2$ and

$$Ra = \frac{g\alpha d^3\Delta T}{\nu\kappa} \quad [224]$$

is the Rayleigh number. The boundary conditions at $z = \pm 1/2$ are obtained from [41] with $\mathcal{T} = 0$, noting that $\mathcal{P}$ is arbitrary to within an additive function of $z$

and/or an additive function $f(x, y)$ satisfying $\nabla_1^2 f = 0$. Impermeability requires $\nabla_1^2 \mathcal{P} = 0$, which can be replaced by the simpler condition $\mathcal{P} = 0$. The no-slip condition at a rigid surface requires $\partial_{13}^2 \mathcal{P} = \partial_{23}^2 \mathcal{P} = 0$, which can be replaced by $\partial_3 \mathcal{P} = 0$. Finally, the vanishing of the tangential stress at a free surface requires $\partial_1 (\nabla_1^2 - \partial_{33}^2) \mathcal{P} = \partial_2 (\nabla_1^2 - \partial_{33}^2) \mathcal{P} = 0$, which can be replaced by $\partial_{33}^2 \mathcal{P} = 0$. In summary, the boundary conditions are

$$\theta = \mathcal{P} = \partial_{33}^2 \mathcal{P} = 0 \text{ (free)} \qquad [225a]$$

or

$$\theta = \mathcal{P} = \partial_3 \mathcal{P} = 0 \text{ (rigid)} \qquad [225b]$$

#### 4.9.2.1 Linear stability analysis

The initial/BVP describing the evolution of small perturbations to the basic state is obtained by linearizing [223] about $(\mathcal{P}, \theta) = (0, 0)$, reducing the resulting equations to a single equation for $\mathcal{P}$ by cross-differentiation, and recasting the boundary conditions $\theta(\pm 1/2) = 0$ in terms of $\mathcal{P}$ with the help of [223a]. For the (analytically simpler) free-boundary case, the result is

$$\left[ \nabla^4 (\nabla^2 - \partial_t) - Ra \nabla_1^2 \right] \mathcal{P} = 0 \qquad [226a]$$

$$\mathcal{P} = \partial_{33}^2 \mathcal{P} = \partial_{3333}^4 \mathcal{P} = 0 \text{ at } z = \pm 1/2 \qquad [226b]$$

Equations [226] admit normal mode solutions of the form

$$\mathcal{P} = P(z) f(x, y) e^{st} \qquad [227]$$

where $\sigma$ is the growth rate and $f(x,y)$ is the planform function satisfying $\nabla_1^2 f = -k^2 f$. Substituting [227] into [226], we obtain

$$\left[ (D^2 - k^2)^2 (D^2 - k^2 - s) + Ra k^2 \right] P = 0 \qquad [228a]$$

$$P = D^2 P = D^4 P = 0 \text{ at } z = \pm 1/2 \qquad [228b]$$

where $D = \mathrm{d}/\mathrm{d}z$. Equations [228] define an eigenvalue problem whose solution is

$$P = \sin n\pi \left( z + \frac{1}{2} \right), \quad s = \frac{Ra\, k^2}{(n^2 \pi^2 + k^2)^2} - n^2 \pi^2 - k^2 \qquad [229]$$

where the index $n$ defines the vertical wavelength of the mode. The growth rate $s$ becomes positive when $Ra$ exceeds a value $Ra_0(k)$ that corresponds to marginal stability. This occurs first for the mode $n = 1$, for which

$$Ra_0 = \frac{(\pi^2 + k^2)^3}{k^2} \qquad [230]$$

The most unstable wave number $k_c$ and the corresponding critical Rayleigh number $Ra_c$ are found by minimizing $Ra_0(k)$, yielding

$$Ra_c = \frac{27\pi^4}{4} \approx 657.5, \quad k_c = \frac{\pi}{\sqrt{2}} \approx 2.22 \qquad [231]$$

The corresponding results for convection between rigid surfaces must be obtained numerically (Chandrasekhar, 1981, pp. 36–42), and are $Ra_c \approx 1707.8$, $k_c \approx 3.117$. The marginally stable Rayleigh number $Ra_0(k)$ is shown for both free and rigid surfaces in **Figure 13(a)**. For future reference, we note that [229] can be written

$$s = (\pi^2 + k^2) \frac{Ra - Ra_0}{Ra_0} \qquad [232]$$

#### 4.9.2.2 Order-parameter equations for finite-amplitude thermal convection

Linear stability analysis predicts the initial growth rate of a normal mode with wave number **k**. Typically, however, many different modes have the same or nearly the same growth rate: examples include convection rolls with the same wave number but different orientations, and rolls with the same orientation but slightly different wave numbers within a narrow band around the most unstable wave number $k_c$. The question therefore arises: among a set of modes with equal (or nearly equal) growth rates, which mode or combination of modes is actually realized for a given (supercritical) Rayleigh number? The answer depends on a combination of two factors: the nonlinear coupling between different modes due to the nonlinear terms in the governing equations, and external biases imposed by the initial and/or boundary conditions. Linear stability analysis is powerless to help us here, and more complicated nonlinear theories are required.

An especially powerful method of this type is to reduce the full 3-D equations governing convection to 2-D equations for one or more order parameters that describe the degree of order or patterning in the system. Newell *et al.* (1993) identified four distinct classes of such equations, depending on whether the degree of supercriticality $\epsilon \equiv [(Ra - Ra_c)/Ra_c]^{1/2}$ is small or of order unity, and whether the horizontal spectrum of the allowable modes is discrete or (quasi-)continuous (**Table 4**). The spectrum is discrete for convection in an infinite layer with a periodic planform characterized by a single fundamental wave number $k$, and also for a layer of finite extent $L = O(d)$ when only a few of the allowable wave

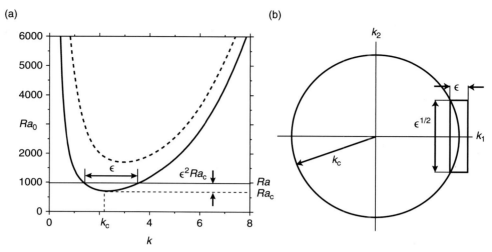

**Figure 13** (a) Marginally stable Rayleigh number $Ra_0$ as a function of wave number $k$ for Rayleigh–Bénard convection between traction-free (solid line) and rigid (dashed line) surfaces. The critical Rayleigh number $Ra_c$ and wave number $k_c$ are indicated for the free-surface case. For a slightly super-critical Rayleigh number $Ra = (1 + \epsilon^2)Ra_c$, a band of wave numbers of width $\sim \epsilon$ is unstable. (b) The most unstable wave vector $\mathbf{k} = (k_1, k_2)$ for Rayleigh–Bénard convection lies on a circle of radius $k_c$. Changing $|\mathbf{k}|$ by an amount $\sim \epsilon$ in the vicinity of the wave vector $(k_c, 0)$ for rolls with axes parallel to the $x_2$-direction corresponds to changing $k_1$ and $k_2$ by amounts $\sim \epsilon$ and $\sim \epsilon^{1/2}$, respectively (rectangle).

**Table 4** Order-parameter equations for Rayleigh–Bénard convection

| Spectrum | $\epsilon \ll 1$ | $\epsilon = O(1)$ |
|---|---|---|
| Discrete | amplitude equations | finite-amplitude convection rolls |
| Continuous | envelope equations | slowly modulated patterns |

numbers $2n\pi/L$ are unstable for a given Rayleigh number. The wave numbers of the unstable modes become ever more closely spaced as $L$ increases, until in the limit $L/d \gg 1$ they can be regarded as forming a quasi-continuous spectrum.

Historically, the first case to be studied (Malkus and Veronis, 1958) was that of a single mode with $\epsilon \ll 1$. Here the order parameter is the amplitude $A(t)$ of the dominant mode, whose temporal evolution is described by a nonlinear Landau equation (Landau, 1944). Subsequently, these weakly nonlinear results were extended to $\epsilon = O(1)$ by Busse (1967a), who used a Galerkin method to obtain numerical solutions for convection rolls and to examine their stability. The more complicated case of a continuous spectrum was first studied by Segel (1969) and Newell and Whitehead (1969), who derived the evolution equation governing the slowly varying (in time and space) amplitude envelope $A(\epsilon x, \epsilon^{1/2}y, \epsilon^2 t)$ of weakly nonlinear ($\epsilon \ll 1$) convection with modes contained in a

narrow band surrounding the critical wave vector $\mathbf{k}_c = (k_c, 0)$ for straight parallel rolls. Finally, Newell *et al.* (1990) extended these results to $\epsilon = O(1)$ by deriving the phase diffusion equation that governs the slowly varying phase $\Phi$ of a convection pattern that locally has the form of straight parallel rolls. We now examine each of these four cases in turn.

#### 4.9.2.3 Amplitude equation for convection rolls

Consider convection between traction-free boundaries in the form of straight rolls with axes parallel to $\mathbf{e}_y$. The equations and boundary conditions satisfied by $\mathcal{P}$ and $\theta$ are [223], where $\mathcal{P}$ and $\theta$ are independent of $y$ and $\partial_2 \equiv 0$.

The basic idea of weakly nonlinear analysis is to expand the dependent variables in powers of a small parameter $\epsilon$ that measures the degree of supercriticality. By substituting these expansions into the governing equations and gathering together the terms proportional to different powers of $\epsilon$, one reduces the original nonlinear problem to a sequence of linear (but inhomogeneous) problems that can be solved sequentially. Thus we write

$$\mathcal{P} = \sum_{n=1}^{\infty} \epsilon^n \mathcal{P}_n, \quad \theta = \sum_{n=1}^{\infty} \epsilon^n \theta_n,$$

$$Ra = Ra_0 + \sum_{n=1}^{\infty} \epsilon^n Ra_n \qquad [233]$$

where the last expansion can be regarded as an implicit definition of the supercriticality $\epsilon(Ra)$. Now because the boundary conditions are the same on both surfaces, the hot and cold portions of the flow are mirror images of each other. The problem is therefore invariant under the transformation $\{\mathcal{P} \rightarrow -\mathcal{P}, \theta \rightarrow -\theta\}$ or (equivalently) $\epsilon \rightarrow -\epsilon$. Application of the latter transformation to the expanded form of [223a] shows that $Ra_n = 0$ for all odd $n$. To lowest order, therefore,

$$Ra - Ra_0 \approx \epsilon^2 Ra_2 \qquad [234]$$

where $Ra_2 = O(1)$ is to be determined. Now comparison of [234] with [232] shows that $s \propto \epsilon^2$. We therefore introduce a slow time $T \equiv \epsilon^2 t$ which is of order unity during the initial stage of exponential growth, whence

$$\partial_t = \epsilon^2 \partial_T \qquad [235]$$

Substituting [233], [234], and [235] into [223] and collecting terms proportional to $\epsilon$, we obtain

$$\nabla^4 \mathcal{P}_1 + Ra_0 \theta_1 = 0, \quad \nabla^2 \theta_1 + \partial_{11}^2 \mathcal{P}_1 = 0 \qquad [236]$$

subject to the boundary conditions $\mathcal{P}_1 = \partial_{33}^2 \mathcal{P}_1 = \theta_1 = 0$ at $z = \pm 1/2$. This is just the linear stability problem, for which the solution is

$$\mathcal{P}_1 = -\frac{C}{k^2} \cos \pi z \cos kx, \quad \theta_1 = \frac{C}{\pi^2 + k^2} \cos \pi z \cos kx,$$
$$Ra_0 = \frac{(\pi^2 + k^2)^3}{k^2} \qquad [237]$$

where the amplitude $C(T)$ of the vertical velocity $w_1 \equiv \partial_{11}^2 \mathcal{P}_1$ remains to be determined. Next, we collect terms proportional to $\epsilon^2$ to obtain

$$\nabla^4 \mathcal{P}_2 + Ra_0 \theta_2 = -Ra_1 \theta_1 \equiv 0 \qquad [238a]$$

$$\nabla^2 \theta_2 + \partial_{11}^2 \mathcal{P}_2 = \mathcal{L}_j \mathcal{P}_1 \partial_j \theta_1 \equiv -\frac{C^2 \pi}{2(\pi^2 + k^2)} \sin 2\pi z \qquad [238b]$$

Now because the inhomogeneous term in [238b] depends only on $z$, we must seek solutions of [238] of the forms

$$\mathcal{P}_2 = \tilde{\mathcal{P}}_2(x, z), \quad \theta_2 = \tilde{\theta}_2(x, z) + \bar{\theta}_2(z) \qquad [239]$$

where $\tilde{\mathcal{P}}_2$ and $\tilde{\theta}_2$ are the fluctuating (periodic in $x$) parts of the solution and $\bar{\theta}_2$ is the mean ($x$-independent) part. The mean part of $\mathcal{P}_2$ is set to zero because the poloidal scalar is arbitrary to within an additive function of $z$, as can be verified by inspection of [41]. Now the equations satisfied by $\tilde{\mathcal{P}}_2$ and $\tilde{\theta}_2$ are just the homogeneous forms of [238], which are identical to the order $\epsilon$ equations [236]. We may therefore set

$\tilde{\mathcal{P}}_2 = \tilde{\theta}_2 = 0$ with no loss of generality. The solution for $\bar{\theta}_2$ is then obtained by integrating [238b] subject to the boundary conditions $\bar{\theta}_2(\pm 1/2) = 0$. We thereby find

$$\mathcal{P}_2 = 0, \quad \theta_2 \equiv \bar{\theta}_2 = \frac{C^2}{8\pi(\pi^2 + k^2)} \sin 2\pi z \qquad [240]$$

Physically, $\bar{\theta}_2(z)$ describes the average heating (cooling) of the upper (lower) half of the layer that is induced by the convection. Despite appearances, [240] is consistent with [238a] because $\mathcal{P}_2$ is arbitrary to within an additive function of $z$.

The parameter $Ra_2$ is still not determined, so we must proceed to order $\epsilon^3$, for which the equations are

$$\nabla^4 \mathcal{P}_3 + Ra_0 \theta_3 = -Ra_2 \theta_1 \qquad [241a]$$

$$\nabla^2 \theta_3 + \partial_{11}^2 \mathcal{P}_3 = \partial_T \theta_1 + \mathcal{L}_j \mathcal{P}_1 \partial_j \theta_2 \qquad [241b]$$

We now evaluate the RHSs of [241], set $\mathcal{P}_3 = \tilde{\mathcal{P}}_3(z, T) \cos kx$ and $\theta_3 = \tilde{\theta}_3(z, T) \cos kx$, and reduce the resulting equations to a single equation for $\tilde{\mathcal{P}}_3$, obtaining

$$\left[ (D^2 - k^2)^3 + k^2 Ra_0 \right] \tilde{\mathcal{P}}_3$$
$$= -\left[ \frac{Ra_0 (C^3 \cos 2\pi z + 4\dot{C})}{4(\pi^2 + k^2)} - Ra_2 C \right] \cos \pi z \qquad [242]$$

where $D = d/dz$ and a superposed dot denotes $d/dT$. Now the homogeneous form of [242] is identical to the eigenvalue problem at order $\epsilon$, and thus has a solution $\propto \cos \pi z$ that satisfies the boundary conditions. The Fredholm alternative theorem then implies that the inhomogeneous equation [242] will have a solution only if the RHS satisfies a solvability condition. In the general case, this condition is found by multiplying the inhomogeneous equation by the solution of the homogeneous adjoint problem and then integrating over the domain of the dependent variable. For our (self-adjoint) problem, the homogeneous adjoint solution is $\propto \cos \pi z$, and the procedure just described yields

$$\int_{-1/2}^{1/2} \left[ (D^2 - k^2)^3 + k^2 Ra_0 \right] \tilde{\mathcal{P}}_3 \cos \pi z \, dz$$
$$= -\frac{\left[ Ra_0 (C^3 + 8\dot{C}) - 8(\pi^2 + k^2) Ra_2 C \right]}{16(\pi^2 + k^2)} \qquad [243]$$

Now the LHS of [243] is identically zero, as one can easily show by integrating repeatedly by parts and applying the boundary conditions on $\tilde{\mathcal{P}}_3$. The RHS of [243] must therefore vanish. Now by [233] and [237], the amplitude of the vertical velocity is $\epsilon C \equiv W$. Because only the product of $\epsilon$ and $C$ is physically meaningful, the definition of $\epsilon$ itself – or, equivalently, of $Ra_2$ – is arbitrary. Making the most convenient choice $Ra_2 = Ra_0$, we find that the vanishing of the RHS of [243] yields the following evolution equation for the physical amplitude $W$:

$$\frac{dW}{dT} = (\pi^2 + k^2) W - \frac{Ra_0}{8(Ra - Ra_0)} W^3 \qquad [244]$$

Initially, $W$ grows exponentially with a growth rate $W^{-1}\dot{W} = \pi^2 + k^2$. At long times, however, $\dot{W} \to 0$ and the amplitude approaches a steady value

$$W(T \to \infty) = \left[\frac{8(\pi^2 + k^2)(Ra - Ra_0)}{Ra_0}\right]^{1/2} \qquad [245]$$

### 4.9.2.4 Finite-amplitude convection rolls and their stability

The extension of the above results to strongly nonlinear rolls ($\epsilon = O(1)$) between rigid surfaces is due to Busse (1967a), whose development we follow here with some changes of notation. The first step is to determine steady roll solutions of [223] using a Galerkin method (Fletcher, 1984) whereby $\theta$ and $\mathcal{P}$ are expanded in orthogonal functions that satisfy the boundary conditions and the coefficients are then chosen to satisfy approximately the governing equations.

We begin by expanding $\theta$ into a complete set of Fourier modes that satisfy the boundary conditions $\theta(x, \pm 1/2) = 0$:

$$\theta = \sum_{m,n} c_{mn} e^{imkx} f_n(z), \quad f_n(z) = \sin n\pi\left(z + \frac{1}{2}\right) \qquad [246]$$

where $2\pi/k$ is the wavelength of the convection rolls, $c_{mn} = \bar{c}_{n-m}$, where the overbar denotes complex conjugation, and the summations are over $-\infty \le m \le \infty$ and $1 \le n \le \infty$. Now substitute [246] into [223a] and solve the resulting equation to obtain

$$\mathcal{P} = -Ra \sum_{m,n} c_{mn} e^{imkx} Q_n(mk, z) \qquad [247a]$$

$$Q_n(r, z) = \frac{f_n(z) + n\pi b_n(r)}{(r^2 + n^2\pi^2)^2} \qquad [247b]$$

$$b_n(r) = \begin{cases} (2SC + r)^{-1}(2zC \sin h\, rz - S\cosh rz) & \text{(n odd)} \\ (2SC - r)^{-1}(C \sin h\, rz - 2zS\cosh rz) & \text{(n even)} \end{cases} \qquad [247c]$$

$$S = \sinh\frac{r}{2}, \quad C = \cosh\frac{r}{2} \qquad [247d]$$

To determine the coefficients $c_{mn}$, we substitute [247] into [223b], multiply the result by $f_q(z)e^{-ipkx}$, and integrate over the fluid layer. By using for $p$ and $q$

all integers in the ranges of $m$ and $n$, respectively, we obtain an infinite set of nonlinear algebraic equations for $c_{mn}(Ra)$ that can be truncated and solved numerically (Busse, 1967a.)

Another advantage of the Galerkin method is that a stability analysis of the solution can easily be performed. The linearized equations governing infinitesimal perturbations $(\tilde{\mathcal{P}}, \tilde{\theta})$ to the steady roll solution $(\mathcal{P}_0, \theta_0)$ are

$$\nabla^4 \tilde{\mathcal{P}} = -Ra\tilde{\theta} \qquad [248a]$$

$$\tilde{\theta}_t + \mathcal{L}_j \mathcal{P}_0 \partial_j \tilde{\theta} + \partial_j \theta_0 \mathcal{L}_j \tilde{\mathcal{P}} - \nabla_1^2 \tilde{\mathcal{P}} = \nabla^2 \tilde{\theta} \qquad [248b]$$

subject to $\tilde{\theta} = \tilde{\mathcal{P}} = \partial_3 \tilde{\mathcal{P}} = 0$ at $z = \pm 1/2$. Now [248] are linear PDEs with coefficients that are periodic in the $x$-direction but independent of both $z$ and $t$. The general solution can be written as a sum of solutions which depend exponentially on $x$, $y$, and $t$, multiplied by a function of $x$ having the same periodicity as the stationary solution (Busse, 1967a.) We therefore write

$$\tilde{\theta} = \left\{\sum_{m,n} \tilde{c}_{mn} e^{imkx} f_n(z)\right\} e^{i(ax+by)+st} \qquad [249]$$

and note further that [248a] and the boundary conditions on $\mathcal{P}$ are satisfied exactly if

$$\tilde{\mathcal{P}} = -Ra\left\{\sum_{m,n} \tilde{c}_{mn} e^{imkx} Q_n\left(\sqrt{(mk+a)^2 + b^2}, z\right)\right\}$$
$$\times e^{i(ax+by)+st} \qquad [250]$$

Substituting [249] and [250] into [248b], multiplying by $f_q(z)e^{-i(pkx+ax+by)-st}$, and averaging over the fluid layer, we obtain an infinite system of linear equations for the coefficients $\tilde{c}_{mn}$. The system is then truncated and the largest eigenvalue $s(Ra, k, a, b)$ is determined numerically, assuming that $s$ is real in view of the fact that $\Im(s) = 0$ at the onset of convection (Busse, 1967a).

The results show that convection rolls are stable ($s < 0$) in an elongate region of the $(Ra, k)$ space often called the Busse balloon (**Figure 14**). Above $Ra \approx 22600$, convection rolls of any wave number are unstable. The lower left edge of the balloon (denoted by $Z$ in **Figure 14**) represents the onset of a zigzag instability in the form of rolls oblique to the original ones. Around the rest of the balloon (denoted by $C$), the stability of rolls is limited by the crossroll instability, which grows in the form of perpendicularly aligned rolls. The results of Busse (1967a) were extended to the case of traction-free boundaries by Straus (1972), Busse and Bolton (1984), and Bolton and Busse (1985).

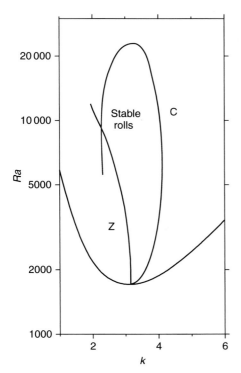

**Figure 14** Region of stability in the Rayleigh number/wave number plane of 2-D convection rolls between rigid isothermal surfaces (Busse balloon). Portions of the boundary of the balloon labeled $Z$ and $C$ correspond to the onset of the zigzag and cross-roll instabilities, respectively.

### 4.9.2.5 Envelope equation for modulated convection rolls

The analyses in Sections 4.9.2.3 and 4.9.2.4 assume convection in the form of straight parallel rolls with a single dominant wave number $k$. In the laboratory, however, rolls often exhibit a more irregular pattern in which both the magnitude and direction of the wave vector $\mathbf{k} = (k_1, k_2)$ vary slowly as functions of time and position. To describe this behavior, Newell and Whitehead (1969) and Segel (1969) derived an envelope equation for modulated convection rolls whose wave vectors form a continuous spectrum within a narrow band centered on the wave vector $(k_c, 0)$ for straight parallel rolls. The envelope equation is derived via a multiscale expansion that accounts in a self-consistent way for both the fast and slow variations of the flow field in time and space. Richter (1973) applied a similar method to convection modulated by long-wavelength variations in the boundary temperatures.

The first step is to determine the scales over which the slow (i.e., long-wavelength) spatial variations occur. Because the marginal stability curve $Ra_0(k)$ is a parabola in the vicinity of its minimum $(k, Ra_0) = (k_c, Ra_c)$, the wave numbers $|\mathbf{k}|$ that become unstable when $Ra$ exceeds $Ra_c$ by an amount $\epsilon^2 Ra_c$ comprise a continuous band of width $\sim \epsilon$ centered on $k_c$ (**Figure 13(a)**). However, the orientation of the rolls, measured by the ratio $k_2/k_1$, may also vary. Now the most unstable wave vector lies on a circle of radius $k_c$ (**Figure 13(b)**). Therefore if we change $|\mathbf{k}|$ by an amount $\sim \epsilon$ in the vicinity of the straight-roll wave vector $(k_c, 0)$, the (maximum) corresponding changes of $k_1$ and $k_2$ are $\sim \epsilon$ and $\sim \epsilon^{1/2}$, respectively. The appropriate slow variables are therefore

$$X = \epsilon x, \quad Y = \epsilon^{1/2} y, \quad T = \epsilon^2 t \qquad [251]$$

where the expression for $T$ derives from the argument preceeding [235].

The essence of the multiscale procedure is to treat the flow fields (here, $\mathcal{P}$ and $\theta$) as functions of both the fast variables $(x, y, t)$ and the slow variables $(X, Y, T)$. Accordingly, the asymptotic expansions analogous to [233] are

$$\mathcal{P} = \sum_{n=1}^{\infty} \epsilon^n \mathcal{P}_n(X, Y, T, x, z)$$
$$\theta = \sum_{n=1}^{\infty} \epsilon^n \theta_n(X, Y, T, x, z) \qquad [252]$$

Because the solution for steady straight rolls is independent of $y$ and $t$, $\mathcal{P}_n$ and $\theta_n$ depend on these variables only through the variables $Y$ and $T$ that measure the slow modulation of the pattern. By the chain rule, derivatives of the expansions [252] with respect to the fast variables transform as

$$\partial_t \rightarrow \epsilon^2 \partial_T, \quad \partial_x \rightarrow \partial_x + \epsilon \partial_X,$$
$$\partial_y \rightarrow \epsilon^{1/2} \partial_Y, \quad \partial_z \rightarrow \partial_z \qquad [253]$$

We now substitute [252] into the governing equations [223] and collect terms proportional to like powers of $\epsilon$, just as in Section 4.9.2.3. The solutions at order $\epsilon$ analogous to [237] are

$$\mathcal{P}_1 = -\frac{1}{2k_c^2} \left[ C e^{ik_c x} + \bar{C} e^{-ik_c x} \right] \cos \pi z \qquad [254a]$$

$$\theta_1 = \frac{1}{2(\pi^2 + k_c^2)} \left[ C e^{ik_c x} + \bar{C} e^{-ik_c x} \right] \cos \pi z \qquad [254b]$$

where $C = C(X, Y, T)$ is the slowly varying envelope of the roll solution and overbars denotes complex conjugation. The solution at order $\epsilon^2$ analogous to [240] is

$$\mathcal{P}_2 = 0, \quad \theta_2 = \frac{C\bar{C}}{8\pi(\pi^2 + k_c^2)} \sin 2\pi z \quad [255]$$

The temperature $\theta_2$ also contains free modes proportional to $\partial_X(C, \bar{C})$ and $\partial_{YY}(C, \bar{C})$, but these do not change the solvability condition at order $\epsilon^3$ and can therefore be neglected. Evaluating this solvability condition as in Section 4.9.2.3 and setting $k_c = \pi/\sqrt{2}$, we obtain the Newell–Whitehead–Segel equation for the envelope $W \equiv \epsilon C$ of the vertical velocity field:

$$\frac{\partial W}{\partial T} - \frac{4}{3}\left(\frac{\partial}{\partial X} - \frac{i}{\sqrt{2\pi}}\frac{\partial^2}{\partial Y^2}\right)^2 W$$
$$= \frac{3\pi^2}{2} W - \frac{Ra_c}{8(Ra - Ra_c)}|W|^2 W \quad [256]$$

Equation [256] differs from [244] by the addition of a diffusion-like term on the LHS, which represents the interaction of neighboring rolls via the buoyant torques they apply to each other. Equations generalizing [256] to $N$ interacting wavepackets are given by Newell and Whitehead (1969).

### 4.9.2.6 Phase diffusion equation for thermal convection

The final case is that of large-amplitude ($\epsilon = O(1)$) convection in the form of modulated rolls, which can be described by a phase diffusion equation. The derivation below follows Newell *et al.* (1990) ≡ NPS90.

The basic assumption is that the flow field consists locally of straight parallel rolls whose wave vector varies slowly over the fluid layer. This wave vector can be written as $\mathbf{k} \equiv \nabla\phi$, where $\phi$ is the phase of the roll pattern. The phase is the crucial dynamical variable far from onset, because the amplitude is an algebraic function of the wave number $k$ and the Rayleigh number and no longer an independent parameter. The amplitude is therefore slaved to the phase gradients, and the phase diffusion equation alone suffices to describe the dynamics.

The problem involves two very different length scales, the layer depth $d$ and the tank width $L \gg d$, and may therefore be solved using a multiscale expansion. Because the convection is strongly nonlinear, however, the relevant small parameter is no longer the degree of supercriticality (as in Section 4.9.2.5), but rather the inverse aspect ratio $d/L \equiv \Gamma^{-1}$. Now the lateral scale of the variation of $\mathbf{k}$ is the tank

width $L \equiv \Gamma d$, and its characteristic timescale is the lateral thermal diffusion time $L^2/\kappa \equiv \Gamma^2 d^2/\kappa$. Accordingly, the slow variables that characterize the modulation of the basic roll pattern are

$$X = \Gamma^{-1}x, \quad Y = \Gamma^{-1}y, \quad T = \Gamma^{-2}t \quad [257]$$

The starting point of the analysis is the Galerkin representation [246]–[247] for large-amplitude steady rolls with axes parallel to the $y$-direction and wave vector $\mathbf{k} = (k, 0)$. Let $A$ be some measure of the amplitude of this solution, which can be calculated via the Galerkin procedure as a function of $Ra$ and $k$. Next, we seek modulated roll solutions for the dependent variables $\mathbf{v} = (\mathcal{P}, \theta)$ in the form

$$\mathbf{v} = \mathbf{F}(\phi = \Gamma\Phi(X, Y, T), z; A(X, Y, T)) \quad [258]$$

In [258], $\Phi$ is a slow phase variable defined such that

$$\mathbf{k} = \nabla_x\phi = \nabla_X\Phi \quad [259]$$

where $\nabla_x$ and $\nabla_X$ are the horizontal gradient operators with respect to the variables $(x, y)$ and $(X, Y)$, respectively. Because derivatives act on functions of $\phi$, $z$, $X$, $Y$, and $T$, the chain rule implies the transformations

$$\partial_z \to \partial_z, \quad \nabla_x \to \mathbf{k}\partial_\phi + \Gamma^{-1}\nabla_X,$$
$$\partial_t \to \Gamma^{-1}\partial_T\Phi\partial_\Theta + \Gamma^{-2}\partial_T,$$
$$\nabla^2 \to k^2\partial_{\phi\phi}^2 + \partial_{zz}^2 + \Gamma^{-1}D\partial_\phi + \Gamma^{-2}\nabla_X^2, \quad [260]$$
$$D = 2\mathbf{k}\cdot\nabla_X + \nabla_X\cdot\mathbf{k}$$

where $\mathbf{k} = (k_1, k_2)$.

Because of the slow modulation, [258] is no longer an exact solution of the Boussinesq equations. We therefore seek solutions in the form of an asymptotic expansion in powers of $\Gamma^{-1}$,

$$\mathbf{v} = \mathbf{v}_0 + \Gamma^{-1}\mathbf{v}_1 + \Gamma^{-2}\mathbf{v}_2 + \cdots \quad [261]$$

where $\mathbf{v}_0$ is the Galerkin solution for steady parallel rolls. Substituting [261] into the Boussinesq equations, using [260], and collecting terms proportional to $\Gamma^{-1}$, we obtain a BVP for $\mathbf{v}_1$ which has a solution only if a solvability condition is satisfied. This condition yields the phase diffusion equation

$$\frac{\partial\Phi}{\partial T} + \frac{1}{\tau(k)}\nabla\cdot[\mathbf{k}B(k)] = 0 \quad [262]$$

where the functions $\tau(k)$ and $B(k)$ are determined numerically. The diffusional character of [262] becomes obvious when one recalls that $\mathbf{k} = \nabla_X\Phi$. In writing [262], we have neglected an additional mean drift term that is nonzero only for finite values of the

Prandtl number (NPS90, eqn. (1.1)). NPS90 show that a linear stability analysis of [262] reproduces closely all the portions of the boundary of the Busse balloon for which the wavelength of the instability is long relative to the layer depth.

### 4.9.3 Convection at High Rayleigh Number

Even though convection in the form of steady 2-D rolls is unstable when $Ra > 22\,600$ (Section 4.9.2.4), much can be learned by studying the structure of such rolls in the limit $Ra \to \infty$. Because temperature variations are now confined to thin TBLs around the edges of the cells, scaling analysis and BL theory are particularly effective tools for this purpose.

#### 4.9.3.1 Scaling analysis

The essential scalings for cellular convection at high Rayleigh number can be determined by a scaling analysis of the governing equations. The following derivation (for the cell geometry shown in **Figure 6**) generalizes the analysis of McKenzie *et al.* (1974) to include both traction-free and rigid-surface boundary conditions. For simplicity, we assume in this subsection only that the aspect ratio $\beta$ does not differ much from unity.

The analysis proceeds by determining six equations relating six unknown quantities: the thicknesses $\delta_p$ of the thermal plumes and $\delta_h$ of the horizontal TBLs, the maximum vertical velocity $v_p$ and the vorticity $\omega$ in the plumes, the vertical velocity $v(\delta_h)$ in the plume at the edge of the horizontal BLs, and the heat flux $q$ across the layer (per unit length along the roll axes). The heat flux carried by an upwelling plume is

$$q \sim \rho c_p v_p \Delta T \delta_p \qquad [263]$$

where $c_p$ is the heat capacity at constant pressure. Because the convection is steady, the flux [263] must equal that lost by conduction through the top horizontal BL, implying

$$q \sim k_c d \Delta T / \delta_h \qquad [264]$$

where $k_c$ is the thermal conductivity. The thickness $\delta_h$ of the horizontal BLs is controlled by the balance $v T_y \sim \kappa T_{yy}$ of advection and diffusion, which implies

$$v(\delta_h) \Delta T / \delta_h \sim \kappa \Delta T / \delta_h^2 \qquad [265]$$

Now $v(\delta_h)$ depends on whether the horizontal surfaces are free or rigid. Because the velocity parallel to the boundary is constant to lowest order across a TBL at a free surface ($n = 0$ say) and varies linearly across a TBL at a rigid surface ($n = 1$),

$$v(\delta_h) \sim v_p (\delta_h/d)^{n+1} \qquad [266]$$

The force balance in the plumes is scaled using the vorticity equation $\nu \nabla^2 \omega = g \alpha T_x$, which implies

$$\omega \sim g \alpha \Delta T \delta_p / \nu \qquad [267]$$

Finally, the vorticity $\omega$ in the plumes is of the same order as the rotation rate of the isothermal core, or

$$\omega \sim v_p / d \qquad [268]$$

The simultaneous solution of [263]–[268] yields scaling laws for each of the six unknown quantities as a function of $R$. **Table 5** shows the results for $v_p$, $\delta_p$, and the Nusselt number $Nu \equiv d/\delta_h$. The free-surface heat transfer law $Nu \sim Ra^{1/3}$ corresponds to a dimensional heat flux $q \equiv k_c \Delta T Nu/d$ that is independent of the layer depth $d$, whereas $q \propto d^{-2/5}$ for rigid surfaces. A revealing check of the results is to note that $v_p d/\kappa$ (column 2) is just the effective Peclet number $Pe$ for the flow. The scaling laws for $Nu$ (column 4) then imply $Nu \sim Pe^{1/(n+2)}$, in agreement with our previous expression [12] for the heat transfer from a hot sphere moving in a viscous fluid.

#### 4.9.3.2 Flow in the isothermal core

Given the fundamental scales of **Table 5**, we can now carry out a more detailed analysis based on BL theory. This approach, pioneered by Turcotte (1967) and Turcotte and Oxburgh (1967) and extended by Roberts (1979), Olson and Corcos (1980), and Jimenez and Zufiria (1987) (henceforth JZ87), is applicable to convection in a layer bounded by either free or rigid surfaces. Here we consider only the former case, following JZ87 with some changes of notation.

The dimensional equations governing the flow are

$$\nu \nabla^4 \psi = -g \alpha T_x \qquad [269a]$$

$$u \theta_x + v \theta_y = \kappa \nabla^2 T \qquad [269b]$$

**Table 5**  Scaling laws for vigorous Rayleigh–Bénard convection

| Boundaries | $v_p d/\kappa$ | $\delta_p/d$ | $Nu \equiv d/\delta_h$ |
|---|---|---|---|
| Free | $Ra^{2/3}$ | $Ra^{-1/3}$ | $Ra^{1/3}$ |
| Rigid | $Ra^{3/5}$ | $Ra^{-2/5}$ | $Ra^{1/5}$ |

where $T$ is the temperature, $\mathbf{u} = u\mathbf{e}_x + v\mathbf{e}_y \equiv -\psi_y\mathbf{e}_x + \psi_x\mathbf{e}_y$ is the velocity, and $\psi$ is the streamfunction. In the limit $Ra \to \infty$, each cell comprises a nearly isothermal core surrounded by thin vertical plumes and horizontal BLs. Because temperature gradients are negligible in the core, the principal agency driving the flow within it is the shear stresses applied to its vertical boundaries by the plumes. Our first task is therefore to derive a boundary condition on the core flow that represents this agency.

Consider for definiteness the upwelling plume at the left boundary $x = -\beta d/2 \equiv x_-$ of the cell (**Figure 6**), and let its half-width be $\delta_p$. The (upward) buoyancy force acting on the right half of this plume must be balanced by a downward shear stress at its edge $x = x_- + \delta_p$ (the shear stress on $x = x_-$ is zero by symmetry). This requires

$$\nu v_x|_{x=x_-+\delta_p} = -g\alpha \int_{x_-}^{x_-+\delta_p} (T - T_c)\mathrm{d}x \quad [270]$$

where $T_c$ ($\equiv 0$) is the temperature in the core. Multiplying [270] by $v_p$, we obtain

$$v_p v_x|_{x=x_-+\delta_p} = -\frac{\alpha g}{\nu\rho c_p}\left[\rho c_p \int_{x_-}^{x_-+\delta_p} v_p(T - T_c)\mathrm{d}x\right] \quad [271]$$

where we have used the fact that $v_p \equiv \psi_x$ is constant to first order across the plumes to take it inside the integral on the RHS. Now the quantity [...] in [270] is just the heat flux carried by the right half of the plume, which is $k_c\Delta T Nu/2$. Moreover, because the plume is thin, $v_x \equiv \psi_{xx}$ can be evaluated at $x = x_-$ rather than at $x = x_- + \delta_p$. Equation [271] then becomes

$$\psi_x\psi_{xx}|_{x=x_-} = -\frac{1}{2}\frac{g\kappa\alpha\Delta T}{\nu}Nu \quad [272]$$

which is a nonlinear and inhomogeneous boundary condition on the core flow. The corresponding condition at $x = +\beta d/2$ is obtained from [272] by reversing the sign of the RHS.

We now invoke the results of the scaling analysis (**Table 5**) to nondimensionalize the equations and boundary conditions for the core flow. We first write the scaling law for $Nu$ as

$$Nu = C(\beta)Ra^{1/3} \quad [273]$$

where $C(\beta)$ (to be determined) measures the dependence of the heat transfer on the aspect ratio. Rewriting the equations and boundary conditions in terms of the dimensionless variables $(x', z') = (x, z)/d$

and $\psi' = \psi/\kappa C(\beta)^{1/2} Ra^{2/3}$ and then dropping the primes, we obtain

$$\nabla^4\psi = 0 \quad [274a]$$

$$\psi(x, \pm 1/2) = \psi_{yy}(x, \pm 1/2) = 0 \quad [274b]$$

$$\psi(\pm\beta/2, y) = \psi_x(\pm\beta/2, y)$$
$$\times \psi_{xx}(\pm\beta/2, y) \mp 1/2 = 0 \quad [274c]$$

The problem [274] can be solved either numerically (JZ87) or using a superposition method (Section 4.5.4.1).

### 4.9.3.3 TBLs and heat transfer

The next step is to determine the temperature in the plumes and the horizontal TBLs using BL theory. The temperature in all these layers is governed by the transformed BL equation [149]. Turcotte and Oxburgh (1967) solved this equation assuming self-similarity, obtaining solutions of the form [151]. However, Roberts (1979) pointed out that this is not correct, because the fluid traveling around the margins of the cell sees a periodic boundary condition which is alternatingly isothermal (along the horizontal boundaries) and insulating (in the plumes). Now because [149] is parabolic, its solution can be written in convolution-integral form in terms of an arbitrary boundary temperature $T_b(\tau)$ and an initial temperature profile $T(\psi, 0)$ at $\tau = 0$, where $\tau = \int U(s)\mathrm{d}s$ is the time-like variable introduced in Section 4.7.1.1 and the velocity $U(s)$ parallel to the boundary is determined (to within the unknown scale factor $C(\beta)$) from the core flow streamfunction $\psi$ that satisfies [274]. However, in the limit $\tau \to \infty$ corresponding to an infinite number of transits around the cell, the convolution integral describing the evolution of the initial profile vanishes, and the solution is (JZ87)

$$T(\psi, \tau) = -\frac{\psi}{2(\pi\kappa)^{1/2}}$$
$$\times \int_0^\infty \frac{T_b(\tau-t)}{t^{3/2}} \exp\left(-\frac{\psi^2}{4\kappa t}\right)\mathrm{d}t \quad [275]$$

where all variables are dimensional. Now $T_b$ is known only on the top and bottom surfaces. Along the (insulating) plume centerlines, it can be found by setting to zero the temperature gradient $\partial_\psi T|_{\psi=0}$ calculated from [275] and solving numerically the resulting integral equation. The temperature everywhere in the plumes and horizontal BLs can then be determined from [275]. The final step is to determine the unknown scale factor $C(\beta)$ by matching the heat flow advected vertically by the plumes to the conductive heat flux across the upper boundary (JZ87, eqn [23]).

The solution of JZ87 shows that the scale factor $C(\beta)$ (and hence the Nusselt number $Nu = CRa^{1/3}$) is maximum for $\beta \approx 0.75$, which may therefore be the preferred cell aspect ratio (Malkus and Veronis, 1958). A particularly interesting result from a geophysical perspective is that the temperature profile in the TBL beneath the cold upper surface exhibits a hot asthenosphere, that is, a range of depths where the temperature exceeds that of the cell's interior.

### 4.9.3.4 Structure of the flow near the corners

The solutions described above break down near the corners of the cell, where the vorticity of the core flow solution becomes infinite and the BL approximation is no longer valid. The (dimensional) characteristic radius $r_c$ of this corner region can be determined using a scaling argument. Consider for definiteness the corner above the upwelling plume, and let $(r, \varphi)$ be polar coordinates with origin at the corner such that $\varphi = 0$ on the upper surface and $\varphi = \pi/2$ on the plume centerline. Close to the corner, the streamfunction has the self-similar form $\psi(r, \varphi) = r^\lambda F(\varphi)$ for some $\lambda$, where $F$ is given by [64]. Near the corner, the dimensional forms of the boundary conditions [274b] and [274c] are

$$\psi(r, 0) = \psi(r, \pi/2) = \psi_{\varphi\varphi}(r, 0) = 0 \qquad [276a]$$

$$r^{-3}\psi_\varphi(r, \pi/2)\psi_{\varphi\varphi}(r, \pi/2) = \frac{1}{2}\frac{g\kappa\alpha\Delta T}{\nu}Nu \qquad [276b]$$

Equation [276b] immediately implies $\lambda = 3/2$, and the streamfunction which satisfies all the conditions [276] is

$$\psi = \frac{\kappa}{2}\left(\frac{r}{d}\right)^{3/2} C(\beta)^{1/2} Ra^{2/3}\left(\sin\frac{3}{2}\varphi - \sin\frac{1}{2}\varphi\right) \qquad [277]$$

Note that the vorticity $\omega \sim -(\kappa/d^2)Ra^{2/3}(r/d)^{-1/2}$ becomes infinite at the corner. Now the balance between viscous forces and buoyancy in the corner region requires $\nu\nabla^4\psi \sim -g\alpha T_x$, or

$$\psi \sim \kappa Ra(r_c/d)^3 \qquad [278]$$

Equating [278] with the scale $\psi \sim \kappa Ra^{2/3}(r_c/d)^{3/2}$ implied by [277], we find (Roberts, 1979)

$$r_c \sim Ra^{-2/9}d \qquad [279]$$

A corrected solution for the flow in the corner region that removes the vorticity singularity was proposed by JZ87.

### 4.9.3.5 Howard's scaling for high-Ra convection

While the above solution for cellular convection is illuminating, we know that rolls between rigid boundaries with $R > 22\,600$ are not stable to small perturbations (Section 4.9.2.4). In reality, convection at high Rayleigh number $Ra > 10^6$ is a quasi-periodic process in which the TBLs grow by thermal diffusion, become unstable, and then empty rapidly into plumes, at which point the cycle begins again. The characteristic timescale of this process can be estimated via a simple scaling argument (Howard, 1964). The thickness $\delta$ of both TBLs initially increases by thermal diffusion as $\delta \approx (\pi\kappa t)^{1/2}$. Instability sets in when the growth rate of R–T instabilities in the layers (see Section 4.9.1) becomes comparable to the thickening rate $\dot{\delta}/\delta$, or (equivalently) when the Rayleigh number $Ra(\delta) \equiv g\alpha\Delta T\delta^3/\nu\kappa$ based on the TBL thickness attains a critical value $Ra_c \approx 10^3$. This occurs after a time

$$t_c \approx \frac{1}{\pi\kappa}\left(\frac{\kappa\nu Ra_c}{g\alpha\Delta T}\right)^{2/3} \qquad [280]$$

that is independent of the layer depth. The scaling law [280] has been amply confirmed by laboratory experiments (e.g., Sparrow et al., 1970.)

### 4.9.4 Thermal Convection in More Realistic Systems

As the simplest and most widely studied example of thermal convection, the R–B configuration is the source of most of what has been learned about thermal convection during the past century. From a geophysical point of view, however, it lacks many crucial features that are important in the Earth's mantle: the variation of viscosity as a function of pressure, temperature, and stress; the presence of solid-state phase changes; internal production of heat by radioactive decay; density variations associated with differences in chemical composition; and a host of other factors such as rheological anisotropy, the effect of volatile content, etc. Because analytical methods thrive on simple model problems, the addition of each new complexity reduces their room for manoeuvre; but many important analytical results have been obtained nevertheless. Here we review briefly some of these, focussing on the effects of variable viscosity, phase transitions, and chemical heterogeneity.

A variety of important analytical results have been obtained for fluids with temperature-dependent viscosity. Stengel *et al.* (1982) performed a linear stability analysis of convection in a fluid whose viscosity $\nu(T)$ varies strongly (by up to a factor of $2 \times 10^4$) with temperature. For fluids for which $\nu(T)$ is an exponential function, the critical Rayleigh number (based on the viscosity at the mean of the boundary temperatures) first increases and then decreases as the total viscosity contrast across the layer increases, reflecting the fact that convection begins first in a low-viscosity sublayer near the hot bottom boundary. A nonlinear stability analysis of convection in a fluid with a weak linear dependence of viscosity on temperature was carried out by Palm (1960), using a method like that described in Section 4.9.2.3. Because the variation of the viscosity breaks the symmetry between upwelling and downwelling motions, the amplitude equation analogous to [244] contains a quadratic term that permits transcritical bifurcation. Subsequent work (e.g., Busse, 1967b, Palm *et al.*, 1967) showed that the stable planforms are hexagons, hexagons and 2D rolls, or 2D rolls alone, depending on the Rayleigh number. Busse and Frick (1985) used a Galerkin method (Section 4.9.2.4) to study convection in a fluid whose viscosity depends strongly (but linearly) on temperature, and found that a square planform becomes stable when the viscosity contrast is sufficiently large.

Convection at high Rayleigh number in fluids with strongly variable viscosity is a particularly difficult analytical challenge, but some noteworthy results have been obtained. Morris and Canright (1984) and Fowler (1985) used BL analyses similar to that described in Section 4.9.3 to determine the Nusselt number in the stagnant lid limit $\Delta T/\Delta T_r \to \infty$ for convection in a fluid whose viscosity depends on temperature as $\nu = \nu_0 \exp[-(T-T_0)/\Delta T_r]$. Both studies find the Nusselt number to be $Nu \sim (\Delta T/\Delta T_r)^{-1} Ra_r^{1/5}$ where $Ra_\gamma = \alpha g \Delta T_r d^3/\nu_0 \kappa$ is the Rayleigh number based on the rheological temperature scale $\Delta T_r$ and the viscosity $\nu_0$ at the hot bottom boundary. Solomatov (1995) presented a comprehensive scaling analysis for convection in fluids with temperature- and stress-dependent viscosity. He found that three different dynamical regimes occur as the total viscosity contrast across the layer increases: quasi-isoviscous convection, a transitional regime with a mobile cold upper BL, and a stagnant-lid regime in which the upper BL is motionless and convection is confined beneath it. Busse *et al.* (2006) generalized

Turcotte and Oxburgh's (1967) BL analysis of steady cellular convection to include thin low-viscosity layers adjoining the top and bottom boundaries, whose presence increases the cell aspect ratio that maximizes the heat transport.

The influence of a phase transition on the stability of convection in a fluid with constant viscosity was analyzed by Schubert and Turcotte (1971). The novel elements here are two matching conditions at the depth of the phase change: one of these equates the energy released by the transformation to the difference in the perturbation heat flux into and out of the phase boundary, while the second requires the phase boundary to lie on the Clapeyron slope. The resulting eigenvalue problem yields the critical Rayleigh number $Ra_c(S, Ra_Q)$, where

$$S = \frac{\Delta\rho/\rho}{\alpha d(\rho g/\gamma - \beta)}, \quad Ra_Q = \frac{\alpha g d^3 Q}{8 c_p \nu \kappa} \quad [281]$$

$\Delta\rho$ is the density difference between the phases, $Q$ is the energy per unit mass required to change the denser phase into the lighter, $\gamma$ is the Clapeyron slope, and $\beta$ is the magnitude of the temperature gradient in the basic state. $Ra_c$ is a decreasing function of $S$ (which measures the destabilizing effect of the density change) but an increasing function of $Ra_Q$ (which measures the stabilizing influence of latent heat). These results were extended to a divariant phase change by Schubert *et al.* (1972).

Convection in a chemically layered mantle was first studied by Richter and Johnson (1974), who performed a linear stability analysis of a system comprising two superposed fluid layers of equal depth and viscosity but different densities. The critical Rayleigh number $Ra_c$ depends on the value of a second Rayleigh number $Ra_\rho = g\Delta\rho d^3/\nu\kappa$ proportional to the magnitude of the stabilizing density difference $\Delta\rho$ between the layers. Three distinct modes of instability are possible: convection over the entire depth of the fluid with advection of the interface; separate convection within each layer; and standing waves on the interface, corresponding to an imaginary growth rate at marginal stability. Busse (1981) extended these results to layers of different thicknesses. Rasenat *et al.* (1989) examined a still larger region of the parameter space, and demonstrated the possibility of an oscillatory two-layer regime with no interface deformation, in which the coupling between the layers oscillates between thermal and mechanical. Le Bars and Davaille (2002) mapped out the linear stability of two-layer convection as a function of the interlayer viscosity contrast $\gamma$ and the layer depth ratio $r$, and showed that the transition from oscillatory to stratified convection occurs at a critical value of the buoyancy ratio $B \equiv Ra_\rho / Ra$ that depends on $\gamma$ and $r$.

For the time being, studies like those discussed above represent the limit of what can be learned about convection in complex systems using analytical methods alone. The impressive additional progress that has been made using experimental and numerical approaches is discussed in Chapters 2 and 4 of this volume.

## Acknowledgment

This is IPGP contribution number 2202.

## References

Albers M and Christensen UR (2001) Channeling of plume flow beneath mid-ocean ridges. *Earth and Planetary Science Letters* 187: 207–220.

Ansari A and Morris S (1985) The effects of a strongly temperature-dependent viscosity on Stokes's drag law: Experiments and theory. *Journal of Fluid Mechanics* 159: 459–476.

Backus GE (1958) A class of self-sustaining dissipative spherical dynamos. *Annals of Physics* 4: 372–447.

Bai Q, Mackwell SJ, and Kohlstedt DL (1991) High-temperature creep of olivine single crystals. Part 1: Mechanical results for buffered samples. *Journal of Geophysical Research* 96: 2441–2463.

Backus GE (1967) Converting vector and tensor equations to scalar equations in spherical co-ordinates. *Geophysical Journal of the Royal Astronomical Society* 13: 71–101.

Barcilon V and Richter FM (1986) Nonlinear waves in compacting media. *Journal of Fluid Mechanics* 164: 429–448.

Barenblatt GI (1996) *Scaling, Self-Similarity, and Intermediate Asymptotics.* Cambridge: Cambridge University Press.

Batchelor GK (1967) *An Introduction to Fluid Dynamics.* Cambridge: Cambridge University Press.

Batchelor GK (1970) Slender-body theory for particles of arbitrary crosssection in Stokes flow. *Journal of Fluid Mechanics* 44: 419–440.

Bercovici D (1994) A theoretical model of cooling viscous gravity currents with temperature-dependent viscosity. *Geophysical Research Letters* 21: 1177–1180.

Bercovici D and Lin J (1996) A gravity current model of cooling mantle plume heads with temperature-dependent buoyancy and viscosity. *Journal of Geophysical Research* 101: 3291–3309.

Biot MA (1954) Theory of stress-strain relations in anisotropic viscoelasticity and relaxation phenomena. *Journal of Applied Physics* 25: 1385–1391.

Bird RB, Armstrong RC, and Hassager O (1987) *Dynamics of Polymeric Liquids, Vol 1: Fluid Mechanics,* 2nd edn. New York, NY: John Wiley & Sons.

Blake JR (1971) A note on the image system for a Stokeslet in a no-slip boundary. *Proceedings of the Cambridge Philosophical Society* 70: 303–310.

Bloor MIG and Wilson MJ (2006) An approximate analytic solution method for the biharmonic problem. *Proceedings of the Royal Society of London A* 462: 1107–1121.

Bolton EW and Busse FH (1985) Stability of convection rolls in a layer with stress-free boundaries. *Journal of Fluid Mechanics* 150: 487–498.

Buckingham E (1914) On physically similar systems; illustrations of the use of dimensional equations. *Physical Review* 4: 345–376.

Buckmaster JD, Nachman A, and Ting L (1975) The buckling and stretching of a viscida. *Journal of Fluid Mechanics* 69: 1–20.

Busse FH (1967a) On the stability of two-dimensional convection in a layer heated from below. *Journal of Mathematical Physics* 46: 140–150.

Busse FH (1967b) The stability of finite amplitude cellular convection and its relation to an extremum principle. *Journal of Fluid Mechanics* 30: 625–649.

Busse FH (1981) On the aspect ratio of two-layer mantle convection. *Physics of the Earth and Planetary Interiors* 24: 320–324.

Busse FH and Bolton EW (1984) Instabilities of convection rolls with stress-free boundaries near threshold. *Journal of Fluid Mechanics* 146: 115–125.

Busse FH and Frick H (1985) Square-pattern convection in fluids with strongly temperature-dependent viscosity. *Journal of Fluid Mechanics* 150: 451–465.

Busse FH, Richards MA, and Lenardic A (2006) A simple model of high Prandtl and high Rayleigh number convection bounded by thin low-viscosity layers. *Geophysical Journal International* 164: 160–167.

Canright DR (1987) *A Finite-Amplitude Analysis of the Buoyant Instability of a Highly Viscous Film over a Less Viscous Half-Space.* PhD Thesis, University of California, Berkeley.

Canright D and Morris S (1993) Buoyant instability of a viscous film over a passive fluid. *Journal of Fluid Mechanics* 255: 349–372.

Chandrasekhar S (1981) *Hydrodynamic and Hydromagnetic Stability.* New York, NY: Dover.

Conrad CP and Molnar P (1997) The growth of Rayleigh–Taylor-type instabilities in the lithosphere for various rheological and density structures. *Geophysical Journal International* 129: 95–112.

Cox RG (1970) The motion of long slender bodies in a viscous fluid. Part 1: General theory. *Journal of Fluid Mechanics* 44: 791–810.

Davaille A (1999) Two-layer thermal convection in miscible viscous fluids. *Journal of Fluid Mechanics* 379: 223–253.

Dziewonski AM and Anderson DL (1981) Preliminary reference Earth model. *Physics of the Earth and Planetary Interiors* 25: 297–356.

England P and McKenzie D (1983) Correction to: A thin viscous sheet model for continental deformation. *Geophysical Journal of the Royal Astronomical Society* 73: 523–532.

Farrell WE (1972) Deformation of the Earth by surface loads. *Review of Geophysics and Space Physics* 10: 761–797.

Feighner M and Richards MA (1995) The fluid dynamics of plume-ridge and plume-plate interactions: An experimental investigation. *Earth and Planetary Science Letters* 129: 171–182.

Fenner RT (1975) On local solutions to non-Newtonian slow viscous flows. *International Journal of Non-Linear Mechanics* 10: 207–214.

Fleitout L and Froidevaux C (1983) Tectonic stresses in the lithosphere. *Tectonics* 2: 315–324.

Fletcher CAJ (1984) *Computational Galerkin Methods.* New York, NY: Springer Verlag.

Forte AM and Peltier WR (1987) Plate tectonics and aspherical Earth structure: The importance of poloidal-toroidal coupling. *Journal of Geophysical Research* 92: 3645–3679.

Fowler AC (1985) Fast thermoviscous convection. *Studies in Applied Mathematics* 72: 189–219.

Gantmacher FR (1960) *Matrix Theory,* 2 vols. Providence, RI: AMS Chelsea Publishing.

Gomilko AM, Malyuga VS, and Meleshko VV (2003) On steady Stokes flow in a trihedral rectangular corner. *Journal of Fluid Mechanics* 476: 159–177.

Gratton J and Minotti F (1990) Self-similar viscous gravity currents: Phase-plane formalism. *Journal of Fluid Mechanics* 210: 155–182.

Griffiths RW (1986) Thermals in extremely viscous fluids, including the effects of temperature-dependent viscosity. *Journal of Fluid Mechanics* 166: 115–138.

Hager BH and O'Connell RJ (1981) A simple global model of plate tectonics and mantle convection. *Journal of Geophysical Research* 86: 4843–4867.

Hager BH, Clayton RW, Richards MA, Comer RP, and Dziewonski AM (1985) Lower mantle heterogeneity, dynamic topography and the geoid. *Nature* 313: 541–545.

Happel J and Brenner H (1991) *Low Reynolds Number Hydrodynamics,* 2nd rev. edn. Dordrecht: Kluwer Academic.

Hauri EH, Whitehead JA, Jr., and Hart SR (1994) Fluid dynamic and geochemical aspects of entrainment in mantle plumes. *Journal of Geophysical Research* 99: 24275–24300.

Hinch EJ (1991) *Perturbation Methods.* Cambridge: Cambridge University Press.

Houseman GA and Molnar P (1997) Gravitational (Rayleigh–Taylor) instability of a layer with non-linear viscosity and convective thinning of continental lithosphere. *Geophysical Journal International* 128: 125–150.

Howard LN (1964) Convection at high Rayleigh number. In: Görtler H (ed.) *Proceedings of the 11th International Congress in Applied Mechanics*, pp. 1109–1115. Berlin: Springer.

Huppert HE (1982) The propagation of two-dimensional and axisymmetric viscous gravity currents over a rigid horizontal surface. *Journal of Fluid Mechanics* 121: 43–58.

Ito G (2001) Reykjanes 'V'-shaped ridges originating from a pulsing and dehydrating mantle plume. *Nature* 411: 681–684.

Jackson JD (1975) *Classical Electrodynamics.* New York, NY: John Wiley & Sons.

Jimenez J and Zufiria JA (1987) A boundary-layer analysis of Rayleigh–Bénard convection at large Rayleigh number. *Journal of Fluid Mechanics* 178: 53–71.

Jeong J-T and Moffatt HK (1992) Free-surface cusps associated with flow at low Reynolds number. *Journal of Fluid Mechanics* 241: 1–22.

Johnson RE (1980) Slender-body theory for slow viscous flow. *Journal of Fluid Mechanics* 75: 705–714.

Katopodes FV, Davis AMJ, and Stone HA (2000) Piston flow in a two-dimensional channel. *Physics of Fluids* 12: 1240–1243.

Keller JB and Rubinow SI (1976) An improved slender-body theory for Stokes flow. *Journal of Fluid Mechanics* 99: 411–431.

Kerr RC and Lister JR (1987) The spread of subducted lithospheric material along the mid-mantle boundary. *Earth and Planetary Science Letters* 85: 241–247.

Kevorkian J and Cole JD (1996) *Multiple Scale and Singular Perturbation Methods.* New York, NY: Springer.

Kim S and Karrila SJ (1991) *Microhydrodynamics: Principles and Selected Applications.* Boston, MA: Butterworth-Heinemann.

Koch DM and Koch DL (1995) Numerical and theoretical solutions for a drop spreading below a free fluid surface. *Journal of Fluid Mechanics* 287: 251–278.

Koch DM and Ribe NM (1989) The effect of lateral viscosity variations on surface observables. *Geophysical Research Letters* 16: 535–538.

Lachenbruch AH and Nathenson M (1974) Rise of a variable viscosity fluid in a steadily spreading wedge-shaped conduit with accreting walls. *Open File Report.* 74–251, 27 pp. US Geological Survey, Menlo Park, CA.

Ladyzhenskaya OA (1963) *The Mathematical Theory of Viscous Incompressible Flow.* New York, NY: Gordon and Breach.

Lamb H (1932) *Hydrodynamics.* Cambridge: Cambridge University Press.

Landau LD (1944) On the problem of turbulence. *Comptes Rendus de l'Academie des Sciences URSS* 44: 311–314.

Landau LD and Lifshitz EM (1986) *Theory of Elasticity,* 3rd edn. Oxford: Pergamon.

Langlois WE (1964) *Slow Viscous Flow.* New York, NY: Macmillan.

Le Bars M and Davaille A (2002) Stability of thermal convection in two superposed miscible viscous fluids. *Journal of Fluid Mechanics* 471: 339–363.

Lee SH and Leal LG (1980) Motion of a sphere in the presence of a plane interface. Part 2: An exact solution in bipolar co-ordinates. *Journal of Fluid Mechanics* 98: 193–224.

Lemery C, Ricard Y, and Sommeria J (2000) A model for the emergence of thermal plumes in Rayleigh-Bénard convection at infinite Prandtl number. *Journal of Fluid Mechanics* 414: 225–250.

Lister JR and Kerr RC (1989a) The effect of geometry on the gravitiational instability of a region of viscous fluid. *Journal of Fluid Mechanics* 202: 577–594.

Lister JR and Kerr RC (1989b) The propagation of two-dimensional and axisymmetric viscous gravity currents at a fluid interface. *Journal of Fluid Mechanics* 203: 215–249.

Longman IM (1962) A Green's function for determining the deformation of the Earth under surface mass loads. *Journal of Geophysical Research* 67: 845–850.

Loper DE and Stacey FD (1983) The dynamical and thermal structure of deep mantle plumes. *Physics of the Earth and Planetary Interiors* 33: 305–317.

Malkus WVR and Veronis G (1958) Finite amplitude cellular convection. *Journal of Fluid Mechanics* 4: 225–260.

Manga M (1996) Mixing of heterogeneities in the mantle: Effect of viscosity differences. *Geophysical Research Letters* 23: 403–406.

Manga M (1997) Interactions between mantle diapirs. *Geophysical Research Letters* 24: 1871–1874.

Manga M and Stone HA (1993) Buoyancy-driven interaction between two deformable viscous drops. *Journal of Fluid Mechanics* 256: 647–683.

Manga M, Stone HA, and O'Connell RJ (1993) The interaction of plume heads with compositional discontinuities in the Earth's mantle. *Journal of Geophysical Research* 98: 19979–19990.

Mangler W (1948) Zusammenhang zwischen ebenen und rotationssymmetrischen Grenzschichten in kompressiblen Flüssigkeiten. *Zeitschrift für Angewandte Mathematik und Mechanik* 28: 97–103.

Marsh BD (1978) On the coolling of ascending andesitic magma. *Philosophical Transactions of the Royal Society of London* 288: 611–625.

Marsh BD and Carmichael ISE (1974) Benioff zone magmatism. *Journal of Geophysical Research* 79: 1196–1206.

McKenzie DP (1969) Speculations on the consequences and causes of plate motions. *Geophysical Journal of the Royal Astronomical Society* 18: 1–32.

McKenzie DP and Bowin C (1976) The relationship between bathymetry and gravity in the Atlantic ocean. *Journal of Geophysical Research* 81: 1903–1915.

McKenzie DP, Roberts JM, and Weiss NO (1974) Convection in the Earth's mantle: Towards a numerical simulation. *Journal of Fluid Mechanics* 62: 465–538.

Meleshko VV (1996) Steady Stokes flow in a rectangular cavity. *Proceedings of the Royal Society of London A* 452: 1999–2022.

Mitrovica JX and Forte AM (2004) A new inference of mantle viscosity based upon joint inversion of convection and glacial isostatic adjustment data. *Earth and Planetary Science Letters* 225: 177–189.

Moffatt HK (1964) Viscous and resistive eddies near a sharp corner. *Journal of Fluid Mechanics* 18: 1–18.

Molnar P and Houseman GA (2004) The effects of buoyant crust on the gravitational instability of thickened mantle lithosphere at zones of intracontinental convergence. *Geophysical Journal International* 158: 1134–1150.

Morris S (1980) *An Asymptotic Method for Determining the Transport of Heat and Matter by Creeping Flows with Strongly Variable Viscosity; Fluid Dynamic Problems Motivated by Island Arc Volcanism*. PhD Thesis, The Johns Hopkins University

Morris S (1982) The effects of a strongly temperature-dependent viscosity on slow flow past a hot sphere. *Journal of Fluid Mechanics* 124: 1–26.

Morris S and Canright D (1984) A boundary-layer analysis of Bénard convection in a fluid of strongly temperature-dependent viscosity. *Physics of the Earth and Planetary Interiors* 36: 355–373.

Muskhelishvili NI (1953) *Some Basic Problems in the Mathematical Theory of Elasticity*. Groningen: P. Noordhoff.

Nayfeh A (1973) *Perturbation Methods*. New York, NY: John Wiley & Sons.

Newell A C, Passot T, and Lega J (1993) Order parameter equations for patterns. *Annual Reviews of Fluid Mechanics* 25: 399–453.

Newell AC, Passot T, and Souli M (1990) The phase diffusion and mean drift equations for convection at finite Rayleigh numbers in large containers. *Journal of Fluid Mechanics* 220: 187–252.

Newell AC and Whitehead JA, Jr. (1969) Finite bandwidth, finite amplitude convection. *Journal of Fluid Mechanics* 38: 279–303.

Niordson FI (1985) *Shell Theory*. Amsterdam: North-Holland.

Novozhilov VV (1959) *Theory of Thin Shells*. Groningen: P. Noordhoff.

Olson P (1990) Hot spots, swells and mantle plumes. In: Ryan MP (ed.) *Magma Transport and Storage*, pp. 33–51. New York, NY: John Wiley.

Olson P and Christensen U (1986) Solitary wave propagation in a fluid conduit within a viscous matrix. *Journal of Geophysical Research* 91: 6367–6374.

Olson P and Corcos GM (1980) A boundary-layer model for mantle convection with surface plates. *Geophysical Journal of the Royal Astronomical Society* 62: 195–219.

Olson P, Schubert G, and Anderson C (1993) Structure of axisymmetric mantle plumes. *Journal of Geophysical Research* 98: 6829–6844.

Olson P and Singer H (1985) Creeping plumes. *Journal of Fluid Mechanics* 158: 511–531.

Palm E (1960) On the tendency towards hexagonal cells in steady convection. *Journal of Fluid Mechanics* 8: 183–192.

Palm E, Ellingsen T, and Gjevik B (1967) On the occurrence of cellular motion in Bénard convection. *Journal of Fluid Mechanics* 30: 651–661.

Parsons B and Daly S (1983) The relationship between surface topography, gravity anomalies, and the temperature structure of convection. *Journal of Geophysical Research* 88: 1129–1144.

Peltier WR (1974) The impulse response of a Maxwell Earth. *Reviews of Geophysics and Space Physics* 12: 649–669.

Peltier WR (1995) Mantle viscosity. In: Peltier WR (ed.) *Mantle Convection: Plate Tectonics and Global Dynamics*, pp. 389–478. New York, NY: Gordon and Breach Science Publishers.

Pozrikidis C (1990) The deformation of a liquid drop moving normal to a plane wall. *Journal of Fluid Mechanics* 215: 331–363.

Pozrikidis C (1992) *Boundary Integral and Singularity Methods for Linearized Viscous Flow*. Cambridge: Cambridge University Press.

Rasenat S, Busse FH, and Rehberg I (1989) A theoretical and experimental study of double-layer convection. *Journal of Fluid Mechanics* 199: 519–540.

Rayleigh L (1922) *Theory of Sound*, vol. II, 313 pp. New York, NY: Dover.

Ribe NM (1992) The dynamics of thin shells with variable viscosity and the origin of toroidal flow in the mantle. *Geophysical Journal International* 110: 537–552.

Ribe NM (1998) Spouting and planform selection in the Rayleigh–Taylor instability of miscible viscous fluids. *Journal of Fluid Mechanics* 234: 315–336.

Ribe NM (2003) Periodic folding of viscous sheets. *Physical Review E* 68: 036305.

Ribe NM and Christensen U (1994) Three-dimensional modeling of plume-lithosphere interaction. *Journal of Geophysical Research* 99: 669–682.

Ribe NM and Christensen U (1999) The dynamical origin of Hawaiian volcanism. *Earth and Planetary Science Letters* 171: 517–531.

Ribe NM, Christensen UR, and Theissing J (1995) The dynamics of plume-ridge interaction. Part 1: Ridge-centered plumes. *Earth and Planetary Science Letters* 134: 155–168.

Ribe NM and Delattre WL (1998) The dynamics of plume-ridge interaction-III. The effects of ridge migration. *Geophysical Journal International* 133: 511–518.

Ribe NM, Stutzmann E, Ren Y, and van der Hilst R (2007) Buckling instabilities of subducted lithosphere beneath the transition zone. *Earth and Planetary Science Letters* 254: 173–179.

Ricard Y, Fleitout L, and Froidevaux C (1984) Geoid heights and lithospheric stresses for a dynamic earth. *Annals of Geophysics* 2: 267–286.

Richards MA and Hager BH (1984) Geoid anomalies in a dynamic Earth. *Journal of Geophysical Research* 89: 5987–6002.

Richards MA, Hager BH, and Sleep NH (1988) Dynamically supported geoid highs over hotspots: Observation and theory. *Journal of Geophysical Research* 893: 7690–7708.

Richter FM (1973) Dynamical models for sea floor spreading. *Reviews of Geophysics and Space Physics* 11: 223–287.

Richter FM and Johnson CE (1974) Stability of a chemically layered mantle. *Journal of Geophysical Research* 79: 1635–1639.

Richter FM and McKenzie DP (1984) Dynamical models for melt segregation from a deformable matrix. *Journal of Geology* 92: 729–740.

Roberts GO (1979) Fast viscous Bénard convection. *Geophysical and Astrophysical Fluid Dynamics* 12: 235–272.

Schubert G, Olson P, Anderson C, and Goldman P (1989) Solitary waves in mantle plumes. *Journal of Geophysical Research* 94: 9523–9532.

Schubert G and Turcotte DL (1971) Phase changes and mantle convection. *Journal of Geophysical Research* 76: 1424–1432.

Schubert G, Yuen DA, and Turcotte DL (1972) Role of phase transitions in a dynamic mantle. *Geophysical Journal of the Royal Astronomical Society* 42: 705–735.

Scott DR and Stevenson DJ (1984) Magma solitons. *Geophysical Research Letters* 11: 1161–1164.

Scott DR, Stevenson DJ, and Whitehead JA, Jr. (1986) Observations of solitary waves in a deformable pipe. *Nature* 319: 759–761.

Segel L (1969) Distant side-walls cause slow amplitude modulation of cellular convection. *Journal of Fluid Mechanics* 38: 203–224.

Selig F (1965) A theoretical prediction of salt dome patterns. *Geophysics* 30: 633–643.

Shankar PN (1993) The eddy structure in Stokes flow in a cavity. *Journal of Fluid Mechanics* 250: 371–383.

Shankar PN (2005) Eigenfunction expansions on arbitrary domains. *Proceedings of the Royal Society of London A* 461: 2121–2133.

Sleep NH (1987) Lithospheric heating by mantle plumes. *Geophysical Journal of the Royal Astronomical Society* 91: 1–11.

Solomatov VS (1995) Scaling of temperature- and stress-dependent viscosity convection. *Physics of Fluids* 7: 266–274.

Sparrow EM, Husar RB, and Goldstein RJ (1970) Observations and other characteristics of thermals. *Journal of Fluid Mechanics* 41: 793–800.

Stengel KC, Oliver DS, and Booker JR (1982) Onset of convection in a variable-viscosity fluid. *Journal of Fluid Mechanics* 120: 411–431.

Stevenson DJ and Turner JS (1977) Angle of subduction. *Nature* 270: 334–336.

Stimson M and Jeffrey GB (1926) The motion of two spheres in a viscous fluid. *Proceedings of the Royal Society of London A* 111: 110–116.

Stokes GG (1845) On the theories of the internal friction of fluids in motion, and of the equilibrium and motion of elastic solids. *Transactions of the Cambridge Philosophical Society* 8: 287–347.

Straus JM (1972) Finite amplitude doubly diffusive convection. *Journal of Fluid Mechanics* 56: 353–374.

Tanimoto T (1998) State of stress within a bending spherical shell and its implications for subducting lithosphere. *Geophysical Journal International* 134: 199–206.

Tovish A, Schubert G, and Luyendyk BP (1978) Mantle flow pressure and the angle of subduction: Non-Newtonian corner flows. *Journal of Geophysical Research* 83: 5892–5898.

Turcotte DL (1967) A boundary-layer theory for cellular convection. *International Journal of Heat and Mass Transfer* 10: 1065–1074.

Turcotte DL (1974) Membrane tectonics. *Geophysical Journal of the Royal Astronomical Society* 36: 33–42.

Turcotte DL and Oxburgh ER (1967) Finite amplitude convection cells and continental drift. *Journal of Fluid Mechanics* 28: 29–42.

Umemura A and Busse FH (1989) Axisymmetric convection at large Rayleigh number and infinite Prandtl number. *Journal of Fluid Mechanics* 208: 459–478.

Van Dyke M (1975) *Perturbation Methods in Fluid Mechanics*. Stanford, CA: Parabolic Press.

Von Mises R (1927) Bemerkungen zur Hydrodynamik. *Zeitschrift für Angewandte Mathematik und Mechanik* 7: 425–431.

Watts AB (1978) An analysis of isostasy in the world's oceans. Part 1: Hawaiian-Emperor Seamount chain. *Journal of Geophysical Research* 83: 5989–6004.

Weinstein SA and Olson PL (1992) Thermal convection with non-Newtonian plates. *Geophysical Journal International* 111: 515–530.

Whitehead JA, Jr., Dick HBJ, and Schouten H (1984) A mechanism for magmatic accretion under spreading centers. *Nature* 312: 146–148.

Whitehead JA, Jr. and Helfrich KR (1986) The Korteweg-de Vries equation from laboratory conduit and magma migration equations. *Geophysical Research Letters* 13: 545–546.

Whitehead JA, Jr. and Helfrich KR (1988) Wave transport of deep mantle material. *Nature* 335: 59–61.

Whitehead JA, Jr. and Luther DS (1975) Dynamics of laboratory diapir and plume models. *Journal of Geophysical Research* 80: 705–717.

Whitham GB (1974) *Linear and Non-Linear Waves*. Sydney: Wiley-Interscience.

Whittaker RJ and Lister JR (2006a) Steady axisymmetric creeping plumes above a planar boundary. Part I: A point source. *Journal of Fluid Mechanics* 567: 361–378.

Whittaker RJ and Lister JR (2006b) Steady axisymmetric creeping plumes above a planar boundary. Part II: A distributed source. *Journal of Fluid Mechanics* 567: 379–397.

Yale MM and Phipps Morgan J (1998) Asthenosphere flow model of hotspot-ridge interactions: A comparison of Iceland and Kerguelen. *Earth and Planetary Science Letters* 161: 45–56.

Yuen DA and Peltier WR (1980) Mantle plumes and the thermal stability of the $D''$ layer. *Geophysical Research Letters* 7: 625–628.

Yuen DA and Schubert G (1976) Mantle plumes: a boundary-layer approach for Newtonian and non-Newtonian temperature-dependent rheologies. *Journal of Geophysical Research* 81: 2499–2510.

# 5 Numerical Methods for Mantle Convection

**S. J. Zhong**, University of Colorado, Boulder, CO, USA

**D. A. Yuen**, University of Minnesota, Minneapolis, MN, USA

**L. N. Moresi**, Monash University, Clayton, VIC, Australia

## 5.1 Introduction

The governing equations for mantle convection are derived from conservation laws of mass, momentum, and energy. The nonlinear nature of mantle rheology with its strong temperature and stress dependence and nonlinear coupling between flow velocity and temperature in the energy equation require that numerical methods be used to solve these governing equations. Understanding the dynamical effects of phase transitions (e.g., olivine-to-spinel phase transition) and multicomponent flow also demands numerical methods. Numerical modeling of mantle convection has a rich history since the late 1960s (e.g., Torrance and Turcotte, 1971; McKenzie *et al.*, 1974). Great progress in computer architecture along with improved numerical techniques has helped advance the field of mantle convection into its own niche in geophysical fluid dynamics (e.g., Yuen *et al.*, 2000).

In this chapter, we will present several commonly used numerical methods in studies of mantle convection with the primary aim of reaching out to students and new researchers in the field. First, we will present the governing equations and the boundary and initial conditions for a given problem in mantle convection, and discuss the general efficient strategy to solve this problem numerically (Section 5.2). We will then briefly discuss finite-difference (FD), finite-volume (FV), and spectral methods in Section 5.3. Since finite elements (FEs) have attained very high popularity in the user community, we will discuss FEs in greater details as the most basic numerical tool (Section 5.4). For simplicity and clarity, we will focus our discussion on homogeneous, incompressible fluids with the

Boussinesq approximation. However, we will also describe methods for more complicated and realistic mantle situations by including non-Newtonian rheology, solid-state phase transitions, and thermo-chemical (i.e., multicomponent) convection (Section 5.5). Finally, in Section 5.6, we will discuss some new developments in computational sciences, such as software development and visualization, which may impact our future studies of mantle convection modeling.

## 5.2   Governing Equations and Initial and Boundary Conditions

The simplest mathematical formulation for mantle convection assumes incompressibility and the Boussinesq approximation (e.g., McKenzie *et al.*, 1974). Under this formulation, the nondimensional conservation equations of the mass, momentum, and energy are (*see* Chapter 2):

$$u_{i,i} = 0 \qquad [1]$$

$$\sigma_{ij,j} + Ra\,T\delta_{i3} = 0 \qquad [2]$$

$$\frac{\partial T}{\partial t} + u_i T_{,i} = (\kappa T_{,i})_{,i} + \gamma \qquad [3]$$

where $u_i$, $\sigma_{ij}$, $T$, and $\gamma$ are the velocity, stress tensor, temperature, and heat-production rate, respectively, $Ra$ is Rayleigh number, and $\delta_{ij}$ is a Kronecker delta function. Repeated indexes denote summation, and $u_{,i}$ represents partial derivative of variable $u$ with respect to coordinate $x_i$. These equations were obtained by using the following characteristic scales: length $D$, time $D^2/\kappa$; and temperature $\Delta T$, where $D$ is often the thickness of the mantle or a fluid layer, $\kappa$ is thermal diffusivity, and $\Delta T$ is the temperature difference across the fluid layer (*see* Chapters 2 and 4 for discussion on nondimensionalization). The stress tensor $\sigma_{ij}$ can be related to strain rate $\dot{\varepsilon}_{ij}$ via the following constitutive equation:

$$\sigma_{ij} = -P\delta_{ij} + 2\eta\dot{\varepsilon}_{ij} = -P\delta_{ij} + \eta(u_{i,j} + u_{j,i}) \qquad [4]$$

where $P$ is the dynamic pressure and $\eta$ is the viscosity.

Substituting eqns [4] into [2] reveals three primary unknown variables: pressure, velocity, and temperature. The three governing eqns [1]–[3] are sufficient to solve for these three unknowns, together with adequate boundary and initial conditions. Initial conditions are only needed for temperature due to the first-order derivative with respect to time in the energy equation. Boundary conditions are in general a combination of prescribed stress and velocity for the momentum equation, and of prescribed heat flux and temperature for the energy equation. The initial and boundary conditions can be expressed as

$$T(r_i, t = 0) = T_{\text{init}}(r_i) \qquad [5]$$

$$u_i = g_i \text{ on } \Gamma_{g_i}, \qquad \sigma_{ij}n_j = b_i \text{ on } \Gamma_{b_i} \qquad [6]$$

$$T = T_{\text{bd}} \text{ on } \Gamma_{T_{\text{bd}}}, \qquad (T_{,i})_n = q \text{ on } \Gamma_q \qquad [7]$$

where $\Gamma_{g_i}$ and $\Gamma_{b_i}$ are the boundaries where $i$th components of velocity and forces are specified to be $g_i$ and $b_i$, respectively, $n_j$ is the normal vector of boundary $\Gamma_{b_i}$, and $\Gamma_{T_{\text{bd}}}$ and $\Gamma_q$ are the boundaries where temperature and heat flux are prescribed to be $T_{\text{bd}}$ and $q$, respectively.

Often free-slip (i.e., zero tangential stresses and zero normal velocities) and isothermal conditions are applied to the surface and bottom boundaries in studies of mantle dynamics, although in some studies surface velocities may be given in consistent with surface plate motions (e.g., Bunge *et al.*, 1998). When steady-state or statistically steady-state solutions are to be sought, as they often are in mantle dynamics, the choice of initial condition can be rather arbitrary and it does not significantly affect final results in a statistical sense.

Although full time-dependent dynamics of thermal convection involves all three governing equations, an important subset of mantle dynamics problems, often termed as instantaneous Stokes flow problem, only require solutions of eqns [1] and [2]. For Stokes flow problem, one may consider the dynamic effects of a given buoyancy field (e.g., one derived from seismic structure) or prescribed surface plate motion on gravity anomalies, deformation rate, and stress at the surface and the interior of the mantle (Hager and O'Connell, 1981; Hager and Richards, 1989, also *see* Chapters 2 and 4).

These governing equations generally require numerical solution procedures for three reasons. (1) The advection of temperature in eqn [3], $u_i T_{,i}$, represents a nonlinear coupling between velocity and temperature. (2) The constitutive law or eqn [4] is often nonlinear in that stress and strain rate follow a power-law relation; that is, the viscosity $\eta$ in eqn [4] can only be considered as effective viscosity that depends on stress or strain rate. (3) Even for the Stokes flow problem with a linear rheology, spatial variability in viscosity can make any analytic solution method difficult and impractical.

Irrespective of numerical methods used for the treatment of the individual governing equations, it is usual to solve the coupled system explicitly in time as follows: (1) At a given time step, solve eqns [1] and [2] (i.e., the instantaneous Stokes flow problem) for flow velocity for given buoyancy or temperature. (2) Update the temperature to next time step from eqn [3], using the new velocity field. (3) Continue this process of time stepping by going back to step (1).

## 5.3 Finite-Difference, Finite-Volume, and Spectral Methods

In this section, we will briefly discuss how finite-difference (FD), finite-volume (FV), and spectral methods are used in studies of mantle convection. These methods have a long history in modeling mantle convection (e.g., Machetel *et al.*, 1986; Gable *et al.*, 1991), and remain important in the field (e.g., Tackley, 2000; Kageyama and Sato, 2004; Stemmer *et al.*, 2006; Harder and Hansen, 2005).

### 5.3.1 Finite Difference

FD methods have a much earlier historical beginning than FE, spectral, or FV methods because they are motivated intuitively by differential calculus and are based on local discretization of the derivative operators based on a Taylor series expansion with an assigned order of accuracy about a given point. The unknowns at each grid point depend on those of the neighboring points from the local Taylor expansion. For examples of FD implementation, the reader is urged to consult an excellent introductory book by Lynch (2005). Additional material of FORTRAN 95 listings of library subprograms using FD methods can be found in Griffiths and Smith (2006). Hence FD formulations are easy to grasp and to program and they have been the leading tool in numerical computations since the 1950s, as evidenced by the pioneering codes written for meteorology and applied weapons research (e.g., Richtmeyer and Morton, 1967). All of the initial codes in mantle convection were written with the FD formulation (Torrance and Turcotte, 1971; Turcotte *et al.*, 1973; McKenzie *et al.*, 1974; Parmentier *et al.*, 1975; Jarvis and McKenzie, 1980). The initial numerical techniques in solving mantle convection in the nonlinear regime were based on second-order FD methods (Torrance and Turcotte, 1971; McKenzie *et al.*, 1974). In the early 1980s, splines with a fourth-

order accuracy were used to solve two-dimensional (2-D) problems with variable viscosity (Christensen, 1984). Malevsky (1996) developed a 3-D mantle convection code based on 3-D splines, which allowed one to reach very high Rayleigh numbers, like $10^8$ (Malevsky and Yuen, 1993). Recently, Kageyama and Sato (2004) have developed a 3-D FD technique, using a baseball-like topological configuration called the 'yin–yang' grid, for solving the 3-D spherical convection problem.

#### 5.3.1.1 FD implementation of the governing equations

For 2-D mantle convection within the Boussinesq approximation and isoviscous flow in FD methods, a commonly used formulation employs stream-function $\Psi$ and vorticity $\omega$ to eliminate the pressure and velocities, and the governing equations are written in terms of stream function $\Psi$, vorticity $\omega$, and temperature $T$ (e.g., McKenzie *et al.*, 1974). They are written in time-dependent form as (also see eqns [1–3])

$$\nabla^2 \Psi = -\omega \qquad [8]$$

$$\nabla^2 \omega = Ra \frac{\partial T}{\partial x} \qquad [9]$$

$$\frac{\partial T}{\partial t} = \nabla^2 T - \left( \frac{\partial \Psi}{\partial x} \frac{\partial T}{\partial z} - \frac{\partial \Psi}{\partial z} \frac{\partial T}{\partial x} \right) + \gamma \qquad [10]$$

where the velocity vector is defined as

$$\mathbf{v} = (v_x, v_z) = \left( \frac{\partial \Psi}{\partial z}, -\frac{\partial \Psi}{\partial x} \right) \qquad [11]$$

which automatically satisfies the continuity, and $\omega$ is the only component left of the vorticity vector $\nabla \times \bar{v}$, $x$ and $z$ are the horizontal and vertical coordinates, respectively, with $z$ vector pointing upward, and $t$ is the time. We note that eqns [8] and [9] are given by a set of coupled second-order partial differential equations. Alternatively, we could have combined them to form a single fourth-order partial differential equation in terms of $\Psi$, called the biharmonic equation, as developed in the numerical scheme of Christensen (1984) using bicubic splines and also by Schott and Schmeling (1998) using FDs. We note that solving the biharmonic solution by FDs takes more time than solving two coupled Laplacian equations. We have restricted ourselves here to constant viscosity for pedagogical purposes. Examples of variable-viscosity convection equations can be found in Christensen and Yuen (1984) and Schubert *et al.* (2000).

Zero tangential stress and impermeable boundary conditions can be expressed as

$$\omega|_\Gamma = \Psi|_\Gamma = 0 \qquad [12]$$

Other boundary conditions are similar to those discussed in Section 5.2.

In 2-D mantle convection problems, one can also utilize the so-called primitive variables (i.e., velocity and dynamical pressure) formulation by solving the coupled equations of mass and momentum conservations at each time step. This procedure may involve more numerical care and computational time because of more unknowns. The examples can be found in Auth and Harder (1999) and Gerya and Yuen (2003, 2007).

For 3-D mantle convection, we can similarly employ the poloidal potential $\Phi$ and a vorticity-like scalar function $\Omega$ (Busse, 1989; Travis *et al.*, 1990). It is to be noted that this potential $\Phi$ is not the same as the stream function $\Psi$ in 2-D (*see* Chapter 4). From the general representation of an arbitrary solenoidal vector field (Busse, 1989), we can write a 3-D velocity vector as

$$\mathbf{v} = \nabla \times \nabla \times (\mathbf{e}_z \Phi) + \nabla \times (\mathbf{e}_z \Theta) \qquad [13]$$

where $\mathbf{e}_z$ is the unit vector in the vertical $z$-direction pointing upward. $\Theta$ is the toroidal potential and is present in problems with lateral variations of viscosity (Zhang and Christensen, 1993; Gable *et al.*, 1991). Thus, for constant viscosity, $\Theta$ is a constant and the velocity vector $\mathbf{v} = (u, v, w)$ involves higher-order derivatives of $\Phi$ in this formulation:

$$u = \frac{\partial^2 \Phi}{\partial y \partial z}, \quad v = \frac{\partial^2 \Phi}{\partial x \partial z}, \quad w = -\left(\frac{\partial^2 \Phi}{\partial x^2} + \frac{\partial^2 \Phi}{\partial y^2}\right) \quad [14]$$

The 3-D momentum equation for constant properties can be written as a system of coupled Poisson equations in 3-D:

$$\nabla^2 \Phi = \Omega \qquad [15]$$

$$\nabla^2 \Omega = Ra\,T \qquad [16]$$

with the energy equation as

$$\frac{\partial T}{\partial t} = \nabla^2 T - \mathbf{v} \cdot \nabla T + \gamma \qquad [17]$$

where all differential operators are 3-D in character, and $\Omega$ is a scalar function playing a role analogous to the vorticity in the 2-D formulation.

We note that in FD and FV methods the primitive variables are now dominant in terms of popularity in 3-D mantle convection with variable viscosity (Tackley, 1994; Albers, 2000; Ogawa *et al.*, 1991; Trompert and Hansen, 1996).

### 5.3.1.2 Approximations of spatial derivatives and solution approaches

In FD methods, spatial derivations in the above equations need to be approximated by values at grid points. There are many algorithms for generating the formulas in the FD approximation of spatial differential operators (e.g., Kowalik and Murty, 1993; Lynch, 2005). The most common algorithm is to use a Taylor expansion. For example, a second-order differential operator (for simplicity, consider the spatial derivative in the $x$-direction) can be expressed as

$$\frac{d^2 f(x)}{dx^2} = [f(x_1) + f(x_3) - 2f(x_2)]/(\Delta x)^2 \qquad [18]$$

where $x_k$ is a grid point while the derivative is computed at a point $x$ which is not necessarily a grid point (**Figure 1**(a)). This approximation is second-order accurate. We will use a recursive algorithm for generating the weights in the FD approximation with high-order accuracy. We will, for simplicity, consider the spatial derivative in the $x$-direction. The $m$th-order differential operator can be approximated by a FD operator as

$$\frac{d^n f(x)}{dx^n} = \sum_{k=1}^{j} W_{n,j,k} f(x_k) \qquad [19]$$

where $W_{n,j,k}$ are the weights to be applied at grid point $x_k$, in order to calculate the $n$th order derivative at the point $x$. The stencil is $j$ grid points wide. An ideal algorithm should accurately and efficiently produce weights for any order of approximation at arbitrarily distributed grid points. Such an algorithm has been developed and tested for simple differential operators by Fornberg (1990, 1995). The cost for generating each weight is just four operations and this can be done simply with a FORTRAN program given in Fornberg's book (Fornberg, 1995). The weights are collected in a 'differentiation matrix'. This results in calculating the derivatives of a vector, which is derived from a matrix–vector multiplicative operation. This algorithm can be easily generalized for determining the spatial derivatives for higher-dimensional operators.

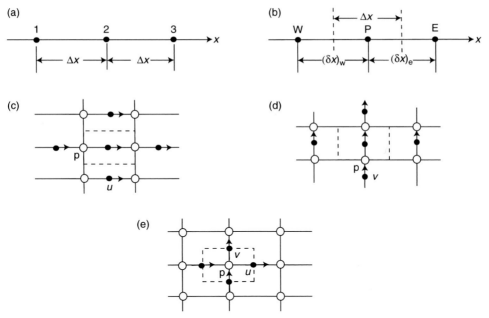

**Figure 1** A simple three-point one-dimensional (1-D) finite-difference stencil (a), a 1-D finite-volume control-cell (b), a staggered 2-D grid for velocity–pressure for horizontal (c) and vertical (d) components of the momentum equation and for the continuity equation (e).

In general, a $p$th-order FD method converges as $O(k - p)$ for $k$ grid points. However, variable coefficients and nonlinearities, such as the heat advection term, greatly complicate the convergence. The reader can find a more detailed discussion of the issues related to high-order FD methods and spectral methods in Fornberg (1995). Mantle convection in both 2-D and 3-D configurations using this variable-grid, high-order FD method has been studied by Larsen *et al.* (1995).

Using the stream-function vorticity formulation for 2-D and the scalar potential approach for 3-D, we can see that we have a large system of algebraic equations to solve at each time step. The partial differential equations posed by eqns [8] and [9] for 2-D and eqns [15] and [16] in 3-D represent one of the difficult aspects in mantle convection problems because of its elliptic nature. The linear algebra arising from this elliptic problem for large systems with varying size of matrix elements due to variable viscosity makes it very difficult to solve accurately. They can be written down symbolically as a large-scale matrix algebra problem,

$$Au = f \qquad [20]$$

where $A$ is a matrix operator involving the second-order spatial derivatives and the associated boundary conditions for the stream function and vorticity, $u$ is the solution vector over the set of 2-D grid points for $(\Phi, \Omega)$, and $f$ is a vector representing the right-hand side of the system from eqns [8] and [9], namely $\Omega$ and $\partial T/\partial x$ values at 2-D grid points from the previous time step.

The matrix equation from 2-D mantle convection problems with variable viscosity can be solved effectively and very accurately with the direct method, such as the Cholesky decomposition, with current computational memory architecture. A high-resolution 2-D problem with around $700 \times 700$ grid points and billions of tracers (Rudolf *et al.*, 2004) can be tackled on today's shared-memory machines with 1–2 TB of RAM memory, such as the S.G.I. Altix at National Center for Supercomputing Applications (NCSA) or the IBM Power-5 series in National Center for Atmospheric Research (NCAR). This tact of using shared-memory architecture has been employed by Gerya *et al.* (2006) to solve high-resolution 2-D convection problems with strongly variable viscosity. The presence of variable viscosity would make the matrix $A$ very singular because of the disparity of values to the matrix elements. The adverse condition of the matrix is further aggravated by the presence of non-Newtonian rheology, which makes the matrix $A$ nonlinear, thus making it necessary to use iterative method for solving the matrix equation. Multigrid iterative method (e.g., Hackbush, 1985; Press *et al.*, 1992; Trottenberg *et al.*, 2001; Yavneh, 2006) is a well-proven way of solving the elliptic problem with strong nonlinearities. Davies (1995) solved variable-viscosity convection problem in 2-D with the FD-based multigrid technique.

For 3-D mantle convection problems, one must still employ the iterative method because of memory requirements associated with a FD grid configuration with over a million unknowns. The multigrid iterative method used for an FD grid has been used by Parmentier and Sotin (2000) for solving 3-D high Rayleigh number mantle convection. We note that for variable-viscosity convection in 3-D one must use the pressure–velocity formulation cast in FDs (e.g., Peyret and Taylor, 1983) for the momentum equation. The resulting FD equations can be solved with the multigrid method. In this connection higher-order FD method (Fornberg, 1995) with variable grid points together with the multigrid technique may bring about a renaissance to the FD method because of its favorable posture in terms of memory requirements over FEs.

Another source of numerical difficulties in the mantle convection is due to the temperature advection term, $\mathbf{v} \cdot \nabla T$. It is well known from linear analysis that numerical oscillations result, if simple FD schemes are used to calculate the spatial derivative due to the interaction between velocity and the FD approximations of $\nabla T$. Excellent discussions of this problem involving numerical dispersion can be found in Kowalik and Murty (1993). Different numerical approximations have been proposed to treat the advection term in the FD: the first is by Spalding (1972), which involves a weighted upwind scheme and is first-order accurate, which has been employed in the early FD codes of mantle convection (Turcotte *et al.*, 1973). A more popular and effective method is an iterated upwind correction scheme that is correct to second order and was proposed by Smolarkiewicz (1983, 1984). This is a scheme based on the positively definite character of the advection term. The solutions, in general, do not change much after one to two correction steps and it is easy to program for parallel computers. This scheme has been implemented by Parmentier and Sotin (2000) and by Tackley (2000) for mantle convection problems. Use of higher-order FD schemes (Fornberg, 1995) will also help to increase the accuracy of the advection of

temperature because this will lead closer to a pseudospectral quality. Other advection schemes include the semi-Lagrangian technique, which is based on tracer characteristics and have been used in mantle convection for the stream-function method (Malevsky and Yuen, 1991) and for primitive variables (Gerya and Yuen, 2003).

### 5.3.2   FV Method

A FV method is often used to solve differential equations (Patankar, 1980). FV methods share some common features with FE and FD methods. In FV methods, discretization of a differential equation results from integrating the equation over a control-volume or control-cell and approximating differential operators of reduced order at cell boundaries (Patankar, 1980). Patankar (1980) suggested that the FV formulation may be considered as a special case of the weighted residual method in FE in which the weighting function is uniformly one within an element and zero outside of the element. The FV method is also similar to FD method in that they both need to approximate differential operators using values at grid points.

We will use a simple example to illustrate the basic idea of the FV method (Patankar, 1980). Consider a 1-D heat conduction equation with heat source $\gamma$.

$$\frac{\mathrm{d}}{\mathrm{d}x}\left(k\frac{\mathrm{d}T}{\mathrm{d}x}\right) + \gamma = 0 \qquad [21]$$

where $k$ is the heat conductivity. This equation is integrated over a control-cell bounded by dashed lines in **Figure 1(b)**, and the resulting equation is

$$\left(k\frac{\mathrm{d}T}{\mathrm{d}x}\right)_{\mathrm{e}} - \left(k\frac{\mathrm{d}T}{\mathrm{d}x}\right)_{\mathrm{w}} + \int_{\mathrm{w}}^{\mathrm{e}} \gamma\,\mathrm{d}x = 0 \qquad [22]$$

where e and w represent the two ends of the control-cell (**Figure 1(b)**). Introducing approximations for the flux at each end and the source term in eqn [22] leads to

$$\frac{k_{\mathrm{e}}(T_{\mathrm{E}} - T_{\mathrm{P}})}{(\delta x)_{\mathrm{e}}} - \frac{k_{\mathrm{w}}(T_{\mathrm{P}} - T_{\mathrm{W}})}{(\delta x)_{\mathrm{w}}} + \bar{\gamma}\Delta x = 0 \qquad [23]$$

where $k_{\mathrm{e}}$ and $k_{\mathrm{w}}$ are the heat conductivity at cell boundaries, $T_{\mathrm{E}}$, $T_{\mathrm{P}}$, and $T_{\mathrm{W}}$ are temperatures at nodal points, $\bar{\gamma}$ is the averaged source in the control-cell, and $\Delta x = [(\delta x)_{\mathrm{e}} + (\delta x)_{\mathrm{w}}]/2$ is the size of the control-cell (**Figure 1(b)**). The eqn [23] is a discrete equation with nodal temperatures as unknown for the given control-cell. Applying this procedure to all control-cells leads to a system of equations with unknown temperatures at all grid points, similar to eqn [20] from a FD method.

FV methods have also been used extensively in modeling mantle convection, possibly starting with Ogawa et al. (1991). Tackley (1994) implemented an FV method coupled with a multigrid solver for 2-D/3-D Cartesian models. Ratcliff et al. (1996) developed a 3-D spherical-shell model of mantle convection using an FV method. Harder and Hansen (2005) and Stemmer et al. (2006) also developed FV mantle convection codes for spherical-shell geometries using a cubed sphere grid. All these mantle convection studies used second-order accurate FV methods, although higher-order formulas can be formulated.

To solve fluid flow problems that are governed by the continuity and momentum equations, such as those in mantle convection, it is often convenient to use the pressure–velocity formulation, similar to FE method. Also, a staggered grid is often used in which pressures and velocities are defined at different locations of a control-cell (Patankar, 1980; Tackley, 1994) (**Figures 1(c)–1(e)**). Such a staggered grid helps remove checkerboard pressure solutions. In the staggered grid, control-cells for the momentum equation are different from those for the continuity equation. For example, for the continuity equation, a pressure node is at the center of a control-cell, while velocities are defined at the cell boundaries such that each velocity component is perpendicular to the corresponding cell boundary (**Figure 1(e)**). For the momentum equation, the velocities are at the center of a cell, while the pressures are defined at cell boundaries (**Figures 1(c)** and **1(d)**).

For a given pressure field, applying the FV procedure leads to discrete equations for velocities. However, solutions to the velocity equations are only accurate if the pressure field is accurate. The continuity equation is used to correct the pressure field. This iterative procedure between the velocity and pressure is implemented efficiently in a SIMPLER algorithm (Patankar, 1980) that is used in the FV convection codes (Tackley, 1993, 1994; Stemmer et al., 2006). A variety of methods can be used to solve the discrete equations of velocities and pressure, similar to that in FD method. They may include successive over-relaxation and Gauss–Seidel iteration (Harder and Hansen, 2005; Stemmer et al., 2006), or a multigrid method (Tackley, 1994), or the alternating-direction implicit (i.e., ADI) method (Monnereau and Yuen, 2002).

Like in FE and FD methods, the FV method also requires special treatment of advection term $\mathbf{v} \cdot \nabla T$ in the energy equation. Either upwind scheme or an iterative correction scheme such as that proposed by Smolarkiewicz (1983, 1984) in Section 5.3.1 on FD can be used in the FV method to treat the advection term.

### 5.3.3 Spectral Methods

The spectral method is a classical method, which is motivated by its analytical popularity and inherent accuracy. It is based on the concept of orthogonal eigenfunction expansion based on the differential operators associated with the Laplace equation and works on orthogonal curvilinear coordinate systems. For mantle convection problems this means that we can express the horizontal dependence by using Fourier expansion for Cartesian geometry and spherical harmonics for the 3-D spherical shell. Symbolically we can write down this expansion as

$$F(x_1, x_2, x_3) = \sum_{ijk} a_{ijk} f(x_3) g(x_1) h(x_2) \quad [24]$$

where $F$ is the field variable being expanded, $a_{ijk}$ is the spectral coefficient, $x_1$ and $x_2$ are the horizontal coordinates, $x_3$ is the vertical or radial coordinate, $g$ and $h$ are the eigenfunctions being employed, and $f$ is a function describing the vertical or radial dependence. We note that $f$ can be determined by solving a two-point boundary value, using propagator matrices, or an orthogonal function expansion, as in the case of Chebychev polynomials, or using the FD (e.g., Cserepes and Rabionowicz, 1985; Cserepes et al., 1988; Gable et al., 1991; Travis et al., 1990; Machetel et al., 1986; Glatzmaier, 1988).

Balachandar and Yuen (1994) developed a spectral-transform 3-D Cartesian code for mantle convection, based on expansion of Chebychev functions for solving the two-point boundary-value problem in the vertical direction and fast Fourier transforms along the two horizontal directions. This method has also been extended to variable-viscosity problems with the help of Krylov subspace iterative technique (Saad and Schultz, 1986) for solving iteratively the momentum equation. The first 3-D codes in spherical-shell convection were developed by Machetel et al. (1986) who used an FD scheme in the radial direction and spherical harmonic decomposition over the longitudinal and latitudinal directions and by Glatzmaier (1988) who employed Chebychev polynomials in solving the radial direction and spherical harmonics for the spherical surface. The Chebychev polynomials allow for very high accuracy near the thermal boundary layers. They can also be expanded about internal boundary layers, such as the one at 670 km discontinuity. Zhang and Yuen (1995 and 1996a) have devised a 3-D code for spherical geometry based on higher-order FD method (Fornberg, 1995) and spectral expansion in the field variables in spherical harmonics along the circumferential directions. They also developed an iterative technique for the momentum equation to solve for variable viscosity with a lateral viscosity contrast up to 200 (Zhang and Yuen, 1996b) in compressible convection. It is important to check also the spectra of the viscosity decay with the degree of the spherical harmonics (Zhang and Yuen, 1996a) in order to ascertain whether Gibbs phenomenon is present.

Spectral methods are very accurate, much more so than FEs and FDs for the same computational efforts, and their efficacy in terms of convergence can be assessed by examining the rate of decay of the energy in the spectrum with increasing number of terms in the spectral expansion in eqn [24] (Peyret and Taylor, 1983). Spectral methods also have a couple of other distinct advantages. First, they are very easy to implement, because they reduce the problems to either an algebraic set or simple weakly coupled ordinary differential equations, as discussed earlier. Second, using fast transform algorithms such as fast Fourier transform, spectral methods can be very efficient and fast on a single-processor computer.

However, spectral methods only work well for constant material properties or depth-dependent properties in simple geometries (Balachandar and Yuen, 1994). Furthermore, spectral techniques do not go as far in terms of viscosity contrasts (less than a factor of 200) for variable-viscosity convection (Balachandar et al., 1995, Zhang and Yuen, 1996b). We note, however, that Schmalholz et al. (2001) showed that in a folding problem their spectral method could go up to a viscosity contrast of around $10^4$. Perhaps a better choice of a preconditioner in the solution of the momentum equation in the spectral expansion of a variable-viscosity problem will enable a larger viscosity contrast than $10^4$. This issue is still open and remains a viable research topic. Spectral methods are also difficult to be used efficiently on parallel computers because of the global basis functions used in these methods. At present, spectral methods, although elegant mathematically, are not as effective as other numerical methods for solving

realistic mantle convection problems. Spectral methods still remain the main driver for geodynamo problems (e.g., Glatzmaier and Roberts, 1996; Kuang and Bloxham, 1999) because of the exclusive choice of constant material properties in solving the set of magneto-hydrodynamic equations. Lastly, we wish to mention briefly a related spectral method called the spectral-element method (Maday and Patera, 1989), which is now being used in geophysics in seismic-wave propagation (Komatitsch and Tromp, 1999). This approach may hold some promise for variable-viscosity mantle convection.

## 5.4 An FE Method

FE methods are effective in solving differential equations with complicated geometry and material properties. FE methods have been widely used in the studies of mantle dynamics (Christensen, 1984; Baumgardner, 1985; King et al., 1990; van den Berg et al., 1993; Moresi and Gurnis, 1996; Bunge et al., 1997; Zhong et al., 2000). This section will go through some of the basic steps in using FE methods in solving governing equations for thermal convection.

The FE formulation for Stokes flow that is described by eqns [1] and [2] is independent from that for the energy equation. Hughes (2000) gave detailed description on a Galerkin weak-form FE formulation for the Stokes flow. Brooks (1981) developed a streamline upwind Petrov–Galerkin formulation (SUPG) for the energy equation involving advection and diffusion. These two formulations remain popular for solving these types of problems (Hughes, 2000) and are employed in mantle convection codes ConMan (King et al., 1990) and Citcom/CitcomS (Moresi and Gurnis, 1996; Zhong et al., 2000). The descriptions presented here are tailored from Brooks (1981), Hughes (2000), and Ramage and Wathen (1994) specifically for thermal convection in an incompressible medium, and they are also closely related to codes ConMan and Citcom.

### 5.4.1 Stokes Flow: A Weak Formulation, Its FE Implementation, and Solution

#### 5.4.1.1 A weak formulation

The Galerkin weak formulation for the Stokes flow can be stated as follows: find the flow velocity $u_j = v_i + g_i$ and pressure $P$, where $g_i$ is the prescribed boundary velocity from eqn [6] and $v_i \in \mathbf{V}$, and $P \in \mathbf{P}$, where $\mathbf{V}$ is a set of functions in which each function,

$w_i$, is equal to zero on $\Gamma_{g}$, and $\mathbf{P}$ is a set of functions $q$, such that for all $w_i \in \mathbf{V}$ and $q \in \mathbf{P}$

$$\int_{\Omega} w_{i,j}\sigma_{ij}\,\mathrm{d}\Omega - \int_{\Omega} qu_{i,i}\,\mathrm{d}\Omega = \int_{\Omega} w_i f_i\,\mathrm{d}\Omega$$
$$+ \sum_{i=1}^{n_{sd}} \int_{\Gamma_{b_i}} w_i b_i\,\mathrm{d}\Gamma \qquad [25]$$

$w_i$ and $q$ are also called weighting functions. Equation [25] is equivalent to eqns [1] and [2] and boundary conditions equation [6], provided that $f_i = RaT\delta_{iz}$ (Hughes, 2000). Equation [25] can be written as

$$\int_{\Omega} w_{i,j}c_{ijkl}v_{k,l}\,\mathrm{d}\Omega - \int_{\Omega} qv_{i,i}\,\mathrm{d}\Omega - \int_{\Omega} w_{i,i}P\,\mathrm{d}\Omega$$
$$= \int_{\Omega} w_i f_i\,\mathrm{d}\Omega + \sum_{i=1}^{n_{sd}} \int_{\Gamma_{b_i}} w_i b_i\,\mathrm{d}\Gamma - \int_{\Omega} w_{i,j}c_{ijkl}g_{k,l}\,\mathrm{d}\Omega \qquad [26]$$

where

$$c_{ijkl} = \eta(\delta_{ik}\delta_{jl} + \delta_{il}\delta_{jk}) \qquad [27]$$

is derived from constitutive eqn [4]. It is often convenient to rewrite

$$w_{i,j}c_{ijkl}v_{k,l} = \varepsilon(\mathbf{w})^{\mathrm{T}}D\varepsilon(\mathbf{v}) \qquad [28]$$

where for 2-D plane strain problems,

$$\varepsilon(\mathbf{v}) = \left\{ \begin{array}{c} v_{1,1} \\ v_{2,2} \\ v_{1,2} + v_{2,1} \end{array} \right\}, \qquad D = \begin{bmatrix} 2\eta & 0 & 0 \\ 0 & 2\eta & 0 \\ 0 & 0 & \eta \end{bmatrix} \qquad [29]$$

(note here the change in the definition of the off-diagonal components of the strain-rate tensor). It is straightforward to write the expressions for other coordinate systems including 3-D Cartesian, axisymmetric (Hughes, 2000), or spherical geometry (Zhong et al., 2000).

Suppose that we discretize the solution domain $\Omega$ using a set of grid points (**Figure 2**) so that the velocity and pressure fields and their weighting functions can be expressed anywhere in the domain in terms of their values at the grid (nodal) points and the so-called 'shape functions' which interpolate the grid points:

$$\mathbf{v} = v_i \mathbf{e}_i = \sum_{A \in \Omega^v - \Gamma_{g_i}^v} N_A v_{iA} \mathbf{e}_i,$$

$$\mathbf{w} = w_i \mathbf{e}_i = \sum_{A \in \Omega^v - \Gamma_{g_i}^v} N_A w_{iA} \mathbf{e}_i,$$

$$\mathbf{g} = \sum_{A \in \Gamma_{g_i}^v} N_A g_{iA} \mathbf{e}_i \qquad [30]$$

$$P = \sum_{B \in \Omega^p} M_B P_B, \qquad q = \sum_{B \in \Omega^p} M_B q_B \qquad [31]$$

where $N_A$ is the shape functions for velocity at node A, $M_B$ is the shape functions for pressure at node B,

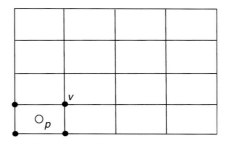

**Figure 2** Finite-element discretization and grid in 2-D. For a four-node bilinear element, the pressure is defined at the center of the element, while velocities are defined at the four corners.

$\Omega^v$ is the set of velocity nodes, $\Omega^p$ is the set of pressure nodes, and $\Gamma_{g}^v$ is the set of velocity nodes along boundary $\Gamma_{g}$. Note that the velocity shape functions and velocity nodes can be (and usually are) different from those for pressure (**Figure 2**).

Substituting eqns [30] into [28] leads to the following equation:

$$\varepsilon(\mathbf{w})^{\mathrm{T}}D\varepsilon(\mathbf{v})$$
$$= \varepsilon\left( \sum_{A \in \Omega^v - \Gamma_{g_i}^v} N_A w_{iA}\mathbf{e}_i \right)^{\mathrm{T}} D\varepsilon\left( \sum_{B \in \Omega^v - \Gamma_{g_j}^v} N_B v_{jB}\mathbf{e}_j \right)$$
$$= \left[ \sum_{A \in \Omega^v - \Gamma_{g_i}^v} \varepsilon(N_A \mathbf{e}_i)^{\mathrm{T}} w_{iA} \right] D\left[ \sum_{B \in \Omega^v - \Gamma_{g_j}^v} \varepsilon(N_B \mathbf{e}_j) v_{jB} \right]$$
$$= \sum_{A \in \Omega^v - \Gamma_{g_i}^v} w_{iA}\left[ \sum_{B \in \Omega^v - \Gamma_{g_j}^v} \mathbf{e}_i^{\mathrm{T}} B_A^{\mathrm{T}} D B_B \mathbf{e}_j v_{jB} \right] \qquad [32]$$

where for 2-D plane strain problems

$$B_A = \begin{bmatrix} N_{A,1} & 0 \\ 0 & N_{A,2} \\ N_{A,2} & N_{A,1} \end{bmatrix} \qquad [33]$$

Substituting eqns [30] and [31] into [26] leads to the following equation:

$$\sum_{A \in \Omega^v - \Gamma_{g_i}^v} w_{iA}\left[ \sum_{B \in \Omega^v - \Gamma_{g_j}^v} \left( \mathbf{e}_i^{\mathrm{T}} \int_{\Omega} B_A^{\mathrm{T}} D B_B\,\mathrm{d}\Omega\,\mathbf{e}_j v_{jB} \right) \right.$$
$$- \sum_{B \in \Omega^p} \left( \mathbf{e}_i \int_{\Omega} N_{A,i} M_B\,\mathrm{d}\Omega\,P_B \right) \Bigg]$$
$$- \sum_{A \in \Omega^p} \left[ q_A \sum_{B \in \Omega^v - \Gamma_{g_j}^v} \left( \int_{\Omega} M_A N_{B,j}\,\mathrm{d}\Omega\,\mathbf{e}_j v_{jB} \right) \right]$$
$$= \sum_{A \in \Omega^v - \Gamma_{g_i}^v} w_{iA}\left[ \int_{\Omega} N_A \mathbf{e}_i f_i\,\mathrm{d}\Omega + \sum_{i=1}^{n_{sd}} \int_{\Gamma_{b_i}} N_A \mathbf{e}_i b_i\,\mathrm{d}\Gamma \right.$$
$$- \sum_{B \in \Gamma_{g_j}^v} \left( \mathbf{e}_i^{\mathrm{T}} \int_{\Omega} B_A^{\mathrm{T}} D B_B\,\mathrm{d}\Omega\,\mathbf{e}_j g_{jB} \right) \Bigg] \qquad [34]$$

Because eqn [34] holds for any weighting functions $w_{iA}$ and $q_A$, it implies the following two equations:

$$\sum_{B \in \Omega^v - \Gamma^v_{g_j}} \left( \mathbf{e}_i^T \int_\Omega B_A^T D B_B \, d\Omega \, \mathbf{e}_j v_{jB} \right)$$

$$- \sum_{B \in \Omega^p} \left( \mathbf{e}_i \int_\Omega N_{A,\,i} M_B \, d\Omega \, P_B \right)$$

$$= \int_\Omega N_A \mathbf{e}_i f_i \, d\Omega + \sum_{i=1}^{n_{sd}} \int_{\Gamma b_i} N_A \mathbf{e}_i h_i \, d\Gamma$$

$$- \sum_{B \in \Gamma^v_{g_j}} \left( \mathbf{e}_i^T \int_\Omega B_A^T D B_B \, d\Omega \, \mathbf{e}_j g_{jB} \right) \qquad [35]$$

$$\sum_{B \in \Omega^v - \Gamma^v_{g_j}} \left( \int_\Omega M_A N_{B,\,j} \, d\Omega \, \mathbf{e}_j v_{jB} \right) = 0 \qquad [36]$$

Combining eqns [35] and [36] into a matrix form leads to

$$\begin{bmatrix} K & G \\ G^T & 0 \end{bmatrix} \begin{Bmatrix} V \\ P \end{Bmatrix} = \begin{Bmatrix} F \\ 0 \end{Bmatrix} \qquad [37]$$

where the vector $V$ contains the velocity at all the nodal points, the vector $P$ is the pressure at all the pressure nodes, the vector $F$ is the total force term resulting from the three terms on the right-hand side of eqn [35], the matrices $K$, $G$, and $G^T$ are the stiffness matrix, discrete gradient operator, and discrete divergence operator, respectively, which are derived from the first and second terms of eqns [35] and [36], respectively. Specifically, the stiffness matrix is given by

$$K_{lm} = \mathbf{e}_i^T \int_\Omega B_A^T D B_B \, d\Omega \, \mathbf{e}_j \qquad [38]$$

where subscripts A and B are the global velocity node numbers as in eqn [30], $i$ and $j$ are the degree of freedom numbers ranging from 1 to $n_{sd}$, and $l$ and $m$ are the global equation numbers for the velocity ranging from 1 to $n_v n_{sd}$ where $n_v$ is the number of velocity nodes.

### 5.4.1.2  An FE implementation

We now present an FE implementation of the Galerkin weak formulation for the Stokes flow and the resulting expressions of different terms in [37]. A key point of the FE approach is that all of the equations to be solved are written in the form of integrals over the solution domain and can, therefore, without approximation, be written as a sum of integrals over convenient subdomains, and the matrix eqn [37] may be decomposed into overlapping sums of contributions from these domains.

Let us first introduce the elements and shape functions. A key feature of the standard FE method is that a local basis function or shape function is used such that the value of a variable within an element depends only on that at nodal points of the element. The shape functions are generally chosen such that they have the value of unity on their parent node and zero at all other nodes and zero outside the boundary of the element. Unless there is a special, known form to the solution, it is usual to choose simple polynomial shape functions and form their products for additional dimensions. In addition, this interpolating requirement constrains the patterns of nodes in an element and which elements can be placed adjacent to each other. Bathe (1996) gives an excellent overview of the issues. (As usual, there are some useful exceptions to this rule, one being the element-free Galerkin method where smooth, overlapping interpolating kernels are used. This does mean, however, that the global problem cannot be trivially decomposed into local elements.) Zienkiewicz et al. (2005) is also an excellent source of reference for these topics.

For simplicity, we consider a 2-D domain with quadrilateral elements. We employ mixed elements in which there are four velocity nodes per element each of which occupies a corner of the element, while the only pressure node is at the center of the element (**Figure 2**). For these quadrilateral elements, the velocity interpolation in each element uses bilinear shape functions, while the pressure is constant for each element.

As a general remark on FE modeling of deformation/flow of incompressible media, it is important to keep interpolation functions (shape functions) for velocities at least 1 order higher than those for pressure, as we did for our quadrilateral elements (Hughes, 2000). Spurious flow solutions may arise sometimes even if this condition is satisfied. The best-known example is the 'mesh locking' that arises from linear triangle elements with constant pressure per element for which incompressibility (i.e., a fixed elemental area) constraint per element demands zero deformation/flow everywhere in the domain (Hughes, 2000).

For any given element e, velocity and pressure within this element can be expressed through the following interpolation:

$$\mathbf{v} = v_i \mathbf{e}_i = \sum_{a=1}^{n_{en}} N_a v_{ia} \mathbf{e}_i, \quad \mathbf{w} = w_i \mathbf{e}_i = \sum_{b=1}^{n_{en}} N_b w_{ib} \mathbf{e}_i,$$

$$\mathbf{g} = g_i \mathbf{e}_i = \sum_{a=1}^{n_{en}} N_a g_{ia} \mathbf{e}_i \qquad [39]$$

$$P = \sum_{a=1}^{n_{ep}} M_a P_a, \quad q = \sum_{a=1}^{n_{ep}} M_a q_a \qquad [40]$$

where $n_{en}$ and $n_{ep}$ are the numbers of velocity and pressure nodes per element, respectively, and $n_{en} = 4$ and $n_{ep} = 1$ for our quadrilateral elements. The shape function $N_a$ for $a = 1, \ldots, n_{en}$ depends on coordinates, and $N_a$ is 1 at node $a$ and linearly decreases to zero at other nodes of the element. The locality of the shape functions greatly simplifies implementation and computational aspects of the Galerkin weak formulation. For example, the integrals in eqns [35] and [36] may be decomposed into sum of integrals from each element and the matrix equation [37] may be decomposed into sums of elemental contributions. Specifically, we may now introduce elemental stiffness matrix, discrete gradient and divergence operators, and force term.

$$k^e = [k_{lm}^e], \quad g^e = [g_{ln}^e], \quad f^e = \{f_l^e\} \qquad [41]$$

where $1 \le l$, $m \le n_{en} n_{sd}$, $1 \le n \le n_{ep}$ (note that for quadrilateral elements, $n_{en} = 4$, $n_{ep} = 1$, and $n_{sd} = 2$), $k^e$ is a square matrix of $n_{en} n_{sd}$ by $n_{en} n_{sd}$, and $g^e$ is a matrix of $n_{en} n_{sd}$ by $n_{ep}$.

$$k_{lm}^e = \mathbf{e}_i^T \int_{\Omega^e} B_a^T D B_b \, \mathrm{d}\Omega \, \mathbf{e}_j \qquad [42]$$

where $l = n_{sd}(a-1) + i$, $m = n_{sd}(b-1) + j$, $a, b = 1, \ldots, n_{en}$, and $i, j = 1, \ldots, n_{sd}$

$$g_{ln}^e = -\mathbf{e}_i \int_{\Omega^e} N_{a,i} M_n \, \mathrm{d}\Omega \qquad [43]$$

where $n = 1, \ldots, n_{ep}$, and the rest of the symbols have the same definitions as before:

$$f_l^e = \int_{\Omega^e} N_a f_i \, \mathrm{d}\Omega + \int_{\Gamma_{b_i}^e} N_a b_i \, \mathrm{d}\Gamma - \sum_{m=1}^{n_{sd} n_{en}} k_{lm}^e g_m^e \qquad [44]$$

Determinations of these elemental matrices and force term require evaluations of integrals over each element with integrands that involve the shape functions and their derivatives. It is often convenient to

use isoparametric elements for which the coordinates and velocities in an element have the interpolation schemes (Hughes, 2000). For example, for 2-D quadrilateral elements that we discussed earlier, the velocity shape functions for node $a$ of an element in a parent domain with coordinates $\xi \in (-1, 1)$ and $\eta \in (-1, 1)$ is given as

$$N_a(\xi, \eta) = 1/4(1 + \xi_a \xi)(1 + \eta_a \eta) \qquad [45]$$

where $(\xi_a, \eta_a)$ is $(-1, -1)$, $(1, -1)$, $(1, 1)$, and $(-1, 1)$ for $a = 1, 2, 3$, and 4, respectively. The pressure shape functions for $M_a$ is 1, as there is only one pressure node per element. Although the integrations in [42]–[44] are in the physical domain (i.e., $x_1$ and $x_2$ coordinates) rather than the parent domain, they can be expressed in the parent domain through coordinate transformation.

In practice, these element integrals are calculated numerically by some form of quadrature rule. In 2-D, Gaussian quadrature rules are optimal and are usually recommended; in 3-D, other rules may be more efficient but are not commonly used for reasons of programming simplicity (Hughes (2000) documents integration procedure in detail). There are a small number of cases where it may be worthwhile using a nonstandard procedure to integrate an element. The most common is where the constitutive parameters (i.e., $D$ in eqn [29]) change within the element; a higher-order quadrature scheme than the standard recommended one can give improved accuracy in computing $k^e$. When the constitutive parameters are strongly history dependent, $D$ is known only at a number of Lagrangian sample points; if these are used directly to integrate $k^e$, the Lagrangian integration point methods such as MPM result (Sulsky *et al.*, 1994; Moresi *et al.*, 2003 for application in geodynamics).

With elemental $k^e$, $g^e$, and $f^e$ determined, it is straightforward to assemble them into global matrix eqn [37]. If an iterative solution method is used to solve [37], one may carry out calculations of the left-hand side of [37] element by element without assembling elemental matrices and force terms into the global matrix equation form [37].

The boundary conditions described in eqn [44] are simple to implement when the boundary is aligned with the coordinate system, and when the boundary condition is purely velocity or purely boundary tractions. The difficulty arises in the case of boundaries which are not aligned with the coordinate system and which specify a mixture of boundary tractions and velocities. This form of boundary condition is common in engineering applications of FE method, designing mechanical components, for example, and in geodynamics it occurs when modeling arbitrarily oriented contact surfaces such as faults, or a spherical boundary in a Cartesian solution domain. Unless treated appropriately, this form of boundary condition leads to a constraint on a linear combination of the degrees of freedom at each constrained node. The solution is straightforward: instead of using the global coordinate system to form the matrix equation, we first transform to a local coordinate system which does conform to the boundary geometry (Desai and Abel, 1972).

Let $R$ be a rotation matrix which transforms the global coordinate system XYZO to a local one X'Y'Z'O' aligned with the boundary at node $n$. The transformed stiffness matrix and force vector contributions for this node then become

$$K'_n = R_n^T K_n R_n \qquad [46]$$

and

$$F'_n = R_n^T F_n \qquad [47]$$

$R$ can be formulated for each node or it may be assembled for an element or for the global matrices but, other than the nodes where the 'skew' boundary conditions occur, the block entries are simply the identity matrix. It is possible to use this procedure at every node in a system with a natural coordinate system (e.g., the spherical domain) so that one can switch freely between a spherical description of forces and velocities where convenient, and an underlying Cartesian formulation of the constitutive relations. This method was used in modeling faults in Zhong *et al.* (1998).

### 5.4.1.3   The Uzawa algorithm for the matrix equation

Similar to FD or FV methods, the FE discretization of the differential equations leads to a matrix equation such as [37]. The remaining question is to solve the matrix equation, which we now discuss in this section. We can use either direct or iterative methods to solve the matrix equation, depending on problems that we are interested in, similar to what we discussed earlier for matrix equation [20] from FD method. Here we will focus on iterative solution methods, because they require significantly less memory and computation than direct solution approaches.

Iterative solution approaches are the only feasible and practical approaches for 3-D problems. Later we will briefly discuss a penalty formulation for the incompressible Stokes flow that requires a direct solution approach and is only effective for 2-D problems.

The system of equations as it stands is singular due to the block of zero entries in the diagonal, but it is symmetric, and the stiffness matrix $K$ is symmetric positive definite and we can use this to our advantage in finding a solution strategy. An efficient method is the Uzawa algorithm which is implemented in Citcom code (Moresi and Solomatov, 1995). In the Uzawa algorithm, the matrix equation [37] is broken into two coupled systems of equations (Atanga and Silvester, 1992; Ramage and Wathen, 1994):

$$KV + GP = F \qquad [48]$$

$$G^T V = 0 \qquad [49]$$

Combining these two equations and eliminating $V$ to form the Schur complement system for pressure (Hughes, 2000)

$$(G^T K^{-1} G)P = G^T K^{-1} F \qquad [50]$$

Notice that matrix $\hat{K} = G^T K^{-1} G$ is symmetric positive definite. Although in practice eqn [51] cannot be directly used to solve for $P$ due to difficulties in obtaining $K^{-1}$, we may use it to build a pressure-correction approach by using a conjugate gradient algorithm which does not require construction of matrix $\hat{K}$ (Ramage and Wathen, 1994). The procedure is presented and discussed in detail as follows.

With the conjugate gradient algorithm, for symmetric positive definite $\hat{K}$, the solution to a linear system of equations $\hat{K}P = H$ can be obtained with the operations in the left column of **Figure 3** (Golub and van Loan, 1989, p. 523).

For equations [48] and [50] for both velocities and pressure, with initial guess pressure $P_0 = 0$, the initial velocity $V_0$ can be obtained from

$$KV_0 = F, \quad \text{or} \quad V_0 = K^{-1} F \qquad [51]$$

and the initial residual for pressure equation [50], $r_0$, and search direction, $s_1$, are $r_0 = s_1 = H = G^T K^{-1} F = G^T V_0$ (see the left column of **Figure 3**). To determine the search step $\alpha_k$ in the conjugate gradient algorithm, we need to compute the product of search direction $s_k$ and $\hat{K}$, $s_k^T \hat{K} s_k$ (**Figure 3**). This product can be evaluated without explicitly constructing $\hat{K}$ for the following reasons.

```
k = 0; P₀ = 0; r₀ = H
while |rₖ| > a given tolerance, ε
    k = k + 1
    if k = 1
        s₁ = r₀
    else
        βₖ = rᵀₖ₋₁rₖ₋₁/rᵀₖ₋₂rₖ₋₂
        sₖ = rₖ₋₁ + βₖsₖ₋₁
    end
    αₖ = rᵀₖ₋₁rₖ₋₁/sᵀₖK̂sₖ
    Pₖ = Pₖ₋₁ + αₖsₖ
    rₖ = rₖ₋₁ + αₖK̂sₖ
end
P = Pₖ
```

```
k = 0; P₀ = 0
Solve KV₀ = F
r₀ = H = GᵀV₀
while |rₖ| > a given tolerance, ε
    k = k + 1
    if k = 1
        s₁ = r₀
    else
        βₖ = rᵀₖ₋₁rₖ₋₁/rᵀₖ₋₂rₖ₋₂
        sₖ = rₖ₋₁ + βₖsₖ₋₁
    end
    Solve Kuₖ = Gsₖ
    αₖ = rᵀₖ₋₁rₖ₋₁/(Gsₖ)ᵀuₖ
    Pₖ = Pₖ₋₁ + αₖsₖ
    Vₖ = Vₖ₋₁ − αₖuₖ
    rₖ = rₖ₋₁ − αₖGᵀuₖ
end
P = Pₖ, V = Vₖ
```

**Figure 3**   The left column is the original conjugate gradient algorithm, and the right column is the Uzawa algorithm for solving eqns [48] and [49].

The product can be written as

$$s_k^T \hat{K} s_k = s_k^T G^T K^{-1} G s_k = (G s_k)^T K^{-1} G s_k \qquad [52]$$

If we define $u_k$, such that

$$K u_k = G s_k, \quad \text{or} \quad u_k = K^{-1} G s_k \qquad [53]$$

then we have

$$s_k^T \hat{K} s_k = (G s_k)^T K^{-1} G s_k = (G s_k)^T u_k \qquad [54]$$

This indicates that if we solve [53] for $u_k$ with $G s_k$ as the force term, the product $s_k^T \hat{K} s_k$ can be obtained without actually forming $\hat{K}$. Similarly, $\hat{K} s_k$ in updating the residual $r_k$ (the left column of **Figure 3**) can be obtained without forming $\hat{K}$, because

$$\hat{K} s_k = G^T K^{-1} G s_k = G^T u_k$$

As the pressure $P$ is updated via $P_k = P_{k-1} + \alpha_k s_k$ from the conjugate gradient algorithm (**Figure 3**), the velocity field can also be updated accordingly via

$$V_k = V_{k-1} - \alpha_k u_k \qquad [55]$$

This can be seen from the following derivation. At iteration step $k-1$, the pressure and velocity are $P_{k-1}$ and $V_{k-1}$, respectively, and they satisfy eqn [48]

$$K V_{k-1} + G P_{k-1} = F \qquad [56]$$

At iteration step $k$, the updated pressure is $P_k$, and the velocity $V_k = V_{k-1} + v$, where $v$ is the unknown increment to be determined. Substituting

$P_k$ and $V_k$ into [48] and considering $P_k = P_{k-1} + \alpha_k s_k$, $V_k = V_{k-1} + v$, and [56] lead to

$$K v + \alpha_k G s_k = 0 \quad \text{or} \quad v = -\alpha_k K^{-1} G s_k \qquad [57]$$

From [53], it is clear that the velocity increment $v = -\alpha_k u_k$, and consequently eqn [55] updates the velocity.

The final algorithm is given in the right column of **Figure 3** (Ramage and Wathen, 1994). The efficiency of this algorithm depends on how efficiently eqn [53] is solved. We note that the choice of conjugate gradient is simply one of a number of possible choices here. Any of the 'standard' Krylov subspace methods including biconjugate gradient, GMRES, can, in principle, be developed in the same way using [52]–[54] wherever matrix-vector products involving the inverse of $K$ are required. Preconditioning for the pressure equation can be of great help in improving the convergence of this iteration, as discussed in Moresi and Solomatov (1995).

It should be pointed out that the Uzawa algorithm outlined above can also be used in connection with other numerical methods including FV and FD methods, provided that the matrix equations for pressure and velocities from these numerical methods have the form of [37].

### 5.4.1.4 Multigrid solution strategies

The stiffness matrix $K$ is symmetric positive definite, and this allows for numerous possible solution approaches. For example, multigrid solvers have been used for solving eqn [53] (Moresi and Solomatov, 1995) as in Citcom code. Newer versions of Citcom including CitcomS/CitcomCU employ full multigrid solvers with a consistent projection scheme for eqn [53] and are even more efficient (Zhong *et al.*, 2000).

The multigrid method works by formulating the FE problem on a number of different scales – usually a set of grids which are nested one within the other sharing common nodes (see **Figure 4(a)**), similar to how multigrid methods are used in FD and FV methods. The solution progresses on all of the grids at the same time with each grid eliminating errors at a different scale. The effect is to propagate information very rapidly between different nodes in the grid which would otherwise be prevented by the local support of the element shape functions. In fact, by a single traverse from fine to coarse grid and back, all nodes in the mesh can be directly connected to every other – allowing nodes which are physically coupled but remote in the mesh to communicate directly during each iteration cycle. This matches the physical structure of the Stokes flow problem in which stresses are transmitted instantaneously to all parts of the system in response to changes anywhere in the buoyancy forces or boundary conditions.

The multigrid effect relies upon using an iterative solver on each of many nested grid resolutions which acts like a smoother on the residual error at the characteristic scale of that particular grid. Gauss–Seidel iteration is a very common choice because it has exactly this property, although its effectiveness depends on the order in which degrees of freedom are visited by the solver and, consequently, it can be difficult to implement in parallel. On the coarsest grid it is possible to use a direct solver because the number of elements is usually very small.

For an elliptic operator such as the stiffness matrix, $K$, of the Stokes flow problem in eqn [53] we write

$$K_h v_h = F_h \qquad [58]$$

where the $h$ subscript indicates that the problem has been discretized to a mesh of fineness $h$. An initial estimate of the velocity on grid $h$, $v_h$, can be improved by a correction $\Delta v_h$ determining the solution to

$$K_h \Delta v_h = F_h - K_h v_h \qquad [59]$$

Suppose we obtain our approximate initial estimate by solving the problem on a coarse grid. The reduction of the number of degrees of freedom leads to a more manageable problem which can be solved quickly. The correction term is therefore

$$K_h \Delta v_h = F_h - K_h R_h^H v_H \qquad [60]$$

where $H$ indicates a coarser level of discretization, and the operator $R_h^H$ is an interpolation from the coarse-level $H$ to the fine-level $h$.

To find $v_H$ we need to solve a coarse approximation to the problem

$$K_H v_H = F_H \qquad [61]$$

where $K_H$ and $F_H$ are the coarse-level equivalent of the stiffness matrix and force vector. One obvious way to define these is to construct them from a coarse representation of the problem on the mesh $H$ exactly as would be done on $h$. An alternative is to define

$$K_H = R_H^h K_h R_h^H \quad \text{and} \quad F_H = R_H^h F_h \qquad [62]$$

where $R_h^H$ is a 'restriction' operator which has the opposite effect to the interpolation operation in that it lumps nodal contributions from the fine mesh onto the coarse mesh.

The power of the algorithm is in a recursive application. The coarse-grid correction is also

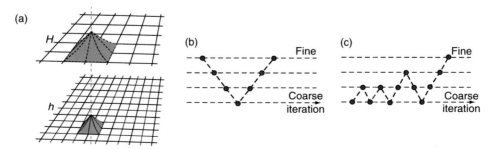

**Figure 4** A nested grids with fine and coarse meshes (a), structure of V cycle multigrid iteration (b) and structure of a full multigrid Iteration (c).

Obtain an approximate solution $v_h$ at the finest grid, $h$

Calculate residual $r_h = F_h - K_h v_h$

Project residual by $N$ levels to $h - N$

   *repeat:* $r_{h-i} = R_{h-i}^{h-(i-1)} r_{h-i-1}$

Solve exactly

   $\Delta v_{h-N} = K_{h-N} r_{h-N}$

Interpolate and improve

   $r_{h-(i-1)} = R_{h-(i-1)}^{h-i} K_{h-i} \Delta v_{h-i}$

   Improve $\Delta v_{h-(i-1)}$

$v_{h-(i-1)} = v_{h-(i-1)} + \Delta v_{h-(i-1)}$

**Figure 5**  A simple sawtooth multigrid algorithm for solving eqn [51].

calculated through the use of a still coarser grid and so on, until the problem is so small that an exact solution can be obtained very rapidly. One very simple, but instructive, algorithm for hierarchical residual reduction is the sawtooth cycle given in **Figure 5**, and its logical layout is same as the V-cycle in **Figure 4(b)**.

The step in which the velocity correction is 'improved' is an iterative method for reducing the residual at the current level which has the property of smoothing the error strongly at the current mesh scale. At each level these smoothing operators reduce the residual most strongly on the scale of the discretization – the hierarchical nesting of different mesh sizes allows the residual to be reduced at each scale very efficiently (see Yavneh (2006) for a more lengthy discussion).

The projection and interpolation operators have to be chosen fairly carefully to avoid poor approximations to the problem at the coarse levels and ineffectual corrections propagated to the fine levels. The interpolation operator is defined naturally from the shape functions at the coarse levels. The projection operator is then defined to complement this choice (the operators should be adjoint).

The sawtooth cycle outlined in this section is the simplest multigrid algorithm. Developments include improving the residual at each level of the restriction as well as the interpolation, known as a 'v-cycle', and cycles in which the residual is interpolated only part way through the hierarchy before being reprojected and subjected to another set of improvements (a 'w-cycle').

The full multigrid algorithm introduces a further level of complexity. Instead of simply casting the problem at a single level and projecting/improving

the residual on a number of grids, the whole problem is defined for all the grids. In this way the initial fine-grid approximation is obtained by interpolating from the solution to the coarsest-grid problem. The solution at each level is still obtained by projecting to the finest level and reducing the residual at each projection step. The resulting cycle is illustrated in **Figure 4(c)**.

A recent overview of multigrid methods is required reading at this point. Yavneh (2006) introduces the method which is then explained for a number of relevant examples by Oosterlee and Gaspar-Lorenz (2006), Bergen *et al.* (2006), and Bastian and Wieners (2006).

## 5.4.2  Stokes Flow: A Penalty Formulation

For 2-D problems, an efficient method to solve the incompressible Stokes flow is a penalty formulation with a reduced and selective integration. This method has been widely used in 2-D thermal convection problems, for example, in ConMan (King *et al.*, 1990). We now briefly discuss this penalty formulation, and detailed descriptions can be found in Hughes (2000).

The key feature in the penalty formulation is to allow for slight compressibility or $u_{k,k} \approx 0$. Here it is helpful to make an analogy to isotropic elasticity. The constitutive equations for both compressible and incompressible isotropic elasticity are given by the following two equations:

$$\sigma_{ij} = -P\delta_{ij} + \eta(u_{i,j} + u_{j,i}) \qquad [63]$$

$$u_{k,k} + P/\lambda = 0 \qquad [64]$$

where $\lambda$ is the Lame constant which is finite for compressible media but infinite for incompressibility media (i.e., to satisfy $u_{k,k} = 0$ for finite $P$). To allow for slight compressibility, $\lambda$ is taken finite but significantly larger than $\eta$, such that the error associated with the slight compressibility is negligibly small. Using words of 64 bit long (i.e., double precision), $\lambda/\eta \sim 10^7$ is effective. For finite $\lambda$, the constitutive equation becomes

$$\sigma_{ij} = \lambda u_{k,k} \delta_{ij} + \eta(u_{i,j} + u_{j,i}) \qquad [65]$$

which replaces eqn [4].

An interesting consequence of this new constitutive equation is that the pressure is no longer needed in the momentum equation, and this simplifies the

FE analysis. The weak form of the resulting Stokes flow problem is

$$\int_{\Omega} w_{i,j} c_{ijkl} v_{k,l} \, d\Omega = \int_{\Omega} w_i f_i \, d\Omega + \sum_{i=1}^{n_{sd}} \int_{\Gamma_{b_i}} w_i b_i \, d\Gamma \\ - \int_{\Omega} w_{i,j} c_{ijkl} g_{k,l} \, d\Omega \qquad [66]$$

where

$$c_{ijkl} = \lambda \delta_{ij} \delta_{kl} + \eta(\delta_{ik} \delta_{jl} + \delta_{il} \delta_{jk}) \qquad [67]$$

The FE implementation of eqn [66] is similar to that in Section 5.4.1. With the pressure excluded as a primary variable, the matrix equation is simply

$$[K]\{V\} = \{F\} \qquad [68]$$

While the elemental force vector is defined the same as that in [44], the elemental stiffness needs some modification in comparison with that in [42]:

$$k_{lm}^e = \mathbf{e}_i^T \left( \int_{\Omega^e} B_a^T D B_b \, d\Omega + \int_{\Omega^e} B_a^T \bar{D} B_b \, d\Omega \right) \mathbf{e}_j \qquad [69]$$

where the first integral is the same as in [42] but the second integral is a new addition with

$$\bar{D} = \begin{bmatrix} \lambda & \lambda & 0 \\ \lambda & \lambda & 0 \\ 0 & 0 & 0 \end{bmatrix} \qquad [70]$$

The matrix eqn [68] only yields correct solution for velocities if a reduced and selective integration scheme is used to evaluate the elemental stiffness matrix (e.g., Hughes, 2000). Specifically, the numerical quadrature scheme for the second integral of eqn [69] needs to be one order lower than that used for the first integral. For example, if for a 2-D problem, a $2 \times 2$ Gaussian quadrature scheme is used to evaluate the first integral, then a one-point Gaussian quadrature scheme is needed for the second integral. Hughes (2000) discussed the equivalence theorem for the mixed elements and the penalty formulation with the reduced and selective integration. Moresi *et al.* (1996) showed that these two formulations yield essentially identical results for the Stokes flow problems by comparing solutions from ConMan code employing a penalty formulation and Citcom which uses a mixed formulation.

Finally, we make two remarks about this penalty formulation.

First, although the pressure is not directly solved from the matrix equation, the pressure can be obtained through postprocessing via $P = -\lambda u_{k,k}$ for each element. Such obtained pressure fields often display a checkerboard pattern. However, a pressure-smoothing scheme (Hughes, 2000) seems to work well. The pressure field is important in many geophysical applications including computing dynamic topography and melt migration.

Second, with $\lambda/\eta \sim 10^7$, the stiffness matrix is not well conditioned and is not suited for any iterative solvers. A direct solver is required for this type of equations, as done in ConMan. This implies that this formulation may not be applicable to 3-D problems because of the memory and computation requirements associated with direct solvers. Reducing $\lambda/\eta$ improves the condition for the stiffness matrix; however, this is not recommended as it results in large errors associated with relaxing the incompressibility constraints.

### 5.4.3   The SUPG Formulation for the Energy Equation

The convective transport of any quantity at high Peclet number (the ratio of advective transport rate to diffusion rate) is challenging in any numerical approach in which the grid does not move with the material deformation. The transfer of quantities from grid points to integration points in order to calculate their updated values introduces a nonphysical additional diffusion term. Furthermore, the advection operator is difficult to stabilize and many different schemes have been proposed to treat grid-based advection in both an accurate and stable fashion. Many of the successful approaches include some attempt to track the flow direction and recognize that the advection operator is not symmetric in the upstream/downstream directions. This section introduces an SUPG formulation and a predictor–multicorrector explicit algorithm for time-dependent energy equation (i.e., eqn [3]). This method was developed by Hughes (2000) and Brooks (1981) some twenty years ago and remains an effective method in FE solutions of the equations with advection and diffusion such as our energy equation. FE mantle convection codes Citcom and ConMan both employ this method for solving the energy equation.

A weak-form formulation for the energy equation [3] and boundary conditions [7] is (Brooks, 1981)

$$\int_{\Omega} w(\dot{T} + u_i T_{,i}) d\Omega + \int_{\Omega} w_{,i}(\kappa T_{,i}) d\Omega$$
$$+ \sum_e \int_{\Omega_e} \bar{w}[\dot{T} + u_i T_{,i} - (\kappa T_{,i})_{,i} - \gamma] d\Omega$$
$$= \int_{\Omega} w\gamma \, d\Omega + \int_{\Gamma_q} wq \, d\Gamma - \int_{\Omega} w_{,i} \kappa g_{,i} d\Omega \quad [71]$$

where $w$ is the regular weighting functions and is zero on $\Gamma_q$, $\dot{T}$ is the time derivative of temperature, and $\bar{w}$ is the streamline upwind contribution to the weighting functions.

The FE implementation of [71] is similar to what was discussed for the Stokes flow in Section 5.4.1.2. While the weighting function $w$ is similar to what was defined in [39] except it is now a scalar, the streamline upwind part $\bar{w}$ is defined through artificial diffusivity $\tilde{\kappa}$ as

$$\bar{w} = \tilde{\kappa} \hat{u}_j w_{,j} / |u| \quad [72]$$

where $|u|$ is the magnitude of flow velocity, $\hat{u}_j = u_j / |u|$ represents the directions of flow velocity, and $\tilde{\kappa}$ is defined as

$$\tilde{\kappa} = \left( \sum_{i=1}^{n_{sd}} \xi_i u_i h_i \right) \Big/ 2 \quad [73]$$

$$\xi_i = \begin{cases} -1 - 1/\alpha_i, & \alpha_i < -1 \\ 0, & -1 \le \alpha_i \le 1, \\ 1 - 1/\alpha_i, & \alpha_i > 1 \end{cases} \quad \text{for } \alpha_i = \frac{u_i h_i}{2\kappa} \quad [74]$$

where $u_i$ and $h_i$ are flow velocity and element lengths in certain directions. It should be pointed out that eqns [72] and [74] are empirical and other forms are possible. Such defined streamline upward weighting function $\bar{w}$ can be thought as adding artificial diffusion to the actual diffusion term to lead to total diffusivity

$$\kappa + \tilde{\kappa} \hat{u}_i \hat{u}_j \quad [75]$$

$\bar{w}$ is discontinuous across elemental boundaries, different from $w$. This is why the integral in the third term of [72] is for each element. $\tilde{w} = w + \bar{w}$ is also sometimes called the Petrov–Galerkin weighting functions which indicates that the shape function used to weight the integrals and the shape function for interpolation are distinct.

A reasonable assumption is the weighted diffusion for an element in the third term of eqn [71], $\bar{w}(\kappa T_{,i})_{,i}$, is negligibly small. Therefore, eqn [71] can be written as

$$\int_{\Omega} w_{,i}(\kappa T_{,i}) d\Omega + \sum_e \int_{\Omega_e} \tilde{w}(\dot{T} + u_i T_{,i} - \gamma) d\Omega$$
$$= \int_{\Gamma_q} wq \, d\Gamma - \int_{\Omega} w_{,i} \kappa g_{,i} \, d\Omega \quad [76]$$

We now present relevant matrices at an element level. The $\dot{T}$ term in [71] implies that a mass matrix is needed and it is given as

$$m_{ab}^e = \int_{\Omega^e} N_a N_b \, d\Omega \quad [77]$$

where $a, b = 1,\ldots,n_{en}$. Elemental stiffness $k^e$ is

$$k_{ab}^e = \int_{\Omega^e} B_a^T \kappa B_b \, d\Omega \quad [78]$$

where for 2-D problems

$$B_a^T = (N_{a,1} \quad N_{a,2}) \quad [79]$$

Elemental force vector $f^e$ is given as

$$f_a^e = \int_{\Omega^e} \tilde{N}_a \gamma \, d\Omega + \int_{\Gamma_q^e} \tilde{N}_a q \, d\Gamma - \sum_{b=1}^{n_{en}} k_{ab}^e g_b^e \quad [80]$$

where $\tilde{N}_a$ is the Petrov–Galerkin shape function.
Elemental advection matrix $c^e$ is given as

$$c_{ab}^e = \int_{\Omega^e} \tilde{N}_a u_i N_{b,i} \, d\Omega \quad [81]$$

The combined matrix equation may be written as

$$M\dot{\Phi} + (K + C)\Phi = F \quad [82]$$

where $\Phi$ is the unknown temperature, and $M$, $K$, $C$, and $F$ are the total mass, stiffness, advection matrices, and force vector assembled from all the elements.

Equation [82] can be solved using a predictor–corrector algorithm (Hughes, 2000) with some initial condition for temperature (e.g., eqn [5]). Suppose that temperature and its time derivative at time step $n$ are given, $\Phi_n$ and $\dot{\Phi}_n$, the solutions at time step $n + 1$ with time increment $\Delta t$ can be obtained with the following algorithm:

1. Predictor:

$$\Phi_{n+1}^0 = \Phi_n + \Delta t(1-\alpha)\dot{\Phi}_n, \quad \dot{\Phi}_{n+1}^0 = 0,$$
$$\text{iteration step } i = 0 \quad [83]$$

2. Solving:

$$M^* \Delta \dot{\Phi}^i_{n+1} = \prod_e (f^e_{n+1} - m^e \dot{\Phi}^i_{n+1} - (k^e + c^e)\Phi^i_{n+1}) \quad [84]$$

3. Corrector:

$$\Phi^{i+1}_{n+1} = \Phi^i_{n+1} + \Delta t \alpha \Delta \dot{\Phi}^i_{n+1}, \dot{\Phi}^{i+1}_{n+1} = \dot{\Phi}^i_{n+1} + \Delta \dot{\Phi}^i_{n+1} \quad [85]$$

4. If needed, set iteration step $i = i + 1$ and go back step 2.

We make four remarks for this algorithm. First, this method is second-order accurate if $\alpha = 0.5$ (Hughes, 2000). Second, typically two iterations are sufficient. Third, in [84] $\prod$ represents the operation of assembling elemental matrix into global matrix, and $M^*$ is the lumped mass matrix which essentially makes this scheme an explicit scheme. Fourth, time increment $\Delta t$ needs to satisfy Courant time-stepping constraints to make the scheme stable (Hughes, 2000).

## 5.5   Incorporating More Realistic Physics

In Section 5.2, we presented the governing equations for thermal convection in a homogeneous incompressible fluid with a Newtonian (linear) rheology and the Boussinesq approximation. However, the Earth's mantle is likely much more complicated with heterogeneous composition and non-Newtonian rheology (see Chapter 2). In addition, non-Boussinesq effects such as solid–solid phase transitions may play an important role in affecting the dynamics of the mantle. In this section, we will discuss the methods that help incorporate these more realistic physics in studies of mantle convection. We will focus on modeling thermochemical convection, solid-state phase transitions, and non-Newtonian rheology.

### 5.5.1   Thermochemical Convection

Thermal convection for a compositionally heterogeneous mantle has gained a lot of interest in recent years (Lenardic and Kaula, 1993; Tackley, 1998a; Davaille, 1999; Kellogg et al., 1999; Chapter 10), with focus on the roles of mantle compositional anomalies and crustal structure in mantle dynamics. This is also called thermochemical convection. Different from purely thermal convection for which the fluid has the same composition, thermochemical convection involves fluids with different compositions. Here we will present governing equations and numerical methods for solving these equations (see also Chapters 2 and 10).

#### 5.5.1.1   Governing equations

Governing equations for thermochemical convection include a transport equation that describes the movement of compositions, in addition to the conservation laws of the mass, momentum, and energy (i.e., eqns [1]–[3]). Suppose that $C$ describes the compositional field, the transport equation is

$$\frac{\partial C}{\partial t} + u_i C_{,i} = 0 \quad [86]$$

This transport equation is similar to the energy equation [3] except that chemical diffusion and source terms are ignored, which is justified given that chemical diffusion is likely extremely small for the length- and timescales that we consider in mantle convection. For a two-component system such as the crust–mantle system or depleted-primordial mantle system, $C$ can be either 0 or 1, representing either component. If the fluids of different compositions have intrinsically different density, then the momentum eqn [2] needs to be modified to take into account the compositional effects on the buoyancy

$$\sigma_{ij,j} + Ra(T - \beta C)\delta_{iz} = 0 \quad [87]$$

where $\beta$ is the buoyancy number (van Keken et al., 1997; Tackley and King, 2003) and is defined as

$$\beta = \Delta\rho/(\rho\Delta T\alpha) \quad [88]$$

where $\Delta\rho$ is the density difference between the two compositions, $\rho$ and $\Delta T$ are the reference density and temperature, and $\alpha$ is the reference coefficient of thermal expansion.

A special class of thermochemical convection problems examine how the mantle compositional heterogeneity is stirred by mantle convection (e.g., Gurnis and Davies, 1986; Christensen, 1989; Kellogg, 1992; van Keken and Zhong, 1999). For these studies on the mixing of the mantle, we may assume that the fluids with different compositions have identical density with $\beta = 0$.

#### 5.5.1.2   Solution approaches

Solving the conservation equations of the mass, momentum, and energy for thermochemical convection is identical to what was introduced in Section 5.3 for purely thermal convection. The additional compositional buoyancy term in the momentum equation [87] does not present any new difficulties numerically, provided that the composition $C$ is given. The new challenge is to solve the transport equation [86] effectively.

A number of techniques have been developed or adapted in solving the transport equation in thermochemical convection studies. They include a field method with a filter (e.g., Hansen and Yuen, 1988; Lenardic and Kaula, 1993), a marker chain method (Christensen and Yuen, 1984; van Keken *et al.*, 1997; Zhong and Hager, 2003), and a particle method (e.g., Weinberg and Schmeling, 1992; Tackley, 1998a; Tackley and King, 2003; Gerya and Yuen, 2003). As reviewed by van Keken *et al.* (1997), while these techniques work to some extent, they also have their limitations, particularly in treating entrainment and numerical diffusion of composition $C$. We will briefly discuss each of these methods with more emphasis on the particle method.

In the particle method, the transport equation for $C$ (i.e., eqn [86]) is not solved directly. Composition $C$ at a given time is represented by a set of particles. This representation requires a mapping from the distribution of particles to compositional field $C$ which is often represented on a numerical mesh. With the mapping, to update $C$, all that is needed is to update the position of each particle to obtain an updated distribution of particles. This effectively solves the transport equation for $C$.

Two different particle methods have been used to map distribution of particles to $C$: absolute and ratio methods (Tackley and King, 2003). In the absolute method, particles are only used to represent one type of composition (e.g., for dense component or with $C = 1$). The population density of particles can be mapped to $C$. For example, $C$ for an element/grid cell with volume $\Omega_e$ and particles $N_e$ can be given as

$$C_e = AN_e/\Omega_e \qquad [89]$$

where the constant $A$ is the reciprocal of initial density of particles for composition $C=1$ (i.e., total number of particles divided by the volume of composition $C=1$). Clearly, the absence of particles in an element/grid cell represents $C=0$. A physically unrealistic situation with $C>1$ may arise due to statistical fluctuations in particle distribution or particle settling. Therefore, for this method to work effectively, a large number of particles are required (Tackley and King, 2003).

In the ratio method, two different types of particles are used to represent the compositional field $C$, type 1 for $C=0$ and type 2 for $C=1$. $C$ for an element/grid cell that includes type 1 particles $N_1$ and type 2 particles $N_2$ is given as

$$C_e = N_2/(N_1 + N_2) \qquad [90]$$

In the ratio method, $C$ can never be greater than 1. Tackley and King (2003) found that the ratio method is particularly effective in modeling thermochemical convection in which the two components occupy similar amount of volumes.

We now discuss briefly procedures to update the positions of particles. One commonly used method is a high-order Runge–Kutta method (e.g., van Keken *et al.*, 1997). Here we present a predictor–corrector scheme for updating the particle positions (e.g., Zhong and Hager, 2003). Suppose that at time $t=t_0$, flow velocity is $\mathbf{u}_0$ and compositional field is $C_0$ that is defined by a set of particles with coordinates, $\mathbf{x}_0^i$, for particle $i$. The algorithm for solving composition at the next time step $t = t_0 + \mathrm{d}t = t_1$, $C_1$, can be summarized as follows:

1. Using a forward Euler scheme, predict the new position for each particle $i$ with $\mathbf{x}_{1p}^i = \mathbf{x}_0^i + \mathbf{u}_0 \mathrm{d}t$ and mapping the particles to compositional field $C_{1p}$ at $t = t_1$.
2. Using the predicted $C_{1p}$, solve the Stokes equation for new velocity $\mathbf{u}_{1p}$.
3. Using a modified Euler scheme with second-order accuracy, compute the position for each particle $i$ with $\mathbf{x}_1^i = \mathbf{x}_0^i + 0.5(\mathbf{u}_0 + \mathbf{u}_{1p})\mathrm{d}t$ and compositional field $C_1$ at $t = t_1$.

The marker chain method is similar to the particle method in a number of ways. In the marker chain method, composition $C$ is defined by an interfacial boundary that separates two different components. The interfacial boundary is a line for 2-D problems or a surface for 3-D. Using the flow velocity, one tracks the evolution of the interfacial boundary and hence composition $C$. Often the interfacial boundary is represented by particles or markers. Therefore, updating the interfacial boundary is essentially the same as updating the particles in the particle method. Composition $C$ on a numerical grid which is desired for solving the momentum and energy equations [87] and [3] can be obtained by projection. As van Keken *et al.* (1997) indicated, the marker chain method is rather effective for compositional anomalies with relatively simple structure and geometry in 2-D.

The field method is probably the most straightforward. By setting diffusivity to be zero, we can employ the same solver for the energy equation (e.g., in

Section 5.4.3) to solve the transport equation for *C*. However, this often introduces numerical artifacts including numerical oscillations and numerical diffusion. Lenardic and Kaula (1993) introduced a filter scheme that removes the numerical oscillations while conserving the total mass of compositional field.

### 5.5.2 Solid-State Phase Transition

Solid-state phase transitions are important phenomena in the mantle. Major phase transitions include olivine–spinel transition at 410 km depth and spinel-to-perovskite and magnesium-to-wustite transitions at 670 km depth, that are associated with significant changes in mantle density and seismic wave speeds. Recently, it was proposed that the D″ discontinuity near the core–mantle boundary is also caused by a phase transition from perovskite to post-perovskite (Murakami, *et al.*, 2004). These phase transitions may affect the dynamics of mantle convection in two ways: (1) on the energetics due to latent heating associated with phase transitions, (2) on the buoyancy due to undulations at phase boundary caused by lateral variations in mantle temperature that affects the pressure at which phase transitions occur (Richter, 1973; Schubert *et al.*, 1975; Christensen and Yuen, 1985; Tackley *et al.*, 1993; Zhong and Gurnis, 1994). In this section, following Richter (1973) and Christensen and Yuen (1985), we present a method to model phase transitions.

The undulations of a phase boundary represent additional buoyancy force that affects the momentum equation. For phase transition *k* with density change $\Delta\rho_k$, the phase boundary undulations for phase transition *k* with density change $\Delta\rho_k$ can be described by a dimensionless phase-change function $\Gamma_k$ that varies from 0 to 1 where regions with $\Gamma_k$ of 0 and 1 represent the two phases separated by this phase-change boundary. The momentum equation can be written as

$$\rho_{ij,j} + (RaT - \sum_k Ra_k\Gamma_k)\delta_{iz} = 0 \qquad [91]$$

where phase-change Rayleigh number $Ra_k$ is

$$Ra_k = \frac{\Delta\rho_k gD^3}{\kappa\eta_0} \qquad [92]$$

The phase-change function $\Gamma_k$ is defined via 'excess pressure'

$$\pi_k = P - P_0 - \gamma_k T \qquad [93]$$

where $\gamma_k$ and $P_0$ are the Clapeyron slope and phase-change pressure at zero-degree temperature for the

*k*th phase transition. After normalizing pressure by $\rho_0 gD$ and Clapeyron slope by $\rho_0 gD/\Delta T$, the nondimensional 'excess pressure' can be written as

$$\pi_k = 1 - d_k - z - \gamma_k(T - T_k) \qquad [94]$$

where $\gamma_k$, $d_k$, and $T_k$ are the nondimensional Clapeyron slope, reference-phase transition depth, and reference-phase transition temperature for the *k*th phase transition, respectively. The dimensionless phase-change function is then given as

$$\Gamma_k = \frac{1}{2}\left(1 + \tanh\frac{\pi_k}{d}\right) \qquad [95]$$

where *d* is dimensionless phase transition width which measures the depth segment over which the phase change occurs. It should be pointed out that the effect of phase transition on buoyancy force can also be modeled with 'effective' coefficient of thermal expansion (Christensen and Yuen, 1985).

The latent heating effect, along with viscous heating and adiabatic heating, can be included in the energy equation (also see eqn [3]) as (Christensen and Yuen, 1985)

$$\left[1 + \sum_k \gamma_k^2 \frac{Ra_k}{Ra}\frac{d\Gamma_k}{d\pi_k}D_i\ (T + T_s)\right]\left(\frac{\partial T}{\partial t} + \mathbf{v}\cdot\nabla T\right)$$
$$+ \left(1 + \sum_k \gamma_k \frac{Ra_k}{Ra}\frac{d\Gamma_k}{d\pi_k}\right)(T + T_s)D_i v_z$$
$$= \nabla^2 T + \frac{D_i}{Ra}\tau_{ij}\frac{\partial v_i}{\partial x_j} + \gamma \qquad [96]$$

where $T_s$, $v_z$, and $\tau_{ij}$ are the surface temperature, vertical velocity, and deviatoric stress, respectively; *k* is phase-change index; $D_i$ is the dissipation number and is defined as

$$D_i = \frac{\alpha gD}{C_p} \qquad [97]$$

where $\alpha$ and $C_p$ are the coefficient of thermal expansion and specific heat (*see also* Chapter 2). Christensen and Yuen (1985) called the effects of latent heating, viscous heating, and adiabatic heating as non-Boussinesq effects and termed this formulation as extended-Boussinesq formulation. They suggest that these effects are all of similar order, proportional to $D_i$, and should be considered simultaneously.

The modified momentum and energy equations [91] and [96] can be solved with the same algorithms such as the Uzawa and SUPG for mantle convection problems with extended-Boussinesq approximations in 3-D (e.g., Zhong, 2006; Kameyama and Yuen, 2006).

### 5.5.3  Non-Newtonian Rheology

Laboratory studies suggest that the deformation of olivine, the main component in the upper mantle, follows a power-law rheology (e.g., Karato and Wu, 1993):

$$\dot{\varepsilon} = A\tau^n \qquad [98]$$

where $\dot{\varepsilon}$ is the strain rate, $\tau$ is the deviatoric stress, the pre-exponent constant $A$ represents other effects such as grain size and water content, and the exponent $n$ is $\sim$3. The nonlinearity in the rheology arises from $n \neq 1$ (*see also* Chapter 2).

The effects of non-Newtonian rheology on mantle convection were first investigated by Parmentier *et al.* (1976) and Christensen (1984). More recent efforts have been focused on how non-Newtonian rheology including viscoplastic rheology may lead to dynamic generation of plate tectonics (King and Hager, 1990; King *et al.*, 2002; Bercovici, 1995; Moresi and Solomatov, 1998, Zhong *et al.*, 1998; Tackley, 1998b; Trompert and Hansen, 1998).

Solutions of nonlinear problems in general require an iterative approach. The power-law rheolgy may be written as an expression for effective viscosity

$$\sigma_{ij} = -P\delta_{ij} + 2\eta_{\text{eff}}\dot{\varepsilon}_{ij} \qquad [99]$$

$$\eta_{\text{eff}} = \tau/\dot{\varepsilon} = \frac{1}{A}\dot{\varepsilon}^{(1-n)/n} \qquad [100]$$

where $\dot{\varepsilon}$ is the second invariant of strain-rate tensor

$$\dot{\varepsilon} = \left(\frac{1}{2}\dot{\varepsilon}_{ij}\dot{\varepsilon}_{ij}\right)^{1/2} \qquad [101]$$

It is clear that the effective viscosity depends on strain rate which in turn depends on flow velocity. Therefore, a general strategy for this problem is (1) starting with some guessed effective viscosity, solve the Stokes flow problem for flow velocities; (2) update the effective viscosity with the newly determined strain rate, and solve the Stokes flow again; (3) keep this iterative process until flow velocities are convergent.

Implementation of this iterative scheme is straightforward. The convergence for this iterative process depends on the exponent $n$. For regular power-law rheology with $n \sim 3$, convergence is usually not a problem. However, for large $n$ (e.g., in case of viscoplastic rheology), the iteration may

converge very slowly or may diverge. Often some forms of damping may help improve convergence significantly (e.g., King and Hager, 1990).

## 5.6  Concluding Remarks and Future Prospects

In this chapter, we have discussed four basic numerical methods for solving mantle convection problems: FE, FD, FV, and spectral methods. We have focused our efforts on FE method, mainly because of its growing popularity in mantle convection studies over the past decade, partially prompted by the easily accessed FE codes from Conman to Citcom. To this end, the discussions on FE method should help readers understand the inner working of these two FE codes. However, our discussions on FD, FV, and spectral methods are rather brief and are meant to give readers a solid basis for understanding the rudiments of these methods and the references with which to delve deeper into the subjects. These three methods have all been widely used in studies of mantle convection and will most likely remain so for years to come. It is our view that each of these methods has its advantages and disadvantages and readers need to find the one that is most suited to their research. This chapter is by no means exhaustive or extremely advanced in character. We did not cover many potentially interesting and powerful numerical techniques, such as spectral elements (e.g., Komatitsch and Tromp, 1999), wavelets (Daubeschies *et al.*, 1985), level-set method (Osher and Fedkiew, 2003), and adaptive grid techniques (Berger and Oliger 1984; Bruegmann and Tichy, 2004), and interested readers can read these references to learn more.

Rapid advancement in computing power has made 3-D modeling of mantle convection practical, although 2-D modeling still plays an essential role. While a variety of solvers with either iterative or direct solution method are available to 2-D models, 3-D modeling requires iterative solution techniques due to both computer memory and computational requirements. A powerful iterative solution approach is the multigrid method that can be used in either FE or FV method. Such methods are already implemented in a number of mantle convection codes including STAG3D (Tackley, 1994), Citcom/CitcomS (Moresi and Solomatov, 1995; Zhong *et al.*, 2000), and Terra (Baumgardner, 1985). We spent some effort in discussing the multigrid method, and more on this topic can be found in Brandt (1977), Trottenberg *et al.* (2001), and Yavneh (2006). The multigrid idea is powerful in that one can generalize this to structures other than grids, for instance multiscale or multilevel techniques (Trottenberg *et al.*, 2001).

Three-dimensional modeling should almost certainly make a good use of parallel computing with widely available parallel computers from PC-clusters to super-parallel computers. Parallel computing technology poses certain limitations on numerical techniques as well. For example, spectral methods, while having some important advantages over other grid-based numerical methods, are much more difficult to implement efficiently in parallel computing. This may severely limit its use to tackle next-generation computing problems. Fortunately, many other numerical methods including FE, FD, and FV methods are very efficient for parallel computing. Many of the codes mentioned earlier for mantle convection studies including STAG3D, Citcom/CitcomS, Terra, and those in Harder and Hansen (2005) and Kageyama and Sato (2004) are fully parallelized and can be used on different parallel computers. These codes can be scaled up to possibly thousands of processors with a great potential yet to be explored in helping understand high-resolution, high-Rayleigh-number mantle convection.

However, there remain many challenges in numerical modeling of mantle convection, both in developing more robust numerical algorithms and in analyzing model results. At least four areas need better numerical algorithms. (1) Thermochemical convection becomes increasingly important in answering a variety of geodynamic questions. We discussed cursorily a Lagrangian technique of advection of tracers in solving thermochemical convection. However, it is clear that most existing techniques do not work well for entrainment in thermochemical convection, as demonstrated by van Keken *et al.* (1997) and Tackley and King (2003). The increasing computing power will help solve this problem by providing significantly high resolution, but better algorithms are certainly needed. (2) The lithosphere is characterized by highly nonlinear rheology including complex shear-localizing feedback mechanisms and history-dependent rheology and plastic deformation (Gurnis *et al.*, 2000; Bercovici, 2003). Convergence deteriorates rapidly when nonlinearity increases. More robust algorithms are needed for solving mantle convection with highly nonlinear rheology. (3) Robust algorithms are needed in order to incorporate compressibility in mantle convection and to better compare with seismic and mineral physics models. 2-D compressible mantle convection models with simple thermodynamics have been formulated (Jarvis and McKenzie, 1980; Ita and King, 1998). We anticipate more developments in the near future. (4) Multiscale physics is an important feature of mantle convection. Earth's mantle convection is of very long wavelengths, for example, $\sim 10\,000$ km for the Pacific Plate. However, mantle convection is also fundamentally controlled by thin thermal boundary layers that lead to thin upwelling plumes ($\sim 100$ km) and downwelling slabs due to high Rayleigh number in mantle convection. Plate boundary processes and entrainment in thermochemical convection also occur over possibly even smaller length scale. Furthermore, material properties are also affected by near-microscopic properties such as grain size, which has not been well incorporated in mantle convection studies. Most existing numerical methods in mantle convection work for largely uniform grids. New methods that work with dynamic adaptive mesh refinement are needed. Finally, it is necessary for all these new methods and algorithms to work efficiently in 2-D/3-D on parallel computers.

Efficient postprocessing and analyses of modeling results are also increasingly becoming an important issue. Mantle convection, along with many other disciplines in the geosciences, now faces an exponential increase in the amount of numerical data generated in large-scale high-resolution 3-D convection. It is currently commonplace to have a few terabytes of data for a single project and this poses significant challenges to conventional ways of data analyses, post-processing, and visualization. Visualization is already a severe problem even in the era of terascale computing. Considering our future quest to the petascale computing, this problem would be greatly exacerbated in the coming decade. It is essential to employ modern visualization tools to confront this challenge. Erlebacher *et al.* (2001) discussed these issues and their solutions, and similar arguments about these problems and their solutions can be found in a report on high-performance computing in geophysics by Cohen (2005). Several potentially useful methodologies such as 2-D/3-D feature extraction, segmentation methods, and flow topology (see Hansen and Johnson, 2005) can help geophysicists understand better the physical structure of time-dependent convection with coherent structures, such as plumes or detached slabs in spherical geometry. 3-D visualization packages, for example, AMIRA or PARAVIEW, can help to alleviate the burden of the researcher in unraveling the model output. Going further into detailed examination of 3-D mantle convection, one would need large-scale display devices such as the PowerWall with more than 12 million pixels or CAVE-like environments. This particular visualization method is described under current operating conditions in the Earth Simulator Center by Ohno *et al.* (2006). Remote visualization of the data under the auspices of Web-services using the client-server paradigm may be a panacea for collaborative projects (see Erlebacher *et al.*, 2006).

Finally, it is vitally important to develop and maintain efficient and robust benchmarks, as in any fields of computational sciences. Benchmark efforts have been made in the past by various groups for different mantle convection problems (e.g., Blankenbach *et al.*, 1989; van Keken *et al.*, 1997; Tackley, 1994; Moresi and Solomatov, 1995; Zhong *et al.*, 2000; Stemmer *et al.*, 2006). As numerical methods and computer codes become more sophisticated and our community moves into tera- and petascale computing era, benchmark efforts become even more important to assure the efficiency and accuracy.

# References

Albers M (2000) A local mesh refinement multigrid method for 3-D convection problems with strongly variable viscosity. *Journal of Computational Physics* 160: 126–150.

Atanga J and Silvester D (1992) Iterative methods for stabilized mixed velocity-pressure finite elements. *International Journal of Numerical Methods in Fluids* 14: 71–81.

Auth C and Harder H (1999) Multigrid solution of convection problems with strongly variable viscosity. *Geophysical Journal International* 137: 793–804.

Balachandar S and Yuen DA (1994) Three-dimensional fully spectral numerical method for mantle convection with depth-dependent properties. *Journal of Computational Physics* 132: 62–74.

Balachandar S, Yuen DA, Reuteler DM, and Lauer G (1995) Viscous dissipation in three dimensional convection with temperature-dependent viscosity. *Science* 267: 1150–1153.

Bastian P and Wieners C (2006) Multigrid methods on adaptively refined grids. *Computing in Science and Engineering* 8(6): 44–55.

Bathe K-J (1996) *Finite Element Procedures*, 1037 pp. Englewood Cliffs, NY: Prentice Hall.

Baumgardner JR (1985) Three dimensional treatment of convectionflow in the Earth's mantle. *Journal of Statistical Physics* 39(5/6): 501–511.

Bercovici D (2003) The generation of plate tectonics from mantle convection. *Earth and Planetary Science Letters* 205: 107–121.

Bercovici D (1995) A source-sink model of the generation of plate tectonics from non-Newtonian mantle flow. *Journal of Geophysical Research* 100: 2013–2030.

Bergen B, Gradl T, Huelsemann F, and Ruede U (2006) A massively parallel multigrid method for finite elements. *Computing in Science and Engineering* 8(6): 56–62.

Berger M and Oliger J (1984) Adaptive mesh refinement for hyperbolic partial differential equations. *Journal of Computational Physics* 53: 484–512.

Blankenbach B, Busse F, Christensen U, et al. (1989) A benchmark comparison of mantle convection codes. *Geophysical Journal International* 98: 23–38.

Bollig EF, Jensen PA, Lyness MD, et al. (in press) VLAB: Web Services, Portlets, and Workflows for enabling cyber-infrastructure in computational mineral physics. *Physics of the Earth and Planetary Interiors*.

Brandt A (1999) Multi-level adaptive solutions to boundary-value problems. *Mathematics of Computation* 31: 333–390.

Briggs WL, Henson VE, and McCormick SF (2000) *A Multigrid Tutorial*, 2nd edn. SIAM Press, 2000.

Brooks AN and Petrov-Galerkin A (1981) *Finite Element Formulation for Convection Dominated Flows*. PhD Thesis, California Institute of Technology, Pasadena, CA.

Bruegmann B and Tichy W (2004) Numerical solutions of orbiting black holes. *Physical Review Letters* 92: 211101.

Bunge HP, Richards MA, Lithgow-Bertelloni C, Baumgardner JR, Grand SP, and Romanowicz BA (1998) Time scales and heterogeneous structure in geodynamic Earth models. *Science* 280: 91–95.

Bunge H-P, Richards MA, and Baumgardner JR (1997) A sensitive study of 3-dimensional spherical mantle convection at 108 Rayleigh number: Effects of depth-dependent viscosity, heating mode, and an endothermic phase change. *Journal of Geophysical Research* 102: 11991–12007.

Busse FH (1989) Fundamentals of thermal convection. In: Peltier WR (ed.) *Mantle Convection, Plate Tectonics and Global Dynamics*, pp. 23–109. New York: Gordon and Breach.

Christensen U (1989) Mixing by time-dependent convection. *Earth and Planetary Science Letters* 95: 382–394.

Christensen UR (1984) Convection with pressure- and temperature-dependent non-Newtonian rheology. *Geophysical Journal of the Royal Astronomical Society* 77: 343–384.

Christensen UR and Yuen DA (1985) Layered convection induced by phase changes. *Journal of Geophysical Research* 90: 1029110300.

Christensen UR and Yuen DA (1984) The interaction of a subducting lithosphere with a chemical or phase boundary. *Journal of Geophysical Research* 89(B6): 4389–4402.

Cohen RE (ed.) (2005) High-Performance Computing Requirements for the Computational Solid-Earth Sciences, 101 pp. (http://www.geoprose.com/computational_SES.html).

Cserepes L, Rabinowicz M, and Rosemberg-Borot C (1988) Three-dimensional infinite Prandtl number convection in one and two layers with implication for the Earth's gravity field. *Journal of Geophysical Research* 93: 12009–12025.

Cserepes L and Rabinowicz M (1985) Gravity and convection in a two-layer mantle. *Earth and Planetary Science Letters* 76: 193–207.

Daubeschies I, Grossmann A, and Meyer Y (1985) Painless nonorthogonal expansions. *Journal of Mathematical Physics* 27: 1271–1283.

Davaille A (1999) Simultaneous generation of hotspots and superswells by convection in a heterogeneous planetary mantle. *Nature* 402: 756–760.

Davies GF (1995) Penetration of plates and plumes through the mantle transition zone. *Earth and Planetary Science Letters* 133: 507–516.

Desai CS and Abel JF (1972) *Introduction to the Finite Element Method. A Numerical Method for Engineering Analysis.* New York: Academic Press.

Erlebacher C, Yuen DA, and Dubuffet F (2001) Current trends and demands in visualization in the geosciences. *Visual Geosciences* 6: 59 (doi:10.1007/s10069-001-1019-y).

Erlebacher G, Yuen DA, Lu Z, Bollig EF, Pierce M, and Pallickara S (2006) A grid framework for visualization services in the Earth Sciences. *Pure and Applied Geophysics* 163: 2467–2483.

Fornberg B (1990) High-order finite differences and the pseudospectral method on staggered grids. *SIAM Journal on Numerical Analysis* 27: 904–918.

Fornberg BA (1995) *A Practical Guide to Pseudospectral Methods.* Cambridge: Cambridge University Press.

Gable CW, O'Connell RJ, and Travis BJ (1991) Convection in three dimensions with surface plates: Generation of toroidal flow. *Journal of Geophysical Research* 96: 8391–8405.

Gerya TV and Yuen DA (in press) Robust characteristics method for modeling multiphase visco-elastoplastic thermo-mechanical problems. *Physics of the Earth and Planetary Interiors.*

Gerya TV, Connolly JAD, Yuen DA, Gorczyk W, and Capel AM (2006) Seismic implications of mantle wedge plumes. *Physics of the Earth and Planetary Interiors* 156: 59–74.

Gerya TV and Yuen DA (2003) Characteristics-based marker method with conservative finite-differences schemes for modeling geological flows with strongly variable transport properties. *Physics of the Earth and Planetary Interiors* 140: 295–320.

Glatzmaier GA (1988) Numerical simulations of mantle convection: Time-dependent, three-dimensional, compressible, spherical shell. *Geophysics and Astrophysics Fluid Dynamics* 43: 223–264.

Glatzmaier GA and Roberts PH (1995) A three-dimensional self-consistent computer simulation of a geomagnetic field reversal. *Nature* 377: 203–209.

Golub GH and van Loan CF (1989) *Matrix Computations*, 642 pp. Baltimore, MD: The Johns Hopkins University Press.

Griffiths DV and Smith IM (2006) *Numerical Methods for Engineers*, 2nd edn. New York: Chapman and Hall.

Gurnis M, Zhong S, and Toth J (2000) On the competing roles of fault reactivation and brittle failure in generating plate tectonics from mantle convection. In: Richards MA, Gordon R, and van der Hilst R (eds.) *The History and Dynamics of Global Plate Motions*, pp. 73–94. Washington, DC: American Geophysical Union.

Gurnis M and Davies GF (1986) Mixing in numerical models of mantle convection incorporating plate kinematics. *Journal of Geophysical Research* 91: 6375–6395.

Hackbush W (1985) *Multigrid Methods and Applications.* Berlin: Springer Verlag.

Hager BH and Richards MA (1989) Long-wavelength variations in Earth's geoid: Physical models and dynamical implications. *Philosophical Transactions of the Royal Society of London A* 328: 309–327.

Hager BH and O'Connell RJ (1981) A simple global model of plate dynamics and mantle convection. *Journal of Geophysical Research* 86: 4843–4878.

Hansen CD and Johnson CR (eds.) (2005) *The Visualization Handbook*, 962 pp. Amsterdam: Elsevier.

Hansen U and Yuen DA (1988) Numerical simulations of thermal chemical instabilities and lateral heterogeneities at the core–mantle boundary. *Nature* 334: 237–240.

Harder H and Hansen U (2005) A finite-volume solution method for thermal convection and dynamo problems in spherical shells. *Geophysical Journal International* 161: 522–537.

Hughes TJR (2000) *The Finite Element Method*, 682 pp. New York: Dover Publications.

Ita JJ and King SD (1998) The influence of thermodynamic formulation on simulations of subduction zone geometry and history. *Geophysical Research Letters* 25: 1463–1466.

Jarvis GT and McKenzie DP (1980) Convection in a compressible fluid with infinite Prandtl number. *Journal of Fluid Mechanics* 96: 515–583.

Kameyama MC and Yuen DA (2006) 3-D convection studies on the thermal state of the lower mantle with post perovskite transition. *Geophysical Research Letters* 33 (doi:10.1029/2006GL025744).

Kageyama A and Sato T (2004) The 'Yin and Yang Grid': An overset grid in spherical geometry. *Geochemistry Geophysics Geosystem* 5(9): Q09005.

Karato S and Wu P (1993) Rheology of the upper mantle: A synthesis. *Science* 260: 771–778.

Kellogg LH, Hager BH, and van der Hilst RD (1999) Compositional stratification in the deep mantle. *Science* 283: 1881–1884.

Kellogg LH (1992) Mixing in the mantle. *Annual Review of Earth and Space Sciences* 20: 365–388.

King SD, Gable CW, and Weinstein SA (1992) Models of convection-driven tectonic plates: A comparison of methods and results. *Geophysical Journal International* 109: 481–487.

King SD, Raefsky A, and Hager BH (1990) ConMan: Vectorizing a finite element code for incompressible two-dimensional convection in the Earth's mantle. *Physics of the Earth and Planetary Interiors* 59: 195–207.

King SD and Hager BH (1990) The relationship between plate velocity and trench viscosity in Newtonian and power-law subduction calculations. *Geophysical Research Letters* 17: 2409–2412.

King SD, Lowman JP, and Gable CW (2002) Episodic tectonic plate reorganizations driven by mantle convection. *Earth and Planetary Science Letters* 203: 83–91.

Komatitsch D and Tromp J (1999) Introduction to the spectral element method for three-dimensional seismic wave propagation. *Geophysical Journal International* 139: 806–822.

Kowalik Z and Murty TS (1993) *Numerical Modeling of Ocean Dynamics*, 481 pp. Toh Tuck Link, Singapore: World Scientific Publishing.

Kuang W and Bloxham J (1999) Numerical modeling of magnetohydrodynamic convection in a rapidly rotating spherical shell: Weak and strong field dynamo action. *Journal of Computational Physics* 153: 51–81.

Larsen TB, Yuen DA, Moser J, and Fornberg B (1997) A higher-order finite-difference method applied to large Rayleigh number mantle convection. *Geophysical and Astrophysical Fluid Dynamics* 84: 53–83.

Lenardic A and Kaula WM (1993) A numerical treatment of geodynamic viscous flow problems involving the advection of material interfaces. *Journal of Geophysical Research* 98: 8243–8269.

Lynch DR (2005) *Numerical Partial Differential Equations for Environmental Scientists and Engineers: A First Practical Course*, 388 pp. Berlin: Springer Verlag.

Machetel Ph, Rabinowicz M, and Bernadet P (1986) Three-dimensional convection in spherical shells. *Geophysical and Astrophysical Fluid Dynamics* 37: 57–84.

Maday Y and Patera A (1989) Spectral -element methods for the incompressible Navier-Stokes equations. In: Noor A and Oden J (eds.) *State of the Art Survey in Computational Mechanics*, pp. 71–143. New York: ASME.

Malevsky AV (1996) Spline-characteristic method for simulation of convective turbulence. *Journal of Computational Physics* 123: 466–475.

Malevsky AV and Yuen DA (1991) Characteristics-based methods applied to infinite Prandtl number thermal convection in the hard turbulent regime. *Physics Fluids A* 3(9): 2105–2115.

Malevsky AV and Yuen DA (1993) Plume structures in the hard-turbulent regime of three-dimensional infinite Prandtl number convection. *Geophysical Research Letters* 20: 383–386.

McKenzie DP, Roberts JM, and Weiss NO (1974) Convection in the Earth's mantle: Towards a numerical solution. *Journal of Fluid Mechanics* 62: 465–538.

McNamara AK and Zhong S (2004) Thermochemical structures within a spherical mantle: Superplumes or piles? *Journal of Geophysical Research* 109: B07402 (doi:10.1029/2003JB002847).

Monnereau M and Yuen DA (2002) How flat is the lower-mantle temperature gradient? *Earth and Planetary Science Letters* 202: 171–183.

Moresi LN, Dufour F, and Muhlhaus H-B (2003) A Lagrangian integration point finite element method for large deformation modeling of viscoelastic geomaterials. *Journal of Computational. Physics* 184: 476–497.

Moresi LN and Solomatov VS (1998) Mantle convection with a brittle lithosphere: Thoughts on the global tectonic styles of the Earth and Venus. *Geophysical Journal International* 133: 669–682.

Moresi LN and Solomatov VS (1995) Numerical investigation of 2D convection with extremely large viscosity variation. *Physics of Fluids* 9: 2154–2164.

Moresi L and Gurnis M (1996) Constraints on the lateral strength of slabs from three-dimensional dynamic flow models. *Earth and Planetary Science Letters* 138: 15–28.

Moresi LN, Zhong S, and Gurnis M (1996) The accuracy of finite element solutions of Stokes' flow with strongly varying viscosity. *Physics of the Earth and Planetary Interiors* 97: 83–94.

Murakami M, Hirose K, Kawamura K, Sata N, and Ohishi Y (2004) Post-perovskite phase transition in MgSiO3. *Science* 304: 855–858.

Ogawa M, Schubert G, and Zebib A (1991) Numerical simulations of 3-dimensional thermal convection in a fluid with strongly temperature-dependent viscosity. *Journal of Fluid Mechanics* 233: 299–328.

Ohno N, Kageyama A, and Kusano K (in press) Virtual reality visualization by CAVE with VFIVE and VTK. *Journal of Plasma Physics*.

Oosterlee CW and Gaspar-Lorenz FJ (2006) Multigrid methods for the Stokes system. *Computing in Science and Engineering* 8(6): 34–43.

Osher S and Fedkiw RP (2003) *Level Set Methods and Dynamic Implicit Surfaces*, 275 pp. Berlin: Springer Verlag.

Pantakar SV (1980) *Numerical Heat Transfer and Fluid Flow*. New York: Hemisphere Publishing Corporation.

Pantakar SV and Spalding DB (1972) A calculation procedure for heat, mass and momentum transfer in three-dimensional parabolic flows. *International Journal of Heat and Mass Transfer* 15: 1787–2000.

Parmentier EM and Sotin C (2000) Three-dimensional numerical experiments on thermal convection in a very viscous fluid: Implications for the dynamics of a thermal boundary layer at high Rayleigh number. *Physics of Fluids* 12: 609–617.

Parmentier EM, Turcotte DL, and Torrance KE (1976) Studies of finite amplitude non-Newtonian thermal convection with application to convection in the Earth's mantle. *Journal of Geophysical Research* 81: 1839–1846.

Parmentier EM, Turcotte DL, and Torrance KE (1975) Numerical experiments on the structure of mantle plumes. *Journal of Geophysical Research* 80: 4417–4425.

Peyret R and Taylor TD (1983) *Computational Methods for Fluid Flow*, 350 pp. Berlin: Springer Verlag.

Press WH, Flannery BP, Teukolsky SA, and Vetterling WA (1992) *Numerical Recipes in FORTRAN*. New York: Cambridge University Press.

Ramage A and Wathen AJ (1994) Iterative solution techniques for the Stokes and Navier-Stokes equations. *International Journal of Numerical Methods in Fluids* 19: 67–83.

Ratcliff JT, Schubert G, and Zebib A (1996) Steady tetrahedral and cubic patterns of spherical-shell convection with temperature-dependent viscosity. *Journal of Geophysical Research* 101: 25 473–25 484.

Richter FM (1973) Finite amplitude convection through a phase boundary. *Geophysical Journal of the Royal Astronomical Society* 35: 265–276.

Richtmeyer RD and Morton KW (1967) *Difference Methods for Initial Value Problems*, 405 pp. New York: Interscience Publishers.

Rudolf M, Gerya TV, Yuen DA, and De Rosier S (2004) Visualization of multiscale dynamics of hydrous cold plumes at subduction zones. *Visual Geosciences* 9: 59–71.

Saad Y and Schultz MH (1986) GMRES: A generalized minimal residual algorithm for solving nonsymmetric linear systems. *SIAM Journal on Scientific Statistical Computing* 7: 856–869.

Schmalholz SM, Podladchikov YY, and Schmid DW (2001) A spectral/finite-difference for simulating large deformations of heterogeneous, viscoelastic materials. *Geophysical Journal International* 145: 199–219.

Schott B and Schmeling H (1998) Delamination and detachment of a lithospheric root. *Tectonophysics* 296: 225–247.

Schubert G, Turcotte DL, and Olson P (2001) *Mantle Convection in the Earth and Planets*. Cambrideg, UK: Cambridge University Press.

Schubert G, Yuen DA, and Turcotte DL (1975) Role of phase transitions in a dynamic mantle. *Geophysical Journal of the Royal Astronomical Society* 42: 705–735.

Smolarkiewicz PK (1983) A simple positive definite advection transport algorithm with small implicit diffusion. *Monthly Weather Review* 111: 479–489.

Smolarkiewicz PK (1984) A fully multidimensional positive definite advection scheme with small implicit diffusion. *Journal of Computational Physics* 54: 325–339.

Spalding DB (1972) A novel finite difference formulation for differential expressions involving both first and second derivatives. *International Journal of Numerical Methods in Engineering* 4: 551–559.

Spiegelman M and Katz RF (2006) A semi-Lagrangian Cranck–Nicolson algorithm for the numerical solution of advection-diffusion problems. *Geochemistry, Geophysics, Geosystems* 7 (doi:10.1029/2005GC001073).

Stemmer K, Harder H, and Hansen U (2006) A new method to simulate convection with strongly temperature-dependent and pressure-dependent viscosity in a spherical shell: Applications to the Earth's mantle. *Physics of the Earth and Planetary Interiors* 157: 223–249.

Sulsky D, Chen Z, and Schreyer HL (1994) A particle method for history-dependent materials. *Computational Methods in Applied Mechanics and Engineering* 118: 179–196.

Tackley PJ (2000) Self-consistent generation of tectonic plates in time-dependent, three-dimensional mantle convection simulations. Part 1: Pseudoplastic yielding. *Geochemistry Geophysics Geosystems* 1(8) (doi:10.1029/2000GC000043).

Tackley PJ (1998a) Three-dimensional simulations of mantle convection with a thermo-chemical basal boundary layer: D″. In: Gurnis M (ed.) *The Core–Mantle Region*, pp. 231–253. Washington, DC: AGU.

Tackley PJ (1998b) Self-consistent generation of tectonic plates in three-dimensional mantle convection. *Earth and Planetary Science Letters* 157: 9–22.

Tackley PJ (1994) *Three-Dimensional Models of Mantle Convection: Influence of Phase Transitions and Temperature-Dependent Viscosity.* PhD Thesis, California Institute of Technology, Pasadena, CA.

Tackley PJ (1993) Effects of strongly temperature-dependent viscosity on time-dependent, 3-dimensional models of mantle convection. *Geophysical Research Letters* 20: 2187–2190.

Tackley PJ and King SD (2003) Testing the tracer ratio method for modeling active compositional fields in mantle convection simulations. *Geochemistry Geophysics Geosystems* 4 (doi:10.1029/2001GC000214).

Tackley PJ, Stevenson DJ, Glatzmeir GA, and Schubert G (1993) Effects of an endothermic phase transition at 670 km depth on spherical mantle convection. *Nature* 361: 137–160.

Torrance KE and Turcotte DL (1971) Thermal convection with large viscosity variations. *Journal of Fluid Mechanics* 47: 113–125.

Travis BJ, Olson P, and Schubert G (1990) The transition from two-dimensional to three-dimensional planforms in infinite-Prandtl-number thermal convection. *Journal of Fluid Mechanics* 216: 71–92.

Trompert RA and Hansen U (1998) Mantle convection simulations with rheologies that generate plate-like behavior. *Nature* 395: 686–688.

Trompert RA and Hansen U (1996) The application of a finite-volume multigrid method to 3-dimensional flow problems in a highly viscous fluid with a variable viscosity. *Geophysical and Astrophysical Fluid Dynamics* 83: 261–291.

Trottenberg U, Oosterlee C, and Schueller A (2001) *Multigrid*, 631 pp. Orlando, FL: Academic Press.

Turcotte DL, Torrance KE, and Hsui AT (1973) Convection in the Earth's mantle in Methods. In: Bolt BA (ed.) *Computational Physics*, vol. 13, pp. 431–454. New York: Academic Press.

van den Berg AP, van Keken PE, and Yuen DA (1993) The effects of a composite non-Newtonian and Newtonian rheology on mantle convection. *Geophysical Journal International* 115: 62–78.

van Keken PE and Zhong S (1999) Mixing in a 3D spherical model of present-day mantle convection. *Earth and Planetary Science Letters* 171: 533–547.

van Keken PE, King SD, Schmeling H, Christensen UR, Neumeister D, and Doin M-P (1997) A comparison of methods for the modeling of thermochemical convection. *Journal of Geophysical Research* 102: 22477–22496.

Verfurth R (1984) A combined conjugate gradient-multigrid algorithm for the numerical solution of the Stokes problem. *IMA Journal of Numerical Analysis* 4: 441–455.

Weinberg RF and Schmeling H (1992) Polydiapirs: Multi-wavelength gravity structures. *Journal of Structural Geology* 14: 425–436.

Yavneh I (2006) Why multigrid methods are so efficient? *Computing in Science and Engineering* 8: 12–23.

Yoshida M and Kageyama A (2004) Application of the Yin-Yang grid to a thermal convection of a Boussinesq fluid with infinite Prandtl number in a three-dimensional spherical shell. *Geophysical Research Letters* 31: doi:10.1029/2004GL019970.

Yuen DA, Balachandar S, and Hansen U (2000) Modelling mantle convection: A significant challenge in geophysical fluid dynamics. In: Fox PA and Kerr RM (eds.) *Geophysical and Astrophysical Convection* pp. 257–294.

Zhang S and Christensen U (1993) Some effects of lateral viscosity variations on geoid and surface velocities induced by density anomalies in the mantle. *Geophysical Jornal International* 114: 531–547.

Zhang S and Yuen DA (1995) The influences of lower mantle viscosity stratification on 3-D spherical-shell mantle convection. *Earth and Planetary Science Letters* 132: 157–166.

Zhang S and Yuen DA (1996a) Various influences on plumes and dynamics in time-dependent, compressible, mantle convection in 3-D spherical shell. *Physics of the Earth and Planetary Interiors* 94: 241–267.

Zhang S and Yuen DA (1996b) Intense local toroidal motion generated by variable viscosity compressible convection in 3-D spherical shell. *Geophysical Research Letters* 23: 3135–3138.

Zhong S (2006) Constraints on thermochemical convection of the mantle from plume heat flux, plume excess temperature and upper mantle temperature. *Journal of Geophysical Research* 111: B04409 (doi:10.1029/2005JB003972).

Zhong S (2005) Dynamics of thermal plumes in 3D isoviscous thermal convection. *Geophysical Journal International* 154(162): 289–300.

Zhong S and Hager BH (2003) Entrainment of a dense layer by thermal plumes. *Geophysical Journal International* 154: 666–676.

Zhong S, Zuber MT, Moresi LN, and Gurnis M (2000) Role of temperature dependent viscosity and surface plates in spherical shell models of mantle convection. *Journal of Geophysical Research* 105: 11063–11082.

Zhong S, Gurnis M, and Moresi LN (1998) The role of faults, nonlinear rheology, and viscosity structure in generating plates from instantaneous mantle flow models. *Journal of Geophysical Research* 103: 15255–15268.

Zhong S and Gurnis M (1994) The role of plates and temperature-dependent viscosity in phase change dynamics. *Journal of Geophysical Research* 99: 15903–15917.

Zienkiewicz OC, Taylor RL, and Zhu JZ (2005) *The Finite Element Method: Its Basis and Fundamentals 6th edn.* Woburn, MA: Butterworth and Heinemann.

## Relevant Websites

http://www.geodynamics.org – Computational Infrastructure for Geodynamics.

http://www.lcse.umn.edu – Laboratory for Computational Science and Engineering, University of Minnesota.

# 6 Temperatures, Heat and Energy in the Mantle of the Earth

**C. Jaupart and S. Labrosse**, Institut de Physique du Globe, Paris, France

**J.-C. Mareschal**, GEOTOP-UQAM-McGill, Montreal, QC, Canada

## 6.1   Introduction

Earth's heat engine works in ways which still elude us after many years of research, both on the fundamental physical aspects of convection in the mantle and on securing new and precise observations. In this context, it is worth remembering that we pursue two different goals in studies of mantle convection. One is to account for specific phenomena, such as hot spots or back-arc spreading centers. The other goal is to go back in time in order to evaluate how the rates of geological processes have changed with time and to decipher processes which are no longer active today. Both goals require a thorough understanding of dynamics but each relies on a different set of constraints. The former can be attained using present-day observations, such as the distributions of seismic velocities and density at depth. The latter goal relies on a global energy balance, which specifies how the convective regime evolved as the Earth cooled down and its energy sources got depleted. From another standpoint, the present-day energy budget of the planet and the distribution of heat flux at the surface are constraints that mantle models must satisfy.

The present-day energy budget reflects how Earth's convective engine has evolved through geological time and hence provides clues on the past. The power of this constraint has motivated a large number of studies. With hind-sight, one may note that the emergence of convection models coincided with the failure of conductive (and radiative) thermal history models to account for the mantle temperature regime and the Earth energy budget (Jacobs and Allan, 1956; Jacobs, 1961; MacDonald, 1959). Convection in the Earth's mantle had become an inescapable conclusion of that failure and was required to explain the observation that the mantle is not molten. The difficulty in running fully consistent dynamical calculations of convection over the whole history of our planet led to so-called 'parametrized' models such that the heat flux out of the Earth is written directly as a function of dimensionless numbers describing the bulk convective system,

such as the Rayleigh number (Schubert *et al.*, 2001). For a given set of initial conditions, the model results were required to match the present-day energy budget, or more precisely the ratio of heat production to heat loss (the Urey ratio). The earliest study of this kind was probably that of McKenzie and Weiss (1975) and was followed by countless others. This approach was used to argue against whole-layer mantle convection (McKenzie and Richter, 1981), to date the emergence of plate tectonics [Peltier and Jarvis, 1982], to derive constraints on the distribution of radiogenic heat sources in the mantle (Schubert *et al.*, 1980), and even to determine the amount of radioactive sources in the Earth's core (Davies, 1980; Breuer and Spohn, 1993). The difficulty in accounting for the wealth of processes that characterize the Earth, such as continental growth as well as degassing and the implied changes of rheological properties, however, has led to disenchantment. Yet, it is clear that the present-day thermal and tectonic regime of the Earth results from several billions of years of convective processes and is best understood within a time-dependent framework.

Determination of Earth's rate of heat loss requires a very large number of heat flux measurements in a variety of geological settings. Local surveys as well as global analyses of large data sets have shown that heat flux varies on a wide range of spatial scales and, in the continents, is not a function of a single variable such as geological age, for example. Heat flux data exhibit large scatter, which has had unfortunate consequences. One has been that few scientists have invested time and energy to sort out the large number of variables and physical processes which come into play. The 1980s saw a rapid decrease in the number of research teams active in that field as well as in the number of measurements carried out at sea and on land. Another consequence has been that, with few notable exceptions (e.g., Pari and Peltier, 1995), the distribution of heat flux is rarely used as a constraint that mantle models must satisfy. Many evolutionary models for convection in the Earth's mantle have thus abandoned the energy budget as a viable constraint and have turned to geochemical data. Here, we will assess the reliability of heat flux measurements and shall demonstrate that the spatial distribution of heat flux provides a key constraint for understanding convection in the Earth.

The last two decades have seen notable advances in the interpretation of heat flux. In the oceans, these include a thorough understanding of hydrothermal circulation through oceanic crust and sediments, as well as detailed and precise heat flux measurements through both very young and very old seafloor. In the continents, sampling of old cratons is now adequate in several areas, heat production of lower-crustal assemblages is better understood, and systematic studies of heat flux and heat production allow strong constraints on the crustal contribution to the surface heat flux. Today, we have a better understanding of the energy sources in the Earth than we did 20 years ago and know how large some of the uncertainties are.

In this chapter, we shall focus on two different, but closely related, problems. One is to evaluate the present-day energy budget of the mantle with emphasis on the associated uncertainties. The other is to evaluate how thermal evolution models must be developed in order to account for this budget. We shall establish the gross thermodynamics of the Earth and shall explain how estimates of heat loss and heat production have been obtained, drawing from recent advances. We shall emphasize the peculiarities of heat loss mechanisms of our planet and in particular the spatial distribution of heat flux. We shall then rely on the heat budget to infer the present-day secular cooling rate of our planet. We shall also evaluate independent constraints on temperature in the Earth's mantle and present a reference temperature profile through the convective mantle. Finally, we shall discuss the thermal evolution of our planet, from the standpoint of both observations and theoretical models. In order to facilitate the reader's task, we give a short summary of major points at the end of each section.

## 6.2 Basic Thermodynamics

### 6.2.1 Breakdown of the Energy Budget

The integral form of the energy balance for the whole planet takes the form

$$\frac{d(U + E_c + E_g)}{dt} = -\int_A \mathbf{q} \cdot \mathbf{n} \, dA + \int_V H \, dV + \int_V \psi \, dV - p_a \frac{dV}{dt} \tag{1}$$

where $U$ is internal energy, $E_c$ is kinetic energy, and $E_g$ is gravitational potential energy. $\mathbf{q}$ is the surface heat flux, $\mathbf{n}$ is the unit normal vector, $A$ is the Earth's outer surface, $H$ is internal heat production per unit volume, $p_a$ is atmospheric pressure, $V$ is the Earth's total volume, and $\psi$ stands for energy transfers to or from external systems, such as tidal dissipation. **Table 1** provides a list of the main symbols used in this chapter. Equation [1] states that the Earth's total energy changes due to heat loss, internal heat generation, energy transfer between our planet interior and its surroundings (atmosphere as well as other celestial bodies), and finally the work of atmospheric pressure as the planet contracts. We have assumed that Earth's surface is stress-free, such that there is no work due to external shear stresses. Dissipation induced by internal convective motions is not included because it is due to internal transfers of energy and does not act to change the total energy of the system. We explain below how these internal energy transfers operate and show that changes of gravitational energy are compensated by changes of strain energy $E_S$, which is the energy required to compress matter to its actual local pressure $p$.

Our main purpose in this chapter is to evaluate the different terms in the energy balance, including changes of strain energy and gravitational energy, and to derive an equation for the average temperature of the Earth. The dominant terms on the right of eqn [1] are the Earth's rate of heat loss and internal heat generation, which are inferred from field measurements and chemical Earth models. The other terms are evaluated theoretically and are shown to be negligible.

The gravitational energy of the Earth is defined as the energy required to bring matter from infinity and, assuming spherical symmetry, can be written as

$$E_g = \int_0^R \rho(r) g(r) r 4\pi r^2 \, dr \tag{2}$$

with $\rho$ and $g$ the spatially varying density and gravity. This energy is negative because the accretion process releases energy. This energy can be computed for the present Earth and an upper limit can be obtained for a sphere with uniform density:

$$E_g = -\frac{3}{5} \frac{GM^2}{R} \tag{3}$$

where $G$ is the gravitational constant and $M$ is the mass of the Earth.

Kinetic energy may be broken down into three different components:

$$E_c = E_{rot} + E_{contr} + E_{conv} \tag{4}$$

**Table 1** Symbols used

| Symbol | Definition | Units (commonly used units/or value) |
|---|---|---|
| $C_p$ | Heat capacity | $J\,kg^{-1}\,K^{-1}$ |
| $C_Q$ | Heat flux/age$^{1/2}$ | $470–510\,mW\,m^{-2}/(My)^{1/2}$ |
| $C_A$ | Seafloor accretion rate | $3.34\,km^2\,My^{-1}$ |
| $D$ | Thickness of convecting layer | m |
| $E_c$ | Kinetic energy | J |
| $E_g$ | Gravitational potential energy | J |
| $E_{rot}$ | Rotational energy | J |
| $E_s$ | Strain energy | J |
| $F_b$ | Buoyancy flux | $W\,m^{-1}$ |
| $g$ | Acceleration of gravity | $m\,s^{-2}$ |
| $G$ | Gravitational constant | $6.67 \times 10^{-11}\,kg\,m^3\,s^{-2}$ |
| $H$ | (volumetric) Heat production | $W\,m^{-3}$ ($\mu W\,m^{-3}$) |
| $I$ | Moment of inertia | $kg\,m^2$ |
| $K$ | Bulk modulus | Pa |
| $k$ | Thermal conductivity | $W\,m^{-1}\,K^{-1}$ |
| $L$ | Length of oceanic plate (length-scale) | m |
| $M$ | Mass of Earth | $5.973 \times 10^{24}\,kg$ |
| $Q, q$ | Heat flux | $W\,m^{-2}$ ($mW\,m^{-2}$) |
| $p$ | Pressure | Pa (MPa, GPa) |
| $R$ | Radius of Earth | 6378 km |
| $Ra$ | Rayleigh number | / |
| $s$ | Entropy per unit mass | $J\,kg^{-1}\,K^{-1}$ |
| $T$ | Temperature | K (°C) |
| $u$ | Internal energy per unit-mass (also used for horizontal velocity component) | $J\,kg^{-1}$ |
| $U$ | Internal energy (also used for horizontal velocity component) | J |
| $Ur$ | Urey ratio | / |
| $V$ | Volume | $m^3$ |
| $v_c$ | Contraction velocity | $m\,s^{-1}$ |
| $w$ | Convective velocity | |
| $\alpha$ | Coefficient of thermal expansion | $K^{-1}$ |
| $\gamma$ | Grüneisen parameter | / |
| $\kappa$ | Thermal diffusivity | $m^2\,s^{-1}$ |
| $\mu$ | Viscosity | $Pa\,s$ |
| $\Omega$ | Angular velocity | $rad\,s^{-1}$ |
| $\phi$ | Heat dissipated by friction | $W\,m^{-3}$ |
| $\psi$ | External energy sources | $W\,m^{-3}$ |
| $\rho$ | Density | $kg\,m^{-3}$ |
| $\Delta S_{cond}$ | Entropy production | $J\,K^{-1}$ |
| $\sigma$ | Deviatoric stress tensor | Pa (MPa) |

corresponding to the bulk rotation of our planet, radial contraction induced by secular cooling, and internal convective motions, respectively. One may easily show that the latter two are small compared to the first one.

**Table 2** lists estimates for gravitational, kinetic, and internal energy components and makes it clear that kinetic energy is very small compared to the other two. A striking result is that the largest component by far is gravitational energy, which is larger than internal energy by at least one order of magnitude. In a constant mass planet, gravitational energy changes are due to thermal contraction, chemical differentiation, and vertical movements of Earth's surface (tectonic processes

and erosion–deposition). These various processes work in different ways and are associated with different energy transport mechanisms, and hence must be dealt with separately.

### 6.2.2 Changes in Gravitational Energy: Contraction due to Secular Cooling

Gravitational energy is the largest component of the budget (**Table 2**) and special care is warranted to evaluate how it gets converted to other forms of energy when the planet contracts. This has been discussed in a series of papers (Lapwood, 1952;

**Table 2**    Numbers – order of magnitude

|  | Value | Units |
|---|---|---|
| Rotational energy | $2.1 \times 10^{29}$ | J |
| Internal energy (for 2500 K average temperature) | $1.7 \times 10^{31}$ | J |
| Gravitational energy (uniform sphere) | $2.2 \times 10^{32}$ | J |
| Rotation angular velocity | $7.292 \times 10^{-5}$ | rad s$^{-1}$ |
| Polar moment of inertia | $8.0363 \times 10^{37}$ | kg m$^2$ |
| Total mass | $5.9737 \times 10^{24}$ | kg |
| Total Volume | $1.08 \times 10^{21}$ | m$^3$ |
| Mass mantle | $\approx 4.0 \times 10^{24}$ | kg |
| Mass crust | $\approx 2.8 \times 10^{22}$ | kg |

Flasar and Birch, 1973). Here, we avoid detailed calculations and throw light on some interesting thermodynamical aspects.

The gravitational energy changes when the planet contracts:

$$\frac{\Delta E_g}{E_g} = -\frac{\Delta R}{R} \quad [5]$$

For uniform cooling by an amount $\Delta T$, we show in Appendix 1 that

$$\frac{\Delta R}{R} \approx \frac{\langle \alpha \rangle \Delta T}{3} \quad [6]$$

where $\langle \alpha \rangle$ is an average value for the coefficient of thermal expansion and $\Delta T$ is negative. Assuming $\langle \alpha \rangle \approx 2 \times 10^{-5}\,\text{K}^{-1}$ and a secular cooling rate of $100\,\text{K Gy}^{-1}$, the contraction velocity is $\text{d}R/\text{d}t \approx -10^{-13}\,\text{m s}^{-1}$, a very small value compared to the typical convective velocity of $\approx 10^{-9}\,\text{m s}^{-1}$. The induced change of gravitational energy, however, is far from being negligible. For the same choice of parameter values, it is $\approx 4\,\text{TW}$, which, as we shall see, corresponds to 10% of the total energy loss of our planet. We shall demonstrate, however, that such changes of gravitational energy are not converted to heat.

Thermal contraction affects the planet's rotation. The moment of inertia $I$ changes:

$$\Delta I / I = 2\Delta R / R \quad [7]$$

and hence

$$\Delta E_{\text{rot}} / E_{\text{rot}} = \Delta \Omega / \Omega = -2\Delta R / R \quad [8]$$

where $\Omega$ is the Earth rotation velocity. Thus, some of the gravitational potential energy goes into the energy of rotation. Rotational energy is much less (three orders of magnitude) than gravitational energy, however, and hence may be neglected in the analysis of energy changes.

To elucidate energy transfer processes, we now consider thermodynamics at the local scale. We focus

on a few specific aspects of interest and refer the reader to the study by Braginsky and Roberts (1995) for a comprehensive analysis. All energies are now written per unit mass with small letters, that is, $e_c$ and $u$ stand for the kinetic energy and internal energy per unit mass, respectively. We begin with the standard form of the first law of thermodynamics (Bird et al., 1960; Schubert et al., 2001, see also Chapter 8):

$$\rho \frac{\text{D}(u + e_c)}{\text{D}t} = -\nabla \cdot \mathbf{q} - \nabla \cdot (p\mathbf{v}) - \nabla \cdot [\boldsymbol{\sigma} \cdot \mathbf{v}] + H + \psi + \rho \mathbf{g} \cdot \mathbf{v} \quad [9]$$

where $\boldsymbol{\sigma}$ is the deviatoric stress tensor, $\mathbf{v}$ velocity, and $\psi$ collects external source terms such as tidal dissipation. From the momentum equation, we get

$$\rho \frac{\text{D}e_c}{\text{D}t} = \rho \frac{\text{D}(v^2/2)}{\text{D}t} = -\mathbf{v} \cdot \nabla p - \mathbf{v} \cdot \nabla \cdot \boldsymbol{\sigma} + \rho \mathbf{g} \cdot \mathbf{v} \quad [10]$$

Subtracting this from the bulk energy balance leads to an equation for the internal energy:

$$\rho \frac{\text{D}u}{\text{D}t} = -\nabla \cdot \mathbf{q} + H + \psi + \phi - p\nabla \cdot \mathbf{v} \quad [11]$$

where $\phi$ stands for viscous dissipation:

$$\phi = -\nabla \cdot [\boldsymbol{\sigma} \cdot \mathbf{v}] + \mathbf{v} \cdot \nabla \cdot \boldsymbol{\sigma} = -\boldsymbol{\sigma} : \nabla \mathbf{v} \quad [12]$$

Equation [11] is thus the usual statement that changes of internal energy $u$ are due to heat gains or losses (which are broken into four contributions) and to the work of pressure (the last term on the right-hand side).

All the equations above stem from standard thermodynamics theory. We now introduce gravitational energy and strain energy. We decompose variables into the sum of the azimuthal average and a perturbation, for example, such as

$$T = \bar{T} + \theta \quad [13]$$

for temperature. The velocity field is decomposed into a component due to contraction, $\mathbf{v}_c$, and a convective component, $\mathbf{w}$. One key difference between these two is that the azimuthal average of the radial convective velocity, $\bar{w}_r$, is zero, in contrast to that of contraction. We may assume that contraction proceeds in conditions close to hydrostatic equilibrium, such that it involves no deviatoric stress and no dissipation. In this case, the azimuthal average of the momentum equation reduces to a hydrostatic balance:

$$0 = -\nabla \bar{p} + \bar{\rho}\mathbf{g} \qquad [14]$$

We now consider separately the effects of contraction, which act on the average density and pressure, and the effects of convection, which involve departures from these averages. We focus on the contraction process with velocity $\mathbf{v}_c$. In the internal energy equation [11], we identify the work done by pressure as a change of strain energy:

$$-\bar{p}\,\nabla \cdot \mathbf{v}_c = \bar{\rho}\frac{De_s}{Dt} \qquad [15]$$

and break internal energy down into heat content $e_T$ and strain energy $e_s$:

$$\bar{\rho}\frac{Du}{Dt} = \bar{\rho}\left(\frac{De_T}{Dt} + \frac{De_s}{Dt}\right) \qquad [16]$$

In the total energy balance [9], the last term on the right-hand side is the work done by the gravity force. By definition, this term can be written as the change of gravitational potential energy when it is carried over to the left-hand side of the balance:

$$\bar{\rho}\frac{De_g}{Dt} = -\bar{\rho}\mathbf{g} \cdot \mathbf{v}_c \qquad [17]$$

This relationship is demonstrated in integral form in Appendix 2.

Collecting all terms, the energy balance [9] is written as

$$\bar{\rho}\frac{D[e_T + e_s + e_g + e_c]}{Dt} = -\nabla \cdot \mathbf{q} + H + \psi$$
$$- \nabla \cdot (\bar{p}\mathbf{v}_c) + (\cdots) \qquad [18]$$

In this equation, terms associated with convective motions are not written explicitly and will be dealt with later. Kinetic energy is also negligible and, by inspection, one may deduce from eqn [18] that

$$\bar{\rho}\frac{D[e_s + e_g]}{Dt} = -\nabla \cdot (\bar{p}\mathbf{v}_c) \qquad [19]$$

This can be demonstrated by recalling the identity

$$\nabla \cdot (\bar{p}\mathbf{v}_c) = \bar{p}\nabla \cdot \mathbf{v}_c + \mathbf{v}_c \cdot \nabla \bar{p} \qquad [20]$$

Using the hydrostatic balance and eqn [17] the right-hand side of this equation can be recast as

$$\bar{p}\nabla \cdot \mathbf{v}_c + \mathbf{v}_c \cdot \nabla \bar{p} = -\bar{\rho}\frac{De_s}{Dt} - \bar{\rho}\frac{De_g}{Dt} \qquad [21]$$

which is indeed eqn [19].

Integrating eqn [19] over the whole-planet volume, we finally obtain

$$\frac{dE_g}{dt} + \frac{dE_s}{dt} = -p_a\frac{dV}{dt} \qquad [22]$$

where $E_s$ is the total strain energy of the planet. The term on the right-hand side is very small and this equation therefore states that the change of gravitational energy is compensated by one of strain energy, so that no heat is generated.

### 6.2.3  Secular Cooling Equation

To derive an equation for temperature, we return to local variables. Introducing variables of state, we write

$$\rho\frac{De_T}{Dt} = \rho T\frac{Ds}{Dt} = \rho C_p\frac{DT}{Dt} - \alpha T\frac{Dp}{Dt} \qquad [23]$$

where $s$, the entropy per unit mass, has been expressed as a function of temperature and pressure and $\alpha$ is the coefficient of thermal expansion. From [11], we deduce that

$$\rho C_p\frac{DT}{Dt} = \alpha T\frac{Dp}{Dt} - \nabla \cdot \mathbf{q} + H + \psi + \phi \qquad [24]$$

By definition,

$$\frac{Dp}{Dt} = \frac{\partial p}{\partial t} + \mathbf{v} \cdot \nabla p = \frac{\partial p}{\partial t} + \mathbf{v}_c \cdot \nabla p + \mathbf{w} \cdot \nabla p \qquad [25]$$

Thus, the first two terms on the right of this equation are responsible for the so-called 'adiabatic heating', which is the only remaining contribution of contraction in eqn [24]. It may be safely neglected because of the small contraction velocity. As we shall see below, the last term on the right is not negligible and, even though $\bar{w}_r = 0$, contributes a key term in the energy budget, the buoyancy flux.

We now focus on the contribution of convective motions to the energy budget and no longer deal with contraction. Thus, time variations of the average density and pressure are neglected. Averaging eqn [24] over a spherical shell of radius $r$ and neglecting second-order terms gives

$$\rho C_p\frac{D\bar{T}}{Dt} = -\rho \alpha g\overline{w_r\theta} - \frac{\partial \bar{q}_r}{\partial r} + H + \psi + \bar{\phi} \qquad [26]$$

where the first term on the right is the buoyancy flux and where $w_r$ is the radial convective velocity component such that $\bar{w}_r = 0$. Integrating over the planet, we finally obtain

$$\int_V \rho C_p \frac{\mathrm{D}\bar{T}}{\mathrm{D}t}\,\mathrm{d}V = M\left\langle C_p\right\rangle \frac{\mathrm{d}\left\langle T\right\rangle}{\mathrm{d}t}$$
$$= -\int_V \rho \alpha g \overline{w_r \theta}\,\mathrm{d}V - \int_A \bar{q}_r\,\mathrm{d}A$$
$$+ \int_V H\,\mathrm{d}V + \int_V \psi\,\mathrm{d}V + \int_V \bar{\phi}\,\mathrm{d}V$$
$$[27]$$

where $M$ is the mass of the Earth and $\left\langle C_p\right\rangle$ and $\left\langle T\right\rangle$ are its average heat capacity and average temperature, respectively. This equation involves viscous dissipation and may be simplified because, as shown in Appendix 3,

$$-\int_V \rho \alpha g \overline{w_r \theta}\,\mathrm{d}V + \int_V \bar{\phi}\,\mathrm{d}V = 0 \qquad [28]$$

This equation states that viscous dissipation is balanced by the bulk buoyancy flux and explains why it does not enter the bulk energy balance (*see also* Chapter 8). This equation is very useful because it allows one to include all the dissipative processes that are active simultaneously. We shall refer to it in Section 6.7, where we discuss the controls on convective motions on Earth.

Subtracting this equation from the bulk energy balance [27], we finally obtain

$$M\left\langle C_p\right\rangle \frac{\mathrm{d}\left\langle T\right\rangle}{\mathrm{d}t} = -\int_A \bar{q}_r\,\mathrm{d}A + \int_V H\,\mathrm{d}V + \int_V \psi\,\mathrm{d}V \quad [29]$$

which is the secular cooling equation.

### 6.2.4 Summary

The thermodynamics of the cooling Earth requires to separate slow contraction from the convective velocities. The gravitational energy from thermal contraction does not enter in the global heat budget because it is stored as strain energy. Similarly, the viscous dissipation can be important but is balanced by the bulk buoyancy flux. The secular cooling is dominated by the imbalance between radiogenic heat production and total heat flow.

## 6.3 Heat Loss through the Seafloor

For the purposes of calculating the rate at which the Earth is losing heat, the most direct and unbiased method is to integrate individual measurements of heat flux over the surface. As we shall see this method fails in the oceans and one has to use theory in order to obtain a reasonable estimate. This approach has drawn a lot of criticisms on the grounds that the procedure is a theoretical one and hence leads to a biased result. We shall discuss this point in detail. Our present understanding of the global thermal budget of the Earth can be summarized in the spherical harmonic representation of the surface heat flux (**Figure 1**). It is important to note that the spherical harmonic coefficients are not simply determined as the best fit to the observations but are adjusted to fit our best model of energy loss by the oceanic lithosphere.

### 6.3.1 Oceanic Heat Flux Data

Heat flow through permeable rock and sediment involves two mechanisms: conduction through the solid, and static, matrix and water flow through pores and fractures into the sea. Measuring the latter directly would be very costly and would require continuous recording over long timescales in order to determine a representative flow rate. It is clear that hydrothermal circulation is a highly efficient heat transport mechanism. It involves large volumes of oceanic rocks as shown by the extent of alteration in ophiolitic massifs (Davis and Elderfield, 2004).

Available measurement methods account only for heat conduction. The vast majority of marine heat flux determinations rely on the probe technique, such that a rigid rod carrying a thermistor chain is shoved into sediments. Obviously, this requires thick sedimentary cover, a systematic bias in the measurement environment that has important consequences discussed below. Another technique is to measure temperatures in deep-sea drillholes. This is clearly the best technique because it relies on measurements over a large depth range through poorly permeable crystalline basement, but it is particularly time consuming. Drilling operations perturb the thermal environment greatly, implying that measurements cannot be done just after drilling is completed and require hole reentry. In addition, the number of deep-sea drillholes is too small to provide a good sampling of the seafloor. The few comparisons that have been made between the two techniques show that the shallow probe technique provides reliable results (Erickson *et al.*, 1975).

High heat flux values near oceanic ridges were one of the decisive observations confirming seafloor spreading and thermal convection as a key geological mechanism (von Herzen, 1959; Langseth *et al.*, 1966).

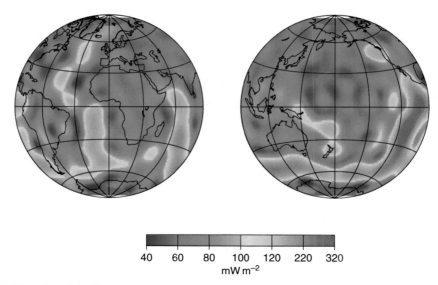

**Figure 1**   Global heat flux of the Earth obtained from the spherical harmonic representation to degree and order 18 (Pollack *et al.*, 1993).

Yet, these early surveys made it very clear that heat flux data exhibit enormous scatter and that heat conduction cannot account for all the observations. Heat flux data from the compilation by Stein and Stein (1992) have been binned in 2 My age intervals and are shown as a function of age in **Figure 2**. It is immediately apparent that there is a lot of scatter, particularly at young ages. Binning data for all oceans by age group is done for statistical reasons, in the hope that measurement errors cancel each other in a large data set. This is not valid if measurement errors are not random, which is the case here. Data reliability depends in part on environmental factors, such as basement roughness and sediment thickness. A rough basement/sediment interface leads to heat refraction effects and focussing of hydrothermal flows. At young ages, seafloor roughness depends in part on spreading velocity and hence varies from ocean to ocean. Also, the thin sedimentary cover implies severe environmental problems. Such effects explain why the data scatter is larger for young ages than for old ages. The first age bin presents a specific problem. It is characterized by the largest heat flux values (as well as the most conspicuous signs of hydrothermal activity) and hence deserves special consideration. Clearly, a proper average for this age bin requires data at very young ages, less than 1 My, say, which are virtually nonexistent. A final issue is that the global data set includes measurements made with different techniques and care. Some early data are associated with larger errors than in recent

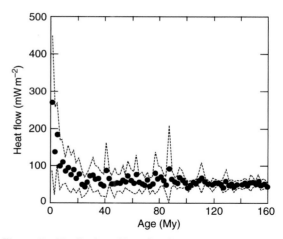

**Figure 2**   Distribution of heat flux data as a function of age from the compilation by Stein and Stein (1992). Dots represent averaged heat flux values in 2 My bins. Dash lines indicate the envelope at one standard deviation.

surveys due to small probe lengths and because thermal conductivity determinations were not made *in situ.*

Accounting for the cooling of oceanic lithosphere at young ages requires detailed understanding of oceanic hydrothermal flows, which have been recently described in Davis and Elderfield (2004). In areas of hydrothermal circulation, sediment ponds are frequently zones of recharge, such that downward advection of cold water lowers the temperature. Recharge tends to occur over wide regions in contrast to discharge, which is usually focussed through basement outcrops. By design, heat flux measurements require sedimentary cover and hence are systematically biased toward anomalously low heat flux areas. This error is systematic and hence cannot be eliminated by a large number of measurements. We shall give a specific example below. Far from oceanic ridges, sediment cover is much thicker, which alleviates most of the measurement problems. With thick hydraulically resistive sediment, hydrothermal circulation occurs in a closed system confined to the crystalline crust. In such conditions, heat flux varies spatially but the integrated value is equal to the heat extracted from the lithosphere. A reliable heat flux determination therefore requires closely spaced stations.

To summarize this description of oceanic heat flux data, we emphasize that systematic errors arise from the measurement environment. In such conditions, error analysis can only be achieved by small-scale local studies. There are very sound reasons that explain why heat flux data underestimates the total heat flux out of the seafloor (Harris and Chapman, 2004). Using the raw data average turns a blind eye to the problem of measuring heat flux through young lithosphere. Furthermore, it pays no attention to the systematics of the scatter. For the purposes of calculating heat loss through the oceans, it is therefore necessary to resort to other methods. We shall rely on detailed heat flux surveys in specific environments to obtain accurate estimates. We shall also show how the topography of the seafloor records the amount of heat that is lost by the oceanic lithosphere.

## 6.3.2 Cooling of the Oceanic Lithosphere

The basic framework for determining the temperature in the oceanic lithosphere is the heat equation:

$$\rho C_p \left( \frac{\partial T}{\partial t} + \mathbf{v} \cdot \nabla T \right) = \nabla \cdot (k \nabla T) \quad [30]$$

where $C_p$ is the heat capacity, $\rho$ is the density of the lithosphere, $k$ is thermal conductivity, and $\mathbf{v}$ is the velocity of the plate. We have neglected radiogenic heat production, which is very small in the oceanic crust and in mantle rocks (see below) and viscous heat dissipation. Over the temperature range of interest here, variations of heat capacity amount to ±20%. Such subtleties will be neglected here for the sake of clarity and simplicity. They must be taken into account, however, for accurate calculations of the thermal structure of oceanic plates (McKenzie et al., 2005). In the upper

boundary layer of a convection cell (see Chapter 3), vertical advective heat transport is negligible. Over the large horizontal distances involved, vertical temperature gradients are much larger than horizontal ones, even in the vicinity of the ridge axis, and hence one may neglect horizontal diffusion of heat. The validity of this standard boundary-layer approximation is easily verified a posteriori. One final simplification is to assume steady-state. A detailed study of oceanic tholeiitic basalts demonstrates that their composition, and hence the temperature of the mantle from which they were derived, remains constant for about 80 My (Humler et al., 1999). Over the lifetime of an oceanic plate, secular cooling may be safely neglected and the plate is in quasi-steady state. We shall neglect temperature variations in the direction parallel to the ridge axis. With these simplifications, the heat equation reduces to

$$\rho C_p u \frac{\partial T}{\partial x} = k \frac{\partial^2 T}{\partial z^2} \quad [31]$$

where $x$ is the distance from the ridge, $z$ is depth, and $u$ is the plate velocity relative to the ridge (i.e., half the spreading rate). Defining age $\tau$ for a constant spreading rate

$$u\tau = x \quad [32]$$

leads to

$$\frac{\partial T}{\partial \tau} = \kappa \frac{\partial^2 T}{\partial z^2} \quad [33]$$

where $\kappa$ is thermal diffusivity. This is the one-dimensional (1-D) heat diffusion equation, whose solution requires a set of initial and boundary conditions. The upper boundary condition is straightforward and robust: a fixed temperature of about 4°C (in practice 0°C for convenience), due to the high efficiency of heat transport in water. All the discussion deals with the validity of the other two. The initial condition requires specification of the thermal structure of an oceanic spreading center. The bottom boundary condition may not account for the complex dynamics of the Earth's upper thermal boundary layer.

### 6.3.2.1 Initial condition: Temperature distribution at the ridge axis
As mantle rises toward the oceanic ridge, it undergoes pressure release and partial melting, which affects temperature. Furthermore, the upwelling is hotter than surrounding mantle and hence loses heat laterally by diffusion. During isentropic ascent

of dry mantle, temperature decreases by about 200 K (McKenzie and Bickle, 1988), which is small relative to the temperature drop at the surface of the ridge. Thus, it is commonly assumed that the axial temperature does not vary with depth and is equal to a constant value noted $T_M$. The validity of this assumption can be assessed by a comparison between model predictions and observations. Recent calculations by McKenzie *et al.* (2005), however, rely on a realistic axial temperature profile.

### 6.3.2.2 Bottom boundary condition

The simplest model has the lower boundary at infinite depth and assumes that temperature remains finite, such that cooling proceeds unhampered over the entire age span of oceanic lithosphere, and has been called the 'half-space' model. The temperature distribution is then

$$T(z, \tau) = T_M \, \text{erf}\left(\frac{z}{2\sqrt{\kappa\tau}}\right)$$
$$= \frac{2 T_M}{\sqrt{\pi}} \int_0^{z/2\sqrt{\kappa\tau}} \exp\left(-\eta^2\right) d\eta \qquad [34]$$

for which the surface heat flux is

$$Q(\tau) = \frac{T_M}{\sqrt{\pi\kappa\tau}} = C_Q \tau^{-1/2} \qquad [35]$$

where $C_Q$ is a constant. One remarkable feature is that the surface heat flux follows the $\tau^{-1/2}$ relationship for arbitrary temperature-dependent physical properties (Carslaw and Jaeger, 1959; Lister, 1977) (Appendix 4). This equation makes the very simple

prediction that heat flux varies as $\tau^{-1/2}$. Lord Kelvin used this property in his attempt to derive the age of the Earth from the present heat loss (Thomson, 1862). Numerical models of mantle convection which are in a plate-like regime conform very well to this relationship. **Figure 3** displays a snapshot of the temperature field and surface observables in such a model, from the study of Grigné *et al.* (2005). One sees clearly that the horizontal velocity at the surface is piece-wise constant, defining plates, and that the heat flux decreases with distance $x$ from ridges as $1/\sqrt{x}$, that is to say $1/\sqrt{\tau}$ for a constant velocity. We shall see that the value for the mantle temperature $T_M$ remains subject to some uncertainty. The value of the constant $C_Q$, however, may be determined empirically from the data, as shown below.

In the other class of models, a boundary condition is applied at some depth, which marks the base of the 'plate'. In principle, one should solve for heat supply from the underlying mantle. However, this requires elaborate physical models of mantle convection relying on specific choices of material properties and simplifying assumptions. For the sake of simplicity, one may consider two simple end-member cases, such that temperature or heat flux is constant at the base of the plate. Both these boundary conditions are approximations. For example, the fixed-temperature boundary condition requires infinite thermal efficiency for heat exchange between the plate and the mantle below.

The choice of the proper basal boundary condition is important because it determines the

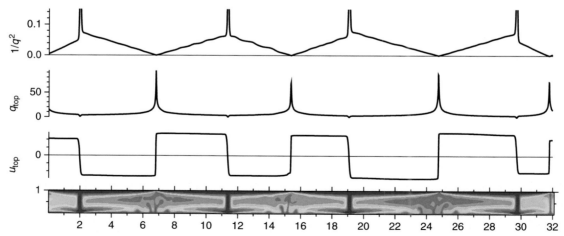

**Figure 3**   Snapshot of temperature, surface velocity ($u_{\text{top}}$), heat flux ($q$), and pseudoage ($1/q^2$) in a numerical convection model with self-consistent plate tectonics (Grigné *et al.*, 2005). See Appendix 8 for details. Note that the pseudoage varies linearly as function of distance to the ridge, which is consistent with the $\tau^{-1/2}$ heat flux law for young oceanic lithosphere.

relationship between the relaxation time and the plate thickness. Specifically, the thermal relaxation time of the plate is four times as long for fixed heat flux at the base than it is for fixed temperature. Alternatively, with fixed temperature at the base the plate must be twice as thick as the plate with fixed heat flux to have the same relaxation time. For very short time, the cooling rate (i.e., the surface heat flux) does not depend on the lower boundary condition and is the same for a plate with fixed heat flux or with fixed temperature at the base: it is the same as the heat flux for the cooling half-space. The details are provided in Appendix 5. Thus, it is better to use a half-space model for young ages because it relies on a reduced set of hypotheses. It does fit the oceanic data, as shown below. Furthermore, it has been tested over and over again and forms the basis for scaling laws of convective heat flux in many different configurations (Howard, 1964; Turcotte and Oxburgh, 1967; Olson, 1987). This simple model breaks down at ages larger than about 80 My for reasons that are still debated. For this reason, it may be wise not to rely on a specific physical model, such as the plate model with fixed basal temperature. At old ages, heat flux data exhibit small scatter and may be used directly.

### 6.3.2.3 Check: Heat flux data

A very detailed heat flux survey on young seafloor near the Juan de Fuca ridge was conducted by Davis *et al.* (1999) with three specific goals: evaluate the intensity and characteristics of hydrothermal circulation, assess local thermal perturbations due to basement irregular topography, and test cooling models for the lithosphere. **Figure 4** shows the salient results. Data between 1 and 3 My conform to the expected $\tau^{-1/2}$ relationship and suggest that constant $C_Q$ in eqn [35] is between 470 and 510 (with heat flux in mW m$^{-2}$ and age in My).

This value can be corroborated independently using heat flux data over a larger age range. To this aim, we use the 'reliable' heat flux data of Sclater *et al.* (1976) in well-sedimented areas younger than 80 My. For those sites, thick sedimentary cover is hydraulically resistive and seals off hydrothermal circulation which may still be effective in the igneous basement. Thus, there are no localized discharge zones and the average heat flux is equal to the rate at which the basement loses energy. Adding the constraint that, according to the half-space model, heat flux tends to zero as age tends to infinity tightens the estimate (Harris and Chapman, 2004). **Figure 5** shows that values for $C_Q$ are between 475 and 500 with the

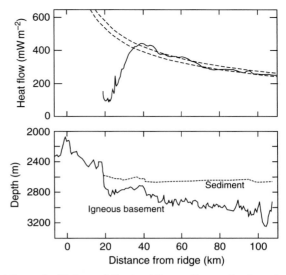

**Figure 4** High-resolution heat flux profile near the Juan de Fuca ridge from Davis *et al.* (1999). Dashed lines stand for two predictions of the half-space cooling model with constant $C_Q$ in the $\tau^{-1/2}$ heat flux–age relationship equal to 470 and 510 (with heat flux in mW m$^{-2}$ and age in My).

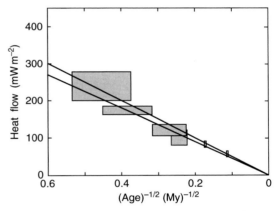

**Figure 5** Averaged heat flux over well-sedimented areas excluding ocean floor older than 80 My. Data through very young ocean floor with very large scatter provide no useful constraints and have been omitted for clarity. Solid lines correspond to values of 475 and 500 for constant $C_Q$.

same units as above, in remarkable agreement with the value deduced from the local Juan de Fuca survey. Combining these two independent determinations, we conclude that $C_Q = 490 \pm 20$, corresponding to an uncertainty of $\pm 4\%$. **Table 3** compares the various estimates that have been used in the past. The heat loss estimate of Pollack *et al.* (1993) was based on $C_Q = 510$, which is clearly an upper bound. This specific value was taken from the analysis of Stein and Stein (1992), which itself was

**Table 3**   Estimates of the continental and oceanic heat flux and global heat loss

|  | Continental (mW m$^{-2}$) | Oceanic (mW m$^{-2}$) | Total (TW) |
|---|---|---|---|
| Williams and von Herzen (1974) | 61 | 93 | 43 |
| Davies (1980) | 55 | 95 | 41 |
| Sclater et al. (1980) | 57 | 99 | 42 |
| Pollack et al. (1993) | 65 | 101 | 44 |
| This study[a] | 65 | 94 | 46 |

[a]The average oceanic heat flux does not include the contribution of hot spots. The total heat loss estimate does include 3 TW from oceanic hot spots.

based on the plate model with constant basal temperature. One feature of this model is that $T_M = 1725$ K, a high value that is not consistent with the average ridge-axis temperature derived from the compositions of mid-ocean ridge basalts (MORBs) (Kinzler and Grove, 1992) (**Table 4**).

#### 6.3.2.4   Depth of the seafloor

An isostatic balance condition leads to a very simple equation for subsidence with respect to the ridge axis (Sclater and Francheteau, 1970):

$$\Delta h(\tau) = h(\tau) - h(0)$$
$$= \frac{1}{\rho_m - \rho_w} \int_0^d (\rho[T(z, \tau)] - \rho[T(z, 0)]) dz \quad [36]$$

where $h(\tau)$ and $\Delta h(\tau)$ are the depth of the ocean floor and subsidence at age $\tau$ and where $\rho_m$ and $\rho_\omega$ denote the densities of mantle rocks at temperature $T_M$ and water, respectively. In this equation, $d$ is some reference depth in the mantle below the thermal boundary layer. This equation neglects the vertical normal stress at depth $d$, which may be significant only above the mantle upwelling structure, that is, near the ridge axis. We are interested in the heat flux out of the seafloor, $q(0, \tau)$. Assuming for

simplicity that the coefficient of thermal expansion $\alpha$ is constant, the equation of state for near-surface conditions is

$$\rho(T) = \rho_m[1 - \alpha(T - T_M)] \quad [37]$$

From the isostatic balance [36], we obtain

$$\frac{db}{d\tau} = \frac{-\alpha \rho_m}{\rho_m - \rho_w} \frac{d}{d\tau} \left[ \int_0^d T(z, \tau) dz \right]$$
$$= \frac{-\alpha}{C_p(\rho_m - \rho_w)} \frac{d}{d\tau} \left[ \int_0^d \rho_m C_p T(z, \tau) dz \right] \quad [38]$$

where we have also assumed that $C_p$ is constant. Heat balance over a vertical column of mantle between $z = 0$ and $z = d$ implies that

$$\frac{db}{d\tau} = \frac{\alpha}{C_p(\rho_m - \rho_w)} [q(0, \tau) - q(d, \tau)] \quad [39]$$

which states that thermal contraction reflects the net heat loss between the surface and depth $d$. Because $q(0, \tau)$ depends on $T_M$, the subsidence rate also depends on the initial temperature at the ridge axis. This equation states that the surface heat flux is the sum of heat flux at depth $d$ and the amount of cooling over vertical extent $d$. Using only the latter therefore leads to an underestimate of the surface heat flux.

Carlson and Johnson (1994) investigated these theoretical predictions using the best data set, basement depths from deep-sea drillholes, which require no correction for sediment thickness. They reached three important conclusions. One is that "no simple plate model has an acceptable degree of systematic misfit over the entire range of ages". They further found that the plate model underestimates depth for ages less than 80 My. Their third conclusion was that the half-space cooling model was consistent with the depth data. Using the best-fit parameters deduced from subsidence data and estimates for the various physical properties of mantle rocks in eqn [39] (i.e., for $\alpha$, $C_p$, and the densities), they predicted heat flux

**Table 4**   Potential temperature of the oceanic upper mantle

|  | Reference | Method |
|---|---|---|
| 1333°C[a] | Parsons and Sclater (1977) | Average depth + heat flux |
| 1450°C[a] | Stein and Stein (1992) | Average depth + heat flux |
| 1300–1370°C[a] | Carlson and Johnson (1994) | True basement depth (DSDP) |
| 1315°C | McKenzie et al. (2005) | Depth + heat flux with $k(T)$, $C_p(T)$ and $\alpha(T)$ |
| 1280°C | McKenzie and Bickle (1988) | Average basalt composition |
| 1315–1475°C | Kinzler and Grove (1992) | Basalt composition |
| 1275–1375°C | Katsura et al. (2004) | Isentropic profile through Ol–Wa phase-change |

[a]Temperature estimate for a cooling model with constant temperature below the ridge axis (i.e., which does not account for isentropic decompression melting).

values that were consistent with reliable heat flux data. For ages less than 80 My, the best-fit depth versus age relationship is

$$b(\tau) = (2600 \pm 20) + (345 \pm 3)\tau^{-1/2} \qquad [40]$$

with $b$ in meters and $\tau$ in My. From this relationship, the predicted heat flux over the same age range is

$$q(0, \tau) = (480 \pm 4)\tau^{-1/2} \qquad [41]$$

with $q$ in mW m$^{-2}$ and $\tau$ in My. We shall return to these estimates below. Using the same physical properties, Carlson and Johnson (1994) find that fitting the depth data with the plate model and half-space model lead to different values for the average ridge-axis temperature $T_M$: 1470°C and 1370°C, respectively. The former value is inconsistent with phase diagram constraints for mantle melting (Kinzler and Grove, 1992) (**Table 4**).

### 6.3.3   Heat Loss through the Ocean Floor

These simple physical considerations demonstrate that seafloor topography and surface heat flux record the same phenomenon, and furthermore that, within uncertainty, the subsidence rate is proportional to surface heat flux. In contrast, the raw oceanic heat flux data set records the conductive heat flux through a thin and heterogeneous superficial permeable layer. Using this data set leaves the seafloor topography unexplained.

The simple plate model with fixed basal temperature leads to an overestimate of the temperature at the ridge axis and hence leads to an overestimate of the oceanic heat flux. No such problem is encountered with the half-space model. To calculate the total rate of oceanic heat loss with as few hypotheses as possible, we break down the data into two different age intervals: 0–80 My, where the simple $t^{-1/2}$ law holds, and older seafloor where heat flux data depart from the half-space model. For seafloor older than 80 My, the heat flux is approximately constant $q_{80} \approx 48$ mW m$^{-2}$ (Lister *et al.*, 1990) (**Figure 6**). Deviations from this value are ±3m W m$^{-2}$ and exhibit no systematic trend as a function of age. The mean is determined with an uncertainty of 1 mW m$^{-2}$, which represents 1% of the average oceanic heat flux. This has no impact on the total heat loss estimate which is dominated by the young seafloor contribution. We also use depth data for the first time interval but not for the second one, because intrinsic basement roughness and volcanic

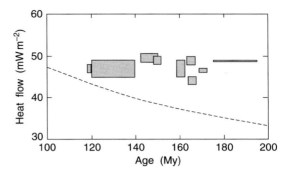

**Figure 6**   Heat flux data and prediction of the half-space cooling model for ages larger than 100 My. From Lister CRB, Sclater JG, Davis EE, Villinger H, and Nagahira S (1990) Heat flow maintained in ocean basins of great age: Investigations in the north equatorial west Pacific. *Geophysical Journal International* 102: 603–630.

constructions obscure age variations. We also add the contribution of marginal basins, whose heat flux conforms to the standard oceanic heat flux model, as demonstrated by Sclater *et al.* (1980).

Heat loss through the ocean floor is equal to

$$Q_0 = \int_0^{\tau_{max}} q(0, \tau)\frac{\mathrm{d}A}{\mathrm{d}\tau}\mathrm{d}\tau \qquad [42]$$

where $A(\tau)$ is the distribution of seafloor with age, which can be deduced from maps of the ocean floor (Sclater *et al.*, 1980; Müller *et al.*, 1997; Royer *et al.*, 1992). Using the most recent global data set of Royer *et al.* (1992), Rowley (2002), and Cogné and Humler (2004) found that a simple linear relationship provides a good fit to the data (**Figure 7**), confirming the earlier result of Sclater *et al.* (1980):

$$\frac{\mathrm{d}A}{\mathrm{d}\tau} = C_A(1 - \tau/\tau_m) \qquad [43]$$

These three different groups of authors agree that $\tau_m = 180$ My but quote slightly different values of the coefficient $C_A$: 3.45 km$^2$ yr$^{-1}$ for Sclater *et al.* (1980), 2.96 km$^2$ yr$^{-1}$ for Rowley (2002) and 2.85 km$^2$ yr$^{-1}$ for Cogné and Humler (2004). The larger estimate of Sclater *et al.* (1980) is due to the inclusion of marginal basins, which contribute ≈0.38 km$^2$ yr$^{-1}$ to the global accretion rate. The small difference of about 0.12 km$^2$ yr$^{-1}$ (4%) between the more recent estimates of Rowley (2002) and Cogné and Humler (2004) arises from the different methods used to fit the data. One independent constraint is brought by the total area of ocean floor, which is sensitive to the exact location of the continent–ocean boundary. A detailed analysis of continental margins leads to a

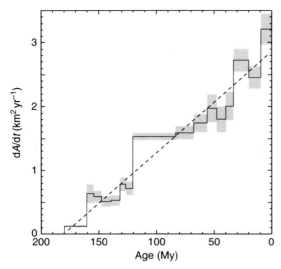

**Figure 7** Distribution of seafloor ages. The dashed line is the best-fit linear function. Modified from Cogné JP and Humler E (2004). Temporal variation of oceanic spreading and crustal production rates during the last 180 My. *Earth and Planetary Science Letters* 227: 427–439.

total continental area of $210 \times 10^6 \, \text{km}^2$ (Cogley, 1984). From this, the total seafloor surface is $300 \times 10^6 \, \text{km}^2$, slightly less than the value used by Sclater *et al.* (1980). For a triangular age distribution with a maximum age $\tau_m = 180 \, \text{My}$, this implies that $C_A = 3.34 \, \text{km}^2 \, \text{yr}^{-1}$. Subtracting the contribution of marginal basins, this corresponds exactly to the Rowley (2002) estimate. Thus, for our purposes, we shall use $C_A = 3.34 \, \text{km}^2 \, \text{yr}^{-1}$. This discussion illustrates that uncertainties may come from unexpected variables, the area of the sea floor in this particular instance. The continental heat flux budget must account for the remaining $210 \times 10^6 \, \text{km}^2$.

Uncertainties come from estimates of the total area of ocean floor, or, more precisely, the total area of continental shelves as well as departures from the simple triangular age distribution. The former is less than 3% and implies a much smaller uncertainty on the global heat loss estimate because the total surface of the Earth is known very precisely: a change in the area of oceans is compensated by an opposite change in the area of continents. Considering the difference between the average oceanic and continental heat flux values, the resulting uncertainty on the global heat loss estimate is only 1%. The impact of departures from the triangular age distribution is best assessed by comparing the heat loss estimate derived from eqn [42] and that obtained by adding the individual contribution of each age group. Parsons (1982) showed that this difference

amounts to about 0.3% of the total, which may be considered negligible. As we shall see later on, however, evaluating the uncertainty on the age distribution must be done over a large timescale and involves consideration of the stability of the convective planform.

Integrating separately seafloor younger and older than 80 My gives

$$Q_{80-} = \int_0^{80} C_Q \, \tau^{-1/2} C_A (1 - \tau/180) \, \mathrm{d}\tau = 24.3 \, \text{TW} \quad [44]$$

$$Q_{80+} = q_{80} \int_{80}^{180} C_A (1 - \tau/180) \, \mathrm{d}t' = 4.4 \, \text{TW} \quad [45]$$

$$Q_{\text{oceans}} = 29 \pm 1 \, \text{TW} \quad [46]$$

where the uncertainty comes mostly from that on coefficient $C_Q$. The present estimate is slightly less than earlier estimates because of the slightly lower ridge temperature (or equivalently, the slightly smaller value of coefficient $C_Q$ in the heat flux vs age relationship) and because of the revised estimate for the mean accretion rate at zero age $C_A$. For $C_Q = 510 \, \text{mW} \, \text{m}^{-2} \, \text{My}^{-1/2}$ and $C_A = 3.45 \, \text{km}^2 \, \text{yr}^{-1}$, the heat loss would be 31 TW.

For small ages, the bathymetry provides a direct measure of the heat lost by the cooling plate. We obtain another heat loss estimate with the following equation:

$$Q(80-) = \frac{C_p}{\alpha} (\rho_m - \rho_w) \times \int_0^{80} \frac{\mathrm{d}b}{\mathrm{d}\tau} \frac{\mathrm{d}A(\tau)}{\mathrm{d}\tau} \, \mathrm{d}\tau \quad [47]$$

$$\approx 24 \, \text{TW} \quad [48]$$

which is formally identical to the heat flux equation above and where $\rho_m = 3300 \, \text{kg} \, \text{m}^{-3}$, $C_p = 10^3 \, \text{J} \, \text{kg}^{-1} \, \text{K}^{-1}$, $\alpha = 3 \times 10^{-5} \, \text{K}^{-1}$. In old basins, where heat flux $\approx 48 \, \text{mW} \, \text{m}^{-2}$, the bathymetry is almost flat and hence cannot be used to estimate the rate of heat loss.

These estimates of oceanic heat loss do not account for the contribution of hot spots which are areas of enhanced heat flux (Bonneville *et al.*, 1997). The heat flux from hot spots can be estimated from the buoyancy of bathymetric swells (Davies, 1988; Sleep, 1990). These estimates are in the range 2–4 TW and must be added to the heat loss due to plate cooling. We discuss below the relationship between the hot-spot component and the core heat loss.

### 6.3.4 Summary

Heat loss through the ocean floor cannot be determined using the raw heat flux data set which includes many measurements that are affected by hydrothermal circulation and irregularities of the sediment cover. Predictions of the 'half-space' theoretical model for the cooling of the lithosphere can be compared successfully to measurements in selected environments where the effects of hydrothermal circulation can be assessed and accounted for. This model also implies values for the mantle temperature beneath mid-ocean ridges that are consistent with independent petrological models for basalt genesis. Finally, it is consistent with the evolution of seafloor bathymetry. Relying on the raw heat flux data would leave bathymetry data unaccounted for. The heat loss estimate requires accurate values for the areal distribution of seafloor ages. Uncertainty in the end result is essentially due to errors on the extent of continental margins. Accounting for the various uncertainties involved, the present-day rate of heat loss through the ocean floor is $32 \pm 2$ TW. This estimate includes the enhanced heat flux over hot spots.

## 6.4 Heat Loss through Continents

### 6.4.1 Average Continental Heat Flux and Heat Loss through Continental Areas

There are more than 10 000 heat flux measurements over the continents and their margins. The raw average of all the continental heat flux values is $80$ mW m$^{-2}$ (Pollack *et al.*, 1993). However, there is a strong bias to high heat flux values because many measurements were made in geothermal areas (e.g., western US, Baikal Rift) and large areas (Antarctica, Greenland, parts of the shields in Brazil and Africa) have almost no data. In the United States, a large fraction of the more than 2000 heat flux measurements belong to the Basin and Range Province. Excluding the values from the United States, the mean continental heat flux is only $66$ mW m$^{-2}$.

Bias in the sampling can be removed by area-weighting the average as demonstrated in **Table 5**. Averaging over $1° \times 1°$ windows yields a mean heat flux of $65.3$ mW m$^{-2}$. Using wider windows does not change this mean value significantly. The histograms of heat flux values or averages over $1° \times 1°$ windows have identical shapes, except for the extremely high values ($>200$ mW m$^{-2}$). Pollack *et al.* (1993) have obtained a mean continental heat flux of $66$ mW m$^{-2}$ by binning heat flux values by tectonic age and weighting by the area. Different methods to estimate the mean continental heat flux consistently yield $63$–$66$ mW m$^{-2}$. Here, uncertainty is due to poor data coverage in several regions (Greenland,

**Table 5** Continental heat flux statistics

|  | $\mu(Q)$[a] (mW m$^{-2}$) | $\sigma(Q)$[b] (mW m$^{-2}$) | N(Q)[c] |
|---|---|---|---|
| *World* |  |  |  |
| All values | 79.7 | 162 | 14123 |
| Averages $1° \times 1°$ | 65.3 | 82.4 | 3024 |
| Averages $2° \times 2°$ | 64.0 | 57.5 | 1562 |
| Averages $3° \times 3°$ | 63.3 | 35.2 | 979 |
| *USA* |  |  |  |
| All values | 112.4 | 288 | 4243 |
| Averages $1° \times 1°$ | 84 | 183 | 532 |
| Averages $2° \times 2°$ | 78.3 | 131.0 | 221 |
| Averages $3° \times 3°$ | 73.5 | 51.7 | 128 |
| *Without USA* |  |  |  |
| All values | 65.7 | 40.4 | 9880 |
| Averages $1° \times 1°$ | 61.1 | 30.6 | 2516 |
| Averages $2° \times 2°$ | 61.6 | 31.6 | 1359 |
| Averages $3° \times 3°$ | 61.3 | 31.3 | 889 |

[a]Mean of the window-averaged heat flux values.
[b]Standard deviation of the window-averaged heat flux values.
[c]Number of windows with heat flux data.

Antarctica, large areas in Africa). Those under-sampled regions account for about 20% of the total continental surface and it would take large departures from the continental heat flux trends to affect the end result significantly.

For a mean continental heat flux value of $65$ mW m$^{-2}$, the contribution of all the continental areas (i.e., $210 \times 10^6$ km$^2$) to the energy loss of the Earth represents $\approx 14$ TW. This number includes the submerged margins and continental areas with active tectonics, where higher than normal heat flux values are associated with thick radiogenic crust and shallow magmatic activity. Uncertainty in this number is due to lack of adequate data coverage in Greenland, Antarctica, and large parts of Africa. To estimate the induced uncertainty, we assume that heat flux in those areas is equal to either the lowest or the highest average heat flux recorded in well-sampled geological provinces (36 and 100 mW m$^{-2}$, respectively). This procedure allows departures of $\pm 1.5$ TW from the estimate of 14 TW. This uncertainty is certainly overestimated because the poorly sampled regions are vast and must encompass geological provinces of various ages and geological histories; for instance, both Antarctica and Greenland are known to include high and low heat flux regions. For the sake of simplicity, we shall retain a final uncertainty estimate of 1 TW only.

## 6.4.2   Various Contributions to the Surface Heat Flux in Continental Areas

Determining the heat loss from the mantle through the continental lithosphere requires accounting for the crustal heat production. In stable continents, for ages greater than about 500 My, continents are near thermal steady-state such that surface heat flux is the sum of heat production in the lithosphere and of the heat supply at the base of the lithosphere. We shall focus on estimating the crustal heat production and shall discuss briefly the contribution of the lithospheric mantle. The average heat flux does not vary significantly for provinces older than 500 My (Sclater *et al.*, 1980) and, only in Archean (i.e. older than 2.5 Gy) provinces, it might be lower than in younger terranes (Morgan, 1985). The number of heat flux determinations in Archean and Pre-Cambrian provinces has increased during the past 20 years. With adequate sampling of heat flux and heat production, and detailed information on geology and crustal structure, the crustal and mantle components of the heat flux can now be determined. It will be shown that, for stable regions, the crustal heat production makes the dominant contribution and the heat flux from the mantle is low.

Recently active regions are in a transient thermal regime and the high surface heat flux reflects cooling of the continental lithosphere. After removing the crustal heat production (which has been determined in stable provinces), it is possible to estimate the transient component of the heat flux, which originates in mantle cooling.

## 6.4.3   Estimating Moho Heat Flux

### 6.4.3.1   *Direct estimates of Moho heat flux*

Early attempts to calculate mantle heat flux relied on an empirical relationship between heat flux and heat production rate, the so-called 'linear heat flow–heat production relationship' (Birch *et al.*, 1968; Roy *et al.*, 1968):

$$Q = Q_r + D \times H \qquad [49]$$

where $Q$ is the local surface heat flux, $H$ is the local surface heat production, and $D$ is a length scale related to the thickness of a shallow layer enriched in radiogenic elements. The intercept $Q_r$ is called the reduced heat flux and represents the contribution of the mantle and crust below the enriched shallow layer. It was suggested that crustal heat production decreases exponentially as a function of depth down to the

Moho (Lachenbruch, 1970). If this were true, values for $D$ ($\approx 10$ km) would imply that the mantle heat flux is equal to $Q_r$. Although it was soon realized than it cannot be so, many studies still rely on heat production that it cannot decrease exponentially with depth.

The significance of the empirical heat flow relationship has been questioned on theoretical grounds (England *et al.*, 1980; Jaupart, 1983). It was shown that, for the rather small wavelengths involved, surface heat flux is only sensitive to shallow heat production contrasts (Jaupart, 1983; Vasseur and Singh, 1986). With more data available, it was found that for many provinces such a linear relationship is not verified (Jaupart *et al.*, 1982; Jaupart and Mareschal, 1999). Secondly, the crustal component of the heat flux can now be estimated from systematic investigations of lower crustal rocks, from both large granulite facies terrains (Fountain and Salisbury, 1981; Ashwal *et al.*, 1987; Fountain *et al.*, 1987) and xenoliths suites (Rudnick and Fountain, 1995). The heat production values, obtained on samples from large exposure of granulite facies terranes in different areas of the Superior Province, are very consistent ($\approx 0.4\,\mu W\,m^{-3}$). They appear to be representative of all granulite facies terranes worldwide, including the Ivrea Zone (Pinet and Jaupart, 1987; Joeleht and Kukkonen, 1998). Thirdly, sampling in superdeep holes (Kola, Russia, and KTB, Germany) demonstrates that heat production shows no systematic variation with depth as would be required by the linear relationship. At Kola, the Proterozoic supracrustal rocks (above 4 km depth) have lower heat production ($0.4\,\mu W\,m^{-3}$) than the Archean basement ($1.47\,\mu W\,m^{-3}$) (Kremenentsky *et al.*, 1989). At KTB, heat production decreases with depth at shallow levels, reaches a minimum between 3 and 8 km and increases again in the deepest parts of the borehole (Clauser *et al.*, 1997). Over a larger depth extent, studies of exposed crustal sections suggest a general trend of decreasing heat production with depth, but this trend is not a monotonic function (Ashwal *et al.*, 1987; Fountain *et al.*, 1987; Ketcham, 1996). Even for the Sierra Nevada batholith where the exponential model had initially been proposed, a recent compilation has shown that the heat production does not decrease exponentially with depth (Brady *et al.*, 2006). In the Sierra Nevada, heat production first increases, then decreases and remains constant in the lower crust beneath 15 km.

Many authors have assumed that the mantle heat flux is $\approx 25\,mW\,m^{-2}$ in stable continental regions, because this was the lowest measured value (Pollack and Chapman, 1977a; Cermak and Bodri, 1986). New heat flux values $\approx 20\ mW\ m^{-2}$ have been obtained at several locations (Chapman and Pollack, 1974; Swanberg *et al.*, 1974; Duchkov, 1991; Mareschal *et al.*, 2000; Mareschal *et al.*, 2005) and the average heat flux over wide areas ($500 \times 500\ km^2$) of the Baltic and Siberian Shields is $<18\ mW\ m^{-2}$. The mantle heat flux cannot be higher than these values. For the mantle heat flux to be equal to such values, the whole crust below specific measurement sites must be completely devoid of heat-producing elements over a large area.

Another approach was to assume that the mantle heat flux is roughly proportional to the average surface heat flux. Pollack and Chapman (1977b) have argued that mantle heat flux is $\approx 40\%$ of the regional average surface heat flux. Their analysis, however, was based on a small data set. In order to detect changes of basal heat flux, small-scale heat flux variations are of little use because they record shallow heat production contrasts and one must work at a minimum scale of about 300 km (Mareschal and Jaupart, 2004). This places stringent constraints on data coverage because heat flux and heat production vary on a typical scale of 10 km due to the heterogeneity of the crust. In North America, there are sufficient heat flux data to sample five well-defined provinces or subprovinces with different geological structures on a scale of about 500 km. For these five large provinces, average values of heat flux and surface heat production are statistically correlated (**Figure 8**). The data are close to a relationship of the form

$$\bar{Q} = Q_{\text{i}} + D\bar{H} \qquad [50]$$

where $\bar{Q}$ and $\bar{H}$ are province-wide-averaged heat flux and heat production. That this relationship takes the same form as the 'local' relationship (eqn [49]) is fortuitous. In northern America, the latter is only valid for relatively small-scale variations (typically 10–50 km) of heat flux and heat production over Appalachian plutons and does not hold in the Pre-Cambrian provinces (Grenville, Trans-Hudson Orogen, Superior Province). The new relationship (eqn [50]) reflects variations of average heat flux on a much larger scale ($>500$ km) and relies on a very large data set. It implies that the average heat flux takes the same value $Q_{\text{i}}$ at some intermediate crustal depth in all provinces. Formally, it is not possible to rule out variations of mantle heat flux between the

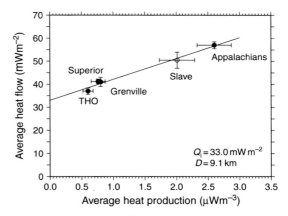

**Figure 8** Averaged heat flux versus average surface heat production for five major geological provinces of North America. The solid line is the best-fit linear relationship of eqn [50].

five provinces but the data require that such variations are exactly compensated by opposite variations of lower-crustal heat production. It is hard to explain how this may be achieved in practice and the most sensible hypothesis is that the mantle heat flux is approximately the same beneath the five provinces. For these provinces, independent geophysical and petrological constraints on crustal structure show indeed that changes of crustal heat production account for the observed heat flux variations (Pinet *et al.*, 1991; Mareschal *et al.*, 1999).

We shall explain how we calculated the crustal heat production by various methods and obtained for the mantle heat flux values that are consistently $\leq 18\ mW\ m^{-2}$.

### 6.4.3.2 Crustal heat production and Moho heat flux

Regions of low surface heat flux provide a strong constraint on the Moho heat flux. In several parts of the Canadian Shield, heat flux values as low as $22\ mW\ m^{-2}$ have been measured (Jaupart and Mareschal, 1999; Mareschal *et al.*, 2000). Similar values have also been reported for the Siberian Shield (Duchkov, 1991), the Norwegian Shield (Swanberg *et al.*, 1974), and western Australia (Cull, 1991). These correspond to areas where the crustal contribution is smallest and hence provide an upper bound to the mantle heat flux. One may refine this estimate further by subtracting some lower bound for crustal heat production. Surface heat flux records a large-scale average of heat production, and hence one should consider a representative crustal assemblage, and not a single rock type, such as gabbro, for example. For no crustal material are heat production estimates lower than $0.1 \, \mu W \, m^{-3}$ (Pinet and Jaupart, 1987; Joeleht and Kukkonen, 1998; Rudnick and Fountain, 1995). Over the average thickness of $\approx 40 \, km$, the contribution of the crust must be at least $4 \, mW \, m^{-2}$, and hence the mantle heat flux must be $< 18 \, mW \, m^{-2}$. In Norway, Swanberg et al. (1974) obtained a heat flux value of $21 \, mW \, m^{-2}$ over an anorthosite body, after estimating the crustal heat production, they concluded that mantle heat flux is about $11 \, mW \, m^{-2}$. The same value of mantle heat flux was obtained from the analysis of all the heat flux and radiogenic heat production data in the Norwegian Shield (Pinet and Jaupart, 1987).

A lower bound on mantle heat flux can be obtained by requiring that melting conditions are not attained in the crust in the absence of tectonic events and magmatic intrusions (Rolandone et al., 2002). In high heat flux areas of the Canadian Shield, crustal rocks are at high temperatures today and were still hotter in the past when radiogenic heat production was higher. The condition of thermal stability provides a lower bound of $11 \, mW \, m^{-2}$ on the mantle heat flux. Combining this result with the independent constraints derived from present-day heat flux values leads to a range of $11-18 \, mW \, m^{-2}$ for the mantle heat flux beneath the Canadian Shield. Arguments different from these have led to the same range of values in other Pre-Cambrian areas (Jones, 1988; Guillou-Frottier et al., 1995; Gupta et al., 1991).

In several regions of the world, a large fraction of the crustal column has been exposed by tectonic processes. Sampling of such exposed cross-sections allows the determination of the vertical distribution of radiogenic elements. If heat flux and seismic data are also available, it is possible to determine the total crustal heat production. For the Kapuskasing structure in the Canadian Shield where the crustal contribution could be determined, the Moho heat flux was calculated to be $13 \, mW \, m^{-2}$ (Ashwal et al., 1987; Pinet et al., 1991). The average crustal heat production can also be estimated in provinces where all crustal levels can be found at the surface. In these provinces, systematic sampling will yield an estimate of the average bulk crustal heat production. In the Grenville Province of the Canadian Shield, the average crustal heat production was determined to be $0.65 \, \mu W \, m^{-3}$ for an average surface heat flux of $41 \, mW \, m^{-2}$. This yields a Moho heat flux of $15 \, mW \, m^{-2}$ (Pinet et al., 1991). Similar results have

been reported for other shields in the world, including South Africa (Nicolaysen et al., 1981) and India (Roy and Rao, 2000), and are listed in **Table 6**.

Other methods have combined heat flux with other geophysical data, mainly long-wavelength Bouguer gravity, to estimate changes in crustal composition. A search for all models consistent with all the available data, including gravity data and bounds on heat production rates for the various rock types involved, leads to a range of $7-15 \, mW \, m^{-2}$ for the mantle heat flux (Guillou et al., 1994).

The above estimates were derived using local geophysical and heat production data in several provinces and rely on knowledge of crustal structure. Independent determinations of the mantle heat flux may be obtained by considering the lithospheric thickness determined by seismic and xenolith studies. Pressure and temperature estimates from mantle xenoliths may be combined to determine a best-fit geotherm consistent with heat transport by conduction. Mantle heat flux estimates obtained in this manner depend on the value assumed for thermal conductivity. Available estimates are consistent with those deduced from crustal models and are listed in **Table 6**.

The estimates of **Table 6** come from Archean and Proterozoic cratons where heat flux values are generally low. Heat flux values tend to be larger in younger stable continental regions. For example, heat flux is higher ($57 \, mW \, m^{-2}$) in the Appalachians than in the Canadian Shield. The crust of the Appalachians contains many young granite intrusions with very high heat production ($> 3 \, \mu W \, m^{-3}$). The elevated heat flux can be accounted for by the contribution of these granites and does not require mantle heat flux to be higher than in the Shield (Pinet et al., 1991; Mareschal et al., 2000). Throughout stable North America, including the Appalachians, variations of the mantle heat flux may not be exactly zero but must be less than departures from the best-fitting relationship (**Figure 8**), or about $\pm 2 \, mW \, m^{-2}$. This estimate is close to the intrinsic uncertainty of heat flux measurements (Jaupart and Mareschal, 1999).

Allowing for the uncertainties and requiring consistency with low heat flux measurements, we retain the range of $15 \pm 3 \, mW \, m^{-2}$ for the mantle heat flux in stable continents. For this range, the differences of average heat flux values between geological provinces cannot be accounted for by changes of mantle heat flux and hence must be attributed to changes of crustal heat production. The ranges of

**Table 6**   Various estimates of the heat flux at Moho in stable continental regions

| Location | Heat flux (mW m$^2$) | Reference |
|---|---|---|
| Norwegian Shield | 11[a] | Swanberg et al. (1974), Pinet and Jaupart (1987) |
| Vredefort (South Africa) | 18[a] | Nicolaysen et al. (1981) |
| Kapuskasing (Canadian Shield) | 11–13[a] | Ashwal et al. (1987), Pinet et al. (1991) |
| Grenville (Canadian Shield) | 13[a] | Pinet et al. (1991) |
| Abitibi (Canadian Shield) | 10–14[a] | Guillou et al. (1994) |
| Siberian craton | 10–12[a] | Duchkov (1991) |
| Dharwar craton (India) | 11[a] | Roy and Rao (2000) |
| Trans-Hudson Orogen (Canadian Shield) | 11–16[a, b] | Rolandone et al. (2002) |
| Slave Province (Canada) | 12–24[c] | Russell et al. (2001) |
| Baltic Shield | 7–15[c] | Kukkonen and Peltonen (1999) |
| Kalahari craton (South Africa) | 17–25[c] | Rudnick and Nyblade (1999) |

[a]Estimated from surface heat flux and crustal heat production.
[b]Estimated from condition of no-melting in the lower crust at the time of stabilization.
[c]Estimated from geothermobarometry on mantle xenoliths.

**Table 7**   Estimates of bulk continental crust heat production from heat flux data (Jaupart and Mareschal, 2003)

| Age group | Heat production ($\mu$W m$^{-3}$) | Total (40 km crust) (mW m$^{-2}$) | %Area[a] |
|---|---|---|---|
| Archean | 0.56–0.73 | 23–30 | 9 |
| Proterozoic | 0.73–0.90 | 30–37 | 56 |
| Phanerozoic | 0.95–1.21 | 37–47 | 35 |
| Total continents | 0.79–0.95 | 32–40 | |

[a]Fraction of total continental surface, from Model 2 in Rudnick and Fountain (1995).

heat flux and heat production values are the same for all provinces between 200 My and 2.5 Gy, with a weak trend of decreasing average heat flux and heat production with age (Perry et al., 2006). The range is narrower in Archean provinces where high heat flux values are not found, possibly because a very radioactive crust would have been too hot to be stabilized (Morgan, 1985). Averaging the heat production of the crust of different ages yields a range of 0.79–0.95 $\mu$W m$^{-3}$ (**Table 7**; Jaupart and Mareschal, 2003).

### 6.4.4   Recently Active Regions and Continental Margins

Submerged and recently active (i.e., during the past 200 My) continental areas cover $92 \times 10^6$ km$^2$, $\approx$45% of the total continental surface (**Table 8**). These regions are not in thermal steady-state and are characterized by higher heat flux than the continental average. Because of the long thermal relaxation time of the continental lithosphere, present surface heat flux samples the inputs of heat from the mantle of the past 100–200 My. The crustal component can

now be calculated from crustal thickness and average heat production. After accounting for crustal heat production, the heat from the mantle (some of which is included in the transient component) can be estimated.

#### 6.4.4.1   Compressional orogens

In compressional orogens, crustal and lithospheric thickening result in reduced temperature gradients and heat flux, but the total heat production in the thick crust is high. These two competing effects lead to a complex transient thermal structure and few generalizations can be made on the surface heat flux. For instance, very high heat flux values (>100 mW m$^{-2}$) have been measured on the Tibetan Plateau (Francheteau et al., 1984; Jaupart et al., 1985; Hu et al., 2000). They have been attributed to shallow magma intrusions and yield little information on the mantle heat flux. In contrast, present surface heat flux remains low in the Alps, and after removing the crustal heat production, heat flux at the Moho is estimated to be as low as 5 mW m$^{-2}$ (Vosteen et al., 2003). After removing the crustal contribution, heat flux from the mantle is also low

**Table 8**   Surface area and heat flux in oceans and continents

|  | Area | Total heat flux |
|---|---|---|
| Oceans |  |  |
| Oceanic | $273 \times 10^6 \, km^2$ |  |
| Marginal basins | $27 \times 10^6 \, km^2$ |  |
| Total oceans | $300 \times 10^6 \, km^2$ | 32 TW |
| Continents |  |  |
| Pre-Cambrian | $95 \times 10^6 \, km^2$ |  |
| Paleozoic | $23 \times 10^6 \, km^2$ |  |
| Stable continents | $118 \times 10^6 \, km^2$ |  |
| Active continental | $30 \times 10^6 \, km^2$ |  |
| Submerged (Margins and basins) | $62 \times 10^6 \, km^2$ |  |
| Total continental | $210 \times 10^6 \, km^2$ | 14 TW |

beneath the North American Cordillera (Brady et al., 2006), and beneath the South American Cordillera, at least where it has not been affected by back-arc extension (Henry and Pollack, 1988).

### 6.4.4.2   Zones of extension and continental margins

In rifts, recently extended regions, continental margins and basins, heat flux is higher than in stable areas because of a large transient component, which ultimately represents additional inputs of heat from the mantle. Crustal extension and lithospheric thinning will instantly result in steepening the temperature gradient and increasing the heat flux. Thermal relaxation from the initial conditions depends on the boundary condition at the base of the lithosphere.

The heat flux is high ($75-125 \, mW \, m^{-2}$) in zones of extension and in continental rifts, significantly more than in stable regions (Morgan, 1983). A striking feature of the zones of extension is that the transition between the region of elevated heat flux and the surrounding is as sharp as the sampling allows one to determine, that is, Colorado Plateau–Basin and Range in North America (Bodell and Chapman, 1982), East African Rift–Tanzanian craton (Nyblade, 1997), Baikal Rift–Siberian craton (Poort and Klerkx, 2004). The absence of lateral diffusion of heat suggests that the enhanced heat flux in the extended area is not due to conductive processes but is the direct result of extension and lithospheric thinning. Where the sampling is sufficient, heat flux also exhibits short-wavelength variations. These variations are partly due to the cooling of shallow magmatic intrusions and to groundwater movement. The actual heat loss is higher than the average conductive heat flux because of the heat transport by hot springs and volcanoes. Lachenbruch and Sass (1978)

have estimated that the heat delivered by volcanoes in rifts and in the Basin and Range is negligible. They have also argued that the integrated effect of heat transport by groundwater is small for the Basin and Range, with the exception of the Yellowstone system, where locally the heat flux is $>40 \, W \, m^{-2}$. However, the total heat loss at Yellowstone remains modest: the conductive and convective heat loss for the entire Yellowstone system has been estimated to be $\approx 5 \, GW$ (Fournier, 1989). It would thus require 200 'Yellowstones' to increase the continental heat loss by 1 TW. The effect of continental hot spots on the budget seems presently negligible. Values for the total heat loss through geothermal systems in the East African Rift are comparable to those of Yellowstone (Crane and O'Connell, 1983). In continental as well as in oceanic rifts, the heat loss is underestimated because of hydrothermal heat transport. However, because continental rifts are narrow and their total surface area is small, the error will not significantly affect the continental heat flux budget. For instance, the total heat loss for the Gregory Rift, in Kenya, is $\approx 20 \, GW$ (Crane and O'Connell, 1983). Similar values have been inferred for Baikal (Poort and Klerkx, 2004). Large igneous provinces testify of periods of enhanced volcanic activity in the continents. Their effect on the heat flow budget is however negligible. In the Deccan, where $500\,000 \, km^3$ of basalts were deposited c. 60 My, there is no heat flow anomaly, suggesting that the lavas did not heat up the lithosphere. Assuming that the lavas were deposited in 1 My, the heat that they carried to the surface contributed less than 0.1 TW to the energy budget.

The contribution of wide regions of extension is more significant than that of rifts. In the Basin and Range Province in the southwestern US, high average heat flux ($105 \, \text{mW m}^{-2}$) has been interpreted to imply an extension of 100% (Lachenbruch and Sass, 1978). This interpretation depends on assumptions on the pre-extensional heat flux and on crustal heat production. Early estimates of the mantle heat loss are probably too large because the crustal heat production was underestimated (Ketcham, 1996). It now appears that the average heat production of the crust is the same as in stable regions and yields a total crustal heat production of $\approx 33 \, \text{mW m}^{-2}$. This implies that the transient component due to cooling and the mantle heat flux in the Basin and Range contribute a total heat flux of $\approx 70 \, \text{mW m}^{-2}$ (Ketcham, 1996). Detailed models to account for this heat loss assume either delamination of the lithospheric mantle or stretching and transport of heat into the lithosphere by intrusions (Lachenbruch and Sass, 1978). Regardless of the mechanism, at least two-thirds of the heat flux in regions of extension comes from the mantle.

Basins and continental margins account for an important fraction ($\approx 30\%$) of the continental surface. Margins are characterized by gradual crustal thinning towards oceanic basins, which implies a lateral variation of the crustal heat flux component. The average heat flux of the margins ($78 \, \text{mW m}^{-2}$) is higher than in stable regions despite the thinner crust. This higher heat flux is explained by the cooling of the stretched lithosphere and is reflected in the thermal subsidence (Vogt and Ostenso, 1967; Sleep, 1971). Where detailed information on crustal thickness is available, the input of heat from the mantle can be calculated.

### 6.4.5 Mantle Heat Loss through Continental Areas

There is a major difference in the thermal regime between stable and active continental regions. In stable continental regions, the mean heat flux is low ($\leq 55 \, \text{mW m}^{-2}$) and mostly comes from crustal heat production. Heat flux from the mantle is $\approx 15 \, \text{mW m}^{-2}$. In extensional regions, the high heat flux ($\geq 75 \, \text{mW m}^{-2}$) includes a contribution from crustal radioactivity ($\approx 30 \, \text{mW m}^{-2}$) and heat from the mantle in the transient component. Despite the thin crust, continental margins also have higher than average heat flux ($\approx 80 \, \text{mW m}^{-2}$) because they are cooling after being extended.

Different methods lead to a value of 14 TW for the integrated heat flux from continental areas. Neglecting geothermal and volcanic transport has

no significant impact on this value. The estimated average heat production of the continental crust ranges between 0.79 and $0.95 \, \mu\text{W m}^{-3}$ (Jaupart and Mareschal, 2003) and the total volume of continental crust is $\approx 0.73 \times 10^{10} \, \text{km}^3$, which gives a total heat production in the crust between 6 and 7 TW.

Little is known about the amounts of radiogenic elements in the lithospheric mantle. Direct estimates rely on a few exposures of peridotite massifs, which are typically depleted (Rudnick *et al.*, 1998), and on mantle xenoliths from kimberlite pipes, which are usually enriched (Russell *et al.*, 2001). Considerations on the thermal stability of continental roots and consistency with heat flux measurements as well as with petrological temperature estimates lead to the conclusion that enrichment must be recent and associated with metasomatic infiltrations (Jaupart and Mareschal, 1999; Russell *et al.*, 2001). This enrichment process is probably limited in both area and volume and our best estimate of radiogenic heat production in the lithospheric mantle comes from peridotite massifs. For the sake of completeness, we take a value of $0.02 \, \mu\text{W m}^{-3}$ from (Rudnick *et al.*, 1998) and consider an average lithosphere thickness of 150 km. The total heat thus generated in the subcontinental lithospheric mantle is about 0.5 TW, which is only accurate within a factor of about two.

Subtracting the contribution of radioactive sources from the total heat loss out of continents, we thus arrive at an estimate of the heat input from the mantle of 6–7 TW, most of which is brought through the tectonically active regions and the continental margins.

### 6.4.6 Summary

Heat flux data are now available for provinces of all ages, including Archean cratons which were poorly sampled 20 years ago. About half of the heat loss through continents is accounted for by crustal radiogenic heat production. Stable continents allow a small heat flux of about $15 \, \text{mW m}^{-2}$ out of the convecting mantle and hence act as insulators at the surface of the Earth.

## 6.5 Heat Sources

### 6.5.1 Radiogenic Sources in the Mantle

The composition of our planet cannot be measured directly for lack of direct samples from the lower mantle and the core, and hence has been estimated using various methods. It had been noted by Birch (1965) that if the Earth had a chondritic composition, its heat production would match what was then thought to be the heat loss (30 TW). This remarkable coincidence did not resist close scrutiny. It was soon noted that the Earth is depleted in K relative to chondrites and this reduces the heat production (Wasserburg *et al.*, 1964). With the same amount of U and Th as in the chondrites, and a terrestrial K/U ratio, the total heat production was estimated to be only 20 TW (Wasserburg *et al.*, 1964). On the other side of the balance, the heat loss is now believed to be much larger than it was then.

All attempts to construct a bulk Earth composition model rely on two different kinds of samples: meteorites, which represent the starting material, and samples of today's upper mantle. Both show rather extensive variations of composition due to their different histories. Processes in the early solar nebula at high temperature contribute one type of compositional variation. Processes within the Earth, which occur at lower temperatures, contribute another type of compositional variation. Stated schematically, one has a range of compositions from the early solar system and a range of compositions for the upper mantle of the Earth, and one must devise the procedure to correct for two different sets of processes.

Chondrites represent samples of undifferentiated silicate material from the solar system prior to melting and metallic core segregation. Their composition derives from the solar composition altered by processes in the early solar nebula which have generated different families of chondrites. Perturbations are essentially brought in the gas state and elemental behavior is classified according to volatility (or condensation temperature). For our present purposes, the important elements are uranium, thorium, and potassium. The first two are associated with very high condensation temperatures and called 'refractory lithophile' elements. That these two elements have the same behavior in the early solar system is demonstrated by the fact that they have the same ratio in all types of chondritic meteorites. Potassium is a 'moderately volatile' element with a lower condensation temperature. The best match with solar concentration ratios is achieved by CI carbonaceous chondrites, which explains why many Earth models have relied on them. However, we do know that CI chondrites have larger amounts of volatiles, including water and $CO_2$, than the Earth. As regards samples from the Earth's mantle, one may establish a

systematic compositional trend through the samples available and identify the most primitive (and least-differentiated) material. With these problems in mind, we review the four main types of approaches that have been used and the resulting estimates of Earth composition.

The first method relies on direct samples from the mantle. Ringwood (1962) argued that basalts and peridotites are complementary rocks, such that the latter is the solid residue of the partial melting event which led to basalt genesis and extraction. Thus, mixing them back together with the appropriate proportions yields the starting material, which was named 'pyrolite'. Clearly, one has to choose the best samples which have not been affected by leaching and low-temperature alteration. Unfortunately, this procedure is not efficient for uranium, which is very mobile.

A second method relies on a choice for the starting material. Many authors (e.g., Hart and Zindler, 1986) used CI chondrites, a particular class of chondrites, and worked their way through the processes that turn these meteorites into Earth-like material: devolatization (loss of water, $CO_2$, and other volatile elements present in very small amounts in the Earth) followed by reduction (loss of oxygen). Errors associated with this obviously come from the mass loss estimates, but also from the starting CI chondrite composition since this group of meteorites is quite heterogeneous. Javoy (1999) argued in favor of a different type of meteorite. His line of reasoning focusses on the oxidation state of the solar nebula as it started to condense. The only meteorites with the right oxidation state are enstatite chondrites, which are therefore close to the material which went into the protoplanets. These chondrites are largely degassed, save for sulfur, so that the volatile loss correction is small.

A third method tries to avoid a specific choice for the starting composition and aims at determining it. Hart and Zindler (1986) defined the compositional trends of chondritic meteorites and peridotites, which are not parallel to one another. Each trend records the effects of the two different sets of processes operating in the primitive solar nebula and in the Earth, and hence the intersection can only be the starting Earth material. In this case, the error comes from the scatter around the two compositional trends.

A fourth method relies on elemental ratios. For refractory lithophile elements, such as uranium and thorium, the concentration ratio is independent of chondrite type and hence is a property of the starting Earth material. Once these ratios have been determined, two procedures can be used to determine primitive abundances from measurements on peridotite samples. One procedure is to study the relationship between abundance and elemental ratios: the primitive abundance is that which corresponds to the chondritic ratio (McDonough and Sun, 1995). In the other procedure, one starts with one specific element for which one can determine a reliable value for the bulk Earth and work sequentially to all the others using elemental ratios. The element of choice is Mg because, although it is not the most refractory element, its behavior during melting and alteration is well understood (Palme and O'Neill, 2003). Uncertainties on the uranium, thorium, and potassium concentrations are large ($\approx$15%, see **Table 9**) (McDonough and Sun, 1995; Palme and O'Neill, 2003). Lyubetskaya and Korenaga (2007) have recently revisited the procedure of McDonough and Sun (1995). The correlation between abundance and elemental ratio can be accounted for by variations in the degree of melting of a peridotite and McDonough and Sun (1995) carried out a linear regression through the data. Depletion effects due to melt extraction, however, are intrinsically nonlinear. With a realistic treatment of depletion effects and statistical analysis of the highly scattered data, Lyubetskaya and Korenaga

(2007) have obtained a model for the bulk silicate Earth (BSE) that is more depleted than previous ones.

As regards the radioactive elements of interest here, the 'pyrolite' method is unreliable and hence was not used. **Table 9** lists estimates obtained by different authors using the other methods. One should note that concentration ratios are constrained more tightly than absolute concentrations. One cannot separate uncertainties due to the starting chemical data from those of the calculation algorithm, because each author uses his own data and method. Save for major modifications in our understanding of early planetary accretion, it may well prove impossible to reduce the spread of results.

Values for U, Th, and K concentrations yield estimates of the radiogenic heat production rate in the Earth. We have used the revised decay constants listed in **Table 10** (Rybach, 1988) (see also *Handbook of Constants*). Those differ slightly from the earlier estimates given by Birch (1965), which are commonly used in the geophysical literature. Heat production values vary within a restricted range, from 3.9 to 5.1 pW kg$^{-1}$. The BSE model of Palme and O'Neill (2003) leads to the highest value and may well be an overestimate. The EH chondrite model does not lead to a major difference for the present-day heat production rate but implies important changes at early

**Table 9** Radioelement concentration and heat production in meteorites, in the bulk silicate Earth, in Earth mantle, and crust

| | U (ppm) | Th (ppm) | K (ppm) | A (pW kg$^{-1}$) |
|---|---|---|---|---|
| *CI Chondrites* | | | | |
| Palme and O'Neill (2003) | 0.0080 | 0.030 | 544 | 3.5 |
| McDonough and Sun (1995) | 0.0070 | 0.029 | 550 | 3.4 |
| *Bulk silicate Earth* | | | | |
| From CI chondrites | | | | |
| Javoy (1999) | 0.020 | 0.069 | 270 | 4.6 |
| From EH Chondrites | | | | |
| Javoy (1999) | 0.013 | 0.0414 | 383 | 3.6 |
| From chondrites and lherzolites trends | | | | |
| Hart and Zindler (1986) | 0.021 | 0.079 | 264 | 4.9 |
| From elemental ratios and refractory lithophile elements abundances | | | | |
| McDonough and Sun (1995) | 0.020 ± 20% | 0.079 ± 15% | 240 ± 20% | 4.8 ± 0.8 |
| Palme and O'Neill (2003) | 0.022 ± 15% | 0.083 ± 15% | 261 ± 15% | 5.1 ± 0.8 |
| Lyubetskaya and Korenaga (2007) | .017 ± 0.003 | .063 ± 0.011 | 190 ± 40 | 3.9 ± 0.7 |
| *Depleted MORB source* | | | | |
| Workman and Hart (2005) | 0.0032 | 0.0079 | 25 | 0.59 |
| *Average MORB mantle source* | | | | |
| Su (2000); Langmuir *et al.* (2005) | 0.013 | 0.040 | 160 | 2.8 |
| *Continental crust* | | | | |
| Rudnick and Gao (2003) | 1.3 | 5.6 | 1.5 10$^4$ | 330 |
| Jaupart and Mareschal (2003) | / | / | / | 293–352 |

**Table 10**    Heat production constants

| Isotope/Element | Natural abundance (%) | Half-life (yr) | Energy per atom ($\times 10^{-12}$ J) | Heat production per unit mass of isotope/element (W kg$^{-1}$) |
|---|---|---|---|---|
| $^{238}$U | 99.27 | $4.46 \times 10^9$ | 7.41 | $9.17 \times 10^{-5}$ |
| $^{235}$U | 0.72 | $7.04 \times 10^8$ | 7.24 | $5.75 \times 10^{-4}$ |
| U | | | | $9.52 \times 10^{-5}$ |
| $^{232}$Th | 100. | $1.40 \times 10^{10}$ | 6.24 | $2.56 \times 10^{-5}$ |
| Th | | | | $2.56 \times 10^{-5}$ |
| $^{40}$K | 0.0117 | $1.26 \times 10^9$ | 0.114 | $2.97 \times 10^{-5}$ |
| K | | | | $3.48 \times 10^{-9}$ |

From Rybach (1988).

ages because its Th/U and K/U ratios differ strongly from the others. One important result is that the uncertainty on each estimate ($\approx 15\%$) is consistent with the range spanned by the various estimates.

Independent constraints can be derived from the compositions of MORBs, which have been sampled comprehensively. MORBs span a rather large compositional range and common practice has been to define end members associated with different chemical reservoirs. According to this framework, depleted MORBs come from a depleted reservoir whose complement is enriched continental crust. Enriched basalts are attributed to primitive mantle tapped by deep mantle plumes or to secondary enrichment processes, for example, infiltrations of low-degree melts and metasomatic fluids in subduction zones (Donnelly et al., 2004). The heat production rate of the depleted MORB mantle source is $\approx 0.6$ pW kg$^{-1}$ (i.e., at this rate, the entire mantle would generate only 2.4 TW). This source, however, does not provide an exact complement of average continental crust (Workman and Hart, 2005). An alternative approach avoids the separation of different mantle reservoirs and determines the average composition of all the mantle that gets tapped by mid-ocean ridges (Su, 2000; Langmuir et al., 2005). Composition of the average MORB mantle source is then derived from a well-constrained melting model. This mantle reservoir is a mixture of different components and is the average mantle lying below oceanic ridges. It is depleted in incompatible elements and represents a complement of continental crust. Thus, it may be interpreted as the mantle reservoir that has been processed to form continents (Langmuir et al., 2005). There may be a volume of primitive mantle lying at depth that has never been sampled by mid-oceanic ridges. Therefore, a lower bound on the total amount of radioelements in the mantle is obtained by assuming that the average MORB source extends through

the whole mantle. From **Table 9**, this leads to a total mantle heat production of 11 TW. Adding radioelements from the continental crust and lithospheric mantle, which contribute 7–8 TW (**Table 11**), we obtain a lower bound of 18 TW for the total rate of heat production in the Earth. This is consistent with the BSE models and their uncertainties.

Various models for the bulk silicate Earth, which includes continental crust, lead to a total rate of heat production of 20 TW, with an uncertainty of 15%. After removing the contributions of the continental crust (6–7 TW) and the lithospheric mantle ($\approx 1$ TW), heat production in the mantle amounts to a total of 13 TW, with an uncertainty of 20%.

### 6.5.2   Heat Flux from the Core

The outer core is made of molten iron and hence has very low viscosity, contrary to the deep mantle which is much more viscous. Thus, the heat flux out of the core is controlled by the efficacy of mantle convection and cannot be considered as an independent input. Nevertheless, the thermal evolution of the core controls the energy sources available to drive the geodynamo and one may thus deduce constraints on the heat flux at the core–mantle boundary (CMB) using thermodynamics. This question is briefly covered here and the interested reader should consult the volume concerning the core and some standard references (Gubbins and Roberts, 1987; Braginsky and Roberts, 1995; Lister and Buffett, 1995; Labrosse, 2005a, 2005b).

An energy balance can be written for the core, in much the same way as it is done above for the mantle. The main differences come from electromagnetic processes and chemical buoyancy due to inner-core crystallization. The low viscosity maintains the convective state very close to the reference (radially symmetric) state. The convective velocity at the

**Table 11** Mantle energy budget: preferred value and range

|  | TW | TW |
|---|---|---|
| Oceanic heat loss ($300 \times 10^6 \, km^2$) | 32 | 30–34 |
| Continental heat loss ($210 \times 10^6 \, km^2$) | 14 | 13–15 |
| Total surface heat loss ($510 \times 10^6 \, km^2$) | 46 | 43–49 |
| Radioactive sources (mantle + crust) | 20 | 17–23 |
| Continental heat production (crust + lith. mantle) | 7 | 6–8 |
| Heat flux from convecting mantle | 39 | 35–43 |
| Radioactive heat sources (convecting mantle) | 13 | 9–17 |
| Heat from core | 8 | 5–10[a] |
| Tidal dissipation in solid earth | 0.1 |  |
| Gravitational energy (differentiation of crust) | 0.3 |  |
| Total input | 21 | 14–27 |
| Net loss (mantle cooling) | 18 | 8–29 |
| Present cooling rate, $K \, Gy^{-1}$ | 118 | 53–190 |
| Present Urey ratio[b] | 0.33 | 0.21–0.49 |

[a]This range includes estimates from core thermodynamics and inference from the perovskite–post-perovskite phase diagram.
[b]Urey ratio for the convecting mantle, leaving out crustal heat sources from both the heat loss and heat production.
The distribution in the range is barely known for most cases and the preferred value is simply the middle one. The cooling rate is computed assuming $C_P = 1200 \, J \, K^{-1} \, kg^{-1}$.

surface of the core, which is of the order of $10^{-4} \, m \, s^{-1}$ (Hulot et al., 2002), suggests relative density fluctuations of order $10^{-9}$ (Braginsky and Roberts, 1995; Labrosse et al., 1997). Such fluctuations correspond to temperature variations $\delta T \sim 10^{-4} \, K$ in the absence of other effects. This is very small compared to the secular temperature decrease, implying a very good separation of scales between the upper boundary layer and the whole outer core. Thus, for the secular cooling of the core, it is sufficient to consider radial profiles of temperature and concentration, which are usually assumed to be isentropic and uniform, respectively. The very thin boundary layer of the outer core is not significant for the bulk energy budget, contrary to that of the mantle.

The energy balance of the core equates the heat flux at the CMB to the sum of secular cooling, $Q_C$, latent heat from inner-core crystallization, $Q_L$, compositional energy due to chemical separation of the inner core (often called gravitational energy, but see Braginsky and Roberts, 1995), $E_\xi$, and, possibly, radiogenic heat generation, $Q_H$. Secular cooling makes the inner core grow, which releases latent heat and compositional energy and the first three energy sources in the balance can be related to the size of the inner core and its growth rate (Braginsky and Roberts, 1995). The current growth rate of the inner core is small (about $300 \, m \, Gy^{-1}$) and cannot be determined by observation. Thus, one has to resort to indirect means. Energy requirements for the

geodynamo do not appear directly in the bulk energy balance for the core because they are accounted for by internal energy transfers, just like viscous dissipation in the mantle. The entropy balance, however, depends explicitly on dissipation ($\Phi_c$), which is achieved mostly in the form of Joule heating (ohmic dissipation). Combining the energy and the entropy balances, an efficiency equation can be written, which is to leading order (Braginsky and Roberts, 1995; Labrosse, 2003; Lister, 2003):

$$\Phi_c + T_\Phi \Delta S_{cond} = \frac{T_\Phi}{T_{CMB}} \left(1 - \frac{T_{CMB}}{T_{ICB}}\right) Q_L$$
$$+ \frac{T_\Phi}{T_{CMB}} \left(1 - \frac{T_{CMB}}{T_C}\right) Q_C$$
$$+ \frac{T_\Phi}{T_{CMB}} \left(1 - \frac{T_{CMB}}{T_H}\right) Q_H + \frac{T_\Phi}{T_{CMB}} E_\xi$$

$$[51]$$

where $T_i$ is the temperature at which heat due to process ($i$) is released and where

$$\Delta S_{cond} \equiv \int k \left(\frac{\nabla T}{T}\right)^2 dV \qquad [52]$$

is the entropy production due to heat conduction (see Chapter 2). The efficiency eqn [51] shows that heat is less efficiently transformed into ohmic dissipation than compositional energy.

All the source terms on the right-hand side of the efficiency eqn [51], save for radiogenic heating, are linked to inner-core growth and are proportional to its growth rate. Therefore, if Ohmic dissipation $\Phi_c$

and radiogenic heat production $Q_H$ can be estimated, one can calculate the inner-core growth rate and the heat flux across the CMB. **Figure 9** shows the different contributions to the energy and entropy budgets as a function of Ohmic dissipation for zero radiogenic heat production from Labrosse (2003). $\Phi_c = 1$ TW implies a heat flux of about 6 TW at the CMB. In addition, about 1 TW is dissipated by conduction along the isentropic temperature gradient.

Ohmic dissipation in the core is dominated by small-scale components of the magnetic field, which cannot be determined directly because they are screened by crustal magnetic sources (Hulot *et al.*, 2002). Using high-resolution numerical models of the geodynamo, Christensen and Tilgner (2004) have recently claimed that $\Phi_c = 0.2$–0.5 TW, which implies a heat flux across the CMB of about 2–4 TW. These models rely on the Boussinesq approximation, and hence neglect the isentropic temperature gradient in both the heat and entropy balances. As a rough correction, one may add an estimate of the associated

dissipation $\Delta S_{cond} \sim 1$ TW, which would bring the total dissipation estimate to $\Phi_c = 1$–2 TW, close to that of Roberts *et al.* (2003). From **Figure 9**, the corresponding heat flux across the CMB would thus be in the range 5–10 TW.

Studies of hot-spot swells have led to estimates of $\approx 2$–4 TW for the heat flux carried by mantle plumes (Sleep, 1990; Davies, 1988). The relevance of this heat flux to the cooling of the core is difficult to assess, in part because some of the plumes may not come from the CMB and in part because plumes account for only part of the core heat loss. The convective heat flux is given by

$$Q_{conv} = \rho C_p \overline{w_r \theta} \qquad [53]$$

where, as above, $\theta$ is the temperature perturbation with respect to the azimuthal temperature average. This may be broken down into two different components due to upwellings and downwellings, for which $\overline{w_r \theta}$ has the same sign. For our present discussion, this states that secular cooling is effected by both mantle plumes and downgoing slabs. In Rayleigh–Bénard convection with no internal heating, each component is equally important. In the Earth, convection is also driven by internal heating and secular cooling and this breaks the symmetry between upwelling and downwelling currents, at the expense of the former. In such a situation, heat transport at the CMB is dominated by the spreading of cold material from subducted slabs (Labrosse, 2002; Zhong, 2005; Mittelstaedt and Tackley, 2006). The contribution of hot spots therefore provides a lower bound to the total core heat loss. As explained below, an independent estimate of $13 \pm 4$ TW was derived by Lay *et al.* (2006) from the post-perovskite phase change and an estimate of thermal conductivity at the base of the mantle.

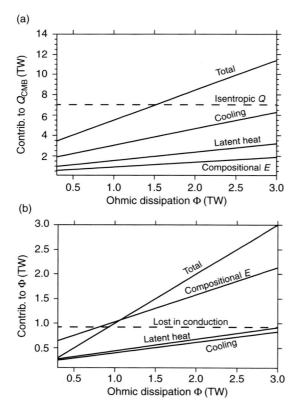

**Figure 9** Contributions to the energy (a) and entropy (b) balance in the core, in the absence of internal heat production. Based on the model of Labrosse (2003).

### 6.5.3 Other Sources: Tidal Heating, Crust–Mantle Differentiation

Earth's rotation is accelerating because of postglacial readjustments, and it is slowing down because of tidal interaction with the moon. The torque exerted on the Moon is due to the lag between the tidal potential and the tidal bulge which is 2.9° ahead of the potential. In the Earth–Moon system, angular momentum is conserved, but there is a net loss of the rotational and gravitational potential energy. This energy is converted into heat by frictional forces. With laser ranging, the changes in the Earth–Moon distance have been measured accurately ($3.7\,cm\,yr^{-1}$) and the slowing down of Earth's rotation due to tidal interaction with the Moon can be calculated exactly. The effect of the solar tidal potential on Earth rotation is $\approx 20\%$ that of the Moon and it must be included in the calculations. The slowing down of Earth's rotation is $5.4 \times 10^{-22}\,rad\,s^{-1}$ leading to $0.024\,ms\,yr^{-1}$ increase in the length of the day. The energy loss has been calculated to be $3\,TW$, which must be accounted by dissipation in the oceans, in the solid earth, and in the Moon. It is commonly assumed that most of the tidal friction comes from dissipation in shallow seas because dissipation in the deep oceans was shown to be small (e.g., Jeffreys, 1962; Munk and MacDonald, 1960). Lambeck (1977) calculated that dissipation in the seas and oceans must account for 90–95% of the energy dissipation. The contribution of the solid Earth tide depends on the quality factor $Q$. For the values of $Q$ in the mantle suggested by seismology, dissipation by the solid earth tide accounts for $<0.1\,TW$ (Zschau, 1986). Such a low value has now found confirmation from satellite observations of the lag between the solid earth tide ($0.16°$) and the lunar potential. This observation implies that the dissipation by the solid earth is $0.083\,TW$ (Ray *et al.*, 1996).

Some gravitational potential energy is also released by the extraction of continental crust out of the mantle. A rough estimate of this energy loss is obtained by considering the potential energy of a differentiated Earth:

$$E_{gm} = -(3/5)GM_m^2/R_1 - GM_m\,M_c/R_1 \qquad [54]$$

where $M_c$ is the mass of crust ($\approx 2.6 \times 10^{22}\,kg$) and residual material with mass $M_m$ is in a sphere of radius $R_1$. Comparing with $E_g$ from eqn [3], we get $\Delta E_g \approx 10^{28}\,J$. For a constant rate of crustal growth during 3 Gy, the contribution to the energy budget is small, $\approx 0.1\,TW$. If the crust differentiated in two or three short episodes, each episode may have added as much as $\approx 1\,TW$ to the mantle budget. In contrast to the change of gravitational energy due to thermal contraction, this energy contributes to the bulk energy balance as compositional energy (see Braginsky and Roberts, 1995).

#### 6.5.3.1 Summary

Various models for the BSE composition lead to results that differ significantly from one another. Uncertainties on the average uranium, thorium, and potassium abundances stem from the propagation of errors through a sequence of elemental ratios and from the correction procedure for depletion effects due to melting in mantle peridotites. A lower bound on the bulk mantle heat production may be derived from the average MORB mantle source. A large uncertainty remains on the heat flux from the core and better constraints will have to come from dynamo theory.

## 6.6 Secular Cooling: Constraints on Mantle Temperatures

In this section, we review evidence for secular cooling, starting with present-day evidence and working backwards in time. The total heat lost by the mantle is more than all the inputs. Our preferred values for the input and output of energy are 21 and $39\,TW$, respectively. The difference $18\,TW$ must be accounted for by the secular cooling of the mantle. Assuming a constant value for the specific heat of $1250\,J\,kg^{-1}\,K^{-1}$, the rate of cooling must be $3.8 \times 10^{-15}\,K\,s^{-1}$ or $120\,K\,Gy^{-1}$, with a range $50–190\,K\,Gy^{-1}$.

### 6.6.1 The Present-Day Mantle Geotherm

The potential temperature of shallow oceanic mantle may be calculated from the composition of MORBs which have not been affected by fractional crystallization and also from heat flux and bathymetry data, as explained above. Such independent determinations are summarized in **Table 4** and are in very good agreement with one another. For temperatures at greater depth, one may use seismic discontinuities and the associated solid-state phase changes. This classical method requires specification of the mantle composition, which is usually taken to be pyrolite (Ringwood, 1962). Well-defined discontinuities at depths of 410 and 660 km have been linked to the olivine–wadsleyite transition and to the dissociation of spinel to perovskite and magnesowustite, the so-called 'post-spinel' transition (**Table 12**). Other seismic discontinuities have been identified, notably at a depth of about 500 km. Some of these are not detected everywhere and seem to have a regional character, and interpretation is still tentative to some extent. Recently, a new phase change relevant to the lowermost mantle has been discovered, from perovskite to post-perovskite (Murakami *et al.*, 2004; Oganov and Ono, 2004).

**Table 12**    Anchor points for the mantle geotherm

| Boundary | Depth (km) | Temperature (K) | Reference |
|---|---|---|---|
| MORB generation | 50 | 1590–1750[a] | Kinzler and Grove (1992) |
| Olivine–Wadsleyite | 410 | $1760 \pm 45$ | Katsura et al. (2004) |
| Post-spinel | 660 | $1870 \pm 50$ | Katsura et al. (2003, 2004) |
| Core–mantle | 2884 | $4080 \pm 130$ | Alfé et al. (2002); Labrosse (2003), this paper |

[a]indicates true range of temperatures in the shallow mantle.

The olivine–wadsleyite phase change has a large Clapeyron slope of $4\,\mathrm{MPa\,K^{-1}}$ and hence provides an accurate temperature estimate (Katsura et al., 2004). Errors on temperature arise from the experimental data as well as from the slight differences in the exact depth of the seismic discontinuity. Accounting for both these errors, the temperature estimate at 410 km is $1760 \pm 45\,\mathrm{K}$ for a pyrolitic upper mantle. Isentropic profiles (**Figure 10**) which pass through these $(P, T)$ values correspond to potential temperatures in the range 1550–1650 K (Katsura et al., 2004), in very good agreement with the independent estimates from the composition of MORBs and heat flux data (**Table 4**).

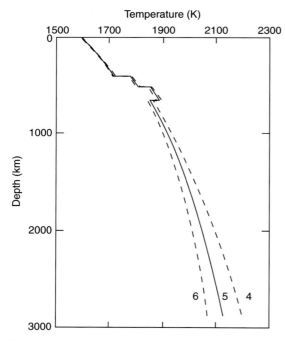

**Figure 10**   Isentropic temperature profiles in the mantle for different values of the Anderson–Grüneisen parameter, as labelled. From Katsura T, Yamada H, Nishikawa O, et al. (2004) Olivine-Wadsleyite transition in the system $(\mathrm{Mg,Fe})_2\mathrm{SiO}_4$. *Journal of Geophysical Research* 109: B02209 (doi:10.1029/2003JB002438).

Recent laboratory studies have cast doubt on the post-spinel transition pressure (Irifune and Isshiki, 1998; Katsura et al., 2003). Furthermore, this transition may have a very small Clapeyron slope (as small as $-0.4\,\mathrm{MPa\,K^{-1}}$, (Katsura et al., 2003), implying that uncertainties on pressure lead to large errors on the transition temperature. For these reasons, one should treat temperature estimates for this transition with caution. Nevertheless, the uncertainty often quoted for this value is similar to that for the 410 km discontinuity, and the present-day 'best' value for the mantle temperature at a depth of 660 km is $1870 \pm 50\,\mathrm{K}$ (Ito and Katsura, 1989; Katsura et al., 2003, 2004).

The other major discontinuity, the CMB, is a chemical boundary and its temperature can be computed from the core side. Once again, the method relies on a phase change, in this case the solidification of iron at the inner-core boundary. Alfé et al. (2002) have recently determined that the melting temperature of pure Fe is in the range 6200–6350 K. Adding the effects of light elements (O, Si, S), the liquidus temperature at the inner-core boundary is lowered by $700 \pm 100\,\mathrm{K}$. The outer core can be assumed to be very close to isentropic (e.g., Braginsky and Roberts, 1995), and the variation in temperature can be linked to the variation in density by (e.g., Poirier, 2000)

$$\left(\frac{\partial T}{\partial \rho}\right)_S = \gamma \frac{T}{\rho} \qquad [55]$$

where $\gamma$ is the Grüneisen parameter (see Chapter 2). According to the theoretical calculations of Vočadlo et al. (2003), $\gamma = 1.5 \pm 0.01$ throughout the core. We may therefore assume that $\gamma$ is constant and integrate eqn [55]:

$$T_{\mathrm{CMB}} = T_{\mathrm{ICB}} \left(\frac{\rho_{\mathrm{CMB}}}{\rho_{\mathrm{ICB}}}\right)^{\gamma} = 0.73\,T_{\mathrm{ICB}} \qquad [56]$$

which relates the temperature at the CMB to that at the ICB and to density structure. Using $\gamma = 1.5$ and density values from the preliminary reference Earth model (PREM) of Dziewonski and Anderson (1981), we obtain a range of 3950–4210 K for the temperature at the top of the core. This does not account for uncertainties in the PREM density values and does not represent the full range of published values, but a full discussion of all the different estimates is outside the scope of this chapter.

The perovskite to post-perovskite phase change (Murakami *et al.*, 2004; Oganov and Ono, 2004) provides a strong constraint on the temperature just above the CMB, if it takes place in the mantle. According to the latest experiments (Hirose *et al.*, 2006; Hirose, 2006), this phase change has a Clapeyron slope of $11.5 \pm 1.4\,MPa\,K^{-1}$ and a temperature of 2400 K at 119 GPa. A key feature is that it occurs very near the boundary layer at the base of the mantle. Assuming an error function temperature profile in this boundary layer, Hernlund *et al.* (2005) predicted that the phase change boundary should be crossed twice in cold regions of the mantle (**Figure 11**), and should not occur in hot regions, which seems to be consistent with seismic observations. As illustrated in **Figure 11**, the fact that the phase boundary is crossed twice implies that the temperature gradient at the base of the mantle is larger than that of the Clapeyron diagram. It also implies that the temperature at the CMB must be larger than the temperature of the phase change at the pressure of the CMB, which is in the range 3800–4200 K. This estimate is consistent with those

obtained from the core side. Lateral variations of the depth of discontinuity that are deduced from seismological observations imply lateral temperature variations of $\approx$1500 K in the lowermost mantle (Hirose, 2006).

Paired seismic discontinuities have also been detected in hot regions near the CMB beneath the Pacific Ocean (Lay *et al.*, 2006). These can be reconciled with the dynamical model of Hernlund *et al.* (2005) if these regions are enriched in iron. Using a dynamical model for the boundary layer above the CMB, Lay *et al.* (2006) have estimated that the heat flux across the CMB is at least $13 \pm 4\,TW$ for a thermal conductivity value of $10\,Wm^{-1}\,K^{-1}$.

### 6.6.2 Temperature versus Time

One may use petrological constraints to investigate past temperatures of the mantle. Continental crustal material was different in the Archean than it is today. Basaltic lavas exhibit systematic compositional trends with time, including a secular decrease in average MgO content. MgO-rich ultramafic lavas named komatiites are common in the Archean and are almost absent from today's rock record. Early workers proposed that komatiites require the mantle source to be at least 300 K hotter than present (Green, 1975; Sleep, 1979) but they considered dry mantle only. The peridotite solidus depends strongly on water content, which in turn depends on the geological setting. If komatiites are generated by

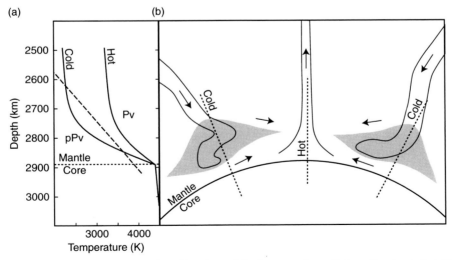

**Figure 11** Sketch of post-perovskite lenses in cold regions of the lowermost mantle from Hernlund *et al.* (2005). (a) Schematic temperature profiles for 'cold' and 'hot' regions (plain curves) superimposed onto the phase boundary. (b) Sketch of thermal boundary layer at the base of the mantle with convective downwellings and upwellings.

mantle plumes, involving mantle that is essentially dry, that is, such that its water content is so small that it does not affect phase boundaries, one deduces that mantle plume temperatures have decreased by about 300 K in 3 Gy (Nisbet *et al.*, 1995). Jarvis and Campbell (1983) suggested that such hot mantle plumes did not require the Archean mantle to be more than 100 K hotter than present on average. According to an alternative hypothesis, komatiites are generated in a subduction environment, involving mantle hydrated by downgoing plates. In that case, one is led to conclude that this part of the mantle was only slightly hotter ($\approx$100 K) in the Archean than it is today (Grove and Parman, 2004). In both cases, komatiites do not sample 'average' mantle and it is not clear how to incorporate these temperature estimates in models for the entire mantle.

Mid-ocean ridge tholeiites are better suited for studies of the mantle's average temperature because they can be sampled over very large areas. They are a compositionally heterogeneous group, however, which translates into a wide temperature range ($\approx$200 K) (Klein and Langmuir, 1987; Kinzler and Grove, 1992). Abbott *et al.* (1994) calculated the liquidus temperature for Phanerozoic MORBs and Archean MORB-like greenstones and determined the maximum and minimum mantle potential temperatures versus time. Although the range of temperatures for each period is wide ($\approx$200 K), the trend is well marked. Abbott *et al.* (1994) concluded that mantle temperatures decreased by $\approx$150 K since 3 Gy (137 K for the mean or 187 K for the maximum temperature recorded in each age bin), which is less than the range of mantle potential temperatures at a given time. Cooling of the mantle by $50\,\text{K}\,\text{Gy}^{-1}$ represents $\approx$8 TW.

### 6.6.3 Early Earth

A full description of Earth's thermal evolution must include the initial conditions. Here, 'initial' refers to the time when the Earth had completed its main phase of core–mantle differentiation and the mantle had solidified to the point where its dynamics can be described as sub-solidus convection. Before reaching that point, a host of processes with different dynamics occurred. They may be separated into three categories: accretion, core formation, and magma ocean crystallization. The process of formation of the Earth brought together matter which was originally dispersed in the proto-solar nebula, thereby releasing

gravitational energy. One may estimate the total energy released by taking the difference between the total gravitational energy before and after. The fate of this energy, however, depends on the way it is dissipated and transformed into another type of energy. The effect of core differentiation is quite different from that of accretion. Most of the processes involved remain speculative to some extent and we restrict our discussion to the points that are directly relevant to the early thermal structure. Several review articles (e.g., Stevenson, 1989; Wetherill, 1990) and two books (e.g., Newsom and Jones, 1990; Canup and Righter, 2000) deal with these issues in detail.

During accretion, the gravitational energy of impactors is first transformed into kinetic energy and then dissipated in the form of heat at the impact. One may define two limit-cases. If no energy is lost to space, the temperature of the whole Earth is raised by an amount equal to

$$\Delta T = \frac{-E_{\text{g}}}{MC_{\text{P}}} \sim 3.75 \times 10^5 \text{ K} \qquad [57]$$

which would be sufficient to vaporize the whole planet. Most of the impact energy, however, was released at shallow levels and lost to space by radiation. Stevenson (1989) estimated that, if all the energy is made available for radiation, accretion would raise the temperature of the Earth by less than 70 K relative to that of the nebula. The actual evolution lies somewhere between these two limiting cases, involving partial dissipation of the impact energy within the planet and radiative heat transfer through the primordial atmosphere. One important factor is the size of the impactors. 'Small' impactors, which are much smaller than the target, account for the vast majority of impacts on Earth after the planetary embryo stage (Melosh and Ivanov, 1999). The depth of energy release increases with the size of the impactor and one key variable is the ratio between the time for energy transport to the surface and the time between two impacts. During accretion, evolution towards larger and fewer impactors has two competing effects: energy gets buried at greater depth while the time between two impacts increases, which enhances heat loss to the atmosphere. Assuming heat transport by diffusion, Stevenson (1989) concluded that typical accretion scenarios lead to significant energy retention within the planet. The extreme case is that of the giant impact thought to be at the origin of Moon formation. Calculations suggest that the whole-Earth temperature was raised to as high as 7000 K (Cameron, 2001; Canup, 2004). In such conditions, the whole Earth melted and parts of it were vaporized to form a thick atmosphere. The question of whether or not previous impacts were able to melt the Earth becomes irrelevant.

The formation of the core also has important energetic implications. Some gravitational energy is released by going from a uniform composition to a stratified core–mantle system. Kinetic energy plays no role in this process, in contrast to the accretionary sequence, and gravitational potential energy is directly dissipated by viscous heating in both the iron and silicate phases. Flasar and Birch (1973) estimated that this process would heat the whole Earth by about 1700 K. This estimate relies on the bulk difference in gravitational energy between the initial and final states and hence gives no information on where energy gets dissipated. We know now that iron-silicate differentiation occurred very early in the solar system and affected planetesimals (Kleine *et al.*, 2002). Clearly, core formation within planetesimals and in the Earth after the giant impact involve different dynamics. Models of the giant impact show that large parts of the cores of the two proto-planets merge without remixing with silicates (Canup, 2004). Emulsification of iron in the molten silicate is a possibility (Rubie *et al.*, 2003), because of the large viscous stresses involved.

Our current understanding of core formation suggests three mechanisms: iron droplets 'raining' through a magma ocean, diapirs generated by Rayleigh–Taylor instability at a rheological interface and interstitial flow across a solid permeable matrix (Stevenson, 1990; Rushmer *et al.*, 2000). All three mechanisms may have been active at different times and have different implications for dissipation. For a Newtonian rheology, the amount of viscous heating is $\phi \sim \mu U^2 / L^2$, where $\mu$ is the viscosity of the fluid phase, which may be silicate or iron, and $U$ and $L$ are the scales for velocity and length. In the case of an iron diapir, the velocity and length scales are the same for the metal and silicate phases, but the viscosity of the former is several orders of magnitudes smaller. Viscous heating is thus concentrated in the silicate phase and little heating of the iron phase results because heat diffusion is not efficient over the descent timescale. This would differentiate a core that is initially colder than the lower mantle. In the case of interstitial flow, the small size of iron veins makes heat diffusion very effective and thermal equilibration with the surrounding silicate phase is likely. In the case of iron droplets raining down

through molten silicate, the droplet size is set by equilibrium between surface tension and viscous drag and is typically 1cm. Again, thermal equilibrium is likely and the core should initially be at the temperature of the lower mantle.

### 6.6.4 Magma Ocean Evolution

Both the giant impact and core-formation processes generated temperatures that were high enough for a magma ocean to include the whole silicate Earth. Cooling and crystallization of such a deep magma ocean involves heat transfer through the primordial atmosphere, convection, rotation, and crystal-melt separation. Available models have been aimed mostly at determining the extent of chemical stratification at the end of crystallization (Abe, 1997; Solomatov, 2000). The pressure effect on the liquidus causes more crystallization at the bottom. The low viscosity of the melt and the size of Earth imply highly turbulent convective flows and rapid cooling, such that the lower parts of the magma ocean solidify in a few kiloyears. Two rheological transitions, from pure magma to slurry and from slurry to mush, affect the convective regime and the cooling rate. One important fact is that, in a convecting region, the isentropic temperature gradient is less than the gradients of liquidus and solidus.

Starting from a superheated magma ocean (i.e., at temperatures above the liquidus), the initial phase has a fully molten upper layer which becomes thinner as cooling proceeds. A first transition occurs when the fully molten layer vanishes. At this stage, the Earth is made of a partially crystallized magma ocean which may lie over already fully solidified mantle. The radial temperature profile is tied to the solidus which is steeper than the isentropic profile, which leads to convective overturn in the solid layer (Elkins-Tanton *et al.*, 2003). The two layers evolve at vastly different timescales because of their different rheologies. The bulk cooling rate is set by the heat loss through Earth's surface, which is controlled by the dynamics of the partially crystallized surficial magma ocean. In this second phase, heat transport occurs mostly by melt–solid separation and solidification proceeds from the bottom up. The fully solidified layer at the base of the magma ocean thickens rapidly and eventually becomes unstable. Convective overturn is slower than the cooling of the magma ocean and may be considered as a separate event which leads to decompression melting and the formation of a secondary magma ocean at the

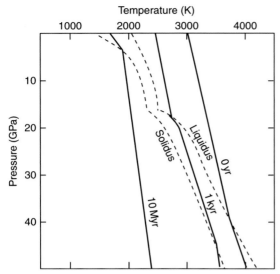

**Figure 12**  Three geotherms at different times in the early Earth. The timescale for the thermal evolution is set by heat loss at the upper boundary which decays rapidly with temperature. At about 10 My, the solid content in the partially molten upper mantle layer reaches the threshold value of 60%, which marks the cessation of liquid behavior. After that time, convection is in the subsolidus regime controlled by solid behavior which still prevails today. Adapted from Abe Y (1997) Thermal and chemical evolution of the terrestrial magma ocean. *Physics of the Earth and Planetary Interiors* 100: 27–39.

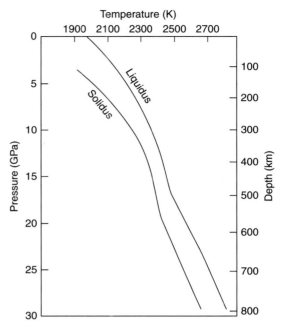

**Figure 13**  Solidus and liquidus for dry pyrolite as a function of pressure. Adapted from Litasov K and Ohtani E (2002) Phase relations and melt compositions in CMAS-pyrolite-$H_2O$ system up to 25 GPa. *Physics of the Earth and Planetary Interiors* 134: 105–127.

surface. The process of cooling and solidification of this magma ocean then repeats itself. This regime prevails until the shallow magma ocean reaches the rheological threshold between liquid and solid behavior, which probably occurs at a crystal fraction of about 60%. At this stage, the shallow partially crystallized layer becomes strongly coupled to the solid mantle below and cooling proceeds through bulk convection everywhere. According to Abe (1993, 1997), this was achieved in a few 10 My (**Figure 12**) and sets the initial conditions for secular cooling models of the solid Earth. From the most recent phase diagram (Herzberg and Zhang, 1996; Litasov and Ohtani, 2002), the final rheological transition corresponds to a potential temperature of about 1800 ± 100 K for a mantle composed of dry pyrolite (**Figure 13**).

Recent advances in geochemistry have confirmed the theoretical estimates of Abe (1997). Caro *et al.* (2003) have found evidence that pushes early crust formation as far back as 4.4 Gy. Others have claimed that liquid water was present on Earth's surface at 4.3 Gy (Mojzsis *et al.*, 2001). These studies cannot

demonstrate that plate tectonics was already active at such early times, but provide some support for a solid upper mantle.

### 6.6.5  Average Secular Cooling Rate

Subsolidus convection began at a mantle potential temperature of about 1800 ± 100 K, which exceeds the present-day temperature by about 200 K. Even if the timing is not known precisely, this constrains the average cooling rate of the Earth to be about 50 K Gy$^{-1}$. The analysis of Phanerozoic MORBs and Archean MORB-like greenstones due to Abbott *et al.* (1994) leads to the same estimate.

The present-day mantle potential temperature is fixed at 1600 K by a fit to heat flux and bathymetry data regardless of the water content of mantle rocks (McKenzie *et al.*, 2005). If the mantle contained significant amounts of water at the end of the magma ocean phase, however, the phase diagram must be shifted to lower temperatures. One consequence is that the starting potential temperature at the beginning of subsolidus convection was less than the

1800 K estimate given above. In this case, the average cooling rate must be even less than $50\,\mathrm{K\,Gy^{-1}}$.

### 6.6.6   Summary

At the end of accretion and core–mantle separation, a magma ocean probably extended through a large fraction of the silicate Earth. Crystallization of the magma ocean was achieved rapidly and led to a stratified mantle with a solid lower layer and a partially crystallized upper layer. At the start of solid-state mantle convection, upper-mantle temperatures were such that the surficial partially molten region had a solid fraction of about 60%. This sets the initial temperature of the solid Earth to a value which is about 200 K higher than the present.

## 6.7   Thermal Evolution Models

### 6.7.1   The Urey Ratio

The decay time of the bulk radiogenic heat production, which is such that heat production decreases by a factor $e$ and which is the weighted average of the individual decay times of the four relevant isotopes (**Table 9**), is 3 Gy. Thus over the Earth's history, heat sources have decreased by a factor of about four. The efficiency of Earth's convective engine in evacuating heat generated by radioactive decay is commonly measured by the Urey ratio, $Ur$, which is the ratio of heat production over heat loss:

$$Ur = \frac{\int_V H\,\mathrm{d}V}{\int_A \mathbf{q}\cdot\mathbf{n}\,\mathrm{d}A} \qquad [58]$$

To calculate this ratio, we do not take continental heat sources into account because they are stored in the continental lithosphere and hence are not involved in mantle convection. Using the data of **Table 11**, we find that $Ur = 0.33$, with a total range of 0.21–0.49.

The heat budget of **Table 11** allows calculation of the present-day cooling rate. Secular variations of basalt compositions and consideration of initial thermal conditions provide constraints on the total temperature drop over Earth's history and hence on the average cooling rate. Thus, physical models are not needed to determine how the Earth has cooled down. Instead, the data allow a test of our understanding of mantle convection processes. Available constraints on the cooling rate can be turned into

one for the rate of heat loss. The global heat balance reads as

$$M\langle C_p\rangle \frac{\mathrm{d}T}{\mathrm{d}t} = -Q + H \qquad [59]$$

where $M$ is the mass of the Earth and $\langle C_p\rangle$ an 'effective' heat capacity which accounts for the isentropic variation of temperature with depth. Integrating over the age of the Earth, one deduces that

$$\frac{\bar{Q} - \bar{H}}{Q - H} = \frac{(\mathrm{d}T/\mathrm{d}t)_{av}}{\mathrm{d}T/\mathrm{d}t} \qquad [60]$$

where $\bar{Q}$ and $\bar{H}$ are the time-averaged values of heat loss and heat production and $(\mathrm{d}T/\mathrm{d}t)_{av}$ is the average cooling rate. The cooling rate has an average value of about $50\,\mathrm{K\,Gy^{-1}}$ and a larger present-day value (about $120\,\mathrm{K\,Gy^{-1}}$). Thus, the ratio in eqn [60] is less than 1 and probably as small as 0.4. This implies that the rate of heat loss varies less rapidly than that of heat production.

### 6.7.2   Parametrized Cooling Models

In steady-state well-mixed homogeneous convective layers with constant physical properties, the heat flux through the top boundary is a function of the Rayleigh number, which itself depends on the temperature difference across the layer (e.g., Schubert and Young, 1976; Sharpe and Peltier, 1978; Davies, 1980). A very robust scaling law relates the dimensionless heat flux (i.e., the Nusselt number) to the Rayleigh number:

$$Nu = \frac{Q/A}{kT/D} = C_1 Ra^{\beta} \qquad [61]$$

where $C_1$ is a proportionality constant, $Q/A$ the heat flux, $T$ the temperature difference across the layer, $D$ the layer thickness, and $Ra$ the Rayleigh number:

$$Ra = \frac{g\alpha TD^3}{\kappa\nu_M} \qquad [62]$$

where $\nu_M = \mu_M/\rho$ is the kinematic viscosity. This relationship can be turned into an equation for heat loss $Q$ of the form:

$$Q = C_2 T^{1+\beta}\nu_M^{-\beta} \qquad [63]$$

where the constants $C_2$ and $\beta$ are obtained from boundary-layer theory as well as laboratory experiments (Howard, 1964; Turcotte and Oxburgh, 1967; Olson, 1987). This relationship is valid if, and only if, instability always occurs in the same conditions, for example, when a Rayleigh number defined locally in the boundary layer exceeds a critical value. Typically, $\beta = 1/3$, such that heat loss is governed solely by local instabilities of the upper boundary layer. The value of constant $C_2$, but not that of exponent $\beta$, is set by the instability threshold and hence depends on the length of the convective cell (Olson, 1987; Grigné et al., 2005). Cooling models of this kind have been termed 'parametrized' because they collapse all the physics of mantle convection into a single equation involving only temperature and two parameters, $C_2$ and $\beta$.

Further developments involved temperature-dependent physical properties. The key principle is that a hot layer evolves much more rapidly than a cold one because of the strong dependence of viscosity on temperature. One can approximate an Arrhenius law for viscosity by an equation of the form $\nu = \nu_0 (T/T_0)^{-n}$, with $n \sim 35$, which is valid for $T \sim T_0$ (Davies, 1980; Christensen, 1985). The thermal evolution equation then takes the following form:

$$M\langle C_p \rangle \frac{dT}{dt} = -Q_0 \left(\frac{T}{T_0}\right)^{1 + \beta(1+n)} + H(t) \quad [64]$$

where $Q_0$ is the heat loss at the reference potential temperature $T_0$. Temperature changes in the Earth are small compared to the absolute temperature (i.e., $\approx 200\,K$ for a present-day temperature of $\approx 1600\,K$). One may thus linearize the above equation by considering the temperature variations around $T_0$:

$$T = T_0 + \Theta \quad \text{with} \quad \Theta \ll T_0 \quad [65]$$

$$M\langle C_p \rangle \frac{d\Theta}{dt} = -Q_0 \left[1 + \frac{\Theta}{T_0}(1 + \beta + \beta n)\right] + H(t) \quad [66]$$

The heat sources can be approximated as decreasing exponentially with time, $H(t) = H_0 \times \exp(-t/\tau_r)$, where $\tau_r \approx 3000$ My. The solution of eqn [66] is

$$\Theta = \Theta_0 \times \exp(-t/\tau_p) + \frac{Q_0 \tau_p}{M\langle C_p \rangle}(\exp(-t/\tau_p) - 1)$$

$$+ \frac{H_0 \tau_p \tau_r}{M\langle C_p \rangle (\tau_r - \tau_p)}(\exp(-t/\tau_r) - \exp(-t/\tau_p)) \quad [67]$$

where the relaxation time constant $\tau_p$ is given by

$$\tau_p = \frac{M\langle C_p \rangle T_0}{(1 + \beta + \beta n)Q_0} \quad [68]$$

Using standard values for the parameters and variables involved, $n = 35$, $\beta = 1/3$, $M = 6 \times 10^{24}$ kg, $Q_0 = 30$ TW, $T_0 = 1300$ K (the temperature jump across the boundary layer is the relevant parameter

here) and $\langle C_p \rangle = 1200\,J\,kg^{-1}K^{-1}$, this thermal adjustment timescale is about 800 My. From eqn [67], the Urey ratio as a function of time can be obtained. For $t \gg \tau_p$, the Urey ratio tends to a constant value $(\tau_r - \tau_p)/\tau_r \approx 0.75$. This is larger than observed.

A key point is that, after about 2 Gy, model predictions are not sensitive to the initial conditions, which has two implications. One is that failure of a model to reproduce the present-day Urey ratio cannot be blamed on the poorly known initial condition. The other implication is that 'backward' thermal calculations starting from the present become unreliable for old ages.

Parametrized models for whole-mantle convection lead to present-day values of the Urey ratio that are larger than 0.7, significantly more than observed. This reveals a fundamental flaw in the model setup. In order to meet the constraint of the Urey ratio, one must increase the adjustment time of mantle convection. One option is to appeal to a layered mantle (McKenzie and Richter, 1981). Another option is to make the bulk rate of heat loss less sensitive to temperature, which may be achieved by decreasing the value of exponent $\beta$ in eqn [68]. According to Christensen (1984, 1985), this may be attributed to the temperature dependence of viscosity. For very large variations of viscosity, however, convection occurs in the stagnant lid regime (e.g., Ogawa et al., 1991; Davaille and Jaupart, 1993; Solomatov and Moresi, 1997). In this case, plate tectonics is shut off, which is not a satisfactory solution. There may be an intermediate regime with subduction of the very viscous lid, such that $\beta = 0.293$ (Solomatov and Moresi, 1997), but this $\beta$-value is too large to meet the constraint of the Urey ratio. Another mechanism that has been invoked for decreasing the exponent $\beta$ is the resistance to bending of the plate at subduction zones which, according to Conrad and Hager (1999), should lead to $\beta \sim 0$. Following a similar idea, Korenaga (2003) added the effect of temperature on the depth of melting, hence the thickness of the crust, as discussed by Sleep (2000). He proposed a negative value of $\beta$. Negative values of $\beta$ would solve the problem at hand but there is no evidence that resistance to bending actually limits subduction on Earth.

This discussion shows that simple convection models cannot account for the observations. We now discuss how to properly account for the behavior of Earth's convective system. We discuss the relevance of heat loss 'parameterizations' to the true

Earth and evaluate whether the present-day heat balance is representative of secular cooling models which, by construction, deal only with long-term temperature changes.

### 6.7.3    The Peculiarities of Mantle Convection: Observations

The Earth's convecting mantle exhibits several features which make it very distinctive. One of them is the triangular age distribution of the seafloor, which occurs because subduction affects all ages with the same probability (**Figure 7**; Parsons, 1982). This is at odds with other convecting systems as well as with the parametrized schemes discussed above. We show in Appendix 8 the age distribution at the upper boundary of several convective systems. None of them resembles that of Earth, which illustrates current limitations in reproducing mantle convection processes.

A few other peculiar features of mantle convection are worth mentioning. Heat loss is unevenly distributed at the surface. The Pacific Ocean alone accounts for almost 50% of the oceanic total, and 34% of the global heat loss of the planet. This is due in part to the large area of this ocean and in part to its high spreading rate. Oceanic plates are transient, such that changes of oceanic heat loss may occur when a new ridge appears or when one gets subducted. For example, the heat flux out of the Atlantic Ocean is about 6 TW, 17% of the oceanic total (Sclater *et al.*, 1980). This ocean has almost no subduction and started opening only at 180 My. At that time, the generation of a new mid-ocean ridge led to an increase of the area of young seafloor at the expense of old seafloor from the other oceans, and hence to enhanced heat loss. The triangular age distribution may well be a consequence of this relatively recent plate reorganization. From the standpoint of the dynamics of convection, the most challenging features of mantle convection are perhaps the large variations of plate speeds and dimensions. With the small number of plates that are present, averaging values of spreading velocity and plate size may well be meaningless.

Earth's surface is partially covered by continents which have two effects on mantle convection. They do not allow large heat fluxes through them and generate boundaries with complicated shapes that constrain mantle flow. The controls they exert on secular cooling have been investigated by few authors (Lenardic *et al.*, 2005).

### 6.7.4    Convection with Plates

Here, we recapitulate the main physical controls on mantle convection. With rigid plates, several physical mechanisms are involved and it is convenient to use the dissipation equation [28].

For the Earth, the buoyancy flux is due almost entirely to subducting plates and their associated downwelling currents. For a plate going down with velocity $U$ at a trench of length $l$, the buoyancy flux $F_b$ is:

$$F_b = \rho \alpha g \overline{w_r \theta} \delta_T l = \frac{1}{2} \rho \alpha g U T_M \delta_T l \qquad [69]$$

where $\delta_T = \sqrt{\kappa \tau_m}$ is the thermal boundary-layer thickness. Here, $\delta_T l$ is the cross-sectional area of the plate that goes down into the mantle and the temperature profile through the plate is taken to be linear. Subducted material gets heated by radioactive sources and goes through mantle that is cooling down. Thus, the buoyancy flux is not conserved during descent. For a simple argument, we neglect these effects. In this case, if the plate remains vertical down to the CMB, the bulk buoyancy flux in eqn (28) is

$$\Phi_b = \int_V \rho \alpha g \overline{w_r \theta}\, \mathrm{d}V = \frac{1}{2} \rho \alpha g U T_M \delta_T l D \qquad [70]$$

where $D$ is the mantle thickness.
Viscous dissipation may be written as

$$\phi_v \propto \sigma \frac{U}{\delta_v} \qquad [71]$$

where $\sigma$ is the deviatoric stress and $\delta_v$ the thickness of the momentum boundary layer. For material with viscosity $\mu_M$,

$$\sigma \propto \mu_M \frac{U}{\delta_v} \qquad [72]$$

which implies that $\phi_v \propto U^2$. For cell length $L$, balancing the total dissipation with the bulk buoyancy flux leads to

$$U \propto \frac{\rho g \alpha T_M \delta_T D}{\mu_M} \frac{\delta_v}{L} \qquad [73]$$

For large values of the Prandtl number, the momentum boundary layer is spread over large distances. For cellular convection, it extends throughout the convecting layer, such that $\delta_v \propto D$. In the Earth, however, convection is driven mostly by cooling from above, such that velocities are highest in downwellings and low in large-scale upward return flow. In this case, viscous dissipation occurs mostly in the vicinity of downwellings, such that $\delta_v$ may not scale with the depth of the layer (Jarvis and McKenzie, 1980).

In the Earth, subduction involves cold and stiff plates, such that bending at subduction zones induces large amounts of dissipation (Conrad and Hager, 1999). Assuming that the plate behaves viscously, the rate of dissipation is

$$\phi_P \propto \mu_P \frac{(U\delta_T)^2}{R_P^2} \quad [74]$$

where $\mu_P$ is the effective viscosity of the plate and $R_P$ its radius of curvature. Note that this dissipation rate is also proportional to $U^2$, as for internal viscous dissipation. For an estimate of the bulk dissipation rate due to bending, one must specify the volume of plate that gets deformed. With this extra dissipation mechanism, one obtains another equation for velocity $U$.

Another equation is required to specify heat loss. For the sake of simplicity, we use a half-space cooling model because the predicted heat flux differs from observations at old ages only, and this difference represents a small fraction of the total: changing heat loss through seafloor older than 80 My by 30% (an overestimate) impacts the total heat loss value by only 6%. Thus, we assume that $q = kT_M/\sqrt{\pi\kappa\tau}$. For a rectangular plate of length $L$ and width $l$, heat loss $Q_P$ is

$$Q_P = \int_A q \, dA = \int_0^{\tau_m} ql U \, d\tau = 2\frac{k}{\sqrt{\pi\kappa}} T_M U l \sqrt{\tau_m} \quad [75]$$

where $\tau_m$ is the maximum plate age. The rate of heat loss over the total area of the plate, $A_P = Ll$, is

$$Q_P = 2A_P \frac{kT_M}{\sqrt{\pi\kappa\tau_m}} \quad [76]$$

Note that this result was derived directly from [75] independently of eqn [73] for velocity $U$, that is, without involving dynamical constraints. All the dynamical information needed is in $\tau_m$, which has not been determined yet.

All the key variables of convection, heat loss $Q_P$, velocity $U$, and plate length $L$, depend on $\tau_m$. To obtain a closure equation for this variable, a widely used argument has been that the thermal boundary layer becomes unstable when a local Rayleigh number exceeds a threshold value such that

$$Ra_\delta = \frac{\rho g \alpha T_M \delta_T^3}{\kappa\mu_P} = Ra_c \quad [77]$$

where $Ra_c$ is some critical value. This leads to the 'parametrized' model equation with $\beta = 1/3$, which

is not satisfactory. A key piece of physics is missing, which prevents us from specifying how $\tau_m$ changes when the mantle cools down. When the Earth was hotter, decompression melting proceeded to larger melt fractions and hence to larger amounts of depletion. This is demonstrated by secular changes of the composition of MORBs (Abbott et al., 1994). Larger degrees of melting enhance dehydration of the residual solid and hence generate stiffer lithospheric plates. $\mu_P$, the plate effective viscosity, increases with increasing temperature and hence was larger in the past than today (Korenaga, 2006). How this affects $\tau_m$ or $L$ is not clear, however. One might imagine for example that higher intrinsic stiffness of lithospheric material gets compensated because subduction involves younger and hence thinner plates such that the total dissipation due to bending did not change much.

### 6.7.5   Cooling through the Ocean Floor

With the physical arguments developed above, one runs into two different types of difficulties when applying them to the Earth. Firstly, one must specify either the plate length $L$ or the maximum plate age $\tau_m$. Secondly, one must go from a single convection cell to several cells with different characteristics. Here, we derive several useful relationships that are independent of dynamical arguments.

A successful physical model must rely on accurate equations for the rate of heat loss and for the bulk buoyancy flux. Let us consider a thermal balance for the volume encompassing the thermal boundary layer below Earth's surface:

$$\int_{A^-} \rho C_p \overline{w_r \theta} \, dA = \int_{A^+} \mathbf{q} \cdot \mathbf{n} \, dA \quad [78]$$

where $A^+$ is the upper surface in contact with the hydrosphere and $A^-$ lies at the base of the thermal boundary layer. This equation states that the subducting plates carry the energy deficit that has been accumulated by cooling to the sea. From this relation, we deduce that the bulk buoyancy flux is proportional to the rate of heat loss:

$$\Phi_b = \frac{\alpha g}{C_p} QD \quad [79]$$

Thus, accurate parametrizations of the buoyancy flux and of the rate of heat loss are equivalent.

We derive a general equation for heat loss on Earth. Because crustal radioactivity accounts for a large fraction of the continental heat flux, the basal heat flux into continents, out of the convecting mantle, is very small. We consider that continental radioactivity plays no dynamic role and thus equate the heat loss of the Earth to that of the oceans. We shall again use the half-space cooling model which is sufficiently accurate. The distribution of sea floor age $f$ in function of dimensionless age $\tau/\tau_m$ is such that

$$\frac{dA}{d\tau} = C_A f\left(\frac{\tau}{\tau_m}\right) \qquad [80]$$

Where $C_A$ is the plate accretion rate and with $f(0) = 1$. Using the half-space expression [35], we obtain the total oceanic heat loss:

$$Q_{oc} = A_o \frac{kT_M}{\sqrt{\pi\kappa\tau_m}} \frac{\int_0^1 \frac{f(u)}{\sqrt{u}} du}{\int_0^1 f(u)du} = A_o \frac{kT_M}{\sqrt{\pi\kappa\tau_m}}\lambda(f) \qquad [81]$$

Where $A_o$ is the total ocean surface and $\lambda(f)$ a coefficient which depends on the dimensionless age distribution. This equation has the same form as eqn [76] derived above but it includes a new variable, the dimensionless age distribution. For the present-day triangular age distribution

$$Q_{oc} = \frac{8A_o}{3}\frac{kT_M}{\sqrt{\pi\kappa\tau_m}} \qquad [82]$$

Even though we have not used any dynamical argument to derive the heat loss equation, we can relate changes of heat loss to changes of spreading rate. The total oceanic area is

$$A_o = C_A\tau_m \int_0^1 f(u)du \qquad [83]$$

If we assume that the oceanic area and the age distribution remain constant, changes of heat loss imply changes of $\tau_m$, which in turn imply changes of $C_A$, the rate of plate generation.

The oceanic heat loss can be equated with little error to the convective heat loss on Earth. Thus eqns [79], [81], and [83] provide the most compact description of the physical controls on cooling. They avoid the problem of defining an 'average' plate with average velocity and length. They say nothing, however, about the maximum plate age $\tau_m$.

## 6.7.6 Vagaries of Seafloor Spreading

In eqn [81], one coefficient that can vary significantly over short time scales is $\lambda(f)$, which depends on the dimensionless distribution of seafloor ages. Cogné and Humler (2004) have attempted to calculate spreading

rates in the past. The last 180 My have seen the closure of the Tethys Ocean and the subduction of several ridges in the paleo-Pacific Ocean (Engebretson et al., 1984). Accounting for those, the total seafloor generation rate did not vary significantly over the last 180 My (**Figure 14**). Fluctuations in seafloor spreading may occur on a larger timescale. Subduction of young seafloor occurs mostly at the edge of continents and may be due to the complex geometry of the ocean–continent boundaries. With all continents assembled in a single landmass, the large continuous oceanic area imposes less constraints on spreading and subduction. In other words, the present-day distribution of subduction zones may be a transient feature associated with the breakup of Gondwana. The assembly and breakup of supercontinents occur over some characteristic time $\tau_W$. Allègre and Jaupart (1985) have related this time to the 'mean free path' of continents at Earth's surface, such that continents sweep the whole surface of the Earth and necessarily run into one another. They obtained $\tau_W \approx 400$ My for present-day spreading rates and distribution of continents, which is less than the thermal adjustment time. $\tau_W$ varies as a function of continental area and drift velocity and has probably been larger in the past when continents accounted for a smaller fraction of Earth's surface. Geological data

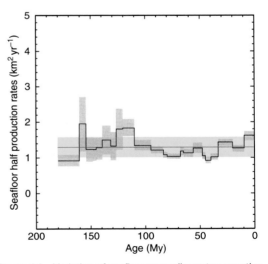

**Figure 14** Variation of seafloor spreading rates over the last 180 My accounting for oceanic ridges that got subducted in the Pacific and Tethys Oceans. The black line is the half spreading rate and the gray area represent the uncertainty. The red line is the average value. Adapted from Cogné J-P and Humler E (2004) Temporal variation of oceanic spreading and crustal production rates during the last 180 My. *Earth and Planetary Science Letters* 227: 427–439.

support such an increase of $\tau_W$ (Hoffman, 1997). Note that this observation runs against the intuitive notion that plates moved faster in the past. If the rate of heat loss of the Earth depends on the distribution of continents, it oscillates on a timescale $\tau_W$ over a long-term decreasing trend. We now attempt to estimate how large these oscillations can be.

Convective systems that are not constrained by lateral boundaries lead to age distributions that are almost rectangular (see Appendix 8). In this case, subduction occurs at the same age everywhere and the parameter $\lambda(f) = 2$, which is 25% less than the value for the triangular distribution (8/3). This changes the estimated oceanic heat loss by about 8 TW, which amounts to about half of the difference between present heat loss and heat production in the mantle (**Table 11**). Assuming for the sake of argument that the life time of a supercontinent is as long as the Wilson cycle and that these cycles are accompanied by changes of the age distribution of seafloor, the time-averaged oceanic heat loss could be 12.5% (i.e., 4 TW) less than the estimate of **Table 11**. A similar result was obtained by Grigné et al. (2005) using numerical models with many interacting convection cells of variable wavelength.

We have so far assumed that the total area of oceans remains constant. Over a few tens of million years, changes of oceanic area that are not related to rigid plate tectonics may occur, however, due to zones of diffuse deformation. Such zones are found in both oceans and continents and presently account for ≈15% of Earth's surface (Gordon, 2000). In continents, extension occurs at the expense of oceans whereas shortening increases the oceanic area. These zones are usually very active and characterized by high heat flux, as in the Basin and Range Province, for example. Assuming that the average heat flux over such zones is equal to that in the Basin and Range (105 mW m$^{-2}$), the net effect on the global heat loss is small because this heat flux is approximately the average oceanic one.

The two types of short-lived transient phenomena that we have identified, changes of the age distribution of seafloor and diffuse deformation, may induce fluctuations of heat loss on the order 4 TW. This is not sufficient to explain the difference between the present and average values of the secular cooling rate.

Yet another type of transient may be due to enhanced or subdued hot-spot activity. Assessing the magnitude of the implied heat loss variations is difficult from the geological record and is best achieved through consideration of thermal boundary layers.

### 6.7.7  Heat Flow Out of the Core

The discussion above has made no distinction between mantle and core. Labrosse (2002) has argued that the integrated heat flux at the CMB is mostly due to the spreading of cold subducted material. If this is correct, the value of this heat flux varies in tandem with the surface heat flux. Let us sketch some implications of such a proposition.

One possibility is that the core cools at the same rate as the mantle. Such a slow variation is compatible with thermal evolution models of the Earth's core, provided that ohmic dissipation increased by a factor of four when the inner core started crystallizing (Labrosse, 2003). It is likely, however, that the core heat loss fluctuates over timescales that are short compared to the age of the Earth, much as the surface heat flux. The core loses heat to an unstable boundary layer which grows at the base of the mantle. Compared to the upper boundary layer of mantle convection, this layer involves a smaller heat flux. For the sake of discussion, let us estimate the energy contained in the D″ layer, taken here as the lower thermal boundary layer of the mantle. The temperature difference across this 200 km thick layer can be estimated to be about $\delta T = 1000$ K (e.g., Lay et al., 1998). Assuming a linear temperature profile, the energy content of this layer is

$$U = \rho C_P 4\pi b^2 h \frac{\delta T}{2} \simeq 7.5 \times 10^{28} \, \text{J} \qquad [84]$$

with $b = 3480$ km the radius of the CMB, $\rho \simeq 5 \times 10^3$ kg m$^{-3}$ density, $h = 200$ km, and $C_P = 1000$ J kg$^{-1}$ K$^{-1}$. This energy is transferred to the mantle when the boundary layer goes unstable. The time scale is that of conductive thickening of the layer, that is $h^2/\pi\kappa \simeq 400$ My for $\kappa = 10^{-6}$ m$^2$ s$^{-1}$. Dividing the total energy given by eqn [84] by this timescale predicts a heat flux variation of 5 TW. This accounts for a large fraction of the total heat flux across the CMB and has important implications for the dynamo.

Further changes of core heat loss may be due to variations of the mass flux of downwellings, which may occur due to plate reorganizations and supercontinent cycles.

### 6.7.8  Summary

Theoretical models for the cooling of the Earth rely on theory and measurements in convective systems that do not possess one key feature of the present plate tectonic regime, which is that ocean floor of all ages gets subducted. The present imbalance between heat loss and heat production may be due in part to a particular convection phase with enhanced heat loss compared to the long-term evolution.

The main stumbling block in developing a realistic secular cooling model is our poor understanding of the subduction process. This hampers our ability to properly parametrize plate velocity and heat loss. Specifically, scaling arguments for mantle convection depend on only one parameter, $\tau_m$, the maximum age of a plate at Earth's surface. Plate velocity $U$ and plate length $L$, as well as the rate of heat loss depend on how this variable changes with time. Alternatively, past mantle temperature changes that are documented in a variety of ways provide powerful constraints for convection models. They imply that, in the past, the bulk rate of heat loss has changed less rapidly than heat production.

## 6.8 Conclusions

Studies of the thermal evolution of the Earth are as old as geophysics and remain central to geology, for they deal with the energy that drives all geological processes. The heat budget of the mantle can be established with a reasonable accuracy ($\approx 20\%$) thanks to tremendous improvements in our knowledge of physical properties and data coverage. The Urey number is the ratio of heat production to heat loss, two imperfectly known quantities whose estimates are summarized in **Table 11**. Our current estimates for Urey number are in the range 0.2–0.5, which rules out early parametrized cooling models. This may be explained by the particular behavior of oceanic plates, which are subducted at all ages. This feature seems to be unique among convective systems of all kinds. It now appears that heat loss does not follow the decay of radiogenic heat production and that the time lag is on the order of the age of the Earth (Section 6.7.1).

The present mantle energy budget implies a secular cooling rate in the range of 50–190 K Gy$^{-1}$. Over long timescale, the average value for the cooling rate estimated from geological constraints appears to be at the very low end of this range (50 K Gy$^{-1}$). There is no reason to assume that the cooling rate has remained constant through time. Both geological data and

physical constraints on the thermal structure of the early Earth indicate that the cooling rate increased as the planet got older (Section 6.7.6). Independent constraints on cooling rate come from considerations of early Earth evolution. Plate tectonics is a regime of mantle convection attainable only in a state of subsolidus rheology, that is with at most 40% melt. The most recent phase diagrams for the mantle (Zerr *et al.*, 1998; Litasov and Ohtani, 2002) indicate that this threshold is reached when the potential temperature is about 200 K higher than the present. If plate tectonics has been operating since the end of the magma ocean, the total amount of cooling is constrained and the average cooling rate cannot be more than about 50 K Gy$^{-1}$.

## Appendix 1: Contraction of Earth due to Secular Cooling

The planet contracts as it cools down. This induces changes of gravity which themselves induce changes of pressure and density. Here, we derive an approximate solution for a homogeneous planet in order to demonstrate that the most important effect is that of temperature.

Assuming spherical symmetry, governing equations for a spherical planet are as follows:

$$\frac{1}{r^2}\frac{d}{dr}\left(r^2 g\right) = 4\pi G\rho \qquad [85]$$

$$\frac{dP}{dr} = -\rho g \qquad [86]$$

where $G$ is the gravitational constant and $P$ pressure, and where all variables depend on radial distance $r$ only. From the equation of state, we deduce that

$$\frac{d\rho}{dr} = \frac{d\rho}{dP}\frac{dP}{dr} = \frac{\rho}{K_s}\frac{dP}{dr} \qquad [87]$$

$$= -\frac{\rho^2 g}{K_s} \qquad [88]$$

where $K_s$ is the isentropic bulk modulus. For simplicity, the bulk modulus is assumed to be constant. In order to make the equations dimensionless, we use radius $R$ as length-scale and density at the top (we shall define it more precisely later) $\rho_T$ as density-scale. From those, we derive a gravity scale

$$[g] = G\rho_T R \qquad [89]$$

Equations can now be written using dimensionless variables:

$$\frac{1}{r^2}\frac{d}{dr}\left(r^2 g\right) = 4\pi\rho \qquad [90]$$

$$\frac{d\rho}{dr} = -\epsilon\rho^2 g \qquad [91]$$

where $\epsilon$ is a dimensionless number which provides a measure of the magnitude of density changes due to pressure:

$$\epsilon = \frac{\rho_T[g]R}{K_s} = \frac{G\rho_T^2 R^2}{K_s} \qquad [92]$$

In this problem, we consider the cooling of the Earth's mantle, where the radial temperature profile is almost isentropic. We neglect the upper thermal boundary layer which does not contribute much to the total mass and energy. Thus, reference density $\rho_T$ is the mantle density at the atmospheric pressure, that is, at the potential temperature of the isentropic radial profile. For what follows, the key point is that $\rho_T$ does not depend on pressure, which varies in the planet interior during contraction. For $\rho_T = 3.3 \times 10^3\,kg\,m^{-3}$, $R = 6370\,km$ and a value of 150 GPa for $K_s$, $\varepsilon \approx 0.2$. This is small and we expand all the variables in series of $\epsilon$:

$$\rho = \rho_o + \epsilon\rho_1 + \cdots \qquad [93]$$

$$g = g_o + \epsilon g_1 + \cdots \qquad [94]$$

We find that

$$\rho_o = 1 \qquad [95]$$

$$\rho_1 = -\frac{4\pi}{6}\rho_o^3\left(r^2 - 1\right) \qquad [96]$$

$$g_o = \frac{4\pi}{3}\rho_o r \qquad [97]$$

$$g_1 = \frac{16\pi^2}{90}\rho_o^3 r\left(5 - 3r^2\right) \qquad [98]$$

We then calculate the mass of the planet to first order in $\epsilon$:

$$M = \frac{4}{3}\pi\rho_T R^3\left(1 + \frac{4\pi}{15}\epsilon\right) = \frac{4}{3}\pi\rho_T R^3\left(1 + \frac{4\pi}{15}\frac{G\rho_T^2 R^2}{K_s}\right) \qquad [99]$$

Writing that mass is conserved when $\rho_T$ and $R$ change, we obtain

$$\frac{\Delta R}{R} = -\frac{1}{3}\frac{\Delta\rho_T}{\rho_T}\left(1 + \frac{16\pi}{45}\frac{G\rho_T^2 R^2}{K_s}\right) \qquad [100]$$

By definition, surface density $\rho_T$ depends on temperature only, such that

$$\frac{\Delta\rho_T}{\rho_T} = -\alpha\Delta T \qquad [101]$$

where $\alpha$ is the thermal expansion coefficient. The end result is therefore

$$\frac{\Delta R}{R} = \frac{\alpha\Delta T}{3}\left(1 + \frac{16\pi}{45}\frac{G\rho_T^2 R^2}{K_s}\right) \qquad [102]$$

This shows that contraction is enhanced by the change in gravity field and also that the correction to the temperature effect is small. An exact calculation involving a pressure-dependent bulk modulus would not alter this conclusion.

## Appendix 2: Gravitational Energy Changes

Here, we present a demonstration for changes of the bulk gravitational energy which was made by Paul Roberts (personal communication). For a sphere of uniform density and radius $R$,

$$E_g = -\frac{16\pi^2}{15}G\rho^2 R^5 \qquad [103]$$

For a contracting object with moving material boundaries, it is convenient to employ Lagrangian variables. Because of mass conservation, the most convenient variable is the mass in a sphere of radius $r$:

$$M(r) = 4\pi\int_0^r \rho(r)r^2\,dr = \int_0^r \rho(r)dV \qquad [104]$$

Thus,

$$g(r) = \frac{-GM(r)}{r^2} \qquad [105]$$

The gravitational energy in a material volume bounded by spheres of radii $r_1$ and $r_2$ is

$$E_{12} = +\int_{r_1}^{r_2}\rho g r\,dV = +\int_{M_1}^{M_2} rg\,dM = +\int_{M_1}^{M_2}\frac{GM}{r}\,dM \qquad [106]$$

The Lagrangian derivative is easily calculated using the variable $M$:

$$\frac{dE_{12}}{dt} = \int_{M_1}^{M_2}\frac{dr}{dt}\frac{GM}{r^2}\,dM \qquad [107]$$

For this calculation, the velocity field is limited to contraction, such that $dr/dt = v_c$. Thus

$$\frac{dE_{12}}{dt} = \int_{M_1}^{M_2} v_c \frac{GM}{r^2} dM = -\int_{r_1}^{r_2} \rho(\mathbf{v_c} \cdot \mathbf{g}) dV \qquad [108]$$

which is the integral form of eqn [17] in the main text.

## Appendix 3: Viscous Dissipation

Here, we derive an equation for viscous dissipation which, by definition, involves deviatoric stresses and hence departures from hydrostatic equilibrium in the momentum equation. For convection in the solid Earth, inertial terms can be neglected in the momentum balance, with the consequence that changes of kinetic energy are also negligible in the mechanical energy balance [10]. Subtracting the hydrostatic balance [14] from the momentum balance, we get

$$0 = -\nabla(\delta p) - \nabla \cdot \boldsymbol{\sigma} + \delta\rho\mathbf{g} \qquad [109]$$

where $\delta_p$ and $\delta_\rho$ are deviations from the azimuthal averages of pressure and density, respectively, and where we have neglected perturbations to the gravity field, which only appear in second-order terms. Introducing the temperature perturbation, $\theta$, the density perturbation is obtained by a Taylor expansion about the mean:

$$\delta\rho = \left(\frac{\partial\rho}{\partial p}\right)_a \delta p + \left(\frac{\partial\rho}{\partial T}\right)_a \theta \qquad [110]$$

where derivatives are taken along the azimuthally averaged density profile. Introducing the thermal expansion coefficient, this can be recast as

$$\delta\rho = \frac{\partial\bar{\rho}/\partial r}{\partial\bar{p}/\partial r} \delta p - \bar{\rho}\alpha\theta \qquad [111]$$

Substituting for the density perturbation in eqn [109] and using the hydrostatic expression for $\partial\bar{p}/\partial r$ (eqn [14] of the main text), we get

$$0 = -\nabla(\delta p) - \frac{\partial\bar{\rho}/\partial r}{\bar{\rho}g} \delta p\mathbf{g} - \nabla \cdot \boldsymbol{\sigma} - \bar{\rho}\alpha\theta\mathbf{g} \qquad [112]$$

This can be recast as

$$0 = -\bar{\rho}\nabla\left(\frac{\delta p}{\bar{\rho}}\right) - \nabla \cdot \boldsymbol{\sigma} - \bar{\rho}\alpha\theta\mathbf{g} \qquad [113]$$

From this, we deduce a modified form of the mechanical energy balance:

$$0 = -\bar{\rho}\mathbf{w} \cdot \nabla\left(\frac{\delta p}{\bar{\rho}}\right) - \mathbf{w} \cdot \nabla \cdot \boldsymbol{\sigma} - \bar{\rho}\alpha\theta\mathbf{w} \cdot \mathbf{g} \qquad [114]$$

Integrating this equation over the planet volume yields three separate integrals which are evaluated separately. The first term on the right-hand side yields

$$\int_V \bar{\rho}\mathbf{w} \cdot \nabla\left(\frac{\delta p}{\bar{\rho}}\right) dV = \int_V \nabla \cdot (\delta p\mathbf{w}) dV - \int_V \frac{\delta p}{\bar{\rho}} \nabla \cdot (\bar{\rho}\mathbf{w}) dV$$
$$= \int_A \delta p\,\mathbf{w} \cdot d\mathbf{A} - \int_V \frac{\delta p}{\bar{\rho}} \nabla \cdot (\bar{\rho}\mathbf{w}) dV$$
$$[115]$$

where the first term on the right is zero because the pressure perturbation $\delta p$ must vanish at Earth's surface where pressure is held constant. From the continuity equation in the anelastic approximation, $\nabla \cdot (\bar{\rho}\mathbf{w}) = 0$ (Braginsky and Roberts, 1995). The second term on the right of the mechanical energy balance [114] yields

$$\int_V \mathbf{w} \cdot \nabla \cdot \boldsymbol{\sigma}\,dV = \int_V (\mathbf{w} \cdot \nabla \cdot \boldsymbol{\sigma} - \nabla \cdot [\boldsymbol{\sigma} \cdot \mathbf{w}]) dV$$
$$+ \int_V \nabla \cdot [\boldsymbol{\sigma} \cdot \mathbf{w}] dV$$
$$= \int_V \phi\,dV + \int_A \mathbf{w} \cdot [\boldsymbol{\sigma} \cdot d\mathbf{A}] \qquad [116]$$

The last term on the right of this equation is the work of shear stresses at Earth's surface, which is negligible. Using these results, the mechanical energy equation may be reduced to

$$-\int_V \rho\alpha g\overline{w_r\theta}\,dV + \int_V \bar{\phi}\,dV = 0 \qquad [117]$$

This equation states that viscous dissipation is balanced by the bulk buoyancy flux and explains why it does not enter the bulk energy balance.

## Appendix 4: Half-Space Cooling Model with Temperature-Dependent Properties

The general form of the 1-D heat equation is

$$\rho C_p \frac{\partial T}{\partial \tau} = \frac{\partial}{\partial z}\left(k\frac{\partial T}{\partial z}\right) \qquad [118]$$

where all properties depend on temperature $T$. In this section, the temperature for the half-space is initially $T_M$ and temperature is set to zero at the surface: $T(z = 0, \tau) = 0$.

For constant physical properties, the temperature for the half-space is given by Carslaw and Jaeger (1959):

$$T(z, t) = T_M \text{erf}\left(\frac{z}{2\sqrt{\kappa\tau}}\right) \qquad [119]$$

and the surface heat flux is

$$q(0, t) = \frac{kT_M}{\sqrt{\pi\kappa\tau}} \qquad [120]$$

For temperature-dependent properties, Carslaw and Jaeger (1959) introduce the following variable:

$$\theta = \frac{1}{k_0} \int_0^T k \, dT \qquad [121]$$

where $k_0$ is thermal conductivity at the reference temperature $T = 0$. The heat equation becomes

$$\frac{1}{\kappa}\frac{\partial\theta}{\partial\tau} = \frac{\partial^2\theta}{\partial z^2} \qquad [122]$$

where $\kappa$ is the temperature-dependent thermal diffusivity. For the boundary and initial conditions of interest here, this equation can be written in terms of a similarity variable $\eta$:

$$\theta(z, t) = \Theta(\eta), \text{ with } \eta = z\tau^{-1/2} \qquad [123]$$

$$-\frac{\eta}{2\kappa}\Theta' = \Theta'' \qquad [124]$$

with the following boundary conditions:

$$\Theta(0) = 0, \quad \Theta(\infty) = \Theta_M, \quad \Theta'(\infty) = \Theta''(\infty) = 0 \qquad [125]$$

The surface heat flux is thus

$$q = k\frac{\partial T}{\partial z}\bigg|_{z=0} = k_0\tau^{-1/2}\Theta'(0) \qquad [126]$$

which is of the form $C_Q \tau^{-1/2}$ regardless of the specific functional form of $\rho$, $C_p$, and $k$. This beautiful result was pointed out first by Lister (1977).

## Appendix 5: Plate Models for the Oceanic Lithosphere

For a plate of thickness $L$ initially at temperature $T_M$, with $T(z = 0, \tau) = 0$ and $T(z = L, \tau) = T_M$, the temperature for $0 < z < L$ is given by

$$\tau(z, \tau) = T_M$$

$$\times \left(\frac{z}{L} + \sum_{n=1}^{\infty}\frac{2}{n\pi}\sin\left(\frac{n\pi z}{L}\right)\exp\left(\frac{-n^2\pi^2\kappa\tau}{L^2}\right)\right) \qquad [127]$$

and the surface heat flux is given by

$$q(0, t) = \frac{kT_M}{L}\left(1 + 2\sum_{n=1}^{\infty}\exp\left(\frac{-n^2\pi^2\kappa\tau}{L^2}\right)\right) \qquad [128]$$

which diverges at $\tau = 0$. For $\tau \ll L^2/\kappa$, the heat flux

$$q(0, \tau) \approx \frac{kT_M}{\sqrt{\pi\kappa\tau}} \qquad [129]$$

With fixed heat flux at the base, $Q(L, \tau) = kT_M/L$, the temperature of the plate is given by

$$T(z, \tau) = \frac{T_M z}{L} - \frac{4T_M}{\pi}\sum_{n=0}^{\infty}\frac{(-1)^n}{(2n+1)}$$

$$\times \cos\left(\frac{(2n+1)\pi(z-L)}{2L}\right)$$

$$\times \exp\left(\frac{-(2n+1)^2\pi^2\kappa\tau}{4L^2}\right) + \frac{8T_M}{\pi^2}\sum_{n=0}^{\infty}\frac{(-1)^n}{(2n+1)^2}$$

$$\times \sin\left(\frac{(2n+1)\pi z}{2L}\right)\exp\left(\frac{-(2n+1)^2\pi^2\kappa\tau}{4L^2}\right)$$

$$[130]$$

and the surface heat flux is given by

$$\frac{q(0, t)}{(kT_M/L)} = 1 + 2\sum_{n=0}^{\infty}\exp\left(\frac{-(2n+1)^2\pi^2\kappa\tau}{4L^2}\right)$$

$$+ \frac{4}{\pi}\sum_{n=0}^{\infty}\frac{(-)^n}{(2n+1)}\exp\left(\frac{-(2n+1)^2\pi^2\kappa\tau}{4L^2}\right) \qquad [131]$$

The surface heat flux for these three different boundary conditions are compared in **Figure 15**, with the

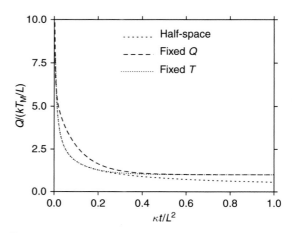

**Figure 15** Normalized heat flux as a function of reduced time for different cooling models for the oceanic lithosphere. Length-scale $L^*$ is equal to $L$, the plate thickness, in the fixed temperature model and to $L/2$ in the fixed flux model. In the half-space model, $L^*$ cancels out. See Appendix 5.

same length scale $L$ for the half-space and the constant-temperature boundary condition and $L/2$ for the heat flux boundary condition.

## Appendix 6: Differences Between Estimates of the Energy Budget

Table 13 compares our construction of the mantle energy budget with those proposed by Stacey (1992) and Davies (1999). Although the total energy budget is almost identical, there are major differences in the breakdown of the budget between different items. These differences originate in the various assumptions made as well as in the objectives that are sought. Our estimate uses a slightly different total oceanic area which reflects our better knowledge of continental margins. It also includes the contribution of hot spots, which is not accounted for by bulk lithosphere cooling. Finally, it allows for a small amount of radioelements in the subcontinental lithospheric mantle and takes advantage of our improved constraints on crustal heat production.

For Stacey (1992), the total heat production (27 TW) is significantly higher than the value of 20 TW for BSE, and almost equal to that of the nondepleted chondritic model of Birch (1965). It seems that Stacey has added the crustal heat production to BSE. The core heat loss is low because it is assumed identical to the heat carried by hot spots. Stacey assumed that all the gravitational energy released by thermal contraction (2.1 TW) goes to heat.

In Davies (1999), the secular cooling of the mantle is assumed to be fixed by petrological data and the lower mantle heat production is the variable that is adjusted to balance the budget when all the other variables are fixed. Core cooling is also assumed identical to the total heat flux from hot spots. Upper mantle heat production is known to be low (from samples of the mantle carried to the surface). Lower mantle is assumed to be a mixture between a depleted chondritic composition, which would give 11 TW, and a MORB-like component, which would yield 27 TW. The ratio of those two components is adjusted to balance the budget. (Note that depleted mantle should give only 3 TW, and the MORB-like component seems to have the same heat production as that of chondrites).

The global energy budget of the Earth was one of the arguments used by Kellogg et al. (1999) to propose that the lowermost mantle forms a distinct reservoir with chondritic concentration in radioelements, the 'abyssal layer'. The heat production in the depleted MORB mantle (i.e., the source of depleted MORBs) is $\approx 0.6$ pW kg$^{-1}$. Assuming that this is representative of the whole mantle, the total mantle heat production would only be 2.5 TW, that is, much less than the 14 TW obtained by removing the crustal heat production from BSE. (Note that Kellogg et al. (1999) used 31 TW for the total heat production in a 'chondritic' Earth).

## Appendix 7: Average Thermal Structure and Temperature Changes in Upwellings and Downwellings

The reference vertical temperature profile is often called 'adiabatic', which is misleading. Here, we recapitulate the definitions and introduce two different

**Table 13** Various estimates of the global budget (TW)

|  | Stacey (1992) | Davies (1999)[a] | This study |
|---|---|---|---|
| Total heat loss | 42 | 41 | 46 |
| Continental heat production | 8 | 5 | 7 |
| Upper mantle |  | 1.3 |  |
| Lower mantle |  | 11–27 |  |
| Mantle heat production | 19 | 12–28[b] | 13 |
| Latent heat – core differentiation | 1.2 | <1 |  |
| Mantle differentiation | 0.6 | 0.3 | 0.3 |
| Gravitational (Thermal contraction) | 2.1 |  |  |
| Tidal dissipation |  | 0.1 | 0.1 |
| Core heat loss | 3 | 5 | 8 |
| Mantle cooling | 10 | 9 | 18 |
| Present Urey ratio | 0.64 | 0.3–0.68 | 0.21–0.49 |

[a]Mantle cooling is fixed.
[b]Lower mantle heat production is variable and calculated to fit the mantle cooling rate.

reference temperature profiles. The equation for the entropy per unit mass, $s$, is

$$\rho T \frac{Ds}{Dt} = \rho C_p \frac{DT}{Dt} - \alpha T \frac{Dp}{Dt} \qquad [132]$$

$$= -\nabla \cdot \mathbf{q} + \phi + H \qquad [133]$$

Note that this shows that entropy is not conserved due to irreversible dissipation and radioactive decay. Density changes due to temperature have a small impact on pressure and dynamic pressure variations are small compared to the hydrostatic pressure. Thus,

$$\frac{Dp}{Dt} \approx -\rho g w \qquad [134]$$

where $w$ is the radial velocity component.

From these equations, we may deduce the isentropic temperature profile, such that $Ds/Dt = 0$. In the interior of the convecting system, far from the upper and lower boundaries, the dominant velocity component is vertical. Assuming steady-state and using eqn [132]:

$$\rho C_p w \frac{dT_S}{dr} = -\alpha T_S \rho g w \qquad [135]$$

where $T_S$ stands for the isentropic temperature profile. Thus,

$$\frac{dT_S}{dr} = -\frac{\alpha g}{C_p} T_S \qquad [136]$$

This is close to the vertical profile of the azimuthally averaged temperature in a steady-state well-mixed convective system with no internal heat production and negligible viscous dissipation.

The isentropic temperature gradient derived above provides a convenient 'reference' profile which illustrates the role of compressibility. However, it is a poor approximation for the temperature path followed by a rising (or sinking) mantle parcel. We may consider for simplicity that such a parcel does not exchange heat with its surroundings, which is a good approximation for the broad return flow away from subduction zones. In this case, we set $\mathbf{q} = 0$, use the same approximation as before for pressure and obtain

$$\rho C_p \frac{DT}{Dt} \approx \rho C_p \left( \frac{\partial T}{\partial t} + w \frac{\partial T}{\partial r} \right) = -(\alpha T)\rho g w + H + \phi$$

$$[137]$$

The material parcel's temperature changes due to radiogenic heat production and dissipation as well

as due to the work of pressure forces. This may be recast as follows:

$$\frac{\partial T}{\partial r} = -\frac{\alpha g}{C_p} T + \frac{1}{w} \left( \frac{\phi + H}{\rho C_p} - \frac{\partial T}{\partial t} \right) \qquad [138]$$

where one should note that secular cooling acts in the same direction as internal heat production. The radial temperature gradient from eqn [138] differs from the isentropic value by about 30%. Here, we see the importance of knowing precisely the secular cooling rate $\partial T/\partial t$.

## Appendix 8: Seafloor Age Distribution as Seen from Models of Mantle Convection

We present here a few examples of temperature fields obtained in numerical models of convection and compute the seafloor age distribution they would provide for comparison with that of Earth today. In all cases presented, convection is set by imposing a destabilizing temperature difference $\Delta T$ between top and bottom and a uniform internal heating rate per unit volume $H$. This defines two dimensionless parameters (e.g., Sotin and Labrosse, 1999), the Rayleigh number $Ra$ and the dimensionless internal heating rate $H^*$:

$$Ra = \frac{g\alpha\Delta T D^3}{\kappa \nu_0} \quad ; \quad H^* = \frac{HD^2}{k\Delta T} \qquad [139]$$

where $D$ is the thickness of the system and $\nu_0$ is the dynamic viscosity at some reference point. The other symbols are the same as in the rest of the chapter. A third dimensionless number is the Prandtl number $Pr = \nu_0/\kappa$, which is of order $10^{23}$ in the mantle and is accordingly set to infinity. The viscosity of mantle material is known to vary with temperature, pressure, mineral phase, etc. To reproduce the plate tectonics regime of mantle convection, a complex rheological law must be used (see Chapters 1 and 2). This introduces yet another set of parameters. For example, the pseudoplastic rheology used by Tackley (2000) is defined by an effective viscosity

$$\eta_{\text{eff}} = \min \left[ \eta(T), \frac{\sigma_y}{2\dot\varepsilon} \right] \qquad [140]$$

where $\sigma_y$ is a yield stress and $\dot\varepsilon = \sqrt{\dot\varepsilon_{i,j}\dot\varepsilon_{i,j}}/2$ the second invariant of the strain-rate tensor. Viscosity is temperature dependent:

$$\eta(T) = \exp\left[\frac{23.03}{T+1} - \frac{23.03}{2}\right] \qquad [141]$$

This specific rheological law was used to obtain the results of **Figure 3** Grigné *et al.*, 2005).

In numerical models of mantle convection, determining the age of each point at the surface requires extensive calculations with markers. To alleviate this difficulty, we use the fact that the surface heat flux is due to conductive cooling and define a pseudoage as follows:

$$q \propto \frac{1}{\sqrt{\tau}} \qquad [142]$$

We thus use heat flux values to determine the age distribution.

The first configuration tested is that of isoviscous Rayleigh–Bénard convection with internal heating. Using a snapshot from Labrosse (2002), we get a distribution of pseudoages that is peaked at low values and exhibits a roughly exponential tail at large ages (**Figure 16**). This reflects that, with a large amount of internal heating, convective flow is dominated by strong cold plumes and passive return flow in most of the interior. There are weak hot plumes in the lower part of the domain but they do not contribute much to the heat flux, in a manner reminiscent of that on Earth. The distribution of surface heat flux is due to the distributed return flow.

A somewhat more realistic model relying on the pseudoplastic rheology has been developed by Grigné *et al.* (2005) and a snapshot is shown in **Figure 3**. The

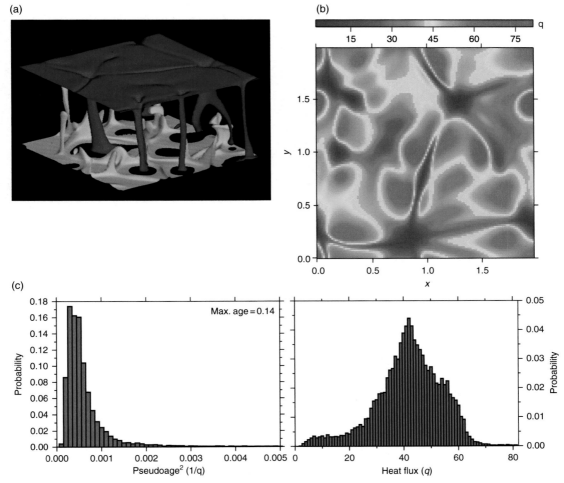

**Figure 16**  Snapshot of temperature (a), surface heat flux (b), and the corresponding pseudoage distribution (c) in a convection system with a Rayleigh number $Ra = 10^7$ and an internal heating rate $H = 20$. Note that the age distribution is far from that of the Earth's oceans, which is triangular. From the model of Labrosse (2002). See Appendix 8 for details.

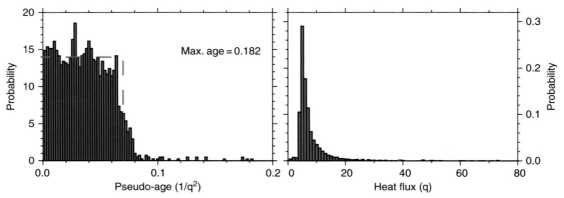

**Figure 17**   Distribution of surface heat flux ($q$, right) and pseudoage ($1/q^2$) for a model with plate-like behavior, corresponding to **Figure 3**. The age distribution is close to a rectangular one (dashed line). See Appendix 8 for details. From Grigné C, Labrosse S, and Tackley PJ (2005) Convective heat transfer as a function of wavelength: Implications for the cooling of the Earth. *Journal of Geophysical Research* 110(B03): 409 (doi:10.1029/2004JB003376).

distribution of surface heat flux and pseudoage for this model is given in **Figure 17**. The plate-like behavior is such that the distribution of heat flux is peaked at low values and the pseudoage distribution is approximately rectangular, reflecting the fact that all plates have similar sizes and velocities.

# References

Abbott D, Burgess L, Longhi J, and Smith WHF (1994) An empirical thermal history of the Earth's upper mantle. *Journal of Geophysical Research* 99: 13835–13850.

Abe Y (1993) Physical state of the very early Earth. *Lithos* 30: 223–235.

Abe Y (1997) Thermal and chemical evolution of the terrestrial magma ocean. *Physics of the Earth and Planetary Interiors* 100: 27–39.

Alfé D, Gillan MJ, and Price GD (2002) *Ab initio* chemical potentials of solid and liquid solutions and the chemistry of the Earth's core. *Journal of Chemical Physics* 116: 7127–7136.

Allègre CJ and Jaupart C (1985) Continental tectonics and continental kinetics. *Earth and Planetary Science Letters* 74: 171–186.

Ashwal LD, Morgan P, Kelley SA, and Percival J (1987) Heat production in an Archean crustal profile and implications for heat flow and mobilization of heat producing elements. *Earth and Planetary Science Letters* 85: 439–450.

Birch F (1965) Speculations on the thermal history of the Earth. *Bulletin of the Geological Society of America* 76: 133–154.

Birch F, Roy RF, and Decker ER (1968) Heat flow and thermal history in New England and New York. In: An-Zen E (ed.) *Studies of Appalachian Geology*, pp. 437–451. New York: Wiley-Interscience.

Bird RB, Stewart WE, and Lightfoot EN (1960) *Transport Phenomena*. New York: Wiley.

Bodell JM and Chapman DS (1982) Heat flow in the North Central Colorado plateau. *Journal of Geophysical Research* 87: 2869–2884.

Bonneville A, Von Herzen RP, and Lucazeau F (1997) Heat flow over Reunion hot spot track: Additional evidence for thermal rejuvenation of oceanic lithosphere. *Journal of Geophysical Research* 102: 22731–22748.

Brady RJ, Ducea MN, Kidder SB, and Saleeby JB (2006) The distribution of radiogenic heat production as a function of depth in the Sierra Nevada batholith, California. *Lithos* 86: 229–244.

Braginsky SI and Roberts PH (1995) Equations governing convection in Earth's core and the geodynamo. *Geophysical and Astrophysical Fluid Dynamics* 79: 1–97.

Breuer D and Spohn T (1993) Cooling of the Earth, Urey ratios, and the problem of potassium in the core. *Geophysical Research Letters* 20: 1655–1658.

Cameron AGW (2001) From interstellar gas to the Earth–Moon system. *Meteoritics and Planetary Science* 36: 9–22.

Canup RM (2004) Simulations of a late lunar-forming impact. *Icarus* 168: 433–456.

Canup RM and Righter K (eds.) (2000) *Origin of the Earth and Moon*. Tucson, AZ: The University of Arizona Press.

Carlson RL and Johnson HP (1994) On modeling the thermal evolution of the oceanic upper mantle: An assessment of the cooling plate model. *Journal of Geophysical Research* 99: 3201–3214.

Caro G, Bourdon B, Birck J-L, and Moorbath S (2003) $^{146}Sm-^{142}Nd$ evidence from Isua metamorphosed sediments for early differentiation of the Earth's mantle. *Nature* 423: 428–432 doi:10.1038/nature01668.

Carslaw HS and Jaeger JC (1959) *Conduction of Heat in Solids* 2nd edn. Oxford (UK): Clarendon Press.

Cermak V and Bodri L (1986) Two-dimensional temperature modelling along five East-European geotraverses. *Journal of Geodynamics* 5: 133–163.

Chapman DS and Pollack HN (1974) 'Cold spot' in West Africa: Anchoring the African plate. *Nature* 250: 477–478.

Christensen UR (1984) Heat transport by variable viscosity convection and implications for the Earth's thermal evolution. *Physics of the Earth and Planetary Interiors* 35: 264–282.

Christensen UR (1985) Thermal evolution models for the Earth. *Journal of Geophysical Research* 90: 2995–3007.

Christensen UR and Tilgner A (2004) Power requirement of the geodynamo from ohmic losses in numerical and laboratory dynamos. *Nature* 429: 169–171.

Clauser C, Giese P, Huenges E, *et al.* (1997) The thermal regime of the crystalline continental crust: Implications from the KTB. *Journal of Geophysical Research* 102: 18417–18441.

Cogley JG (1984) Continental margins and the extent and number of continents. *Reviews Geophysics and Space Physics* 22: 101–122.

Cogné J-P and Humler E (2004) Temporal variation of oceanic spreading and crustal production rates during the last 180 My. *Earth and Planetary Science Letters* 227: 427–439.

Conrad CP and Hager BH (1999) The thermal evolution of an Earth with strong subduction zones. *Geophysical Research Letters* 26: 3041–3044.

Crane K and O'Connell S (1983) The distribution and implications of the heat flow from the Gregory rift in Kenya. *Tectonophysics* 94: 253–272.

Cull JP (1991) Heat flow and regional geophysics in Australia. In: Cermak V and Rybach L (eds.) *Terrestrial Heat Flow and the Lithosphere Structure*, pp. 486–500. Berlin: Springer Verlag.

Davaille A and Jaupart C (1993) Transient high-Rayleigh-number thermal convection with large viscosity variations. *Journal of Fluid Mechanics* 253: 141–166.

Davies GF (1980) Thermal histories of convective Earth models and constraints on radiogenic heat production in the Earth. *Journal of Geophysical Research* 85: 2517–2530.

Davies GF (1980) Review of oceanic and global heat flow estimates. *Reviews of Geophysics and Space Physics* 18: 718–722.

Davies GF (1988) Ocean bathymetry and mantle convection, Part 1: large-scale flow and hotspots. *Journal of Geophysical Research* 93: 10467–10480.

Davies GF (1999) *Dynamic Earth: Plates, Plumes, and Mantle Convection*. Cambridge: Cambridge University Press.

Davis EE, Chapman DS, Wang K, et al. (1999) Regional heat flow variations across the sedimented Juan de Fuca ridge eastern flank: Constraints on lithospheric cooling and lateral hydrothermal heat transport. *Journal of Geophysical Research* 104: 17675–17688.

Davis EE and Elderfield H (2004) *Hydrogeology of the oceanic lithosphere*. Cambridge: Cambridge University Press.

Donnelly KE, Goldstein SL, Langmuir CH, and Spiegelman M (2004) Origin of enriched ocean ridge basalts and implications for mantle dynamics. *Earth and Planetary Science Letters* 226: 347–366.

Duchkov AD (1991) Review of Siberian heat flow data. In: Cermak V and Rybach L (eds.) *Terrestrial Heat Flow and the Lithosphere Structure*, pp. 426–443. Berlin: Springer-Verlag.

Dziewonski AM and Anderson DL (1981) Preliminary reference Earth model. *Physics of the Earth and Planetary Interiors* 25: 297–356.

Elkins-Tanton LT, Parmentier EM, and Hess PC (2003) Magma ocean fractional crystallization and cumulate overturn in terrestrial planets: Implications for Mars. *Meteoritics and Planetary Science* 38: 1753–1771.

Engebretson DC, Cox A, and Gordon RG (1984) Relative motions between oceanic plates of the Pacific basin. *Journal of Geophysical Research* 89: 10291–10310.

England PC, Oxburg ER, and Richardson SW (1980) Heat refraction in and around granites in north–east England. *Geophysical Journal of the Royal Astronomical Society* 62: 439–455.

Erickson AJ, Von Herzen RP, Sclater JG, Girdler RW, Marshall BV, and Hyndman R (1975) Geothermal measurements in deep-sea drill holes. *Journal of Geophysical Research* 80: 2515–2528.

Flasar FM and Birch F (1973) Energetics of core formation: A correction. *Journal of Geophysical Research* 78: 6101–6103.

Fountain DM and Salisbury MH (1981) Exposed cross-sections through the continental crust: Implications for crustal structure, petrology, and evolution. *Earth and Planetary Science Letters* 56: 263–277.

Fountain DM, Salisbury MH, and Furlong KP (1987) Heat production and thermal conductivity of rocks from the Pikwitonei–Sachigo continental cross-section, central Manitoba: Implications for the thermal structure of Archean crust. *Canadian Journal of Earth Sciences* 24: 1583–1594.

Fournier RO (1989) Geochemistry and dynamics of the Yellowstone National Park hydrothermal system. *Annual Review of Earth and Planetary Sciences* 17: 13–53.

Francheteau J, Jaupart C, Jie SX, et al. (1984) High heat flow in southern Tibet. *Nature* 307: 32–36.

Gordon RG (2000) Diffuse oceanic plate boundaries: Strain rates, vertically averaged rheology, and comparisons with narrow plate boundaries and stable plate interiors. In: Richards MA, Gordon RG, and van der Hilst RD (eds.) *Geophysical Monograph Series* 121: *The History and Dynamics of global plate Motions*, pp. 143–159. Washington, DC: American Geophysical Union.

Green DH (1975) Genesis of Archean peridotitic magmas and constraints on Archean geothermal gradients and tectonics. *Geology* 3: 15–18.

Grigné C, Labrosse S, and Tackley PJ (2005) Convective heat transfer as a function of wavelength: Implications for the cooling of the Earth. *Journal of Geophysical Research* 110(B03): 409 (doi:10.1029/2004JB003376).

Grove TL and Parman SW (2004) Thermal evolution of the Earth as recorded by komatiites. *Earth and Planetary Science Letters* 219: 173–187.

Gubbins D and Roberts PH (1987) Magnetohydrodynamics of the Earth's core. In: Jacobs JA (ed.) *Geomagnetism*, vol. 2, pp. 1–183. London: Academic.

Guillou L, Mareschal JC, Jaupart C, Gariépy C, Bienfait G, and Lapointe R (1994) Heat flow and gravity structure of the Abitibi belt, Superior Province, Canada. *Earth and Planetary Science Letters* 122: 447–460.

Guillou-Frottier L, Mareschal J-C, Jaupart C, Gariepy C, Lapointe R, and Bienfait G (1995) Heat flow variations in the Grenville Province, Canada. *Earth and Planetary Science Letters* 136: 447–460.

Gupta ML, Sharma SR, and Sundar A (1991) Heat flow and heat generation in the Archean Dharwar craton and implication for the southern Indian Shield geotherm. *Tectonophys* 194: 107–122.

Harris RN and Chapman DS (2004) Deep-seated oceanic heat flow, heat deficits and hydrothermal circulation. In: Davis E and Elderfield H (eds.) *Hydrogeology of the Oceanic Lithosphere*, pp. 311–336. Cambridge: Cambridge University Press.

Hart SR and Zindler A (1986) In search of a bulk-Earth composition. *Chemical Geology* 57: 247–267.

Henry SG and Pollack HN (1988) Terrestrial heat flow above the Andean subduction zone in Bolivia and Peru. *Journal of Geophysical Research* 93: 15153–15162.

Hernlund JW, Thomas C, and Tackley PJ (2005) Phase boundary double crossing and the structure of Earth's deep mantle. *Nature* 434: 882–886 (doi:10.1038/nature03472).

Herzberg C and Zhang J (1996) Melting experiments on anhydrous peridotite KLB-1: Compositions of magmas in the upper mantle and transition zone. *Journal of Geophysical Research* 101: 8271–8296.

Hirose K (2006) Postperovskite phase transition and its geophysical implications. *Reviews of Geophysics* 44: RG3001 (doi:10.1029/2005RG000186).

Hirose K, Sinmyo R, Sata N, and Ohishi Y (2006) Determination of post-perovskite phase transition boundary in $MgSiO_3$ using Au and MgO pressure standards. *Geophysical Research Letters* 33: L01310 (doi:10.1029/2005GL024468).

Hoffman PF (1997) Tectonic geology of North America. In: van der Pluijm B and Marshak S (eds.) *Earth Structure: An*

*Introduction to Structural Geology and Tectonics*, pp. 459–464. New York: McGraw Hill.

Howard LN (1964) Convection at high Rayleigh number. In: Gortler H (ed.) *Proceedings of the Eleventh International Congress of Applied Mechanics*, pp. 1109–1115. New York: Springer-Verlag.

Hu S, He L, and Wang J (2000) Heat flow in the continental area of China: A new data set. *Earth and Planetary Science Letters* 179: 407–419.

Hulot G, Eymin C, Langlais B, Mandea M, and Olsen N (2002) Small-scale structure of the geodynamo inferred from Oersted and Magsat satellite data. *Nature* 416: 620–623.

Humler E, Langmuir C, and Daux V (1999) Depth versus age: New perspectives from the chemical compositions of ancient crust. *Earth and Planetary Science Letters* 173: 7–23.

Irifune T and Isshiki M (1998) Iron partitioning in a pyrolite mantle and the nature of the 410-km seismic discontinuity. *Nature* 392: 702–705.

Ito E and Katsura T (1989) A temperature profile of the mantle transition zone. *Geophysical Research Letters* 16: 425–428.

Jacobs J and Allan DW (1956) The thermal history of the Earth. *Nature* 177: 155–157.

Jacobs JA (1961) *Some Aspects of the Thermal History of the Earth*, 267p, The Earth Today, 1961.

Jarvis GT and Campbell IH (1983) Archean komatiites and geotherms – solution to an apparent contradiction. *Geophysical Research Letters* 10: 1133–1136.

Jarvis GT and McKenzie DP (1980) Convection in a compressible fluid with infinite Prandtl number. *Journal of Fluid Mechanics* 96: 515–583.

Jaupart C (1983) Horizontal heat transfer due to radioactivity contrasts: Causes and consequences of the linear heat flow-heat production relationship. *Geophysical Journal of the Royal Astronomical Society* 75: 411–435.

Jaupart C and Mareschal J-C (1999) The thermal structure of continental roots. *Lithos* 48: 93–114.

Jaupart C and Mareschal JC (2003) Constraints on crustal heat production from heat flow data. In: Rudnick RL (ed.) *Treatise on Geochemistry, The Crust*, vol. 3, pp. 65–84. New York: Permagon.

Jaupart C, Mann JR, and Simmons G (1982) A detailed study of the distribution of heat flow and radioactivity in New Hampshire (USA). *Earth and Planetary Science Letters* 59: 267–287.

Jaupart C, Francheteau J, and Shen X-J (1985) On the thermal structure of the southern Tibetan crust. *Geophysical Journal of the Royal Astronomical Society* 81: 131–155.

Javoy M (1999) Chemical Earth models. *Comptes Rendus de l'Académie des Sciences* 329: 537–555.

Jeffreys H (1962) *The Earth: Its Origin, History, and Physical Constitution*, 5th edn. Cambridge, UK: Cambridge University Press.

Joeleht TH and Kukkonen IT (1998) Thermal properties of granulite facies rocks in the Precambrian basement of Finland and Estonia. *Tectonophysics* 291: 195–203.

Jones MQW (1988) Heat flow in the Witwatersrand Basin and environs and its significance for the South African shield geotherm and lithosphere thickness. *Journal of Geophysical Research* 93: 3243–3260.

Katsura T, Yamada H, Shinmei T, *et al.* (2003) Post-spinel transition in $Mg_2SiO_4$ determined by *in-situ* X-ray diffractometry. *Physics of the Earth and Planetary Interiors* 136: 11–24.

Katsura T, Yamada H, Nishikawa O, *et al.* (2004) Olivine-Wadsleyite transition in the system $(Mg,Fe)SiO_4$. *Journal of Geophysical Research* 109: B02209 (doi:10.1029/2003JB002438).

Kellogg LH, Hager BH, and van der Hilst RD (1999) Compositional stratification in the deep mantle. *Science* 283: 1881–1884.

Ketcham RA (1996) Distribution of heat-producing elements in the upper and middle crust of southern and west central Arizona: Evidence from the core complexes. *Journal of Geophysical Research* 101: 13611–13632.

Kinzler RJ and Grove TL (1992) Primary magmas of mid-ocean ridge basalts. Part 2: Applications. *Journal of Geophysical Research* 97: 6907–6926.

Klein EM and Langmuir CH (1987) Global correlations of ocean ridge basalt chemistry with axial depth and crustal thickness. *Journal of Geophysical Research* 92: 8089–8115.

Kleine T, Münker C, Mezger K, and Palme H (2002) Rapid accretion and early core formation on asteroids and the terrestrial planets from Hf-W chronometry. *Nature* 418: 952–955.

Korenaga J (2003) Energetics of mantle convection and the fate of fossil heat. *Geophysical Research Letters* 30, doi:10.1029/2003GL016982.

Korenaga J (2006) Archean geodynamics and the thermal evolution of the Earth. In: Benn K, Mareschal JC, and Condie KC (eds.) *Geophysical Monograph Series: Archean Geodynamics and Environments*, vol. 164, pp. 7–32. Washington, DC: American Geophysical Union.

Kremenentsky AA, Milanovsky SY, and Ovchinnikov LN (1989) A heat generation model for the continental crust based on deep drilling in the Baltic shield. *Tectonophysics* 159: 231–246.

Kukkonen IT and Peltonen P (1999) Xenolith-controlled geotherm for the central Fennoscandian shield: Implications for lithosphere–asthenosphere relations. *Tectonophysics* 304: 301–315.

Labrosse S (2002) Hotspots, mantle plumes and core heat loss. *Earth and Planetary Science Letters* 199: 147–156.

Labrosse S (2003) Thermal and magnetic evolution of the Earth's core. *Physics of the Earth and Planetary Interiors* 140: 127–143.

Labrosse S (2005a) Heat flow across the core–mantle boundary. In: Gubbins D and Herrero-Bervera E (eds.) *Encyclopedia of Geomagnetism and Paleomagnetism*. New York: Springer.

Labrosse S (2005b) Energy source for the geodynamo. In: Gubbins D and Herrero-Bervera E (eds.) *Encyclopedia of Geomagnetism and Paleomagnetism*. New York: Springer.

Labrosse S, Poirier J-P, and Le Mouël J-L (1997) On cooling of the Earth's core. *Physics of the Earth and Planetary Interiors* 99: 1–17.

Lachenbruch AH (1970) Crustal temperature and heat production: Implications of the linear heat flow heat production relationship. *Journal of Geophysical Research* 73: 3292–3300.

Lachenbruch AH and Sass JH (1978) Models of an extending lithosphere and heat flow in the Basin and Range province. *Memoirs of the Geological Society of America* 152: 209–258.

Lambeck K (1977) Tidal dissipation in the oceans: Astronomical, geophysical, and oceanographic consequences. *Philosophical Transactions of the Royal Society of London Series A* 287: 545–594.

Langmuir CH, Goldstein SL, Donnelly K, and Su YK (2005) Origins of enriched and depleted mantle reservoirs. In: *Eos Transactions*, Fall Meeting Supplement, vol. 86, p. 1, American Geophysical Union, Abstract V23D–02.

Langseth MG, Le Pichon X, and Ewing M (1966) Crustal structure of the mid-ocean ridges, 5, heat flow through the Atlantic Ocean floor and convection currents. *Journal of Geophysical Research* 71: 5321–5355.

Lapwood ER (1952) The effect of contraction in the cooling of a gravitating sphere, with special reference to the Earth. *Monthly Notices of the Royal Astronomical Society Geophysics* (supplement 6): 402–407.

Lay T, Hernlund J, Garnero EJ, and Thorne MS (2006) A Post-Perovskite lens and D" heat flux beneath the central Pacific. *Science* 314: 1272–1276.

Lay T, Williams Q, and Garnero EJ (1998) The core–mantle boundary layer and deep Earth dynamics. *Nature* 392: 461–468 (doi:10.1038/33083).

Lenardic A, Moresi L-N, Jellinek M, and Manga M (2005) Continental insulation, mantle cooling, and the surface area of oceans and continents. *Earth and Planetary Science Letters* 234: 317–333.

Lister CRB (1977) Estimators for heat flow and deep rock properties based on boundary layer theory. *Tectonophysics* 41: 157–171.

Lister CRB, Sclater JG, Davis EE, Villinger H, and Nagahira S (1990) Heat flow maintained in ocean basins of great age: Investigations in the north equatorial west Pacific. *Geophysical Journal International* 102: 603–630.

Lister JR (2003) Expressions for the dissipation driven by convection in the Earth's core. *Physics of the Earth and Planetary Interiors* 140: 855–158.

Lister JR and Buffett BA (1995) The strength and efficiency of thermal and compositional convection in the geodynamo. *Physics of the Earth and Planetary Interiors* 91: 17–30.

Litasov K and Ohtani E (2002) Phase relations and melt compositions in CMAS-pyrolite-$H_2O$ system up to 25 GPa. *Physics of the Earth and Planetary Interiors* 134: 105–127.

Lyubetskaya T and Korenaga J (2007) Chemical composition of Earth's primitive mantle and its variance. Part 1: Method and results. *Journal of Geophysical Research* 112: B03211 (doi:10.1029/2005JB004223).

MacDonald GJF (1959) Calculations on the thermal history of the Earth. *Journal of Geophysical Research* 64: 1967–2000.

Mareschal JC and Jaupart C (2004) Variations of surface heat flow and lithospheric thermal structure beneath the North American craton. *Earth and Planetary Science Letters* 223: 65–77.

Mareschal JC, Jaupart C, Cheng LZ, et al. (1999) Heat flow in the trans-Hudson Orogen of the Canadian Shield: Implications for proterozoic continental growth. *Journal of Geophysical Research* 104: 29007–29024.

Mareschal JC, Jaupart C, Gariépy C, et al. (2000) Heat flow and deep thermal structure near the southeastern edge of the Canadian Shield. *Canadian Journal of Earth Sciences* 37: 399–414.

Mareschal JC, Poirier A, Rolandone F, et al. (2000) Low mantle heat flow at the edge of the North American continent Voisey Bay, Labrador. *Geophysical Research Letters* 27: 823–826.

Mareschal JC, Jaupart C, Rolandone F, et al. (2005) Heat flow, thermal regime, and rheology of the lithosphere in the Trans-Hudson Orogen. *Canadian Journal of Earth Sciences* 42: 517–532.

McDonough WF and Sun SS (1995) The composition of the Earth. *Chemical Geology* 120: 223–253.

McKenzie D and Bickle MJ (1988) The volume and composition of melt generated by extension of the lithosphere. *Journal of Petrology* 29: 625–679.

McKenzie DP and Richter FM (1981) Parameterized thermal convection in a layered region and the thermal history of the Earth. *Journal of Geophysical Research* 86: 11667–11680.

McKenzie DP and Weiss N (1975) Speculations on the thermal and tectonic history of the Earth. *Geophysical Journal of the Royal Astronomical Society* 42: 131–174.

McKenzie DP, Jackson J, and Priestley K (2005) Thermal structure of oceanic and continental lithosphere. *Earth and Planetary Science Letters* 233: 337–349.

Melosh HJ and Ivanov BA (1999) Impact crater collapse. *Annual Review of Earth and Planetary Sciences* 27: 385–415.

Mittlestaedt E and Tackley PJ (2006) Plume heat flow is much lower than CMB heat flow. *Earth and Planetary Science Letters* 241: 202–210.

Mojzsis SJ, Harrison TM, and Pidgeon RT (2001) Oxygen-isotope evidence from ancient zircons for liquid water at the Earth's surface 4,300 Myr ago. *Nature* 409: 178–181.

Morgan P (1983) Constraints on rift thermal processes from heat flow and uplift. *Tectonophysics* 94: 277–298.

Morgan P (1985) Crustal radiogenic heat production and the selective survival of ancient continental crust. *Journal of Geophysical Research* (Supplement 90): C561–C570.

Müller RD, Roest WR, Royer J-Y, Gahagan LM, and Sclater JG (1997) Digital isochrons of the world's ocean floor. *Journal of Geophysical Research* 102: 3211–3214.

Munk WH and MacDonald GJF (1960) *The Rotation of the Earth.* Cambridge, UK: Cambridge University Press.

Murakami M, Hirose K, Kawamura K, Sata N, and Ohishi Y (2004) Post-perovskite phase transition in $MgSiO_3$. *Science* 304: 855–858.

Newsom HE and Jones JH (eds.) (1990) *Origin of the Earth.* Oxford: Oxford University Press.

Nicolaysen LO, Hart RJ, and Gale NH (1981) The Vredefort radioelement profile extended to supracrustal rocks at Carletonville, with implications for continental heat flow. *Journal of Geophysical Research* 86: 10653–10661.

Nisbet EG, Cheadle MJ, Arndt NT, and Bickle MJ (1995) Constraining the potential temperature of the Archean mantle: A review of the evidence from komatiites. *Lithos* 30: 291–307.

Nyblade AA (1997) Heat flow across the east African Plateau. *Geophysical Research Letters* 24: 2083–2086.

Oganov AR and Ono S (2004) Theoretical and experimental evidence for a post-perovskite phase of $MgSiO_3$ in Earth's D" layer. *Nature* 430: 445–448.

Ogawa M, Schubert G, and Zebib A (1991) Numerical simulations of three-dimensional thermal convection in a fluid with strongly temperature dependent viscosity. *Journal of Fluid Mechanics* 233: 299–328.

Olson PA (1987) A comparison of heat transfer laws for mantle convection at very high Rayleigh numbers. *Physics of the Earth and Planetary Interiors* 48: 153–160.

Palme H and O'Neill HSC (2003) Cosmochemical estimates of mantle composition. In: Carlson RW (ed.) *Treatise on Geochemistry,* Vol 2: Mantle and Core, pp. 1–38. Amsterdam: Elsevier.

Pari G and Peltier WR (1995) The heat flow constraint on mantle tomography-based convection models: Towards a geodynamically self-consistent inference of mantle viscosity. *Journal of Geophysical Research* 100: 12731–12752.

Parsons B (1982) Causes and consequences of the relation between area and age of the ocean floor. *Journal of Geophysical Research* 87: 289–302.

Parsons B and Sclater JG (1977) An analysis of the variation of the ocean floor bathymetry and heat flow with age. *Journal of Geophysical Research* 82: 803–827.

Peltier WR and Jarvis GT (1982) Whole mantle convection and the thermal evolution of the Earth. *Physics of the Earth and Planetary Interiors* 29: 281–304.

Perry HKC, Jaupart C, Mareschal JC, and Bienfait G (2006) Crustal heat production in the Superior Province, Canadian Shield, and in North America inferred from heat flow data. *Journal of Geophysical Research* 111: B04401 (doi:10.1029/2005JB003893).

Pinet C and Jaupart C (1987) The vertical distribution of radiogenic heat production in the Precambrian crust of Norway and Sweden: Geothermal implications. *Geophysical Research Letters* 14: 260–263.

Pinet C, Jaupart C, Mareschal JC, Gariépy C, Bienfait G, and Lapointe R (1991) Heat flow and structure of the lithosphere in the Eastern Canadian shield. *Journal of Geophysical Research* 96: 19941–19963.

Poirier J-P (2000) *Introduction to the Physics of the Earth's Interior, 2 edn, Cambridge: Cambridge University Press.*

Pollack HN and Chapman DS (1977a) On the regional variations of heat flow, geotherms, and lithospheric thickness. *Tectonophysics* 38: 279–296.

Pollack HN and Chapman DS (1977b) Mantle heat flow. *Earth and Planetary Science Letters* 34: 174–184.

Pollack HN, Hurter SJ, and Johnston JR (1993) Heat flow from the Earth's interior: Analysis of the global data set. *Reviews of Geophysics* 31: 267–280.

Poort J and Klerkx J (2004) Absence of a regional surface thermal high in the Baikal rift; new insights from detailed contouring of heat flow anomalies. *Tectonophysics* 383: 217–241.

Ray RD, Eanes RJ, and Chao BF (1996) Detection of tidal dissipation in the solid Earth by satellite tracking and altimetry. *Nature* 381: 595–597.

Ringwood AE (1962) A model for the upper mantle. *Journal of Geophysical Research* 67: 857–867.

Roberts PH, Jones CA, and Calderwood AR (2003) Energy fluxes and ohmic dissipation in the Earth's core. In: Jones CA, Soward AM, and Zhang K (eds.) *Earth's Core and Lower Mantle*, pp. 100–129. London: Taylor & Francis.

Rolandone F, Jaupart C, Mareschal JC, et al. (2002) Surface heat flow, crustal temperatures and mantle heat flow in the Proterozoic Trans-Hudson Orogen, Canadian Shield. *Journal of Geophysical Research* 107: 2314 (doi:10.1029/2001JB000,698).

Rowley DB (2002) Rate of plate creation and destruction; 180 Ma to present. *Geological Society of America Bulletin* 114: 927–933.

Roy RF, Decker ER, Blackwell DD, and Birch F (1968) Heat Flow in the United States. *Journal of Geophysical Research* 73: 5207–5221.

Roy S and Rao RUM (2000) Heat flow in the Indian shield. *Journal of Geophysical Research* 105: 25587–25604.

Royer J, Müller R, Gahagan L, et al. (1992) A global isochron chart. *Technical Report* 117, University of Texas Institute for Geophysics, Austin, 1992.

Rubie DC, Melosh HJ, Reid JE, Liebske C, and Righter K (2003) Mechanisms of metal-silicate equilibration in the terrestrial magma ocean. *Earth and Planetary Science Letters* 205: 239–255.

Rudnick RL and Fountain DM (1995) Nature and composition of the continental crust: A lower crustal perspective. *Reviews of Geophysics* 33: 267–309.

Rudnick RL and Gao S (2003) Composition of the continental crust. In: Rudnick RL (ed.) *Treatise on Geochemistry, The Crust*, vol. 3, pp. 1–64. New York: Permagon.

Rudnick RL and Nyblade AA (1999) The thickness of Archean lithosphere: Constraints from xenolith thermobarometry and surface heat flow. In: Fei Y, Bertka CM, and Mysen BO (eds.) *Mantle Petrology: Field Observations and high pressure Experimentation: A Tribute to Francis R. (Joe) Boyd*, pp. 3–11. Houston, TX: Geochemical Society.

Rudnick RL, McDonough WF, and O'Connell RJ (1998) Thermal structure, thickness and composition of continental lithosphere. *Chemical Geology* 145: 395–411.

Rushmer T, Minarik WG, and Taylor GJ (2000) Physical processes of core formation. In: Canup RM and Drake K

(eds.) *Origin of the Earth and Moon*, pp. 227–243. Tucson, AZ: The University of Arizona Press.

Russell JK, Dipple GM, and Kopylova MG (2001) Heat production and heat flow in the mantle lithosphere, Slave Craton, Canada. *Physics of the Earth and Planetary Interiors* 123: 27–44.

Rybach L (1988) Determination of heat production rate. In: Haenel R, Rybach L, and Stegena L (eds.) *Handbook of Terrestrial Heat-Flow Density Determinations*, pp. 125–142. Amsterdam: Kluwer.

Schubert G, Stevenson D, and Cassen P (1980) Whole planet cooling and the radiogenic heat source contents of the Earth and Moon. *Journal of Geophysical Research* 85: 2531–2538.

Schubert G, Turcotte DL, and Olson P (2001) *Mantle Convection in the Earth and Planets*. Cambridge, UK: Cambridge University Press.

Schubert G and Young RE (1976) Cooling the Earth by whole mantle subsolidus convection: A constraint on the viscosity of the lower mantle. *Tectonophysics* 35: 201–214.

Sclater JG and Francheteau J (1970) Implications of terrestrial heat flow observations on current tectonic and geochemical models of crust and upper mantle of Earth. *Geophysical Journal of the Royal Astronomical Society* 20: 509–542.

Sclater JG, Crowe J, and Anderson RN (1976) On the reliability of oceanic heat flow averages. *Journal of Geophysical Research* 81: 2997–3006.

Sclater JG, Jaupart C, and Galson D (1980) The heat flow through oceanic and continental crust and the heat loss of the Earth. *Reviews of Geophysics and Space Physics* 18: 269–312.

Sharpe HN and Peltier WR (1978) Parameterized mantle convection and the Earth's thermal history. *Geophysical Research Letters* 5: 737–740.

Sleep NH (1971) Thermal effects of the formation of Atlantic continental margins by continental breakup. *Geophysical Journal of the Royal Astronomical Society* 24: 325–350.

Sleep NH (1979) Thermal history and degasing of the Earth: Some simple calculations. *Journal of Geology* 87: 671–686.

Sleep NH (1990) Hotspots and mantle plumes: Some phenomenology. *Journal of Geophysical Research* 95: 6715–6736.

Sleep NH (2000) Evolution of the mode of convection within terrestrial planets. *Journal of Geophysical Research* 105: 17563–17578.

Solomatov VS (2000) Fluid dynamics of a terrestrial magma ocean. In: Canup RM and Righter K (eds.) *Origin of the Earth and Moon*, pp. 323–338. Tucson, AZ: The University of Arizona Press.

Solomatov VS and Moresi LN (1997) Three regimes of mantle convection with non-newtonian viscosity and stagnant lid convection on the terrestrial planets. *Geophysical Research Letters* 24: 1907–1910.

Sotin C and Labrosse S (1999) Three-dimensional thermal convection of an isoviscous, infinite–Prandtl–number fluid heated from within and from below: Applications to heat transfer in planetary mantles. *Physics of the Earth and Planetary Interiors* 112: 171–190.

Stacey FD (1992) *Physics of the Earth, 2nd edn.*, Brisbane, Qld: Brookfield Press.

Stein CA and Stein S (1992) A model for the global variation in oceanic depth and heat flow with lithospheric age. *Nature* 359: 123–129.

Stevenson DJ (1989) Formation and early evolution of the Earth. In: Peltier W (ed.) *Mantle Convection*, pp. 817–873. NY: Gordon and Breach.

Stevenson DJ (1990) Fluid dynamics of core formation. In: Newsom HE and Jones JH (eds.) *Origin of the Earth*, pp. 231–249. New York: Oxford University Press.

Su YJ (2000) Mid-ocean ridge basalt trace element systematics: Constraints from database management, ICPMS analysis, global data compilation and petrologic modeling. Unpub. Ph.D. thesis, Columbia University, New York, 1, 569p, 2000.

Swanberg CA, Chessman MD, Simmons G, Smithson SB, Gronlie G, and Heier KS (1974) Heat-flow heat-generation studies in Norway. *Tectonophysics* 23: 31–48.

Tackley PJ (2000) Self-consistent generation of tectonic plates in time-dependent, three-dimensional mantle convection simulations. Part 1: pseudoplastic yielding. *Geochemistry Geophysics Geosystems* 1, doi:10.1029/2000GC000036.

Thomson W (Lord Kelvin) (1862) On the secular cooling of the Earth. *Transactions of the Royal Society of Edinburgh* 23: 295–311.

Turcotte DL and Oxburgh ER (1967) Finite amplitude convective cells and continental drift. *Journal of Fluid Mechanics* 28: 29–42.

Vasseur G and Singh RN (1986) Effects of random horizontal variations in radiogenic heat source distribution on its relationship with heat flow. *Journal of Geophysical Research* 91: 10397–10404.

Vogt P and Ostenso N (1967) Steady state crustal spreading. *Nature* 215: 810–817.

von Herzen RP (1959) Heat flow values from the southwestern Pacific. *Nature* 183: 882–883.

Vosteen HD, Rath V, Clauser C, and Lammerer B (2003) The thermal regime of the Eastern Alps from inverse analyses along the TRANSALP profile. *Physics and Chemistry of the Earth* 28: 393–405.

Vočadlo L, Alfè D, Gillan MJ, and Price GD (2003) The properties of iron under core conditions from first principles calculations. *Physics of the Earth and Planetary Interiors* 140: 101–125.

Wasserburg G, McDonald GJF, Hoyle F, and Fowler (1964) Relative contributions of uranium, thorium, and potassium to heat production in the Earth. *Science* 143: 465–467.

Wetherill GW (1990) Formation of the Earth. *Annual Review of Earth and Planetary Sciences* 18: 205–256.

Williams DL and von Herzen RP (1974) Heat loss from the Earth: New estimate. *Geology* 2: 327–328.

Workman RK and Hart SR (2005) Major and trace element composition of the depleted MORB mantle (DMM). *Earth and Planetary Science Letters* 231: 53–72.

Zerr A, Diegeler A, and Boehler R (1998) Solidus of Earth's deep mantle. *Nature* 281: 243–246.

Zhong SJ (2005) Dynamics of thermal plumes in three-dimensional isoviscous thermal convection. *Geophysical Journal International* 162: 289–300.

Zschau J (1986) Constraints from the Chandler wobble period. In: Anderson AJ and Cazenave A (eds.) *Space Geodesy and Geodynamics*, pp. 315–344. New York: Academic Press.

# 7 The Dynamics and Convective Evolution of the Upper Mantle

**E. M. Parmentier**, Brown University, Providence, RI, USA

## 7.1 Introduction

The thermal structure of the oceanic upper mantle plays an important role in the larger-scale dynamics of the mantle. Fortunately, this region of the Earth is one that is accessible to observation by a variety of geophysical tools and so can provide a basis for understanding solid-state convective instability, with implications for even more general settings than the Earth's upper mantle. It can be argued that the Earth's upper mantle is the best natural laboratory available to us to study the convective evolution of planetary interiors.

The goal of this chapter is to summarize current understanding of the evolution of the upper mantle based on observations and theoretical predictions, beginning with some of the earliest observations on how cooling the mantle from above would be expressed in seafloor depth, heat flow, and geoid height as functions of age. It is not possible to consider the role of convective instability in the upper mantle in isolation from other forms of larger-scale mantle

flow. While no general agreement has yet emerged on how various scales of convection may affect the upper mantle, small-scale thermal convective instability generated within the upper mantle provides a viable explanation for many important observations. Convection in the upper mantle may be driven by buoyancy resulting both from cooling at the surface and by melting. Mantle cooling is expressed in oceanic heat flow and in geoid height and seafloor depth through the effect of cooling on mantle density.

Buoyancy due to melting may be expressed in intraplate volcanism. Only a few long-lived, age progressive, linear volcanic chains meet all the criteria of the deep-mantle plume-fixed hot-spot hypothesis. Decompression melting in small-scale thermally driven convective upwellings has been one favored explanation for volcanic ridges aligned with plate motion. Seafloor volcanism may also reflect spontaneously generated instability driven by decompression melting, which has been termed 'magmatic convective storms'. Buoyant decompression melting beneath moving plates is a possible mechanism for the abundant short-lived island chains and volcanic ridges identified on the Earth's seafloor.

Finally, the role of mantle upwelling and melting beneath spreading centers is important to consider, in part because it is responsible for the generation of the oceanic crust and is a factor controlling spreading center structure. However, melting and melt extraction under spreading centers leave behind residual mantle that is transported away from the spreading center. Compositional changes are expected to affect viscosity or creep rate and create a stable chemical stratification, thus fundamentally influencing convective instability everywhere in the upper mantle.

## 7.2    Cooling the Mantle from Above

### 7.2.1    A Historical Perspective

Turcotte and Oxburgh (1967) identified the moving lithospheric plates as the conductive thermal boundary layer at the top of mantle convection cells, thus establishing a physical relationship between upper-mantle thermal structure and mantle dynamics. During the emergence of plate tectonics, measurements of seafloor depth and heat flow in the oceans provided the first direct evidence on the thermal structure of the upper mantle. Heat flow generally decreased with crustal age but showed great variability, particularly at young ages. This variability is now understood to be due primarily to hydrothermal circulation of seawater through permeable crustal rocks. In contrast, seafloor depths showed a much more systemic variation with seafloor age. The age dependence of isostatic seafloor depth, which depends on the depth-averaged temperature, was recognized as a stronger observational constraint than heat flow on upper-mantle thermal structure (Langseth *et al.*, 1966; McKenzie, 1967; Vogt and Ostenso, 1967; McKenzie and Sclater, 1969; Sleep, 1969).

The thermal structure that develops due to conductive cooling should be, at least to first order, only a function of crustal age, rather than an independent function of both distance from spreading center and plate velocity. Observations of seafloor depth and age generally confirmed this age dependence. The Turcotte and Oxburgh boundary layer hypothesis, in the simple form suggested originally, indicated that the thermal boundary layer should continue to thicken as the square root of age so that old seafloor would continue to subside. This did

not appear to explain observations showing a relatively uniform depth of old seafloor which required that the thermal boundary layer evolve to a nearly constant thickness. Models for upper-mantle thermal structure that were consistent with average depth of seafloor as a function of age from the growing collection of observations treated the conductive cooling of horizontally moving upper mantle, as in the simple boundary layer theory, but with the added assumption of a uniform temperature at a prescribed depth (Langseth *et al.*, 1966; McKenzie, 1967). A relatively uniform seafloor depth at old ages requires a mechanism to transport heat to bottom of the thermal boundary layer, thus reducing the rate at which it thickens. This set the stage for research in the following several decades and is still a source of continuing study and debate.

### 7.2.2    Convective Instability versus Heating by Hot Spots – Depth–Age and Heat flow

Both small-scale convective instability of the cool thermal boundary and heating at hot spots (*see* Chapter 9) may play a role in transporting heat to the bottom of the thermal boundary layer, but no general consensus has yet emerged on a single mechanism of heating that can explain all available observations. Ideally, observed variations in seafloor depth might constrain the relative importance of these two mechanisms, but such interpretations are limited by the relatively small amount of old seafloor with a well-determined age and by the uncertainty in correcting for the thickness and load of sediments. Simply averaging all depth measurements at a given age indicates that old seafloor is, on average, shallower than would be predicted by conductive cooling alone. Johnson and Carlson (1992; see also Carlson and Johnson (1994)) report a recent assessment of this averaged depth–age relationship using data at sites where drilling has reached basement and for which the age is well determined. As shown in **Figure 1**, old seafloor appears, on average, to be at least several hundred meters shallower than the depth predicted by the purely conductive cooling models that fits very well at relatively young ages. So the question is why old seafloor is often shallower than predicted by conductive cooling and how we might infer the physical processes responsible.

Heating due to hot spots should correlate with the length of time that a given piece of seafloor has spent in the proximity of a hot spot (Crough, 1975).

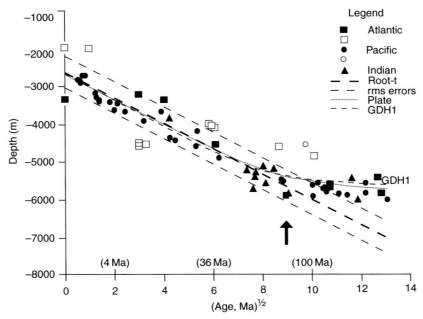

**Figure 1** Depth–age for various oceans, based on sediment and crustal thickness corrected seafloor depths (Johnson and Carlson, 1992), compared to conductive half-space, and plate models with 120 km thick plate and 1350°C mantle temperature (the plate model) and a 95 km thick plate and 1450°C mantle (the GDH1 model).

Heestand and Crough (1981) sorted the depth of seafloor in the North Atlantic by distance from the nearest hot-spot track. Comparing depth–age for seafloor in constant distance ranges indicated no flattening at old ages. In the South Atlantic, where hot-spot influences should be less significant, Hayes (1988) noted that depth followed a purely conductive cooling curve to ages approaching 120 My, but noted the complication of a persistent asymmetry in apparent subsidence rate across the spreading axis. An asymmetry in apparent subsidence rate was also found in the Southeast Indian Ocean (Hayes, 1988) and across the East Pacific Rise (EPR) (Eberle and Forsyth, 1995). One possibility may be that this asymmetry reflects the dynamic topography of larger-scale mantle flow on which the depth–age subsidence is superimposed.

The Pacific Plate contains not only relatively large areas of old seafloor but also numerous seamounts and hot-spot tracks (e.g., Wessel and Lyons, 1997). Schroeder (1984) compiled depth–age data for the Pacific. Eliminating data within 800 km of hot-spot tracks resulted in a good correlation of depth with square root of age for ages less than 80 My. All seafloor older than 80 My was shallower than predicted by conductive cooling alone and was within 800 km of hot spots or hot-spot tracks. Renkin and

Sclater (1988) analyzed the effect of uncertainties in basement age and sediment thickness on depth–age correlations in the North Pacific. Arguing that volcanic constructs always result in shallower seafloor, they proposed that modal seafloor depth, or more precisely a range enclosing two-thirds of the measured depths at any age, provides an estimate of subsidence associated with cooling least biased by volcanism. Modal depths increase with age more slowly than predicted by conductive cooling for ages exceeding about 80 My. Plotting modal depths with age along a corridor of Pacific seafloor containing several swell-like features, Renkin and Sclater argue that not even the deepest depths fall along a conductive cooling model.

Small-scale convective instability of the thermal boundary layer, in addition to hot spots, may advectively transport heat to the bottom of the conductive boundary layer. The possible importance of small-scale convection was first suggested on the basis of averaged depth–age curves that deviate from purely conductive cooling at a seafloor age of about 70–80 My, which was presumed to correspond to the onset of convective instability (Parsons and McKenzie, 1978). Beyond this age, convective heat transfer was thought to maintain the nearly constant thermal boundary layer thickness implied by the plate model (Davis and Lister, 1974; Parsons and Sclater, 1977; Stein and Stein, 1992). Plate thickness must be consistent with the heat flow measured on old seafloor (Davis *et al.*, 1984; Lister *et al.*, 1990; Nagihara *et al.*, 1996), as well as its depth. **Figure 2** from Nagihara *et al.* (1996) shows heat flow values for old seafloor in the northwestern Pacific and in the western North Atlantic, both areas where high-quality heat flow and data needed to correct depth for sediment loading and crustal thickness variations are available. Also shown are predictions from two versions of the plate model and purely conductive cooling, all of which are thought to provide reasonable fits to the depth–age relationship of younger seafloor.

Heat flow is higher and seafloor is consistently shallower than predicted by conductive cooling. The plate model of parameters of Parsons and Sclater (1977) with a plate thickness 125 km and a mantle temperature 1350°C fits the depths well but underestimates the heat flow. In contrast, the hotter and thinner plate model of Stein and Stein (1992) with a plate thickness 95 km and a mantle temperature 1450°C fits the heat flow well but underestimates old seafloor depth. It is worth noting that this mantle temperature is significantly higher than that estimated from melting in adiabatically upwelling mantle beneath presently active spreading centers, frequently inferred to be 1325–1350°C.

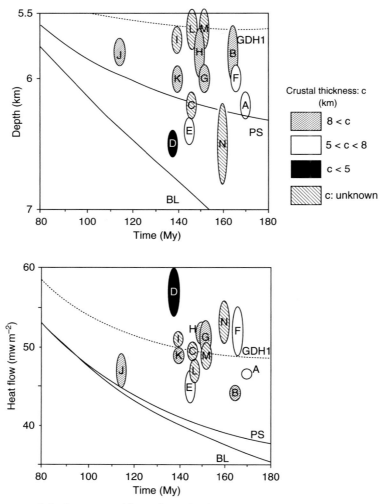

**Figure 2**  Seafloor basement depth, corrected for sediment load and crustal thickness where available (top) and heat flow (bottom) at sites in the western North Pacific (A–F) and northwestern Atlantic (G–N) as functions of age from Nagihara *et al.* (1996). The location of specific sites is identified in table 1 and figure 1 of Nagihara, *et al.* Model curves for the plate models (PS and GDH1) and half-space cooling (BL).

### 7.2.3 Geoid Height as a Measure of Upper-Mantle Thermal Structure

As discussed above, the earliest interpretations of the thermal evolution of the upper mantle relied largely on heat flow and isostatic seafloor depth, with the isostatic assumption justified by the absence of free-air gravity anomalies correlated with the depth–age variation. Following the advent of sea-surface elevation measurements from satellite altimetry, the long wavelength geoid and gravity provided further constraint on upper-mantle thermal structure (Chase, 1985; Smith, 1998). Isostatic seafloor depth measures the density averaged over depth in the mantle; geoid height measures the first moment of the density distribution with depth (Haxby and Turcotte, 1978). Conductive cooling predicts that geoid height decreases linearly with age (Haxby and Turcotte, 1978) as they observed along a Geos3 altimetry track in the North Atlantic. Also using early Geos3 altimetry data, Sandwell and Schubert (1980) found that geoid height decreases approximately linearly with seafloor age for ages less than about 80 My in the Atlantic and southeast Indian spreading centers with geoid slopes in the range $-0.131$ to $-0.149 \, \text{m My}^{-1}$, generally consistent with a conductive cooling thermal boundary layer. In constrast, the geoid height predicted by the plate model should flatten over old seafloor. Sandwell and Schubert (1980) found that geoid height in the southeast Pacific was not consistent with a simple linear geoid–age relationship, but decreased rapidly with age. The decrease was much more rapid than predicted by the plate model used to explain flattening of old seafloor. This is further discussed below.

The geoid height is sensitive to density variations over much larger distances than gravity anomalies. Therefore, density variations throughout the Earth, not just in the upper mantle, contribute to geoid height variations. The difficulty in using geoid as a constraint on the age dependence of the upper mantle is to separate variations in geoid height with seafloor age from other contributions, for example, from oceanic swells or hot spots. One approach to filtering out the long wavelength contributions is to measure the change in geoid height across fracture zones which juxtapose seafloor of different ages (Crough, 1979a; Detrick, 1981). If geoid height decreases linearly with age as expected for conductive cooling, the change in geoid height across a fracture zone of constant age offset should remain constant as seafloor on each side of the fracture zone ages. In constrast, the plate model

should show a change in geoid height across the fracture zone that decreases with age and which vanishes once thermal boundary layer thickness becomes constant. Observations suggest that geoid slope decreases with age, but it appears to do so at a much earlier age than that at which the seafloor flattens. Studies of variations of geoid height with age include Cazenave (1984), Driscoll and Parsons (1988), and Freedman and Parsons (1990). Geoid height as a function of age from Cazenave (1984), plotted as geoid–age slope, is shown in **Figure 3**.

Determining the geoid height change with age using measurements across fracture zones is elegant in concept but difficult in practice. Mechanical behavior of lithosphere, both plate flexure associated with differential subsidence (Sandwell and Schubert, 1982a, 1982b) and thermal bending (Parmentier and Haxby, 1985; Wessel and Haxby, 1990), create uncompensated seafloor topography over distances of 50–100 km adjacent to fracture zones. The resulting geoid anomalies may obscure the identification of the change in isostatic geoid height. Even given these

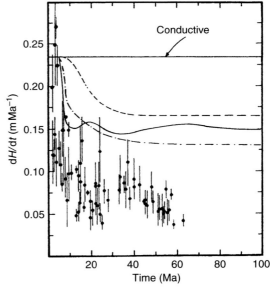

**Figure 3** Geoid–age slope as a function of age with data from Cazenave (1984). Pure conductive cooling leads to the constant geoid–age slope as shown. The observations indicate a rapid decline with age. Curves show model predictions from Buck and Parmentier (1986) for convective cooling with an activation energy of $420 \, \text{kJ mol}^{-1}$ and an activation volume of $10 \, \text{cm}^3 \, \text{mol}^{-1}$. Dashed and long-short dashed curves are for a model reference viscosity of $10^{18}$ and $5 \times 10^{18} \, \text{Pa s}$, respectively. The solid curve is also for a $10^{18} \, \text{Pa s}$ viscosity but in a wider and more finely discretized model domain.

uncertainties, geoid height variation with age is not explained by either the half-space cooling or the plate model mantle thermal structure.

### 7.2.4   Seismic Velocity and Attenuation as Measures of Upper-Mantle Structure and Flow

The age dependence of the apparent elastic thickness of the oceanic lithosphere might also constrain the variation of mantle thermal structure with age. While the focal depths of oceanic intraplate earthquakes increase with age following a 650–700°C isotherm in a conductive cooling model (Bergman and Solomon, 1984; Weins and Stein, 1984), most of the seismicity reported occurs at ages less than 70 My. Flexure of lithosphere at trenches should also reflect the mechanical thickness. Levitt and Sandwell (1995) indicate that plate thickness increases with age of subducting lithosphere and that the plate in the Stein and Stein (1992) thermal model may be too hot and thin. However, the data do not resolve the difference between simple conductive cooling and a plate model with a thicker plate.

Seismic velocities and attenuation in the upper mantle also constrain thermal structure and its age evolution. Surface wave dispersion has been used to infer the velocity variation with age and depth (Nishimura and Forsyth, 1989; Ritzwoller *et al.*, 2004; Priestley and McKenzie, 2006; Maggi *et al*, 2006; Weeraratne *et al.*, 2007; Zhou *et al.*, 2006). As shown in **Figure 4**, seismic shear-wave velocities in the upper mantle generally decrease with depth at a given age and increase with age at a given depth. Seismic velocities continue to change with age at depths exceeding 150 km. This change, if attributable to temperature, clearly contradicts the plate model which implicitly assumes that mantle temperature at depths greater than the plate thickness do not change with age.

Seismic velocity distributions like those in **Figure 4** they may also constrain the depth to which convective instability has cooled the mantle. Early studies like that of Nishimura and Forsyth (1989) probably do not resolve variations deeper than ~200 km. More recent studies using longer period (and consequently longer wavelength) surface waves (Maggi *et al*, 2006; Zhou *et al.*, 2006), which should resolve velocity variations to greater depth, indicate variations deeper than 200 km. However, the extent to which these variations correlate with plate age has not been established.

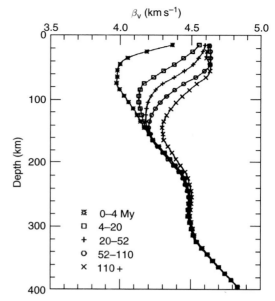

**Figure 4**   Vertically polarized shear-wave velocity in the Pacific as a function of depth in various age intervals determined from Rayleigh wave dispersion (Nishimura and Forsyth, 1989). Note that shear-wave velocity changes with age at depths greater than would be consistent with a plate model that fits the depth–age relationship (≤125 km). Changes with age continue beneath seafloor with ages exceeding 110 My.

Since the pressure, or depth, of phase changes in the upper mantle depends on temperature, temperature variations should be reflected in variations in the thickness of the mantle transition zone. Using travel times of converted phases and SS precursors, the results of Lawrence and Shearer (2006; figure 6b) show systematic westward thickening of the transition zone in the Pacific, correlating with increasing plate age. This could imply temperature variations associated with convective instabilities beneath cooling plates may extend to depths of at least the top of the transition zone, ~440 km deep.

The origin of the seismically defined low-velocity, high-attenutation asthenosphere is not fully resolved. The possible effects of small amounts of melt have been frequently invoked but recent studies indicate that the presence of melt may not be required (Stixrude and Lithgow-Bertelloni, 2006; Priestley and McKenzie, 2006), based on new understanding of the factors controlling shear-wave velocities at seismic frequencies now emerging. Stress relaxation at grain boundaries is thought to be an important anharmonic effect that introduces a frequency–grain size variation and a temperature dependence to the elastic moduli in addition to that of high-frequency (harmonic) variations with temperature and pressure (Gribb and Cooper, 1998; Faul and Jackson, 2005). Grain-size variations could thus contribute to the velocity variation with depth like those shown in **Figure 4**. Using the parametrizations of Priestley and McKenzie (2006) as a first attempt to assess the role of temperature alone, the velocity change of $0.1\,\mathrm{km\,s^{-1}}$ at 150 km depth between 52 and 110 My and >110 My would require cooling in the range of 100–200°C.

Seismic anisotropy in the upper mantle also provides a possible indication of mantle flow. Azimuthal anisotropy has been long recognized (cf. Nishimura and Forsyth, 1989), but its depth variation and the relative contributions from the lithosphere and asthenosphere are not yet resolved. In a simple unidirectional parallel flow, the seismic fast direction is usually assumed to lie close to the shear plane and in the flow direction. In more complex flows, seismically fast direction has been assumed to coincide with the local direction of maximum accumulated elongation. However, the strength and local direction of anisotropy will be determined by the rates at which preferred orientation of mineral grains is created and destroyed by shear and dynamic recrystallization, respectively (Kaminski and Ribe, 2001, 2002; Wenk and Tomé, 1999; Chastel et al., 1993). Several recent global tomography studies have also reported radial anisotropy that is more pronounced beneath the Pacific Plate than elsewhere in the oceanic upper mantle (e.g., Ekstrom and Dziewonski, 1998; Gaboret et al., 2003). Could this radial anisotropy be a consequence of seismically fast directions vertically aligned in the upwellings and downwellings of convective motion? Hopefully, questions of this type can be resolved by continuing study.

### 7.2.5 Upper-Mantle Electrical Conductivity–Water Content and Temperature

Long-period magnetotelluric studies of mantle electrical structure beneath the eastern North Pacific Ocean can detect electrical conductivity structure between 150 and 1000 km (Lizarralde et al., 1995). Interpretation of this data reveals a conductive zone between 150 and 400 km depth with continuously decreasing conductivity at greater depth. Mantle conductivity in this region is comparable to that in the Basin and Range Province and much higher than

that in the Canadian Shield. High conductivities could be explained by the presence of gravitationally stable partial melt. Alternatively, conductivity estimates on measurements of hydrogen solubility and diffusivity in olivine can explain the high conductivities observed. This reinforces the possible role of water in controlling the physical properties, and particularly the rheology, of the upper mantle (Hirth and Kohlsedt, 1996; Hirth et al., 2000). Several recent studies measuring the electrical conductivity of wet olivine (Yoshino et al., 2006; Wang et al., 2006) reach opposing conclusions on whether intragranular water (H-defects) alone can explain observed conductivities or whether partial melt is required.

### 7.2.6 The Age of Thermal Convective Instability beneath Oceanic Lithosphere

In global gravity anomaly maps derived from Seasat altimetry data, Haxby noticed gravity lineations in the Pacific with 150–200 km wavelength aligned in the direction of plate motion as shown in **Figure 5**. Haxby and Weissel (1986) suggested that these gravity lineations were a consequence of mantle convection currents organized by plate motions. In a sheared fluid layer with an initial linear temperature gradient, corresponding to heating from below and cooling from above, Richter (1973) and Richter and Parsons (1975) showed that convective instability should take the form of convective rolls aligned with shearing due to plate motion. The presence of gravity lineations due to convective instability beneath lithosphere only a few million years old contrasted with the view that convective instability at ages ~70 My explained the flattening of old seafloor.

Motivated by the observed gravity anomalies, numerical solutions for steady-state finite amplitude thermal convection showed that convective instability at young ages could be consistent with mantle thermally activated creep rheology (Buck, 1985) and estimated the convective heat fluxes required to explain the observed gravity anomalies (Lin and Parmentier, 1985). Numerical solutions for the development of thermal convection in a fluid layer cooled from above, with a plausible range of rheological parameters, predicted the effect of convective instability on geophysical observables (Buck and Parmentier, 1986). This study assumed that stresses generated by cool mantle sinking from the bottom of the unstable thermal boundary layer did not contribute to seafloor topography and resulting geoid

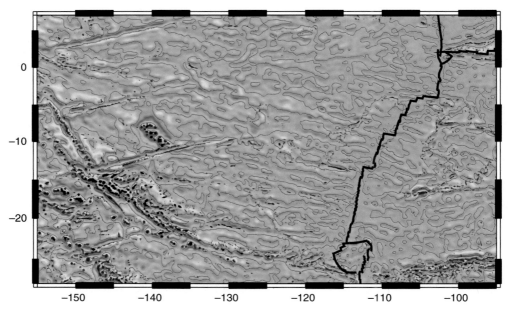

**Figure 5**   Haxby gravity lineations dipicted in this satellite gravity image by Sandwell and Fialko (2004). Band-pass filtered gravity anomaly (80 km < $\lambda$ < 600 km) derived from retracked satellite altimeter data. Color scale saturates at $\pm$15 mGal. Gravity lineaments with 140 km wavelength develop between the ridge axis and 6 Ma and are oriented in the direction of absolute plate motion. Lineaments on older seafloor have somewhat longer wavelength ($\sim$180 km) and cross the grain of the seafloor spreading fabric. Gravity lineaments also occur on the plate to the east of the East Pacific Rise.

anomalies, implying that this negatively buoyant mantle was supported by higher-viscosity mantle at depth. Robinson and Parsons (1988) showed that this is a reasonable approximation if mantle viscosity increases sufficiently with depth. If this were not so, convective instability, which increases the rate of cooling of mantle columns, would cause more rapid seafloor subsidence rather than reduced subsidence as previously assumed (cf. O'Connell and Hager, 1980). The calculated geoid anomaly showed a rapid decline in the geoid–age slope as convective instability developed and comparing qualitatively with observed variations in geoid slope with age shown in **Figure 3**. Other more recent studies argue that convective motions that develop at young ages may explain depth and geoid–age data better than plate model advocates have assumed. Doin and Fleitout (1996) showed that a uniform heat flux supplied to the bottom of the thermal boundary layer, presumably by convective motions, explains depth–age, as well as a plate model, and also the rapid decline in the geoid–age slope at young ages.

Haxby and Weissel (1986) suggested the presence of longer wavelengths of small-scale convection beneath older regions of the Pacific. Wessel *et al.* (1996) identified gravity undulations with wavelengths of $\sim$280 and 1000 km that form small circles about the current pole of Pacific Plate motion. The shorter wavelengths would correspond to the Haxby gravity lineations. The longer wavelengths could reflect the spacing of hot-spot tracks or a larger scale of thermal boundary convective instability corresponding to that usually invoked to explain the plate model. The earlier study of Cazenave *et al.* (1995) had also indicated a wavelength of about 1000 km. Mantle seismic tomography using ray paths along a profile connecting Tonga and Hawaii identified seismic velocity anomalies with a spacing of about 1500 km that correlated with variations in gravity and bathymetry (Katzman *et al.*, 1998). As the depth of a convectively cooled mantle layer increases with time, simple dimensional and scaling arguments suggest that the wavelength of convective motions should increase, a behavior seen in early studies of convective instability (Buck and Parmentier, 1986), but more recent studies, summarized below, further address this.

The Haxby gravity lineations are subtle features of the gravity field, generally much smaller than gravity anomalies due to fracture zones, for example. Current Pacific and Nazca Plate motion is oblique to fracture zones so that lineations aligned with plate motion cut obliquely across them. For other plates that move nearly parallel to fracture zones within them, gravity lineations of comparable amplitude to those in the Pacific would be difficult to detect, so as to identify the possibility convective instability at young ages requires other evidence. If convective instability occurs beneath the fast moving Nazca and Pacific Plates created at the fast spreading EPR does it also occur beneath more slowly moving plates created by slower spreading? As discussed above, Sandwell and Schubert (1980) found that geoid height decreases approximately linearly with age for ages less than about 80 My for the Atlantic and southeast Indian spreading centers with geoid–age slopes comparable to that expected for purely conductive cooling, suggesting that small-scale convective instability at young ages does not occur beneath these more slowly spreading plates.

Previous studies have often treated either hot spots or small-scale convective instability as the mechanism of lithospheric heating in all ocean basins. However, one single mechanism of heat transfer need not be responsible for heating the lithosphere everywhere. With due caution concerning the nonlinearity of thermal convection (meaning that simply adding heat fluxes due to hot spots and small-scale convection may not be valid), it may even be reasonable to think that both mechanisms operate simultaneously but to different degrees beneath different ocean basins. Gravity lineations at young ages are visible only on the Pacific and Nazca Plates, suggesting that convective instability at young ages may develop under these fast moving plates. In the South Atlantic, where hot-spot tracks are sparse and gravity lineations are not detectable, root-age subsidence to ages exceeding 100 My would be consistent with purely conductive mantle cooling beneath this slower moving plate. Seafloor flattening at old ages in the North Atlantic could be due to heating by numerous hot spots.

### 7.2.7 Implications from Recent Theoretical and Experimental Studies of Convective Instability

Convective instability of a thermal boundary layer has been examined in numerous studies using a range of analytical methods (see for example, Yuen and Fleitout (1984) and Marquart et al. (1999)) as well as both laboratory and numerical experiments (see Chapters 3 and 5), the latter taking advantage of advances in computer speed and memory, to better

understand the development of convective instability in a time-dependent basic state (conductive cooling) and temperature-dependent viscosity. The strong temperature dependence of mantle viscosity presents a significant challenge for numerical experiments and emphasizes the important continuing role of laboratory experiments. Davaille and Jaupart (1994) derived scaling laws for convective onset time (age) and heat flow in a viscous fluid with strongly temperature-dependent viscosity cooled from above.

A convective heat flux $f$ independent of the depth of the convecting fluid, implying that the boundary thickness is small compared to the fluid depth, is given by

$$f = Ck\left(\frac{\rho\alpha g}{\mu\kappa}\right)^{1/3}\Delta T_c^{4/3}$$

where k, $\kappa$, $\alpha$, and $\rho$ are the thermal conductivity, thermal diffusivity, thermal expansion coefficient, and density, respectively, and $C = 0.16$ is a dimensionless constant determined from laboratory or numerical experiments. This expression for the heat flux follows directly from dimensional analysis. Davaille and Jaupart (1994) showed that the convecting thermal boundary layer coincided with a 10-fold increase in viscosity, corresponding to temperature decrease $\Delta T_c$. For thermally activated creep

$$\mu = \mu_m\exp\left[\frac{Q}{R}\left(\frac{1}{T}-\frac{1}{T_m}\right)\right]$$

where $\mu_m$ and $T_m$ correspond to the temperature and viscosity beneath the thermal boundary layer. Then

$$\Delta T_c = 2.24\frac{RT_m^2}{Q}$$

This heat flux scaling and the values of $C$ is also confirmed by numerical experiments (Grasset and Parmentier, 1998). Heatflux as a function of $\mu_m$ and $Q$ is shown in **Figure 6**. In the mantle, a heat flux due to small scale convection comparable to that which would be required by the plate model ($\sim$40 mW m$^{-2}$) and creep activation energies from $Q = 250$ to $600$ kJ mol$^{-1}$ would require asthenosphere viscosities between $3\times10^{18}$ and $4\times10^{17}$ Pa s.

The small scale of thermal structures and convective motions in the mantle predicted by laboratory and numerical experiments is important to appreciate. For a convective heat flux beneath old lithosphere of $\sim$40 mW m$^{-2}$ and $\Delta T_c \sim 100°C$, the convecting thermal boundary layer thickness

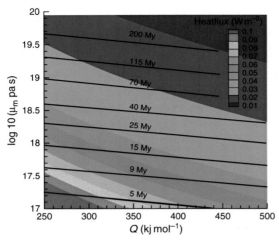

**Figure 6** Contours of mantle heat flux beneath old oceanic lithosphere (based on heat flux scaling of Davaille and Jaupart (1994)) and onset time (age) of convective instability (from Zaranek and Parmentier, 2004) as a function of mantle viscosity and activation energy for thermally activated creep. For mantle heat flow below old oceanic crust of $\sim$50 mW m$^{-2}$ (see **Figure 2**), onset of convection at young ages should occur if the creep activation energy is sufficiently high and mantle viscosity sufficiently low.

$\delta_c \sim k\Delta T_c/f$ is less than 10 km, and convective instability of the boundary layer generates cold downwellings of comparable width.

The onset time of convective instability can also be determined from both laboratory and numerical experiments. For $Q = 250\,\mathrm{kJ\,mol^{-1}}$ and a mantle temperature of 1300°C, Davaille and Jaupart predicted the onset time from their laboratory-derived scaling to be in the range 52–65 My. High-resolution numerical experiments examining the onset of convection in a fluid with strongly temperature-dependent viscosity have been reported in several recent studies (Huang et al., 2003; Korenaga and Jordan, 2003a; Zaranek and Parmentier, 2004). These studies predict shorter onset times than earlier studies. Since convective motions, even those restricted to the upper mantle, must allow for scales as large as 1000 km, the small scales of convective instability indicated above would be difficult to resolve numerically. An understanding of the scaling of onset time with temperature dependence is needed to extrapolate well-resolved numerical experiments to the very stronger temperature dependences expected in the mantle. Recent numerical experiments, which are all in good agreement with each other, indicate a scaling for onset time that differed from that proposed by

Davaille and Jaupart (2004). The onset time scaling of Zaranek and Parmentier (2004), for example, leads to the predictions shown in **Figure 6**.

In **Figure 6**, contours of heat flux provided to the base of the conductive lithosphere by small-scale convection in color shading and onset time in solid contours are plotted as a function of the creep activation energy $Q$ and mantle viscosity $\mu_m$ beneath the thermal boundary layer. The upper-mantle viscosity mantle beneath spreading centers and other convectively active areas is not expected to exceed about $10^{19}$ Pa s (Bills, 1994; Passey, 1981; Sigmundsson, 1991). Estimates based on laboratory rheological measurements suggest an even lower viscosity of $10^{18}$ Pa s (Hirth and Kohlstedt, 1996). If small-scale convection provides all the heat flux to old seafloor ($\sim$45–50 mW m$^{-2}$, see **Figure 2**) and if an upper-mantle viscosity of $10^{18}$ Pa s is assumed, convective instability would occur at ages of $\sim$15 My for $Q = 350$–400 kJ mol$^{-1}$, values intermediate between diffusion ($\sim$300 kJ mol$^{-1}$) and dislocation creep ($\sim$500 kJ mol$^{-1}$) in olivine.

The onset of convection at ages $\leq$10 My would require $\mu_m$ below $10^{18}$ Pa s and $Q$ near the value for dislocation creep in olivine. For convective instability at 70 My with a convective heat flux of 40 mW m$^{-2}$, values of $\mu_m$ higher than those cited above would be required with a creep activation energy lower than 300 kJ mol$^{-1}$. For $Q$ within the 300–500 kJ mol$^{-1}$ range, onset of convection in the 50–70 My age range would be possible only with an upper-mantle viscosity of $10^{19}$ Pa s, and then only if the convective heat flux due to small-scale convection beneath old seafloor were $\leq$30 mW m$^{-2}$. This may be possible if small-scale convection transports only a fraction of the heat needed to explain the surface heatflow and bathymetry of old seafloor, the remainder being due to heating of the lithosphere at hot spots, as discussed earlier.

Mantle viscosity ($\mu_m$ introduced above) may differ near a spreading center and beneath older seafloor. Near spreading centers, the presence of melt may reduce viscosity. Conversely, since the presence of intragranular water in nominally anhydrous minerals reduces creep viscosity and water is a relatively incompatible element, the extraction of melt may dehydrate the mantle, thus increasing viscosity (Karato, 1986; Hirth and Kohlstedt, 1996). As discussed below, the latter effect may be the more dominant one, depending on the sensitivity of creep rates to water content and the water content of the mantle prior to melting.

As indicated above, convective instability at young ages may be present only beneath the Nazca and Pacific Plates, which are both fast moving and formed at a high spreading rate. Why might convective instability at young ages be restricted to fast spreading or fast moving plates? Is spreading rate or plate velocity the determining factor? Slower spreading rates should lead to a thicker thermal boundary layer beneath the spreading axis thus promoting convective instability at younger ages (Sparks and Parmentier, 1993). Other possible effects of spreading rate are discussed later. The linear stability analysis of Korenaga and Jordan (2003b) indicates that convective rolls aligned with plate motion may be stable only in the presence of strong shearing beneath fast moving plates. Rheological effects might also be important. Higher strain rates beneath fast moving plates could create smaller grain size resulting in more rapid diffusion creep or higher stresses resulting in deformation dominated by dislocation creep with higher $Q$. As shown in **Figure 6**, both effects would favor convective instability at younger ages. Rising mantle that is enriched in water relative to that in other regions might also explain a lower-mantle viscosity beneath the EPR, as discussed below, and promote early onset times.

The scale or wavelength of small-scale convection appears to increase to $\geq 1000$ km beneath the old Pacific seafloor. Korenaga and Jordan (2004) recently examined how the horizontal scale of convective motions in a fluid with temperature and depth-dependent viscosity would increase with age, including in particular the effect of an endothermic phase transition at the base of the upper mantle. They suggested that convection initiated at small scales beneath moving plates could eventually penetrate the stable phase transition and evolve into whole-mantle convection.

All of the studies discussed above treat the development and evolution of convective motions as two-dimensional (2-D) and time dependent, most closely approximating convective motions in a vertical plane orthogonal to plate motion as a function of plate age. Despite significant advances in both digital technologies and numerical methodologies, fully 3-D treatments of thermal convection with the very strongly temperature-dependent viscosities believed to characterize mantle flow remain challenging. Numerical solutions must simultaneously resolve 10 km thick thermal boundary layer layers and cold plumes embedded in several thousand kilometer scale flow.

van Hunen *et al.* (2003) studied 3-D convective instability beneath a moving plate. In agreement with earlier analysis, plate motion enhanced the development of longitudinal convective rolls relative to motion-perpendicular transverse rolls. They found that the onset age of fully 3-D convective motions was similar to that in earlier 2-D studies. It would be interesting to more closely examine the evolution of convective motions with age, in particular how the scale convective motions increase as cooling proceeds. A first impression based on van Hunen *et al.* (2003, figure 1) suggests that the scale of convective motions does not increase as rapidly as suggested by earlier 2-D studies.

## 7.3 Convective Instability due to Melting in the Upper Mantle

### 7.3.1 Intraplate Volcanism as an Indicator of Upper-Mantle Convective Activity

Volcanism away from plate boundaries (*see* Chapter 9) is generally thought to be a consequence of decompression melting due to convective motions that arise from deeper in the mantle or from instability generated in the upper mantle. Linear, long-lived, and age-progressive volcanic chains have been explained as the manifestation of fixed hot spots, possibly generated by buoyant plumes of rising material originating deep in the mantle (Morgan, 1971). While the Wilson–Morgan fixed hot-spot model has been successfully applied to volcanic chains in all ocean basins, important disagreements have been recognized between geochronologic observations and the simplest predictions of the model. For example, Okal and Batiza (1987) and McNutt (1998) outline numerous examples where the fixed hot-spot model fails to explain: (1) observed departures from linearity of individual volcanic chains and inconsistent orientations among multiple chains which lie on the same plate, (2) short-lived chains and ones which fluctuate in size, and (3) violations of predicted along-chain age versus distance behavior. For example, the Cook–Austral Chain (Turner and Jarrard, 1982) and the Line Islands (Schlanger *et al.*, 1984), exhibit complex age progressions with volcanic activity occuring along multiple lineaments and with volcanoes of distinctly different ages in close proximity to one another. Furthermore, some linear volcanic ridges in the Pacific form much more rapidly than would be predicted by fixed hot-spot model (Bonatti *et al.*, 1977; Sandwell *et al.*, 1995).

The creation of large igneous provinces or oceanic plateaux by partial melting of the starting plume heads is a corollary of the mantle plume hypothesis for the origin of hot spots (Richards *et al.*, 1989). Of 14 possible Pacific hot-spot tracks studied by Clouard and Bonneville (2001), only three (Louisville, Easter, and Marquesas) can be traced to an oceanic plateau. Seven hot spots have short tracks <35 My and clearly cannot be traced to an oceanic plateau. Koppers *et al.* (2003) found that linear volcanic chains in the Western Pacific and South Pacific Superswell region typically display intermittent volcanic activity with longevities shorter than 40 My, superposed volcanism, and motion relative to other longer-lived hot spots. Finally, study of marine satellite gravity data in the Pacific (Wessel and Lyons, 1997) shows the presence of many volcanoes with a range of sizes in a variety of geologic settings. Most of these volcanoes do not clearly align in chains or ridges, and most seem too small to be explained by deep-mantle plumes. While the long-lived age progressive volcanic chains that fit the fixed hot-spot model are remarkable features, other mechanisms of intraplate volcanism must also be active.

### 7.3.2    Other Possible Mechanisms for Intraplate Volcanism

Diffuse plate extension (Sandwell *et al.*, 1995) has been frequently cited as a possible mechanism of intraplate volcanism. Proposed partly on morphological grounds, this mechanism was thought to explain rapid propagation of volcanic ridges and the formation of volcanic ridges in troughs of topography associated with the Haxby gravity lineations. Volcanism was thought to be attributable to either allowing melt already present in the upper mantle access to the surface or to decompression melting in upwellings associated with lithospheric boundinage. Dunbar and Sandwell (1988) calculated that 10% extension would be required for this mechanism. However, the amount of extension that can be allowed in the Pacific is less than 1% (Goodwillie and Parsons, 1992; Gans *et al.*, 2003), seemingly much too small for significant boudinage. Thus, agreement seems to be emerging that the boudinage hypothesis does not satisfy available observations. An interesting alternative is cracking of the lithosphere under the action of thermal contraction bending moments (Gans *et al.*, 2003; Sandwell and Fialko, 2004). In this hypothesis, cracking of the

plate allows lithosphere between cracks to bend, thus relieving thermal bending moments. Cracks form topographic troughs, and magma already existing beneath the lithosphere exploits these regions to reach the seafloor. Mantle seismic velocities beneath volcanic ridges formed in this way should be higher than in adjacent mantle, which does not seem to be the case, at least beneath the few ridges where regional seismic data is available. Weeraratne *et al.* (2006) find lower seismic velocities, implying higher temperatures and/or more melt beneath ridges. However, the ridges examined in this study are all near the EPR spreading center on seafloor younger than 3 Ma. Seismic velocities beneath recently active volcanic ridges further from the spreading axis have not yet been studied in comparable ways.

### 7.3.3    Melting in Convectively Driven Upwellings

Volcanism due to decompression melting in small-scale convective upwellings is perhaps the favored popular hypothesis for the origin of volcanic ridges aligned with plate motion. Melting in convective upwellings would be consistent with lower seismic velocities in the mantle beneath ridges. Buoyant convective upwellings should elevate the seafloor so that the resulting volcanism should occur on topographic highs. However, volcanic ridges were observed to lie in lows of the gravity lineations. If gravity lows correspond to topographic troughs, this would be inconsistent with a convective origin (Sandwell *et al.*, 1995). However, seafloor topography with 100 m amplitudes over horizontal scales of several hundred kilometers are subtle features compared to the topography of volcanic constructs and the flexure caused by this loading. After removing lithospheric flexure due to the topographic loading, Harmon *et al.* (2006) found that volcanic ridges near the EPR actually lie on topographic highs. Topographic highs correspond to negative residual mantle Bouguer gravity anomalies which were obtained from the observed free air anomaly by subtracting the effect of crustal thickness variations and the attraction of topography. If this relationship holds for other volcanic ridges, particularly those further from the EPR axis, then a convective origin for volcanic ridges and gravity lineations would be indicated.

## 7.3.4 Buoyant Decompression Melting

In addition to decompression melting in thermally driven convection, intraplate volcanism may be a product of decompression melting in convective upwellings that result from the buoyancy associated with melting itself (Tackley and Stevenson, 1993; Raddick *et al.*, 2002; Hernlund and Tackley, 2003). This 'buoyant decompression melting' in a layer that is initially at its melting temperature may organize from small, initially random perturbations and thus might be termed 'magmatic convective storms'. Melt extraction leaves behind a buoyant and creep resistant residual mantle. The accumulation of this relatively immobile and infertile mantle ultimately limits the amount of melt that can be produced by the decompression melting mechanism (Raddick *et al.*, 2002). This results in an inverse correlation between the rate of melt production and the duration of melting, which may be diagnostic of the buoyant decompression melting process.

Buoyant decompression melting could occur in a number of geologic settings; however, if the oceanic upper mantle has previously melted beneath a spreading center and subsequently cooled, spontaneous buoyant decompression melting may be possible only in regions where a large-scale mantle upwelling can counteract conductive cooling, keeping the mantle at its solidus temperature over some depth range. The South Pacific Superswell has been interpreted as a region of large-scale mantle upwelling (McNutt, 1998; Gaboret *et al.*, 2003), and buoyant decompression melting may explain the abundance of volcanism associated with it. Alternatively, to balance the downward flux associated with sinking lithosphere, a small upward velocity should be present in large areas of the upper mantle away from convergent plate boundaries. Since transition zone mineral phases like the $\beta$- and $\gamma$-olivine have a larger water storage capacity than their lower pressure isomorph, upwelling mantle may dehydrate forming a water-rich melt above the transition zone. During the upward percolation of buoyant melt, as envisioned in an early study by Frank (1968), buoyant instabilities may lead to localization of melting, upwelling, and surface volcanism.

In the absence of a large-scale upwelling, where mantle is slightly cooler than its melting temperature, buoyant melting may not occur spontaneously but may be triggered by some initial upwelling due to relief on the bottom of the lithosphere, for example, across oceanic fracture zones that move obliquely across the mantle. This may provide a physical explanation for intraplate volcanism that is controlled by lithosphere structure.

## 7.4 Upwelling and Melting beneath Oceanic Spreading Centers

### 7.4.1 Melting, Melt Extraction, and the Chemical Lithosphere

Mantle upwelling and decompression melting beneath spreading centers is expected to have important consequences for the development of convective motions as the lithosphere ages. Two effects may be particularly important: (1) melting and melt extraction are expected to affect both rheology and density and (2) convective instability beneath the spreading axis due to density variations associated with the presence of melt and its extraction from residual mantle may influence the scale and evolution of off-axis convective instability. Forsyth (1992) provides a relatively recent review of mantle flow beneath spreading centers and the mantle electromagnetic and tomography (MELT) experiment (Forsyth *et al.*, 1998) provides the best geophysical evidence on the amount and distribution of melt in the mantle beneath a very fast spreading section of the EPR. Geochemistry provides evidence that garnet was present during melting, thus providing a minimum estimate of ~70 km for the beginning of melting in upwelling mantle.

Hirth and Kohlstedt (1996) and Phipps Morgan (1997) pointed out that melting beneath spreading centers should produce a compositional lithosphere that is both more viscous, as mentioned above, and compositionally buoyant. Intragranular water that enhances the creep rate of nominally anhydrous mantle minerals behaves as a highly incompatible element during melting. The extraction of melt during fractional melting would leave behind a dry residual mantle. Removing essentially all the water present at concentrations inferred for upper mantle that melts beneath spreading centers may increase its viscosity by more than a factor of 100 (Hirth and Kohlstedt, 1996). The thickness of the residual layer should be comparable to the maximum depth of melting beneath a spreading center. Evans *et al.* (2005), using electrical conductivity, inferred a 60 km dehydration depth beneath a fast spreading section of the EPR (the MELT area at 17°S). The dehydrated residual mantle layer generated at a spreading center might be thought to correspond to the plate in the plate model seafloor evolution. However, this depth of dehydration appears significantly thinner than estimates of plate thickness in the range of 95–125 km discussed above. One possibility is that small amounts of wet melt form at the even greater depths comparable to the plate thickness. Interpretations of seismic data from the MELT experiment do suggest small amounts of melting at depths exceeding 100 km (Forsyth *et al.*, 1998).

## 7.4.2   Influence of Melt Extraction Mechanism

The rheological consequences of melt extraction should depend on the mechanism of melt migration. Does melt percolating upward remain distributed along mineral grain edges maintaining equilibrium with solid mantle as it goes? Or does it localize into larger channels after only small amounts of melt form? Melt that rapidly localizes into larger channels approximates ideal fractional melting which removes incompatible elements more effectively than equilibrium transport.

A fundamental observation is that the chemical composition of basalt erupted at spreading centers is not in equilibrium with residual mantle at low pressure (e.g., O'Hara, 1965; Stolper, 1980; Elthon and Scarfe, 1984). Melts must rise from depths of at least 30 km to the surface without extensive re-equilibration with surrounding mantle at shallow depth in order to preserve this deep geochemical signature. Focused melt flow through high permeability conduits or channels is one possible mechanism. Several lines of evidence suggest that the porosity structure in the melt generation and extraction region of the mantle is heterogeneous consisting of interconnected high-porosity dunite channels embedded in a low-porosity harzburgite or lherzolite matrix (Kelemen *et al.* (1997) and references therein). Remnant dunite channels have been observed as veins, tabular or sometimes irregular shaped bodies in ophiolites. These dunite dikes or veins make up 5–15% of the mantle in the Oman ophiolite with widths ranging from tens of millimeters to ~200 m and lengths from tens of meters to at least 10 km (Kelemen *et al.*, 1997). Although their spatial distribution in the mantle is still not well constrained, the presence of a high porosity, interconnected, coalescing network of dunite channels present above or within the melting region, has been envisioned. Porous flow through dunite channels is capable of producing observed Uranium-series disequilibria and significant trace

element fractionations during mantle melting and melt extraction (e.g., Spiegelman and Elliot, 1993; Kelemen *et al.*, 1997; Spiegelman and Kelemen, 2003; Lundstrom, 2003).

At fast spreading rates, mantle upwelling and melting appear to occur over a several hundred kilometer wide region (e.g., Forsyth *et al.*, 1998) but the oceanic crust is emplaced within a few kilometers of the spreading axis. Thus, a second major constraint on the mechanism of melt migration is that it must be capable of focusing melt to the spreading axis. If the viscosity of upwelling mantle is sufficiently high ($\geq 10^{20}$ Pa s) pressure gradients in the mantle flow may be sufficient to drive melt to the spreading axis (Spiegelman and McKenzie, 1987; Phipps Morgan, 1987). Alternatively, melt may migrate vertically to collect in a decompaction boundary layer that develops as melt begins to freeze. This should create a high-porosity melt channel that slopes away from the spreading axis providing a conduit for melt flow toward the axis (Sparks and Parmentier, 1991; Spiegelman, 1993). Rabinowitz and Ceuleneer (2005) suggest, in fact, that dunites in the Oman ophiolite may be the preserved remnants of this decompaction boundary layer. If melt in the decompaction layer is isolated in dunite channels, then some signature of high pressures will be preserved in the melts that accumulate near the spreading axis. However, melting need not be perfectly fractional so that mantle need not be fully dehydrated.

No seismic or electromagnetic evidence on the distribution of melting and upwelling in the mantle comparable to that for southern EPR is available for slower spreading centers. If buoyancy related to melting is important, then 'active' upwelling could lead to more localized melting at slow spreading rates. Active upwelling would produce higher, more uniform degrees of melting creating more strongly and uniformly dehydrated residual mantle less prone to convective instability. This is illustrated in **Figure 7** from Braun *et al.* (2000).

## 7.4.3   2-D versus 3-D Upwelling and the Spreading Rate Dependence of Seafloor Structure

At slow spreading rates, buoyancy related to melting may also result in localized columns of upwelling and melting beneath the spreading axis (Parmentier and Phipps Morgan, 1990; Choblet and Parmentier, 2001). At high spreading rates, upwelling remains sheet like along the spreading axis. This is one possible

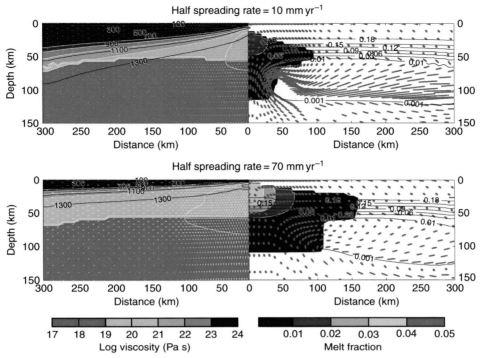

**Figure 7** Mantle upwelling and melting beneath fast and slow spreading centers from Braun, *et al.* (2000). Solid contours in the left panel show the mantle viscosity, and black lines depict isotherms (at 200°C intervals) for a potential temperature of 1350°C and an adiabatic gradient of 0.3°C km$^{-1}$. Ellipses (red ovals) representing accumulated finite strain are plotted along solid flow streamlines corresponding to the velocity field shown by white vectors in the left panel. Solid contours on the right panel show the steady-state melt fraction (>0.1%) overlain by isodepletion contours (black lines). Rheology includes estimates of the effects of dehydration, melt, and grain boundary sliding with a reference viscosity of $5 \times 10^{18}$ Pa s. Since the effects of dehydration dominate those of melt, the viscosity and velocity structures resemble more passive-like plate spreading flow above the dry solidus. Even with lower reference viscosities, buoyant flow is restricted to depths beneath the dry solidus. Buoyant localization of solid flow and melt generation at depth increases as the half-spreading rate decreases. The magnitude and localization of strain also increases with decreasing spreading rate.

explanation for the difference in seafloor morphology and structure between fast and slow spreading centers, and may also explain the strongly lineated morphology of seafloor produced at slow spreading rates (Phipps Morgan and Parmentier, 1995). At high spreading rates, buoyant decompression melting may produce off-axis upwelling columns that may explain near-axis volcanic ridges (Jha *et al.*, 1997) providing a possible alternative to melting in thermally driven upwellings as discussed above.

## 7.5 Summary

The deviation of the age dependence of seafloor depth from predictions of a simple conductive thermal boundary layer (half-space cooling model) need not be explained by only a single mechanism. Both heating by hot spots and convective instability due to

cooling from above may affect thermal evolution but differ in importance in different settings, for example, the North Atlantic and Pacific, respectively. Discussion has sometimes focused on which mechanism is the correct one rather than on assessing which may be more important and why. The relative importance of convective instability should depend on spreading rate and mantle composition, perhaps particularly through the effect of water on rheology.

The seafloor age at which convective instability begins clearly depends strongly on spreading rate and particularly mantle rheology. Convective instability at young ages remains a preferred explanation for gravity lineations on the Pacific and Nazca Plates, but is not required beneath other plates. Beneath thicker, colder lithosphere at slow spreading centers, small-scale convection may be present but less visible. If small-scale convection is present beneath the fast spreading EPR but not beneath slower spreading centers, a number of explanations for this possible difference in behavior can be envisioned. The higher shear rates beneath rapid moving plates may organize convection into a lineated structure that is more visible in gravity data than other less-organized patterns. Higher shear rates may also result in lower effective viscosity of the upper mantle beneath faster moving plates; therefore yielding earlier onset times.

Intraplate volcanism not associated with hot spots on Pacific and Nazca Plates is much more abundant than in other ocean basins. Does this reflect an influence of deeper-mantle processes or convective instability that develops within the upper mantle? The latter possibility would be broadly consistent with a lower-viscosity (higher strain rate, smaller grain size, and/or wetter) mantle beneath the Pacific and Nazca Plates than elsewhere in the upper mantle.

# References

Bergman EA and Solomon SC (1984) Source mechanisms of earthquakes near mid-ocean ridges from body waveform inversion; implications for the early evolution of oceanic lithosphere. *Journal of Geophysical Research* 89: 11415–11441.

Bills BG (1994) Viscosity estimates for the crust and upper mantle from patterns of lacustrine shoreline deformation in the eastern Great Basin. *Journal of Geophysical Research* 99: 22059–22086.

Bonatti E, Harrison CGA, Fisher DE, et al. (1977) Easter volcanic chain (southeast Pacific): A mantle hotline. *Journal of Geophysical Research* 82: 2437–2478.

Braun M, Hirth G, and Parmentier EM (2000) Effects of deep damp melting on mantle flow and melt generation beneath mid-ocean ridges. *Earth and Planetary Science Letters* 176: 339–356.

Buck WR (1985) When does small scale convection begin beneath oceanic lithosphere? *Nature* 313: 775–777.

Buck WR (1987) Analysis of the cooling of a variable-viscosity fluid with applications to the Earth. *Geophysical Journal of the Royal Astronomical Society* 89: 549–577.

Buck WR and Parmentier EM (1986) Convection beneath young oceanic lithosphere: Implications for thermal structure and gravity. *Journal of Geophysical Research* 91: 1961–1974.

Carlson RL and Johnson HP (1994) On modeling the thermal evolution of the oceanic upper mantle: An assessment of the cooling plate model. *Journal of Geophysical Research* 99: 3201–3214 (doi: 10.1029/93JB02696).

Cazenave A (1984) Thermal cooling of the oceanic lithosphere: New constraints from geoid height data. *Earth and Planetary Science Letters* 70: 395–406.

Cazenave A, Lago B, and Dominh K (1982) Geoid anomalies over the north-east Pacific fracture zones from satellite altimeter data. *Geophysical Journal* 69: 15–31.

Cazenave A, Lago B, and Dominh K (1983) Thermal parameters of the oceanic lithosphere estimated from geoid height data. *Journal of Geophysical Research* 88: 1105–1118.

Cazenave A, Parsons B, and Calcagno P (1995) Geoid lineations of 1000-km wavelength over the central Pacific. *Geophysical Research Letters* 22: 97–100.

Chase CG (1985) The geological significance of the geoid. *Annual Review of Earth and Planetary Sciences* 13: 97–117.

Chastel YB, Dawson PR, Wenk H-R, and Bennett K (1993) Anisotropic convection with implications for the upper mantle. *Journal of Geophysical Research* 98: 17757–17772 (doi:10.1029/93JB01161).

Choblet G and Parmentier EM (2001) Mantle upwelling and melting beneath slow spreading centers: Effects of variable rheology and melt productivity. *Earth and Planetary Science Letters* 184: 589–604.

Clouard V and Bonneville A (2001) How many Pacific hotspots are fed by deep-mantle plumes? *Geology* 29: 695–698.

Colin P and Fleitout L (1990) Topography of the ocean floor: Thermal evolution of the lithosphere and interaction of the deep mantle heterogeneities with the lithosphere. *Geophysical Research Letters* 17: 1961–1964.

Craig CH and McKenzie DP (1986) The existence of a thin low-viscosity layer beneath the lithosphere. *Earth and Planetary Science Letters* 78: 420–426.

Crough ST (1978) Thermal origin of mid-plate hot-spot swells. *Geophysical Journal of the Royal Astronomical Society* 55: 451–469.

Crough ST (1975) Thermal model of oceanic lithosphere. *Nature* 256: 388–390.

Crough ST (1977) Approximate solutions for the formation of the lithosphere. *Earth and Planetary Science Letters* 14: 365–377.

Crough ST (1979a) Geoid anomalies across fracture zones and the thickness of the lithosphere. *Earth and Planetary Science Letters* 44: 224–230.

Crough ST (1979b) Hotspot epeirogeny. *Tectonophysics* 61: 321–333.

Davaille A and Jaupart C (1994) One onset of thermal convection in fluids with temperature-dependent viscosity: Application to the ocean mantle. *Journal of Geophysical Research* 99: 19853–19866.

Davis EE and Lister CRB (1974) Fundamentals of ridge crest topography. *Earth and Planetary Science Letters* 21: 405–413.

Davis EE, Lister CRB, and Sclater JG (1984) Towards determining the thermal state of the thermal state of old oceanic lithosphere; heat flow measurements from the Blake-Bahama outer ridge, northwestern Atlantic. *Geophysical Journal of the Royal Astronomical Society* 78: 707–1545.

Detrick RS (1981) An analysis of geoid anomalies across the Mendocino fracture zone: Implications for thermal models of the lithosphere. *Journal of Geophysical Research* 86: 11751–11762.

Detrick RS and Crough ST (1978) Island subsidence, hot spots and lithosphere thinning. *Journal of Geophysical Research* 83: 1236–1244.

Doin MP and Fleitout L (1996) Thermal evolution of the oceanic lithosphere: An alternative view. *Earth and Planetary Science Letters* 142: 121–136.

Doin M-P, Fleitout L, and McKenzie D (1996) Geoid anomalies and the structure of continental and oceanic lithospheres. *Journal of Geophysical Research* 101: 16119–16136 (doi:10.1029/96JB00640).

Driscoll ML and Parsons B (1988) Cooling of the oceanic lithosphere – evidence from geoid anomalies across the Udintsev and Eltanin fracture zones. *Earth and Planetary Science Letters* 88: 289–307.

Dunbar J and Sandwell D (1988) A boudinage model for cross-grain lineations (abs.). *EOS Transactions of the American Geophysical Union* 69: 1429.

Eberle MA and Forsyth DW (1995) Regional viscosity variations, small-scale convection and the slope of the depth-age $\frac{1}{2}$ curve. *Geophysical Research Letters* 22: 473–476.

Ekstrom G and Dziewonski AM (1998) The unique anisotropy of the Pacific upper mantle. *Nature* 394: 168–172.

Elthon D and Scarfe CM (1984) High-pressure phase equilibria of a high-magnesia basalt and the genesis of primary oceanic basalts. *American Mineralogist* 69: 1–15.

Evans RL, Hirth G, Baba K, Forsyth D, Chave A, and Mackie R (2005) Geophysical evidence from the MELT area for compositional controls on oceanic plates. *Nature* 437: 249–252.

Faul U and Jackson I (2005) The seismological signature of temperature and grain size variations in the upper mantle. *Earth and Planetary Science Letters* 234: 119–234.

Fleitiout L and Yuen DA (1984) Secondary convection and the growth of the oceanic lithosphere. *Physics of the Earth and Planetary Interiors* 36: 181–212.

Fleitout L and Moriceau C (1992) Short-wavelength geoid, bathymetry and convective pattern beneath the Pacific Ocean. *Geophysical Journal International* 110: 6–28.

Forsyth DW (1992) Geophysical constraints on mantle flow and melt generation beneath mid-ocean ridges. In: Phipps MJ, Blackman DK, and Sinton JM (eds.) *Geophysical Monograph Series 71: Evolution of Mid Ocean Ridges*, pp. 1–66. Washington, DC: American Geophysical Union.

Forsyth DW, and The MELT Seismic Team (1998) Imaging the deep seismic structure beneath a mid-ocean ridge: The MELT experiment. *Science* 280: 1215–1218.

Frank FC (1968) Two component model for convection in the Earth's upper mantle. *Nature* 220: 350–352.

Freedman AP and Parsons B (1990) Geoid anomalies over two South Atlantic fracture zones. *Earth and Planetary Science Letters* 100: 18–41.

Gaboret C, Forte AM, and Montagner JP (2003) The unique dynamics of the Pacific hemisphere mantle and its signature on seismic anisotropy. *Earth and Planetary Science Letters* 208: 219–233.

Gans KD, Wilson DS, and Macdonald KC (2003) Pacific plate gravity lineaments: Diffuse extension or thermal contraction? *Geochemistry Geophysics Geosystems* 4(9): 1074 (doi:10.1029/2002GC000465).

Goodwillie AM and Parsons B (1992) Placing bounds on lithospheric deformation in the central Pacific Ocean. *Earth and Planetary Science Letters* 111: 123–139.

Grasset O and Parmentier EM (1998) Thermal convection in a volumetrically heated, infinite Prandtl number fluid with strongly temperature-dependent viscosity: Implications for planetary thermal evolution. *Journal of Geophysical Research* 103: 18171–18181.

Gribb TT and Cooper RF (1998) Low-frequency shear wave attenuation in polycrystalline olivine: Grain boundary diffusion and the physical significance of the Andrade model for viscoelastic rheology. *Journal of Geophysical Research* 103: 27267–27279.

Harmon N, Forsyth DW, and Scheirer DS (2006) Analysis of gravity and topography in the GLIMPSE study region: Isostatic compensation and uplift of the Sojourn and Hotu Matua Ridge systems. *Journal of Geophysical Research* 111: B11406 (doi: 10.1029/2005JB004071).

Haxby WF and Turcotte DL (1978) Isostatic geoid anomalies. *Journal of Geophysical Research* 83: 5473–5478.

Haxby WF and Weissel JK (1986) Evidence for small scale convection for SEASAT altimeter data. *Journal of Geophysical Research* 91: 3507–3520.

Hayes DE (1988) Age-depth relationships and depth anomalies in the southeast Indian Ocean and south Atlantic Ocean. *Journal of Geophysical Research* 93: 2937–2954.

Heestand RL and Crough ST (1981) The effect of hot spots on the oceanic age-depth relation. *Journal of Geophysical Research* 86: 6107–6114.

Hernlund JW and Tackley PJ (2003) Post-laramide volcanism and upper mantle dynamics in the Western US: Role of small-scale convection, *Eos Transactions AGU*, Fall Meeting Supplement 84(46):, Abstract V31F-07.

Hirth G, Evans RL, and Chave AD (2000) Comparison of continental and oceanic mantle electrical conductivity: Is the Archean lithosphere dry? *Geochemistry Geophysics Geosystems* 1: (doi:10.1029/2000GC000048).

Hirth G and Kohlstedt DL (1996) Water in the oceanic upper mantle: Implications for rheology, melt extraction, and the evolution of the lithosphere. *Earth and Planetary Science Letters* 144: 93–108.

Huang J and Zhong S (2005) Sublithospheric small-scale convection and its implications for the residual topography at old ocean basins and the plate model. *Journal of Geophysical Research* 110: B05404 (doi:10.1029/2004JB003153).

Huang J, Zhong S, and van Hunen J (2003) Controls on sublithospheric small-scale convection. *Journal of Geophysical Research* 108: 2405 (doi:10.1029/2003JB002456).

Jha K, Parmentier EM, and Sparks DW (1997) Buoyant mantle upwelling and crustal production at oceanic spreading centers: On-axis segmentation and off-axis melting. *Journal of Geophysical Research* 102: 11979–11989.

Johnson HP and Carlson RL (1992) Variation of sea floor depth with age: A test of model based on drilling results. *Geophysical Research Letters* 19: 1971–1974.

Kaminski É and Ribe NM (2001) A kinematic model for recrystallization and texture development in olivine polycrystals. *Earth and Planetetary Science Letters* 189: 253–267.

Kaminski É and Ribe NM (2002) Timescales for the evolution of seismic anisotropy in mantle flow. *Geochemistry Geophysics Geosystems* 3: (doi:10.1029/2001GC000222).

Karato S (1986) Does partial melting reduce the creep strength of the upper mantle. *Nature* 319: 309–310.

Katzman R, Zhao L, and Jordan TH (1998) High-resolution, two-dimensional vertical tomography of the central Pacific mantle using ScS reverberations and frequency-dependent travel times. *Journal of Geophysical Research* 103: 17933–17972 (doi:10.1029/98JB00504).

Kelemen PB, Hirth G, Shimizu N, Spiegelman M, and Dick HJB (1997) A review of melt migration processes in the adiabatically upwelling mantle beneath oceanic spreading ridges. *Philosophical Transactions of the Royal Society of London A* 355: 282–318.

Koppers AAP, Staudigel H, Pringle MS, and Wijbrans JR (2003) Short-lived and discontinuous intraplate volcanism in the south Pacific: Hot spots or extensional volcanism? *Geochemistry Geophysics Geosystems* 4: 1089 (doi:10.1029/2003GC000533).

Korenaga J and Jordan TH (2003a) Linear stability analysis of Richter rolls. *Geophysical Research Letters* 30(22): 2157 (doi:10.1029/2003GL018337).

Korenaga J and Jordan TH (2003b) Physics of multiscale convection in Earth's mantle: Onset of sublithospheric convection. *Journal of Geophysical Research* 108: 2333 (doi:10.1029/2002JB001760).

Korenaga J and Jordan TH (2004) Physics of multiscale convection in Earth's mantle: Evolution of sublithospheric convection. *Journal of Geophysical Research* 109: B01405 (doi:10.1029/2003JB002464).

Langseth MG, LePichon X, and Ewing M (1966) Crustal structure of mid-ocean ridges. Part 5: Heat flow through the Atlantic Ocean and convection currents. *Journal of Geophysical Research* 71: 5321–5355.

Lawrence JF and Shearer PM (2006) A global study of transition zone thickness using receiver functions. *Journal of Geophysical Research* 111: B06307 (doi:10.1029/2005JB003973).

Levitt DA and Sandwell DT (1995) Lithospheric bending at subduction zones based on depth soundings and satellite gravity. *Journal of Geophysical Research* 100: 379–400.

Lin J and Parmentier EM (1985) Surface topography due to convection in a variable viscosity fluid: Application to short wavelength gravity anomalies in the central Pacific Ocean. *Geophysical Research Letters* 12: 357–360.

Lister CRB, Sclater JG, Davis EE, Villinger H, and Nagihara S (1990) Heat flow maintained in ocean basins of great age: Investigations in the north-equatorial west Pacific. *Geophysical Journal International* 102: 603–630.

Lizarralde D, Chave A, Hirth G, and Schultz A (1995) Northeastern Pacific mantle conductivity profile from long-period magnetotelluric sounding using Hawaii-to-California submarine cable data. *Journal of Geophysical Research* 100: 17837–17854.

Lundstrom CC (2003) Uranium-series disequilibria in mid-ocean ridge basalts: Observations and models of basalt genesis. In: Bourdon B, Henderson GM, Lundstrom CC, and Turner SP (eds.) *Reviews in Mineralogy and Geochemistry 52: Uranium-Series Geochemistry*, pp. 175–214. Chantilly, VA: Mineralogical Society of America.

Lynch MA (1999) Linear ridge groups: evidence for tensional cracking in the Pacific Plate. *Journal of Geophysical Research* 104: 29321–29333.

Maggi A, Debayle E, Priestly K, and Barruol G (2006) Multimode surface waveform tomography of the Pacific ocean: A close look at the lithospheric cooling signature. *Geophysical Journal International* 166: 1384–1397.

Marquart G, Schmeling H, and Braun A (1999) Small-scale instabilities below the cooling oceanic lithosphere. *Geophysical Journal International* 138: 655–666.

Marty J-C and Casanave A (1989) Regional variations in subsidence rate of oceanic plates: A global analysis. *Earth and Planetary Science Letters* 94: 301–315.

Marty J-C and Cazenave A (1988) Thermal evolution of the lithosphere beneath fracture zones inferred from geoid anomalies. *Geophysical Research Letters* 15: 593–596.

Marty J-C, Cazenave A, and Lago B (1988) Geoid anomalies across Pacific fracture zones. *Geophysical Journal* 93: 1–23.

McKenzie DP (1967) Some remarks on heat flow and gravity anomalies. *Journal of Geophysical Research* 72: 6261–6273.

McKenzie DP and Sclater JG (1969) Heat flow in the eastern Pacific and sea-floor spreading. *Bulletin of Volcanology* 33: 101–118.

McNutt MK (1998) Superswells. *Reviews of Geophysics* 36: 211–244.

Morgan WJ (1968) Rises, trenches, great faults, and crustal blocks. *Journal of Geophysical Research* 73: 1959.

Morgan WJ (1971) Convection plumes in the lower mantle. *Nature* 230: 42–43.

Morgan WJ (1972) Deep mantle convection plumes and plate motions. *American Association of Petroleum Geologist Bulletin* 56: 203–213.

Nagihara S, Lister CRB, and Sclater JG (1996) Reheating of old oceanic lithosphere: Deductions from observations. *Earth and Planetary Science Letters* 139: 91–104.

Nishimura CE and Forsyth DW (1989) The anisotropic structure of the upper mantle in the Pacific. *Geophysical Journal* 96: 203–229.

O'Connell RJ and Hager BH (1980) On the thermal state of the Earth. In: Dziewonski A and Boschi E (eds.) *Physics of the Earth's Interior*, pp. 270–317. Amsterdam: Elsevier.

O'Hara MJ (1965) Primary magmas and the origin of basalts. *Scottish Journal of Geology* 1: 19–40.

Okal EA and Batiza R (1987) Hotspots: The first 25 years. In: Keating BH, Fryer P, Batiza R, and Boehlert GW (eds.) *Seamounts, Islands, and Atolls*, pp. 1–11, Geophysical Monograph 43. Washington, DC: AGU.

Parmentier EM and Haxby WF (1986) Thermal stresses in the oceanic lithosphere: Evidence from geoid anomalies at fracture zones. *Journal of Geophysical Research* 91: 7193–7204.

Parmentier EM and Phipps MJ (1990) The spreading rate dependence of three-dimensional oceanic spreading center structure. *Nature* 348: 325–328.

Parsons B and McKenzie D (1978) Mantle convection and the thermal structure of the plates. *Journal of Geophysical Research* 83: 4485–4496.

Parsons B and Sclater JG (1977) An analysis of the variation of ocean floor bathymetry and heat flow with age. *Journal of Geophysical Research* 82: 803–827.

Passey QR (1981) Upper mantle viscosity derived from the difference in rebound of the Provo and Bonneville shorelines; Lake Bonneville Basin, Utah. *Journal of Geophysical Research* 86: 11701–11708.

Phipps MJ (1987) Melt migration beneath mid-ocean spreading centers. *Geophysical Research Letters* 14: 1238–1241.

Phipps MJ (1997) The generation of a compositional lithosphere by mid-ocean ridge melting and its effect on subsequent off-axis hotspot upwelling and melting. *Earth and Planetary Science Letters* 146: 213–232.

Phipps MJ and Parmentier EM (1995) Crenulated seafloor: Evidence for spreading rate dependent structure of mantle upwelling and melting beneath a mid-oceanic spreading center. *Earth and Planetary Science Letters* 129: 73–84.

Priestley D and McKenzie DM (2006) The thermal structure of the lithosphere from shear velocities. *Earth and Planetary Science Letters* 244: 285–301.

Rabinowicz M and Ceuleneer G (2005) The effect of sloped isotherms on melt flow in the shallow mantle: a physical and numerical model based on observations in the Oman ophiolite. *Earth and Planetary Science Letters* 229: 231–246.

Raddick MJ, Parmentier EM, and Scheirer DS (2002) Buoyant decompression melting: A possible mechanism for intraplate volcanism. *Journal of Geophysical Research* 107(B10): 2228 (doi:10.1029/2001JB000617).

Renkin ML and Sclater JG (1988) Depth and age in the North Pacific. *Journal of Geophysical Research* 93: 2910–2935.

Richards MA, Duncan RA, and Vincent EC (1989) Flood basalts and hot-spot tracks: Plume heads and tails. *Science* 246: 103–107.

Richter FM (1973) Convection and the large scale circulation of the mantle. *Journal of Geophysical Research* 74: 8735–8745.

Richter FM and Parsons B (1975) On the interaction of two scales of convection in the mantle. *Journal of Geophysical Research* 80: 2529–2541.

Ritzwoller MH, Shapiro NM, and Zhong S (2004) Cooling history of the Pacific lithosphere. *Earth and Planetary Science Letters* 226: 69–84.

Robinson EM and Parsons B (1988) Effect of a shallow low-viscosity zone on small-scale instabilities under the cooling oceanic plates. *Journal of Geophysical Research* 93: 3469–3479.

Sandwell DT (1984) Thermomechanical evolution of oceanic fracture zones. *Journal of Geophysical Research* 89: 11401–11413.

Sandwell DT and McAdoo DC (1988) Marine gravity of the southern ocean and Antarctic margin from Geosat. *Journal of Geophysical Research* 93: 10389–10396.

Sandwell D and Fialko Y (2004) Warping and cracking of the Pacific plate by thermal contraction. *Journal of Geophysical Research* 109: B10411 (doi:10.1029/2004JB003091).

Sandwell D and Schubert G (1980) Geoid height versus age for symmetric spreading ridges. *Journal of Geophysical Research* 85: 7235–7241.

Sandwell DT and Schubert G (1982a) Geoid height-age relation from SEASAT altimeter profiles across the Mendocino fracture zone. *Journal of Geophysical Research* 87: 3949–3958.

Sandwell DT and Schubert G (1982b) Lithospheric flexure at fracture zones. *Journal of Geophysical Research* 87: 4657–4667.

Sandwell DT and Smith WHF (1997) Marine gravity anomaly from Geosat and ERS 1 satellite altimetry. *Journal of Geophysical Research* 102: 10039–10054 (doi:10.1029/96JB03223).

Sandwell DT, Winterer EL, Mammerickx J, *et al.* (1995) Evidence for diffuse extension of the Pacific plate from PukaPuka Ridges and Cross-Grain gravity lineations. *Journal of Geophysical Research* 100: 15087–15099.

Schlanger SO, Garcia MO, Keating BH, *et al.* (1984) Geology and geochronology of the Line Islands. *Journal of Geophysical Research* 89: 11261–11272.

Schroeder W (1984) The empirical age-depth relation and depth anomalies in the Pacific Ocean. *Journal of Geophysical Research* 89: 9873–9884.

Schubert G, Froidevaux C, and Yuen DA (1976) Oceanic lithosphere and asthenosphere: Thermal and mechanical structure. *Journal of Geophysical Research* 81: 3525–3540.

Schubert G and Turcotte DL (1972) One dimensional model of shallow mantle convection. *Journal of Geophysical Research* 77: 945–951.

Schubert G, Yuen DA, Froidevaux C, Fleitout L, and Souriau M (1978) Mantle circulation with partial shallow return flow. *Journal of Geophysical Research* 83: 745–758.

Sclater JG, Lawver LA, and Parsons B (1975) Comparison of long-wavelength residual elevation and free air gravity anomalies in the north Atlantic and possible implications for thickness of the lithosperic plate. *Journal of Geophysical Research* 80: 1031–1052.

Sigmundsson F (1991) Post-glacial rebound and asthenosphere viscosity in Iceland. *Geophysical Research Letters* 18: 1131–1134 (doi:10.1029/91GL01342).

Sleep NH (1969) Sensitivity of heat flow and gravity to the mechanism of seafloor spreading. *Journal of Geophysical Research* 74: 542–549.

Smith WHF (1998) Seafloor tectonic fabric from satellite altimetry. *Annual Review of Earth and Planetary Science* 26: 697–738.

Sparks DW and Parmentier EM (1991) Melt extraction from the mantle beneath spreading centers. *Earth and Planetary Science Letters* 105: 368–377.

Sparks DW and Parmentier EM (1993) The structure of three-dimensional convection beneath spreading centers. *Geophysical Jounal International* 112: 81–91.

Spiegelman M (1993) Physics of melt extraction: Theory, implications and applications. *Philosophical Transactions of the Royal Society of London A* 342: 23–41.

Spiegelman M and Elliot T (1993) Consequences of melt transport for Uranium series disequilibrium. *Earth and Planetary Science Letters* 118: 1–20.

Spiegelman M and Kelemen PB (2003) Extreme chemical variability as a consequence of channelized melt transport. *Geochemistry Geophysics Geosystems* 4(7): 1055 (doi: 1029/2002GC000336).

Spiegelman M and McKenzie D (1987) Simple 2-D models for melt extraction at mid-ocean ridges and island arcs. *Earth and Planetary Science Letters* 83: 137–152.

Stein CA and Stein S (1992) A model for the global variation in oceanic depth and heat flow with lithospheric age. *Nature* 359: 123–129.

Stein CA and Stein S (1993) Contraints on Pacific midplate swells from global depth–age and heat flow-age models. In: Pringle MS, Sager WW, Sliter WV, and Stein S (eds.) *Geophysical Monograph Series, vol. 77: The Mesozoic Pacific: Geology, Tectonics, and Volcanism*, pp. 53–76. Washinhton, DC: AGU.

Stixrude L and Lithgow-Bertelloni C (2005) Mineralogy and elasticity of the oceanic upper mantle: origin of the low-velocity zone. *Journal of Geophysical Research* 110: B03204 (doi:10.1029/2004JB002965).

Stolper E (1980) A phase diagram for mid-ocean ridge basalts: Preliminary results and implications for petrogenesis. *Contributions to Mineralogy and Petrology* 74: 13–27.

Tackley PJ and Stevenson DJ (1993) A mechanism for spontaneous self-perpetuating volcanism on the terrestrial planets. In: Stone DB and Runcorn SK (eds.) *Flow and Creep in the Solar System: Observations, Modeling and Theory*, pp. 307–321. Norwell, MA: Kluwer.

Turcotte DL and Oxburgh ER (1967) Finite amplitude convective cells and continental drift. *Journal of Fluid Mechanics* 28: 24–42.

Turner DL and Jarrard RD (1982) K-Ar dating of the Cook-Austral island chain: A test of the hot-spot hypothesis. *Journal of Volcanology and Geothermal Research* 12: 187–220.

van Hunen J, Huang J, and Zhong S (2003) The effect of shearing on the onset and vigor of small-scale convection in a Newtonian rheology. *Geophysical Research Letters* 30(19): 1991 (doi:10.1029/2003GL018101).

Vogt PR and Ostenso NA (1967) Steady state crustal spreading. *Nature* 215: 811–817.

Wang D, Mookerjee M, Xu Y, and Karato S (2006) The effect of water on the electrical conductivity of olivine. *Nature* 443: 977–980 (doi:10.1038/nature05256).

Wenk H-R and Tomé CN (1999) Modeling dynamic recrystallization of olivine aggregates deformed in simple shear. *Journal of Geophysical Research* 104: 25513–25528 (doi:10.1029/1999JB900261).

Weeraratne DS, Forsyth DW, Yang Y, and Webb SC (2007) Rayleigh wave tomography beneath intraplate volcanic ridges in the South Pacific. *Journal of Geophysical Research* 112: B06303 (doi: 10.1029/2006JB004403).

Wessel P (2001) Global distribution of seamounts inferred from gridded Geosat/ERS-1 altimetry. *Journal of Geophysical Research* 106: 19431–19441.

Wessel P and Haxby WF (1990) Thermal-stresses, differential subsidence, and flexure at oceanic fracture zones. *Journal of Geophysical Research* 95: 375–391.

Wessel P, Kroenke LW, and Bercovici D (1996) Pacific Plate motion and undulations in geoid and bathymetry. *Earth and Planetary Science Letters* 140: 53–66.

Wessel P and Lyons S (1997) Distribution of large Pacific seamounts from Geosat/ERS-1: Implications for the history of intraplate volcanism. *Journal of Geophysical Research* 102: 22459–22475.

Wiens DA and Stein S (1983) Age dependence of oceanic intraplate seismicity and implications for lithospheric evolution. *Journal of Geophysical Research* 88: 6455–6468.

Yoshino T, Matsuzaki T, Yamashita S, and Katsura T (2006) Hydrous olivine unable to account for conductivity anomaly at the top of the asthenosphere. *Nature* 443: 973–976 (doi:10.1038/nature05223).

Yuen DA and Fleitout L (1984) Stability of the oceanic lithosphere with variable viscosity: An initial-value approach. *Physics of the Earth and Planetary Interiors* 343: 173–185.

Zaranek SE and Parmentier EM (2004) Convective instability of a fluid with temperature-dependent viscosity cooled from above. *Earth and Planetary Science Letters* 224: 371–386.

Zhou Y, Nolet G, Dahlen FA, and Laske G (2006) Global upper-mantle structure from finite-frequency surface-wave tomography. *Journal of Geophysical Research* 111: B04304 (doi:10.1029/2005JB003677).

# 8 Mantle Downwellings and the Fate of Subducting Slabs: Constraints from Seismology, Geoid Topography, Geochemistry, and Petrology

**S. D. King**, Virginia Tech, Blacksburg, VA, USA

## 8.1 Introduction

Subduction zones are a major component of the dynamic-plate, convecting-mantle system, forming the downwelling limb of the mantle thermal boundary layer at Earth's surface. Subducting slabs are the dominant source of gravitational potential energy (GPE), that is, the driving force for mantle convection and plate tectonics. It is through subducted slabs that oceanic crust and sediments, including volatile compounds such as water are recycled back into the mantle. Earth is the only terrestrial planet with active subduction zones and it is possible that it is the only planet where the process of subduction as we know it ever occurred. It is conceivable that without subduction, Earth would not have been a suitable place for life to develop. Without the recycling of volatile compounds back into the mantle at subduction zones, it is possible that Earth would have developed a runaway greenhouse similar to Venus.

There have been a number of recent reviews of subduction zones (e.g., Bebout *et al.*, 1996; King, 2001; Stern, 2002; Eiler, 2003; van Keken and King, 2005) with some specifically focusing on seismology (Lay, 1994; van der Hilst *et al.*, 1997), the mantle wedge (van Keken, 2003), phase changes and the fate of slabs (Christensen, 1995), and water in the mantle (Hirschmann, 2006). In an attempt to provide more than a summary of these reviews and recent papers, we will focus this chapter around a broad theme that emerges from recent observations and models of convection. Through a variety of observations (e.g., petrology, careful relocation of earthquake hypocenters, seismic tomography, shear-wave splitting; van der Hilst *et al.*, 1997) it is clear that there is significant three-dimensional (3-D) structure and complexity within subduction zones. Not only do subducted slabs change dip along the length of an arc, studies increasingly recognize slab tears, slab windows, and other forms of slab deformation. While such complexity was noted in some of the earliest subduction zone studies (e.g., Isacks and Molnar, 1971), most slab thermal structure models assume no deformation of the downgoing slab. There is a continuing effort to document and understand changes in petrology along arcs as well, and in many cases these changes cannot be correlated with the traditional parameters (convergence rate and slab age) that control simple slab models. Complementary to the increasing complexity of the observations, convection modeling tools have increased numerical resolution and sophistication in terms of the complexity of rheology formulations, and physics of melting, dehydration, and flow. These models are now attempting to model specific subduction zones using constraints that were not possible to consider with earlier generations of models.

Our understanding of the geometry and composition of subducted slabs, as well as the physical process of subduction, comes from indirect observations including seismology (e.g., earthquake hypocenters, seismic structure, anisotropy, and attenuation), gravity, topography, geochemistry, and petrology of island arc lavas. Numerical calculations of the dynamics of subducted slabs and slab thermal models also play an important role, enabling us to test hypotheses. There have been many advances in all these areas in the past decade, and it would be impossible to do justice to all the important contributions that have been made, some of which are chapters of other parts of this volume. An attempt has been made in this chapter to point to other review papers along the way that highlight areas which cannot be given proper attention here.

The development of the worldwide seismographic network enabled the accurate location of earthquake hypocenters. Gutenberg and Richter (1939) first classified earthquakes by focal depth and compiled the distribution of global earthquake hypocenters clearly showing that intermediate- and deep-focus earthquakes occurred in narrow zones that are consistent with descending plate material beneath trenches (cf. Isacks *et al.*, 1968). In addition, it was recognized that earthquake focal mechanisms of the deepest earthquakes are aligned with the axis of compression along the dip of the plane delineated by the hypocenters (cf. Isacks and Molnar, 1971) and this was taken to be evidence that the downgoing lithosphere meets some form of resistance with depth (e.g., Richter, 1973; Vassiliou *et al.*, 1984). It was also recognized that the focal mechanisms of intermediate-focus earthquakes were aligned with the axis of extension along the dip of the plane delineated by the hypocenters in areas where there are no deep-focus earthquakes. The newly recognized process of seafloor spreading required the consumption of lithosphere to maintain a constant volume of crust (Morgan, 1968; Le Pichon, 1968) and Le Pichon (1968) calculated the rate of seafloor consumption based on the rate of formation of new seafloor. Le Pichon's estimate agrees with the predicted rate of destruction of seafloor at trenches from the seismically determined slab geometry (e.g., Isacks *et al.*, 1968).

The realization that cold plate was sinking into the mantle (e.g., Vening Meinesz, 1954) quickly led to the development of the first slab thermal models (McKenzie, 1969; Minear and Töksoz, 1970), recognizing that the thermal structure of the descending slab would impact surface heat flow, gravity, and seismic-wave traveltimes in the area surrounding the trench. It was apparent from these models that the temperature of the slab was controlled by the rate of seafloor spreading and thermal properties of the slab (McKenzie, 1969). Some of these early models predicted that for the slowest moving plates ($\approx 1$ cm $yr^{-1}$), the slab would be in thermal equilibrium with the mantle by a depth of 600 km (Minear and Töksoz, 1970). Turcotte and Schubert (1973) explained the spatial relationship between the newly discovered downgoing slabs and the linear island chains associated with trenches by frictional heating associated with the motion of the descending slab. Using a 1-D analytical treatment of narrow shear zones, Yuen *et al.* (1978) showed that the mantle deforms readily enough that shear zones are insufficient to produce melting. The relative importance of frictional heating has been debated (Turcotte and Schubert, 1973; Scholz, 1990; Molnar and England, 1990; Peacock, 1992; 1993; Tichelaar and Ruff, 1993; Hyndman and Wang, 1993), and we will return to this topic when we examine thermal models below.

### 8.1.1 Subducted Slabs and GPE Driven Plate Tectonics

We begin by focusing on the GPE associated with subducted slabs, the primary energy source driving mantle convection (e.g., Turcotte and Oxburgh, 1967; Forsyth and Uyeda, 1975; Richter, 1977; Davies and Richards, 1992). First, we estimate the thermal structure (hence buoyancy) that is coming into the trench as a result of the cooling of the oceanic plate. The thickness of an oceanic plate, $d$, is proportional to the square root of the time the plate spends at the surface. The constant of proportionality is the square root of the thermal diffusivity, $\kappa$. This time can be estimated by the length of the plate, $L_p$, divided by the plate velocity $v_p$. Hence

$$d = \sqrt{\frac{\kappa L_p}{v_p}} \qquad [1]$$

(cf. Chapter 4). As the plate subducts, the cold material is more dense than the warmer mantle surrounding it. Thus the density difference between the slab and mantle will be $g\rho\alpha\overline{\Delta T}$ where, $\alpha$ is the coefficient of thermal expansion, $\rho$ is the density, $g$ is the gravitational acceleration, and $\overline{\Delta T}$ is the average temperature difference between the interior mantle and surface. This buoyancy must be balanced by resistance of the mantle surrounding the slab, which has the form $\eta D_{ij}$, where $\eta$ is the coefficient of dynamic viscosity and $D_{ij}$ is the deformation-rate tensor (*see* Chapter 2). A crude estimate of the deformation rate is the velocity gradient, which we will assume goes from $v_p$ at the plate to zero at some point near the middle of the mantle. Assuming the length of the plate, $L_p$, is approximately the same as the depth of the mantle, $D$, and balancing the buoyancy force and viscous resistance, one obtains

$$v = D\left[\frac{g\rho\alpha\overline{\Delta T}\kappa^{1/2}}{2\eta}\right]^{2/3} \qquad [2]$$

With a reasonable choice of values for these parameters, one obtains a value for $v_p$ on the order of $100\,\text{mm yr}^{-1}$. This value is remarkably close to the velocity of oceanic plates with subduction zones (cf. Forsyth and Uyeda, 1975), especially when one considers the simplicity of the analysis.

For each of the major plates, Forsyth and Uyeda (1975) compiled plate velocity, area of plate, length of ridge, trench, and transform fault along the circumference of the plate boundary and calculated the relative importance of each contribution, concluding that the subducting slabs dominate the force balance. Taking their data (from table 1 of Forsyth and Uyeda, 1975), we plot the covariance matrix (**Figure 1**) illustrating that the percent of plate boundary dominated by trench (i.e., effective trench boundary – as trenches at opposite sides of the plate will presumably cancel their contribution) has the strongest correlation with plate velocity (correlation coefficient of 0.92). The solid colors indicate a strong correlation (green, positive; pink, negative) and the more washed out the color, the weaker the correlation. The percentage of continental area is next (correlation coefficient of 0.64) and the percent of effective ridge along the plate boundary is third (correlation coefficient of 0.46). The total area of the plate and the percentage of the boundary that is transform fault (as well as the other observations tabulated by Forsyth and Uyeda (1975)) are uncorrelated with plate velocity. (This figure is a compact representation of the graphs shown in figures 5–9 of Forsyth and Uyeda (1975).)

## 8.2 Slab Geometry

Historically, the geometry of subducted slabs has been constrained by earthquake hypocenters and focal mechanisms (e.g., Isacks *et al.*, 1968; Isacks and Molnar, 1969). **Figure 2** illustrates a global summary of the distribution of down-dip compressive stresses in subduction zones (Isacks and Molnar, 1971). While slab tomography is playing an increasingly important role in understanding slab structure, it does not provide information on the stress field. Slab geometry is variable, and significant attention has been given to correlating slab dip with a variety of observables related to subduction zones. A seminal compilation of 26 parameters measured or estimated for 39 modern subduction zones was undertaken by Jarrard (1986). Using these observations Jarrard performed a multistep correlation analysis, finding that three independent variables – the age of the oceanic plate at the trench, the convergence rate of the plates at the trench (based on global plate reconstructions, e.g., Chase, 1978; Minster and Jordan, 1978), and intermediate slab dip (the dip angle between the trench and the slab at 100 km depth) – correlate significantly with the other observations. There is an excellent correlation between the slab length and the product of the convergence rate and the age of the downgoing slab, consistent with the predictions of some slab thermal models (Molnar *et al.*, 1979). This product

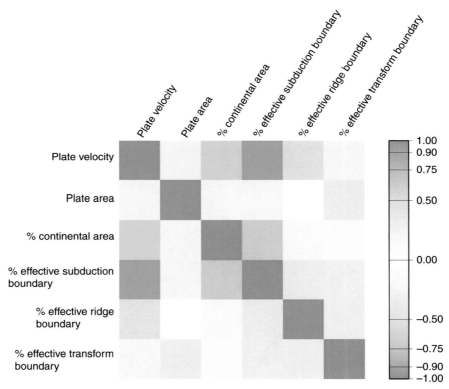

**Figure 1**    Color shading of the covariance matrix for selected plate geometry characteristics from Forsyth and Uyeda (1975) (see table 1 of Forsyth and Uyeda). The percentage of subduction zone, ridge, and transform fault boundary are the effective percentages (as defined by Forsyth and Uyeda), where the effects of boundaries that occur on opposing sides of the plate are assumed to cancel out and are not included in the percentage of length of that type along the boundary. Green is strong positive correlation, red is strong negative correlation.

has become known as the slab thermal parameter (Kirby *et al.*, 1991, 1996).

We summarize the subduction zone parameter correlations by plotting a covariance matrix with nine of the best-constrained observations from Jarrard's compilation plus the slab thermal parameter (slab age times convergence velocity) in **Figure 3**. There is also a strong positive correlation between the age of the plate at the trench and the age calculated at the slab tip, which is expected if the age of the slab tip is a linear function of the age of the incoming plate. This would be the case if the slab was rigid and the fact that this correlation is high is one indicator that slab deformation is limited. The other striking correlation is that between the age of the arc and the intermediate slab dip (trench to 100 km depth) and the deep slab dip (the average of the slab dip between 100 and 400 km). Given the attention focused on trench migration, it is worth pointing out that the analysis shows at best a very weak correlation between slab dip and trench migration.

The relationship between trench rollback and slab age has been examined in a number of studies with conflicting results. Jarrard (1986) finds as many advancing as retreating trenches while Garfunkel *et al.* (1986) find that trench rollback, or retreat, predominates. Slab dynamics and back-arc deformation style have recently been re-examined (Heuret and Lallemand, 2005) using relocated hypocenters (Engdahl *et al.*, 1998), updated plate ages (Mueller *et al.*, 1997), and an updated global plate velocity model (Gripp and Gordon, 2002). After excluding regions where continental lithosphere, volcanic arc crust, or oceanic plateaus are being subducted, they divide the oceanic subduction zones into 159 segments with a uniform interval of two degrees of trench length. Heuret and Lallemand (2005) find as many advancing as retreating trenches with slab motion occurring in two distinct populations: one with essentially zero trench velocity and one with a significant ($-50 \, \text{mm} \, \text{yr}^{-1}$) component of retreat. As one might expect, they find back-arc extension when

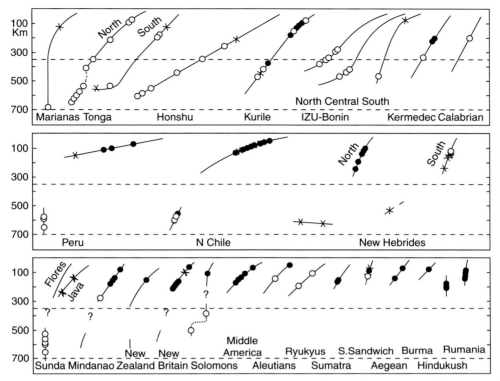

**Figure 2** A global summary of down-dip compressive stresses calculated from focal mechanisms. The open circle is for down-dip compressive or *P* axis roughly parallel to the slab dip and the filled circles are down-dip *T* or tensional axis parallel to slab dip. The line represents the dip and length of the seismicity in the subduction zone. Reproduced from Isacks B and Molnar P (1971) Distribution of stresses in the descending lithosphere from a global survey of focal-mechanism solutions of mantle earthquakes. *Reviews of Geophysics and Space Physics* 9: 103–174. With permission from American Geophysical Union.

the overriding plate is retreating from the trench and back-arc compression when the overriding plate is advancing. The correlation between the absolute velocity of the overriding plate and the style of back-arc deformation from this new compilation is consistent with Jarrard (1986). However, when correlating slab motion and age, Heuret and Lallemand's result is opposite of the result seen in many numerical models; on Earth older subducting plates are observed to be advancing, not retreating as is the case in many dynamic subduction calculations. Heuret and Lallemand (2005) suggest that slabs are partially anchored because of the slabs' resistance to large-scale lateral flow (e.g., Gurnis and Hager, 1988; Griffiths *et al.*, 1995; Scholz and Campos, 1995; Becker *et al.*, 1999). The degree of variability in the data suggests that local flow effects, such as small-scale flow in the vicinity of retreating slab edges or tears or more regional-scale flow associated with the shrinking of the Pacific plate area (e.g., Garfunkel

*et al.*, 1986) may significantly impact slab–trench dynamics.

Following a slightly different approach, Cruciani *et al.* (2005) re-examine the global correlation of slab dip with subduction zone observations using the regionalized upper mantle (RUM) seismic model (Gudmundsson and Sambridge, 1998) to map out the slab geometry rather than using relocated earthquake hypocenters, as both Jarrard (1986) and Heuret and Lallemand (2005) did. Cruciani *et al.* (2005) construct 164 trench-perpendicular cross-sections through the RUM model using a digital seafloor age model for the age of the incoming slab (Mueller *et al.*, 1997) and a global plate velocity model for the convergence velocity (DeMets *et al.*, 1994). There is more scatter in the Cruciani *et al.* (2005) compilation than the original compilation by Jarrard (1986), possibly because of the use of a seismic model as opposed to the more direct focal mechanism data, but more likely reflecting the actual complexity of slab geometries. The product

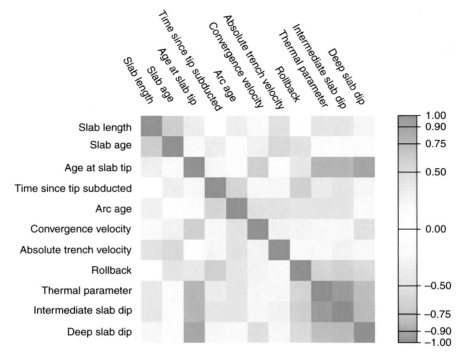

**Figure 3** Covariance matrix for nine observations compiled from 39 modern subduction zones (Jarrard, 1986). The color scale represents the value of the correlation coefficient, with +1.0 being strongly correlated and −1.0 being strongly anticorrelated. Rollback is the rate of trench migration estimated from the global plate models (Minster and Jordan, 1978). Velocity is the velocity of the plate at the trench from the same global plate models. Slab age is the age of the plate entering the trench, and slab-tip age is calculated using the length of the slab and rate of change of the plate age for the plate entering the trench. The intermediate dip is calculated from the trench to the 100 km depth of the slab. The deep dip is calculated from the 100–400 km depth of the slab. The thermal parameter is the product of the slab age and the length of the slab.

of slab age and plate convergence rate (thermal parameter) correlates with slab dip better than either slab age or plate convergence rate individually, although the correlation is weak (correlation coefficient of 0.45), and suggests that forces beyond the local subduction environment may be important. Hager and O'Connell (1981) show that they could produce a reasonably good fit to slab dips simply with global plate motions and large-scale return mantle flow, without a detailed model of the deformation in the subduction zone, suggesting that global-scale mantle flow may play an important role in slab geometry.

As one example where deep mantle flow is almost certainly impacting the deformation of a slab, Gurnis *et al.* (2000a) show that the Tonga slab is being deformed by a large-scale upwelling associated with the Pacific superplume. There is no substantial aseismic penetration into the lower mantle beneath Tonga, consistent with initiation of subduction during the Eocene. Both the pattern and amount of deformation and the seismic energy release in the

Tonga slab show that it is deforming faster and has accumulated more deformation than any other slab. Gurnis *et al.* (2000a) argue that the strong deformation of the Tonga–Kermadec slab in the transition zone is a result of the short subduction history (40 My) and the upward flow in the lower mantle from a broad upwelling associated with the Pacific superplume. The lack of deep aseismic extension of the Tonga slab into the lower mantle is consistent with the short subduction history and the deep mantle flow. While recognizing the importance of regional plate motions, especially the opening of the Lau basin, Gurnis *et al.* argue that the structure and geometry of the Tonga–Kermadec slab is best explained with controls from deep and shallow forces.

### 8.2.1 Slab Structure

In addition to studies of subducted slab dip, numerous studies illustrate that slabs bend, kink, and thicken. These include the location of Wadati–Benioff zones

(WBZs) as mapped through earthquake hypocenters (Isacks and Barazangi, 1977; Giardini and Woodhouse, 1984, 1986; Fischer et al., 1991; Chen et al., 2001; Brudzinski and Chen, 2003a), the study of earthquake focal mechanisms and moment tensors (Isacks and Molnar, 1969, 1971; Vassiliou, 1984; Giardini and Woodhouse, 1986; Bevis, 1988; Fischer and Jordan, 1991; Holt, 1995), and tomography (Zhou and Clayton, 1990; Van der Hilst et al., 1991, 1993, 1997; Chiu et al., 1991; Fukao et al. 1992; van der Hilst and Seno, 1993; Zhao et al., 1994; Van der Hilst and Mann, 1994; Van der Hilst, 1995; Gudmundsson and Sambridge, 1998; Widiyantoro et al., 1999; Miller et al., 2004).

The rate of accumulation of seismic moment in WBZs can be used to estimate the average down-dip strain rate in subducting slabs (Bevis, 1986; Fischer and Jordan, 1991; Holt, 1995). This assumes that the amount of aseismic deformation is small and that the viscous deformation is parallel with the brittle layer with the same deformation-rate tensor. Thus, the seismic deformation rate probably represents a minimum estimate of slab deformation. Because slabs are generally in down-dip compression or down-dip extension, the average change in the length of a slab due to the cumulative seismicity can be related to the sum of the seismic moments of the individual seismic events (Bevis, 1988). Even after accounting for the scatter in focal mechanisms, the average strain rate in subducting slabs is estimated to be $10^{-15}\,\mathrm{s}^{-1}$ in the depth range of 75–175 km (Bevis, 1988). This compares with a characteristic asthenospheric strain rate of $3 \times 10^{-14}\,\mathrm{s}^{-1}$ (e.g., take a characteristic plate velocity of $5\,\mathrm{cm}\,\mathrm{yr}^{-1}$ divided by a thickness of 200 km), leading to the rather surprising suggestion that the difference in deformation rates between the asthenosphere and subducting slabs is no more than a factor of 3. Taking the average descent rate of slabs this deformation rate corresponds to a total accumulated strain in the depth range of 75–175 km of 5%. Fischer and Jordan (1991) find that the seismic data require a thickening factor of 1.5 or more in the seismogenic core of the slab in central Tonga and complex deformation in northern and southern Tonga. Holt (1995) argues that the average seismic strain rate in Tonga represents as much as 60% of the total relative vertical motion between the surface and 670 km.

The geometry of WBZs also indicates that slabs are significantly deformed. Subducting slabs bend to accommodate overriding plates as they descend into the mantle; they also unbend, otherwise WBZs would curve back on themselves (Lliboutry, 1969). The dip

of a subducting slab changes over a narrow depth interval, from 10–20° in the interplate thrust zone to 30–65° at depths near 75 km (Isacks and Barazangi, 1977). This change in shallow dip is a feature that is not well represented by any slab model. In addition to deformation as the slab descends into the mantle, there is also deformation in the horizontal plane at subduction zones. Because oceanic plates themselves are spherical caps, the degree of misfit between subducting slab and the best-fit spherical cap provides an estimate of the amount of slab deformation (Bevis, 1986). Based on this analysis, the Alaska–Aleution, Sumatra–Java–Flores, Caribbean, Scotia, Ryukyu, and Mariana arcs undergo a minimum of 10% strain (Bevis, 1986; Giardini and Woodhouse, 1986).

There is also a difference in the average dip angle between the populations of slabs where the overriding plate is continental or oceanic plate (Furlong et al., 1982; Jarrard, 1986). While the average of the two populations is distinctly different (40° vs 66° from Jarrard's data), there are notable exceptions including the shallow west-dipping Japan slab and the steep eastward-dipping Solomon and New Hebrides slabs. Studies of supercontinent breakup observe shallow dipping slabs under continents (e.g., Lowman and Jarvis, 1996). Because we are currently in a phase where continents are generally moving away from the site of the former supercontinent and overriding oceanic plates, it is not clear whether this correlation once again reflects the effect of trench migration or mechanical differences in the overriding plate.

In both numerical and laboratory experiments, young buoyancy-driven slabs steepen with age from the time of the initiation of subduction until the slab reaches the transition zone (Gurnis and Hager, 1988; Griffiths et al., 1995, Becker et al., 1999). Once slabs reach the top of the lower mantle, slab-dip angles become progressively shallower if there is oceanward trench migration because the deep slab becomes anchored in the higher-viscosity lower mantle. The numerical and laboratory experiments observe trench rollback, while compilations of slab geometry show that older slabs are advancing, not retreating (e.g., Heuret and Lallemand, 2005). It is interesting to note that the strongest and only significant correlation in the data set compiled by Jarrard (1986) is the correlation of the duration of subduction with dip of the deepest part of the slab, defined by Jarrard as 100–400 km. While it is not possible to isolate this from the other factors that influence slab geometry, this correlation is consistent with the assumption that subducted slabs are evolving, time-dependent features that are not at steady-state equilibrium.

## 8.2.2 Slab Structure Seen from Refracted and Scattered Waves

At high frequencies (0.5–10 Hz) seismic waves are particularly sensitive to the presence and state of subducted oceanic crust. Both S and P waves that travel long distances along the top of the slab appear dispersed (e.g., Abers and Sarker, 1996; Abers, 2000) and the dispersion is consistent with a constant-velocity waveguide of constant thickness. Observations in all North Pacific subduction zones suggest that crust remains distinct to depths of 150–250 km. The signals show waveguide behavior at the scale of a few kilometers: short-wavelength, high-frequency energy ($\geq 3$ Hz) is delayed 5–7% relative to that of low frequencies ($\geq 1$ Hz), systematically. Velocities in a low-velocity layer 1–7 km thick, likely subducted crust, must remain seismically slow relative to surrounding mantle at these depths to explain the observations. The inferred velocities are similar to those estimated for blueschists, suggesting that hydrous assemblages persist past the volcanic front.

Abers (2005) solves for traveltimes for waves traveling through a low-velocity channel with variable velocity with depth. Using both S and P waves from events deeper than 100 km recorded at Global Seismic Network (GSN) stations between 1992 and 2001, Abers compiles a data set of 210 P waves and 156 S waves for study and finds that the data from seven circum-Pacific arcs (Aleutian, Alaska, Hokkaido-S. Kurile, N. Honshu, Mariana, Nicaragua, and Kurile–Kamchatka) are consistent with a waveguide extending to greater than 150 km depth with a low-velocity channel 28 km thick with anomalies as large as $d\ln V_{\mathrm{p}} = 14\%$ compared with surrounding mantle, although there is substantial variability with depth. Below 150 km the waveguide velocity is close to the surrounding mantle velocity. The low velocities are consistent with either low-temperature hydrated mafic rocks or metastable gabbros. The waveguide velocities also vary substantially from arc to arc, correlating with slab dip but not with other subduction-related parameters (e.g., plate age, convergence velocity, thermal parameter). Abers favors fluids or hydrous minerals as opposed to variable metastability of anhydrous gabbro as an explanation pointing out that the variability with dip is consistent with fluids from shallow slabs

being driven into the mantle wedge while for steeper slabs the fluid preferentially migrates up the sediment layer.

## 8.2.3 Slab Structure Seen from Seismic Tomography

With an increasing number of tomographic studies in subduction zones (Zhou and Clayton, 1990; Van der Hilst et al., 1991, 1993; Chiu et al., 1991; Fukao et al., 1992; van der Hilst and Seno, 1993; Van der Hilst, 1995; Gudmundsson and Sambridge, 1998; Widiyantoro et al., 1999; Miller et al., 2004; 2005) it is becoming clear that subducted slabs are not 2-D tabular features. For example, slab dip varies along the length of many arcs. In the case of the Cocos slab, the dip angle changes along the Middle American Trench from about 80° at the northern end, with earthquakes occurring to depths of about 200 km, to 60° in the central part, with earthquakes occurring to depths of about 125 km, and even shallower (dip and maximum earthquakes) to the south (Colombo et al., 1997). The age of the Cocos Plate, the relative velocity between the Cocos and North/South American Plates, the heat flow, and even the seamount distribution change along the length of the arc and all of these are known to impact slab thermal structure.

Seismic tomography has been used to image the slab beneath the Mariana arc penetrating vertically into the lower mantle, whereas beneath the Izu–Bonin arc the slab appears to be deflected horizontally on top of the 660 km discontinuity (Zhou and Clayton, 1990; Chiu et al., 1991; Fukao et al., 1992; van der Hilst and Seno, 1993; Gudmundsson and Sambridge, 1998; Widiyantoro et al., 1999; Miller et al., 2005). Recent analysis of tomographic slab models (Miller et al., 2005) shows a distinct change in the seismic velocity in the Pacific slab beneath the Izu–Bonin arc at 300–450 km depth, in a position north and west of the Ogasawara Plateau at the southern end of the Izu–Bonin arc. The change in morphology of the slab could be interpreted as a 'slab tear'. Beneath northern Tonga, a complex morphology is observed with a seismically fast anomaly lying subhorizontally above the 660 km discontinuity for almost 1000 km before (apparently) descending into the lower mantle (Van der Hilst, 1995). Flat-lying fast seismic-velocity anomalies at the base of the transition zone have been found beneath Japan, Java, and Izu–Bonin trenches (van der Hilst et al., 1991, 1993), and these are interpreted to be the extension (often aseismic) of ongoing subducting slabs.

The pattern of deep seismicity beneath Tonga provides yet another illustration of the complexity in subduction zones (Hamburger and Isacks, 1987; Fischer *et al.*, 1991; van der Hilst, 1995; Chen and Brudzinski, 2001, 2003; Brudzinski and Chen, 2003a, 2003b). There is a group of well-located deep events that lies about 200 km above the deepest end of the Tonga slab, making it difficult to connect this cluster of earthquakes with the actively subducting lithosphere. If indeed this cluster of earthquakes represents a remnant slab, it is not clear that it is attached to the current Tonga slab. A possible explanation for the origin of a detached slab is subduction of the Pacific Plate along the fossil Vitiaz trench where subduction ceased about 5–8 Ma (Hamburger and Isacks, 1987; Fischer *et al.*, 1991; Okal and Kirby, 1998; Chen and Brudzinski, 2001). Although another interpretation is that this cluster of earthquakes represents a subhorizontal extension of the Tonga slab along the subduction zone (Fukao *et al.*, 1992; van der Hilst *et al.*, 1993). The interaction of the Tonga slab with a mantle upwelling (Gurnis *et al.*, 2000a) is likely to be a significant factor (perhaps in combination with the other events). With either explanation, it is clear that the interaction of subducted material with the transition zone can yield complex planforms of subducted material.

Another region of slab complexity is the South American slab, in the vicinity of the 1994 deep Bolivian earthquake (Engdahl *et al.*, 1995). Tomographic images of the deep structure of the Andean subduction zone beneath western South America indicate that the Nazca slab is continuous both horizontally and with depth over most regions; in contrast, the Nazca slab penetrates the lower mantle beneath central South America, but is partly deflected to the south. There are two regions of deep earthquakes (a northern and southern zone) separated by the region containing the 1994 deep Bolivia main shock. The region between the north and south deep earthquake zones is modeled as a northwest-striking and steeply northeast-dipping slab structure.

A summary of tomographic cross-sections across Pacific subduction zones (**Figure 4**) from Karason (2002) illustrates both a variety of slab planforms and the general broadening of slabs in lower mantle. It is possible that the general appearance of thinning of the slab at the top of the transition zone is a result of the seismic properties of the mantle in this region, rather than reflecting a thinning of the slab itself. Without independently solving for the topography on the phase boundary itself, traveltime anomalies due to the deflection of the phase boundary will be mapped into volume perturbations in the 3-D seismic

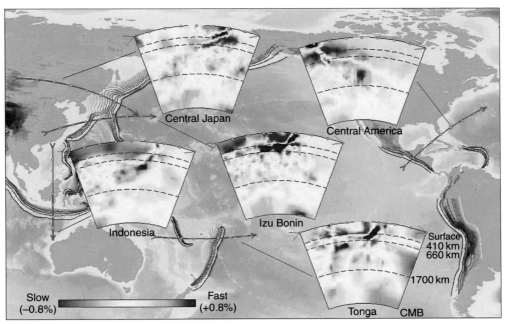

**Figure 4**  Summary of seismic tomography P-wave cross-sections through Pacific subduction zones. Reproduced from Albarède F and van der Hilst RD (2002) *Zoned mantle convection. Philosophical Transactions of the Royal Society of London A* 360: 2569–2592, with permission from Royal Society.

model. Because the jump in velocity at the phase boundary is on the order of 5%, this is a significant contribution to the total traveltime anomaly. Furthermore, a difference between the reference model and the average velocity with depth will lead to the apparent thickening or thinning of vertical anomalies when plotting residual anomaly cross-sections. Thus, especially near phase boundaries, caution must be exercised when interpreting seismic cross-sections. It is also important to bear in mind that this kind of global compilation does not capture the along-strike variability within a single arc.

Deal and Nolet (1999) construct a high-resolution 3-D P-wave velocity model beneath the northwest Pacific. Assuming the positive velocity deviations in the subducting lithosphere are to first order due to a temperature anomaly, they construct a theoretical slab temperature profile using an approach based on the appendix of Davies and Stevenson (1992) and convert this to a synthetic slab velocity model using $dV_p/dT \approx 4.8 \times 10^{-4} \, km \, s^{-1} \, °C^{-1}$. They then invert this model with the observed tomographic model, to estimate the slab thickness and mantle potential temperature, obtaining thickness estimates of $84–88 \, km \pm 8 \, km$ for the Izu-Bonin, Japan, and Kuril slabs. The uncertainty in the thickness is dominated by the uncertainty in the mantle temperature and the estimated variability of 8 km is a result of their estimated 100° uncertainty in mantle temperature. The initial tomography result is then modified to closely resemble the synthetic slab tomogram with a method that guarantees that the new tomography model will satisfy the original seismic delay times. The modified slab images show continuous and narrow slabs compared to the initial tomographic results.

Deal et al. (1999) use the same approach in Tonga and the theoretical temperature model gives an optimal slab thickness of 82 km for a region near 29° S in Tonga. The final image shows a very narrow and continuous slab with maximum velocity anomalies of the order of 6–7%; many of the gaps within the slab, as well as artifacts around the slab which were present in the minimum-norm solution, are absent in the biased image. One can deduce from these experiments that some of the gaps seen in seismic tomography images are not required by the seismic data and that a more continuous slab model is equally acceptable. A different and slightly less rigorous approach was taken by Zhao (2001) who used a slab model *a priori* as a part of the starting model. Because the final model retained the signature of the original slab starting model, one can conclude that the traveltime data were consistent

with the starting model, or at least that there was insufficient information in the seismic data to require a modification to the starting model.

### 8.2.4 Shear-Wave Splitting in Subduction Zones

Perhaps the most challenging seismic observations to reconcile with flow models are the estimates of seismic anisotropy, as measured by shear-wave splitting observations (Savage, 1999). Seismic anisotropy in the mantle is generally assumed to be the result of lattice-preferred orientation (LPO) of mantle minerals such as olivine. In the absence of water, olivine *a*-axes (the seismically fastest axis) are assumed to be aligned with the direction of flow in the dislocation creep regime (e.g., Karato and Wu, 1993). However, recent experimental studies have suggested that the olivine slip system changes under high stress in hydrated systems (Jung and Karato, 2001). Clearly, subduction zones are likely to be hydrated and these new laboratory results complicate the interpretation of the shear-wave splitting observations.

Shear-wave splitting observations include a fast polarization direction, reflecting the orientation of fabric, and a splitting time, reflecting the organization and strength of the fabric. As is the case with the other seismic observations, subduction zones exhibit a variety of shear-wave splitting orientations along a single arc. These have been summarized in global compilations, such as **Figure 5** taken from Lassak et al. (2006).

A number of studies have used mantle flow models to predict the LPO development and resulting shear-wave splitting (e.g., Fischer et al., 2000; Fouch et al., 2000; Hall et al. 2000; Kaminski and Ribe, 2001; Lassak et al., 2006). These studies map the LPO for mantle mineralogies by either orienting crystallographic axes to the local flow direction or calculating finite-strain directions and using them as a guide for the orientation of crystallographic axes. Kaminiski and Ribe (2001) have developed a model for the evolution of LPO in olivine aggregates that accounts for intercrystalline slip and dynamic recrystallization.

In the Tonga subduction zone (Bowman and Ando, 1987; Fischer et al., 2000; Smith et al., 2001), shear-wave splitting results vary from trench-parallel fast directions near the trench to convergence-parallel fast directions closer to the backarc. Jung and Karato (2001) suggest that hydration of olivine may influence the mantle close to the trench and modeling of anhydrous–hydrous transition predicts a fast polarization direction pattern similar to the observed

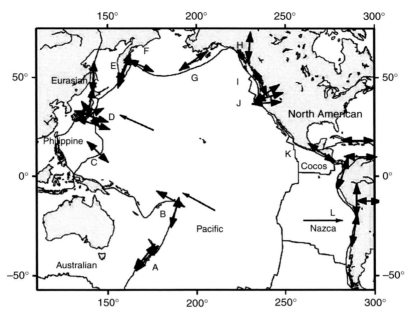

**Figure 5** Summary of shear-wave splitting observations around the Pacific Rim illustrating the variety of observed fast directions near subduction zones from Lassak et al. (2006). Double-headed vectors indicate the average regional orientation of the fast polarization direction. Many regions exhibit fast direction populations with orientations both parallel and orthogonal to local trench strike. For details see Lassak et al. (2006). From Lassak TM, Fouch MJ, Hall CE, and Kamiknsi E (2006) Seismic characterization of mantle flow in subduction systems: Can we resolve a hydrated mantle wedge? *Earth and Planetary Science Letters* 243: 632–649.

range for Tonga, but are not able to match the amplitude (Lassak *et al.*, 2006). The mantle wedge flow associated with the Tonga slab is complex, due to complexity in the slab that is interpreted as a slab tear in the north and the possible influence of the Samoan plume (Smith *et al.*, 2001).

In the Japanese subduction zone, shear-wave splitting observations also range from trench-parallel near the trench to convergence-parallel in the back-arc (Hiramatsu and Ando, 1996). The Japan subduction system is complex with two triple junctions and multiple subducting slabs. In addition, crustal anisotropy may be responsible for some local variability of splitting variations (e.g., Fouch and Fischer, 1996); however, there is agreement that mantle anisotropy is required to explain the majority of the shear-wave splitting results across the region (e.g., Fouch and Fischer, 1996; Nakajima and Hasegawa, 2004; Long and van der Hilst, 2005). Near Hokkaido, Nakajima and Hasegawa (2004) observe splitting variations that are consistent with north–south shear in the over-riding plate or the presence of a hydrated mantle wedge. Further south near Honshu and Ryukyu, Fouch and Fischer (1996) and Long and van der Hilst (2005) observe similar shear-wave splitting

directions with fast directions that rotate from trench-parallel near the trench to trench-orthogonal into the back-arc. The splitting directions are more consistent than the splitting times, with no clear systematics from the splitting times. While most studies conclude that the Japan subduction zone shear-wave-splitting results are consistent with a hydrated wedge (Kneller *et al.*, 2005), the modeling leaves plenty of room for alternatives.

In the Kamchatka subduction zones, shear-wave splitting results once again range from convergence-parallel near the trench to trench-parallel toward the back-arc (Peyton *et al.*, 2001; Levin *et al.*, 2004). Levin *et al.* (2004) interpret these measurements as the result of a complex flow regime with mantle flow moving around the northern bending edge of the downgoing slab due to slab loss/detachment, possibly combined with deformation of material in the wedge. Lassak *et al.* (2006) suggest that hydration of the wedge may also be a possible explanation. South America also exhibits a range of shear-wave splitting directions (Russo *et al.*, 1994; Polet *et al.*, 2000; Anderson *et al.*, 2004).

Yang *et al.* (1995) inferred possible strain geometries in the mantle beneath the eastern Aleutians by mapping seismic anisotropy with shear-wave splitting. Their observed splitting parameters are well matched by predicted values for olivine-rich mantle wedge models with 1% SV anisotropy where the olivine *b*-axis is arc-orthogonal and the *a*-axis is vertical (no arc-parallel strain) or arc-parallel (no vertical strain). These results are consistent with mantle wedge strain models in which arc-normal compression is accompanied by arc-parallel or vertical shearing or extension. Mehl *et al.* (2003) describe residual mantle exposures in the accreted Talkeetna arc, Alaska, that provide the first rock analog for the arc-parallel flow that is inferred from seismic anisotropy at several modern arcs. Stretching lineations and olivine [100] slip directions are subparallel to the Talkeetna arc for over 200 km indicating that mantle flow was parallel to the arc axis. Slip occurred chiefly on the (001) [100] slip system, rarely the dominant slip system observed in olivine, and the alignment of the olivine [100] axes yields a calculated S-wave anisotropy with the fast polarization direction parallel to the arc.

The key point to be made here is that the shear-wave splitting observations, like many of the other seismic observations discussed above, appear to point to a flow environment in subduction zones that is more complex than the simple corner flow model that has been envisaged for the past 30 years. It is possible that some of the interpretation of the shear-wave splitting observations may actually be better explained by less complexity in the flow model and recognition that shape-preferred orientation, due to lenses or veins of melt (e.g., Spiegelman, 2003) or hydration (e.g., Jung and Karato, 2001; Lassak *et al.*, 2006), can also give rise to shear-wave splitting.

Given the complexity of the slab geometry and the petrologic variations introduced by the phase transformations (and possible structure due to nonequilibrium phase transformations), the interpretation of these measurements is difficult.

Seismic studies in subduction zones are extensively reviewed in this series.

## 8.3  Slab Modeling and Thermal Structure

Slab thermal modeling takes a complex region of the Earth where the downgoing plate descends into the mantle and attempts to capture this in a mathematically tractable manner. Typically slab thermal models assume a thermal structure for the incoming plate, some form of coupling between the subducting and overriding plates in the 'seismogenic zone', and some assumptions about the flow in the mantle wedge (**Figure 6**). In most slab thermal models the slab is assumed to be rigid and descends with a constant dip. Thermal models differ as to assumptions regarding flow in the mantle wedge; however, the largest difference comes from a variety of assumptions (sometimes not clearly stated) regarding shear heating and mechanical coupling at the slab and overriding-plate interface are used in slab thermal models. Because of the limited domain of slab thermal models (both horizontally and vertically), it is not possible to calculate geoid or dynamic topography anomalies from slab thermal models. This is one motivation for multiscale models, which use a coarse global model and a refined model near the subduction zone (e.g., Billen *et al.*, 2003).

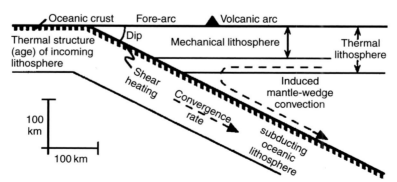

**Figure 6**  Cartoon cross-section of a subduction zone showing the major features in slab thermal models. Reproduced from Peacock SM (1996) Thermal and petrologic structure of subduction zones. In: Bebout G, Scholl D, Kirby S, and Platt J (eds.) *American Geophysical Union Geophysical Monograph 96: Subduction Top to Bottom*, pp. 119–133. Washington, DC: American Geophysical Union, with permission from American Geophysical Union.

There are two end-members for modeling subducted slabs: kinematic and dynamic subduction models. Most present-day kinematic models make the assumption that slabs are (largely) 2-D and rigid and are generally used for detailed slab thermal structure calculations. They are quite often used in seismic, petrologic, or mineral physics investigations because it is possible to control the slab geometry *a priori* (e.g., Minear and Toksöz, 1970; Molnar *et al.*, 1979; Helffrich *et al.*, 1989; Peacock, 1991, 1996; Staudigel and King, 1992; Peacock *et al.*, 1994; Kirby *et al.*, 1996; Bina, 1996; 1997; Marton *et al.*, 1999, 2005). When using a kinematic model, the velocity of the slab is imposed and it is only the careful use of initial and boundary conditions that ensures that the slab velocity is consistent with the buoyancy that provides the driving force for slab motion. The majority of the *P–T–t* (pressure–temperature–time) curves used by petrologists are taken from kinetic slab models (e.g., Peacock 1996, 2003). While kinematic models have been used to predict the lateral variability of phase transformations in slabs, these models cannot address the dynamic impact of phase transformations (*see* Chapter 2) on slabs (e.g., do phase changes retard slab motion?) because the motion of the slab is prescribed. In addition, the deformation that occurs as a plate bends, breaks, and generally deforms as it passes through a subduction zone (e.g., Conrad and Hager, 1999; Buffett and Rowley, 2006) is not accounted for in kinematic models, except in so far as the model prescribes deformation by imposing a curved slab or variation in velocity. With the clear recognition of changes in slab dip, thickening of the slab, possible tears in the slab, and along-slab flow, it seems that there is an increasing need for models with the resolution of kinematic models and the ability for the slab to deform without *a priori* specification. One important result of subduction zone thermal structure studies is that the temperature structure that develops in the upper 50–100 km of the slab is one of the primary controls of the structure of the deep slab, where we take 'deep slab' to mean 100–400 km as in the definition used by Jarrard (1986); that is, whatever else happens to the slab as it subducts, the thermal structure that develops in the upper 100 km is the most important controlling factor. The assumed rate of shear heating has a profound effect on the slab thermal structure, with temperatures on the slab–wedge interface varying by several hundred degrees at 100 km depth (cf. Peacock, 1992, 1993, 1996).

Dynamic subduction zone models often use mantle convection codes, in which the thermal (and sometimes chemical) buoyancy of the slab drive the motion of the slab and the larger-scale mantle flow. These models can include viscous faults, weak zones, complex, nonlinear rheologies, and compositional variation within the slab. In dynamic slab calculations, the plate velocity, slab velocity, and slab dip are not input controls, making it more difficult to set up a calculation with a geometry that resembles a specific subduction zone. Dynamic slab models are also generally lower resolution than kinematic models and cover a larger region of the mantle by necessity. The numerical resolution and the difficulty in reproducing specific subduction zone geometries have inhibited the use of dynamic slab models in seismic, petrologic, or mineral physics investigations where detailed slab thermal structure models are needed.

Dynamic slab models have been used to address the mechanisms of slab dip (Gurnis and Hager, 1988), the interaction of slabs with phase changes (Chrsitensen and Yuen, 1984; Zhong and Gurnis, 1994: Christensen, 1995; 1996; Ita and King, 1998), the geoid and topographic profiles over subduction zones (e.g., Davies, 1981, 1984, 1986; King and Hager, 1994; Zhong and Gurnis, 1995; Chen and King, 1998; Zhong and Davies, 1999), and the mass transfer between the upper and lower mantle (e.g., Davies 1995, Christensen, 1996; Kincaid and Olson, 1987; Kincaid and Sacks, 1997; King *et al.* 1997; Ita and King, 1998). There has been some effort to compare the thermal structures from buoyancy-driven slabs and kinematic slab models (Peacock, 1996; van Keken *et al.*, in preparation).

## 8.3.1 Kinematic Models and Slab Thermal Structure

The first slab thermal structure model, presented by McKenzie (1969), solves for the thermal structure within the slab assuming a hot, uniform-temperature mantle surrounding the slab. Beginning with the conservation of energy (*see* Chapter 2), the temperature within the slab is given by

$$\rho C_p\left(\frac{\partial T}{\partial t} + v \cdot \nabla T\right) = \nabla(k\nabla T) + H \qquad [3]$$

where $\rho$ is the density of the slab, $C_p$ is the coefficient of specific heat, $T$ is the temperature, $v$ is the slab velocity, $k$ is the thermal conductivity, and $H$ is the rate of internal heat generation. Ignoring radioactivity in the slab and assuming that the slab is in thermal equilibrium with the surrounding mantle, eqn [1] reduces to

$$\rho C_p v \cdot \nabla T = \nabla (k \nabla T) + H \qquad [4]$$

Substituting characteristic values for the temperature difference between the mantle and slab, $T_o$, and the length scale of the problem, $l$ (e.g., the width of the slab) (*see* Chapter 2)

$$T = T_o T'; \quad x = l x'; \quad z = l z' \qquad [5]$$

then aligning the coordinate system such that the *x*-axis is along the down-dip slab direction and the *z*-axis is perpendicular, we can write eqn [4] as

$$\frac{\partial^2 T'}{\partial x'^2} - 2R \frac{\partial T'}{\partial x'} + \frac{\partial^2 T'}{\partial z'^2} = 0 \qquad [6]$$

where $R$ is the Thermal Reynolds number, given by $R = (\rho C_p v_x l)/2k$. Note that $v$ becomes $v_x$ in this coordinate system and $v_z$ is zero because the assumption is that the slab is 2-D and does not deform. If we assume the temperature at the top and bottom of the slab maintained at a uniform temperature by the surrounding mantle, the solution can be simplified to

$$T' \approx 1 - \frac{2}{\pi} \exp \left( - \frac{\pi^2 x'}{2R} \right) \sin \pi z' \qquad [7]$$

There are several generalizations of kinematic slab thermal models that are illustrated quite clearly by this derivation. First, the mantle surrounding the slab is assumed to be at a uniform temperature and is not locally impacted by the slab. Second, the slab is assumed to descend at a uniform rate that is always parallel to the dip of the slab everywhere (e.g., no slab buckling or internal strain). Finally, the slab does not exchange mass with the surrounding mantle (e.g., no thermal erosion, no melting or dehydration).

Using this model, McKenzie concluded that the maximum depth of the deepest earthquakes in the Tonga–Fiji–Kermadec arc coincide with the depth of the 680° C isotherm in the model. While he avoided attaching a significance to the exact value of the isotherm, noting the simplicity of the calculations, he suggested a thermal control to the maximum depth extent of deep earthquakes, a suggestion which has been subsequently questioned (e.g., Chen *et al.*, 2004). In addition, he uses this the thermal structure to calculate a buoyancy force due to the cold subducting slab which he estimated at $f = 2.5 \sin \phi$ kbars.

Some of the limiting assumptions in the McKenzie slab model are addressed in the numerical model by Minear and Toksöz (1970). In the Minear and Toksöz model, the slab material moves with a fixed velocity relative to a stationary mantle and the computational scheme consists of translating the temperature field along the slab-dip direction and allowing the slab to equilibrate over a fixed time interval. As they state in the text, "in essence, we have assumed the dynamics and have computed the temperature field given this motion field." An appealing aspect of the Minear and Toksöz model is that the slab geometry, age of the incoming ocean lithosphere, and slab velocity are specified, making it a fairly straightforward exercise to model observed subduction zone geometries. In addition, the effect of additional sources of energy such as adiabatic compression, phase transformations, and shear heating could be incorporated in the slab thermal structure through the imposed heat source term, $H$, in eqn [3]. Based on their resolution analysis, Minear and Toksöz (1970) estimate that the temperatures resulting from these computations "are probably in error by less than 10%." Because this model enables researchers to calculate the thermal structure of a subducting slab using geophysically observable constraints such as slab dip, incoming plate age, and plate velocity, the Minear and Toksöz model has been extensively used (e.g., Spiegelman and McKenzie, 1987; Helffrich *et al.*, 1989; Ita and Stixrude, 1992; Stein and Stein, 1996; Kirby *et al.*, 1996; Bina, 1996, 1997; Marton *et al.*, 1999, 2005; Wiens, 2001).

Another kinematic approach to the subduction problem is to use the corner flow solution (e.g., Batchelor, 1967; Stevenson and Turner, 1977; Turcotte and Schubert, 1982; Peacock *et al.*, 1994) for the velocity field of the subducting slab. This is sometimes referred to as the 'Batchelor solution' after the original reference (i.e., Batchelor, 1967). With the corner flow solution, the velocity field of the slab and the surrounding mantle, including the mantle wedge, satisfy the equations of motion for a constant-viscosity fluid. This solution uses the stream-function formulation (*see* Chapter 4) to solve the equations of motion assuming a uniform velocity, rigid slab and plate. As unlikely as it may seem, it is possible to write the stream function, $\psi$, in the form

$$\psi = Ax + Bz + (Cx + Dz) \arctan \left( \frac{y}{x} \right) \qquad [8]$$

where $A$, $B$, $C$, and $D$ are constants to be determined by the boundary conditions. Because of the geometry of the descending slab, there are two distinct stream functions, that is, two sets of constants to be solved for in this problem: one above and one below the slab. The general problem leads to quite complicated integrals; however, a slab dip of $\pi/4$ allows easy evaluation of the integrals. Solving for the solution in the mantle wedge, we have the following boundary conditions

$$u = v = 0 \quad \text{on} \quad z = 0, \ x > 0 \text{ or } \arctan\frac{z}{x} = 0 \quad [9]$$

$$u = v = \frac{U\sqrt{2}}{2} \quad \text{on} \quad z = x \text{ or } \arctan\frac{z}{x} = \frac{\pi}{4} \quad [10]$$

By straightforward substitution, it is possible to show that the pressure in the arc corner is

$$p = \frac{\mu U\sqrt{2}[\pi x + (4-\pi)z]}{(2 - \pi^2/4)(x^2 + z^2)} \quad [11]$$

We can evaluate this expression using the fact that $x = z = \sqrt{2}/2$ where $r$ is the length along the slab and we find that

$$p = \frac{4\mu U}{(2 - \pi^2/4)r} \approx \frac{-8\mu U}{r} \quad [12]$$

The negative pressure means the back-arc region flow tends to lift the slab against the force of gravity. Notice that the pressure varies as $1/r$ along the slab, with a singularity at the trench. The total lifting torque on the slab will be $r \times P$, where $P$ is a constant.

For the ocean side of the trench, the boundary conditions are

$$u = U, \ v = 0 \quad \text{on} \quad z = 0, \ x < 0 \text{ or } \arctan\frac{z}{x} = \pi \quad [13]$$

$$u = v = \frac{U\sqrt{2}}{2} \quad \text{on} \quad z = x \text{ or } \arctan\frac{z}{x} = \frac{\pi}{4} \quad [14]$$

Again by straightforward substitution, the pressure on the bottom side of the slab is given by

$$p = \frac{\mu U}{r}\left(\frac{3\pi\sqrt{2}-4}{9\pi^2/4-2}\right) \approx \frac{0.5\mu U}{r} \quad [15]$$

Given the geometry and the signs of the pressures, this flow also exerts a lifting torque on the slab, that is about a factor of 20 smaller than the torque on the slab from the arc flow (Stevenson and Turner, 1977). The corner flow solution does not overcome the ridge slab approximation nor does it allow for a slab velocity that is consistent with the slab driving force. As is shown below, the effect of a temperature-dependent viscosity in the mantle wedge part of the flow has a significant effect on the flow near the

corner and a significant effect on the slab thermal structure. This has reduced the difference between estimates of slab temperature from flow calculations and petrology.

While the stream-function solution solves the equations of motion, a solution to the energy equation is also needed and several different solution strategies have been employed cf.(Peacock, 1996). Because of the discontinuity in the velocity boundary conditions at the corner of the wedge, there is a singularity in the pressure field that presents problems for fluid/melt flow models (e.g., Spiegelman and McKenzie, 1987).

**Figures 7** and **8** illustrate results of two different kinematic slab thermal models: the first based on the method described in Minear and Toksöz (1970) and the analytic corner flow solution described above plus the Streamline Upwind Petrov–Galerkin solver from ConMan (King *et al.*, 1990) for the temperature (energy) equation. The problem is a 50-Myr-old slab descending $45°$ at $5 \text{ cm yr}^{-1}$ and the complete list of parameters for these calculations is provided in **Table 1**. There are two reasons for the significant difference between the solutions shown in **Figures 7** and **8**. First, in the Minear and Toksöz (1970) thermal model the slab curves (bends) as it enters the subduction zone, whereas for the corner flow solution, the slab simply descends at $45°$ at the left edge of the box. This accounts for the difference between the two solutions seen in the top 50 km. Second, the Minear and Toksöz (1970) thermal model does not have flow in the wedge, whereas this is a significant component of the corner flow solution with the ConMan result. This accounts for the majority of the difference in the region below 50 km. In the Minear and Toksöz (1970) thermal model, perpendicular to the slab heat is transferred by conduction only. (There is an induced flow approximation in the Minear and Toksöz code, but this does not improve the agreement between the solutions.) The importance of the induced wedge flow on the slab thermal structure will become more significant in the discussion of temperature-dependent rheology. Returning to the Minear and Toksöz claim that the solution is in error by less than 10% (due to the grid) it is important to recognize that this is a statement of the numerical solution on the grid which they used in their paper relative to the solution with the same basic underlying physical assumptions in their model on a more highly refined grid. From **Figure 7**, the difference between the Minear and Toksöz solution and the Batchelor solution plus advection is much more

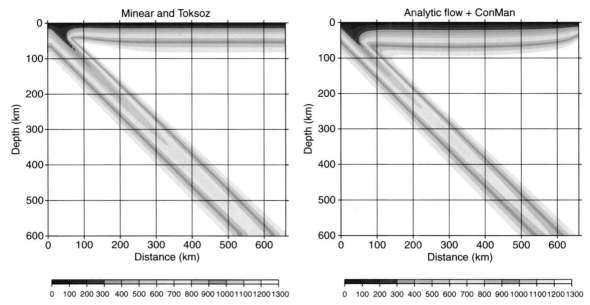

**Figure 7**   A comparison of slab thermal structures from a 50-Myr-old, 45° dipping slab descending at 5 cm yr$^{-1}$ with kinematic slab approximations (left) (Minear and Toksöz, (1970)) and (right) the analytic corner flow solution for flow and ConMan's temperature solver. Below 50 km depth, the Minear and Toksöz slab is colder than the corner flow plus ConMan solution.

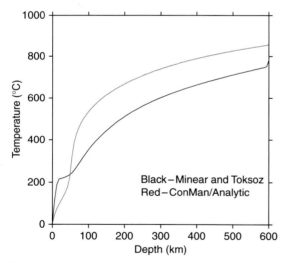

**Figure 8**   Temperature along the line $x = z$ (i.e., the slab interface) for the kinematic slab thermal models shown above. The difference of approximately 150° C is largely due to the lack of corner flow in the Minear and Toksöz model.

**Table 1**   Parameters used in subduction zone thermal models shown in **Figures 7** and **8**.

| Parameter | Symbol | Value |
|---|---|---|
| Slab velocity | $v$ | 5 cm yr$^{-1}$ |
| Slab dip | $\Theta$ | 45° |
| Box depth | $d$ | 600 km |
| Box width | $w$ | 660 km |
| Slab age | $T_{slab}$ | 50 My |
| Thickness of overriding plate | $d_{plate}$ | 50 km |
| Density | $\rho$ | 3300 kg m$^{-3}$ |
| Thermal conductivity | $k$ | 3.0 W m$^{-1}$ K$^{-1}$ |
| Specific heat | $c_p$ | 1250 J kg$^{-1}$ K$^{-1}$ |
| Thermal diffusivity | $\kappa = k/(\rho c_p)$ | 7.272 × 10$^{-5}$ m$^2$ s$^{-1}$ |
| Reference temperature | $T_o$ | 1300° C |

than 10% for the problem described above and as we will see when we move to temperature-dependent rheology the difference can be even greater still. In the ConMan solution, we use a grid spacing on the order of 1 km in the corner of the wedge. We find that grid spacing on this order is necessary to resolve the flow near the pressure singularity near the corner. In the real Earth, it is likely that because of temperature-dependent rheology and serpentinization of the region near the corner of the wedge (e.g., Bostock *et al.*, 2002) that this singularity does not present the problem it does in trying to match the corner flow solution. However, for this problem, the resolution is important because the temperature of the slab is controlled by what happens in the upper 50–100 km.

A series of calculations described by Peacock *et al.* (1994) and Peacock (1996) solve the energy equation (eqn [3]) for the thermal structure of a subduction zone using a finite-difference formulation described in Peacock (1991). Peacock focuses on the petrologic implications of the slab thermal structure, as opposed to slab deformation and driving mechanisms of plate motions. The velocity of the slab is imposed and the analytic 'Batchelor solution' (Batchelor, 1967) is used for the corner flow in the mantle wedge. Beneath the volcanic arc, the thermal models predict temperatures in the oceanic crust in the range of 500–700°C. Peacock (1996) concludes that with the exception of very rare circumstances, the slab does not melt; however, dehydration of Na-amphibole, lawsonite, and chlorite in the subducted oceanic crust can trigger partial melting in the slab wedge. We will return to the petrology of slabs as well as the role of water later.

Kinematic slab models have been used to map the mineralogy of the slab (e.g., Ita and Stixrude, 1992; Bina, 1996, 1997; Marton *et al.*, 1999, 2005). These mineralogical models produce significant differences in slab density, and changes in slab density lead to significant differences in the plate and slab velocities (e.g., King and Ita, 1995; Ita and King, 1998); however, the change in slab density is not accounted for in the flow field in the kinematic models because buoyancy forces do not drive slab flow in the kinematic slab approximation. While kinematic models have provided an estimate of where the transitions in mineralogy might occur, they do not have the predictive ability to follow the implications of the mineralogical models as the slabs evolve dynamically (e.g., Ita and King, 1998). We would like to be able, for example, to calculate whether metastable olivine would have an observable impact on subduction velocity, dip, and slab deformation. This kind of study has been carried out in larger-scale fluid models (e.g., Gurnis and Hager, 1988; Zhong and Gurnis, 1994; Christensen, 1996; Zhong and Gurnis, 1997; Ita and King, 1998). However, those calculations are not able to resolve detailed slab thermal structure with the same resolution as the kinematic models. In most dynamic models, the grid resolution is on the order of 10 km and the mechanics of the deformation near the trench are often grossly approximated by a fault zone (sometimes extending to depths of 100 km or more) or large, prescribed mechanically weak zone.

A kinematic description of slab motion remains popular in part because the input parameters are easily observed properties of subduction zones (i.e., plate age, plate velocity, and slab dip) and because

dynamic slab models require a careful choice of boundary conditions in order to incorporate the effects of global mantle flow (cf. Billen and Gurnis, 2001; Billen *et al.*, 2003). Until recently (Billen and Gurnis, 2001; Billen *et al.*, 2003), convection models have not attempted to reproduce the major geophysical characteristics (e.g., geoid, topography, heat flow) of specific subduction zones, so kinematic models have been the only tools available to researchers. Physical experiments have shed new light on slab morphology and dynamics (e.g., Kincaid and Olson, 1987; Griffiths *et al.*, 1995; Guillou-Frottier *et al.*, 1995; Becker *et al.*, 1999); however, this kind of modeling cannot provide detailed thermal structure information. A new hybrid form of thermal structure calculations (e.g., Peacock and Wang, 1999; van Keken *et al.*, 2002; Peacock *et al.*, 2005) has done an excellent job of matching the geometry and heat flow from specific subducting slab environments; however, these calculations are not able to make use of geoid or dynamic topograpy observations because they lack large-scale dynamic flow.

## 8.3.2 Hybrid Kinematic–Dynamic Slab–Wedge Models

A relatively new approach to subduction zone modeling is a combination of the kinematic and dynamic flow models. In this approach, the slab velocity is imposed (kinematic) and the energy equation (eqn [3]) as well as the equations for incompressible viscous flow are solved numerically (e.g., Davies and Stevenson, 1992; Furukawa, 1993; van Keken *et al.*, 2002; Conder *et al.*, 2002; Kelemen *et al.*, 2003; Arcay *et al.*, 2005; Manea *et al.*, 2005; Peacock *et al.*, 2005. While the slab motion is still imposed, this formulation enables greater flexibility in modeling the flow in the mantle wedge, specifically including the effect of temperature-dependent and/or stress-dependent rheology. Even with this new more dynamic approach, the coupling of the overriding plate with the subducting plate (i.e., the seismogenic zone) remains problematic.

Of particular importance has been the recognition that with the effect of temperature-dependent rheology, flow in the wedge corner is significantly stronger than that predicted by the isoviscous corner flow model, leading to warmer temperatures at the slab–wedge interface in the shallow slab (e.g., Furukawa, 1993; van Keken *et al.*, 2002; Kelemen *et al.*, 2003; Arcay *et al.*, 2005; Manea *et al.*, 2005; Peacock *et al.*, 2005). **Figure 9** illustrates the temperature field from

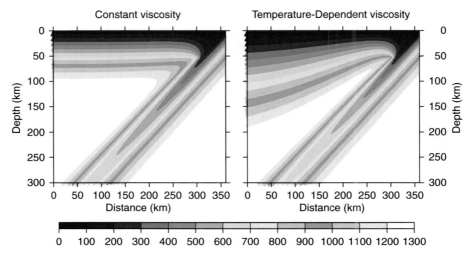

**Figure 9** Thermal structure from two slab calculations using ConMan (King *et al.*, 1990). Both calculations have a slab descending at 45° at a fixed velocity of 5 cm yr$^{-1}$. The incoming plate material is 50 My old. The left image is a constant-viscosity mantle wedge and the right image is a temperature-dependent viscosity wedge, with an activation energy of 335 kJ mol$^{-1}$.

two slab calculations using the convection code ConMan (King *et al.*, 1990). The mesh is nonuniform with significant refinement near the corner of the wedge with a grid spacing of approximately 1 km. The grid extends to 660 km × 600 km and the boundary conditions along the inflow/outflow sides of the wedge are natural boundary conditions (i.e., normal and tangential components of total stress are zero). Both calculations have a slab descending at 45° at a fixed velocity of 5 cm yr$^{-1}$. The incoming plate material is 50 My old and the velocity of the overriding plate is set to zero to a depth of 50 km. The left image is a constant-viscosity mantle wedge and the right image is a temperature-dependent viscosity wedge, with an activation energy of 335 kJ mol$^{-1}$. The flow in the wedge in the temperature-dependent viscosity case extends upward along the slab to a depth of almost 50 km (the depth of the imposed zero-velocity plate) while the constant-viscosity flow has subhorizontal isothermal below the overriding plate. This difference has a significant effect on the thermal structure of the downgoing slab and the heat flow near the trench. The heat flow near the trench is elevated by 10% in the temperature-dependent case. Compared with the constant-viscosity calculation. The temperature in the downgoing slab differs by 98° at the slab–wedge interface at 150 km depth.

The higher temperatures near the slab–wedge interface warm the sediments and oceanic crust at the top of the slab and this may explain the

conflicting geochemical evidence for high temperatures at the top of the slab based on sediment melting (Johnson and Plank, 1999) and low temperatures required by the Boron measurements in arc lavas (Leeman, 1996) as discussed in van Keken *et al.* (2002). The low viscosities in the mantle wedge (due to temperature, volatile content, and high strain rates) suggest that buoyancy in the mantle wedge may be dynamically important (e.g., Kelemen *et al.*, 2003), although in the calculations above, we have not included the effect of buoyancy on flow in the mantle wedge. Buoyancy in the mantle wedge presents a challenge because of the nature of the stress-free boundary conditions at the wedge boundary and the stability of the overriding plate.

Because of our limited understanding of the processes in the seismogenic zone and the variety of approximations used in the seismogenic zone, it is useful to compare the temperature along the seismogenic zone from the hybrid kinematic–dynamic models with other estimates of the temperature in this zone. Molnar and England (1990) develop analytic expressions for the steady-state temperature in the top 50 km of the subduction zone. Their expressions (eqns [16] and [23]) are

$$T = \frac{(Q_0 + Q_{SH})z_f/k}{S} \qquad [16]$$

where $Q_0$ is the basal heatflux (in W m$^{-2}$), $Q_{SH}$ is the rate of shear heating (in W m$^{-2}$), $z_f$ is the depth to the fault (m), $k$ is the thermal conductivity (in $W\,m^{-1}\,K$) and $S$ is given by

$$S = 1 + b\sqrt{Vz_f \sin\delta/\kappa} \qquad [17]$$

where $b$ is a constant of order 1, $V$ is the convergence velocity, $\delta$ is the angle of subduction, and $\kappa$ is the thermal diffusivity ($m^2\,s^{-1}$). Using this expression with values from **Table 1** and $Q_0$ as $35\,mW\,m^{-2}$, and plotting this with the temperatures from the temperature-dependent ConMan result in **Figure 10**, there is a good agreement with the analytic result with a shear stress of $40\,MPa$. The two ConMan results are for calculations with a 'fault' at the seismogenic zone, that is, velocities from the overriding plate and the slab are explicitly decoupled (black line), and a calculation where the velocity across the seismogenic zone varies continuously with the finite-element grid (red line). The difference is less than $10°$, showing that the seismogenic zone formulation in this calculation does not have an overwhelming influence on the solution. To match the Molnar and England solution with the ConMan result without including shear heating requires an unrealistically high $Q_0$ value of $95\,mW\,m^{-2}$. It is worth emphasizing that the ConMan focmulation does not have an explicit shear heating term in the energy equation, so the need for a shear heating term

in the Molnar and England (1990) formulation to match the ConMan solution illustrates that there is a difference between the solutions that has a similar effect as shear heating.

Manea *et al.* (2005) investigate the Guerrero subduction zone in Mexico, characterized by flat subduction with a distant volcanic arc. Using a temperature-dependent rheology to model the temperature structure in the slab and wedge, they find significantly higher temperatures below the volcanic arc compared with previous studies using an isoviscous mantle wedge (e.g., Currie *et al.*, 2001). In Manea *et al.*'s model, temperatures at the slab–wedge interface are also sufficient to allow for a slab contribution to the arc volcanism from the melting of hydrated sediment and oceanic crust, in agreement with geochemical observations of magmas in the Central Mexican Volcanic belt. Manea *et al.* (2005) also investigate the effect of buoyant melt in the wedge using the predicted wedge flow structures. Using the location of the Popocatepetl volcano they estimate that magmatic blobs need to be larger than $10\,km$ in diameter to have reasonable trajectories in the mantle wedge with realistic estimates for viscosity.

Peacock *et al.* (2005) construct high-resolution 2-D finite-element models using the geometry along four profiles perpendicular to the Central American subduction zone in Nicaragua and Costa Rica. Realistic stress- and temperature-dependent rheology in the mantle wedge is necessary to have sufficiently high temperature in the mantle wedge below the volcanic arc to produce melting. This subduction zone is characterized by rapid convergence of young oceanic lithosphere with minor changes in the convergence speed and age of the lithosphere along strike but significant variations in the depth of the seismicity and the geochemistry of arc volcanism. The small along-strike variations in the convergence speed (increasing to the southwest from 79 to $88\,mm\,yr^{-1}$) and age (decreasing from 24 to 15 My) nearly offset and the predicted slab surface temperatures are remarkably similar throughout the subduction zone. Comparing the thermal structures predicted with phase diagrams for mafic (crust) and ultramafic (uppermost mantle) compositions (Hacker *et al.*, 2003) suggests that the predicted temperature in the uppermost $500\,m$ of the subducting oceanic crust is high enough to generate partial melting if the oceanic crust and sediments are hydrated. The thermal models predict that dehydration is nearly complete beneath southern Costa Rica, but that a significant

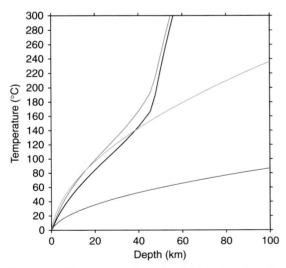

**Figure 10** Temperature along the slab interface from the temperature-dependent ConMan result in **Figure 9** with a discontinuity at the seismogenic zone (black), with no discontinuity at the seismigenic zone (red), the Molnar and England (1990) analytic solution with zero shear heating (blue) and with $60\,mW\,m^{-2}$ (green) – corresponding to a shear stress of $40\,MPa$. See text for further description.

amount of water can be transported into the deep mantle below Nicaragua and NW Costa Rica. However, the predicted variations in the thermal structure are insufficient to explain the large along-strike variations in seismicity and arc geochemistry, leading to the conclusion that these variations are due to regional variations in sediment subduction, crustal structure, and the distribution of hydrous minerals in the incoming lithosphere (e.g., Rüpke *et al.*, 2004).

One of the remaining trouble spots for slab thermal modeling is the coupling of the subducting and overriding plates. There are two traditional approaches to model the decoupling. The first assumes that both the overriding plate and wedge have viscous rheology and decoupling occurs through a weak zone, or a slipping fault. The second approach defines the overriding plate as a rigid zone below which the subducting slab couples with the viscously defined mantle wedge. The depth of decoupling, type of weak-zone or slipping-fault parameters are not well constrained by observations and this can result in significant variations in the predicted temperatures in the wedge and subducting slab. Conder (2005) suggests an approach that incorporates new rheological criteria in which ambient temperature and strain rate define the brittle–ductile transition, which in turn defines the depth at which the boundary condition between slab and overriding plate changes from fully decoupled to fully coupled. The slab surface temperatures in Conder's models are generally higher than observed in other models, and may provide an explanation for the geochemical signatures of sediment melting in some subduction zone.

Van Keken (2003) provides an excellent review of the structure and dynamics of the mantle wedge.

### 8.3.3 Dynamic Models and Mantle Flow

There are many questions that cannot be addressed with subduction models with imposed slab velocities, chief among them the question of the driving forces of plate motions. Dynamic subduction models are generally an attempt to produce a subduction-like geometry within a mantle convection calculation. Some of the earliest calculations have downwellings along the edge of the domain (e.g., Christensen and Yuen, 1984) These calculations address the effect of phase transformations on subduction. In tank experiments with fluids stratified by density and frozen 'plates' atop the fluid, Kincaid and Olson (1984)

produce slabs with morphologies similar to those in Christensen and Yuen (1984) (see also Chapter 3 for more on tank experiments).

Attempts to produce asymmetric downwellings in dynamic calculations include velocity boundary conditions (e.g., Stevenson and Turner, 1977; Davies, 1989; Christensen, 1996), *a priori* specified mechanically weak zones (e.g., Kopitzke, 1979; Gurnis and Hager, 1988; King and Hager, 1994; Chen and King, 1998), and/or dipping viscous faults (e.g., Zhong and Gurnis, 1992, 1994, 1995; Toth and Gurnis, 1998; Ita and King, 1998). King *et al.* (1992), Zhong *et al.* (1998), Gurnis *et al.* (1999), Bercovici *et al.* (2000), and Bercovici (2003) and references therein discuss and compare various plate-generation methods. The influence of the vertical sidewalls in 2-D is a significant limitation to subduction models because the 2-D geometry requires a strong 'return flow' which has a tendency for the slab to steepen with depth, and in many cases dipping more than 90° something not observed in subduction zones on Earth. Attempts to circumvent this problem have included the use of periodic boundary conditions (e.g., Gurnis and Hager, 1988; Lowman and Jarvis, 1996; Chen and King, 1998; Han and Gurnis, 1999) or a cylindrical geometry (Puster *et al.*, 1995; Zhong and Gurnis, 1995; Ita and King, 1998).

It has been recognized that there are limitations to the planform of convection in 2-D and there was a hope that subduction-like geometries might be easier to achieve in 3-D calculations. Yet the model geometry alone is insufficient to produce subduction zone geometries; 3-D spherical geometries fail to produce the slab geometries observed in modern subduction zones (e.g., Bercovici *et al.*, 1989; Tackley *et al.*, 1993; 1994; Ratcliff *et al.*, 1995, 1996; Bunge *et al.*, 1996, 1997). In 3-D constant-viscosity calculations (e.g., Bercovici *et al.*, 1989) the global planform of the flow could be described as linear, sheet-like downwellings that break off into isolated drips and cylindrical, plume-like upwellings. Although these features do not resemble plumes or slabs in any detail, the results were encouraging. When temperature-dependent rheology is included in 3-D spherical calculations, the planform changes, that is, cylindrical downwellings at the pole and sheet-like upwellings (Ratcliff *et al.*, 1995, 1996) or stagnant-lid convection (Moresi and Solomatov, 1995; Solomatov and Moresi, 1996).

Weak zones and faults are difficult to implement numerically, require additional implementation challenges if the fault or weak zone move and/or evolve with the flow, and implicitly assume some heterogeneity (either in strength or composition) pre-exists before subduction begins. This might be entirely realistic; the initiation of subduction may require some level of pre-existing heterogeneity (cf. Toth and Gurnis, 1998; Gurnis *et al.*, 2000b). Some investigators have attempted to produce dynamic, mobile plates using complex constitutive equations (e.g., Bercovici, 1993, 1995; Tackley, 1998; Trompert and Hansen 1998). A constitutive equation is a relationship between two or more variables that is needed to close the equations of motion. Constitutive equations are generally not based on conservation laws, but are empirically derived. For mantle convection (i.e., creeping viscous flow), this is a relationship between stress and strain rate. The simplest relationship is a linear relationship between stress and strain rate; in this case, the constant of proportionality is the kinematic viscosity, which can be a constant or function of temperature, pressure, or composition (*see* Chapter 2). If the relationship between stress and strain rate is linear, this is referred to as a Newtonian viscosity. If the relationship between stress and strain rate is nonlinear, it is called a non-Newtonian or power-law fluid.

A series of investigations of convection with temperature-dependent Newtonian and/or non-Newtonian rheologies documented a transition from the free-slip planform found in constant-viscosity calculations through a sluggish-lid planform to a stagnant-lid planform of convection as the stiffness of the boundary layer increases (Christensen, 1985; Solomatov, 1995; Moresi and Solomatov, 1995; Solomatov and Moresi, 1996). In stagnant- and sluggish-lid convection, most of the top boundary layer is rigid and remains at the surface and only the weakest bottom part of the boundary layer participates in the active flow. Thus, most of the cold boundary layer is never recycled into the interior of the fluid and is quite different from our picture of subduction, where most, if not all, of the subducting plate descends into the interior of the fluid. In the stagnant-lid planform, a significant amount of the negative buoyancy in the top thermal boundary layer (i.e., the plate) does not participate in the downwelling (Moresi and Solomatov, 1995; Conrad and Hager, 1999). This has important implications for the plate driving force due to subducted slabs and for the calculation of gravity anomalies over subduction zones because the thermal structure is related to the density structure through the coefficient of thermal expansion. Furthermore, the surface heat flow from

stagnant-lid convection calculations is significantly smaller than the heat flow from models with temperature-dependent rheology when a plate formulation is included, which has a significant effect on the cooling of the Earth through time (i.e., thermal history calculations, cf. Gurnis, 1989). The sluggish-lid and stagnant-lid convective planforms are inconsistent with piecewise-uniform surface velocities, and to date constitutive theory models have yet to produce realistic subduction zone geometries.

The addition of a strain-weakening (non-Newtonian) component to the rheology weakens parts of the cold thermal boundary layer, especially at regions of high stress (such as in the corners of the computational domain). When coupled with temperature-dependent rheology, this can produce nearly uniform surface velocities (van den Berg *et al.*, 1991; Weinstein and Olson, 1992; King *et al.*, 1992); however, power-law rheology calculations are typically unstable and the plate-like behavior quickly breaks down as these calculations evolve away from the carefully chosen initial conditions. Rheologies based solely on the properties of mantle minerals require the imposition of a yield stress (Tackley, 1998, 2000a, 2000b) or damage theory (Bercovici *et al.*, 2001; Bercovici, 2003; Bercovici and Richard, 2005) and have not been used for subduction modeling experiments. Some researchers argue that subduction forms at pre-existing zones of weakness in the lithosphere (e.g., Kemp and Stevenson, 1996; Schubert and Zhang, 1997; Toth and Gurnis, 1998; Regenauer-Lieb and Yuen, 2000; Regenauer-Lieb *et al.*, 2001; Regenauer-Lieb and Yuen, 2003); hence, heterogeneous properties of the lithosphere are important for dynamic subduction zone models.

A criticism of the calculations discussed above is that they are limited, for the most part, to 2-D geometries, or 3-D geometries with reflecting sidewall boundary conditions. Symmetric, near-90° dipping downwellings are also the observed planform in 3-D spherical convection models (e.g., Tackley *et al.*, 1994; Bunge *et al.*, 1997). Even when temperature-dependent rheology is included in spherical calculations, dipping slab-like features are not observed (Ratcliff *et al.*, 1995, 1996). The use of periodic boundary conditions in 2-D eliminates the effect of the sidewalls and 2-D calculations can be formulated in such a way that they are formally equivalent to calculations that explicitly allow the trench to move relative to the grid (cf. Han and Gurnis, 1999).

Trench migration (roll-back) has been one of the most studied mechanisms for producing dipping slabs

(Gurnis and Hager, 1988; Zhong and Gurnis, 1995; Griffiths *et al.*, 1995; Christensen, 1996a; Ita and King, 1998). It is interesting to note that while numerical calculations and tank experiments have demonstrated that trench migration is an effective mechanism for generating shallow dipping slabs in models, compilations of data from various subduction zones (Jarrard, 1986) show almost no correlation between the dip of a slab in the 100–400 km depth range and the rate of trench migration (**Figure 11**). The data from the recent compilations (e.g., Heuret and Lallemand, 2005; Cruciani *et al.*, 2005) still do not resolve this problem.

A number of researchers have explored the fate of compositionally defined lithosphere as it encounters a viscosity increase at the boundary between the upper and lower mantle (e.g., Christensen, 1996; Richards and Davies, 1989; Gaherty and Hager, 1994; van Keken *et al.*, 1996). In these calculations, the subducted lithosphere is represented as a layered composite of denser eclogite underlain by more buoyant harzburgite. The slabs thicken and fold strongly as they penetrate the more viscous lower mantle with a factor of 2 thickening occurring just above the discontinuity. Slab evolution is largely controlled by the thermal buoyancy and whether or not there is separation of the compositional slab components at the discontinuity, depends on the rheology structure of the slab. It is unlikely that

compositional buoyancy leads to separation of the slab components. Furthermore, these calculations demonstrate that a significant amount of slab deformation is expected in the transition zone. This is explored in tank experiments by Guilloufrottier *et al.* (1995) showing a wide range of slab deformation styles (e.g., sinking slab, stagnant slab, spreading slab, sinking pile, and stagnant pile) when slight changes in the parameters are imposed. These deformation modes are governed by two velocity ratios, which characterize the horizontal and the vertical components of the slab velocity near the interface.

There are several cases where 3-D models have been employed to study specific regions with considerable success matching seismic structure (van der Hilst and Seno, 1993; Moresi and Gurnis, 1996). In these cases, the focus of attention was in the western Pacific, where slabs are old and steeply dipping, and a significant amount of trench migration has been documented. It is possible to reconcile the apparent lack of correlation of the global trench migration observations with slab geometry and the success of the regional subduction studies by admitting that subduction is a time-dependent phenomenon and the shape of subducting slabs evolves with time. Both numerical and tank experiments show that buoyancy-driven subduction is a time-dependent phenomenon (Gurnis and Hager, 1988; Griffiths *et al.*, 1995, Becker *et al.*, 1999). It is important to keep in mind that the geometries of present-day subduction zones provide a single snapshot of a time-dependent phenomenon and that each subduction zone is at a different stage in this process.

There are several observations that have not been fully exploited in convection modeling. The first of these is the difference in dip between eastward-dipping and westward-dipping slabs (e.g., Le Pichon, 1968; Ricard *et al.*, 1991; Doglioni, 1993; Marotta and Mongelli, 1998; Doglioni *et al.*, 1999). The difference between Chilean style subduction zones and Mariana style subduction zones has been recognized for some time (e.g., Uyeda and Kanamori, 1979). It is difficult to separate the effects of dip direction from the fact that most of the eastward-dipping slabs are being overridden by the North and South American Plates (i.e., continents) while many of the slabs in the Pacific are being overridden by island arcs or oceanic plates.

Yet, a series of flexure calculations that include the effect of motion of the slab–trench system relative to the underlying mantle fit slab geometries remarkably well (Marotta and Mongelli, 1998). One of the reasons that this observation has not been

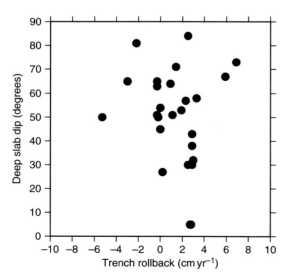

**Figure 11** Deep slab dip, as measured by the shape of the Wadati–Benioff zone in the 100–400 km depth range vs trench rollback using the Minster and Jordan plate motion model. Data taken from Jarrard RD (1986) Relations among subduction parameters. *Reviews of Geophysics* 24: 217–284.

explored is that many of the techniques used to generate subduction features in 2D are not practical to implement in a complete sphere in 3-D calculations. Furthermore, the grid resolution that is required to study subduction problems exceeds what can be practically achieved in a 3-D spherical shell at present.

The area where the asymmetry of subduction is most apparent is the shallow-dipping slabs, such as the subduction of the Pacific Plate beneath Alaska, the Juan de Fuca Plate under Washington and Oregon, and the Cocos and Nasca Plates under Central and South America. The subduction of young lithosphere (in most cases the slab is significantly younger than the slabs in the western Pacific) is not driven by the negative buoyancy of the slab (cf. van Hunen *et al.*, 2000), which is the underlying assumption in most convection subduction studies. Convection models have focused on the case of older slabs where the negative buoyancy of the slab itself provides the driving motion for the plate system (e.g., slab pull). There are a few examples of studies of the subduction of young lithosphere (e.g., England and Wortel, 1980; Vlaar, 1983; van Hunen *et al.*, 2000, 2001, 2002, 2004). Young slabs have less negative buoyancy than older slabs, and thus, younger slabs should have shallower dip angles than older slabs, all other things being equal. However, van Hunen *et al.* (2000, 2001) find that weak crustal rheology, a 'subduction fault' between the two plates, nonlinear rheology, and viscous shear heating all play a role in weakening the shallow crust and mantle and enabling shallow subduction. The subduction of a buoyant oceanic plateau has been suggested as a possible cause for shallow subduction; however, this does not appear to be a significant factor in the case of Peru (van Hunen *et al.*, 2004).

### 8.3.4 3-D Numerical Models of Regional Subduction

Billen and Gurnis (2001) and Billen *et al.* (2003) examine 3-D, dynamic subduction calculations using a slab geometry appropriate for the Tonga–Kermadec and Aleutian slabs. By varying the slab and wedge rheology as well as the slab density structure, they are able to match the observed geoid, topography, and the state of stress in the slab. There are several important restrictions to note in the details of their model development: (1) these are instantaneous snapshots of the flow field (i.e., the slab morphology, plate/slab velocities, and slab thermal structure do not evolve with the flow field); (2) the element size ranges from 2.5 km within the wedge to 100 km at the size boundaries (compare this with the results from the 2-D slab thermal models discussed above); and (3) the thermal structure of the slab (i.e., the initial density structure) is derived from a kinematic thermal structure model. Nonetheless, this work represents an important and significant step forward in understanding subduction zone dynamics.

Although the work of Billen and colleagues (Billen and Gurnis, 2001; Billen *et al.*, 2003) does not focus on slab thermal structure, one of their results motivates the need for more detailed studies of this kind. They find that they need to increase the buoyancy in their slabs by 30% more than predicted by a reasonable equation-of-state model in order to drive the plate and slab at the appropriate observed velocity and match the geoid and topographic profiles. It is reasonable to ask whether the problem is the result of buoyancy estimate from the equation of state, the dissipation of energy in the 'trench region' due to their fault and/or rheology (cf. Conrad and Hager, 2001), or whether the slab thermal structure which they use as the initial condition for their model is wrong. Billen and Gurnis choose to vary the buoyancy; however, previous work has shown that there is a tradeoff between slab buoyancy and slab strength. For example, Griffiths *et al.* (1995) and Houseman and Gubbins (1997) have shown that the deformation of a slab depends on the slab stiffness and slab buoyancy. Griffiths *et al.* define a slab stiffness (the viscosity of the slab divided by the viscosity of the mantle) and find a relationship between the slab deformation and slab stiffness. There are additional observational constraints that can be used to address this question. For example, Bevis (1986, 1988) used slab geometry and earthquake sources to estimate slab strength. The distribution of earthquakes in Benioff zones (Oliver and Isacks, 1967; Lundgren and Giardini, 1992), regional tomography (van der Hilst, 1995), and scattering experiments all constrain the morphology of the slab in the upper mantle.

### 8.3.5 Phase Transformations: Dynamics

The eventual fate of deep slabs has been a topic of debate for several decades (see Lay, 1994; Christensen, 1995; *see* Chapter 10). One of the original arguments for a barrier to convection at 670 km depth was the cessation of earthquakes at this depth. The amount of slab penetration into the lower mantle controls the rate of heat and mass transfer between the upper and the lower mantle (e.g., Lay, 1994; Silver *et al.*, 1988) and understanding slab penetration is critical to understanding thermal convection and chemical mixing of the mantle (e.g., Albarède and van der Hilst, 1999; Kellogg *et al.*, 1999). Proposed hindrances to slab penetration into the lower mantle include the phase transformation of wadsleyite to perovskite plus magnesiowustite (Schubert and Turcotte, 1971; Schubert *et al.*, 1975; Ringwood and Irifune, 1988), a change in chemical composition (e.g., Richter, 1977; Anderson, 1979), and an increase in viscosity (e.g., Hager, 1984). A compositional barrier to convection at around 660 km is not considered likely by most and the topic of viscosity will be covered in a later section. Here we focus on the effect of phase transformations.

A phase change, in and of itself, will not necessarily impact convective flow. Because convection is driven by horizontal density gradients, the idealized phase transformation that happens at one pressure, $P$, for all temperatures will, to first order, have no impact on the flow. There is an effect due to the latent heat release due to the phase change and the small effect of dynamic pressure, which varies horizontally, but is generally on the order of 1% of the hydrostatic pressure. Phase transformations primarily affect convective flow in the mantle because of the temperature variation of the pressure of the phase transformation. This is expressed by the Clausius–Clapeyron relationship,

$$\frac{\partial p}{\partial T} = \gamma = -\frac{\Delta S}{\Delta V} = \frac{\Delta S \rho^2}{\Delta \rho} = \frac{Q_L \rho^2}{\Delta \rho T} \qquad [18]$$

where $T$ is the absolute temperature, $\Delta S$ is the entropy change, $\Delta V$ is the specific volume change, $Q_L$ is the latent heat, and $\rho$ is the mean density of the two phases (*see* Chapter 2). A positive Clapeyron slope, such as the transformation of olivine to wadsleyite, means the denser phase will occur at shallower depth in the slab, with the effect of increasing the negative buoyancy in the slab, enhancing convection. A negative Clapeyron slope, such as the transformation of spinel (ringwoodite) to perovskite plus ferropericlase, means the lighter phase will be present to greater depths in the cold slab, as compared to the surrounding mantle, decreasing the net negative buoyancy of the slab and retarding convective motion (see **Figure 12**).

A simple comparison of densities is sufficient to illustrate the importance of phase transformations on buoyancy. The coefficient of thermal expansion for

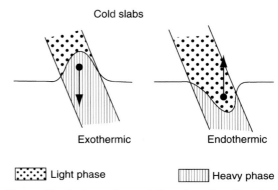

**Figure 12** Cartoon of an endothermic and exothermic phase transformation within a cold slab. Reproduced from Christensen UR (1995) Effects of phase transitions on mantle convection. *Annual Review of Earth and Planetary Sciences* 23: 65–88, with permission from Annual Reviews.

mantle minerals is on the order of $2.0 \times 10^{-5}\,\mathrm{K}^{-1}$. If we assume a 400 km column of slab material, $D$, that is 500 °C colder than its surrounding material, $\Delta T$, this slab will have an integrated buoyancy anomaly, $\rho g \alpha \Delta T D$, of $132\,\mathrm{N\,m}^{-2}$. Comparing this with a phase transformation with a change in density of 6% that has a topography on the phase boundary of 30 km, the phase-change topography will have a buoyancy anomaly of $0.06 \times \rho g b_{pc}$ of $59.4\,\mathrm{N\,m}^{-2}$. Thus, a 30 km deflection of a phase boundary can have a comparable buoyancy anomaly to a 400 km column of 500 °C anomalously cold slab.

It is important to recognize that the mantle is not a single-component system. Because of the nature of the transformations from olivine to wadsleyite and ringwoodite (spinel) to perovskite plus ferropericlase and the close association of the pressure of these phase transformations to the 410- and 660-km seismic discontinuities, most studies of convection with phase transformations have included only the olivine system in their calculations. For a simplified phase diagram see **Figure 13**. It is generally assumed, because the pyroxene-to-garnet transitions are spread out over a larger pressure range, that the effect of these phase transitions will be small and the overall effect of the combine pyrolite system will be that the dynamic effect of the olivine-to-wadsleyite and ringwoodite-to-perovskite plus ferropericlase phase changes will be weaker than in a pure olivine system, because the pyroxene/garnet component makes up about 40% of the composition.

Because deep earthquakes are restricted to subduction zones, the mechanisms of deep earthquakes are assumed to be related to the relatively cold

(a)
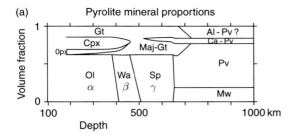

(b)
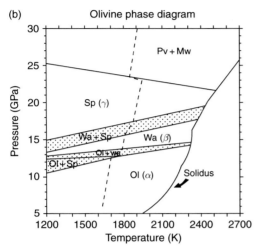

**Figure 13** Phase relationship for a pyrolite upper mantle composition. (a) Volume fraction of minerals as a function of pressure. (b) Phase boundaries for the olivine component of the mantle with the dashed line as the 1573 K adiabat. Reproduced from Ita J and Stixrude L (1992). Petrology, elasticity and composition of the mantle transition zone. *Journal of Geophysical Research* 97: 6849–6866, with permission from American Geophysical Union.

temperature of the interiors of slabs. The specific mechanisms that have been proposed for deep earthquakes include transformational faulting of metastable olivine (e.g., Green and Burnley, 1989; Kirby *et al.*, 1991) and dehydration or amorphization of hydrous phases (e.g., Meade and Jeanloz, 1991). Both hypotheses involve the presence of compositional or mineralogical anomalies that are characterized by low seismic-wave speeds in the mantle transition zone and in principle, this could be detected by seismology.

Global traveltime tomography models have moderate traveltime anomalies implying fast seismic-wave speeds in the lower mantle along the Tonga subduction zone (e.g., Grand *et al.*, 1997; van der Hilst *et al.*, 1997) that are consistent with earlier analysis of traveltimes that indicate an anomaly extending down to no more than 900 km in depth (Fischer *et al.*, 1991).

This has led to a variety of hypotheses on what hinders the cold Tonga slab, including subhorizontal deflection of the slab (e.g., van der Hilst *et al.*, 1995) or the collection of slab material at the base of the transition zone to form a large megolith (Ringwood and Irifune, 1988; Irifune, 1993). The complexity of the Tonga subduction zone and the absence of a deep fast anomaly beneath Tonga point to some form of hindrance to slabs at the base of the transition zone.

It is important to recognize that while the overriding assumption made in almost all subduction zone studies is that buoyancy within a slab is dominated by the effect of temperature, phase transformations have significant effects locally (e.g., Daessler and Yuen, 1996; Bina 1996; 1997; Christensen, 1996b, 1985; Ita and King, 1998; Schmeling *et al.*, 1999; Marton *et al.*, 1999). The impact of phase transformations on buoyancy is most pronounced if the olivine-to-spinel phase transformation is kinetically hindered in cold slabs (see Kirby *et al.* (1996) and references therein). The presence of metastable olivine can reduce slab descent velocities by as much as 30% (Kirby *et al.*, 1996; Schmeling *et al.*, 1999; Marton *et al.*, 1999) and will have the greatest effect on the oldest and coldest slabs. In addition, the pattern of stresses induced by the differential buoyancy are consistent with the pattern of deep earthquakes (Bina, 1997).

Early attempts to model subduction zones using dynamic mantle convection models used a non-Newtonian fluid with an endothermic phase transformation. This approach produced stiff downwelling limbs that descended at a 90° dip angle as a direct consequence of the free-slip (reflecting) boundary conditions on the sides of the domain (Christensen and Yuen, 1984). The symmetry of the downwelling (i.e., material from both sides of the thermal boundary layer at the downwelling) is the major shortcoming of this approach. Christensen and Yuen identified three planforms of deep slabs at a phase boundary: slab penetration, slab stagnation, and partial slab penetration. Fluid experiments with corn syrup which used a setup designed to produce an asymmetric downwelling confirm these basic planforms (Kincaid and Olson, 1987; see also Chapter 2). The tank experiments induce an asymmetric downwelling by placing two cold sheets of concentrated sucrose solution on the top surface of the fluid in the tank. The larger of the two sheets is introduced with a shallow-dipping bend at its leading edge; this provides the instability that allows this plate to subduct under the other plate. The agreement between the tank and numerical experiments (**Figure 14**) suggests that the three

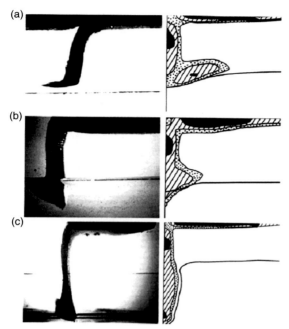

**Figure 14** Experimentally observed styles of slab penetration through a density discontinuity (left) compared with 2-D calculation by Christensen and Yuen (1984) (right). (a) Slab deflection with $R \approx -0.2$. (b) Partial slab penetration with $R \approx 0.0$. (c) Complete slab penetration with $R \approx 0.5$. Reproduced from Kincaid C and Olson PL (1987). An experimental study of subduction and slab migration. *Journal of Geophysical Research* 92: 13832–13840, with permission from American Geophysical Union.

planforms discussed above are robust features of cold downwellings interacting with phase transformations and/or chemical boundaries. Subsequent numerical (Zhong and Gurnis, 1994; King and Ita, 1995; Ita and King, 1998) and tank (Griffiths *et al.*, 1995; Guillou-Frottier *et al.*, 1995) experiments have confirmed the basic results and added additional insight into the interaction between subducting slabs and phase transformations.

There is a general consensus that has emerged from both 2-D and 3-D calculations in both Cartesian and spherical geometries. The evolution of the flow can be described by three distinct stages: (1) the initial impingement of the leading edge of the subducting slab on the spinel to perovskite plus ferropericlase phase boundary; (2) a period of increased trench rollback and draping of the slab on the phase boundary; and (3) virtual cessation of trench rollback as the slab penetrates into the lower mantle.

Studies have clearly demonstrated that the effect of a phase transformation on the pattern of flow in a

convecting fluid is sensitive to the Rayleigh number, initial conditions, boundary conditions, and the equation-of-state approximation (e.g., Ita and King, 1994, 1998; King *et al.*, 1997). For example, Ita and King (1998) show that a depth-dependent coefficient of thermal expansion approximation develops a stagnant slab and layered convection while the slab penetrates into the lower mantle with a calculation with a self-consistent equation of state based on Ita and Stixrude (1992) and otherwise identical parameters. King *et al.* (1997) show that calculations including an exothermic olivine-to-spinel phase transformation decreases layering. Ita and King (1994) show that at Rayleigh numbers of order $10^5$, the initial flow pattern imposed on the system can persist for 30–40 transit times or more and reflecting sidewall boundary conditions used in a Cartesian, box-like system tend to produce layered flows more easily than those with flow-through conditions. Ita and King (1994) go on to show a decrease in layering as the thermodynamic consistency is increased (i.e., Bousinessq calculations are more likely to be layered than compressible calculations with a self-consistent equation of state). The effect of latent heat on the topography of phase boundaries can be important (cf. Christensen, 1998) and is not well accounted for in most convection calculations.

### 8.3.6 Shear Heating

There is a large degree of uncertainty regarding the relative importance of shear heating in the thermal structure of a subduction zone, in large part because of a disagreement on the magnitude of the subduction zone shear stress with some estimates ranging from 100 MPa (e.g., Scholz, 1990; Molnar and England, 1990) and others an order of magnitude lower (e.g., Peacock, 1992; Tichelaar and Ruff, 1993; Hyndman and Wang, 1993). The rate of shear heating is given by the product of the shear stress, $\tau$, and the convergence velocity, $v$ (*see* Chapter 6),

$$Q_{\mathrm{sh}} = \tau v \qquad [19]$$

While the convergence rate is well constrained (e.g., Jarrard, 1986), the magnitude of the shear stress in subduction zones is estimated based on surface heat flow measurements and/or petrologic arguments and this quantity is at best loosely constrained. In a fully compressible convection formulation, the shear stress can be calculated consistently in the viscous flow formulation (e.g., Yuen *et al.*, 1978). However, even in such a formulation there are uncertainties because the deformation mechanism within the slab is poorly understood and perhaps more importantly, because of the uncertainty and difficulty in representing the deformation between the slab and overriding plate. The fore-arc heat flow places a strong constraint on the amount of shear heating that can be added in the upper 50 km. Early kinematic subduction zone thermal models required a high shear stress in order to match the observed heat flow (e.g., Turcotte and Schubert, 1973). With temperature-dependent rheology, the wedge corner flow brings more hot material close to the surface (and the slab) and a lower rate of shear heating is needed (e.g., van Keken et al., 2002). The uncertainty in the amount of shear heating, particularly below 50 km depth, leads to several hundred degrees difference in the temperature of the slab (Peacock, 1996).

### 8.3.7 Elastic/Plastic Subduction Calculations

The discussion thus far has focused on calculations with either imposed velocities or viscous flow; however, there is another important group of calculations that include the effect of elastic strength and brittle faults (e.g., Kemp and Stevenson, 1996; Schubert and Zhang, 1997; Toth and Gurnis, 1998; Gurnis et al., 2000a, 2004; Regenauer-Lieb and Yuen, 2000; Regenauer-Lieb et al., 2001; Regenauer-Lieb, 2003; Hall et al. 2003). The work of Vassiliou et al. (1984) showed that viscous flow was able to reproduce the patterns of stress attributed to elastic bending of the plate as it entered the trench; hence, viscous flow has been thought to capture the broad pattern of deformation at a subduction zone. While this is probably true, the recognition that the deformation in the seismogenic zone (the upper 50 km or so of the boundary between the subducting and overriding plate) impacts the deeper slab structure by altering the slab thermal structure, more work in this area is needed. A useful overview of the role of pre-existing faults and subduction can be found in Gurnis et al. (2000a).

## 8.4 Petrology, Geochemistry, and Arc Volcanics

Perhaps one of the most interesting subduction zone puzzles is that subduction zones are where cold surface material returns to the mantle, yet island arcs, a

significant source of volcanic activity, are associated with subduction zones. Subduction zones are cool, relative to the average upper-mantle temperature, so why is there any volcanic activity associated with subduction zones? The early postplate tectonic view is that island arc volcanism was the result of slab melting (e.g., Ringwood, 1975) which required temperatures approaching the melting point of basalt. The only way to achieve such high temperatures in a subduction environment was with large rates of frictional heating (e.g., Oxburgh and Turcotte, 1968). However, other constraints on the rate of frictional heating (cf. Peacock, 1996), such as the fore-arc heat flow, produce slabs that are significantly cooler than necessary for basalt melting (e.g., Peacock, 1991; Davies and Stevenson, 1992).

There are several ways to resolve this apparent inconsistency: volatiles, mainly water from the oceanic sediments and crust, reduce the melting point of mantle minerals leading to melting at lower temperatures; low melting point sediments could be the major source of arc volcanics; frictional shear heating could significantly increase the temperature near the slab–wedge interface; and induced flow from the downgoing slab could advect additional heat into the mantle wedge, warming the wedge relative to average upper mantle. Each of these mechanisms will be examined below and in reality some combination of these factors is likely to be important.

While most of the subducting slab is not melted and may eventually wind up at the core–mantle boundary, subduction zones are the primary source of continental crust formation and the major mechanism for recycling volatiles into the mantle. The subducting slab is comprised of layers of compositionally distinct components: sediments, oceanic crust, and oceanic lithosphere. Often these components are considered as a single entity, the 'slab', and most convection models do not have the resolution to track the individual components separately. Given that oceanic lithosphere makes up more than 90% of the mass (volume) of the subducting slab, it is not surprising that this controls the mechanics of subduction. The mass (volume) of oceanic crust and sediment will be insufficient to play the major role in the buoyancy or strength of the subducting slab. Possible examples of where oceanic crust may play a significant role include the subduction of oceanic plateaus. Another caveat is that sediments, crust, and/or volatiles coming off the slab may play an influential role in the mechanics of the seismogenic zone. Even though the mechanics of subduction is dominated by the oceanic lithosphere, it is important to keep in mind that that the crust and sediments play important geochemical roles in subduction zone processes.

The emerging view is that the geochemistry points to both volatile-induced melting and sediment melting (Elliott, 2003). There are extensive reviews by Elliott (2003) and Gaetani and Grove (2003) that cover work that is beyond the scope of this review. While there is a vast amount of work on arc petrology and geochemistry, much of this work has only begun to be integrated in subduction zone thermal modeling (e.g., van Keken *et al.*, 2002; Kelemen *et al.*, 2003; Peacock, 2003). There are several reasons for this, the primary is that until the 'rediscovery' of Furukawa's (1993) work on the impact of temperature-dependent rheology on flow (and hence temperature) in the slab corner, there was an inconsistency in slab thermal models – the only way to achieve the temperatures needed for sediment melting and significant dehydration was a high rate of shear heating. However, the high shear stress required was inconsistent with fore-arc heat flow (cf. Peacock, 1996). Another reason that the geochemical and petrological constraints have not been more closely linked with subduction zone models is that the scale of subduction zone models, particularly dynamic models, has been too coarse to study the region of interest. This is changing with the emergence of the new hybrid kinematic–dynamic models (e.g., van Keken *et al.*, 2002; Kelemen *et al.*, 2003; Peacock *et al.*, 2005).

## 8.4.1 Subduction and Volatiles

Oceanic crust contains a large fraction of the water carried into subduction zones (Wallmann, 2001). Mid-ocean ridge basalt (MORB) is depleted in incompatible trace elements and contains 50–200 ppm water (Dixon *et al.*, 2002). Therefore, water, $CO_2$, and incompatible trace elements in the crust are the result of hydrothermal alteration and seafloor weathering. Staudigel *et al.* (1996) infer that most of the water, $CO_2$, and incompatible trace elements are concentrated the upper 500 m of basalts. Formation of amphibolites in the lower crust stores water but little else (Carlson, 2001). Serpentine may be the most important mineral for water transport because it is stable to 7 GPa and can carry an order of magnitude more water than hydrated crust (Pawley and Holloway, 1993).

The role of water in melting at arcs has been studied extensively, dating back to Green and Ringwood (1967) and recent reviews have covered the topic in a depth not allowed in this chapter (Poli and Schmidt, 2002; Eiler, 2003; Elliott, 2003; Hirschmann, 2006). Water plays several important roles in arc geochemistry and dynamics by reducing the melting temperature (Hirth and Kohlstedt, 1996; Asimow *et al.*, 2004) and impacting the rheology (Hirth and Kohlstedt, 2003).

The subducted slab plays an important role in the water cycle, connecting the surface to deep mantle. The extent to which water (and other volatiles) may recycle back to the deep mantle is an intriguing problem. Peacock (2001) suggests that the lower line of earthquakes in double Bennioff zones may be related to dehydration. The water-storage capacity for olivine increases significantly with depth (Kohlstedt *et al.*, 1996) and the transition zone phases of wadsleyite and ringwoodite have large water-storage capacities (Kohlstedt *et al.*, 1996; Smyth *et al.*, 1997). It is possible that if water can get to the transition zone, the transition zone could contain as much water as is present on the surface (Smyth, 1994; Bercovici and Karato, 2003). While subducted oceanic crust is thought to be dehydrated before reaching the transition zone (e.g., Schmidt and Poli, 1998), peridotite could carry water into the transition zone, particularly by water that has percolated to depth in the slab along extensional faults formed at the outer rise (e.g., Peacock, 2001; Rüpke *et al.*, 2004). It is possible that hydrated phases may only be stable to the transition zone along unusually cold adiabats (e.g., Staudigel and King, 1992), requiring old ocean lithosphere or very high convergence rates, or both.

Arcay *et al.* (2005) model subduction zone thermal structure with a dynamic mantle convection code and include dehydration reactions and a rheology that depends on pressure, temperature, strain rate (i.e., non-Newtonian), and water content. Consistent with other recent studies (e.g., van Keken *et al.*, 2002; Conder *et al.*, 2002, Kelemen *et al.*, 2003) they find that non-Newtonian rheology leads to a hotter slab surface temperature and greater erosion of the overlying plate than in isoviscous corner flow models. At high convergence rates, serpentine can be transformed into hydrated phase A, leading to recycling of water to significant depth. Even in cases where a significant amount of water is carried to depth, the mantle wedge is significantly hydrated within 250 km of the trench, effectively broadening the low-viscosity zone near the tip of the slab wedge. The hydrated weak wedge region may contribute to back-arc deformation.

## 8.4.2 Melting of Sedimentary Material

The most compelling evidence for a 'slab' component in arc magmas comes from the trace elements, B, Be, Th, and Pb (e.g., Morris *et al.*, 1990; Hawkesworth, *et al.*, 1993; Plank and Langmuir, 1993; Ishikawa and Nakamura, 1994; Smith *et al.*, 1995; Elliott *et al.*, 1997; Wunder *et al.*, 2005; George *et al.*, 2005; Plank, 2005). The presence of $^{10}$Be, an isotope of Be only produced in the upper atmosphere with a half-life of 1.6 My, is strong evidence that some oceanic sediments are recycled though subduction zones, travel through the mantle wedge, and wind up in the source region of arc lavas (Morris *et al.*, 1990). Furthermore, the short half-life places a tight constraint on the transport of material through the subduction/arc system.

The volume and composition of sediment subducted into various trenches varies considerably (e.g., von Huene and Scholl, 1991). This might represent an along-trench variation apart from slab geometry (dip and convergence velocity) that could explain along-arc variations in subduction zones. There is pretty clear evidence now that the geochemistry identified two distinct components (e.g., Elliott, 2003): a melting of sedimentary material component and an aqueous fluid-derived component.

## 8.4.3 Phase Transformation in Slabs

Phase transformations within slabs can have a significant effect on subduction dynamics (e.g., Daessler and Yuen, 1993, 1996; Bina, 1996, 1997; Christensen, 1996b, 1998; Ita and King, 1998; Schmeling *et al.*, 1999; Marton *et al.*, 1999). The impact of phase transformations on buoyancy is most pronounced if the olivine-to-wadsleyite phase transformation is kinetically hindered in cold slabs (see Kirby *et al.* (1996) and references therein). The presence of metastable olivine may reduce slab descent velocities by as much as 30% (Kirby *et al.*, 1996; Schmeling *et al.*, 1999; Marton *et al.*, 1999) and will have the greatest effect on the oldest and coldest slabs. The stresses induced by the differential buoyancy are consistent with the pattern of deep earthquakes (Bina, 1997).

Marton *et al.* (2005) use a thermokinetic model of subduction zone thermal structure with a thermal conductivity that depends on pressure, temperature, and mineralogy. These calculations do not consider variable rheology. The thermal conductivity decreases exponentially with increasing temperature and the effect of pressure is much smaller, especially over the depth range of the upper mantle.

Considering the effect of radiative heat transfer, they conclude that thermal conductivity of the slab could increase by more than 50%, leading to a 50–100° increase in the temperature of the interior of the slab compared to that in a standard model with uniform thermal conductivity, independent of the slab geometry. Marton *et al.* (2005) use these models to address the question of olivine metastability and the ability of a metastable wedge to explain the pattern of deep seismicity. They conclude that the deepest earthquakes occur in regions that have already transformed to wadsleyite or ringwoodite, even when taking into account all of the uncertainties in their parametrization. In the case of Tonga the deepest earthquakes extend 140 km beyond the olivine metastability region.

## 8.5 Geoid and Topography

The association between local maxima in the long-wavelength component of the Earth's gravitational potential field (geoid) and subduction zones has been recognized by many authors (Runcorn, 1967; Kaula, 1972; McKenzie, 1977; Chase, 1979; Crough and Jurdy, 1980; Davies, 1981; Hager, 1984). The geoid is the surface of constant potential energy (i.e., gravitational energy plus centrifugal potential energy) that coincides (almost) with mean sea level over the oceans. (Mean sea level is not quite a surface of constant potential, due to dynamic processes within the ocean; however, for illustrative purposes it will suffice.) The geoid is actually the height of the potential surface referenced to an oblate spheroid resulting from a hydrostatic spinning Earth (Nakiboglu, 1982).

In a static mantle, with no dynamic flow, the geoid over a positive mass anomaly is positive (**Figure 15**, left panel). In an isoviscous fluid (**Figure 15**, middle panel), the gravitational potential (or geoid) over a positive mass anomaly, such as a subducting slab, is negative (Richards and Hager, 1984). This is because the flow driven by the mass anomaly (blue in **Figure 15**) deforms the surface (and core mantle boundary) creating the red, negative mass anomalies (i.e., long-wavelength depressions). The sum of the contributions from the positive and negative mass anomalies in the isoviscous fluid is negative (and small) (cf. Richards and Hager, 1984; Ricard *et al.*, 1984). With a layered viscosity increase with depth (**Figure 15**, right panel), the amount of surface deformation from the same positive mass anomaly is

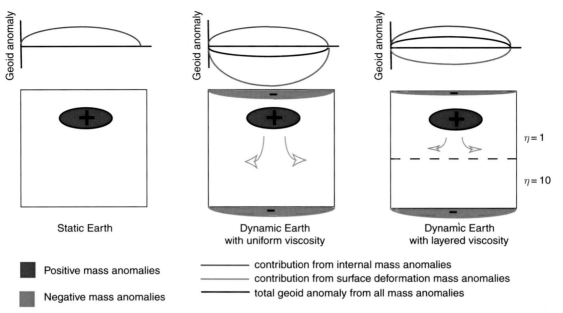

**Figure 15** Schematic representation of geoid anomalies over a positive mass anomaly for a static (left), isoviscous (center) and layered viscosity (right) mantle. The blue line is the component of the anomaly due to the internal mass, the red line is the component of the anomaly due to the surface deformation, and the black line is the sum of the components.

reduced (i.e., the mass anomaly is supported by the stiff layer below) and the resulting anomaly can change sign depending on the viscosity structure (cf. Richards and Hager, 1984; Ricard *et al.*, 1984; Chapters 2 and 4).

The long-wavelength geoid associated with subduction zones requires that subducting slabs encounter a resistance to flow, modeled as an increase in effective viscosity by a factor of 30, or more, at a depth of 670 km (Hager, 1984; Hager and Richards, 1989; Ricard *et al.*, 1989; Hager and Clayton, 1989; Zhong and Gurnis, 1992; King and Hager, 1994). Because the equation for the gravitational potential is linear, we can decompose the problem; the total geoid anomaly is the sum of the contribution from each density anomaly within the Earth. This includes density anomalies that drive mantle convection, such as dense slabs or buoyant plumes, and density anomalies that result from deformed boundaries as a result of convection. This problem was first addressed by Pekeris (1935) but has been expanded on by others (e.g., Runcorn, 1967; Morgan, 1965; McKenzie, 1977; Richards and Hager, 1984; Ricard *et al.*, 1984). Focusing on subduction zones for purposes of illustration, the mass excess due to the dense, sinking slab contributes a positive term to the total geoid while the mass deficit, due to the down-warping of the surface above the

slab, contributes a negative term to the total geoid. These contributions are similar in magnitude and opposite in sign. The resulting total geoid is sensitive to variations in either the density structure of the slab or the deformation of the surface. The magnitude of the down-warping of the surface is a strong function of the viscosity structure of the medium (cf. Richards and Hager, 1984; Ricard *et al.*, 1984). (It should be noted that the boundary between the mantle and core is also deformed as a result of the flow driven by the sinking slab; however, at the wavelengths appropriate to subduction zone problems, this is not a significant contribution even though it is included in most geoid calculations (Richards and Hager, 1984).)

While the density structure of a subducting slab is a complex function of temperature, chemical variations, and phase transformations, there is little doubt that subducting slabs are more dense than surrounding mantle. In order to create a positive geoid anomaly over a subducting slab in an isoviscous fluid, it would require that estimates of slab thermal structure underpredict the density anomaly due to the slab by approximately a factor of 4 (Davies, 1981) or that approximately 300 km of slab material is piled in the transition zone (Hager, 1984). This is much greater than the combined uncertainties in the thermal structure, thermal properties of the slab, and the errors made by neglecting the compositional component of the slab. Thus, some form of resistance to the downward motion of the slab is necessary to reduce the associated surface deformation. In the majority of studies on subduction, it has been assumed that the resistance to the slab is provided by an increase in viscosity at some depth (usually taken to be between the upper and lower mantle) and King (2002) has shown that the phase transformation from ringwoodite to perovskite plus ferropericlase is insufficient to change the sign of the geoid. With a sufficient reduction in the surface deformation, provided by the increase in viscosity of the lower mantle, the total geoid anomaly over the downwelling slab can be positive. In this way, the association of local maxima in the geoid with subduction zones provides a constraint on the viscosity of the mantle.

Based on the analysis described above, an increase in mantle viscosity with depth by a factor of 30–100 is required to produce local geoid maxima over subducting slabs. The low end of this range is consistent with studies of other geophysical observations that constrain mantle rheology including postglacial rebound (cf. Mitrovica, 1996; Mitrovica and Forte, 1997; Forte and Mitrovica, 2001), plate velocities (cf. Lithgow-Bertelloni and Richards, 1995) and rotation. The high end of the viscosity increase required by the

subduction zone geoid observation is more difficult to reconcile with most mantle viscosity models from postglacial rebound studies; although a few postglacial rebound studies have predicted a viscosity increase this large (e.g., Lambeck *et al.*, 1990). It is also possible that both observations are compatible because the depth to which the postglacial rebound observation and the subduction zone observation are sensitive are different. For example, some models of mantle viscosity predict an increasing viscosity with depth throughout much of the lower mantle (e.g., Ricard and Wuming, 1991). If these viscosity models are correct, then it is possible to reconcile a higher value of lower-mantle viscosity predicted by the subduction zone observation because the depth to which the subducting slab is sensitive is greater than the depth to which the postglacial rebound observation is sensitive.

The power spectrum of the geoid is dominated by the longest wavelengths and the power decreases at approximately the fourth power of the spherical harmonic degree (Kaula, 1968). This observation has become known as Kaula's rule. As a result, the analysis of all but the longest components of the geoid requires spatial filtering. **Figures 16** and **17** compare the global geoid from the NIMA/NASA EGS96 geopotential model (Lemoine *et al.*, 1997) that has

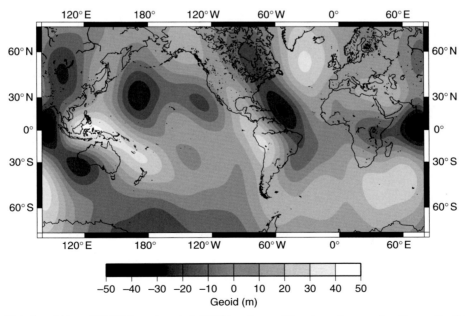

**Figure 16**  Global geoid from EGM96 (Lemoine *et al.*, 1997) band-pass-filtered passing wavelengths smaller than 15 000 km and cutting wavelengths greater than 5000 km. These parameters are comparable to the degree 4–9 'slab' geoid used by Hager (1984). Routines in the GMT software package were used to filter the data (Wessel and Smith, 1991). Reproduced from King SD (2002) Geoid and topography over subduction zones: The effect of phase transformations. *Journal of Geophysical Research* 107 (doi:10.1029/2000JB000141).

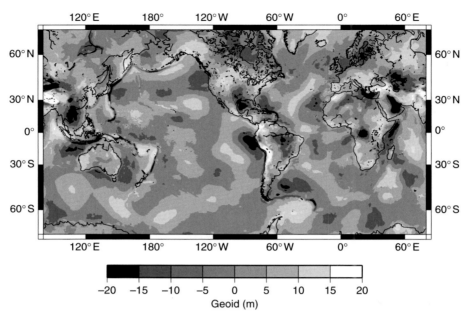

**Figure 17** Global geoid from EGM96 (Lemoine *et al.*, 1997) band-pass-filtered passing wavelengths smaller than 5000 km. Reproduced from King SD (2002) Geoid and topography over subduction zones: The effect of phase transformations. *Journal of Geophysical Research* 107 (doi:10.1029/2000JB000141).

been band-pass filtered using two bands. The first image (**Figure 16**) cuts wavelengths greater than 15 000 km and smaller than 4000 km and passes wavelengths between 10 000 and 5000 km. The second image (**Figure 17**) cuts wavelengths greater than 5000 km and passes wavelengths smaller than 4000 km. **Figure 16** is comparable to the spherical harmonic degree 4–9 geoid that has been shown to correlate well with subduction zones (Hager, 1984) and the correlation between geoid highs and subduction zones is clearly evident in with the Aleutian arc being a notable exception. Short-wavelength geoid anomalies, like those presented in **Figure 17**, have been presented in trench-perpendicular profiles (e.g., Davies, 1981; Zhong and Gurnis, 1992; King and Hager, 1994; Billen *et al.*, 2003) and the geoid and topography at this wavelength has been used to constrain dynamic subduction zone models (Zhong and Gurnis, 1992, 1994, 1995; King and Hager, 1994; Chen and King, 1998; King, 2002). Richards and Engebretson (2002) point out that the long wavelength geoid correlates with the general pattern of fast seismic velocities in the Pacific.

Focusing on short-wavelength geoid (**Figure 17**), there is significant structure in the geoid that appears to be related to subduction zones but is obscured by the longer-wavelength components. Specifically, there

is a narrow trough in the geoid (a local geoid minimum) that is that approximately 200 km in width and spatially coincident with the trench (**Figure 17**). There are local geoid maxima in the back-arc region of almost every subduction zone in the short-wavelength geoid, and the geometry of the geoid maxima matches the geometry of the associated arcs. At this scale, there are geoid local maxima associated with subduction zones even at locations where the association fails at the longer wavelengths (e.g., the Aleutian, Kamchatkan, and Middle-America subduction zones.)

## 8.6 Discussion

Having reviewed the major observations, we turn to several outstanding questions that remain unresolved in subduction zone dynamics.

### 8.6.1 Are Slabs Strong or Weak?

While convection models can reproduce many subduction zone observations with both strong and weak slabs (Tao and O'Connell, 1993), there are several indications from convection studies that slabs are weak. Houseman and Gubbins (1997) use a dynamic model of the lithosphere that assumes that the properties of the slab are uniform throughout the entirety of the slab; the slab is both more viscous and more dense than the surrounding mantle. They find that the shape of the deformed slab is a strong function of the viscosity of the slab. They also observe a buckling mode that produces slab geometries similar to the Tonga slab. The effective viscosity of the slab needed to produce this slab geometry is $2.5 \times 10^{22}$ Pa s, no more than 200 times the upper-mantle viscosity and maybe only a factor of 2 greater than the lower-mantle viscosity (cf. King, 1995).

The regional gravitational potential, or geoid, high over subduction zones has an important constraint on mantle rheology (Hager, 1984). There have been a number of efforts to improve the uniform-viscosity flow model used by Hager by including both depth-dependent and temperature-dependent viscosity, the effect of phase transformations, and plate formulations with temperature-dependent rheology (Richards and Hager, 1989; Ritzert and Jacoby, 1992; Zhong and Gurnis, 1992; King and Hager, 1994; Moresi and Gurnis, 1996; Chen and King, 1998; Zhong and Davies, 1999). The surprising result from this work is that weak slabs provide a much better fit to the geoid than strong (high-viscosity) slabs. King (2002) shows that dynamic support of the slab from an endothermic phase transformation is not enough to explain the geoid highs over subduction zones.

Creager et al. (1995) show that the concave ocean-ward bend in the Bolivian Orocline (South American slab) forces along-strike compression in the subducting slab. Their calculations of membrane strain rates in an assumed continuous slab demonstrate that this can be accommodated by either 10% along-strike compressive strain or geometric buckling. They present evidence for complex deformation in the deep slab that could contribute to the conditions for very large earthquakes by dramatically increasing deformation rate, anomalous advective thickening of a proposed overdriven olivine wedge. This provides a mechanism for local thickening of the olivine wedge and allows transformation faulting to occur over a much larger volume than for normal subduction.

As discussed in the slab structure section, the rate of accumulation of seismic moment in WBZs can be used to estimate the average down-dip strain rate in subducting slabs (Bevis, 1986; Fischer and Jordan, 1991; Holt, 1995), assuming that the amount of aseismic deformation is small and that the viscous deformation is parallel with the brittle layer with the same deformation-rate tensor. In the depth range of 75–175 km the average strain rate in subducting slabs is estimated to be $10^{-15}$ s$^{-1}$ (Bevis, 1988). This compares with a characteristic asthenospheric strain rate of $3 \times 10^{-14}$ s$^{-1}$ (Turcotte and Schubert, 1982; Hager and O'Connell, 1978), leading to the rather surprising suggestion that the difference in deformation rates between the asthenosphere and subducting slabs is no more than a factor of 3. Taking the average descent rate of slabs this deformation rate corresponds to a total accumulated strain in the depth range of 75–175 km of 5%. Fischer and Jordan (1991) find that the seismic data require a thickening factor of 1.5 or more in the seismogenic core of the slab in central Tonga and complex deformation in northern and southern Tonga. Holt (1995) argues that the average seismic strain rate in Tonga represents as much as 60% of the total relative vertical motion between the surface and 670 km.

It would be easy to dismiss any one of these observations taken by itself; however, the number and diversity of the observations that slabs are not significantly stronger than ambient mantle suggests that this should receive serious consideration. A weak slab is not necessarily inconsistent with the laboratory observations. While the viscosity of olivine is a strong function of temperature, it is also a strong function of grain size (Riedel and Karato, 1997). Grain-size reduction resulting from recrystallization of minerals at phase boundaries may counter-balance the effects of temperature on viscosity. In addition, the brittle component of the slab may be mechanically weakened by the faulting and thus the single-crystal measurements of viscosity may not correctly describe the deformation of the slab.

It is important to remember that assumptions regarding slab viscosity are built into some slab thermal models. For example, the kinematic velocity field used to advect the temperatures is uniform except in the corner. Because the slab has a uniform velocity, there is zero strain, and hence zero deformation. The evidence reviewed above suggests that slabs thicken with depth. Because the interior of a slab warms primarily by diffusion, this suggests that the interior of thickening slabs will be colder than the kinematic models predict, all other things being equal. Because there have been no direct comparisons between kinematic models, which do not include slab deformation and dynamic models, which can include slab deformation, it is impossible to make a quantitative statement.

A consequence of the kinematic slab approximation is that there is no guarantee that the resulting thermal structure models are dynamically consistent. The results of Billen and Gurnis (2001) illustrate the problem, even though high-resolution thermal slab structure was not the focus of their work. In order to produce the plate/slab velocities that give rise to the initial slab thermal structure, they have to modify the buoyancy in their slab in a way that is inconsistent with their assumed equation of state. (This is not meant to take away from their work, which is an important step forward; rather, their study illustrates the potential that is available to include more realistic equations of state and attempts to model realistic subduction geometries.) This has rather significant implications for studies that have used kinematic models to map out the mineralogy of the slab (e.g., Kirby et al., 1991, 1996; Ita and Stixrude, 1992; Bina, 1996; Marton et al., 1999). These mineralogical models produce significant differences in slab density. Changes in slab density lead to significant differences in the plate and slab velocities (e.g., Christensen, 1996b; Ita and King, 1998); however, the change in slab density is not accounted for in the flow field because buoyancy forces do not drive slab flow in the kinematic slab approximation. While kinematic models have provided an estimate of where the transitions in mineralogy might occur, they do not have the predictive ability to follow the implications of the mineralogical models as the slabs evolve dynamically (e.g., Ita and King, 1998). For example, we would like to be able to calculate whether metastable olivine and/or an increase in rheology would have an observable impact on subduction velocity, dip and slab deformation. Similar studies have been carried out in dynamic models (e.g., Gurnis and Hager, 1988; Zhong and Gurnis, 1992, 1997; King and Hager, 1994; Christensen, 1996b; Kincaid and Sacks, 1997; Ita and King, 1998; Chen and King, 1998). However, none of those calculations were able to resolve slab thermal structure with the resolution of the kinematic models. It is now possible to use dynamic models with the kind of resolution needed for thermal structure calculations (e.g., van Keken et al., 2002; van Keken, 2003).

### 8.6.2 Are Deep Earthquakes the Result of a Metastable Transformation of Olivine to Wadsleyite?

Transformational faulting of metastable olivine into wadsleyite or ringwoodite arose as an intriguing possible mechanism for the source of deep earthquakes, following the Green and Kirby laboratory

experiments (cf. Green and Burnley, 1989; Kirby et al., 1991, 1996; Stein and Rubie, 1999). However, the rupture areas of the two big, deep 1994 events in Bolivia and Fiji seem to be quite large, based on finite fault modeling and aftershock studies (Wiens et al., 1994; Silver et al., 1995) while thermokinetic mineralogical models of subducting slabs (Daessler and Yuen, 1993; 1996; Daessler et al., 1996; Devaux et al., 2000) imply relatively thin 'wedges' of olivine at the depths of the two large deep earthquakes. The combination of the rupture paths and the thermokinetic modeling appears to rule out transformational faulting as a mechanism for deep earthquakes (cf. Wiens, 2001). However it is important to note that the work of Daessler and colleagues has been performed with kinematic slab models and the Tonga–Fiji region is a complex zone with multiple interpretations (e.g., Gurnis et al., 2000a; Chen and Brudzinski, 2001, 2003; Burdzinski and Chen, 2003a, 2003b), exactly the kind or region where a kinematic slab approximation is most suspect.

The seismic observations supporting metastable phase transformations are weak at best. In addition to the large rupture zones of the deep earthquakes of 1994 and 1995 (Wiens et al., 1994; Silver et al., 1995), Collier et al. (2001) find no evidence for nonequilibrium phase transformations based on the elevation of the 410 km discontinuity at subduction zones, while Chen and Brudzinski (2003) find a patten of mantle anisotropy beneath Tonga–Fiji that could be consistent with a metastable olivine wedge. Sandvol and Ni (1997) combined shear-wave splitting parameters of local and teleseismic S waves from intermediate and deep earthquakes in the southern Kurile and Japan subduction zones with splitting parameters obtained from SKS and SKKS waves to determine depth variation in shear-wave splitting both above and below the earthquake. The shear-wave splitting lag times indicate significant variation in azimuthal anisotropy below 350 km depth. This inference is consistent with the source-side splitting of 0.08 s lag time observed from deep teleseismic S waves that traverse the upper-mantle/lower-mantle boundary. Metastable olivine in the flattened and broadened southern Japan slab would be consistent with these observations; although, the presence of an anisotropic layer composed primarily of highly anisotropic $\gamma$-spinel at the base of the 410 km discontinuity would also be consistent with the splitting observations.

Koper et al. (1998) present the results of a 3-month deployment of a 1000 km seismic line of 23 ocean-bottom seismometers (OBSs) and island broadband seismic stations extending from the Lau back-arc basin, across the Tonga trench and onto the Pacific Plate. The traveltime anomalies recorded on their line are well fit by standard slab thermal models. The presence of a metastable olivine wedge has a subtle effect on the traveltimes because the first arriving waves avoid the low-velocity region. Wedge velocity models provide a slightly better fit to the data than equilibrium models, but F tests indicate the improvement is not significant at the 95% level.

Increasingly the role of water in the transition zone is gaining attention because studies suggest that water may be carried to greater depths in subduction zones than previously thought (e.g., Peacock and Wang, 1999; Peacock, 2003). Wood (1995) suggests that the width of the 410 km seismic discontinuity is strongly impacted by the presence of water and uses this to place a bound on the amount of water in upper-mantle olivine of about 200 ppm. The idea that deep earthquakes could be caused by brittle fracture that results from the release of water from minerals at these high pressures had been around for some time (cf., Meade and Jeanloz, 1991). With the recognition of the potential water-storage capacity of the transition zone (cf., Bercovici and Karato, 2003; Hirschmann, 2006), it is perhaps not surprising that this mechanism is receiving new attention. However, it is difficult to explain the large fault widths and orientations of some deep earthquakes with dehydration fracture, and the earthquakes do not have the orientations expected if they are occurring along reactivated shallow faults (Wiens, 2001). In addition, dehydration mechanisms provide no explanation for the dependence of deep earthquake properties on slab temperature.

### 8.6.3  How Much Water Is Carried into the Transition Zone?

Subduction zones play a key role in the recycling of water into the mantle. By simple mass balance, oceanic crust contains a large fraction of the water carried into subduction zones (Wallmann, 2001). Olivine is an important host of hydrogen in the Earth's upper mantle, and the $OH^-$ abundance in olivine determines many important physical properties of the planet's interior. The water-storage capacity for olivine increases significantly with depth (Kohlstedt et al., 1996) and the transition zone phases of wadsleyite and ringwoodite have large water-storage capacities (Kohlstedt et al., 1996; Smyth et al., 1997).

Natural and experimentally hydrated olivines have been typically analyzed using uncalibrated spectroscopic methods with large uncertainties in accuracy. Bell et al. (2003) determined the hydrogen content of natural olivines using a new technique and indicate that an upward revision of some previous determinations by factors of between 2 and 4 is necessary. This could impact the calculations that have been performed to date. It is possible that if water can be carried into the transition zone, the transition zone could contain as much water as is present on the surface (Smyth, 1994; Bercovici and Karato, 2003). The question remains, "how much water can be carried into the transition zone?" Equally interesting, this raises the question, "how much water can leave through the bottom of the transition zone?" (see for instance, Karato et al., 2006).

There are several studies that address this question; however, the results are not yet clear. Davies and Stevenson (1992) suggest that a significant amount of water can be carried into the wedge via a mechanism of lateral transport via solid matrix flow and vertical transport of the aqueous phase. However, because their calculations assumed a constant-viscosity wedge, more recent results suggest that the $P–T–t$ paths will be quite different from temperature-dependent wedge results. Arcay et al. (2005) show that recycling of water to significant depth is possible in their subduction calculations. Even in cases where a significant amount of water is carried to depth, the mantle wedge is significantly hydrated within 250 km of the trench, effectively broadening the low-viscosity zone near the tip of the slab wedge, consistent with the geoid and topography observations, that require a low-viscosity wedge (e.g., Billen et al., 2003). The idea that volatiles may only be preserved in the slab at high convergence rates (and perhaps only in unusually high convergence rates) was put forward by Staudigel and King (1992) to explain isolated geochemical anomalies such as the Dupal anomaly.

Serpentine may be the most important mineral for water transport (Pawley and Holloway, 1993) and is the most abundant of the hydrous minerals in ultramafic rocks under 500°C (Hacker et al., 2003). There have been a number of slab models that suggest that the mantle wedge is significantly (up to 50%) serpentinized (Peacock and Hyndman, 1999; Peacock, 2001; Hyndman and Peacock, 2003). Hyndman and Peacock (2003) find a low-density, low-viscosity wedge that is 20–50% serpentenized, which again would be consistent with the geoid and topography observations (e.g., Billen *et al.*, 2003).

In contrast to the studies above, Iwamori (1998) uses a corner flow (constant-viscosity) model with a porous flow model for the migration of the aqueous fluid. He concludes that the fluid coming off the slab is absorbed in a serpentinite layer nearly parallel to the slab, rather than the entire wedge. This sepentinite layer may extend to 150 km below the depth of the dehydration, advected by the induced mantle wedge flow. Bostock *et al.* (2002) interpret their seismic image of the southern Cascadia subduction zone as a highly hydrated and serpentinized fore-arc region, which they point out is consistent with thermal and petrological models of the fore-arc mantle wedge. Bostock *et al.* (2002) further recognize that serpentinized material is thought to have low strength and go on to suggest that this may control the large-scale flow in the mantle wedge.

While the thermal models above all suggest the mantle wedge is highly serpentinized, they provide little constraint on how much water gets into the transition zone. Subducted oceanic crust may become dehydrated before reaching the transition zone (e.g., Schmidt and Poli, 1998); however, peridotite could carry water into the transition zone, particularly by water that has percolated to depth in the slab along extensional faults formed at the outer rise (e.g., Peacock, 2001; Rüpke *et al.*, 2004). So perhaps the best argument to be put forward is based on direct observations. Typical MORB contains 50–200 ppm water (Dixon *et al.*, 2002). Wood (1995) shows that the seismically determined width of the 410 km phase change is consistent with ≤200 ppm water at the transition zone, suggesting that the upper bound might be that the transition zone is at most as wet as the MORB source region. Measurement of water in the transition zone via electrical conductivity could provide another constraint (Huang *et al.*, 2005).

We have said little about water leaving the transition zone including whether water can be trapped in the transition zone (cf. Bercovici and Karato, 2003; Karato *et al.*, 2006; Dasgupta *et al.*, 2005; Hirschmann, 2006). Without attempting to evaluate the transition zone water filter model, it is important to recognize that the transport of water and the fate of the subducted slab may be significantly decoupled (Ricard *et al.*, 2006). While many slabs appear to penetrate into the lower mantle, some slabs may stagnate in the transition zone for a significant period of time. In addition, there is also

evidence that components of the slab may separate, depending on the rheology of the slab (e.g., Christensen, 1998; Richards and Davies, 1989; Gaherty and Hager, 1994; van Keken *et al.*, 1996). We will leave the fate of water and the fate of slabs beyond the transition zone to other reviews (cf. Chapter 10).

### 8.6.4   How Can We Resolve the Nature of the Deformation in the Seismogenic Zone?

New geophysical measurements are helping to provide additional needed constraints on the deformation of the upper 50–100 km of subduction zones. For example, residual mantle exposures in the accreted Talkeetna arc, Alaska provide rock analogs for the arc-parallel flow that is inferred from seismic anisotropy at several modern arcs (Mehl *et al.*, 2003).

Bevis *et al.* (2001) interpret the interseismic crustal velocity field of the central Andes using a simple three-plate model in which the Andean mountain belt is treated as a rigid microplate located between the Nazca and South American Plates. They obtain our best fit to the geodetic velocities if the main plate boundary is fully (100%) located between depths of 10 and 50 km and 8.5% of Nazca–South American Plate convergence is achieved in the back arc (by underthrusting of the Brazilian Shield beneath the Subandean zone).

Cassidy and Bostock (1996) observe shear-wave splitting in three-component broadband recordings of local earthquakes near southern Vancouver Island, British Columbia. These measurements constraint shear-wave anisotropy in the continental crust above the subducting Juan de Fuca Plate and support weak coupling between the downgoing Juan de Fuca Plate and the overlying North America Plate, because the principal stress is perpendicular to the direction of subduction, and the S-wave splitting above the subducting plate is nearly perpendicular to the SKS splitting direction in the upper mantle beneath the subducting plate.

## 8.7   Summary

Subduction zones are complex regions of the Earth with challenging physical and chemical processes that impact global geodynamics, geochemistry, and thermal evolution of the planet. While we have a fairly good general understanding about the processes occurring in subduction zones, many details (and a few major pieces) remain elusive. It is increasingly evident that future progress requires collaboration and communication among a variety of specialists. So perhaps topping any list of future directions or outstanding questions is the need for better communication between researchers of various backgrounds and expertise. While this has become a common statement in our scientific lexicon, when it comes to understanding subduction zones, it is probably also true. Aside from this, we provide our own list of most interesting outstanding questions:

1. Why does Earth have plate tectonics while the other terrestrial planets do not? While this has not really been a topic discussed in this review, it is one of the foremost questions related to our understanding of global planetary geodynamics. It is hard to imagine that we really understand the process of subduction if we cannot explain why Earth is the only terrestrial planet with active subduction.

2. What is the nature of deformation in the seismogenic zone? Are there observations in addition to fore-arc heat flow that can be used to constrain the deformation and shear heating in this region? Our understanding of the deformation and heat generation in the top 50–100 km of a subduction zone is limited. Unfortunately, the thermal structure acquired in the top 100 km is largely translated down slab dip, so our ignorance of the processes in this region is propagated along with the slab.

3. How strong are slabs and does slab deformation significantly effect the thermal structure of the slab? For many problems of interest, slab strength may not be the major controlling factor, but for understanding the interaction of slabs with phase transformations in the transition zone, understanding the role of slab buoyancy in plate dynamics, and understanding whether slab components are able to separate (which could be significant in the chemical evolution of the mantle) slab rheology is almost certainly critical.

4. How much water is carried into (and out of) the transition zone by slabs? The recognition that wadsleyite has a large water storage capacity has led to the interesting possibility that a large amount of water could reside in the transition zone. With the evidence for a weak mantle wedge thought to be due to dehydration of the slab and/or serpentinization of the mantle wedge, it is by no means clear that slabs retain any water by the time they reach the transition zone. The fate of water at the base of the transition zone is even more obscure. It is important to

remember that this may not be a steady-state process. Water may only be carried into the transition zone during relatively unusual periods of fast subduction.

5. How does subduction begin? While there have been some interesting developments in this area related to preexisting regions of weakness and/or lithospheric discontinuities, there is a lot more to understand. It seems the easiest way to make a new subduction zone assume preexisting subduction. This begs the question as to how the first subduction zone started. When subduction first began and what subduction looked like in the early Earth are also open questions.

## Acknowledgments

This work was supported by NSF EAR-0408005. A number of the figures were produced with the GMT software package (Wessel and Smith, 1991).

## References

Abers GA (2000) Hydrated subducted crust at 100–250 km depth. *Earth and Planetary Science Letters* 176: 323–330.

Abers GA (2005) Seismic low-velocity layer at the top of subducting slabs: Observations, predictions, and systematics. *Physics of the Earth and Planetary Interiors* 249: 7–30.

Abers GA and Sarker G (1996) Dispersion of regional body waves at 100–150 km depth beneath Alaska: *In situ* constraints on metamorphism of subducted crust. *Geophysical Research Letters* 23: 1171–1174.

Albarède F and van der Hilst RD (1999) New mantle convection models reconcile conflicting evidence. *EOS Transactions of the American Geophysical Union* 45: 535–539.

Albarède F and van der Hilst RD (2002) Zoned mantle convection. *Philosophical Transactions of the Royal Society of London A* 360: 2569–2592.

Anderson DL (1979) Chemical stratification of the mantle. *Journal of Geophysical Research* 84: 6297–6298.

Anderson ML, Zandt G, Triep E, Fouch M, and Beck S (2004) Anisotropy and mantle flow in the Chile–Argentina subduction zone from shear wave splitting analysis. *Geophysical Research Letters* 31: L23608.

Arcay D, Tric E, and Doin M-P (2005) Numerical simulations of subduction zones: Effect of slab dehydration on the mantle wedge dynamics. *Physics of the Earth and Planetary Interiors* 149: 133–154.

Asimow PD, Dixon JE, and Langmuir CH (2004) A hydrous melting and fractionation model for mid-ocean ridge basalts: Application to the Mid-Atlantic Ridge near the Azores. *Geochemistry Geophysics Geosystems* 5: Q01E16 (doi:10.1029/2003GC000568).

Batchelor GK (1967) *An Introduction to Fluid Dynamics.* Cambridge: Cambridge University Press.

Bebout G, Scholl D, Kirby S, and Platt J (eds.) (1996) *Subduction: Top to Bottom. Am. Geophys. Un. Geophys. Mono. 96*, Washington.

Becker TW, Faccenna C, OConnell RJ, and Giardini D (1999) The development of slabs in the upper mantle: Insights from numerical and laboratory experiments. *Journal of Geophysical Research* 104: 15207–15226.

Bell DR, Rossman GR, Maldener J, Endisch D, and Rauch F (2003) Hydroxide in olivine: A quantitative determination of the absolute amount and calibration of the IR spectrum. *Journal of Geophysical Research* 108: 2105 (doi:10.1029/2001JB000679).

Bercovici D (2003) The generation of plate tectonics from mantle convection. *Earth and Planetary Science Letters* 205: 107–121.

Bercovici D (1995) A source-sink model of the generation of plate tectonics from non-Newtonian mantle flow. *Journal of Geophysical Research* 100: 2013–2030.

Bercovici D (1993) A simple model of plate generation from mantle flow. *Geophysical Journal International* 114: 635–650.

Bercovici D and Karato S-I (2003) Whole-mantle convection and the transition-zone water filter. *Nature* 425: 39–44.

Bercovici D and Ricard Y (2005) Tectonic plate generation and two? phase damage: Void growth versus grainsize reduction. *Journal of Geophysical Research* 110: B03401 (doi:10.1029/2004JB003181).

Bercovici D, Ricard Y, and Schubert G (2001) A two-phase model for compaction and damage. Part 1: General theory. *Journal of Geophysical Research* 106: 8887–8906.

Bercovici D, Ricard Y, and Richards MA (2000) The relation between mantle dynamics and plate tectonics: A primer. In: Richards MA, Gordon R, and Van der Hilst R (eds.) *AGU Monograph Series: History and Dynamics of Global Plate Motions*, pp. 5–46. Washington, DC: American Geophysical Union.

Bercovici D, Schubert G, and Glatzmaier GA (1989) Three-dimensional spherical models of convection in the Earth's mantle. *Science* 244: 950–955.

Bevis M (1988) Seismic slip and down-dip strain rates in Wadati-Benioff zones. *Science* 240: 1317–1319.

Bevis M (1986) The curvature of Wadati-Benioff zones and the torsional rigidity of subducting plates. *Nature* 323: 52–53.

Bevis M, Kendrick E, Smalley R, Jr., Brooks B, Allmendinger R, and Isacks B (2001) On the strength of interplate coupling and the rate of back arc convergence in the central Andes: An analysis of the interseismic velocity field. *Geochemistry Geophysics Geosystems* 2: (doi:10.1029/2001GC000198).

Billen MI and Gurnis M (2001) A low viscosity wedge in subduction zones. *Earth and Planetary Science Letters* 193: 22736.

Billen MI, Gurnis M, and Simons M (2003) Multiscale dynamics of the Tonga–Kermadec subduction zone. *Geophysical Journal International* 153: 359–388.

Bina CR (1996) Phase transition buoyancy contributions to stresses in subducting lithosphere. *Geophysical Research Letters* 23: 3563–3566.

Bina CR (1997) Patterns of deep seismicity reflect buoyancy stresses due to phase transitions. *Geophysical Research Letters* 24: 3301–3304.

Bostock MG, Hyndman RD, Rondenay SJ, and Peacock S (2002) An inverted continental Moho and the serpentinization of the forearc mantle. *Nature* 417: 536–568.

Bowman JR and Ando M (1987) Shear-wave splitting in the upper-mantle wedge above the Tonga subduction zone. *Geophysical Journal of the Royal Astronomical Society* 88: 25–41.

Brudzinski MR and Chen W-P (2003a) Visualization of seismicity along subduction zones: Toward a physical basis. *Seismological Research Letters* 74: 731–738.

Brudzinski MR and Chen W-P (2003b) A petrologic anomaly accompanying outboard earthquakes beneath Figi-Tonga:

Corresponding evidence from broadband P and S waveforms. *Journal of Geophysical Research* 108(B6): (doi:10.1029/2002JB002012).

Brudzinski MR and Chen W-P (2005) Earthquakes and strain in subhorizontal slabs. *Journal of Geophysical Research* 110: B08303 (doi:10.1029/2004JB003470).

Buffett B and Rowley DB (2006) Plate bending at subduction zones: Consequences for the direction of plate motions. *Earth and Planetary Science Letters* 245: 359–364.

Bunge HP, Richards MA, and Baumgardner JR (1996) The effect of depth-dependent viscosity on the planform of mantle convection. *Nature* 379: 436–438.

Bunge HP, Richards MA, and Baumgardner JR (1997) A sensitivity study of 3-dimensional spherical mantle convection at 108 Rayleigh number – Effects of depth-dependent viscosity, heating mode, and an endothermic phase-change. *Journal of Geophysical Research* 102: 11991–12007.

Carlson RL (2001) The abundance of ultramafic rocks in Atlantic Ocean crust. *Geophysical Journal International* 144: 37–48.

Cassidy JF and Bostock MG (1996) Shear-wave splitting above the subducting Juan de Fuca Plate. *Geophysical Research Letters* 23: 941–944.

Chase CG (1978) Plate kinematics: The Americas, east Africa and the rest of the world. *Earth and Planetary Science Letters* 37: 355–368.

Chase CG (1979) Subduction, the geoid, and lower mantle convection. *Nature* 282: 464–468.

Chen J and King SD (1998) The influence of temperature and depth dependent viscosity on geoid and topography profiles from models of mantle convection. *Physics of the Earth and Planetary Interiors* 106: 75–92.

Chen P-F, Bina CR, and Okal E (2004) A global survey of stress orientations in subducting slabs as revealed by intermediate-depth earthquakes. *Geophysical Journal International* 159: 721–733.

Chen P-F, Bina CR, and Okal E (2001) Variations in slab dip along the subducting Nazca Plate, as related to stress patterns and moment release of intermediate-depth seismicity and to surface volcanism. *Geochemistry Geophysics Geosystems* 2: 12 (doi:10.1029/2001GC000153).

Chen W-P and Brudzinski MR (2001) Evidence for a large-scale remnant of subducted lithosphere. *Science* 292: 2475–2479.

Chen W-P and Brudzinski MR (2003) Seismic anisotropy in the mantle transition zone beneath Figi–Tonga. *Geophysical Research Letters* 30: 13 (doi:10.1029/2002GL016330).

Chase CG (1979) Subduction, the geoid, and lower mantle convection. *Nature* 282: 464–468.

Chiu J-M, Isacks BL, and Cardwell RK (1991) 3-D configuration of subducted lithosphere in the Western Pacific. *Geophysical Journal International* 106: 99–111.

Christensen UR (1985) Thermal evolution models for Earth. *Journal of Geophysical Research* 90: 2995–3007.

Christensen UR (1995) Effects of phase transitions on mantle convection. *Annual Review of Earth and Planetary Sciences* 23: 65–88.

Christensen UR (1996a) The influence of trench migration on slab penetration into the lower mantle. *Earth and Planetary Science Letters* 10: 27–39.

Christensen UR (1996b) Influence of chemical buoyancy on the dynamics of slabs in the transition zone. *Journal of Geophysical Research* 102: 22435–22443.

Christensen UR (1998) Dynamic phase boundary topography by latent heat effects. *Earth and Planetary Science Letters* 154: 295–306.

Christensen UR and Yuen DA (1984) The interaction of a subducting lithospheric slab with a chemical or phase boundary. *Journal of Geophysical Research* 89: 4389.

Creager KC, Chiao L-Y, Winchester JP, Jr., and Engdahl ER (1995) Membrane strain rates in the subducting plate beneath South America. *Geophysical Research Letters* 22: 2321–2324.

Collier JD, Helffrich GR, and Wood BJ (2001) Seismic discontinuities and subduction zones. *Physics of the Earth and Planetary Interiors* 127: 35–49.

Colombo D, Cimini GB, and de Franco R (1997) Three-dimensional velocity structure of the upper mantle beneath Costa Rica from a teleseismic tomography study. *Geophysical Journal International* 131: 189–208.

Conder JA, Weins DA, and Morris J (2002) On the decompression melting structure at volcanic arcs and back-arc spreading centers. *Geophysical Research Letters* 29: (doi:10.1029/2002GL015390).

Conder JA (2005) A case for hot slab surface temperatures in numerical viscous flow models of subduction zones with an improved fault zone parameterization. *Physics of the Earth and Planetary Interiors* 149: 155–164.

Conrad CP and Hager BH (1999) Effects of plate bending and fault strength at subduction zones on plate dynamics. *Journal of Geophysical Research* 104: 17551–17572.

Conrad CP and Hager BH (2001) Mantle convection with strong subduction zones. *Geophysical Journal International* 144: 271–288.

Crough ST and Jurdy DM (1980) Subducted lithosphere, hotspots, and the geoid. *Earth and Planetary Science Letters* 48: 15–22.

Cruciani C, Carminati E, and Doglioni C (2005) Slab dip vs. lithosphere age: No direct function. *Earth and Planetary Science Letters* 238: 298–310.

Currie CA, Cassidy JF, and Hyndman RD (2001) A regional study of shear wave splitting above the Cascadia subduction zone: Margin-parallel crustal stress. *Geophysical Research Letters* 28: 659–662.

Currie CA, Cassidy JF, Hyndman RD, and Bostock MG (2004) Shear wave anisotropy beneath the Cascadia subduction zone and Western North American craton. *Geophysical Journal International* 157: 341–353.

Dasgupta R, Hirschmann MM, and Dellas N (2005) The effect of bulk composition on the solidus of carbonated eclogite from partial melting experiments at 3 GPa. *Contributions to Mineralogy and Petrology* 149: 288–305.

Daessler R and Yuen DA (1993) The effects of phase transition kinetics on subducting slabs. *Geophysical Research Letters* 20: 2603–2606.

Daessler R and Yuen DA (1996) The metastable olivine wedge in fast subducting slabs; constraints from thermo-kinetic coupling. *Earth and Planetary Science Letters* 137: 109–118.

Daessler R, Yuen DA, Karato SI, and Riedel MR (1996) Two-dimensional thermo-kinetic model for the olivine-spinel phase transition in subducting slabs. *Physics of the Earth and Planetary Interiors* 94: 217–239.

Davies GF (1977) Whole mantle convection and plate tectonics. *Geophysical Journal of the Royal Astronomical Society* 49: 459–486.

Davies GF (1981) Regional compensation of subducted lithosphere: Effects on geoid, gravity and topography from a preliminary model. *Earth and Planetary Science Letters* 54: 431–441.

Davies GF (1984) Lagging mantle convection, the geoid and mantle structure. *Earth and Planetary Science Letters* 69: 187–194.

Davies GF (1986) Mantle convection under simulated plates; effects of heating modes and ridge and trench migration, and implications for the core–mantle boundary, bathymetry, the geoid and Benioff zones. *Geophysical Journal of the Royal Astronomical Society* 84: 153–183.

Davies GF (1989) Mantle convection model with a dynamic plate: Topography, heat-flow and gravity anomalies. *Geophysical Journal* 98: 461–464.

Davies GF (1995) Penetration of plates and plumes through the mantle transition zone. *Earth and Planetary Science Letters* 133: 507–516.

Davies GF and Richards MA (1992) Mantle convection. *Journal of Geology* 100: 151–206.

Davies JH (1999) The role of hydraulic fractures and intermediate-depth earthquakes in generating subduction-zone magmatism. *Nature* 398: 142–145.

Davies JH and Stevenson DJ (1992) Physical models of source region of subduction zone volcanics. *Journal of Geophysical Research* 97: 2037–2070.

Deal MM and Nolet G (1999) Slab temperature and thickness from seismic tomography. Part 2: Izu-Bonin, Japan, and Kuril subduction zones. *Journal of Geophysical Research* 104: 28803–28812.

Deal MM, Nolet G, and van der Hilst RD (1999) Slab temperature and thickness from seismic tomography. Part 1: Method and application to Tonga. *Journal of Geophysical Research* 104: 28789–28802.

Devaux JP, Fleitout L, Schubert G, and Anderson C (2000) Stresses in a subducting slab in the presence of a metastable olivine wedge. *Journal of Geophysical Research* 105: 13365–13373.

DeMets C, Gordon RG, Argus DF, and Stein S (1994) Effect of recent revisions to the geomagnetic reversal time scale on estimates of current plate motions. *Geophysical Research Letters* 21: 2191–2194.

Dixon JE, Dixon TH, Bell DR, and Malservisi R (2004) Lateral variation in upper mantle viscosity: Role of water. *Earth and Planetary Science Letters* 222: 451–467.

Dixon JE, Leist L, Langmuir C, and Schilling J-G (2002) Recycled dehydrated lithosphere observed in plume-influenced mid-ocen-ridge basalt. *Nature* 420: 385–389.

Doglioni C (1993) Geological evidence for a global tectonic polarity. *Journal of the Geological Society of London* 150: 991–1002.

Doglioni C, Harabaglia P, Merlini S, Mongelli F, Peccerillo A, and Piromallo C (1999) Orogens and slabs vs. their direction of subduction. *Earth-Science Reviews* 45: 167–208.

Eiler J (ed.) *Inside the Subduction Factory*, vol. 138, 311 pp. Washington, DC: American Geophysical Union.

Elliott T (2003) Tracers of the slab. In: Eiler J (ed.) *Inside the Subduction Factory*, vol. 138, pp. 23–43. Washington, DC: American Geophysical Union.

Elliott T, Plank T, Zindler A, White W, and Bourdon B (1997) Element transport from slab to volcanic front at the Mariana arc. *Journal of Geophysical Research* 102: 14991–15019.

Engdahl RD, van der Hilst RD, and Berrocal J (1995) Imaging of subducted lithosphere beneath South America. *Geophysical Research Letters* 22: 2317–2320.

Engdahl RD, van der Hilst RD, and Buland RP (1998) Global teleseismic earthquake relocation with improved travel times and procedures for depth determination. *Bulletin of the Seismological Society of America* 88: 722–743.

Engdahl R, van der Hilst R, and Buund R (1998) Global teleseismic earthquake relocation with improved travel times and procedures for depth determination. *Bulletin of the Seismological Society of America* 88: 722–743.

England P and Wortel R (1980) Some consequences of the subduction of young slabs. *Earth and Planetary Science Letters* 47: 403–415.

Fischer KM, Creager KC, and Jordan TH (1991) Mapping the Tonga slab. *Journal of Geophysical Research* 96: 14403–14427.

Fischer KM and Jordan TH (1991) Seismic strain rate and deep slab deformation in Tonga. *Journal of Geophysical Research* 96: 14403–14427.

Fischer KM, Parmentier EM, Stine AR, and Wolf ER (2000) Modeling anisotropy and plate-driven flow in the Tonga subduction zone back arc. *Journal of Geophysical Research* 105: 16181–16191.

Fischer KM and Wiens DA (1996) The depth distribution of mantle anisotropy beneath the Tonga subduction zone. *Earth and Planetary Science Letters* 142: 253–260.

Forsyth DW and Uyeda S (1975) On the relative importance of the driving forces of plate motion. *Geophysical Journal of the Royal Astronomical Society* 43: 163–200.

Forte AM and Woodward RL (1997) Seismic-geodynamic constraints on three-dimensional structure, vertical flow, and heat transfer in the mantle. *Journal of Geophysical Research* 102: 17981–17994.

Forte AM and Mitrovica JX (2001) Deep-mantle high-viscosity flow and thermochemical structure inferred from seismic and geodynamic data. *Nature* 410: 1049–1056.

Fouch MJ and Fischer KM (1996) Mantle anisotropy beneath northwest Pacific subduction zones. *Journal of Geophysical Research* 101: 15987–16002.

Fouch MJ and Fischer KM (1998) Shear wave anisotropy in the Mariana subduction zone. *Journal of Geophysical Research* 25: 1221–1224.

Fouch MJ, Fischer KM, Wysession ME, and Clarke TJ (2000) Shear wave splitting, continental keels, and patterns of mantle flow. *Journal of Geophysical Research* 105: 6255–6276.

Fukao Y, Obayashi M, Inoue H, and Nenbai M (1992) Subducting slabs stagnant in the mantle transition zone. *Journal of Geophysical Research* 97: 4809–4822.

Furukawa Y (1993) Depth of the decoupling plate interface and thermal structure under arcs. *Journal of Geophysical Research* 98: 20005–20013.

Furlong KP, Chapman DS, and Alfeld PW (1982) Thermal modeling of the geometry of subduction with implications for the tectonics of the overriding plate. *Journal of Geophysical Research* 87: 1786–1802.

Gaetani GA and Grove TL (2003) Experimental constraints on melt generation in the mantle wedge. In: Eiler J (ed.) *Inside the Subduction Factory*, vol.138, pp. 107–134. Washington, DC: American Geophysical Union.

Gaherty JB and Hager BH (1994) Compositional vs thermal buoyancy and the evolution of subducted lithosphere. *Geophysical Research Letters* 21: 141–144.

Garfunkel Z, Anderson CA, and Schubert G (1986) Mantle circulation and the lateral migration of subducted slabs. *Journal of Geophysical Research* 91: 7205–7223.

George R, Turner S, Morris J, Plank T, Hawkesworth C, and Ryan J (2005) Pressure-temperature-time paths of sediment recycling beneath the Tonga-Kennadec arc. *Earth and Planetary Science Letters* 233: 195–211.

Giardini D and Woodhouse JH (1984) Deep seismicity and modes of deformation in the Tonga subduction zone. *Nature* 307: 505–509.

Giardini D and Woodhouse JH (1986) Horizontal shear flow in the mantle beneath the Tonga arc. *Nature* 319: 551–555.

Grand SP, van der Hilst RD, and Widiyantoro S (1997) Global seismic tomography: A snapshot of convection in the Earth. *GSA Today* 7: 1–7.

Green HW, II and Burnley PC (1989) A new self-organizing mechanism for deep-focus earthquakes. *Nature* 341: 733–737.

Green TH and Ringwood AE (1967) Crystallization of basalt and andesite under high pressure hydrous conditions. *Earth and Planetary Science Letters* 3: 481–489.

Griffiths RW, Hackney RI, and van der Hilst RD (1995) A laboratory investigation of effects of trench migration on the descent of subducted slabs. *Earth and Planetary Science Letters* 133: 1–17.

Gripp AE and Gordon RG (2002) Young tracks of hotspots and current plate velocities. *Geophysical Journal International* 150: 321–361.

Gudmundsson O and Sambridge M (1998) A regionalized upper mantle (RUM) seismic model. *Journal of Geophysical Research* 103: 7121–7136.

Guillou-Grottier L, Buttles J, and Olson P (1995) Laboratory experiments on the structure of subducted lithosphere. *Earth and Planetary Science Letters* 133: 19–34.

Gurnis M (1989) A reassessment of the heat transport by variable viscosity convection with plates and lids. *Geophysical Research Letters* 16: 179–182.

Gurnis M and Hager BH (1988) Controls of the structure of subducted slabs. *Nature* 335: 317–321.

Gurnis M, Hall C, and Lavier LL (2004) Evolving force balance during inipient subduction. *Geochemistry Geophysics Geosystems* 5: Q07001 (doi:10.1029/2003GC000681).

Gurnis M, Ritsema J, van-Heijst HJ, and Zhong S (2000a) Tonga slab deformation; the influence of a lower mantle upwelling on a slab in a young subduction zone. *Geophysical Research Letters* 27: 2373–2376.

Gurnis M, Zhong S, and Toth J (2000b) On the competing roles of fault reactivation and brittle failure in generating plate tectonics from mantle convection. In: Richards MA, Gordon R, and Van der Hilst R (eds.) *AGU Monograph Series: History and Dynamics of Global Plate Motions* pp. 73–94. Washington, DC.

Gutenberg B and Richter CF (1939) Depth and geographical distribution of deep-focus earthquakes. *Geological Society of America Bulletin* 50: 1511–1528.

Hacker BR, Abers GA, and Peacock SM (2003) Sunduction factory. Part 1: Theoretical mineralogy, density seismic wave speeds and $H_2O$ content. *Journal of Geophysical Research* 108: 2029.

Hager BH (1984) Subducted slabs and the geoid; constraints on mantle rheology and flow. *Journal of Geophysical Research* 89: 6003–6015.

Hager BH and O'Connell RJ (1978) Subduction zone dip angles and flow driven by plate motion. *Tectonophysics* 50: 111–133.

Hager BH and O'Connell RJ (1981) A simple global model of plate dynamics and mantle convection. *Journal of Geophysical Research* 86: 4843–4867.

Hager BH and Clayton RW (1989) Constraints on the structure of mantle convection using seismic observations, flow models, and the geoid. In: Peltier WR (ed.) *Mantle Convection* pp. 658–758. Pergamon Press.

Hager BH, Clayton RW, Richards MA, Comer RP, and Dziewonski AM (1985) Lower mantle heterogeneity, dynamic topography and the geoid. *Nature* 313: 541–545.

Hager BH and Richards MA (1989) Long-wavelength variations in Earths geoid; physical models and dynamical implications. *Philosophical Transactions of the Royal Society of London Series A* 328: 309–327.

Hall C, Gurnis M, Sdrolias M, Lavier LL, and Müller RD (2002) Catastrophic initiation of subduction following forced convergence across fracture zones. *Earth and Planetary Science Letters* 212: 15–30.

Hall CE, Fischer KM, Parmentier EM, and Blackman DK (2000) The influence of plate motions on three-dimensional back arc mantle flow and shear wave splitting. *Journal of Geophysical Research* 105: 28009–28033.

Hamburger MW and Isacks BL (1987) Deep earthquakes in the southwest Pacific: A tectonic interpretation. *Journal of Geophysical Research* 92: 13841–13854.

Han L and Gurnis M (1999) How valid are dynamic models of subduction and convection when plate motions are prescribed? *Physics of the Earth and Planetary Interiors* 110: 235–246.

Hawkesworth CJ, Gallagher K, Hergt JM, and McDermott F (1993) Mantle and slab contributions in arc magmas. *Annual Review of Earth and Planetary Sciences* 21: 175–204.

Hawkesworth CJ, Turner SP, McDermott F, Peate DW, and van Calstern P (1997) U–Th isotopes in arc magmas: Implications for element transfer from the subducted crust. *Science* 276: 551–555.

Helffrich G (1996) Subducted lithospheric slab velocity structure: Observations and mineralogical inferences. In: Bebout G, Scholl D, Kirby S, and Platt J (eds.) *American Geophysical Union Monograph 96: Subduction Top to Bottom*, pp. 215–222. Washington, DC: American Geophysical Union.

Helffrich G and Abers GA (1997) Slab low-velocity layer in eastern Aleutian subduction zone. *Geophysical Journal International* 130: 640–648.

Helffrich G, Stein S, and Wood BJ (1989) Subduction zone thermal structure and mineralogy and their relationship to seismic wave reflections and conversions at the slab/ mantle interface. *Journal of Geophysical Research* 94: 753–763.

Heurut A and Lallenmand S (2005) Plate motions, slab dynamics and back-arc deformation. *Physics of the Earth and Planetary Interiors* 149: 31–52.

Hiramatsu Y and Ando M (1996) Seismic anisotropy near source region in subduction zone around Japan. *Physics of the Earth and Planetary Interiors* 95: 237–250.

Hirschmann MM (2006) Water, melting, and the deep Earth H2O cycle. *Annual Review of Earth and Planetary Sciences* 34: 629–653.

Hirth G and Kohlstedt DL (1995) Experimental constraints on the dynamics of the partially molten upper mantle. Part 2: Deformation in the dislocation creep regime. *Journal of Geophysical Research* 100: 15441–15449.

Hirth G and Kohlsted D (2003) Rheology of the upper mantle and mantle wedge: A view from the experimentalists. In: Eiler JM (ed.) *Geophysical Monograph, 138: Inside the Subduction Factory*, pp. 83–105. Washington, DC: American Geophysical Union.

Holt WE (1995) Flow fields within the Tonga slab determined from the moment tensors of deep earthquakes. *Geophysical Research Letters* 22: 989–992.

Houseman GA and Gubbins D (1997) Deformation of subducted oceanic lithosphere. *Geophysical Journal International* 131: 535–551.

Huang X, Xu Y, and Karato S (2005) Water content in the transition zone from electrical conductivity of wadsleyite and ringwoodite. *Nature* 434: 746–749.

Hyndman RD and Peacock SM (2003) Serpentinization of the forearc mantle. *Earth and Planetary Science Letters* 212: 417–432.

Hyndman RD and Wang K (1993) Thermal constraints on zones of major thrust earthquake failure: the Cascadia subduction zone. *Journal of Geophysical Research* 98: 2039–2060.

Hyndman RD, Yamano M, and Oleskevich DA (1997) The seismogenic zone of subduction thrust faults. *Island Arch* 6(3): 244–260.

Iidaka T and Obara K (1995) Shear-wave polarization anisotropy in the mantle wedge above the subducting Pacific plate. *Tectonophysics* 249: 53–68.

Irifune T (1993) Phase transformations in the Earths mantle and subducting slabs: Implications for their compositions, seismic velocity and density structures and dynamics. *Island Arch* 2: 55–71.

Isacks BL and Barazangi M (1977) Geometry of Benioff zones: Lateral segmentation and downwards bending of the subducted lithosphere. In: Talwani M and Pitmann WC, III (eds.) *Island Arcs, Deep Sea Trenches and Back Arc Basins*, pp. 99–114. Washington, DC: American Geophysical Union.

Isacks B and Molnar P (1969) Mantle earthquake mechanisms and the sinking of the lithosphere. *Nature* 233: 1121–1124.

Isacks B and Molnar P (1971) Distribution of stresses in the descending lithosphere from a global survey of focal-mechanism solutions of mantle earthquakes. *Reviews of Geophysics and Space Physics* 9: 103–174.

Isacks B, Oliver J, and Sykes LR (1968) Seismology and the new global tectonics. *Journal of Geophysical Research* 73: 5855–5899.

Ishikawa T and Nakamura E (1994) Origin of the slab component in arc lavas from across-arc variations of B and Pb isotopes. *Nature* 370: 205–208.

Ita JJ and King SD (1994) Sensitivity of convection with an endothermic phase change to the form of governing equations, initial conditions, boundary conditions and equation of state. *Journal of Geophysical Research* 99: 15919–15938.

Ita JJ and King SD (1998) The influence of thermodynamic formualation on simulations of subduction zone geometry and history. *Geophysical Research Letters* 25: 1463–1466.

Ita J and Stixrude L (1993) Density and elasticity of model upper mantle composition and their implications for whole mantle structure. In: Takahaski E, Jeanloz R, and Rubie D (eds.) *Evolution of Earth and Planets*, pp. 131–141. Washington, DC: American Geophysical Union.

Ita J and Stixrude L (1992) Petrology, elasticity and composition of the mantle transition zone. *Journal of Geophysical Research* 97: 6849–6866.

Iwamori H (1998) Transportation of $H_2O$ and melting in subduction zones. *Earth and Planetary Science Letters* 160: 65–80.

Jarrard RD (1986) Relations among subduction parameters. *Reviews of Geophysics* 24: 217–284.

Johnson MC and Plank T (1999) Dehydration and melting experiments constrain the fate of subducted sediments. *Geochemistry Geophysics Geosystems* 1: (doi:10.1029/ 199GC000014).

Jung H and Karato S (2001) Water-induced fabric transitions in olivine. *Science* 293: 1460–1463.

Kaminski E and Ribe NM (2001) A kinematic model for recrystallization and texture development in olivine polycrystals. *Earth and Planetary Science Letters* 189: 253–267.

Karason H (2002) *Constraints on Mantle Convection from Seismic Tomography and Flow Modeling.* PhD Thesis, MIT.

Karato S, Bercovici D, Leahy G, Richard G, and Jing Z (2006) The transition zone water filter model for global material circulation: Where do we stand? In: Jacobsen SD and van der Lee SC (eds.) *AGU Monograph Series, 168: Earths Deep Water Cycle*, 289. Washington, DC: American Geophysical Union.

Karato S-I and Jung H (1998) Water, partial melting and the origin of the seismic low velocity and high attenuation zone in the upper mantle. *Earth and Planetary Science Letters* 157: 193–207.

Karato S and Wu P (1993) Rheology of the upper mantle: A synthesis. *Science* 260: 771–778.

Kaula WM (1968) *An Introduction to Planetary Physics, the Terrestrial Planets.* New York: John Wiley & Son.

Kaula WM (1972) Global gravity and tectonics. In: Robertson EC (ed.) *The Nature of the Solid Earth*, pp. 386–405. New York: McGraw-Hill.

Kelemen PB, Rilling JL, Parmentier EM, Mehl L, and Hacker BR (2003) Thermal structure due to solid-state flow in the mantle wedge beneath arcs. In: Eiler J (ed.) *Inside the Subduction Factory*, pp. 293–311. Washington, DC: American Geophysical Union.

Kellogg LH, Hager BH, and van der Hilst RD (1999) Compositional stratification in the deep mantle. *Science* 283: 1881–1884.

Kemp DV and Stevenson DA (1996) A tensile flexural model for the initiation of subduction. *Geophysical Journal International* 125: 73–94.

Kincaid C and Olson PL (1984) An experimental study of subduction and slab migration. *Journal of Geophysical Research* 92: 13832–13840.

Kincaid C and Sacks IS (1997) Thermal and dynamical evolution of the upper mantle in subduction zones. *Journal of Geophysical Research* 102(B6): 12295–12315.

King SD (2002) Geoid and topography over subduction zones: The effect of phase transformations. *Journal of Geophysical Research* 107 (doi:10.1029/2000JB000141).

King SD (2001) Subduction: Observations and geodynamic models. *Physics of the Earth and Planetary Interiors* 127: 9–24.

King SD (1995) Models of mantle viscosity. In: Ahrens TJ (ed.) *AGU Reference Shelf 2: Mineral Physics and Crystallography: Handbook of Physical Constants*, pp. 227–236. Washington, DC: American Geophysical Union.

King SD, Balachandar S, and Ita JJ (1997) Using eigenfunctions of the two-point correlation function to study convection with multiple phase transformations. *Geophysical Research Letters* 24: 703–706.

King SD, Gable C, and Weinstein S (1992) Models of convection driven tectonic plates: A comparison of methods and results. *Geophysical Journal International* 109: 481–487.

King SD and Hager BH (1994) Subducted slabs and the geoid. Part 1: Numerical experiments with temperature-dependent viscosity. *Journal of Geophysical Research* 99: 19843–19852.

King SD and Ita JJ (1995) The effect of slab rheology on mass transport across a phase transition boundary. *Journal of Geophysical Research* 100: 20211–20222.

King SD and Masters G (1992) An inversion for radial viscosity structure using seismic tomography. *Geophysical Research Letters* 19: 1551–1554.

King SD, Raefsky A, and Hager BH (1990) ConMan: Vectorizing a finite element code for incompressible two-dimensional convection in the Earth's mantle. *Physics of the Earth and Planetary Interiors* 59: 196–208.

Kirby SH, Durham WB, and Stern LA (1991) Mantle phase changes and deep-earthquake faulting in subducting lithosphere. *Science* 252: 216–225.

Kirby SH, Stein S, Okal EA, and Rubie DC (1996) Metastable mantle phase transformations and deep earthquakes in subducting oceanic lithosphere. *Reviews of Geophysics* 34: 261–306.

Kneller EA, van Keken PE, Karato S, and Park J (2005) B-type olivine fabric in the mantle wedge; insightsmor high resolution non-Newtonian subduction models. *Earth and Planetary Science Letters* 237: 781–797.

Kohlstedt DL, Keppler H, and Rubie DC (1996) Solubility of water in the $\alpha$, $\beta$ and $\gamma$ phases of (Mg,Fe)2SiO4. *Contributions to Mineralogy and Petrology* 123: 345–357.

Koper KD, Wiens DA, Dorman LM, Hildebrand JA, and Webb SC (1998) Modeling the Tonga slab: Can travel time data resolve a metastable olivine wedge? *Journal of Geophysical Research* 103: 30079–30100.

Kopitzke U (1979) Finite element convection models: Comparison of shallow and deep mantle convection and temperatures in the mantle. *Journal of Geophysics* 46: 97–121.

Lambeck K, Johnston P, and Nakada M (1990) Holocene glacial rebound and sea-level change in NW Europe. *Geophysical Journal International* 103: 451–468.

Lassak TM, Fouch MJ, Hall CE, and Kamiknsi E (2006) Seismic characterization of mantle flow in subduction systems: Can we resolve a hydrated mantle wedge? *Earth and Planetary Science Letters* 243: 632–649.

Lay T (1994) Seismological constraints on the velocity structure and fate of subducting slabs: 25 years of progress. *Advances in Geophysics* 35: 1–180.

Lay T, Kanamori H, Ammon CJ, *et al.* (2005) The Great Sumatra–Andaman earthquake of December 26, 2004. *Science* 308: 1127–1133.

Leeman WP (1996) Boron and other fluid-mobile elements in volcanic arc lavas: Implications for subduction processes. In: Bebout GE, Scholl DW, Kirby SH, and Platt JP (eds.) *AGU Geophysical Monograph 96: Subduction: Top to Bottom*, pp. 269–276. Washington, DC: American Geophysical Union.

Lemoine FG, Smith DE, Kunz L, *et al.* (1997) The development of the NASA GSFC and NIMA joint geopotential model. In: Segawa J, Fujimoto H, and Okubo S (eds.) *International Association of Geodesy Symposia: Gravity, Geoid and Marine Geodesy*, pp. 461–469. Berlin: Springer-Verlag.

Le Pichon X (1968) Sea floor spreading and continental drift. *Journal of Geophysical Research* 73: 3661–3697.

Levin V, Droznin D, Park J, and Gordeev E (2004) Detailed mapping of seismic anisotropy with local shear waves in southeastern Kamchatka. *Geophysical Journal International* 158: 1009–1023.

Lithgow-Bertelloni C and Richards MA (1995) Cenozoic plate driving forces. *Geophysical Research Letters* 22: 1317–1320.

Lliboutry L (1969) Seafloor spreading continental drift and lithosphere sinking with an asthenosphere at melting point. *Journal of Geophysical Research* 74: 6525–6540.

Long MD and van der Hilst RD (2005) Upper mantle anisotropy beneath Japan from shear wave splitting. *Physics of the Earth and Planetary Interiors* 151: 206–222.

Lowman JP and Jarvis GT (1996) Continental collisions in wide aspect ratio and high Rayleigh number two-dimensional mantle convection models. *Journal of Geophysical Research* 101: 25485–25497.

Lundgren PR and Giardini D (1990) Lateral structure of the subducting Pacific plate beneath the Hokkaido corner from intermediate and deep earthquakes. *Pure and Applied Geophysics* 134: 385–404.

Lundgren PR and Giardini D (1992) Seismicity, shear failure and modes of deformation in deep subduction zones. *Physics of the Earth and Planetary Interiors* 74: 63–74.

Manea VC, Manea M, Kostoglodov V, and Sewell GS (2005) Thermo-mechanical model of the mantle wedge in the central Mexican subduction zone and a blob tracing approach for magma transport. *Physics of the Earth and Planetary Interiors* 149: 165–186.

Marton F, Bina CR, Stein S, and Rubie DC (1999) Effects of slab mineralogy on subduction rates. *Geophysical Research Letters* 26: 119–122.

Marton FC, Shankland TJ, Rubie DC, and Xu Y (2005) Effects of variable thermal conductivity on the mineralogy of subducting slabs and implications for mechanisms of deep earthquakes. *Physics of the Earth and Planetary Interiors* 149: 5364.

Marotta AM and Mongelli F (1998) Flexure of subducted slabs. *Geophysical Journal International* 132: 701–711.

McKenzie DP (1969) Speculations on the consequences and causes of plate motions. *Geophysical Journal of the Royal Astronomical Society* 18: 1–32.

McKenzie DP (1977) Surface deformation, gravity anomalies and convection. *Geophysical Journal of the Royal Astronomical Society* 48: 211–238.

Meade C and Jeanloz R (1991) Deep-focus earthquakes and recycling of water into the Earth's mantle. *Science* 252: 68–72.

Mehl L, Hacker BR, Hirth G, and Kelemen PB (2003) Arc-parallel flow within the mantle wedge: Evidence from the accreted

Talkeetna arc, south central Alaska. *Journal of Geophysical Research* 108: 2375.

Mei S and Kohlstedt DL (2000) Influence of water on plastic deformation of olivine aggregates. Part 1: Diffusion creep regime. *Journal of Geophysical Research* 105: 21471–81. The creep rate of olivine is enhanced significantly with the presence of water. At a water fugacity of ?300 MPa, samples crept ?56 times faster than those deformed under anhydrous conditions at similar differential stresses and temperatures.

Mei S and Kohlstedt DL (2000) Influence of water on plastic deformation of olivine aggregates. Part 2: Dislocation creep regime. *Journal of Geophysical Research* 105: 21457–21470. As a consequence, for olivine aggregates the concentration of silicon interstitials, the rate of silicon diffusion, and therefore the rate of diffusion creep increase systematically with increasing water fugacity (i.e., OH concentration).

Miller MS, Gorbatov A, and Kennett BLN (2005) Heterogeneity within the subducting Pacific slab beneath the Izu-Bonin-Mariana arc: Evidence from tomography using 3D ray-tracing inversion techniques. *Earth and Planetary Science Letters* 235: 331–342.

Miller MS, Gorbatov A, Kennett BLN, and Lister GS (2004) Imaging changes in morphology, geometry, and physical properties of the subducting Pacific Plate along the Izu-Bonin-Mariana arc. *Earth and Planetary Science Letters* 224: 363–370.

Minear JW and Toksöz NM (1970) Thermal regime of a downgloing slab and new global tectonics. *Journal of Geophysical Research* 75: 1397–1419.

Minster JB and Jordan TH (1978) Present-day plate motions. *Journal of Geophysical Research* 83: 5331–5354.

Mitrovica JX (1996). Haskell (1935) revisited. *Journal of Geophysical Research* 101: 555–569.

Mitrovica JX and Forte AM (1997) Radial profile of mantle viscosity: Results from the joint inversion of convection and postglacial rebound observables. *Journal of Geophysical Research* 102: 2751–2769.

Molnar P and England P (1990) Temperatures, heat flux, and frictional stress near major thrust faults. *Journal of Geophysical Research* 95: 4833–4856.

Molnar P, Freedman D, and Shih JSF (1979) Lengths of intermediate and deep seismic zones and temperatures in downgoing slabs of lithosphere. *Geophysical Journal of the Royal Astronomical Society* 56: 41–54.

Moresi LN and Solomatov VS (1995) Numerical investigation of 2D convection with extremely large viscosity variations. *Physics of Fluids* 7: 2154–2162.

Moresi L and Gurnis M (1996) Constraints on the lateral strength of slabs from three-dimensional dynamic flow models. *Earth and Planetary Science Letters* 138: 15–28.

Morgan WJ (1968) Rises, trenches, great faults and crustal blocks. *Journal of Geophysical Research* 73: 1959–1982.

Morgan WJ (1965) Gravity anomalies and convection currents. Part 1: A sphere and cylinder sinking beneath the surface of a viscous fluid. *Journal of Geophysical Research* 70: 6175–6187.

Morris JD, Leeman WP, and Tera F (1990) The subducted component in island arc lavas: constraints from Be isotopes and B–Be systematics. *Nature* 344: 31–36.

Mueller RD, Roest WR, Royer JY, Gahagan LM, and Sclater JG (1997) Digital isochrons of the world's ocean floor. *Journal of Geophysical Research* 102: 3211–3214.

Nakajima J and Hasegawa A (2004) Shear-wave polarization anisotropy and subduction-induced flow in the mantle wedge of northeastern Japan. *Earth and Planetary Science Letters* 225: 365–377.

Nakiboglu SM (1982) Hydrostatic theory of the Earth and its mechanical implications. *Physics of the Earth and Planetary Interiors* 28: 302–311.

Oliver J and Isacks B (1967) Deep earthquake zone, anomalous structures in the upper mantle, and the lithosphere. *Journal of Geophysical Research* 72: 4259–4275.

Okal EA and Kirby SH (1998) Deep earthquakes beneath the north and south Fiji basins, SW Pacific: Earth's most Intense deep seismicity in stagnant slabs. *Physics of the Earth and Planetary Interiors* 109: 25–63.

Oxburgh ER and Turcotte DL (1968) Problems of high heat flow and volcanism with zones of descending mantle convective flow. *Nature* 216: 1041–1043.

Pawley AR and Holloway JR (1993) Water sources for subduction zone volcanism: New experimental constraints. *Science* 260: 664–667.

Peacock SM (1990) Numerical simulation of metamorphic pressure–temperature–time paths and fluid production in subducting slabs. *Tectonics* 9: 1197–1211.

Peacock SM (1991) Numerical-simulation of subduction zone pressure temperature time paths: Constraints on fluid production and arc magmatism. *Philosophical Transactions of the Royal Society of London A* 335: 341–353.

Peacock SM (1992) Blueschist-facies metamorphism, shear heating, and P–T–t paths in subduction shear zones. *Journal of Geophysical Research* 97: 17693–17707.

Peacock SM (1993) Large-scale hydration of the lithosphere above subducting slabs. *Chemical Geology* 108: 49–59.

Peacock SM (1996) Thermal and petrologic structure of subduction zones. In: Bebout G, Scholl D, Kirby S, and Platt J (eds.) *American Geophysical Union Geophysical Monograph 96: Subduction Top to Bottom*, pp. 119–133. Washington, DC: American Geophysical Union.

Peacock SM (2001) Are the lower planes of double seismic zones caused by serpentine dehydration in subducting oceanic mantle? *Geology* 29: 299–302.

Peacock SM (2003) Thermal structure and metamorphic evolution of subducting slabs. In: Eiler JM (ed.) *Geophysical Monograph, 138: Inside the Subduction Factory*, pp. 7–22. Washington, DC: American Geophysical Union.

Peacock SM and Hyndman RD (1999) Hydrous minerals in the mantle wedge and the maximum depth of subduction thrust earthquakes. *Geophysical Research Letters* 26: 2517–2520.

Peacock SM, Rushmer T, and Thompson AB (1994) Partial melting of subducting oceanic crust. *Earth and Planetary Science Letters* 121: 227–244.

Peacock SM, van Keken PE, Holloway SD, Hacker BR, Abers G, and Fergason RL (2005) Thermal structure of the Costa Rica – Nicaragua subduction zone: Slab metamorphism, seismicity and arc magmatism. *Physics of the Earth and Planetary Interiors* 149: 187–200.

Peacock SM and Wang K (1999) Seismic consequences of warm versus cool subduction metamorphism, examples from Southwest and Northeast Japan. *Science* 286: 937–939.

Pekeris CL (1935) Thermal convection in the interior of the Earth. *Monthly Notices of the Royal Astronomical Society Geophysics Supplement* 3: 343–367.

Peyton V, Levin V, Park J, et al. (2001) Mantle flow at a slab edge: Seismic anisotropy in the kamchatka region. *Geophysical Research Letters* 28: 379–382.

Plank T (2005) Constraints from Th/La sediment recycling at subduction zones and the evolution of the continents. *Journal of Petrology* 46: 921–944.

Plank T and Langmuir CH (1993) Tracing trace elements from sediment input to volcanic output at subduction zones. *Nature* 362: 739–742.

Plank T and Langmuir CH (1998) The chemical composition of subducting sediment and its consequence for the crust and mantle. *Chemical Geology* 145: 325–394.

Polet J, Silver PG, Beck SL, et al. (2000) Shear wave anisotropy beneath the Andes from the BANJO, SEDA, and PISCO experiments. *Journal of Geophysical Research* 105: 6287–6304.

Poli S and Schmidt MW (2002) Petrology of subducted slabs. *Annual Review of Earth and Planetary Sciences* 30: 207–235.

Puster P, Hager BH, and Jordan TH (1995) Mantle convection experiments with evolving plates. *Geophysical Research Letters* 22: 2223–2226.

Ratcliff JT, Schubert G, and Zebib A (1995) Three-dimensional variable viscosity convection of an infinite Prandtl number Bousinessq fluid in a spherical shell. *Geophysical Research Letters* 22: 2227–2230.

Ratcliff JT, Schubert G, and Zebib A (1996) Effects of temperature-dependent viscosity on thermal convection in a spherical shell. *Physica D* 97: 242–252.

Regenauer-Lieb K and Yuen DA (2003) Modeling shear zones in geological and planetary sciences: Solid- and fluid- thermal–mechanical approaches. *Earth-Science Reviews* 63: 295–349.

Regenauer-Lieb K and Yuen DA (2000) Fast mechanisms for the formation of new plate boundaries. *Tectonophysics* 322: 53–67.

Regenauer-Lieb K, Yuen DA, and Branlund J (2001) The initiation of subduction: Criticality by addition of water? *Science* 294: 578–580.

Ricard Y, Doglioni C, and Sabadini R (1991) Differential rotation between lithosphere and mantle: A consequence of lateral mantle viscosity. *Journal of Geophysical Research* 96: 8407–8415.

Ricard Y, Fleitout L, and Froidevaux C (1984) Geoid heights and lithospheric stresses for a dynamic Earth. *Annales of Geophysics* 2: 267–286.

Ricard Y, Vigny C, and Froidevaux C (1989) Mantle heterogeneities, geoid and plate motion: A Monte Carlo inversion. *Journal of Geophysical Research* 94: 13739–13754.

Ricard Y and Wuming B (1991) Inferring viscosity and the 3-D density structure of the mantle from geoid, topography and plate velocities. *Geophysical Journal International* 105: 561–572.

Richard G, Bercovici D, and Karato S-i (2006) Hydration of the mantle transition zone by subducting slabs. *Earth and Planetary Science Letters* 251: 156–167.

Richard G, Monnereau M, and Ingrin J (2002) Is the transition zone an empty water reservoir? Inferences from numerical model of mantle dynamics. *Earth and Planetary Science Letters* 205: 37–51.

Richards MA and Davies GF (1989) On the separation of relatively buoyant components from subducted lithosphere. *Geophysical Research Letters* 16: 831–834.

Richards MA and Engebretson DC (1992) Large-scale mantle convection and the history of subduction. *Nature* 355: 437–440.

Richards MA and Hager BH (1989) Effects of lateral viscosity variations on long-wavelength geoid anomalies and topography. *Journal of Geophysical Research* 94: 10299–10313.

Richards MA and Hager BH (1984) Geoid anomalies in a dynamic Earth. *Journal of Geophysical Research* 89: 5987–6002.

Richter FM (1979) Focal mechanisms and seismic energy release of deep and intermediate earthquakes in the Tonga–Kermadec region and their bearing on the depth extent of mantle flow. *Journal of Geophysical Research* 84: 6783–6795.

Richter FM (1977) On the driving mechanism of plate tectonics. *Tectonophysics* 38: 61–88.

Richter FM (1973) Dynamical models for sea floor spreading. *Reviews of Geophysics and Space Physics* 11: 223–287.

Riedel MR and Karato S (1997) Grain-size evolution in subducted oceanic lithosphere associated with the olivine-spineal

transformation and its effect on rheology. *Earth and Planetary Science Letters* 148: 27–43.

Ringwood AE (1975) *Composition and Petrology of the Earth's Mantle*, 618 pp. New York: Mc Graw-Hill.

Ringwood AE and Irifune T (1988) Nature of the 650 km seismic discontinuity: Implications for mantle dynamics and differentiation. *Nature* 331: 131–136.

Ritzert M and Jacoby WR (1992) Geoid effects in a convecting system with lateral viscosity variations. *Geophysical Research Letters* 19: 1547–1550.

Runcorn SK (1967) Flow in the mantle inferred from the low degree harmonics of the geopotential. *Geophysical Journal of the Royal Astronomical Society* 42: 375–384.

Russo RM and Silver PG (1994) Trench-parallel flow beneath the Nazca plate from seismic anisotropy. *Science* 263: 1105–1111.

Ruff LJ and Tichelaar BW (1996) What controls the seismogenic plate interface in subduction zones? In: Bebout GE, Scholl DW, Kirby SH, and Platt P (eds.) *AGU Geophysics Monograph Series vol.96: Subduction: Top to Bottom*, pp. 105–111. Washington, DC: American Geophysical Union.

Rüpke LH, Morgan JP, Hort M, and Connolly JAD (2004) Serpentine and the subduction zone water cycle. *Earth and Planetary Science Letters* 223: 17–34.

Sandvol E and Ni J (1997) Deep azimuthal seismic anisotropy in the southern Kurile and Japan subduction zones. *Journal of Geophysical Research* 102: 9911–9922.

Savage MK (1999) Seismic anisotropy and mantle deformation: What have we learned from shear wave splitting? *Reviews of Geophysics* 37: 65–106.

Schmeling H, Monz R, and Rubie DC (1999) The influence of olivine metastability on the dynamics of subduction. *Earth and Planetary Science Letters* 165: 55–66.

Schmidt MW and Poli S (1998) Experimentally based water budgets for dehydrating slabs and consequences for arc magma generation. *Earth and Planetary Science Letters* 163: 361–379.

Scholz CH (1990) *The Mechanics of Earthquakes and Faulting*, 439 pp. Cambridge: Cambridge University Press.

Scholz CH and Campos J (1995) On the mechanism of seismic decoupling and back arc spreading at subduction zones. *Journal of Geophysical Research* 100: 22103–22116.

Schubert G and Turcotte DL (1971) Phase changes and mantle convection. *Journal of Geophysical Research* 76: 1424–1432.

Schubert G, Yuen DA, and Turcotte DL (1975) Role of phase transitions in a dynamic mantle. *Geophysical Journal of the Royal Astronomical Society* 42: 705–735.

Schubert G and Zhang K (1997) Foundering of the lithosphere at the onset of subduction. *Geophysical Research Letters* 24: 1527–1529.

Silver PG, Beck SL, Wallace TC, *et al.* (1995) Rupture characteristics of the deep Bolivian earthquakes of 9 June 1994 and the mechanism of deep-focus earthquakes. *Science* 268: 69–73.

Silver PG, Carlson RW, and Olson P (1998) Deep slabs, geochemical heterogeneity and the large-scale structure of mantle convection: Investigation of an enduring paradox. *Annual Reviews of Earth and Planetary Science* 16: 477–541.

Smith HJ, Spivack AJ, Staudigel H, and Hart SR (1995) The boron isotopic composition of altered oceanic crust. *Chemical Geology* 126: 119–135.

Smith GP, Wiens DA, Fischer KM, Dorman LM, Webb SC, and Hildebrand JA (2001) A complex pattern of mantle flow in the Lau Basin. *Science* 292: 713–716.

Smyth JR (1994) A crystallographic model for hydrous wadsleyite ($\beta$-Mg2SiO4): An ocean in the Earths interior? *American Mineralogist* 79: 1021–1024.

Smyth JR, Kawamoto T, Jacobesen SD, Swope RJ, Hervig RL, and Holloway JR (1997) Crystal structure of monoclinic hydrous wadsleyite. *American Mineralogist* 82: 270–275.

Solomatov VS (1995) Scaling of temperature- and stress-dependent viscosity convection. *Physics of Fluids* 7: 266–274.

Solomatov VS and Moresi LN (1996) Stagnant lid convection on Venus. *Journal of Geophysical Research* 101: 4737–4753.

Spiegelman M (2003) Linear analysis of melt band formation by simple shear. *Geochemistry Geophysics Geosystems* 4: 8615.

Spiegelman M and McKenzie D (1987) Simple 2-D models for melt extraction at mid-ocean ridges and island arcs. *Earth and Planetary Science Letters* 83: 137–152.

Staudigel H and King SD (1992) Ultrafast subduction: The key to slab recycling efficiency and mantle differentiation?. *Earth and Planetary Science Letters* 109: 517–530.

Staudigel H, Plank T, White B, and Schmincke H-U (1996) Geochemical fluxes during seafloor alteration of the basaltic upper oceanic crust: DSDP sites 417 and 418. In: Bebout GE, Scholl DW, Kirby SH, and Platt P (eds.) *AGU Geophysical Monograph Series vol. 96: Subduction: Top to Bottom*, pp. 19–38. Washington, DC: American Geophysical Union.

Stein SA and Rubie DC (1999) Deep earthquakes in real slabs. *Science* 286: 909–910.

Stein S and Stein CA (1996) Thermo-mechanical evolution of oceanic lithosphere: Implications for the subduction process and deep earthquakes. In: Bebout GE, Scholl DW, Kirby SH, and Platt P (eds.) *AGU Geophysical Monograph Series*, vol. 96: *Subduction: Top to Bottom*, pp. 1–17. Washington, DC: American Geophysical Union.

Stern RJ (2002) Subduction zones. *Reviews of Geophysics* 40: 3-1–3-38.

Stevenson DJ and Turner JS (1977) Angle of subduction. *Nature* 270: 334–336.

Tackley PJ (2000b) Self-consistent generation of tectonic plates in time?dependent, three?dimensional mantle convection simulations. Part 2: Strain-weakening. *Geochemistry Geophysics Geosystems* 1 (doi:10.1029/2000GC000043).

Tackley PJ (2000a) Self-consistent generation of tectonic plates in time?dependent, three?dimensional mantle convection simulations. Part 1: Pseudoplastic yielding. *Geochemistry Geophysics Geosystems* 1 (doi:10.1029/2000GC000036).

Tackley PJ (1998) Self-consistent generation of tectonic plates in three-dimensional mantle convection. *Earth and Planetary Science Letters* 157: 9–22.

Tackley PJ, Stevenson DJ, Glatzmaier GA, and Schubert G (1993) Effects of an endothermic phase transition at 670 km depth in a spherical model of convection in the Earth's mantle. *Nature* 361: 699–704.

Tackley PJ, Stevenson DJ, Glatzmaier GA, and Schubert G (1994) Effects of multiple phase transitions in a 3-D spherical model of convection in the Earths mantle. *Journal of Geophysical Research* 99: 15877–15901.

Tao W and O'Connell RJ (1993) Deformation of a weak subducted slab and variation of seismicity with depth. *Nature* 361: 626–628.

Tichelaar BW and Ruff LJ (1993) Depth of seismic coupling along subduction zones. *Journal of Geophysical Research* 98(B2): 2017–2037.

Tommasi A (1998) Forward modeling of the development of seismic anisotropy in the upper mantle. *Earth and Planetary Science Letters* 160: 1–13.

Toth J and Gurnis M (1998) Dynamics of subduction initiation at preexisting fault zones. *Journal of Geophysical Research* 103: 18053–18067.

Turcotte DL and Oxburgh ER (1967) Finite amplitude convective cells and continental drift. *Journal of Fluid Mechanics* 28: 29–42.

Turcotte DL and Schubert G (1982) *Geodynamics: Applications of Continuum Physics of Geological Problems.* New York, NY: John Wiley and Sons.

Turcotte DL and Schubert G (1973) Frictional heating of the descending lithosphere. *Journal of Geophysical Research* 78: 5876–5886.

Trompert R and Hansen U (1998) Mantle convection simulations with rheologies that generate plate-like behavior. *Nature* 395: 686–689.

Uyeda S and Kanamori H (1979) Back arc opening and the mode of subduction. *Journal of Geophysical Research* 84: 1049–1061.

van der Hilst RD (1995) Complex morphology of subducted lithosphere in the mantle beneath the Tonga Trench. *Nature* 374: 154–157.

van der Hilst RD, Engdahl ER, and Spakman W (1993) Tomographic inversion of P and pP data for aspherical mantle structure below the northwest Pacific region. *Geophysical Journal International* 115: 264–302.

van der Hilst RD, Engdahl R, Spakman W, and Nolet G (1991) Tomographic imaging of subducted lithosphere below Northwest Pacific island arcs. *Nature* 353: 37–43.

van der Hilst R and Mann P (1994) Tectonic implications of tomographic images of subducted lithosphere beneath northwestern South America. *Geology* 22: 451–454.

van der Hilst R and Seno T (1993) Effects of relative plate motion on the deep structure and penetration depth of slabs below the Izu-Bonin and Mariana island arcs. *Earth and Planetary Science Letters* 120: 395–407.

van der Hilst RD, Widiyantoro S, and Engdahl ER (1997) Evidence for deep mantle circulation from global tomography. *Nature* 386: 578–584.

van der Berg AP, Yuen DA, and van Keken PE (1991) Effects of depth-variations in creep laws on the formation of plates in mantle dynamics. *Geophysical Research Letters* 18: 2197–2200.

van Hunen J, van den Berg AP, and Vlaar NJ (2004) Various mechanisms to induce present-day shallow flat subduction and implications for the younger Earth; a numerical parameter study. *Physics of the Earth and Planetary Interiors* 146: 179–194.

van Hunen J, van den Berg AP, and Vlaar NJ (2002) On the role of subducting oceanic plateaus in the development of shallow flat subduction. *Tectonophysics* 352: 317–333.

van Hunen J, van den Berg AP, and Vlaar NJ (2001) Latent heat effects of the major mantle phase transitions on low-angle subduction. *Earth and Planetary Science Letters* 190: 125–135.

van Hunen J, van den Berg AP, and Vlaar NJ (2000) A thermo-mechanical model of horizontal subduction below an overriding plate. *Earth and Planetary Science Letters* 182: 157–169.

van Keken PE (2003) The structure and dynamics of the mantle wedge. *Earth and Planetary Science Letters* 215: 323–338.

van Keken PE, Karato S, and Yuen DA (1996) Rheological control of oceanic crust separation in the transition zone. *Geophysical Research Letters* 23: 1821–1824.

van Keken PE, Kiefer B, and Peacock SM (2002) High-resolution models of subduction zones; implications for mineral dehydration reactions and the transport of water into the deep mantle. *Geochemistry, Geophysics, Geosystems* 3(10): 1056 (doi:10.1029/2001GC000256).

van Keken PE and King SD (2005) Thermal structure and dynamics of subduction zones: Insights from observations and modeling. *Physics of the Earth and Planetary Interiors* 149: 1–6.

Vassiliou MS (1984) The state of stress in subducting slabs as revealed by earthquakes analyzed by moment tensor inversion. *Earth and Planetary Science Letters* 69: 195–202.

Vassiliou MS and Hager BH (1988) Subduction zone earthquakes and stress in slabs. *Pure and Applied Geophysics* 128: 547–624.

Vassiliou MS, Hager BH, and Raefsky A (1984) The distribution of earthquakes with depth and stress in subduction slabs. *Journal of Geodynamics* 1: 11–28.

Venig Meinesz FA (1954) Indonesian archipelago; a geophysical study. *Geological Society of America Bulletin* 65: 143–164.

Vlaar NJ (1983) Thermal anomalies and magmatism due to lithospheric doubling and shifting. *Earth and Planetary Science Letters* 65: 322–333.

von Huene R and Scholl DW (1991) Observations at convergent margins concerning sediment subduction, subduction erosion, and the growth of continental crust. *Reviews of Geophysics* 29: 279–316.

Wallamann K (2001) The geological water cycle and the evolution of marine $\delta^{18}O$ values. *Geochemica Cosmochemica Acta* 65: 2469–2485.

Weinstein S and Olson P (1992) Thermal convection with non-Newtonian plates. *Geophysical Journal International* 111: 515–530.

Wessel P and Smith WHF (1991) Free software helps map and display data. *EOS Transactions of the American Geophysical Union* 72: 441–446.

Widiyantoro S, Kennett BLN, and van der Hilst RD (1999) Seismic tomography with P and S data reveals lateral variations in the rigidity of deep slabs. *Earth and Planetary Science Letters* 173: 91–100.

Wiemer S, Tytgat G, Wyss M, and Duenkel U (1999) Evidence for shear-wave anisotropy in the mantle wedge beneath south central Alaska. *Bulletin of the Seismological Society of America* 89: 1313–1322.

Wiens DA (2001) Seismological constraints on the mechanism of deep earthquakes: Temperature dependence of deep earthquake source properties. *Physics of the Earth and Planetary Interiors* 127: 145–163.

Wiens DA, McGuire JJ, Shore JP, *et al.* (1994) A deep earthquake aftershock sequence and implications for the rupture mechanism of deep earthquakes. *Nature* 372: 540–543.

Wood BJ (1995) The effect of H2O on the 410-kilometer seismic discontinuity. *Science* 268: 74–76.

Wunder B, Meixner A, Romer RL, Wirth R, and Heinrich W (2005) The geochemical cycle of boron: Constraints from boron isotope partitioning experiments between mica and fluid. *Lithos* 84: 206–216.

Yang X, Fischer KM, and Abers GA (1995) Seismic anisotropy beneath the Shumagin Islands segment of the Aleutian–Alaska subduction zone. *Journal of Geophysical Research* 100: 18165–181677.

Young TE, Green HW, Hofmeister AM, and Walker D (1993) Infrared spectroscopic investigation of hydroxyl in $\beta$-(Mg, Fe)$_2$SiO$_4$ and coexisting olivine: Implications for mantle evolution and dynamics. *Physics and Chemistry of Minerals* 19: 409–422.

Yuen DA, Fleitout L, Schubert G, and Froidevaux C (1978) Shear deformation zones along major transform faults and subducting slabs. *Geophysical Journal of the Royal Astronomical Society* 54: 93–119.

Zhao D (2001) Seismological structure of subduction zones and its implications for arc magmatism and dynamics. *Physics of the Earth and Planetary Interiors* 127: 197–214.

Zhao D, Hasegawa A, and Kanamori H (1994) Deep structure of Japan subduction zone as derived from local, regional, and teleseismic event. *Journal of Geophysical Research* 99(B11): 22313–22329.

Zhao D, Xu Y, Wiens D, Dorman LM, Hildebrand J, and Webb SC (1997) Depth extent of the Lau back-arc spreading center and its relation to subduction processes. *Science* 278: 245–247.

Zhong S and Davies GF (1999) Effects of plate and slab viscosities on the geoid. *Earth and Planetary Science Letters* 170: 487–496.

Zhong S and Gurnis M (1992) Viscous flow model of a subduction zone with a faulted lithosphere; long and short wavelength topography, gravity and geoid. *Geophysical Research Letters* 19: 1891–1894.

Zhong S and Gurnis M (1994) Controls on trench topography from dynamic models of subducted slabs. *Journal of Geophysical Research* 99: 15683–15695.

Zhong S and Gurnis M (1995) Mantle convection with plates and mobile, faulted plate margins. *Science* 267: 838–843.

Zhong S and Gurnis M (1995) Towards a realistic simulation of plate margins in models of mantle convection. *Geophysical Research Letters* 22: 981–984.

Zhong S and Gurnis M (1996) Interaction of weak faults and non-Newtonian rheology produces plate-tectonics in a 3D model of mantle flow. *Nature* 383: 245–247.

Zhong S and Gurnis M (1997) Dynamic interaction between tectonic plates, subducting slabs, and the mantle. *Earth Interactions* 1.

Zhong S, Gurnis M, and Moresi L (1998) Role of faults, nonlinear rheology, and viscosity structure in generating plates from instantaneous mantle flow models. *Journal of Geophysical Research* 103: 15255–15268.

Zhou HW and Clayton (1990) P and S wave travel time inversions for the subducting slab under the island arcs of the northwest pacific. *Journal of Geophysical Research* 95: 6829–6851.

# 9    Hot Spots and Melting Anomalies

**G. Ito**, University of Hawaii, Honolulu, HI, USA

**P. E. van Keken**, University of Michigan, Ann Arbor, MI, USA

## Nomenclature

| | | | |
|---|---|---|---|
| $g$ | acceleration of gravity ($m\,s^{-2}$) | $Ra$ | thermal Rayleigh number |
| $\bar{h}$ | average swell height (m) | $Ra_c$ | critical Rayleigh number |
| $q_p$ | plume heat flux associated with swell buoyancy flux | $T$ | temperature (K) |
| | | $U_p, U$ | plate speed, seafloor spreading rate ($m\,s^{-1}$) |
| $s$ | hot-spot swell volume flux | | |
| $t$ | time (s) | $V$ | volume ($m^3$) |
| | | $\bar{W}$ | average intraplate swell width (m), or steady-state ridge-axis swell depth (m) |
| $x$ | horizontal dimension (m) | | |
| $x_r$ | distance between plume source and ridge axis (m) | $W$ | swell width (m) |
| | | $W_0$ | characteristic width scale (m) |
| $B$ | buoyancy flux ($kg\,s^{-1}$) | $\alpha$ | thermal expansivity ($K^{-1}$) |
| $C$ | composition | $\delta$ | thickness of boundary layer (m) |
| $C_1, C_2,$ | constants used in scaling of swell width | $\kappa$ | thermal diffusivity ($m^2\,s^{-1}$) |
| $C_3$ | | $\tau$ | characteristic growth time (s) |
| $E$ | equation of an ellipse | $\eta, \mu$ | viscosity (Pa s) |
| $F$ | fraction partial melting | $\rho$ | density ($kg\,m^{-3}$) |
| $H$ | thickness of fluid | $\rho_c$ | crustal density ($kg\,m^{-3}$) |
| $L_0$ | characteristic length scale (m) | $\rho_m$ | mantle density ($kg\,m^{-3}$) |
| $M$ | volumetric rate of melt generation ($m^3\,s^{-1}$) | $\rho_w$ | density of sea water ($kg\,m^{-3}$) |
| | | $\Delta T$ | Temperature (contrast) (K) |
| $P$ | pressure (Pa) | $\Delta\rho$ | density difference between buoyant and normal mantle ($kg\,m^{-3}$) |
| $Q$ | volume flux of buoyant material ($m^3\,s^{-1}$) | | |

## 9.1 Introduction

The original work by Wilson (1963, 1973), Morgan (1971, 1972), and Crough (1978) established the concept of 'hot spot' as a broad swelling of topography capped by volcanism, which, combined with plate motion, generates volcanoes aligned in a chain and with ages that progress monotonically. In some cases, these chains project back to massive volcanic plateaus, or large igneous provinces (LIPs), suggesting that hot-spot activity began with some of the largest magmatic outbursts evident in the geologic record (Morgan, 1972; Richards *et al.*, 1989; Duncan and Richards, 1991). Hot-spot volcanism is dominantly basaltic and therefore largely involves melting of mantle peridotite, a process that also produces mid-oceanic ridge volcanism. Yet mid-ocean ridge basalts (MORBs) and hot-spot basalts typically have distinct radiogenic isotope characteristics (Hart *et al.*, 1973; Schilling, 1973). These differences indicate that the two forms of magmatism come from mantle materials that have preserved distinct chemical identities for hundreds of millions of years.

The above characteristics suggest that hot-spot volcanism has an origin that is at least partly decoupled from plate processes. A straightforward explanation is that hot spots are generated by convective upwellings, or plumes of unusually hot, buoyant mantle, which rise from the lower mantle (Wilson, 1963, 1973; Morgan, 1971, 1972; Whitehead and Luther, 1975) possibly through a chemically stratified mantle (e.g., Richter and McKenzie, 1981). The large mushroom-shaped head of an initiating mantle plume and the trailing, more narrow plume stem has become a popular explanation for the formation of a LIP followed by a hot-spot track (e.g., Richards *et al.*, 1989; Campbell and Griffiths, 1990).

Studies of hot spots have flourished over the past few decades. Recent articles and textbooks have reviewed some of the classic connections between hot spots and mantle plumes (e.g., Jackson, 1998; Davies, 1999; Condie, 2001; Schubert *et al.*, 2001), the role of mantle plumes in deep-mantle convection and chemical transport (Jellinek and Manga, 2004), and oceanic hot spots (e.g., Ito *et al.*, 2003; Hekinian *et al.*, 2004). Alternative mechanisms, which emphasize processes in the asthenosphere and lithosphere, are being re-evaluated and some new ones proposed (Foulger *et al.*, 2005). It has become clear that few hot spots confidently show all of the above characteristics of the classic description. The term hot spot itself implies a localized region of anomalously high mantle temperature, but some features that were originally called hot spots may involve mantle with little or no excess heat, volcanoes spanning large distances of a chain with similar ages, or both. Thus, the terms 'magmatic anomaly' or 'melting anomaly' may be more general and appropriate to describe the topic of this chapter.

Progress made in the last decade on studies of hot spots and melting anomalies is emphasized here. We summarize the recent observations and discuss the major dynamical processes that have been explored and evaluate their ability to explain the main characteristics. Mechanisms involving hot mantle plumes have seen the most extensive quantitative testing, but the recent observations compel the exploration and rigorous testing of other mechanisms. We summarize the main observations, outline mechanisms that have been proposed, and pose questions that need quantitative answers.

## 9.2 Characteristics

Guided by the classical description of hot spots, we examine four main characteristics: (1) geographic age progression along volcano chains, (2) initiation by massive flood basalt volcanism, (3) anomalously shallow topography surrounding volcanoes (i.e., a hot-spot swell), and (4) basaltic volcanism with geochemical distinction from MORBs. Given the marked progress in seismic methods over the past decade, we also summarize the findings of mantle seismic structure beneath hot spots and surface melt anomalies. **Table 1** summarizes what we have compiled about the above characteristics for 69 hot spots and melting anomalies. **Figure 1** shows a global map of their locations with abbreviations and the main large igneous provinces that we will discuss.

### 9.2.1 Volcano Chains and Age Progression

#### 9.2.1.1 Long-lived age-progressive volcanism

At least 13 hot-spot chains record volcanism lasting >50 My (**Table 1**). The Hawaiian–Emperor and the Louisville chains, for example, span thousands of kilometers across the Pacific basin (∼6000 and >4000 km, respectively), record volcanism for >75 My (Duncan and Clague, 1985; Watts et al., 1988; Duncan and Keller, 2004; Koppers et al., 2004), and were among the first chains that led to the establishment of the hot-spot

concept. As both chains terminate at subduction zones, the existing volcanoes likely record only part of the activities of these hot spots. The Galápagos is the other Pacific hot spot with a similar duration. Its interaction with the Galápagos Spreading Center has produced two chains: the Galápagos Archipelago–Carnegie Ridge on the Nazca Plate (Sinton et al., 1996) and the Cocos Ridge on the Cocos Plate. The Cocos Ridge records oceanic volcanism for ∼14.5 My (Werner et al., 1999) and projects toward the Caribbean LIP (Duncan and Hargraves, 1984), which has $^{40}Ar/^{39}Ar$ dates of 69–139 Ma (e.g., Sinton et al., 1997; Hoernle et al., 2004). The geochemical similarity of these lavas with the Galápagos Archipelago is compelling evidence for a ∼139 My life span for the Galápagos hot spot (Hoernle et al., 2002, 2004).

In the Indian Ocean, Müller et al.'s (1993b) compilation of ages associates the Réunion hot spot with volcanism on the Mascarene Plateau at 45 Ma (Duncan et al., 1990), the Cocos–Laccadive Plateau ∼60 Ma (Duncan, 1978, 1991), and finally the Deccan flood basalts in India, which are dated at 65–66 Ma (see also Sheth (2005)). The Comoros hot spot can be linked to volcanism around the Seychelles islands dated at 63 Ma (Emerick and Duncan, 1982; Müller et al., 1993b). Volcanism associated with the Marion hot-spot projects from Marion island (<0.5 Ma (McDougall et al., 2001)) along a volcanic ridge to Madagascar. While geologic dating is sparse, Storey et al. (1997) infer an age progression along this track back to ∼88 Ma. The Kerguelen hot spot is linked to Broken Ridge and Ninetyeast Ridge on the Australian Plate, as well as multiple stages of volcanism on the Kerguelen Plateau dating to 114 Ma (Frey et al., 2000; Nicolaysen et al., 2000) (**Figure 2**).

In the Atlantic Ocean, the Tristan–Gough and St. Helena chains record volcanism on the African Plate for ∼80 My (O'Connor and Roex, 1992; O'Connor and Duncan, 1990; O'Connor et al., 1999). The connection of Tristan–Gough to the Paraná flood basalts in South America and the Etendeka basalts in Namibia suggests a duration for Tristan–Gough of ∼130 My (see Peate (1997) and references therein). The Trindade–Martin Vaz chain (Fodor and Hanan, 2000) extends eastward from Brazil to where the Alto Paraniba and Poxoreu volcanic provinces erupted ∼85 Ma (Gibson et al., 1997). In the North Atlantic, the Madiera and Canaries chains have recorded age-progressive volcanism for nearly 70 My (Guillou et al., 2004; Geldmacher et al., 2005). The Canaries are unusual in that single volcanoes often remain active for tens of millions of years (**Figure 3**) (Geldmacher et al., 2005).

**Table 1** Global compilation of hot spots with their geophysical and geochemical characteristics

| Name (abbreviation) | Hot spot E. Long., N. Lat. | Age progression? | Age range | Swell?/width (km) | Connection to LIP? | Geoch. distinct from MORB |
|---|---|---|---|---|---|---|
| *Pacific* | | | | | | |
| Austral (AU) | −140.0, −29.37 | No | 0–58.1 Ma | Yes/600 | No | $^{206}Pb/^{204}Pb$ |
| Baja (BAJ) | −113, 27 | — | — | No | No | — |
| Bowie-Kodiak (BOW) | −130, 49.5 | Ok | 0.1–23.8 Ma | Yes/250 | No | May be $^{206}Pb/^{204}Pb$ |
| Caroline (CAR) | −197, 5.3 | Weak | 1.4 Ma (east) to 4.7–13.9 Ma (west) | — | No | No |
| Cobb (COB) | −128.7, 43.6 | Good | 1.5–29.2 Ma | Yes/370 | No | No |
| Cook (CK) | −149.5, −23.5 | No | 0.2–19.4 Ma | Yes/500 | No | $^{206}Pb/^{204}Pb$ |
| Easter (EAS) | −109, −27 | Good | 0–25.6 Ma | Yes/580 | May be Tuamotu and Mid-Pacs | $^{206}Pb/^{204}Pb$ |
| Foundation (FOU) | −111, −39 | Good | 2.1–21 Ma | Yes/250 | No | $^{206}Pb/^{204}Pb$ |
| Galápagos (GAL) | −91.6, −0.4 | Yes | 0–14.5 Ma offshore; 69–139 Ma, Caribbean LIP | Yes/300 | Caribbean LIP | $^{206}Pb/^{204}Pb$ |
| Geologist (GEO) | −157, 19 | No | 82.7–84.6 Ma | — | No | — |
| Guadalupe (GUA) | −118, 29 | — | <3.4 to ~20.3 Ma | May be/? | No | — |
| Hawaiian-Emperor (HAW) | −155.3 18.9 | Good | 0–75.8 Ma | Yes/920 | No | $^{3}He/^{4}He$ and $^{87}Sr/^{86}Sr$ for Islands but not Emperors $^{206}Pb/^{204}Pb$ |
| Japanese-Wake (JWK) | — | No | 78.6–119.7 Ma | No | No | |
| Juan Fernandez (JFE) | −79, −34 | Weak | 1–4 Ma (2 volcanoes dated) | Yes/? | No | $^{3}He/^{4}He$ and $^{87}Sr/^{86}Sr$ |
| Line Islands (LIN)— | No | 35.5–91.2 My | Partially/? | May be Mid-Pacs | — | |
| Louisville (LOU) | −141.2, −53.55 | Good | 1.1–77.3 Ma | Yes/540 | Doubtfully OJP | $^{206}Pb/^{204}Pb$, may be $^{87}Sr/^{86}Sr$ |
| Magellan Seamounts (MAG) | — | No | 87–18.6 Ma | No | No | $^{87}Sr/^{86}Sr$ and $^{206}Pb/^{204}Pb$ |
| Marquesas (MQS) | −138.5, −11 | Ok | 0.8–5.5 Ma | Yes/850 | May be Shatsky or Hess | $^{87}Sr/^{86}Sr$, may be $^{206}Pb/^{204}Pb$ |
| Marshall Islands (MI) | −153.5, −21.0 | No | 68–138 Ma | May be/? | No | $^{206}Pb/^{204}Pb$ |
| Mid-Pacific | — | No | 73.5–128 Ma | No | It could be a LIP | — |

| | | | | | | |
|---|---|---|---|---|---|---|
| Mountains (MPM) | | | | | | |
| Musician (MUS) | — | Ok | 65.5–95.8 Ma | No | No | — |
| Pitcairn (PIT) | −129.4, −25.2 | Good | 0–11.1 Ma | Yes/570 | No | $^{87}Sr/^{86}Sr$ |
| Puka-Puka (PUK) | −165.5, −10.5 | Ok | 5.6–27.5 Ma | Yes/? | No | $^{206}Pb/^{204}Pb$ |
| Samoa (SAM) | −169, −14.3 | Weak | 0–23 Ma | Yes/396 | No | $^{87}Sr/^{86}Sr$, $^3He/^4He$, and $^{206}Pb/^{204}Pb$ |
| San Felix (SF) | −80, −26 | — | — | Yes/? | No | no |
| Shatsky (SHA) | — | Yes | 128–145 Ma | No | It is a LIP | $^{87}Sr/^{86}Sr$ and |
| Society (SOC) | −148, −18 | Good | 0.01–4.2 Ma | Yes/? | No | $^{206}Pb/^{204}Pb$ |
| Socorro (SCR) | −111, 19 | — | — | Yes/? | No | — |
| Tarava (TAR) | 173, 3 | Weak | 35.9 Ma and 43.5 Ma | Yes/? | No | — |
| Tuamotu (TUA) | — | — | ? | Yes/? | It could be a LIP | — |
| *North America* | | | | | | |
| Yellowstone (YEL) | −111, 44.8 | Yes | 16–17 Ma | Yes/600 | May be Columbia River Basalts | — |
| *Australia* | | | | | | |
| Balleny (BAL) | 164.7, −67.4 | Weak | — | — | May be Lord Howe rise | $^{206}Pb/^{204}Pb$ (2 analyses) |
| East Australia (AUS) | 143, −38 | — | — | — | — | — |
| Lord Howe (LHO) | 159, −31 | — | — | — | It could be LIP | — |
| Tasmantid (TAS) | 153, −41.2 | Yes | — | Yes/290 | May be Lord Howe rise | — |
| *Atlantic* | | | | | | |
| Ascension/ Circe (ASC) | −14, −8 | — | <1 Ma (Ascension) and 6 Ma (Circe) | 820 | No | $^{206}Pb/^{204}Pb$ |
| Azores (AZO) | −28, 38 | Seafloor spreading | 0–20 Ma, possibly ~85 Ma | Yes/2300 | No | $^{87}Sr/^{86}Sr$ and $^{206}Pb/^{204}Pb$ |
| Bermuda (BER) | −65, 32 | — | — | Yes/500 × 700 (parallel × perp to plate motion) | No | — |
| Bouvet (BOU) | 3.4, −54.4 | — | ? | Yes/900 | — | $^{206}Pb/^{204}Pb$, may be $^{87}Sr/^{86}Sr$ and $^3He/^4He$ |

(*Continued*)

Table 1 (Continued)

| Name (abbreviation) | Hot spot E. Long., N. Lat. | Age progression? | Age range | Swell?/width (km) | Connection to LIP? | Geoch. distinct from MORB |
|---|---|---|---|---|---|---|
| Cameroon (CAM) | 6, −1 | No | 1–32 Ma | Yes/500–600 | No | $^{206}Pb/^{204}Pb$ |
| Canaries (CAN) | −17, 28 | Ok | 0–68 Ma | No | No | $^{206}Pb/^{204}Pb$ |
| Cape Verde (CAP) | −24, 15 | No | Neogene | Yes/800 | No | $^{87}Sr/^{86}Sr$ and $^{206}Pb/^{204}Pb$ |
| Discovery (DIS) | −6.45, −44.45 | — | 25 Ma | Yes/600 | No | — |
| Fernando Do Norona (FER) | −32, −4 | — | — | Yes/200–300 | — | $^{87}Sr/^{86}Sr$ and $^{206}Pb/^{204}Pb$ |
| Great Meteor (GM) | −28.5, 31 | — | — | Yes/800 | — | — |
| Iceland (ICE) | −17.58, 64.64 | Yes | 0–62 Ma | Yes/2700 | N. Atlantic LIP | $^{3}He/^{4}He$ $^{87}Sr/^{86}Sr$ |
| Jan Mayen (JM) | −8, 71.17 | — | — | Yes | N. Atlantic LIP? | |
| Madeira (MAD) | −17.5, 32.7 | Yes | 0–67 Ma | No | No | $^{206}Pb/^{204}Pb$ |
| New England (NEW) | −57.5, 35 | Yes | 81–103, 122–124 Ma | No | No | $^{206}Pb/^{204}Pb$, may be $^{87}Sr/^{86}Sr$ |
| Shona (SHO) | −4, −52 | — | not dated | Yes/~900 | — | — |
| Sierra Leone (SL) | −29, 1 | — | not dated | — | It could be a LIP | — |
| St. Helena (SHE) | −10, −17 | Yes | 3–81 Ma | Yes/720 | No | $^{206}Pb/^{204}Pb$ |
| Trindade-Martin Vaz (TRN) | −12.2, −37.5 | Probably | <1 Ma to ~85 Ma | Yes/1330 | Small eruptions north of Parana flood basalts and onto Brazilian margin | $^{87}Sr/^{86}Sr$ and $^{206}Pb/^{204}Pb$ |
| Tristan-Gough (TRI) | −9.9, −40.4 (Gough); −12.2, −37.5 (Tristan) | Yes | 0.5–80 Ma and 130 Ma | Yes/850 | Rio Grande-Walvis and Parana-Entendeka | $^{87}Sr/^{86}Sr$ |
| Vema (VEM) | 16, −32 | — | >11 Ma | Yes/200–300 | No | $^{206}Pb/^{204}Pb$ |
| *Indian* | | | | | | |
| Amsterdam-St. Paul (AMS) | 77, −37 | No | — | Yes/300–500 | May be Kerguelen | may be $^{87}Sr/^{86}Sr$ |

| | | | | | | |
|---|---|---|---|---|---|---|
| Comores (COM) | 44, −12 | Yes | 0–5.4 Ma on island chain and ~50 Ma (Seychilles) | Yes/700–800 | — | $^{87}Sr/^{86}Sr$ and $^{206}Pb/^{204}Pb$ |
| Conrad (CON) | 48, −54 | — | Not dated | Half-width 400 south of seamounts | It could be a LIP | — |
| Crozet (CRO) | 50, −46 | — | Not dated | Yes/1120 | May be Madagascar | $^{206}Pb/^{204}Pb$, may be $^{87}Sr/^{86}Sr$ (but few samples) |
| Kerguelen (KER) | 63, −49 | Yes | 0.1–114 Ma (Kerg) and 38–82 Ma (Ninety east-Broken Ridge) | Yes/1310 | It is a LIP | $^{87}Sr/^{86}Sr$ |
| Marion (MAR) | 37.75, −46.75 | Weak | <0.5 Ma (Marion) and 88 Ma (Madagascar) | Half-width 500 or Along-axis >1700 | Madagascar Plateau and Madagascar Island flood basalts | may be $^{87}Sr/^{86}Sr$ |
| Réunion (REU) | 55.5, −21 | Yes | 0–66 Ma | Yes/1380 | Mascarene, Chagos-Lacc. and Deccan 30–70 My | $^{3}He/^{4}He$ and $^{87}Sr/^{86}Sr$ |
| *Africa* | | | | | | |
| Afar (AF) | 42, 12 | No | — | — | — | — |
| East Africa/ Lake Victoria (EAF) | 34, 6 | No | — | — | — | — |
| Darfur (DAR) | 24, 13 | No | — | — | — | — |
| Ahaggar (AHA) | 6, 23 | No | — | — | — | — |
| Tibesti (TIB) | 17, 21 | No | — | — | — | — |
| *Eurasia* | | | | | | |
| Eifel (EIF) | 7, 50 | No | — | — | — | — |

Horizontal bars indicate that there are no data, where we did not find any data, or where available data are inconclusive.

This is significantly longer than, for example, the activity of Hawaiian volcanoes which have a main stage lasting ~1 My (e.g., Clague and Dalrymple, 1987; Ozawa *et al.*, 2005). The long life span of some of the Canary volcanoes has contributed to uncertainty in defining a geographic age progression, but is consistent with a slow propagation rate of age-progressive volcanism (**Figure 3**).

Iceland is often cited as a classic hot spot (**Figure 4**). The thickest magmatic crust occurs along the Greenland– and Faeroe–Iceland volcanic ridges extending NW and SE from Iceland. Anomalously thick oceanic crust immediately adjacent to these ridges shows datable magnetic lineations (Macnab *et al.*, 1995; White, 1997; Jones *et al.*, 2002b). Extrapolating the ages of these lineations onto the

Greenland– and Faeroe–Iceland Ridges reveals age-progressive volcanism with seafloor spreading, which is most easily explained by the Iceland hot spot causing excess magmatism very near to or at the Mid-Atlantic Ridge (MAR) since the time of continental breakup (Wilson, 1973; White, 1988, 1997). Earlier volcanism occurs as flood basalts along the continental margins of Greenland, the British Isles, and Norway ~56 Ma, and even further away from Iceland in Baffin Island, West and East Greenland, and the British Islands beginning ~62 Ma (e.g., Lawver and Mueller, 1994; Saunders *et al.*, 1997). The latter date provides a minimum estimate for the age of the Icelandic hot spot.

A few volcano chains fail to record volcanism for longer than a few tens of millions of years but the initiation of volcanism or the connection to older

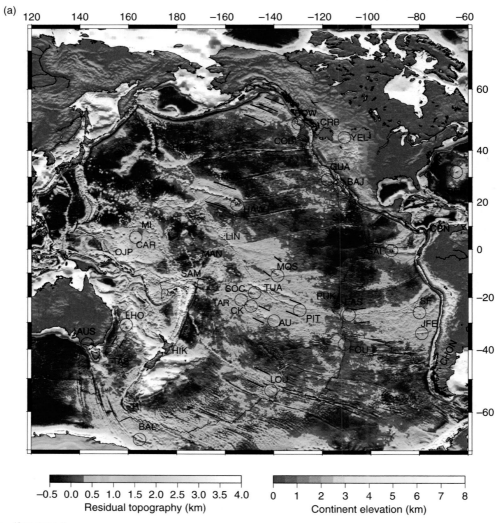

**Figure 1** *(Continued)*

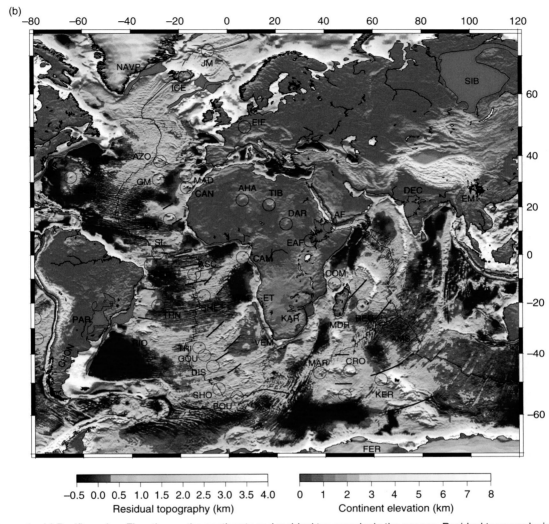

**Figure 1** (a) Pacific region. Elevation on the continents and residual topography in the oceans. Residual topography is the predicted bathymetry grid of Smith and Sandwell (1997) corrected for sediment loading and thicknesses (Laske and Masters, 1997), and for seafloor subsidence with age (Stein and Stein, 1992) using seafloor ages, updated from Müller et al. (1993a) (areas without ages are interpolated using cubic splines). Grid processing and display was done using GMT (Wessel and Smith, 1995). Color change from blue to turquoise is at 300 m and delineates the approximate boundaries of anomalously shallow seafloor. Circles mark estimated locations of most recent (hot spot) volcanism. Pairs of lines are used to measure the widths of some of the hot-spot swells (**Table 1**). Flood basalt provinces on the continents and continental margins are red; abbreviations are identified in Section 9.2.3.2. Axes are in degrees latitude and east longitude. (b). Atlantic and Indian oceans.

volcanic provinces is somewhat unclear. For example, the New England seamount chain records ~20 My of oceanic volcanism (Duncan, 1984), but an extrapolation to the volcanic provinces in New England could extend the duration another 20 My (see O'Neill et al. (2005)). The duration of activity at the Azores hot spot is not clear. Gente et al. (2003) hypothesize that the Azores hot spot formed the

Great Meteor and Corner seamounts as conjugate features ~85 Ma. Yet, age constraints of these edifices are poor and a geochemical association with the Azores group is yet to be tested. The most robust feature of this hot spot is its sudden influence on the MAR starting ~20 Ma as seen in geophysical surveys and dated using interpolations of seafloor isochrons (Cannat et al., 1999; Gente et al., 2003).

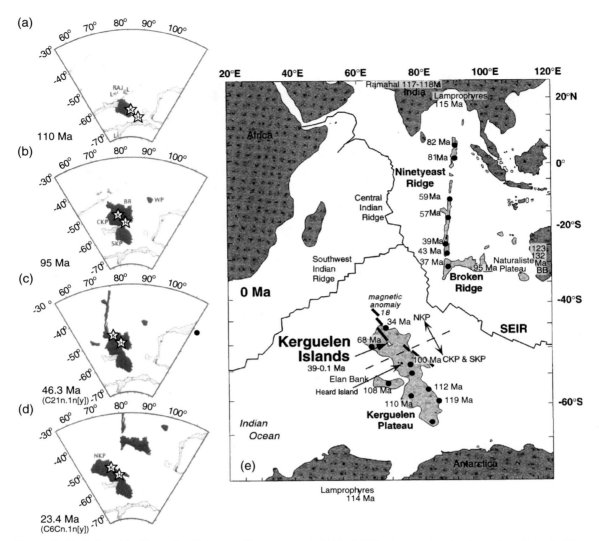

**Figure 2** Evolution of the Kerguelen Plateau and hot-spot track. (a) The initial pulse of volcanism formed the Rajmahal Traps (RAJ), Lamprophreys (L) and the Southern Kerguelen Plateau (SKP) from ~120 to ~110 Ma. Stars mark reconstructed positions of the hot spot (Müller et al., 1993b) assuming a location at Kerguelen Archiplago (K) and Heard Island (H). By 110 Ma, seafloor spreading between India, Antarctica, and Australia is well underway. Formation of the Rajmahal Traps by the Kerguelen hot spot requires the hot spot to have moved by ~10° relative to the Earth's spin-axis (Kent et al., 2002) as consistent with a model of a mantle plume rising through a convecting mantle (Steinberger and O'Connell, 1998) (see also Section 9.3.5). (b) By 95 Ma, Central Kerguelen Plateau (CKP) and Broken Ridge (BR) have formed. (c) At 46.3 Ma, northward migration of the Indian plate forms the Ninetyeast Ridge along a transform fault in the mid-ocean ridge system. (d) The Southeast Indian Ridge has propagated northwestward to separate CKP and BR ~42 Ma, which continue drifting apart at 23.4 Ma. (e) Current configuration as shown by Nicolaysen et al. (2000), but with ages from Coffin et al. (2002). Bold line along NE margin of the plateau marks magnetic anomaly 18 (41.3–42.7 Ma). Reproduced from Coffin MF, Pringle MS, Duncan RA, et al. (2002) Kerguelen hotspot magma output since 130 Ma. *Journal of Petrology* 43: 1121–1139, by permission of Oxford University Press.

On the Pacific Plate (e.g., see Clouard and Bonneville (2005) and references therein), the Cobb (1.5–29.2 Ma (Turner et al., 1980; Desonie and Duncan, 1990) and Bowie–Kodiak chains (0.2–23.8 Ma (Turner et al., 1980)) terminate at the subduction zone south of Alaska. The Easter chain on the Nazca Plate extends to 25.6 Ma (Clouard and Bonneville, 2005) but the record on the Pacific Plate may extend further into the past, perhaps to the Tuamotu Plateau (Ito et al., 1995; Clouard and Bonneville, 2001). This tentative connection certainly warrants more age dating.

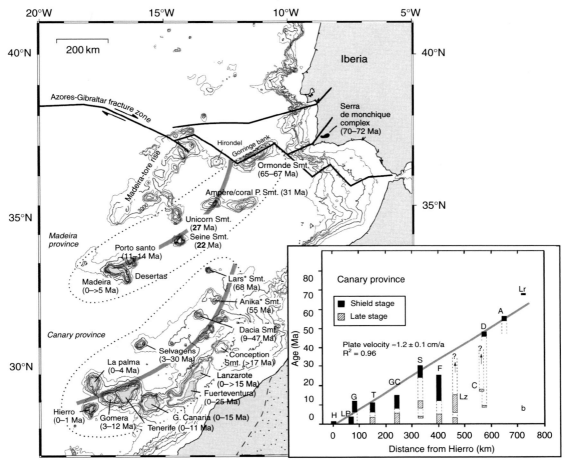

**Figure 3** Map showing bathymetry around the Madeira and Canary group (only contours above 3500 m are shown). Stippled areas separate the two provinces based on geochemical distinctions. Thick gray lines mark possible hot-spot tracks. Lower-right plot shows the age range of the labeled volcanoes vs distance from Hierro. The oldest dates follow a trend of increasing age with distance at an average rate of ~12 km My⁻¹. Reproduced from Geldmacher J, Hoernle K, Van der Bogaard P, Duggen S, and Werner R (2005) New Ar-40/Ar-39 age and geochemical data from seamounts in the Canary and Madeira volcanic provinces: Support for the mantle plume hypothesis. *Earth And Planetary Science Letters* 237: 85–101.

### 9.2.1.2  Short-lived age-progressive volcanism

At least eight volcanic provinces show age-progressive volcanism lasting <22 My (e.g., Clouard and Bonneville, 2005). The Society and Marquesas (Caroff *et al.*, 1999) islands represent voluminous volcanism but for geologically brief durations of 4.2 and 5.5 My, respectively. Clouard and Bonneville (2001) use geometrical considerations to argue that the Marquesas hot spot could have formed the Line Islands and Hess Rise. If this interpretation is correct, the large gap in volcanism between the three provinces suggests a strongly time-varying mechanism. Durations of 10–20 My occur along the Pitcairn (0–11 Ma), Caroline (1.4–13.9 Ma), Foundation

(2.1–21 Ma), Tarava (35.9–43.5 Ma), and Pukapuka (5.6–27.5 Ma) chains in the Pacific, as well as the Tasmandid chain (7–24.3 Ma) (McDougall and Duncan, 1988; Müller *et al.*, 1993b) near Australia.

An intriguing, yet enigmatic form of age-progressive volcanism is represented by the Pukapuka and Sojourn Ridges, which extend NW away from the East Pacific Rise. With respect to its geographic trend and duration, the Pukapuka Ridge resembles some of the other volcano chains in the region, such as the Foundation chain (**Figure 5**) (O'Connor *et al.*, 2002, 2004). The Pukapuka Ridge, however, stands out because of its smaller volcano volumes and the more rapid and variable rate of age progression (Janney *et al.*, 2000). It is thus unclear at this point whether these

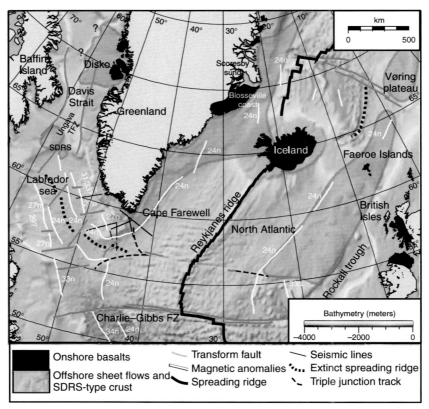

**Figure 4** Distribution of Paleogene flood basalts in the North Atlantic Volcanic Province. Selective seafloor magnetic lineations, major transform faults, active and extinct spreading ridges, and seismic lines are marked as labeled. From Nielsen TK, Larsen HC, and Hopper JR (2002) Contrasting rifted margin styles south of Greenland; implications for mantle plume dynamics. *Earth and Planetary Science Letters* 200: 271–286.

differences indicate deviant behaviors of a common mechanism or a distinct mechanism entirely. Batiza (1982) proposed a distinction between hot-spot volcanoes and smaller (and more numerous) 'non-hot-spot' volcanoes. The association of such seamounts with near-axis, mid-ocean ridge volcanism is good reason to consider a volcano group non-hot-spot, but such a characterization is less straightforward for Pukapuka. While it projects to the region of the young Rano–Rani seamounts near the East Pacific Rise, most of Pukapuka formed on older seafloor (**Figure 5**). The above ambiguities blur the distinction between hot-spot and non-hot-spot volcanism.

### 9.2.1.3 No age-progressive volcanism

An important form of oceanic volcanism does not involve simple geographic age progressions. Amsterdam–St. Paul and Cape Verde are two examples that represent opposite extremes in terms of size and duration. Amsterdam–St. Paul is on the Southeast Indian Ridge, just NE of Kerguelen. Geochemical

distinctions between lavas at Amsterdam–St. Paul and on Kerguelen suggest that they come from separate sources in the mantle (Doucet *et al.*, 2004; Graham *et al.*, 1999). With this interpretation, the Amsterdam–St. Paul hot spot represents a relatively small and short-lived ($\leq 5$ My) melting anomaly. Its small size and duration, as well as its location on a mid-ocean ridge likely contribute to the lack of an identifiable age progression. The Cape Verde volcanoes, on the other hand, are larger and likely to have existed since early Neogene (e.g., McNutt, 1988). The islands do not show a monotonic age progression, but this is reasonably well explained with a hot spot occurring very close to the Euler pole that describes the motion of the African Plate relative to a hot-spot reference frame (McNutt, 1988).

Other examples of volcanoes with complex age–space relations are not well explained within the hot-spot framework. These include older provinces in the Pacific such as the Geologist seamounts, the Japanese and Markus–Wake seamounts, the Marshal Islands,

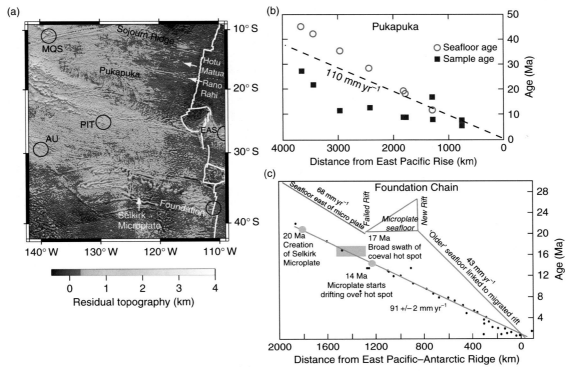

**Figure 5** (a) Residual topography just east of the South Pacific Superswell. (b) Ages of samples seamounts along the Pukapuka Ridge and adjacent seafloor. (c) Figure showing ages along the Foundation Chain between 110° and 131° W. (b) Reproduced from Janney PE, Macdougall JD, Natland JH, and Lynch Ma (2000) Geochemical evidence from the Pukapuka volcanic ridge system for a shallow enriched mantle domain beneath the South Pacific Superswell. *Earth and Planetary Science Letters* 181: 47–60. (c) From O'Connor JM, Stoffers P, and Wijbrans JR (2004) The Foundation Chain: inferring hot-spot-plate interaction from a weak seamount trail. In: Hekinian R, Stoffers P, and Cheminee J-L (eds.) *Oceanic Hotspots*. Berlin: Springer.

the Mid-Pacific Mountains, and the Magellan Rise (e.g., Clouard and Bonneville (2005) and references therein). The lack of modern dating methods applied to samples of many of these provinces leads to significant age uncertainties, and the large number of volcanoes scattered throughout the western Pacific makes it difficult to even define volcano groups. Three notable examples show nonprogressive volcanism with ages confirmed with modern dating methods. The Line Islands and Cook–Austral groups are oceanic chains that both involve volcanism over tens of millions of years with synchronous or near-synchronous events spanning distances >2000 km (**Figure 6**) (Schlanger *et al.*, 1984; McNutt *et al.*, 1997; Davis *et al.*, 2002; Devey and Haase, 2004; Bonneville *et al.*, 2006). The Cameroon line, which extends from Africa to the SW on to the Atlantic seafloor (**Figure 7**) (Marzoli *et al.*, 2000), may represent a continental analog to the former oceanic cases.

The remaining oceanic hot spots in **Table 1** do not have sufficient chronological data to test for time–space relations. This list includes cases that are geographically localized (e.g., Discovery, Ascension/Circe, Sierra Leone, Conrad, Bermuda), have spatial distributions due to interactions with spreading centers (e.g., Bouvet, Shona, Balleny), or have elongated geographic trends but have simply not been adequately dated (Socorro, Tuamotu, Great Meteor).

### 9.2.1.4 Continental hot spots

In addition to the intraplate oceanic volcanism there are a number of volcanic regions in the continents that are not directly associated with present-day subduction or continental rifting. In **Figure 1** we included a number of these areas such as the Yellowstone (YEL)–Snake River volcanic progression, the European Eifel hot spot (EIF), and African hot spots such as expressed by the volcanic

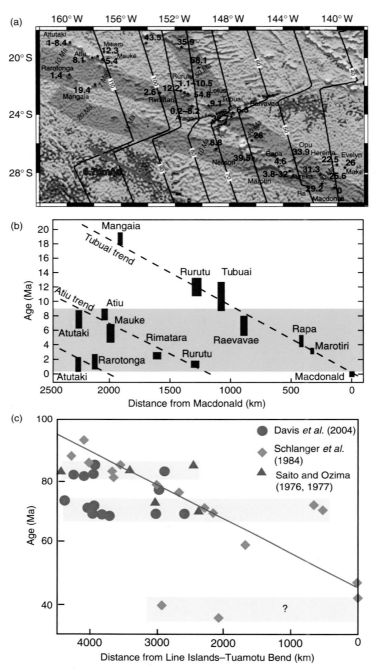

**Figure 6**    (a) Map of Cook–Austral group with age dates marked near sample locations. Shaded bands show possible hot-spot tracks using stage poles for absolute Pacific Plate motion of Wessel and Kroenke (1997). (b) A subset of the above dates vs distance from Macdonald seamount . This plot omits the ages >20 Ma along the northernmost hot-spot track shown in (a). Dashed lines have slopes of 110 km My$^{-1}$. (c) Age vs distance along the Line Islands. Diagonal red line represents a volcanic propagation rate of 96 km My$^{-1}$ as proposed by Schlanger *et al.* (1984). Yellow bands show at least two and possibly three episodes of nearly sychronous volcanism. (a) Reproduced from Clouard V and Bonneville A (2005) Ages of seamounts, islands, and plateaus on the Pacific Plate. In: Foulger G., Natland JH, Presnall DC, and Anderson DL (eds.) *Plumes, Plates, and Paradigms,* pp. 71–90. Boulder, CO: GSA, with permission from GSA. (b) From Devey CW and Haase KM (2004) The sources for hotspot volcanism in the South Pacific Ocean. In: Hekinian R, Stoffers, P and Cheminee J-L (eds.) *Oceanic Hotspots*, pp. 253–280. New York: Springer. (c) Reproduced from Davis AS, Gray LB, Clague DA, and Hein JR (2002) The Line Islands revisited: New $^{40}$Ar/$^{39}$Ar geochronologic evidence for episodes of volcanism due to lithospheric extension. *Geochemistry Geophysics Geosystems* 3(3), 10.1029/2001GC000190, with permission from AGU.

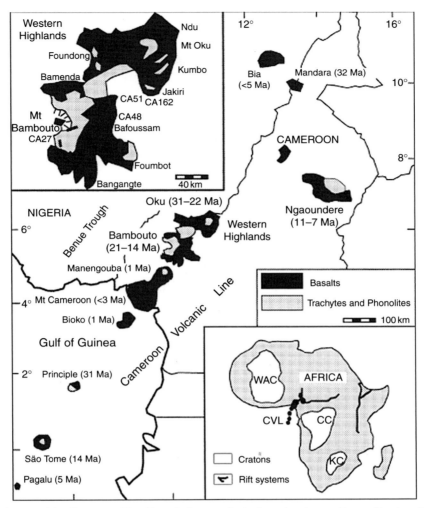

**Figure 7** Sketch map of the Cameroon Line. Reported ages refer to the volcanism on the continent and to the onset of basaltic volcanism on the ocean islands. Inset, top left: sketch map of Western Cameroon Highlands showing location of basaltic samples used for $^{40}Ar/^{39}Ar$ dating. Inset, bottom right: West African Craton (WAC), Congo Craton (CC) and Kalahari Craton (KC). Reproduced from Marzoli A, Piccirillo EM, Renne PR, Bellieni G, Iacumin M, Nyobe JB, and Tongwa AT (2000) The Cameroon volcanic line revisited; petrogenesis of continental basaltic magmas from lithospheric and asthenospheric mantle sources. *Journal of Petrology* 41: 87–109, by permission of Oxford University Press.

mountains of Jebel Mara (DAR), Tibesti (TIB), and Ahaggar (AHA) (Burke, 1996). The Yellowstone hot spot is the only one with a clear age progression. The lack of age progression of the other hot spots could indicate very slow motion of the African and European Plates.

The Yellowstone hot spot is centered on the caldera in Yellowstone National Park. Its signatures include a topographic bulge that is 600 m high and approximately 600 km wide, high heat flow, extensive hydrothermal activity, and a 10–12 m positive geoid anomaly (Smith and Braile, 1994). The trace of the Yellowstone hot spot is recorded by silicic-caldera-

forming events starting at 16–17 My at the Oregon–Nevada border, ~700 km WSW of the hot spot (**Figure 8**). The original rhyolitic volcanism is followed by long-lived basaltic volcanism that now forms the Snake River Plain. The effective speed of the hot-spot track is 4.5 cm yr$^{-1}$, which is interpreted to include a component of the present-day plate motion (2.5 cm yr$^{-1}$) and a component caused by the Basin and Range extension. The excess topography of the Snake River Plain decays systematically and is consistent to that of a cooling and thermally contracting lithosphere following the progression of the American Plate over a hot spot (Smith and Braile, 1994).

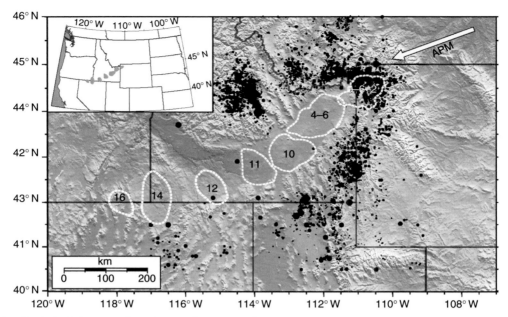

**Figure 8** Shaded relief topography, seismicity (black circles), and calderas (white dashed lines with age indication in Ma) of the Yellowstone–Snake River province. Reproduced from Waite GP, Smith RB, and Allen RM (2006) V–P and V–S structure of the Yellowstone hot spot from teleseismic tomography: Evidence for an upper mantle plume. *Journal of Geophysical Research* 111: B04303, with permission from AGU.

### 9.2.1.5 The hot-spot reference frame

The existence of long-lived volcano chains with clear age progression led Morgan (1971, 1972) to suggest that hot spots remain stationary relative to one another and therefore define a global kinematic reference frame separate from the plates (Morgan, 1983; Duncan and Clague, 1985). However, ongoing studies have established that hot spots do move relative to the Earth's spin-axis and that there is motion between the Pacific and Indo-Atlantic hot spots with speeds comparable to the average plate speed (Molnar and Stock, 1987; Acton and Gordon, 1994; Tarduno and Gee, 1995; DiVenere and Kent, 1999; Raymond et al., 2000; Torsvik et al., 2002). Paleomagnetic evidence (Tarduno et al., 2003; Pares and Moore, 2005) suggests rapid southward motion of the Kerguelen hot spot in the past 100 My and of the Hawaiian hot spot prior to the Hawaiian–Emperor Bend at 50 Ma (Sharpe and Clague, 2006). Southward motion of the Hawaiian hot spot appears to have slowed or nearly ceased since this time (Sager et al., 2005). Combined with a possible shift in plate motion (Norton, 2000; Sharpe and Clague, 2006) this change most likely contributes to the sharpness of the Hawaiian–Emperor Bend (Richards and Lithgow-Bertelloni, 1996). Near present-day (i.e., <4–7 Ma)

motion between hot spots globally is unresolvable or at least much slower than it has been in the geologic past (Wang and Wang, 2001; Gripp and Gordon, 2002).

While past rapid motion between groups of hot spots in different oceans is likely, rapid motion between hot spots on single plates is not. Geometric analyses of volcano locations suggest no relative motion, to within error, between some of the larger Pacific hot spots (Harada and Hamano, 2000; Wessel and Kroenke, 1997). Such geometric methods are independent of volcano ages and therefore have the advantage of using complete data sets of known volcano locations, unhindered by sparse age dating of variable quality. Koppers et al. (2001), however, argue that geometric analyses alone are incomplete and that when age constraints are considered as well, motion between the large Hawaiian and Louisville hot spots is required. But Wessel et al. (2006) show that the hot-spot track predicted by Koppers et al. (2001) misses large sections of both chains and present an improved plate-motion model derived independently of age dating. Thus, at this point, there appears to be little or no motion between the prominent Hawaiian and Louisville hot spots. Shorter chains in the Pacific do appear to move relative to larger chains (Koppers et al., 1998;

Koppers and Staudigel, 2005) but as noted above, the existence of age progression along many short-lived chains is questionable. One enigma, in particular, is the apparent motion required between Iceland and both the Pacific and Atlantic hot spots (Norton, 2000; Raymond *et al.*, 2000). For a more complete description of the methods and above issues regarding absolute plate motion.

### 9.2.2 Topographic Swells

**Figure 1** illustrates that most oceanic hot spots and melt anomalies show anomalously shallow topography extending several hundred kilometers beyond the area of excess volcanism. The prominence of hot-spot swells as established in the 1970s was initially attributed to heat anomalies in the mantle (Crough, 1978, 1983; Detrick and Crough, 1978). The evidence for their cause is somewhat ambiguous: for one, heat flow data fail to show evidence for heated or thinned lithosphere (Stein and Stein, 1993, 2003; DeLaughter *et al.*, 2005). However, it is likely that hydrothermal circulation associated with volcanic topography obscures the deep lithospheric signal (McNutt, 2002; Harris and McNutt, 2007). Seismic studies are limited but provide some clues. Rayleigh-wave dispersion shows evidence for a lithosphere of normal thickness beneath the Pitcairn hot spot (Yoshida and Suetsugu, 2004), whereas an S-wave receiver function study argues for substantial lithospheric thinning a few hundred kilometers up the Hawaiian chain from the hot spot (Li *et al.*, 2004).

Shallower-than-normal topography along the axes of hot-spot-influenced ridges indicates another form of hot-spot swell. Gravity and crustal seismic evidence indicate that topography along the ridges interacting with the Galápagos (Canales *et al.*, 2002) and Iceland (White *et al.*, 1995; Hooft *et al.*, 2006) hot spots are largely caused by thickened oceanic crust. In the case of the Galápagos spreading center, a resolvable contribution to topography likely comes from the mantle (Canales *et al.*, 2002), but for Iceland, it appears that crustal thickness alone can explain the observed topography (Hooft *et al.*, 2006).

An example of a much larger-scale swell is the South Pacific Superswell (McNutt and Fischer, 1987) which spans a geographic extent of ~3000 km and encompasses the hot spots in French Polynesia. Other swells of comparable size include the ancient Darwin Rise in the far northwestern Pacific (McNutt *et al.*, 1990) and the African Superswell (e.g., Nyblade and Robinson, 1994), which encompasses the

southern portion of Africa and the South Atlantic down to the Bouvet Triple Junction. Both the South Pacific and African Superswells involve clusters of individual hot spots. The hot spots in the South Pacific (French Polynesia) are mostly short lived, including some without simple age progressions (see Section 9.2.1) (McNutt, 1998; McNutt *et al.*, 1997). This region is also known to have anomalously low seismic wave speeds in the mantle (e.g., Hager and Clayton, 1989; Ritsema and Allen, 2003) and relatively high geoid (McNutt *et al.*, 1990) suggesting an origin involving anomalously low densities but without substantially thinned lithosphere (e.g., McNutt, 1998). The African Superswell, on the other hand, does not show a seismic anomaly in the upper mantle but rather a broad columnar zone of slow velocities in the lower mantle (e.g., Dziewonski and Woodhouse, 1987; Li and Romanowicz, 1996; Ritsema *et al.*, 1999).

For individual hot spots, we identify the presence of a swell if residual topography exceeds an arbitrarily chosen value of 300 m and extends appreciably (>~100 km) away from volcanic topography (**Figure 1**). We find that such hot-spot swells are very common (**Table 1**). Swells are even apparent on chains with very small volcanoes such as the Tasmandid tracks and the Pukapuka (**Figure 5**) and Sojourn Ridges (Harmon *et al.*, 2007).

One characteristic of hot-spot swells is that they are not present around extremely old volcanoes (**Table 1**). The Hawaiian Swell for example is prominent in the youngest part of the chain but then begins to fade near ~178° W, disappears near the Hawaiian–Emperor Bend (50 Ma), and is absent around the Emperor chain (**Figure 1**). Swells also appear to fade along the Louisville chain (near a volcano age of ~34 Ma), along the Tristan chain near Walvis Ridge (62–79 Ma), and possibly along the Kerguelen track as evident from the shallow seafloor that extends away from the Kerguelen Plateau on the Antarctic Plate but that is absent around the southernmost portion of Broken Ridge (~43 Ma) on the Indian Plate. A swell appears to be present over the whole length of the St Helena chain out to ~80 Ma, but it is not possible with the current data to separate the swell around the oldest portion of this chain with that around the Cameroon line. The SE portion of the Line Islands (ages ranging from 35 to 91 Ma) most likely has a swell but the NE portion (55–128 Ma) may not. Given the diversity of possible times of volcanism in the Line Islands, however, it is not clear which episode is associated with the current swell. Old chains that lack swells in our analysis include the Japanese–Wake seamounts (>70 Ma), the Magellan Seamounts (>70 Ma), the Mid-Pacs (>80 Ma), and the Musician Seamounts (>65.6 Ma). Overall, it thus appears that if a swell forms at a hot spot, it decreases in height with time until it can no longer be detected at an maximum age of 80 Ma, but often even after <50 Ma.

While swells are very prominent even around small or short-lived volcano chains, the Madeira and Canary hot spots are two cases that break the rule. The lack of obvious swells around these large and long-lived volcano chains is indeed very puzzling.

### 9.2.3  Flood Basalt Volcanism

LIPs provide further constraints on the nature of hot spots and mantle dynamics. In this chapter we use the term LIP to represent continental flood basalt provinces, oceanic plateaus, and volcanic passive margins, which are typified by massive outpourings and intrusions of basaltic lava, often occurring within a couple of million years. Reviews of the nature and possible origin of LIPs are provided by Richards *et al.* (1989); Coffin and Eldholm (1994); Mahoney and Coffin (1997); and Courtillot and Renne (2003). As characterized by Coffin and Eldholm (1994), continental LIPs, such as the Siberian Traps or the Columbia River Basalts, often form by fissure eruptions and horizontal flows of massive tholeiitic basalts. Volcanic passive margins, such as those of the North Atlantic Volcanic Province or Etendeka–Paraná, form generally just before continental rifting. The initial pulse is rapid and can be followed by a longer period of excess

oceanic crust production and long-term generation of a seamount chain. Oceanic LIPs form broad, flat-topped features of thickened oceanic crust with some eruptions being subaerial (Kerguelen oceanic plateau) and others seeming to be confined to below sea level (e.g., Ontong Java Plateau (OJP) and Shatsky Rise). This section reviews some basic geophysical and geological observations of LIPs that formed since the Permian. For information on older LIPs we refer the reader to other reviews (Ernst and Buchan, 2001, 2003). **Figure 1** shows locations (abbreviations defined in the following text), and **Figure 9** summarizes the areas and volumes of the provinces described as follows.

#### 9.2.3.1  Continental LIPs

*Columbia River Basalts (CRBs).* The CRBs erupted a volume of ~0.17 Mkm$^3$ between 16.6–15.3 Ma (Courtillot and Renne, 2003). Excellent exposures provide insights into flow structures and relationships to feeder dikes. Individual eruptions have volumes in excess of 2000 km$^3$ and flow over distances up to 600 km (Hooper, 1997). Lack of collapse structures suggests that large amounts of magma were rapidly derived from a deep source without being stored in the crust (Hooper, 1997).

*Emeishan (EM).* The EM province in west China is estimated to have spanned an area of at least 2 Mkm$^2$ and volume of 1 Mkm$^3$ when it first formed (Zhou *et al.*, 2002). Eruption ages are dated at 251–255 Ma and ~258 Ma (e.g., see Courtillot and Renne (2003), and Zhang *et al.* (2006) and references therein), with a date of ~259 Ma being confirmed by a Zircon U–Pb dating (Ali *et al.*, 2005). A rapid, kilometer scale uplift appears to have preceded the basalt eruption by

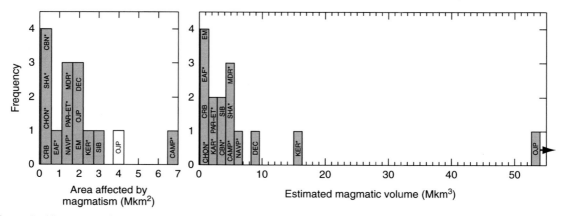

**Figure 9** Histograms of estimated areas affected by magmatism and magmatic volumes of the large igneous provinces discussed in the text. The OJP–Manihiki–Hikurangi LIP would have the area shown by white bar and an even larger volume than shown, as indicated by arrow. Asterisks indicate provinces with voluminous magmatism that could have endured for >3 My to a few tens of millions of years; no asterisks indicate cases that mostly likely erupted in <3 My.

~3 My (Xu *et al.*, 2004; He *et al.*, 2003). Basalts erupted rapidly and were accompanied by high MgO basalts (He *et al.*, 2003; Zhang *et al.*, 2006). Since, the Emeishan traps have been fragmented and eroded, they currently encompass an area of only ~0.3 Mkm$^2$ (Xu *et al.*, 2001).

*Siberia (SIB).* The Siberian Traps are presently exposed over an area of only 0.4 Mkm$^2$ and have an average thickness of 1 km (Sharma, 1997). There are strong indications that the volcanics extend below sedimentary cover and into the West Siberian Basin (Reichow *et al.*, 2005). Additional dikes and kimberlites suggest a maximum extent of 3–4 Mkm$^2$ with a possible extrusive volume of >3 Mkm$^3$. Lack of significant sedimentary rocks or paleosols between flows suggests rapid extrusion (Sharma, 1997). Most of the province probably erupted within ~1 My coinciding with the Permo-Triassic boundary at 250 Ma (Courtillot and Renne, 2003). Individual flows can be as thick as 150 m and can be traced over lengths of hundreds of kilometers. Use of industry seismic and borehole data in the West Siberian Basin indicates that the basin elevation remained high during rifting, suggesting dynamic mantle support (Saunders *et al.*, 2005).

*Yemen/Ethiopia/East Africa Rift System (EAF).* An early volcanic episode in southernmost Ethiopia starting ~45 Ma was followed by widespread flood basalt volcanism in Northwest Ethiopia ~30 Ma and in Yemen starting 31–29 Ma. The 30 Ma Ethiopia event consisted of tholeiites and ignimbrites (Pik *et al.*, 1998) that erupted within 1–2 My (Hofmann *et al.*, 1997; Ayalew *et al.*, 2002). In Yemen, 0.35–1.2 Mkm$^3$ of mafic magmas were produced, followed by less voluminous silicic volcanism starting ~29 Ma (Menzies *et al.*, 1997). Flood volcanism appears to have occurred several million years prior to the onset of extension along the EAR ~23 Ma (Morley *et al.*, 1992; Hendrie *et al.*, 1994) and in the Gulf of Aden ~26 Ma (Menzies *et al.*, 1997). The volcanism is bimodal with shield volcanoes forming on top of tholeiitic basalts (Courtillot and Renne, 2003; Kieffer *et al.*, 2004). Currently, the Ethiopian and Kenyan Rift Systems are on an area of elevated topography ~1000 km in diameter. A negative Bouguer gravity anomaly suggests this topography is dynamically supported in the mantle (Ebinger *et al.*, 1989).

*Older continental flood basalts.* These are often more difficult to detect in the geological record due to the effects of surface uplift and erosion. A general characteristic that is attributed to continental LIPs is

radiating dike swarms (Mege and Korme, 2004; Mayborn and Lesher, 2004). These dike swarms provide the main pathways for basaltic magmas vertically from the mantle, as well as laterally over distances up to 2500 km, as suggested for the 1270-My-old Mackenzie dike swarm in N. America (Lecheminant and Heaman, 1989; Ernst and Baragar, 1992). The longest dikes usually extend well beyond the original boundaries of the main lava field.

### 9.2.3.2 LIPs near or on continental margins

*Central Atlantic Magmatic Province (CAMP).* The CAMP is primarily delineated by giant dike swarms and is associated with the early breakup of Gondwana between North Africa, North America, and Central South America. Widely separated eruptions and dike swarms are present over an area of ~7 Mkm$^2$ with an estimated magmatic volume of ~2 Mkm$^3$ (Marzoli *et al.*, 1999), while seismic and magnetic studies on the eastern margin of North America suggest that this offshore portion of the CAMP could have a volume as large as 3 Mkm$^3$ (Holbrook and Kelemen, 1993). These estimates bring the total volume to near 5 Mkm$^3$. $^{40}$Ar/$^{39}$Ar dates spanning 197–202 Ma suggests an emplacement episode lasting ~5 My (Hames *et al.*, 2000; Marzoli *et al.*, 1999; Courtillot and Renne, 2003). Only the offshore portion is mapped in **Figure 1**.

*Chon Aike (CHON).* In contrast to the other large igneous provinces discussed in the chapter, the Chon Aike Province in Patagonia is primarily silicic with rhyolites dominating over minor mafic and intermediate lavas. The rhyolites may have formed due to intrusion of basalts into crust that was susceptible to melting. The province is relatively small with an area of 0.1 Mkm$^2$ and total volume of 0.235 Mkm$^3$ (Pankhurst *et al.*, 1998). Chon Aike had an extended and punctuated eruptive history from Early Jurassic through Early Cretaceous (184–140 Ma). Pankhurst *et al.* (2000) recognize episodic eruptions with the first coinciding with the Karoo and Ferrar LIPs. The province potentially extends into present-day West Antarctica.

*Deccan (DEC).* The Deccan Traps provide one of the most impressive examples of continental flood basalts. It formed by primarily tholeiitic magmatism over Archean crust interspersed over an area of ~1.5 Mkm$^2$, with an estimated volume of 8.2 Mkm$^3$ (Coffin and Eldholm, 1993). Eruptions straddle the magnetic chrons C30n, C29r, C29n within 1 My around the K–T boundary, as confirmed by $^{40}Ar$–$^{39}Ar$ and Re–Os dating (Allégre *et al.*, 1999; Courtillot *et al.*, 2000; Hofmann *et al.*, 2000). An iridium anomaly embedded between flows suggests that the Chicxulub impact happened while the Deccan Traps were active (Courtillot and Renne, 2003). Seafloor spreading between India and the Seychelles started a few million years after the major Deccan event, ~63 Ma (Vandamme *et al.*, 1991; Dyment, 1998). Unlike older continental flood basalts associated with the breakup of Gondwana, the Deccan basalts that are least contaminated by continental lithosphere closely resemble hot-spot basalts in oceanic areas and the major element contents agree with predictions for high-temperature melting (Hawkesworth *et al.*, 1999).

*Karoo–Ferrar* (*KAR–FER*). The Karoo province in Africa and Ferrar basalts in Antarctica record a volume of 2.5 Mkm$^3$ which erupted at ~184 Ma (Encarnacion *et al.*, 1996; Minor and Mukasa, 1997), possibly followed by a minor event at 180 Ma (Courtillot and Renne, 2003). The short (<1 My) duration is questioned by Jourdan *et al.* (2004, 2005) who obtained ages of ~179 Ma for the northern Okavango dike swarm in Botswana and consequently prefer a longer-lived initial activity that propagated from the south to the north. In Africa, tholeiitic basalts dominate but some picrites and some rhyolites occur (Cox, 1988). The triple-junction pattern of the radiating dike swarm that supplied the Karoo basalt was likely controlled by pre-existing lithospheric discontinuities that include the Kaapvaal and Zimbabwe Craton boundaries and the Limpopo mobile belt (Jourdan *et al.*, 2006). The Ferrar Province spans an area of ~0.35 Mkm$^2$ (Elliot and Fleming, 2004) in a linear belt along the Transantarctic Mountains. The two provinces were split by continental rifting and then seafloor spreading ~156 Ma.

*Madagascar* (*MDR*). Wide-spread voluminous basaltic flows and dikes occurred near the northwestern and southeastern coasts of Madagascar during its rifting from India around 88 Ma (Storey *et al.*, 1997). The oldest seafloor magnetic anomaly to form is chron 34 (84 Ma). Flood volcanism was probably prolonged as it continued to form the Madagascar Plateau to the south, perhaps 10–20 My later as inferred from the reconstructed positions of Marion hot spot.

*North Atlantic Volcanic Province* (*NAVP*). The NAVP covers ~1.3 Mkm$^2$ (Saunders *et al.*, 1997) with an estimated volume of 6.6 Mkm$^3$ (Coffin and Eldholm, 1993) and is closely linked to continental rifting and oceanic spreading (e.g., Nielsen *et al.*, 2002) (**Figure 2**). Prior to the main pulse of flood volcanism, seafloor spreading was active south of the Charlie–Gibbs fracture zone at 94 Ma and propagated northward into the Rockall Trough, which stopped in the late Cretaceous near or prior to the earliest eruptions of the NAVP. The early NAVP eruptions occurred as large picritic lavas in West Greenland and Baffin Island (Gill *et al.*, 1992, 1995; Holm *et al.*, 1993; Kent *et al.*, 2004) soon followed by massive tholeiitic eruptions in West and Southeast Greenland, British Isles, and Baffin Island at 61 Ma (2 Mkm$^3$) and in East Greenland and the Faeroes at 56 Ma (>2 Mkm$^3$) (Courtillot and Renne, 2003). The initial episodes were followed by rifting between Greenland and Europe recorded by Chron 24, 56–52 Ma, continental margin volcanisms, and ocean crust formation, which included the formation of thick seaward-dipping seismic reflection sequences. Spreading slowed in Labrador Sea ~50 Ma, stopped altogether at 36 Ma, but continued further to the west on the Aegir Ridge and eventually along the Kolbeinsey Ridge at ~25 Ma, where it has persisted since. This provides an intriguing suggestion that the presence of hot spots can guide the location of seafloor spreading following continental breakup.

Many of the volcanic margin sequences erupted subaerially or at shallow depths, suggesting widespread regional uplift during emplacement (Clift and Turner, 1995; Hopper *et al.*, 2003). Uplift in the Early Tertiary is documented by extensive erosion and changes in the depositional environments as far as the North Sea Basin (e.g., see Nadin *et al.* (1997) and Mackay *et al.* (2005) and references therein). Reconstructions from drill cores show that uplift was rapid and synchronous and preceded the earliest volcanism by >1 My (Clift *et al.*, 1998).

*Paraná–Etendeka* (*PAR–ET*). Paraná and Etendeka are conjugate volcanic fields split by the breakup of South America and Africa. The Paraná field in South America covers 1.2 Mkm$^2$ with estimated average thickness of 0.7 km (Peate, 1997). Extensive dike swarms surrounding the provinces suggest the original extent could have been even larger (Trumbull *et al.*, 2004). Volcanism is bimodal with dominating tholeiitic lavas and rhyolites. $^{40}Ar$–$^{39}Ar$ dates suggest a peak of eruption ~133–130 Ma (Turner and al., 1994; Renne *et al.*, 1996; Courtillot and Renne, 2003), preceded by minor eruptions in the northwest of the Paraná basin at 135–138 Ma (Stewart *et al.*, 1996). Younger magmatism persisted along the coast

(128–120 Ma) and into the Atlantic Ocean, subsequently forming the Rio Grande (RIO) and Walvis (WAL) oceanic plateaus.

The Etendeka province covers 0.08 Mkm$^2$ and is very similar to the Paraná flood basalts in terms of eruptive history, petrology, and geochemistry (Renne *et al.*, 1996; Ewart *et al.*, 2004). Seafloor spreading in the South Atlantic progressed northward, with the oldest magnetic anomalies near Cape Town (137 or 130 Ma). Earliest magnetic anomaly near Paraná is ~127 Ma. The formation of onshore and offshore basins suggests a protracted period of rifting well in advance of the formation of oceanic crust and the emplacement of the Paraná basalts (Chang *et al.*, 1992).

### 9.2.3.3   Oceanic LIPs

*Caribbean* (*CBN*). The Caribbean LIP is a Late Cretaceous plateau, which is now partly accreted in Colombia and Equador. Its present area is 0.6 Mkm$^2$ with thickness of oceanic crust ranging from 8–20 km. A volume of 4 Mkm$^3$ of extrusives erupted in discrete events from 91–88 Ma (Courtillot and Renne, 2003). The full range of $^{40}$Ar/$^{39}$Ar dates of 69–139 Ma (e.g., Sinton *et al.*, 1997; Hoernle *et al.*, 2004), however, suggests a protracted volcanic history that is poorly understood. The cause volcanism has been attributed to the Galápagos hot spot, which is currently in the Eastern Pacific.

*Kerguelen* (*KER*). The Kerguelen hot spot has a fascinating history of continental and oceanic flood eruptions and rifting, as well as prolonged volcanism (**Figure 2**). The breakup of India, Australia, and Antarctica coincided closely in time with the eruption of the Bunbury basalts in southwest Australia, dated at 123 and 132 Ma (Coffin *et al.*, 2002). The first massive volcanic episode formed the Southern Kerguelen Plateau at 119 Ma, the Rajmahal Traps in India at 117–118 Ma followed by lamprophyres in India and Antarctica at 114–115 Ma (Coffin *et al.*, 2002; Kent *et al.*, 2002). The Central Kerguelen Plateau formed by ~110 Ma and Broken Ridge formed by 95 Ma (Coffin *et al.*, 2002; Frey *et al.*, 2000; Duncan, 2002). The above edifices represent the most active period of volcanism with a volcanic area and volume of 2.3 Mkm$^2$ and 15–24 Mkm$^3$, respectively (Coffin and Eldholm, 1993), but unlike many other flood basalt provinces, it spanned tens of millions of years. Another unusual aspect is the presence of continental blocks as suggested from wide-angle seismics (Operto and Charvis, 1996) and more directly from trace-element and isotopic data of

xenoliths and basalts from Southern Kerguelen Plateau and Broken Ridge (Mahoney *et al.*, 1995; Neal *et al.*, 2002; Frey *et al.*, 2002). During 82 to 43 Ma northward motion of the Indian Plate formed the Ninetyeast Ridge on young oceanic lithosphere, suggesting the Kerguelen hot spot stayed close to the Indian–Antarctic spreading center (Kent *et al.*, 1997). At ~40 Ma, the Southeast Indian Ridge formed and separated Kerguelen and Broken Ridge. Volcanism has since persisted on the Northern Kerguelen Plateau until 0.1 Ma (Nicolaysen *et al.*, 2000). Dynamic uplift of the plateau in the Cretaceous is indicated by evidence for subaerial environment, but subsidence since then is not much different than normal oceanic subsidence (Coffin, 1992).

*Ontong Java Plateau* (*OJP*). A recent set of overview papers on the origin and evolution of the OJP is provided by Fitton *et al.* (2004) and Neal *et al.* (1997). This plateau extends across ~2 Mkm$^2$, has crust as thick as 36 km, and has volumes estimated at 44 Mkm$^3$ and 57 Mkm$^3$ for accretion off, and on a mid-ocean ridge, respectively (Gladczenko *et al.*, 1997). The majority of the basalts were erupted in a relatively short time (~1–2 My) near 122 Ma as revealed by $^{40}$Ar–$^{39}$Ar (Tejada *et al.*, 1996), Re–Os isotope (Parkinson *et al.*, 2002), and paleomagnetic (Tarduno *et al.*, 1991) studies. Basalts recovered from Site 803, and the Santa Isabel and Ramos Islands indicate a smaller episode of volcanism at 90 Ma (Neal *et al.*, 1997). Age contrasts between the surrounding seafloor and OJP suggest it erupted near a mid-ocean ridge. Furthermore, rifting is evident on some of its boundaries and Taylor (2006) suggests that the OJP is only a fragment of what was originally an even larger edifice that included Manihiki (MAN) and Hikurangi (HIK) Plateaus. The inferred original size of the edifice makes it, by far, the largest of any flood basalt province in the geologic record, approaching an order of magnitude more voluminous than any continental flood basalt.

The plateau has been sampled on the tectonically uplifted portions in the Solomon Islands (e.g., Tejada *et al.*, 1996, 2002; Petterson, 2004) and with eight ocean drill holes (most recently during Ocean Drilling Program Leg 192; Mahoney *et al.* (2001)). The volcanics are dominated by massive flows of low-K tholeiitic basalts. Petrological modeling is consistent with the primary magmas formed by 30% melting of a peridotitic source (Fitton and Godard, 2004; Herzberg, 2004b; Chazey III and Neal, 2004), which would require a hot (1500 °C) mantle under thin lithosphere. The low volatile content of volcanic glasses (Roberge *et al.*, 2004, 2005), as well as the range and limited variability of Pb, Sr, Hf, and Nd isotope compositions (Tejada *et al.*, 2004) resemble characteristics of many MORB.

Besides its gigantic volume, the other aspect that makes OJP extremely enigmatic is that it appears to have erupted below sea level with little evidence for hot-spot-like uplift (Roberge *et al.*, 2005; Korenaga, 2005b; Ingle and Coffin, 2004; Coffin, 1992; Ito and Clift, 1998). These aspects must be explained by any successful model for the origin of this flood basalt province.

*Shatsky–Hess Rise (SHA–HES).* Shatsky Rise is one of the large Pacific oceanic plateaus with an area of 0.48 Mkm$^2$ and volume of 4.3 Mkm$^3$. The initial eruption is associated with the jump of a triple junction between the Pacific, Izanagi, and Farallon Plates toward the plateau (Nakanishi *et al.*, 1999; Sager, 2005) near 145 Ma (Mahoney *et al.*, 2005). Subsequently, volcanism progressed northeast together with the triple junction (which migrated with repeated jumps as indicated by seafloor magnetic lineation) until ~128 Ma. The most voluminous portion of the plateau is the central-southwest portion. Magmatism appears to have diminished toward the northeast (**Figure 1**). Thus Shatsky appears to show a short-lived age-progressive volcanism on timescales much like many smaller volcano groups (e.g., Cook–Austral and Pukapuka). Volcanism, however, may have continued with a renewed pulse starting some 10–20 My later with the formation of the Hess Rise, which is comparable in area to Shatsky. Age constraints on Hess Rise are poor because of the lack of sampling and its location on Cretaceous Quite Zone seafloor. The possible coincidence of both plateaus at a mid-ocean ridge suggests a dynamic linkage between their formation and seafloor spreading. Another notable aspect is that Pb and Nd isotope compositions for Shatsky Rise are indistinguishable from those of the present-day East Pacific Rise (Mahoney *et al.*, 2005).

### 9.2.3.4  Connections to hot spots

The possible links between hot spots and LIPs are important for testing the origin of both phenomena, with particular regard to the concept of a starting mantle plume head and trailing, narrower plume stem. While linkages are clear for some cases, a number of proposed connections are obscured by ridge migrations or breakup of the original LIP. Below we list the connections of hot spots to LIPs, in approximate order of decreasing reliability.

At least six examples have strong geographical, geochronological, and geochemical connections between hot-spot volcanism and flood basalt provinces. These are: (1) Iceland and the North Atlantic Volcanic Province, including Greenland, Baffin Island, Great Britain volcanics, Greenland–UK (Faeroe) Ridge (Saunders *et al.*, 1997; Smallwood and White, 2002); (2) Kerguelen, and Bunbury, Naturaliste, Rajmahal (E. India), Broken Ridge, and Ninetyeast Ridge (Kent *et al.*, 1997); (3) Réunion and Deccan (Roy, 2003), W. Indian, Chagos–Laccadive, Mascarenas, Mauritius; (4) Marion and Madagascar (Storey *et al.*, 1997); (5) Tristan da Cunha and Paraná, Etendeka, Rio Grande, Walvis Ridge (Peate, 1997); (6) Galapagos and Caribbean (Feigenson *et al.*, 2004; Hoernle *et al.*, 2004).

In addition to these six examples, a tentative link exists between Yellowstone and the early eruptive sequence of the CRBs based on geochemistry (Dodson *et al.*, 1997), but the paleogeographical connection is somewhat indirect since the main CRB eruption occurred up to 500 km north of the hot-spot track. This may suggest a mechanism for the formation of the flood basalts that is independent of the Yellowstone hot spot (Hales *et al.*, 2005). Yet, the earliest manifestation of the CRBs is possibly the Steens Mountain basalt (Oregon) which is located well to the south of the main eruptive sequences (Hooper, 1997; Hooper *et al.*, 2002) and closer to the proposed location of the Yellowstone hot spot. This suggests a south-to-northward propagation of the basalts and supports the suggestion that flood basalts may have been forced sideways from their mantle source by more competent continental lithosphere toward a weaker 'thin spot' in the lithosphere (Thompson and Gibson, 1991).

The links are less clear – in part due to lack of data – for other flood basalt provinces. The Bouvet hot spot has been linked to the Karoo–Ferrar; the Balleny hot spot may be connected to the Tasmanian Province or to Lord Howe Rise (Lanyon *et al.*, 1993); and the Fernando hot spot has been linked to the Central Atlantic Magmatic Province. Only three Pacific hot spots possibly link back to LIPs: Louisville–OJP, Easter–Mid-Pac, and Marquesas–Hess-Shatsky (Clouard and Bonneville, 2001). The Louisville–OJP connection is doubtful at best: kinematic arguments against such a link have been made (Antretter *et al.*, 2004) and geochemical distinctions between the oldest Louisville seamounts and OJP would require distinct geochemistry between plume head and tail or a difference in melting conditions (Mahoney *et al.*,

1993; Neal *et al.*, 1997). Finally, some hot spots such as Hawaii, Bowie–Kodiak, and Cobb terminate at subduction zones, so any record of a possible LIP has been destroyed.

It is interesting to note that the strongest connections between hot-spot chains and LIPs involve flood basalt provinces near continental margins. The possible exception is Kerguelen, for which most of the magmatism occurred away from any continental margins. The presence of continents probably plays a key role in the origin of many flood basalt provinces (Anderson, 1994b).

## 9.2.4 Geochemical Heterogeneity and Distinctions from MORB

The geochemistry of MORBs and basalts from hot spots and melting anomalies has played a key role in our understanding of mantle dynamics. Isotope ratios of key trace elements reflect long-term ($10^2$–$10^3$ My) concentration ratios between parent and daughter elements and therefore have been used to fingerprint the mantle from which different lavas arise. We will focus on three commonly used ratios: $^{87}Sr/^{86}Sr$, $^{206}Pb/^{204}Pb$, and $^{3}He/^{4}He$. High (or low) $^{87}Sr/^{86}Sr$ is associated with mantle material that is enriched (or depleted) in highly incompatible elements (i.e., those that partition mostly into melt when in equilibrium with the solid during partial melting) relative to moderately incompatible elements (e.g., Rb/Sr). High (or low) $^{206}Pb/^{204}Pb$ ratios are associated with mantle with a long-term high (or low) U/Pb. The $^{3}He/^{4}He$ ratio is a measure of the amount of primordial helium ($^{3}He$, which was present in the presolar nebula and is only lost from the Earth to space via degassing) relative to $^{4}He$, which is primarily generated by radioactive decay of U and Th.

MORB is well characterized by low and small variability in the above ratios. **Figure 10** shows frequency distributions of basalts from the three major oceans; the data are from http://www.petdb.org/ and absolutely no data (e.g., Iceland) are excluded. One standard deviation about the median for all three datasets combined defines the MORB range for $^{87}Sr/^{86}Sr$ of 0.7025–0.7033, for $^{206}Pb/^{204}Pb$ of 17.98–18.89, and for $^{3}He/^{4}He$ of 7.08–10.21 (where $^{3}He/^{4}He$ is given in multiples of the atmosphere ratio, $R_a$). Significantly, while the median values from the three major spreading systems vary individually, each fall within the above ranges for all three isotope ratios. In contrast, ocean-island basalts (OIBs) have much larger variability,

extending from MORB values to much higher values (**Figure 10**). By comparing the median hot-spot compositions (solid bar) with the MORB ranges, it becomes clear that, with few exceptions, the hot spots and melting anomalies have compositions that are distinguishable from MORB by at least one of the three isotope ratios (see also **Table 1**). This result appears to be independent of the duration of age-progression (e.g., Tristan vs. Marquesas), the presence or absence of a swell (Hawaii vs. Canaries), or even volcano size (Kerguelen vs. Pukapuka (Janney *et al.* (2000)) and Foundation (Maia *et al.* (2000)).

There are at least four possible exceptions: Cobb, Bowie–Kodiak, the Caroline seamounts, and the Shatsky Rise. Helium isotopes are not yet available at these locations but $^{87}Sr/^{86}Sr$ and $^{206}Pb/^{204}Pb$ compositions for each case fall within or very near to the MORB range. These examples span a wide range of forms, from a small, short-lived seamount chain (Caroline), to longer-lived, age-progressive volcanism (Cobb, Bowie–Kodiak), to an oceanic LIP (Shatsky). The possibility that these cases are geochemically indistinguishable from MORB has far-reaching implications about mantle processes and chemical structure; it clearly needs testing with further sampling.

## 9.2.5 Mantle Seismic Anomalies

### 9.2.5.1 Global seismic studies

Seismic wave propagation is generally slowed by elevated temperature, volatile content, the presence of melt, and mafic (primarily garnet) content of the mantle (e.g., Anderson, 1989). Seismology is therefore the primary geophysical tool for probing the mantle signature of hot spots and melting anomalies. Seismic tomography has become a popular method of 'imaging' the mantle. While it has provided important insights into the deep transport of subducting slabs (e.g., Grand *et al.*, 1997; Bijwaard *et al.*, 1998), seismic tomography has yet to produce robust images with sufficient spatial resolution in the deep mantle beneath hot spots (Nataf, 2000; Ritsema and Allen, 2003). The lack of methods that can probe the lower mantle with sufficient resolution makes it particularly difficult to address questions regarding the putative lower-mantle source for hot spots.

Global seismic models provide a first attempt to trace the surface expressions of hot spots to seismic anomalies into the mantle (Niu *et al.*, 2002; Zhao, 2001; DePaolo and Manga, 2003; Montelli *et al.*, 2004). The most robust features are the voluminous low-velocity

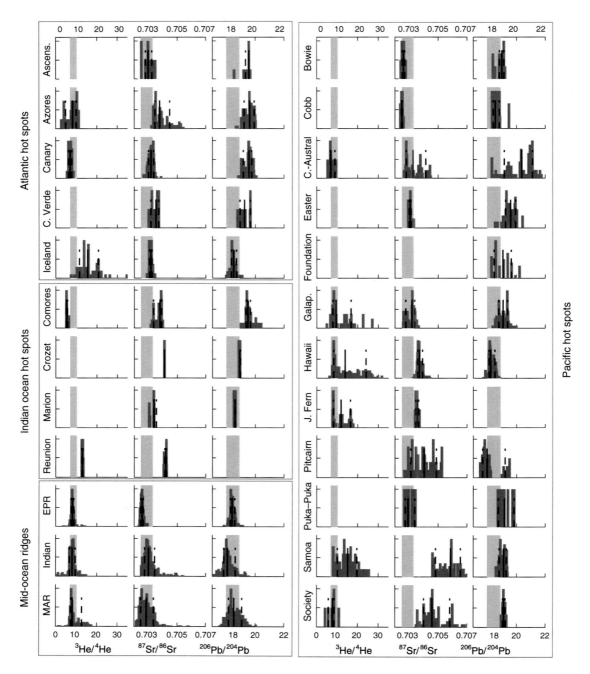

**Figure 10** Frequency distributions (dark gray, normalized by maximum frequency for each case, so the peaks are at 1.0) of isotope measurements taken from the shown oceanic hot spots and mid-ocean ridges (lower left). Solid lines mark median values and dashed lines encompass 68% (i.e., one standard deviation of a normal distribution) of the samples. Light gray bars denote the range of values encompassing 68% of all of the MORB measurements (sum of the three ridges shown). Most of these data are from the GEOROC database with key references for $^3$He/$^4$He data given in Ito and Mahoney (2006). Data for the Puka–Puka are from Janney et al. (2000) and for the Foundation chain from Maia et al. (2000).

anomalies in the lowermost mantle below the South Pacific and Africa Superswell regions (Breger et al., 2001; Ni et al., 2005; van der Hilst and Karason, 1999; Trampert et al., 2004). The sharp edges of these anomalies (Ni et al., 2002; To et al., 2005) and their reproduction in dynamical models (Tan and Gurnis, 2005; McNamara and Zhong, 2005) suggest both a compositional and thermal origin of these anomalies. It is more difficult to identify low-velocity anomalies at smaller spatial scales and shallower depths. For example, Ritsema and Allen (2003) investigated the correlation between seismic low-velocity regions in the global S-wave model S20RTS (Ritsema et al., 1999) and hot spots from a comprehensive list (Sleep, 1990). They confidently detected anomalously low seismic wave speeds in the upper mantle for only a small numbers of hot spots. Stronger correlations between a number of hot spots and deep-mantle anomalies were found using finite-frequency P- and S-wave tomography (Montelli et al., 2004), but the ability of this method to actually improve the resolution with the available data is debated (de Hoop and van der Hilst, 2005; van der Hilst and de Hoop, 2005; Boschi et al., 2006).

The topography of seismic discontinuities in the transition zone may also reflect heterogeneity. A hotter anomaly causes the exothermic phase change at 410 km to occur deeper and the endothermic phase change at 660 km to occur shallower, assuming the olivine system dominates the phase changes (Ito and Takahashi, 1989; Helffrich, 2000). Global observations of the transition zone thickness provide some support for hotter-than-normal mantle below some hot spots (e.g., Helffrich, 2002; Li et al., 2003a, 2003b), although other observations suggest that global seismic data can resolve strong correlations between topography of the 410 km discontinuity and seismic velocities only at wavelengths larger than those of individual hot spots (Chambers et al., 2005).

### 9.2.5.2 Local seismic studies of major hot spots

Improved insights into the upper mantle beneath hot spots can be obtained using regional or array studies, including the use of surface waves (e.g., Pilidou et al., 2005).

*Iceland.* Regional seismic studies have confidently imaged a body of anomalously slow seismic wave speeds in the upper mantle beneath Iceland. Conventional ray theory was used to first image the anomaly (Allen et al., 1999a, 2002; Foulger et al., 2001; Wolfe et al., 1997) but improved finite-frequency techniques (Allen and Tromp, 2005) resolve the feature to be roughly columnar with lateral dimension of 250–300 km and peak P- and S-wave anomalies of −2.1% and −4.2%, respectively (Hung et al., 2004) (**Figure 11**). Recent studies using Rayleigh waves and local earthquakes confirm these high amplitudes, which most likely require a combination of excess temperature and melt (Yang and Shen, 2005; Li and Detrick, 2006).

In addition, studies of surface waves and shear-wave splitting reveal significant seismic anisotropy in the Icelandic upper mantle (Li and Detrick, 2003; Bjarnason et al., 2002; Xue and Allen, 2005). Overall, they find that the fast S-wave propagation directions are mostly NNE–SSW in central Iceland with stronger E–W components to the west and east. The anisotropy in west and eastern Iceland deviates significantly from the directions of motion of the two plates and thus could indicate a large-scale mantle flow in the region (Bjarnason et al., 2002). However, in central Iceland, the strong rift-parallel anisotropy near the active rift zones is interpreted to indicate ridge-parallel flow associated with a mantle plume (Li and Detrick, 2003; Xue and Allen, 2005).

The anomalous seismic structure extends well below 410 km as evidenced by the thinning of the transition zone beneath Iceland (Shen et al., 1998) (**Figure 12**). While Shen et al. (1998) show evidence for both a deepening 410- and shoaling 660-discontinuity, Du et al. (2006) argue that the discontinuity at 660 km is instead flat. The precise nature of both discontinuities is important in determining whether the Iceland anomaly initiates in the upper mantle or is present below 660 km (Shen et al. 1998). One global tomography model suggests that the anomaly extends into the lower mantle (Bijwaard and Spakman, 1999) and another study identifies an ultralow-velocity zone near the core–mantle boundary (CMB) below Iceland (Helmberger et al., 1998). The available seismic data, however, leave a lower mantle origin open to debate (e.g., Foulger et al., 2001) and a robust test awaits improved regional seismic experiments.

*Other Atlantic hotspots.* Slow surface-wave speed anomalies extend to 200 km below the Azores hot spot as part of an along-strike perturbation of the velocity structure beneath the MAR (Pilidou et al., 2004). Receiver function analysis at Cape Verde (Lodge and Helffrich, 2006) indicates a thickened crust (∼15 km) and a high-velocity, low-density zone to a depth of ∼90 km. The oldest volcanoes sit on top of the thickest parts of the crust and the high-velocity layer. Such high velocities in the shallow mantle beneath active hot spots are unusual. They

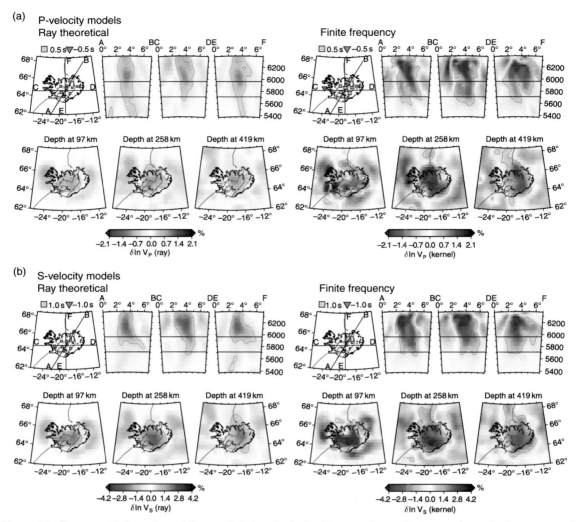

**Figure 11** Tomographic inversions of the mantle below the Iceland hot spot imaged by suggesting major improvement in the resolving power using a finite frequency approach (right) compared to traditional ray based methods (left) in both P- (top) and S- (bottom) wave models. Reproduced from Hung SH, Shen Y, and Chiao LY (2004) Imaging seismic velocity structure beneath the Iceland hot spot: A finite frequency approach. *Journal of Geophysical Research* 109: B08305 (doi:10.1029/2003JB002889), with permission from AGU.

suggest major-element heterogeneity (e.g., due to melt depletion) dominate over other effects such as the presence of volatiles, melt, or excess temperature.

*Hawaii.* Anomalously low seismic wave speeds beneath the Hawaiian hot spot have been found from preliminary surface wave (Laske *et al.*, 1999) and tomographic (Wolfe *et al.*, 2002) studies. The anomaly imaged by tomography appears broader and higher in amplitude for S-waves (200 km diameter and up to −1.8%) than it does for P-waves (100 km diameter and −0.7%). A significant low S-wave speed (<4 km s⁻¹) is observed below 130 km suggesting the presence of partial melt below this

depth (Li *et al.*, 2000). Additional evidence for an upper-mantle melt anomaly is provided by a seafloor magnetotelluric study (Constable and Heinson, 2004) which suggested a columnar zone of 5–10% partial melting with a radius <100 km and a depth extent of 150 km. Insights into the deeper structure using seismic tomography are currently hindered by poor coverage of stations and earthquake sources, but resolution tests indicate that the anomaly is unlikely to be restricted to the lithosphere (Wolfe *et al.*, 2002). A deep origin is suggested by evidence for a thinning of the transition zone by 40–50 km (Li *et al.*, 2000, 2004; Collins *et al.*, 2002). These studies, as well as a

P660s–P410s (s;IASP91-observed)

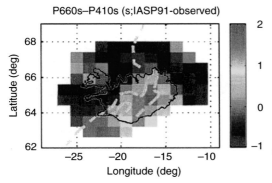

**Figure 12** Difference in travel time between PS conversions at depths of 660 and 410 km with respect to the IASPEI91 (Kennett and Engdahl, 1991) reference model in seconds. The sharp decrease in velocity in a focused zone below Iceland suggests a strong thinning of the transition zone which is best explained by a significant increase in local mantle temperature. Reprinted by permission from Macmillan Publishers Ltd: (Nature) Shen Y, Solomon SC, Bjarnason IT, and Wolfe C 1998) Seismic evidence for a lower mantle origin of the Iceland plume. *Nature* 395: 62–65, copyright (1998).

combined seismic and electromagnetic inversion show that the transition-zone anomaly is consistent with an excess temperature of 200–300 K (Fukao *et al.*, 2004). A still deeper origin, possibly to the CMB, is suggested by a recent tomographic study that incorporates core phases (Lei and Zhao, 2006).

*Galápagos.* The Galápagos hot spot is part of a broad region in the Nazca and Cocos Basins with significantly reduced long-period Love and Rayleigh wave speeds (Vdovin *et al.*, 1999; Heintz *et al.*, 2005). The transition zone surrounding the Galápagos has similar thickness to that of the Pacific Basin except for a narrow region of ∼100 km in radius slightly to the west of the Galápagos archipelago where it is thinned by ∼18 km. This amount of thinning suggests an excess temperature of 130 K (Hooft *et al.*, 2003). Preliminary results of a regional tomography study indicate a low-velocity feature of comparable dimension, extending above the transition zone into the shallow upper mantle (Toomey *et al.*, 2001). The western edge of the archipelago shows shear-wave splitting of up to 1s with a direction consistent with E–W plate direction. The anisotropy disappears beneath the archipelago where the upper-mantle wave speeds are anomalously low, suggesting that melt or complex flow beneath the hot spot destroy the plate-motion derived anisotropy (Fontaine *et al.*, 2005).

*Yellowstone.* Early tomographic studies revealed a complex velocity structure in the upper mantle beneath the Snake River Plain, southwest of the

Yellowstone hot spot. This structure was interpreted to represent compositional variability restricted to the upper mantle associated with melting (Saltzer and Humphreys, 1997). More recent work suggests that a narrow, low-velocity feature extends from the upper mantle into the top of the transition zone (Waite *et al.*, 2006; Yuan and Dueker, 2005). The shallow upper-mantle anomaly is present over a distance of more than 400 km, spanning from the northeastern extent of the Snake River Plain to Yellowstone National Park, including a short segment to the northeast of the Yellowstone caldera. The anomaly is strongest at depths 50–200 km with peak anomalies of −2.3% for Vp and −5.5% for Vs (Waite *et al.*, 2006). The velocity reductions are interpreted to represent 1% partial melt at a temperature of 200 K above normal (Schutt and Humphreys, 2004). Initial transition-zone studies showed significant topography of the 410 discontinuity throughout the region (Dueker and Sheehan, 1997). More recent studies show that the 410 discontinuity deepens by 12 km near the intersection of the low-velocity anomaly identified by Waite *et al.* (2006) and Yuan and Dueker (2005), but interestingly, the 660 km discontinuity appears flat in this area (Fee and Dueker, 2004). Shear-wave splitting measurements around the Yellowstone–Snake River Plain show fast S-wave speeds primarily aligning with apparent plate motion, except for two stations in the Yellowstone caldera, perhaps due to local melt effects (Waite *et al.*, 2005).

*Eifel.* The Eifel region in Western Germany is characterized by numerous but small volcanic eruptions with contemporaneous uplift by 250 m in the last 1 My. Tomographic imaging indicates a mantle low-velocity anomaly extending to depths of at least 200 km (Passier and Snieder, 1996; Pilidou *et al.*, 2005). Inversions using a high-resolution local array study indicate a fairly narrow (100 km) P-wave anomaly of −2% that possibly extends to a depth of 400 km (Ritter *et al.*, 2001; Keyser *et al.*, 2002). The connection with the deeper mantle is unclear but has been suggested to include the low-velocity structure in the lower mantle below central Europe (Goes *et al.*, 1999) (**Figure 13**). Shear-wave splitting measurements show the largest split times for S-waves polarized in the direction of absolute plate motion, but the pattern is overprinted by complex orientations, suggestive of parabolic mantle flow around the hot spot (Walker *et al.*, 2005). A comparison of the seismic anomaly structure below the Eifel, Iceland, and Yellowstone is provided in **Figure 14** (from Waite *et al.*, 2006).

*East Africa.* Body and surface-wave studies indicate a strong regional low-velocity anomaly in the mantle

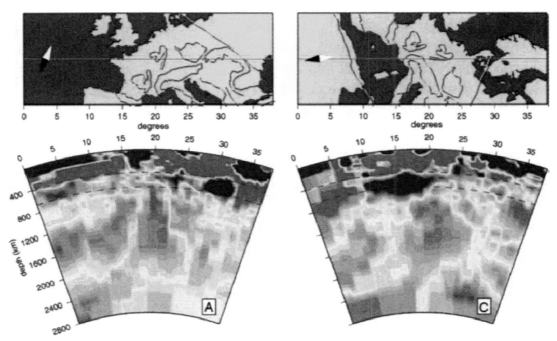

**Figure 13** Whole-mantle P tomography below Europe showing possible connections of mid-European volcanism to deep seated low-velocity anomalies in the lower mantle. Reprinted with permission from Goes S, Spakman W, and Bijwaard H (1999) A lower mantle source for central European volcanism. *Science* 286: 1928–1931. Copyright (1999) AAAS.the AAAS.

below the East Africa Rift System. Below Afar, surface-wave studies image anomalous velocities ($\leq-6\%$ in vertically polarized S-waves) extending to a depth of $\sim$200 km (Sebai *et al.*, 2006). In the North Ethiopian Rift, a narrow (75–100 km) tabular feature extends to depths >300 km and is slightly broader in the northern portion which trends toward Afar. Between the flanks the rift zones, P-wave speeds are reduced by 2.5% and S-wave speeds by 5.5% (Bastow *et al.*, 2005). Independent P-wave tomography confirms the general trend and amplitude of the anomaly with a tentative suggestion that it trends to the south and west (Benoit *et al.*, 2006) toward the broad lower-mantle seismic anomaly below Africa (Ritsema *et al.*, 1999; Grand, 2002). The Tanzania Craton, to the northwest, is imaged as having high P- and S-wave speeds to depths of at least 200 km (Ritsema *et al.*, 1998), is thinner than other African cratons, and is surrounded by the slow seismic-wave speeds associated with the East African Rift System (Sebai *et al.*, 2006). The transition zone shows complicated variations but is generally thinner below the Eastern Rift by 30–40 km compared to the more normal thickness under areas of the Tanzania Craton (Owens *et al.*, 2000; Nyblade *et al.*, 2000). Combined, these results suggest that the mantle below the rifts is

hotter than normal mantle by 200–300 K with partial melt in the shallow upper mantle.

Seismic anisotropy is dominantly parallel to the main Ethiopian Rift, in an area that extends to nearly 500 km away from the ridge axis, which likely rules out simple extension-driven asthenospheric flow (Gashawbeza *et al.*, 2004). The regional anisotropy is most likely caused by pre-existing features in the late Proterozoic Mozambique Belt but may be locally enhanced by aligned melt in the Ethiopian and Kenyan Rifts (Gashawbeza *et al.*, 2004; Walker *et al.*, 2004; Kendall *et al.*, 2005).

*South Pacific Superswell.* Recordings of French nuclear explosions in French Polynesia provide evidence for seismically fast velocities in the shallow mantle which suggest compositional heterogeneity without evidence for excess temperature (Rost and Williams, 2003). Rayleigh-wave dispersion measurements across the Pitcairn hot-spot trail suggest an absence of lithospheric thinning (Yoshida and Suetsugu, 2004). Underside reflections of S-waves at the 410- and 660-discontinuities show normal thickness of the mantle transition zone, except in a 500-km-wide area beneath the Society hot spot (Niu *et al.*, 2002). Seismic anisotropy in French Polynesia generally aligns with apparent plate motion, although

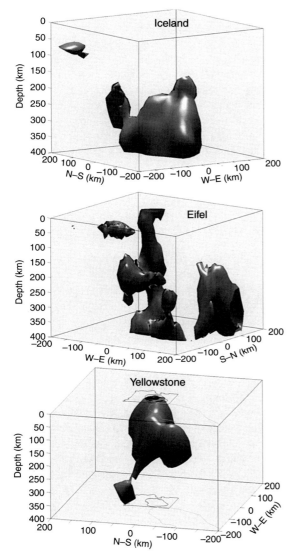

**Figure 14** Contours of −1% P-wave velocity anomaly below the Eifel, Iceland, and Yellowstone hot spots. Reproduced from Waite GP, Smith RB, and Allen RM (2006) V–P and V–S structure of the Yellowstone hot spot from teleseismic tomography: Evidence for an upper mantle plume. *Journal of Geophysical Research* 111: B04303, with permission from AGU.

the absence of anisotropy beneath Tahiti and minor deviations beneath other islands could indicate the presence of local flow or magma (Russo and Okal, 1998).

*Bowie.* The presence of a narrow, low-velocity zone near the base of the transition zone below the Bowie hot spot is based on delays in seismic records of Alaskan earthquakes measured in the NW United States (Nataf and VanDecar, 1993). The delays are

consistent with a zone of 150 km diameter with an excess temperature of 200 K.

*LIPs.* Beneath the Deccan Traps, seismic speeds in the shallow upper-mantle are anomalously low to a depth of at least 200 km (Kennett and Widiyantoro, 1999). The anomaly appears to be absent at depths near the transition zone (Kumar and Mohan, 2005). The Paraná Province is underlain by a distinct region of low seismic-wave speeds in the upper mantle (VanDecar *et al.*, 1995; Schimmel *et al.*, 2003). The OJP has an upper-mantle velocity anomaly of −5% with respect to Preliminary Reference Earth Model (PREM) (Dziewonski and Anderson, 1981) with a maximum depth extent of 300 km (Richardson *et al.*, 2000), whereas ScS reverberations show that this region is less attenuating than 'normal' Pacific asthenosphere (Gomer and Okal, 2003). Seismic anisotropy beneath the OJP is weak, which is interpreted to indicate that the residual mantle root has remained largely undeformed since it formed ~120 Ma (Klosko *et al.*, 2001). The presence of anomalously slow mantle beneath such old flood basalts is enigmatic in that any thermal anomaly is expected to have diffused away.

### 9.2.6 Summary of Observations

Long-lived (>50 My) age-progressive volcanism occurs in 13 hot spots. At present day, these hot spots define a kinematic reference frame that is deforming at rates lower than average plate velocities. Over geologic time, however, there has been significant motion between the Indo-Atlantic hot spots, the Pacific hot spots, and Iceland. Short-lived (≤22 My) age progressions occur in at least eight volcano chains. The directions and rates of age progression in the short-lived chains suggest relative motion between these hot spots, even on the same plate. Finally, a number of volcano groups, which sometimes align in chains (e.g., Cook–Austral, Cameroon), fail to show simple age–distance relations but instead show episodic volcanism over tens of millions of years.

Anomalously shallow topographic swells are very common among hot spots and melting anomalies. These swells are centered by the volcanoes and span geographic widths of hundreds of kilometers to >1000 km. Swells appear to diminish with time; they are usually present around volcanoes with ages <50 Ma and are typically absent around volcanoes older than ~70 My. Conspicuously, the Madeira and Canary hot spots are two active volcano chains without substantial swells.

Large igneous provinces represent the largest out-pourings of magma but also show a huge range in magmatic volumes and durations. They can be as large as 50 Mkm$^3$ (OJP) to <2 Mkm$^3$ (CRB) (see **Figure 9**). Voluminous magmatism can occur in dramatic short bursts, lasting 1–2 My (CAMP, CRB, EM, OJP, SIB) or can be prolonged over tens of millions of years (e.g., CHON, CBN, KER, NAVP, SHA). Main eruptive products are tholeiitic basalts which, on the continents, typically are transported through radiating dike swarms. High-MgO basalts or picrites are found in a number of provinces (NAVP, OJP, CAR) which indicate high degrees of melting. Smaller rhyolitic eruptions in continental flood basalts (KAR, PAR, CHON, EARS) indicate melting of continental crust. Dynamic topographic uplift is evident around the main eruptive stages of some LIPs (EM, NAVP, SIB, KER) but may not have occurred at some oceanic plateaus (OJP, SHA). While an appreciable amount of the geologic record is lost to subduction, a half-dozen recently active hot-spot chains are confidently linked to LIPs with the remainder having tenuous or non-existent links. Those volcano chains that clearly backtrack to LIPs involve those at the centers or margins of continents; Kerguelen is one of the possible exception. Also, most of the LIPs we have examined are associated with rifting, either between continents (PAR, KAR, CAMP, NAVP, DEC, MDR) or at mid-ocean ridges (OJP–MAN–HIK, KER, SHA–HES). The above characteristics compel substantial revisions and/or alternatives to the hypothesis of an isolated head of a starting mantle plume as the only origin of LIPs.

Basalts from hot spots and other melting anomalies, for the most part, are more heavily influenced by mantle materials that are distinct in terms of $^{87}$Sr/$^{86}$Sr, $^{206}$Pb/$^{204}$Pb, and/or $^3$He/$^4$He ratios from the MORB source. Four possible exceptions, which show MORB-like $^{87}$Sr/$^{86}$Sr and $^{206}$Pb/$^{204}$Pb compositions (but lack constraints from $^3$He/$^4$He) are the Shatsky Rise and the Bowie–Kodiak, Cobb, and Caroline chains.

Most hot spots are associated with anomalously low seismic-wave speeds below the lithosphere and in the upper mantle. The transition zone below hot spots is often thinned by tens of kilometers. The above findings are consistent with elevated mantle temperature by 150–200 K and with excess partial melt in the shallow upper mantle. Improved understanding of mineral physics at appropriately high pressure and temperature are needed to better constrain the magnitude of the possible temperature anomalies and to quantify the potential contribution

of compositional heterogeneity. Finally, while there are hints of seismic anomalies extending into the lower mantle and even to the CMB, definitive tests of a deep origin for some melting anomalies require more extensive regional seismic experiments and modern methods of interpretation.

The key characteristics described above provide information needed to test various proposed dynamical mechanisms for the formation of hot spots and melting anomalies. Some trends and generalities are apparent but substantial deviations likely reflect a range of interacting processes. In other words, it is very unlikely that a single overarching mechanism applies to all cases.

## 9.3  Dynamical Mechanisms

This section reviews the mechanisms proposed to generate hot spots and melting anomalies. We begin with a summary of methods used to quantitatively explore the mechanisms (Section 9.3.1). We then discuss the shallower processes of melting (Section 9.3.2) and swell formation (Section 9.3.3) before addressing the possible links to the deeper mantle, with specific focus on the extensive literature on mantle plumes (Section 9.3.4). In the context of whole mantle convection, we discuss possible causes of volcano age progressions and the inferred approximately coherent motion among hot spots on the same (Indo-Atlantic and the Pacific) plates (Section 9.3.5). Proposed mechanisms for generating LIPs and their possible connection to hot spots are explored in Section 9.3.6. The diversity of observations of hot spots and LIPs requires important modifications to the thermal plume hypothesis, as well as alternative possibilities as presented in Section 9.3.7. In light of these possibilities we discuss possible causes for the differences in geochemistry between hot-spot basalts and MORB in Section 9.3.8.

### 9.3.1  Methods

The origin and evolution of hot spots and melting anomalies can be constrained by studying the transport of energy, mass, and momentum in the solid and partially molten mantle (see Chapter 2). Key parts of the above processes can be described mathematically by the governing equations and solved with analytical or numerical approaches, or can be studied by simulation in laboratory experiments using analog materials.

Analytical approaches provide approximate solutions and scaling laws that reveal the main relationships between phenomena and key para-meters (*see* Chapter 4). Relevant examples of applications include boundary layer analysis, which is important to quantifying the time and length scales of convection, and lubrication theory, which has been instrumental in understanding hot-spot swell forma-tion. A separate approach uses experimental methods (*see* Chapter 3), involving analog materials such as corn syrup and viscous oils to simulate mantle dynamics at laboratory timescales. Analog methods were instrumental in inspiring the now-classic images of plumes, with voluminous spherical heads followed by narrow columnar tails. Numerical tech-niques are necessary to solve the coupling between, and non-linearities within the equations such as those caused by strongly varying (e.g., temperature, pressure, and composition-dependent) rheology (*see* Chapter 5), phase transitions, and chemical reactions. Numerical models are therefore well suited for simulating more-or-less realistic conditions and for allowing quantita-tive comparisons between predictions and observations. In this way, numerical techniques pro-vide a means to directly test conceptual ideas against the basic laws of physics and to delineate the condi-tions under which a proposed mechanism is likely to work or has to be rejected based on observations.

Unfortunately, fully consistent modeling is diffi-cult to achieve and is hampered by multiple factors. First, the constitution of the Earth's mantle can only be approximated. Information about material properties such as density, rheology, and thermal conductivity, are essential for quantitative modeling but is not precisely known and becomes increasingly imprecise with depth. Second, the transition from the brittle lithosphere to the viscous asthenosphere involves a rapid temperature increase; the common assumption that the deformation of the Earth's mantle can be approximated as creeping viscous flow is only correct at high temperatures and the details of the lithosphere–asthenosphere interaction depend on poorly known processes that are difficult to model self-consistently. Finally, the problem is multiscale, involving processes occurring from scales as small as individual grains (e.g., fluid–solid interac-tion, chemical transport), to as large as the whole mantle (*see* Chapter 2). Addressing these challenges will require careful comparisons between the differ-ent techniques, adjustments according to improved insights from experimental and observational work, and smart use of increasing computing technology.

### 9.3.2 Generating the Melt

Understanding of the causes for excess melt genera-tion is essential for our understanding of the dynamics of hot spots and melting anomalies. The rate that an infinitesimal bulk quantity of mantle melts to a fraction $F$ can be described by

$$\begin{aligned} DF/Dt = {} & (\partial F/\partial T)_{P,C}\,(DT/Dt) \\ & + (\partial F/\partial C)_{P,T}\,(DC/Dt) \\ & + (-\partial F/\partial P)_S\,(-DP/Dt) \end{aligned} \qquad [1]$$

The first two terms on the right-hand side describe nonisentropic processes. The first term describes the melt produced by heating and is pro-portional to the change in $F$ with temperature $T$ at constant pressure $P$ (i.e., isobaric productivity); melt-ing by this mechanism may occur in a variety of settings but is likely to be comparatively small and thus has not been a focus of study. The second term describes melt generated by the open-system change in composition. This term may be important behind subduction zones where the addition of fluids into the mantle wedge causes 'flux' melting and the for-mation of arc and back-arc volcanism. Finally, the third term describes isentropic decompression melt-ing, which is perhaps the most dominant process of melt generation at mid-ocean ridges, hot spots, and other melting anomalies.

For decompression melting, the rate of melt gen-eration is controlled by the melt productivity $-\partial F/\partial P$, which is positive if temperature exceeds the solidus. Both the solidus and value of $\partial F/\partial P$ (e.g., McKenzie, 1984; Hirschmann *et al.*, 1999; Phipps Morgan, 2001) depend on the equilibrium composition of the solid and liquid at a given pressure. The rate of decom-pression $-DP/Dt$ is controlled by mantle dynamics and is primarily proportional to the rate of mantle upwelling. The total volumetric rate of melt genera-tion is approximately proportional to $DF/Dt$ integrated over the volume $V$ of the melting zone,

$$\dot{M} = \frac{\rho_m}{\rho_c} \int_V (\partial F/\partial P)_S (DP/Dt)\, dV \qquad [2]$$

where $\rho_m$ is mantle density and $\rho_c$ is igneous crustal density. Melting anomalies thus require one or more of the following conditions: excess temperature, pre-sence of more fusible or fertile material, and mantle upwelling. Higher temperatures increase $V$ by increasing the pressure at which the solidus is inter-sected, more fusible mantle can change both $(-\partial F/\partial P)_S$ and increase $V$, and both factors may influence $DP/Dt$ through their effects on mantle buoyancy.

#### 9.3.2.1 Temperature

Melting caused by an increase in temperature has been a major focus of previous studies. One way to estimate mantle temperature is based on comparisons between predicted and observed melt-production rates at different settings. Another way involves using compositions of primitive lavas. Given an assumed starting source composition and a model for melt–solid interaction as the melt migrates to the surface (e.g., 'batch', 'fractional', 'continuous' melting), the liquid concentrations of key oxides (e.g., $Na_2O$, $CaO$, $Al_2O_3$, $SiO_2$, $FeO$, $MgO$) as well as incompatible trace elements can be predicted based on their dependence on $P$, $T$, and $F$, which are related through solid mineralogy and liquid composition. The above relationships can be established by thermodynamic theory and constrained with laboratory experiments.

Estimates of temperature variations at mid-ocean ridges are aided by the relative simplicity of lava compositions, relatively straightforward measurements of magma production rate (e.g., crustal thickness times spreading rate), and the ability to correlate variations in the above observational parameters in space (i.e., along mid-ocean ridges). In addition, spreading rate provides a constraint on the rate of mantle upwelling, for example, if one assumes, to first order, that mantle upwelling is a passive (kinematic) response to plate spreading. Based on the conditions needed for a lherzolitic mantle to yield observed crustal thicknesses and major element variations near Iceland, excess mantle temperatures beneath Iceland relative to normal mid-ocean ridges are estimated to range from about 100°C to >250°C (e.g., Klein and Langmuir, 1987; McKenzie and Bickle, 1988; Langmuir *et al.*, 1992; Shen and Forsyth, 1995; Presnall *et al.*, 2002; Herzberg and O'Hara, 2002). Excess temperature estimates based on inversions of crustal thickness and incompatible trace-element compositions also fall within the above range for Iceland (White *et al.*, 1995; Maclennan *et al.*, 2001) and other hot spots (White *et al.*, 1992).

Beneath Hawaii, the maximum mantle potential temperatures (i.e., temperature at zero pressure after removing the effects of adiabatic decompression) is estimated at 1500–1600°C based on predicted melt production rates from numerical models of mantle upwelling, driven by thermal buoyancy (Ribe and Christensen, 1999; Watson and McKenzie, 1991). This temperature range is 200–300°C higher than the estimated potential temperature of 1280°C

beneath normal mid-ocean ridges using the same melting model (McKenzie and Bickle, 1988).

Another method of estimating mantle temperatures is based on Fe–Mg content of primary melts and the olivine phenocrysts with which they equilibrate. This method depends on experimentally constrained partitioning of Fe and Mg between liquid and olivine, measured forsterite content of olivine crystals, and estimated Fe and Mg content of primary magmas (i.e., magmas that segregated from the mantle melting zone and have not been further modified by shallow processes such as crystal fractionation or accumulation). One group suggests that the mantle is no hotter beneath Hawaii than beneath many mid-ocean ridges (Green *et al.*, 2001). Other groups, however, suggest elevated temperatures of 50–100°C (Herzberg and O'Hara, 2002; Herzberg, 2004a) and 100–200°C (Putirka, 2005) beneath Hawaii and Iceland. As essentially all sampled lavas have evolved to varying degrees after they left the mantle source, an important uncertainty is the MgO content of the primary liquids. Putirka (2005) argues, for example, that the lower MgO contents derived by Green *et al.* (2001) for Hawaii could lead to an underestimate of temperature. A recent critical evaluation of the criteria for determining mantle potential temperatures below ridges and hot spots is provided by Herzberg *et al.* (in press).

#### 9.3.2.2 Composition

A major source of uncertainty for all of the above temperature estimates is the composition of the mantle source. Water and $CO_2$, for example, can dramatically reduce melting temperatures even in the small proportions (i.e., well below saturation) likely to be present in the MORB source (Asimow and Langmuir, 2003; Dasgupta and Hirschmann, 2006). While such small concentrations of volatiles are not likely to increase the total extent of melting significantly, they can enhance the amount of melt produced for a given temperature by appreciably expanding the volume of the melting zone. As the mantle beneath hot spots is likely to be more volatile rich, temperature estimates based on dry peridotite may be too high. For example, excess temperatures beneath the hot-spot-influenced Galápagos spreading center may be reduced from ~50°C for anhydrous melting models to <40°C when water is considered (Cushman *et al.*, 2004; Asimow and Langmuir, 2003). Similarly, estimates for the mantle excess temperature beneath Azores have been revised from ~75°C to ~55°C (Asimow and Langmuir, 2003). Hydrous melting models have yet to be explored in detail for the larger Iceland and Hawaii hot spots.

The mantle beneath hot spots may also contain more fusible, mafic lithologies, such as those generated by the recycling of subducted oceanic crust. The presence of such 'fertile' mantle has been suggested for hot spots such as Hawaii (Hauri, 1996; Takahashi, 2002), Iceland (Korenaga and Kelemen, 2000), the CRBs (Takahashi *et al.*, 1998), Galápagos (Sallares *et al.*, 2005), and others (Hofmann, 1997). Pyroxenite lithologies have both a lower solidus and greater productivity ($\partial F/\partial P$) than peridotite (Pertermann and Hirschmann, 2003) and therefore require much lower temperatures to produce the same volume of melt as peridotite.

Some have argued that fertile mantle melting could generate many melting anomalies with very small or zero excess temperatures (Korenaga, 2005b). An important difficulty with this hypothesis is that mafic materials will tend to form eclogite, which is significantly denser than lherzolite in the upper mantle (Irifune *et al.*, 1986). This material must therefore

produce substantial melt at depths where upwelling (i.e., $-DP/Dt$, eqn [2]) is not appreciably impeded by negative buoyancy. It has been suggested that eclogite becomes neutrally buoyant near the base of the upper mantle (~660 km) (Hirose *et al.*, 1999; Ringwood and Irifune, 1988). To reach the solidus and initiate melting at this depth most likely requires temperatures >300 K higher than normal (e.g., Hirose and Fei, 2002). Alternatively, it has been suggested that rapid upwelling driven by shallow thermal convection (Korenaga, 2004) (**Figure 15**) or fast seafloor spreading (Korenaga, 2005b) could entrain eclogite upward from a neutrally buoyant layer at 660 km and cause melting in the upper mantle without usually hot mantle. More recent experiments, however, suggest that subducted basalt is actually denser than peridotite throughout the upper mantle (Aoki and Takahashi, 2004) and thus would unlikely accumulate near 660 km. It is clear that a more complete understanding of the properties and phase relations of different lithologies at a range of mantle pressures and temperatures is needed to test the importance of fertile mantle melting.

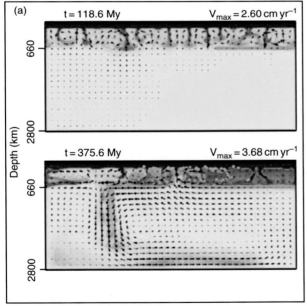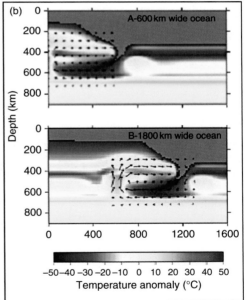

**Figure 15** Two forms of small-scale convection in the upper mantle. (a) Solutions of 2-D numerical models in which sublithospheric thermal instabilities drive convection (arrows show mantle flow) in the upper mantle at two time steps as labeled. Colors show temperature with blue being cold (lithosphere) and light yellow being hottest. Green tracers of subducted crustal fragments first form a layer at 660 km; some eventually sink into the lower mantle and some are drawn upward where they could melt. (b) Predictions of 2-D numerical models in which small-scale convection is driven by the edge of thick lithosphere (gray). Colors show temperature contrast from the mantle adiabat. (a) From Korenaga J (2004) Mantle mixing and continental breakup magmatism. *Earth and Planetary Science Letters* 218: 463–473. (b) From King S and Anderson DL (1998) Edge-driven convection. *Earth and Planetary Science Letters* 160: 289–296.

### 9.3.2.3 Mantle flow

The final major factor that can lead to melting anomalies is enhanced mantle upwelling ($-DP/Dt$). Thermal buoyancy can cause rapid upwelling and dramatically enhance melt production (e.g., Ito *et al.*, 1996; Ribe *et al.*, 1995). Compositional buoyancy could also enhance upwellings (Green *et al.*, 2001). However, compositionally lighter material such as those with less iron and dense minerals (i.e., garnet), perhaps due to prior melt depletion (Oxburgh and Parmentier, 1977; Jordan, 1979), are likely to be less fusible than undepleted or fertile mantle. Behaving in complimentary fashion to fertile mantle, depleted mantle must be light enough such that the associated increase in upwelling rate ($-DP/Dt$) overcomes the reduction in fusability ($V$ and $(-\partial F/\partial P)_S$ in eqn [2]).

Buoyancy-driven upwelling, however, only requires enhanced lateral variations in density. Besides hot, plume-like upwellings from the deep mantle, large density variations can occur near the lithospheric thermal boundary layer (TBL). Sublithospheric boundary layer instabilities can drive small-scale convection in the upper mantle (e.g., Richter, 1973; Korenaga and Jordan, 2003; Huang *et al.*, 2003; Buck and Parmentier, 1986) (**Figure 15(a)**). A related form of small-scale convection can occur where there are large variations in lithospheric thicknesses such as that near rifted continental margins (Buck, 1986; King and Anderson, 1998) (**Figure 15(b)**). While the physics of convection in such situations have been explored to some degree, the volumes, timescales, and compositions of magmatism that could be produced have not.

The process of thinning the continental lithosphere can also cause rapid passive upwelling in the underlying asthenosphere. Thinning could occur by the foundering and delamination of the lower lithosphere or by continental rifting. Both mechanisms have been proposed to form flood volcanism on continents or continental margins without elevated mantle temperatures (Hales *et al.*, 2005; van Wijk *et al.*, 2001) (see Section 9.3.6).

Another form of enhanced mantle upwelling can occur in response to melting itself. Partial melting reduces the density of the solid residue (discussed above) and generates intergranular melt. Both factors can reduce the bulk density of the partially molten mantle and drive buoyant decompression melting (**Figure 16**). Buoyant decompression melting has been shown to be possible beneath mid-ocean ridges but could also occur away from mid-ocean ridges (Tackley and Stevenson, 1993; Raddick *et al.*, 2002). The key requirements for buoyant decompression melting to occur mid-plate is the presence of an appreciable thickness of mantle to be very near its

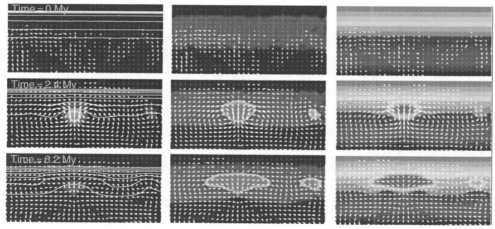

**Figure 16** Predictions of 2-D numerical models which simulate buoyant decompression melting for three time steps as labeled. Left column shows fractional melting rate (red = 0.0021 My$^{-1}$, blue = 0), middle column shows melt fraction retained in the mantle $\phi$ (red = 0.02), and right column shows volume fraction of melt extracted $\xi$ (red = 0.108). Density decreases as linear functions of both $\phi$ and $\xi$, and lateral density variations are what drive upwelling. Melting is therefore limited by the accumulated layer low-density residue. Reproduced from Raddick MJ, Parmentier EM, and Scheirer DS (2002) Buoyant decompression melting: A possible mechanisms for intra plate volcanism. *Journal of Geophysical Research* 107: 2228 (doi: 1029/2001JB000617), with permission from AGU.

solidus and a perturbation to initiate upwelling and melting. Perturbations could be caused by rising small bodies of hotter or more fusible mantle, the flow of mantle from beneath the old and thick side of a fracture zone to the young and thin side (Raddick *et al.*, 2002), or even sublithospheric convection.

Lastly, mantle upwelling can be driven by vertical motion of the lithosphere, such as that due to lithospheric flexure. The flexural arch around large intraplate volcanoes, for example, is caused by the growth of volcanoes which not only pushes the underlying lithosphere downward but causes upward flexing in a donut-shaped zone around the volcano. Flexural arching also occurs on the seaward side of subduction zones. Decompression melting can occur beneath flexural arches, again, if an appreciable thickness of asthenosphere is near or at its solidus (Bianco *et al.*, 2005).

### 9.3.3 Swells

One of the more prominent characteristics of hot spots and melting anomalies is the presence of broad seafloor swells. One indication of the possible origin is an apparent dependence of swell size with the rate of plate motion. We measure swell widths using the maps of residual topography (**Figure 1**) with the criterion that the swell is defined by the area that exceeds a height of +300 m in the direction perpendicular to volcano chains. This direction usually corresponds to the direction in which the swell is least wide (except for Trindade). The parallel bars in **Figure 1** mark the widths we measure for different hot spots. **Figure 17(a)** shows how swell width $\bar{W}$ varies with the plate speed $U_p$ at the hot spot, relative to the hot-spot reference frame (compiled by Kerr and Mériaux (2004)). For plate speeds <80 km My$^{-1}$, $\bar{W}$ appears to decrease with increasing plate speed. The prominent Pacific hot spots (Hawaii, Marquesas, Easter, Pitcairn, and Louisville) break the trend and have widths comparable to many of the swells in the Atlantic. We now test whether the observations can be explained by buoyant, asthenospheric material ponding beneath the lithosphere.

#### 9.3.3.1 Generating swells: Lubrication theory

Lubrication theory is a simplifying approach of solving the equations governing fluid flow against a solid interface. The method eliminates partial derivatives with respect to depth by assuming that fluid

layers are thin compared to their lateral dimension, and therefore describes fluid-layer thickness in map view. This theory was first applied to the formation of hot-spot swells by Sleep (1987) and Olson (1990) (*see* Chapter 4). Here, buoyant asthenosphere is introduced at the base of a moving, rigid (lithospheric) plate. The buoyant material is dragged laterally by plate motion and expands by self-gravitational spreading away from the source such that the extent $W$ perpendicular to plate motion increases with distance $x$ from the source (**Figure 18(a)**). Without plate motion the material would expand axisymmetrically like a pancake.

As confirmed with laboratory experiments and numerical models, the width of the buoyant material and swell far from the source can be approximated by the dimensionless equation (**Figure 18(a)**):

$$W/L_0 = C_1 (x/L_0)^{1/5} \qquad [3]$$

where $C_1$ is a constant ($\sim$3.70) and $L_0$ is the characteristic length scale of the problem defined as (Ribe and Christensen, 1994)

$$L_0 = \left( \frac{B^3 g}{96\pi^3 \Delta\rho^2 \eta U_p^4} \right)^{1/4} \qquad [4]$$

This length scale contains information about the key parameters controlling the width of the flow: buoyancy flux $B$ (kg s$^{-1}$), acceleration of gravity $g$, density contrast between the buoyant and normal mantle $\Delta\rho$, viscosity of the buoyant mantle $\eta$, and plate speed $U_p$. Alternatively, if the buoyant material is introduced along a semi-infinite line, then $W$ will increase more rapidly with distance. Using the same reasoning as Kerr and Mériaux (2004), this case can be directly compared to [3], using the length scale $L_0$

$$W/L_0 = C_2 L_0^{3/5} (x/L_0)^{4/5} \qquad [5]$$

where $C_2$ is a constant (in **Figure 18(a)** $C_2 L_0^{3/5} = 1.23$).

In either case, $W$ is predicted to be of the same order as $L_0 \propto B^{3/4}/U_p$. This relation predicts $L_0$ and thus $W$ to be inversely proportional to $U_p$ because faster plates allow less time for the layer to expand while it is dragged a distance $x$. The above relations also show an important dependence on $B$, which is not considered in our initial plot in **Figure 17(a)**. Larger buoyancy fluxes (proportional to $Q$) lead to larger $W$ primarily by enhancing the rate of gravitational expansion.

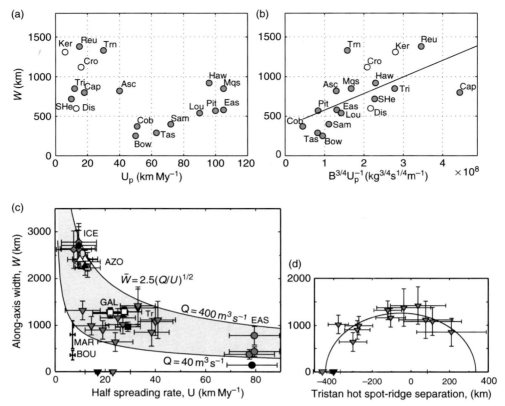

**Figure 17** (a) Average widths $\bar{W}$ of intraplate hot-spot swells measured by lines shown in **Figure 1** versus the rate $U_p$ of plate motion relative to the hot spots. (b) Lubrication theory of buoyant mantle expanding beneath the lithosphere predicts a swell width scale $L_0 \propto B^{-3/4} / U_p$, where $B$ is buoyancy flux. Line shows best fitting regression to the data, but excluding Cape Verde. (c) Widths $W$ for hot-spot–ridge interaction are the total along-isochron span of positive residual topography: diamonds, Iceland; white triangles, Azores; white squares, Galápagos; inverted triangles, Tristan; circles, Easter. Plate rate $U$ is the half spreading rate of the ridge during times corresponding to isochron ages (Ito and Lin, 1995). Black symbols mark present-day ridge-axis anomalies. Bold error bars show along-axis mantle-Bouguer gravity anomaly widths along the Southwest Indian Ridge near the Bouvet (black) and Marion (purple) hot spots (Georgen et al., 2001). Curves show predictions of scaling laws based on lubrication theory for a range of volume fluxes of buoyant mantle $Q$. (d) Along-isochron widths of residual bathymetric anomalies vs plume-ridge separation distance at times corresponding to isochron ages for the Tristan–MAR system. Curve is best fitting elliptical function $E(x/W_0)$ in eqn [7]. (c, d) Reproduced from Ito G, Lin J, and Graham D (2003) Observational and theoretical studies of the dynamics of mantle plume-mid-ocean ridge interaction. *Review of Geophysics* 41: 1017 (doi:10.1029/2002RG000117), with permission from AGU.

Buoyancy flux can be estimated based on the volumetric rate of swell creation, (Davies, 1988; Sleep, 1990)

$$B = \bar{h}\bar{W}U_p(\rho_m - \rho_w) \qquad [6]$$

where $\bar{h}$ and $\bar{W}$ are averages of swell height and width, respectively, and $\rho_m - \rho_w$ is the density contrast between the mantle and water.

Using estimated values for $B$ and $U_p$ (Kerr and Meriaux, 2004), a plot of $\bar{W}$ versus $B^{3/4}/U_p$ indeed shows a positive correlation (**Figure 17(b)**). Some of the scatter could be due to errors in $B$, perhaps due to uncertainties in swell height (Cserepes et al., 2000) as

well as the oversimplifying assumption in [6] that the buoyant material flows at the same speed as the plate (Ribe and Christensen, 1999). Other sources of scatter could be differences in $\Delta\rho$ and $\eta$ between hot spots.

For hot spots interacting with mid-ocean ridges, swell widths are the extent that positive residual topography extends in the direction parallel to ridges (Ito et al., 2003). Swell widths along hot-spot-influenced ridges and nearby seafloor isochrons appear to depend on the full spreading rate $U$ as well as hot-spot–ridge separation (**Figures 17(c)** and **17(d)**). Lubrication theory predicts that along-axis 'waist' widths $\bar{W}$ will reach a steady state when the volume flux of buoyant

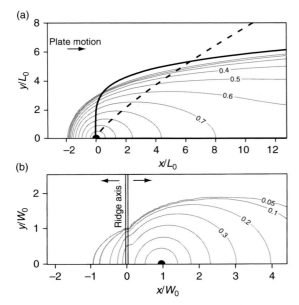

**Figure 18** (a) Plan view of predicted thickness (gray contours) of buoyant material expanding by gravitational spreading beneath a moving plate. Horizontal distances (x,y) are normalized by the width scale $L_0$ given by [4]. Buoyant material is introduced at a point at the origin. Bold curve is the analytical solution for the total width $W$ of the layer far from the source (eqn [3]). Dashed curve shows predicted $W$ for buoyant material emanating from a uniform line source (bold horizontal line) that starts at $x/L_0 = 0$ (eqn [5]). (b) Predicted thickness of buoyant material expanding from a source near two spreading plates. Distances are normalized by the width scale $W_0$ (eqn [8]). The width along the ridge axis reaches a steady-state value of $\bar{W}$. The numerical model predicts the material to expand away from the source due to its own buoyancy, to be dragged away from the ridge by plate motion, to thin in response to ambient mantle upwelling and lithospheric accretion, and to be channeled toward the ridge by the ridge-ward sloping lithosphere. Reproduced from Ribe NM and Christensen UR (1994) Three-dimensional modelling of plume-lithosphere interaction. *Journal of Geophysical Research* 99: 669–682, with permission from AGU.

material at the source $Q$ is balanced by the sinks associated with lithospheric accretion near the mid-ocean ridge (**Figure 18(b)**). Results of numerical models are well explained by the scaling law (Feighner and Richards, 1995; Ribe, 1996; Ito *et al.*, 1997; Albers and Christensen, 2001; Ribe *et al.*, 2007):

$$\bar{W}/W_0 = C_3 \left( \frac{Q\, \Delta \rho g}{48 \eta U^2} \right)^c E(x_r / W_0) \qquad [7]$$

where $C_3 \sim 2$, $c \sim 0.07$, and $E$ is an equation for an ellipse in terms of the normalized distance $x_r / W_0$ between the source and ridge axis. The characteristic width scale is

$$W_0 = (Q/U)^{1/2} \qquad [8]$$

The curves in **Figure 17(c)** show widths predicting by [7] and [8] for seven cases of hot-spot–ridge interaction. The general inverse relationship between $\bar{W}$ and $U^{1/2}$ explains the data reasonably well. Dispersion of $\bar{W}$ at a given $U$ can be caused by differences in $Q$ and in hot-spot-ridge separation $x_r$ (**Figure 17(d)**).

Overall, the apparent correlations between hot-spot widths, fluxes, and plate rates can be well explained by buoyant material being introduced at the base of the lithosphere. Compositionally or thermally buoyant upwellings rising from below the asthenosphere are possible sources and have been widely explored in context of the mantle-plume hypothesis (see also below). Buoyant mantle could also be generated near the base of the lithosphere, perhaps due to buoyant decompression melting. Such a mechanism for swell generation may be an alternative to deep-seated thermal upwellings.

### 9.3.3.2 Generating swells: Thermal upwellings and intraplate hot spots

Hot mantle plumes provide a straightforward mechanism to explain both the swells and excess volcanism associated with some hot spots. Three dimensional (3-D) numerical models that solve the governing equations of mass, energy, and momentum equilibrium of a viscous fluid have quantified the physics of plume-generated swells (Ribe and Christensen, 1994; Zhong and Watts, 2002; van Hunen and Zhong, 2003). They have, for example, successfully predicted the shape and uplift history of the Hawaiian swell. They also predict the eventual waning of swell topography to occur as a result of the thinning (see also **Figure 18(a)**) and cooling plume material beneath the lithosphere. Such a prediction provides a simple explanation for the disappearance of hot-spot swells along the Hawaiian and Louisville chain, as well as the lack of swells around very old portions of other volcano chains. Ribe and Christensen (1994) also predict minimal thinning of the lithosphere; therefore, the predicted elevation in heat flow is smaller than the variability that can be caused by local crustal or topographic effects (DeLaughter *et al.*, 2005; Harris and McNutt, 2007).

A similar model but with melting calculations defined the range of lithospheric thicknesses, potential temperatures, and buoyancy fluxes needed to generate the Hawaiian magma fluxes, swell width, and swell height (Ribe and Christensen, 1999)

(**Figure** 19). For reference, a plume composed of anhydrous lherzolite requires high potential temperatures of 1500–1600°C to roughly match the observations. In addition to a main (shield) melting phase, this model also predicts a secondary zone of upwelling and melting substantially down the mantle 'wind' of the plume. The location of this melting zone away from the main zone is consistent with it contributing to part of the rejuvenated stages on some Hawaiian Islands.

### 9.3.3.3 Generating swells: Thermal upwellings and hot-spot–ridge interaction

Another series of numerical modeling studies help define the conditions for hot upwelling plumes to explain swells and melting anomalies along hot-spot influenced ridges. Initially, studies showed that the swell width and the crustal thickness along the MAR near Iceland required a very broad upwelling (radius ∼300 km) of only modest excess temperature (<100°C above an ambient of 1350° C) (Ribe *et al.*,

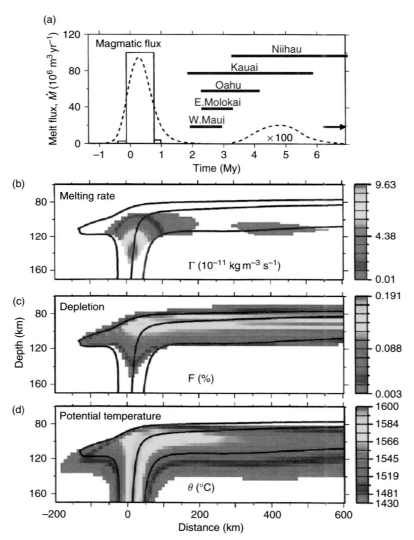

**Figure 19** Predictions of a 3-D numerical model for the generation of the Hawaiian swell and volcanism. A thermally buoyant mantle plume is introduced at a model depth of 400 km; it rises, interacts with a moving lithosphere, and melts. (a) Predicted duration and flux of volcanism (eqn [2]) (dashed) is comparable to the pre-shield, shield, and post-shield volcanic stages (three boxes). Blue lines show the spans of rejuvenated volcanism on the five labeled islands at distances from the concurrently active shield volcanoes (Bianco *et al.*, 2005). Cross-sections through the center of the plume show (b) melting rate $\Gamma = \rho_m DF/Dt$, (c) depletion *F*, and (d) potential temperature. Bold curves are streamlines. From Ribe NM and Christensen UR (1994) Three-dimensional modelling of plume-lithosphere interaction. *Journal of Geophysical Research* 99: 669–682.

1995; Ito *et al.*, 1996). These characteristics appeared to be inconsistent with the evidence from seismic tomography for a much narrower and hotter body (Wolfe *et al.*, 1997; Allen *et al.*, 1999b, 2002). However, calculations involving such a narrow and higher temperature upwelling predict crustal thicknesses that are several times that measured on Iceland – even assuming dry lherzolite as the source material.

One solution is provided by the likelihood that water that was initially dissolved in the unmelted plume material is extracted at the onset of partial

melting, and this dramatically increases the viscosity of the residue (Hirth and Kohlstedt, 1996). Numerical models that simulate this effect predict that the lateral expansion of the plume material occurs beneath the dry solidus and generates the observed swell width along the MAR (Ito *et al.*, 1999). Above the dry solidus, where most of the melt is produced, the mantle rises slowly enough to generate crustal thicknesses comparable to those at Iceland and along the MAR (**Figure 20**). A similar model, but with a variable flux of material rising in the plume produces

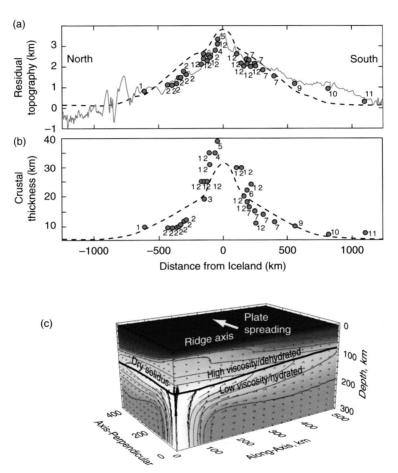

**Figure 20** (a)–(c) Comparisons between observations and predictions of a 3-D numerical model of a hot mantle plume rising into the upper mantle, interacting with two spreading plates, and melting. (a) Observed (light blue from gridded bathymetry and dots from refraction experiments) residual topography along the MAR, compared to the predictions, which assumes isostatic topography due to the thickened crust and low-density mantle. Dots show height above the seismically determined crustal thicknesses that are shown in (b). Dashed curve in (b) is predicted crustal thickness. (c) Perspective view of potential temperatures (white > ~1500° C, orange = 1350° C) within the 3-D model. The vertical cross-sections are along (left) and perpendicular (left) to the ridge. Viscosity decreases with temperature and increases at the dry solidus by $10^2$ because water is extracted from the solid with partial melting (Ito *et al.*, 1999). (a, b) Reproduced from Hooft EEE, Brandsdóttir B, Mjelde R, Shimamura H, and Murai Y (2006) Asymmetric plume-ridge interaction around Iceland: The Kolbeinsey Ridge Iceland Seismic Experiment. *Geophysics, Geochemistry, and Geosystems* 7: Q05015 (doi:10.1029/2005GC001123), from AGU. (c) Reproduced from Ito G. Shen Y, Hirth G, and Wolfe CJ (1999) Mantle flow, melting, and dehydration of the Iceland mantle plume. *Earth and Planetary Science Letters* 165: 81–96.

fluctuations in crustal production at the ridge that propagates away from the plume source along the ridge axis (Ito, 2001). This behavior was shown to explain V-shaped ridges that straddle the MAR north and south of Iceland for hundreds of kilometers (Vogt, 1971; Jones *et al.*, 2002b). A north–south asymmetry in crustal thickness documented by Hooft *et al.* (2006) is not predicted by the above models and could hold clues to larger-scale mantle flow, heterogeneity, or both.

### 9.3.4   Dynamics of Buoyant Upwellings

Our discussion of hot spots and melting anomalies is set amidst the background of larger-scale processes of plate tectonics and mantle convection. The cooling of the oceanic lithosphere is the main driving force for plate motion and the associated convection in the mantle. While the volume of mantle that participates in the plate-tectonic cycle is still debated, a consensus model has emerged of moderated whole-mantle convection, with significant material exchange between the upper mantle with relatively low viscosity and the more sluggishly convecting lower mantle. Observational evidence for whole-mantle convection is based on geoid and topography (e.g., Richards and Hager, 1984; Davies, 1998), geodynamic inversions (Mitrovica and Forte, 1997), seismological observations of slab extensions in to the lower mantle (Creager and Jordan, 1986), and seismic tomography (Grand, 1994; Su *et al.*, 1994, Grand *et al.*, 1997). Geodynamic models of whole-mantle convection show reasonable agreement with seismic tomography (Lithgow-Bertelloni and Richards, 1998; McNamara and Zhong, 2005), seismic anisotropy in the lower mantle (McNamara *et al.*, 2002), as well as surface heat flux and plate motions (e.g., van Keken and Ballentine, 1998). Hot spots and melting anomalies represent smaller-scale processes that likely involve mantle plumes.

The lack of detailed knowledge about lower-mantle properties, such as rheology, and thermal conductivity and expansivity, provides speculative opportunities about dynamical behavior; but numerical calculations can be used to map out the likely range of outcomes. For example, whole-mantle convection models with reasonable degrees of internal heating can satisfactorily explain both the average surface heat flow and plate velocities, but only when a higher-viscosity mantle is assumed (e.g., van Keken and Ballentine, 1998). Mineral physics provide strong suggestions for a reduction in

thermal expansivity and increase in thermal conductivity with pressure. The combined effects will reduce convective vigor, but models that incorporate reasonable depth variations of these properties predict that this does not render the lower mantle immobile (e.g., van Keken and Ballentine, 1999; van den Berg *et al.*, 2005; Matyska and Yuen, 2006a). Dynamical theory provides an essential stimulus for the mantle-plume hypothesis, since thermal plumes form naturally from hot boundary layers in a convecting system. The CMB is the main candidate to have a significant TBL (e.g., Boehler, 2000), but other boundary layers may exist at locations where sharp transitions in material properties or composition occur, such as the bottom of the transition zone at 670 km depth and the proposed thermochemical layer at the base of the mantle. We will first summarize the fluid dynamics of plumes rising from a TBL before addressing the consequences of chemical buoyancy forces and depth-dependent mantle properties.

### 9.3.4.1   TBL instabilities

In it simplest form, the growth of an upwelling instability from a hot boundary layer can be approximated as a Rayleigh–Taylor instability with the onset time and growth rate controlled by the local (or boundary) Rayleigh number,

$$Ra_\delta = \frac{\rho g \alpha \Delta T \delta^3}{\mu \kappa} \qquad [9]$$

The instability is enhanced by thermal expansivity $\alpha$ ($\rho$ is density and $g$ is acceleration of gravity), the temperature jump across the boundary layer $\Delta T$, and the layer thickness $\delta$, and is hampered by viscosity $\mu$, and thermal diffusivity $\kappa$ (including potential radiative effects in the deep mantle). Large-scale mantle flow tends to suppress the growth of instabilities but the temperature dependence of rheology will enhance its growth. For more specifics on governing equations for boundary layer instabilities and examples of their modeling with laboratory and numerical techniques see chapters 6 and 11 of Schubert *et al.* (2001). Indeed, analytical methods provide important insights to the rate of formation of the instability and the dependence on ambient conditions (see e.g., Whitehead and Luther, 1975; Ribe and de Valpine, 1994). The growth of a diapir to a full plume can be understood with nonlinear theory; for example, Bercovici and Kelly (1997) show that growth is retarded by draining of the source layer and the diapir can temporarily stall. Experimental and numerical investigations confirm and expand these predictions and, quite importantly, provide direct verification of model predictions made by independent approaches (e.g., Olson *et al.*, 1988; Ribe *et al.*, 2007). In general, most studies find that for reasonable lower-mantle conditions, the boundary layer instabilities will grow on the order of 10–100 My (e.g., Christensen, 1984; Olson *et al.*, 1987; Ribe and de Valpine, 1994).

As the diapir rises it will generally be followed by a tail of hot material that traces back to the boundary layer. The rise speed of the diapir is proportional to its buoyancy, the square of its radius, and is inversely proportional to viscosity. The morphology of the plume head and tail are controlled by the viscosity contrast between the hot and ambient fluid. A more viscous plume will tend to form a head approximately the same width as the tail (a 'spout' morphology) whereas a lower-viscosity plume will tend to form a voluminous plume head much wider than the tail (a 'mushroom' or 'balloon' geometry following terminology by Kellogg and King (1997)). Since mantle viscosity is a strong function of temperature it is generally expected that the mushroom/balloon geometry should dominate, but Korenaga (2005a) proposes an interesting counter argument for grain-size controlled, high-viscosity plumes. As we will discuss below, chemical buoyancy may have significant control on the shape of the plume as well.

Dynamical experiments without large-scale flow generally demonstrate that a boundary layer will become unstable with many simultaneous plumes that interact with each other as they rise through the fluid (e.g., Whitehead and Luther, 1975; Olson *et al.*, 1987; Kelly and Bercovici, 1997; Manga, 1997; Lithgow-Bertelloni *et al.*, 2001). To study the dynamics of a single plume, it has become common to use a more narrow or point source of heat, which in laboratory experiments can be achieved by inserting a small patch heater at the base of the tank (e.g., Kaminski and Jaupart, 2003; Davaille and Vatteville, 2005) or to inject hot fluid through a small hole in the base of the tank (Griffiths and Campbell, 1990). The latter work showed that with strongly temperature-dependent viscosity the plume head entrains ambient fluid, forming a characteristic mushroom-shaped head. Interestingly, this same shape was observed also by Whitehead and Luther (1975) but for mixing of fluids with similar viscosity (their figure 9). Van Keken's (1997) replication of Griffiths and

Campbell's (1990) laboratory experiment also showed that this form of plume is retained when it originates from a TBL or when olivine, rather than corn syrup rheology, was assumed. Other relevant numerical experiments are provided by Davies (1995) and Kellogg and King (1997).

### 9.3.4.2 *Thermochemical instabilities*

Studies of thermal plumes originating from TBLs have guided much of the classic descriptions of mantle upwellings and represent a logical starting point for understanding them. The Earth, however, is more complex since density is likely to be controlled by composition, as well as temperature. The seismic structure in the deep mantle beneath the African and South Pacific Superswell regions provides evidence for such deep compositional heterogeneity. Mantle convection models suggest that dense layers are likely to form distinct large blobs or piles that are away from areas of active downwellings (Tackley, 2002; McNamara and Zhong, 2005). Due to the spatial and temporal interaction between chemical and thermal buoyancy forces, the upwellings that form from a thermochemical boundary layer can be dramatically different from the classical thermal plume and interaction with the lower-viscosity upper mantle can significantly alter their shape (Farnetani and Samuel, 2005). The stable topography of high-density layers could provide an anchoring point above which thermal plumes can rise and thus define a fixed reference frame for different hot-spot groups (Davaille *et al.*, 2002; Jellinek and Manga, 2002).

A compelling cause for compositional heterogeneity is the recycling of oceanic crust in subduction zones (e.g., Christensen and Hofmann, 1994). The density of the mafic (eclogitic) crust likely remains higher than that of the ambient mantle through most of the lower mantle (Ono *et al.*, 2001). A layer generated by oceanic crust recycling is likely to remain stable if its density is in the range of 1–6% greater than that of the ambient mantle (Sleep, 1988; Montague and Kellogg, 2000; Zhong and Hager, 2003; Brandenburg and van Keken, in press). Entrainment of this layer by plumes provides a straightforward explanation for the geochemically observed oceanic crust component in OIBs (Shirey and Walker, 1998; Eiler *et al.*, 2000). The potential for entrainment of a deep chemical boundary was studied systematically by Lin and van Keken (2006a, 2006b) who found that with strongly temperature-dependent viscosity the entrainment would become episodic under a large range of conditions. The style of entrainment ranged from nearly stagnant large plumes in the lower mantle to fast episodic pulsations traveling up the pre-existing plume conduit, which provides an explanation for the pulses of LIP volcanism (Lin and van Keken, 2005). Oscillatory instabilities in starting plumes can be caused by the competing effects of thermal and chemical buoyancy with particularly interesting effects where the effective buoyancy is close to zero (Samuel and Bercovici, 2006).

### 9.3.4.3 Effects of variable mantle properties

Large variations in material properties can lead to complex forms and time dependence of buoyant upwellings. For example, the combination of increasing ambient viscosity, thermal conductivity, and decreasing thermal expansivity with depth will likely cause plumes to be relatively broad in the deep mantle but become thinner when migrating upward (Albers and Christensen, 1996). Changes in rheology fundamentally alter plume dynamics. The sharp decrease of ambient viscosity for a plume rising from the lower, to the upper mantle will cause a rapid increase in speed and resulting drop in plume width (van Keken and Gable, 1995). Such necking may also cause the formation of a second boundary layer with episodic diapirism with timescales on the order of 1–10 My (van Keken et al., 1992, 1993). The viscosity change can also completely break apart a starting mantle-plume head into multiple diapirs, perhaps contributing to multiple flood basalt episodes (Bercovici and Mahoney, 1994). An important aspect of mantle rheology is the non-Newtonian behavior, which is characterized by a viscosity that has a strong, nonlinear function of stress in addition to temperature and pressure. The strong stress dependence can dramatically enhance the deformation rate of boundary layer instabilities and lead to much higher rise speeds than is observed in Newtonian fluids (where strain rate and stress are linearly related) (Larsen and Yuen, 1997; Van Keken, 1997; Larsen et al., 1999). Such behavior can cause starting plume heads to rise sufficiently fast to almost completely separate from the smaller tail, thus providing an alternative explanation for the observed LIP episodicity (Van Keken, 1997).

The transition zone is also characterized by major phase changes in the upper-mantle mineral assemblages, dominated by the exothermic 400 km discontinuity and the endothermic 670 km

discontinuity. The phase changes provide a dynamical influence that can strongly modify plume flow, with predictions for more episodic or faster plume flow in the upper mantle (Nakakuki et al., 1997; Brunet and Yuen, 2000). One model provides a source for plumes even just below the 660 km discontinuity (Cserepes and Yuen, 2000), an intriguing possibility for hot spots with sources above the CMB.

The sluggish nature of the lower mantle may be enhanced if heat radiation becomes efficient at high temperature. This has been explored for mantle plumes by Matyska et al. (1994) who suggest that the radiative components will strongly enhance the stability and size of large 'super'plume regions in the lower mantle, even without chemical stabilization. The plume stability may be enhanced by the possible post-perovskite transition at the base of the lower mantle (Matyska and Yuen, 2005, 2006a).

Thus, the concept of a cylindrical plume, rising vertically from the CMB to the lithosphere, is probably far too simple. Upwellings are likely to take on complex shapes, have a wide range of sizes, be strongly time dependent, and originate from different depths in the mantle. We will revisit such issues in Section 9.3.7.

### 9.3.4.4 Plume buoyancy flux and excess temperature

A long-standing question concerns the efficiency of heat transport in mantle plumes. If hot spots are dynamically supported by plumes that rise from the CMB, then we can use surface observations to estimate the heat from the core. The topography of hotspot swells provides a fundamental constraint on the buoyancy flux of plumes. Davies (1988) estimated, from swell heights of 26 hot spots provided by Crough (1983), that new topography is being generated at a rate $S = 17.5 \, \mathrm{m^3 \, s^{-1}}$ (in comparison to $300 \, \mathrm{m^3 \, s^{-1}}$ for the total mid-oceanic ridge system). Iceland was excluded from this compilation and the overall flux was dominated by the Pacific hot spots and in particular by Hawaii, Society, and the Marquesas. The total buoyancy flux follows by multiplication with the density difference between mantle and sea water, $B = (\rho_m - \rho_w)S = 40 \, \mathrm{Mg \, s^{-1}}$, assuming $(\rho_m - \rho_w) = 2300 \, \mathrm{kg \, m^{-3}}$. The estimated heat flux carried by the plumes $q_p$ follows from $q_p = \rho_m C_p S / \alpha$, which is around 2 TW, assuming $C_p = 1000 \, \mathrm{J \, kg^{-1} \, K^{-1}}$ and $\alpha = 3 \times 10^{-5} \, \mathrm{K^{-1}}$. Sleep (1990) provided a similar analysis for 34 hot spots, including Iceland, and found a slightly larger value of $55 \, \mathrm{Mg \, s^{-1}}$ for the total buoyancy flux, implying a core heat loss of ~2.7 TW. The plume heat flux is therefore significantly smaller than the total heat flux of the Earth of 44 TW (Pollack *et al.*, 1993).

The low estimates for plume heat flux are suggestive of only a minor contribution of plumes to the cooling of the Earth, and potentially a small contribution of core cooling, under the assumption that hot-spot heat is related to the heat from the core. However, the role of compressible convection, internal heating modes, and interaction with the large-scale flow and plate tectonics may be important in masking heat rising from the core. For example, a large-scale mantle circulation associated with plate tectonics could be drawing heat from the CMB toward mid-ocean ridges rather than allowing it to rise in vertical plumes to form mid-plate hot spots (Gonnermann *et al.*, 2004). Statistical arguments for the power-law distribution of plumes have been used to debate that many small plumes are entrained in the large-scale flow and do not express themselves as hot spots (Malamud and Turcotte, 1999). This idea has been supported by mantle convection models that simulate a wide range of heating modes (Labrosse, 2002; Mittelstaedt and Tackley, 2006).

The temperature increase across the TBL at the CMB can be estimated based on the adiabatic extrapolation of upper-mantle temperatures and mineral physics constraints on the temperature of the core. An estimate exceeding 1000 K (Boehler, 2000) is much greater than the temperature anomalies expected in the upper mantle beneath the major hot spots of only 200–300 K (see Section 9.3.2.1). If plumes rise from the CMB, what then is the mechanism for reducing the plume temperature by the time it reaches the base of the lithosphere? While entrainment of, and diffusive heat loss to the surrounding cooler mantle will reduce the plume excess temperature, most calculations based on the classical plume model suggest only modest reductions (e.g., Leitch *et al.*, 1996; Van Keken, 1997).

In our view, the likely important role of thermochemical convection and the variable properties of the mantle (see Sections 9.3.4.2 and 9.3.4.3) can provide a self-consistent resolution to the above discrepancy. For example, Farnetani (1997) showed that if a compositionally dense layer stabilizes at the base of D″, plumes will tend to rise from only the top of the TBL, which is substantially less hot than the CMB. In addition, decompression and subadiabaticity can enhance the cooling of plumes as they rise through the mantle and further reduce their surface expression (Zhong, 2006).

### 9.3.5 Chains, Age Progressions, and the Hot-spot Reference Frame

Thermal plumes rising from below the upper mantle to the lithosphere provide a reasonably straightforward explanation for a source of at least some hot-spot swells and age-progressive volcano chains. An origin residing below the asthenosphere allows for the possibility of a kinematic reference frame that is distinct from the plates. On the other hand, if thermal plumes rise from the deep, convecting mantle, it should be intuitive that hot spots are not stationary (Section 9.2).

A series of studies initiated by Steinberger (Steinberger and O'Connell, 1998; Steinberger, 2000) have used numerical models to simulate plumes rising in a convecting mantle. Their calculations of a spherical Earth assume that mantle flow is driven kinematically by the motion of the plates with realistic geometries, and dynamically by internal density variations estimated from different seismic tomography models. The viscosity structure includes high viscosities in the lower mantle ($\sim 10^{21} - 10^{23}$ Pa s) and lower viscosities in the upper mantle ($\sim 10^{20} - 10^{21}$ Pa s). A plume is simulated by inserting a vertical conduit in the mantle at a specified time in the past. Velocities at each point along the conduit are the vector sum of the ambient mantle velocity and the buoyant rise speed of the conduit, which is computed based on scaling laws derived from theory and laboratory experiments. For simplicity, the ambient mantle flow is not influenced by the plumes. Plume conduits therefore deform with time and where they intersect the base of the lithosphere defines the location of the hot spots (**Figure 21**).

This method was applied to examine the evolution of the Hawaiian, Louisville, and Easter hot spots in the Pacific ocean (Steinberger, 2002; Koppers *et al.*, 2004). A mantle flow model was found to optimize fits between predicted and observed age progressions along the whole lengths of all three chains. For the Hawaiian hot spot, the models predict absolute southward motion that was rapid (average $\sim 40 \text{ km My}^{-1}$) during 50–80 Ma (prior to the Hawaiian–Emperor Bend) and slower ($< 20 \text{ km My}^{-1}$) since 50 Ma, thereby providing an explanation for the paleomagnetic evidence (Tarduno *et al.*, 2003; Pares and Moore, 2005; Sager *et al.*, 2005). The models also predict slow eastward motion of Louisville, consistent with the observed nonlinear age progression (Koppers *et al.*, 2004), as well as WSW motion of the Easter hot spot at rates of $\sim 20 \text{ km My}^{-1}$.

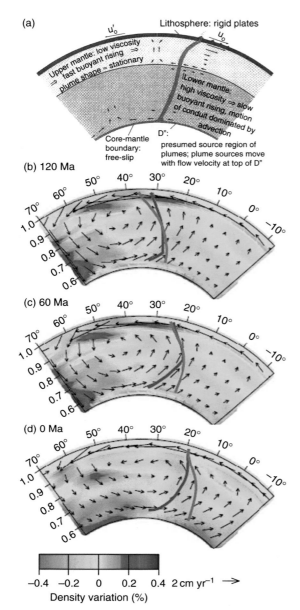

**Figure 21** (a) Cartoon illustrating the rise and deformation of plumes through a flowing mantle with layered viscosity and surface plate motion. (b)–(d) show predicted mantle flow (arrows), density variations (colors), and deformation of the Hawaiian plume rising from the CMB at a location that is fixed (purple, initiated 150 Ma, or moving with the mantle (red, initiated 170 Ma). Times in the past are labeled. (a) Reproduced from Steinberger B and O'Connell RJ (2000) Effects of mantle flow on hotspot motion. *Geophysical Monograph* 121: 377–398, with permission from AGU. (b)–(d) Reprinted by permission from Macmillan Publishers Ltd: (Nature) Steinberger B, Sutherland R, and O'Connell RJ (2004) Prediction of Emperor-Hawaii seamount locations from a revised model of global plate motion and mantle flow. *Nature* 430: 167–173, copyright (2004).

The above studies addressed one group of hot spots but a key challenge is to explain the age progressions of both the Pacific and Indo-Atlantic hot spots in a single whole-mantle flow model. Steinberger *et al.* (2004) started this by considering two kinematic circuits to define the relative motions between the Pacific and African Plates: (1) through Antarctica, south of New Zealand and (2) through the Lord Howe Rise, north of New Zealand, Australia, and then Antarctica. The reference case of fixed hot spots predicts hot-spot tracks that deviate substantially from observed locations (dotted curves in **Figure 22**) for both plate circuits. Considering moving hot spots derived from mantle flow simulations with plate circuit (1) yields reasonable matches to the tracks for ages <50 Ma but predicts a track too-far west of the Emperor chain for ages >50 Ma. Finally, predicted tracks with moving hot spots using plate circuit (2) provide the closest match to the observed tracks. This model successfully predicts the geographic age progressions along most of the Tristan, Réunion, and Louisville tracks, and for the Hawaiian track since 50 Ma, including a bend between the Hawaiian and Emperor seamounts. But the bend is not sharp enough: the models still predict a trajectory for the Emperor seamounts too far west.

The above studies illustrate that models of plumes rising in a geophysically constrained, mantle flow field can explain many key aspects of apparent hot-spot motion. The studies, however, underscore the importance of uncertainties in defining relative plate motions, particularly in the presence of diffuse plate boundaries – for example, that near Lord Howe Rise. Still more uncertainties are associated with the locations of volcanism in time and in paleomagnetic latitudes. The models are sensitive to a number of properties such as mantle viscosity structure, the choice of seismic tomography model, the mapping between seismic velocities and density, as well as the buoyancy and dimensions (which control the rise speed) of the mantle plumes. A recent study has just begun to quantify the observational uncertainties and to use them to define statistically robust mantle flow solutions (O'Neill *et al.*, 2003). But many observations remain poorly understood, including the location and trend of the older portion of the Emperor chain.

### 9.3.6    Large Igneous Provinces

The rapid and massive magmatic production of many LIPs, combined with their strong connection to continental breakup, but inconsistent connection to

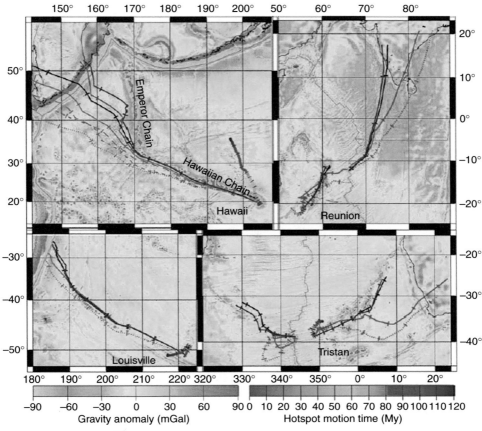

**Figure 22** Computed hot-spot motion (rainbow colored bands with color indicating position in time according to scale on the lower right) and tracks overlain on gravity maps (left color scale). Tracks are plotted on all plates regardless of whether a hot spot was actually on the plates during those times. Ticks along tracks are every 10 My. Dotted (red and green) lines are solutions assuming fixed hot spots. Solid red lines (plate circuit (1) shown for Hawaii and Louisville) and purple lines (plate circuit (2) shown for Hawaii) are for moving hot spots in mantle flowing in response to absolute plate motions that optimize fits to only the Tristan and Réunion hot-spot tracks. Green lines (shown for Réunion and Tristan) are best fit solutions to only the Hawaii and Louisville tracks using plate circuit (1). Black (plate circuit (1)) and blue (plate circuit (2)) are solutions that optimize fits to all four tracks. A least-squares method is used to optimize the fit to locations and radiometric ages of seamounts. Reprinted by permission from Macmillan Publishers Ltd: (Nature) Steinberger B, Sutherland R, and O'Connell RJ (2004) Prediction of Emperor-Hawaii seamount locations from a revised model of global plate motion and mantle flow. *Nature* 430: 167–173, copyright (2004).

present-day hot-spot volcanism are challenging to understand. Moreover, the wide range of eruptive volumes (**Figure 8**) and durations suggest that there may not be just one overarching mechanism.

The observation of large plume heads followed by thin tails in fluid dynamical experiments has traditionally been used to explain the LIP-hotspot connection (Richards *et al.*, 1989) and remains, because of its simplicity and plausibility, an attractive base model for the formation of many LIPs. Its strengths include that: (1) it is supported by fluid dynamics for increasingly realistic assumptions

about mantle composition and rheology (see Section 9.3.4), in fact these modifications to the base model allow for an explanation of some of the diversity seen in the geological record (**Figure 23**); (2) it offers a dynamical cause for the common disconnect between LIP and hotspot volcanism (Farnetani and Samuel, 2005) (i.e., some plume heads rise to the surface without plume tails and some upwellings form narrow plume tails without heads, **Figure 23**). (3) It predicts the hottest material of rising plume heads will erupt first (Farnetani and Richards, 1995) which explains high MgO basalts early in the LIP record; and (4) it

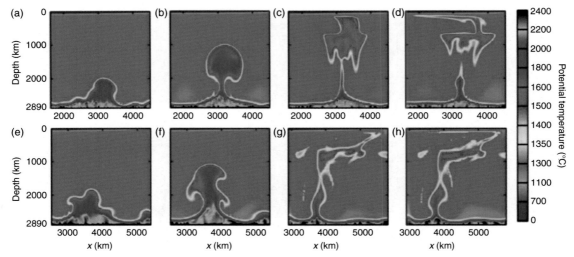

**Figure 23** Temperatures of upwelling instabilities rising from the CMB predicted by 3-D numerical models. The CMB is heated from below and is blanked by a layer initially 200 km in thickness and containing material with an intrinsic, compositional density that is 1% greater than that of the surrounding lower mantle. Upwelling through a depth of 670 km is inhibited by an endothermic phase change with a Clapeyron slope of −2.5 MPa K$^{-1}$. In the top row a large, roughly spherical plume head rapidly rises and could generate flood volcanism at the surface. The head detaches from the stem and this may delay or prevent the formation of a chain of volcanoes extending away from the flood basalts. In the bottom row, only a narrow upwelling rises into the upper mantle. Reproduced from Farnetani CG and Samuel H (2005) Beyond the thermal plume hypothesis. *Geophysical Research Letters* 32: L07311 (doi:10.1029/2005GL022360), with permission from AGU.

predicts that the arrival of the plume at the surface leads to uplift and extension which is observed in the geological record of many LIPs (see Section 9.2.3). Plume-based models, however, have yet to adequately explain the strong correlation between LIPS and continental breakup and the lack of uplift during the OJP eruptions (d'Acremont *et al.*, 2003).

There are a number of alternative mechanisms that address the above issues, which include shallow, sublithospheric processes or metorite impacts. While the plume model has received significant attention and quantitative hypothesis testing, the majority of the alternatives are currently still in rather qualitative form.

The first alternative to thermal plumes pertains to LIPs formed on continental margins. Anderson (1994b) proposes that excess heat can build below continents during tectonic quiescence and/or supercontinent formation, which then causes the massive eruptions during continental breakup. This hypothesis addresses the correlation between LIPs and continental breakup and the lack of connection of some continental LIPs to hot-spot trails. One aspect not addressed specifically is why the volcanism is typically not margin-wide but, instead, is more restricted in total extent. Nevertheless, the correlation between the LIPs and continental breakup is intriguing and it is quite likely that regional variations in the composition and strength of the lithosphere have an important control on the location of magma eruption.

Second, delamination of continental lithosphere and secondary convection at rifted margins (**Figure 15(b)**) have been forwarded to generate LIPs near continents (King and Anderson, 1995, 1998; van Wijk *et al.*, 2001; Hales *et al.*, 2005; Anderson, 2005). Like the above concept of subcontinental mantle incubation, these models can explain continental LIPs without hot-spot tracks, but have yet to show how they could form LIPs with hot-spot tracks. Indeed, more quantitative modeling of shallow mantle processes needs to be done.

A third alternative, which could apply to LIPs that form near sites of continental or oceanic rifting, is that compositional, rather than thermal effects cause excess melting, for example, more fertile mantle such as eclogite and/or water in the source (e.g., Anderson, 1994a, 2005; Cordery *et al.*, 1997; Korenaga, 2004; 2005b) (**Figure 15(a)**). The strengths of this possibility include that some compositional effects are expected and these can strongly enhance melt production (see Section 9.3.2.2); the lack of uplift of OJP could be explained if dense eclogite occurs in the source; and it can explain the formation of LIPs not connected to long-lived hot-spot tracks. Again, a key dynamic weakness of models invoking eclogite in the source is that eclogite is dense and requires some mechanism to stay near, or be brought back up to the surface.

Fourth, the enigmatic nature of the OJP has led to the suggestion that a meteorite impact could be responsible for the emplacement of LIPs (Rogers, 1982; Jones *et al.*, 2002a; Tejada *et al.*, 2004; Ingle and Coffin, 2004). The strengths are that the decompression of mantle following impact may generate extensive melting (although perhaps with some qualifications (Ivanov and Melosh, 2003)) with less uplift than expected from a hot plume head and without a connection to a hot-spot track. It is rare, however, to find direct evidence for meteorite impact during LIP emplacement. One of the few convincing observations is the iridium anomaly embedded in the Deccan Traps, but if this impact signal is related to the Chixculub impact, it post-dates the start of volcanism (Courtillot and Renne, 2003). Also, it is statistically unlikely that the majority of Phanerozoic LIPs can be explained by impacts (Ivanov and Melosh, 2003; Elkins-Tanton and Hager, 2005). Finally, a cursory investigation of planetary impact cratering suggests that many large craters on the Moon, Mars, and Venus form without contemporaneous volcanism. An interesting recent speculation is that the formation of the lunar impact basins and (much) later volcanic activity that formed the maria are indeed related and caused by the long term storage of heat in the lunar mantle (Elkins-Tanton *et al.*, 2004).

### 9.3.7 Hot Spots: Modifications and Alternatives

#### 9.3.7.1 Variable hot-spot durations from transient thermal plumes

Section 9.2.1 showed that the duration of hot-spot activity varies significantly from <10 to >100 My with numerous cases falling near both extremes. A broad range of timescales is generally consistent with the variable scales of flows that characterize mantle convecting at high Rayleigh numbers ($\geq 10^7$). Thermal upwellings, for example, are, in general, transient phenomena as shown by laboratory experiments of basally heated viscous fluids. An excellent example is provided by Davaille and Vatteville (2005); they find

that well-developed plumes in the laboratory often detach from their bases on timescales comparable to, but less than, that needed to initiate the instability from the TBL (**Figure 24(a)**). Laboratory experiments show that this timescale is approximately (e.g., Sparrow *et al.*, 1970; Manga *et al.*, 2001)

$$\tau = \frac{H^2}{\pi\kappa}\left(\frac{Ra_c}{Ra}\right)^{2/3} \quad [10]$$

where $H$ is the thickness of the fluid, $Ra_c$ is the critical Rayleigh number (i.e., the minimum needed to grow an instability and is of order $10^3$), and $Ra$ is the Rayleigh number of the system. This is essentially the time it takes heat diffusion to sufficiently thicken the TBL (to a thickness $\delta = \sqrt{\pi\kappa\tau_c}$) such that the local Rayleigh number (eqn [8]) equals $Ra_c$. If the TBL is at the CMB, then $H = 2800$ km, $Ra = 10^7$, and plume life times are expected to be $\tau \sim 200$ My, which is sufficient to maintain long-lived hot-spot chains. If the TBL is near the base of the upper mantle ($H = 660$ km, $Ra = 1e7$), then $\tau \sim 10$ My is comparable to the duration of many short-lived volcano chains.

There are a few ways to initiate plumes from boundary layers internal to the mantle. High-resolution 2-D convection simulations by Matyska and Yuen (2006b) predict large-scale ($10^3 - 10^4$ km), superplume-like upwellings as a result of relatively low local Rayleigh numbers caused by high viscosities, low thermal expansivity, and radiative heat transfer in the lower mantle (**Figure 24(b)** and **24(d)**). As superplumes rise through the lower mantle, further rise can be inhibited by the endothermic phase change at 660 km. The hot tops of the superplumes can generate a TBL from which smaller scale ($10^2$ km) upper-mantle upwellings can originate. Another possible surface for a mid-mantle TBL is the top of a chemically dense layer in the lower mantle (see Section 9.3.4.2). Laboratory experiments show that when an initially chemically stratified system is heated from below and cooled from above, a variety of forms of upwellings and downwellings occur, depending on the ratio of chemical-to-thermal buoyancy (Davaille, 1999). When the negative chemical buoyancy of the lower layer is ~0.35–0.55 times the positive buoyancy due to the basal heating, the two layers remain separate but the surface between them undulates to form broad downwellings and superplume upwellings. Above the upwellings, smaller instabilities can rise into the upper layer and to the surface (**Figure 24(e)**). In both of the two examples described above, smaller upper-mantle plumes are shown to rise from the top of broad

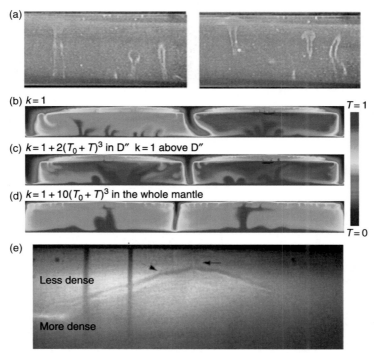

**Figure 24** (a) Photographs, the left taken prior to the right, of a laboratory experiment involving sucrose solution heated from the base. Isotherms (white streaks) outline the hot upwellings that detach from the base, as imaged using suspended liquid crystals and glass particles. (b) 2-D numerical simulations of whole-mantle convection in which the mantle is heated internally, at its base, and is cooled at the surface. Viscosity is temperature- and pressure-dependent. Calculations also include an endothermic phase change at 670 km and an exothermic phase change at 2650 km. Coefficient of thermal expansion decreases by >10 times from the surface to the CMB. (b) A thermal conductivity $k = 1$ corresponds to that due to heat diffusion, whereas (c) and (d) include an additional term (dependent on temperature) for radiative heat transfer. (e) Laboratory experiments involving a chemically dense and more viscous fluid underneath a chemically less dense and less viscous fluid. The system is heated from below and cooled from above. Dark and light bands show the interface between the two fluids, which bows upward and resembles the upper surface of a mantle superplume. Smaller plumes are imaged to be rising from this surface into the upper layer. (a) Reproduced from Davaille A and Vatteville (2005) On the transient nature of mantle plumes. *Geophysical Research Letters* 32, doi:1029/2005GL023029, with permission from AGU. (b)–(d) Reproduced from Matyska C and Yuen DA (2006b) Upper-mantle versus lower-mantle plumes: are they the same? In: Foulger GR and Jurdy DM (eds.) *The Origins of Melting Anomalies*: *Plates, Plumes, and Planetary Processes*. GSA, with permission from GSA. (e) Reprinted by permission from Macmillan Publishers Ltd: (Nature) Davaille A (1999) Simultaneous generation of hotspots and superswells by convection in a heterogeneous planetary mantle. *Nature* 402: 756–760, copyright (1999).

superplumes. Such a situation could provide an explanation for the large frequency of short-lived hot spots in the superswell regions of the South Pacific and Africa (e.g., Courtillot *et al.*, 2003; Koppers *et al.*, 2003).

### 9.3.7.2  *Forming melting anomalies by upper-mantle processes*

Buoyant upwellings have been the focus of a large number of studies but are probably not the only phenomena giving rise to melting anomalies. Upper-mantle processes that are largely decoupled from the lower mantle undoubtedly contribute to the magmatism in various ways.

Small-scale, sublithospheric convection, as introduced in Section 9.3.2, is one possible mechanism. Sublithospheric convection could be evidenced by a number of observations: it could limit the maximum thickness of the lithosphere and slow the subsidence of old seafloor (e.g., Huang *et al.*, 2003); it could give rise to the prominent gravity lineations over the Pacific seafloor (Haxby and Weissel, 1986); and it could explain the periodic fluctuations in upper-mantle seismic structure as imaged perpendicular to hot-spot swells in the Pacific (Katzman *et al.*, 1998). Small-scale sublithospheric convection may explain magmatism along lineaments parallel to plate motion (e.g., Richter, 1973; Huang *et al.*, 2003), but perhaps without a systematic geographic age progression. Quantitative studies of melting have yet to be done but are needed to quantify the rates and time dependence of magma production, as well as whether small-scale convection could generate seafloor swells.

Another mechanism that has been recently proposed is fingering instabilities of low-viscosity asthenosphere (Weeraratne *et al.*, 2003; Harmon *et al.*, 2006). When two fluids are contained in a thin layer (Hele–Shaw cell in the laboratory and possibly the asthenosphere in the upper mantle) and one fluid laterally displaces a more viscous fluid, the boundary between the two becomes unstable and undulates with increasing amplitude (Saffman and Taylor, 1958). Fingers of the low-viscosity fluid lengthen and penetrate the high-viscosity fluid. Perhaps hot mantle rising beneath the South Pacific Superswell area is supplying hot, low-viscosity asthenosphere that is fingering beneath the Pacific Plate and generating some of the volcanic lineaments such as Pukapuka (Harmon *et al.*, 2006; Weeraratne *et al.*, in press). Weeraratne and Parmentier (2003) explore this possibility using laboratory experiments that simulate asthenospheric conditions.

Finally, numerous studies have suggested that heterogeneity in lithospheric stresses or structure can allow magma that is already present in the asthenosphere to erupt at the surface (e.g., Anguita and Hernan, 2000; Clouard *et al.*, 2003). Lithospheric stress associated with regional (e.g., Sandwell *et al.*, 1995) or local (e.g., Mittelstaedt and Ito, 2005) tectonics, as well as thermal contraction (Gans *et al.*, 2003) could initiate fissures that can propagate and cause volcanic lineaments over a range of scales. Sandwell and Fialko (2004), for example, demonstrate that top-down cooling of the lithosphere generates thermoelastic tensile stress, which is optimally released by local zones of fracturing with spacing comparable to the flexural wavelength of the lithosphere and to the distances between the Pukapuka, Sojourn, and Hotu–Matua Ridges in the South Pacific. Natland and Winterer (2005) propose a lithospheric fissure origin for most or all of the Pacific hot spots, but such a hypothesis has yet to be tested quantitatively.

Another form of stress-influenced magma penetration could redistribute magma from diffuse volumes in the asthenosphere to discrete, localized eruption sites at the surface. The weight of a volcano can draw further magmatism if lithospheric stresses due to loading focuses magma-filled cracks toward the volcano

(Muller *et al.*, 2001) or if damage that is related to volcano loading enhances permeability in the lithosphere beneath it (Hieronymus and Bercovici, 1999). Parametrized models of damage-enhanced lithospheric permeability predict volcano chains to form either from a plate moving over a hot-spot-like source of magma (Hieronymus and Bercovici, 1999) or without a hot-spot source, but with nonlithostatic, horizontal tension in the lithosphere related to plate tectonics (Hieronymus and Bercovici, 2000) (**Figure 25**). The latter result provides another

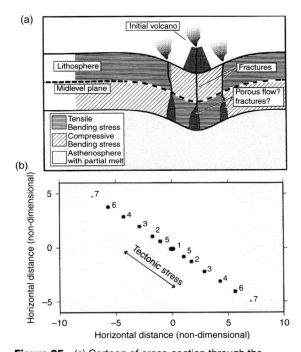

**Figure 25** (a) Cartoon of cross-section through the lithosphere, which is being flexed by a growing volcano. Nonlithostatic stresses due to bending are shaded dark for tension and light for compression. The asthenosphere is assumed to be partially molten. Damage-enhanced permeability is high beneath the volcano where bending stresses are largest. New volcanoes are predicted to form where the tension in the upper half of the plate is maximum. Remote, tectonic stresses can cause horizontal, nonlithostatic tension, which is depth independent and can cause new eruptions to occur in lines. (b) Predicted pattern of volcanoes resulting from flexural stresses and nonlithostatic, tectonic stress that is tensile parallel to the volcano lineament (arrows). The location of volcano number 1 was imposed and subsequent volcanoes formed in numerical order as labeled. The model predicts new volcanism to propagate in both directions away from the initial volcano. From Hieronymus CF and Bercovici D (2000) Non-hotspot formation of volcanic chains; control of tectonic and flexural stresses on magma transport. *Earth and Planetary Science Letters* 181: 539–554.

plausible mechanism for forming lineaments without a monotonic age progression, but instead, with testable progressions of decreasing volcano age in both directions along the lineament. Another test is provided by the relationship between tectonic stress and the direction of volcano propagation. Models based on fracture mechanics predict propagation of volcanism perpendicular to the tensile direction, while Hieronymus and Bercovici (2000) predict volcanism to propagate along the tensile direction by the interaction of point-load flexural and background stresses (**Figure 25**).

### 9.3.8 Geochemistry of Hotspots and Melting Anomalies Vs MORB

The geochemical differences between basalts erupted at hot spots and melting anomalies versus MORB provide a vital constraint on the causal mechanisms (*see also* Chapter 10). We have focused on $^{87}Sr/^{86}Sr$, $^{206}Pb/^{204}Pb$, $^{3}He/^{4}He$ ratios because they are key to tracing at least five flavors of mantle materials with different, time-averaged, chemical histories (Hart *et al.*, 1992; Hanan and Graham, 1996; Zindler and Hart, 1986). Lavas with low $^{3}He/^{4}He$ and minimal $^{87}Sr/^{86}Sr$ and $^{206}Pb/^{204}Pb$ probably come from depleted mantle material (DM), referring to a long-term ($>10^{2}$ My) depletion in incompatible elements. Enriched (EM1 or EM2, i.e., high $^{87}Sr/^{86}Sr$) and HIMU (i.e., high $^{206}Pb/^{204}Pb$) mantle are thought to be influenced by subducted and recycled material, the former by old oceanic sediments or metasomatized lithosphere, and the latter perhaps by oceanic crust that has been hydrothermally altered and then devolatized during subduction (e.g., Cohen and O'Nions, 1982; Hofmann and White, 1982; Hart *et al.*, 1992; Zindler and Hart, 1986; Hofmann, 1997). High $^{3}He/^{4}He$, moderately low $^{87}Sr/^{86}Sr$, and intermediate-to-high $^{206}Pb/^{204}Pb$ compositions mark the fifth geochemical material; it has been identified with various names and we will refer to it as FOZO (for FOcal ZOne (Hart *et al.*, 1992)). Its origin, however, is not well understood. The 'standard' hypothesis is that $^{3}He/^{4}He$ measures the primordial nature of the source material with low ratios reflecting material that has experienced substantial degassing of primordial $^{3}He$ and the high ratios indicating relatively undegassed mantle. But more recent evidence weakens the standard hypothesis and instead suggests that FOZO, in fact, has been depleted in highly incompatible elements. In this scenario, the high $^{3}He/^{4}He$ ratio could reflect a low $^{4}He$ concentration as a result of low U and Th

content (Coltice and Ricard, 1999; Stuart *et al.*, 2003; Meibom *et al.*, 2005; Parman *et al.*, 2005).

The key issue is that MORB appears to be heavily influenced by DM and minimally influenced by subducted materials and FOZO, whereas hot spots and melting anomalies appear to be influenced substantially by all five components (albeit to different degrees for different volcano groups). One possibility is that the pressure/temperature dependence of mantle viscosity and mineralogy, as well as density differences between the different mantle materials promotes large-scale layering in mantle geochemistry. DM is likely to be compositionally light and may tend to concentrate in the upper mantle where it is sampled by mid-ocean ridge magmatism. Mantle plumes, which feed hot spots, rise from deeper levels in the mantle and incorporate the other materials in addition to DM.

The formation of the different geochemical components, as well as the possibility of large-scale layering in the presence of vigorous, whole-mantle convection is actively being studied with both computational and laboratory methods (e.g., Christensen and Hofmann, 1994; van Keken and Ballentine, 1999; Davaille, 1999; Ferrachat and Ricard, 2001; Xie and Tackley, 2004). On the one hand, such studies have successfully predicted the formation of deep layers that are concentrated in dense subducted mafic material, which, if entrained in upwelling plumes could explain some of the elevated $^{206}Pb/^{204}Pb$ ratios in hot-spot basalts. On the other hand, it remains to be seen how it is possible to generate and physically separate two (or more) different components that may be depleted of mafic components: one with low $^{3}He/^{4}He$ that is prominent in MORB and the other with high-$^{3}He/^{4}He$ that is weakly expressed in MORB and more prominently expressed in some hot-spot lavas. Another challenge is to reconcile the geochemical character of hot spots/melting anomalies with the possibility that some could be caused by plumes originating very deep in the mantle, some by plumes originating from shallower in the mantle, and others from shallow mechanisms completely unrelated to plumes. Finally, the small heat flux of mantle plumes implied by observations of swell buoyancy flux (Section 9.3.4.4), as well as constraints on excess temperatures of plumes in the upper mantle (Zhong, 2006) require that incompatible-element-rich materials are present both above and below the source layer of most mantle plumes.

One key process to consider in addressing the above issues is the chemical extraction of the different components by melting (Phipps Morgan, 1999). Geochemical evidence indicates that heterogeneity is likely to be present over a range of spatial scales, including scales much smaller than the size of upper-mantle melting zones (e.g., Niu *et al.*, 1996; Phipps Morgan, 1999; Saal *et al.*, 1998; Reiners, 2002; Salters and Dick, 2002; Stracke *et al.*, 2003; Kogiso *et al.*, 2004; Ellam and Stuart, 2004). The likelihood that different materials begin melting at different depths for a given mantle temperature makes it probable that differences in lithospheric thickness, as well as the rate of mantle flow through the melting zone can influence the relative proportions of incompatible elements that are extracted from the different components. Mid-ocean ridge magmatism could most substantially melt the refractory component (DM) because the thin lithosphere allows for the greatest amount of decompression melting. Magmatism away from mid-ocean ridges could be less influenced by DM and proportionally more by the other, perhaps less refractory components owing to the thicker lithosphere. Melting of a buoyant upwelling – like a mantle plume – can also emphasize the least refractory components, even beneath relatively thin lithosphere, because the buoyancy pushes mantle through the deepest portions of the melting zone more rapidly than in the shallowest portions (Ito and Mahoney, 2005, 2006). Unraveling the above clues provided by magma geochemistry will thus require integrated geochemical, geophysical, and geodynamic investigations of the character of the mantle source, as well as the mantle convection, melting, and melt extraction.

## 9.4  Conclusions and Outlook

The rich diversity of observations and dynamical behavior makes it likely that a variety of mechanisms cause hot spots and melting anomalies. Our future task is to design observational and theoretical tests of which mechanisms can and cannot explain individual systems. The variety of observational techniques will lead to improved constraints. These observations include volume and durations of volcanism, the nature and depth extent of mantle seismic anomalies, and presence or absence of four key characteristics: swells, age progressions, connections with LIPS, and geochemical distinctions from MORB. We will close this chapter with a short outlook in the form of a wish list for future work.

1. Origin of melting anomalies: We need to explore different mechanisms of mantle flow and melting that include increasingly realistic dynamics (non-Newtonian, time-dependent, and 3-D) and lithologic variability, and we need to test model predictions against observed volumes and durations of volcanism. We need to do this to understand the relative importance of temperature and composition, which are both coupled to the upper-mantle dynamics. Example processes that are relatively less well understood include sublithospheric convection, fertile mantle melting, viscous fingering in the asthenosphere, as well as the role of the lithosphere in controlling magmatism on the surface.

2. Origin of swells: Previous work has shown that plumes can explain many observations of swells but future work is needed to explore whether nonplume mechanisms can cause swells, and in particularly, how they vary with plate speed and buoyancy flux. We are also faced by explaining the presence of melting anomalies without swells, such as the prominent Canaries and Madeira hot spots. Perhaps such systems are dominated by fertile mantle melting.

3. Age progressions and lack thereof: For long-lived age progressions, future challenges involve reducing observational uncertainties with further geophysical studies, more accurate and precise dating, and improving geodynamic models of mantle flow and evolution. The latter will require improved methods of tracking mantle flow further into the past, and in defining the ranges of allowable mantle density and viscosity structures from seismology and mineral physics. For hot spots with short-lived age progressions models will need to consider the possibility of them originating from upwellings from boundary layers above the CMB or nonplume sources. For hot spots without simple age progressions, plume may be unlikely and thus other mechanisms should be explored. In fact some of these mechanisms (e.g., propagating fracture) may predict age progressions and quantitative models are needed to explore these possibilities.

4. LIPs are the most dramatic but potentially the least understood dynamical processes on the planet. It is critical to evaluate how the evolving plume theory can self-consistently address the formation of some LIPs and to which cases alternatives are required.

5. Geochemistry: Differences in isotope geochemistry and in particular the distinctions from MORB of most OIBs require a chemically layered mantle, differences in melting of a nonlayered, heterogeneous mantle, or some combination of above.

6. Seismology: The key challenge is to confidently resolve if any hot spots have seismic anomalies extending into the lower mantle and if so, which ones do and do not. Such information will be critical for evaluating plume versus nonplume mechanisms. Combined with geochemical observations, such information could be the key in addressing the possibility or nature of geochemical layering.

7. Integrated and interdisciplinary work: We need to meet our capabilities of simulating increasingly complex dynamic behaviors with increasing quality of geophysical and geochemical data.

## Acknowledgments

This work is partly supported by the National Science Foundation (OCE-0351234 and EAR-0440365 to Garrett Ito and EAR-0229962 to Peter van Keken).

## References

Acton GD and Gordon RG (1994) Paleomagnetic tests of Pacific plate reconstructions and implications for motion between hotspots. *Science* 263: 1246–1254.

Albers M and Christensen UR (1996) The excess temperature of plumes rising from the core–mantle boundary. *Geophysical Research Letters* 23: 3567–3570.

Albers M and Christensen UR (2001) Channeling of plume flow beneath mid-ocean ridges. *Earth and Planetary Science Letters* 187: 207–220.

Ali JR, Thompson GM, Zhou MF, and Song XY (2005) Emeishan large igneous province, SW China. *Lithos* 79: 475–489.

Allégre CJ, Birck JL, Capmas F, and Courtillot V (1999) Age of the Deccan traps using Re-187–Os-187 systematics. *Earth and Planetary Science Letters* 170: 197–204.

Allen RM, Nolet G, Morgan WJ, *et al.* (1999a) The thin hot plume beneath Iceland. *Geophysical Journal International* 137: 51–63.

Allen RM, Nolet G, Morgan WJ, *et al.* (1999b) The thin hot plume beneath Iceland. *Geophysical Journal International* 137: 51–63.

Allen RM, Nolet G, Morgan WJ, *et al.* (2002) Imaging the mantle beneath Iceland using integrated seismological techniques. *Journal of Geophysical Research* 107: 10.1029/2001JB000595.

Allen RM and Tromp J (2005) Resolution of regional seismic models: Squeezing the Iceland anomaly. *Geophysical Journal International* 161: 373–386.

Anderson DL (1989) *Theory of the Earth.* Brookline Village, MA: Blackwell.

Anderson DL (1994a) The sublithospheric mantle as the source of continental flood basalts: the case against the continental
lithosphere and plume head reservoirs. *Earth and Planetary Science Letters* 123: 269–280.

Anderson DL (1994b) Superplumes or supercontinents? *Geology* 22: 39–42.

Anderson DL (2005) Large igneous provinces, delamination, and fertile mantle. *Elements* 1(5): 271–275.

Anguita F and Hernan F (2000) The Canary Islands origin; a unifying model. *Journal of Volcanology and Geothermal Research* 103: 1–26.

Antretter M, Riisager P, Hall S, Zhao X, and Steinberger B (2004) Modelled paleolatitudes for the Louisville hot spot and the Ontong Java Plateau. *GSA, Special Publication* 299: 21–30.

Aoki I and Takahashi E (2004) Density for MORB eclogite in the upper mantle. *Physics of the Earth and Planetary Interiors* 143–144: 129–143.

Asimow P and Langmuir CH (2003) The importance of water to oceanic melting regimes. *Nature* 421: 815–820.

Ayalew D, Barbey P, Marty B, Reisberg L, Yirgu G, and Pik R (2002) Source, genesis, and timing of giant ignimbrite deposits associated with Ethiopian continental flood basalts. *Geochimica et Cosmochimica Acta* 66: 1429–1448.

Bastow ID, Stuart GW, Kendall JM, and Ebinger CJ (2005) Upper-mantle seismic structure in a region of incipient continental breakup: Northern Ethiopian rift. *Geophysical Journal International* 162: 479–493.

Batiza R (1982) Abundances, distribution and sizes of volcanoes in the Pacific Ocean and implications for the origin of non-hotspot volcanoes. *Earth and Planetary Science Letters* 60: 195–206.

Benoit MH, Nyblade AA, and Vandecar JC (2006) Upper mantle P-wave speed variations beneath Ethiopia and the origin of the Afar hotspot. *Geology* 34: 329–332.

Bercovici D and Kelly A (1997) The non-linear initiation of diapirs and plume heads. *Physics of the Earth and Planetary Interiors* 101: 119–130.

Bercovici D and Mahoney J (1994) Double flood basalts and plume head separation at the 660-kilometer discontinuity. *Science* 266: 1367–1369.

Bianco TA, Ito G, Becker JM, and Garcia MO (2005) Secondary Hawaiian volcanism formed by flexural arch decompression. *Geochemistry Geophysics Geosystems* 6: Q08009 (doi:10.1029/2005GC000945).

Bijwaard H and Spakman W (1999) Tomographic evidence for a narrow whole mantle plume below Iceland. *Earth and Planetary Science Letters* 166: 121–126.

Bijwaard H, Spakman W, and Engdahl ER (1998) Closing the gap between regional and global travel time tomography. *Journal of Geophysical Research* 103: 30055–30078.

Bjarnason IT, Silver PG, Rumpker G, and Solomon SC (2002) Shear wave splitting across the Iceland hot spot: Results from the ICEMELT experiment. *Journal of Geophysical Research* 107: 2382 (doi:10.1029/2001JB000916).

Boehler R (2000) High-pressure experiments and the phase diagram of lower mantle and core materials. *Reviews of Geophysics* 38: 221–245.

Bonneville A, Dosso L, and Hildenbrand A (2006) Temporal evolution and geochemical variability of the South Pacific superplume activity. *Earth and Planetary Science Letters* 244: 251–269.

Boschi L, Becker TW, Soldati G, and Dziewonski AM (2006) On the relevance of Born theory in global seismic tomography. *Geophysical Research Letters* 33: L06302.

Brandenburg JP and van Keken PE (in press) Preservation of oceanic crust in a vigorously convecting mantle. *Journal of Geophysical Research.*

Breger L, Romanowicz B, and Ng C (2001) The Pacific plume as seen by S, ScS, and SKS. *Geophysical Research Letters* 28: 1859–1862.

Brunet D and Yuen DA (2000) Mantle plumes pinched in the transition zone. *Earth and Planetary Science Letters* 178: 13–27.

Buck WR (1986) Small-scale convection induced by passive-rifting: The cause for uplift of rift shoulders. *Earth Planetary and Science Letters* 77: 362–372.

Buck WR and Parmentier EM (1986) Convection beneath young oceanic lithosphere: Implications for thermal structure and gravity. *Journal of Geophysical Research* 91: 1961–1974.

Burke K (1996) The African Plate. *South African Journal of Geology* 99: 339–409.

Campbell IH and Griffiths RW (1990) Implications of mantle plume structure for the evolution of flood basalts. *Earth and Planetary Science Letters* 99: 79–93.

Canales JP, Ito G, Detrick RS, and Sinton J (2002) Crustal thickness along the western Galapagos spreading center and the compensation of the Galapagos hotspot swell. *Earth and Planetary Science Letters* 203: 311–327.

Cannat M, Brias A, Deplus C, et al. (1999) Mid-Atlantic Ridge-Azores hotspot interactions: Along-axis migration of a hotspot-derived event of enhanced magmatism 10 to 4 Ma ago. *Earth and Planetary Science Letters* 173: 257–269.

Caroff M, Guillou H, Lamiaux M, Maury RC, Guille G, and Cotten J (1999) Assimilation of ocean crust by hawaiitic and mugearitic magmas: And example from Eiao (Marquesas). *Lithos* 46: 235–258.

Chambers K, Woodhouse JH, and Deuss A (2005) Topography of the 410-km discontinuity from PP and SS precursors. *Earth and Planetary Science Letters* 235: 610–622.

Chang HK, Kowsmann RO, Figueiredo AMF, and Bender AA (1992) Tectonics And stratigraphy of the East Brazil Rift system – An overview. *Tectonophysics* 213: 97–138.

Chazey WJ, III and Neal CR (2004) Large igneous province magma petrogenesis from source to surface: Platinum-group element evidence from Ontong Java Plateau basalts recovered during ODP legs 130 and 192. In: Fitton JG, Mahoney JJ, Wallace PJ, and Saunders AD (eds.) *Origin and Evolution of the Ontong Java Plateau*, pp. 449–484. London: Geological Society.

Christensen U (1984) Instability of a hot boundary-layer and initiation of thermo-chemical plumes. *Annales Geophysicae* 2: 311–319.

Christensen UR and Hofmann AW (1994) Segregation of subducted oceanic crust and the convecting mantle. *Journal of Geophysical Research* 99: 19867–19884.

Clague DA and Dalrymple GB (1987) The Hawaiian-Emperor volcanic chain. In: Decker RW, Wright TL, and Stauffer PH (eds.) *Volcanism in Hawaii*, pp. 5–54. Honolulu: USGC.

Clift PD, Carter A, and Hurford AJ (1998) The erosional and uplift history of NE Atlantic passive margins; constrains on a passing plume. *Journal of the Geological Society of London* 155: 787–800.

Clift PD and Turner J (1995) Dynamic support by the Icelandic Plume and vertical tectonics of the Northeast Atlantic continental margins. *Journal of Geophysical Research* 100: 24473–24486.

Clouard V and Bonneville A (2001) How many Pacific hotspots are fed by deep-mantle plumes? *Geology* 29: 695–698.

Clouard V and Bonneville A (2005) Ages of Seamounts, islands, and plateaus on the Pacific plate. In: Foulger G, Natland JH, Presnall DC, and Anderson DL (eds.) *Plumes, Plates, and Paradigms*, pp. 71–90. Boulder, CO: GSA.

Clouard V, Bonneville A, and Gillot P-Y (2003) The Tarava Seamounts; a newly characterized hotspot chain on the South Pacific Superswell. *Earth and Planetary Science Letters* 207: 117–130.

COFFIN MF (1992) Emplacement and subsidence of Indian Ocean plateaus and submarine ridges. In: Duncan RA, Rea DK, Kidd RB, Rad UV and Weissel JK (eds.) *Synthesis of Results from Scientific Drilling in the Indian Ocean, Geophysical Monograph 70*; pp. 115–125. Washington, DC: American Geophysical Union.

Coffin MF and Eldholm O (1993) Scratching the surface: Estimating dimensions of large igneous provinces. *Geology* 21: 515–518.

Coffin MF and Eldholm O (1994) Large igneous provinces: Crustal structure, dimensions, and external consequences. *Reviews of Geophysics* 32: 1–36.

Coffin MF, Pringle MS, Duncan RA, et al. (2002) Kerguelen hotspot magma output since 130 Ma. *Journal Of Petrology* 43: 1121–1139.

Cohen RS and O'nions RK (1982) Identification of recycled continental material in the mantle from Sr, Nd, and Pb isotope investigations. *Earth and Planetary Science Letters* 61: 73–84.

Collins JA, Vernon FL, Orcutt JA, and Stephen RA (2002) Upper mantle structure beneath the Hawaiian Swell; constraints from the ocean seismic network pilot experiment. *Geophysical Research Letters* 29: 1522 (doi:10.1029/2001GL013302).

Coltice N and Ricard Y (1999) Geochemical observations and one layer mantle convection. *Earth and Planetary Science Letters* 174: 125–137.

Condie KC (2001) *Mantle Plumes and Their Record in Earth History*. Cambridge: Cambridge University Press.

Constable S and Heinson G (2004) Hawaiian hot-spot swell structure from seafloor MT sounding. *Tectonophysics* 389: 111–124.

Cordery MJ, Davies GF, and Cambell IH (1997) Genesis of flood basalts from eclogite-bearing mantle plumes. *Journal of Geophysical Research* 102: 20179–20197.

Courtillot V, Davaille A, Besse J, and Stock J (2003) Three distinct types of hotspots in the Earth's mantle. *Earth and Planetary Science Letters* 205: 295–308.

Courtillot V, Gallet Y, Rocchia R, et al. (2000) Cosmic markers, Ar-40/Ar-39 dating and paleomagnetism of the KT sections in the Anjar Area of the Deccan large igneous province. *Earth and Planetary Science Letters* 182: 137–156.

Courtillot VE and Renne PR (2003) On the ages of flood basalt events. *Comptes Rendus Geoscience* 335: 113–140.

Cox KG (1988) The Karoo province. In: Macdougall JD (ed.) *Continental Flood Basalts*, 35 p. Dordrecht: Kluwer Academic Publishers.

Creager KC and Jordan TH (1986) Slab penetration into the lower mantle below the Mariana and other island arcs of the Northwest Pacific. *Journal of Geophysical Research* 91: 3573–3589.

Crough ST (1978) Thermal origin of mid-plate hot-spot swells. *Geophysical Journal of the Royal Astronomical Society* 55: 451–469.

Crough ST (1983) Hotspot swells. *Annual Reviews of Earth and Planetary Science* 11: 165–193.

Cserepes L, Christensen UR, and Ribe NM (2000) Geoid height versus topography for a plume model of the Hawaiian swell. *Earth and Planetary Science Letters* 178: 29–38.

Cserepes L and Yuen DA (2000) On the possibility of a second kind of mantle plume. *Earth and Planetary Science Letters* 183: 61–71.

Cushman B, Sinton J, Ito G, and Dixon JE (2004) Glass compositions, plume-ridge interaction, and hydrous melting along the Galapagos spreading center, 90.5 degrees W to 98 degrees W. *Geochemistry, Geophysics, and Geosystems* 5: Q08E17 (doi:10.1029/2004GC000709).

D'acremont E, Leroy S, and Burov EB (2003) Numerical modelling of a mantle plume; the plume head-lithosphere interaction in the formation of an oceanic large igneous province. *Earth and Planetary Science Letters* 206: 379–396.

Dasgupta R and Hirschmann MM (2006) Melting in the Earth's deep upper mantle caused by carbon dioxide. *Nature* 440: 659–662.

Davaille A (1999) Simultaneous generation of hotspots and superswells by convection in a heterogeneous planetary mantle. *Nature* 402: 756–760.

Davaille A, Girard F, and Lebars M (2002) How to anchor hotspots in a convecting mantle? *Earth and Planetary Science Letters* 203: 621–634.

Davaille A and Vatteville (2005) On the transient nature of mantle plumes. *Geophysical Research Letters* 32: (doi:1029/2005GL023029).

Davies GF (1988) Ocean bathymetry and mantle convection 1. Large-scale flow and hotspots. *Journal of Geophysical Research* 93: 10467–10480.

Davies GF (1995) Penetration of plates and plumes through the lower mantle transition zone. *Earth and Planetary Science Letters* 133: 507–516.

Davies GF (1998) Topography: A robust constraint on mantle fluxes. *Chemical Geology* 145: 479–489.

Davies GF (1999) *Dynamic Earth*. Cambridge, UK: Cambridge University Press.

Davis AS, Gray LB, Clague DA, and Hein JR (2002) The Line Islands revisited: New $^{40}Ar/^{39}Ar$ geochronologic evidence for episodes of volcanism due to lithospheric extension. *Geochemistry, Geophysics and Geosystems* 3(3): (doi:10.1029/2001GC000190).

De Hoop MV and van Der Hilst RD (2005) On sensitivity kernels for 'wave-equation' transmission tomography. *Geophysical Journal International* 160: 621–633.

Delaughter JE, Stein CA, and Stein S (2005) Hotspots: A view from the swells. *GSA, Special Publication* 388: 881.

Depaolo DJ and Manga M (2003) Deep origin of hotspots – The mantle plume model. *Science* 300: 920–921.

Desonie DL and Duncan RA (1990) The Cobb–Eickelberg seamount chain: hotspot volcanism with mid-ocean ridge basalt affinity. *Journal of Geophysical Research, B, Solid Earth and Planets* 95: 12697–12711.

Detrick RS and Crough ST (1978) Island subsidence, hot spots, and lithospheric thinning. *Journal of Geophysical Research* 83: 1236–1244.

Devey CW and Haase KM (2004) The sources for hotspot volcanism in the South Pacific Ocean. In: Hekinian R, Stoffers P, and Cheminée J-L (eds.) *Oceanic Hotspots*, pp. 253–280. New York: Springer.

Divenere V and Kent DV (1999) Are the Pacific and Indo-Atlantic hotspots fixed? Testing the plate circuit through Antarctica. *Earth and Planetary Science Letters* 170: 105–117.

Dodson A, Kennedy BM, and Depaolo DJ (1997) Helium and neon isotopes in the Imnaha Basalt, Columbia River Basalt Group; evidence for a Yellowstone plume source. *Earth and Planetary Science Letters* 150: 443–451.

Doucet S, Weis D, Scoates JS, Debaille V, and Giret A (2004) Geochemical and Hf-Pb-Sr-Nd isotopic constraints on the origin of the Amsterdam-St. Paul (Indian Ocean) hotspot basalts. *Earth and Planetary Science Letters* 218: 179–195.

Du Z, Vinnik LP, and Foulger GR (2006) Evidence from P-to-S mantle converted waves for a flat '660-km' discontinuity beneath Iceland. *Earth and Planetary Science Letters* 241: 271–280.

Dueker KG and Sheehan AF (1997) Mantle discontinuity structure from midpoint stacks of converted P to S waves across the Yellowstone hotspot track. *Journal of Geophysical Research, B, Solid Earth and Planets* 102: 8313–8327.

Duncan RA (1978) Geochronology of basalts from the Ninetyeast Ridge and continental dispersion in the Eastern Indian Ocean. *Journal of Volcanology and Geothermal Research* 4: 283–305.

Duncan RA (1984) Age progressive volcanism in the New England Seamounts and the opening of the Central Atlantic Ocean. *Journal of Geophysical Research* 89: 9980–9990.

Duncan RA (1991) Age distribution of volcanism along aseismic ridges in the eastern Indian Ocean. *Proceedings of the Ocean Drilling Program, Scientific results*. College Station, TX, USA, Ocean Drilling Program.

Duncan RA (2002) A time frame for construction of the Kerguelen Plateau and Broken Ridge. *Journal of Petrology* 43: 1109–1119.

Duncan RA, Backman J, Peterson LC, et al. (1990) The volcanic record of the Reunion hotspot. *Proceedings of the Ocean Drilling Program, Scientific Results* 115: 3–10.

Duncan RA and Clague DA (1985) Pacific plate motion recorded by linear volcanic chains. In: Nairn AEM, Stehli FG, and Uyeda S (eds.) *The Ocean Basins and Margins*, pp. 89–121. New York: Plenum Press.

Duncan RA and Hargraves RB (1984) Caribbean region in the mantle reference frame. In: Bonini W, et al. (ed.) *The Caribbean-South American Plate Boundary and Regional Tectonic*, pp. 81–93. Vancouver: Geological Society of America.

Duncan RA and Keller R (2004) Radiometric ages for basement rocks from the Emperor Seamounts, ODP Leg 197. *Geochemistry, Geophysics and Geosystems* 5: Q08L03 (doi:10.1029/2004GC000704).

Duncan RA and Richards MA (1991) Hotspots, mantle plumes, flood basalts, and true polar wander. *Reviews of Geophysics* 29: 31–50.

Dyment J (1998) Evolution of the Carlsberg Ridge between 60 and 45 Ma: Ridge propagation, spreading asymmetry, and the Deccan-reunion hotspot. *Journal of Geophysical Research* 103: 24067–24084.

Dziewonski AM and Anderson DL (1981) Preliminary reference Earth model. *Physics of the Earth and Planetary Interiors* 25: 297–356.

Dziewonski AM and Woodhouse JH (1987) Global images of the Earth's interior. *Science* 236: 37–48.

Ebinger CJ, Bechtel TD, Forsyth DW, and Bowin CO (1989) Effective elastic plate thickness beneath the East African and Afar Plateaus and dynamic compensation of the uplifts. *Journal of Geophysical Research* 94: 2883–2901.

Eiler JM, Schiano P, Kitchen N, and Stolper EM (2000) Oxygen isotope evidence for the origin of chemical variations in the sources of mid-ocean ridge basalts. *Nature* 403: 530–534.

Elkins-Tanton LT and Hager BH (2005) Giant meteoroid impacts can cause volcanism. *Earth and Planetary Science Letters* 239: 219–232.

Elkins-Tanton LT, Hager BH, and Grove TL (2004) Magmatic effects of the lunar heavy bombardment. *Earth and Planetary Science Letters* 222: 17–27.

Ellam RM and Stuart FM (2004) Coherent He-Nd-Sr isotope trends in high $^3He/^4He$ basalts: Implications for a common reservoir, mantle heterogeneity and convection. *Earth and Planetary Science Letters* 228: 511–523.

Elliot DH and Fleming TH (2004) Occurrence and dispersal of magmas in the Jurassic Ferrar Large igneous province, Antarctica. *Gondwana Research* 7: 223–237.

Emerick CM and Duncan RA (1982) Age progressive volcanism in the Comores Archipelago, western Indian Ocean and implications for Somali plate tectonics. *Earth and Planetary Science Letters* 60: 415–428.

Encarnacion J, Fleming TH, Elliot DH, and Eales HV (1996) Synchronous emplacement of Ferrar and Karoo dolerites and the early breakup of Gondwana. *Geology* 24: 535–538.

Ernst RE and Baragar WRA (1992) Evidence from magnetic fabric for the flow pattern of magma in the mackenzie giant radiating dyke swarm. *Nature* 356: 511–513.

Ernst RE and Buchan KL (eds.) (2001) *Mantle Plumes: Their Identification Through Time*. Boulder, CO: Geological Society of America.

Ernst RE and Buchan KL (2003) Recognizing mantle plumes in the geological record. *Annual Review of Earth and Planetary Sciences* 31: 469–523.

Ewart A, Marsh JS, Milner SC, *et al.* (2004) Petrology and geochemistry of Early Cretaceous bimodal continental flood volcanism of the NW Etendeka, Namibia; part 1, introduction, mafic lavas and re-evaluation of mantle source components. *Journal of Petrology* 45: 59–105.

Farnetani CG (1997) Excess temperature of mantle plumes: The role of chemical stratification of D″. *Geophysical Research Letters* 24: 1583–1586.

Farnetani CG and Richards MA (1995) Thermal entrainment and melting in mantle plumes. *Earth and Planetary Science Letters* 133: 251–267.

Farnetani CG and Samuel H (2005) Beyond the thermal plume hypothesis. *Geophysical Research Letters* 32: L07311 (doi:10.1029/2005GL022360).

Fee D and Dueker K (2004) Mantle transition zone topography and structure beneath the Yellowstone hotspot. *Geophysical Research Letters* 31: L18603 (doi:10.1029/2004GL020636).

Feigenson MD, Carr MJ, Maharaj SV, Juliano S, and Bolge LL (2004) Lead isotope composition of Central American volcanoes; influence of the Galapagos plume. *Geochemistry, Geophysics and Geosystems* 5: Q06001 (doi:10.1029/2003GC000621).

Feighner MA and Richards MA (1995) The fluid dynamics of plume-ridge and plume-plate interactions: An experimental investigation. *Earth and Planetary Science Letters* 129: 171–182.

Ferrachat S and Ricard Y (2001) Mixing properties in the Earth's mantle: Effects of viscosity stratification and of oceanic crust segregation. *Geochemistry, Geophysics and Geosystems* 2 (doi:10.1029/2000GC000092).

Fitton JG and Godard M (2004) Origin and evolution of magmas on the Ontong Java Plateau. In: Fitton JG, Mahoney JJ, Wallace PJ, and Saunders AD (eds.) *Origin and Evolution of the Ontong Java Plateau*, pp. 151–178. London: Geological Society.

Fitton JG, Mahoney JJ, Wallace PJ, and Saunders AD (eds.) (2004) *Origin and Evolution of the Ontong Java Plateau*. London: Geological Society.

Fodor RV and Hanan BB (2000) Geochemical evidence for the Trindade hotspot trace; Columbia Seamount ankaramite. *Lithos* 51: 293–304.

Fontaine FR, Hooft EEE, Burkett PG, Toomey DR, Solomon SC, and Silver PG (2005) Shear-wave splitting beneath the Galapagos archipelago. *Geophysical Research Letters* 32: 121308.

Foulger GR, Natland JH, and Anderson DL (2005) A source for Icelandic magmas in remelted Iapetus crust. *Journal of Volcanology and Geothermal Research* 141: 23–44.

Foulger GR, Pritchard MJ, Julian BR, *et al.* (2001) Seismic tomography shows that upwelling beneath Iceland is confined to the upper mantle. *Geophysical Journal International* 146: 504–530.

Frey FA, Coffin MF, Wallace PJ, *et al.* (2000) Origin and evolution of a submarine large igneous province; the Kerguelen Plateau and Broken Ridge, southern Indian Ocean. *Earth and Planetary Science Letters* 176: 73–89.

Frey FA, Weis D, Borisova AY, and XU G (2002) Involvement of continental crust in the formation of the Cretaceous Kerguelen Plateau: New perspectives from ODP Leg 120 sites. *Journal Of Petrology* 43: 1207–1239.

Fukao Y, Koyama T, Obayashi M, and Utada H (2004) Trans-Pacific temperature field in the mantle transition region derived from seismic and electromagnetic tomography. *Earth and Planetary Science Letters* 217: 425–434.

Gans KD, Wilson DS, and Macdonald KC (2003) Pacific Plate gravity lineaments: Diffuse extension or thermal contraction?

*Geochemistry, Geophysics and Geosystems* 4: 1074 (doi:10.1029/2002GC000465).

Gashawbeza EM, Klemperer SL, Nyblade AA, Walker KT, and Keranen KM (2004) Shear-wave splitting in Ethiopia: Precambrian mantle anisotropy locally modified by Neogene rifting. *Geophysical Research Letters* 31: L18602 (doi:10.1029/2004GL020471).

Geldmacher J, Hoernle K, van Der Bogaard P, Duggen S, and Werner R (2005) New Ar-40/Ar-39 age and geochemical data from seamounts in the Canary and Madeira volcanic provinces: Support for the mantle plume hypothesis. *Earth and Planetary Science Letters* 237: 85–101.

Gente P, Dyment J, Maia M, and Goslin J (2003) Interaction between the Mid-Atlantic Ridge and the Azores hot spot during the last 85 Myr: Emplacement and rifting of the hot spot-derived plateaus. *Geochemistry Geophysics Geosystems* 4: 8514 (doi:10.1029/2003GC000527).

Georgen JE, Lin J, and Dick HJB (2001) Evidence from gravity anomalies for interactions of the Marion and Bouvet hotspots with the Southwest Indian Ridge: Effects of transform offsets. *Earth and Planetary Science Letters* 187: 283–300.

Gibson SA, Thompson RN, Weska R, Dickin AP, and Leonardos OH (1997) Late Cretaceous rift-related upwelling and melting of the Trindade starting mantle plume head beneath western Brazil. *Contributions to Mineralogy and Petrology* 126: 303–314.

GILL RCO, Holm PM, and Nielsen TFD (1995) Was a short-lived Baffin Bay plume active prior to initiation of the present Icelandic plume? Clues from the high-Mg picrites of West Greenland. *Lithos* 34: 27–39.

Gill RCO, Pedersen AK, and Larsen JG (1992) Tertiary picrites from West Greenland: Melting at the periphery of a plume? In: Storey BC, Alabaster T, and Pankhurst RJ (eds.) *Magmatism and the Causes of Continental Break-Up*, pp. 335–348. London: Geological Society.

Gladczenko TP, Coffin MF, and Eldholm O (1997) Crustal structure of the Ontong Java Plateau: Modeling of new gravity and existing seismic data. *Journal of Geophysical Research* 102: 22711–22729.

Goes S, Spakman W, and Bijwaard H (1999) A lower mantle source for central European volcanism. *Science* 286: 1928–1931.

Gomer BM and Okal EA (2003) Multiple ScS probing of the Ontong Java Plateau. *Physics of the Earth and Planetary Interiors* 138: 317–331.

Gonnermann HM, Jellinek AM, Richards MA, and Manga M (2004) Modulation of mantle plumes and heat flow at the core mantle boundary by plate-scale flow: Results from laboratory experiments. *Earth and Planetary Science Letters* 226: 53–67.

Graham DW, Johnson KTM, Priebe LD, and Lupton JE (1999) Hotspot-ridge interaction along the Southeast Indian Ridge near Amsterdam and St. Paul islands; helium isotope evidence. *Earth and Planetary Science Letters* 167: 297–310.

Grand SP (1994) Mantle shear structure beneath the Americas and surrounding oceans. *Journal of Geophysical Research* 99: 11591–11621.

Grand SP (2002) Mantle shear-wave tomography and the fate of subducted slabs. *Philosophical Transactions of the Royal Society of London Series A, Mathematical Physical and Engineering Sciences* 360: 2475–2491.

Grand SP, van Der Hilst RD, and Widiyantoro S (1997) Global seismic tomography: A snapshot of convection in the mantle. *GSA Today* 7: 1–7.

Green DH, Falloon TJ, Eggins SM, and Yaxley GM (2001) Primary magmas and mantle temperatures. *European Journal of Mineralogy* 13: 437–451.

Griffiths RW and Campbell IH (1990) Stirring and structure in mantle starting plumes. *Earth and Planetary Science Letters* 99: 66–78.

Gripp AE and Gordon RG (2002) Young tracks of hotspots and current plate velocities. *Geophysical Journal International* 150: 321–361.

Guillou H, Carracedo JC, Paris R, and Torrado FJP (2004) Implications for the early shield-stage evolution of Tenerife from K/Ar ages and magnetic stratigraphy. *Earth and Planetary Science Letters* 222: 599–614.

Hager BH and Clayton RW (1989) Constraints on the structure of mantle convection using seismic observations, flow models, and the geoid. In: Peltier WR (ed.) *Mantle Convection*, pp. 98–201. New York: Gordon and Breach.

Hales TC, Abt DL, Humphreys ED, and Roering JU (2005) A lithospheric instability origin for Columbia River flood basalts and Wallowa Mountains uplift in northeast Oregon. *Nature* 438: 842–845.

Hames WE, Renne PR, and Ruppel C (2000) New evidence for geologically instantaneous emplacement of earliest Jurassic Central Atlantic magmatic province basalts on the North American margin. *Geology* 28: 859–862.

Hanan BB and Graham DW (1996) Lead and helium isotope evidence from oceanic basalts for a common deep source of mantle plumes. *Science* 272: 991–995.

Harada Y and Hamano Y (2000) Recent progress on the plate motion relative to hotspots. In: Richards MA, Gordon RG, and Vanderhilst RD (eds.) *AGU Geophysics Monograph Series: The History and Dynamics of Global Plate Motions*, pp. 327–338. Washington, DC: AGU.

Harmon N, Forsyth D, and Scheirer D (2007) Analysis of gravity and topography in the GLIMPSE study region: Isostatic compensation and uplift of the Sojourn and Hotu Matua ridge systems. *Journal of Geophysical Research* 111: B11406, doi: 10.1029/2005JB004071.

Harris RN and Mcnutt MK (2007) Heat flow on hotspot swells: Evidence for fluid flow. *Journal of Geophysical Research* 112(B3): B03407, doi: 10.1029/2006JB004299.

Hart SR, Hauri EH, Oschmann LA, and Whitehead JA (1992) Mantle plumes and entrainment: Isotopic evidence. *Science* 256: 517–520.

Hart SR, Schilling J-G, and Powell JL (1973) Basalts from Iceland and along the Reykjanes Ridge: Sr isotope geochemistry. *Nature* 246: 104–107.

Hauri E (1996) Major-element variability in the Hawaiian mantle plume. *Nature* 382: 415–419.

Hawkesworth C, Kelley S, Turner S, Le Roex A, and Storey B (1999) Mantle processes during Gondwana break-up and dispersal. *Journal of African Earth Sciences* 28: 239–261.

Haxby WF and Weissel JK (1986) Evidence for small-scale mantle convection from Seasat altimeter data. *Journal of Geophysical Research* 91: 3507–3520.

He B, Xu YG, Chung SL, Xiao L, and Wang Y (2003) Sedimentary evidence for a rapid, kilometer-scale crustal doming prior to the eruption of the Emeishan flood basalts. *Earth and Planetary Science Letters* 213: 391–405.

Heintz M, Debayle E, and Vauchez A (2005) Upper mantle structure of the South American continent and neighboring oceans from surface wave tomography. *Tectonophysics* 406: 115–139.

Hekinian R, Stoffers P, and Cheminee J-L (eds.) (2004) *Oceanic Hotspots*. New York: Springer.

Helffrich G (2000) Topography of the transition zone seismic discontinuities. *Reviews of Geophysics* 38: 141–158.

Helffrich G (2002) Thermal variations in the mantle inferred from 660 km discontinuity topography and tomographic wave speed variations. *Geophysical Journal International* 151: 935–943.

Helmberger DV, Wen L, and Ding X (1998) Seismic evidence that the source of the Iceland hotspot lies at the core–mantle boundary. *Nature* 396: 248–251.

Hendrie DB, Kusznir NJ, Morley CK, and Ebinger CJ (1994) Cenozoic extension in Northern Kenya – A quantitative model of rift basin development in the Turkana Region. *Tectonophysics* 236: 409–438.

Herzberg C (2004a) Geodynamic information in peridotite petrology. *Journal of Petrology* 45: 2507–2530.

Herzberg C (2004b) Partial melting below the Ontong Java Plateau. In: Fitton JG, Mahoney JJ, Wallace PJ, and Saunders AD (eds.) *Origin and Evolution of the Ontong Java Plateau*, pp. 179–184. London: Geological Society.

Herzberg C, Asimow P, Arndt P, et al. (in press) Temperatures in ambient mantle and plumes: Constraints from basalts, picrites and komatiites. *Geochemistry, Geophysics and Geosystems* 8: Q02006, doi: 10.1029/2006GC001390.

Herzberg C and O'hara MJ (2002) Plume-associated ultramafic magmas of Phanerozoic age. *Journal of Petrology* 43: 1857–1883.

Hieronymus CF and Bercovici D (1999) Discrete alternating hotspot islands formed by interaction of magma transport and lithospheric flexure. *Nature* 397: 604–607.

Hieronymus CF and Bercovici D (2000) Non-hotspot formation of volcanic chains; control of tectonic and flexural stresses on magma transport. *Earth and Planetary Science Letters* 181: 539–554.

Hirose K and Fei Y (2002) Subsolidus and melting phase relations of basaltic composition in the uppermost lower mantle. *Geochimica etCosmochimica Acta* 66: 2099–2108.

Hirose K, Fei Y, Ma Y, and Ho-Kwang M (1999) The fate of subducted basaltic crust in the Earth's lower mantle. *Nature* 387: 53–56.

Hirschmann MM, Asimow PD, Ghiorso MS, and Stolper EM (1999) Calculation of peridotite partial melting from thermodynamic models of minerals and melts. III. Controls on isobaric melt production and the effect of water on melt production. *Journal of Petrology* 40: 831–851.

Hirth G and Kohlstedt DL (1996) Water in the oceanic upper mantle: Implications for rheology, melt extraction, and the evolution of the lithosphere. *Earth and Planetary Science Letters* 144: 93–108.

Hoernle K, Hauff F, and van Den Bogaard P (2004) 70 my. history (139-69 Ma) for the Caribbean large igneous province. *Geology* 32: 697–700.

Hoernle K, van Den Bogaard P, Werner R, Lissina BHF, Alvarado G, and Garbe-Schönberg D (2002) Missing history (16-71 Ma) of the Galápagos hotspot: Implications for the tectonic and biological evolution of the Americas. *Geology* 30: 795–798.

Hofmann AW (1997) Mantle geochemistry: The message from oceanic volcanism. *Nature* 385: 219–229.

Hofmann AW and White WM (1982) Mantle plumes from ancient oceanic crust. *Earth and Planetary Science Letters* 57: 421–436.

Hofmann C, Courtillot V, Feraud G, et al. (1997) Timing of the Ethiopian flood basalt event and implications for plume birth and global change. *Nature* 389: 838–841.

Hofmann C, Feraud G, and Courtillot V (2000) Ar-40/Ar-39 dating of mineral separates and whole rocks from the Western Ghats lava pile: Further constraints on duration and age of the Deccan traps. *Earth and Planetary Science Letters* 180: 13–27.

Holbrook WS and Kelemen PB (1993) Large igneous province on the US Atlantic margin and implications for magmatism during continental breakup. *Nature* 364: 433–436.

Holm PM, Gill RCO, Pedersen AK, et al. (1993) The Tertiary picrites of West Greenland; contributions from 'Icelandic' and other sources. *Earth and Planetary Science Letters* 115: 227–244.

Hooft EEE, Brandsdóttir B, Mjelde R, Shimamura H, and Murai Y (2006) Asymmetric plume-ridge interaction around Iceland: The Kolbeinsey Ridge Iceland Seismic Experiment. *Geophysics Geochemistry and Geosystems* 7: Q05015 (doi:10.1029/2005GC001123).

Hooft EEE, Toomey DR, and Solomon SC (2003) Anomalously thin transition zone beneath the Galapagos hotspot. *Earth and Planetary Science Letters* 216: 55–64.

Hooper PR (1997) The Columbia River flood basalt province: Current status. In: Mahoney JJ and Coffin MF (eds.) *Large Igneous Provinces: Continental, Oceanic, and Planetary Flood Volcanism*, pp. 1–27. Washington, DC: AGU.

Hooper PR, Binger GB, and Lees KR (2002) Ages of the Steens and Columbia River flood basalts and their relationship to extension-related calc-alkalic volcanism in eastern Oregon. *Geological Society of America Bulletin* 114: 43–50.

Hopper JR, Dahl-Jensen T, Holbrook WS, *et al.* (2003) Structure of the SE Greenland margin from seismic reflection and refraction data: Implications for nascent spreading center subsidence and asymmetric crustal accretion during North Atlantic opening. *Journal of Geophysical Research* 108: 2269.

Huang J, Zhong S, and van Hunen J (2003) Controls on sublithospheric small-scale convection. *Journal of Geophysical Research* 108: (doi:10.1029/2003JB002456).

Hung SH, Shen Y, and Chiao LY (2004) Imaging seismic velocity structure beneath the Iceland hot spot: A finite frequency approach. *Journal of Geophysical Research* 109: B08305 (doi:10.1029/2003JB002889).

Ingle S and Coffin MF (2004) Impact origin for the greater Ontong Java Plateau? *Earth and Planetary Science Letters* 218: 123–134.

Irifune T, Sekine T, Ringwood AE, and Hibberson WO (1986) The eclogite–garnetite transformations at high pressure and some geophysical implications. *Earth and Planetary Science Letters* 77: 245–256.

Ito E and Takahashi E (1989) Postspinel transformations in the system $Mg_2SiO_4$–$Fe_2SiO_4$ and some geophysical implications. *Journal of Geophysical Research* 94: 10637–10646.

Ito G (2001) Reykjanes 'V'-shaped ridges originating from a pulsing and dehydrating mantle plume. *Nature* 411: 681–684.

Ito G and Clift P (1998) Subsidence and growth of Pacific Cretaceous plateaus. *Earth and Planetary Science Letters* 161: 85–100.

Ito G and Lin J (1995) Oceanic spreading center-hotspot interactions: Constraints from along-isochron bathymetric and gravity anomalies. *Geology* 23: 657–660.

Ito G, Lin J, and Gable C (1997) Interaction of mantle plumes and migrating mid-ocean ridges: Implications for the Galápagos plume-ridge system. *Journal of Geophysical Research* 102: 15403–15417.

Ito G, Lin J, and Gable CW (1996) Dynamics of mantle flow and melting at a ridge-centered hotspot: Iceland and the Mid-Atlantic Ridge. *Earth and Planetary Science Letters* 144: 53–74.

Ito G, Lin J, and Graham D (2003) Observational and theoretical studies of the dynamics of mantle plume–mid-ocean ridge interaction. *Review of Geophysics* 41: 1017 (doi:10.1029/2002RG000117).

Ito G and Mahoney J (2006) Melting a high $^3He/^4He$ source in a heterogeneous mantle. *Geochemistry, Geophysics and Geosystems* 7: Q05010 (doi:10.1029/2005GC001158).

Ito G and Mahoney JJ (2005) Flow and melting of a heterogeneous mantle: 1. Method and importance to the geochemistry of ocean island and mid-ocean ridge basalts. *Earth And Planetary Science Letters* 230: 29–46.

Ito G, Mcnutt M, and Gibson RL (1995) Crustal structure of the Tuamotu Plateau, 15°S, and implications for its origin. *Journal of Geophysical Research* 100: 8097–8114.

Ito G, Shen Y, Hirth G, and Wolfe CJ (1999) Mantle flow, melting, and dehydration of the Iceland mantle plume. *Earth and Planetary Science Letters* 165: 81–96.

Ivanov BA and Melosh HJ (2003) Impacts do not initiate volcanic eruptions: Eruptions close to the crater. *Geology* 31: 869–872.

Jackson I (ed.) (1998) *The Earth's Mantle, Composition, Structure, and Evolution*. Cambridge: Cambridge University Press.

Janney PE, Macdougall JD, Natland JH, and Lynch MA (2000) Geochemical evidence from the Pukapuka volcanic ridge system for a shallow enriched mantle domain beneath the South Pacific Superswell. *Earth and Planetary Science Letters* 181: 47–60.

Jellinek AM and Manga M (2002) The influence of the chemical boundary layer on the fixity, spacing and lifetime of mantle plumes. *Nature* 418: 760–763.

Jellinek AM and Manga M (2004) Links between long-lived hot spots, mantle plumes, D″, and plate tectonics. *Reviews of Geophyiscs* 42: RG3002 (doi:10.1029/2003RG000144).

Jha K, Parmentier EM, and Morgan JP (1994) The role of mantle depletion and melt-retention buoyancy in spreading-center segmentation. *Earth and Planetary Science Letters* 125: 221–234.

Jones AP, Price GD, Price NJ, Decarli PS, and Clegg RA (2002a) Impact induced melting and the development of large igneous provinces. *Earth and Planetary Science Letters* 202: 551–561.

Jones SM, White N, and Maclennan J (2002b) V-shaped ridges around Iceland: Implications for spatial and temporal patterns of mantle convection. *Geochemistry, Geophysics and Geosystems* 3: 1059 (doi:10.1029/2002GC000361).

Jordan TH (1979) Mineralogies, densities and seismic velocities of garnet lherzolites and their geophysical implications. In: Boyd FR and Meyer HOA (eds.) *Mantle Sample: Inclusions in Kimberlites and Other Volcanics*, pp. 1–14. Washington, DC: American Geophysical Union.

Jourdan F, Feraud G, Bertrand H, *et al.* (2004) The Karoo triple junction questioned: Evidence from Jurassic and Proterozoic Ar-40/Ar-39 ages and geochemistry of the giant Okavango dyke swarm (Botswana). *Earth and Planetary Science Letters* 222: 989–1006.

Jourdan F, Feraud G, Bertrand H, *et al.* (2005) Karoo large igneous province: Brevity, origin, and relation to mass extinction questioned by new Ar-40/Ar-39 age data. *Geology* 33: 745–748.

Jourdan F, Feraud G, Bertrand H, Watkeys MK, Kampunzu AB, and Le Gall B (2006) Basement control on dyke distribution in Large Igneous Provinces: Case study of the Karoo triple junction. *Earth and Planetary Science Letters* 241: 307–322.

Kaminski E and Jaupart C (2003) Laminar starting plumes in high-Prandtl-number fluids. *Journal of Fluid Mechanics* 478: 287–298.

Katzman R, Zhao L, and Jordan TH (1998) High-resolution, two-dimensional vertical tomography of the central Pacific mantle using ScS reverberations and frequency-dependent travel times. *Journal of Geophysical Research* 103: 17933–17971.

Kellogg LH and King SD (1997) The effect of temperature dependent viscosity on the structure of new plumes in the mantle; results of a finite element model in a spherical, axisymmetric shell. *Earth and Planetary Science Letters* 148: 13–26.

Kelly A and Bercovici D (1997) The clustering of rising diapirs and plume heads. *Geophysical Research Letters* 24: 201–204.

Kendall JM, Stuart GW, Ebinger CJ, Bastow ID, and Keir D (2005) Magma-assisted rifting in Ethiopia. *Nature* 433: 146–148.

Kennett BLN and Engdahl ER (1991) Traveltimes for global earthquake location and phase identification. *Geophysical Journal International* 105: 429–465.

Kennett BLN and Widiyantoro S (1999) A low seismic wavespeed anomaly beneath northwestern India; a seismic signature of the Deccan Plume? *Earth and Planetary Science Letters* 165: 145–155.

Kent AJR, Stolper EM, Francis D, Woodhead J, Frei R, and Eiler J (2004) Mantle heterogeneity during the formation of the North Atlantic Igneous Province: Constraints from trace element and Sr-Nd-Os-O isotope systematics of Baffin Island picrites. *Geochemistry, Geophysics and Geosystems* 5: Q11004 (doi:10.1029/2004GC000743).

Kent RW, Pringle MS, Müller RD, Sauders AD, and Ghose NC (2002) $^{40}Ar/^{39}Ar$ geochronology of the Rajmahal basalts, India, and their relationship to the Kerguelen Plateau. *Journal of Petrology* 43: 1141–1153.

Kent W, Saunders AD, Kempton PD, and Ghose NC (1997) Rajmahal Basalts, Eastern India: Mantle sources and melt distribution at a volcanic rifted margin. In: Mahoney JJ and Coffin MF (eds.) *Large Igneous Provinces: Continental, Oceanic and Planetary Flood Volcanism*, pp. 183–216. Washington, DC: American Geophysical Union.

Kerr RC and Meriaux C (2004) Structure and dynamics of sheared mantle plumes. *Geochemistry, Geophysics and Geosystems* 5: Q12009 (doi:10.1029/2004GC000749).

Keyser M, Ritter JRR, and Jordan M (2002) 3D shear-wave velocity structure of the Eifel Plume, Germany. *Earth and Planetary Science Letters* 203: 59–82.

Kieffer B, Arndt N, Lapierre H, *et al.* (2004) Flood and shield basalts from Ethiopia: Magmas from the African superswell. *Journal of Petrology* 45: 793–834.

King S and Anderson DL (1998) Edge-driven convection. *Earth and Planetary Science Letters* 160: 289–296.

King SD and Anderson DL (1995) An alternative mechanism of flood basalt formation. *Earth and Planetary Science Letters* 136: 269–279.

Klein EM and Langmuir CH (1987) Global correlations of ocean ridge basalt chemistry with axial depth and crustal thickness. *Journal of Geophysical Research* 92: 8089–8115.

Klosko ER, Russo RM, Okal EA, and Richardson WP (2001) Evidence for a rheologically strong chemical mantle root beneath the Ontong-Java Plateau. *Earth and Planetary Science Letters* 186: 347–361.

Kogiso T, Hirschmann MM, and Reiners PW (2004) Length scales of mantle heterogeneities and their relationship to ocean island basalt geochemistry. *Geochemica et Cosmochimica Acta* 68: 345–360.

Koppers AA, Staudigel H, Pringle MS, and Wijbrans JR (2003) Short-lived and discontinuous intraplate volcanism in the South Pacific: Hot spots or extensional volcanism. *Geochemistry, Geophysics and Geosystems* 4: 1089 (doi:10.1029/2003GC000533).

Koppers AAP, Duncan RA, and Steinberger B (2004) Implications of a nonlinear $^{40}Ar/^{39}Ar$ age progression along the Louisville seamount trail for models of fixed and moving hot spots. *Geochemistry, Geophysics and Geosystems* 5: Q06L02 (doi:10.1029/2003GC000671).

Koppers AAP, Phipps Morgan J, Morgan JW, and Staudigel H (2001) Testing the fixed hotspot hypothesis using (super 40) Ar/ (super 39) Ar age progressions along seamount trails. *Earth and Planetary Science Letters* 185: 237–252.

Koppers AAP and Staudigel H (2005) Asynchroneous bends in Pacific seamounts trails: A case for extensional volcanism? *Science* 307: 904–907.

Koppers AAP, Staudigel H, Wijbrans JR, and Pringle MS (1998) The Magellan Seamount Trail; implications for Cretaceous hotspot volcanism and absolute Pacific Plate motion. *Earth and Planetary Science Letters* 163: 53–68.

Korenaga J (2004) Mantle mixing and continental breakup magmatism. *Earth and Planetary Science Letters* 218: 463–473.

Korenaga J (2005a) Firm mantle plumes and the nature of the core–mantle boundary region. *Earth And Planetary Science Letters* 232: 29–37.

Korenaga J (2005b) Why did not the Ontong Java Plateau form subaerially? *Earth and Planetary Science Letters* 234: 385–399.

Korenaga J and Jordan TH (2003) Physics of multiscale convection in Earth's mantle: Onset of sublithospheric convection. *Journal of Geophysical Research* 108(B7): (doi:10.1029/2002JB001760).

Korenaga J and Kelemen PB (2000) Major element heterogeneity in the mantle source of the North Atlantic igneous province. *Earth and Planetary Science Letters* 184: 251–268.

Kumar AR and Mohan G (2005) Mantle discontinuities beneath the Deccan volcanic province. *Earth and Planetary Science Letters* 237: 252–263.

Labrosse S (2002) Hotspots, mantle plumes and core heat loss. *Earth and Planetary Science Letters* 199: 147–156.

Langmuir CH, Klein EM, and Plank T (1992) Petrological systematics of mid-ocean ridge basalts: Constraints on melt generation beneath ocean ridges. In: Phipps Morgan J, Blackman DK, and Sinton JM (eds.) *Mantle Flow and Melt Generation at Mid-Ocean Ridges*, pp. 1–66. Washington, DC: American Geophysical Union.

Lanyon R, Varne R, and Crawford AJ (1993) Tasmanian Tertiary basalts, the Balleny plume, and opening of the Tasman Sea (Southwest Pacific Ocean). *Geology* 21: 555–558.

Larsen TB and Yuen DA (1997) Ultrafast upwelling bursting through the upper mantle. *Earth and Planetary Science Letters* 146: 393–399.

Larsen TB, Yuen DA, and Storey M (1999) Ultrafast mantle plumes and implications for flood basalt volcanism in the Northern Atlantic region. *Tectonophysics* 311: 31–43.

Laske G and Masters G (1997) A Global digital map of sediment thickness. *EOS, Transactions American Geophysical Union, Fall Meeting Supplement* 78: F483.

Laske G, Morgan JP, and Orcutt JA (1999) First results from the Hawaiian SWELL pilot experiment. *Geophysical Research Letters* 26: 3397–3400.

Lawver LA and Mueller RD (1994) Iceland hotspot track. *Geology* 22: 311–314.

Lecheminant AN and Heaman LM (1989) Mackenzie igneous events, Canada – Middle Proterozoic hotspot magmatism associated with ocean opening. *Earth and Planetary Science Letters* 96: 38–48.

Lei JS and Zhao DP (2006) A new insight into the Hawaiian plume. *Earth And Planetary Science Letters* 241: 438–453.

Leitch AM, Steinbach V, and Yuen DA (1996) Centerline temperature of mantle plumes in various geometries: Incompressible flow. *Journal of Geophysical Research* 101: 21829–21846.

Li A and Detrick RS (2003) Azimuthal anisotropy and phase velocity beneath Iceland: Implication for plume–ridge interaction. *Earth and Planetary Science Letters* 214: 153–165.

Li A and Detrick RS (2006) Seismic structure of Iceland from Rayleigh wave inversions and geodynamic implications. *Earth and Planetary Science Letters* 241: 901–912.

Li X, Kind R, Priestley K, Sobolev SV, Tilmann F, Yuan X, and Weber M (2000) Mapping the Hawaiian plume conduit with converted seismic waves. *Nature* 405: 938–941.

Li X, Kind R, Yuan X, *et al.* (2003a) Seismic observation of narrow plumes in the oceanic upper mantle. *Geophysical Research Letters* 30: 1334 (doi:10.1029/2002GL015411).

Li X, Kind R, Yuan X, Woelbern I, and Hanka W (2004) Rejuvenation of the lithosphere by the Hawaiian Plume. *Nature* 427: 827–829.

Li XD and Romanowicz B (1996) Global mantle shear velocity model developed using nonlinear asymptotic coupling theory. *Journal of Geophysical Research* 101: 22245–22272.

Li XQ, Kind R, and Yuan XH (2003b) Seismic study of upper mantle and transition zone beneath hotspots. *Physics of the Earth and Planetary Interiors* 136: 79–92.

Lin S-C and Van Keken PE (2006a) Dynamics of thermochemical plumes: 1. Plume formation and entrainment of a dense layer. *Geochemistry, Geophysics and Geosystems* 7: Q02006 (doi:10.1029/2005GC001071).

Lin SC and Van Keken PE (2005) Multiple volcanic episodes of flood basalts caused by thermochemical mantle plumes. *Nature* 436: 250–252.

Lin SC and Van Keken PE (2006b) Dynamics of thermochemical plumes: 2. Complexity of plume structures and its implications for mapping mantle plumes. *Geochemistry, Geophysics and Geosystems* 7: Q03003 (doi:10.1029/2005GC001072).

Lithgow-Bertelloni C and Richards MA (1998) The dynamics of Cenozoic and Mesozoic plate motions. *Reviews of Geophysics* 36: 27–78.

Lithgow-Bertelloni C, Richards MA, Conrad CP, and Griffiths RW (2001) Plume generation in natural thermal convection at high Rayleigh and Prandtl numbers. *Journal of Fluid Mechanics* 434: 1–21.

Lodge A and Helffrich G (2006) Depleted swell root beneath the Cape Verde Islands. *Geology* 34: 449–452.

Mackay LM, Turner J, Jones SM, and White NJ (2005) Cenozoic vertical motions in the Moray Firth Basin associated with initiation of the Iceland Plume. *Tectonics* 24: TC5004.

Maclennan J, Mckenzie D, and Gronvold K (2001) Plume-driven upwelling under central Iceland. *Earth and Planetary Science Letters* 194: 67–82.

Macnab R, Verhoef J, Roest W, and Arkani-Hamed J (1995) New database documents the magnetic character of the Arctic and North Atlantic. *EOS Transactions of American Geophysical Union* 76(45): 449.

Mahoney JJ and Coffin MF (eds.) (1997) *Large Igneous Provinces: Continental, Oceanic and Planetary Flood Volcanism*. Washington, DC: American Geophysical Union.

Mahoney JJ, Duncan RA, Tejada MLG, Sager WW, and Bralower TJ (2005) Jurassic–Cretaceous boundary age and mid-oceanic-ridge-type mantle source for Shatsky Rise. *Geology* 33: 185–188.

Mahoney JJ, Fitton JG, Wallace PJ, et al. (2001) *Proceedings of the Ocean Drilling Program, Initial Reports, 192* (online).

Mahoney JJ, Jones WB, Frey FA, Salters VJ, Pyle DG, and Davies HL (1995) Geochemical characteristics of lavas from Broken Ridge, the Naturaliste Plateau and southernmost Kerguelen plateau: Cretaceous plateau volcanism in the SE Indian Ocean. *Chemical Geology* 120: 315–345.

Mahoney JJ, Storey M, Duncan RA, Spencer KJ, and Pringle M (1993) Geochemistry and age of the Ontong Java Plateau. In: Pringle M, Sager W, Sliter W, and Stein S (eds.) *Geophysical Monograph Series: The Mesozoic Pacific: Geology, Tectonics, and Volcanism*, pp. 233–261. Washington, DC: AGU.

Maia M, Ackermand D, Dehghani GA, et al. (2000) The Pacific–Antarctic ridge-foundation hotspot interaction; a case study of a ridge approaching a hotspot. *Marine Geology* 167: 61–84.

Malamud BD and Turcotte DL (1999) How many plumes are there?. *Earth and Planetary Science Letters* 174: 113–124.

Manga M (1997) Interactions between mantle diapirs. *Geophysical Research Letters* 24: 1871–1874.

Manga M, Weeraratne D, and Morris SJS (2001) Boundary-layer thickness and instabilities in Bénard convection of a liquid with a temperature-dependent viscosity. *Physics of Fluids* 13: 802–805.

Marzoli A, Piccirillo EM, Renne PR, et al. (2000) The Cameroon volcanic line revisited; petrogenesis of continental basaltic magmas from lithospheric and asthenospheric mantle sources. *Journal of Petrology* 41: 87–109.

Marzoli A, Renne PR, Piccirillo EM, Ernesto M, Bellieni G, and de Min A (1999) Extensive 200-million-year-old continental flood basalts of the Central Atlantic magmatic province. *Science* 284: 616–618.

Matyska C, Moser J, and Yuen DA (1994) The potential influence of radiative heat transfer on the formation of megaplumes in the lower mantle. *Earth and Planetary Science Letters* 125: 255–266.

Matyska C and Yuen DA (2005) The importance of radiative heat transfer on superplumes in the lower mantle with the new post-perovskite phase change. *Earth and Planetary Science Letters* 234: 71–81.

Matyska C and Yuen DA (2006a) Lower mantle dynamics with the post-perovskite phase change, radiative thermal conductivity, temperature- and depth-dependent viscosity. *Physics of the Earth and Planetary Interiors* 154: 196–207.

Matyska C and Yuen DA (2006b) Upper-mantle versus lower-mantle plumes: Are they the same? In: Foulger GR and Jurdy DM (eds.) *The Origins of Melting Anomalies: Plates, Plumes, and Planetary Processes*. Vancouver, BC: GSA.

Mayborn KR and Lesher CE (2004) Paleoproterozoic mafic dike swarms of northeast Laurentia: Products of plumes or ambient mantle? *Earth and Planetary Science Letters* 225: 305–317.

Mcdougall I and Duncan RA (1988) Age progressive volcanism in the Tasmantid Seamounts. *Earth and Planetary Science Letters* 89: 207–220.

Mcdougall I, Verwoerd W, and Chevallier L (2001) K–Ar geochronology of Marion Island, Southern Ocean. *Geological Magazine* 138: 1–17.

Mckenzie D (1984) The generation and compaction of partially molten rock. *Journal of Petrology* 25: 713–765.

Mckenzie D and Bickle MJ (1988) The volume and composition of melt generated by extension of the lithosphere. *Journal of Petrology* 29: 625–679.

Mcnamara AK, Van Keken PE, and Karato SI (2002) Development of anisotropic structure in the Earth's lower mantle by solid-state convection. *Nature* 416: 310–314.

Mcnamara AK and Zhong SJ (2005) Thermochemical structures beneath Africa and the Pacific Ocean. *Nature* 437: 1136–1139.

Mcnutt M (1988) Thermal and mechanical properties of the Cape Verde Rise. *Journal of Geophysical Research* 93: 2784–2794.

Mcnutt MK (1998) Superswells. *Reviews of Geophysics* 36: 211–244.

Mcnutt MK (2002) Heat flow variations over Hawaiian swell controlled by near-surface processes, not plume properties. In: Takahashi E (ed.) *Hawaiian Volcanoes: Deep Underwater Presepectives, Geophysics Monograph 128*. Washington, DC: American Geophysical Union.

Mcnutt MK, Caress DW, Reynolds J, Jordahl KA, and Duncan RA (1997) Failure of plume theory to explain midplate volcanism in the southern Austral Island. *Nature* 389: 479–482.

Mcnutt MK and Fischer KM (1987) The South Pacific superswell. In: Keating GH, Fryer P, Batiza R, and Boehlert GW (eds.) *Seamounts, Islands, and Atolls*, pp. 25–34. Washington, DC: AGU.

Mcnutt MK, Winterer EL, Sager WW, Natland JH, and Ito G (1990) The Darwin Rise: A Cretacous superswell? *Geophysical Research Letters* 17: 1101–1104.

Mege D and Korme T (2004) Dyke swarm emplacement in the Ethiopian large igneous province: Not only a matter of stress. *Journal of Volcanology and Geothermal Research* 132: 283–310.

Meibom A, Sleep NH, Zahnle K, and Anderson DL (2005) Models for nobel gasses in mantle geochemistry: Some observations and alternatives. *Plumes, Plates, and Paradigms: Geological Society of America Special Paper 388*. GSA.

Menzies M, Baker J, Chazot G, and Al'kadasi M (1997) Evolution of the Red Sea volcanic margin, Western Yemen. In: Mahoney JJ and Coffin MF (eds.) *Large Igneous Provinces: Continental, Oceanic, and Planetary Flood Volcanism*. Washington, DC: American Geophysical Union.

Minor DR and Mukasa SB (1997) Zircon U-Pb and hornblende Ar-40-Ar-39 ages for the Dufek layered mafic intrusion, Antarctica: Implications for the age of the Ferrar large igneous province. *Geochimica et Cosmochimica Acta* 61: 2497–2504.

Mitrovica JX and Forte AM (1997) Radial profile of mantle viscosity: Results from the joint inversion of convection and postglacial rebound observables. *Journal of Geophysical Research* 102: 2751–2769.

Mittelstaedt E and Tackley PJ (2006) Plume heat flow is much lower than CMB heat flow. *Earth and Planetary Science Letters* 241: 202–210.

Mittelstaedt EL and Ito G (2005) Plume–ridge interaction, lithospheric stresses, and origin of near-ridge volcanic linements. *Geochemistry, Geophysics and Geosystems* 6: Q06002 (doi:10.1029/2004GC000860).

Molnar P and Stock J (1987) Relative motions of hotspots in the Pacif, Atlantic and Indian Oceans since late Cretaceous time. *Nature* 327: 587–591.

Montague NL and Kellogg LH (2000) Numerical models for a dense layer at the base of the mantle and implications for the geodynamics of D″. *Journal of Geophysical Research* 105: 11101–11114.

Montelli R, Nolet G, Dahlen FA, Masters G, Engdahl ER, and Hung SH (2004) Finite-frequency tomography reveals a variety of plumes in the mantle. *Science* 303: 338–343.

Morgan WJ (1971) Convection plumes in the lower mantle. *Nature* 230: 42–43.

Morgan WJ (1972) Plate motions and deep mantle convection. *The Geological Society of America Memoir* 132: 7–22.

Morgan WJ (1983) Hotspot tracks and the early rifting of the Atlantic. *Tectonophysics* 94: 123–139.

Morley CK, Wescott WA, Stone DM, Harper RM, Wigger ST, and Karanja FM (1992) Tectonic evolution of the Northern Kenyan Rift. *Journal of the Geological Society* 149: 333–348.

Muller JR, Ito G, and Martel SJ (2001) Effects of volcano loading on dike propagation in an elastic half-space. *Journal of Geophysical Research* 106: 11101–11113.

Müller RD, Roest WR, Royer J-Y, Gahagan LM, and Sclater JG (1993a) A digital age map of the ocean floor. 93-30 ed., Scripps Institution of Oceanography Reference Series.

Müller RD, Royer J-Y, and Lawver LA (1993b) Revised plate motions relative to the hotspots from combined Atlantic and Indian Ocean hotspot tracks. *Geology* 21: 275–278.

Nadin PA, Kusznir NJ, and Cheadle MJ (1997) Early Tertiary plume uplift of the North Sea and Faeroe-Shetland basins. *Earth and Planetary Science Letters* 148: 109–127.

Nakakuki T, Yuen DA, and Honda S (1997) The interaction of plumes with the transition zone under continents and oceans. *Earth and Planetary Science Letters* 146: 379–391.

Nakanishi M, Sager WW, and Klaus A (1999) Magnetic lineations within Shatsky Rise, northwest Pacific Ocean: Implications for hot spot-triple junction interaction and oceanic plateau formation. *Journal of Geophysical Research* 104: 7539–7556.

Nataf HC (2000) Seismic imaging of mantle plumes. *Annual Review of Earth and Planetary Sciences* 28: 391–417.

Nataf HC and Vandecar J (1993) Seismological detection of a mantle plume? *Nature* 364: 115–120.

Natland JH and Winterer EL (2005) Fissure control on volcanic action in the Pacific. In: Foulger GR, Natland JH, Presnall DC, and Anderson DL (eds.) *Plumes, Plates, and Paradigms*, pp. 687–710. Vancouver: Geological Society of America.

Neal CR, Mahoney JJ, and Chazey WJ (2002) Mantle sources and the highly variable role of continental lithosphere in basalt petrogenesis of the Kerguelen Plateau and Broken Ridge LIP: Results from ODP Leg 183. *Journal of Petrology* 43: 1177–1205.

Neal CR, Mahoney JJ, Kroenke LW, Duncan RA, and Petterson MG (1997) The Ontong Java Plateau. In: Mahoney JJ and Coffin MF (eds.) *Large Igneous Provinces: Continental, Oceanic, and Planetary Flood Volcanism*, pp. 183–216. Washington, DC: American Geophysical Union.

Ni SD, Helmberger DV, and Tromp J (2005) Three-dimensional structure of the African superplume from waveform modelling. *Geophysical Journal International* 161: 283–294.

Ni SD, Tan E, Gurnis M, and Helmberger D (2002) Sharp sides to the African superplume. *Science* 296: 1850–1852.

Nicolaysen K, Frey FA, Hodges KV, Weis D, and Giret A (2000) $^{40}Ar/^{39}Ar$ geochronology of flood basalts from the Kerguelen Archipelago, southern Indian Ocean; implications for Cenozoic eruption rates of the Kerguelen plume. *Earth and Planetary Science Letters* 174: 313–328.

Nielsen TK, Larsen HC, and Hopper JR (2002) Contrasting rifted margin styles south of Greenland; implications for mantle plume dynamics. *Earth and Planetary Science Letters* 200: 271–286.

Niu F, Solomon SC, Silver PG, Suetsugu D, and Inoue H (2002) Mantle transition-zone structure beneath the South Pacific Superswell and evidence for a mantle plume underlying the Society Hotspot. *Earth and Planetary Science Letters* 198: 371–380.

Niu Y, Waggoner G, Sinton JM, and Mahoney JJ (1996) Mantle source heterogeneity and melting processes beneath seafloor spreading centers: The East Pacific Rise 18°–19° S. *Journal of Geophysical Research* 101: 27711–27733.

Norton IO (2000) Global hotspot reference frames and plate motion. In: Richards MA, Gordon RG, and Vanderhilst RD (eds.) *The History and Dynamics of Global Plate Motions*. Washington, DC: AGU.

Nyblade AA, Owens TJ, Gurrola H, Ritsema J, and Langston CA (2000) Seismic evidence for a deep upper mantle thermal anomaly beneath East Africa. *Geology* 28: 599–602.

Nyblade AA and Robinson SW (1994) The African superswell. *Geophysical Research Letters* 21: 765–768.

O'connor JM and Duncan RA (1990) Evolution of the Walvis Ridge-Rio Grande Rise hot spot system: Implications for African and South American plate motions over plumes. *Journal of Geophysical Research* 95: 17475–17502.

O'connor JM and Roex APL (1992) South Atlantic hot spot-plume systems: 1. Distribution of volcanism in time and space. *Earth and Planetary Science Letters* 113: 343–364.

O'connor JM, Stoffers P, van Den Bogaard P, and Mcwilliams M (1999) First seamount age evidence for significantly slower African Plate motion since 19 to 30 Ma. *Earth and Planetary Science Letters* 171: 575–589.

O'connor JM, Stoffers P, and Wijbrans JR (2002) Pulsing of a focused mantle plume; evidence from the distribution of foundation chain hotspot volcanism. *Geophysical Research Letters* 29(9): 4.

O'connor JM, Stoffers P, and Wijbrans JR (2004) The foundation chain; inferring hotspot–plate interaction from a weak seamount trail. In: Hekinian R, Stoffers P, and Cheminee J-L (eds.) *Oceanic Hotspots*, pp. 349–372. Berlin: Springer.

O'neill C, Mueller D, and Steinberger B (2003) Geodynamic implications of moving Indian Ocean hotspots. *Earth and Planetary Science Letters* 215: 151–168.

O'neill C, Müller D, and Steinberger B (2005) On the uncertainties in hot spot reconstructions and the significance of moving hot spot reference frames. *Geochemistry, Geophysics and Geosystems* 6: Q04003 (doi:10.1029/2004GC000784).

Olson P (1990) Hot spots, swells and mantle plumes. In: Ryan MP (ed.) *Magma Transport and Storage*, pp. 33–51. New York: John Wiley and Sons.

Olson P, Schubert G, and Anderson C (1987) Plume formation in the D″-layer and the roughness of the core–mantle boundary. *Nature* 327: 409–413.

Olson P, Schubert G, Anderson C, and Goldman P (1988) Plume formation and lithosphere erosion: A comparison of laboratory and numerical experiments. *Journal of Geophysical Research* 93: 15065–15084.

Ono S, Ito E, and Katsura T (2001) Mineralogy of subduction basaltic crust (MORB) for 27 to 37 GPa and the chemical heterogeneity of the lower mantle. *Earth and Planetary Science Letters* 190: 57–63.

Operto S and Charvis P (1996) Deep structure of the southern Kerguelen Plateau (southern Indian Ocean) from ocean bottom seismometer wide-angle seismic data. *Journal of Geophysical Research* 101: 25077–25103.

Owens TJ, Nyblade AA, Gurrola H, and Langston CA (2000) Mantle transition zone structure beneath Tanzania, East Africa. *Geophysical Research Letters* 27: 827–830.

Oxburgh ER and Parmentier EM (1977) Compositional and density stratification in oceanic lithosphere – Causes and consequences. *Geological Society of London* 133: 343–355.

Ozawa A, Tagami T, and Garcia MO (2005) Unspiked K–Ar dating of the Honolulu rejuvenated and Koolau shield volcanism on O'ahu, Hawaii. *Earth and Planetary Science Letters* 232: 1–11.

Pankhurst RJ, Leat PT, Sruoga P, et al. (1998) The Chon Aike province of Patagonia and related rocks in West Antarctica: A silicic large igneous province. *Journal of Volcanology and Geothermal Research* 81: 113–136.

Pankhurst RJ, Riley TR, Fanning CM, and Kelley SP (2000) Episodic silicic volcanism in Patagonia and the Antarctic Peninsula: Chronology of magmatism associated with the break-up of Gondwana. *Journal of Petrology* 41: 605–625.

Pares JM and Moore TC (2005) New evidence for the Hawaiian hotspot plume motion since the Eocene. *Earth and Planetary Science Letters* 237: 951–959.

Parkinson IJ, Shaefer BF, and Arculus RJ (2002) A lower mantle origin for the world's biggest LIP? A high precision Os isotope isochron from Ontong Java Plateau basalts drilled on ODP Leg 192. *Geochimica et Cosmochimica Acta* 66: A580.

Parman SW, Kurz MD, Hart SR, and Grove TL (2005) Helium solubility in olivine and implications for high $^3He/^4He$ in ocean island basalts. *Nature* 437: 1140–1143 (doi:10.1038/nature04215).

Passier ML and Snieder RK (1996) Correlation between shear wave upper mantle structure and tectonic surface expressions: Application to central and southern Germany. *Journal of Geophysical Research* 101: 25293–25304.

Peate DW (1997) The Paraná–Etendeka provinces. In: Mahoney JJ and Coffin MF (eds.) *Large Igneous Provinces: Continental, Oceanic, and Planetary Flood Volcanism*, pp. 145–182. Washington, DC: American Geophysical Union.

Pertermann M and Hirschmann MM (2003) Partial melting experiments on a MORB-like pyroxenite between 2 and 3 GPa: Constraints on the presence of pyroxenite in basalt source regions from solidus locations and melting rate. *Journal of Geophysical Research* 108(B2): (doi:10.1029/2000JB000118).

Petterson MG (2004) The geology of north and central Malaita, Solomon Islands: The thickest and most accessible part of the world's largest (Ontong Java) oceanic plateau. In: Fitton JG, Mahoney JJ, Wallace PJ, and Saunders AD (eds.) *Origin and Evolution of the Ontong Java Plateau*, pp. 63–81. London: Geological Society.

Phipps Morgan J (1999) Isotope topology of individual hotspot basalt arrays: Mixing curves or melt extraction trajectories. *Geochemistry, Geophysics and Geosystems* 1: (doi:10.1029/1999GC000004).

Phipps Morgan J (2001) Thermodynamics of pressure release melting of a veined plum pudding mantle. *Geochemistry, Geophysics and Geosystems* 2: (doi:10.1029/2000GC000049).

Pik R, Deniel C, Coulon C, Yirgu G, Hofmann C, and Ayalew D (1998) The northwestern Ethiopian Plateau flood basalts. Classification and spatial distribution of magma types. *Journal of Volcanology and Geothermal Research* 81: 91–111.

Pilidou S, Priestley K, Debayle E, and Gudmundsson O (2005) Rayleigh wave tomography in the North Atlantic: High resolution images of the Iceland, Azores and Eifel mantle plumes. *Lithos* 79: 453–474.

Pilidou S, Priestley K, Gudnmundsson O, and Debayle E (2004) Upper mantle S-wave speed heterogeneity and anisotropy beneath the North Atlantic from regional surface wave tomography: The Iceland and Azores plumes. *Geophysical Journal International* 159: 1057–1076.

Pollack HN, Hurter SJ, and Johnson JR (1993) Heat-flow from the earths interior - analysis of the global data set. *Reviews of Geophysics* 31: 267–280.

Presnall DC, Gudfinnsson GH, and Walter MJ (2002) Generation of mid-ocean ridge basalts at pressures from 1 to 7 GPa. *Geochimica et Cosmochimica Acta* 66: 2073–2090.

Putirka KD (2005) Mantle potential temperatures at Hawaii, Iceland, and the mid-ocean ridge system, as inferred from olivine phenocrysts: Evidence for thermally driven mantle plumes. *Geochemistry, Geophysics and Geosystems* 6: Q05L08 (doi:10.1029/2005GC000915).

Raddick MJ, Parmentier EM, and Scheirer DS (2002) Buoyant decompression melting: A possible mechanisms for intra plate volcanism. *Journal of Geophysical Research* 107: 2228 (doi: 1029/2001JB000617).

Raymond C, Stock JM, and Cande SC (2000) Fast Paleogene motion of the Pacific hotspots from revised global plate circuit constraints. *The History and Dynamics of Global Plate Motions, AGU Geophysical Monographs* 121: 359–375.

Reichow MK, Saunders AD, White RV, Al'mukhamedov AL, and Medvedev AY (2005) Geochemistry and petrogenesis of basalts from the West Siberian basin: An extension of the Permo-Triassic Siberian Traps, Russia. *Lithos* 79: 425–452.

Reiners PW (2002) Temporal-compositional trends in intraplate basalt eruptions: Implications for mantle heterogeneity and melting processes. *Geochemistry, Geophysics and Geosystems* 3: 1011 (doi:10.1029/2002GC000250).

Renne PR, Deckart K, Ernesto M, Feraud G, and Piccirillo EM (1996) Age of the Ponta Grossa dike swarm (Brazil), and implications to Parana flood volcanism. *Earth and Planetary Science Letters* 144: 199–211.

Ribe N (1996) The dynamics of plume–ridge interaction 2. Off-ridge plumes. *Journal of Geophysical Research* 101: 16195–16204.

Ribe N, Christensen UR, and Theissing J (1995) The dynamics of plume–ridge interaction, 1: Ridge-centered plumes. *Earth and Planetary Science Letters* 134: 155–168.

Ribe NM and Christensen UR (1994) Three-dimensional modelling of plume–lithosphere interaction. *Journal of Geophysical Research* 99: 669–682.

Ribe NM and Christensen UR (1999) The dynamical origin of Hawaiian volcanism. *Earth and Planetary Science Letters* 171: 517–531.

Ribe NM, Davaille A, and Christensen UR (2007) Fluid dynamics of mantle plumes. In: Ritter J and Christensen U (eds.) *Mantle Plumes – A Multidisciplinary Approach*. Springer.

Ribe NM and de Valpine DP (1994) The global hotspot distribution and instability of D″. *Geophysical Research Letters* 21: 1507–1510.

Richards MA, Duncan RA, and Courtillot VE (1989) Flood basalts and hotspot tracks: Plume heads and tails. *Science* 246: 103–107.

Richards MA and Hager BH (1984) Geoid anomalies in a dynamic Earth. *Journal of Geophysical Research* 89: 5987–6002.

Richards MA and Lithgow-Bertelloni C (1996) Plate motion changes, the Hawaiian-Emperor bend, and the apprent success and failure of geodynamic models. *Earth and Planetary Science Letters* 137: 19–27.

Richardson WP, Okal EA, and van der Lee S (2000) Rayleigh-wave tomography of the Ontong-Java Plateau. *Physics of the Earth and Planetary Interiors* 118: 29–61.

Richter FM (1973) Convection and the large-scale circulation of the mantle. *Journal of Geophysical Research* 78: 8735–8745.

Richter FM and Mckenzie DP (1981) On some consequences and possible causes of layered mantle convection. *Journal of Geophysical Research* 86: 6133–6142.

Ringwood AE and Irifune T (1988) Nature of the 650-km seismic discontinuity: Implications for mantle dynamics and differentiation. *Nature* 331: 131–136.

Ritsema J and Allen RM (2003) The elusive mantle plume. *Earth and Planetary Science Letters* 207: 1–12.

Ritsema J, Nyblade AA, Owens TJ, Langston CA, and Vandecar JC (1998) Upper mantle seismic velocity structure beneath Tanzania, east Africa: Implications for the stability of cratonic lithosphere. *Journal of Geophysical Research* 103: 21201–21213.

Ritsema J, van Heijst HJ, and Woodhouse JH (1999) Complex shear velocity structure beneath Africa and Island. *Science* 286: 1925–1928.

Ritter JRR, Jordan M, Christensen UR, and Achauer U (2001) A mantle plume below the Eifel volcanic fields, Germany. *Earth and Planetary Science Letters* 186: 7–14.

Roberge J, Wallace PJ, White RV, and Coffin MF (2005) Anomalous uplift and subsidence of the Ontong Java Plateau inferred from $CO_2$ contents of submarine basaltic glasses. *Geology* 33: 501–504.

Roberge J, White RV, and Wallace PJ (2004) Volatiles in submarine basaltic glasses from the Ontong Java Plateau (ODP Leg 192): Implications for magmatic processes and source region compositions. In: Fitton JG, Mahoney JJ, Wallace PJ, and Saunders AD (eds.) *Origin and Evolution of the Ontong Java Plateau*, pp. 339–342. London: Geological Society.

Rogers GC (1982) Oceanic plateaus as meteorite impact signatures. *Nature* 299: 341–342.

Rost S and Williams Q (2003) Seismic detection of sublithospheric plume head residue beneath the Pitcairn hot-spot chain. *Earth and Planetary Science Letters* 209: 71–83.

Roy AB (2003) Geological and geophysical manifestations of the Reunion Plume–Indian lithosphere interactions; evidence from northwest India. *Gondwana Research* 6: 487–500.

Russo RM and Okal EA (1998) Shear wave splitting and upper mantle deformation in French Polynesia; evidence from small-scale heterogeneity related to the Society Hotspot. *Journal of Geophysical Research, B, Solid Earth and Planets* 103: 15089–15107.

Saal AE, Hart SR, Shimizu N, Hauri EH, and Layne GD (1998) Pb isotopic variability in melt inclusions from oceanic island basalt, Polynesia. *Science* 282: 1481–1484.

Saffman PG and Taylor GI (1958) The penetration of a fluid into a porous medium of Hele-Shaw cell containing a more viscous liquid. *Proceedings of the Royal Society of London A* 245: 312–329.

Sager WW (2005) What built Shatsky Rise, a mantle plume or ridge tectonics? In: Foulger G, Natland JH, Presnall DC, and Anderson DL (eds.) *Plumes, Plates, and Paradigms*, pp. 721–733. Boulder, CO: Geological Society of America.

Sager WW, Lamarch AJ, and Kopp C (2005) Paleomagnetic modeling of seamounts near the Hawaiian-Emperor bend. *Tectonophysics* 405: 121–140.

Sallares V, Charvis P, Flueh ER, and Bialas J (2005) Seismic structure of the Carnegie ridge and the nature of the Galapagos hotspot. *Geophysical Journal International* 161: 763–788.

Salters VJM and Dick HJB (2002) Mineralogy of the mid-ocean-ridge basalt source from neodymium and isotopic composition of abyssal peridotites. *Nature* 418: 68–72.

Saltzer RL and Humphreys ED (1997) Upper mantle P wave velocity structure of the eastern Snake River Plain and its relationship to geodynamic models of the region. *Journal of Geophysical Research* 102: 11829–11841.

Samuel H and Bercovici D (2006) Oscillating and stagnating plumes in the Earth's lower mantle. *Earth and Planetary Science Letters* 248: 90–105.

Sandwell D and Fialko Y (2004) Warping and cracking of the Pacific plate by thermal contraction. *Journal of Geophysical Research* 109: B10411 (doi:10.1029/2004JB0003091).

Sandwell DT, Winterer EL, Mammerickx J, *et al.* (1995) Evidence for diffuse extension of the Pacific plate from Pukapuka ridges and cross-grain gravity lineations. *Journal of Geophysical Research* 100: 15087–15099.

Saunders AD, England RW, Relchow MK, and White RV (2005) A mantle plume origin for the Siberian traps: Uplift and extension in the West Siberian Basin, Russia. *Lithos* 79: 407–424.

Saunders AD, Fitton JG, Kerr AC, Norry MJ, and Kent RW (1997) The North Atlantic igneous province. In: Mahoney JJ and Coffin MF (eds.) *Large Igneous Provinces: Continental, Oceanic, and Planetary Flood Volcanism*, pp. 45–93. Washington, DC: American Geophysical Union.

Schilling J-G (1973) Iceland mantle plume: Geochemical study of Reykjanes Ridge. *Nature* 242: 565–571.

Schimmel M, Assumpcao M, and Vandecar JC (2003) Seismic velocity anomalies beneath SE Brazil from P and S wave travel time inversions. *Journal of Geophysical Research* 108: 2191 (doi:10.1029/2001JB000187).

Schlanger SO and Al E (1984) Geology and geochronology of the Line Islands. *Journal of Geophysical Research* 89: 11261–11272.

Schubert G, Turcotte DL, and Olson P (2001) *Mantle Convection in the Earth and Planets*. Cambridge: Cambridge University Press.

Schutt DL and Humphreys ED (2004) P and S wave velocity and V-P/V-S in the wake of the Yellowstone hot spot. *Journal of Geophysical Research, Solid Earth* 109: B01305.

Scott DR and Stevenson DJ (1989) A self-consistent model for melting, magma migration and buoyancy-driven circulation beneath mid-ocean ridges. *Journal of Geophysical Research* 94: 2973–2988.

Sebai A, Stutzmann E, Montaner J-P, Sicilia D, and Beucler E (2006) Anisotropic structure of the African upper mantle from Rayleigh and Love wave tomography. *Physics of the Earth and Planetary Interiors* 155: 48–62.

Sharma M (1997) Siberian Traps. In: Mahoney JJ and Coffin MF (eds.) *Large Igneous Provinces: Continental, Oceanic, and Planetary Flood volcanism*. Washington, DC: American Geophysical Union.

Sharpe WD and Clague DA (2006) 50-Ma initiation of Hawaiian-Emperor bend records major change in Pacific plate motion. *Science* 313: 1281–1284.

Shen Y and Forsyth DW (1995) Geochemical constraints on initial and final depths of melting beneath mid-ocean ridges. *Journal of Geophysical Research* 100: 2211–2237.

Shen Y, Solomon SC, Bjarnason IT, and Wolfe C (1998) Seismic evidence for a lower mantle origin of the Iceland plume. *Nature* 395: 62–65.

Sheth HC (2005) From Deccan to Réunion: No trace of a mantle plume. In: Foulger G, Natland JH, Presnall DC, and Anderson DL (eds.) *Plumes, Plates, and Paradigms*, pp. 477–501. Vancouver: Geological Society of America.

Shirey SB and Walker RJ (1998) The Re-Os isotope system in cosmochemistry and high-temperature geochemistry. *Annual Reviews of the Earth and Planetary Sciences* 26: 423–500.

Sinton CW, Christie DM, and Duncan RA (1996) Geochronology of the Galápagos seamounts. *Journal of Geophysical Research* 101: 13689–13700.

Sinton CW, Duncan RA, and Denyer P (1997) Nicoya Peninsula, Costa Rica: A single suite of Caribbean oceanic plateau magmas. *Journal of Geophysical Research* 102: 15507–15520.

Sleep NH (1987) Lithospheric heating by mantle plumes. *Geophysical Journal of the Royal Astronomical Society* 91: 1–11.

Sleep NH (1988) Gradual entrainment of a chemical layer at the base of the mantle by overlying convection. *Geophysical Journal International* 95: 437–447.

Sleep NH (1990) Hotspots and mantle plumes: Some phenomenology. *Journal of Geophysical Research* 95: 6715–6736.

Smallwood JR and White RS (2002) Ridge-plume interaction in the North Atlantic and its influence on continental breakup and seafloor spreading. *Geological Society Special Publications* 197: 15–37.

Smith RB and Braile LW (1994) The yellowstone hotspot. *Journal of Volcanology and Geothermal Research* 61: 121–187.

Smith WHF and Sandwell DT (1997) Global seafloor topography from satellite altimetry and ship depth soundings. *Science* 277: 1957–1962.

Sparks DW, Parmentier EM, and Morgan JP (1993) Three-dimensional mantle convection beneath a segmented spreading center: Implications for along-axis variations in crustal thickness and gravity. *Journal of Geophysical Research* 98: 21977–21995.

Sparrow EM, Husar RB, and Goldstein RJ (1970) Observations and other characteristics of thermals. *Journal of Fluid Mechanics* 41: 793–800.

Stein CA and Stein S (1992) A model for the global variation in oceanic depth and heat flow with lithospheric age. *Nature* 359: 123–129.

Stein CA and Stein S (1993) Constraints on Pacific midplate swells from global depth-age and heat flow-age models. In: Pringle M, Sager W, Sliter W, and Stein S (eds.) *Geophysical Monograph 76: The Mesozoic Pacific*, pp. 53–76. Washington, DC: American Geophysical Union.

Stein CA and Stein S (2003) Mantle plumes: Heat-flow near Iceland. *Astronomy and Geophysics* 44: 8–10.

Steinberger B (2000) Plumes in a convecting mantle: Models and observations for individual hotspots. *Journal of Geophysical Research* 105: 11127–11152.

Steinberger B (2002) Motion of the Easter Hot Spot relative to Hawaii and Louisville hot spots. *Geochemistry, Geophysics and Geosystems* 3: 8503 (doi:10.1029/2002GC000334).

Steinberger B and O'connell RJ (1998) Advection of plumes in mantle flow; implications for hot spot motion, mantle viscosity and plume distribution. *Geophysical Journal International* 132: 412–434.

Steinberger B and O'connell RJ (2000) Effects of mantle flow on hotspot motion. *Geophysical Monograph* 121: 377–398.

Steinberger B, Sutherland R, and O'connell RJ (2004) Prediction of Emperor-Hawaii seamount locations from a revised model of global plate motion and mantle flow. *Nature* 430: 167–173.

Stewart K, Turner S, Kelley S, Hawkesworth C, Kirstein L, and Mantovani M (1996) 3-D, (super 40) Ar- (super 39) Ar geochronology in the Parana continental flood basalt province. *Earth and Planetary Science Letters* 143: 95–109.

Storey M, Mahoney JJ, and Saunders AD (1997) Cretaceous basalts in Madagascar and the transition between plume and continental lithosphere mantle sources. In: Mahoney JJ and Coffin MF (eds.) *Large Igneous Provinces: Continental, Oceanic, and Planetary Flood Volcanism*, pp. 95–122. Washington, DC: American Geophysical Union.

Stracke A, Zindler A, Salters VJM, et al. (2003) Theistareykir revisited. *Geochemistry, Geophysics and Geosystems* 4: 8507 (doi:10.1029/2001GC000201).

Stuart FM, Lass-Evans S, Fitton JG, and Ellam RM (2003) High $^3$He/$^4$He ratios in pictritic basalts from Baffin Island and the role of a mixed reservoir in mantle plumes. *Nature* 424: 57–59.

Su W-J, Woodward RL, and Dziewonski AM (1994) Degree 12 model of shear velocity heterogeneity in the mantle. *Journal of Geophysical Research* 99: 6945–6980.

Tackley PJ (2002) Strong heterogeneity caused by deep mantle layering. *Geochemistry, Geophysics and Geosystems* 3 (doi:10.1029/2001GC000167).

Tackley PJ and Stevenson DJ (1993) A mechanism for spontaneous self-perpetuating volcanism on terrestrial planets. In: Stone DB and Runcorn SK (eds.) *Flow and Creep in the Solar System: Observations, Modeling and Theory*, pp. 307–322. Norwell, MA: Kluwer.

Takahashi E (2002) The Hawaiian plume and magma genesis. *Geophysical Monograph* 128: 347–348.

Takahashi E, Nakajima K, and Wright TL (1998) Origin of the Columbia Rivers basalts: Melting model of a heterogeneous plume head. *Earth and Planetary Science Letters* 162: 63–80.

Tan E and Gurnis M (2005) Metastable superplumes and mantle compressibility. *Geophysical Research Letters* 32: L20307 (doi:10.1029/2005GL024190).

Tarduno JA, Duncan RA, Scholl DW, et al. (2003) The Emperor Seamounts; southward motion of the Hawaiian Hotspot plume in Earth's mantle. *Science* 301: 1064–1069.

Tarduno JA and Gee J (1995) Large-scale motion between Pacific and Atlantic hotspots. *Nature* 378: 477–480.

Tarduno JA, Sliter WV, Kroenke LW, et al. (1991) Rapid formation of Ontong Java Plateau by Aptian mantle plume volcanism. *Science* 254: 399–403.

Taylor B (2006) The single largest oceanic plateau: Ontong Java-Manihiki-Hikurangi. *Earth and Planetary Science Letters* 241: 372–380.

Tejada MLG, Mahoney JJ, Castillo PR, Ingle SP, Sheth HC, and Weis D (2004) Pin-pricking the elephant: Evidence on the origin of the Ontong Java Plateau from Pb-Sr-Hf-Nd isotopic characteristics of ODP Leg 192. In: Fitton JG, Mahoney JJ, Wallace PJ, and Saunders AD (eds.) *Origin and Evolution of the Ontong Java Plateau*, pp. 133–150. London: Geological Society.

Tejada MLG, Mahoney JJ, Duncan RA, and Hawkins MP (1996) Age and geochemistry of basement and alkalic rocks of Malaita and Santa Isabel, Solomon Islands, southern margin of Ontong Java Plateau. *Journal of Petrology* 37: 361–394.

Tejada MLG, Mahoney JJ, Neal CR, Duncan RA, and Petterson MG (2002) Basement geochemistry and geochronology of central Malaita and Santa Isabel, Solomon Islands, with implications for the origin and evolution of the Ontong Java Plateau. *Journal of Petrology* 37: 449–484.

Thompson RN and Gibson SA (1991) Subcontinental mantle plumes, hotspots and preexisting thinspots. *Journal of the Geological Society* 148: 973–977.

To A, Romanowicz B, Capdeville Y, and Takeuchi N (2005) 3D effects of sharp boundaries at the borders of the African and Pacific Superplumes: Observation and modeling. *Earth and Planetary Science Letters* 233: 137–153.

Toomey DR, Hooft EEE, Solomon SC, James DE, and Hall ML (2001) Upper mantle structure beneath the Galapagos Archipelago from body wave data. *EOS Transactions of the American Geophysical Union, Fall Meeting Supplement* 82: F1205.

Torsvik TH, van der Voo R, and Redfield TF (2002) Relative hotspot motions versus true polar wander. *Earth and Planetary Science Letters* 202: 185–200.

Trampert J, Deschamps F, Resovsky J, and Yuen D (2004) Probabilistic tomography maps chemical heterogeneities throughout the lower mantle. *Science* 306: 853–856.

Trumbull RB, Vietor T, Hahne K, Wackerle R, and Ledru P (2004) Aeromagnetic mapping and reconnaissance geochemistry of the Early Cretaceous Henties Bay-Outjo dike swarm, Etendeka Igneous Province, Namibia. *Journal of African Earth Sciences* 40: 17–29.

Turner DL, Jarrard RD, and Forbes RB (1980) Geochronology and origin of the Pratt-Welker seamount chain, Gulf of Alaska: A new pole of rotation for the Pacific plate. *Journal of Geophysical Research* 85: 6547–6556.

Turner S and Al E (1994) Magmatism and continental bread-up in the South Atlantic: High precision $^{40}Ar$–$^{39}Ar$ geochronology. *Earth and Planetary Science Letters* 121: 333–348.

Van Den Berg AP, Rainey ESG, and Yuen DA (2005) The combined influences of variable thermal conductivity, temperature- and pressure-dependent viscosity and core–mantle coupling on thermal evolution. *Physics of the Earth and Planetary Interiors* 149: 259–278.

Van Der Hilst RD and De Hoop MV (2005) Banana-doughnut kernels and mantle tomography. *Geophysical Journal International* 163: 956–961.

Van Der Hilst RD and Karason H (1999) Compositional heterogeneity in the bottom 1000 kilometers of Earth's mantle: Toward a hybrid convection model. *Science* 283: 1885–1888.

van Hunen J and Zhong S (2003) New insight in the Hawaiian plume swell dynamics from scaling laws. *Geophysical Research Letters* 30: 1785 (doi:10.1029/2003GL017646).

van Keken PE (1997) Evolution of starting mantle plumes: A comparison between numerical and laboratory models. *Earth and Planetary Science Letters* 148: 1–11.

van Keken PE and Ballentine CJ (1998) Whole-mantle versus layered mantle convection and the role of a high-viscosity lower mantle in terrestrial volatile evolution. *Earth And Planetary Science Letters* 156: 19–32.

van Keken PE and Ballentine CJ (1999) Dynamical models of mantle volatile evolution and the role of phase transitions and temperature-dependent rheology. *Journal of Geophysical Research* 104: 7137–7151.

van Keken PE and Gable CW (1995) The interaction of a plume with a rheological boundary: A comparision between two- and three-dimensional models. *Journal of Geophysical Research* 100: 20291–20302.

van Keken PE, Yuen DA, and Vandenberg AP (1992) Pulsating diapiric flows: Consequences of vertical variations in mantle creep laws. *Earth and Planetary Science Letters* 112: 179–194.

van Keken PE, Yuen DA, and Vandenberg AP (1993) The effects of shallow rheological boundaries in the upper mantle on inducing shorter time scales of diapiric flows. *Geophysical Research Letters* 20: 1927–1930.

Van Wijk JW, Huismans RS, ter Voorde M, and Cloetingh S (2001) Melt generation at volcanic continental margins: No need for a mantle plume? *Geophysical Research Letters* 28: 3995–3998.

Vandamme D, Courtillot V, Besse J, and Montinny R (1991) Paleomagnetism and age determinations of the Deccan Traps (India): Results of a Nagpur–Bombay traverse and review of earlier work. *Reviews of Geophysics* 29: 159–190.

Vandecar JC, James DE, and Assumpcao M (1995) Seismic evidence for a fossil mantle plume beneath South America and implications for plate driving forces. *Nature* 378: 25–31.

Vdovin O, Rial JA, Levshin AL, and Ritzwoller MH (1999) Group-velocity tomography of South America and the surrounding oceans. *Geophysical Journal International* 136: 324–340.

Vogt PR (1971) Asthenosphere motion recorded by the ocean floor south of Iceland. *Earth and Planetary Science Letters* 13: 153–160.

Waite GP, Schutt DL, and Smith RB (2005) Models of lithosphere and asthenosphere anisotropic structure of the Yellowstone hot spot from shear wave splitting. *Journal of Geophysical Research* 110: B11304.

Waite GP, Smith RB, and Allen RM (2006) V-P and V-S structure of the Yellowstone hot spot from teleseismic tomography: Evidence for an upper mantle plume. *Journal of Geophysical Research* 111: B04303.

Walker KT, Bokelmann GHR, Klemperer SL, and Bock G (2005) Shear-wave splitting around the Eifel hotspot: Evidence for a mantle upwelling. *Geophysical Journal International* 163: 962–980.

Walker KT, Nyblade AA, Klemperer SL, Bokelmann GHR, and Owens TJ (2004) On the relationship between extension and anisotropy: Constraints from shear wave splitting across the East African Plateau. *Journal of Geophysical Research* 109.

Wang S and Wang R (2001) Current plate velocities relative to hotspots; implications for hotspot motion, mantle viscosity and global reference frame. *Earth and Planetary Science Letters* 189: 133–140.

Watson S and Mckenzie D (1991) Melt generation by plumes – A study of Hawaiian volcanism. *Journal of Petrology* 32: 501–537.

Watts AB, Weissel JK, Duncan RA, and Larson RL (1988) Origin of the Louisville ridge and its relationship to the Eltanin fracture zone system. *Journal of Geophysical Research* 93: 3051–3077.

Weeraratne DS, Forsyth DW, and Yang Y (in press) Rayleigh wave tomography of the oceanic mantle beneath intraplate seamount chains in the South Pacific. *Journal of Geophysical Research*.

Weeraratne DS, Parmentier EM, and Forsyth DW (2003) Viscous fingering of miscible fluids in laboratory experiments and the oceanic mantle asthenosphere. *AGU Fall Meeting Supplement*, abstract V21B-03.

Werner R, Hoernle K, van den Bogaard P, Ranero C, von Huene R, and Korich D (1999) Drowned 14-my.-old Galapagos Archipelago off the coast of Costa Rica; implications for tectonic and evolutionary models. *Geology* 27: 499–502.

Wessel P, Harada Y, and Kroenke LW (2006) Toward a self-consistent, high resolution absolute plate motion model for the Pacific. *Geochemistry, Geophysics and Geosystems* 7: Q03L12.

Wessel P and Kroenke LW (1997) A geometric technique for relocating hotspots and refining absolute plate motions. *Nature* 387: 365–369.

Wessel P and Smith WHF (1995) New version of the generic mapping tools released. *EOS Transactions of American Geophysical Union* 76: 329.

White RS (1988) A hot-spot model for early Tertiary volcanism in the N Atlantic. *Geological Society Special Publications* 39: 3–13.

White RS (1997) Rift–plume interaction in the North Atlantic. *Philosophical Transactions of the Royal Society of London Series A* 355: 319–339.

White RS, Bown JW, and Smallwood JR (1995) The temperature of the Iceland plume and origin of outward propagating V-shaped ridges. *Journal of the Geological Society of London* 152: 1039–1045.

White RS, Mckenzie D, and O'nions RK (1992) Oceanic crustal thickness from seismic measurements and rare earth element inversions. *Journal of Geophysical Research* 97: 19683–19715.

Whitehead JA and Luther DS (1975) Dynamics of laboratory diapir and plume models. *Journal of Geophysical Research* 80: 705–717.

Wilson JT (1963) A possible origin of the Hawaiin Islands. *Canadian Journal of Physics* 41: 863–870.

Wilson JT (1973) Mantle plumes and plate motions. *Tectonophysics* 19: 149–164.

Wolfe CJ, Bjarnason IT, Vandecar JC, and Solomon SC (1997) Seismic structure of the Iceland mantle plume. *Nature* 385: 245–247.

Wolfe CJ, Solomon SC, Silver PG, Vandecar JC, and Russo RM (2002) Inversion of body-wave delay times for mantle structure beneath the Hawaiian Islands; results from the PELENET experiment. *Earth and Planetary Science Letters* 198: 129–145.

Xie S and Tackley PJ (2004) Evolution of helium and argon isotopes in a convecting mantle. *Earth and Planetary Science Letters* 146: 417–439.

Xu Y, Chung S-L, Jahn B-M, and WU G (2001) Petrologic and geochemical constraints on the petrogenesis of Permian-Triassic Emeishan flood basalts in southwestern China. *Lithos* 58: 145–168.

Xu Y, He B, Chung S-L, Menzies MA, and Frey FA (2004) Geologic, geochemical, and geophysical consequences of plume involvement in the Emeishan flood-basalt province. *Geology* 32: 917–920.

Xue M and Allen RM (2005) Asthenospheric channeling of the Icelandic upwelling: Evidence from seismic anisotropy. *Earth and Planetary Science Letters* 235: 167–182.

Yang T and Shen Y (2005) P-wave velocity structure of the crust and uppermost mantle beneath Iceland from local earthquake tomography. *Earth and Planetary Science Letters* 235: 597–609.

Yoshida Y and Suetsugu D (2004) Lithospheric thickness beneath the Pitcairn hot spot trail as inferred from Rayleigh wave dispersion. *Physics of the Earth and Planetary Interiors* 146: 75–85.

Yuan HY and Dueker K (2005) Teleseismic P-wave tomogram of the Yellowstone plume. *Geophysical Research Letters* 32: L07304 (doi:10.1029/2004GL022056).

Zhang ZC, Mahoney JJ, Mao JW, and Wang FH (2006) Geochemistry of picritic and associated basalt flows of the western Emeishan flood basalt province, China. *Journal of Petrology* 47: 1997–2019.

Zhao D (2001) Seismic structure and origin of hotspots and mantle plumes. *Earth and Planetary Science Letters* 192: 251–265.

Zhong SJ (2006) Constraints on thermochemical convection of the mantle from plume heat flux, plume excess temperature, and upper mantle temperature. *Journal of Geophysical Research* 111: B04409 (doi:10.1029/2005JB003972).

Zhong SJ and Hager BH (2003) Entrainment of a dense layer by thermal plumes. *Geophysical Journal International* 154: 666–676.

Zhong SJ and Watts AB (2002) Constraints on the dynamics of mantle plumes from uplift of the Hawaiian Islands. *Earth and Planetary Science Letters* 203: 105–116.

Zhou Z, Malpas J, Song XY, *et al.* (2002) A temporal link between the Emeishan large igneous province (SW China) and the end-Guadalupian mass extinction. *Earth and Planetary Science Letters* 196: 113–122.

Zindler A and Hart S (1986) Chemical geodynamics. *Annual Reviews of Earth and Planetary Sciences* 14: 493–571.

## Relevant Websites

http://www.mantleplumes.org – Discussing the Origin of 'Hot Spot' Volcanism.

http://www.georoc.mpch-mainz.gwdg.de – Geochemistry of the Rocks of the Oceans and Continents.

http://www.petdb.org – Petrological Database of the Ocean Floor.

# 10  Mantle Geochemical Geodynamics

**P. J. Tackley**, Institut für Geophysik, ETH Zurich, Switzerland

## 10.1 Observations and the Origin of Heterogeneity

### 10.1.1 Introduction

Geochemical constraints on mantle heterogeneity are thoroughly reviewed in several papers (e.g., Graham, 2002; Hilton and Porcelli, 2003; Hofmann, 2003), but as they are so important for understanding this chapter, it is appropriate to briefly summarize here the basic observations and 'traditional' interpretations. New interpretations are discussed Section 10.5. It is noted that with modern analysis techniques, particularly multi-

collector inductively coupled mass spectrometer (MC-ICPMS) machines, the number of isotopes that geochemists are measuring has increased, and while many isotope systems are consistent in their behavior, additional complexity in interpretation is introduced by isotope systems that are affected by additional processes.

This chapter starts with models of the bulk-silicate Earth (BSE) composition, then constraints on the variations of composition, which fall into three main categories: (1) mass-balance arguments based on the inferred composition of the mid-ocean ridge basalt (MORB) source and continental crust,

(2) noble gas constraints, and (3) end-member components identified by analysis of the trace-element isotopic compositions of ocean island basalts (OIBs) and MORB. This section also reviews geophysical constraints on chemical heterogeneity and mantle structure, and the major types of mantle conceptual model that have been proposed to explain geochemical and/or geophysical observations.

## 10.1.2 Bulk Earth and BSE models

The bulk composition of the whole Earth (i.e., core + mantle + crust) is estimated using a variety of constraints as reviewed in Palme and O'Neill (2003), particularly the composition of the Sun, meteorites, and upper-mantle-derived rocks. The relative abundance of refractory elements is thought to be basically chondritic (i.e., solar), while nonrefractory to volatile elements show increasing degrees of depletion. The composition of the silicate part, that is, BSE is further influenced by core extraction, which enriched the mantle in nonsiderophile elements but depleted it in siderophile elements. A commonly used compositional BSE model is that of McDonough and Sun (1995), but a more depleted model was recently proposed by Lyubetskaya and Korenaga (2007a). Knowledge of the trace-element content of the BSE is important in making 'mass-balance' arguments summarized later, as is knowledge of the composition of the continental crust (e.g., Rudnick, 1995; Rudnick and Fountain, 1995; Taylor and McLennan, 1995). A recent review concludes, however, that Earth's composition is unlike any meteorite (Drake and Righter, 2002). The reader is referred to the above cited literature for further detailed arguments.

## 10.1.3 Fractionation of Major and Trace Elements

When solid and liquid coexist in equilibrium with each other in a multicomponent material such as mantle silicates, the composition of the solid and liquid components is different in both major and trace elements. The two main situations for solid–liquid equilibration are pressure-release partial melting and the crystallization of a magma ocean.

A crude way of viewing major-element fractionation is that different minerals have different melting temperatures – for example, in the shallow mantle garnet and clinopyroxene melt at a lower temperature than olivine. Thus, partial melting followed by melt segregation and eruption causes both the crust

that forms from the resulting melt, and the residual solid, to have a different composition from the original rock. In a cooling magma ocean, this results in the solid crystals that form having a different composition from the remaining liquid, which might allow them to settle to the top or bottom due to related density differences. Indeed, density differences that result from differences in major-element composition may well play an important role in mantle dynamics, as discussed in later sections.

Trace-element partitioning between solid and liquid is expressed in terms of partition coefficients $d$, which describe the equilibrium concentration of the element in the solid divided by the concentration in the liquid:

$$d = \frac{C_{\text{solid}}}{C_{\text{liquid}}} \qquad [1]$$

where $C_{\text{solid}}$ and $C_{\text{liquid}}$ are the concentration of the element in the solid and liquid, respectively. A partition coefficient $\ll 1$ means that the element preferentially enters the melt, and is known as an incompatible element. Conversely, a partition coefficient $\gg 1$ indicates that the element is compatible and stays in the solid. The partition coefficient depends on the element (because different elements have difference sizes and electronic properties), which results in trace elements being fractionated during melting, that is, the ratio of their concentration changes. Partition coefficients may also depend on pressure (depth), oxygen fugacity, and, of course, the minerals present. Different isotopes of the same element have partition coefficients that are almost the same, and can be treated as identical for the situations discussed in this chapter. For more information see Righter and Drake (2003), Wood and Blundy (2003).

This fractionation of trace elements plays a key role in the evolution of isotope ratios of interest in mantle geochemistry. In a simple radioactive decay system in which a parent isotope radioactively decays to a daughter that is a different element, the parent to daughter ratio (P/D) will change as a result of melting and crustal formation, whereas the ratio of the daughter to a stable isotope of the same element (D/D$_\text{S}$) will not change. This change in (P/D) affects the future rate of change of (D/D$_\text{S}$). By analyzing the (D/D$_\text{S}$) ratio in two or more samples that have experienced fractionation in a single event, it is possible to derive the elapsed time since the fractionation event, and this is the principle of radiogenic dating. The specific example of Pb–Pb dating is more complicated because it combines two decay systems, but this also makes it more robust (for details see, e.g., Stacey (1992)).

The concentration of trace elements in the melt depends not only on the partition coefficient but also on the melt fraction. For example, a highly incompatible element essentially all enters the melt, which means that its concentration in the melt is inversely proportional to melt fraction. In the more general case, it is straightforward to derive an equation for the equilibrium concentration of a trace element in the melt (Shaw, 1970):

$$C_{\text{liquid}} = \frac{C_0}{F + d(1 - F)} \qquad [2]$$

where $F =$ melt fraction and $C_0 =$ concentration in the total (bulk) system. Thus the enrichment of a magmatic product in trace elements relative to its source can be used to estimate the melt fraction.

## 10.1.4   Trace-Element Budgets of the Crust and Mantle

The trace-element composition of the continental crust and the inferred composition of the MORB source are often compared to the composition of the BSE discussed earlier. This comparison indicates that the continental crust is strongly enriched in incompatible trace elements by a factor of $\sim$50–100 in the most incompatible elements, whereas the MORB source is depleted. Furthermore, the patterns of enrichment and depletion (i.e., as a function of elements) appear to be complementary. The enriched continental crust and the depleted MORB source region have thus long been thought to be complementary in their trace-element signature, implying that the MORB source region is the residue from melting that produced the continental crust (e.g., Hofmann (1988)). Mass-balance calculations have been used to estimate the fraction of the mantle that must be depleted to make the continental crust (e.g., Jacobsen and Wasserburg, 1979; O'Nions et al., 1979; Davies, 1981; Allegre et al., 1983a; Hofmann, 1988) with estimated depleted reservoir size ranging between 30% and 97% of the mantle and the rest of the mantle being primitive, that is, roughly chondritic. Even with the latest constraints there is a considerable range in the estimated depleted reservoir size: 50–80% of the mantle (Hofmann, 2003). A curious feature of this calculation is that the degree of melting required to produce continental crust is

small, for example, $\sim$1% (Hofmann, 2003) – about an order of magnitude lower than the 8–10% melting thought to produce present-day MORB. This indicates that incompatible trace elements must be scavenged from a large volume and concentrated into a small volume in order to produce the concentrations observed in continental crust, as discussed in Section 10.1.5.2.

### 10.1.4.1   Reservoir of 'missing' elements

The above mass-balance calculation implies the existence of a primordial reservoir of primitive (i.e., BSE) composition containing the trace elements that are 'missing' from the continental crust plus depleted (MORB source) mantle. This inferred reservoir of volume 20–50% of the mantle is often argued to be important in explaining the mantle's heat budget, as it would contain a large amount of heat-producing elements, and the concentration of heat-producing elements in the MORB source appears to be too low to explain the mantle's heat budget. The reservoir appears to be 'hidden', in the sense that it produces no observed geochemical signature, except perhaps that of high $^3$He/$^4$He. As an alternative to a primitive reservoir, it has also been proposed that the 'missing' trace elements are contained in a smaller, enriched reservoir (e.g., Coltice and Ricard, 1999; Tolstikhin and Hofmann, 2005; Boyet and Carlson, 2006; Tolstikhin et al., 2006).

### 10.1.5   Origin of Heterogeneities

Chemical variations inside the mantle could arise either from primordial differentiation of the Earth, through subsequent magmatism, or through possible minor reactions with the core, and these are discussed in this section. Quantitative models of either process require a knowledge of the bulk composition of the mantle, which was discussed in Section 10.1.2.

### 10.1.5.1   Primordial (fractionation of magma ocean, core formation, early atmosphere)

The extent to which the mantle might have started off layered due to fractional crystallization of a magma ocean has been hotly debated over the years. Key issues are whether a deep magma ocean existed, and if so whether its cooling would result in a fractionation due to crystal settling, or whether it would instead solidify into a chemically homogeneous mantle. This depends on various factors such as type of transient atmosphere present, grain size, viscosities, convection mode, etc. (Solomatov and Stevenson, 1993a, 1993b; Abe, 1997). The existence of a deep magma ocean has previously been argued against because the geochemical signature of the mantle is inconsistent with wholesale fractionation (e.g., Ringwood (1990a)), but if magma ocean convection is vigorous enough that crystals do not settle (Tonks and Melosh, 1990; Solomatov *et al.*, 1993; Solomatov and Stevenson, 1993b), then this difficulty vanishes. Furthermore, models of Earth formation predict collisions with planetesimals the size of the Moon or even Mars (the latter of which offers the most popular explanation for the formation of the moon), which would inevitably result in large-scale melting and a deep terrestrial magma ocean (e.g., Benz and Cameron, 1990; Melosh, 1990). Another possibly important aspect of magma ocean evolution is its interaction with the primordial atmosphere, through which it might have absorbed volatile elements (e.g., Ballentine, 2002), although Ballentine *et al.* (2005) suggest that this was not important for primitive volatile (particularly He/Ne) acquisition.

While there is therefore much uncertainty about magma ocean evolution, a recent synthesis by Solomatov (2000) offers a scenario that is consistent with both geochemical and dynamical constraints. In this, following a giant impact the lower mantle solidifies rapidly (i.e., <1000 years) and undifferentiatedly, leaving a shallow magma ocean in which fractionation and much slower cooling occur. This is consistent with the earlier work of Abe (1997), who also found that following a giant impact, lower-mantle differentiation is uncertain whereas upper-mantle differentiation is likely, and a shallow magma ocean could remain for 100–200 My.

Another chemically important process is the separation of the metal core from the remaining silicate. Metal–silicate separation is likely to have occurred in (at least) two environments (Stevenson, 1990; Karato and Murthy, 1997): small metal droplets falling through a magma ocean, where separation would occur rapidly (Tonks and Melosh, 1990; Solomatov, 2000; Rubie *et al.*, 2003; Hoink *et al.*, 2006); and large metal diapirs sinking through solid silicate or primitive material. The details of these strongly influence the partitioning of trace elements between the mantle and core as well as the extent to which the core and mantle are chemically equilibrated in general. Specifically, chemical equilibration is efficient in a magma ocean, but is ineffective for large sinking diapirs, leading to the view that metal–silicate equilibration last happened at

temperatures and pressures relevant to the base of the magma ocean (Murthy and Karato, 1997). Such mechanisms are relevant to questions such as whether the core contains potassium or noble gases, as well as whether reactions between the mantle and core take place.

In summary, at the present moment the consensus from geochemical constraints combined with dynamical considerations is that primordial chemical stratification in the deep mantle is unlikely, whereas primordial upper-mantle chemical stratification is likely. Partitioning of elements between the mantle and core is sensitive to poorly known processes that occurred during core–mantle separation.

### 10.1.5.2   Formation of continental and oceanic crust

It is clear that much differentiation has taken place due to melting subsequent to Earth's formation. The most obvious manifestations of this are (1) the formation of the continental crust, which is strongly enriched (by a factor of up to 50–100) in incompatible trace elements compared to BSE models and can survive over billions of years, and (2) the formation of the oceanic crust, which is less enriched and gets subducted back into the mantle after 100–200 My possibly to get mixed again (see later discussion). A thorough review of crustal formation and its geochemical consequences was given in Hofmann (2003); here the main points relevant to the discussion in this chapter are summarized.

### 10.1.5.2.(i)   Continental crust   As discussed above, the (enriched) continental crust is typically thought to be complementary to the depleted mantle, with an implied degree of melting of ~1%. This is very low, and suggests that continental crust is produced by a more complex process than simple partial melting (by ~1%) and eruption – indeed, melting experiments carried out on peridotite produce magmas that do not resemble (i.e., are less 'evolved' than) the continental crust in their major-element composition (Hirose and Kushiro, 1993; Rudnick, 1995). Various possibilities have been proposed, as reviewed by Rudnick (1995). Modern continental crust is produced either by the intraplate volcanism (i.e., flood basalts on continents, or accretion of oceanic plateaus) or at convergent margins, for which an important process is dehydration of subducted crust and removal and transport of incompatible elements by the resulting fluid in the subduction wedge, followed by metasomatism and melting to produce new

crust. Processes generating continental crust may have been different in the Archaean (e.g., Rudnick, 1995; Taylor and McLennan, 1995): possibilities include melting of subducted oceanic crust, weathering of surface rocks leading to depletion in MgO. Furthermore, the growth history of continents is highly uncertain, and reflects a balance between processes of production of new continents and entrainment of delamination of old continental material, with several different models proposed (e.g., Reymer and Schubert, 1984; Taylor and McLennan, 1995). Recent zircon data imply that continents already existed 4.4–4.5 Gy before present and were being rapidly recycled into the mantle (Harrison *et al.*, 2005). According to Rudnick (1995), anywhere between 40% and 100% of the present mass of the continents may have existed since the Archaean. New material is still being added.

***10.1.5.2.(ii)  Oceanic crust*** While continental crust formation is generally considered to be the dominant influence on the trace-element abundances in the mantle (the present-day oceanic crust is much less enriched and smaller in volume), a large fraction of the mantle has passed through the melting ('processing') zone beneath mid-ocean ridges, and thus the production and recycling of oceanic crust has had a major influence on mantle heterogeneity. The rate at which this happens is somewhat uncertain and is discussed in detail in the next section. Regardless of the exact rate of processing, subducted MORB is clearly a volumetrically significant component in the mantle and should not be ignored in calculating mantle geochemical evolution.

### 10.1.5.3  Processing rate of mid-ocean ridge melting

The exact volume of material that has been processed by MOR melting is quite uncertain, because of differing estimates of the present-day rate of processing and uncertainty about how this changed in the past, over billions of years. One measure of the processing rate is the 'residence time', which for random sampling of the mantle is equal to the length of time taken for the mass of the mantle to pass through the processing zone. At present rates of seafloor spreading, estimates of this processing time or residence time have varied widely, from 2.9 to 10 Ga. Example estimates are 4 Gy (O'Connell and Hager, EOS 61, 373, 1980, as referenced in Davies (1981)), 10 Gy (Gurnis and Davies, 1986b), 5.7 Gy (Kellogg and Wasserburg, 1990), 3–6 Gy (Davies, 2002),

9.5 Gy (Phipps Morgan, 1998), 4.5 Gy (Xie and Tackley, 2004b), or 2.9 Gy (Coltice, 2005). It is worth analyzing why there is such a large variation in these estimates. The processing rate, expressed as the volume of mantle processed per year, is given by

$$P = Ad\rho \qquad [3]$$

where $A$ is the area of oceanic seafloor produced per year, $d$ is the depth of onset of melting beneath spreading centers, and $\rho$ is the density of the processed material. Observations of the age distribution of ocean floor (e.g., summarized in Phipps Morgan (1998)) indicate that $A = 2.7\,\mathrm{km^2\,yr^{-1}}$ for seafloor produced in the last 1 Ma, or $A = 2.9\,\mathrm{km^2\,yr^{-1}}$ based on an average over the last 100 Ma, the latter being similar to the $3.0\,\mathrm{km^2\,yr^{-1}}$ estimated by Stacey (1992). Other authors have preferred to use 'back of the envelope' estimates, for example, Davies (2002) assumed a ridge length of 40 000 km and spreading velocity of 5–10 cm yr$^{-1}$ (giving $A = 2$–$4\,\mathrm{km^2\,yr^{-1}}$) and Coltice (2005) assumed a ridge length of 60 000 km and plate velocity of 5 cm yr$^{-1}$ ($A = 6\,\mathrm{km^2\,yr^{-1}}$). Regarding onset depth of melting $d$ Phipps Morgan (1998) assumed that the oceanic crustal thickness of 6.5 km was produced with 10% melt ($d = 65\rho_{\mathrm{crust}}/\rho_{\mathrm{mantle}}$ km), whereas Davies (2002) assumed $d = 100$ km and Coltice (2005) assumed $d = 150$ km. Using the observationally constrained $A = 2.9\,\mathrm{km\,yr^{-1}}$ together with, for example, $d = 80$ km and peridotite density $= 3000\,\mathrm{kg\,m^{-3}}$ leads to $P = 6.96 \times 10^{14}\,\mathrm{kg\,yr^{-1}}$, which gives a timescale of 5.74 Gy to process the mantle mass of $4 \times 10^{24}$ kg. Thus, estimates of processing time as low as 3 Gy should be regarded as extreme lower limits. Estimates of the present-day processing time based on geochemical fluxes (Albarède, 2005b) range from 4 to 9 Gy, bracketing the geophysical value discussed above; this estimate is further discussed in Section 10.2.5.

It is likely that the rate of seafloor spreading and hence crustal production was considerably higher in the past when the mantle was hotter, which would greatly reduce the processing/residence time in the past (Gurnis and Davies, 1986b). In addition, the depth of melting would have been larger due to the hotter mantle, as considered by Tajika and Matsui (1992, 1993), and Davies (2002), further increasing processing rate. The appropriate increase of seafloor spreading in the past is not known: in simple, internally heated convection in secular equilibrium the surface velocity scales as radiogenic heat production[2]. Davies (1980) and Phipps Morgan (1998) gave detailed arguments why this also applies to plate tectonics. It is, however, likely that the early Earth did not have plate tectonics as it exists today because the thicker oceanic crust and younger oceanic plates (when they reach subduction zones) would make plates difficult to subduct (Vlaar, 1985; Davies, 1990, 1992), and perhaps because the mechanical oceanic lithosphere was thicker because water depletion, which causes an increase in viscosity, occurred deeper (Korenaga, 2003). Thus, the most likely increase in crustal production rate lies between constant and $H^2$, where $H$ = radiogenic heat production rate.

To give some quantitative examples of the influence of faster processing in the past: (1) Xie and Tackley (2004b) calculated the volume of oceanic crust produced over the age of the Earth to be ~10% of the mantle mass if the present-day production rate applied to whole Earth history, 54% if the $H^2$ law of Davies (1980) applied, or 82% if the increase in depth of melting due to higher mantle temperatures were additionally taken into account. In the latter two cases, this clearly implies that the same material melts two or more times, as it is not possible to produce this amount of MORB from the residue of continental crust production. (2) Davies (2002) calculated that the $H^2$ law leads to four times the amount of processing over geological time, which he expressed as being equivalent to 18 Gy of constant-rate processing. This means that 98.6% of the mantle is processed if the present-day residence/processing time is 4.2 Gy.

### 10.1.5.4 Reactions with the core
Some researchers have studied the possibility of an iron-rich layer forming above the core–mantle boundary (CMB) caused by reactions with the core. There is some geochemical evidence (based on the Os–Pt system) for mass exchange between the mantle and core (Walker et al., 1995; Brandon et al., 1998, 2003; Humayun et al., 2004), but this is controversial (Schersten et al., 2004; Brandon and Walker, 2005).

Experiments indicate that iron reacts with silicate perovskite to form iron oxide and silica (Knittle and Jeanloz, 1989, 1991). Experiments with liquid iron and solid silicate show that the silicate becomes depleted in Fe and O (Goarant et al., 1992). Liquid iron also reacts with $Al_2O_3$ at CMB pressures (Dubrovinsky et al., 2001) but not with silica (Dubrovinsky et al., 2003). Using thermodynamic calculations, Song and Ahrens (1994) determined that

the observed reactions are thermodynamically possible. Thus, core–mantle reactions are expected from an experimental standpoint, although it is possible that the presence of water inhibits them (Boehler et al., 1995; Poirier et al., 1998).

It appears difficult to generate significant mantle heterogeneity this way, because the thickness of the reaction zone is relatively small. Poirier and Le Mouel (1992) and Poirier (1993) find that liquid Fe can penetrate 1–100 m into the mantle along grain boundaries if it reacts with the silicate. Knittle and Jeanloz (1991) proposed that the Fe ascents by a capillary action driven by surface tension. However, a much more efficient mechanism is a suction mechanism in areas of downwelling (where the CMB is depressed) which may entrain Fe of order ~1 km into the mantle with fractions of up to 10%, depending on parameters such as the effective bulk viscosity (Stevenson, 1988; Kanda and Stevenson, 2006). Petford et al. (2005) also proposed a dynamically assisted Fe transfer mechanism assuming that the D″ region behaves as a poro-viscoelastic granular material with dilatant properties, but the required strain rates are rather high ($10^{-12}\,s^{-1}$). Thus, it appears that Fe might be able to infiltrate up to ~1 km at significant volume fractions, but this needs further study.

If Fe does infiltrate the mantle in some regions in a thin ($\leq 1$ km) layer and react to form dense products, the question is then whether the continual flux of new mantle material through this reaction zone is sufficient, over geological time, to affect a volumetrically significant amount of mantle material. Knittle and Jeanloz (1991) suggested that the material would not be raised far into the mantle because of its increased density, and could accumulate into a few 100-km-thick D″ layer, but this needs to be tested using detailed modeling. Numerical models of this performed to date have parametrized core reactions as an effective chemical diffusivity, which may not correctly capture the process. The most advanced model is that of Kellogg (1997), who parametrized the infiltration of Fe using a chemical diffusivity 100–1000 times smaller than the thermal diffusivity, and found that large density contrasts ($B = 10$) result in a thin (<100 km) layer that slows further growth, whereas lower density contrast ($B = 1$) result in an undulating layer several hundreds of kilometers thick that piles up under upwellings, as is observed in earlier calculations (Hansen and Yuen, 1988). If a metal-bearing layer does build up, it could have a thermal conductivity an order of magnitude higher than that of silicates (Manga and Jeanloz, 1996), which could further influence mantle convection.

It has been proposed that silicate melt (partial melt fractions up to 30%) exists in some areas at the CMB as an explanation of seismically observed ultralow velocity zones (ULVZs) (Williams and Garnero, 1996; Lay *et al.*, 2004). Such areas of partial melt could have interesting geochemical consequences. Firstly, they might facilitate reactions between the mantle and the core due to the larger flux of mantle material that comes into contact with the core. Secondly, incompatible trace elements would be redistributed between melt and different solid phases with different partition coefficients. If they were still incompatible then the melt could act as a reservoir of such elements; however, Hirose *et al.* (2004) found that large ion lithophile (LIL) elements partition into the solid calcium–perovskite phase rather than the melt, which should result in measurable geochemical signatures for plumes. Lateral variations in the occurrence and/or thickness of partially molten zones could be due to lateral variations in bulk composition, or possibly to the interplay between upwellings, downwellings, and melt migration. In the past, the CMB temperature was higher and so such regions of partially molten silicate would have been larger and/or more prevalent, or at least, more likely.

### 10.1.6 MORB and OIB Trace-Element Signatures

#### 10.1.6.1 *Noble gases*
The most common noble gases considered in mantle geochemistry are the isotopes of He and Ar, with Ne and Xe also sometimes considered.

#### 10.1.6.1.(i) *$^3$He/$^4$He ratios* The two isotopes of Helium are $^3$He, which is primordial, and $^4$He, which is produced by the radioactive decay of U and Th. Histograms of the ratio of $^3$He/$^4$He for mid-MORBs and OIBs show clear differences. MORBs have a relatively constant $^3$He/$^4$He ratio with a mean value of $1.16 \times 10^{-5}$, which is a factor of 8.4 higher than the atmospheric value (Kurz *et al.*, 1983; Allegre *et al.*, 1995; Graham, 2002). The range of $^3$He/$^4$He ratios between 7 and 9 is thus taken as representative of the depleted upper mantle. OIB $^3$He/$^4$He ratios are, in contrast, extremely heterogeneous, extending to both higher and lower values than MORB, with the highest ratios measured (at Baffin Island) about 50 times atmospheric (Stuart *et al.*, 2003).

There are several possible explanations for the high $^3$He/$^4$He material. The 'traditional' interpretation is that this component comes from an undegassed, 'primitive' source, because noble gases are highly incompatible and volatile, and are thus expected to enter any melt and be heavily out gassed prior to or during eruption (Allegre *et al.*, 1983b; Farley *et al.*, 1992; Porcelli and Wasserburg, 1995) (note that outgassing requires $CO_2$ bubbling, which occurs at depths less than 60 km). This component has traditionally been identified as the lower mantle, consistent with the need to keep it intact for billions of years, with other arguments for the lower mantle being primitive, and with the idea of plumes sampling the lower mantle and spreading centers sampling the upper mantle. However, the actual volume and location of this component is not constrained. If the 'primitive' interpretation is correct, one might expect the absolute concentration of He in OIB to be higher than that in MORB, but this is not observed, perhaps because magma chamber processes cause loss of He (Farley and Neroda, 1998).

The leading alternative explanation for high $^3$He/$^4$He is that it is associated with recycled depleted oceanic lithosphere (Albarède, 1998; Anderson, 1998; Coltice and Ricard, 1999; Ferrachat and Ricard, 2001; Coltice and Ricard, 2002). This is based on the fact that as high $^3$He/$^4$He simply requires high $^3$He/(U+Th) because decay of U and Th produces $^4$He. Therefore, an alterative to high [$^3$He] is low [U+Th], which is found in material that has been depleted in incompatible trace elements due to melting. The evolution of He ratios is illustrated in **Figure 1**. Note that $^3$He/$^4$He decreases with time in all 'reservoirs' due to the radiogenic ingrowth of $^4$He. If the ratio $^3$He/(U+Th) is changed in different components due to fractionation (during which $^3$He/$^4$He remains the same), then with time, it will evolve $^3$He/$^4$He that is higher or lower than the original material. If He is less incompatible than U and Th (e.g., Graham *et al.*, 1990), and as found in recent laboratory experiments (Parman *et al.*, 2005), then $^3$He/(U+Th) is higher in the depleted residue and after some time its $^3$He/$^4$He will be higher than that of the original material. However, Anderson (1998), Coltice and Ricard (1999), Ferrachat and Ricard (2001), Coltice and Ricard (2002) showed that it is not necessary for He to be less incompatible if the residuum is stored for substantial time periods while the rest of the mantle becomes increasingly outgassed. To understand this, note that in the crust, $^3$He/(U+Th) is extremely low

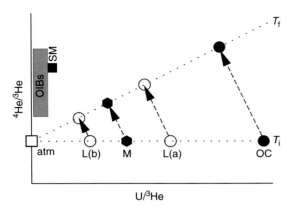

**Figure 1** Schematic diagram showing the Helium isotope evolution of different mantle components. M, mantle; L, lithosphere; atm, atmosphere; OC, oceanic crust. When M melts it produces oceanic crust, lithosphere (a) or (b) depending on the relative partition coeffients of U and He, and some He is lost to the atmosphere. Between melting events, U decays and produces $^4$He, so U/$^3$He decreases and $^4$He/$^3$He increases. The shallow mantle (SM) is a mixture of recycled lithosphere and oceanic crust, so recycled lithosphere will always have lower $^4$He/$^3$He (high $^3$He/$^4$He) than the average, regardless of the relative partition coefficients. With successive melting events, the mantle average ratios move to the right due to He outgassing. From Coltice N and Ricard Y (2002) On the origin of noble gases in mantle plumes. *Philosophical Transactions of the Royal Society of London Series A-Mathematical Physical and Engineering Sciences* 360: 2633–2648.

due to degassing of He and concentration of U and Th, so the crust will have much lower $^3$He/$^4$He regardless of the exact relative He and U partitioning. If the mantle has almost all differentiated, such that to first order it is made of a mixture of subducted crust and depleted residue, then the average mantle $^3$He/$^4$He will inevitably be in between the ratios in residue and subducted crust, making residue $^3$He/$^4$He appear to be high.

Another proposed explanation of the 'primordial' helium component is that it comes from the core. However, the only measurements of noble gas partitioning between silicate melt and iron melt under pressures up to 100 kbar indicate that the partition coefficients are much less than unity and that they decrease systematically with increasing pressure (Matsuda *et al.*, 1993). These results suggest that Earth's core contains only negligible amounts of noble gases if core separation took place under equilibrium conditions at high pressure, although the uncertainties are too great to rule out the core as a potential mantle $^3$He source (Porcelli and Halliday, 2001).

Finally, it has been proposed that a large amount of $^3$He was delivered to the Earth's surface by cosmic dust, and then recycled to the mantle (Allegre *et al.*, 1993; Anderson, 1993). Several arguments have been made against this proposal (Farley and Neroda, 1998); in particular, the dust flux is several orders of magnitude too low to account for the $^3$He flux out of the mantle, and $^3$He is expected to be degassed from the dust particles during subduction.

***10.1.6.1.(ii) Heat–Helium imbalance*** The present-day flux of $^4$He out of the mid-ocean ridge system has been estimated by measuring the concentration of $^4$He in ocean water. O' Nions and Oxburgh (1983) pointed out that this $^4$He flux is only a small fraction of what would be expected from the concentration of $^{238}$U, $^{235}$U, and $^{232}$Th in MORB and even lower than U and Th in BSE, which has led to the term 'heat–Helium imbalance'. The 'traditional' explanation of this is that the radiogenic sources are trapped in the lower mantle, from which heat can escape by conduction across a thermal boundary layer, but $^4$He cannot. It has been suggested that the concentration of $^4$He might be misleading because this is influenced by short-timescale processes, whereas the concentration of U and Th is a long-timescale process. Possible solutions to this are discussed in later Sections 10.4.1.1 and 10.5.1.

***10.1.6.1.(iii) Argon outgassing*** Argon has three stable isotopes, the two of most interest being 'primitive' $^{36}$Ar, and 'radiogenic' $^{40}$Ar. There is virtually no primordial $^{40}$Ar in the Earth (Ozima and Kudo, 1972), it being entirely produced by the decay of $^{40}$K. As Ar does not escape from the atmosphere (unlike He) due to its large atomic mass, the amount of $^{40}$Ar in the Earth's atmosphere represents the total amount of argon degassed from the Earth's interior over the age of the Earth. This atmospheric $^{40}$Ar budget, when combined with an estimate of the total $^{40}$K content of the Earth, implies that approximately half of all $^{40}$Ar produced within the Earth since its formation is retained within the solid Earth (Allegre *et al.*, 1996; O' Nions and Tolstikhin, 1996), which gives a constraint on the fraction of the mantle that has been processed through mid-ocean ridge or other melting environments. There is some uncertainty in this figure due to uncertainty in the amount of $^{40}$K in the mantle and a fairly small uncertainty in how much of the $^{40}$Ar in continental crust has actually outgassed. The uncertainty in total amount of K arises because in the planetary formation process K is moderately volatile, so the total amount of K is usually calculated from the total amount of (refractory) U assuming some a particular K/U ratio, estimates of which range from a widely accepted value of 12 700 (Jochum *et al.*, 1983) to as low as 2800 (Stacey, 1992). The continental crust contains ~1/3 to 1/2 of the K budget (Rudnick and Fountain, 1995) and for the above mass-balance calculation is usually assumed to be completely outgassed in the resulting $^{40}$Ar, which may not be the case (see, e.g., discussion in Phipps Morgan (1998)). The linkage between K assimilation by the continental crust and $^{40}$Ar crust and mantle outgassing can be used to constrain the relative amounts of K in the crust and Ar in the atmosphere (Coltice *et al.*, 2000a). In any case this '50% outgassing' constraint is commonly misinterpreted to mean that 50% of the mantle is undegassed – in fact, because Earth's $^{40}$Ar was continuously produced over geological time from a negligible initial concentration, the mantle could have almost entirely degassed early in its history, and still match the Ar constraint. In other words, early or accretionary degassing events do not influence $^{40}$Ar outgassing. Quantitatively, for a constant convective vigor van Keken and Ballentine (1999) show that outgassing 50% radiogenic $^{40}$Ar implies outgassing 72% of primordial, nonradiogenic species. If the convective vigor were higher in the past then the latter fraction can be much higher. Phipps Morgan (1998) showed that the outgassing constraint is perfectly compatible with whole-mantle convection, even if the processing rate is much higher in the past: for example, assuming that the present-day processing time is 9.6 Gy and was faster according to (heating rate)$^2$ in the past, one obtains 93% outgassing of primitive $^{36}$Ar but only 58% outgassing of radiogenic $^{40}$Ar, or 98% primitive outgassed while retaining 31% of radiogenic Ar. A better expression of this constraint is '50% of the mantle has been undegassed since the $^{40}$Ar was produced in it' (Phipps Morgan, 1998).

### 10.1.6.1.(iv)   *The Argon paradox*   A more problematic constraint arises from the concentration of Ar in the MORB source region, which is much lower than what would be expected if the $^{40}$Ar remaining in the mantle were evenly distributed (Turner, 1989). This has traditionally been interpreted to mean that the additional, 'missing' $^{40}$Ar must be 'hidden' in the lower mantle, which should have about 50 times the $^{40}$Ar concentration of the upper mantle

(Allegre *et al.*, 1996). It has been argued that this difficulty could be resolved if the amount of $^{40}$K in the mantle were lower than commonly estimated (Albarède, 1998; Davies, 1999). Another reconciliation is that this, as well as the heat–Helium imbalance and the apparent need for a deep $^3$He reservoir, could be reconciled if the estimate of $^{40}$Ar in the shallow mantle, which comes from measured short-term $^3$He fluxes into the oceans, were too low by a factor of 3.5 (Ballentine *et al.*, 2002). It is important to note that the proposal of K in the core does not help with this paradox, because it is the imbalance between mantle Ar and K that is the problem.

### 10.1.6.1.(v)   *Neon, Xenon, Krypton*   Far less attention has been paid to isotopes of these elements, largely because of atmospheric contamination problems (Farley and Neroda, 1998; Ballentine and Barford, 2000; Ballentine, 2002), but some recent work has important implications for mantle processes. Neon has three isotopes (atomic weights 20, 21, 22), of which $^{21}$Ne is radiogenic and $^{20}$Ne and $^{22}$Ne are primordial. Recent measurements of $^{20}$Ne/$^{22}$Ne in magmatic $CO_2$ well gases (Ballentine *et al.*, 2005) provide the first unambiguous value for the convecting mantle noble gas abundance and isotopic composition, and indicate that the upper mantle has a $^{20}$Ne/$^{22}$Ne similar to that of solar wind-irradiated meteorites (~12.5, so-called Neon-B), which is significantly lower than $^{20}$Ne/$^{22}$Ne of the solar nebula (13.6–13.8). This Neon-B value is reproduced in samples from Iceland and Hawaii (Trieloff *et al.*, 2000) and Reunion (Hopp and Trieloff, 2005). Some studies (e.g., Honda *et al.*, 1999; Dixon *et al.*, 2000; Moreira *et al.*, 2001) have argued for a solar component based on extrapolating plots of $^{20}$Ne/$^{22}$Ne versus $^{21}$Ne/$^{22}$Ne but their measured $^{20}$Ne/$^{22}$Ne values are not higher than the Neon-B value. However, some samples from Iceland (Harrison *et al.*, 1999) and the Kola peninsular (Yokochi and Marty, 2004) appear to have $^{20}$Ne/$^{22}$Ne > 12.5, which if verified would represent a minor component.

These measurements imply that the convecting mantle is dominated by Neon gained by solar wind irradiation of accreting material (Neon-B), but that another 'solar' component might exist, sometimes sampled by plumes, associated with early accreted material with a solar nebula composition. The latter might be in the deep mantle or core (Ballentine *et al.*, 2005). Alteration by recycled material appears to have played a minimal role. While measured Xe and Kr ratios generally appear air-like (Ballentine,

2002), evidence for 'solar' Xe has also been detected in $CO_2$ well gases by Caffee *et al.* (1999). A recent analysis suggests that the mantle content of 'heavy' noble gases (i.e., Xe, Kr, Ar) is dominated by ocean water subduction (Holland and Ballentine, 2006), consistent with the finding of Sarda *et al.* (1999) that a significant amount of mantle Ar is recycled.

### 10.1.6.2   The end-member 'zoo'
Analysis of the varying trace-element isotopic compositions of OIBs and MORBs, particularly those of Sr, Nd, Hf, and Pb suggest that they are the result of mixing between several 'end members', which are typically referred to as DMM, HIMU, EM1, and EM2 (White, 1985; Zindler and Hart, 1986; Carlson, 1994; Hofmann, 2003). Additionally considering He data leads to a further component, commonly referred to as C or FOZO. The common interpretations of these are as follows:

#### 10.1.6.2.(i)   DMM (depleted MORB mantle)
This end-member represents the most depleted MORB. Most MORBs are not as depleted as DMM, but plot on the DMM–HIMU trend.

#### 10.1.6.2.(ii)   HIMU (high-$\mu$)
$\mu$ is the ratio $^{238}U/^{204}Pb$. As U and Th decay into $^{207}Pb$ and $^{206}Pb$ with time, high $\mu$ leads to high ratio of radiogenic to primordial lead, that is, $^{206}Pb/^{204}Pb$ – a 'radiogenic' signature. This component is typically interpreted to be recycled oceanic crust (Chase, 1981; Hofmann and White, 1982; Allegre and Turcotte, 1986; Hofmann, 1997). However, as U and Pb are approximately equally compatible (or U is slightly more incompatible) when making oceanic crust, some additional mechanisms must be invoked to fractionate U and Pb to make the HIMU signature. The invoked mechanisms are hydrothermal removal of Pb from the subducted slab into subduction-related magmas (Hofmann, 1988) and enrichment of the oceanic crust in continental U carried by eroded sediments.

#### 10.1.6.2.(iii)   EM (enriched mantle) 1 and 2
These signatures are observed in OIBs, but their origin remains controversial. Leading explanations are by Carlson (1994) and Hofmann (2003): EM1 may originate from either recycling of delaminated subcontinental lithosphere or recycling of subducted ancient pelagic sediment, while EM2 may be recycled oceanic crust with a small amount of sediment, melt-impregnated oceanic lithosphere, or

sediment contamination of plume-derived magmas as they pass through the crust.

#### 10.1.6.2.(iv)   FOZO (Focus zone)
Components that are essentially the same as FOZO (Hart *et al.*, 1992, Hilton *et al.*, 1999) have also been named C (Hanan and Graham, 1996), PHEM (Farley *et al.*, 1992), PREMA (Zindler and Hart, 1986), and HRDM (Stuart *et al.*, 2003). This corresponds to the high $^3He/^4He$ component, and in other isotopes has a composition on the DMM–HIMU trend, although in the definition of (Stracke *et al.*, 2005) it lies off the DMM–HIMU trend when considering the Sr–Rb system. Samples from many OIB associations appear to form linear trends that radiate from FOZO to various enriched compositions.

#### 10.1.6.2.(v)   Isotopic 'Ages'
As discussed earlier, isotope ratios evolve along separate paths in different materials due to fractionation of radiogenic parent and daughter elements during melting and subsequent radiogenic ingrowth of daughter isotopes. The resulting slopes on isotope diagrams indicate an age of fractionation and are widely used to date geological processes. Even though the mantle has undergone continuous differentiation rather than a single differentiation event, plots of $^{207}Pb/^{204}Pb$–$^{206}Pb/^{204}Pb$ from MORB and OIB samples display a coherent line, or 'pseudo-isochron', from which an 'effective age' can be calculated. The Pb system has the advantage that Pb isotopes do not fractionate from each other on melting, which is a disadvantage with, for example, the Sm–Nd system. Pb–Pb plots display a trend from DMM to the highly radiogenic HIMU end-member. The Pb–Pb pseudo-isochron ages for various MORB and OIB groups correspond to effective ages of 1.5–2 Gy (Hofmann, 1997), although the exact meaning of these ages is not straightforwardly interpretable (Hofmann, 1997; Albarède, 2001). A recent mathematical analysis reveals that for ongoing differentiation the Pb–Pb 'pseudo-isochron' age is generally much larger than the average age of crustal differentiation (Rudge, 2006).

#### 10.1.6.2.(vi)   Lead paradoxes
Another constraint is that all oceanic basalts and most MORB plot to the high $^{206}Pb/^{204}Pb$ side of the geochron, implying that the mantle has experienced a net increase in $\mu$, and implying that there is a hidden reservoir of unradiogenic (low $\mu$) material. This problem is still unresolved and is discussed in more

detail in Hofmann (2003) and Murphy *et al.* (2003). A second lead paradox, often called the kappa conundrum, is related to the discrepancy between measured $^{208}Pb/^{206}Pb$ ratios and the ratio of their parents, $\kappa = {}^{232}Th/^{238}U$. A layered-mantle 'steady-state' solution to this was proposed by Galer and O'Nions (1986), with a more recent layered solution proposed by Turcotte *et al.* (2001). Other recent work has focused on reconciling it with whole-mantle convection; a reasonable resolution is starting to recycle uranium into the mantle $\sim 2.5$ billion years ago (Kramers and Tolstikhin, 1997; Elliott *et al.*, 1999).

### 10.1.7 Length scales: Geochemical and Seismological

Of primary interest for understanding mantle-mixing processes and the preservation of chemical endmembers is the length-scale distribution of chemical heterogeneity in the mantle, so this section reviews constraints from geochemical and seismological observations. In general, it is observed that heterogeneity exists at all scales, but quantitative constraints on the spectrum are minimal.

#### 10.1.7.1  Geochemical

Chemical heterogeneity is observed at all scales (Carlson, 1987, 1994; Meibom and Anderson, 2004), even within the relatively uniform MORB source. For example, at the smallest scale (i.e., melt inclusions in olivine phenocrysts), Pb isotope ratios in OIB span a large fraction of the global range (Sobolev and Shimizu, 1993; Sobolev, 1996; Saal *et al.*, 1998; Sobolev *et al.*, 2000; Hauri, 2002), while Nd ratios in the Ronda peridotite massif in Spain exceed the global OIB Nd ratio variation (Reisberg and Zindler, 1986). According to Sobolev *et al.* (2000), extreme heterogeneity of trace-element patterns in olivine phenocrysts at Hawaii indicate that the isotopic signature is maintained up to crustal magma chambers, limiting the scale of heterogeneities in the source region to a few kilometers or less. At the centimeter to meter scale, mantle rocks outcropping in peridotite massifs have been observed to contain centimeter-to-decimeter-thick isotopically enriched pyroxenite veins embedded within a mostly depleted peridotite matrix (Polve and Allegre, 1980; Reisberg and Zindler, 1986; Suen and Frey, 1987; Pearson *et al.*, 1993). Such variations have inspired the proposal of a 'marble cake' or 'plum pudding' style of chemical variation in the mantle, in which small volumes of enriched material reside within a matrix of depleted

MORB mantle (Allegre and Turcotte, 1986), a concept that has been the topic of much recent investigation as discussed in later sections. At the longest scale, there is also abundant evidence for compositional heterogeneity at scales of >1000 km in the form of the Dupal anomaly (Schilling, 1973; Dupre and Allegre, 1983; Hart, 1984; Klein *et al.*, 1988; Hart, 1984; Castillo, 1988), a $\sim 10^8 \, km^2$ region centered in the Indian Ocean that stretches in an almost continuous belt around the Southern Hemisphere between the equator and 60° S and is characterized by its anomalous Sr and Pb ratios.

Limited information exists about the spectrum of geochemical heterogeneity from sampling variations in isotope ratios along mid-ocean ridges. An early compilation of $^{87}Sr/^{86}Sr$ data from MORB and OIBs (Gurnis, 1986b) seemed consistent with a flat or 'white' spectrum, but the wealth of data that has been gathered since then reveals more complex patterns. The spectrum of $^3He/^4He$ for 5800 km of the southeast Indian ridge displays periodicities of 150 and 400 km (Graham *et al.*, 2001), which was interpreted to be indicative of upper-mantle convection. Using new high-resolution MC-ICPMS data, Agranier *et al.* (2005) plotted the spectra of various isotope ratios along the mid-Atlantic ridge, finding two distinct signals: a 6–10° length-scale signal visible in Pb, Sr, and He ratios and the principle component 1 and associated with the Iceland hot spot, and a power-law spectrum with a slope of −1 visible in Nd, Hf ratios and principle component 2, perhaps indicative of dynamic stretching processes discussed later. Graham *et al.* (2006) found a bimodal distibution of hafnium isotope ratios along the southest Indian ridge, which appears to indicate compositional striations with an average thickness of ~40 km in the upper mantle.

#### 10.1.7.2  Seismological

The long-wavelength spectrum of heterogeneity is revealed by global seismic tomographic models, but much of the heterogeneity at this wavelength is thermal. There have, however, been attempts to separate thermal and chemical components to the seismic velocity variations (e.g., Ishii and Tromp, 1999; Forte and Mitrovica, 2001; Romanowicz, 2001; Trampert *et al.*, 2004). Particularly notable are those studies that use normal-mode splitting data because these are sensitive to density variations, providing an additional constraint. In particular, at the longest wavelengths of thousands of kilometers, Ishii and Tromp (1999) and Trampert *et al.* (2004) found that a substantial fraction of the shear-wave heterogeneity in the deepest mantle is due to compositional variations, and that the two low-velocity 'megaplumes' seen in global tomographic models seem to have dense material at their bases. Short-wavelength heterogeneity is constrained by scattering: short-period precursors to PKP reveal the presence of small ~8 km, weak (1% rms) velocity perturbations throughout the mantle (Hedlin et al., 1997). There are strong regional differences in scattering strength (Hedlin and Shearer, 2000), with strong scattering beneath South and Central America, eastern Europe, and Indonesia, and some correlation with large-scale anomalies revealed by seismic tomography including the African superplume and Tethys trench. Other types of seismic observation find structure with a gradient in velocity that is too large to be explained by diffusive thermal structures, including a dipping low-velocity layer in the mid-lower mantle (Kaneshima and Helffrich, 1999) and sharp sides to the deep-mantle 'superplumes' beneath Africa and the Pacific (Ni et al., 2002). In summary, while there is ample seismic evidence for chemical heterogeneity in the mantle, it is not sufficient at this point to constrain the spectrum over all wavelengths, and in particular not to the subkilometer wavelengths to which subducted MORB is expected to be stretched.

The spectrum of heterogeneity expected from stirring theory is discussed in the later in Section 10.2.7.

## 10.1.8 Geophysical Constraints in Favor of Whole-Mantle Convection

A number of types of geophysical evidence argue that mantle convection is not completely layered at the 660 km discontinuity, which is in conflict with the models based on mass balance of Ar and other trace elements discussed earlier.

First, seismological observations have long favored penetration of slabs into the lower mantle. While arguments initially revolved around interpreting traveltimes (Creager and Jordan, 1984; Creager and Jordan, 1986), subsequent 3-D regional or global seismic tomographic models (Fukao, 1992; Fukao et al., 1992; Grand, 1994; Van der Hilst et al., 1997; Masters et al., 2000) clearly show slabs penetrating immediately in some areas, or penetrating after a period of stagnation in other areas, although it has also been argued that thermal coupling in a two-layered convection could produce similar features (e.g., Cizkova et al., 1999). Significant flow

stratification should lead to a strong decorrelation in global tomographic models across 660 km depth, which is not observed (Jordan et al., 1993; Puster and Jordan, 1994, 1997; Puster et al., 1995). Recent tomographic models have also revealed some upwelling plumes that appear to be continuous across the whole depth of the mantle (Montelli et al., 2004), although there is an active debate on this topic.

Several types of geodynamic evidence also favor penetration of flow into the lower mantle. First, if the 660 km discontinuity were a total barrier to convection, then there would a strong thermal boundary layer at the base of the upper mantle, and all of the heat coming from the lower mantle would be carried by strong upwellings in the upper mantle. As demonstrated by Davies (1988), such pervasive strong hot upwellings would violate constraints based on the topography and geoid of the ocean floors, which instead appear to be passively spreading. Identifiable hot-spot swells assumed to be caused by upwelling plumes only account for 6–10% of the global heat flow (Davies, 1988; Sleep, 1990). Second, modeling the global geoid and dynamic topography using driving forces derived from seismic tomography indicates a better fit to observations with whole-mantle flow than with layered flow (Hager et al., 1985; Ricard et al., 1988, 1989; Hager and Richards, 1989), although some successful fits are possible with layered models (Wen and Anderson, 1997). A third line of evidence comes from correlating global tomographic models with the spatial locations of subduction zones in the past 140–180 Ma (Richards and Engebretson, 1992; Scrivner and Anderson, 1992; Kyvalova et al., 1995): past subduction appears to match well with lower-mantle tomography, implying that the subducted slabs are in the lower mantle. Extending this approach, 3-D mantle structure models have been constructed by combining historical plate reconstructions with a global flow solver, and the results show a reasonable match to fast structures observed in global tomographic models (Deparis et al., 1995; Megnin et al., 1997; Bunge et al., 1998). Finally, numerical convection models indicate that the endothermic phase transition at 660 km depth is too weak to completely layer the flow at present-day convective vigor – only intermittent regional layering is plausible (Christensen and Yuen, 1985; Machetel and Weber, 1991; Peltier and Solheim, 1992; Tackley et al., 1994; Bunge et al., 1997), although this does not rule out chemical layering per se, as chemical layering with a large-enough density contrast is additive to the phase-change buoyancy effect (Christensen and Yuen, 1984).

While complete layering at 660 km depth seems ruled out, there is evidence in favor of chemical layering in the lowest ~few hundred kilometers of the mantle, perhaps localized in certain regions. First, the D″ region has long been found to be seismically highly heterogeneous (e.g., Lay and Garnero, 2004), which might require compositional variations to explain. Second, long-wavelength seismic tomographic models robustly find two large regions of low seismic velocity underneath Africa and Pacific. While these might plausibly be smeared-out groups of narrow plumes (Schubert *et al.*, 2004), seismic evidence indicates that these are not purely thermal in origin, including (1) 'anomalous' scaling between P-wave and S-wave velocities (Kennett *et al.*, 1998; Masters *et al.*, 2000; Bolton and Masters, 2001), (2) a positive density anomaly, according to some models that use normal-mode splitting data (Ishii and Tromp, 1999; Trampert *et al.*, 2004), (3) sharp sides, which are difficult to maintain with a diffusive thermal field (Luo *et al.*, 2001; Wen, 2001, 2002). Some thermochemical convection models discussed later produce structures that are consistent with these seismological observations.

It has also been argued that geophysical evidence favors a ~1300 km deep, highly undulating layer in the deep mantle (Kellogg *et al.*, 1999; van der Hilst and Karason, 1999). The boundary of this should, however, produce strong observable signatures (heterogeneity and scattering) that have not presently been found (Vidale *et al.*, 2001; Tackley, 2002; Castle and van der Hilst, 2003a, 2003b). At the present time, the evidence favors a deeper layer that is probably not global.

### 10.1.9 Overview of Geochemical Mantle Models

Before discussing the geophysical findings regarding thermochemical convection, it is useful to give an overview of the major models that have been proposed to explain geochemical observations and reconcile geochemical and geophysical observations. Some of these are discussed in more detail in the final section. Rather than discuss a large number of proposed models, some of which have similarities, it is useful to discuss the two main characteristics that differ among models, and which tend to be the aspects that models focus on. **Figure 2** schematically represents the main models.

#### 10.1.9.1 Vertical layering
##### 10.1.9.1.(i) Layered at 660 km
The traditional layered model is chemically and dynamically stratified at the 660 km discontinuity (e.g., slabs do not penetrate), with the upper mantle being a depleted, degassed region that is the complement of the continental crust, and the lower mantle being primitive – the repository for 50% of the BSEs Argon, for the high $^3$He/$^4$He component, and for other trace elements that are missing from the MORB source + continental crust according to mass-balance calculations (Allegre *et al.*, 1996). Variations on this model include intermittent breakdown of layering (Stein and Hofmann, 1994), the idea that this layering broke down in recent history (Allegre, 1997), and the model of Anderson (1995), in which the upper mantle is divided into an enriched shallow layer (the source of OIBs) and a depleted transition zone (the source of MORB). This model is contradicted by a number of geophysical observations, as discussed in Section 10.1.8.

##### 10.1.9.1.(ii) Deep, highly undulating layer (hidden anomalous layer)
This model, proposed by Kellogg *et al.* (1999), is conceptually similar to the traditional layered model except that the layer boundary is moved downwards to approximately 1300 km above the CMB, undulating by ~1000 km. Recycled components reside on top of the layer. The layer is 'hidden' because the seismic-wave velocity anomalies caused by the different composition and higher temperature cancel out. While this proposal has been thought provoking and has stimulated increasing progress toward findings a unified model, it should produce a seismic signature that is not found, as discussed in Section 10.1.8.

##### 10.1.9.1.(iii) Thin layer (D″ region)
A few hundred kilometers thick layer above the CMB, perhaps highly undulating and/or localized into certain areas, and which could potentially contain primitive material, recycled material, or some mixture of the two, is a possibility. It has long been proposed that such a layer could contain recycled oceanic crust that provides the HIMU signature (Hofmann and White, 1982). Such oceanic crust contains a higher concentration of incompatible trace elements than primitive material so could also account for the 'missing' trace elements (about one-third of the total inventory of heat-producing elements according to Coltice and Ricard (1999)). Whether such a layer could exist depends on the density contrast of MORB at these

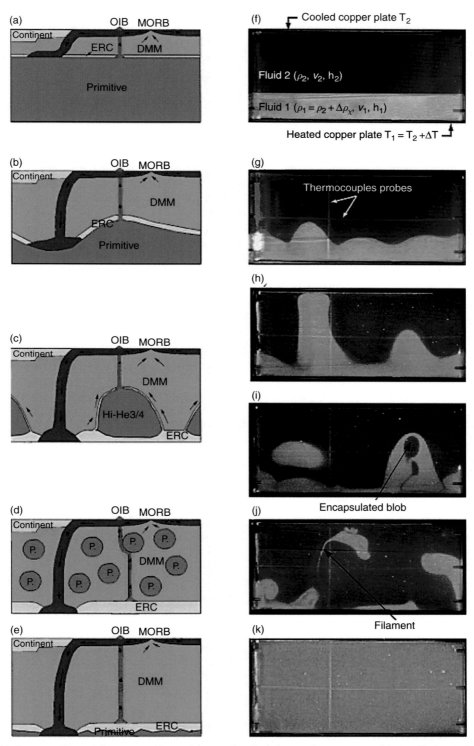

**Figure 2** Various mantle models proposed to explain geochemical observations compared to results from a laboratory experiment at different times in its evolution. From Le Bars and Davaille (2004b). Cartoons in the left column are from Tackley (2000). (a) Layered at the 660 km discontinuity; (b) deep, high-topography hidden layer (Kellogg et al., 1999); (c) 'piles' (Tackley, 1998) or 'domes' (Davaille, 1999a); (d) primitive blobs (Davies, 1984; Manga, 1996; Becker et al., 1999; du Vignaux and Fleitout, 2001); (e) whole-mantle convection with thin dense layer at the base.

high pressures and temperatures, as discussed in Section 10.3.3.2. It has recently been argued that this enriched layer may have formed shortly after Earth's accretion process (Tolstikhin and Hofmann, 2005; Tolstikhin et al., 2006). Another proposal is that the layer preferentially contains oceanic crust containing oceanic plateaus, with normal oceanic crust not reaching this depth (Albarède and van der Hilst, 2002). Seismological observations support the existence of a thin and perhaps intermittent layer, as discussed in Section 10.1.8.

### 10.1.9.2 Distributed heterogeneity

Instead of, or as well as, radial stratification, lateral compositional heterogeneity is typically thought to exist. An early justification for this was that mantle rocks outcropping in peridotite massifs have been observed to contain centimeter to decimeter thick isotopically enriched pyroxenite veins embedded within a mostly depleted peridotite matrix (Polve and Allegre, 1980; Reisberg and Zindler, 1986; Suen and Frey, 1987; Pearson et al., 1993). Such variations have inspired the proposal of a 'marble cake' or 'plum pudding' style of chemical variation in the mantle, in which small volumes of enriched material reside within a matrix of depleted MORB mantle (Allegre and Turcotte, 1986), a concept that has been the topic of much recent investigation as discussed in later sections, with several variations and refinements proposed. At the larger scale, it has been proposed that very large ($\sim$100 km scale) blobs may exist (Becker et al., 1999). Seismic scattering supports distributed heterogeneity, as discussed in Section 10.1.7. Dynamically, key issues are how rapidly distributed heterogeneity gets mixed (see Section 10.2), which will be influenced by any composition-dependence of viscosity (Section 10.2.3.5).

### 10.1.10  Mantle–Exosphere (Biosphere) Interactions

Cycling of volatiles (particularly water and carbonate) between the interior and fluid envelope of a terrestrial planet could play a major role in the evolution of the fluid envelope, and hence the long-term habitability of the surface environment. The mantle influences the fluid envelope by volatile outgassing associated with volcanism, and by volatile recycling back into the interior. Indeed, a long-term cycle associated with carbonate recycling and reoutgassing may provide a critical feedback mechanism that maintains surface temperature in a habitable range

(Kasting et al., 1993b; Sleep and Zahnle, 2001). Volatile recycling also affects the redox state of the mantle, which, through outgassing, influences the redox state of the atmosphere (Kasting et al., 1993a; Lecuyer and Ricard, 1999; Delano, 2001), and may have been responsible for the rapid rise in atmospheric oxygen $\sim$2 Gy ago (Kump et al., 2001). Crucial to recycling is the existence of plate tectonics, through which carbonate, water, and other volatiles are subducted into the mantle (e.g., Ruepke et al., 2004); note, however, that until recently it was widely accepted that a 'subduction barrier' existed that prevented volatiles from being recycled into the mantle (Staudacher and Allegre, 1989) and even recent mass-balance calculations (Hilton et al., 2002) indicate that the mass of water erupted by arc volcanics far exceeds the unbound water subducted. Thus, the existence of plate tectonics on Earth may be important for long-term habitability. Multiple feedbacks exist in the fluid envelope–mantle–core system. Water strongly affects the viscosity of the mantle (Hirth and Kohlstedt, 1996), so water exchange between the mantle and exterior may have had important influences on the evolution of the coupled system. Furthermore, water greatly reduces the strength of the lithosphere through its effect on faults and the yield strength, and is thus commonly thought to be necessary for plate tectonics to exist. The convective regime and heat flow out of the mantle are strongly affected by the fluid envelope through the influence of water on mantle viscosity and plate tectonics. Recent analysis indicates that subducted seawater accounts for about 50% of the unbound water presently in the convecting mantle (Holland and Ballentine, 2006).

## 10.2   Mantle Stirring and Mixing of Passive Heterogeneities

### 10.2.1   Stirring, Stretching, Mixing, and Dispersal

It is important to distinguish between different concepts that are all sometimes loosely referred to as 'mixing'. Strictly speaking, 'mixing' involves homogenization of the chemical differences by a combination of stirring and chemical diffusion. Stirring involves stretching, which is the process by which an initial 'blob' of chemically distinct material is elongated into a thin tendril or lamella by strain caused by the convective flow. This greatly increases the surface area of the heterogeneity and decreases its width, facilitating chemical diffusion. Chemical diffusion of most elements is extremely slow, acting over length scales of centimeters to meters over geological time; thus, it is necessary to stretch heterogeneities to this scale before they can be mixed. As discussed later, however, observed MORB or OIB compositions probably involve averaging over a volume that is larger than this, so stretching to diffusion scales may not be necessary to eliminate the signature of an individual heterogeneity during melting.

'Dispersal' refers to the dispersal of initially close heterogeneities evenly around the mantle. While this occurs partly by similar processes to stretching (i.e., the increasing separation of two initially very close points), effective dispersal involves material transport across convective 'cells'. It is possible in some flows for rapid stretching to occur within a cell, but for inter-cell transfer to be slow. Conversely, as discussed later, high-viscosity blobs undergo very slow stretching, but may get rapidly dispersed around the domain.

Many of the concepts introduced in this section are illuminatingly illustrated in the review of van Keken *et al.* (2003). *See also* Chapter 2.

## 10.2.2 Stretching Theory: Laminar and Turbulent Regimes

### 10.2.2.1 Finite deformation tensor

The mathematical description of finite deformation was presented by McKenzie (1979) following Malvern (1969). We are interested in the deformation tensor relating the vector $\mathbf{y}'(t)$ which joins two particles in a fluid element at positions $\mathbf{x}'_1(t)$ and $\mathbf{x}'_2(t)$ to $\mathbf{y}$, the vector joining the same two particles at $t = 0$:

$$\mathbf{y}'(t) = \mathbf{F}(t)\mathbf{y} \qquad [4]$$

The $\mathbf{F}$ matrix is initially the unit matrix and evolves according to

$$D_t F_{ij} = L_{ik} F_{kj} \qquad [5]$$

where $D_t$ is the Lagrangian time derivative and $\mathbf{L}$ is the velocity gradient tensor:

$$L_{ij} \equiv \frac{\partial v_i}{\partial x_j} \qquad [6]$$

(can also be written in the material frame (Farnetani *et al.*, 2002)) which is a combination of the strain-rate tensor (symmetric part of $\mathbf{L}$) and the vorticity tensor (antisymmetric part of $\mathbf{L}$).

As the $\mathbf{F}$ tensor has nine components in three dimensions, or four components in two dimensions, it is desirable to extract a simpler quantity to describe the stretching. In two dimensions a convenient scalar measure is the logarithm of the aspect ratio of the strain ellipse (McKenzie, 1979), that is,

$$f(t) = \log_{10}(a/b) \qquad [7]$$

where $a$ and $b$ are the major and minor axes of the ellipse, respectively, and $a/b$ can be found from the second invariant of the $\mathbf{F}$ tensor, that is,

$$a/b = \gamma + (\gamma^2 - 1)^{1/2}, \quad \gamma = \frac{1}{2} F_{ij} F_{ij} \qquad [8]$$

Some authors prefer to measure the change in the semimajor or semiminor axis of the strain ellipse, which is the square root (or one over the square root) of $(a/b)$. In two dimensions, Kellogg and Turcotte (1990) show how a complete description of the total deformation of a heterogeneity can be obtained by tracking only two quantities for each tracer, the aspect ratio $(a/b \equiv e)$, and orientation $\theta$ of the strain ellipse. In three dimensions, the strain ellipse becomes an ellipsoid and a method of tracking that is simpler than the above equation has not yet been proposed.

### 10.2.2.2 Asymptotic stretching rate

Two types of shear flow, simple shear and pure shear, result in two different asymptotic stretching behaviors termed laminar and turbulent, respectively (see **Figure 3**). The theory is described by Olson *et al.* (1984a, 1984b) based on previous theory by Batchelor (1952, 1959) and Corrsin (1961), and is briefly summarized here; *see also* Chapter 2.

Laminar mixing (**Figure 3(a)**) occurs in a simple shear flow and involves a linear increase in the cumulative strain $\varepsilon$ with time $t$:

$$\varepsilon = \dot{\varepsilon} t \qquad [9]$$

where $\dot{\varepsilon}$ is the average strain rate experienced by the heterogeneity. A linear decrease with time occurs for the component of the wave-number vector that is perpendicular to the velocity:

$$k_x(t) = k_x(0), \quad k_z(t) = k_z(0) + k(0)\dot{\varepsilon} t \qquad [10]$$

for a shear flow described as $\mathbf{u} = -\dot{\varepsilon} z \hat{x}$. This effectively describes a cascade of heterogeneity to shorter wavelengths, with 'blobs' being stretched into thin strips aligned with the direction of flow velocity. In the absence of diffusion, the Fourier amplitude of the

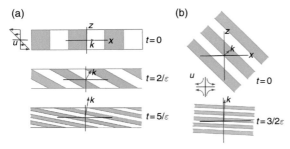

**Figure 3** Diagrams illustrating the evolution of a single Fourier component of a passive anomaly for (a) uniform laminar shear flow, and (b) stagnation point flow, based on **Figures 4** and **5** of Olson et al. (1984b). The velocity field **u** is sketched for each case. In a uniform shear flow (a), the Fourier component is rotated into the direction of shear while its wavelength decreases, such that the wave-number magnitude $|\mathbf{k}|$ grows at a rate $\dot{\varepsilon}$, the Lagrangian strain rate. Near a stagnation point (b) the Fourier components also become rotated, this time into the direction of compression, but the wave-number magnitude grows at a rate that is proportional to time, that is, $\dot{\varepsilon}t$.

heterogeneity does not change, it simply moves to higher $k$. For the case with diffusion see below.

Turbulent mixing is associated with pure shear (also known as normal strains (Olson et al., 1984a, 1984b)), which occurs at stagnation points in convective flows (**Figure 3(b)**). In this case, the velocity is proportional to the distance from the stagnation point ($\mathbf{u} = \dot{\varepsilon}x\hat{\mathbf{x}} - \dot{\varepsilon}z\hat{\mathbf{z}}$), so the cumulative strain increases exponentially with time:

$$\varepsilon = \exp(\dot{\varepsilon}t) \qquad [11]$$

with wave numbers evolving exponentially in both directions:

$$k_z(t) = k_z(0)\exp(\dot{\varepsilon}t), \quad k_x(t) = k_x(0)\exp(-\dot{\varepsilon}t) \quad [12]$$

As turbulent mixing is far more rapid than laminar mixing, it is important to establish which regime is most appropriate to characterize flows associated with mantle convection, so this has been a major focus of research. This issue has been investigated theoretically and for kinematically driven flows and for convectively driven flows. The general findings are that (1) the type of mixing can be quite sensitive to exact flow, although greater time-dependence results in a greater tendency towards turbulent mixing, and (2) different 'blobs' may undergo substantially different rates of stretching in the same flow depending on their location in that flow. This is discussed in the next section.

Some semantic issues are worth noting. First, the very high Prandtl number flow that the mantle undergoes should actually be termed laminar due to the unimportance of intertial terms; nevertheless, it can lead to the turbulent stretching regime. Second, turbulent mixing is often referred to as chaotic mixing, and laminar mixing is often referred to as regular mixing.

### 10.2.2.3 Measures of stretching

Most measures of stretching are based on the elongation of an infinitesimal strain ellipse, that is, the ratio of the major axis to its original value ($a/a_0$), calculated using the equations in Section 10.2.2.1, and tracked using infinitesimal tracer particles distributed throughout the domain. The width of the strain ellipse is the inverse of this value. Values of ($a/a_0$) exhibit a considerable spread, so it is important to consider the range in values as well as the average. When the time-averaged stretching rate increases exponentially with time, the appropriate average to take is the median (Christensen, 1989a) or the geometric mean, which is related to the mean in logarithmic space. In this approach, the strain ellipse is always considered to be infinitesimal in size even if it reaches $10^{10}$ or larger, so these measures are not appropriate for estimating the separation of two initially closely spaced points, which cannot get larger than the size of the domain.

Other than the median or geometric mean stretching, it is useful to consider quantities that convey information about the distribution of strains. The statistical distribution of stretching can be plotted in various ways; for example, Kellogg and Turcotte (1990) plot the probability distribution of strain ($a/a_0$ vs number of tracers), the cumulative distribution ($a/a_0$ vs fraction of particles exceeding ($a/a_0$)), and the time-dependence of strain evolution (time vs $a/a_0$ with different lines for the percentage of tracers exceeding $a/a_0$). An example of the latter quantity is the size reduction exceeded by 90% of tracers plotted by Christensen (1989a).

Often the goal is to determine what fraction of heterogeneities have 'mixed', in the sense of being stretched to less than the diffusive or sampling length scale. Kellogg and Turcotte (1990) plotted as a function of time the fraction of tracers that have been completely homogenized. Gurnis and Davies (1986b) were interested in heterogeneities that remained relatively intact, so plotted as a function of time the fraction that have strain less than 5.

**10.2.2.3.(i)  Lyapunov exponents**  The Lyapunov exponent is the exponential coefficient for the rate at which two initially close heterogeneities are separated. This can be written as

$$\lambda(P, \vec{y}) = \lim_{\substack{t \to 0 \\ y \to 0}} \left[ \frac{1}{t} \ln \left( \frac{y(t)}{y} \right) \right] \qquad [13]$$

where $y$ is the length of separation vector '**y**' located at position $P$, and $y(t)$ is its length after a time $t$. This is zero for laminar stretching and positive or negative for exponential stretching. The value depends on the direction of '$y$' relative to the velocity gradient tensor, so $\lambda$ has maximum and minimum values. Kellogg and Turcotte (1990) use the maximum Lyapunov exponent and the minimum Lyapunov exponent to quantify the stretching (Wolf, 1986), and these are straightforwardly related to $\bar{\alpha}_e$, the average strain rate measured from the numerical experiments, by

$$\lambda_{max} = \frac{\bar{\alpha}_e}{\ln 2} \qquad [14]$$

where

$$\bar{\alpha}_e = \frac{1}{\tau} \ln \left( \frac{a}{a_0} \right) \quad \text{or} \quad a = a_0 \exp(\bar{\alpha}_e \tau) \qquad [15]$$

In numerical applications it is often more convenient to calculate the Lyapunov exponent using finite rather than infinitesimal time difference $t$, or finite separation $y$, for example by considering the motion of nearby tracers. Lyapunov exponents calculated using these finite quantities are referred to as finite-time or finite-size Lyapunov exponents, respectively, and have been used in the mantle dynamics literature (Ferrachat and Ricard, 1998, 2001; Farnetani *et al.*, 2002; Farnetani and Samuel, 2003).

In general, for finite quantities we can write using the equations above that:

$$\frac{y(t)}{y(0)} = \frac{|\mathbf{F}(t)\mathbf{y}|}{|\mathbf{y}|} = \left( \frac{\mathbf{y}^T \mathbf{F}^T(t)\mathbf{F}(t)\mathbf{y}}{\mathbf{y}^T \mathbf{y}} \right)^{1/2} \qquad [16]$$

The maximum and minimum eigenvalues of the right Cauchy–Green strain tensor $\mathbf{F}^T\mathbf{F}$ can then be used to calculate the maximum and minimum finite-time Lyapunov exponents (e.g., Farnetani and Samuel, 2003):

$$\lambda_{max} = \frac{1}{t} \ln \sigma_{max}, \qquad \lambda_{min} = \frac{1}{t} \ln \sigma_{min} \qquad [17]$$

In 2-D incompressible flow, $\lambda_{min} = -\lambda_{max}$.

The hyperbolic persistence time method is a method to analyze stable or unstable material trajectories (unstable meaning that nearby particle trajectories separate from it), and was introduced to the mantle community by Farnetani and Samuel (2003). Readers are referred to this paper for more details.

### 10.2.3  Stretching: Numerical Results

#### 10.2.3.1  General discussion

A convective flow includes both stagnation points (e.g., at the corners of a 'cell'), which will be associated with pure shear and exponential stretching, and laminar regions (away from corners), which will be associated with simple shear and linear stretching. It is therefore likely that a heterogeneity will experience linear stretching most of the time, with the occasional episode of exponential stretching. Which regime dominates the long-term behavior depends on the details of the flow. If the flow is steady-state and 2-D, then most of the heterogeneities never pass close to a stagnation point, so the average stretching rate follows a linear stretching law (McKenzie, 1979; Olson *et al.*, 1984b). If the flow is highly unsteady, then the stagnation points move around and new ones are generated, allowing essentially all heterogeneities to pass close to stagnation point at some point, and giving on average an exponential stretching rate. The overall behavior is, however, highly sensitive to the details of the time dependence, and this has been studied by a variety of models ranging from purely kinematic flows to purely convective flows (e.g., Christensen, 1989a). As all heterogeneities are not stretched at the same rate, it has proved to be important to study the range of stretching rates as well as the average. Most of the research has been done in two dimensions, so 3-D studies are discussed in a later section.

#### 10.2.3.2  Steady-state 2-D flows

Steady-state 2-D Bénard convection cells follow the laminar stretching law, whether kinematic (Olson *et al.*, 1984b) or convective (McKenzie, 1979). The fit to theory is reasonable for both nondiffusive and diffusive heterogeneity (Olson *et al.*, 1984b). It is instructive to understand the resulting stretching regime as a baseline for more interpreting more complex flows.

Heterogeneities are stretched into spirals of lamellae that are locally oriented subparallel to the streamlines, because different streamlines are rotating at different rates. For example, near the top of the domain the lamellae are locally subhorizontal. In a completely heated-from below steady-state Bénard convection cell the interior of the cell (far from the boundaries) undergoes almost rigid-body rotation with an associated low stretching rate, but near the boundaries, where heterogeneities pass close to stagnation points in the corners, the cumulative strain can increase substantially (e.g., the aspect ratio of the strain ellipse ($a/b$) by a factor of 1000) over one rotation (McKenzie, 1979). It is interesting to note that strain does not increase monotonically, but rather as large oscillations with a superimposed linear trend (McKenzie, 1979). The parts of these oscillations where the strain decreases have been termed 'unmixing'. The addition of internal heating makes the situation more complex as there is no uniformly-rotating area in the middle, but again the cumulative strain ($a/b$) undergoes considerable oscillation and near the edge of the 'cell' can reach a factor of 100 over one rotation (McKenzie, 1979). The highest strain rates (hence velocity gradients) occur near the downwelling. For finite-sized bodies (initially cylinders), Hoffman and McKenzie (1985) found that for pure basal heating the amount of stretching was rapid for a blob initially situated near the edge of the flow, but with internal heating the blob became deformed but not greatly stretched in one revolution.

Steady corner flows relevant to mid-ocean ridges and subduction zones have also been studied (McKenzie, 1979). The most severe strain ($a/b \sim 30$) was experienced in the corner above a subducting slab, with the principle axis of the strain ellipse aligned subparallel to the streamline. Again, strain does not necessarily increase monotonically with time.

### 10.2.3.3  Time-dependent 2-D flows

Various types of 2-D flows have been studied, including purely kinematic flows, purely convective flows, and convective flows with kinematic upper boundary conditions. This has led to the identification of three main stirring regimes (Christensen, 1989a): a regime in which all particles are undergoing exponential stirring, a 'slow' regime similar to the laminar regime, and a hybrid regime in which some particles/regions are undergoing turbulent stretching while some are undergoing slow laminar stretching. Fully time-dependent convection with no kinematic constraints has always been found to give turbulent mixing, whereas controlling the flow with a kinematic upper boundary condition or fully kinematic description may lead to the 'slow' or hybrid regimes, though some kinematic flows lead to the turbulent mixing regime (e.g., Kellogg and Turcotte, 1990).

More discussion of the factors that lead to these regimes now follows.

*10.2.3.3.(i)  Slow stirring*  In this regime, most heterogeneities are stretched into long streaks, but clumps exist where the streaks get folded, leading to the simultaneous existence of tendrils with average stretching rate and blobs with very slow stretching (Gurnis, 1986c) which can survive for up to tens of transit times. This regime has been observed in kinematically prescribed flows consisting of two counter-rotating cells with a boundary that moves smoothly and periodically with time (Gurnis, 1986c; Gurnis and Davies, 1986b; Christensen, 1989a). While visually the system may look reasonably well mixed, the average stretching rate is slow compared to the other regimes, and lies somewhere between exponential and linear with time (Gurnis, 1986c; Christensen, 1989a). Gurnis and Davies (1986b) also obtained a regime visually similar to this with a combination of convection and kinematic plates, but the asymptotic stretching rate was not analyzed.

*10.2.3.3.(ii)  Exponential  stirring*  In  this regime, all heterogeneities experience exponential stretching, although at different rates. There are no unmixed islands or persistent large blobs. All fully time-dependent purely convective (i.e., with no kinematic control) 2-D flows in the mantle convection literature that have been suitably analyzed are in this regime, including those in Hoffman and McKenzie (1985) and Christensen (1989a). So are kinematic flows in which there is an oscillation between a single-cell and three-cell structure, like those of Kellogg and Turcotte (1990) and simplified versions in Christensen (1989a). A characteristic that this kinematic flow shares with time-dependent convection is the creation of new cell boundaries, which can occur due to plume formation, sublithosphere convection or delamination, or new subduction zone formation. In some cases it can also be obtained by two-cell kinematic flow with the boundary motion containing more than one frequency component (Christensen, 1989a).

It is important to note that the effective stretching rate that goes into the exponential stretching law (sometimes called the Lagrangian strain rate (Coltice, 2005)) is much lower than the mean strain rate of the flow – approximately 10–30 lower according to Christensen (1989a). In 2-D convective flows with internal heating up to $Ra = 10^{10}$ and with basal heating up to $Ra = 10^9$ Coltice (2005) found that the

Lagrangian strain rate scales in proportion to the convective velocity, that is, $Ra^{1/2}$ and $Ra^{2/3}$, respectively, even though the higher $Ra$ flows are more time-dependent, implying that the stretching rate is simply proportional to the rate at which material is moving, provided the flow is time dependent. This finding also applied to cases with a 100- or 1000-fold viscosity jump at mid-depth.

### 10.2.3.3.(iii) Hybrid stirring
In this regime, most of the flow is exponentially stirred but there exist unmixed regions or 'islands' in the flow; the histogram of strain distributions can exhibit a bimodal distribution with one peak at very low strains (Christensen, 1989a). In the experiments of Christensen (1989a) this was found in relatively few cases – in two of the kinematic flows with more than one driving frequency. Nevertheless, it is a mode that is commonly observed in mixing experiments performed outside the solid Earth science community (see Metcalfe *et al.* (1995) and references therein). The wide range of stretching rates observed in this mode and the 'slow stirring' mode above indicate the importance of considering the range rather than only the median or mean.

The statistical range of heterogeneity stretching in the turbulent mixing regime was studied by Kellogg and Turcotte (1990) using a chaotic kinematic flow with time dependence driven by the Lorenz equations. Specifically, the flow had two cells, and the boundary between moved smoothly but chaotically such that sometimes there was only one cell. Into this they injected a large number of particles with different starting positions, in order to obtain robust statistics regarding their stretching. Naturally, exponential (turbulent) stretching was obtained, but the amount of strain experienced by different tracers varied very widely, such that after, for example, five overturns, the semimajor axis of the strain ellipse divided by its initial radius ($a/a_0$) averaged $10^4$ but varied between $\sim10^1$ and $\sim10^7$ (1st to 99th percentile) or $\sim10^2$–$\sim10^6$ (10th to 90th percentile). In other words, the (logarithmic) variation in cumulative strain is as large as the average cumulative strain. It is thus not meaningful to talk about precise quantities such as 'mixing time' based on average stretching; everything must be qualified in terms of probability distributions. The effective strain rate for this stretching was found to be 13% of the maximum strain rate that occurred at the stagnation point in the corner of the cell.

### 10.2.3.4 3-D flows
It is expected that mixing in three dimensions is quantitatively different from that in two dimensions, but so far studies have been few and appear contradictory about the sign of the change. Two key concepts are relevant.

### 10.2.3.4.(i) Toroidal motion
Three-dimensional flows can have toroidal motion, the component of motion associated with rotation about a vertical axis and strike-slip motion, whereas 2-D flows (in a vertical plane) cannot. In steady-state 2-D flows the streamlines (particle paths) are always closed, but in 3-D the presence of toroidal motion can cause chaotic, space-filling particle paths, even if the flow is stationary in time (Ottino, 1989). One way of viewing this, as discussed by Ferrachat and Ricard (1998), is that toroidal motion increases by one the number of degrees of freedom of the Hamiltonian, for which one degree of freedom gives regular mixing and two degrees of freedom gives chaotic flow. For steady-state flows driven by an upper boundary condition mimicking a spreading center with a transform offset this effect was demonstrated by Ferrachat and Ricard (1998). Areas of rapid, turbulent mixing, as well as islands of laminar mixing, were identified based on Poincare sections and maps of finite-time Lyapunov exponents. Quantitatively, a 3-D flow with zero toroidal component was found to give similar Lyapunov exponent to 2-D flow, while the addition of a toroidal component increased the exponents by a factor of $\sim2$. A large lower-mantle viscosity tends to confine toroidal motion to the upper mantle (Gable *et al.*, 1991; Ferrachat and Ricard, 1998).

### 10.2.3.4.(ii) Plumes versus sheets
Because 3-D upwellings (and downwellings if slabs are not included) can be columnar plumes rather than infinite sheets, they have a smaller effect on disrupting the flow pattern and causing inter-cell mixing. This was apparent in time-dependent, constant-viscosity convection at Rayleigh numbers up to $8 \times 10^5$ calculated by Schmalzl *et al.* (1996): after several overturns tracers had dispersed efficiently within individual cells (as indicated by histograms of tracer vertical positions), but dispersal between cells was much slower in 3-D than in equivalent calculations in 2-D. A recent suite of calculations comparing 2-D and 3-D (poloidal only) flows up to much higher Rayleigh numbers found, however, that 2-D and

3-D flows have approximately the same Lagrangian strain rate (Coltice and Schmalzl, 2006).

In summary, if the results of Coltice and Schmalzl (2006) are validated for flows with more complex physics, then overall mixing efficiency in 3-D is likely to be similar to (if poloidal-only) or greater than (if toroidal motion is present) 2-D flows with the same parameters.

It is worth considering the particle trajectories in 3-D flows. In steady-state 3-D convection with zero toroidal component, inter-cell transfer is zero because individual flow lines lie on a 2-D distorted (but closed) toroidal surface (Schmalzl *et al.*, 1995). Individual streamlines are generally not closed in small number of orbits around the torus, but rather the surface is progressively filled by the trace of a single streamline. Extreme shear strains can occur but mixing is in 2-D. This is illustrated in **Figure 4**. With the addition of time dependence, boundary layer instabilities perturb these 'toroidal' structures, allowing for cross-cell mixing (Schmalzl *et al.*, 1996). Here it is important to note that the 'toroidal paths' followed by particles in a steady-state flow are completely different to the 'toroidal motion' caused by surface rotation and strike-slip motion – indeed toroidal motion breaks up the tori found with purely poloidal motion (Ferrachat and Ricard, 1998). The only relevant laboratory experiments were the steady-state experiments of Richter *et al.* (1982) in which for convective rolls blobs were smeared out into spiral sheets, or corkscrews if a superimposed lateral motion, with more complicated topologies for spoke pattern convection, but still within individual cells and not cross-cell.

The analysis of mixing in steady-state 3-D flows was extended to spherical geometry by van Keken and Zhong (1999), using a flow intended to be similar to the flow in the present-day mantle, generated by buoyancy from a slab model combined with an approximate plate rheology. As with the simple flow of Ferrachat and Ricard (1998), only limited regions exhibited laminar mixing, with most other regions containing corkscrew-like particle tracks and probably chaotic mixing: the areas with highest stretching rates are those with high toroidal motion. Long-range dispersal was also studied, as discussed in a later section.

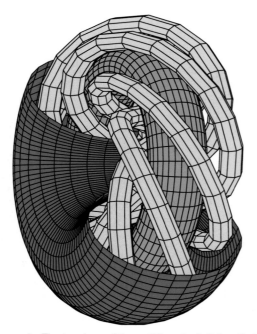

**Figure 4** The topology of the pathlines for 3-D flow that is purely poloidal. Individual tracer paths are restricted to tubes or tori, which they spiral round. Adding increasing amounts of toroidal motion breaks up the tubes, then the tori. Extracted from a figure from Ferrachat S and Ricard Y (1998) Regular vs. chaotic mantle mixing. *Earth and Planetary Science Letters* 155: 75–86.

### 10.2.3.5  *Effect of viscosity variations*

*10.2.3.5.(i)  Depth-dependent viscosity*  For some time it was thought that if the lower mantle has a substantially higher viscosity, then motion would be so sluggish that chemically distinct material could be retained for long periods of time (e.g., Davies, 1983). Numerical studies have, however, found that the effect of higher lower-mantle viscosity on stirring and residence times is relatively small, and that it cannot provide an explanation for long-lived heterogeneities *per se*. These numerical studies, which all study passive tracers, are discussed in this section.

Somewhat confusingly, the earliest study to address this issue (Gurnis and Davies, 1986a) did appear to show very slow stirring in a 100-times higher-viscosity lower mantle, even though the heterogeneities were passive, that is, had no density anomaly to keep them in the lower mantle. The circulation in these cases was however extremely sluggish, and was arguably driven mainly by a prescribed plate-like boundary condition rather than convection. (Although the authors were careful to make the prescribed surface velocities consistent with convection for constant-viscosity cases, this was not the case for depth-dependent viscosity cases (Christensen, 1989a, 1990)).

With a much higher convective vigor more relevant to the Earth and free-slip upper boundary, a lower-mantle viscosity up to a factor of 100 higher than the upper-mantle viscosity does not prevent large-scale mixing of the mantle (van Keken and Ballentine, 1998). The latter's calculations tracked helium degassing and ingrowth by radioactive decay, and it was found that the horizontally averaged, %degassed, and $^3$He/$^4$He ratio were approximately constant with radius even for high-viscosity lower-mantle cases, indicating good vertical mixing, although lateral heterogeneity was observed. Curiously, higher lower-mantle viscosity seemed to cause higher outgassing rates – this was because the viscosity of the upper mantle was decreased in order to maintain the same surface heat flux in all cases, which led to higher velocities in the shallow mantle, hence more rapid outgassing. Subsequent models (van Keken and Ballentine, 1999) demonstrated that temperature-dependent viscosity and an endothermic phase change at 660 km depth also do not result in chemical stratification, even when the mantle starts out layered in trace-element content, although again, there was a significant amount of lateral heterogeneity (i.e., $^3$He/$^4$He varies from 0 to a maximum value of 36–72).

Regarding stretching rate, a 100-fold viscosity jump was found by Coltice (2005) to make no difference to the average Lagrangian strain rate for internally heated convection, but for basally heated convection it decreased strain rate by a factor of ~4, probably because plumes from the lower boundary are affected by the viscosity jump, but instabilities from the upper boundary are not. In these calculations, the upper-mantle viscosity and Rayleigh number were held constant.

Processing (residence) time was found to be insensitive to a viscosity jump, regardless of heating mode (Coltice, 2005). Consistent with random sampling and the He ratio results of van Keken and Ballentine (1998), Hunt and Kellogg (2001) found that the a 100-fold viscosity jump at 660 km depth did not cause any age stratification. Seemingly inconsistent with Coltice (2005), Hunt and Kellogg (2001) found that the mean particle age increased by a factor of ~2, but this could be because they removed particles not only when they were sampled by melting, but also when they became widely dispersed from other particles that were introduced at the same time, which means that the mean age cannot be interpreted

as a simple residence time. The viscosity jump does, however, make a substantial difference to the ability of tracers to migrate laterally from their starting positions, because the flow is more steady, and this is something that should be studied further.

A non-convective, steady-state flow, driven by top boundary forcing and assumed idealized density anomalies due to downwelling slabs was studied by Ferrachat and Ricard (2001), who found that a 100-fold viscosity jump at mid-depth increased the lower mantle mixing time (using a modified version of Olson et al. (1984a) definition based on even dispersion across boxes) by a factor of ~5, and that the finite-time Lyapunov exponents increased by a factor ~2–3. It is not clear how much such an idealized flow applies to the real Earth, however.

### 10.2.3.5.(ii) Composition-dependent viscosity
If a blob of compositionally distinct material has a higher viscosity than its surroundings, then the rate at which it gets stretched is considerably reduced. Simple theory predicts that in pure shear, the deformation rate is proportional to $1/(\lambda + 1)$, where $\lambda$ is the viscosity contrast (Spence et al., 1988; Manga, 1996). In a steady-state 2-D convective flow, Manga (1996) found that the deformation rate of extremely large (diameter ~1000 km) blobs was dramatically reduced with only a factor of ~10 viscosity increase, whereas if the viscosity of the blob is lower than its surroundings then it rapidly gets stretched into tendrils. Manga (1996) also found that these large high-viscosity blobs tend to aggregate, raising the possibility that they could combine into even larger blobs.

For time-dependent convection (with a prescribed time-dependent top velocity condition) du Vignaux and Fleitout (2001) showed that stretching rate is more highly dependent on viscosity ratio than in simple shear, and that the stretching is exponential with time, consistent with the turbulent mixing regime. Specifically, they found that

$$e_m \equiv \frac{d \ln(a^*/a_0)}{dt} = \frac{C}{(1+\lambda)^2} \quad [18]$$

where $e_m$ is the mixing efficiency, $a^*$ is the mean distance between two points initially at separation $a_0$, and $C$ is a constant. The quantity $a^*$ is commonly used in mixing studies as a measure of stretching (Ottino, 1989) and is related to, but not the same as, the semimajor axis of the strain ellipse $a$ (du Vignaux and Fleitout, 2001). This equation is valid for viscosity contrast $\lambda$ of ~an order of magnitude or more. This scaling can be explained using an analytic theory of a blob experiencing repeated pure shear events in random directions (du Vignaux and Fleitout, 2001). While one might intuitively expect repeated shear in random directions to average out, or at least add like a random walk, the fact that when the major axis of the ellipse is at an angle like 45° to the pure shear direction the stretching increases regardless of its orientation, means that the overall effect is one of exponentially increasing strain.

Thus, differing viscosity of components can have a dramatic effect on the rate at which they get stretched and hence mixed. It is quite likely that chemically different components have different viscosities. Subducted basaltic crust in the upper mantle has a large proportion of garnet, which has a relatively high viscosity (Karato *et al.*, 1995). The presence of water can reduce viscosity by up to ~two orders of magnitude (Hirth and Kohlstedt, 1996), which will increase the viscosity of depleted residue and possibly decrease the viscosity of primitive material, which is rich in volatiles. In the lower mantle, magnesiowüstite likely has a higher viscosity than perovskite (Yamazaki and Karato, 2001), so assemblages that have different proportions of these minerals will have differing viscosity.

### 10.2.3.5.(iii)   Non-Newtonian viscosity   Mineral
physics results suggest that a substantial part of the upper mantle is undergoing non-Newtonian dislocation creep (Karato and Wu, 1993), yet very few studies have addressed the influence of non-Newtonian power-law rheology on mixing. In a series of papers, (Ten *et al.*, 1996, 1997, 1998) compared the mixing properties of highly time-dependent Newtonian and non-Newtonian flows of similar vigor, in 2-D models that also had temperature- and pressure-dependent viscosity. Their basic finding is that mixing is less efficient in non-Newtonian flows, with unmixed 'islands' persisting for long time periods, both in plumes (Ten *et al.*, 1996) and global convection (Ten *et al.*, 1997, 1998). The local patterns of mixing were quite different between the different rheologies, with a greater richness in the scales of spatial heterogeneities in the non-Newtonian case (Ten *et al.*, 1997). Non-Newtonian rheology causes chemical heterogeneity to persist for longer, implying that original chemical heterogeneities can remain unmixed for a long period and may be concentrated at certain depths (Ten *et al.*, 1998). These results suggest that Newtonian models underestimate the

mixing time in the upper mantle, something that warrants further study.

### 10.2.4   Dispersal

#### 10.2.4.1   Measures of dispersal
##### 10.2.4.1.(i)   Based on sampling cells   One
class of measure for measuring the dispersal of initially close particles through the domain is based on dividing the domain into sampling cells and using statistical measures based on the number of particles in each cell. Initially, all the tracers will be in one cell, but when they are completely dispersed, the distribution of number of particles per cell will be indistinguishable from a random distribution.

Measures that have been used include the fraction of sampling bins with 0 tracers (Christensen, 1989a) or the fraction of sampling bins with >5 times the average number of tracers (Christensen, 1989a) (both measures decrease with time), or the fraction of cells with >1 tracer, which increases with time (Schmalzl *et al.*, 1996). Schmalzl *et al.* (1996) plot statistics for each of several initial blocks, and derive a 'rate of mixing' from the slope.

The statistics of number of particles/call can also be used to directly calculate a 'mixing time' (Olson *et al.*, 1984a), which is the decay time for reduction of the variance, as described in Section 10.2.6.1.

##### 10.2.4.1.(ii)   Based on separation   Given a distribution of particles, the 'two-particle correlation function' $H(r)$ gives the fraction of pairs of particles with separation less than $r$ (Schmalzl and Hansen, 1994). To calculate $H(r)$, all pairs of particles, that is, $N(N-1)/2$ combinations, must be considered. $H(0)$ is zero, then $H(r)$ increases with $r$, reaching a maximum at either the size of the domain or the size of the tracer cloud (if smaller than the domain). The initial increase can generally be approximated as $H(r) \sim r^{\alpha}$, where $\alpha$ is the 'correlation dimension', and gives information about the dimensionality of the tracer cloud. In particlar, $\alpha = 1$ corresponds to tracers spread into a line, $\alpha = 2$ to well-mixed 2-D convection, and $\alpha = 3$ to well-mixed 3-D convection. In spherical geometry, a random distribution of heterogeneities has a correlation dimension that depends on length scale, rising to a maximum of 2.8 at ~4000 km, falling monotonically to ~2 at ~7500 km then plunging to 0 by 10 000 km (Stegman *et al.*, 2002).

The 'root-mean square dispersion index' was introduced by Hoffman and McKenzie (1985) to quantify long-wavelength dispersion. It is defined as the square root of the second moment of the auto-correlation of the density distribution, the general form of which is

$$R_2 = \frac{\left( \iint r^2 \, dV_1 \, dV_2 \right)^{1/2}}{\left( \iint dV_1 \, dV_2 \right)^{1/2}} \qquad [19]$$

where $r$ is the separation of two volume elements $dV_1$ and $dV_2$ of the same body. To study the 1-D horizontal dispersion of a body represented by $N$ tracers in a 2-D domain this is simplified to

$$R_2 H = \frac{1}{N} \left( \sum_{i=1}^{N} \sum_{j=1}^{N} x_{ij}^2 \right)^{1/2} \qquad [20]$$

there $x_{ij}$ is the horizontal separation of two particles $i$ and $j$, and this measure reaches an asymptotic value (for perfectly random mixing) of

$$R_2 H(\infty) = \left[ 1/6(N^2 - 1) \right]^{1/2} \qquad [21]$$

***10.2.4.1.(iii) Others*** Other quantities that have been used are the time spectrum of a tracer's vertical position, which gives the 'overturn time' of the convection (Schmalzl *et al.*, 1996), and Poincaré sections (Schmalzl *et al.*, 1995; Ferrachat and Ricard, 1998), that is, 2-D slices of a 3-D volume showing where the particle paths intersect it. These are useful for identifying whether motion is restricted to closed 2-D surfaces as in Schmalzl *et al.* (1995), or is chaotic (Ferrachat and Ricard, 1998).

### 10.2.4.2 Numerical results

Many discussions of 'mixing' actually refer to the dispersal of heterogeneities around the domain, rather than the process of stretching and diffusion previously discussed. This dispersal is often studied by placing an array of tracers in one part of the domain, and studying how rapidly they spread out and become evenly distributed. A common observation in laboratory and numerical experiments (e.g., Richter *et al.*, 1982; Schmalzl and Hansen, 1994) is that dispersal (and mixing) occurs rapidly inside a convective cell, but much more slowly between one cell and another, that is, long-range lateral dispersal can be a relatively slow process. Another way of expressing this is that 'mixing' properties depend on spatial scale. The rate of cross-cell dispersal depends on how time dependent the flow is (e.g.,

how time-dependent upwellings and downwellings at the cell boundaries are) and whether and how often the cell pattern reorganizes. In the limit of steady-state flow, inter-cell dispersion is zero in 2-D flows and in 3-D flows with only poloidal motion, but nonzero if toroidal motion is present, as discussed later.

***10.2.4.2.(i) Intra-cell dispersal*** The dispersal of tracers within a cell scales similarly to stretching. In the experiments of Christensen (1989a) using a variety of kinematic and convective flows as discussed earlier, stretching-related diagnostics and dispersion-related diagnostics were compared, and found to obey similar behaviors, including the delineation of three regimes of mixing.

***10.2.4.2.(ii) Inter-cell dispersal*** For time-dependent 2-D flows, several numerical studies have studied the rate of lateral dispersion as a function of parameters. Convection that is completely internally heated is typically highly time dependent with no long-lived cells, and Richter *et al.* (1982) found that at $Ra = 1.4 \times 10^6$ tracers were fairly rapidly (e.g., in 0.1 diffusion times) dispersed across a box of aspect ratio 8 in what visually resembles a diffusion process. Using an analytic time-dependent flow they argued that rapid long-wavelength dispersal requires that the convective pattern change on a timescale comparable to the overturn time. Basally heated convection tends to have a steadier cellular pattern. Hoffman and McKenzie (1985) presented a series of calculations, also in an $8 \times 1$ box at $Ra = 1.4 \times 10^6$, with either 100% internal heating or 50% basal and 50% internal, and found that the flows with basal heating produced slower lateral dispersal, although it was still quite rapid. Completely heated from below flows were investigated by Schmalzl and Hansen (1994), who found that at range $Ra = 10^6 - 10^8$ heterogeneities within one cell are destroyed rapidly while two adjacent cells can remain unmixed for substantially longer, although at $Ra = 3 \times 10^6$ particles were well-dispersed by nondimensional time 0.02. They found the particle correlation function $H(r)$ and corresponding correlation dimension to be effective measures of long-wavelength dispersion, with a value of 2 corresponding to a perfectly dispersed field. Higher Rayleigh number flows are more time dependent such that the increase in the rate of laterally dispersion with $Ra$ is higher than the increase in velocity with $Ra$ (Schmalzl and Hansen, 1994).

*10.2.4.2.(iii) 3-D spherical results* Results regarding intercell dispersal in Cartesian 3-D flows were mentioned in a previous section. Long-range dispersal in steady-state 3-D spherical flows was studied by van Keken and Zhong (1999), using a flow intended to be similar to the flow in the present-day mantle, generated by buoyancy from a slab model combined with an approximate plate rheology. They found that particles could be efficiently transported far from their source: after 4 billions years of motion in this steady-state flow, pairs of particles initially 10 km apart were dispersed around the mantle with an almost flat probability distribution (i.e., number vs separation). This suggests that dispersal in the actual mantle is relatively efficient. Stegman *et al.* (2002) extended such 3-D spherical modeling in order to determine whether dispersal is reduced by higher lower-mantle viscosity (see **Figure 5**). Using the correlation dimension to measure dispersion, they found that in a model with 100 times higher lower-mantle viscosity and driven by a slab model and present-day plate motions, the correlation dimension reached after a few billion years is only of order 1, which is normally indicative of stretching into a line. A representative structure can be seen in **Figure 5**. They also found that in a model with 100 times higher lower-mantle viscosity and driven by a slab model and present-day plate motions, the upper mantle became mixed about 60% faster than the lower mantle, where mixing is determined by the correlation dimension. They conclude that this is insufficient differential mixing to explain the difference between the relatively high heterogeneity of OIBs and relatively low heterogeneity of MORBs, although it could be argued that the latter is related to the small-scale structures in the melting region, whereas the correlation dimension is quantifying variations at thousands of kilometers length scales.

### 10.2.4.3  Eddy diffusivity?

In the atmospheric community, long-wavelength lateral dispersal is commonly treated using an 'eddy diffusivity', although it has been found to be dependent on the initial length scale (separation) being considered: larger separations lead to larger estimates of effective eddy diffusivity (Richardson, 1926; Richter *et al.*, 1982). It is interesting to know whether this concept could be applied to the mantle. Olson *et al.* (1984b) studied whether an effective diffusion could be used to parametrize mixing within a convecting cell (an 'eddy'), and concluded that it could not, because the effective diffusion coefficient would need to be larger for larger heterogeneity size. However, the two studies that have attempted to fit lateral dispersal across convective cells with an effective 'eddy' viscosity do appear to have had reasonable success.

For internally heated convection at $Ra = 1.4 \times 10^6$, Richter *et al.* (1982) use the distance-neighbor approach of Richardson, (1926) to estimate an eddy diffusivity for the large-scale dispersal of heterogeneities at a scale length several times that of the depth, and estimated a value of 30 times the thermal diffusivity, that is, $\sim 0.003\ \mathrm{m^2\,s^{-1}}$, independent of initial separation. In similar experiments but

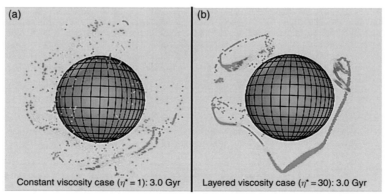

| (a) | (b) |
|---|---|
| Constant viscosity case ($\eta^* = 1$): 3.0 Gyr | Layered viscosity case ($\eta^* = 30$): 3.0 Gyr |

**Figure 5**  Dispersal of an initial 125 km radius cluster of tracers around the mantle after 3 Gy. Mantle flow is driven entirely by plate-like boundary conditions. Comparing the two panels show the effect of a 30* viscosity increase at the 660 km discontinuity. In the lower mantle the heterogeneity is stretched into a ribbon-like structure, but this is more rapidly destroyed in the upper mantle, or in an isoviscous mantle, where toroidal flow is more prominent. Reproduced from Stegman DR, Richards MA, and Baumgardner JR (2002) Effects of depth-dependent viscosity and plate motions on maintaining a relatively uniform mid-ocean ridge basalt reservoir in whole mantle flow. *Journal of Geophysical Research* 107 (doi:10.1029/2001JB000192).

with varying heating modes, Hoffman and McKenzie (1985) estimated a value of 25 times the thermal diffusivity for internally heated convection, and 20 times the thermal diffusivity for 50% internal heated convection, by fitting the time evolution of the horizontal rms dispersion index with a diffusion curve, that is,

$$R_2 H = \sqrt{\kappa t} \qquad [22]$$

The slight difference from the almost identical experiment of Richter *et al.* (1982) is probably due to a difference in estimation technique and is not meaningful. They find that their numerical experiment agrees with this square root of time curve, so that eddy diffusivity is a good approximation for large scales of over three times the layer depth, and they get 20 times the thermal diffusion coefficient rather than 30 times. Based on these two studies, an eddy viscosity seems to offer a reasonable parametrization of lateral dispersal over distances greater than ~3 times the layer depth in a simple 2-D convective system, but the application to more complex situations and as a function of convective parameters remains to be determined.

## 10.2.5  Residence Time

The concept of residence time plays a central role in geochemical theory (e.g., Albarède, 1998), and is something that can be investigated in numerical calculations of stirring and mixing processes. Residence time is defined as the length of time a heterogeneity (tracer) stays in the specified reservoir or other container. In geochemical box modeling, it may be applied to how long (on average) an isotope stays in a reservoir such as the continental crust, lower mantle, etc. In numerical modeling of mantle evolution it usually means the average length of time before a tracer is sampled by melting, either from the beginning of the calculation or, if tracers are being continually introduced at subduction zones, the average length of time before they are next sampled (melted). For whole-mantle convection in statistical equilibrium, this is equal to the time taken for one mantle mass to pass through the mid-ocean ridge processing zone, the quantitative value of which was discussed earlier in Section 10.1.5.3, and is probably around 6 Gy at the present processing rate, but could have been much faster (i.e., lower) in the past, which Phipps Morgan (1998) showed can easily

match the '50% Argon outgassing constraint' as discussed in Section 10.1.6.1.

From a chemical perspective, the residence time for lithophile elements was calculated by Albarède (2005b) by dividing the amount of the element in a mean mantle model (i.e., BSE – continental crust) by the current rate at which the element is extracted into the ocean crust. Residence times in the range 4–9 Gy were obtained, which bracket the geophysically constrained value of ~6 Gy discussed in Section 10.1.5.3. Again, it is likely that processing was substantially faster in the past, as is required to match the Argon outgassing constraint (Phipps Morgan, 1998).

In geochemical box modeling it is commonly assumed that a reservoir is randomly sampled, that is, the probability of melting each atom or tracer in the reservoir is equal regardless of how long they have been in the reservoir, in which case the overall process is analogous to radioactive decay. Thus, if a reservoir starts off with $N_0$ atoms or tracers of a particular isotope, the rate of sampling is given by

$$\frac{dN}{dt} = \frac{N}{\tau} \qquad [23]$$

where $\tau$ is the residence time. Hence the number left after time $t$ is given by

$$N = N_0 \exp(-t/\tau) \qquad [24]$$

The random sampling assumption is valid if tracers are effectively dispersed throughout the domain in question by time-dependent flows and, as discussed in Section 10.2.3.3, fully time-dependent free convection (i.e., without kinematic constraints) experiments have always displayed chaotic mixing, which effectively disperses tracers. In steady-state 2-D flows, tracers are instead restricted to individual streamlines, so only streamlines that intersect the sampling area are sampled and the sampling increases linearly with time until all tracers on those streamlines have been sampled (Coltice, 2005). Viscosity stratification has only a minimal effect, as discussed elsewhere. If heterogeneities are anomalously dense they may preferentially reside near the base of the domain, reducing the number in the sampling area which can increase average residence time.

The earliest study to check residence times was by Gurnis and Davies (1986b), who investigated a low-$Ra$ flow driven by internally heated convection at $Ra = 10^5$ and 'organized by' a kinematic plate-like surface velocity condition. Tracers were introduced at the simulated moving trench, and it was found that the length of time tracers spend in the box before being sampled was easily predicted by the applied volumetric sampling rate and the rate of introduction of new tracers, consistent with the random sampling assumption. For free convection, Coltice (2005) confirmed that random sampling applies when the mixing is chaotic.

### 10.2.5.1  Effect of chemical density
In a related study (Gurnis, 1986a), the effect of the heterogeneity being denser than surrounding material was studied, and found to have a fairly small effect, despite chemical buoyancy ratios as large as 2. However, with a more Earth-like model setup, this conclusion is modified (Christensen and Hofmann, 1994; Davies, 2002). Davies (2002) adopted a similar model setup, but at the correct convective vigor for Earth, and with a chemical buoyancy ratios of up to 0.8, and also obtained residence times consistent with random sampling. While the mean age of tracers was only slightly increased (from a scaled 2.42 to 2.52 Ga), some dense basaltic tracers settled into piles at the base, and their mean age increased by 32% from 2.57 to 3.38 Ga. If tracers are truly randomly sampled, then the distribution of ages should have a particular probability distribution that depends on the source function. Davies (2002) showed that tracers settling the base of the domain have a probability density function (PDF) that is dramatically different from those in the MORB sampling region, which implies that random sampling does not apply to all tracers even though the average residence time may not change much. An appropriate description is probably that a different residence time applies to material that settles above the CMB, and Christensen and Hofmann (1994) calculated such a quantity for their calculations in which somewhat more basalt settling was obtained. They obtained residence times in the dense basal layer of between 0.36 and 1.62 Ga depending on parameters. For the whole system, they did not calculate the 'mean age' but rather the lead–lead pseudo-isochron age; changing the buoyancy ratio from 1.0 to 2.25 while keeping other parameters constant changed the Pb–Pb age from 1.37 to 2.66 Ga. Thus, if density contrasts cause a layer to form, it is important to study residence times in the layer, and not just the average of the whole system.

### 10.2.5.2  Effect of viscosity stratification
The effect of viscosity stratification and heating mode on residence time was determined by Coltice (2005). When the $Ra$ is high enough for the flow to effectively disperse the tracers, the residence timescales with the theoretical prediction based on fraction of the mantle sampled per unit time, independent of viscosity contrast. This is consistent with works that have studied the effect of viscosity stratification on the depth distribution of mean age (van Keken and Ballentine, 1998; Hunt and Kellogg, 2001) – the effect is minimal.

## 10.2.6  Mixing Time
### 10.2.6.1  Background
Several definitions of 'mixing time' have been used, which, together with sensitivity of stirring rate to the exact type of assumed flow, have led to a wide variety of estimates for mixing time in the mantle. Some estimates have been based on stretching, while others have been based on dispersal. As discussed previously, 'mixing' (dispersal) across several convective cells can be far slower than 'mixing' in a single convective cell. In any case, a key quantity is the length scale of interest, which could be the diffusion length scale, or sampling length scale.

The strictest definition of mixing time is the time required for the heterogeneities to get stretched thin enough to be homogenized by diffusion, which requires stretching down to submeter length scales for silicate rocks. It is questionable, however, whether such small-scale variations can be observed in geochemical observations (see later discussion) so an important concept is that of sampling volume (or, in 2-D, sampling area) (Richter and Ribe, 1979), which can be re-expressed as a sampling length scale by taking the cubed (or square) root (Olson et al., 1984a, 1984b). This sampling volume describes the region over which heterogeneities are homogenized by the sampling process, such as partial melting and extraction and mixing of magmas with each other and with the rock through which they are passing. While this has often been assumed to be ~tens of kilometers, some of the geochemical observations discussed in Section 10.1.7 indicate that shorter-length-scale heterogeneity can be preserved in the resulting rocks. When calculating compositional heterogeneity in models, heterogeneity should be integrated (averaged) over the sample volume. In summary the mixing time is the time required for heterogeneity to get stretched to the sampling size or the diffusion length scale, whichever is larger.

It has been usual to scale estimates of mixing time to either upper-mantle convection or whole-mantle convection, which generally results in rather small estimated timescales for upper-mantle mixing. These are probably irrelevant to the present-day mantle, as it is now clear that the upper mantle is not convecting independently as a simple heated-from-below Rayleigh–Bénard system, but instead the convective regime is dominated by whole-mantle flow and the plates impose a long-wavelength organization to the upper mantle. It is possible that small-scale convection below the plates exists in the upper mantle, but this would be less vigorous than assumed in the 'upper mantle' mixing estimates and its effect on mixing has not yet been evaluated.

Some important statistical measures of heterogeneity are now introduced, following Olson *et al.* (1984a). The two-point correlation function is given by

$$R(\mathbf{r}, t) = \langle \theta(\mathbf{x'})\theta(\mathbf{x'} + \mathbf{r}) \rangle \qquad [25]$$

where $\theta(\mathbf{x})$ is the compositional field, $\mathbf{x}$ is the position vector, $\mathbf{r}$ is the separation vector, and the angle brackets denote the volume average. Its Fourier transform, the energy spectral density (Olson *et al.*, 1984a) in $n$ spatial dimensions is given by

$$E(\mathbf{k}, t) = \left(\frac{1}{\pi}\right)^n \int_0^\infty R(\mathbf{r}, t)\exp(-i\mathbf{k} \cdot \mathbf{r})\mathbf{dr} \qquad [26]$$

and the related variance of composition $\sigma^2$ is given by

$$\sigma^2 = R(0, t) = \int_0^\infty E(\mathbf{k}, \mathbf{t}) \, \mathbf{dk} \qquad [27a]$$

If heterogeneity is not created or destroyed but only changes wavelength, the variance remains constant with time (*see* Chapter 2). However, if there is a sampling wavelength below which heterogeneity can no longer be detected, then it is appropriate to use the corresponding wave number as the upper limit for the integration, leading to the sampling variance:

$$\bar{\sigma}^2 = \int_0^{k_s} E(\mathbf{k}, \mathbf{t}) \, \mathbf{dk} \qquad [27b]$$

which will decrease with time due the cascade of heterogeneity to shorter wavelengths if heterogeneity is not replenished at long wavelengths. This has led to one definition of the mixing time (Olson *et al.*, 1984a)

$$\tau_{\text{mix}} = \int_0^\infty \frac{\bar{\sigma}^2(t)}{\bar{\sigma}^2(t_0)} \, dt \qquad [28]$$

the physical meaning of which is the time needed to reduce the sample variance of the composition significantly. A modification of this should be made when calculating $\bar{\sigma}^2$ numerically using a finite number of sampling cells containing a finite number of tracer particles, as described by Ferrachat and Ricard (2001). In this situation, $\bar{\sigma}^2$ does not tend asympotically to zero, so the integral does not converge. Instead, the statistical limit is

$$\lim_{t \to \infty} \frac{\bar{\sigma}^2(t)}{\bar{\sigma}^2(t_0)} = \frac{s - 1}{n} \qquad [29]$$

where $s$ is the number of sampling cells and $n$ is the number of particles. Thus, Ferrachat and Ricard (2001) introduce a definition for mixing time that includes this and does converge:

$$\tau_{\text{mix}} = \int_0^\infty \left(\frac{\bar{\sigma}^2(t)}{\bar{\sigma}^2(t_0)} - \frac{s-1}{n}[1 - \exp(-t/\tau_{\text{mix}})]\right) dt \qquad [30]$$

The mixing time is expected to depend on sampling length scale, as discussed elsewhere. To test this, the number of sample cells $s$ can be varied, as was done by Olson *et al.* (1984a) and Ferrachat and Ricard (2001). Ferrachat and Ricard (2001) found that the mixing time increases regularly with 1/length scale, with a moderate increase in mixing time for decreasing cell size – approximately a factor of 4–5 for factor 32 decrease in length scale.

### 10.2.6.2 Laminar flows

A heterogeneity is stretched until it is either homogenized by diffusion or has a width of less than the sampling length scale. The balance between diffusion and sampling length scale in controlling the mixing time was analyzed by Olson *et al.* (1984b) for idealized flows, then compared to numerical results of a steady-state Rayleigh–Bénard convection cell in which the stretching rate is laminar with time. Substituting the spectral expansion of a heterogeneity into the above equation led to the mixing time for nondiffusive laminar stirring being proportional to initial length scale / sampling length scale, that is,

$$\tau_{\text{mix}} \approx \frac{\pi}{\dot{\varepsilon}k_0\Delta z} = \frac{\lambda_0}{2\dot{\varepsilon}\Delta z} \qquad [31]$$

where $\lambda_0$ is the initial wavelength of heterogeneity (if a localized heterogeneity, twice its size) and $\Delta z$ is the sampling length scale. If, however, diffusion is rapid enough to homogenize the heterogeneity before it reaches the sampling length scale, a different expression is obtained:

$$\tau_{\text{mix}} \approx \left( \frac{\lambda_0^2}{4\kappa\dot{\varepsilon}^2} \right)^{1/3} \qquad [32]$$

where $\kappa$ is the diffusivity. There is a smooth transition between these types of behavior.

Olson et al. (1984b) tested these equations using a single Bénard convection cell and found that the results fitted reasonably well for both nondiffusive and diffusive heterogeneity. Scaled to the Earth with an assumed $\dot{\varepsilon} = 3 \times 10^{-15}\,\text{s}^{-1}$, an initial anomaly wavelength equal to the depth and a sampling length scale of 5 km, this leads to mixing time of 2.3 Gy for upper-mantle convection or 10 Gy for whole-mantle convection.

In the convective + kinematic internally heated ($Ra = 10^5$) flows of Gurnis and Davies (1986b), long-lived clumps (folded strips of crust) that persisted over tens of transit times were observed, even though most tracers were well dispersed. Based on qualitative measures the authors estimated the survival time of the clumps to be 2.4 Gy, or 1.4 Gy if faster processing in the past is taken into account.

### 10.2.6.3 Turbulent regime

For turbulent mixing, the equivalent result for nondiffusive flow is logarithmic and thus more rapid (Olson et al., 1984b):

$$\tau_{\text{mix}} \approx \frac{1}{\dot{\varepsilon}} \ln\left( \frac{2\lambda_0}{\Delta z} \right) \qquad [33]$$

An important point to note is that effective strain rate $\dot{\varepsilon}$ that goes into this equation can be more than an order of magnitude smaller than the average (e.g., root mean square) strain rate of the flow and depends on the details of the flow. For the convective flows investigated by Christensen (1989a) at Rayleigh numbers in the range $10^5$–$10^6$, the effective strain rate was a factor of 10–30 less than the mean strain rate of the flow. The effective strain rate scales in proportion to convective velocity over a wide range of Rayleigh number for both basally heated and internally heated flows (Coltice, 2005).

As for mixing time, Christensen (1989a) assumed the mantle well mixed when >90% of the heterogeneities had been reduced to less than 1/50th of their original size, and scaled his numerical results to the Earth using the transit time. This led to an estimate of 200–300 Ma for upper-mantle

convection, or 1500–2000 My for whole-mantle convection with a reasonable viscosity increase with depth. There are two points to note. This mixing criterion involves stretching oceanic crust down to 100 m thickness, which is larger than what other estimates have assumed: if instead the mixing is taken to meter scale, this timescale estimate is ~doubled to 3–4 Gy for whole-mantle convection. Secondly, there is a wide distribution in stretching rates, so that even when the 'mixed' criterion is met, there are still heterogeneities that are substantially larger and could influence isotopic signatures.

Hoffman and McKenzie (1985) also estimate mixing time from the scaling relationships that they found for small-scale stretching and large-scale dispersion. They estimated that 'any reasonable body' experiencing upper-mantle convection will be thinned to $10^{-3}$ layer depths (700 m) in 400 My and $10^{-6}$ layer depths (0.7 m) in 700 My; for whole-mantle convection these timescales are 1.0 Gy (for 3 km) or 1.5 Gy (for 3 m). However, lateral dispersion takes much longer: for half the circumference of the Earth and whole-mantle convection the timescale is 4.6 Gy, similar to the age of the Earth.

The mixing time for subducted oceanic crust in a kinematic chaotic flow was calculated by Kellogg and Turcotte (1990). They assumed that a heterogeneity must be stretched thin enough for diffusive homogenization to take place based on the model of Kellogg and Turcotte (1986), which gives an equation for mixing time of

$$t_b = \frac{12b}{\pi u_{\text{sm}} \bar{\alpha}_e} \ln\left[ 4b_0 \left( \frac{\pi u_{\text{sm}} \bar{\alpha}_e}{12Db} \right)^{1/2} \right] \qquad [34]$$

where $D$ is the chemical diffusivity (assumed to be $10^{-19}\,\text{m}^2\,\text{s}^{-1}$ based on Hofmann and Hart (1978) and Sneeringer et al. (1984), $b$ is the depth of the convecting layer, $b_0$ is the initial heterogeneity size (assumed to be 6 km), $u_{\text{sm}}$ is the average surface velocity (assumed to be 10 cm/yr$^{-1}$), and $\bar{\alpha}_e$ is the effective strain rate from the numerical experiments. With these assumptions $t_b$ was calculated to be 240 and 960 My for upper-mantle convection or whole-mantle convection, respectively, which is the time for 50% of the heterogeneities to be homogenized by diffusion and corresponds to nine overturns. The diffusive homogenization length scale for this is less than 1 cm for both cases, and the amount of thinning is 6–7 orders of magnitude. Due to idealizations in their model, however, these times must be interpreted as lower limits. One idealization is that this is not a real convective flow, and it is not clear how representative the effective strain rate will be of the real mantle. For example, in the real mantle the viscosity increases with depth, which will reduce the strain rate in the deep mantle and increase the 'overturn time', a concept which itself is not well defined for flow in the actual mantle. An obvious quantitative issue is that the assumed average surface velocity of 10 cm yr$^{-1}$ is approximately three times larger than Earth's actual mean poloidal surface velocity, the poloidal component of which is around 3.5 cm yr$^{-1}$ (Lithgow-Bertelloni *et al.*, 1993). If the times are recalculated with this value, they change to 660 and 2630 My, respectively, the latter of which may well be consistent with geochemical evidence for survival of geochemical heterogeneity over 2 billion year timescales.

In the actual mantle, heterogeneity is continuously introduced at the same time as 'old' heterogeneity is destroyed by stirring and diffusion, so Kellogg and Turcotte (1990) also consider the fraction of the mantle that is completely homogenized by the present day when oceanic crust is continuously introduced into the mantle over geological time. For this it is first necessary to assume an equation for crustal production versus time, and they assume that oceanic crustal production in the past scales with radiogenic heat production in the mantle with a $1/e$ decay time of 2.5 Gy (Turcotte and Schubert, 1982), that the present rate of production is 2.8 km$^2$/year, and that the processing depth under mid-ocean ridges in 60 km, assumptions that lead to half of the mantle being processed in 800 My for layered convection or 2200 My for whole-mantle convection. Combining this melting model with their mixing model, they derive an expression for $F(d)$, the fraction of particles narrower than $d$. Hence setting $d$ to the diffusion length scale of $10^{-7}$, they obtain a curve for the fraction of particles that have been completely homogenized, in terms of the mixing time. For their estimated whole-mantle mixing time of 960 My, 67% of the mantle is

perfectly mixed, but for the adjusted mixing time of 2.6 Gy as discussed above, this fraction is 28%, that is, over 70% of the present-day mantle is still in the form of strips of varying thickness. Thus, they predict substantial chemical heterogeneity even with the relatively rapid mixing in their experiments.

Kellogg and Stewart (1991) estimate the effective strain rate and mixing time for a chaotic flow using the same method, but this time the flow is the non-linear interaction of a single convection roll with a single spatial subharmonic and the obtained mixing times are about a factor of 2 smaller.

### 10.2.6.4 Summary of different 'mixing time' estimates

**Table 1** summarizes the various estimates of mixing time for whole-mantle convection, as discussed above. As far as the author is aware these are all the published estimates given in dimensional time units, although it might be possible to obtain additional estimates by scaling nondimensional results from other studies. Estimates vary by an order of magnitude, from 1 to 10 Gy, because of different assumptions about the length scale to which heterogeneities should be stretched, the type of flow, and the flow velocity. Gurnis and Davies (1986b) were concerned with the survival of large clumps, so their estimate is not directly comparable to the other estimates. The estimate of Olson *et al.* (1984b) is large because they assumed laminar flow, whereas it is now established that time-dependent convective flows tend to give turbulent mixing, as occurred in the last three estimates. The estimate of Kellogg and Turcotte (1990) should be at least tripled because their assumed velocity was too large, as discussed above. The estimate of Christensen (1989a) refers to stretching oceanic crust to ~100 m thickness, and should be at least doubled for comparison with the Kellogg and Turcotte (1990) estimate. The estimate of Hoffman and McKenzie (1985) was made before it was established that the lower mantle is probably a

**Table 1** Summary of mixing time estimates

| Study | $\tau_{mix}$ (Ga) | Length scale (inner) | Notes | Rescaled $\tau_{mix}$ (Gyr) |
|---|---|---|---|---|
| (Gurnis and Davies, 1986b) | 2.4 | Visible clumps | Survival time | |
| (Olson *et al.*, 1984b) | 10 | Diffusion | Laminar stretching | |
| (Christensen, 1989a) | 1.5–2.0 | 1/50 (~100 m) | 90% reduced to this size | 3–4 |
| (Hoffman and McKenzie, 1985) | 1.0 | 3 km | Only 3* viscosity | >3 |
| | 1.5 | 3 m | Increase with depth | |
| (Kellogg and Turcotte, 1990) | 0.96 | Diffusive | 50% homogenized | ~3 |

factor of ~30 more viscous than the upper mantle (they assumed a factor of 3), so should be at least doubled when this is taken into account. When such scaling is done, estimates are in better agreement, in the 3–4 Gy range for stretching crust to the submeter lenthscale. If an equivalent table were constructed for upper-mantle convection estimates then the times would be much shorter, but as argued earlier, geophysical constraints indicate that the dominant mode of convection in the present Earth is whole-mantle, and the effect of possible 'small-scale' convection beneath the oceanic plates has not yet been evaluated.

### 10.2.7 Spectrum of Chemical Heterogeneity

As was discussed in a Section 10.1.7, our knowledge of the spectrum of chemical heterogeneity in the mantle is far from perfect but some useful information does exist. Geochemical observations show the existence of heterogeneity at all length scales, from the scale of melt inclusions to thousands of kilometers, and quantitative information regarding the spectrum is becoming available. As also discussed earlier, seismological information points to the existence of scatterers throughout the lower mantle, with the amount of scattering varying from place to place. The longest wavelengths are constrained by seismic tomography but separating thermal from compositional contributions has major uncertainties. This section reviews the dynamical theory about the expected length scales of heterogeneity.

Mantle heterogeneity is introduced at long wavelengths by subduction (and possibly, primordial layering). If subducted material segregates into a layer then this acts to increase its wavelength, but if it remains in the convective flow then it is progressively stretched until it reaches small-enough length scales to be homogenized by diffusion. Thus, as discussed by Olson *et al.* (1984a), the spectrum of mantle heterogeneity undergoing stirring is expected to display three regions with increasing wave numbers (decreasing wavelength): a 'source' region corresponding to the spectral characteristics of the source, a 'cascade' region in which the heterogeneities are being stretched to shorter wavelengths, and a 'dissipation' region, in which chemical diffusion acts to eliminate heterogeneity at the shortest scales. It is useful to consider the spectrum that is obtained in statistical equilibrium, in which heterogeneity is introduced at a constant rate and the total spectrum

does not change with time. This depends on the stretching regime in the cascade region: For laminar stirring the wave number $k$ of an individual heterogeneity increases linearly with time, which leads to a flat ('white') spectrum for all heterogeneities in statistical equilibrium (Olson *et al.*, 1984a). In contrast, for turbulent stirring the wave vector of an individual heterogeneity increases exponentially with time, that is, the rate of increase is proportional to its present value, which leads to a statistical equilibrium spectrum that scales as $1/k$. This at first seems inconsistent with the with the $-5/3$ slope obtained by Metcalfe *et al.* (1995) for a kinematic chaotic stirring of a laboratory fluid, but as this system was not in statistical equilibrium the slope reflects a transient regime.

Geochemical observations, as discussed in Section 10.1.7, can be compared to these theoretical slopes. The early analysis of Gurnis (1986b) using $^{87}Sr/^{86}Sr$ data from MORB and OIBs appeared to be consistent with a 'white' laminar mixing spectrum, but the more highly constrained spectrum of isotope ratios along the mid-Atlantic ridge obtained by Agranier *et al.* (2005) instead indicates the turbulent, $1/k$ regime, consistent with expectations from mantle-mixing theory discussed earlier.

Such theory also successfully predicts the smaller scales of heterogeneity observed in peridotite massifs in Beni Bousera (Kellogg and Turcotte, 1990). They derive that the probability of finding a strip of material larger than size $d$ is given by $1/d(1 - F(d))$, which, when evaluated for scales between 1 and 100 cm, gives a slope close to $-1$, indicating that the $(1/d)$ term is dominating and $(1 - F(d))$ does not change very much over this range. Field samples match this slope reasonably well.

In any case, it is clear that because thermal heterogeneity diffuses rapidly whereas chemical heterogeneity does not, short-wavelength mantle heterogeneity is dominantly chemical in origin, while thermal heterogeneity should be important, and perhaps dominant, at long wavelengths, although recent seismological studies argue for a large chemical component (Trampert *et al.*, 2004; Deschamps *et al.*, 2007). These expectations are confirmed in the thermochemical convection calculations of Nakagawa and Tackley (2005b), in which the slope of the chemical heterogeneity at long wavelengths also appears to be consistent with a $1/k$ scaling, although they do not study how this scales at short wavelengths.

## 10.3    Mantle Convection with Active (Buoyant) Chemical Heterogeneity

This section considers thermochemical mantle convection in which the composition affects density. Studies have typically considered what happens with a preexisting dense layer, or the development of a layer over time by settling of dense material or reactions with the core. Because the chemical diffusivity in the mantle is orders of magnitude lower than thermal diffusivity, it is typically assumed that the chemical field is nondiffusive, although if the field is being treated as a continuous field then some numerical diffusion is typically introduced in order to stabilize the advection. The Lewis number gives the ratio of chemical to thermal diffusivity:

$$Le = \frac{\kappa_T}{\kappa_C} \qquad [35]$$

Its value is quoted as $\sim 10^8$ by Hansen and Yuen (1989), but the chemical diffusivity quoted by Kellogg and Turcotte (1986) would lead to $Le = 10^{13}$.

Attention has generally focused on two cases: compositional layering at the 660 km interface, or a relatively thin composition layer above the CMB corresponding to D″.

### 10.3.1    Stability and Dynamics of Chemical Layering in a Convecting Mantle

#### 10.3.1.1    The balance between chemical and thermal buoyancy

If a layer of dense material is present, the parameter that most strongly influences its dynamics and evolution is the ratio of chemical density contrast to thermal density contrast, that is, the buoyancy ratio, here written as $B$:

$$B = \frac{\Delta\rho_C}{\Delta\rho_T} = \frac{\Delta\rho_C}{\rho_0 \alpha \Delta T} \qquad [36]$$

where $\Delta\rho_c$ is the density contrast due to composition, $\Delta\rho_T$ is that due to temperature, $\rho_0$ is the density, $\alpha$ is the thermal expansivity, and $\Delta T$ is the temperature contrast. If $\Delta T$ represents the total thermal contrast available in the system, then it has been found that the layering displays long-term stability with a fairly flat interface if $B$ is greater than some critical value of order 1, and undergoes either large-topography layering or various shorter-lived behaviors that end in mixing and whole-mantle convection, if $B < 1$ (Richter and Johnson, 1974; Richter and McKenzie, 1981; Olson and Kincaid, 1991;

Montague et al., 1998; Davaille, 1999a; Montague and Kellogg, 2000; Le Bars and Davaille, 2004a). The behaviors of short-lived modes were systematically mapped out as a function of $B$, viscosity contrast, and layer thickness using laboratory studies (Davaille, 1999a, 1999b; Le Bars and Davaille, 2004a, 2004b) supplemented by mathematical analysis (Le Bars and Davaille, 2002), and are generally consistent with earlier laboratory and numerical studies, as follows, and as illustrated in the domain diagram of **Figure 6**:

1. $B > 1$ stable stratification, both layers convecting with slow entrainment,
2. $B > 1$ but thin, nonconvecting lower layer,
3. $1 > B > 0.5$: layered but with large topography,
4. $0.3 < B < 0.5$, layer thickness between 0.2 and 0.7: doming, for and, number of pulsations increases rapidly with viscosity contrast, and
5. $B < 0.2$: rapid overturn and mixing of layers.

If the two layers initially have the same temperature, then additional transient behavior occurs as a temperature difference develops across the layers, thereby decreasing the total (i.e., thermal + chemical) density contrast. In this case, layers that should be unstable according to the above criteria may initially

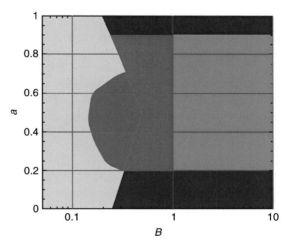

**Figure 6**    Domain diagram for the behavior of thermochemical convection as a function of buoyancy ratio $B$ and initial depth of dense layer a, as determined by laboratory experiments (Davaille, 1999b; Davaille et al., 2002; Le Bars and Davaille, 2002, 2004a, 2004b). Green, flat layers; red, large topography layers; pink, transient oscillating domes; blue, thin noninternally convecting layer progressively entrained by thermochemical plumes; yellow, rapid overturn. Reproduced from Le Bars M and Davaille A (2004b) Whole layer convection in a heterogeneous planetary mantle. *Journal of Geophysical Research* 109: 23.

be stable, as in, for example, the calculations of Samuel and Farnetani (2003). *B* may decrease with time due to the entrainment of each layer into the other layer, as discussed below. Indeed, in laboratory experiments of Le Bars and Davaille (2004b) (**Figure 2**), the experiment evolved through several different regimes as the density difference between the layers decreased. Another initial condition that causes transient behavior is a linear compositional gradient throughout the domain and constant temperature, as studied by Hansen and Yuen (2000), who found that after some time the system developed a stable but large-topography layer. After that, the evolution discussed in the remainder of this section applies.

In the mantle, the relevant physical properties ($\alpha$, $\rho_0$) vary with depth, and the above criteria are most valid if values of the physical properties close to the chemical boundary area are considered, although some studies have used, for example, reference or 'surface' values and thus require careful interpretation. Scaling to the Earth, the appropriate $\Delta T$ is the superadiabatic temperature drop between surface and CMB, which might be 2500 K. For layering at around 660 km depth, $\alpha \sim 2.5 \times 10^{-5}$, so the critical $\Delta\rho_c/\rho_0 \sim 6.25\%$, whereas at the CMB $\alpha \sim 1.0 \times 10^{-5}$, so $\Delta\rho_c/\rho_0 \sim 2.5\%$. These are rough estimates, as it has been found that other characteristics such as the viscosity contrast between the two layers (a combination of intrinsic viscosity contrast and temperature-dependent contrast), the thickness of the dense layer, and the thermal conductivity of the dense layer also play a role, as discussed later.

The detailed dynamics have been investigated variously by laboratory and numerical experiments with some mathematical analyses, with laboratory experiments having the advantage of physical correctness, which is particularly important because entrainment rates are difficult to accurately measure numerically (van Keken *et al.*, 1997; Tackley and King, 2003), whereas numerical experiments have the advantage that the variation of physical properties (viscosity, thermal expansivity, conductivity) with pressure and temperature can be more realistically ascertained and full information about the various fields is available for analysis. Experiments (both numerical and laboratory) with layering are often initialized in a way that is far from a steady state (e.g., temperature is isothermal instead of higher in the lower layer), which results in a transient behavior. Furthermore, layered experiments are intrinsically transient in that the layering evolves with time, and this may result in the final state

retaining a memory of the initial conditions. Further discussion of laboratory experiments can be found in Chapter 3.

### 10.3.1.2 Initially stable layering: pattern and dynamics

#### 10.3.1.2.(i) Flat, thick, internally convecting layers
For thick, convecting layers, layered convection occurs and a thermal boundary layer builds up over the interface, with heat transport accomplished by thermal diffusion, substantially reducing the global heat flow and affecting Earth's thermal evolution (Montague and Kellogg, 2000). For layers of approximately equal thickness and viscosity, as in the laboratory and numerical experiments of Richter and McKenzie (1981), the temperature difference between the layers is $\Delta T/2$ (Richter and McKenzie, 1981), but this varies as the viscosity contrast and thicknesses are changed. The pattern of convection in the upper and lower layers are coupled. When the layers are of similar viscosity, mechanical coupling dominates such that upwellings in one layer match with downwellings in the other layer (Richter and McKenzie, 1981; Cserepes and Rabinowicz, 1985). When there is a large viscosity contrast exceeding a factor of $\sim 100$ (Cserepes and Rabinowicz, 1985), however, thermal coupling dominates, in which upwellings in the lower layer underlie upwellings in the upper layer. Intermediate viscosity contrasts, such as 25–30, result in a mixture of thermal and mechanical (viscous) coupling, as show by 3-D calculations (Cserepes *et al.*, 1988; Glatzmaier and Schubert, 1993). In this mixed regime, the flow direction of rolls in the upper layer was observed to become perpendicular to those in the lower layer by (Cserepes *et al.*, 1988). Entrainment occurs at cusps in the interface, as discussed later.

If one layer is substantially more viscous, the form of upwellings or downwellings forming at the interface is also different (Davaille, 1999b), with 2-D sheets forming in the more-viscous layer but 3-D plumes forming in the less-viscous layer, which have different entrainment rates as discussed below.

#### 10.3.1.2.(ii) Thin, nonconvecting lower layer
In this case the lower layer is embedded within the lower thermal boundary layer of the main convection, and the dense material gets swept by the convection, forming a spoke pattern of ridges on top, with plumes at the junctions of the ridges, and entrainment takes place there, as shown by 3-D numerical experiments (Tackley, 1998) and

laboratory experiments (Olson and Kincaid, 1991; Davaille *et al.*, 2002). The numerical experiments show that this dense layer acts like a rigid boundary, reducing the horizontal wavelength of convective planform relative to a free-slip lower boundary (Tackley, 1998), and that if the layer internally convects (Montague *et al.*, 1998; Montague and Kellogg, 2000), as would be the case if its viscosity is low enough (Solomatov and Moresi, 2002; Schott and Yuen, 2004), small-wavelength thermal structure is generated within the layer. These numerical experiments also show that even in 100% internally heated cases, entrainment exists at cusps anticorrelated with where the downwellings are, and larger amounts of internal heating reduce the buoyancy ratio needed for stability. Temperature-dependent viscosity decreases the density contrast needed for stable layering (Tackley, 1998; Schott *et al.*, 2002), and when the layer is marginally stable, can lead to complicated schlieren structures of entrained material, and complicated layer topography (Schott *et al.*, 2002). For reasonable Earth parameters Tackley (1998) found for an internally heated case that ~1% density contrast was sufficient to maintain a layer over reasonable timescales.

### 10.3.1.2.(iii) Thick, large-topography lower layer

With a lowish buoyancy ratio like 0.6 (Montague *et al.*, 1998), the topography on the layer interface is large (Davaille, 1999a; Kellogg *et al.*, 1999), which may result in discontinuous 'piles' or 'hills' if the layer thickness is less than the topography. In 2-D calculations at low Rayleigh number, this results in wide stable 'hills' below the upwelling, which cause strong (~several kilometers) negative topography on the CMB (Davies and Gurnis, 1986; Hansen and Yuen, 1989), with temperature-dependent viscosity decreasing the magnitude of this negative topography (Hansen and Yuen, 1989). The topography on a layer interface decreases with increasing Rayleigh number, approximately as $Ra^{-1/3}$ (Davies and Gurnis, 1986; Tackley, 1998), which predicts relatively little topography at the Earth's convective vigor (Montague and Kellogg, 2000). If, however, the increase of viscosity and thermal conductivity and decrease of thermal expansivity with depth is taken into account, they lead to a low effective $Ra$ for the deep mantle, so large interface fluctuations can still occur, as shown in 3-D (Tackley, 1998, 2002). It has been argued that such structures may account for the superplumes observed in seismic tomography, discussed in Section 10.1.8.

An important point shown by van Thienen *et al.*, (2005) is that any layering around 660 km induced by the post-spinel phase transition makes the layer more stable over long time periods, presumably because it reduces the temperature drop across the layer and the convective velocities.

Regarding the 3-D structure of a strongly undulating layer, Cartesian calculations (Tackley, 1998, 2002), either Boussinesq or compressible with either constant, depth-dependent or depth- and temperature-dependent viscosity, found that the long-term solution of piles tend to be ridges/spokes. In spherical geometry Boussinesq calculations (Oldham and Davies, 2004) there appears to be a greater tendency for the piles to be isolated. A viscosity contrast between the layers makes a major difference to the planform: McNamara and Zhong (2004b) show that temperature-dependent viscosity leads to linear piles that are swept around and spread through the entire lower mantle, whereas compositional-dependent viscosity in which the dense material is more viscous leads to isolated round 'superplumes', particularly with contrasts as large as factor 500. This seems to be consistent with the 'domes' in the laboratory results of Davaille (1999a), except that they also observe domes when the dense layer is less viscous. The reconciliation of these is that such domes are transient features – for long-lived features domes are only observed when the lower layer is more viscous (McNamara and Zhong, 2004b).

Applying the specific geometry and motion of Earth's plates to a spherical thermochemical convection model, McNamara and Zhong (2005) found that structures similar to those seismologically observed under Africa and the Pacific were generated. Specifically, Earth's subduction history tends to focus dense material into a ridge-like pile beneath Africa and a relatively more rounded pile under the Pacific Ocean, which they argue is consistent with seismic observations.

### 10.3.1.3 Influence of a dense layer on plume dynamics

#### 10.3.1.3.(i) Temperature
The presence of a dense layer reduces the temperature anomaly of plumes (Farnetani, 1997; Zhong, 2006), which may help to explain why the inferred temperature anomaly of plumes close to the lithosphere is much lower than the estimated temperature drop across the lower thermal boundary layer (TBL) for whole-mantle convection.

***10.3.1.3.(ii) Stability*** Laboratory experiments find that plumes rising from cusps of a dense layer are stable over longer time periods than plumes rising from the lower TBL of whole-mantle convection. In particular, Davaille *et al.* (2002) find that the dense layer acts as a 'floating anchor', reducing the lateral migration of plumes, which then have a scaled drift velocity of $1-2$ mm yr$^{-1}$ – larger than simple scalings predict for thermal plumes. In addition the dense layer enables them to survive for hundreds of million years. The plume spacing is proportional to the stratified boundary layer thickness. In the laboratory experiments of Jellinek and Manga (2002), the dense, low-viscosity material forms a network of ridges (embayment and divides) and they delineate the tradeoff between viscosity contrast and layer topography in the stabilization of plumes. They found that if the dense material has a similar viscosity, the topography required to stabilize plumes is $>2-3$ times the thermal boundary layer thickness (to overcome viscous stresses), but this required topography decreases as the layer viscosity decreases, because the lower layer then acts as a free-slip boundary. Indeed, with $>2.5$ orders of magnitude viscosity contrast, only minimal topography is needed. With no dense layer present, upwellings are in the form of intermittent thermals.

This laboratory finding that a dense layer stabilizes plumes is, however, seemingly contradicted by numerical experiments (McNamara and Zhong, 2004a). In these numerical experiments, plume stability was quantified as the ratio of horizontal plume velocity to average surface velocity. McNamara and Zhong (2004a) found this to be about the same in layered calculations as it is in isochemical calculations with a rigid lower boundary, regardless of the viscosity contrast of the dense material. They also found that layer topography does not influence fixity, but can lead to longer-lived plume conduits that follow the topographic peaks around. The reconciliation of these apparently contradictory findings is not clear. One possibility is that the dense layer reduces the convective vigor and that the laboratory experiments did not take this into account. Another is that there is a difference in the dynamics of downwellings (e.g., they might be stronger in the numerical experiments), which push the lower layer and associated hot plumes around. More study is needed to reconcile laboratory and numerical results on this issue.

***10.3.1.3.(iii) Starting plumes*** Despite the effect that a dense layer can have on increasing plume stability as discussed above, some studies have found that for starting plumes, the presence of a dense layer can add complexity and modes of time-dependence that are not present with purely thermal plumes. Lin and van Keken (2006a, 2006b) investigated axisymmetric models set up to produce a plume at the axis, various temperature dependences of viscosity, and a dense layer with various thicknesses, buoyancy ratios, and viscosity contrasts. They found that entrainment commonly becomes a transient process for models with variable viscosity. In models where a large fraction of the dense layer was entrained, variable viscosity could lead to secondary instabilities developing after the initial plume head, whereas in models with only a thin filament entrained, variable viscosity could lead to an initial pulse of entrainment. In general, the dense material led to a wider range of plume morphologies than observed with purely thermal plumes. As with results discussed elsewhere, they found that decreasing the viscosity of the layer made it more stable.

Even greater morphological variability and complexity was found in the 3-D models of Farnetani and Samuel (2005), in which thermochemical plume formation was modeled in a compressible mantle with no imposed symmetry, an initially 200-km-thick layer with buoyancy ratio $B = 0.1$, temperature- and depth-dependent viscosity. Plume morphologies were much less regular than 'classical' plumes, and included plumes with blob-like or irregular heads, tilted plumes, $\sim 1000$ km wide 'megaplumes', broad plumes without heads, as well as 'classical' head plus narrow tail plumes, often with several types coexisting. The buoyancy flux of conduits varied widely, and some structures ponded under the 660 km discontinuity. Plume conduits were founds to be laterally heterogeneous (i.e., without concentric zonation), but as with previous results (Farnetani *et al.*, 2002) the only material hot enough to melt underneath the lithosphere is from the source region.

### 10.3.1.4 Entrainment of surrounding material by mantle plumes

As discussed above, plumes clearly entrain material from a dense layer at their base. A long-standing question is how much this composition is diluted by material entrained from the upper layer as they rise. The initial plume head and subsequent conduit have different structures and so are expected to be different in their entrainment characteristics.

Depending on the viscosity contrast, rising hot diapirs, analogous to plume heads, thermally entrain surrounding material as they rise, forming a folded structure (Griffiths, 1986). It was demonstrated that this structure also occurs in plume conduits if they are sheared by the surrounding material (Richards and Griffiths, 1989), to such an extent that "small plumes or narrow plume conduits may consist of only a small fraction of original source material" (Richards and Griffiths, 1989). Hauri *et al.* (1994) presented analytic solutions for plume conduits, finding that they entrain significant amounts of ambient mantle due to lateral thermal conduction from the plume. They found a huge range of possible entrainment, depending on the buoyancy flux, with most of the entrained material coming from the lower half of the layer they are passing through.

Although plumes may entrain a significant amount of surrounding material, Farnetani and Richards (1995) found that the entrained material does not contribute significantly to the melt produced by the plume, with >90% of the primary plume magmas being composed of primary plume material. The first-order reason for this is that the entrained material has a much lower temperature than the core of the original plume, and so is much less likely to melt underneath the lithosphere. Reinforcing and extending this finding, Farnetani *et al.* (2002) analyzed entrainment by a thermal plume with temperature-dependent viscosity contrast 100, and found that the material that melts is from the thermal boundary layer at the base of the mantle. The material in the plume head was found to be much more extensively stretched and stirred than that in the plume conduit, implying greater homogenization of geochemical anomalies. At that viscosity contrast, the plume head did not entrain surrounding material, although a substantial amount of lower-mantle material was advected upwards with the plume into the upper mantle.

Entrainment of the dense layer by plumes is further discussed in the next section.

## 10.3.2   Long-Term Entrainment of Stable Layers

In a system with stable compositional layering, the system nevertheless evolves with time because the thermal convection in each layer causes entrainment of material from the other layer, in the form of thin schlieren, thereby diminishing the density contrast (Olson, 1984). There are two possible outcomes to this evolution: (1) If the entrainment is asymmetric (typically due to the lower layer being thinner than

the other, e.g., Samuel and Farnetani (2003)) then one layer may eventually disappear due to complete entrainment into the other layer. Indeed, it is likely that in the long term there is substantial entrainment of any dense material residing above the CMB, which may be balanced by addition of material by segregation or core reactions, as discussed later. (2) The density contrast decreases to the point where the layering is no longer stable and large-scale overturn and stirring occurs. Numerical calculations and laboratory experiments have been useful in elucidating the details of these processes, while mathematical analysis has been useful in developing scaling laws for entrainment.

### 10.3.2.1   Comparison of entrainment laws and estimates

Several authors have obtained expressions for the entrainment rate as a function of parameters such as density contrast, viscosity contrast, etc. Studies usually focus on one of two situations: a thin layer at the base of the mantle, or two layers of substantial thickness. While the same laws might apply, the former case obviously has much greater asymmetry and the lower layer might not be convecting.

**10.3.2.1.(i)   Laboratory   experiments**   From studying the entrainment in laboratory experiments with two layers of equal depths, Olson (1984) found that the mixing rate of the two layers, defined as the rate of change of their density divided by the density difference (i.e., the fractional volume flux), scales as

$$\dot{M}/\dot{\varepsilon} \approx 0.046 Ri^{-1} \qquad [37]$$

where the Richardson $Ri = \Delta\rho g D / \tau$ with $\tau$ being the average viscous stress within each layer. Eliminating $Ri$ leads to

$$\dot{M}/\dot{\varepsilon} \approx 0.046 \frac{\tau}{\Delta\rho g D} \qquad [38]$$

For two thick stable layers where the lower layer is more viscous by a factor of $1.0$–$6.4 \times 10^4$, Davaille (1999b) found that the scaling for entrainment by linear sheets (typically found in the more viscous layer) is different from that for axisymmetric plumes (found in the less-viscous layer). For entrainment by linear sheets

$$Q_{21} = C_1 \frac{\kappa b_0}{B} Ra^{1/5} \qquad [39]$$

whereas for columnar plumes

$$Q_{12} = C_2 \frac{\kappa h_0}{B^2} Ra^{1/3} \qquad [40]$$

where $Q$ is the flux of material, $h_0$ is the initial thickness of lower layer, $\kappa$ is the thermal diffusivity, and $C_1$ and $C_2$ are two experimentally determined constants. The 'plume' law has the same $1/B^2$ form as the one proposed by Sleep (1988) (discussed below). When the lower layer is substantially more viscous than the upper layer, it is difficult for the upper layer to entrain it, requiring an additional factor (Davaille, 1999a):

$$Q = C_1 \frac{\kappa H}{B^2} Ra^{1/3} \frac{1}{1 + \Delta\eta/B} \qquad [41]$$

where $\Delta\eta$ is the viscosity contrast. This additional factor applies for $\Delta\eta > 1$, that is, for entrainment of the more-viscous layer into the less-viscous layer. If the layer being entrained is substantially less viscous than the entraining layer then $Q$ could also be reduced due to draining of the less-viscous material back into the layer it came from, but this was not observed by Davaille (1999a) and Davaille et al. (2002) at $\Delta\eta$ as low as 0.01. This entrainment law was subsequently verified for the case of a thin layer with plume entrainment (Davaille et al., 2002). Depending on parameters, a mantle plume would entrain chemically dense material at a rate of 0.1–15% of its total volume flux. Davaille et al. (2002) present graphs of layer survival time versus thickness and density contrast, based on thermal expansivity $10^{-5} K^{-1}$ and 'temperature heterogeneity' of 500 K. As an example, entrainment of 100 km over 4.5 Gy corresponds to a 3.5% density contrast if the layers are of the same viscosity or $\sim$2% density contrast if the dense material is 10 times more viscous.

Several other authors have conducted laboratory experiments to study entrainment and long-term evolution. In experiments with an initial layer thickness of 0.2 (600 km) and similar viscosity materials, Gonnermann et al. (2002) find vigorous convection in both layers and entrainment consistent with the scaling laws of Davaille (1999a). They estimate that an initially 300-km-thick layer can survive over Earth history if it has a 2% density contrast, but if initially thicker then it can survive initial density contrast < 2%. In some cases the dynamics changed from stable layering to doming.

The strong influence of viscosity contrast on entrainment of a thin layer was demonstrated by Namiki (2003) using viscosity ratios from 0.1 to 400.

It was found that when the lower layer is more viscous it grows in size due to asymmetric entrainment of the upper layer. When the lower layer is less viscous, an interfacial layer developed and the volumes of the layers did not change because plumes entrained the interfacial layer, not the lower layer. This interfacial layer is a novel finding not observed in any other mantle-related studies, so it is worth considering what causes it. The layer might develop because of incomplete entrainment of the lower, less-viscous layer by the upper layer due to reduced viscous coupling. It might form as a result of double-diffusive convection which helps to transport material across the boundary (M. Manga, personal communication). If the latter explanation is correct, it would not form in Earth because the chemical diffusion rate is very low. In any case, by considering stresses as in Olson (1984) Namiki (2003) derives an entrainment law that is consistent with those discussed above:

$$\dot{E}_i \sim \frac{\kappa_i}{B_i^2 \delta_{th_i}} \qquad [42]$$

where $\dot{E}_i$ is the entrainment rate of layer $i$, $\delta$ is the thermal boundary layer thickness, related to the Rayleigh number as $\delta_{th} \sim 5 L_i Ra^{-1/3}$ leading to an overall scaling of

$$\dot{E}_i \sim 0.2 \frac{\kappa_i}{B_i^2 L_i} Ra^{1/3} \qquad [43]$$

In this parametrization the effect of a viscosity difference is to alter the $\Delta T$ for each layer, hence $B$ and $Ra$. The survival time of a 200-km-thick layer was calculated as a function of its density difference, assuming the total $\Delta T = 4000$ K, the viscosity of D″ is $10^{19}$ Pa s, the viscosity of the mantle is $10^{21}$ or $10^{22}$ Pa s, and thermal expansivity is $2 \times 10^{-5} K^{-1}$. It was estimated that a density difference of at least 300 kg m$^{-2}$ (5.5%) is needed for the layer to survive over geological time (ignoring the possible presence of an interfacial layer). With a more realistic thermal expansivity of $1 \times 10^{-5} K^{-1}$, this would be reduced to 2.75% contrast, comparable with other estimates.

A scaling law that explicitly includes viscosity contrast was proposed by Jellinek and Manga (2002) on the basis of laboratory experiments:

$$q = 0.49 \frac{\kappa}{H} \frac{Ra^{1/3}}{\Delta\eta B} \qquad [44]$$

where $H$ is the depth of tank and $\Delta\eta > 1$ is the ratio of upper- to lower-layer viscosity. This is similar to the expression of Davaille (1999a) given above, in the limit of $\Delta\eta \gg B$. It is interesting that in this limit, the dependence on $B$ is $1/B$ rather than $1/B^2$.

### 10.3.2.1.(ii) Mathematical or numerical approaches

Gradual entrainment of thin chemical layer at the base of the mantle by either 2-D upwelling sheets or axisymmetric plumes was analyzed by Sleep (1988) to determine the fraction of entrained material in the upwelling as a function of chemical density contrast. The analysis did not consider the details of the entrainment process, but rather the carrying capacity of the plume conduit. Thus, a 1-D section across the upwelling was considered, with the entrained material in the form of a thin filament in the center of the upwelling. To determine the thickness of this filament, it was assumed that the thickness adjusts so as to maximize entrainment rate (e.g., if it is too thick then the upwelling velocity decreases). The models included temperature-dependent viscosity. While a general scaling law for all parameter combinations was not explicitly stated, the general form was found to be

$$F = C/B^2 \qquad [45]$$

where $C$ is a constant and $B$ is the chemical buoyancy ratio. Based on these models, Sleep (1988) estimated that a chemical density contrast of 6% is necessary for a 100-km-thick, nonconvecting layer to survive over the age of the Earth if the thermal expansivity is $2 \times 10^{-5}\,\mathrm{K}^{-1}$. Subsequent mineral physics data (e.g., Anderson *et al.*, 1992; Chopelas, 1996) indicate a thermal expansivity closer to $1 \times 10^{-5}\,\mathrm{K}^{-1}$ at the CMB, so this estimate of density contrast should be adjusted to 3%. Additionally, if the initial layer thickness were larger, then a higher entrainment rate could be tolerated. In any case, this density contrast corresponds to $B \sim 3$ based on the temperature drop over the lower boundary layer, or $B \sim 1.5$ using the temperature drop over the entire system – both substantially larger than that required to stabilize the layer against rapid overturn. Sleep (1988) also considered the case of an internally convecting layer and concluded that the above estimate is still reasonable.

The entrainment process at the base of the plume (not considered by Sleep (1988)) was included in the high-resolution 2-D numerical models of Zhong and Hager (2003). They found that treating the lateral flow at the bottom of the plume makes some significant differences compared to the 1-D analysis of Sleep (1988), with the 1-D model overestimating entrainment by up to more than a factor of 4 compared to their 2-D

calculations. They found that the entrainment rate $Q$ is mainly controlled by the buoyancy ratio $B$ and the radius of the thermal plume $r_{\mathrm{T}}$ but insensitive to the thickness of the dense layer, with a relationship

$$Q \sim B^{-2.48} r_{\mathrm{T}}^{3.80} \qquad [46]$$

Because their model was set up to produce a plume with prescribed parameters, they did not obtain a scaling law for a full convective system. They did, however, apply the scaling to estimate the survival time of a 1000-km-thick layer, and it was estimated that more than 90% of a 1000-km-thick dense layer can survive for 4.5 Gy if the dense material has a net negative buoyancy of $\sim 1\%$ . It is important to note that this is the 'net' buoyancy (thermal + chemical) and not the chemical buoyancy! For a thermal expansivity of $2.5 \times 10^{-5}\,\mathrm{K}^{-1}$ and the dense material 800 K hotter, the chemical density contrast would have to be 3%. This could be reduced by a factor of $\sim 2$ using a more realistic deep-mantle thermal expansivity.

Recently, Lin and van Keken (2006a, 2006b) revisited the transient dynamics of a plume forming from a boundary layer with an included dense chemical layer, first studied by Christensen (1984). The dynamics are discussed in Section 10.3.1.3. Although the short-lived (250 Ma) nature of their simulations was not intended to study long-term entrainment, they did compare short-term entrainment with existing laws and also found some interesting behavior that is worth noting here. In particular, they found two behaviors that are not accounted for in standard entrainment theory: (1) a very thin layer, that is, much thinner than the thermal boundary layer, could be entrained very rapidly despite having $B = 2$ or 3, (2) for $B \sim 1$, a new starting plume head caused a 'pulse' in entrainment, bringing into question the applicability of steady-state entrainment laws for a dynamic, time-dependent mantle. As expected, temperature-dependent viscosity resulted in much lower entrainment, for example with a factor of 1000, even a layer with $B < 0.5$ was stable. Regarding entrainment rate, their results were consistent with $F = C/B^2$ found by Sleep (1988) with $C = 0.07$, and the scaling law of Davaille (1999b) with $C = 1.1$, although there was some deviation from the latter when $B > 1$, which they suggested is due to the effective $B$ and $Ra$ adapting with time in the laboratory experiments. They estimate that a layer with temperature-dependent viscosity can survive if it is thicker than 75 km and its compositional density contrast is larger than 1.5% assuming thermal expansivity is $3 \times 10^{-5}$, which would be reduced to ~0.5% with an Earth-like thermal expansivity.

### 10.3.2.1.(iii) Summary of layer survival estimates

As several studies discussed above have estimated the density contrast required for a dense layer to survive over geological time in the face of continuous entrainment, it is useful to consider whether these estimates are consistent with each other. In general, the published estimates assume a deep-mantle thermal expansivity at least a factor of 2 higher than realistic, so in **Table 2** they have been rescaled if necessary to a thermal expansivity of $1 \times 10^{-5} \, \text{K}^{-1}$.

In general, the rescaled estimates are consistent, with thinner layers increasing the required density contrast and a viscosity contrast (either higher or lower) reducing the required density contrast. Perhaps the exception is Lin and van Keken (2006a), whose density estimate is very low. This could be because they have a larger viscosity contrast than any other study, and this is certainly worth exploring further. Another factor might be that they consider a single plume with $\Delta T = 750 \, \text{K}$ rather than a full convective system as in the laboratory studies, and convective downwellings could help to destabilize the layer. On the other hand, Sleep (1988) and Zhong and Hager (2003) also consider isolated plumes. In summary, this issue requires investigation in a fully convecting system with large viscosity contrast.

### 10.3.3 Evolution of a Mantle with Ongoing Differentiation

#### 10.3.3.1 General evolution

So far the discussion has focused on the evolution of a system with pre-existing layering. However, while the possibility of primordial layering exists, most of the observed heterogeneity appears to be related to

recycled material, so several models have studied the evolution of a mantle in which heterogeneity is continuously introduced over time, generally to simulate the differentiated components in subducted slabs (Gurnis, 1986a; Ogawa, 1988, 1993, 1997, 2000b, 2003; Christensen, 1989b; Christensen and Hofmann, 1994; Kameyama et al., 1996; Ogawa and Nakamura, 1998; Walzer and Hendel, 1999; Davies, 2002; Tackley and Xie, 2002; Ogawa, 2003; Nakagawa and Tackley, 2004b; van Thienen et al., 2004; Xie and Tackley, 2004b; Nakagawa and Tackley, 2005a, 2005b; Tackley et al., 2005). Some of these have already been discussed to a certain extent, in the context of the effect on residence time for example.

In this section, models in which active heterogeneity is introduced are discussed. Some of the models track trace elements as well as crust–residue differentiation; their trace-element aspects are discussed in Section 10.4. One of the main differences between studies is how they treat the secular evolution of the planet, in which the mantle and core cools and heat-producing elements produce less heat. While some studies attempt to directly include this (e.g., Ogawa, 2003; Xie and Tackley, 2004b; Nakagawa and Tackley, 2005a), others have parametrized the faster earlier evolution as a longer run time at a constant convective vigor (Davies, 2002). Other differences are how crustal production is treated (self-consistent melting criterion or assumed thickness crust) and how plate tectonics is treated (kinematic plates or using a yield-stress approach).

While there are substantial variations in the assumptions and results among different calculations, it is possible to generalize the typical evolution.

1. Initial condition: chemically homogeneous.
2. Vigorous early melting occurs. If subducted crust and residue components are able to separate at some depth, then depleted residue rises to the top and

**Table 2** Scaled estimates of density contrast required for long-term survival

| Study | Thickness (km) | $\Delta \rho$ | $\eta$ |
|---|---|---|---|
| (Sleep, 1988) | 100 | 3.0% | 1 |
| (Davaille et al., 2002) | 100 | 3.5% | 1 |
| (Davaille et al., 2002) | 100 | 2.0% | 10 |
| (Gonnermann et al., 2002) | 300 | 2.0% | 1 |
| (Namiki, 2003) | 200 | 2.75% | 0.01 |
| (Zhong and Hager, 2003) | 1000 | 1.5% | 1 |
| (Lin and Van Keken, 2006a) | >75 | >0.5% | 0.001 |

crust sinks to the bottom, forming a depleted upper mantle and layer of subducted crust at the bottom. Rapid formation of a depleted upper mantle was observed in some of the calculations of Ogawa and Nakamura, (1998), Davies (2002), Ogawa (2003), and Nakagawa and Tackley (2005b), but particularly in the calculation presented by Davies (2006) because of its high convective vigor (internally heated $Ra = 7.7 \times 10^{11}$) designed to represent the early Earth. Davies (2006) argues that this depletion helps plate tectonics to operate early on.

3. Increasing chemical stratification (depending on parameters), possibly including above the CMB and around 660 km (see later sections for discussion), increasing heterogeneity as more recycled material gets subducted and stirred. It is possible that the phase changes around 660 km depth play a much larger role than they do now, because of the basic scaling with convective vigor (Christensen and Yuen, 1985; Ogawa, 2003).

4. Transition to a slowly evolving long-term state, in which a dynamic equilibrium may be set up between segregation and entrainment, and between the generation (by melting) and destruction (by mixing and/or remelting) of heterogeneity. The mantle is highly heterogeneous, consisting of a large fraction of recycled material. At this point most of the mantle may be depleted relative to the start condition if subducted crust is able to form a layer in the deep mantle (e.g., Ogawa, 1997; Xie and Tackley, 2004a).

5. Possible switches in the regime can occur, such as remixing of previously formed layers (Ogawa, 1994, 1997). Other possible changes are a reduction in stratification at 660 km depth (Ogawa, 2003), or changes in the plate tectonic regime (e.g., as proposed by Vlaar (1986), Davies (1992), Sleep (2000)), although the latter possibility has not been observed in models yet.

It is worth noting some findings of such calculations. First, layers that form from segregation and settling of dense material (as discussed in the next section) tend not to have sharp boundaries like layers that modelers introduce *a priori* (**Figure 7**). This affects the ability of plumes to form at their interfaces and should also effect entrainment. Second, in models that include secular cooling, melting can be a very important transporter of heat early on. For example, in the calculations of Xie and Tackley (2004b), heat transported by melting, which is the sum of energy transport due to latent heat and energy loss to rapid

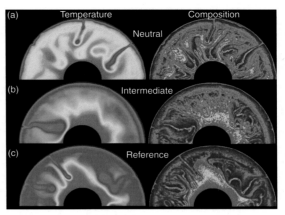

**Figure 7** Laterally heterogeneous mantle and 'messy' layering resulting from continuous oceanic differentiation and recycling of MORB and residue for billions of years. MORB, red; harzburgite, blue. All models include composition-dependent depth of the transition to perovskite (deeper for MORB) and the same compositional density contrasts in the upper mantle and top of lower mantle. The density contrast of MORB relative to pyrolite at the CMB is either (a) zero, (b) 1.1%, or (c) 2.2%. When MORB is dense at the CMB, some of it segregates into a layer, which is intermittent in the intermediate-density case. In all cases, the transition zone is enriched and the top of the upper mantle is depleted, due to the different depths of the perovskite phase transition. From Nakagawa T and Tackley PJ (2005a) Deep mantle heat flow and thermal evolution of Earth's core in thermo-chemical multiphase models of mantle convection. *Geochemistry Geophysics Geosystems* 6 (doi:10.1029/2005GC000967).

cooling of the newly formed crust at the surface, is about equally important as conductive heat loss early in the calculation, but later on subsides to a small fraction of the total heat transport as is the case on present-day Earth. The vigorous melting that occurs early on was found in the calculations of Ogawa (1997) and Ogawa and Nakamura (1998) to cause a compositionally layered internal structure, whereas when the melting rate decreased due to cooling the internal structure became well mixed. The early, layered mode occurred when the average temperature no longer intersected the solidus. Dupeyrat *et al.* (1995) also obtained a switch between layered and well-mixed modes, from a homogeneous start.

Shallow mantle depletion can cause some dynamical effects, as found by Dupeyrat *et al.* (1995) modeling upper-mantle convection. A layer of depleted residue building up under the lithosphere reduces convective velocity, and increases the wavelength of convective cells. When plates are included, they carry the buoyant residue down to the bottom where it can subsequently rise in the form of diapirs, which increases mixing rate.

### 10.3.3.2    Development of a dense layer by gravitational settling of subducted crust

The idea that subducted oceanic crust might settle at the CMB was initially proposed by Hofmann and White (1982) as a way of separating the subducted crust from convective mixing for long enough for it to develop the isotopic age observed in OIBs. Long-term storage of subducted oceanic crust may also help to explain other geochemical signatures – for example, the positive Nd anomaly of Archaean rocks (Chase and Patchett, 1988). This section discusses constraints on the plausibility of this process. The possibility of a layer building up over time due to reactions with the core is discussed in another section.

#### 10.3.3.2.(i)    MORB density contrast    A key parameter in this process is the density contrast of the MORB assemblage at the pressure and temperature range of the CMB. Unfortunately, this is still uncertain due to uncertainties in the properties of the mineral phases present in MORB at these pressures, although there are ongoing improvements in our knowledge of these. Historically, Ringwood (1990b) proposed that basalt remains 2–4% denser than peridotite throughout the whole mantle, with the exception of the depth range 660–720 km, where it is less dense, the consequences of which are discussed in Section 10.3.3.3. Other mineral physicists have, however, calculated that MORB might actually be less dense than pyrolite at CMB pressures, in particular Kesson *et al.* (1998) and Ono *et al.* (2001). According to Kesson *et al.* (1998), MORB is intrinsically 0.9% denser than pyrolite at 1100 km depth but $-0.5\%$ ($28\,kg\,m^{-3}$) less dense than pyrolite at the CMB, while depleted peridotite is less dense than pyrolite at all depths: $-0.8\%$ and $-0.7\%$ at 1100 km and the CMB, respectively. At face value this implies that MORB would be positively buoyant at the CMB, but what may matter is the density relative to residue, rather than relative to pyrolite, which is $+0.2\%$ ($11\,kg\,m^{-3}$). This is because it seems likely that 95–99% of the mantle has been processed through MOR melting, and thus has been differentiated into residue and MORB, with very little pyrolite remaining. In any case, $11\,kg\,m^{-3}$ density difference is not enough to have a significant dynamical effect. More recently, Ono *et al.* (2001) calculated that although MORB is $\sim100\,kg\,m^{-3}$ denser at the top of the lower mantle, its density profile intersects an average mantle density at around 1500–2000 km depth in the lower mantle, making MORB neutrally buoyant at $\sim1600$ km depth and positively buoyant at greater depths.

The most recent estimates of MORB density, however, appear to be moving back in the direction of it being more dense than pyrolite at CMB pressures. From *in situ* X-ray observations of peridotite and NMORB and K-rich basalts at up to CMB pressures and temperatures, Ono *et al.* (2005b) estimated that MORB is still $40–80\,kg\,m^{-3}$ denser than the average mantle at a depth of 1800 km, at which depth he had previously estimated MORB to have $\sim$zero density anomaly (Ono *et al.*, 2001). Guignot and Andrault (2004) refined the equation of state of phases containing Na, K, and Al, and calculated MORB to be denser than pyrolite by 100–$160\,kg\,m^{-3}$ at the top of the lower mantle reducing to 25–$95\,kg\,m^{-3}$ at the CMB, which corresponds to 0.45%–1.7% of the density at the CMB, and results in a buoyancy parameter $B$ of 0.2–0.7 (assuming a thermal expansivity of $1.0 \times 10^{-5}\,K^{-1}$ and superadiabatic temperature contrast of 2500 K), or 0.4–1.4 (assuming a local temperature contrast of 1200 K). Most recently, Hirose *et al.* (2005) measured physical properties of MORB samples at pressure and temperatures up to those near the CMB, and found that MORB is substantially denser than PREM throughout the lower mantle and by about $200\,kg\,m^{-3}$ (3.6%) near the CMB, corresponding to a buoyancy ratio $B$ of 1.4 ($\Delta T = 2500$ K) or 3.0 ($\Delta T = 1200$ K). Results from several studies can be used to estimate whether settling is possible at these density contrasts.

#### 10.3.3.2.(ii)    Modeling studies    The earliest results on this issue (Gurnis, 1986a) found that, with $B = 2$, although basaltic tracers that were already separated from the residue component of the slab could aggregate into piles under upwellings for completely internally heated flows, with basal heating they got rapidly swept back into the main flow, and if accompanied by complementary depleted tracers did not settle at all. One way of understanding this result is in terms of the Stokes velocity of a sinking blob of crust: at high lower-mantle viscosities this is extremely low, for example, $0.1\,mm\,yr^{-1}$ for a 5 km blob with mantle viscosity $= 10^{21}\,Pa\,s$. If, however, the strong temperature-dependence of viscosity of mantle rocks is taken into account, then relatively low viscosities may be present in the hot lower thermal boundary layer, which allow the crust to settle from the residue fast enough, and several studies that have included this have found significant crustal settling is possible.

The laboratory study of Olson and Kincaid (1991) is particularly enlightening because there are no questions of resolution, artificial diffusion, or other issues that plague numerical treatments of composition. They introduced compositionally stratified slabs with equally thick layers of 1% denser and 1% less dense material (B $\sim\pm0.6$) and initially 3.5 orders of magnitude higher viscosity to the bottom of their tank, and found that after a period of warming, the buoyant component of the slab underwent a Rayleigh–Taylor instability in which the low-density 'depleted' material rose through the high-density 'crustal' material, which then remained above the CMB in a 'spoke' pattern and subsequently got entrained over time, as discussed later. The scaled time from slab incidence to dense spokes was 1.8 Ga. In this experiment, which has a buoyancy ratio in the range discussed above, almost all of the dense 'crust' settled to the CMB, but this may be exaggerated because the 'crust' and 'residue' layers were assumed to be equally thick, whereas in reality the crustal layer is much thinner.

A similar process has been obtained in numerical experiments, but with a smaller fraction of the crustal material remaining at the CMB. Christensen (1989b) included temperature-dependent viscosity as well as a decrease in thermal expansivity with depth in a calculation with the continuous introduction of slabs containing crust and residue layers, taking care to scale the thickness of the compositional layers in proportion to the thermal thickness of the slab. He found that for a buoyancy ratio $\sim$1, between 5% and 25% of the basaltic crust settled at the CMB. Again, the observed process was a Rayleigh–Taylor instability of the residue component, leaving behind some fraction of the crust. With constant viscosity, no segregation occurred, consistent with Gurnis (1986a). Interestingly, if a layer of dense crustal material already existed, a larger fraction of the incoming basalt was able to segregate and settle. These calculations were greatly developed in Christensen and Hofmann (1994) in which it was found that for $B = 1.5$, $\sim$14% of the total amount of crust in the mantle was in 'pools' (similar to piles) above the CMB, implying that $\sim$1/6 of the subducted crust initially settled into a layer at bottom. Higher Rayleigh number decreased the size of the bottom 'pools', but higher $B$ increased them (e.g., 26% of the total crust for $B = 2.25$). Making the viscosity less temperature-dependent reduced the amount of settling, as with previous results. Interestingly, the bottom pools were not 100% crust – rather their crustal content was $\sim$60%.

The first models of the discussed type that are at an Earth-like convective vigor were presented by Davies (2002), who assumed a density difference of up to 100 kg m$^{-3}$ but a constant $\alpha = 2 \times 10^{-5}$ K$^{-1}$, about a factor of 2 higher than realistic at the CMB. In contrast to the models of Christensen and Hofmann (1994), the model was heated entirely from within, with no hot thermal boundary layer and hence no low-viscosity region at the bottom (the fraction of basal heating was arguably higher than realistic in the models of Christensen and Hofmann (1994)). Nevertheless, despite the absence of a low-viscosity TBL and high $\alpha$, a significant amount of basalt settled above the CMB, although less than obtained by Christensen and Hofmann (1994). Davies (2002) suggested other mechanisms that assist crustal settling, such as thickening of the crust due to slab buckling (the Stokes velocity increases as size$^2$), and the fact that the vertical velocity approaches zero towards the lower boundary (making it more difficult for the convective flow to entrain sinking blobs). All these mechanisms may play a role, but as Christensen (1989b) and Christensen and Hofmann (1994) clearly demonstrated the strong influence of temperature-dependent viscosity on crustal settling above the CMB, it is likely that Davies (2002) would have obtained more crustal settling, if the lower boundary region were less viscous.

The studies discussed in this section so far have controlled the surface velocities (to make them plate-like) and the rate at which compositionally distinct material is injected. A number of other studies (e.g., Ogawa, 2000a, 2003; Xie and Tackley, 2004a; Nakagawa and Tackley, 2005c) attempted to treat these aspects self-consistently, with the crustal production rate determined by melting induced when the temperature reaches the solidus, and the surface velocities determined by solution of the governing equations, often using a rheology that facilitates a 'plate-like' behavior (e.g., Moresi and Solomatov, 1998; Yoshida and Ogawa, 2004). It is not clear that this makes the resulting calculations more representative of Earth, but they are different from the previously discussed studies in other ways that make it instructive to discuss the results here.

The 2-D box calculations of (Ogawa, 2003), building on a series of earlier papers (Ogawa, 1988, 1993, 1994, 1997, 2000a, 2000b; Kameyama et al., 1996; Ogawa and Nakamura, 1998), include self-consistent differentiation, a plate treatment involving hysteresis (Yoshida and Ogawa, 2004), and a cooling mantle with decaying heat sources. With MORB having a density anomaly of $120\,kg\,m^{-3}$ and a somewhat high constant thermal expansivity of $3 \times 10^{-5}\,K^{-1}$, subducted MORB settles into a layer above the CMB that accumulates with time, while subducted residue undergoes diapir-like instabilities raising it back through the mantle. The basaltic layer has large undulations and typically has a thickness of $\sim 1/4$ of the domain depth, which scales to 500 km.

The 2-D cylindrical calculations of Tackley and Xie (2002), Xie and Tackley (2004a, 2004b) Nakagawa and Tackley (2004b, 2005a, 2005c) include self-consistent magmatism, temperature- and depth-dependent viscosity and yielding-induced plate tectonics, and decreasing convective vigor due to decaying heat-producing elements and a cooling core. The material properties such as thermal expansivity have reasonable depth dependences. Various density profiles for the MORB component relative to the residue component in the lower mantle are considered, and the density of the average composition ('pyrolite') is assumed to be a weighted average of the densities of the two end members, consistent with the small-scale structure being a heterogeneous mixture of the two end members, but not primitive never-melted pyrolite. These calculations show that even for a relatively small density contrast for MORB in the deep mantle (e.g., $62\,kg\,m^{-3}$; Nakagawa and Tackley (2005a)), subducted MORB settles at the base and builds up into large discontinuous 'piles' of material extending up to $\sim 700\,km$ into the mantle. With double this density contrast ($123\,kg\,m^{-3}$; Xie and Tackley (2004a, 2004b); Nakagawa and Tackley (2005a)), a continuous undulating layer of dense material is formed, again up to 500–700 km thick. When basalt is neutrally buoyant at CMB pressures, the CMB region is still heterogeneous because subducted crust and residue reside at the bottom for some period while they heat up sufficiently to be stirred back in to the convective flow (Xie and Tackley, 2004a). If there is crossover in density, with MORB positively buoyant at the CMB, then a layer of depleted residue can build up above the CMB, with blobs of MORB residing in the mid lower mantle where it is neutrally buoyant (Xie and Tackley, 2004a, 2004b).

In these calculations (Ogawa, 2003; Tackley *et al.*, 2005) the layer of dense material can become much thicker than observed in the calculations of Christensen and Hofmann (1994) and Davies

(2002), a major reason being that the total amount of MORB that can be produced by melting of primitive material is higher. Christensen and Hofmann (1994) and Davies (2002) limit the total amount of MORB in the system to 10%, whereas the other authors assume, based on the fraction of garnet + pyroxene in pyrolite being 30–40%, that up to 30% or 40% basalt fraction can be generated by melting. The true situation is probably between these extremes: melting by more than 10% probably did occur in the hotter, early Earth, and remelting of once-depleted pyrolite could produce additional basalt, but petrological considerations indicate that the bulk composition changes depending on the degree of melting (e.g., McKenzie and Onions (1992); Asimow *et al.*, 2001).

A recent factor that decreases the stability of a deep, dense layer is the recently discovered post-perovskite phase transition (Murakami *et al.*, 2004; Oganov and Ono, 2004). The positive Clapeyron slope of this transition acts to slightly destabilize the lower thermal boundary layer (Nakagawa and Tackley, 2004a), and any resulting chemical layering (Nakagawa and Tackley, 2005b). The destabilization effect is however relatively small, and seems to be lower than the uncertainties in MORB density discussed earlier, and uncertainties due to composition dependence of the PPV transition (e.g., Murakami *et al.*, 2005; Ono *et al.*, 2005a). In any case, because dense material tends to accumulate under hot upwellings, and the positive Clapeyron slope leads to an anticorrelation between the phase transition depth and temperature, the occurrence of a thick PPV layer is anticorrelated with the occurrence of thick pile of dense material.

An uncertain factor in all of the above calculations is the influence of 2-D geometry on quantitative entrainment rates, and hence long-term stability. As discussed in Section 10.3.2.1, in the laboratory experiments of Davaille (1999b), different scaling laws were found (i.e., with different exponents for $B$ and $Ra$) for entrainment by 2-D sheets and 3-D plumes. This is something that will have to be addressed in the future.

### 10.3.3.3 Chemical layering induced by phase transitions

The intriguing possibility exists that one or more phase transitions near or below 660 km depth could enforce or enhance chemical layering. Some research has focused on the effect of a single phase transition – the endothermic spinel to perovskite + magnesio-wüstite transition at around 660 km depth, which, according to dynamical models that assume this transition applies to all mantle material, can cause partial layering of thermal convection, with the degree of flow stratification very sensitive to the value of the Clapeyron slope (e.g., Christensen and Yuen, 1985; Machetel and Weber, 1991; Peltier and Solheim, 1992; Weinstein, 1993; Tackley *et al.*, 1994). In reality, however, olivine makes up only ~60% of the mantle assemblage, and the phase changes in the other, pyroxene–garnet system, occur at different pressures. In particular, the transition to the perovskite phase occurs with a positive Clapeyron slope and probably at a greater depth such as 750 km. This has two effects: (1) the effect of the combined phase transitions on purely thermal convection is much reduced relative to what would occur if the olivine transitions applied to all mantle material, and (2) subducted oceanic crust might be positively buoyant (less dense than other compositions) in a region tens of kilometers thick below the 660 km discontinuity, whereas it is negatively buoyant (more dense) at other pressures (Irifune and Ringwood, 1993). This has been suggested to cause the separation of slab components in the vicinity of 660 km, with the accumulation of 'megaliths' (Ringwood and Irifune, 1988; Ringwood, 1991).

Estimates of depth extent over which this anomalous buoyancy occurs have varied widely: a recent study of Ono *et al.* (2001) puts it at 660–720 km. A possible intermediate transition to ilmenite (Chudinovskikh and Boehler, 2002) would reverse this picture and increase the layering (van den Berg *et al.*, 2002), but it is doubtful whether ilmenite occurs in a pyrolite composition (S. Ono, personal communication).

### 10.3.3.3.(i) A single endothermic phase transition

If a chemical boundary already exists at a similar depth to the 660 km phase transition, Christensen and Yuen (1984) showed that their effects on deflecting a subducting slab and therefore causing flow layering can be added linearly. For the parameters they assumed, flow layering required a 4.5% chemical contrast or a Clapeyron slope of $-5\,\mathrm{MPa\,K^{-1}}$ or a linear combination of the two. If chemical stratification does not, however, already exist, can an endothermic phase transition dynamically induce it? For a long-wavelength initial chemical heterogeneity, Weinstein (1992) showed that it can for certain combinations of chemical buoyancy ratio and Clapeyron slope. The endothermic phase was shown to act as a filter, trapping

chemically dense material below it and chemically less-dense material above it, because the total buoyancy (thermal + chemical) of these types of material is insufficient to overcome the phase change buoyancy. For this mechanism to operate, the required chemical buoyancy contrasts were in the range 1.1–1.5% (assuming $\alpha = 2 \times 10^{-5}\,\mathrm{K^{-1}}$) – much lower than those required to cause layering on their own. The required Clapeyron slope was, however, larger than realistic, in the range $-5$ to $-8.6\,\mathrm{MPa\,K^{-1}}$, though this is partly because the calculations were performed at a low convective vigor ($Ra = 10^5$), and would be substantially reduced if an Earth-like vigor could be reached (Christensen and Yuen, 1985). For small-scale heterogeneities introduced by the continuous injection of compositionally layered slabs, Mambole and Fleitout (2002) showed that small but measurable chemical stratification can be introduced by Clapeyron slopes as weak as $-1\,\mathrm{MPa\,K^{-1}}$, with the amount of chemical stratification increasing steadily with the magnitude of the Clapeyron slope. With only the endothermic transition active, the dense component, which is the enriched, subducted MORB, is more prevalent in the lower mantle, while depleted, relatively buoyant material is more prevalent in the upper mantle. This result was confirmed by Nakagawa and Buffett (2005) in the case where the compositional anomalies were generated by melting-indiced differentiation. If the full system of phase transitions is considered the sense of layering is reversed, as discussed below.

### 10.3.3.3.(ii) Composition-dependent phase transitions and subducted slabs

An important question is whether the enriched basaltic crust and depleted harzburgitic residue can separate out due to the density inversions that occur around and below 660 km depth, as suggested by Ringwood (1994). Three studies have demonstrated that this does not occur for slabs with a simple rheology, due to the strong viscous coupling between crust and rest of the slab (Richards and Davies, 1989; Gaherty and Hager, 1994; Christensen, 1997), whereas van Keken *et al.* (1996) showed that if a weak layer separates the crust from the rest of the slab, crustal separation may be possible. While Richards and Davies (1989) investigated constant-viscosity slabs, Gaherty and Hager (1994) studied slabs with temperature-dependent viscosity falling vertically onto a viscosity increase at 660 km depth and a local density inversion below this depth, but their model did not include density anomalies due to phase-change deflection or net compositional buoyancy of the slab below 660 km, which was an important component of the proposal of (Ringwood and Irifune, 1988). Thus, Christensen (1997) included these effects as well as a more realistic slab angle associated with trench migration at the subduction zone, with basaltic rock buoyant between 680 and 900 km, but still found that compositional effects are negligible for 100 My-old plates, and although they could cause younger plates to stagnate at lower trench velocity they did not prevent eventual slab penetration into the lower mantle. The rheology of the slab is more complicated than simple temperature dependence. van Keken *et al.* (1996) considered a slab with an idealized rheological sandwich containing a soft (low-viscosity) layer between the 'hard garnet' crust and the cold viscous slab interior, and showed that the crust can separate once the slab passes below 660 km depth, if most of the slab is 10 times more viscous than the surrounding material and the weak peridotitic layer is 10 or 100 times less viscous than the rest of the slab. This model underestimates the viscosity of the slab based on simple viscosity laws, but once the peridotitic part of the slab passes through the spinel to perovskite + magnesiowüstite transition, it may be considerably weakened due to transformational superplasticity associated with grain size reducing by the phase transition (Karato, 1995), whereas the crustal component would not. In summary, with 'conventional' rheology, compositional effects play only a minor role in slab interaction with the 660–800 km region and separation of slab components at 660 km depth is not possible, but if there is substantial weakening of the subcrustal layer due to transformation superplasticity or other effects, then separation of the basaltic crust might occur.

### 10.3.3.3.(iii) Composition-dependent phase transitions in a convecting system

Regardless of the viability of slab components separating at around 660 km depth, several studies have shown that once slabs reach the bottom thermal boundary layer and heat up, the compositionally distinct components can indeed separate (Olson and Kincaid, 1991; Christensen and Hofmann, 1994); thus, the question of how separated crust and residue components interact with the phase transitions is of great relevance. The stability of a transition layer (consisting of an inseparable mixture of eclogite and harzburgite) that has already formed in the 660 region to thermal convection was studied by Christensen (1988), who found that it could only

survive for a few hundred million years, but this model did not allow for separate basalt and harzburgite components, nor the possibility of recharge of these components. Models in which differentiated components are constantly introduced by subduction (Fleitout *et al.*, 2000; Ogawa, 2000a) show the development of a strong compositional gradient across the 660, with depleted harzburgitic material accumulating at the top of the lower mantle, enriched basaltic material accumulating at the bottom of the upper mantle, and a chemical gradient present throughout the entire mantle, as evident in the later calculations of (Tackley and Xie, 2002; Xie and Tackley, 2004a, 2004b). This entrapment of distinct components typically does not occur when a downwelling first encounters the region, but rather after the crust and residue components have separated in the deep mantle and are subsequently circulating as 'blobs'. That the chemical and flow stratification is due mainly to the different depths of the perovskite transition in garnet and olivine systems rather than Clapeyron slope effects was demonstrated by Tackley *et al.* (2005). Nakagawa and Buffett (2005) found that slab penetration into the lower mantle is accompanied by an upwelling counter flow of depleted material in the area surrounding the slab, which is a mechanism for generating depleted regions in the upper mantle.

Multicomponent phase transitions is not the only mechanism that can cause an accumulation of subducted basalt above 660 km: Davies (2002, 2006) showed that a low-viscosity (high Rayleigh number) upper mantle in combination with a viscosity jump at 660 km depth can also be cause this.

In summary, it is quite likely that local chemical stratification exists across the 660 km interface caused by differing depths of the perovskite transition in the olivine and pyroxene–garnet systems. There is, however, still uncertainty as to these depths. This local chemical stratification around 660–720 km is a prediction that could be tested seismically, which would help to determine whether the currently estimated depths of these phase transitions are indeed correct. For additional information on phase transitions.

### 10.3.3.4 The effect of chemical layering on core heat flow and planetary thermal evolution

As was noted early on (e.g., Christensen and Hofmann, 1994), the presence of a compositionally dense layer above the CMB reduces the heat flow out of the core. This reduction can be dramatic if the dense layer completely covers the core, and should be taken into account in thermal evolution calculations (e.g., McNamara and Van Keken, 2000). Unfortunately, when the layer undulates greatly and is discontinuous, such that some patches of the core are not covered, the appropriate parametrization of core heat flux is unknown, so direction numerical simulation becomes the preferred approach.

A long-standing paradox in modeling Earth's thermal evolution is explaining how the heat flux out of the core can have stayed large enough over billions of years to maintain the geodynamo, without growing the inner core to a much larger size than observed. Parametrized models of core–mantle evolution in which no complexities are present predict a final inner-core size that is much larger than that observed (e.g., Labrosse et al., 1997; Labrosse, 2003; Nimmo et al., 2004). As the heat conducted out of the core is determined by what is happening in the mantle, the time history of CMB heat flow is an important quantity to investigate using numerical models, both from the point of view of understanding core evolution and in providing an additional constraint on mantle models.

Nakagawa and Tackley (2004b) used a thermochemical convection model similar to those already presented but with a parametrized core heat balance based on Buffett et al. (1992, 1996) to show that the presence of a layer of dense material above the CMB, either primordial or arising from segregated crust, substantially reduces the CMB heat flow hence inner-core growth. A problem is, CMB heat flow can be reduced so much that it is insufficient to facilitate dynamo action, and can even become zero or negative. Because of this, a discontinuous layer, rather than a global layer, was found to be the most promising scenario. Nakagawa and Tackley (2005a) subsequently found that it is difficult to match the constraints of correct inner-core size and enough heat flow out of the core, unless there is radioactive potassium in the core (e.g., Nimmo et al., 2004), in that case viable evolution solutions were obtained, with a tradeoff found between crustal density and core K content.

As was discussed in an earlier section, chemical layering has a strong influence on plume temperature and dynamics. A systematic study of the effect of chemical layering on plume heat flux, plume excess temperature, and upper-mantle temperature was conducted by Zhong (2006) in 3-D spherical geometry, assuming a sharp chemical interface and a global layer, and moderately temperature-dependent viscosity. The possible range of parameters (layer depth, internal heating rates) was constrained by comparing to observations. The preferred model is one in which the dense layer was relatively thin ($< 350 \, km$) and the internal heating rate in the upper layer is $\sim 3$ times higher than that inferred for the MORB source region, which, curiously, is similar to the heating deficit inherent in the 'heat-helium paradox' (Ballentine et al., 2002).

### 10.3.4 On the Accuracy of Numerical Thermochemical Convection Calculations

A fundamental problem with treating chemical variations numerically is the negligible diffusivity of chemical variations, which leads to the possibility of sharp interfaces and very narrow lamella (e.g., less than one grid spacing), both of which are very difficult to treat numerically. Two major classes of method are used in the mantle dynamics community to track compositional variations: field-based approaches on a fixed (Eulerian) grid, and Lagrangian particles/tracers that are advected through the grid. Various benchmark tests of these have been performed in the mantle convection community (van Keken et al., 1997; Tackley and King, 2003). Further discussion can be found in Chapter 5.

Field-based approaches have the advantage of computational speed and little additional computational complexity, but the disadvantages of numerical diffusion, which smears out sharp features, and numerical dispersion, which causes artifacts like overshoots and ripples. Other research communities that employ computational fluid dynamics techniques (e.g., the atmospheric science community) face similar problems and there is much ongoing research into improving advection methods, a small sampling of which is Muller (1992), Xu-Dong and Osher (1996), Smolarkiewicz and Margolin (1998), and Yabe et al. (2002). In the mantle dynamics community, the relatively straight-forward interface-sharpening scheme of Lenardic and Kaula (1993) can yield a dramatic improvement in the advection of interfaces (e.g., Tackley and King, 2003).

Tracer-based approaches have the advantage of minimal numerical diffusion (a small amount does exist because of finite error in updating each tracer's position), thus the apparent ability to represent discontinuities and sub-grid-scale features, but the disadvantage that one needs 5–50 times as many tracers as grid points to prevent statistical noise from unduly influencing the solution (e.g., Christensen and Hofmann, 1994; Tackley and King, 2003), which adds a considerable computational burden. This noise arises because in order to calculate the buoyancy caused by compositional variations, some averaging must be done over tracers in the vicinity of each grid or nodal point. It has been found that taking a local average of the $C$ value of each tracer (as exemplified by the 'ratio method') is far less noisy than methods that rely on a number count of tracers (the so-called 'absolute method'), regardless of the underlying numerical scheme (Tackley and King, 2003). Another type of tracer method tracks only the interface between two compositions, and is thus far more computationally efficient than methods where tracers exist everywhere (e.g., Schmalzl and Loddoch, 2003; Lin and Van Keken, 2006a), though it is not suited for continuously varying chemical fields, and the computational needs increase up to exponentially with time as the interface becomes more complex, unless interface simplification algorithms are applied (Schmalzl and Loddoch, 2003).

Benchmarks of these various methods for cases with an active compositional layer (van Keken et al., 1997; Tackley and King, 2003), indicate a sensitivity of the results to the details of the method being used, including a strong sensitivity of the entrainment rate of a stable compositional layer to resolution (both grid and number of tracers) (Tackley and King, 2003). This means that the various numerical results reviewed in this section may not be quantitatively correct in the amount of entrainment that exists as a function of density contrast except in calculations where the grid is greatly refined in the region of entrainment (e.g., Zhong and Hager, 2003; Lin and Van Keken, 2006a), although convergence tests suggest that they are qualitatively correct in the general planforms and features that arise. This is an issue that will require more attention in the future, particularly as more calculations are performed in 3-D.

# 10.4 Convection Models Tracking Trace-Element Evolution

## 10.4.1 Passive Heterogeneity

### 10.4.1.1 Helium and argon

The evolution of noble gases in a mantle that included outgassing and radiogenic ingrowth but not major-element differentiation was studied by van Keken and Ballentine (1998, 1999). These studies investigated various proposed mechanisms for maintaining a relatively undegassed lower mantle over billions of years, finding that neither high deep-mantle viscosity, a strongly endothermic phase transition at 660 km depth, or temperature-dependent viscosity, are capable of causing a relatively undegassed deep mantle, in contrast to earlier modeling at lower convective vigor (Gurnis and Davies, 1986a). They did find that the observed total amount of $^{40}Ar$ outgassing is consistent with whole-mantle convection over geological history, although the convective vigor in their models was constant with time. In that case, 50% degassing of radiogenic $^{40}Ar$ implies ~70% degassing of primitive isotopes.

van Keken et al. (2001) extended the analysis to the heat-helium imbalance, tracking the relationship between the rate of helium outgassing and the heat flux. Large fluctuations in the helium outgassing were obtained, but only the most extreme fluctuations would explain the 'heat-helium' issue as it is currently understood, that is, with a $^{4}He$ concentration 3.5 times higher than current estimates (Ballentine et al., 2002).

The helium ratio evolution caused by subduction of differentiated oceanic plates was studied by Ferrachat and Ricard (2001), who tracked the evolution of oceanic crust and residue and showed that if oceanic crust segregates at the CMB, a large region of recycled oceanic lithosphere with high $^{3}He/^{4}He$ can form above it. However, in their model compositional variations were also purely passive, not influencing the flow, whereas models that account for the density anomalies of differentiated components have found that depleted material tends to rise.

### 10.4.1.2 Continent formation

The melting-induced differentiation of the continental crust and depleted mantle from primitive mantle was the focus of models by Walzer and Hendel (1997a, 1997b, 1999). In Walzer and Hendel (1997a, 1997b) the 670 km discontinuity was impermeable, and upon melting, trace-element ratios in the crust and depleted mantle were taken to be those in Hofmann (1988) rather than calculated using partition coefficients. As a result of these model assumptions, the lower mantle remained primitive while the upper mantle rapidly differentiated into continental crust and depleted mantle with exactly the trace-element ratios of Hofmann (1988). This approach was extended in Walzer and Hendel (1999) to include a permeable 660 km discontinuity with phase changes at 410 and 660 and a plate model that allowed continents to aggregate. Curiously, despite the lack of compositional density contrasts their models typically developed an average (but very irregular) vertical stratification with primitive material in the deepest mantle and depleted material above it. This could be related to extremely high viscosities in their model mid-lower mantle, typically in the range $10^{24}$–$10^{26}$ Pa s, much higher than is assumed in other models or obtained in inversions of postglacial rebound or geoid and topography.

### 10.4.2 Active Heterogeneity

#### 10.4.2.1 U–Pb and HIMU

The hypothesis that segregation and sequestering of subducted MORB is responsible for the HIMU component and related Pb–Pb isochron (Hofmann and White, 1982) was investigated in pioneering calculations by Christensen and Hofmann (1994) using a numerical mantle convection model that included tracking of crust–residue differentiation and isotopes of trace elements U, Pb, Sm, and Nd, which partitioned between crust and residue on melting. The model was partly heated from within and partly from below and had a kinematic time-dependent plate-like upper boundary condition. As discussed earlier, a fraction of the subducted crust segregated into a layer above the CMB, and with some parameter combinations, realistic Pb–Pb isotope diagrams and 'ages' were obtained. The isotope age depended on the length of time subducted MORB resided near the CMB. However, they found it necessary to run the calculations for only 3.6 Gy instead of the full age of the Earth, otherwise the age was too high, contrary to previous expectations from mixing time estimates (reviewed in Section 10.2.6) that it would be difficult to obtain a large-enough age. This would be appropriate if the mantle were completely mixed 3.6 Gy before present. Furthermore, their model mantle had a constant convective vigor (e.g., velocities, rate of melting), rather than, as discussed earlier, a vigor that decreases with time as the mantle temperature and radiogenic heat production decrease.

The calculations of Xie and Tackley (2004b) followed a similar approach to those of Christensen and Hofmann (1994), but with a convective vigor that decreases with time due to secular cooling and the decay of heat-producing elements, and the investigation of 4.5 Gy run times as well as 3.6 Gy. Technical differences were the use of a viscoplastic rheology to allow plate-like behavior of the lithosphere (Moresi and Solomatov, 1998) rather than kinematic velocity

boundary conditions, and a more self-consistent treatment of melting and crustal generation based on comparing the temperature to a solidus. The higher convective vigor early in the calculations plus the longer evolution resulted in a lot of differentiation early in the modeled evolution, which resulted in Pb–Pb ages that are much too large (3.39 Gy for the reference case). This large age is not surprising, because the rate of melting is much higher earlier in the evolution, so early differentiating material dominates the isotopic signature. Additionally, early differentiated material has developed a more radiogenic signature so influences the least-squares fit more than recently differentiated material. This finding is consistent with previous attempts at modeling lead isotope ratios starting from 4.5 Gy ago that obtained unrealistically high slopes in $^{207}Pb/^{204}Pb$–$^{206}Pb/^{204}Pb$ space (Armstrong and Hein, 1973; Allegre et al., 1980). It is also consistent with the numerical models of Davies (2002), in which although Pb–Pb ages were not calculated, found mean ages (since melting) of ~2.7 Gy for MORB. Christensen and Hofmann (1994) attributed this 'too old' problem to the rapid decay of $^{235}U$ into $^{207}Pb$ in the early stage of the model, also finding that the early differentiation has a strong effect on the isotopic signatures at the end of the run. Thus they chose to start their models 3.6 Gy ago with uniform Pb composition, which assumes that any earlier differentiation in the mantle has been completely homogenized by the convection.

Several possible explanations for the discrepancy between models and data were investigated by Xie and Tackley (2004b). These are that HIMU material may not have been subducted into the mantle prior to 2.0–2.5 Gy BP, that melting is not rapid enough in the model (i.e., too low processing rate meaning too high residence time), that stretching of heterogeneities is inadequately represented, that the 'sampling volume' over which isotopic ratios are calculated is incorrect, that the deep-mantle density of crustal material is incorrect, or that the Pb–Pb slope is made artificially large by noise associated with tracer discretization. The most successful and straightforward explanation was found to be the first, which is discussed in more detail below, but it is worth briefly discussing the others. Regarding the melting-induced processing rate, this was found to be within the range of possibilities for the Earth (see discussion in Section 10.1.5.3). A parametrized box model indicated that only the most extreme assumptions would lead to a low-enough residence time. Inadequate treatment of stretching highlights a problem with tracer treatments: whereas in reality a tracer should get exponentially stretched into a long tendril that extends over an area much wider than the sampling volume thereby reducing its contribution to the isotopic signature of a melt, in fact it remains as a localized entity. Xie and Tackley (2004b) tested one approximation of this: tracking the integrated strain as in McKenzie (1979) and Kellogg and Turcotte (1990), then ignoring high-strain tracers when calculating the isotopic signature of a sampling volume. This was found to reduce the Pb–Pb somewhat, but not enough to match observations. Changing the sampling volume over which heterogeneities are averaged does not change the Pb–Pb slope, only the amount of scatter.

One possible reason for the ∼1.8 Ga Pb–Pb age is that HIMU material did not enter the mantle before 2.0–2.5 Gy ago, for which there are two possible explanations. One of these is that higher subduction zone temperatures in the past may have caused essentially all Pb, U, and other trace elements to be stripped from the subducted crust by melting (Martin, 1986). Supporting this explanation is eclogite xenoliths from the upper mantle under western Africa that are highly depleted in incompatible elements and have been interpreted as the melting residue of Archaean subducted crust (Barth *et al.*, 2001). Another possible explanation is that the HIMU end-member was not produced prior to the rise in atmospheric oxygen about 2.2 Gy ago. In this latter scenario, HIMU is produced by addition of continental U to oceanic crust by hydrothermal circulation at ridges (Michard and Albarède, 1985; Elliott *et al.*, 1999), with efficient stripping of U from continents being accomplished by an oxygen-rich atmosphere oxidizing U to its 6+ state (Holland, 1984, 1994). This is not the only process that has been proposed to produce HIMU: the other is preferential stripping of Pb (relative to U) from the subducted crust to the overlying mantle in subduction zones (Newsom *et al.*, 1986; Hofmann, 1988; Chauvel *et al.*, 1995). It is possible that both of these processes may operate. In the numerical model of Xie and Tackley (2004b), as in that of Christensen and Hofmann (1994), HIMU was produced by artificially changing the partition coefficients of U and Pb so that they fractionate on melting. In order to evaluate the effect of it not being produced prior to some point in the past, Xie and Tackley (2004b) ran cases in which the U and Pb partition coefficients are set equal for the first part of the model run, then changed to their

default values at a specified point in the past. Setting the transition to 2.5 Gy before present caused a Pb–Pb isotopic age of 1.75 Gy, whereas setting it to 2.0 Gy before present caused an isotopic age of 1.4 Gy. Thus, from a modeling perspective a lack of HIMU production prior to 2–2.5 Gyr ago is an appealing mechanism for explaining the effective age of the Pb–Pb isotope diagrams, although it needs to be shown that this fits other isotope systems too.

A mathematical analysis of the relationship between the probability distribution of heterogeneity age (i.e., time since last melting), and the pseudo-isochron age, was conducted by Rudge (2006), superceding an earlier box model approach by Allegre and Lewin (1995). He obtained simple relationships between mean remelting time and isochron age for a constant melting rate, which closely fit the Pb–Pb pseudo-isochron ages in the numerical simulations of Christensen and Hofmann (1994). His more general theory was also able to fit the isochron ages of the time-dependent melting histories in the simulations and calculations of Xie and Tackley (2004b). An important finding is that for constant melting rate, the Pb–Pb pseudo-isochron age is substantially higher than the mean time before remelting – in particular fitting a 2.0 Gyr Pb–Pb pseudo-isochron age with a constant melting rate over Earth's history requires a time interval of only 0.5 Gy before material remelts, much shorter than seems reasonable.

### 10.4.2.2 $^3$He/$^4$He

The origin of the high $^3$He/$^4$He end-member is discussed in Section 10.1.6.1. The two most commonly cited origins have both been the foci of numerical investigations: Samuel and Farnetani (2003) demonstrated the dynamical viability of the 'primitive' explanation, while Ferrachat and Ricard (2001) and Xie and Tackley (2004a) demonstrated the viability of the 'recycled' explanation. In any case, it is important to recall that the ratio of $^3$He/$^4$He constantly decreases with time due to radiogenic ingrowth of $^4$He from U and Th decay. Elemental fractionation between He and (U + Th) upon melting, and outgassing of He to the atmosphere influence He/(U + Th) in the crust and residue, which over time can then develop $^3$He/$^4$He different from primitive material. Thus the present-day distribution of He isotope ratios reflects a combination of ingrowth, fractionation, and outgassing.

*10.4.2.2.(i) Primitive* Samuel and Farnetani (2003) used a fully dynamical convection model (e.g., with no imposed surface velocities) to demonstrate that a thick basal layer of primitive material, consistent with the proposal of Kellogg *et al.* (1999) and van der Hilst and Karason (1999), is capable of generating $^3$He/$^4$He histograms similar to those observed for MORB and OIB. Isotopes of U, Th, K, and He were tracked in three different components: primitive, subducted MORB, and subducted residue, with the primitive layer being the 'mass-balance' reservoir of U, Th, and K as well as the high $^3$He/$^4$He component. Partitioning of trace elements between crust and residue was based on an assumed melt fraction. With a compositional density contrast of 2.4% (100 kg m$^{-3}$), depth-dependent thermal expansivity, temperature-dependent viscosity, and a primitive layer making up 25% of the mantle volume, it was found that $^3$He/$^4$He values in MORB cluster in the observed range, whereas OIB samples, which were associated with plumes forming from the layer interface, display a much wider spread because they contain a heterogeneous mixture of the three different components. Heterogeneity dispersed more rapidly in the upper layer than in the lower layer due to the higher convective vigor there. Lower layer material was entrained into the upper layer much more rapidly than upper layer into the lower layer: after 2.1 Gy the percentages were 25% and 3%, respectively.

*10.4.2.2.(ii) Recycled* As discussed in Section 10.4.1.1, Ferrachat and Ricard (2001) found that a high $^3$He/$^4$He repository of subducted residue built up in the deep mantle in a steady-state flow with no compositional buoyancy. Studies that do include compositional buoyancy find, however, that the depleted residue tends to rise to the shallow mantle, as in the preliminary calculations of Tackley and Xie (2002), in which high $^3$He/$^4$He residue aggregates near the top due to its buoyancy, where it is sampled by mid-ocean ridge volcanism. Thus, Xie and Tackley (2004a) investigated the possibility that active (buoyant) compositional variations caused by crustal production may play an important role in the evolution of noble gas isotopes and the generation and maintenance of distinct end-members, using numerical convection models with a self-consistent melting criterion and plate-like dynamics. The distribution of $^3$He/$^4$He ratios in the modeled MORB source region was found to be highly dependent on the density contrasts between the different

components (primitive, residue, and basalt) and compatibility of He relative to U and Th. While in most cases the $^3$He/$^4$He was scattered to too high values, two parameter combinations were identified that led to a 'realistic' MORB-like distribution of $^3$He/$^4$He (i.e., clustered around 8–10 times atmospheric): (1) He less compatible than U, Th, and crust dense in the deep mantle. In this case, the shallow mantle contained a large proportion of depleted residue, which had MORB-like $^3$He/$^4$He, and the high $^3$He/$^4$He component was primitive material. (2) He equally compatible as U, Th, and crust buoyant in the deep mantle. In this case, subducted crust in the shallow mantle brought the $^3$He/$^4$He into the correct range; both primitive material and residue had the same 'primitive' $^3$He/$^4$He.

These calculations demonstrate that a MORB-like shallow mantle $^3$He/$^4$He and the existence of a high $^3$He/$^4$He end-member are consistent with whole-mantle convection from a homogeneous start, but with the ambiguity of two alternative explanations. Indeed there may be a continuum of 'successful' solutions in (partition coefficient, buoyancy) space, which would require the consideration of additional constraints to constrain further. These calculations did not attempt to distinguish between MORB and OIB melting signatures. Thus, more investigations are necessary, preferably in 3-D.

## 10.5 New Concepts and Future Outlook

### 10.5.1 Revisiting Noble Gas Constraints

#### 10.5.1.1 Ar and He budget

Noble gas constraints were reviewed in Section 10.1.6.1. As discussed in that section, the '50% $^{40}$Ar degassing' constraint is straightforward to match with whole-mantle convection, even with substantially higher processing rates in the past. A much more troubling constraint from the point of view of whole-mantle convection models, is that the Argon concentration in the MORB source is much lower than it would be if the Ar thought to be in the mantle were evenly distributed, which has been taken as evidence of layered convection with the lower layer containing the 'missing' argon. Along similar lines is the 'heat-helium imbalance', that is, the concentration of He in the MORB source region seems to be too low relative to concentrations of U and Th. Recently, Ballentine *et al.* (2002) noted that both of these constraints are based on fluxes of $^3$He through the oceans that represent an average over 1000 years, and if the true long-term average were a factor of 3.5 higher, both constraints would be removed. He argues that two other indicators of $^3$He concentration in the MORB source – the concentration in the 'popping rock' and the estimate of carbon concentration in MORBs based on graphite-melt equilibrium, give estimates of $^3$He concentration up to several times higher than obtained from the He flux through the oceans. On the other hand, Saal *et al.* (2002) obtain rather lower $^3$He concentrations by considering carbon concentrations in undegassed melt inclusions. Numerical models of convection that track the relevant isotopes (van Keken *et al.*, 2001) obtain considerable variation in the outgassing rate over timescales much longer than 1000 years.

The 'standard' reconciliation of the heat-helium imbalance invokes the concept that He and heat have different transport mechanisms, that is, heat is transported across a thermal boundary layer at 660 km depth, whereas He is not. Castro *et al.* (2005) found that a similar concept occurs to He transport in continental aquifers, which, when applied to the oceans, suggests that the He flux into the ocean may be much lower than the flux out of the mantle, due to the action of seawater circulating through the oceanic crust (see Albarède (2005a) for further discussion).

The 'missing $^{40}$Ar' constraint would also be eased if the total amount of $^{40}$K in the mantle were lower than commonly assumed, which is based on BSE abundances (e.g., McDonough and Sun, 1995) combined with the K/U ratio observed in MORB (Jochum *et al.*, 1983). It has been shown that this difficulty could be resolved if the amount of $^{40}$K in the mantle were a factor $\sim 2$ than commonly estimated (Albarède, 1998; Davies, 1999), which would be the case if the K/U ratio were lower than that observed in MORB, perhaps due to fractionation of K from U (Albarède, 1998). Lassiter (2004) suggests that K/U could be significantly lower than normally assumed, if the K depletion of oceanic crust in subduction zones is taken into account. Then, subducted MORB in the mantle would have a relatively low K/U, lowering the mantle average.

As a result of these various arguments, it seems plausible that the 'missing $^{40}$Ar' and 'heat-helium imbalance' constraints are less robust than previously thought.

### 10.5.1.2   High $^3$He/$^4$He

Although the high $^3$He/$^4$He end member (C or FOZO) is often assumed to be primitive and undegassed, several authors (e.g., Albarède, 1998;

Anderson, 1998; Coltice and Ricard, 1999, 2002; Coltice *et al.*, 2000b; Ferrachat and Ricard, 2001) have suggested that it could instead be generated by recycling, as was discussed in detail in Section 10.1.6.1 and investigated by the numerical models of Ferrachat and Ricard (2001) and Xie and Tackley (2004a) discussed in Section 10.4.2.2. If high $^3$He/$^4$He is caused by low [U+Th] rather than high [$^3$He], then an inverse correlation is expected between [U+Th] and $^3$He/$^4$He, and such a correlation was found in the data plotted by Coltice and Ricard (1999). Meibom *et al.* (2003) point out that the unfiltered $^3$He/$^4$He data set for MORB has a Gaussian distribution similar to that for $^{187}$Os/$^{188}$Os, and propose that both are due to mixing of radiogenic and nonradiogenic components (associated with basalt and residue, respectively) with varying amounts of partial melting. They propose that residue can retain substantial He due to the capture of He-rich bubbles in magma chambers. Further reinforcing this interpretation is a recent compilation of OIB isotope data which shows that high $^3$He source has the same isotopic and major-element composition as depleted mantle, therefore represents already-melted material (Class and Goldstein, 2005). Hauri *et al.* (1994) had earlier concluded that the FOZO represented differentiated material. Class and Goldstein (2005) find an inverse correlation between $^3$He/$^4$He and Th abundance consistent with He being a mixture of a small amount ($\sim 2\%$) of primitive material plus ingrown $^4$He. Based on the difference between plumes and OIBs they infer that amount of time the high $^3$He/$^4$He material has remained separate from the rest of the mantle is $\sim 1$–2 Gy.

## 10.5.2   Improved Recipes for Marble Cakes and Plum Puddings

### 10.5.2.1   Two recipes

Extreme compositional heterogeneity of the mantle, building on the 'marble cake' proposal of Allegre and Turcotte (1986), is a common theme in recent attempts to explain mantle geochemistry, and contrasts with earlier views of a relatively homogeneous upper mantle, with the necessary OIB heterogeneity supplied by plumes from the deep mantle. The proposed mantle assemblage typically consists of a mostly depleted residue with $\sim$few % primitive material, $\sim 10$–20% recycled MORB and $\sim$few % recycled sediment (derived from continental crust) and possibly continental lithosphere and/or OIB.

One possible recipe is that the assemblage in the MORB source region is different from that in the plume (hence OIB) source region, but both assemblages are derived mostly by recycling processes. This is consistent with the proposal of Hofmann and White (1982) and Christensen and Hofmann (1994) that HIMU and perhaps EM components arise from subducted crust that has segregated into a layer above the CMB. This recycled deep layer can also be the location for storing incompatible elements that are 'missing' according to mass budget calculations (Coltice and Ricard, 1999), and have higher concentrations in MORB than in primitive material, requiring a smaller storage area. The high $^3He/^4He$ component might be caused by strips of former oceanic lithosphere (Coltice and Ricard, 1999, 2002) as discussed in the previous section. In the marble cake of Coltice and Ricard (1999, 2002), this subducted lithosphere forms a layer above D″, accounting for different MORB and OIB signatures, but it is not clear that such a layer would be stable. A deep layer of residue was obtained in the numerical calculations of Ferrachat and Ricard (2001), but compositional density contrasts were not accounted for in those calculations, and when they are, residue tends to rise into the upper part of the mantle (e.g., Tackley and Xie, 2002; Ogawa, 2003; Xie and Tackley, 2004a), although this depends on uncertain density relations in the deep mantle, as discussed earlier.

Another possible recipe is that OIB and MORB result from the same statistical distribution with their geochemical differences due entirely to melting processes, with the details of the melting processes varying in different proposals. In the original proposal (Allegre and Turcotte, 1986), enriched plums exist in a depleted matrix, and the higher heterogeneity of OIBs was ascribed to smaller sampling region and lower degrees of melt. More recent proposals tend to emphasize the heterogeneity of the entire mixture, with no ∼uniform residue. Based on modeling the evolution of the distribution of trace-element ratios in the differentiating continental crust-depleted mantle system, Kellogg (2004) proposed something of an inversion of this model, with depleted 'plums' consisting the residue from continental crust extraction, and a 'matrix' with a C/FOZO composition consisting of small-scale strips of everything else. Meibom et al. (2003) and Meibom and Anderson (2004), proposed that MORB and OIB derive from the same source region (SUMA – the Statistical Upper Mantle Assemblage; the composition of the lower mantle was not specified) with

the geochemical differences due to different statistical sampling of this highly heterogeneous source region – essentially, smaller melt fractions. A whole mantle consisting mainly of residue and strips of subducted crust was also favored by Helffrich and Wood (2001), who used various constraints to estimate plausible fractions of the different components that satisfy global heat budget constraints: for the mantle their model contained between 10 and 24% recycled oceanic crust, with the rest mostly sterile mantle (i.e., zero heat-producing elements) and recycled continental material making up to 0.4%. The viability of this class of model is strongly supported by the recent modeling of Ito and Mahoney, 2005a 2005b, 2006) who calculated the melting of a heterogeneous four or five-component mantle, and found that in most cases OIBs and MORBs could be drawn from the same source distribution, although different OIBs require different proportions of the components. This analysis is further discussed below.

A further proposed recipe is that large primitive 'blobs' exist throughout the mantle (Becker et al., 1999). The proposed blobs are larger than the sampling length scale and are not sampled by ridge volcanism. The dynamical feasibility of this model has not been established and it will not be further discussed here.

### 10.5.2.2 Two-stage melting

It was proposed by Phipps Morgan and Morgan, 1999), that the OIB and MORB signatures can be explained by two-stage melting of a heterogeneous 'plum pudding' source region. In this model, the OIB source is 'normal' mantle that includes lots of enriched basalt–pyroxenite veins, while the MORB source is produced from this by depletion associated with hot-spot melting. Thus, the initial stage of melting is associated with plumes producing OIBs with low-degree melting. The once-depleted plume material fills the asthenosphere, constituting the MORB source; subsequent second-stage melting underneath spreading centers produces the depleted MORB signature. A parametrized box model is presented to demonstrate the success of this model in explaining geochemical data. While the literal topology of this two-stage model is not widely accepted, it is useful to summarize the outcome of the parametrized model because the heterogeneous mixture of different components could have wider applicability. The model mantle contains the following components: PRIM (primitive material), ORES (residue from OIB melting), MRES (residue from MORB melting), recycled components ZOIB, ZMORB and ZCONT, and the continental crust. The rate of oceanic floor production is $2.7\,km^2\,yr^{-1}$ at the present day and increases in the past as (heat production)$^2$. By the end of the calculation, 95% of mantle has passed through MOR processing, and the proportions of the components are: PRIM = 5%, ZCONT + ZOIB = 2%, ZMORB = 20%, ORES = 5%, MRES = 62.5%, with continental crust being 0.5%. A Rb–Sr pseudo-age of 1.3 Ga is obtained, which is much younger than whole-mantle pseudo-isochron of 3.5 Ga because it represents the average time between OIB melting and MORB melting. Phipps Morgan and Morgan (1999) also present an explanation for the Pb kappa conundrum (Th/U of MORB is 2.5, whereas lead ratios imply it should be 4) based on the different compatibilities of Th and U, and for the heat-He imbalance based on He being outgassed mainly by hot spots and continental flood basalts (as previously suggested by Kellogg and Wasserburg, 1990), which leads to the prediction that atmospheric He concentrations should be elevated after flood basalt events.

### 10.5.2.3 Source statistics and component fractions

A series of papers (Anderson, 2000, 2001; Meibom et al., 2002, 2003, 2005; Meibom and Anderson, 2004) studied the statistical distributions of isotope ratios in MORB and OIB environments, and found that they support the idea that the upper mantle consists of a multiple-length-scale mixture of different 'plums' with no uniform matrix, from long-term recycling of crustal and sedimentary components. Helium ratio distributions for the 'unfiltered' MORB data set were plotted by Anderson (2000, 2001) and found to be drawn from the same statistical distribution as OIB helium ratios but with smaller standard deviation, which was argued to be related to averaging over larger sampling regions and degrees of melting consistent with the central limit theorem in which "independent averaging of samples from any given distribution will approach a Gaussian distribution with a variance that becomes smaller as the sample volume increases" (Meibom et al., 2002). This random sampling model was subsequently found to be consistent with the observed Gaussian distribution of $^{187}Os/^{188}Os$ in mantle-derived grains, implying random sampling between ancient radiogenic and unradiogenic mantle with relatively high degrees of partial melting (Meibom et al., 2002). The similarity between Os ratio distributions and He ratio

distributions was discussed in Meibom et al. (2003) and argued to be evidence for the same processes acting on both systems, with Meibom and Anderson (2004) further analyzing ratios of Pb, Sr, Nd, pointing out the Gaussian shape reflects the dominance of the homogenization during sampling process by partial melting and magma chamber mixing, rather than the nature of the source region. They argue that the dominant length scales are in the range 100 m–100 km, rather than the centimeter to meter scale implicit in the original marble cake proposal (Allegre and Turcotte, 1986), and that therefore the efficiency of convective stirring is limited, with the dominant homogenization processes occurring during the melting and magmatic stages. There are also larger-scale variations related to different plate tectonic histories.

A length-scale constraint comes from the Os analysis of mantle-derived grains in Meibom et al. (2002): as a single grain corresponds to about $1\,m^3$ of mantle, this is the minimum length scale of heterogeneity, and it is consistent with observations of Os variations in abyssal peridotites on length scales 10–100 m (Parkinson et al., 1998; Brandon et al., 2000). An age constraint comes from the least radiogenic Os ratio they found, which indicates a depletion age of 2.6 Gy.

Quantitative modeling of the distribution of isotope ratios by sampling a heterogeneous system was performed by Kellogg et al. (2002) using a new type of 'box model' that allows the distribution of isotope ratios to be calculated in addition to the mean values (Allegre and Lewin, 1995), as a function of melt fraction, sampling zone size, and evolution. Essentially, each melting event produces new 'sub-reservoirs', which are stretched with time, evolve isotopically and can be statistically sampled at a particular length scale. The results showed that indeed, as the sampling length scale or melt fraction is increased, isotope ratio distributions tends towards Gaussian, as discussed above. The method was used to successfully match Rb–Sr and Sm–Nd isotope ratio diagrams by differentiation and recycling in the upper mantle–continental crust system, and subsequently extended to include the U–Th–Pb system (Kellogg, 2004), and later applied to the whole-mantle system including the oceanic differentiation cycle (Kellogg, 2004).

A simpler mathematical model of mantle isotopic variability was developed by Rudge et al. (2005). Their model considered the mantle to be a single reservoir that experiences ongoing melting events with a certain average time between remelting.

They found that the best fit to the isotopic scatter was obtained with remelting timescale of 1.4–2.4 Ga and melt fraction of around 0.5%, the latter of which is similar to Kellogg *et al.* (2002) and Kellogg (2004), although the latter study was intended to represent continental formation melting. This did not, however, fit Pb isotopic data.

### 10.5.2.4   Can OIB and MORB be produced by the same statistical distribution?

To answer this question, it is necessary to understand the composition that is obtained when a heterogeneous mixture is partially melted, and when it is melted in different environments – in particular, underneath spreading centers (MORB) versus underneath the lithosphere (OIBs), and this is far from straightforward. Ito and van Keken (Chapter 9) present a review of some of the relevant observation and explanations.

When a source region that is heterogeneous in major-element chemistry is partially melted, it is likely that the enriched pyroxenite material melts at a lower temperature than depleted residue, which means that low degrees of melting will preferentially sample the enriched material, while high degrees of melting will average over the whole assemblage (Sleep, 1984). Whether this can quantitatively account for the differences between MORB and OIB compositions has not, until recently, been quantitatively assessed. Note that this is preferential sampling of enriched components is different from the proposal of Anderson (2001) and Meibom and Anderson (2004), in which the same material would be melted in MORB and OIB environments but with smaller melt fraction producing greater variability with the same mean. Based on quantitative modeling of trace-element distributions, Kellogg *et al.* (2002) find that OIB and MORB cannot be produced from the same distribution simply by differences in sample volume (with smaller samples being more heterogeneous): differences in the melting processes and/or source regions are necessary.

If the depleted matrix (DM) and pyroxenite (PX) veins are at small-enough length scales to be in thermal equilibrium, Phipps Morgan (2001) found that melting of the PX is greatly enhanced by heat flowing from the surrounding matrix. The details depend on the temperature–pressure gradient of PX solidus relative to DM solidus, consistent with other calculations (Hirschmann *et al.*, 1999) and confirming the prediction of Sleep (1984).

Complicating interpretations of hot-spot melting is that different basalts from the same hot-spot do not generate the same composition, but rather elongate tube-like fields (Hart *et al.*, 1992), a traditional interpretation of which is a mixing line between two end-member components present in the source region. Phipps Morgan (1999) finds, however, that such linear arrays are naturally generated by progressive melting of a heterogeneous source mixture, and remain linear independent of the number of distinct isotopic components in the initial source mixture.

Calculations with up to five components were performed by Ito and Mahoney, 2005a, 2005b, 2006), who quantitatively modeled the different trace-element compositions that arise as a result of differing OIB and MORB environments. Two key differences arise between OIB and MORB melting environments as a result of the differing mantle flow. Whereas flow beneath a spreading center rises at a roughly constant velocity to the surface, the vertical flow caused by plumes impinging on the base of a rigid lithosphere decreases towards the base of the lithosphere tens of kilometers deep. Hence the total amount of decompression melting that takes place in OIB environments is typically less, but even if the temperature is high enough to make it the same, the depth-distribution of OIB melting is concentrated towards greater depths whereas that of MORB melting is constant with depth. As the enriched components are assumed to melt at lower temperature (greater depth) due to their different lithologies and (sometimes) enrichment in volatiles, they make up a larger proportion of OIB-type melts, whereas MORB melting is diluted to a much larger extent by the depleted component, leading to a more enriched signature for OIBs. The details of the trace element are sensitive to the exact composition of enriched components in the melting zone, particularly for OIB melting.

Ito and Mahoney (2005a) study a heterogeneous mantle consisting of three components: DM, depleted mantle (similar to the DMM of Hart *et al.* (1992)); PX, pyroxenite (essentially subducted MORB, and carrying the HIMU signature); and EM, enriched mantle (similar to EM1 and EM2). They calculate the trace-element signature (La/Sm, Sr, Nd, Pb) of the resulting melt assuming perfect melt homogenization, as a function of the proportions of PX and EM (keeping DM fixed at 90%), the thickness of lithosphere, and the potential temperature. Their results reproduce many of the features observed in MORB and OIB data, particularly the difference between OIBs erupting over old (thick) seafloor and those erupting over young seafloor, which are that over young seafloor the compositions of hot-spot islands are less variable and overlap with MORB observations, whereas over old seafloor, trace-element compositions are more enriched and highly variable due to sensitivity to exact proportions of PX and EM, and temperature, in the source region. The results are also able to reproduce some features of 'arrays' in ratio–ratio space, the trajectories of which are often curved and have corners due to the transition from melting one component to melting another. Because the observed isotope ratios of each hot-spot typically cover a wide range (an elongated 'array'), the 'fit' must focus on fitting the scatter rather than precise values. Incomplete magma mixing during ascent, which is not treated, may cause melt batches to have more extreme isotopic signatures. Evidence for pressure being a most important sampling parameter is in correlated major-element, trace-element, and isotope compositions in postglacial Icelandic basalts (Stracke et al., 2003).

This analysis is then applied to specific island chains in Ito and Mahoney (2005b), with the goal of answering the question of whether two distinct sources are needed to explain OIB and MORB. The authors use a simple grid search scheme to determine the range of mantle sources that can fit 13 specific hot-spot island chains based on correlations in Sr, Nd, Pb isotopes space, then calculate the isotopic composition you would get if melting each one below mid-ocean ridges. The variables are mantle temperature excess, mass fractions of components, and isotopic compositions of EM and PX. Most model OIB compositions produce MORB that is indistinguishable from observed MORB compositions, suggesting that the whole mantle could be made of the same statistical assemblage. The proportion of different components was different for different OIBs, indicating large-scale heterogeneity. However, some predicted MORB compositions lie outside the range of observed normal MORB data. The fit to MORB is optimal when DM fractions are 85–95%, with the rest being PX and EM. Regarding abundances of enriched components, on average PX is slightly more abundant than EM (by a factor 1.17). Regarding the global budget issue, they calculated that if the upper bound for U and Th in continents and upper bound for U and Th in model are assumed, an additional primitive or otherwise enriched source is not necessary, though in that case only 30% of the mantle + core heat budget is radiogenic.

This approach was extended to include He in Ito and Mahoney (2006), with the addition of a high $^3He/^4He$ component corresponding to C (Hanan and Graham, 1996) or FOZO (Hart et al., 1992; Hilton et al., 1999) and basically the same as HRDM (Stuart et al., 2003). It was found that OIB and MORB can arise out of the same heterogeneous source region if three conditions are met: (1) the high $^3He/^4He$ component starts melting deeper than DM, (2) DM is $>= 85\%$ of the mantle, and (3) the concentration of He in C/FOZO is comparable to that in DM. The models are successful in matching the low variation in MORB $^3He/^4He$.

The requirement for C/FOZO to have the same He concentration as DM implies that it is not primitive but has already melted reinforces the findings of Class and Goldstein (2005) and earlier proposals and evidence discussed earlier (Albarède, 1998; Coltice and Ricard, 1999): it keeps a high He ratio due to strong depletion in U and Th while retaining relatively high $^3He$, perhaps in olivine-rich lithologies (Meibom et al., 2005; Parman et al., 2005) or by mixing with primitive material, with the difference that Ito and Mahoney (2006) remove the requirement for this component to be distributed differently (e.g., located in the deep mantle and brought up by plumes or depleted by a two-stage melting process (Phipps Morgan and Morgan, 1999) – it can be part of the general upper-mantle assemblage as proposed by Meibom and Anderson (2004). Still, C/FOZO and DM are distinct components despite their similarity in major and trace elements other than He, so it needs to be established how such similar material can evolve two quite different He ratios.

### 10.5.3 Transition Zone Water Filter Concept

As mentioned in Section 10.1.10, water is a potentially very important constituent in Earth's mantle due to its effect on rheology (e.g., Hirth and Kohlstedt, 1996), and is cycled between the interior and the biosphere/exosphere. Despite this importance, only one study has attempted to track its transport around the mantle (Richard et al., 2002). Water is also thought to play an important role in transporting trace elements from the subducting slab through the mantle wedge region, and several studies have tried to calculate how much water leaves the slab and/or its influence on the mantle wedge region (e.g., van Keken et al., 2002; Gerya and Yuen, 2003; Ruepke et al., 2004; Arcay et al., 2005).

It has also been proposed that water plays an important role in trace-element transport in the transition zone region. Due to different solubilities of water in different minerals, the transition zone between 410 and 660 km depth can contain 10–30 times higher water concentrations than the mantle above or below it. This has led to the idea (Bercovici and Karato, 2003) that some water may be trapped in the transition zone, and with it, incompatible trace elements. If so, this provides a way of maintaining a mantle that is layered in trace elements but still undergoing whole-mantle convection.

The high water content of enriched transition zone material rising through the 410 km discontinuity would cause it to partially melt, and the incompatible trace elements would enter the melt, leaving a depleted solid in the MORB source region. A requirement for this melt to be trapped is that it is denser than the surrounding solid, in which case a melt layer would build up above 410 km, with the production of new melt being balanced by entrainment and solidification of melt near downwelling slabs. This mechanism applies to the slow passive upwelling return flow characteristic of a mainly internally heated system. Plumes from the water-poor lower mantle could, however, escape being stripped of their trace elements because they have a relatively low water content and move through the transition zone too quickly for water to diffuse into them. Thus, plumes in the upper mantle appear enriched relative to the filtered MORB. Material traveling in the opposite direction and entering the lower mantle across the 660 km discontinuity would likely be too cold to melt but water could enter a free phase at grain boundaries and percolate back into the transition zone.

While appealing in its simplicity and ability to reconcile geochemical layering and whole-mantle convection, several aspects of the proposal require testing, as is presently happening. The existence of a partial melt layer above 410 km should be testable seismically, and some studies have indeed indicated such a zone (e.g., Revenaugh and Sipkin, 1994; Song et al., 2004). Mineral properties such as the density of melt above 410 km depth, the partition coefficients or relevant trace elements at this pressure, and the diffusivity of hydrogen need to be established. The detailed dynamical and geochemical consequences need to be worked out. A geochemical prediction of the model is that at steady state the average isotopic composition of the convecting upper mantle must be

almost the same as that of the lower mantle, with modifications for radiogenic ingrowth, which implies that noble gas residence times in the upper mantle must be very long (C. J. Ballentine, personal communication, 2007).

Some of this research is in progress, as summarized in Karato et al. (2006). For example, Leahy and Bercovici (2004) calculate the convective consequences of concentrating heat-producing elements in the transition zone and/or lower mantle, and find that the exact distribution has a minor influence on the flow except in the extreme case where all heat-producing elements are concentrated in the transition zone. An analysis of electrical conductivity of the relevant phases compared to that inferred in the mantle implies the predicted amount of water in the transition zone (Huang et al., 2005, 2006), although this has been doubted (Hirschmann, 2006a). Further discussions of water transport in the mantle can be found in Hirschmann (2006b).

### 10.5.4 Outlook and Future Directions

#### 10.5.4.1 Can geochemical and geophysical observations be reconciled with current paradigms?

Recent advances in paradigms and modeling bring us closer to an integrated model that is consistent with both geochemical and geophysical constraints, but several major questions remain.

Likely characteristics of an integrated geodynamical–geochemical model are

- The turbulent (exponential) mixing regime, with a whole-mantle mixing time (to $\sim$ tens of centimeter scale lengths) $\sim$3 Gy as discussed in Section 10.2.6.4.
- The mantle is compositionally heterogeneous at all scales, from thousands of kilometers to centimeters.
- The upper mantle that is not much better mixed than the lower mantle, as demonstrated by recent numerical results (Hunt and Kellogg, 2001; Stegman et al., 2002). ('Upper mantle' mixing time estimates are meaningless in whole-mantle convection, as discussed earlier).
- Almost all of the mantle (i.e., 95–99%) has been processed through mid-ocean ridge melting and so presumably been degassed of primordial noble gases. This is consistent with the constraint of only $\sim$50% degassing of radiogenic $^{40}$Ar.

Less certain, but looking increasingly feasible are

- Many OIBs can arise from a statistically similar heterogeneous source as MORB, through differences in melting processes. It is also feasible, however, that plume-related volcanism could tap a different statistical assemblage.
- There may be some vertical stratification of incompatible trace elements due either to gravitational stratification of major elements (crust settling above the CMB, perhaps some of it generated very early on, and/or the composition-dependence of the depth of the perovskite phase transition, or effects of water (the transition zone water filter) or melting (possible entrapment in melt zones at the CMB).

Many geochemical issues remain to be explained, although hypotheses exist. These include the radiogenic heat budget of the mantle (where are the heat-producing elements missing from the MORB source region? – although recent silicate Earth models (Lyubetskaya and Korenaga, 2007a, 2007b) suggest that there are no missing elements), the correct explanation of high $^3$He/$^4$He ratios (the 'recycled' hypothesis has been gaining ground on the 'primitive' explanation recently), and the Pb paradox. In some cases, uncertainties arise because of uncertain physical properties such as partition coefficients and densities.

### 10.5.4.2 Quantitative modeling approaches

Recent years have seen the improvement of several modeling approaches that are capable of generating model geochemical data to compare to observations. These include improved 'box models' that treat the statistical distribution of isotope ratios as well as their mean values (Kellogg *et al.*, 2002; Kellogg and Tackley, 2004; Rudge *et al.*, 2005; Rudge, 2006), improved methods for calculating the trace-element signatures that arise from melting in different environments (Phipps Morgan, 1999; Ito and Mahoney, 2005a, 2005b, 2006), and numerical thermochemical convection models that track trace elements and make quantitative predictions about trace-element distributions and evolution, as well as the seismic structure of the mantle. These different methods are complementary and can be combined to enhance understanding. Several modeling challenges remain in the future. The global numerical models do not include continent formation, because the relevant processes are poorly understood; the complex mass transport occurring in subduction zones is an important part of this. Global numerical models cannot treat the smaller scales, which must therefore be done using some statistical approach.

## Acknowledgments

The author thanks Yanick Ricard for providing **Figure 4**.

## References

Abe Y (1997) Thermal and chemical evolution of the terrestrial magma ocean. *Physics of the Earth and Planetary Interiors* 100: 27–39.

Agranier A, Blichert-Toft J, Graham DW, Debaille V, Schiano P, and Albarède F (2005) The spectra of isotopic heterogeneities along the Mid-Atlantic ridge. *Earth and Planetary Science Letters* 238: 96–109.

Albarède F (1998) Time-dependent models of U–Th–He and K–Ar evolution and the layering of mantle convection. *Chemical Geology* 145: 413–429.

Albarède F (2001) Radiogenic ingrowth in systems with multiple reservoirs: Applications to the differentiation of the mantle–crust system. *Earth and Planetary Science Letters* 189: 59–73.

Albarède F (2005a) Helium feels the heat in Earth's mantle. *Science* 310: 1777–1778.

Albarède F (2005b) The survival of mantle geochemical heterogeneities. In: Van Der Hilst RD, Bass JD, Matas J, and Trampert J (eds.) *Earth's Deep Mantle: Structure, Composition, and Evolution*, pp. 27–46. Washington, DC: American Geophysical Union.

Albarède F and van der Hilst RD (2002) Zoned mantle convection. *Philosophical Transactions of the Royal Society of London A* 360: 2569–2592.

Allegre CJ (1997) Limitation on the mass exchange between the upper and lower mantle: The evolving convection regime of the Earth. *Earth and Planetary Science Letters* 150: 1–6.

Allegre CJ, Brevart O, Dupre B, and Minster JF (1980) Isotopic and chemical effects produced in a continuously differentiating convecting Earth mantle. *Philosophical Transactions of the Royal Society of London A* 297: 447–477.

Allegre CJ, Hart SR, and Minster JF (1983a) Chemical structure and evolution of the mantle and continents determined by inversion of Nd and Sr isotopic data. Part I: Theoretical methods. *Earth and Planetary Science Letters* 66: 177–190.

Allegre CJ, Hofmann A, and O'Nions K (1996) The argon constraints on mantle structure. *Geophysical Research Letters* 23: 3555–3557.

Allegre CJ and Lewin E (1995) Isotopic systems and stirring times of the Earths mantle. *Earth and Planetary Science Letters* 136: 629–646.

Allegre CJ, Moreira M, and Staudacher T (1995) $^4$He/$^3$He dispersion and mantle convection. *Geophysical Research Letters* 22: 2325–2328.

Allegre CJ, Sarda P, and Staudacher T (1993) Speculations about the cosmic origin of He and Ne in the interior of the Earth. *Earth and Planetary Science Letters* 117: 229–233.

Allegre CJ, Staudacher T, Sarda P, and Kurz MD (1983b) Constraints on evolution of Earth's mantle from rare gas systematics. *Nature* 303: 762–766.

Allegre CJ and Turcotte DL (1986) Implications of a 2-component marble-cake mantle. *Nature* 323: 123–127.

Anderson DL (1993) He-3 From the mantle – Primordial signal or cosmic dust. *Science* 261: 170–176.

Anderson DL (1995) Lithosphere, asthenosphere, and perisphere. *Reviews of Geophysics* 33: 125–149.

Anderson DL (1998) A model to explain the various paradoxes associated with mantle noble gas geochemistry. *Proceedings of the National Academy of Sciences of the United States of America* 95: 9087–9092.

Anderson DL (2000) The statistics of helium isotopes along the global spreading ridge system and the central limit theorum. *Geophysical Research Letters* 27: 77–82.

Anderson DL (2001) A statistical test of the two reservoir model for helium isotopes. *Earth and Planetary Science Letters* 193: 77–82.

Anderson OL, Oda H, and Isaak D (1992) A model for the computation of thermal expansivity at high compression and high temperatures – MgO as an example. *Geophysical Research Letters* 19: 1987–1990.

Arcay D, Tric E, and Doin M-P (2005) Numerical simulations of subduction zones. Effect of slab dehydration on the mantle wedge dynamics. *Physics of the Earth and Planetary Interiors* 149: 133–153.

Armstrong RL and Hein SM (1973) Computer simulation of Pb and Sr isotope evolution of the Earth's crust and upper mantle. *Geochemica Cosmochimica Acta* 37: 1–18.

Asimow PD, Hirschmann MM, and Stolper EM (2001) Calculation of peridotite partial melting from thermodynamic models of minerals and melts. Part IV: Adiabatic decompression and the composition and mean properties of mid-ocean ridge basalts. *Journal of Petrology* 42: 963–998.

Ballentine CJ (2002) Geochemistry – Tiny tracers tell tall tales. *Science* 296: 1247–1248.

Ballentine CJ and Barford DN (2000) The origin of air-like noble gases in MORB and OIB. *Earth and Planetary Science Letters* 180: 39–48.

Ballentine CJ, Marty B, Lollar BS, and Cassidy M (2005) Nean isotopes constrain convection and volatile origin in the Earth's mantle. *Nature* 433: 33–38.

Ballentine CJ, van Keken PE, Porcelli D, and Hauri EH (2002) Numerical models, geochemistry and the zero-paradox noble-gas mantle. *Philosophical Transactions of the Royal Society of London Series A-Mathematical Physical and Engineering Sciences* 360: 2611–2631.

Barth MG, Rudnick RL, Horn I, et al. (2001) Geochemistry of xenolithic eclogites from west Africa. Part I: A link between low MgO eclogites and Archean crust formation. *Geochemica Cosmochimica Acta* 65: 1499–1527.

Batchelor GK (1952) The effect of homogeneous turbulence on material lines and surfaces. *Proceedings of the Royal Society A* 213: 349–366.

Batchelor GK (1959) Small-scale variation of convected quantities like temperature in a turbulent fluid. *Journal of Fluid Mechanics* 5: 113–133.

Becker TW, Kellogg JB, and O'Connell RJ (1999) Thermal constraints on the survival of primitive blobs in the lower mantle. *Earth and Planetary Science Letters* 171: 351–365.

Benz W and Cameron AGW (1990) Terrestrial effects of the giant impact. In: Newsom HE and Jones JH (eds.) *Origin of the Earth*, pp. 61–68. New York: Oxford University Press.

Bercovici D and Karato S (2003) Whole-mantle convection and the transition-zone water filter. *Nature* 425: 39–44.

Boehler R, Chopelas A, and Zerr A (1995) Temperature and chemistry of the core–mantle boundary. *Chemical Geology* 120: 199–205.

Bolton H and Masters G (2001) Travel times of P and S from the global digital seismic networks: Implications for the relative variation of P and S velocity in the mantle. *Journal of Geophysical Research-Solid Earth* 106: 13527–13540.

Boyet M and Carlson RW (2006) A new geochemical model for the Earth's mantle inferred from $^{146}$Sm–$^{142}$Nd systematics. *Earth and Planetary Science Letters* 250: 254–268.

Brandon A, Snow JE, Walker RJ, Morgan JW, and Mock TD (2000) $^{190}$Pt–$^{186}$Os and $^{187}$Re–$^{187}$Os systematics of abyssal peridotites. *Earth and Planetary Science Letters* 177: 319–335.

Brandon AD and Walker RJ (2005) The debate over core–mantle interaction. *Earth and Planetary Science Letters* 232: 211–225.

Brandon AD, Walker RJ, Morgan JW, Norman MD, and Prichard HM (1998) Coupled 186Os and 187Os evidence for core–mantle interaction. *Science* 280: 1570–1573.

Brandon AD, Walker RJ, Puchtel IS, Becker H, Humayun M, and Revillon S (2003) 186Os–187Os systematics of Gorgona Island komatiites: Implications for early growth of the inner core. *Earth and Planetary Science Letters* 206: 411–426.

Buffett BA, Huppert HE, Lister JR, and Woods AW (1992) Analytical model for solidification of the Earth's core. *Nature* 356: 329–331.

Buffett BA, Huppert HE, Lister JR, and Woods AW (1996) On the thermal evolution of the Earth's core. *Journal of Geophysical Research* 101: 7989–8006.

Bunge HP, Richards MA, and Baumgardner JR (1997) A sensitivity study of 3-dimensional spherical mantle convection at $10^8$ Rayleigh number – Effects of depth-dependent viscosity, heating mode, and an endothermic phase change. *Journal of Geophysical Research* 102: 11991–12007.

Bunge HP, Richards MA, Lithgowbertelloni C, Baumgardner JR, Grand SP, and Romanowicz BA (1998) Time scales and heterogeneous structure in geodynamic Earth models. *Science* 280: 91–95.

Caffee MW, Hudson GB, Velsko C, Huss GR, Alexander EC, and Chivas AR (1999) Primordial noble gases from Earth's mantle: Identification of a primitive volatile component. *Science* 285: 2115–2118.

Carlson RW (1987) Geochemical evolution of the crust and mantle. *Reviews of Geophysics* 25: 1011–1020.

Carlson RW (1994) Mechanisms of Earth differentiation – Consequences for the chemical-structure of the mantle. *Reviews of Geophysics* 32: 337–361.

Castillo P (1988) The Dupal anomaly as a trace of the upwelling lower mantle. *Nature* 336: 667–670.

Castle JC and van der Hilst RD (2003a) Searching for seismic scattering off mantle interfaces between 800 km and 2000 km depth. *Journal of Geophysical Research* 108 (doi:10.1029/2001JB000286).

Castle JC and van der Hilst RD (2003b) Using ScP precursors to search for mantle structures beneath 1800 km depth. *Geophysical Research Letters* 30 (doi:10.1029/2002GL016023).

Castro MC, Patriarche D, and Goblet P (2005) 2-D numerical simulations of groundwater flow, heat transfer and 4He transport – implications for the He terrestrial budget and the mantle helium-heat imbalance. *Earth and Planetary Science Letters* 237: 893–910.

Chase CG (1981) Oceanic island Pb: Two-stage histories and mantle evolution. *Earth and Planetary Science Letters* 52: 277–284.

Chase CG and Patchett PJ (1988) Stored mafic/ultramafic crust and early Archean mantle depletion. *Earth and Planetary Science Letters* 91: 66–72.

Chauvel CS, Goldstein SL, and Hofmann AW (1995) Hydration and dehydration of oceanic crust controls Pb evolution in the mantle. *Chemical Geology* 126: 65–75.

Chopelas A (1996) Thermal expansivity of lower mantle phases Mgo and Mgsio3 perovskite at high-pressure derived from vibrational spectroscopy. *Physics of the Earth and Planetary Interiors* 98: 3–15.

Christensen U (1984) Instability of a hot boundary layer and initiation of thermo-chemical plumes. *Annales Geophysicae* 2: 311–319.

Christensen U (1988) Is subducted lithosphere trapped at the 670-km discontinuity? *Nature* 336: 462–463.

Christensen U (1989a) Mixing by time-dependent convection. *Earth and Planetary Science Letters* 95: 382–394.

Christensen U (1990) Mixing by time-dependent mantle convection – Reply. *Earth and Planetary Science Letters* 98: 408–410.

Christensen UR (1989b) Models of mantle convection – one or several layers. *Philosophical Transactions of the Royal Society of London A* 328: 417–424.

Christensen UR (1997) Influence of chemical buoyancy on the dynamics of slabs in the transition zone. *Journal of Geophysical Research* 102: 22435–22444.

Christensen UR and Hofmann AW (1994) Segregation of subducted oceanic crust in the convecting mantle. *Journal of Geophysical Research* 99: 19867–19884.

Christensen UR and Yuen DA (1984) The interaction of a subducting lithospheric slab with a chemical or phase-boundary. *Journal of Geophysical Research* 89: 4389–4402.

Christensen UR and Yuen DA (1985) Layered convection induced by phase transitions. *Journal of Geophysical Research* 90: 10291–10300.

Chudinovskikh L and Boehler R (2002) The MgSiO3 ilmenite-perovskite phase boundary: Evidence for strongly negative Clapeyron slope.

Cizkova H, Cadek O, Van Den Berg AP, and Vlaar NJ (1999) Can lower mantle slab-like seismic anomalies be explained by thermal coupling between the upper and lower mantles? *Geophysical Research Letters* 26: 1501–1504.

Class C and Goldstein SL (2005) Evolution of helium isotopes in the Earth's mantle. *Nature* 436: 1107–1112.

Coltice N (2005) The role of convective mixing in degassing the Earth's mantle. *Earth and Planetary Science Letters* 234: 15–25.

Coltice N, Albarède F, and Gillet P (2000a) $^{40}K-^{40}Ar$ constraints on recycling continental crust into the mantle. *Science* 288: 845–847.

Coltice N, Ferrachat S, and Ricard Y (2000b) Box modeling the chemical evolution of geophysical systems: Case study of the Earth's mantle. *Geophysical Research Letters* 27: 1579–1582.

Coltice N and Ricard Y (1999) Geochemical observations and one layer mantle convection. *Earth and Planetary Science Letters* 174: 125–137.

Coltice N and Ricard Y (2002) On the origin of noble gases in mantle plumes. *Philosophical Transactions of the Royal Society of London Series A-Mathematical Physical and Engineering Sciences* 360: 2633–2648.

Coltice N and Schmalzl J (2006) Mixing times in the mantle of the early Earth derived from 2-D and 3-D numerical simulations of convection. *Geophysical Research Letters* 33 (doi:10.1029/2006GL027707).

Corrsin S (1961) The reactant concentration spectrum in turbulent mixing with a first order reaction. *Journal of Fluid Mechanics* 11: 407–416.

Creager KC and Jordan TH (1984) Slab Penetration Into the lower mantle. *Journal of Geophysical Research* 89: 3031–3049.

Creager KC and Jordan TH (1986) Slab penetration into the lower mantle beneath the Mariana and other island arcs of the northwest Pacific. *Journal of Geophysical Research Solid Earth and Planets* 91: 3573–3589.

Cserepes L and Rabinowicz M (1985) Gravity and convection in a 2-layer mantle. *Earth and Planetary Science Letters* 76: 193–207.

Cserepes L, Rabinowicz M, and Rosembergborot C (1988) 3-Dimensional infinite Prandtl number convection in one and 2 layers with implications for the Earths gravity-field. *Journal of Geophysical Research Solid Earth and Planets* 93: 12009–12025.

Davaille A (1999a) Simultaneous generation of hotspots and superswells by convection in a heterogeneous planetary mantle. *Nature* 402: 756–760.

Davaille A (1999b) Two-layer thermal convection in miscible viscous fluids. *Journal of Fluid Mechanics* 379: 223–253.

Davaille A, Girard F, and Le BM (2002) How to anchor hotspots in a convecting mantle? *Earth and Planetary Science Letters* 203: 621–634.

Davies GF (1980) Thermal histories of convective Earth models and constraints on radiogenic heat production in the Earth. *Journal of Geophysical Research* 85: 2517–2530.

Davies GF (1981) Earth's neodymium budget and structure and evolution of the mantle. *Nature* 290: 208–213.

Davies GF (1983) Viscosity structure of a layered convecting mantle. *Nature* 301: 592–594.

Davies GF (1984) Geophysical and isotopic constraints on mantle convection – An interim synthesis. *Journal of Geophysical Research* 89: 6017–6040.

Davies GF (1988) Ocean bathymetry and mantle convection. Part: Large-scale flow and hotspots. *Journal of Geophysical Research Solid Earth and Planets* 93: 10467–10480.

Davies GF (1990) Heat and mass transport in the early Earth. In: Newsome HE and Jones JH (eds.) *Origin of the Earth*, pp. 175–194. New York: Oxford University Press.

Davies GF (1992) On the emergence of plate-tectonics. *Geology* 20: 963–966.

Davies GF (1999) Geophysically constrained mantle mass flows and the $^{40}Ar$ budget: A degassed lower mantle? *Earth and Planetary Science Letters* 166: 149–162.

Davies GF (2002) Stirring geochemistry in mantle convection models with stiff plates and slabs. *Geochemica Cosmochimica Acta* 66: 3125–3142.

Davies GF (2006) Depletion of the ealy Earth's upper mantle and the viability of early plate tectonics. *Earth and Planetary Science Letters* 243: 376–382.

Davies GF and Gurnis M (1986) Interaction of mantle dregs with convection – lateral heterogeneity at the core-mantle boundary. *Geophysical Research Letters* 13: 1517–1520.

Delano JW (2001) Redox history of the Earth's interior since similar to 3900 Ma: Implications for prebiotic molecules (Review). *Origins of Life and Evolution of the Biosphere* 31: 311–341.

Deparis V, Legros H, and Ricard Y (1995) Mass anomalies due to subducted slabs and simulations of plate motion since 200 My. *Physics of the Earth and Planetary Interiors* 89: 271–280.

Deschamps F, Trampert J, and Tackley PJ (2007) Thermo-chemical structure of the lower mantle: Seismological evidence and consequences for geodynamics. In: Yuen DA, Maruyama S, Karato SI, and Windley BF (eds.) *Superplume: Beyond Plate Tectonics.* Springer (in press).

Dixon ET, Honda M, McDougall I, Campbell IH, and Sigurdsson I (2000) Preservation of near-solar isotopic ratios in Icelandic basalts. *Earth and Planetary Science Letters* 180: 309–324.

Drake MJ and Righter K (2002) Determining the composition of the Earth (Review). *Nature* 416: 39–44.

Dubrovinsky L, Annersten H, Dubrovinskaia N, *et al.* (2001) Chemical interaction of Fe and $Al_2O_3$ as a source of heteroeneity at the Earth's core–mantle boundary. *Nature* 412: 527–529.

Dubrovinsky L, Dubrovinskaia N, Langenhorst F, *et al.* (2003) Iron–silica interaction at extreme conditions and the electrically conducting layer at the base of Earth's mantle. *Nature* 421: 58–61.

Dupeyrat L, Sotin C, and Parmentier EM (1995) Thermal and chemical convection in planetary mantles. *Journal of Geophysical Research, B, Solid Earth and Planets* 100: 497–520.

Dupre B and Allegre CJ (1983) Pb–Sr variation in Indian Ocean basalts and mixing phenomena. *Nature* 303: 142–146.

du Vignaux NM and Fleitout L (2001) Stretching and mixing of viscous blobs in Earth's mantle. *Journal of Geophysical Research-Solid Earth* 106: 30893–30908.

Elliott T, Zindler A, and Bourdon B (1999) Exploring the kappa conundrum: The role of recycling in the lead isotope evolution of the mantle. *Earth and Planetary Science Letters* 169: 129–145.

Farley KA, Natland JH, and Craig H (1992) Binary mixing of enriched and undegassed (primitive?) mantle components (He, Sr, Nd, Pb) in Samoan lavas. *Earth and Planetary Science Letters* 111: 183–199.

Farley KA and Neroda E (1998) Noble gases in the Earth's mantle. *Annual Review of Earth and Planetary Sciences* 26: 189–218.

Farnetani CG (1997) Excess temperature of mantle plumes – The role of chemical stratification across D″. *Geophysical Research Letters* 24: 1583–1586.

Farnetani CG, Legras B, and Tackley PJ (2002) Mixing and deformations in mantle plumes. *Earth and Planetary Science Letters* 196: 1–15.

Farnetani CG and Samuel H (2003) Lagrangian structures and stirring in the Earth's mantle. *Earth and Planetary Science Letters* 206: 335–348.

Farnetani CG and Samuel H (2005) Beyond the thermal plume paradigm. *Geophysical Research Letters* 32 (doi:10.1029/2005GL022360).

Farnetani DG and Richards MA (1995) Thermal entrainment and melting in mantle plumes. *Earth and Planetary Science Letters* 136: 251–267.

Ferrachat S and Ricard Y (1998) Regular vs. chaotic mantle mixing. *Earth and Planetary Science Letters* 155: 75–86.

Ferrachat S and Ricard Y (2001) Mixing properties in the Earth's mantle: Effects of the viscosity stratification and of oceanic crust segregation. *Geochemistry Geophysics Geosystems* 2, doi:10.1029/2000GC000092.

Fleitout L, Mambole A, and Christensen U (2000) Phase changes around 670 km depth and segregation in the Earth's mantle. *EOS Transactions AGU, Fall Meeting Supplement* 81: Abstract T12E–11.

Forte AM and Mitrovica JX (2001) Deep-mantle high-viscosity flow and thermochemical structure inferred from seismic and geodynamic data. *Nature* 410: 1049–1056.

Fukao Y (1992) Seismic tomogram of the Earths mantle – Geodynamic implications. *Science* 258: 625–630.

Fukao Y, Obayashi M, Inoue H, and Nenbai M (1992) Subducting slabs stagnant in the mantle transition zone. *Journal of Geophysical Research-Solid Earth* 97: 4809–4822.

Gable CW, O'Connell RJ, and Travis BJ (1991) Convection in 3 dimensions with surface plates – Generation of toroidal flow. *Journal of Geophysical Research* 96: 8391–8405.

Gaherty JB and Hager BH (1994) Compositional vs thermal buoyancy and the evolution of subducted lithosphere. *Geophysical Research Letters* 21: 141–144.

Galer SJG and O'Nions RK (1986) Magmagenesis and the mapping of chemical and isotopic variations in the mantle. *Chemical Geology* 56: 45–61.

Gerya TV and Yuen DA (2003) Rayleigh–Taylor instabilities from hydration and melting propel 'cold plumes' at subduction zones. *Earth and Planetary Science Letters* 212: 47–62.

Glatzmaier GA and Schubert G (1993) 3-Dimensional spherical-models of layered and whole mantle convection. *Journal of Geophysical Research* 98: 21969–21976.

Goarant F, Guyot F, Peyronneau J, and Poirier JP (1992) High-pressure and high-temperature reactions between silicates and liquid iron alloys, in the diamond anvil cell, studied by analytical electron microscopy. *Journal of Geophysical Research* 97: 4477–4487.

Gonnermann HM, Manga M, and Jellinek AM (2002) Dynamics and longevity of an initially stratified mantle. *Geophysical Research Letters* 29: 1399.

Graham D, Lupton J, Albarède F, and Condomines M (1990) Extreme temporal homogeneity of Helium isotopes at Piton de la Fournaise, Reunion Island. *Nature* 347: 545–548.

Graham DW (2002) Noble gas isotope geochemistry of mid-ocean ridge and ocean island basalts; characterizazion of mantle source reservoirs. In: Porcelli D, Wieler R, and Ballentine CJ (eds.) *Reviews in Mineralogy and Geochemistry: Noble Gases in Geochemistry and Cosmochemistry*, pp. 247–318. Washington, DC: Mineralogical Society of America.

Graham DW, Blichert-Toft J, Russo CJ, Rubin KH, and Albarède F (2006) Cryptic striations in the upper mantle revealed by hafnium isotopes in southest Indian ridge basalts. *Nature* 440: 199–202.

Graham DW, Lupton JE, Spera FJ, and Christle DM (2001) Upper-mantle dynamics revealed by helium isotope variations along the Southeast Indian Ridge. *Nature* 409: 701–703.

Grand SP (1994) Mantle shear structure beneath the America and surrounding oceans. *Journal of Geophysical Research-Solid Earth* 99: 11591–11621.

Griffiths RW (1986) Dynamics of mantle thermals with constant buoyancy or anomalous internal heating. *Earth and Planetary Science Letters* 78: 435–446.

Guignot N and Andrault D (2004) Equations of state of Na–K–Al host phases and implications for MORB density in the lower mantle. *Physics of the Earth and Planetary Interiors* 143–144: 107–128.

Gurnis M (1986a) The effects of chemical density differences on convective mixing in the Earth's mantle. *Journal of Geophysical Research* 91: 1407–1419.

Gurnis M (1986b) Quantitative bounds on the size spectrum of isotopic heterogeneity within the mantle. *Nature* 323: 317–320.

Gurnis M (1986c) Stirring and mixing in the mantle by plate-scale flow – Large persistent blobs and long tendrils coexist. *Geophysical Research Letters* 13: 1474–1477.

Gurnis M and Davies GF (1986a) The effect of depth-dependent viscosity on convective mixing in the mantle and the possible survival of primitive mantle. *Geophysical Research Letters* 13: 541–544.

Gurnis M and Davies GF (1986b) Mixing in numerical-models of mantle convection incorporating plate kinematics. *Journal of Geophysical Research Solid Earth and Planets* 91: 6375–6395.

Hager BH, Clayton RW, Richards MA, Comer RP, and Dziewonski AM (1985) Lower mantle heterogeneity, dynamic topography and the geoid. *Nature* 313: 541–546.

Hager BH and Richards MA (1989) Long-wavelength variations in Earths geoid – Physical models and dynamical implications. *Philosophical Transactions of the Royal Society of London A* 328: 309–327.

Hanan BB and Graham DW (1996) Lead and helium isotope evidence from oceanic basalts for a common deep source of mantle plumes. *Science* 272: 991–995.

Hansen U and Yuen DA (1988) Numerical simulations of thermal–chemical instabilities at the core mantle boundary. *Nature* 334: 237–240.

Hansen U and Yuen DA (1989) Dynamical influences from thermal–chemical instabilities at the core–mantle boundary. *Geophysical Research Letters* 16: 629–632.

Hansen U and Yuen DA (2000) Extended-Boussinesq thermal–chemical convection with moving heat sources and variable viscosity. *Earth and Planetary Science Letters* 176: 401–411.

Harrison D, Burnard P, and Turner G (1999) Noble gas behaviour and composition in the mantle: Constraints from the Iceland plume. *Earth and Planetary Science Letters* 171: 199–207.

Harrison TM, Blichert-Toft J, Mueller W, Albarade F, Holden P, and Mojzsis SJ (2005) Heterogeneous hadean hafnium: Evidence of continental crust at 4.4 to 4.5 Gyr. *Science* 310: 1947–1950.

Hart SR (1984) A large-scale isotope anomaly in the Southern Hemisphere mantle. *Nature* 309: 753–757.

Hart SR, Hauri EH, Oschmann JA, and Whitehead JA (1992) Mantle plumes and entrainment: Isotopic evidence. *Science* 256: 517–520.

Hauri E (2002) SIMS analysis of volatiles in silicate glasses. Part 2: Isotopes and abundances in Hawaiian melt inclusions. *Chemical Geology* 183: 115–141.

Hauri EH, Whitehead JA, and Hart SR (1994) Fluid dynamic and geochemical aspects of entrainment in mantle plumes. *Journal of Geophysical Research Solid Earth* 99: 24275–24300.

Hedlin MAH and Shearer PM (2000) An analysis of large-scale variations in small-scale mantle heterogeneity using global seismographic network recordings of precursors to PKP. *Journal of Geophysical Research* 105: 13655–13673.

Hedlin MAH, Shearer PM, and Earle PS (1997) Seismic evidence for small-scale heterogeneity throughout the Earth's mantle. *Nature* 387: 145–150.

Helffrich GR and Wood BJ (2001) The Earth's mantle (Review). *Nature* 412: 501–507.

Hilton DR, Fischer TP, and Marty B (2002) Noble gases and volatile recycling at subduction zones. *Reviews in Mineralogy and Geochemistry* 47: 319–370.

Hilton DR, Gronvold K, Macpherson CG, and Castillo PR (1999) Extreme 3He/4He ratios in northwest Iceland: Constraining the common component in mantle plumes. *Earth and Planetary Science Letters* 173: 53–60.

Hilton DR and Porcelli D (2003) Noble gases as mantle tracers. In: Carlson RW (ed.) *Treatise on Geochemistry, vol. 2: The Mantle and Core*, pp. 277–318. Amsterdam: Elsevier.

Hirose K and Kushiro I (1993) Partial melting of dry peridotites at high pressures: Determination of compositions of melts segregated from peridotite using aggregates of diamond. *Earth and Planetary Science Letters* 114: 477–489.

Hirose K, Shimizu N, van Westeren W, and Fei Y (2004) Trace element partitioning in Earth's lower mantle and implications for geochemical consequences of partial melting at the core–mantle boundary. *Physics of the Earth and Planetary Interiors* 146: 249–260.

Hirose K, Takafuji N, Sata N, and Ohishi Y (2005) Phase transition and density of subducted MORB crust in the lower mantle. *Earth and Planetary Science Letters* 237: 239–251.

Hirschmann M (2006a) A wet mantle conductor? *Nature* 439: E3.

Hirschmann MM (2006b) Water, melting, and the deep Earth $H_2O$ cycle. *Annual Review of Earth and Planetary Sciences* 34: 629–653.

Hirschmann MM, Asimow PD, Ghiorso MS, and Stolper EM (1999) Calculation of peridotite partial melting from thermodynamic models of minerals and melts. Part III: Controls on isobaric melt production and the effect of water on melt production. *Journal of Petrology* 40: 831–851.

Hirth G and Kohlstedt DL (1996) Water in the oceanic upper-mantle – Implications for rheology, melt extraction and the evolution of the lithosphere. *Earth and Planetary Science Letters* 144: 93–108.

Hoffman NRA and McKenzie DP (1985) The destruction of geochemical heterogeneities by differential fluid motions during mantle convection. *Geophysical Journal of the Royal Astronomical Society* 82: 163–206.

Hofmann AW (1988) Chemical differentiation of the Earth: The relationship between mantle, continental crust, and oceanic crust. *Earth and Planetary Science Letters* 90: 297–314.

Hofmann AW (1997) Mantle geochemistry: The message from oceanic volcanism. *Nature* 385: 219–229.

Hofmann AW (2003) Sampling mantle heterogeneity through oceanic basalts: Isotopes and trace elements. In: Carlson RW (ed.) *Treatise on Geochemistry*, pp. 61–101. Amsterdam: Elsevier.

Hofmann AW and Hart SR (1978) An assessment of local and regional isotopic equilibrium in the mantle. *Earth and Planetary Science Letters* 38: 44–62.

Hofmann AW and White WM (1982) Mantle plumes from ancient oceanic-crust. *Earth and Planetary Science Letters* 57: 421–436.

Hoink T, Schmalzl J, and Hansen U (2006) Dynamics of metal-silicate separation in a terrestrial magma ocean. *Geochemistry Geophysics Geosystems* 7 (doi:10.1029/2006GC001268).

Holland G and Ballentine CJ (2006) Seawater subduction controls the heavy noble gas composition of the mantle. *Nature* 441: 186–191.

Holland HD (1984) *The Chemical Evolution of the Atmosphere and Oceans*. Princeton, NJ: Princeton University Press.

Holland HD (1994) Early proterozoic atmospheric change. In: Bengtson S (ed.) *Early Life on Earth*, pp. 237–244. New York: Columbia University Press.

Honda M, McDougall I, Patterson DB, Doulgeris A, and Clague DA (1999) Noble gases in submarine pillow basalt glasses from Loihi and Kilauea, Hawaii: A solar component in the Earth. *Geochemica Cosmochimica Acta* 57: 859–874.

Hopp J and Trieloff M (2005) Refining the noble gas record of the Reunion mantle plume source: Implications on mantle geochemistry. *Earth and Planetary Science Letters* 240: 573–588.

Huang X, Xu Y, and Karato S (2005) Water content in the transition zone from electrical conductivity of wadsleyite and ringwoodite. *Nature* 434: 746–749.

Huang X, Xu Y, and Karato S (2006) A wet mantle conductor? Reply. *Nature* 439: E3–E4.

Humayun M, Qin L, and Norman MD (2004) Geochemical evidence for excess iron in the mantle beneath Hawaii. *Science* 306: 91–94.

Hunt DL and Kellogg LH (2001) Quantifying mixing and age variations of heterogeneities in models of mantle convection: Role of depth-dependent viscosity. *Journal of Geophysical Research* 106: 6747–6760.

Irifune T and Ringwood AE (1993) Phase-transformations in subducted oceanic-crust and buoyancy relationships at depths of 600–800 km in the mantle. *Earth and Planetary Science Letters* 117: 101–110.

Ishii M and Tromp J (1999) Normal-mode and free-air gravity constraints on lateral variations in velocity and density of Earth's mantle. *Science* 285: 1231–1235.

Ito G and Mahoney JJ (2005a) Flow and melting of a heterogeneous mantle. Part 1: Method and importance to the geochemistry of ocean island and mid-ocean ridge basalts. *Earth and Planetary Science Letters* 230: 29–46.

Ito G and Mahoney JJ (2005b) Flow and melting of a heterogeneous mantle. Part 2: Implications for a chemically non-layered mantle. *Earth and Planetary Science Letters* 230: 47–63.

Ito G and Mahoney JJ (2006) Melting a high $^3$He/$^4$He source in a heterogeneous mantle. *Geochemistry Geophysics Geosystems* 7: Q05010 (doi:10.1029/2005GC001158).

Jacobsen SB and Wasserburg GJ (1979) The mean age of mantle and crustal reservoirs. *Journal of Geophysical Research* 84: 7411–7427.

Jellinek AM and Manga M (2002) The influence of a chemical boundary layer on the fixity and lifetime of mantle plumes. *Nature* 418: 760–763.

Jochum K, Hofmann AW, Ito E, Seufert HM, and White WM (1983) K, U and Th in mid-ocean ridge basalt glasses and heat production, K/U and K/Rb in the mantle. *Nature* 306: 431–436.

Jordan TH, Puster P, Glatzmaier GA, and Tackley PJ (1993) Comparisons between seismic Earth structures and mantle flow models based on radial correlation functions. *Science* 261: 1427–1431.

Kameyama M, Fujimoto H, and Ogawa M (1996) A thermo-chemical regime in the upper mantle in the early Earth inferred from a numerical model of magma-migration in a convecting upper mantle. *Physics of the Earth and Planetary Interiors* 94: 187–215.

Kanda RVS and Stevenson DJ (2006) Suction mechanism for iron entrainment into the lower mantle. *Geophysical Research Letters* 33 (doi:10.1029/2005GL025009).

Kaneshima S and Helffrich G (1999) Dipping low-velocity layer in the mid-lower mantle:Evidence for geochemical heterogeneity. *Science* 283: 1888–1891.

Karato S (1995) Interaction of chemically stratified subducted oceanic lithosphere with the 660 km discontinuity. *Proceedings of the Japan Academy Series B Physical and Biological Sciences* 71: 203–207.

Karato S, Bercovici D, Leahy G, Richard G, and Jing Z (2006) The transition zone water filter model for global material circulation: Where do we stand? In: Jacobsen SD and Van Der Lee S (eds.) *AGU Monograph Series, 168: Earth's Deep Water Cycle*, pp. 289–314. Washington, DC: American Geophysical Union.

Karato S and Murthy VR (1997) Core formation and chemical-equilibrium in the earth. Part 1: Physical considerations. *Physics of the Earth and Planetary Interiors* 100: 61–79.

Karato S, Wang ZC, Liu B, and Fujino K (1995) Plastic-deformation of garnets – Systematics and implications for the rheology of the mantle transition zone. *Earth and Planetary Science Letters* 130: 13–30.

Karato S and Wu P (1993) Rheology of the upper mantle – A synthesis. *Science* 260: 771–778.

Kasting JF, Eggler DH, and Raeburn SP (1993a) Mantle redox evolution and the oxidation state of the Archean atmosphere. *Journal of Geology* 101: 245–257.

Kasting JF, Whitmore DP, and Reynolds RT (1993b) Habitable zones around main sequence stars. *Icarus* 101: 108–128.

Kellogg JB (2004) *Towards and Understanding of Chemical and Isotopic Heterogeneity in the Earth's Mantle*. Cambridge, MA: Harvard University.

Kellogg JB, Jacobsen SB, and O'Connell RJ (2002) Modeling the distribution of isotopic ratios in geochemical reservoirs. *Earth and Planetary Science Letters* 204: 183–202.

Kellogg JB and Tackley PJ (2004) A comparison of methods for modeling geochemical variability in the Earth's mantle. *EOS Transactions AGU, Fall Meeting Supplement* 85, Abstract U41A-0728.

Kellogg LH (1997) Growing the Earth's D′ layer: Effect of density variations at the core–mantle boundary. *Geophysical Research Letters* 24: 2749–2752.

Kellogg LH, Hager BH, and van der Hilst RD (1999) Compositional stratification in the deep mantle. *Science* 283: 1881–1884.

Kellogg LH and Stewart CA (1991) Mixing by chaotic convection in an infinite Prandtl number fluid and implications for mantle convection. *Physics of Fluids a Fluid Dynamics* 3: 1374–1378.

Kellogg LH and Turcotte DL (1986) Homogenization of the mantle by convective mixing and diffusion. *Earth and Planetary Science Letters* 81: 371–378.

Kellogg LH and Turcotte DL (1990) Mixing and the distribution of heterogeneities in a chaotically convecting mantle. *Journal of Geophysical Research Solid Earth and Planets* 95: 421–432.

Kellogg LH and Wasserburg GJ (1990) The role of plumes in mantle helium fluxes. *Earth and Planetary Science Letters* 99: 276–289.

Kennett BLN, Widiyantoro S, and Vanderhilst RD (1998) Joint seismic tomography for bulk sound and shear-wave speed in the Earths mantle. *Journal of Geophysical Research Solid Earth* 103: 12469–12493.

Kesson SE, Gerald JDF, and Shelley JM (1998) Mineralogy and dynamics of a pyrolite lower mantle. *Nature* 393: 252–255.

Klein EM, Langmuir CH, Zindler A, Staudigel H, and Hamelin H (1988) Isotope evidence of a mantle convection boundary at the Australian-Antarctic discordance. *Nature* 333: 623–629.

Knittle E and Jeanloz R (1989) Simulating the core-mantle boundary – An experimental study of high-pressure reactions between silicates and liquid iron. *Geophysical Research Letters* 16: 609–612.

Knittle E and Jeanloz R (1991) Earths core–mantle boundary – Results of experiments at high-pressures and temperatures. *Science* 251: 1438–1443.

Korenaga J (2003) Energetics of mantle convection and the fate of fossil heat. *Geophysical Research Letters* 30: 1437.

Kramers JD and Tolstikhin IN (1997) Two terrestrial lead isotope paradoxes, forward transport modelling, core formation and the history of the continental crust. *Chemical Geology* 139: 75–110.

Kump LR, Kasting JF, and Barley ME (2001) Rise of atmospheric oxygen and the 'upside-down' Archean mantle. *Geochemistry Geophysics Geosystems* 2 (doi:10.1029/2000GC000114).

Kurz MD, Jenkins WJ, Hart SR, and Clague D (1983) Helium isotopic variations in volcanic rocks from Loihi seamount and the island of Hawaii. *Earth and Planetary Science Letters* 66: 388–406.

Kyvalova H, Cadek O, and Yuen DA (1995) Correlation analysis between subduction in the last 180 Myr and lateral seismic structure of the lower mantle: Geodynamical implications. *Geophysical Research Letters* 22: 1281–1284.

Labrosse S (2003) Thermal and magnetic evolution of the Earth's core. *Physics of the Earth and Planetary Interiors* 140: 127–143.

Labrosse S, Poirier JP, and Le Mouel JL (1997) On cooling of the Earth's core. *Physics of the Earth and Planetary Interiors* 99: 1–17.

Lassiter JC (2004) Role of recycled oceanic crust in the potassium and argon budget of the Earth: Toward a resolution of the 'missing argon' problem. *Geochemistry Geophysics Geosystems* 5: Q11012 (doi:10.1029/2004GC000711).

Lay T and Garnero EJ (2004) Core–mantle boundary structures and processes. In: Sparks RSJ and Hawkesworth CJ (eds.) *The State of the Planet: Frontiers and Challenges in Geophysics*, (doi:10.1029/150GM04). pp. 25–41. Washington, DC: AGU.

Lay T, Garnero EJ, and Williams Q (2004) Partial melting in a thermo-chemical boundary layer at the base of the mantle. *Physics of the Earth and Planetary Interiors* 146: 441–467.

Le Bars M and Davaille A (2002) Stability of thermal convection in two superimposed miscible viscous fluids. *Journal of Fluid Mechanics* 471: 339–363.

Le Bars M and Davaille A (2004a) Large interface deformation in two-layer thermal convection of miscible viscous fluids. *Journal of Fluid Mechanics* 499: 75–110.

Le Bars M and Davaille A (2004b) Whole layer convection in a heterogeneous planetary mantle. *Journal of Geophysical Research* 109: 23.

Leahy GM and Bercovici D (2004) The influence of the transition zone water filter on convective circulation in the mantle. *Geophysical Research Letters* 31 (doi:10.1029/2004GL021206).

Lecuyer C and Ricard Y (1999) Long-term fluxes and budget of ferric iron: Implication for the redox states of the Earth's mantle and atmosphere. *Earth and Planetary Science Letters* 165: 197–211.

Lenardic A and Kaula WM (1993) A numerical treatment of geodynamic viscous-flow problems involving the advection of material interfaces. *Journal of Geophysical Research-Solid Earth* 98: 8243–8260.

Lin S-C and van Keken PE (2006a) Dynamics of thermochemical plumes. Part 1: Plume formation and entrainment of a dense layer. *Geochemistry Geophysics Geosystems* 7 (doi:10.1029/2005GC001071).

Lin S-C and van Keken PE (2006b) Dynamics of thermochemical plumes. Part 2: Complexity of plume structures and its implications for mapping of mantle plumes. *Geochemistry Geophysics Geosystems* 7 (doi:10.1029/2005GC001072).

Lithgow-Bertelloni C, Richards M A, Ricard Y, O'Connell R J, and Engebretson DC (1993) Toroidal-poloidal partitioning of plate motions since 120 Ma. *Geophysical Research Letters* 20: 375–378.

Luo S-N, Ni S, and Helmberger DV (2001) Evidence for a sharp lateral variation of velocity at the core-mantle boundary from multipathed PKPab. *Earth and Planetary Science Letters* 189: 155–164.

Lyubetskaya T and Korenaga J (2007a) Chemical composition of Earth's primitive mantle and its variance. Part 1: Method and results. *Journal of Geophysical Research* 112(B3): B03211, doi:10.1029/2005JB004223.

Lyubetskaya T and Korenaga J (2007b) Chemical composition of Earth's primitive mantle and its variance. Part 2: Implications for global geodynamics. *Journal of Geophysical Research* 112(B3): B03212, doi:10.1029/2005JB004224.

Machetel P and Weber P (1991) Intermittent layered convection in a model mantle with an endothermic phase-change at 670 km. *Nature* 350: 55–57.

Malvern LE (1969) *Introduction to the Mechanics of a Continuous Medium.* Englewood Cliffs: NJ Prentice Hall.

Mambole A and Fleitout L (2002) Petrological layering induced by an endothermic phase transition in the Earth's mantle. *Geophysical Research Letters* 29: 2044.

Manga M (1996) Mixing of heterogeneities in the mantle – Effect of viscosity differences. *Geophysical Research Letters* 23: 403–406.

Manga M and Jeanloz R (1996) Implications of a metal-bearing chemical-boundary layer in D″ for mantle dynamics. *Geophysical Research Letters* 23: 3091–3094.

Martin H (1986) Effect of steeper archean geothermal gradient on geochemistry of subduction zone magmas. *Geology* 14: 753–756.

Masters G, Laske G, Bolton H, and Dziewonski A (2000) The relative behavior of shear velocity, bulk sound speed, and compressional velocity in the mantle: Implications for

chemical and thermal structure. In: Karato S, Forte a M, Liebermann RC, Masters G, and Stixrude L (eds.) *Geophysical Monograph on Mineral Physics and Seismic Tomography fom the Atomic to the Global Scale,* pp. 63–87. Washington, DC: American Geophysical Union.

Matsuda JMS, Ozima M, Ito K, Ohtaka O, and Ito E (1993) Noble gas partitioning between metal and silicate under high pressures. *Science* 259: 788–790.

McDonough WF and Sun SS (1995) The composition of the Earth. *Chemical Geology* 120: 223–253.

McKenzie D (1979) Finite deformation during fluid flow. *Geophysical Journal of the Royal Astronomical Society* 58: 689–715.

McKenzie D and Onions RK (1992) Partial melt distributions from inversion of rare-earth element concentrations. *Journal of Petrology* 32: 1021–1091.

McNamara AK and Van Keken PE (2000) Cooling of the Earth: A parameterized convection study of whole versus layered models. *Geochemistry Geophysics Geosystems* 1 (doi:10.1029/2000GC000045).

McNamara AK and Zhong S (2004a) The influence of thermochemical convection on the fixity of mantle plumes. *Earth and Planetary Science Letters* 222: 485–500.

McNamara AK and Zhong S (2004b) Thermochemical structures within a spherical mantle: Superplumes or piles? *Journal of Geophysical Research* 109 (doi:10.1029/2003JB00287).

McNamara AK and Zhong S (2005) Thermochemical piles beneath Africa and the Pacific Ocean. *Nature* 437: 1136–1139.

Megnin C, Bunge HP, Romanowicz B, and Richards MA (1997) Imaging 3-D spherical convection models – What can seismic tomography tell us about mantle dynamics. *Geophysical Research Letters* 24: 1299–1302.

Meibom A and Anderson DL (2004) The statistical upper mantle assemblage. *Earth and Planetary Science Letters* 217: 123–139.

Meibom A, Anderson DL, Sleep NH, et al. (2003) Are high He-3/He-4 ratios in oceanic basalts an indicator of deep-mantle plume components? *Earth and Planetary Science Letters* 208: 197–204.

Meibom A, Sleep NH, Chamberlain CP, et al. (2002) Re–Os isotopic evidence for long-lived heterogeneity and equilibration processes in the Earth's upper mantle. *Nature* 419: 705–708.

Meibom A, Sleep NH, Zahnle K, and Anderson DL (2005) Models for noble gases in mantle geochemistry: Some observations and alternatives. *Plumes, Plates, and Paradigms, Special Paper Geological Society of America* 388: 347–363.

Melosh HJ (1990) Giant impacts and the thermal state of the early Earth. In: Newsom HE and Jones JH (eds.) *Origin of the Earth,* pp. 69–83. New York: Oxford University Press.

Metcalfe G, Bina CR, and Ottino JM (1995) Kinematic considerations for mantle mixing. *Geophysical Research Letters* 22: 743–746.

Michard A and Albarède F (1985) Hydrothermal uranium uptake at ridge crests. *Nature* 317: 61–88.

Montague NL and Kellogg LH (2000) Numerical models for a dense layer at the base of the mantle and implications for the geodynamics of D″. *Journal of Geophysical Research* 105: 11101–11114.

Montague NL, Kellogg LH, and Manga M (1998) High Rayleigh number thermo-chemical models of a dense boundary layer in D′. *Geophysical Research Letters* 25: 2345–2348.

Montelli R, Nolet G, Dahlen FA, Masters G, Engdahl ER, and Hung SH (2004) Finite-frequency tomography reveals a variety of plumes in the mantle. *Science* 303: 338–343.

Moreira M, Breddam K, Curtice J, and Kurz MD (2001) Solar nean in the Icelandic mantle: New evidence for an undegassed lower mantle. *Earth and Planetary Science Letters* 185: 15–23.

Moresi L and Solomatov V (1998) Mantle convection with a brittle lithosphere – Thoughts on the global tectonic styles of the Earth and Venus. *Geophysical Journal International* 133: 669–682.

Muller R (1992) The performance of classical versus modern finite-volume advection schemes for atmospheric modeling in a one-dimensional test-bed. *Monthly Weather Review* 120: 1407–1415.

Murakami M, Hirose K, Kawamura K, Sata N, and Ohishi Y (2004) Post-perovskite phase transition in $MgSiO_3$. *Science* 304: 855–858.

Murakami M, Hirose K, Sata N, and Ohishi Y (2005) Post-perovskite phase transition and mineral chemistry in the pyrolitic lowermost mantle. *Geophysical Research Letters* 32 (doi:10.1029/2004GL021956).

Murphy DT, Kamber BS, and Collerson KD (2003) A refined solution to the first terrestrial Pb-isotope paradox. *Journal of Petrology* 44: 39–53.

Murthy VR and Karato S (1997) Core formation and chemical-equilibrium in the earth. Part 2: Chemical consequences for the mantle and core. *Physics of the Earth and Planetary Interiors* 100: 81–95.

Nakagawa T and Buffett BA (2005) Mass transport mechanism between the upper and lower mantle in numerical simulations of thermochemical mantle convection with multicomponent phase changes. *Earth and Planetary Science Letters* 230: 11–27.

Nakagawa T and Tackley PJ (2004a) Effects of a perovskite-post perovskite phase change near the core–mantle boundary on compressible mantle convection. *Geophysical Research Letters* 31: L16611 (doi:10.1029/2004GL020648).

Nakagawa T and Tackley PJ (2004b) Effects of thermo-chemical mantle convection on the thermal evolution of the Earth's core. *Earth and Planetary Science Letters* 220: 107–119.

Nakagawa T and Tackley PJ (2005a) Deep mantle heat flow and thermal evolution of Earth's core in thermo-chemical multiphase models of mantle convection. *Geochemistry Geophysics Geosystems* 6 (doi:10.1029/2005GC000967).

Nakagawa T and Tackley PJ (2005b) The interaction between the post-perovskite phase change and a thermo-chemical boundary layer near the core–mantle boundary. *Earth and Planetary Science Letters* 238: 204–216.

Nakagawa T and Tackley PJ (2005c) Three-dimensional numerical simulations of thermo-chemical multiphase convection in Earth's mantle. *Proceedings of the Third MIT Conference on Computational Fluid and Solid Mechanics*.

Namiki A (2003) Can the mantle entrain D″? *Journal of Geophysical Research* 108 (doi:10.1029/2002JB002315).

Newsom HE, White WM, Jochum KP, and Hofmann AW (1986) Siderophile and chalcophile element abundances in oceanic basalts. *Earth and Planetary Science Letters* 80: 299–313.

Ni S, Tan E, Gurnis M, and Helmberger DV (2002) Sharp sides to the African superplume. *Science* 296: 1850–1852.

Nimmo F, Price GD, Brodholt J, and Gubbins D (2004) The influence of potassium on core and geodynamo evolution. *Geophysical Journal International* 156: 363–376.

O'Nions RK, Evensen NM, and Hamilton PJ (1979) Geochemical modeling of mantle differentiation and crustal growth. *Journal of Geophysical Research* 84: 6091–6101.

O'Nions RK and Oxburgh ER (1983) Heat and helium in the Earth. *Nature* 306: 429.

O'Nions RK and Tolstikhin IN (1996) Limits on the mass flux between lower and upper mantle and stability of layering. *Earth and Planetary Science Letters* 139: 213–222.

Oganov AR and Ono S (2004) Theoretical and experimental evidence for a post-perovskite phase of $MgSiO_3$ in Earth's D' layer. *Nature* 430: 445–448.

Ogawa M (1988) Numerical experiments on coupled magmatism-mantle convection system – implications for mantle evolution and archean continental crusts. *Journal of Geophysical Research Solid Earth and Planets* 93: 15119–15134.

Ogawa M (1993) A numerical-model of a coupled magmatism mantle convection system in Venus and the Earth's mantle beneath archean continental crusts. *Icarus* 102: 40–61.

Ogawa M (1994) Effects of chemical fractionation of heat-producing elements on mantle evolution inferred from a numerical-model of coupled magmatism mantle convection system. *Physics of the Earth and Planetary Interiors* 83: 101–127.

Ogawa M (1997) A bifurcation in the coupled magmatism-mantle convection system and its implications for the evolution of the Earth's upper mantle. *Physics of the Earth and Planetary Interiors* 102: 259–276.

Ogawa M (2000a) Coupled magmatism-mantle convection system with variable viscosity. *Tectonophysics* 322: 1–18.

Ogawa M (2000b) Numerical models of magmatism in convecting mantle with temperature-dependent viscosity and their implications for Venus and Earth. *Journal of Geophysical Research* 105: 6997–7012.

Ogawa M (2003) Chemical stratification in a two-dimensional convecting mantle with magmatism and moving plates. *Journal of Geophysical Research* 108, doi:10.1029/2002JB002205.

Ogawa M and Nakamura H (1998) Thermochemical regime of the early mantle inferred from numerical models of the coupled magmatism-mantle convection system with the solid-solid phase transitions at depths around 660 km. *Journal of Geophysical Research* 103: 12161–12180.

Oldham D and Davies JH (2004) Numerical investigation of layered convection in a three-dimensional shell with application to planetary mantles. *Geochemistry Geophysics Geosystems* 5, doi:10.1029/2003GC000603.

Olson P (1984) An experimental approach to thermal-convection in a 2-layered mantle. *Journal of Geophysical Research* 89: 1293–1301.

Olson P and Kincaid C (1991) Experiments on the interaction of thermal convection and compositional layering at the base of the mantle. *Journal of Geophysical Research* 96: 4347–4354.

Olson P, Yuen DA, and Balsiger D (1984a) Convective mixing and the fine-structure of mantle heterogeneity. *Physics of the Earth and Planetary Interiors* 36: 291–304.

Olson P, Yuen DA, and Balsiger D (1984b) Mixing of passive heterogeneties by mantle convection. *Journal of Geophysical Research* 89: 425–436.

Ono S, Ito E, and Katsura T (2001) Mineralogy of subducted basaltic crust (MORB) from 25 to 37 GPa, and chemical heterogeneity of the lower mantle. *Earth and Planetary Science Letters* 190: 57–63.

Ono S, Oganov AR, and Ohishi Y (2005a) *In situ* observations of phase transition between perovskite and $CaIrO_3$-type phase in $MgSiO_3$ and pyrolitic mantle composition. *Earth and Planetary Science Letters* 236: 914–932.

Ono S, Oshishi Y, Isshiki M, and Watanuki T (2005b) *In situ* X-ray observations of phase assemblages in peridotite and basalt compositions at lower mantle conditions: Implications for density of subducted oceanic plate. *Journal of Geophysical Research* 110 (doi:10.1029/2004JB0003196).

Ottino JM (1989) *The Kinematics of Mixing: Stretching, Chaos, and Transport*. Cambridge: Cambridge University Press.

Ozima M and Kudo K (1972) Excess argon in submarine basalts and an Earth-atmosphere evolution model. *Nature* 239: 23–24.

Palme H and O'Neill HSC (2003) Cosmochemical esimates of mantle composition. In: Carlson RW (ed.) *Treatise on Geochemistry*, pp. 1–38. Amsterdam: Elsevier.

Parkinson IJ, Hawkesworth CJ, and Cohen AS (1998) Ancient mantle in a modern arc: Osmium isotopes in Izu-Bonin-Mariana forearc peridotites. *Science* 281: 2011–2013.

Parman SW, Kurz MD, Hart SR, and Grove TL (2005) Helium solubility in olivine and implications for high 3He/4He in ocean island basalts. *Nature* 437: 1140–1143.

Pearson DG, Davies GR, and Nixon PH (1993) Geochemical constraints on the petrogenesis of diamond facies pyroxenites from the Beni Bousera peridotite massif, north Morocco. *Journal of Petrology* 34: 125–172.

Peltier WR and Solheim LP (1992) Mantle phase-transitions and layered chaotic convection. *Geophysical Research Letters* 19: 321–324.

Petford N, Yuen D, Rushmer T, Brodholt J, and Stackhouse S (2005) Shear-induced material transfer across the core–mantle boundary aided by the post-perovskite phase transition. *Earth Planets Space* 57: 459–464.

Phipps Morgan J (1998) Thermal and rare gas evolution of the mantle. *Chemical Geology* 145: 431–445.

Phipps Morgan J (1999) Isotope topology of individual hotspot basalt arrays: Mixing curves or melt extraction trajectories? *Geochemistry Geophysics Geosystems* 1, doi:10.1029/1999GC000004.

Phipps Morgan J (2001) Thermodynamics of pressure release melting of a veined plum pudding mantle. *Geochemistry Geophysics Geosystems* 2, doi:10.1029/2000GC000049.

Phipps Morgan J and Morgan J (1999) Two-stage melting and the geochemical evolution of the mantle: A recipe for mantle plum-pudding. *Earth and Planetary Science Letters* 170: 215–239.

Poirier JP (1993) Core-infiltrated mantle and the nature of the D″ layer. *Journal of Geomagnetism and Geoelectricity* 45: 1221–1227.

Poirier JP and Le Mouel JL (1992) Does infiltration of core material into the lower mantle affect the observed geomagnetic field? *Physics of the Earth and Planetary Interiors* 73: 29–37.

Poirier JP, Malavergne V, and Le Mouel JL (1998) Is there a thin electrically conducting layer at the base of the mantle? In: Gurnis M, Wysession ME, Knittle E, and Buffett BA (eds.) *The Core–Mantle Boundary Region*, pp. 131–137. Washington, DC: American Geophysical Union.

Polve M and Allegre CJ (1980) Orogenic lherzolite complexes studies by 87Rb–87Sr: A clue to understand the mantle convection processes? *Earth and Planetary Science Letters* 51: 71–93.

Porcelli D and Halliday AN (2001) The core as a possible source of mantle helium. *Earth and Planetary Science Letters* 192: 45–56.

Porcelli D and Wasserburg GJ (1995) Mass transfer of helium, neon, argon, and xenon through a steady-state upper mantle. *Geochimica et Cosmochimica Acta* 59: 4921–4937.

Puster P and Jordan TH (1994) Stochastic-analysis of mantle convection experiments using 2-point correlation-functions. *Geophysical Research Letters* 21: 305–308.

Puster P and Jordan TH (1997) How stratified is mantle convection? *Journal of Geophysical Research* 102: 7625–7646.

Puster P, Jordan TH, and Hager BH (1995) Characterization of mantle convection experiments using 2-point correlation-functions. *Journal of Geophysical Research Solid Earth* 100: 6351–6365.

Reisberg l and Zindler A (1986) Extreme isotopic variability in the upper mantle: Evidence from Ronda. *Earth and Planetary Science Letters* 81: 29–45.

Revenaugh J and Sipkin SA (1994) Seismic evidence for silicate melt atop the 410-km mantle discontinuity. *Nature* 369: 474–476.

Reymer A and Schubert G (1984) Phanerozioc addition rates to the continental crust and crustal growth. *Tectonics* 3: 63–77.

Ricard Y, Froidevaux C, and Fleitout L (1988) Global plate motion and the geoid – A physical model. *Geophysical Journal Oxford* 93: 477–484.

Ricard Y, Vigny C, and Froidevaux C (1989) Mantle heterogeneities, geoid, and plate motion – A Monte-Carlo inversion. *Journal of Geophysical Research* 94: 13739–13754.

Richard G, Monnereau M, and Ingrin J (2002) Is the transition zone an empty water reservoir? Inferences from numerical models of mantle dynamics. *Earth and Planetary Science Letters* 205: 37–51.

Richards MA and Davies GF (1989) On the separation of relatively buoyant components from subducted lithosphere. *Geophysical Research Letters* 16: 831–834.

Richards MA and Engebretson DC (1992) Large-scale mantle convection and the history of subduction. *Nature* 355: 437–440.

Richards MA and Griffiths RW (1989) Thermal entrainment by deflected mantle plumes. *Nature* 342: 900–902.

Richardson LF (1926) Atmospheric diffusion shown on a distance-neighbor graph. *Proceedings of the Royal Society of London* 110: 709–737.

Richter FM, Daly SF, and Nataf HC (1982) A parameterized model for the evolution of isotopic heterogeneities in a convecting system. *Earth and Planetary Science Letters* 60: 178–194.

Richter FM and Johnson CE (1974) Stability of a chemically layered mantle. *Journal of Geophysical Research* 79: 1635–1639.

Richter FM and McKenzie D (1981) On some consequences and possible causes of layered mantle convection. *Journal of Geophysical Research* 86: 6133–6142.

Richter FM and Ribe NM (1979) On the importance of advection in determining the local isotopic composition of the mantle. *Earth and Planetary Science Letters* 43: 212–222.

Righter K and Drake MJ (2003) Partition coefficients at high pressure and temperature. In: Carlson CW (ed.) *Treatise on Geochemistry, vol. 2: The Mantle and Core*, pp. 425–449. Amsterdam: Elsevier.

Ringwood AE (1990a) Earliest history of the Earth–Moon system. In: Newsom HE and Jones JH (eds.) *Origin of the Earth*, pp. 101–134. New York: Oxford University Press.

Ringwood AE (1990b) Slab-mantleinteractions. Part 3: Petrogenesis of intraplate magmas and structure of the upper mantle. *Chemical Geology* 82: 187–207.

Ringwood AE (1991) Phase-transformations and their bearing on the constitution and dynamics of the mantle. *Geochimica et Cosmochimica Acta* 55: 2083–2110.

Ringwood AE (1994) Role of the transition zone and 660 km discontinuity in mantle dynamics. *Physics of the Earth and Planetary Interiors* 86: 5–24.

Ringwood AE and Irifune T (1988) Nature of the 650-km seismic discontinuity – Implications for mantle dynamics and differentiation. *Nature* 331: 131–136.

Romanowicz B (2001) Can we resolve 3D density heterogeneity in the lower mantle? *Geophysical Research Letters* 28: 1107–1110.

Rubie DC, Melosh HJ, Reid Liebske C, and Righter K (2003) Mechanisms of metal-silicate equilibration in the terrestrial magma ocean. *Earth and Planetary Science Letters* 205: 239–255.

Rudge JF (2006) Mantle pseudo-isochrons revisited. *Earth and Planetary Science Letters* 249: 494–513.

Rudge JF, McKenzie D, and Haynes PH (2005) A theoretical approach to understanding the isotopic heterogeneity of mid-ocean ridge basalt. *Geochemica Cosmochimica Acta* 69: 3873–3887.

Rudnick RL (1995) Making continental crust. *Nature* 378: 571–578.

Rudnick RL and Fountain DM (1995) Nature and composition of the continental crust; a lower crustal perspective. *Reviews of Geophysics* 33: 267–309.

Ruepke L, Phipps Morgan J, Hort M, and Connolly JAD (2004) Serpentine and the subduction zone water cycle. *Earth and Planetary Science Letters* 223: 17–34.

Saal AE, Hart SR, Shimizu N, Hauri EH, and Layne GD (1998) Pb isotopic variability in melt inclusions from oceanic island basalts, Polynesia. *Science* 282: 1481–1484.

Saal AE, Hauri EH, Langmuir CH, and Perfit MR (2002) Vapour undersaturation in primitive mid-ocean-ridge basalt and the volatile content of Earth's upper mantle. *Nature* 419: 451–455.

Samuel H and Farnetani CG (2003) Thermochemical convection and helium concentrations in mantle plumes. *Earth and Planetary Science Letters* 207: 39–56.

Sarda P, Moreira M, and Staudacher T (1999) Argon-lead isotopic correlation in mid-Atlantic ridge basalts. *Science* 283: 666–668.

Schersten A, Elliott T, Hawkesworth C, and Norman M (2004) Tungsten isotope evidence that mantle plumes contain no contribution from the Earth's core. *Nature* 427: 234–237.

Schilling J-G (1973) Iceland mantle plume: Geochemical study of Reykjanes ridge. *Nature* 242: 565–571.

Schmalzl J and Hansen U (1994) Mixing the Earth's mantle by thermal-convection - A scale-dependent phenomenon. *Geophysical Research Letters* 21: 987–990.

Schmalzl J, Houseman GA, and Hansen U (1995) Mixing properties of three-dimensional (3-D) stationary convection. *Physics of Fluids* 7: 1027–1033.

Schmalzl J, Houseman GA, and Hansen U (1996) Mixing in vigorous, time-dependent three-dimensional convection and application to Earth's mantle. *Journal of Geophysical Research* 101: 21847–21858.

Schmalzl J and Loddoch A (2003) Using subdivision surfaces and adaptive surface simplification algorithms for modeling chemical heterogeneities in geophysical flows. *Geochemistry Geophysics Geosystems* 4 (doi:10.1029/2003GC000578).

Schott B and Yuen DA (2004) Influences of dissipation and rheology on mantle plumes coming from the D'-layer. *Physics of the Earth and Planetary Interiors* 146: 139–145.

Schott B, Yuen DA, and Braun A (2002) The influences of composition-and temperature- dependent rheology in thermal-chemical convection on entrainment of the D'-layer. *Physics of the Earth and Planetary Interiors* 129: 43–65.

Schubert G, Masters G, Olson P, and Tackley P (2004) Superplumes or plume clusters? *Physics of the Earth and Planetary Interiors* 146: 147–162.

Scrivner C and Anderson DL (1992) The effect of post Pangea subduction on global mantle tomography and convection. *Geophysical Research Letters* 19: 1053–1056.

Shaw DM (1970) Trace element fractionation during anatexis. *Geochemistry Geophysics Geosystems* 34: 237–242.

Sleep NH (1984) Tapping of magmas from ubiquitous mantle heterogeneities – An Alternative to mantle plumes. *Journal of Geophysical Research* 89: 10029–10041.

Sleep NH (1988) Gradual entrainment of a chemical layer at the base of the mantle by overlying convection. *Geophysical Journal (Oxford)* 95: 437–447.

Sleep NH (1990) Hotspots and mantle plumes – Some Phenomenology. *Journal of Geophysical Research Solid Earth and Planets* 95: 6715–6736.

Sleep NH (2000) Evolution of the mode of convection within terrestrial planets. *Journal of Geophysical Research* 105: 17563–17578.

Sleep NH and Zahnle K (2001) Carbon dioxide cycling and implications for climate on ancient Earth. *Journal of Geophysical Research* 106: 1373–1399.

Smolarkiewicz PK and Margolin LG (1998) MPDATA: A finite-difference solver for geophysical flows. *Journal of Computational Physics* 140: 459–480.

Sneeringer M, Hart SR, and Shimizu N (1984) Strontium and samarium diffusion in diopside. *Geochemica Cosmochimica Acta* 48: 1589–1608.

Sobolev AV (1996) Melt inclusions in minerals as a source of principal petrological information. *Petrology* 4: 209–220.

Sobolev AV, Hofmann AW, and Nikogosian IK (2000) Recycled oceanic crust observed in 'ghost plagioclase' within the source of Mauna Loa lavas. *Nature* 404: 986–990.

Sobolev AV and Shimizu N (1993) Ultra-depleted primary melt included in an olivine from the Mid-Atlantic Ridge. *Nature* 363: 151–154.

Solomatov VS (2000) Fluid dynamics of a terrestrial magma ocean. In: Canup RM and Righter K (eds.) *Origin of the Earth and Moon*, pp. 323–338. Tucson, AZ: The University of Arizona Press.

Solomatov VS and Moresi LN (2002) Small-scale convection in the D' layer. *Journal of Geophysical Research* 107, doi:10.1029/2000JB000063.

Solomatov VS, Olson P, and Stevenson DJ (1993) Entrainment from a bed of particles by thermal-convection. *Earth and Planetary Science Letters* 120: 387–393.

Solomatov VS and Stevenson DJ (1993a) Nonfractional crystallization of a terrestrial magma Ocean. *Journal of Geophysical Research-Planets* 98: 5391–5406.

Solomatov VS and Stevenson DJ (1993b) Suspension in convective layers and style of differentiation of a terrestrial magma ocean. *Journal of Geophysical Research-Planets* 98: 5375–5390.

Song T-RA, Helmberger DV, and Grand SP (2004) Low-velocity zone atop the 410-km seismic discontinuity in the northwestern United States. *Nature* 427: 530–533.

Song X and Ahrens TJ (1994) Pressure-temperature range of reactions between liquid-iron in the outer core and mantle silicates. *Geophysical Research Letters* 21: 153–156.

Spence DA, Ockendon JR, Wilmott P, Turcotte DL, and Kellogg LH (1988) Convective mixing in the mantle: The role of viscosity differences. *Geophysical Journal* 95: 79–86.

Stacey FD (1992) *Physics of the Earth*. Kenmore, QLD: Brookfield Press.

Staudacher T and Allegre CJ (1989) Recycling of oceanic crust and sediments: The noble gas subduction barrier. *Earth and Planetary Science Letters* 89: 173–183.

Stegman DR, Richards MA, and Baumgardner JR (2002) Effects of depth-dependent viscosity and plate motions on maintaining a relatively uniform mid-ocean ridge basalt reservoir in whole mantle flow. *Journal of Geophysical Research* 107 (doi:10.1029/2001JB000192).

Stein M and Hofmann AW (1994) Mantle plumes and episodic crustal growth. *Nature* 372: 63–68.

Stevenson DJ (1988) Infiltration, dissolution, and underplating: Rules for mixing core–mantle cocktails. *EOS AGU, Fall Meeting Supplement* 69: 1404.

Stevenson DJ (1990) Fluid dynamics of core formation. In: Newman HE and Jones JH (eds.) *Origin of the Earth*, pp. 231–249. New York: Oxford University Press.

Stracke A, Hofmann AW, and Hart S (2005) FOZO, HIMU, and the rest of the mantle zoo. *Geochemistry Geophysics Geosystems* 6, doi:10.1029/2004GC000824.

Stracke A, Zindler A, Salters VJM, *et al.* (2003) Theistareykir revisited. *Geochemistry Geophysics Geosystems* 4(2) (doi:10.1029/2001GC000201).

Stuart FM, Lass-Evans S, Fitton JG, and Ellam RM (2003) High 3He/4He ratios in picritic basalts from Baffin Island and the role of a mixed reservoir in mantle plumes. *Nature* 424: 57–59.

Suen CJ and Frey FA (1987) Origins of mafic and ultramafic rocks in the Ronda peridotite. *Earth and Planetary Science Letters* 85: 183–202.

Tackley PJ (1998) Three-dimensional simulations of mantle convection with a thermochemical CMB boundary layer: D″? In: Gurnis M, Wysession ME, Knittle E, and Buffett BA (eds.) *The Core-Mantle Boundary Region*, pp. 231–253. Washington, DC: American Geophysical Union.

Tackley PJ (2000) Mantle convection and plate tectonics: Towards an integrated physical and chemical theory. *Science* 288: 2002.

Tackley PJ (2002) Strong heterogeneity caused by deep mantle layering. *Geochemistry Geophysics Geosystems* 3, doi:10.1029/2001GC000167.

Tackley PJ and King SD (2003) Testing the tracer ratio method for modeling active compositional fields in mantle convection simulations. *Geochemistry Geophysics Geosystems* 4, doi:10.1029/2001GC000214.

Tackley PJ, Stevenson DJ, Glatzmaier GA, and Schubert G (1994) Effects of multiple phase transitions in a 3-dimensional spherical model of convection in Earth's mantle. *Journal of Geophysical Research* 99: 15877–15901.

Tackley PJ and Xie S (2002) The thermo-chemical structure and evolution of Earth's mantle: Constraints and numerical models. *Philosophical Transactions of the Royal Society of London A* 360: 2593–2609.

Tackley PJ, Xie S, Nakagawa T, and Hernlund JW (2005) Numerical and laboratory studies of mantle convection: Philosophy, accomplishments and thermo-chemical structure and evolution. In: Van Der Hilst RD, Bass J, Matas J, and Trampert J (eds.) *AGU Geophysical Monograph: Earth's Deep Mantle: Structure, Composition and Evolution*, pp. 85–102. Washington, DC: AGU.

Tajika E and Matsui T (1992) Evolution of terrestrial proto-Co2 atmosphere coupled with thermal history of the Earth. *Earth and Planetary Science Letters* 113: 251–266.

Tajika E and Matsui T (1993) Evolution of sea-floor spreading rate-based on Ar-40 degassing history. *Geophysical Research Letters* 20: 851–854.

Taylor SR and McLennan SM (1995) The geochemical evolution of the continental crust. *Reviews of Geophysics* 33: 241–265.

Ten A, Yuen DA, Larsen TB, and Malevsky AV (1996) The evolution of material-surfaces in convection with variable viscosity as monitored by a characteristics-based method. *Geophysical Research Letters* 23: 2001–2004.

Ten A, Yuen DA, Podladchikov YY, Larsen TB, Pachepsky E, and Malevsky AV (1997) Fractal features in mixing of non-Newtonian and Newtonian mantle convection. *Earth and Planetary Science Letters* 146: 401–414.

Ten AA, Podladchikov YY, Yuen DA, Larsen TB, and Malevsky AV (1998) Comparison of mixing properties in convection with the particle-line method. *Geophysical Research Letters* 25: 3205–3208.

Tolstikhin IN and Hofmann AW (2005) Early crust on top of the Earth's core. *Physics of the Earth and Planetary Interiors* 148: 109–130.

Tolstikhin IN, Kramers JD, and Hofmann AW (2006) A chemical Earth model with whole mantle convection: The importance of a core-mantle boundary layer (D″) and its early formation. *Chemical Geology* 226: 79–99.

Tonks WB and Melosh HJ (1990) The physics of crystal settling and suspension in a turbulent magma ocean. In: Newsom HE and Jones JH (eds.) *Origin of the Earth*, pp. 151–174. New York: Oxford University Press.

Trampert J, Deschamps F, Resovsky JS, and Yuen D (2004) Probabilistic tomography maps significant chemical heterogeneities in the lower mantle. *Science* 306: 853–856.

Trieloff M, Kunz J, Clague DA, Harrison D, and Allegre CJ (2000) The nature of pristine noble gases in mantle plumes. *Science* 288: 1036–1038.

Turcotte DL, Paul D, and White WM (2001) Thorium–uranium systematics require layered mantle convection. *Journal of Geophysical Research* 106: 4265–4276.

Turcotte DL and Schubert G (1982) *Geodynamics: Applications of Continuum Physics to Geological Problems*. New York: Wiley.

Turner G (1989) The outgassing history of the Earth's atmosphere. *Journal of the Geological Society of London* 146: 147–154.

van den Berg AP, Jacobs MH, and de Jong BH (2002) Numerical models of mantle convection based on thermodynamic data for the MgOSiO2 olivine-pyroxene system. *EOS Transactions AGU, Fall Meeting Supplement* 83: Abstract MR72B–1041.

van der Hilst RD and Karason H (1999) Compositional heterogeneity in the bottom 1000 kilometers of Earth's mantle: Toward a hybrid convection model. *Science* 283: 1885–1888.

Van der Hilst RD, Widlyantoro S, and Engdahl ER (1997) Evidence for deep mantle circulation from global tomography. *Nature* 386: 578–584.

van Keken P and Zhong S (1999) Mixing in a 3D spherical model of present-day mantle convection. *Earth and Planetary Science Letters* 171: 533–547.

van Keken PE and Ballentine CJ (1998) Whole-mantle versus layered mantle convection and the role of a high-viscosity lower mantle in terrestrial volatile evolution. *Earth and Planetary Science Letters* 156: 19–32.

van Keken PE and Ballentine CJ (1999) Dynamical models of mantle volatile evolution and the role of phase transitions and temperature-dependent rheology. *Journal of Geophysical Research* 104: 7137–7151.

van Keken PE, Ballentine CJ, and Hauri EH (2003) Convective mixing in the Earth's mantle. In: Carlson RW (ed.) *Treatise on Geochemistry*, pp. 471–491. Amsterdam: Elsevier.

van Keken PE, Ballentine CJ, and Porcelli D (2001) A dynamical investigation of the heat and helium imbalance. *Earth and Planetary Science Letters* 188: 421–434.

van Keken PE, Karato S, and Yuen DA (1996) Rheological control of oceanic crust separation in the transition zone. *Geophysical Research Letters* 23: 1821–1824.

van Keken PE, Kiefer B, and Peacock SM (2002) High-resolution models of subduction zones: Implications for mineral hydration reactions and the transport of water into the deep mantle. *Geochemistry Geophysics Geosystems* 3, doi:10.1029/2001GC000256.

van Keken PE, King SD, Schmeling H, Christensen UR, Neumeister D, and Doin MP (1997) A comparison of methods for the modeling of thermochemical convection. *Journal of Geophysical Research* 102: 22477–22495.

van Thienen P, van den Berg AP, and Vlaar NJ (2004) Production and recycling of oceanic crust in the early Earth. *Tectonophysics* 386: 41–65.

van Thienen P, van Summeren J, van der Hilst RD, van den Berg AP, and Vlaar NJ (2005) Numerical study of the origin and stability of chemically distinct reservoirs deep in Earth's mantle. In: Van Der Hilst RD, Bass J, Matas J, and Trampert J (eds.) *AGU Geophysical Monograph: The Structure, Evolution and Composition of Earth's Mantle*, pp. 117–136. Washington, DC: AGU.

Vidale JE, Schubert G, and Earle PS (2001) Unsuccessful initial search for a midmantle chemical boundary layer with seismic arrays. *Geophysical Research Letters* 28: 859–862.

Vlaar NJ (1985) Precambrian geodynamical constraints. In: Tobi a C and Touret JLR (eds.) *The Deep Proterozoic Crust in the North Atlantic Provinces*, pp. 3–20. Dordrecht: D. Riedel Publishing Company.

Vlaar NJ (1986) Archaean global dynamics. *Dutch contributions to the International Lithosphere Program; Meeting* 65: 91–101.

Walker RJ, Morgan JW, and Horan MF (1995) 187Os enrichment in some mantle plume sources: Evidence for core–mantle interation? *Science* 269: 819–822.

Walzer U and Hendel R (1997a) Tectonic episodicity and convective feedback mechanisms. *Physics of the Earth and Planetary Interiors* 100: 167–188.

Walzer U and Hendel R (1997b) Time-dependent thermal convection, mantle differentiation and continental-crust growth. *Geophysical Journal International* 130: 303–325.

Walzer U and Hendel R (1999) A new convection-fractionation model for the evolution of the principal geochemical reservoirs of the Earth's mantle. *Physics of the Earth and Planetary Interiors* 112: 211–256.

Weinstein SA (1992) Induced compositional layering in a convecting fluid layer by an endothermic phase-transition. *Earth and Planetary Science Letters* 113: 23–39.

Weinstein SA (1993) Catastrophic overturn of the Earth's mantle driven by multiple phase-changes and internal heat-generation. *Geophysical Research Letters* 20: 101–104.

Wen L (2001) Seismic evidence for a rapidly-varying compositional anomaly at the base of the Earth's mantle beneath the Indian ocean. *Earth and Planetary Science Letters* 194: 83–95.

Wen LX (2002) An SH hybrid method and shear velocity structures in the lowermost mantle beneath the central Pacific and South Atlantic Oceans. *Journal of Geophysical Research-Solid Earth* 107: 2055.

Wen LX and Anderson DL (1997) Layered Mantle Convection – A Model for geoid and topography. *Earth and Planetary Science Letters* 146: 367–377.

White WM (1985) Sources of oceanic basalts: radiogenic isotope evidence. *Earth and Planetary Science Letters* 115: 211–226.

Williams Q and Garnero EJ (1996) Seismic evidence for partial melt at the base of Earth's mantle. *Science* 273: 1528–1530.

Wolf A (1986) Quantifying chaos with Lyapunov exponents. In: Holden AV (ed.) *Chaos*, pp. 273–290. Princeton, NJ: Princeton University Press.

Wood BJ and Blundy JD (2003) Trace element partitioning under crustal and uppermost mantle conditions: The influences of ionic radius, cation charge, pressure, and temperature. In: Carlson RW (ed.) *Treatise on Geochemistry, vol. 2: The Mantle and Core*, pp. 395–424. Amsterdam: Elsevier.

Xie S and Tackley PJ (2004a) Evolution of helium and argon isotopes in a convecting mantle. *Physics of the Earth and Planetary Interiors* 146: 417–439.

Xie S and Tackley PJ (2004b) Evolution of U–Pb and Sm–Nd systems in numerical models of mantle convection. *Journal of Geophysical Research* 109: B11204 (doi:10.1029/2004JB003176).

Xu-Dong L and Osher S (1996) Nonoscillatory high order accurate self-similar maximum principle satisfying shock capturing schemes. Part I:. *SIAM Journal on Numerical Analysis* 33: 760–779.

Yabe T, Ogata Y, Takizawa K, Kawai T, Segawa A, and Sakurai K (2002) The next generation CIP as a conservative semi-Lagrangian solver for solid, liquid and gas. *Journal of Computational and Applied Mathematics* 149: 267–277.

Yamazaki D and Karato S (2001) Some mineral physics constraints on the rheology and geothermal structure of Earth's lower mantle. *American Mineralogist* 86: 385–391.

Yokochi R and Marty B (2004) A determination of the neon isotopic composition of the deep mantle. *Earth and Planetary Science Letters* 225: 77–88.

Yoshida M and Ogawa M (2004) The role of hot uprising plumes in the initiation of plate-like regime of three-dimensional mantle convection. *Geophysical Research Letters* 31: 5607.

Zhong S (2006) Constraints on thermochemical convection of the mantle from plume heat flux, plume excess temperature, and upper mantle temperature. *Journal of Geophysical Research* 111 (doi:10.1029/2005JB003972).

Zhong SJ and Hager BH (2003) Entrainment of a dense layer by thermal plumes. *Geophysical Journal International* 154: 666–676.

Zindler A and Hart S (1986) Geochemical geodynamics. *Earth and Planetary Science Letters* 14: 493–571.

Printed in the United States
By Bookmasters